GENETIC PROGRAMMING

Complex Adaptive Systems
John H. Holland, Christopher G. Langton, and Stewart W. Wilson, advisors

Adaptation in Natural and Artificial Systems: An Introductory Analysis with Applications to Biology, Control, and Artificial Intelligence, John H. Holland

Toward a Practice of Autonomous Systems: Proceedings of the First European Conference on Artificial Life, edited by Francisco J. Varela and Paul Bourgine

Genetic Programming: On the Programming of Computers by Means of Natural Selection, John R. Koza

Genetic Programming: The Movie, John R. Koza and James P. Rice

From Animals to Animats 2: Proceedings of the Second International Conference on Simulation of Adaptive Behavior, edited by Jean-Arcady Meyer, Herbert L. Roitblat, and Stewart W. Wilson

Intelligent Behavior in Animals and Robots, David McFarland and Thomas Bösser

Advances in Genetic Programming, edited by Kenneth E. Kinnear, Jr.

Genetic Programming II: Automatic Discovery of Reusable Programs, John R. Koza

Genetic Programming II Video: The Next Generation, John R. Koza

Turtles, Termites, and Traffic Jams: Explorations in Massively Parallel Microworlds, Mitchel Resnick

From Animals to Animats 3: Proceedings of the Third International Conference on Simulation of Adaptive Behavior, edited by Dave Cliff, Philip Husbands, Jean-Arcady Meyer, and Stewart W. Wilson

Artificial Life IV: Proceedings of the Fourth International Workshop on the Synthesis and Simulation of Living Systems, edited by Rodney A. Brooks and Pattie Maes

Comparative Approaches to Cognitive Science, edited by Herbert L. Roitblat and Jean-Arcady Meyer

Artificial Life: An Overview, edited by Christopher G. Langton

Evolutionary Programming IV: Proceedings of the Fourth Annual Conference on Evolutionary Programming, edited by John R. McDonnell, Robert G. Reynolds, and David B. Fogel

An Introduction to Genetic Algorithms, Melanie Mitchell

Catching Ourselves in the Act: Situated Activity, Interactive Emergence, and Human Thought, Horst Hendriks-Jansen

Toward a Science of Consciousness: The First Tucson Discussions and Debates, edited by Stuart R. Hameroff, Alfred W. Kaszniak, and Alwyn C. Scott

Genetic Programming: Proceedings of the First Annual Conference, edited by John R. Koza, David E. Goldberg, David B. Fogel, and Rick L. Riolo

Genetic Programming

Proceedings of the First Annual Conference
1996

edited by John R. Koza, David E. Goldberg, David B. Fogel, and Rick L. Riolo

A Bradford Book The MIT Press Cambridge, Massachusetts London, England

This book was printed and bound in the United States of America.

This publication may be ordered from The MIT Press, 55 Hayward Street, Cambridge, MA 02142 USA; by phone 1-800-356-0343 (toll-free) or 617-625-8569; by FAX to 617-625-6660; by e-mail to mitpress-orders@mit.edu or from The MIT Press's WWW page at http://www-mitpress.mit.edu/.

ISSN 1088-4750
ISBN 0-262-61127-9

Contents

Short Genetic Programming Papers

Genetic Algorithms Poster Papers

Evolutionary Programming and Evolution Strategies Papers

Classifier Systems Papers

Preface

This book of proceedings contains the papers presented at the first genetic programming conference, held at Stanford University on July 28–31, 1996.

Genetic programming is an automatic programming technique for evolving computer programs that solve (or approximately solve) problems. Starting with a primordial ooze of thousands of randomly created computer programs composed of appropriate programmatic ingredients, a population of computers programs is progressively evolved over many generations using the Darwinian principle of survival of the fittest, a sexual recombination operation, and occasional mutation.

Acknowledgments

The editors of these proceedings and the organizers of this conference gratefully acknowledge the assistance and contributions of the following people and organizations:

- Stanford University,
- the cooperating professional societies:
    - Association for Computing Machinery (ACM)
    - Special Interest Group in Artificial Intelligence (SIGART) of the ACM,
    - IEEE Neural Network Council, and
    - American Association for Artificial Intelligence (AAAI),
- the American Association for Artificial Intelligence (AAAI) for its grant for student travel,
- the Release 1.0 newsletter and its editor, Esther Dyson, for in-kind contributions,
- Applied AI Systems Inc. of Kanata, Ontario and its president, Takashi Gomi, for in-kind contributions,
- the conference's honorary chair, John Holland of the University of Michigan,
- the conference's publicity chair, Patrick Tufts of Brandeis University,
- David Andre and Forrest H Bennett III for serving on the executive committee,
- the Stanford University Conference Office and, in particular, Susan Nunan,
- the American Association for Artificial Intelligence (AAAI) for conference administration services, and special thanks to Julia Bowen of the AAAI for her exceptionally professional work on this conference,
- the MIT Press: in particular, Harry Stanton, Jerry Weinstein, and Deborah Cantor-Adams,
- the conference's two invited speakers, John Holland of the University of Michigan and David E. Goldberg of the University of Illinois at Urbana-Champaign,
- the tutorial speakers (listed on following pages),
- the considerable time and effort of the members of the four program committees (listed on following pages) for genetic programming, genetic algorithms, classifier systems, and evolutionary programming and evolution strategies, who reviewed the submitted papers, and
- the authors who submitted papers for consideration by the conference.

Services were supplied to the conference by Jerry Jager of Jager Communications of San Carlos, California (who did the logotypes for the conference); the Prodigy Press of Palo Alto, California; the Stanford Bookstore; Bagetelos and Fadem; Deloitte and Touche; IEEE Computer Society; SIAM; ACM; Kluwer Academic Publishers; Rubin Response Services Inc.; the IIS Corporation of Mountain View, California; and the Intelligence newsletter. This conference was organized by Genetic Programming Conferences Inc., a California not-for-profit corporation.

The Peer Review Process

Out of an original pool of 120 papers, these proceedings contain 38 long papers, 35 short papers, and 17 poster papers that were selected by a peer review process. Papers were publicly solicited by a "Call for Papers" (with a submission deadline of January 10, 1996) that was widely circulated in the genetic programming, genetic algorithms, evolutionary programming, evolution strategies, artificial intelligence, artificial life, neural network, and machine learning communities. The review and selection of the papers was done separately for the four categories of papers at the conference—genetic programming, genetic algorithms, evolutionary programming and evolutionary strategies, and classifier systems.

David E. Goldberg of the University of Illinois at Urbana-Champaign chaired the reviewing process for the 23 papers submitted in the genetic algorithms category. Each paper was reviewed by him and two other members of the special committee for genetic algorithms appointed by him.

David B. Fogel of Natural Selection, Inc. in La Jolla, California chaired the reviewing process for the papers submitted on evolutionary programming and evolutionary strategies.

Rick L. Riolo of the Program for Study of Complex Systems at the University of Michigan chaired the reviewing process for the papers submitted on classifier systems. The chairs of the special committees on evolutionary programming and evolution strategies and classifier systems invited several papers in their particular subject areas in addition to the submitted papers; there are no invited genetic programming or genetic algorithms papers in these proceedings.

For the genetic programming papers, the peer reviewing of the papers was done by an intentionally large and broad-based program committee consisting of 88 members. All authors of any previously published scientific paper on genetic programming were solicited and encouraged to be members of this program committee. Each member of the program committee was given only about five papers to read and evaluate. Each reviewer completed a detailed 16-part review form for each paper. Since there were 88 submitted papers on genetic programming, each author received about five detailed reviews for his or her paper. There were a total of about 440 peer reviews received for the submitted papers on genetic programming. The tabulation of the results of the peer reviewing process by the program committee was done by an executive committee consisting of David Andre and Forrest H Bennett III and the GP-96 General Chair (John Koza), none of whom reviewed any papers. One of the 16 parts of the peer review form asked the reviewer to allocate a fixed budget of 100 points among the five papers that he reviewed (with an appropriate adjustment being made if the reviewer did not review precisely five papers). A peer reviewer could allocate a maximum of 40 points to any one paper (with the total of all of his allocated points being 100 for his five papers). This "applause meter" forced the reviewer to make sometimes-difficult "relative" overall choices among his papers. Another part of the peer review form asked the reviewer to make the more traditional "absolute" recommendation as to whether he thought that the paper should (definitely or probably) be accepted or rejected (with the peer reviewer being able to recommend acceptance of all, some, or none of his papers). An algorithm combining these "relative" and "absolute" scores was then used to rank the papers. The highest ranked 30 papers were selected to become the long papers. The 29 papers with the next highest rankings were selected to become the short papers. An additional group of papers with the next highest rankings were accepted as one-page poster papers. The remaining submissions were rejected. As it happened, the same 30 papers would have become the long papers if only the "relative" factor had been used in the algorithm. The executive committee and the General Chair briefly read many of the papers prior to the time of the final decision (particularly those that were at the likely boundaries between long papers, short papers, poster papers, and rejections and those that had received highly divergent rankings from the peer reviewers) and satisfied themselves as to the overall reasonableness of the outcome produced by the algorithm combining the peer reviewers' "relative" and "absolute" scores.

The authors of the long papers were allocated nine pages each in the proceedings, and these papers are grouped together in the proceedings (since several authors actually used as few as six pages in their camera-ready papers). The authors of the short papers were allocated six pages each in the proceedings and these papers are also grouped together in the proceedings. One author of two accepted short papers was permitted to unequally reallocate his 12 pages among his two papers; one short paper (whose author was injured just prior to the final deadline for camera-ready papers of April 12, 1996) is published here in its unshortened form. Each of peer reviewers affirmatively answered the following question as part of his 16-part review form:

I have read this paper. I am not at the same institution as any of the authors. I am not reviewing more than one paper for this conference by this author. I do not have an especially close relationship (or an antagonistic relationship) with the author(s). I am not re-reviewing this work (i.e., I have not, within the past year, reviewed substantially the same work by the same author). This review is my own opinion and work and has not been delegated to anyone (specifically including students or subordinates). I have not discussed my review of this paper with any other peer reviewer for this conference. I will regard any new intellectual content of this paper as confidential and will not disclose or use it until such time as it gets published under the author's name or otherwise becomes publicly known (and, if I discussed this review with anyone else at all, they have also explicitly agreed to maintain the author's work as confidential and I notified the General Chair of their agreement to do so). I do not feel that I am unqualified to review this paper.

The members of the executive committee and the General Chair (all of whom had themselves submitted papers to the conference) do not know the identity of the peer reviewers of any specific paper at the conference (including their own). This information was known only by the American Association for Artificial Intelligence (who handled the distribution of the papers to the peer reviewers based on subject matter).

Late-Breaking Papers

In order to provide conference attendees with information about research that was initiated, enhanced, improved, or completed after the original paper submission deadline of January 10, 1996, a rapidly-printed book containing papers describing late-breaking developments in the field of genetic programming is to be distributed to all attendees in addition to these official proceedings. The deadline for late-breaking papers is July 3, 1996. Late-breaking papers are to be briefly examined for minimum standards of acceptability and relevance, but are not peer reviewed. Late-breaking papers are to be presented during a poster or workshop session to be held during the conference. Arrangements will be made to enable others to purchase this book from the Stanford University Book Store after the conference.

Pre-Distribution of Conference Proceedings to Attendees

In order to enable conference attendees to use the conference for in-depth follow-up on the papers presented at the conference, copies of these proceedings are to be mailed to all attendees one or two weeks prior to the conference. This experimental effort is being limited to attendees with U. S. mailing addresses for this year's conference.

Use of Discussants for Each Paper at the Conference

In order to stimulate discussion of the papers at the conference, a discussant has been appointed to study each individual long and short paper prior to the conference. At the conference, each paper is to be presented in a 15-minute period (in either a plenary session or a parallel session) by one of the paper's authors. After the author has orally presented his paper, the discussant will present his prepared comments on the paper and discuss the paper with the presenting author during a 5-minute period. This will be followed by a 5-minute period in which the audience asks questions of the presenting author.

John R. Koza
Stanford University

David E. Goldberg
University of Illinois

David B. Fogel
Natural Selection, Inc.

Rick L. Riolo
University of Michigan

April 15, 1996

Chairs and Committees

Honorary Chair

John Holland, University of Michigan

Invited Speakers

John Holland, University of Michigan
David E. Goldberg, University of Illinois

General Chair

John R. Koza, Stanford University

Publicity Chair

Patrick Tufts, Brandeis University

Executive Committee

David Andre, Stanford University
Forrest H Bennett III, Stanford University

Tutorial Speakers

- Genetic Algorithms—David E. Goldberg, University of Illinois at Urbana-Champaign
- Machine Language Genetic Programming—Peter Nordin, University of Dortmund, Germany
- Genetic Programming using Mathematica—Robert Nachbar—Merck Research Laboratories
- Introduction to Genetic Programming—John Koza, Stanford University
- Classifier Systems—Robert Elliott Smith, University of Alabama
- Evolutionary Computation for Constraint Optimization—Zbigniew Michalewicz, University of North Carolina
- Advanced Genetic Programming—John Koza, Stanford University
- Evolutionary Programming and Evolution Strategies—David Fogel, University of California, San Diego
- Cellular Encoding—Frédéric Gruau, Stanford University (via videotape) and David Andre, Stanford University (in person)
- Genetic Programming with Linear Genomes—Wolfgang Banzhaf, University of Dortmund, Germany
- ECHO—Terry Jones, Santa Fe Institute
- Neural Networks—David E. Rumelhart, Stanford University
- Machine Learning—Pat Langley, Stanford University
- Molecular Biology for Computer Scientists—Russ B. Altman, Stanford University
- Evolvable Hardware—Hugo De Garis, ATR, Nara, Japan and Adrian Thompson, University of Sussex

Program Committee

Russell J. Abbott	California State University, Los Angeles and The Aerospace Corporation
Hojjat Adeli	Ohio State University

Dennis Allison	Stanford University
Lee Altenberg	Hawaii Institute of Geophysics and Planetology
David Andre	Stanford University
Peter J. Angeline	Loral Federal Systems
Wolfgang Banzhaf	University of Dortmund, Germany
Rik Belew	University of California at San Diego
Samy Bengio	INRS Telecommunications
Forrest H Bennett III	Stanford University
Tobias Blickle	Swiss Federal Institute of Technology
Scott Brave	Stanford University
Bill P. Buckles	Tulane University
Walter Cedeno	Primavera Systems Inc.
K. Govinda Char	University of Glasgow
Nichael Lynn Cramer	BBN System and Technologies
Mark J. Crosbie	Purdue University
Jason Daida	University of Michigan
Judith E. Devaney	NIST
Patrik D'haeseleer	University of New Mexico
Marco Dorigo	Université Libre de Bruxelles
Dimitris C. Dracopoulos	Brunel University
Bertrand Daniel Duñay	System Dynamics International
Garett Dworman	The Wharton School
Andrew N. Edmonds	Science in Finance Ltd., UK
H.H. Ehrenburg	CWI, The Netherlands
Gabe Ferrer	University of Virginia
Frank D. Francone	FRISEC
Adam P. Fraser	University of Salford
Alex Fukunaga	University of California, Los Angeles
Frédéric Gruau	Stanford University
Richard J. Hampo	Ford Motor Company
Simon Handley	Stanford University
Thomas D. Haynes	The University of Tulsa
Hitoshi Hemmi	ATR, Kyoto, Japan
Vasant Honavar	Iowa State University
Thomas Huang	University of Illinois
Hitoshi Iba	Electrotechnical Laboratory, Japan
Christian Andrew Johnson	Department of Economics, University of Santiago
Hugues Juille	Brandeis University
Martin A. Keane	Econometrics
Mike Keith	Allen Bradely Controls
Maarten Keijzer	Cap Volmac, Adaptive Systems bv.
Steven O. Kimbrough	The Wharton School
Kenneth E. Kinnear, Jr.	Adaptive Computing Technology
James D. Laing	The Wharton School
W. B. Langdon	University College, London
David Levine	Argonne National Laboratory
Brij Masand	GTE Laboratories
Kenneth Marko	Ford Motor Company
Martin C. Martin	Carnegie Mellon University
Sidney R. Maxwell III	Applied Microsystems, Inc

Nicholas Freitag McPhee	University of Minnesota, Morris
Justin D. Miller	Interactive Television Program
David Montana	BBN System and Technologies
Heinz Muehlenbein	GMD Research Center, Germany
Robert B. Nachbar	Merck Research Laboratories
Peter Nordin	University of Dortmund, Germany
Howard Oakley	Institute of Naval Medicine, UK
Franz Oppacher	Carleton University, Ottawa
Una-May O'Reilly	Carleton University, Ottawa
Michael Papka	Argonne National Laboratory
Timothy Perkis	n/a
Frederick E. Petry	Tulane University
Peter Olsen	Department of Defense
Jordan Pollack	Brandeis University
Bill Punch	Michigan State University
Simon Raik	Monash University, Melbourne
Justinian P. Rosca	University of Rochester
Conor Ryan	University College Cork, Ireland
Dale A. Schoenefeld	University of Tulsa
Gregory Seront	Free University of Brussels
Malcolm Shute	University of Brighton
Eric V. Siegel	Columbia University
Karl Sims	n/a
Andrew Singleton	Creation Mechanics
Lee Spector	Hampshire College
Walter Alden Tackett	Neuromedia
Astro Teller	Carnegie Mellon University
Marco Tomassini	Ecole Polytechnique Federale de Lausanne
Patrick Tufts	Brandeis University
V. Rao Vemuri	University of California at Davis
Peter A. Whigham	Australia
Darrell Whitley	Colorado State University
Man Leung Wong	Chinese University of Hong Kong
Alden H. Wright	University of Montana
Byoung-Tak Zhang	GMD, Germany
Afra Zomorodian	Stanford University

Program Committee for Genetic Algorithms

David E. Goldberg, Chair	University of Illinois—Urbana-Champaign
Kalyanmoy Deb	Indian Institute of Technology—Kanpur
Charles L. Karr	University of Alabama
Sushil J. Louis	University of Nevada—Reno
Laurence D. Merkle	Air Force Institute of Technology
Liwei Wang	University of Illinois—Urbana-Champaign

Program Committee for Classifier Systems

Rick L. Riolo, Chair	University of Michigan

Program Committee for Evolutionary Programming and Evolution Strategies

David B. Fogel, Chair	Natural Selection, Inc., La Jolla
Peter J. Angeline	Loral Federal Systems
Charles Clelland	Deakin University, Australia
John McDonnell	San Diego State University
Don Waagen	TRW

GENETIC PROGRAMMING

Long Genetic Programming Papers

Discovery by Genetic Programming of a Cellular Automata Rule that is Better than any Known Rule for the Majority Classification Problem

David Andre
Visiting Scholar
Computer Science Dept.
Stanford University
860 Live Oak Ave, #4
Menlo Park, CA 94025 USA
andre@flamingo.stanford.edu

Forrest H Bennett III
Visiting Scholar
Computer Science Dept.
Stanford University
Stanford, California 94305
fhb3@slip.net

John R. Koza
Computer Science Dept.
258 Gates Building
Stanford University
Stanford, California 94305
koza@cs.stanford.edu

ABSTRACT

It is difficult to program cellular automata. This is especially true when the desired computation requires global communication and global integration of information across great distances in the cellular space. Various human-written algorithms have appeared in the past two decades for the vexatious majority classification task for one-dimensional two-state cellular automata. This paper describes how genetic programming with automatically defined functions evolved a rule for this task with an accuracy of 82.326%. This level of accuracy exceeds that of the original 1978 Gacs-Kurdyumov-Levin (GKL) rule, all other known human-written rules, and all other known rules produced by automated methods. The rule evolved by genetic programming is qualitatively different from all previous rules in that it employs a larger and more intricate repertoire of domains and particles to represent and communicate information across the cellular space.

1. Introduction

Massively parallel computation is potentially attractive for solving problems. However, it has proved difficult to program massively parallel computing devices.

This paper describes our research on using genetic programming to evolve programs for cellular automata, which are one type of massively parallel computing system. Specifically, this paper focuses on the well-studied majority classification task for one-dimensional two-state cellular automata. For this problem, genetic programming evolved a rule whose performance exceeds that of the original Gacs-Kurdyumov-Levin (GKL) rule, all other known subsequent human-written rules, and all known rules produced by automated approaches for this problem (table 1).

The result in this paper is that genetic programming has produced a result that is slightly better than human performance for this particular problem. This result joins several similar results recently obtained by both our group and other researchers in applying genetic programming to non-trivial problem domains. The transmembrane segment identification problem is another problem where genetic programming is competitive with human performance. The goal in that problem is to classify a given protein segment (i.e., a subsequence of amino acid residues from a protein sequence) as being a transmembrane domain or non-transmembrane area of the protein. The goal is to perform the classification without using the biochemical foreknowledge concerning hydrophobicity that is typically used in human-written algorithms for this problem. Four different versions of genetic programming have been applied to this problem (Koza 1994a; Koza and Andre 1996a, 1996b). The performance of all four versions using genetic programming is slightly superior to that of algorithms written by knowledgeable human investigators.

A second example involves the problem of automatically discovering patterns in databases of DNA and protein sequences. Genetic programming successfully evolved motifs for detecting the D-E-A-D box family of proteins and for detecting the manganese superoxide dismutase family. Both motifs were evolved without prespecifying their length. Both relied on automatically defined functions to capture common subexpressions.

Table 1 Out-of-sample performance of various rules for the majority classification task.

Rule	Accuracy	Test cases
Best rule evolved by bit-string genetic algorithm (Das, Mitchell, and Crutchfield 1994)	76.9%	10^6
GKL 1978 human-written	81.6%	10^6
Davis 1995 human-written	81.8%	10^6
Das 1995 human-written	82.178%	10^7
Best rule evolved by genetic programming (in this paper)	82.326%	10^7

The two genetically evolved consensus motifs detect the two families either as well as, or slightly better than, the comparable human-written motifs found in the PROSITE database (Koza and Andre 1996c).

A third instance involves the automated discovery of an analog electrical circuit for a difficult-to-design asymmetric bandpass filter (Koza, Bennett, Andre, and Keane 1996 in this volume). Other instances have been reported in Kinnear 1994 and Angeline and Kinnear 1996.

Section 2 of this paper provides background on cellular automata. Section 3 presents the majority classification task and previous work on it. Section 4 details the steps required to use genetic programming on the majority classification problem. Section 5 examines the results of our most successful run using genetic programming. In section 6, we examine, in detail, the intricate communication patterns employed by the best rule evolved by genetic programming for this problem.

2. Cellular Automata

Local rules govern the important interactions of many animate and inanimate entities. It has been frequently observed that the simultaneous execution of a single relatively simple rule at many local sites leads to the emergence of interesting and complex global behavior (Langton 1989). This global behavior often occurs when entities that have access only to information about their immediate local environment engage in long-distance communication and long-distance integration of information.

Cellular automata are an abstract way of studying and analyzing the simultaneous execution of local rules (Burks 1970; Farmer, Toffoli, and Wolfram 1983; Wolfram 1986; Gutowitz 1991; Wuensche and Lesser 1992).

A cellular space is a uniform array of cells arranged in a certain topological arrangement in a certain number of dimensions. In a cellular automaton (CA), each cell in a cellular space is occupied by an identical automaton (each with its own current state). The next state of each individual automaton in the cellular space depends on its own current state and on the current states of the other automata in a specified local neighborhood around the individual automaton. The state of each automaton at time 0 is called its *initial condition.*

The best-known cellular automata system is John Conway's game of *Life* (Berlekamp, Conway, and Guy 1985). It involves a two-dimensional arrangement of identical two-state automata. In the game of *Life*, the next state of each automaton depends on its current state and on the states of its eight immediately adjacent neighbors in the two-dimensional cellular space.

For a one-dimensional cellular automata, the cellular space is a linear arrangement of identical automata. The next state of each individual automaton might depend on the current state of that automaton and the current states of its neighbors at a particular specified distance. For example, if the distance is three, we denote the seven states that determine the next state of a given automaton as `X` (for the automaton at the center), `W` (the adjacent automaton to the west), `E` (the adjacent automaton to the east), `WW` (the automaton at distance 2 to the west), `EE`, `WWW`, and `EEE`. A cellular space is said to have *periodic boundary conditions* when the cellular space is toroidal. If the automaton located in each cell has only two states, the state-transition function of the automaton is a Boolean function of its own current state and the states of its neighbors at the specified distance.

Cellular automata are dynamical systems that are discrete in time, in space (the cells), and in site value (the states). Cellular automata are the discrete counterparts of continuous dynamical systems defined by partial differential equations and the physicist's concept of field (Toffoli and Margolus 1987). Simulations of fluid flow and other complex processes that depend only on local states are well suited to cellular automata.

It is extremely difficult, in general, to design a single state-transition rule that, when it operates in each cell of the cellular space, produces a desired global behavior.

Genetic algorithms (Holland 1975) operating on fixed-length character strings have been previously used to evolve the initial conditions and state-transition rules for cellular automata (Meyer, Richards, and Packard 1991). Genetic programming (Koza 1992, 1994a, 1994b; Koza and Rice 1992) has been previously used for automatic programming of cellular automata (Koza 1992).

3. The Majority Classification Problem

The majority classification problem is one vehicle for exploring how complex calculations can be performed over large expanses of time and space using rules of interaction that operate over a relatively small distance.

In one frequently studied version of this vexatious problem, there is a one-dimensional linear arrangement of 149 two-state automata whose update rule operates on information within a distance of three. The initial states of all 149 automata (called the *initial configuration*) are the inputs to the calculation. If all 149 automata relax to a common state (0 or 1) after a certain amount of time, the common state is considered to be the binary output of the calculation. For the majority classification task, a "0" is correct if a majority of the 149 initial bits are 0 and a "1" is correct if a majority of the bits are 1. Thus, to solve this problem, a seven-argument Boolean transition rule must be found such that when it is situated at all 149 cells of a one-dimensional cellular automata system, the automata converge to the correct configuration of 149 0's or 149 1's (after, say, 600 time steps). The fitness of a rule for this task is measured by its ability to correctly classify any arbitrary random initial state.

It is difficult to construct a two-state, seven-neighborhood cellular automata rule that performs this majority classification task on a 149-bit input configuration. Seven bits cannot store an integer as large as 149 (if, for example, one were storing and transmitting a signal communicating a locally observed surplus or deficit of 0's or 1's). It is unknown whether a perfect solution to this problem exists.

The difficulty that human programmers have had with this task is indicated by its history. In 1978, Gacs, Kurdyumov, and Levin developed a two-state, seven-neighbor rule for the purpose of studying reliable computation under random perturbations. This Gacs-Kurdyumov-Levin (GKL) rule performs the majority classification task reasonably well. The GKL rule is successful on 81.6% of the inputs consisting of a configuration of 149 bits (each chosen independently with 50% probability). Gonzaga de Sa and Maes (1992) showed that this system does indeed relax to a common state.

The majority classification task has been the subject of extensive study (Mitchell, Hraber, and Crutchfield 1993; Das, Mitchell, and Crutchfield 1994; Mitchell, Crutchfield, and Hraber 1994; Crutchfield and Mitchell 1995; Das, Crutchfield, Mitchell, and Hanson 1995; Mitchell 1996).

In 1993, Lawrence Davis (1995) cleverly modified the GKL rule and created a rule that achieved an accuracy of about 81.8%. Rajarshi Das(1995) created another clever rule that achieved slightly better accuracy than the GKL rule and the Davis rule. The Das rule has an accuracy of about 82.178%. Table 2 shows the GKL rule, the Davis rule, the Das rule, and the rule evolved by genetic programming in this paper. The 128 bits are presented in the natural order that they would appear in a state transition table starting with state 0000000 and ending at state 1111111.

Das, Mitchell, and Crutchfield (1994) evolved rules using a version of the genetic algorithm operating on fixed-length strings. These rules sometimes exhibited qualitatively the same behavior as the GKL rule. However, none of the rules evolved using the genetic algorithm operating on fixed-length strings were as accurate as the original GKL rule. As Das, Mitchell, and Crutchfield (1994) reported, the genetic algorithm usually found only relatively uninteresting block-expanding rules that score in the range of 65 – 70% accuracy. The best result from the genetic algorithm operating on fixed-length strings had 76.9% accuracy. This reported performance may have been the consequence of factors such as the small population size employed.

Land and Belew (1995) suggested that evolution may actually be hindered by the standard representation of the cellular automata rule as a chromosome string of length 128 used in previous work using the genetic algorithm. They suggest that higher-level representations (such as condition-action pairs) may aid the evolutionary process because of the higher degree of locality in the condition-action pairs. Given Land and Belew's (1995) view, the tree representation employed by genetic programming seems well suited for this task. The size and shape of the ultimate solution is free to undergo evolution in genetic programming. The search strategy employed by genetic programming (and the genetic algorithm) is also important for this problem.

Although many Boolean problems can be trivially solved by hillclimbing methods in a variety of representations, such methods work only when the fitness cases (i.e., the 2^7 lines in the truth table) are independent. When a Boolean function is used for a cellular automata rule, there is no correct answer for a given set of inputs that is independent from the answers for the other inputs. Also, genetic programming supports automatically defined functions, whereas the conventional genetic algorithm operating on fixed-length strings does not have a similar facility for exploiting regularities, symmetries, homogeneities, and modularities of the problem environment.

4. Preparatory Steps

The runs reported here used standard genetic programming with automatically defined functions (Koza 1994a) as summarized in table 3. The problem (coded in ANSI C) was run on a medium-grained parallel Parsytec computer system consisting of 64 80-MHz Power PC 601 processors arranged in a toroidal mesh with a host PC Pentium type computer. The so-called *distributed genetic algorithm* or *island model* for parallelization was used. That is, subpopulations (*demes*) were situated at the processing nodes of the system. Population size was Q = 800 at each of the D = 64 demes for a total population size of 51,200. The initial random subpopulations were created locally at each processing node. Generations were run asynchronously on each node. After a generation of genetic operations was performed locally on each node, four boatloads, each consisting of B = 3% (the migration rate) of the subpopulation (selected on the basis of fitness) were dispatched to each of the four toroidally adjacent nodes. Details of the parallel implementation of genetic programming can be found in Andre and Koza 1996.

Table 2 The succession of "best" cellular automata rules for the majority classification task, presented in truth table order from 0000000 to 1111111 (i.e. 0 to 127).

Rule	State Transitions
GKL 1978 human-written	00000000 01011111 00000000 01011111 00000000 01011111 00000000 01011111 00000000 01011111 11111111 01011111 00000000 01011111 11111111 01011111
Davis 1995 human-written	00000000 00101111 00000011 01011111 00000000 00011111 11001111 00011111 00000000 00101111 11111100 01011111 00000000 00011111 11111111 00011111
Das (1995) human-written	00000111 00000000 00000111 11111111 00001111 00000000 00001111 11111111 00001111 00000000 00000111 11111111 00001111 00110001 00001111 11111111
Best rule evolved by genetic programming (in this paper)	00000101 00000000 01010101 00000101 00000101 00000000 01010101 00000101 01010101 11111111 01010101 11111111 01010101 11111111 01010101 11111111

Table 3 **Tableau for the majority classification problem for one-dimensional cellular automata.**

Objective:	Find a seven argument Boolean function that performs the majority classification problem for a 149-width one-dimensional cellular automata system.
Architecture of the overall program with ADFs:	One result-producing branch and one two-argument automatically defined function, ADF0, and one three-argument automatically defined function, ADF1. ADF1 can refer to ADF0.
Terminal set for the RPB:	X, E, EE, EEE, W, WW, and WWW.
Function set for the RPB:	AND, OR, NAND, NOR, NOT, IF, XOR, ADF0, and ADF1.
Terminal set for ADF0:	X, E, EE, EEE, W, WW, WWW, ARG0, and ARG1.
Function set for ADF0:	AND, OR, NAND, NOR, NOT, IF, and XOR.
Terminal set for ADF1:	X, E, EE, EEE, W, WW, WWW, ARG0, ARG1, and ARG2.
Function set for ADF1:	AND, OR, NAND, NOR, NOT, IF, XOR, and ADF0.
Fitness cases:	• 1,000 149-bit initial configurations were used as in-sample fitness cases. These initial configurations were created randomly, with no bias (i.e., 0 and 1 each have an independent 50% probability of being chosen). Thus, the distribution of densities of the initial state vectors was a binomial distribution centered at 0.50. • The out-of-sample fitness cases consisted of 1,000,000 (and later 10,000,000 and 15,000,000) similarly created initial configurations.
Raw fitness:	1,000 minus the number of fitness cases for which the system relaxes to the correct configuration after 600 time steps.
Standardized fitness:	1,000 minus raw fitness.
Hits:	Raw fitness.
Parameters:	• Population size, *M*, is 51,200 (64 times 800). • Maximum number of generations to be run, *G*, is 51. • 89% crossovers, 10% reproductions, and 1% mutations were used on each generation. • A maximum of 500 points (functions and terminals) for the result-producing branch and 250 points for each automatically defined function was allowed. • Structure-preserving crossover with branch typing was used. • The other parameters for controlling the runs of genetic programming were the default values specified in Koza (1994a).

5. Emergent Computation in the Most Successful Run

We made five runs of genetic programming on this problem. Each run produced numerous individuals that were reminiscent of the GKL rule. Each of the runs produced numerous individuals that scored well above the best accuracy of 76.9% produced by the genetic algorithm operating on fixed-length character strings. The best rule evolved by genetic programming scores better than any known human-written or machine-produced rule (table 1).

An examination of the most successful run illustrates the emergence of many interesting entities prior to the creation of the best individual.

The best individual program from the initial random generation indiscriminately classifies all fitness cases as having a majority of 1's. Since 525 of the 1,000 randomly created fitness cases on this particular run happened to have a majority of 1's, this best of generation 0 scores 525 hits.

The best individual of generation 1 scores 650 hits. The activity of a one-dimensional cellular automata system can be presented as a two dimensional grid in which the top horizontal row contains the states (0 or 1) of the 149 automata at a time 0 (i.e., the initial configuration of the system) and in which each successive row represents the states of the 149 automata at successive time steps. Figure 1 shows a small part of such a diagram for this individual. In the space-time diagram, we see large areas dominated by solid blocks of repeated simple regular patterns, called *domains*, where a given domain is specified by a regular expression. The two domains shown in figure 1 are the domain denoted by the regular expression 1^* and the domain denoted by the expression 0^*. The space-time diagrams in this paper display 1's as black, and 0's as white. Thus we call the 1^* domain black (designated by (**B**)) and the 0^* domain white (designated by (**W**)).

The domains of this generation 1 individual interact in various ways. When a domain consisting of a solid block of 1's is to the left of a domain consisting of a solid block of 0's, the interface consists of ...111000... and the corresponding positions at the next time step also consist of ...111000... . Because these particular two domains do not move left or right over time, this particular domain interface is said to have a velocity of zero.

Following Das, Mitchell, and Crutchfield (1994), we describe such an interface between two domains as a particle, denoted P(xy), where x is the domain on the left and y the domain on the right. Particles are one of the ways that information is communicated across large distances in time and space in a cellular space. In figure 1, line **a** shows the path of the P(BW) particle with a velocity of 0 and line **b** shows the path of the P(WB) particle with a velocity of 1/2.

```
...1111|00000000000000000\11111111111...
...1111|0000000000000001 0\1111111111...
...1111|000000000000000000\111111111...
...1111|0000000000000000101\11111111...
...1111|0000000000000000000\11111111...
...1111|00000000000000000100\1111111...
...1111|000000000000000000000\1111111...
...1111|000000000000000000100\1111111...
...1111|0000000000000000000000\111111...
...1111|00000000000000000001 00\111111...
...1111|00000000000000000000000\11111...
        a                       b
```

Figure 1 Eleven time steps of the partial space-time behavior of the best individual of generation 1 on one fitness case showing the spread of the zero domain.

This evolved rule from generation 1 overpredicts a majority of 1's and underpredicts a majority of 0's. This individual is an example of what Mitchell, Hraber, and Crutchfield (1993) called a *block-expanding rule* (i.e. a rule that converges to a state unless a sufficiently large block of adjacent or nearly adjacent instances of the opposite state exists in the input).

The best program of generation 6 scores 706 hits on the 1,000 in-sample fitness cases. A portion of its space-time behavior is shown in figure 2. The behavior of this individual is similar to the individual of generation 1. However, it scores slightly better because it has modified the conditions under which it expands a block. The interaction between the domain of zeros on the left and ones on the right is quite complex in this individual, and the interface between the two domains is quite large. The particle $P_{(WB)}$ travels at a velocity of 2/7 in this individual.

The best individual of generation 15 scores 815 hits. Its behavior (see figure 3) is somewhat similar to that of the GKL rule, although it scores less than 80% on out-of-sample fitness cases. Like the GKL rule (figure 7), this rule separates white on the left from black on the right with a gray domain that grows into both black and white. When black is to the left of white, a zero-velocity particle characterizes their interaction. The combination of these two concepts yields the basic mechanism that both GKL and this generation 15 individual use to compute global measures of density. If the domain of all white is larger, it will win out over black, and vice versa.

Figure 2 Partial space-time behavior of the best individual of generation 6. Black represents 1s, white 0s.

Although similar to GKL, the rule of generation 15 has several new gray domains that are not quite the same as the gray domain in the GKL rule's behavior. The new domains interact, as can be seen in figure 4. In this example, the zero and one domains do not enter into the computation until the very end – all of the primary computation is performed by the new gray domains. At this point, however, the new gray domains do not interact particularly well – the rule fails often when it computes with the new gray domains.

The best individual of the run emerged on generation 17. The 36-point result-producing branch is

```
(nand (xor (or (adf0 X (not WWW)) (not WWW))
(adf0 (adf0 E X) (or E W))) (or (nand (if EEE E
X) (if EEE W (and EE WWW))) (adf1 (nand W E)
(xor WWW W) E))).
```

The 10-point automatically defined function ADF0 is

```
(if (nor (or X X) (nor arg1 arg0)) EEE WWW).
```

The 44-point automatically defined function ADF1 is

```
(xor W (xor (adf0 (adf0 (if (or (not W) (adf0
WWW arg2)) (xor (adf0 WWW X) (nor WW WWW)) (nand
(xor X X) (or X WW))) (nor (or (if arg2 EEE
arg2) (nor arg2 arg1)) (not (xor W E)))) (xor W
(not arg2))) EEE)).
```

This rule scores 824 hits on the 1,000 in-sample fitness cases, and scores 82.4% accuracy over 100,000 out-of-sample cases. It scores 82.326% accuracy over 10,000,000 additional cases.

This accuracy is slightly better than the score of the GKL rule. We are aware of two human-coded rules that also are better than the GKL rule – the Das rule (1995) and the Davis rule (1995) (table 2). We tested the best individual evolved by genetic programming against each of the other three rules using the same fitness cases (10^6 10^7 fitness cases). Table 1 shows the results. As can be seen, the best rule evolved by genetic programming is slightly better than all the other rules. Non-parametric $\chi 2$ tests with one degree of freedom were performed; the probability(p) that the

Figure 3 Space-time behavior of the best individual of generation 15 on a typical initial state vector.

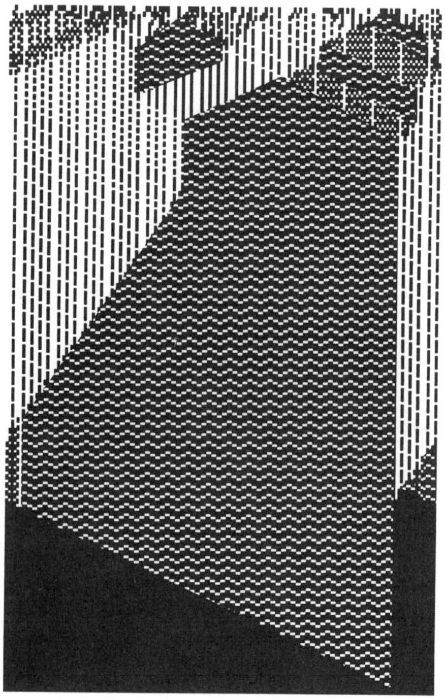

Figure 4 Space-time behavior of the best individual of generation 15 where there are no all black or all white domains until the end.

pairwise differences between the best rule evolved by genetic programming and each of the other three rules were attributable to chance was less than 0.001 (i.e. the differences are statistically significant, $p < 0.001$).

6. The Best Rule Evolved by Genetic Programming

Why does the best rule evolved by genetic programming score better than the previously known (human-written) rules? First, it has more domains than the GKL rule. The 11 domains of the rule evolved by genetic programming (table 4) classify the density of 1's into finer levels of gray than do the black, white, and checkerboard gray domains of the GKL rule (analyzed in detail in Das, Mitchell, and Crutchfield 1994). The checkerboard gray domain of the GKL rule corresponds to the regular expression $(10)^* \cup (01)^*$ (Das, Mitchell, and Crutchfield 1994). In addition, the rule evolved by genetic programming employs a larger number of particles. The rule evolved by genetic programming employs 10 particles that are identical to the particles (Das, Mitchell, and Crutchfield 1994) in the GKL rule; however, it also discovered at least 40 additional particles that involve the new gray domains. All the new particles have velocity 0, +3, or -3.

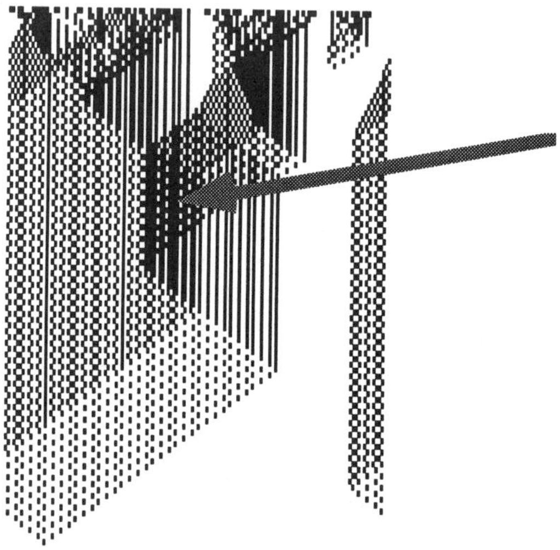

Figure 5 Space-time diagram showing the behavior of the best rule evolved by genetic programming. The differences between the shades of gray and the area highlighted by the arrow are important to convergence.

More domains would be purely a superficial difference if the new domains did not participate in the computation. However, the new gray domains are crucial to the computations being performed by the genetically evolved rule. In the behavior shown in figure 5, for example, the new grays are critical to the success of the rule. If all the grays acted identically, then the automata would not converge in this example. The interactions of the new gray domains in the area indicated by the arrow allow the white domain to escape and encroach to the left.

In addition, there are periods (as shown in figure 6) during the behavior of the best rule evolved by genetic programming where none of the GKL domains exist. In fact, there are periods in which the computation is carried out entirely by the new gray domains.

Even though the new gray domains take part in the

Figure 6 The behavior of the best rule evolved by genetic programming on one set of initial states.

computation, one might question whether they account for the difference in score between the best rule evolved by genetic programming and the GKL rule.

However, the behavior of both rules shown in figure 7 indicates that the finer levels of gray do confer an advantage on the best rule evolved by genetic programming. On the left of the initial state vector, there are two fairly large regions of white (0s) separated by a region of black (1s) marked in the diagrams by the arrows. Under the GKL rule, the black area is eliminated by the two adjacent white domains. In the best rule evolved by genetic programming, these two white areas are classified as light gray domains and the black area becomes a dark gray domain (all shifted to the right slightly). Because it can make use of extra domains, the best rule evolved by genetic programming retains the information that there is a significant density of black (1s) in the vicinity of the arrow until additional global information can resolve the local dispute.

The basic mechanism of best rule evolved by genetic programming is similar to that of the GKL rule in that there is a race among the black, white, and gray domains. Figure 8 shows a space-time diagram of the behavior of the best rule evolved by genetic programming.

In the best rule evolved by genetic programming, a white domain (**W**) is separated on the right from a black domain (**B**) by a growing domain of checkerboard gray (**6**). The white domain (**W**) is separated on the left from the black domain (**B**) by one or both of the separator domains, (**3**) and (**8**), which are domains whose interactions with either white or black have zero velocity. In the case shown in figure 8, the (**B**) domain is larger, and the gray (**6**) domain cuts off the white domain (**W**), allowing the black domain (**B**) to break through. In this case, the extra gray domains of (**7**) and (**9**) are not important, but they play a role in other fitness cases.

The mechanism discussed above is sufficient for many cases, but fails to handle the case when the black and white domains are of approximately equal size. In this circumstance, the gray (**6**) domain cuts off both the white and black domains. In the behavior of the GKL rule, when there is such a 'tie', after a brief transition, the black and white domains seem to 'swap' places, separated by a new growing gray domain (figure 9).

In the behavior of the best rule evolved by genetic programming such collisions occur slightly differently and yield different behavior after the collision. Under this rule, the black (**B**) and white (**W**) domains are separated by a

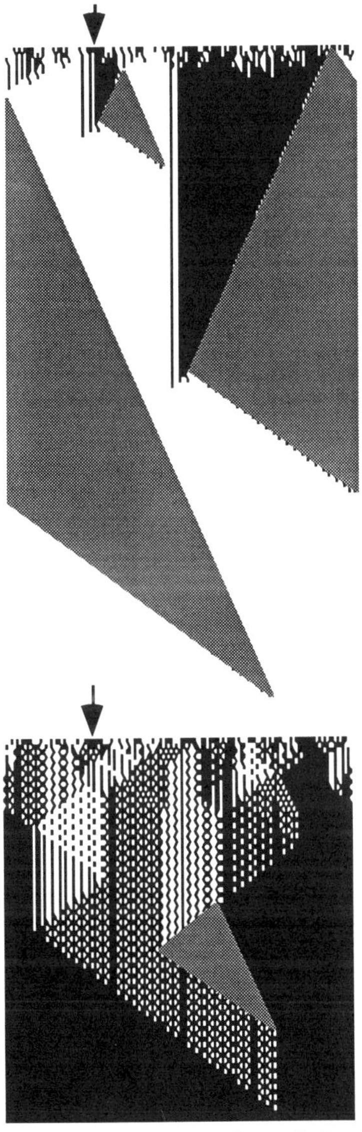

Figure 7 The space-time behavior of the best rule evolved by genetic programming (bottom) and the GKL rule (top) on the same initial state vector. The ability of the evolved rule to classify domains of the input as intermediate gray values allows it to avoid the 'round-off' error that the GKL makes.

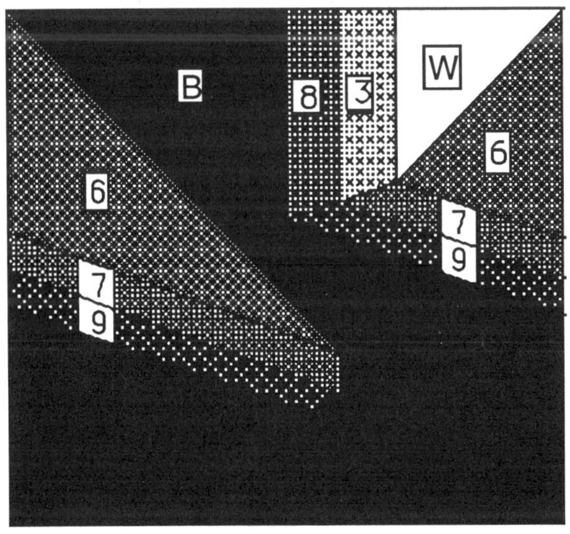

Figure 8 Abstracted Space-Time domain diagram for the best rule evolved by genetic programming showing its typical behavior. The darkness of each domain corresponds to the density of ones.

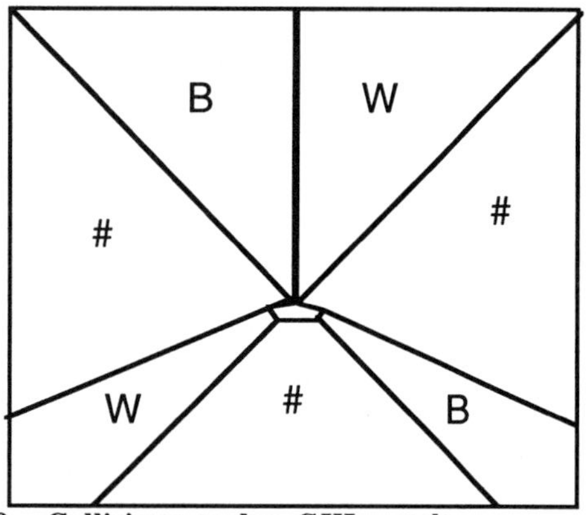

Figure 9 Collisions under GKL rule.

potentially larger distance by one or both of the separator domains (**8**) and (**3**). If there is a tie, new domains emerge after the collision instead of the black and white domains, and there is no immediate gray separator domain. The new gray domains of (**7**) and (**9**) appear to represent the continuation of the black (**B**) domain, whereas the (**2**) and (**1**) domains appear to represent the continuation of the white (**W**) domain. Thus, when there is a tie in the best rule evolved by genetic programming; the original domains of black and white disappear completely; and the computation is carried out entirely by the gray domains. The abstract schematic for this interaction is shown in figure 10.

One problem that a rule utilizing zero-velocity particles can have is that the initial state vector could either be in a state or relax to a state where convergence does not take place. If the automata either starts in or reaches a state where the automata is filled with domains that have zero-velocity interactions with the adjacent patterns, the automata cannot converge. Thus, one potential downside to the best rule evolved by genetic programming is that because it utilizes more domains in its computation that have zero-velocity particles, it may be more likely to be trapped in a non-convergent steady-state. We found only one such input pattern in our testing (see figure 11). Given that the best rule evolved by genetic programming scores better than any other known rule, however, it is possible that the occasional

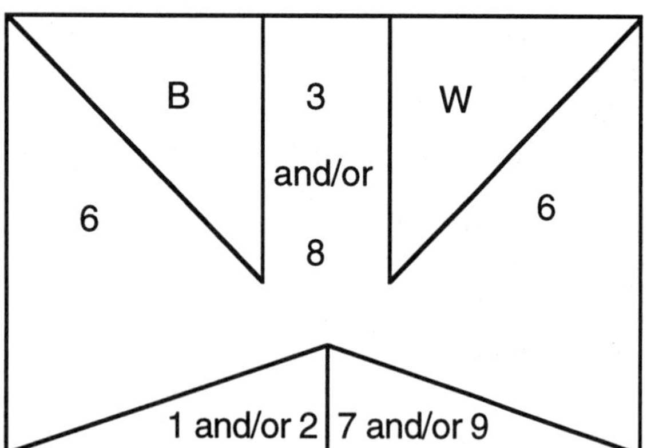

Figure 10 Collisions under the best rule evolved by genetic programming.

Figure 11 The behavior of the best rule evolved by genetic programming on an initial state vector where the automata does not converge.

non-convergence represents a strategic tradeoff rather than a defect.

The asymmetry of the best rule evolved by genetic programming suggests that there may be other rules that are better yet at performing the majority classification task.

7. Conclusions

A cellular automata rule for the majority classification task was evolved using genetic programming with automatically defined functions. This genetically evolved rule has an accuracy of 82.326%. This accuracy exceeds that of the Gacs-Kurdyumov-Levin (GKL) rule, all other known human-written rules, and all other known rules produced by previous automated approaches. The best rule evolved by genetic programming differs qualitatively from these other rules. It utilizes a very fine internal representation of density information; it employs a large number of different domains and particles; and it uses an intricate set of signals for communicating information over large distances in time and space.

Acknowledgements

The authors gratefully thank Lawrence Davis, Melanie Mitchell, and James Crutchfield for helpful discussions and for reviewing a draft of this paper. Simon Handley and Scott Brave made helpful comments on drafts of this paper.

Bibliography

Andre, David and Koza, John R. 1996. Parallel genetic programming: A scalable implementation using the transputer architecture. In Angeline, Peter J. and Kinnear, Kenneth E. Jr. (editors). *Advances in Genetic Programming 2*. Cambridge, MA: The MIT Press.

Angeline, Peter J. and Kinnear, Kenneth E. Jr. (editors). *Advances in Genetic Programming 2*. Cambridge, MA: The MIT Press.

Burks, Arthur W. 1970. *Essays on Cellular Automata*. Urbana, IL: University of Illinois Press.

Crutchfield, J. P. and Mitchell, Melanie 1995. The evolution of emergent computation. *Proceedings of the National Academy of Sciences, USA*. 92 (23).

Das, Rajarshi, Mitchell, Melanie, and Crutchfield, J. P. 1994. A genetic algorithm discovers particle-based computation in cellular automata. In Davidor, Yuval, Schwefel, Hans-Paul, and Maenner, Reinhard (editors). 1994. *Parallel Problem Solving from Nature - PPSN III*. (Lecture Notes in Computer Science, Volume 866). Berlin: Springer-Verlag. 344-353.

Das, Rajarshi, Crutchfield, J. P., Mitchell, Melanie, and Hanson, J. E. 1995. Evolving globally synchronized cellular automata. In Eshelman, Larry J. (editor). *Proceedings of the Sixth International Conference on Genetic Algorithms*. San Francisco, CA: Morgan Kaufmann Publishers.

Das, Rajarshi. 1995. Personal communication.

Davis, Lawrence. 1995. Personal communication.

Farmer, Doyne, Toffoli, Tommaso, and Wolfram, Stephen (editors). 1983. *Cellular Automata: Proceeding of an Interdisciplinary Workshop, Los Alamos, New Mexico, March 7-11, 1983*. Amsterdam: North-Holland Physics Publishing. Also in *Physica D*, volume 10.

Gacs, P., Kurdyumov, G. L., and Levin, L. A. 1978. One dimensional uniform arrays that wash out finite islands. *Problemy Peredachi Informatsii*. 12(1978) 92 – 98.

Gardner, Martin 1970. The fantastic combinations of John Conway's new solitaire game "Life." *Scientific American*. 223(April): 120–123.

Gonzaga de Sa, P. and Maes. C. 1992. The Gacs-Kurdyumov-Levin automaton revisited. *Journal of Statistical Physics*. 67(3/4) 507–522.

Gutowitz, Howard (editor). 1991. *Cellular Automata: Theory and Experiment*. Cambridge, MA: The MIT Press. Also in *Physica D* 1990.

Holland, John H. 1975. *Adaptation in Natural and Artificial System*. Ann Arbor, MI: University of Michigan Press. The 1992 second edition was published by The MIT Press.

Kinnear, Kenneth E. Jr. (editor). *Advances in Genetic Programming*. Cambridge, MA: The MIT Press 1994.

Koza, John R. 1992. *Genetic Programming: On the Programming of Computers by Means of Natural Selection*. Cambridge, MA: The MIT Press.

Koza, John R. 1994a. Genetic Programming II: Automatic Discovery of Reusable Programs. Cambridge, MA: The MIT Press.

Koza, John R. 1994b. *Genetic Programming II Videotape: The Next Generation*. Cambridge, MA: The MIT Press.

Koza, John R. and Andre, David. 1996a. Classifying protein segments as transmembrane domains using architecture-altering operations in genetic programming. In Angeline, Peter J. and Kinnear, Kenneth E. Jr. (editors). 1996. *Advances in Genetic Programming II*. Cambridge, MA: MIT Press. In Press.

Koza, John R. and Andre, David. 1996b. Evolution of iteration in genetic programming. In *Evolutionary Programming V: Proceedings of the Fifth Annual Conference on Evolutionary Programming*. Cambridge, MA: MIT Press. In Press.

Koza, John R. and Andre, David. 1996c. Automatic discovery of protein motifs using genetic programming. In Yao, Xin (editor). 1996. *Evolutionary Computation: Theory and Applications*. Singapore: World Scientific. In Press.

Koza, John R., Bennett III, Forrest H, Andre, David, and Keane, Martin A. 1996. Automated WYWIWYG design of both the topology and component values of analog electrical circuits using genetic programming. In Koza, John R., Goldberg, David E., Fogel, David B., and Riolo, Rick L. (editors). *Genetic Programming 1996: Proceedings of the First Annual Conference, July 28-31, 1996, Stanford University*. Cambridge, MA: MIT Press. In this volume.

Koza, John R., and Rice, James P. 1992. *Genetic Programming: The Movie*. Cambridge, MA: MIT Press.

Land, Mark and Belew, Richard K. 1995. In McDonnell, John R., Reynolds, Robert G., and Fogel, David B. (editors). *Evolutionary Programming IV: Proceedings of the Fourth Annual Conference on Evolutionary Programming*. Cambridge, MA: The MIT Press. Pages 403–434.

Langton, Christopher G. (editor). 1989. *Artificial Life, Santa Fe Institute Studies in the Sciences of Complexity*. Volume VI. Redwood City, CA: Addison-Wesley.

Meyer, Thomas P., Richards, Fred C., and Packard, Norman H. 1989. A learning algorithm for modeling. *Physical Review Letters*. 63(16) 1735-1738. October 16.

Mitchell, Melanie. 1996. *An Introduction to Genetic Algorithms*. Cambridge, MA: The MIT Press.

Mitchell, Melanie, Crutchfield, J. P., and Hraber, P. T. 1994. Dynamics, computation, and the "edge of chaos": A re-examination. In G. Cowan, D. Pines, and D. Melzner (editors). *Complexity: Metaphors, Models, and Reality*. Santa Fe Institute Studies in the Sciences of Complexity, Proceedings, Volume 19. Reading, MA: Addison-Wesley.

Mitchell, Melanie, Hraber, P. T., and Crutchfield, J. P. 1993. Revisiting the edge of chaos: Evolving cellular automata to perform computations. *Complex Systems*. (7) 89-130.

Toffoli, T. and Margolus, N. 1987. *Cellular Automata Machines*. Cambridge, MA: The MIT Press.

Wuensche, Andrew and Lesser, Mike. 1992. The Global Dynamics of Cellular Automata. Santa Fe Institute Studies in the Sciences of Complexity. Reference Volume I. Reading, MA: Addison-Wesley.

Wolfram (editor). 1986. *Theory and Applications of Cellular Automata*. Singapore: World Scientific.

Table 4 The 11 domains of the best rule evolved by genetic programming.

Regular Domain	Domain Name	Color
0*	W	White
(000001)*∪(100000)*∪(010000)*∪(001000)*∪(000100)*∪(000010)*	1	Very Light Gray
(000101)*∪(100010)*∪(010001)*∪(101000)*∪(010100)*∪(001010)*	2	Light Gray
(001)*∪(100)*∪(010)*	3	Light Gray
(001101)*∪(100110)*∪(010011)*∪(101001)*∪(110100)*∪(011010)*	4	Gray
(001011)*∪(100101)*∪(110010)*∪(011001)*∪(101100)*∪(010110)*	5	Gray
(01)*∪(10)*	6	Gray
(011101)*∪(101110)*∪(010111)*∪(101011)*∪(110101)*∪(111010)*	7	Dark Gray
(011)*∪(101)*∪(110)*	8	Dark Gray
(011111)*∪(101111)*∪(110111)*∪(111011)*∪(111101)*∪(111110)*	9	Very Dark Gray
1*	B	Black

A Study in Program Response and the Negative Effects of Introns in Genetic Programming

David Andre
andre@flamingo.stanford.edu
(415) 941-9137
Visiting Scholar - Dept. of CS
Stanford University
860 Live Oak, #4 Menlo Park, CA 94025

Astro Teller
astro@cs.cmu.edu
(412) 268-7123
Dept. of Computer Science
Carnegie Mellon University
Pittsburgh, PA 15213-3890

ABSTRACT

The standard method of obtaining a response in tree-based genetic programming is to take the value returned by the root node. In non-tree representations, alternate methods have been explored. One alternative is to treat a specific location in indexed memory as the response value when the program terminates. The purpose of this paper is to explore the applicability of this technique to tree-structured programs and to explore the intron effects that these studies bring to light. This paper's experimental results support the finding that this memory-based program response technique is an improvement for some, but not all, problems. In addition, this paper's experimental results support the finding that, contrary to past research and speculation, the addition or even facilitation of introns can seriously degrade the search performance of genetic programming.

1 Introduction

The traditional method for extracting the response from each individual in GP was inspired by the functional programming of LISP. The most common way to get a program's response is to take the value returned by the root node. The tree whose value is taken to be the answer, especially when an individual consists of multiple trees (such as automatically defined functions (ADFs)), is usually called the Result Producing Branch (RPB). We label this traditional method Result Producing Branch Response, or RPBR. While using the return value of the RPB is the simplest thing to do, it gives exponentially decreasing importance to nodes further down the tree. In other words, nodes lower in the tree have exponentially less of an effect on the outcome in RPBR. We conjecture that this situation can increase the complexity of the search.

The genesis of this paper was the idea of a Memory-Based Program Response (MBPR) paradigm. This idea has already been suggested and found to be successful in non-tree representations (e.g., [Teller and Veloso, 1995a, Teller and Veloso, 1995b]). The hypothesis that this paper addresses is that the advantages of MBPR outweigh its disadvantages for many problems. This paper includes a description of the MBPR mechanism, presentation of experiments designed to test the hypothesis, and analysis of those experimental results.

Section 2 describes an alternative to RPBR. Section 3 presents a set of runs on a classic GP benchmark on which MBPR does well. Then sections 4 and 5 present two other benchmark domains for which the MBPR costs are negligible and high, respectively. The following two sections (6 and 7) describe the important effects of introns in these reported results and comment on the implications. Finally, section 8 concludes with a summary and future research directions.

2 Introduction to the Memory Based Programming Response Method

The idea of MBPR is that one or more memory elements be considered to be the "Answer Location" in memory. After the program tree has been evaluated the value of this memory location or locations is taken to be the program's response. For this paper we will take indexed memory [Teller, 1994] as our memory model and use cell 0 of memory to be the answer location. Thus, in this version of MBPR, the tree is executed and then `Memory[0]` is taken to be the program's response.

2.1 MBPR: Non Tree Representations

Teller and Veloso have previously reported success using the MBPR paradigm in GP with representations that are not tree-structured [Teller and Veloso, 1995a, Teller and Veloso, 1995b]. While this past research is a good justification for careful investigations into the actual effectiveness of MBPR in alternate program representations, we will mention here a few persuasive reasons for the use of MBPR in non-standard program response situations.

The first non-standard response situation that is becoming increasingly common in the GP field is to have a single program return more than one value. Having separate trees that evolve together is one solution to this problem, but this is really a group of programs that share a fitness measure, not a single program with multiple outputs (e.g., [Langdon, 1995]). A simple way to achieve this effect is to use MBPR and use multiple memory locations for the multiple return values. For example, if we want our programs to return three values, we can treat `Memory[0]`, `Memory[1]`, and `Memory[2]` as these three values upon program completion.

A second non-standard response situation is one in which the programs must report a value periodically while running continuously or in which the programs may run for a long or infinite amount of time while the answer is required after a fixed or at least limited amount of time. In such cases, there is no way to force the programs to output a value from some special function node (e.g., the root node of a tree) when desired. If the MBPR paradigm is used, the value of `Memory[0]`, for example, can be polled as needed without interruption of the program or concern for its state.

A third circumstance is one in which the representation itself is not isomorphic to a tree-structured representation. In such cases, as long as memory is being used, the MBPR paradigm is trivial to implement. Conversely, if an RPBR strategy is to be used, then you have to pick a special unit of the program to treat as the answer and have some reassurance that that unit will be activated within the time constraints of the problem. In the MBPR method, the program becomes responsible, not for returning a best value, but for updating its memory with its best guess.

The three reasons just outlined, and others, clearly show MBPR to be more flexible than RPBR. Thus, only a high computational overhead should prevent the use of the MBPR paradigm.

2.2 MBPR: Tree Representations

The following two definitions will be useful in discussing the relative merits of the RPBR and MBPR paradigms.

Vertical Node Interference is the blocking of a subtree result by one of its ancestors. This phenomena is aggravated by representations in which there is an non-uniform distribution of control among the nodes. The tree representation is such a representation. The tree representation provides an inverse exponential distribution of control among the nodes as a function of depth. The root controls the return value of all N nodes. If the root has arity A then each of its children has, on average, control over N/A nodes.

Sibling Coordination Interference is the blocking of a subtree's result by a node that is *not* an ancestor. This is the search difficulty brought about by representations in which nodes that are not hierarchically related (e.g., ancestor-descendant relation) to each other can still affect each other's results through side-effects. The ant problem representation in [Koza, 1992] suffers from this. In fact, any representation that includes memory usage (e.g., indexed memory) potentially suffers from this effect.

The argument for MBPR *in the tree-structured GP representation* follows. If a program has some portion of it that computes an accurate or perfect answer, it must (independent of the representation and response collection method) rely on the rest of the program not to interfere with that answer until the environment collects it. For example, in tree-GP using RPBR, if a subtree computes a good answer, it must depend on the path from the root of that subtree to the root of the tree to be non-destructive. This is the vertical node interference problem mentioned in the previous section.

Clearly, MBPR suffers from an analogous problem. If a program is allowed to update its "answer" memory cell in any part of its program, then clearly a "destructive" program fragment can overwrite a correct answer that the "good" part of the program had already written to the "answer" location in memory. This is the sibling coordination interference problem.

Because the MBPR approach is largely (although not entirely) position independent, it seems to have a natural advantage over RPBR. A response collection policy is position independent to the extent that a subtree that computes "the answer" for a given individual can be moved without affecting the subtree's ability to produce "the answer" in the new individual. In the MBPR paradigm, a subtree that includes a WRITE to the answer cell in memory can be moved to a new individual program and that subtree will still be able to store its value into the answer cell in memory. Obviously, the high degree of vertical node interference means no such guarantee can be made when RPBR is used. MBPR is not entirely position independent in the sense that while a transplanted subtree can compute an answer and store it in `Memory[0]`, that value could later be overwritten by another subtree.

To believe that the MBPR paradigm will be better than RPBR paradigm for a particular problem, we must believe that searching with high sibling coordination interference will be easier than searching with high vertical node interference. It would seem that the a priori argument for MBPR is that, in the worst case, any tree that is a solution in RPBR space, by adding a WRITE to the top, becomes a solution in the MBPR space. Even if there are other WRITEs to the answer location in the tree, this root node, since it is last, cannot be overwritten. It seems that this "capping" procedure would be simple to evolve if it were necessary. However, the side-effects of introns introduced by memory (sibling coordination interference) complicates this analysis.

All these reasons contribute to our belief that MBPR is a more flexible strategy for program response collection. Now, experiments will help us to understand when, if ever, the costs associated with MBPR outweigh the arguments for using it.

3 Positive Results with MBPR

All of the experiments in this paper refer to four different experimental conditions that were run together for comparisons. Below is a list of the four types of experiments as they are annotated in the tables and an explanation of what they mean.

- *ProblemType*-**RPBR** indicates that the problem was run with the usual function set and the result was taken from the return value of the root node of the tree.
- *ProblemType*-**RPBR-X** (N) indicates that the problem was run with the usual function set plus a perversion of the indexed memory functions in which (READ V) and (WRITE U V) both return their last argument (V). N indicates the number of indexed memory cells (see below for the effect $N < 2$ on these run types). In this case the result is still taken from the return value of the root node of the tree.
- *ProblemType*-**RPBR-M** (N) indicates that the problem was run with the usual function set plus the standard indexed memory functions (READ *index*) and (WRITE *index new-value*) with N possible indexed memory cells. In this case the result is still taken from the return value of the root node of the tree.
- *ProblemType*-**MBPR** (N) indicates that the problem was run with the usual function set plus the indexed memory functions (READ *index*) and (WRITE *index new-value*) and N cells. In this case the result is *not* taken from the return value of the root node of the tree. Instead, after the tree is evaluated, the current value of `Memory[0]` is taken to be the program's result.

When $N = 1$, no indexing is necessary, and READ becomes a terminal and WRITE becomes a single arity function. When $N = 0$, READ is no longer included in the function set. Single arity WRITE is left in the function set to save values, but they can not be referenced later by the program. The *ProblemType*-**RPBR-M** runs were only performed when $N > 0$, because when $N = 0$, the memory functions have no effect on the output in RPBR case, and are therefore equivalent to their inert counterparts used in the *ProblemType*-**RPBR-X** runs.

Table 1: The 5-Parity problem

Basic 5-Parity Function Set

functions	and	or	nand	nor	
Terminals	d0	d1	d2	d3	d4

5-Parity Problem Parameters
Max Generations = 75
Max Tree Size = 500
Pop Size = 32,000

5-Parity Problem Results

Problem Type	Comp. Effort	#Solved	#Run
5Parity-RPBR	3,435,185	26	27
5Parity-RPBR-X (2)	17,820,172	12	27
5Parity-RPBR-M (2)	1,640,000	26	26
5Parity-MBPR (2)	1,189,000	26	26

All the results presented in this paper had several methodological points in common. Where the methodologies differed, the differences will be mentioned when the data is presented. For all of the experiments, we used tournament selection with sample size of 8. Unless otherwise noted, the maximum number of generations was 101. All the runs discussed in this paper with a maximum node limit of 2000 or above had an initial maximum depth of 10. In addition, the runs with the Gudermannian[1] function on the $N = 0$ case with a maximum node limit of 1000 also used an initial maximum depth of 10. All other runs used had an initial maximum depth of 8. The recombination percentages for all experiments were: 10% copy, 1% mutation, 89% crossover (10% leaves, 79% internal nodes).

The first problem on which we tested the MBPR paradigm was the 5-Parity problem. There were two motivations for choosing this problem. The first was that, as the most classic GP benchmark, it is one that most readers are familiar with. The second reason for choosing a parity problem was the desire for a simple, but not trivial, problem where a subtree has often been observed to express a correct calculation.

This problem was run on a medium-grained parallel Parsytec computer system consisting of 64 80 MHz Power

[1] This paper's regression problem is called the "inverse Gudermannian", referred to here simply as the "Gudermannian". The function is $(log((1/cos(X)) + tan(X)))$.

PC 601 processors arranged in a toroidal mesh with a host PC Pentium type computer. The so-called distributed genetic algorithm for parallelization was used with a population size of $Q = 500$ at each of the $D = 64$ demes. On each generation, four boatloads of emigrants, each consisting of $B = 5\%$ (the migration rate) of the node's subpopulation (fitness-based selection) were dispatched to the four toroidally adjacent processing nodes [Andre and Koza, 1996]. This hardware was only used on the Boolean runs. All other runs were performed on a Pentium.

The results on the 5-parity problem are presented in table 1. Not only does the MBPR take 33% of the computational effort of RPBR on the 5-parity problem, but the MBPR style of response actually solves 100% of the time instead of the 96% shown here for RPBR. This large improvement for MBPR may be due to either one or both of the differences between MBPR and RPBR: the addition of memory or the memory based response. In the case where memory is added (5-parity-RPBR-M), 100% of the runs found a perfect solution, and it requires only 48% of the computational effort of the 5-parity RPBR. Taking the answer from memory helps, but just adding memory accounts for more than half of the difference in computational expense.

Clearly, the most dramatic aspect of the data in table 1 is the large amount of computational effort and the low number of solved runs for the 5-Parity-RPBR-X problem. When the inert READ and WRITE operators are added, computational effort increases by a factor of five and solutions occur less than half the time. This is surprising. Why would adding READ and WRITE functions that just passed back their last argument cause such a problem? This question will be addressed in sections 6 and 7. The results from the parity problem indicate that the MBPR approach works well on the most classic GP benchmark.

4 Minimal Cost Results with MBPR

The next benchmark problem we chose for further validation of the MBPR results shown in the previous section was the lawn-mower problem. The lawn-mower problem was chosen for two important reasons. The first is that, as will be described, this version of the lawn-mower problem requires the use of memory. This required use of memory makes it possible to at least partly separate the MBPR paradigm from the general helpfulness of memory (table 1). The second reason was to pick a problem with more complexity than the 5-Parity problem and with a much richer function set.

A version of the lawn-mower problem is described in [Koza, 1992]. The basic concept is that there is a lawn-mower in an 8x8 toroidal grid world. In our version of the problem, the mower only moves once per execution of the tree, and is allowed 200 time steps. If the value returned is less than zero, then the mower turns left. If the value is greater than zero, then it moves forward and mows. If the return value is zero, it does nothing. The lawn-mower's goal is to mow as much of the world as possible during those 200 time steps. Since it receives no sensory input, it must rely on its memory to help it create a path that hits most or all of the grid positions. This description (and the function set shown in table 2) should make it clear that this problem only shares surface level details with the lawn-mower problem in [Koza, 1992]. The main differences are: no state except indexed memory (i.e., no progn), one move at a time (i.e. one move per tree evaluation), and no jumping allowed.

Table 2: The Lawn-mower Problem

Basic Lawn-mower Function Set

functions	or	and	if > 0	not	if $= 0$
	pdiv	add	sub		mult	write
	read	incm[2]	dprog2		
Terminals	move	left	-10...10		

Lawn-mower Problem Parameters

Max Generations = 101
Max Tree Size = 500
Pop Size = 2,000

Lawn-mower Problem Results

Section 1: 20 Indexed Memory Cells

Problem Type	Comp. Effort	#Solved	#Run
Lawn-RPBR-M (20)	15,099	143	143
Lawn-MBPR (20)	24,372	136	136

Section 2: 20 IM Cells and 1 dedicated *answer* cell

Problem Type	Comp. Effort	#Solved	#Run
Lawn-MBPR (20/1)	15,516	209	209

The lawn-mower problem (results for which are shown in table 2:1) was set up so that memory was required[3]. We see that the MBPR takes 1.62 times as much computational effort as RPBR. This is surprising given that the programs are using their memory anyway. One potential explanation for this is that because this is actually a memory critical problem, and zero is a very likely value to occur[4], different program pieces may initially "fight" for control of memory cell 0: some to store the answer, and others to keep track of the mowing movements.

[2](INCM(X) adds 1 to `Memory[X]` and then returns new value of `Memory[X]`)

[3]Lawn-RPBR (no memory) and Lawn-RPBR-X (useless memory functions) are not shown because memory is required for the problem.

[4]Zero is especially likely to occur with a function set such as this where aspects of it, like boolean functions and the initally all zeroed valued memory, contribute to the likliness of zero as a value

The above hypothesis suggests the following experiment: continue to use 20 indexed memory cells, but introduce one new dedicated memory cell for the MBPR "answer." To dedicate this memory cell, a new function was introduced into the function set: WRITE-ANSWER. WRITE-ANSWER takes one parameter, the value, and stores that value in the dedicated memory cell. The results of this experiment are shown in table 2:2. The addition of this dedicated memory cell has successfully reduced the computational cost of the MBPR runs to almost exactly that of the RPBR runs. This change in computational cost demonstrates that contention over the "answer" position in indexed memory (memory cell 0) was indeed the expensive aspect of the MBPR runs shown in table 2:1. In short, MBPR does not reduce computational effort in this domain, but does not increase it either.

5 Negative Results with MBPR

Symbolic Regression was chosen for a third benchmark on which to test the MBPR paradigm both because it is a traditional GP benchmark problem, and more importantly, because it has characteristics that suggest that MBPR will not help. Table 3 shows the basic function set that all the regression problems discussed in this paper used (except where explicitly noted).

The relative simplicity of the function set was an important factor in the decision to choose regression as our third problem domain. The algebraic primitives do not lead directly to intron activity (this is discussed more fully in the next section) because 0 and 1 are not provided as terminals. In addition, the algebraic primitives for regression tend to lead to most or all of the tree being used. This set of functions is relatively homogeneous. This homogeneity leads us to suspect that MBPR will hamper rather than help the search process.

Table 3:1 presents the results of the first set of regression runs with the Gudermannian function and 20 indexed memory units. Not only does the MBPR paradigm not help, it is extremely detrimental to performance. In fourteen runs, it never finds a solution. While this is in the direction that was predicted, it is so negative that it is puzzling. Again, we see that adding useless READ and WRITE functions (the Reg-RPBR-X runs) causes a factor of five rise in computational cost and a 28% reduction in number of solutions. It is at this point that we may begin to suspect that the MBPR effects are being swamped by some other effect.

In an effort to find the cause of the performance drop, several experiments were run in which the memory size was reduced. The rationale is that, since the the MBPR paradigm did well on the 5-Parity and had only two memory cells whereas the MBPR method did less well on the lawn-mower problem where there were 20 memory cells, perhaps reducing the indexed memory size will, for this problem, lower the cost of using MBPR. Table 3:2-4 shows the results.

It has been repeatedly suggested, though never demonstrated, that unnecessarily large indexed memory sizes can degrade performance. Table 3:2-4 indicates that reducing memory size does not seem to explain the poor performance of the RPBR-X, RPBR-M, and MBPR runs.

The last set of runs in table 3:4 (0 cells for indexed memory) shows runs in which there is no indexed memory to confuse the program while it is computing an answer. Since (Write-Answer x) returns X, it basically has no effect except to save off the answer. In this situation, any tree that has WRITE-ANSWER as its root is *exactly* the same as a tree using RPBR from the point of view of the return value. This is true even if there are other WRITE-ANSWERs in the tree, since the root is evaluated last. It is hard to believe that the

Table 3: Regression

Basic Regression Function Set

functions	Add Sub Mult Div
Terminals	X Random Ephemeral Constants

Regression Problem Parameters

Function = $(log((1/cos(X)) + tan(X)))$.

Max Tree Size = 1000

Regression Problem Results

Section 1: 20 Indexed Memory Cells

Problem Type	Comp. Effort	# Solved	# Run
Reg-RPBR	35558	28	29
Reg-RPBR-X (20)	183380	19	28
Reg-RPBR-M (20)	129946	10	15
Reg-MBPR (20)	NA	0	14

Section 2: 2 Indexed Memory Cells

Problem Type	Comp. Effort	# Solved	# Run
Reg-RPBR	35558	28	29
Reg-RPBR-X (2)	183380	19	28
Reg-RPBR-M (2)	142710	6	16
Reg-MBPR (2)	1975913	1	17

Section 3: 1 Indexed Memory Cells

Problem Type	Comp. Effort	# Solved	# Run
Reg-RPBR	22589	32	32
Reg-RPBR-X (1)	178732	22	31
Reg-RPBR-M (1)	114506	25	33
Reg-MBPR (1)	540994	12	32

Section 4: 0 Indexed Memory Cells

Problem Type	Comp. Effort	# Solved	# Run
Reg-RPBR	34408	57	57
Reg-RPBR-X (0)	57550	112	114
Reg-MBPR (0)	212157	43	57

problem is eight times easier than "Solve the Gudermannian with Add, Subtract, etc and put a WRITE-ANSWER at the root node."

The runs in table 3:4 have two main points of interest. Notice that while the computational effort of the RPBR-X runs is worse than that of RPBR runs, it is much closer to that of the RPBR runs that to that of the MBPR runs. The former ratio is 1.67; the later is 3.68. This means two things. First, even inert WRITE-ANSWER alone (RPBR-X) is enough to cause a performance degradation. Second, and at least as importantly, MBPR is causing more of a problem than the the intron effect caused by inert WRITE-ANSWER in this case.

6 The Mystery Cause Unmasked

Introns are almost always possible. Whenever a subtree's return value is ignored, that subtree is an intron. There are, however, ways of making introns easier to generate, and therefore more prevalent in programs. For example, in Regression-RPBR-X (20), (WRITE X Y) not only has no effect on memory, but in simply returning Y, it makes the X subtree an intron. The (READ X) function in the RPBR-X (20) runs does not create an entire intron subtree, but merely is an intron itself.

We see the following two ways in which introns can be detrimental to the search process:

1. They can hamper search by providing locations where crossover can happen and have no effect on behavior.
2. They can take up space in the tree and, if tree space is limited, may cramp the actual space for solutions.

In the experiments reported in this section, we attempt to differentiate between these two types of intron effects. Allowing larger maximum tree sizes should only alleviate the cramping effect of introns and not the search degradation effect. These experiments directly compare the effects of increasing maximum tree size on two different types of introns-those created by WRITE-ANSWER in the RPBR-X (0) case, and those created by sibling coordination interference in the MBPR (0) case. We assume (though there is disagreement about this) that the first of these two problems does occur; in our opinion, a wasted search step can do nothing but slow the search process.

The runs in table 4:1 show the results of the Gudermannian problem runs, where the tree sizes are limited to 2000 nodes per tree (as opposed to 1000 nodes per tree in table 3). Notice that the computational effort of the RPBR and MBPR runs in table 3:4 and table 4:1 increases, but their ratio does not change significantly.

The ratio of the computational effort of the RPBR-X runs to the computational effort of the RPBR runs drops from 1.67 to 1.46 when the maximum tree size is increased from 1000 to 2000. Since nothing but the maximum tree-size has changed, and the control for the experiment (RPBR) takes approximately the same amount of time, this means that larger maximum tree sizes made it easier to solve the problem. The RPBR-X runs shown in table 4 differ from the RPBR runs only in that there is a single, useless function. The inert WRITE-ANSWER, because it passes its argument back up, does not even cause introns below it. This is very significant. Adding a single function can cause a run to take 1.67 times as long to complete (table 4:4) and increasing the maximum tree-size can reduce the penalty to 1.46 (2000 node limit). The question arises, "What is the floor for this ratio?"

We see in table 4:2 that moving the maximum tree size to 5000 again has little effect on the RPBR runs or on the MBPR runs. The computational effort of the RPBR-X runs drops to 1.41 times that of the RPBR runs. This confirms our conclusion that the useless functions take up space and force the useful program fragments into a smaller space. By increasing the maximum tree size, this problem is greatly reduced. In the runs with 10000 node maximum tree size, the computational effort of the RPBR-X runs is 1.36 times the computational effort of the RPBR runs. While this is only slightly lower than the 1.41 just mentioned, it is close to the theoretical penalty for adding one useless function (X+1/X

Table 4: Regression with increased tree size limits

Regression. $(log((1/cos(X)) + tan(X)))$.

Section 1: Max Tree Size (MTS): 2000

Problem	Comp. Effort	# Solved	# Run
Reg-RPBR	32564	47	51
Reg-RPBR-X (0)	47520	101	103
Reg-MBPR (0)	230615	37	52

Section 2: Max Tree Size (MTS): 5000

Problem	Comp. Effort	# Solved	# Run
Reg-RPBR	37525	43	43
Reg-RPBR-X (0)	53001	85	86
Reg-MBPR (0)	208426	30	43

Section 3: Max Tree Size (MTS): 10000

Problem	Comp. Effort	# Solved	# Run
Reg-RPBR	44116	121	121
Reg-RPBR-X (0)	60064	239	243
Reg-MBPR (0)	304157	67	121

Section 4: Increasing MTS for Reg-RPBR-X (0)

MTS	Comp. Effort	solved/runs	Ratio to RPBR
1000	57550	112/114	1.67
2000	47520	101/103	1.46
5000	53001	085/086	1.41
10000	60064	239/243	1.36

where X is the number of useful functions) which in this case is 5/4 or 1.25. Table 4:4 summarizes this trend.

Comparing the computational effort of the RPBR and the RPBR-X runs for the experiments shown in this section, we see a gradual reduction of the difference between the two as the maximum allowed tree sizes increases. In fact, it seems asymptotic to the theoretical penalty for adding a useless function to the function set.

The computational effort using the MBPR paradigm does not improve noticeably as the maximum tree size increases (table 3:4 and table 4). Thus, the negative effects caused by using the MBPR paradigm are not the cramping effects. This leaves only the "search degradation" intron effect as the explanation for MBPR's higher computational expense. This corresponds with the picture we have painted of the trade-offs between RPBR and MBPR. As the tree grows, the relative importance of vertical node interference and sibling coordination interference does not change.

7 Discussion

7.1 MBPR: Tree Representations

In the previous two sections, we showed three things about the MBPR paradigm:

- For the boolean 5 parity problem, MBPR works well, but much of its advantage was because of the introduction of memory.
- For the iterative lawn-mower problem, a problem that already required memory, MBPR was neutral with respect to computational effort.
- For the regression problems, problems that by their nature (homogeneity of function set) are unlikely to benefit from memory, MBPR was very expensive. Even when WRITE-ANSWER was used instead of indexed memory, the computational effort increase was always at least a factor of 2.

The explanation for MBPR's performance on the 5-parity problem is relatively simple. There is a large amount of redundancy in the parity problems that can be exploited through a technique like ADFs [Koza, 1992]. Since ADFs were not used, the introduction of memory served the same purpose. Values can be stored once in memory and then retrieved in several different places.

The performance improvement observed when using MBPR on the 5-parity problem provides evidence that the vertical node interference is more harmful for the 5-parity problem than is sibling coordination interference. Given that the runs with indexed memory (RPBR-M) differ only from the MBPR runs only in that the representation in the RPBR-M runs allows more vertical node interference, the improvement when using MBPR must be due to that difference.

Another way of understanding the success of MBPR on the 5 parity problem is by investigating its function set. In the 5-Parity problem, all four of the functions allow intron subtrees to form if the other subtree always returns a one or a zero. For example, if a subtree to an AND always returns a 0, the result of the other subtree is ignored. Since the 5-Parity is a boolean problem, constant values of 0 or 1 occur quite often. In contrast, in the regression problem, only a return value of 0 (out of infinitely many possible values) can cause a TIMES to have an intron subtree. Because it is relatively easy to make intron subtrees in the Boolean problem, vertical node interference is much more of a problem than is sibling coordination interference. This explains the success of MBPR on this problem

In the lawn-mower problem, it seems to be that the vertical node interference and the sibling coordination interference are approximately equal; moving from RPBR to MBPR does not make a significant difference. We suggest that this approximate equality of the two interference measures exists because memory is already being used in the RPBR lawn-mower problem. As shown in section 4, the moderate increase in computational effort was caused by contention over use of the `Memory[0]` when a dedicated memory cell is not provided. The lawn-mower problem uses memory in a very different way than does the 5-Parity boolean problem. The lawn-mower must use memory to communicate with itself on a later time step, whereas the 5-Parity problem only uses memory to pass values around. Using memory simulates ADFs for the 5-Parity problem, but not for the lawn-mower problem. This point may help to explain some of the difference between MBPR's effect on the 5-Parity and lawn-mower problems. `Memory[0]` could be a point of contention in either case (possibly more so in the 5-Parity problem when there are only 2 cells). However, these different memory requirements may change the dynamics of memory use in ways that we do not yet fully understand.

Most of the results shown in section 5 used WRITE-ANSWER instead of full indexed memory in a effort to disentangle the issues. One of the aspects of this data that we can not yet fully explain is why WRITE-ANSWER, even when it is not used for MBPR, causes such a noticeable jump in computational effort. Section 7.2 gives a plausible explanation for the underlying cause.

7.2 Introns

It should be clear from the previous section that the MBPR results were being confused by (arguably even obscured by) another effect: introns. This paper has suggested that introns can affect performance in a very negative way. The purpose of this section is to summarize the experimental evidence for introns and to present a partial explanation for their effects. We should mention here that our definition of intron is a behavioral definition. That is, we do not care where the genotype/phenotype line is drawn; an uncalled subtree is an intron, as is a subtree that is called and whose result is ignored.

Before we explain our findings, however, we will summarize what we believe to be the contemporary view of introns in the GP field. Introns have recently been often described as a beneficial aspect of the genetic search that serves at least two important functions. The first "benefit" of introns is often stated to be their ability to protect the program of which they are a part from "destructive" crossover. The second "benefit" of introns is to allow the population as a whole to protect and thereby preserve the best "building blocks" in the population. Efforts to insert extra introns into the population as a search aid [Nordin and Banzhaf, 1995] indicate the current favorable standing of introns in GP.

Many researchers (e.g., [Nordin and Banzhaf, 1995, McPhee and Miller, 1995, Rosca and Ballard, 1995]) have argued that introns are helpful, in that they prevent destructive crossover. However, their argument assumes that the best of the current generation are in the right part of the search space (i.e, that destructive crossover is necessarily bad). While destructive crossover may be bad for natural evolution, the goal of GP is often optimization of a single program, not the survival of any particular individual of individuals.

There are at least three types of introns associated with functions. There are also introns associated with useless terminals, but since this paper had only one experiment with a change in the terminal set ((READ) returning the value of memory for $N = 1$), we will concentrate on the following three function intron types:

- **Local:** Useless functions that simply pass on the values they are passed.
- **Hierarchical:** functions that force one or more of their subtree arguments to be ignored. WRITE in the RPBR-X ($N > 0$) runs done in this paper is an example of this intron type.
- **Sibling (horizontal):** functions that may cause the effects of a sibling subtree to become moot at a later point in the program. An example of this intron type is when a subtree has its contributions to memory overwritten by a subtree that is executed later (as can happen in the RPBR-M and MBPR runs in this paper).

Table 5 summarizes the effect that adding inert versions of READ and WRITE has on the computational effort required to solve various problems. In all of the runs in table 5, READ is useless (because it simply passes its one argument up the tree) (causing Local introns). WRITE actually creates an intron by ignoring its first argument and passing its second argument along (causing Hierarchical introns). Measured by either computational effort or by the likelihood that a perfect solution will be found in the required number of generations, the addition of this useless function and this intron causing function have a very strong negative impact on the search process. Since these two functions are not actually acting as memory conduits, the extra computational expense must be caused by the intron effects.

Table 5: Result Comparisons

Run results: Inert Memory (RPBR-X) relative to RPBR
MTS is Max Tree Size.

Section 1: using READ and WRITE

MTS	Problem	Comp. Effort Δ	Solution Rate Δ
1000	5Parity (2)	5.18	2.27
1000	Reg (20)	5.23	1.43
1000	Reg (2)	5.23	1.43

Section 2: using READ and WRITE

MTS	Problem	Comp. Effort Δ	Solution Rate Δ
1000	Reg (1)	7.91	0.70

Section 3: using WRITE-ANSWER

MTS	Problem	Comp. Effort Δ	Solution Rate Δ
1000	Reg (0)	1.67	1.02
2000	Reg (0)	1.46	0.94
5000	Reg (0)	1.41	1.01
10000	Reg (0)	1.36	1.02

In table 5:2 we see that when READ becomes a 0 argument function (i.e. a terminal) and WRITE becomes a single argument function (in the $N = 1$ case), and so both cause only local introns, the performance degradation remains. However, when we remove READ entirely and are left with only WRITE-ANSWER (causing only local introns) ($N = 0$) the performance degradation is lessened, although still negative. Further, with only WRITE-ANSWER we no longer have a function that explicitly causes whole subtrees to become introns; WRITE-ANSWER is itself simply an intron since it does nothing useful. This suggests that the addition of one useless function can have a detrimental effect on the GP search process.

The zero arity READ (table 5:2) returns a zero since it has no value to pass along. Since the presence of (READ) in Reg (1) is the only thing that distinguished it from the Reg (0) runs, the presence of zero as a terminal obviously has a large negative effect. The reason for this can again be explained by intron effects. With random ephemeral constants and X, making an intron is possible using the algebraic primitives, but not as easy as when 0 is also readily available. It is the difference between the likelihood of (* Y 0) and (* Y (- Z Z)). This may not seem like a large difference, but it has a large effect in this case, as shown in the difference between table 5:2 and the first line of table 5:3.

Table 5:3 shows the effect of the useless function (WRITE-ANSWER) as the maximum size allowed for a tree is raised from 1000 to 2000, 5000, and then 10000. In all three cases, the expense of local introns has been reduced to a modest cost. The conclusion is that for populations with sufficiently large members, this particular intron effect (addition of inert functions) ceases to be expensive. This says something important about tree size. The prevailing wisdom seems to be that as long as the tree size maximum is large enough for a non-compacted solution to fit, the search will go fine. Table 5:3 implies that the necessary tree size ceiling is also a function of the ease with which the functions can form introns.

It is important to notice that the conclusion of the previous paragraph is made in the context of $N = 0$. By increasing the maximum tree size, we alleviate this intron "cramping effect". In the case where $N > 0$ and/or MBPR is used, the intron expense incurred from the search degradation effect dominates (as entire subtrees can be intron-ized). In this, the $N > 0$, case, increasing the maximum tree size limit has little effect, as is shown in table 4. Now we have an answer to the question "Why won't the 'capping procedure' (proposed in section 3) guarantee a minimal cost increase for MBPR?" The answer is, the 'capping procedure' presupposes a correct solution to cap. The search degradation caused by introns appears to seriously impede the evolution of such solutions.

8 Conclusions

The hypothesis of this paper is that the greater flexibility and reduction of vertical node interference in MBPR outweighs the possible negative effects of increased sibling coordination interference and the possible cost of introducing memory.

The experimental results presented show that the intron effects from MBPR are too damaging for those problems (e.g., regression) that are very homogeneous. This correlates well with whether or not those problems benefit from the introduction of memory; When introducing memory to a problem does not help or even hurts, MBPR is unlikely to provide a benefit since memory must be added for MBPR to be accomplished.

When memory is necessary, the experimental results indicated that, for the test problem, MBPR did not have a dramatic performance impact. Given the negligible cost, the flexibility and generality of MBPR certainly makes it worth consideration for these problem types. However, if the problem is more "horizontal", the likelihood of introns is already present, and vertical node interference is a frequent occurrence then (e.g., the Parity problem) MBPR is likely to give a boost to the search process.

The take home messages are two fold. First, MBPR does not seem to be a problem independent improvement for tree representations, although it seems to be an advantage when memory use is simple, helpful, or required. In addition, the MBPR paradigm, as described in section 2.1, has several other points in its favor. To verify these conclusions, for both tree-GP and non-tree-GP representations, further experiments should be done with more complex problems.

The second conclusion of this paper is that introns are probably damaging. This goes against conventional wisdom, but in the data presented in this paper, the control experiments identify intron effects as the only possible cause of significant performance degradation under a variety of conditions and on a variety of problems. Results were presented that showed this change, to varying degrees, with local, hierarchical, and sibling introns.

The corollary to this conclusion is that simple changes in the function set can have large effects on computational effort. In particular, the addition of one intron causing function can make it difficult or impossible to solve what otherwise would have been an easy problem in genetic programming.

Acknowledgements

This work is supported in part by the Hertz Foundation whom Astro thanks for their kind support. The support and comments of John Koza and Forrest Bennett are greatly appreciated by both David and Astro.

References

[Andre and Koza, 1996] D. Andre and J. Koza. Parallel genetic programming: A scalable implementation using the transputer architecture. In P. Angeline and K. Kinnear, editors, *Advances In Genetic Programming II*. MIT Press, 1996.

[Koza, 1992] John Koza. *Genetic Programming*. MIT Press, 1992.

[Langdon, 1995] William Langdon. Evolving data structures with genetic programming. In Stephanie Forrest, editor, *Proceedings of the Sixth International Conference on Genetic Algorithms*. Morgan Kauffman, 1995.

[McPhee and Miller, 1995] N. McPhee and J. Miller. Accurate replication in genetic programming. In L. Eshelman, editor, *Proceedings of the Fifth International Conference on Genetic Algorithms*, pages 303–309. Morgan Kauffman, 1995.

[Nordin and Banzhaf, 1995] P. Nordin and W. Banzhaf. Complexity compression and evolution. In L. Eshelman, editor, *Proceedings of the Fifth International Conference on Genetic Algorithms*, pages 310–317. Morgan Kauffman, 1995.

[Rosca and Ballard, 1995] J. Rosca and D. Ballard. Causality in genetic programming. In L. Eshelman, editor, *Proceedings of the Fifth International Conference on Genetic Algorithms*, pages 256–263. Morgan Kauffman, 1995.

[Teller and Veloso, 1995a] Astro Teller and Manuela Veloso. PADO: A new learning architecture for object recognition. In Katsushi Ikeuchi and Manuela Veloso, editors, *Symbolic Visual Learning*, pages 81–116. Oxford University Press, 1995.

[Teller and Veloso, 1995b] Astro Teller and Manuela Veloso. Program evolution for data mining. In Sushil Louis, editor, *The International Journal of Expert Systems. Third Quarter. Special Issue on Genetic Algorithms and Knowledge Bases*. JAI Press, 1995.

[Teller, 1994] Astro Teller. The evolution of mental models. In Kenneth E. Kinnear, editor, *Advances In Genetic Programming*, pages 199–220. MIT Press, 1994.

An Investigation into the Sensitivity of Genetic Programming to the Frequency of Leaf Selection During Subtree Crossover

Peter J. Angeline

Loral Federal Systems
1801 State Route 17C
Owego, NY 13823
pja@lfs.loral.com

ABSTRACT

In genetic programming, crossover swaps randomly selected subtrees between parents. Typically, the probability of selecting a leaf as the subtree to be swapped is reduced, supposedly to allow larger structures on average. This paper reports on a study to determine the effect of modifying the leaf selection frequency for subtree crossover on the performance of a non-standard genetic program. Both a variety of constant values and dynamic update methods are investigated. It is shown that the performance of the genetic program is impacted by the manipulation of the leaf selection frequency and often can be improved using a random process rather than a constant value.

1.0 Introduction

The chief genetic operator used in genetic programming is subtree crossover. Subtree crossover randomly selects a subtree from a parent and swaps it with a randomly selected subtree from a second parent. Often, the selection of a subtree from the parent tree is biased against selecting a leaf as determined by the *leaf selection frequency.* Koza (1992) suggests a value of 0.1 for this parameter. Experiments in Angeline (1996) using self-adaptive crossover mechanisms suggest that a leaf frequency value of 0.1 may not be optimal for many problems.

This paper describes a study that investigates the sensitivity of genetic programs that use subtree crossover to various values for the leaf selection frequency. In addition, a self-adaptive method for selecting leaf frequency is compared with a similar pseudo-adaptive random method.

This study shows that the optimal leaf selection frequency during crossover is problem dependent and probably a priori unpredictable without significant understanding of the problem domain. The results also suggest that the most generally effective method for determining appropriate leaf selection frequency may be a simple pseudo-adaptive random process. An analysis of why this is true is provided.

2.0 Background

There are currently three major forms of Evolutionary Computation: genetic algorithms (Holland 1992; Goldberg 1989), evolutionary programming (Fogel Owens and Walsh 1966; Fogel 1995) and evolution strategies (Rechenberg 1971; Schwefel 1981). Genetic programming (Koza 1992) is a variant of genetic algorithms that evolves tree representations.

As with most genetic algorithms, genetic programming relies mainly on a form of crossover to manipulate offspring. Unlike genetic algorithms, genetic programming's *subtree crossover* is by definition sensitive to the underlying organization of the representation. Subtree crossover randomly selects and swaps complete subtrees between parents to create children. Cramer (1985) first used subtree crossover in a genetic programming-like system. Koza (1992) augmented and popularized this technique while applying it to a number of interesting mainstream engineering problems.

When one of the parents in a crossover operation is a full tree, about 50% of the subtrees that can be selected are simple leaf nodes. Koza (1992) determined that because of this it was important to bias selection of crossover points away from leaves and towards internal nodes. Koza (1992) associated a parameter with subtree crossover, the *leaf selection frequency* or *leaf frequency* for short, to implement this bias. Koza (1992) suggests that an appropriate value for leaf frequency is 0.1 and supplied the following argument:

> This distribution promotes the recombination of larger structures whereas a uniform probability distribution over all points would do an inordinate amount of mere swapping of terminals from tree to tree in a manner more akin to point mutation than to

recombining of small substructures or building blocks. (Koza 1992, p. 114)

While the need to bias subtree crossover away from selecting leaves seems plausible, it was not empirically validated.

Angeline (1996) investigated a self-adaptive method termed Selective Self-Adaptive Crossover (SSAC) that associated an adaptive parameter with each component of evolving genetic programs. When performing crossover, these parameters were used in a roulette wheel selection to choose the actual crossover point in the parent trees. After crossover, the adaptive parameters were updated using a Gaussian random variable. Angeline (1996) showed that this and one other self-adaptive crossover method performed as well and often better than standard non-adaptive subtree crossover. A post experimental analysis of the values of the adaptive parameters showed that the best-of-generation individuals on average would chose leaf nodes for crossover about 50% of the time for the three problems investigated.

The discrepancy between the standard assumption of a leaf frequency of 0.1 against the larger probability for leaf selection preferred by the self-adaptive method suggests that further investigation into the relationship between leaf frequency and the problem solving ability of a genetic program is necessary. The purpose of this study is to determine how sensitive subtree crossover is to different leaf selection frequencies and supply some guidance into selecting appropriate leaf frequency values or methods for unknown problems.

3.0 A Non-Standard Genetic Program

The genetic program used in this study has a number of deviations from the genetic program introduced in Koza (1992). Most of the modifications are standard in evolutionary programming (Fogel, Owens and Walsh 1966; Fogel 1995). While the philosophical underpinnings of evolutionary programming discourages recombination of population members, relying instead on representation specific mutation operations, there is nothing that prevents using recombination as a form of mutation. The genetic program studied here is essentially a standard evolutionary program as described in Fogel (1995) that uses tree-based crossover as one form of mutation. The remainder of this section provides details of the algorithm.

3.1 Parent Selection

The type of selection performed in this study is a form of tournament selection modified to minimize over-selection (i.e. multiple reproductions given to a single high-fitness parent) (Fogel 1995). First, a tournament score is associated with each member of the population based on its fitness relative to other population members. The algorithm assigns an individual's tournament score by randomly selecting k additional population members and summing the number with worse fitness. The population is then sorted by the tournament score and the best 50% is reserved as parents for the next generation. As a result, any given population member can be selected only once as a parent for the next generation.

Unlike the typical tournament selection algorithm used in genetic algorithms, this scheme forces both an individual and its offspring to have above average fitness in order to increase the representation of the lineage in the population. It thus places additional selection pressure on *mutable* fit population members rather than just those with high fitness. While this difference is subtle, it may hold advantages in complex environments.

3.2 Offspring Reproduction

Creation of offspring occurs in three stages. First, each parent selected from the last population is copied to create its offspring. Each parent is copied exactly once for the next generation. Next, crossover is applied to all children in the population. When a crossover operation is executed, a second "child" tree is selected from the population that has yet to undergo crossover. Once crossover is complete, both of the resulting children are placed in the population. Finally, each child is mutated with a user defined probability. The number of mutations performed on the child is given by a Poisson random variable with a user specified rate resampled for each child. Given that k mutations are chosen to be executed on a child, k mutation operations are selected uniformly with replacement from the five mutation operators described below. The chosen set of mutations are applied to the child in the order selected. Any mutation or crossover operation that results in a child with a number of nodes outside a user-defined range is returned to its original form. For the experiments below, a viable tree had more than three nodes and less than 50.

This algorithm employs five separate mutations for evolving trees: *grow*, *shrink*, *cycle*, *switch*, and a *numerical terminal mutation*. *Grow* randomly selects a leaf node of the tree and replaces it with a randomly generated tree. The depth for the generated tree is limited by a user-defined constant set to seven in the experiments below. The *shrink* mutation selects a random subtree in the program and replaces it with a randomly selected terminal. *Cycle* chooses a function symbol in the program and replaces it with another function symbol that requires the same number of parameters. Only the name of the function at a particular node in the tree is changed in this mutation; the children (i.e., parameters) of the node are unmodified. A *switch* mutation selects two sibling subtrees of the same parent node and switches their positions in the parent's function call. Finally, the *numerical terminal mutation* is defined for what Koza (1992) called *ephemeral random constants* and here are called *numerical terminals*. This mutation operation selects a single real-valued numerical terminal in the tree and adds to it Gaussian noise with a particular variance. In this study, the variance of the Gaussian noise for this mutation was always set to 0.1. This mutation is especially important since small changes in a numerical constant within an evolving program will typically correspond to small changes in the behavior of the program. Therefore, even after crossover becomes ineffective due to the emergence of introns (Angeline 1994), manipulation of numerical constants may still improve offspring.

Angeline (1996) demonstrates that this form of genetic program is competitive with the standard genetic program described in Koza (1992). The operational differences between this genetic program and a standard genetic program should not affect the applicability of the results of this study to any genetic program that uses subtree crossover.

4.0 Methods for Investigating Leaf Frequency

In order to fully access the effects of manipulating leaf selection frequency, two separate experimental approaches were taken. The first method compares the performance of the genetic program over multiple constant values for leaf frequency while the second method looks at the effect of two simple dynamic methods for setting the leaf frequency.

4.1 Testing Multiple Values for Leaf Frequency

In order to test how the performance of a genetic program changes with different values for leaf frequency, this study compares runs using several different leaf frequencies. Leaf frequencies used in the study are 0.1, 0.25, 0.5, and 0.75. The values 0.1 and 0.5 are the respective values from Koza (1992) and Angeline (1996). The other two values were selected to balance out the study and to supply more of a feel for how modification to this parameter affects the performance of the genetic program.

4.2 Dynamically Adjusting Leaf Frequency

An alternative to associating a constant global value to leaf frequency is to try to dynamically adjust the parameter's value during the run. Evolutionary computations that dynamically adjust operational parameters are termed *adaptive*.

Angeline (1995) identifies a two dimensional categorization for adaptive evolutionary computations. The first dimension describes at what level of the evolutionary computation the adaptive parameter affects the processing. *Population-level* adaptations modify population-wide parameters, often with heuristics computed over the current or past populations. Such methods include updating the global frequency of operator application (Rechenberg 1973; Davis 1989), and dynamically adjusting the interpretation of the representation (Shaefer 1987; Schraudolph and Belew 1992; Whitley, Mathias and Fitzhorn 1991). *Individual-level* adaptations associate parameters with each individual that determine how the algorithm manipulates the individual. Examples include evolving crossover positions in genetic algorithms (Rosenberg 1967; Schaffer and Morishima 1987) and adapting the relative probability of mutating components of finite state machines in an evolutionary program (Fogel, Fogel and Angeline 1994). *Component-level* adaptations associate adaptive parameters with each component of an evolving individual that determine how each component is modified during reproduction. These methods include evolving the variance of the mutational noise applied to each component of real-valued fixed-length vector representations (Schwefel 1981; Bäch and Schwefel 1993; Fogel et al. 1991; Fogel et al. 1992) and evolving the absolute probability of mutating particular components in finite state machines (Fogel, Fogel and Angeline 1994).

Adaptive evolutionary computations are evolutionary computations that modify the values for certain operational parameters while solving a problem. The ultimate efficiency of an adaptive evolutionary computation follows from the method used to update the adaptive parameters. (Angeline 1995) identifies two distinct types of adaptive parameter update rules. *Absolute update rules* compute a predetermined function over a set of generations or populations and use the changes in this heuristic to determine when and how to modify the algorithm's adaptive parameters. *Empirical update rules* use the same variation and selection process evolving the problem solutions to also modify the adaptive parameters. When an adaptive evolutionary computation evolves the values for its adaptive parameters (i.e. uses an empirical update rule), then it is termed *self-adaptive*. Self-adaptive methods are more flexible and can more readily adjust to a broader range of adaptive situations than absolute update rules (Angeline 1995). Typically, individual-level and component-level adaptive parameters are self-adaptive.

This study implements an individual-level self-adaptive method for determining leaf frequency. A single floating point parameter is associated with each individual in the population along with its program tree. This value is interpreted during crossover as the probability of selecting a leaf in the individual's tree as the crossover point. Consequently, when using this method, each individual will have a unique leaf frequency.

Once crossover has been performed on an individual, random noise is added to the individual's adaptive parameter. The form of random noise added to the adaptive parameter σ_i is given by:

$$\sigma_i' = \sigma_i + \alpha \cdot \sigma_i \cdot N(0, 1)$$

where σ_i' is the new value of the parameter, N(0, 1) is a Gaussian random variable with mean 0 and variance 1 and α is a scaling constant set equal to 0.1 in the experiments described below. Parameter values that wander below 0 or above 1 are reset to the closest value within the range. Initial parameter values for all adaptive parameters are generated using a uniform random variable over the interval (0..1).

In order to determine the effectiveness of the self-adaptive leaf frequency parameter, this study also investigated a pseudo-adaptive technique. In the pseudo-adaptive technique, the adaptive leaf frequency parameter associated with each individual is updated with a uniform random variable rather than the Gaussian form above. Consequently, the parameter's subsequent value has no relationship to its previous value. If the adaptive method is adapting to appropriate values for leaf frequency, then the pseudo-adaptive method should not perform as well.

5.0 Experiments

5.1 Problems

The first problem investigated was the boolean 6-multiplexer function (6-Mux) as originally defined in [Koza 1992]. The function set for this problem is {*not, and, or, if*} and the terminal set is {*a0, a1, d0, d1, d2, d3*} where *d0, d1, d2* and *d3* are the data inputs and *a0* and *a1* are the selector bits. The fitness of an individual is the number of correct responses for the 64 possible inputs.

The interlocking spirals problem is the second problem investigated. In this problem, a function must be learned that separates two classes of data that form interlocking spirals in the coordinate plane. Figure 1a shows the two data sets for the problem. Koza (1992) also investigated this problem and the form of this experiment follows the experiment described there. The function set is defined to be {+, -, *, /, *iflte, sin, cos*} and the terminal set is defined as {*x, y, RNT*} where *x* and *y* are the coordinates of the point to be classified and RNT is a *random numerical terminal*, which is identical to an ephemeral random constant Koza (1992) except it may be mutated with the numerical terminal mutation described above. The fitness of an individual for this problem is the number of points from the 194 points in the two data sets correctly classified.

The third problem used in the experiments is a time series modeling problem. Since the year 1700, the average number of sunspots observed each month have been recorded. The collection of these values are referred to as the Wolfe Sunspot Data and are available on-line.[1] Figure 1b shows the average number of sunspots per year for the time period from 1700 to 1993. This data is a naturally occurring chaotic time series which makes it more appealing than artificial chaotic time series, such as the Mackey-Glass equation, for experimentation since the underlying generating model of the series is unknown. The object of this problem is to predict the number of sunspots that will be observed in the following year given a minimal number of previous data points. Following (Oakley 1994), a function set and terminal set were selected that fit a time series prediction problem. The functions used were {+, -, *, %, *sin, cos*} and the terminal set was {*x1, x2, x4, x8, RNT*} where *x1* is the number of sunspots observed in the year before the predicted year, *x2* is the number of sunspots observed two years before the predicted year and so on. The available terminals were selected to be powers of two for no particular reason other than it gave a good spread over the previous data. Other terminal sets may produce better results than those reported below. It should be noted that the data was not separated into training and test sets making this a time series modeling task rather than a time series prediction task. Given that the object is to minimize the prediction error of the evolved function, the fitness function is the sum of the squared prediction error for each of the data points (years 1708 to 1993). The actual fitness function used negates the error to make this a maximization problem.

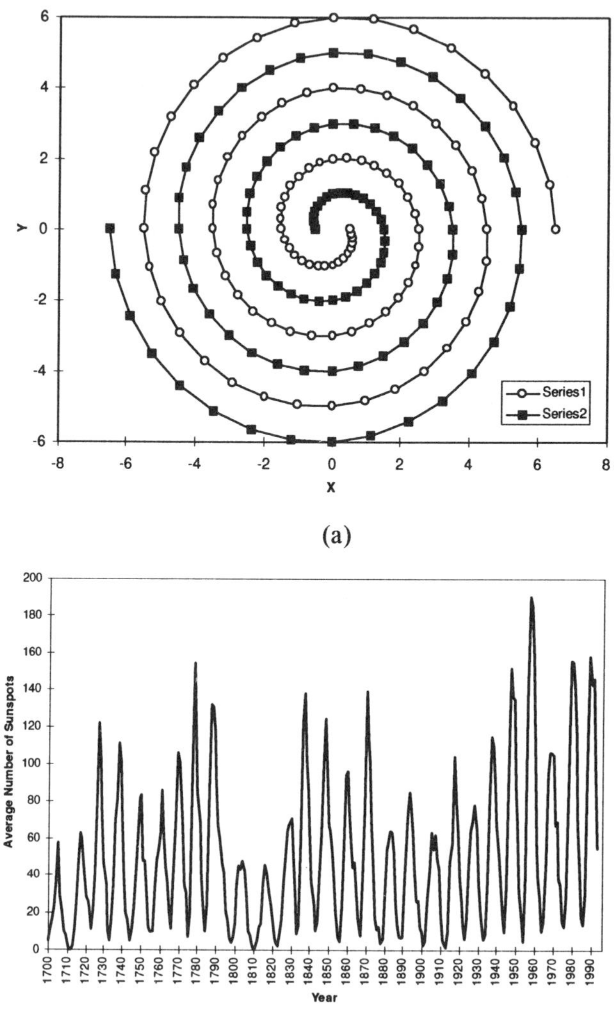

(b)

Figure 1: Data sets for two of the problems investigated in the study. (a) Data sets of the intertwined spiral problem. (b) Data set for the Sunspot prediction problem.

5.2 Experimental Setup

Several experiments were run for each of the three problems. Four experiments evolved solutions to the problems with the global leaf frequency set to a constant from the set {0.1, 0.25, 0.5 0.75}. These experiments are identified respectively by the keys Leaf 0.1, Leaf 0.25, Leaf 0.5 and Leaf 0.75 in the graphs below. Two other experiments were performed using the adaptive and pseudo-adaptive techniques described

1. The World Wide Web address is http://www.ngdc.noaa.gov/stp/SOLAR/SSN/ssn.html.

above and are identified respectively as Adapt and Rand in the graphs. Each experiment consisted of 30 trials allowed to run for at most 1000 generations. The population size was set to 250 with 50% replacement making the population turnover 125 individuals each generation. The maximum number of nodes allowed for a tree was set to be 50 nodes. No restriction was placed on the depth of the trees.

5.3 Measuring Relative Performance

Koza (1992) and Koza (1994) rely on a statistic termed *computational effort* for comparisons between different genetic programs. As defined in Koza (1992), computational effort is a point statistic with no associated confidence interval. Without a confidence interval, the significance of the point statistic cannot be determined and comparisons are inconclusive. This study uses different statistics along with confidence values for all comparisons.

When a series of independent runs (trials) are executed with an evolutionary computation, often it is convenient to use the mean score of the best-of-run individual taken over all trials to summarize performance. As long as the trials are independent their mean will be a reliable statistic. Student's T-test (Press et al. 1992) is a standard statistical technique used to determine if two independent data sets have means that are statistically significantly different. Confidence values of $p<0.05$ suggest a significant (reliable) difference in the means while confidence values of $p<0.01$ suggest a difference that is highly significant. Confidence values of $p>0.05$ are generally considered too unreliable to permit conclusions. Large confidence values suggest there is no significant difference between the mean of the two data sets and consequently no significant difference in the two processes giving rise to the data sets. This study uses this method for analyzing the results for both the spiral and sunspot problems.

For the 6-Mux problem, a large number of the runs actually solve the problem prior to the end of the trial. Consequently, the means of the different experiments are fairly close. In this case the mean generation at which a trial solves the problem is used as the statistic for comparison. Again, since the trials within each experiment are independent, this mean is an appropriate statistic. Student's T-test is again used to calculate a confidence value for the significance of a difference in mean performance between methods.

6.0 Results

Figure 1a graphs the average best-of-generation score for the various constant leaf frequencies on the 6-Mux problem. While there appears to be some separation between the various constant values around generation 100, these differences all but disappear by the end of the runs. Around generation 100, Leaf 0.75 appears to be performing the best while Leaf 0.1 appears to be the worst. All of the graphs approach the maximum of 64 hits by generation 1000. Figure 1b graphs the same results for the two adaptive methods compared with

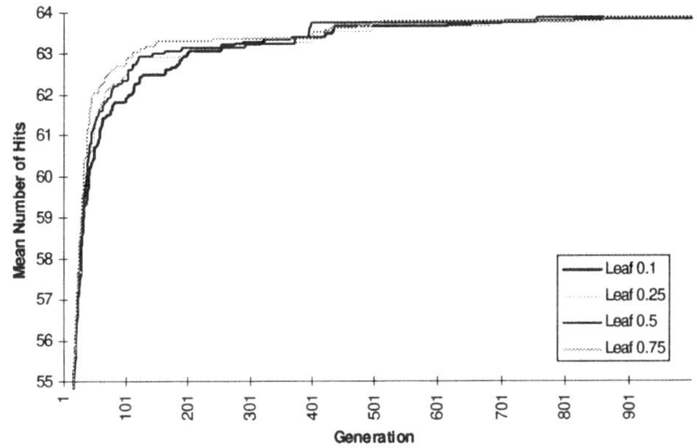

(a)

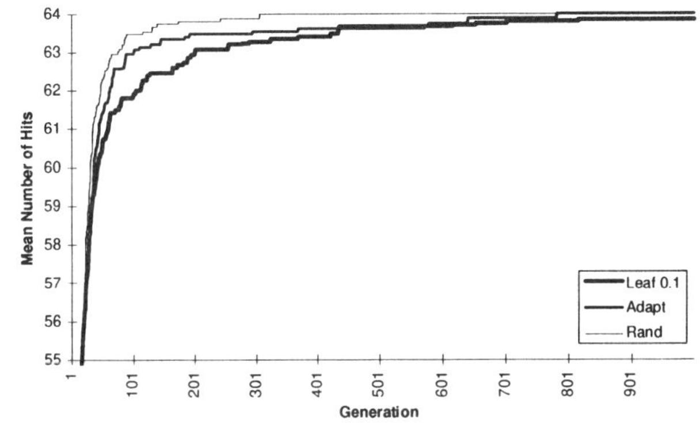

(b)

Figure 2: Results for the leaf frequency experiments on the 6-Mux Problem starting at generation 15. (a) Comparison of different constant leaf frequency values. (b) Comparison of adaptive, pseudo-adaptive (Rand) and traditional constant leaf frequency value of 0.1.

the run using a leaf frequency of 0.1. Again there is separation between the techniques around generation 100. At that generation, both Adapt and Rand have higher mean scores than Leaf 0.1, with Rand performing slightly better than Adapt. Again by generation 1000 all of the mean score approach the maximum score of 64.

Table 1 shows the mean generation of completion over

Table 1: Mean Generation of Completion for the 6-Mux Problem

Method	Mean Generation
Leaf 0.1	189.53
Leaf 0.25	159.67
Leaf 0.5	151.87
Leaf 0.75	160.40
Adapt	142.33
Rand	71.23

the 30 trials for each of the different methods investigated. The table shows that Rand had a significantly lower mean

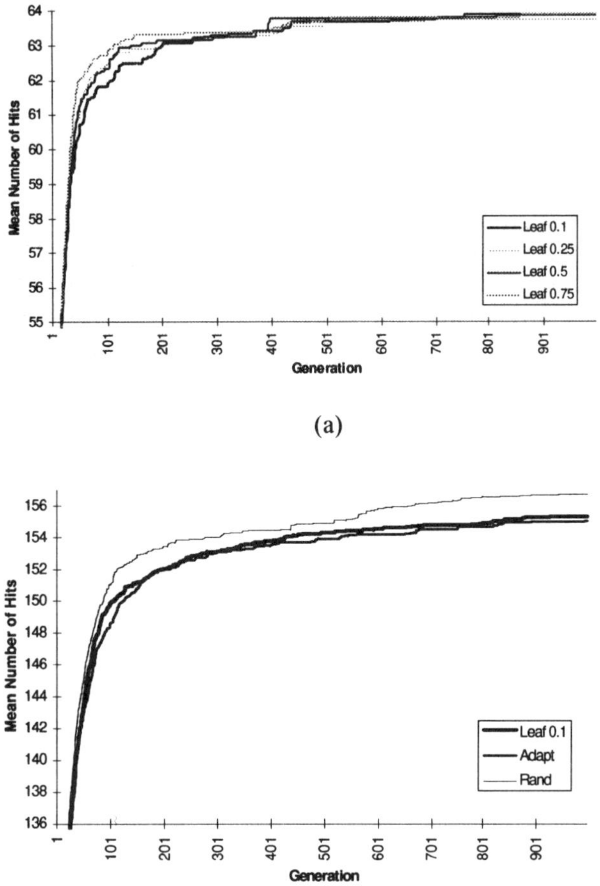

(a)

(b)

Figure 3: Results for the leaf frequency experiments on the Spiral Problem starting at generation 25. (a) Comparison of the different constant leaf frequency values over 30 trials. (b) Comparison of the performance between the adaptive and pseudo-adaptive to the traditional constant leaf frequency value of 0.1.

generation of completion that the rest of the techniques. Rand and Adapt were the only techniques in which all 30 trials found the correct solution. The maximum generation at which Rand found the solution was 306 while for Adapt it was 782.

A t-test between all pairings of techniques was performed to determine which if any of the difference in means for the generation of completion were significant. The performance of Rand was very nearly significant against all other methods but the current data only found a significant difference between Rand vs. Leaf 0.1 ($p<0.05$) and Rand vs. Leaf 0.25 ($p<0.05$). Additional runs may prove that Rand is statistically significantly better than all of the other methods on this problem. No other differences between the means were found to be significant.

Figure 2a graphs the average best-of-generation score for the various constant leaf frequencies on the Spiral problem. Once again there is a clear separation between the various techniques around generation 100 that in this problem continues until the end of the run. On this problem Leaf 0.75 performs slightly worse than the other techniques while Leaf 0.5 appears to perform slightly better on average until late in the run. Figure 2b graphs the results for the two dynamic leaf frequency methods against the performance of Leaf 0.1. As in the 6-Mux problem, Rand appear to perform better on average than both Adapt and Leaf 0.1. For this problem, the performance of Leaf 0.1 and Adapt are essentially identical.

Table 2 shows the mean best-of-run score over the 30 trials for the Spiral Problem for each of the methods. The mean scores are all very close after 1000 generations. Consequently, t-tests performed between all pairings of methods showed no statistical significance for the differences in the means.

Table 2: Mean Best-Of-Run Score for the Spiral Problem

Method	Mean Score
Leaf 0.1	155.27
Leaf 0.25	154.93
Leaf 0.5	155.30
Leaf 0.75	153.10
Adapt	155.00
Rand	156.70

Figure 4a graphs the mean prediction error for the best-of-generation individual for the various constant leaf frequencies on the Sunspot problem. In this problem the runs using constant leaf frequencies separate much earlier in their performance, with Leaf 0.75 starting to fall behind in performance around generation 35. For this problem, Leaf 0.25 appears to perform the best of the constant values. Figure 4b graphs the performance of the dynamic methods against the performance of Leaf 0.1. The performance of both dynamic methods are fairly similar to the performance of Leaf 0.1 for the entire run.

Table 3 lists the mean total prediction error received by the best-of-generation individuals for each of the methods investigated. T-test performed between each pairing of methods showed that there was a highly significant difference in the means of Leaf 0.25 vs. Leaf 0.75 with no significant difference found in all other pairings. Additional runs may also

Table 3: Mean Best-Of-Run Error for Sunspot Problem

Method	Mean Total Error
Leaf 0.1	-80347.098
Leaf 0.25	-75542.528
Leaf 0.5	-77821.334
Leaf 0.75	-84210.293
Adapt	-81926.267
Rand	-81971.782

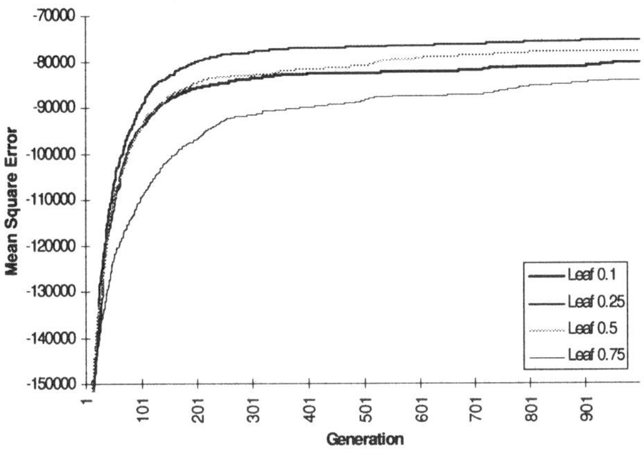

(a)

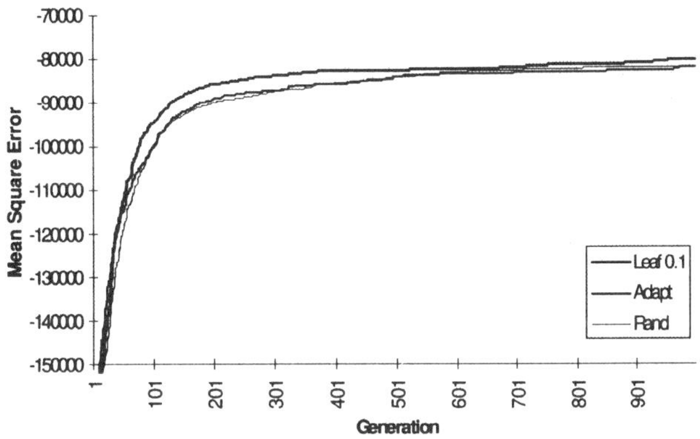

(b)

Figure 4: Results for the leaf frequency experiments on the Sunspot Problem starting at generation 10. (a) Comparison of the different constant leaf frequency values. (b) Comparison of adaptive and pseudo-adaptive method to the traditional leaf frequency value of 0.1.

show a significant difference between Leaf 0.5 vs. Leaf 0.75 as well.

In order to better understand the performance of the Adaptive method and compare it to the self-adaptive parametric results obtained in Angeline (1996) for leaf frequency, the average leaf frequency for each population was recorded for each of the trials in the study. Figure 5 graphs the mean average leaf frequency by generation over all trials for each of the three problems. The resemblance between the graphs is striking. In each, the values all begin around 0.5, as would be expected, and then all slowly rise to about 0.8 over about 100 generations and then stay at about that level for the rest of the run.

7.0 Discussion

There are several interesting issues raised by the results of this study. First and foremost it should be clear from the results above that no one constant value for leaf frequency is

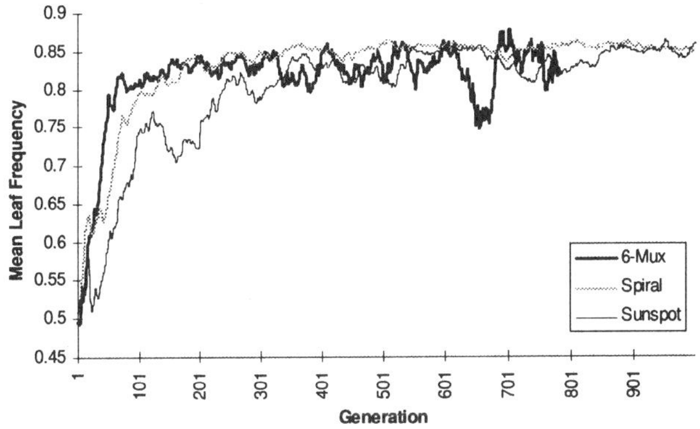

Figure 5: Graph showing the mean individual leaf frequency taken on by the population when using the adaptive leaf frequency. Results are averaged over 30 trials for each problem.

optimal for every problem. Further, the above results show that for some problems a leaf frequency as high as 0.75 may be a more beneficial value than a value close to zero. There is nothing apparent in the problems investigated here that suggest a priori when larger values would be more beneficial.

In general the two dynamic methods performed at least as well or better than most of the constant values for leaf frequency. And, counter to the initial intuition of the investigator, the pseudo-adaptive random method appears to occasionally outperform the self-adaptive method for adjusting leaf frequency.

One reason why the random method may have performed so well is that it allows a large number of different leaf frequencies in the population at the same time. Each of the constant methods, both parents will always have the same leaf frequency. The adaptive method will also place limits on the variety of leaf frequencies in the population since as certain population members do better, more population members with leaf frequency values close to their values will be introduced into the population. This is evidenced by the graphs in Figure 5. Because the Rand method allows a more diverse collection of leaf frequencies in the population. the distribution of possible crossovers is much more uniform on average. This may improve the probability of generating improved offspring over a range of problems.

The Adaptive leaf frequencies shown in Figure 5 suggest that the natural dynamic of leaf frequency when associated with individuals is to climb to 0.8 during the time that the most progress is being made on the problem and then stay there for the remainder of the run. This may be because as the generations pass crossover at internal nodes are more likely to create non-viable offspring while crossover performed at the leaves are generally less harmful. Thus the natural dynamics of this individual-level self-adaptive method is to select leaf frequency values that more consistently produce better children. Unfortunately, it may be that by reducing the number of internal points later in the run inhibits the potential performance of the Adaptive method. This is suggested by both the random pseudo-adaptive method and medium to lower constant values out performing the adaptive method on occasion. The component-level self-adaptive method investigated in Angeline (1996) generally gravitated to values closer to 0.5, suggesting that it was doing a better job of finding advantageous relative probabilities for selecting leaves over internal points. Direct comparisons between these two genetic programs is not possible since they place different restrictions on the legal depth of evolving trees.

8.0 Conclusions

The experiments and discussion above suggest that the most prudent method for selecting a leaf frequency for a genetic program may be to select a random leaf frequency for each parent just before selecting a crossover point. This procedure ensures a good mix of leaf frequencies and may distribute crossover operations more uniformly over the possible offspring.

Finally, the results of this study also suggest that simply removing the leaf frequency parameter from subtree crossover may actually be a prudent choice for many problems. Further experiments comparing the results here with runs that do not use a frequency are needed to determine if this is indeed the case.

Acknowledgments

Thanks to David Fogel for answering my statistical questions so completely and confirming my suspicions. Thanks also to everyone in the Advanced Technology Department at Loral Federal Systems - Owego who continue to allow me to usurp all of the department's machines during the off hours for experiments.

References

Angeline, P. J. (1993). *Evolutionary Algorithms and Emergent Intelligence.* Doctoral Dissertation. The Ohio State University, Computer and Information Science Department.

Angeline, P. J. (1994). Genetic Programming and Emergent Intelligence. In *Advances in Genetic Programming*, K. Kinnear (ed.), Cambridge, MA: MIT Press, pp. 75-96.

Angeline, P. J. (1995). Adaptive and Self-Adaptive Evolutionary Computations. In *Computational Intelligence: A Dynamic System Perspective*, M. Palaniswami, Y. Attikiouzel, R. Marks, D. Fogel and T. Fukuda (eds.), Piscataway, NJ: IEEE Press, pp. 152-163.

Angeline, P. J. (1996). Two Self-Adaptive Crossover Operations for Genetic Programming. In *Advances in Genetic Programming: Volume II*, P. Angeline and K. Kinnear (eds.), Cambridge MA: MIT Press. In Press.

Angeline, P. J. and Pollack, J. B. (1992). The Evolutionary Induction of Subroutines, In *Proceedings of the 14th Annual Conference of the Cognitive Science Society*, Hillsdale, NJ: Lawrence Erlbaum Associates, pp. 236-241.

Angeline, P. J. and Pollack, J. B. (1993). Evolutionary Module Acquisition. In *Proceedings of the Second Annual Conference on Evolutionary Programming*, D.B. Fogel and W. Atmar (eds.), La Jolla, CA: Evolutionary Programming Society, pp. 154-163.

Angeline, P. J. and Pollack, J. B. (1994). Coevolving High-Level Representations. In *Artificial Life III*, C.G. Langton (ed.), Addison-Wesley, Reading, MA, pp. 55-71.

Bäck, T. and Schwefel, H.-P. (1993). An Overview of Evolutionary Algorithms for Parameter Optimization, *Evolutionary Computation*, 1 (1), pp. 1-24.

Cramer, N. (1985) A Representation for the Adaptive Generation of Simple Sequential Programs. In Proceedings of an International Conference on Genetic Algorithms and their Applications, J. Grefensttete (ed.), Hillsdale, NJ: Lawrence Erlbaum.

Davis, L. (1989). Adapting Probabilities in Genetic Algorithms, *Proc. of the Second International Conference on Genetic Algorithms*, J.J. Grefenstette (ed.), Hillsdale, NJ: Lawrence Erlbaum, pp 61 - 69.

Fogel, D.B. (1995). *Evolutionary Computation: Toward a New Philosophy of Machine Intelligence*, Piscataway, NJ: IEEE Press.

Fogel, D.B., Fogel, L.J. and Atmar, J.W. (1991). Meta-Evolutionary Programming. *Proc. of the 25th Asilomar Conference on Signals, Systems and Computers*, R.R. Chen (ed.), San Jose, CA: Maple Press, pp. 540-545.

Fogel, D.B., Fogel, L.J., Atmar, J.W. and Fogel, G.B. (1992). Hierarchic Methods of Evolutionary Programming, *Proceedings of the First Annual Conference on Evolutionary Programming*, D.B. Fogel and W. Atmar (eds.), La Jolla, CA: Evolutionary Programming Society, pp. 175-182.

Fogel, L.J., Owens, A.J. and Walsh, M.J. (1966). *Artificial Intelligence through Simulated Evolution*, New York: John Wiley.

Fogel, L.J., Fogel, D.B. and Angeline, P.J. (1994). A Preliminary Investigation.on Extending Evolutionary Programming to Include Self-adaptation on Finite State Machines. *Informatica*, **18**, pp. 387-398.

Goldberg, D.E. (1989) Genetic Algorithms in Search, Optimization and Machine Learning, Reading,MA: Addison Wesley.

Holland, J.H. (1992). *Adaptation in Natural and Artificial Systems*, Ann Arbor, MI: University of Michigan Press.

Koza, J.R. (1992). *Genetic Programming: On the programming of computers by means of natural selection.* Cambridge, MA: MIT Press.

Koza, J. R. (1994). *Genetic Programming II: Automatic Discovery of Reusable Subprograms.* Cambridge, MA: MIT Press.

Oakley, E. H. N., (1994). Two scientific applications of genetic programming: stack filters and non-linear equation fitting to chaotic data. In *Advances in Genetic Programming*, K. Kinnear (ed.), Cambridge, MA: MIT Press, pp. 369-390.

Press, W. H., Teukolsky, S. A., Vetterking, W. T. and Flannery, B. P. (1992) *Numerical Recipes in C: The Art of Scientific Computing*, Second Edition.

Rechenberg, I. (1973). *Evolutionsstrategie: Optimierung Technischer Systeme nach Prinzipien der Biologischen Evolution*, Stuttgart: Frommann-Holzberg Verlag.

Rosenberg, R.S. (1967). *Simulation of genetic populations with biochemical properties*. Doctoral dissertation, University of Michigan, *Dissertation Abstracts International, 28 (7) 2732B*. (University Microfilms No. 67-17,836)

Schaffer, J.D. and Morishima, A. (1987). An Adaptive Crossover Distribution Mechanism for Genetic Algorithms. *Proc. of the Second International Conference on Genetic Algorithms*, J.J. Grefenstette (ed.), Hillsdale, NJ: Lawrence Erlbaum, pp. 36-40.

Schaudolph, N.N., and Belew, R.K. (1992). Dynamic parameter encoding for genetic algorithms. *Machine Learning,* **9** (1), pp 9-22.

Schwefel, H.-P. (1981). *Numerical Optimization of Computer Models*. Chichester, U.K: John Wiley.

Shaefer, C.G. (1987) The ARGOT System: Adaptive Representation Genetic Optimizing Technique. *Proc. of the Second International Conference on Genetic Algorithms*, J. Grefenstette (ed.), Hillsdale, NJ: Lawrence Erlbaum.

Whitley, D., Mathias, K. and Fitzhorn, P. (1991) Delta coding: an iterative search strategy for genetic algorithms. *Proc. of the Fourth International Conference on Genetic Algorithms*, R. K. Belew and L. B. Booker (eds.), San Mateo, CA: Morgan Kaufmann, pp 77-84.

Automatic Creation of an Efficient Multi-Agent Architecture Using Genetic Programming with Architecture-Altering Operations

Forrest H Bennett III
Visiting Scholar
Computer Science Department
Stanford University
Stanford, California 94305 USA
EMAIL: fhb3@slip.net
PHONE: 415-424-0948

ABSTRACT

Previous work in multi-agent systems has required the human designer to make up-front decisions about the multi-agent architecture, including the number of agents to employ and the specific tasks to be performed by each agent. This paper describes the automatic evolution of these decisions during a run of genetic programming using architecture-altering operations.

Genetic programming is extended to the discovery of multi-agent solutions for a central-place foraging problem for an ant colony. In this problem each individual ant is controlled by a set of agents, where *agent* is used in the sense of Minsky's *Society of Mind.*

We describe the simultaneous evolution of the number of agents needed to solve the problem and the work performing steps of each agent.

Genetic programming was able to evolve time-efficient solutions to this problem by distributing the functions and terminals across successively more agents in such a way as to reduce the maximum number of functions executed per agent. The other source of time-efficiency in the evolved solution was the cooperation that emerged among the ants in the ant colony.

1. Introduction

The thesis that minds are architecturally organized into *agents* has been presented in the literature of several different fields. Minsky (1985) presents the following notion of agents:

> "I'll call 'Society of Mind' this scheme in which each mind is made of many smaller processes. These we'll call *agents*. Each mental agent by itself can only do some simple thing that needs no mind or thought at all. Yet when we join these agents in societies -- in certain very special ways -- this leads to true intelligence."
>
> "To explain the mind, we have to show how minds are built from mindless stuff, from parts that are much smaller and simpler than anything we'd consider smart. Unless we can explain the mind in terms of things that have no thoughts or feelings of their own, we'll only have gone around in circles."

In cognitive science, Dennett 1991 argues that "there is no central Headquarters" in the mind, but instead says that "specialist circuits" operate in "parallel pandemoniums."

In psychology, Baars 1988 summarizes a "gathering consensus" in his field that minds are implemented as a "distributed society of specialists."

In this paper we use the term *agents* in the sense of these references -- that is, the narrow specialist modules that operate in parallel to create an effective overall problem solving organism.

Herein, an agent corresponds to a single branch of the program tree for an individual in the population of the genetic programming system.

Are the minds of real ants organized into agents? This is not known. However, it is known that a typical ant species has as many as 24 to 40 different categories of behaviors; and some species seem to have as many as 50 categories of behaviors (Oster and Wilson 1978). It is easy to envision that each of these behaviors corresponds to one or more agents.

Multi-agent systems are also attractive because of the potential gain in efficiency from their inherent parallelism.

1.1. Central-Place Foraging in an Ant Colony

Social insects such as ants, wasps, bees, and termites dominate the insect fauna and together constitute more than 75% of the insect biomass (Fittkau and Klinge 1973). The success of these insects is due in large part to the survival advantage accruing from their cooperation in a complex social organization and to the efficiencies of specialization that this cooperation makes possible. Of the social insects, ants have the most complex social organizations (Holldobler and Wilson 1990).

The problem presented in this paper is a single aspect of an ant colony's cooperative behavior, namely an abstraction of the central-place foraging problem faced by worker ants. In the version of the problem presented here, as in nature, a single worker ant can perform food foraging competently (Holldobler and Wilson 1990). However, cooperation yields more efficient foraging.

Food sources are initially located by individual *Solenopsis invicta* ants using random search (Holldobler and Wilson 1990). Once a food source is located, the ants lay an alpha-farnesene recruitment pheromone on the ground while returning to the nest, thereby creating a trail of pheromones connecting the food source and the nest. When the odor of this pheromone is detected by other worker ants, they respond by following the pheromone trail. Thus, other ants in the colony are able to go much more directly to a food source that has already been found without having to search for the food. The pheromone-dropping behavior together with the pheromone trail-following behavior is how the colony cooperates to perform food foraging in a more efficient manner than could be accomplished by each ant performing an independent random search. The cooperative behavior is accomplished with each ant responding only to locally available information.

In this paper, the behavior of each individual ant is controlled by a multi-agent algorithm (multi-branch program) where each agent (branch) is executed in parallel. A solution to an ant colony food foraging problem where each ant is running a single sequential algorithm was written by Resnick (1991), and another sequential algorithm was evolved using genetic programming in Koza 1992 (Chapter 12).

The problem in this paper is a harder version of the problem described in Koza 1992 because the function and terminal set used by Koza was much more powerful than the primitives used here (see sections 4.1 and 4.2).

1.2. Previous Genetic Programming Multi-Agent Work

Previous work in multi-agent systems has required the human designer to make up-front decisions about the multi-agent architecture, including the number of agents to employ and the specific tasks to be performed by each agent. This paper describes the automatic evolution of these multi-agent architectures during a run of genetic programming using architecture-altering operations.

In Andre 1995, agents were evolved where the number of agents (two) and the task of each agent was predetermined. In Ryan 1995, the conditions under which each agent would activate was preset, as were the number of agents. In Haynes, et al 1995 the number of agents was preset to four. These were not Minsky-style agents, but rather members of a cooperating team that each controlled its own body. Each agent in Crosbie and Spafford 1995 was generated in a separate run of genetic programming; they were not co-evolved.

1.3. Previous Genetic Programming Work with New Programming Elements

Genetic programming has previously been extended to automatic definition and manipulation of well known programming elements while simultaneously solving a problem. Examples include subroutines with parameters (Koza 1994), program architecture (Koza 1995), generic control structures such as macros (Spector 1996), and iteration (Koza and Andre 1996).

This paper extends genetic programming to the automatic definition and manipulation of parallel algorithms and parallel architectures.

2. The Foraging Problem

Genetic programming is applied to the problem of finding a multi-agent parallel algorithm that, when executed by each ant in a colony, causes efficient central-place foraging behavior in the ant colony. The behavior of each individual ant in this problem is controlled by multiple agents.

The fitness function used in this problem penalizes the ant colony for the time it uses to find and transport each food pellet to the nest and penalizes the colony heavily for each food pellet it does not transport to the nest. Thus there is selection pressure for transporting all the food and for doing so efficiently. A colony in which each ant randomly searches for each food pellet will take a long time to gather the food and thus be less fit. The ant colony can be more competitive by discovering how to cooperate.

The second way that the ant colony can become more efficient (and thus more competitive) is by distributing the functions and terminals in the program tree into as many agents as possible, while still solving the problem. The time-efficiency of a multi-agent algorithm is measured in terms of the greatest number of functions executed in any one of the agents. Thus an algorithm that executes 100 functions in 4 agents (with 25 functions per agent) will be twice as time-efficient as an algorithm that executes 100 functions in 2 agents (with 50 functions per agent). This will be discussed in greater detail in section 4.3.

2.1. The Simulated World

Each of the 20 ants in the colony are controlled by the same multi-agent algorithm (i.e., an individual multi-branch

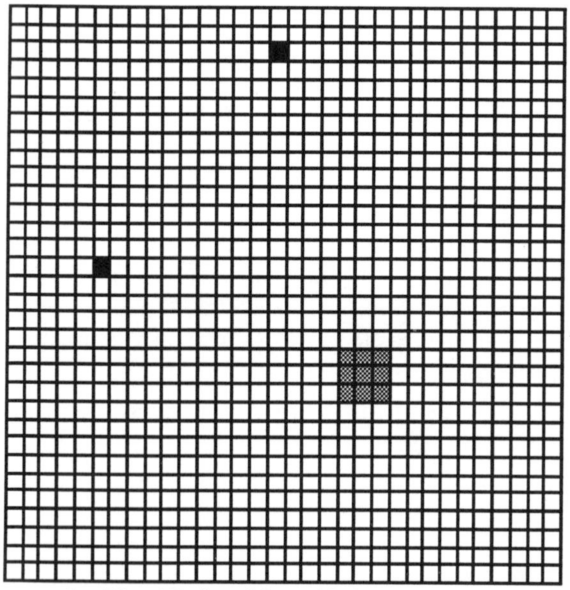

Figure 1 The 32 by 32 grid for the toroidal ant colony world. The two food locations with 72 food pellets each are shown in black. The nine grid locations of the nest are shown in gray. (Location (0, 0) is in the lower left corner.)

program tree in the genetic programming system's population).

The ant colony's world consists of a 32 by 32 toroidal grid shown in figure 1. Each of the colony's ants begins each fitness evaluation at a random grid location within 5 squares of the center of a 3 by 3 nest that is centered at grid location (21, 11).

Each ant is initially facing in a random direction. There are a total of 144 food pellets located in two separate grid locations that each have 72 pellets. One food area is at (5, 17) and the other at (15, 29).

The ant colony is simulated until either all the food pellets have been transported to the nest, or the maximum number of allotted time steps (4,000) is reached.

On each time step of the simulation, the evolved multi-agent algorithm for the colony is evaluated once for each of the 20 ants in the colony. Evaluating the multi-agent algorithm for a single ant involves evaluating each of the agents of the algorithm in a manner that simulates parallel execution.

The result of each agent evaluation is a single action (designated by a terminal) to be performed in the world. The action (result terminal) of each agent is not executed until all of the agents have been evaluated. This delay of the actions insures that all the functions (conditional tests) are evaluated as if they were in parallel, with the world and the ant in the same state. Thus the order of the agents (branches) in the overall program tree is irrelevant.

After each of the agents (branches) of the program tree have been evaluated, the designated terminal actions are executed to change the state of the ant and the world. Each specific behavior (corresponding to a terminal) is only executed one time per ant per time step even if more than one agent evaluates to that action. Regardless of the order of the agents, the actions are always executed in the following order: GRAB-FOOD, UNCONDITIONAL-DROP-PHEROMONE, TURN-LEFT, TURN-RIGHT, MOVE-RANDOM, and MOVE-FORWARD. The two movement actions are executed last to insure that all the actions take place in the ant's current location before any movement occurs.

Figure 2 shows an example program tree with three agents. Suppose this program tree is evaluated for ant on a grid location containing food and that the ant is not currently carrying food. Since IF-FOOD-HERE is true the result of the first (leftmost) agent will be the GRAB-FOOD terminal. The result of the second (middle) agent will always be the MOVE-FORWARD terminal. The result of the third (rightmost) agent will be the MOVE-RANDOM terminal, since the ant is not currently carrying food. So the results of evaluating all the agents is the set of terminals {GRAB-FOOD, MOVE-FORWARD, MOVE-RANDOM}. Now that we have the terminals from each of the agents, the three actions are executed. First, GRAB_FOOD removes one food pellet from the current grid location and sets the state of the ant to indicate that it is carrying food. Second, the MOVE_FORWARD changes the ant's grid location by one square in the direction the ant is currently facing. Third, the MOVE-RANDOM is executed. (See the details on the set of terminals below.)

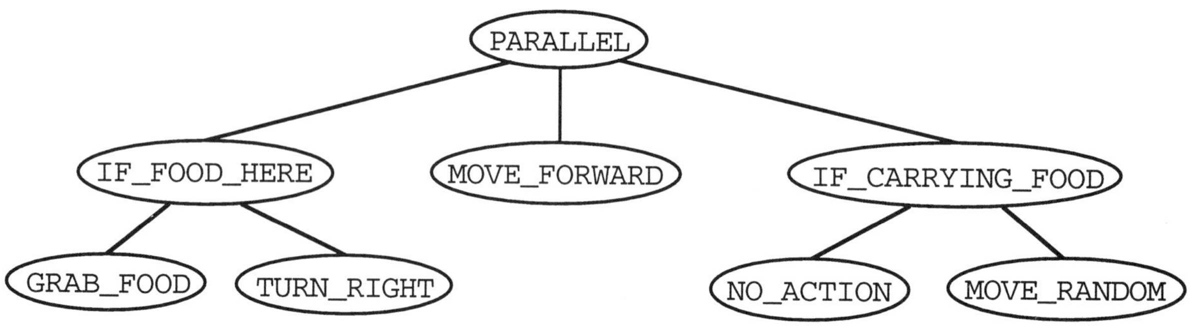

Figure 2 An example program tree showing an individual with three agents. Each agent is represented by one of the three branches connected by the PARALLEL function.

3. Architecture-Altering Operations

Two operations, branch duplication and branch deletion, are used to create the multi-agent architecture.

3.1. Branch Duplication Operation

Branch duplication involves creating a duplicate copy of one of the branches (agents) in an individual multiple agent parallel program. This operation was inspired by naturally occurring gene duplication (Ohno 1970).

The branch duplication operation used here is a simpler version of the one presented in Koza (1995). Koza's version of branch duplication takes into account modifications to hierarchically referenced branches. However, the branches used for this problem do not have any hierarchical references. Thus the steps involved to handle these references during branch duplication are not required.

This operation is applied to 3% of the population in each generation. The branch duplication operation used here works in the following way:

(1) Select an individual from the population based on fitness using tournament selection with a tournament size of 7.

(2) Randomly pick one of the branches in this individual as the branch-to-be-duplicated.

(3) For the branch picked, make a duplicate copy of that branch, and insert it into the individual. This increases by one the number of branches in the individual. If the individual selected already has the maximum number of allowed branches, then do nothing.

Initially this new duplicate branch has no effect, since it does the same thing as the branch from which it was copied. However, the two branches may diverge in future generations because of crossover or mutation.

3.2. Branch Deletion Operation

Branch deletion involves deleting one of the branches (agents) in an individual multiple agent parallel program.

The branch deletion operation used here is also simpler than Koza's because there are no hierarchical references.

This operation is applied to 1% of the population in each generation. The branch deletion operation used here works in the following way:

(1) Select an individual from the population in the same way as was done for branch duplication in the previous section.

(2) Randomly pick one of the branches in this individual as the branch-to-be-deleted.

(3) For the branch picked, delete it from the selected individual. This decreases by one the number of branches in the individual. If the individual selected has only a single branch, then do nothing.

This operation will affect the behavior of the individual unless the deleted branch is a duplicate copy of another branch, the deleted branch was not doing anything, or the effects of the deleted branch are the same as the effects of some combination of one or more of the other branches.

4. Preparatory Steps for the Genetic Programming Run

All the programs in the initial population of generation 0 are created with a single branch (agent). Thus the run begins with all programs in the initial population having the minimum architectural structure.

All the functions used in this problem correspond to percepts and all the terminals used correspond to actions that change the state of the world or the ant.

The parameters for this problem are shown in Table 1.

4.1. Function Set

All the functions in this problem are two-argument conditional branches that have no side effects. The function set consists of the eight functions that are all available to every branch (agent). The effects of evaluating the various functions are as follows:

- `IF-FOOD-HERE` is a two-argument conditional branching function that executes its first branch if there is a food pellet at the same location on the grid as the ant, but otherwise executes its second branch.
- `IF-FOOD-FORWARD` is a two-argument conditional branching function that executes its first branch if there is a food pellet in the grid location adjacent to the ant's location in the direction the ant is facing, but otherwise executes its second branch.
- `IF-CARRYING-FOOD` is a two-argument conditional branching function that executes its first branch if the ant is carrying a food pellet, but otherwise executes its second branch. This conditional will be true if the ant has grabbed a food pellet and the ant has not yet returned to the nest.
- `IF-SMELL-FOOD` is a two-argument conditional branching function that executes its first branch if there is at least one food pellet on one or more of the four (north, east, south, and west) grid locations adjacent to the ant's current location, but otherwise executes its second branch.
- `IF-NEST-HERE` is a two-argument conditional branching function that executes its first branch if the ant is currently in a grid location that is one of the grid locations of the nest, but otherwise executes its second branch. When an ant carrying a food pellet enters the nest, that food pellet is implicitly taken from the ant and dropped into the nest.
- `IF-FACING-NEST` is a two-argument conditional branching function that executes its

Table 1 Tableau for the central-place foraging problem for an ant colony.

Objective:	Find a multi-agent algorithm that, when executed by each ant in a colony of 20 ants, causes efficient central-place foraging behavior in the ant colony.
Terminal set:	MOVE_FORWARD, TURN_RIGHT, TURN_LEFT, MOVE_RANDOM, GRAB_FOOD, UNCONDITIONAL_DROP_PHEROMONE, and NO_ACTION.
Function set:	IF_FOOD_HERE, IF_FOOD_FORWARD, IF_CARRYING_FOOD, IF_NEST_HERE, IF_FACING_NEST, IF_SMELL_FOOD, IF_SMELL_PHEROMONE, and IF_PHEROMONE_FORWARD.
Fitness cases:	The ant colony's world consists of a 32 by 32 toroidal grid. There are a total of 144 food pellets located in two separate grid locations that each have 72 pellets. One food area is at (5, 17) and the other at (15, 29).
Raw fitness:	For each food pellet that the colony transports to the nest, add to the fitness the product of the following: • The number of time steps elapsed when the food pellet arrived at the nest. • The number of sequential IF functions executed by the ant who delivered the food pellet. For each food pellet that the colony does not transport to the nest, add to the fitness the product of the following: • The manhattan distance between the food pellet and the nest. • The maximum number of sequential IF functions that an ant can execute in the allotted number of time steps. • The maximum allotted time steps. The fitness calculated above is divided by 1,000,000.
Standardized Fitness:	Equals raw fitness.
Hits:	The number of food pellets that the colony has transported back to the nest within the allotted time steps. The maximum number of hits is 144.
Parameters:	• Population size, M = 64,000. • The percentages for the genetic operations were: 75% for internal point crossover, 10% for leaf point crossover, 10% for reproduction, 3% for branch duplication, 1% for branch deletion, and 1% for mutation. • A maximum size of 100 points was allowed for each agent. A maximum of 7 agents were allowed per individual, so the maximum size of an individual was 700 points. • The other parameters for controlling the runs of genetic programming were the default values specified in Koza (1994).
Success predicate:	None.

first branch if the ant is currently facing in the direction of the nest, but otherwise executes its second branch. The ant is considered to be facing in the direction of the nest if and only if the execution of a MOVE-FORWARD terminal would not increase the ant's manhattan distance from the center of the nest. Real ants are able to navigate back to the nest using the polarized light in the sky (Holldobler and Wilson 1990).

• IF-SMELL-PHEROMONE is a two-argument conditional branching function that executes its first branch if there is at least one pheromone on one or more of the four (north, east, south, and west) grid locations adjacent to the ant's current location, but otherwise executes its second branch.

• IF-PHEROMONE-FORWARD is a two-argument conditional branching function that executes its first branch if there is a pheromone in the grid location adjacent to the ant's location in the direction the ant is facing, but otherwise executes its second branch.

The function set used by Koza (1992) in his version of the central-place foraging problem for an ant colony was powerful. Indeed, some of the built-in functions provided to the genetic programming system were already as complex as the kind of agents that we wanted to evolve in the experiment described here. For example, Koza's function MOVE-TO-ADJACENT-FOOD-ELSE combines a test of all the adjacent positions in the ants environment, with a move into the location with the food pellet in it. Thus the evolutionary process did not have to learn which location contained the food pellet, and did not have to link this information with the action of moving into that location. Most importantly, it did not have to discover that linking that information with that behavior was a useful thing to do for this problem. In the version of the problem presented here, genetic programming has to discover an equivalent to Koza's MOVE-TO-ADJACENT-FOOD-ELSE function by using some combination of: IF-SMELL-FOOD, TURN-LEFT, TURN-RIGHT, MOVE-RANDOM, IF-FOOD-FORWARD, and MOVE-FORWARD. The same point applies to Koza's rather powerful function MOVE-TO-ADJACENT-PHEROMONE-ELSE.

Koza also uses a PROGN function that allows a list of arguments to be evaluated in series. This function, in effect, provides for a kind of state information to be used. For example, in the branch (PROGN (MOVE-FORWARD) (TURN-LEFT)), the branch is able to execute a TURN-LEFT in the state in which a MOVE-FORWARD has just occurred. Omitting PROGN prevents the use of this kind of implicit state information, and prevents solutions that merely encode a route to the food with a sequence of moves and turns.

4.2. Terminal Set

All the terminals in this problem are actions that have effects on the state of the world or the ant. The terminal set consists of the following seven terminals that are all available to every branch (agent). The effects of evaluating the various terminals are as follows:

- MOVE-FORWARD changes the ant's location to the adjacent grid location in the direction that the ant is currently facing.
- TURN-RIGHT changes the direction the ant is facing by 90 degrees clockwise.
- TURN-LEFT changes the direction the ant is facing by 90 degrees counter-clockwise.
- MOVE-RANDOM changes the direction the ant is facing to one of the four possible directions (north, east, south, and west) with equal probability and then moves the ant two squares forward in that direction.
- GRAB-FOOD takes one food pellet from the grid location where the ant is currently located, if and only if there is a food pellet at that location and the ant is no already carrying a food pellet.
- UNCONDITIONAL-DROP-PHEROMONE drops a pheromone cloud into a 3 by 3 square centered at the current location of the ant. The pheromones decay linearly such that they are no longer present on the grid locations 20 simulation time steps after they are dropped.
- NO-ACTION does nothing, and was included so that an agent could decide that no action was appropriate for it in a given time step based on the percepts it had evaluated.

The DROP-PHEROMONE terminal used in Koza 1992 is a powerful operator that only executes when the ant is carrying food. Again, the discovery of the utility of only using the DROP-PHEROMONE terminal under certain conditions is an example of the kind of agent that we wanted to evolve. The UNCONDITIONAL-DROP-PHEROMONE terminal used here always drops pheromones.

The MOVE-TO-NEST terminal used by Koza combines the perception of the direction of the nest with the action of movement towards the nest. Again, we seek to evolve an agent that combines the percept and the action. This experiment uses movement terminals that are separate from the IF-FACING-NEST percept function.

4.3. Fitness and Hits

The colony's hits are the number of food pellets that it transports to the nest.

The fitness measure combines the amount of time used by the colony to transport food to the nest and the number of functions executed to do so; and it includes a large penalty for each food pellet not transported to the nest. The fitness formula is shown in figure 3.

In figure 3, the number of sequential IF functions executed by an ant to transport a food pellet to the nest is determined as follows. When the agents in the program tree are evaluated for a single time step for a single ant, the maximum number of functions evaluated by one of the

$$\frac{\Sigma^n(t_{food} * f_{food}) + \Sigma^m(t_{max} * f_{max} * d_{food})}{1{,}000{,}000}$$

where:

n	= *Number of food pellets transported to the nest.*
t_{food}	= *Number of time steps elapsed when the food pellet arrived at the nest.*
f_{food}	= *Number of sequential IF functions (as defined below) executed by the ant who transported the food pellet.*
m	= *Number of food pellets not transported to nest.*
t_{max}	= *Maximum allotted time steps = 4,000.*
f_{max}	= *Maximum possible value of* f_{food} = $t_{max} * p_{max}$ = *400,000.*
d_{food}	= *Manhattan distance between food pellet and nest.*
p_{max}	= *Maximum number of points per agent = 100.*

Figure 3 The equation for the fitness function.

agents is counted as the number of sequential IF functions executed by that ant for that time step. A running total of this count is stored for each ant, and the count for an ant is incorporated into the fitness as noted above and then cleared each time the ant arrives at the nest with a food pellet.

Thus, any food pellets not transported to the nest are penalized as though they were brought the remaining distance to the nest by the largest size agent possible that also takes the maximum number of time steps to transport the food to the nest.

The colony's fitness is calculated for a single fitness case, that is for a single world configuration of nest location, food pellet locations, and total number of food pellets. However, note that each fitness evaluation is in two minor senses a distinct fitness case. First, the successive executions of the MOVE-RANDOM terminal will move in different directions in different fitness evaluations. Second, the starting positions of the ants and the directions the ants are facing are different each time the colony is executed. Therefore, an algorithm could not rely on a particular sequence of MOVE-RANDOM executions across generations.

Even though the food is located in the same places in each fitness evaluation, the colony has no way with the given function and terminal set to encode the location of the food (not even implicitly). Therefore, the colony must actually search for the food in each fitness evaluation.

4.4. Termination Criterion and Results Designation

There was no termination criterion for this problem since a fitness of zero was not possible, and there was no way of knowing in advance how good of a fitness might evolve. The run was terminated manually after the fitness value of the best-of-run individual had appeared to plateau.

4.5. Parallel Computer System

The hardware used for this problem is a 64 node parallel computer where each node consists of an 80 MHz PowerPC 601 microprocessor with 32M bytes of RAM. The inter-process communication is handled by an INMOS Transputer at each node and runs at 10M bits/sec. The 64 nodes are connected in a toroidal grid so that each node communicates only with its four nearest neighbors.

Each node runs a distinct copy of the genetic programming system with a population of 10,000 individuals. On each generation, 2000 individuals are selected based on fitness to migrate to each of the four neighboring nodes. See Andre and Koza 1996.

The four distinct levels of parallelism employed in this problem are: the physical parallelism of the 64 processors of the actual hardware used, the conceptual parallelism of the individuals in the population of programs in the genetic programming system, the simulated parallelism of the 20 ants within one ant colony, and the simulated parallelism of the multiple agents controlling a single ant.

5. Results

The best individual of generation 0 has a fitness value of 4,258,013, scored 26 hits, and has a single agent. (All the individuals in generation 0 have a single agent.) The effect of this agent is to wander randomly until it finds food, grab any food that it finds, and randomly wander back to the nest. It does not use smell, orient to the nest, or cooperate in any way with the other ants in the colony. The program tree for this individual is shown here in simplified form:

```
(IF_FOOD_HERE
   (IF_CARRYING_FOOD
      (MOVE_FORWARD) (GRAB_FOOD))
   (MOVE_RANDOM)).
```

The first best-of-generation individual to score all 144 possible hits occurred in generation 7, had a fitness value of 568, and contained two agents. This individual has 79 points in the first agent and 93 points in the second agent.

The best fitness drops by 3 orders of magnitude in generation 7 (see Figure 4). This large drop in fitness is because the best-of-generation colony in generation 7 transported all the food to the nest, and thus does not incur the large fitness penalty for un-transported food. From this generation forward, all the best-of-generation individuals score 144 hits. However, the best fitness continues to decrease (or remain the same) in every generation because the best-of-generation individuals were able to get more efficient between generations 7 and 90. Figure 4 shows that between generations 7 and 90, the number of agents in the best-of-generation individual increases as the best-of-generation fitness decreases.

In generation 87, the best-of-generation individual had 7 agents. This was the only generation in which the best individual had the maximum number of agents.

The best individual of the run appeared in generation 90, had a fitness value of 7.4, and scored 144 hits. This individual has 6 agents, with a total of 380 points (i.e., functions and terminals) in all 6 agents combined. In order to determine the behavior of each of the agents in the best-of-run individual, each agent was isolated and allowed to control the colony by itself. The qualitative behavior of the colony when controlled by each of the individual agents alone are as follows:

- *Agent 0:* The colony divides into two groups. One group moves only upwards on the grid, and the other group moves only to the right on the grid. None of the ants pick up any food or drop pheromones.
- *Agent 1:* Each of the ants in the colony wanders randomly until it finds food. Then it stops next to that food and does nothing else.
- *Agent 2:* The colony divides into three groups. One group moves only upwards, another group

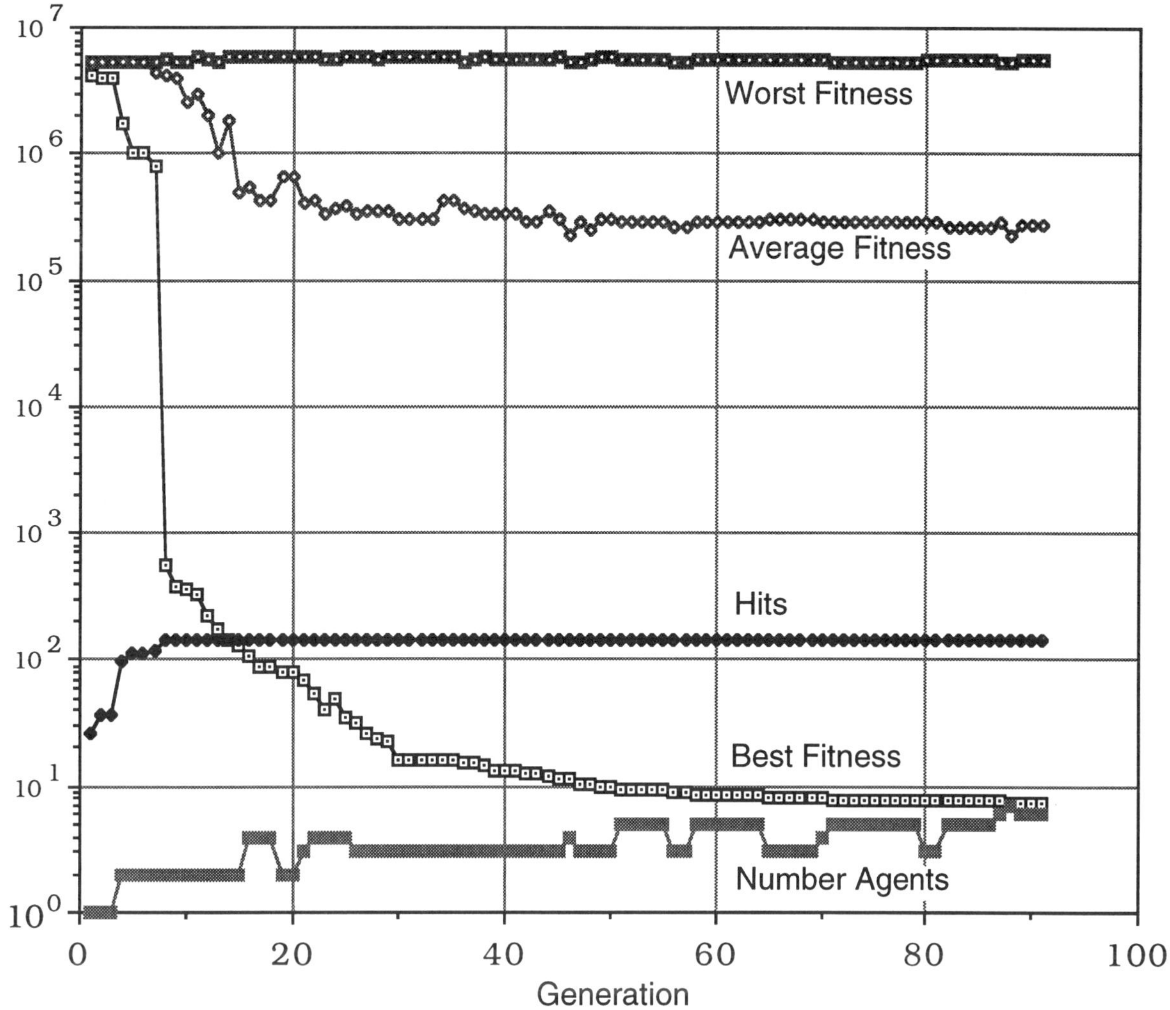

Figure 4 Log plot by generation of Worst-fitness, Average-fitness, Hits for best-of-generation, Best-fitness, and Number of agents for best-of-generation.

moves only to the right, and the last group just stands still next to the nest.

• *Agent 3:* Each of the ants in the colony wanders randomly until it finds food. Then it stops on top of the grid location containing the food and does nothing else.

• *Agent 4:* Apparently the same behavior as agent 0, but not clear since not all of an agent's program tree necessarily gets executed when an agent is run in isolation. This is because (for example) one agent may be depending on another to drop pheromones.

• *Agent 5:* Behaves the same as agent 3.

When both agent 0 and agent 1 are run together to control the colony, the ants perform random foraging, and grab the food and transport it back to the nest, but do not drop pheromones or cooperate in any way.

When the three agents 0, 1, and 2 are all run together to control the colony, the ants perform the full cooperative pheromone-dropping and pheromone-trail-following behavior, but still do not perform as efficiently as all 6 of the agents running together.

When all 6 agents are run together and viewed in an animation, one can observe how the colony solves the problem. First, each ant wanders randomly until it finds food, whereupon it grabs a food pellet and proceeds directly back to the nest while dropping pheromones along the way. Other randomly wandering ants that come upon this pheromone trail follow it to the food and join the food carrying and pheromone dropping. The recruitment, carrying, and pheromone dropping continue until the food source is exhausted. Then the ants resume random wandering and the pheromone trail drys up. Sometimes the second food source is discovered while the first food source is still being exploited. In this case, the colony works on both food sources at the same time, and when one of the food sources is exhausted all of the recruited ants focus on the remaining food source.

The best individual was tested 10,000 times to

determine how well it would generalize over many different random sequences of MOVE-RANDOM. For each of these 10,000 trials this individual scored the maximum possible of 144 hits. The average fitness in these 10,000 trials was 14.6 (compared to 7.4 during the run) and the standard deviation was 6.3.

This run produced many individuals that scored the maximum possible 144 hits. Individuals that scored 144 hits were found that contained every number of agents from two to the maximum allowed (seven).

6. Conclusions

Genetic programming was successfully used to discover multi-agent algorithms that solve a central-place foraging problem for an ant colony. Genetic programming was able to simultaneously evolve time-efficient multi-agent algorithms, multi-agent architectures, and emergent cooperation within the colony of ants.

Time-efficient solutions to this problem were evolved by distributing the functions and terminals across successively more agents in such a way as to reduce the maximum number of functions executed per agent. The other source of time efficiency in the evolved solution was the cooperation that emerged in the ant colony among the ants.

Acknowledgements

The author gratefully acknowledges David Andre for the use of his genetic programming code and for his help in using it, John Koza for his encouragement to undertake this project, Mark Roulo for his discussions during the project, and Simon Handley and Scott Brave for reading and critiquing drafts of this paper.

Bibliography

Andre, David. 1995. The evolution of agents that build mental models and create simple plans. In Eshelman, Larry J. (editor). *Proceedings of the Sixth International Conference on Genetic Algorithms.* San Francisco, CA: Morgan Kaufmann.

Andre, David and Koza, John R. 1996. Parallel genetic programming: A scalable implementation using the transputer architecture. In Angeline, Peter J. and Kinnear, Kenneth E. Jr. (editors). 1996. *Advances in Genetic Programming 2.* Cambridge, MA: The MIT Press.

Baars, B. 1988. *A Cognitive Theory of Consciousness.* Cambridge: Cambridge University Press.

Crosbie, Mark and Spafford, Eugene H. 1995. Applying genetic programming to intrusion detection. *Proceedings of the AAAI-95 Fall Symposium Series on Genetic Programming.*

Dennett, Daniel C. 1991. *Consciousness Explained.* Boston: Little, Brown.

Fittkau, E. J., and H. Klinge. 1973. On biomass and trophic structure of the central Amazonian rain forest ecosystem. *Biotropica,* 5(1): 2-14.

Goldberg, David E. 1989. *Genetic Algorithms in Search, Optimization, and Machine Learning.* Reading, MA: Addison-Wesley.

Haynes, Thomas, Sen, Sandip, Schoenefeld, Dale, and Wainwright, Roger. 1995. Evolving a team. *Proceedings of the AAAI-95 Fall Symposium Series on Genetic Programming.*

Holldobler, Bert, and Wilson, Edward O. 1990. *The Ants.* Cambridge, MA: The Belknap Press of Harvard University Press.

Kinnear, Kenneth E. Jr. (editor). 1994. *Advances in Genetic Programming.* Cambridge, MA: The MIT Press.

Koza, John R. 1992. *Genetic Programming: On the Programming of Computers by Means of Natural Selection.* Cambridge, MA: The MIT Press.

Koza, John R. 1994. *Genetic Programming II: Automatic Discovery of Reusable Programs.* Cambridge, MA: The MIT Press.

Koza, John R. 1995. Gene duplication to enable genetic programming to concurrently evolve both the architecture and the work performing steps of a computer program. *Proceedings of the 1995 International Joint Conference in Artificial Intelligence.*

Koza, John R., and Andre, David. 1996. Evolution of iteration in genetic programming. In *Evolutionary Programming V: Proceedings of the Fifth Annual Conference on Evolutionary Programming.* Cambridge, MA: MIT Press. In Press.

Minsky, Marvin. 1985. *The Society of Mind.* New York: Simon & Schuster.

Ohno, Susumu. 1970. *Evolution by Gene Duplication.* New York: Springer-Verlag.

Oster, G. F., and Wilson, E. O. 1978 *Caste and ecology in the social insects.* (Monographs in Population Biology, no. 12). Princeton University Press, Princeton, N.J.

Resnick, Mitchel. 1991. Animal simulations with *Logo: Massive parallelism for the masses. In Meyer, Jean-Arcady, and Wilson, Stewart W. (editors), *From Animals to Animtats: Proceedings of the First International Conference on Simulation of Adaptive Behavior.* MIT Press.

Ryan, Conor. 1995. GP robots and GP teams - competition, co-evolution and co-operation in genetic programming. *Proceedings of the AAAI-95 Fall Symposium Series on Genetic Programming.*

Spector, Lee. 1996. Simultaneous evolution of programs and their control structures. In Angeline, P. and Kinnear, K. (editors) 1996. *Advances in Genetic Programming 2.* Cambridge, MA: MIT Press.

Evolving Deterministic Finite Automata Using Cellular Encoding

Scott Brave
Computer Science Dept.
Stanford University
Stanford, California 94305
brave@cs.stanford.edu

ABSTRACT

This paper presents a method for the evolution of deterministic finite automata that combines genetic programming and cellular encoding. Programs are evolved that specify actions for the incremental growth of a deterministic finite automata from an initial single-state zygote. The results show that, given a test bed of positive and negative samples, the proposed method is successful at inducing automata to recognize several different languages.

1. Introduction

The automatic creation of finite automata has long been a goal of the evolutionary computation community. Fogel et. al. [1966] was the first to propose the generation of deterministic finite automata (DFAs) by means of an evolutionary process, and the possibility of inferring languages from examples was initially established by Gold [1967]. Since then, much work has been done in the induction of DFAs for language recognition. Tomita [1982] showed that hill-climbing in the space of nine-state automata was both successful and superior to enumerative search at inducing a diverse set of seven regular languages. This group of seven languages, seen in table 1, has become a benchmark for work in language induction and will be the main languages investigated in this paper. The automata produced by Tomita's method, however, rarely generalized to out-of-sample sentences from the desired target language.

Pollack [1991] was able to produce DFAs with better out-of-sample performance using a recurrent high-order back-propagation network, but had difficulties with convergence on Tomita's languages TL2 and TL6. Other work has applied neural networks to the induction of higher order Chomsky languages (e.g. context-free and Turing) with varying success ([Giles et. al. 1990]; [Giles et. al. 1992]; [Waltrous and Kuhn 1992]; [Williams and Zisper 1988]). The majority of these methods, however, require some part of the automata structure to be specified by the user before being applied. In particular, the user almost always has to specify the number of states.

Genetic programming (GP) [Koza 1992] provides the flexibility to evolve the structure and function of automata in their entirety. Dunay et. al. [1994] describes a genetic programming encoding which successfully produces generalizing automata for several languages, including all of the Tomita regular languages.

This paper presents a method for applying cellular encoding, described by Gruau [1994], to the evolution of DFAs. The genetic programming tree is interpreted as a coding for the growth of an automata from an initial single-state zygote. Mimicking the process of cell division in nature, execution of a GP tree causes the progressive division and development of cells in the embryo to form the fully grown automata. The encoding ensures that the GP trees are always valid, specify automata that contain no orphan or sink states, and can be crossed over normally. The encoding also allows for back pointers (cycles) to be generated easily. The results show that the method is successful in inducing several target languages, including all but one of the Tomita regular languages (TL6).

Table 1 Tomita's benchmarks.

Language	Description	Example Sentence in the Language	Example Sentence NOT in the Language
TL1	b*	bbbbb	bbaba
TL2	(ba)*	babababa	abbaaab
TL3	no odd a strings after odd b strings	aaabaabb	abbbaaaaa
TL4	no aaa substrings	bbabbaa	abbaaaba
TL5	pairwise, an even sum of ab's and ba's	aabbaaaba	aabbaaab
TL6	number of b's - number of a's = 3n (multiple of 3)	abbaabbbb	bbabaaa
TL7	a*b*a*b*	aababbb	abbaabba

2. Cellular Encoding of Deterministic Finite Automata

Gruau [1994] presents a genetic programming technique, called cellular encoding, that allows for the concurrent evolution of the architecture, weights, and thresholds of a neural network. Each tree in the population represents a *cellular code* that specifies the development of a neural network from a single neuron (*ancestor cell*). Developmental operations include neuron (*cell*) division, link modification, and threshold specification. The method has shown to be successful at producing neural networks to solve a variety of problems. This paper applies cellular encoding to the evolution of deterministic finite automata.

2.1. Function Set

The functions that make up the GP tree are of two types: those which cause cell (state) division, and those which modify properties of existing states. Deterministic finite automata were chosen, instead of non-deterministic finite automata to make the fitness evaluation as fast as possible. This decision, however, makes the choice of functions somewhat more constrained since each operation on the growing embryonic automata must preserve the determinism. The most intuitive way to implement the cell division described above on a serial machine would be to maintain a pointer to the currently active cell (state) [Gruau 1994]. Functions in the GP tree would then cause the active cell to divide, modify the active cell, or change which cell is active. It turns out, however, that in dealing with deterministic finite automata, it is easier to think of a certain arc being active at a given time, instead of a cell. So, during execution of the GP tree there is always a single arc which is active. Functions in the tree either cause the cell at the head of the active arc to divide, modify where the active arc points, or change which arc is active. The zygote(ancestor) DFA, which classifies all inputs identically, is shown in figure 1.

There are two functions that cause new states to be created in the automata. PARALLEL and PARALLEL-REJ are similar to the PARALLEL function in Gruau's work. PARALLEL is a two argument function that creates a new accepting node to which the active arc will now point. PARALLEL-REJ creates a new rejecting node. The output arcs of the newly-created node are copied from the old node to which the active arc pointed. PARALLEL then makes the output a arc of the new node active and evaluates its first child branch. Next, output b is made active and the second child branch is evaluated. Figure 2 demonstrates the behavior of the PARALLEL function on a section of an automata. If the output arc b of node 1, seen in figure 2a, is active, calling PARALLEL will create the new node labeled N as shown in figure 2b (bold type indicates the active arc and double circles signify accepting states). Notice that if node 2 and node N are both accepting, or both

Figure 1 Single state zygote DFA.

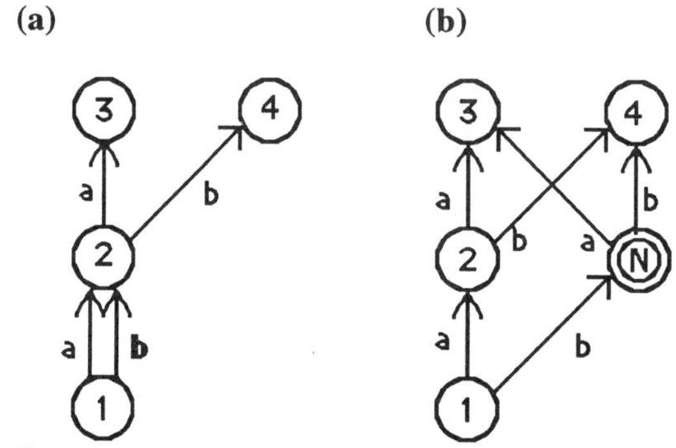

Figure 2 Effect of PARALLEL function. (a) Before calling PARALLEL, and (b) After calling PARALLEL.

rejecting, the behavior of the automata will not be immediately affected. The possibility for specialization has been created, however, since subsequent operations can modify the output arcs of node N.

The second class of functions allow for modification of the active arc, but create no new nodes. RETRACT(arg1) simply makes the active arc point to the node it leaves, creating a self-loop, and then evaluates its argument. WRITE-NODE(arg1) pushes the number of the node at the tail of the active arc onto a stack which is maintained during growth. TO-NODE(arg1) makes the active arc point to the node whose number is on the top of the stack and then pops the stack. If the stack is empty, TO-NODE leaves the active arc unchanged. Both WRITE-NODE and TO-NODE evaluate their argument after performing their operations. DONE is the only terminal, and simply indicates that the active arc is finished being modified.

2.2. Example Encoding

To gain a better understanding of the encoding, we will step through the growth of a simple automata that recognizes the language, (a|b)*bab(a|b)* (all strings that contain the subsequence bab). The example GP tree used for this demonstration is shown in figure 3.

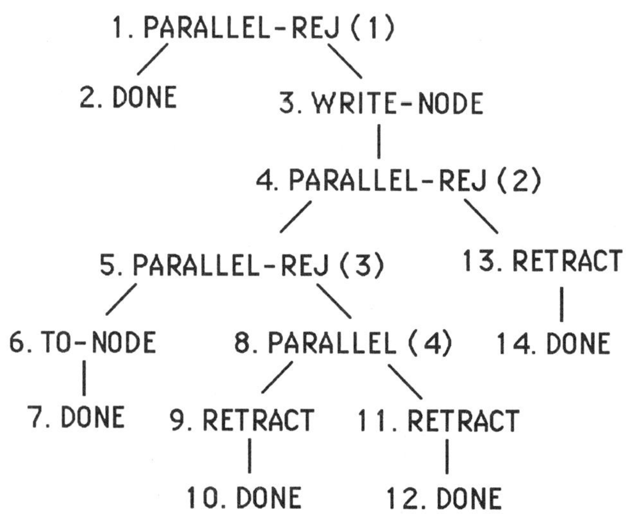

Figure 3 Example GP tree that encodes an automata to recognize (a|b)*bab(a|b)*.

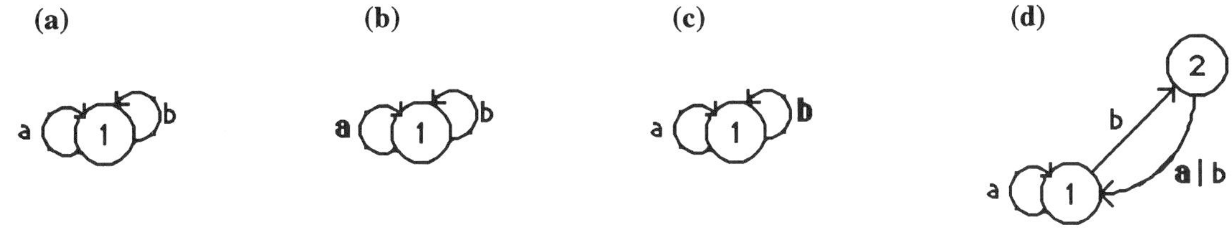

Figure 4 Example growth of an automata (initial steps).

To avoid confusion in this discussion we will refer to nodes in the genetic programming tree as "nodes" and nodes in the finite automata as "states". Nodes in the tree are numbered at the left to indicate the order of evaluation (pre-order) and numbers in parentheses denote creation of a new state in the automata. The program begins with a call to PARALLEL-REJ at node 1. The first call to PARALLEL or PARALLEL-REJ in the GP tree is used to specify the zygote as an accepting or rejecting state (all functions, necessarily of single-arity, before such a call are ignored). Therefore, execution of node 1 causes the initial zygote (start-state) to be a rejecting state as shown in figure 4a. This state is the first created and thus receives the label "1" as indicated in parenthesis in figure 3. The output arc a, as indicated in figure 4b, is now activated and the left subtree of node 1 is evaluated. The function, DONE, at tree node 2 is executed and no modifications to the active arc occur. The control now moves to the right subtree and arc b is activated (figure 4c). Node 3, WRITE-NODE, causes a "1" to be pushed onto the stack and node 4, PARALLEL-REJ, creates a new rejecting state as seen in figure 4d. Program execution now continues by activating arc a of state 2 and evaluating the left subtree of node 4. The function call at node 5 causes another cell (state) division as shown in figure 5a; state 3 is created and the previous output arcs of state 1 are copied (remember that we copy the output arcs from the state to which the active arc previously pointed). The active arc now becomes output arc a of state 3 and nodes 6 and 7 are executed. This causes the "1" to be popped off the stack and the active arc is made to point to state 1. In this case, however, the arc was already pointing to state 1, so the TO-NODE function was redundant. Output arc b of state 3 is now activated and a final cell division at node 8 causes a new accepting state (state 4) to be created as shown in figure 5b.

The two subtrees of node 8 cause both output arcs of state 4 to retract (figure 5c). The output arc b of state 2 now becomes active as the right subtree of node 4 is evaluated. Figure 5d shows the final "mature" automata that correctly recognizes the target language, (a|b)*bab(a|b)*.

Note that, although not illustrated in the above example, the arguments to the arc moving functions, TO-NODE and RETRACT, can be arbitrarily complex. The decision to make these two functions single-arity instead of zero-arity (terminals) allows for arc movements to affect subsequent cell divisions by changing which state the output arcs are copied from.

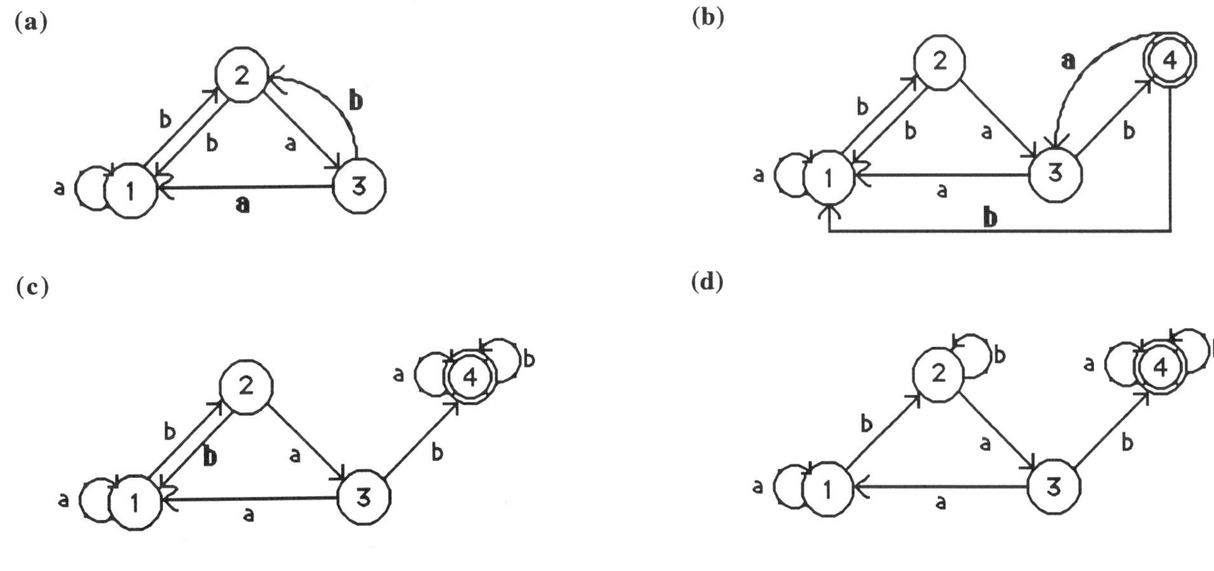

Figure 5 Example growth of an automata (final steps).

Table 2 Tableau for the DFA induction problem.

Objective:	Find a program which encodes a DFA that correctly classifies all strings in the sample set (fitness cases), given the target language.
Function set for the main branch:	`AND(arg1,arg2), OR(arg1,arg2), NOT(arg1,arg2), DFA1, DFA2, DFA3`
Function set for the three DFA branches (DFA1, DFA2, DFA3):	`PARALLEL(arg1,arg2),PARALLEL-REJ(arg1,arg2), RETRACT(arg1), WRITE-NODE(arg1), TO-NODE(arg1), DONE`
Fitness cases:	1,000 example sentences (500 positive and 500 negative)
Fitness:	The number of incorrectly classified sentences.
Parameters:	Population size (*M*): 10,000 Maximum number of generations (*G*): 100
Maximum number of nodes per individual:	Main branch: 50 Each DFA branch: 300
Success predicate:	A fitness of 0 (all sentences in the sample set correctly classified).

3. Structure of the Genetic Program

The genetic program for the language induction problem has four branches: one main branch, and three automata branches. Each automata branch encodes one DFA in the method described above. The main branch has access to these automata in addition to the boolean functions, `AND`, `OR`, and `NOT`. The main branch thus serves as a way of combining the information returned by the three created automata. The main motivation behind this approach was early difficulty in solving the dual parity problem (even number of a's and b's). Although encodings quickly emerged that recognized even a's only or even b's only, these encodings could not be easily combined with crossover to create a solution. To combat this problem, the boolean functions were added in hopes that the greater flexibility would increase building block potential. Since a single automata equivalent to any and/or/not combination of three sub-automata can be trivially created, the GP tree can still be considered an encoding of a single DFA (recall that regular languages are closed under conjunction, disjunction, and complementation). For each sentence to be classified, each of the three DFA branches are evaluated and the resulting automata simulated with the sentence given as input. If a DFA accepts the sentence, all occurrences of that DFA are replaced by `TRUE` in the main branch. If the sentence is not accepted, all occurrences are replaced by `FALSE`. Once these substitutions are made, the main branch is evaluated. If the main branch returns `TRUE`, the sentence has been classified as a member of the language by the genetic program. If the main branch returns `FALSE`, the sentence has been classified as not in the language. For example, one possible solution to the dual parity problem would be to build one automata to recognize and *even* number of a's (DFA1) and one to recognize an *odd* number of b's (DFA2) and then combine them as (AND DFA1 (NOT DFA2)).

4. Results and Discussion

The described method was tested on several languages using 500 positive and 500 negative cases with a population size of 10,000. A tableau summarizing the parameters is given as table 2.

All runs were executed using Dave's Genetic Programming Code in C (DGPC) on a Pentium. The crossover rate was set to 0.8 for internal functions and 0.1 for leaves, the maximum new tree depth was 6, and no mutation was used. Program size was limited to 50 nodes for the main branch and 300 nodes for each DFA branch. Fitness was measured as the number of incorrectly classified strings. The first languages tested were the seven benchmarks described in Tomita [1982] (shown in figure 1). Of these tests, all but TL6 (#1's - #0's = 3n) solved in less than 100 generations, producing automata with 100% success in classifying out of sample strings (500 positive and 500 negative). Furthermore, inspection of the evolved automata showed that all solutions were fully generalizable. Figure 6 shows one such evolved solution to TL3 and figure 7 shows a solution to TL7 (both of these solutions made use of only one ADF). TL6 was run 10 times without success. The method was also successful in producing correct and general automata for the dual parity problem (even number of a's and b's) and the language: `(a|b)*aaaa(a|b)*bbbb(a|b)*`.

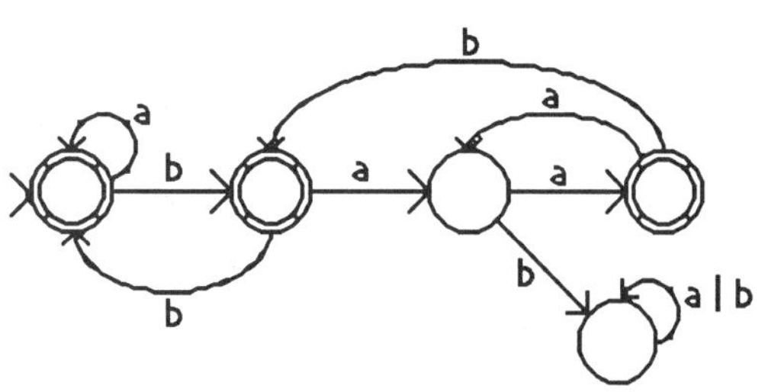

Figure 6 Example evolved solution to TL3.

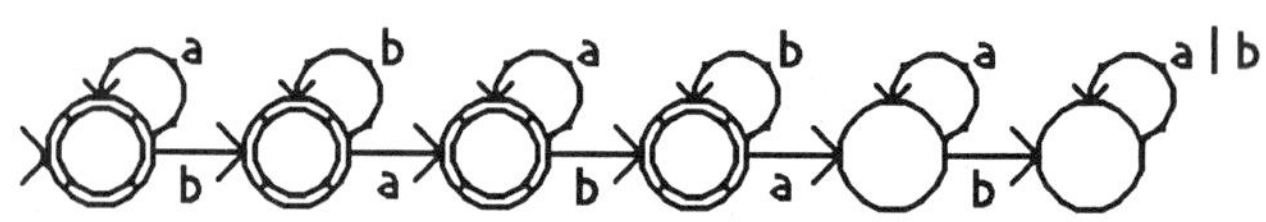

Figure 7 Example evolved solution to TL7.

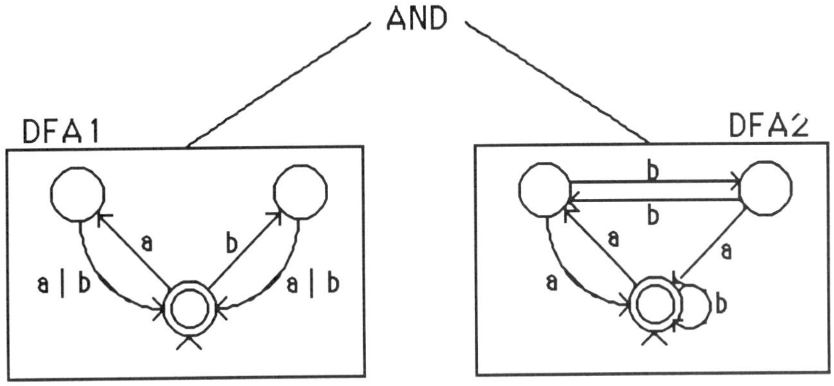

Figure 8 Example solution to the dual parity problem.

An interesting, evolved solution to the dual parity problem can be seen in figure 8. The main branch (not shown) simplified to an AND of DFA1 and DFA2. DFA1 recognizes even length strings and DFA2 recognizes an even number of a's.

The reason why no solution was found to TL6 is still under investigation. Initial observations of the unsuccessful runs, however, suggest that the programs were attempting to enumerate the example sentences. This explanation seems likely given that 1) Any increase in fitness tended to be in small increments 2) The size of the program trees grew more quickly than on runs of other languages 3) Improvements in fitness ceased before the maximum number of generations was reached 4) The function set allows for "easy" enumeration with PARALLEL and PARALLEL-REJ. Such enumerating programs plateau when limitations on program size prevent further enumeration. The reason that the TL6 evolution degenerated to enumeration, while other languages did not, could be due to lack of useful building blocks with the described encoding or even the absence of a possible encoding for the necessary automata. Obviously, all encodings have corresponding automata; however, it is not clear that all automata have an encoding or even that all classes of equivalent automata have at least one encoding. The stack structure used by the WRITE-NODE and TO-NODE functions makes automata like the one in figure 9 impossible to create (though there may exist an encoding for an equivalent automata). The difficulty occurs with the output arcs of state 4. Because arc a is always active before arc b, the creation of this automata would require state 1 to be above state 2 on the stack. This is not possible with the described encoding since state 2 is created after state 1. The possibility also exists, however, that TL6 is simply a more difficult problem, requiring a larger population size or more than 10 runs to find a solution.

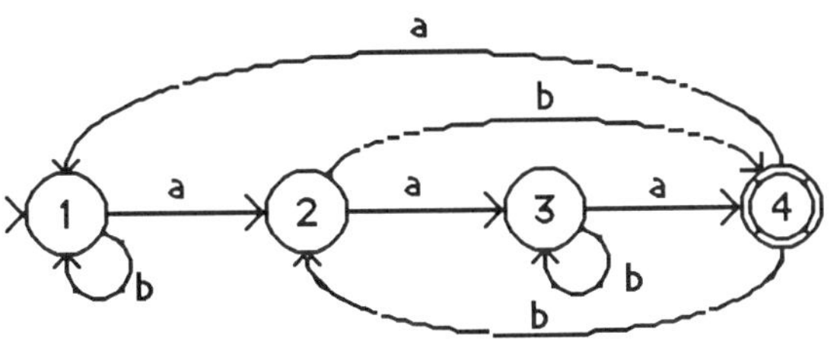

Figure 9 Example automata which is impossible to create with the described encoding.

5. Conclusion

This paper has presented a method for applying cellular encoding to the evolution of deterministic finite automata. The described encoding has the advantage that it 1) maps to only valid automata, 2) employs normal crossover, and 3) provides mechanisms for the easy creation of back pointers. These advantages, along with the successful evolution (induction) of 8 out of the 9 languages tested, indicates that the cellular encoding of DFAs is a method worth further investigation.

6. Future Work

Future work will focus on the improvement of the current function set, in hopes of creating a encoding that can successfully solve all seven of the Tomita regular languages. Memory structures other than the single stack (e.g. two stack, indexed) will be investigated, along with the utility of adding new state-creating functions. In addition, a theoretical analysis will be sought, in an attempt to find and prove an encoding which is able to produce any DFA.

Acknowledgments

Discussions with John Koza, Frederic Gruau, David Andre, and Simon Handley have been very helpful. Many thanks to David Andre for the use of Dave's Genetic Programming Code in C (DGPC).

Bibliography

Dunay, B. D., Petry, F. E., and Buckles, W. P. (1994). Regular language induction with genetic programming, *Proceedings of the First IEEE Conference on Evolutionary Computation.* IEEE Press. Volume I. Pages 396-400.

Fogel, L. J., Owens, A. J., and Walsh, M.J. (1966). *Artificial Intelligence Through Simulated Evolution.* New York, NY: John Wiley and Sons.

Giles, C. L., Sun, G. Z., Chen, H.H, Lee, Y.C., and Chen, D. (1990). Higher order recurrent networks and grammatical inference, *Advances in Neural Information Processing*. D.S. Touretzky, ed. San Mateo, CA: Morgan Kaufmann. Volume 2. Pages 380-387.

Giles, C. L., Miller, C. B., Chen D., Chen, H. H., Sun, G., and Lee, Y. C. (1992). Learning and extracting finite automata with second-order recurrent neural networks, *Neural Computation*.Volume 4. Pages 393-405.

Gold, E. M. (1967). Language identification in the limit, *Inform. Contr. Volume* 10. Pages 447-474.

Gruau, Frederic. (1994). Genetic micro programming of neural networks. *Advances in Genetic Programming*. K. E. Kinnear Jr., ed. Cambridge, MA: The MIT Press. Pages 495–518.

Koza, J. R. (1992). *Genetic Programming: On the Programming of Computers by Means of Natural Selection.* Cambridge, MA: The MIT Press.

Pollack, J. B. (1991). Language induction by phase transition in dynamical recognizers, *Advances in Neural Information Processing*. R. P. Lipmann, J. E. Moody, D. S. Touretzky, ed. San Mateo, CA: Morgan Kaufmann. Volume 3. Pages 619-626.

Tomita, M. (1982). Dynamic construction of finite-state automata from examples using hill climbing, *Proceedings of the Fourth Annual Cognitive Science Conference*. Ann Arbor, MI. Pages 105-108.

Waltrous, R. L. and Kuhn, G. M. (1992). Induction of finite-state languages using second-order recurrent networks, *Neural Computation*. Volume 4. Pages 404-414.

Williams, R. J. and Zipser, D. (1988). A learning algorithm for continually running fully recurrent neural networks, *Institute for Cognitive Science Report* 8805. La Jolla, CA: University of California at San Diego.

Genetic Programming and the Efficient Market Hypothesis

Shu-Heng Chen
AI-ECON Research Group
Department of Economics
National Chengchi University
Taipei, Taiwan 11623
E-mail: chchen@cc.nccu.edu.tw

Chia-Hsuan Yeh
AI-ECON Research Group
Department of Economics
National Chengchi University
Taipei, Taiwan 11623
E-mail: g3258501@grad.cc.nccu.edu.tw

ABSTRACT

While search plays an important role in the efficient market hypothesis (EMH), the traditional formalization of the EMH, based on probabilistic independence, fails to capture it. Due to this failure, recent findings of nonlinear tests misled us into concluding that the EMH is rejected. Even though most economists are reluctant to make this conclusion, the traditional formalization leaves us no other choice. This paper reformalizes the EMH with a biologically-based search program, i.e., genetic programming (GP). The GP-based search enables us to model search in the EMH explicitly. Through this, serach cost as well as search intensity can be measured objectively, and the notion of predictability and profitability can then be formalized. The GP-based notion of the EMH will be exemplified by testing the EMH with a small, medium and large sample of the S&P 500 stock index.

1 Introduction

> Thus, investors cannot device an investment strategy to yield abnormal profits on the basis of an analysis of past price patterns. (Malkiel, 1987, p.127)

In his survey article, Malkiel gave the intuitive notion of the *efficient market hypothesis* cited above. Based on this intuitive notion, to formalize the EMH, two technical notions are needed, namely, a technical notion of *predictability* and a technical notion of *profitability*. However, despite its long history, the formalization of the EMH has been built solely on a technical notion of *predictability* and it proceeded as if an appropriate technical notion of unpredictability would automatically defy profitability in any sense. In the history of the EMH, there is only one technical notion of predictability been formalized and it was constructed in terms of *probabilistic independence* in probability theory. Mathematically speaking, considering the rate of returns R_t a random variable (function) defined in the L^2 probabilistic Hilbert space and Ω_{t-1} the σ-algebra generated by the history of rate of returns $\{R_j\}_{j=-\infty}^{t-1}$, then the EMH simply says that R_t is independent of any random variables in the Ω_{t-1}. Based on the different information sets Ω_{t-1}, one could have different versions of the EMH (Malkiel, 1987). The one introduced above is called *weak-form efficiency*. Furthermore, considering the expected rate of returns the conditional expectation $E(R_t \mid \Omega_{t-1})$, then the EMH also implies that

$$E(R_t \mid \Omega_{t-1}) = 0. \qquad (1)$$

Since Equation (1) is also a consequence of the *random walk* defined in a discrete-time stochastic process, the EMH is often used interchangeably with the random walk hypothesis in finance, even though the former is far from equivalent to the latter.

While this formalization is precise, Chen and Yeh (1995) argued that *it is not computable* in the sense that we can not test it with an effective algorithm. Nevertheless, this problem had not been fully recognized until recently when a series of *nonlinear tests* failed this hypothesis, e.g., Brock, Dechert and Scheinkman (1987), Savit (1988, 1989), Hinich and Patterson (1989), Hsieh (1989), Frank, Gencay and Stengos (1988), Scheinkman and LeBaron (1989), Peters (1991), and Willey (1992). These tests either tell us that there exists nonlinear dependence between R_t and Ω_{t-1} or that $\{R_i\}_{i=-\infty}^{t}$ is a chaotic time series which looks random but is in fact deterministic.

What do these tests imply? If we take Equation (1) as the formalization of the EMH, then the rejection of the EMH naturally follows from the statistical rejection of Equation (1). Unfortunately, this is a certainly a very negative result for the mainstream economists and an unacceptable one too. However, there may be hope of solving this conundrum if we can interpret these results in a different way. For example, Fogler (1995) had the following remarks:

> ...whether a chaotic process or a complex nonlinear process generates a pattern is irrelevant unless one knows the exact functional form generating the process. Even if tests indicate a high embedding dimension, forecasting is impossible unless a specific function form is assumed. (p.16)

Accordingly, the rejection of Equation (1) based on any nonlinearity test does not authorize us to say that the market is not efficient *unless we are able to show the underlying data generating process.* Fogler's remarks powerfully separated the nonlinearity tests from the efficient market hypothesis. But, that also means Equation (1) has to be given up and a new technical notion of unpredictability is needed. To be consistent with Fogler's remarks, the new notion must be *search oriented.* More precisely, to formalize Fogler's notion of unpredictability, a *search algorithm* must be included so that the decision of predictability can be made based on search results. Moreover, since we are searching for the potential regularity in L^2 space and the cardinality of this space is *infinite*, the design of this search algorithm cannot be limited to only a finite number of functions or any special subclass of functions. However, the theoretical technology to design the search algorithm of this sort was not in the classical toolbox of economists. That also partially explains why the search-based notion of unpredictability was not formalized over the last 90 years, while it was already captured by economists' intuition. e.g., Malkiel (1987).

Given the introduction above, the contribution which genetic programming can make to the EMH is clear. Using the novel idea of incorporating the evolutionary operation into the automatic search over the space of programs (L^2 space), genetic programming provides financial economists with a theoretical technology to design a search algorithm and to formalize the notion of unpredictability. This formalization also enables us to define *search intensity* and to give us an objective measurement of *search costs* and *the chance of success in the search.* Based on that, *the expected profits of search* can be derived and the other aspect of the EMH, i.e., the profitability aspect, can also be formalized.

This contribution also distinguishes the application of genetic programming from other applications of biologically inspired computational models to finance. For example, artificial neural nets are mainly used in finance as a problem-solving technique for the nonlinear estimation and forecasting of the asset price or for the computation of option pricing. Regardless, ANNs per se provide little to change or advance our understanding of the fundamental concepts in finance. In contrast, genetic programming not only enriches the finance toolbox, but also enhances our understanding of finance theory from its very root.

Chen and Yeh (1995) conducted a pioneer study to demonstrate the GP-based notion of unpredictability. This formalization of unpredictability was illustrated with examples of predicting chaotic dynamic systems. They applied GP to predict the time series generated by three chaotic dynamic systems which are represented by the *LISP S-Expression* and depicted as a rooted, pointed-labeled tree (GP-tree). They have shown that if the rates of return are generated by a *simple* deterministic chaotic dynamic system, then GP may actually discover it. Here, "*simple*" refers to the *depth* of the GP-tree and has nothing to do with the embedding dimension. Their study illustrated how genetic programming can provide us with an explicit search program upon which an objective measure for predictability can be constructed. Also, it indicated that the use of *Lyapunov exponents* to estimate the predictability of a chaotic time series might not be appropriate.

In the same study, the EMH based on the GP-based notion of unpredictability was exemplified by an application to a mini-size sample of S&P 500 index (sample size = 50). It was found that, while linear models cannot predict better than the random walk, a GP-based search can beat random walk by 50%. However, Chen and Yeh (1995) did not explicitly show how the GP-based notion of unpredictability can be related to the other aspect of the EMH, i.e., *the notion of profitability.* While they could show to Fogler how to find the *underlying data generating process* given the existence of *nonlinear dependence*, they certainly have not answered the essential question *whether predictability implies profitability.* Therefore, their work, at best, only formalized the first half of Malkiel's intuitive notion of the EMH.

This paper will extend Chen and Yeh (1995) to the second half. To do this, we exemplify the notion of profitability by comparing long-range forecasting with large samples and short-range forecasting with small samples. Since different forecasting strategies involve different search cost, i.e., direct *computational cost* and indirect *risk cost*, the GP-based notion of profitability enables us to check whether the better prediction by a certain search strategy was made at high search cost. In other words, we can test the *no-free-lunch property* in Malkiel's intuition of the EMH. This work will be done in Sections 2 and 3. Section 2 will describe the choice of data. Section 3 will present the results of the GP-based search over different sizes of sample along with the analysis.

Section 4, which is motivated by Koza (1992) (pp.245-255), provides an *knowledge-discovery perspective* to study the GP-based formalization of the EMH. In this section, the EMH represented by Equation (1) is regarded as a *benchmark*. We then evaluate the competitiveness of this benchmark by observing how well it can fight against its potential competitors. If, during the evolution, it is beaten by a competitor, then this benchmark is considered not competitive; otherwise, we say it is competitive. It is this sense that we will show that *the EMH is not competitive in the small sample but is quite competitive in the large sample.* Concluding remarks are given in Section 5.

2 The Design of the Data Environment

2.1 The Sample Size

The data environment in which the GP-based search is implemented concerns the daily rate of return of the S&P 500 Index, from 1/2/62 to 9/6/95. There are totally 8,478 observations in this dataset. This dataset is rich enough for us to test the EMH *in the long run* as well as *in the short run*. Distinguishing between short-run tests and long-run tests is motivated by the recent studies of the time series of stock prices which seem to indicate that *even though nonlinear regularities might exist, they are not stable over time.* For example, Peters (1991) made an excellent note on this.

> For too long, we seem to have been divided between "technicians" who believe that the market follows a regular cycle and "quants" who believe there is no cycle at all. The truth lies somewhere in between. There are cycles, but they are not regular, hence may be invisible to standard statistical techniques. (p.62)

LeBaron (1992) also came up with a similar observation.

> Some out-of-sample forecast improvements are demonstrated for the weekly S&P 500 series. The evidence shows that these improvements are difficult to detect since they only occur during certain time periods. ... Although forecast improvements appear significant, they are extremely small and occur only for a fraction of the weeks tested. Stock returns remain, as they should, a relatively difficult series to forecast. (p.381)

Therefore, the application of GP to different sizes of sample does require lots of thought. Suppose that the stock market encounters a sequence of short-term time-variant nonlinear relations, a large sample size may average out all these relations. In this case, a smaller sample size is desirable, whereas for the time-invariant long-term nonlinear relation, a large sample size is needed. In an earlier study, Chen and Yeh (1995) conducted large-scale experiments for a sample with sample-size 50 (about ten weeks). In this paper, we extend the study by testing the EMH with a short-term (sample size=200), medium-term (1000) and long-term (2000) sample of S&P 500. The short-term sample corresponds to data about 10 months long, the medium-term about 4 years and the long-term about 8 years. It will be interesting to see how the validity of the EMH will be affected by the sample size.

2.2 Selecting Data with the MDL

In a data set with 8478 observations, there are many subsets associated with different sample sizes. To test all of them seems to be time-consuming. The alternative is to test a few of them which are representative. If these representative datasets can reject the GP-based test of the EMH, so can the rest. For this purpose, the subsets considered are the ones with the highest complexity. Since there is no unified definition or criterion for *complexity* (Horgan, 1995), our adherence to any particular style is, more or less, arbitrary. Of course, this weakness can be consolidated by simultaneously using different styles of complexity measure. However, in this paper, we only consider the most popular style used in statistics and econometrics, i.e., *Rissanen's minimum description length* (**MDL**).

Rissanen's MDL (minimum description length) is an approximation for *Kolmogorov complexity* which measures the complexity of a set of data by the length of the shortest universal Turing machine program that will generate the data. The measure is well-defined, but not practically computable. The MDL developed by Rissanen (1982) is a way to approximate this uncomputable measure by replacing the universal Turing machine with a class of probabilistic models.

A detailed description of the MDL algorithm used in this paper can be found in Chen and Tan (1996). Briefly speaking, we first transform the original sequence of $\{R_t\}$ from 1/2/62 to 9/6/95 into a 0-and-1 sequence based on the sign of R_t. Then the MDL is computed for each of the 200, 1000 and 2000 consecutive observations in the 0-and-1 sequence by choosing the *Bernoulli class* and *Markov class* as our model classes. The MDL(x_i^{2000}) of S&P 500 is given in Figure 1. From Figure 1, we can see that, in terms of the 2000-day MDL, the second half of the time series is more complex than the first half. It is consistent with what we have been told in finance that complexity has a tendency to increase with time. Based on the maximum MDL criterion, the periods for different sizes of chosen samples are given in Table 1. The S&P 500 time series of the chosen period for the sample with sample size 1000 and 2000 are drawn in Figures 2 and 3.

Once the in-sample data series are given, the next issue is how to determine the length of the post-sample series. In this paper, we arbitrarily set the in-sample data to be ten times the size of the post-sample data. The periods

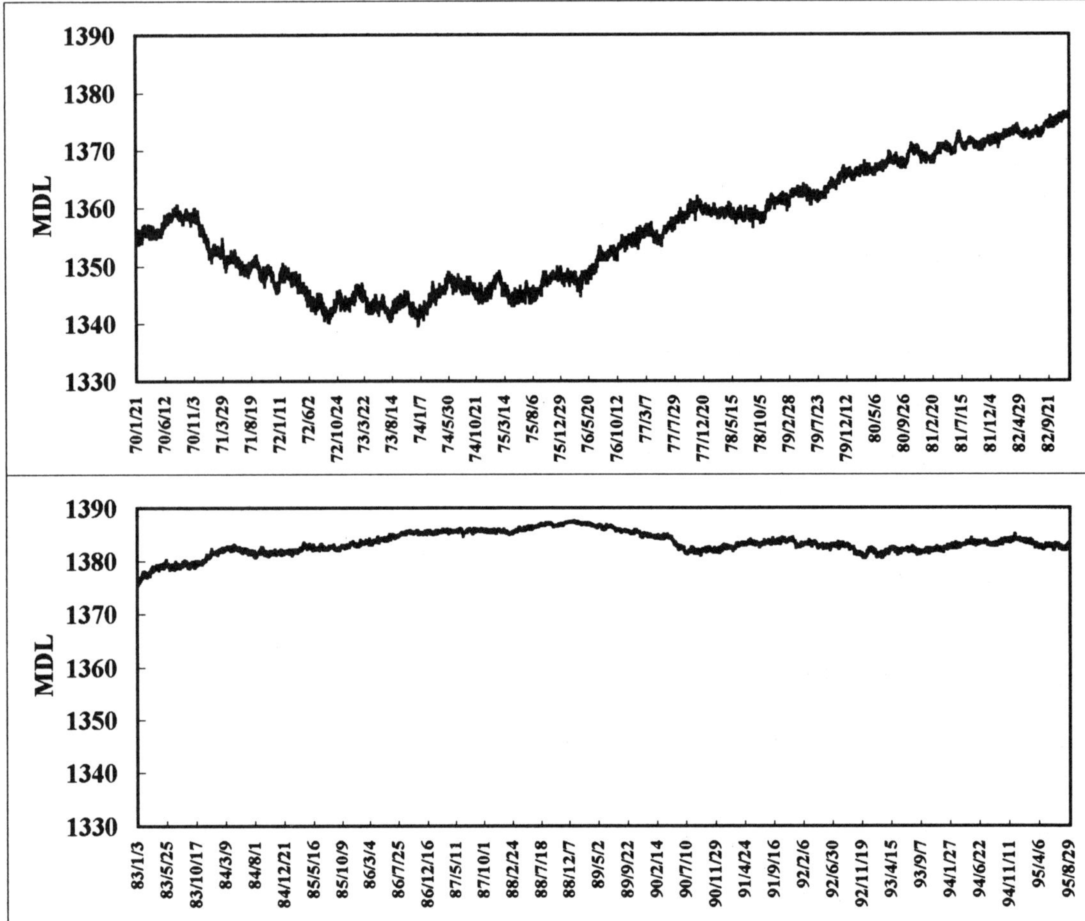

Figure 1 : The MDL of S&P 500 (1962.1.1-1995.9.6) (2000 Daily Returns)

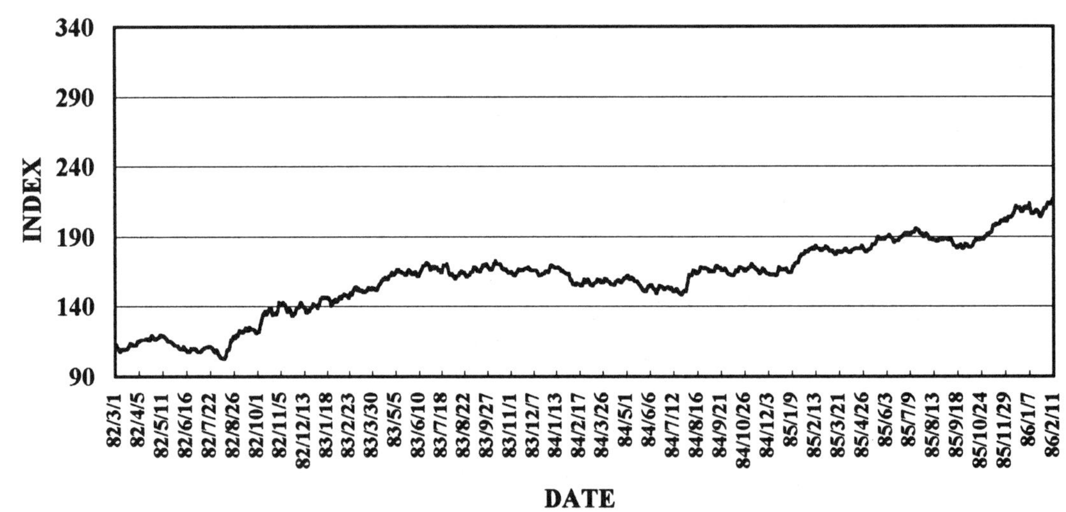

Figure 2 : The Stock Price Indices Corresponding to the Maximum MDL S&P 500 (1000 Daily Returns)

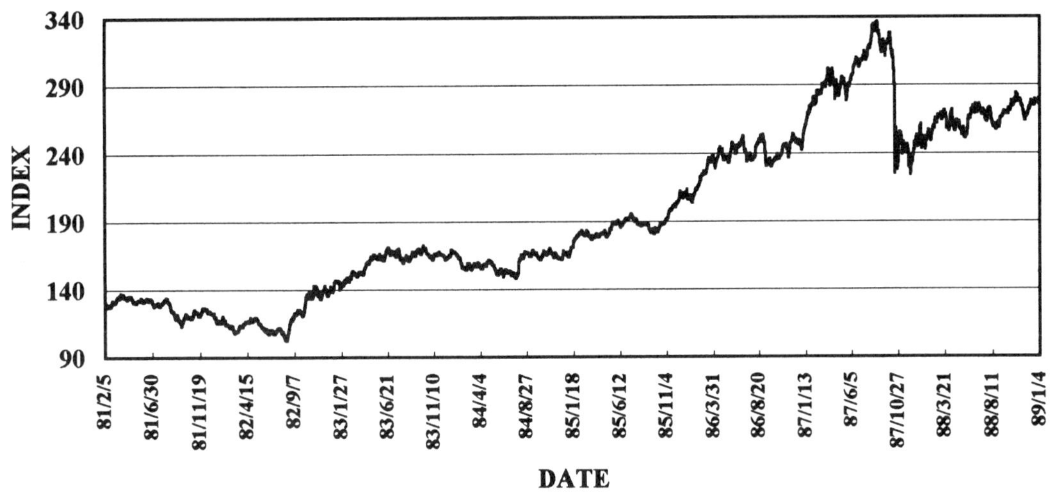

Figure 3 : The Stock Price Indices Corresponding to the Maximum MDL S&P 500 (2000 Daily Returns)

of post-sample series are also given in Table 1.

Table 1: MDL-Based Data Selection: S&P 500

Sample size	Whole Sample Period (Post-Sample Period)	MDL
200	1/3/92 - 10/16/92 (9/21/92 -)	142.472
1000	3/1/82 - 2/11/86 (9/19/85 -)	697.794
2000	2/5/81 - 1/4/89 (88/3/22 -)	1387.579

3 The Empirical Results

To implement genetic programming, the program GP-Pascal is written in Pascal 4.0 by following the instructions given in Koza (1992). A detailed description of this program can be found in Chen, Lin and Yeh (1995). The chosen parameters to run GP-Pascal are given in Table 2. % and *RLOG* appearing in the function set are the protected division function and the protected natural logarithm function (Koza, 1992, pp.82-83). In this paper, all simulations conducted are based on the terminal set, which includes the ephemeral random floating-point constant R ranging over the interval [-9.99, 9.99] and the rate of return lagging up 10 periods, i.e., $R_{t-1}, \ldots R_{t-10}$.

MAPE (*Mean Absolute Percentage Error*) is defined as follows:

$$MAPE = \sum_{i=1}^{m} \frac{\mid \hat{R}_i - R_i \mid}{\mid R_i \mid}, \qquad (2)$$

where m is the sample size and $\hat{R}_i$ is the prediction value of R_i. The choice of the *mean absolute percentage error* as the fitness function is attributed to Makridakis (1993), who suggested a modified form of MAPEs as the most appropriate measure satisfying both theoretical and practical concerns while allowing meaningful relative comparisons. Since the prediction made by the random walk hypothesis is always 0, the corresponding MAPE is always 1. So, *GP is said to beat the RW if its MAPE is less than 1.* In addition, we define $\pi_1(n)$ as the probability that GP can beat the RW in Generation n in the in-sample data and $\pi_2(n)$ the probability that GP can beat the RW in the holdout sample.

Table 2: Tableau for Predicting Rates of Return

Population size	500
Number of trees created by complete growth	50
Number of trees created by partial growth	50
Function set	$\{+, -, \times, \%, Sin, Cos, EXP, RLOG\}$
Terminal set	$\{R_{t-1}, R_{t-2}, \cdots, R_{t-10}, R\}$
Number of trees generated by reproduction	50
Number of new lives (immigrants)	50
Number of trees generated by mutation	100
Probability of mutation	0.2
Maximum length of the tree	17
Probability of leaf selection under crossover	0.5
Number of generations	200
Maximum number in the domain of Exp	1700
Criterion of fitness	MAPE

Based on those control parameters, multiple runs of simulations were executed for each sample. For each of

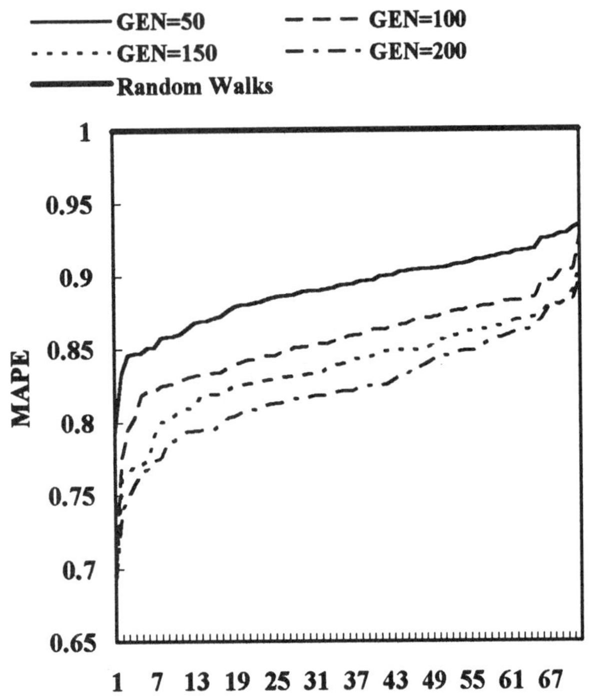

Figure 4 : In-Sample MAPEs (S&P 500)

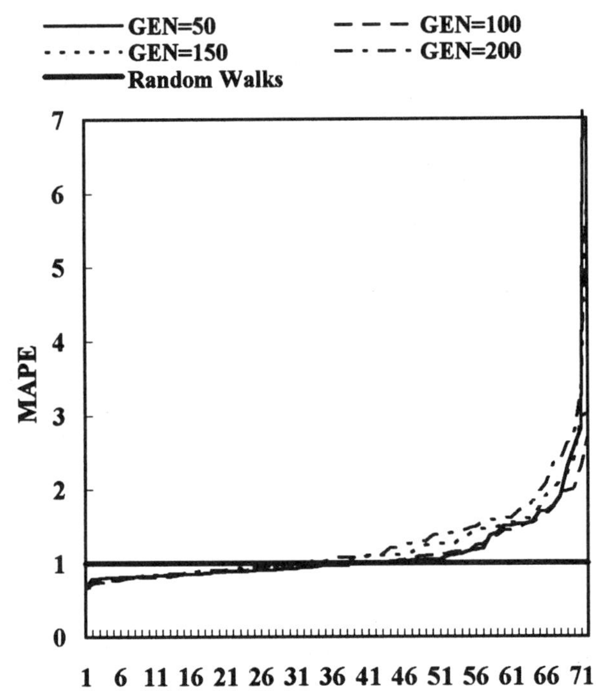

Figure 5 : Post-Sample MAPEs (S&P 500)

the simulations, the MAPE is calculated for the in-sample period and the post-sample period. Statistics show that the sample $\pi_1(n)$ are 100% in Generations 50, 100, 150 and 200 for the small and medium sample (sample size=200 and 1000). Moreover, when evolution takes longer, limited improvement can always be made. For example, in the small sample, the MAPE in the best and worst case of these simulations in Generation 50 is 0.9698 and 0.9904 respectively, and in Generation 200, it is improved to 0.9452 and 0.9842. These results are also reported in Table 3.

Table 3: The In-Sample MAPE of the Best Case and the Worst Case

Sample Size/Gen	50	100	150	200
Mini-Best	0.7925	0.6979	0.6955	0.6919
Small-Best	0.9698	0.9613	0.9567	0.9452
Medium-Best	0.9970	0.9964	0.9964	0.9964
Large-Best	0.9983	0.9980	0.9980	0.9974
Mini-Worst	0.9348	0.9268	0.9077	0.8985
Small-Worst	0.9904	0.9904	0.9870	0.9842
Medium-Worst	0.9999	0.9977	0.9977	0.9977
Large-Worst	1.0000	1.0000	1.0000	1.0000

From Table 3, there are a couple of points worth noticing. First of all, as the sample size gets larger, the difference in the in-sample MAPE between GP-based search and the random walk hypothesis becomes more and more negligible. The difference is particularly striking when we compare the in-sample MAPE performance of the mini sample (sample size = 50) with that of larger samples. To get a more general flavor of the MAPE performance under the GP-based search, the best in-sample and post-sample MAPEs of the mini sample of each simulation in Gen (generation) 50, 100, 150, and 200 are ranked from the lowest to the highest and are shown in Figures 4 and 5 (See Chen and Yeh (1995) for details).

Secondly, the range of the MAPE, i.e., the difference in the MAPE bewteen the best and the worst search tends to narrow as the sample size increases. It may be reasonable to summarize these results by the following *asymptotic-like hypothesis*,

$$MAPE_{gp}(n, m) \rightarrow 1, \quad as \quad m \rightarrow \infty, \tag{3}$$

where n is the number of evolution and m is the size of the training sample.

This hypothesis clarifies some confusing points around the EMH, in particular, the *time horizon* of the EMH. Starting with the pessimistic side, it says that *large sample training is not going to be very helpful.* In fact, our simulations show that it is easier to search for patterns from a smaller sample than from a larger one. This is consistent with the old truism in economics that "if there is a dollar bill on the ground, it is soon picked up.". Some econometricians such as White (1988) suggest that large sample training with recurrent neural nets might give the EMH a real challenge. Our hypothesis says exactly the opposite: it is the large sample that makes the EMH easily acceptable. Turning to the optimistic side, this hypothesis does imply that there is room for profits for *short-range forecasting with the small sample*; but then it also implies that the rule used to forecast is preferred to be updated on a weekly basis rather than on a monthly or yearly basis. As a result, the computational cost associated with this highly frequent revision must be higher than that for *long-range forecasting with the large sample*, and it may be so high that, in terms of net profits, there is no significant difference between long-range and short-range forecasting.

The computational cost is the physical cost. Another important cost concerning forecasting is the *risk cost* which is not directly derived from the physical side but from the utility (psychological) side. In economics and finance, investors are assumed to be *risk aversers*. This being the case, risk can be counted as another cost for them. So, let us turn to the risk side of forecasting strategies, and use the difference in the post-sample MAPE between the best GP-based search and the worst one as a measure of risk. Table 4 summarizes the results on the risk of different styles of the GP-based search.

Table 4: The Post-Sample MAPE of the Best Case and the Worst Case

Sample Size/Gen	50	100	150	200
Mini-Best	0.6712	0.6680	0.6680	0.6680
Small-Best	0.9987	0.9987	1.0000	0.9919
Medium-Best	0.9962	0.9952	0.9951	0.9951
Large-Best	1.0000	1.0000	1.0000	1.0000
Mini-Worst	2.8223	2.7109	3.0470	5.9289
Small-Worst	1.2021	1.0391	1.0383	1.0389
Medium-Worst	1.0239	1.0383	1.0337	1.0333
Large-Worst	1.0019	1.0019	1.0019	1.0075

From Table 4, we can see that the risk under long-range forecasting with large-sample training is much less than that under short-range forecasting with short-sample training. In particular, if we examine the mini sample studied by Chen and Yeh (1995), the risk can be five times higher than that with the large sample. Thus, even though GP-based data mining can be useful for the short-term data, after taking into account the computational cost and the risk cost, its advantages may not be that great.

4 On The Competitiveness of the Random Walk Hypothesis

To get a better grasp of the GP-based notion of the EMH, we propose in this section another perspective to study the EMH by raising the following issues:

- If the efficient market hypothesis is true, can genetic programming automatically discover this truth? In fact, we are asking whether genetic programming can be used to prove the EMH.
- Can the random walk hypothesis survive well in the competitive environment generated by genetic programming?

As to these questions, the major result of our simulations is that *when the sample size is large enough, e.g., sample size = 2000, genetic programming can actually be used to prove the EMH by discovering the random walk hypothesis as the most competitive model.* We shall illustrate this based on Simulation 2000-1, which is the first simulation for the 2000-observation sample. The best model chosen from Gen. 0 is:

$$F^{0}_{best} = Log(R_{t-1} - R_{t-1}) \tag{4}$$

Since $Log(0)$ is defined as 0 in the program, the best model found by GP in the initial generation is exactly the random walk hypothesis. In other words, the RW hypothesis was discovered at the very beginning of the evolution. Still, if the random walk hypothesis is not competitive, then during the evolution, one could expect that other better models will be discovered and replace the RW. However, this was not the case observed in Simulation 2000-1. The RW hypothesis kept on dominating the evolution till the end while it appeared in different styles. For example, the best model from Gen. 16 is:

$$F^{16}_{best} = (((R_{t-10} - R_{t-10})\%ExpR_{t-3})\% - 8.163083) \tag{5}$$

Since R_{t-10} cancels itself in Equation (2), F^{16}_{best} is in fact the zero function. The evolution might generate seemingly complicated functions. After cancellation and reduction, however, they can always be simplified to a zero function. For example, the best model from Gen. 195 is:

$$F^{195}_{best} = Log(((ExpR_{t-1} - (R_{t-8}\%R_{t-10})) - (ExpR_{t-3} \\ -(R_{t-8}\%R_{t-10})))\%Log((R_{t-10} - Cos(R_{t-1} - R_{t-1})) \\ \%(((R_{t-1} - R_{t-1})\%R_{t-1})\%(R_{t-1} - R_{t-10})))) \tag{6}$$

Considering the zero function an equivalent class, then we can see from the evolution in the large sample that *the members of the class of zero function dominated the whole evolution.* In other words, the random walk hypothesis is very competitive in the large sample. GP could not easily find anything better than the RW hypothesis. However, this result does not hold for the small sample. In particular, when the sample size is 200, the random walk hypothesis can be the best only in the initial generation (Gen. 0). It would be quickly replaced by other models once the evolution started. For example, the best model from Gen. 0 in Simulations 200-2 and 200-3 are given in Equations (4) and (5) respectively.

$$G^{0}_{best} = (R_{t-7} * (R_{t-6} - R_{t-6})) \tag{7}$$

$$\mathtt{H}^{0}_{\mathtt{best}} = \mathtt{Sin}((\mathtt{R}_{\mathtt{t-3}} + \mathtt{R}_{\mathtt{t-2}}) * (\mathtt{R}_{\mathtt{t-4}} - \mathtt{R}_{\mathtt{t-4}})) \quad (8)$$

Clearly, both G^0_{best} and H^0_{best} are the members of the equivalent class of zero function, hence they represent the random walk hypothesis. However, in Gen. 1, they were defeated by the G^1_{best} and H^1_{best} (given below) separately.

$$\mathtt{G}^{1}_{\mathtt{best}} = (\mathtt{R}_{\mathtt{t-5}} * (\mathtt{R}_{\mathtt{t-7}} * \mathtt{R}_{\mathtt{t-1}})) \quad (9)$$

$$\mathtt{H}^{1}_{\mathtt{best}} = \mathtt{Sin}(\mathtt{Sin}(\mathtt{CosLogR}_{\mathtt{t-8}} * (\mathtt{R}_{\mathtt{t-1}} - \mathtt{R}_{\mathtt{t-6}}))\%((\mathtt{Log}\ \mathtt{R}_{\mathtt{t-4}} - (4.522217\%\mathtt{R}_{\mathtt{t-3}})) + \mathtt{CosExpR}_{\mathtt{t-2}})) \quad (10)$$

Therefore, the random walk hypothesis is not competitive in the small sample and combined with the finding of Chen and Yeh (1995), we can conclude that *the learnability of the financial data (S&P 500) is only restricted to the small sample.* When the sample gets large, nothing can be learned from it except that it is random.

This conclusion can also be justified by *the complexity of the best discovered model.* That is, instead of asking what rules are discovered by GP, we are inquiring how complex those rules are which are discovered by GP. The length of the *LISP program* is used to measure complexity. The complexity of the best model chosen by GP in Gen 50, 100, 150 and 200 in a few selected simulations is given in Table 5. It is interesting to note that when the sample size is small, in-sample fitness can always be improved by searching for a bigger LISP tree. For example, when the sample size is 200, the length of the best LISP program has a positive relation to the number of generation of the evolution. However, when the sample size is large, the relation disappears.

Table 5: The Length of the Best Model (LISP Program)

Simulation/Gen	50	100	150	200
200-2	75	243	303	315
200-3	24	24	53	126
200-4	48	54	210	465
200-7	210	300	333	339
1000-1	309	387	330	360
1000-3	198	108	111	117
1000-5	114	186	162	198
2000-1	111	69	191	168
2000-3	24	42	153	84

5 Concluding Remarks

> The attempts to formalize the EMH as a consistent, analytical economic theory have met with less success than the empirical tests of the hypothesis. (Ross, 1987, p.7)

Needless to say, at this moment, any conclusion about the validity of the EMH is premature and this paper should not leave readers any impression that we are pro or against the EMH. After all, the validity of the EMH is not the major concern of this paper. However, the paper does have a purpose. It attempts to convince readers that over the last 90 years the EMH has not been formalized in an appropriate way, and that biologically inspired approaches, in particular, genetic programming, hopefully can make it right. Right not only in terms of formalization but also about questions and analysis.

For example, the right question is not whether there is evidence of nonlinear independence but whether such nonlinear patterns can be discovered at reasonable cost. The right analysis is not to test the large sample with asymptotic econometrics but to do a cost-effective analysis of forecasting different ranges of market with different sizes of sample. Similarly, the right question is not whether there will be a sequence of advanced techniques which can beat the random walk but whether the retraining or updating price of these techniques is competitive; the right analysis is not to test whether the random walk can be beaten by one recurrent neural net once but to check whether the random walk can be beaten by *evolutionary neural nets* continuously with reasonable cost. All the right questions with their corresponding analyses have an *economic sense.* Malkiel's intuition of the EMH is also an economic one rather than a statistical one. It is the absence of this economic sense in the traditional analysis that has divided us between *technicians* and *quants.* To capture the economic sense, we need the theoretical technology which, to some extent, can standardize the search activity so that search intensity and search cost can be communicated among different searchers. Genetic programming is just the candidate we are looking for.

This concludes the summary of what this paper intends to do. The purpose is very limited and the attempt is far from mature. There is little doubt that there are lots of imperfections to be fixed and details to be figured out in the follow-up research. But, if the relevance of genetic programming to the EMH is clear, the next non-trivial issue, which is motivated by Maarten Keijzer, is *what is the relationship between different styles of GP with different computing environments and the validity of the EMH.* For example, will the EMH be rejected if the computer power is doubled? Will the conclusion about the EMH differ if we are using supercomputer system to find the pattern instead of a single processor machine? Will the EMH be rejected if it is tested by parallel genetic programming with tick-by-tick data? List of questions of this sort can go on and on. But what are the answers? At the end of his survey article, Malkiel seemed to offer some insight into such an intriguing issue.

> Undoubtedly with the passage of time and with the increasing sophistication of our data bases and empirical techniques, we will document further departures from efficiency and understand their causes more fully. But I suspect that the end result will not be an abandonment of the profes-

sion's belief that stock market is remarkably efficient in its utilization of information. (p.133)

Acknowledgments

The authors are indebted for the research support from NSC grant No.85-2415-H-004-001. This paper was revised from its original version by taking the advice and comments of Christian Johnson, Maarten Keijzer and three other anonymous referees in the **GP'96** program committee. The authors are very grateful for their painstaking review of this paper. Of course, all remaining errors are the authors' sole responsibility.

Bibliography

Brock, W., W. Dechert, and J. Scheinkman 1987. A Test for Indepedence Based on the Correlation Dimension. Working Paper, University of Wisconsin at Madsion, University of Houston, and University of Chicago.

Chen, S., C. Lin and C. Yeh 1995. On the Model Selection and its Stability of the Natural Vacancy Rate of Housing: An Application of Genetic Programming. *Journal of Housing Studies*. 3. Pages 73-98.

Chen, S. and C. Yeh 1995. Toward a Computable Approach to the Efficient Market Hypothesis: An Application of Genetic Programming. Working Paper Series, No. 9502. Department of Economics, National Chengchi University.

Chen, S. and C. Tan 1996. Measuring Stock Market Efficiency by Rissanen's MDL. Paper presented at the *Pacific Rim Allied Economic Organizations 2nd Biennial Conference* coordinated by *Western Economic Association International*. Jan. 10-15, 1996. Hong Kong.

Diebold, F. X. and J. A. Nason 1990. Nonparametric Exchange Rate Prediction?. *Journal of International Economics*. 28. Pages 315-332.

Fogler, H. R. 1995, Investment Analysis and New Quantitative Tools. Paper presented at the *1995 Eastern Finance Association Meeting*. Hilton Head Island, South Carolina.

Frank, M. Z., R. Gencay, and T. Stengos 1988. International Chaos?. *European Economic Review*. 32. Pages 1569-1584.

Hinich, M. J. and D. M. Patterson 1985. Evidence of Nonlinearity in Daily Stock Returns. *Journal of Business and Economic Statistics*. Vol. 3, No. 1. Pages 69-77.

Horgan, J. 1995, From Complexity to Perplexity. *Scientific American*. June, Pages 104-109.

Hsieh, D. A. 1989. Testing for Nonlinear Dependence in Daily Foreign Exchange Rates. *Journal of Business*. 62. Pages 339-368.

Koza, J. 1992. *Genetic Programming: On the Programming of Computers by Means of Natural Selection*. Cambridge, MA: The MIT Press.

LeBaron, B. 1992. Nonlinear Forecasts for the S&P Stock Index. In M. Casdagli and S. Eubank (eds.). *Nonlinear Modeling and Forecasting*. Pages 381-393. Addison-Wesley Press.

Makridakis, S. 1993, Accuracy Measure: Theoretical and Practical Concerns. *International Journal of Forecasting*. 9. Pages 527-529.

Malkiel, B. G. 1987. Efficient Market Hypothesis, in J. Eatwell, M. Milgate and P. Newman (eds.). *The New Palgrave: Finance*. Pages 127-134. Norton.

Peters, E. E. 1991. A Chaotic Attratcors for the S&P 500. *Financial Analysis Journal*. March-April. Pages 55-62.

Rissanen, J. 1982. A Universal Prior for Integers and Estimation by Minimum Description Length. *Annals of Statistics*. 11. Pages 416-431.

Ross, S. 1987. Finance. In J. Eatwell, M. Milgate and P. Newman (eds.). *The New Palgrave: Finance*. Pages 1-34. Norton.

Savit, R. 1988. When Random Is Not Random: An Introduction to Chaos in Market Prices. *Journal of Futures Markets*. 8. Pages 271-290.

Savit, R. 1989. Nonlinearities and Chaotic Effects in Option Prices. *Journal of Futures Markets*. 9. Pages 507-518.

Scheinkman, J. A. and B. LeBaron 1989. Nonlinear Dynamics and Stock Returns. *Journal of Business*. 3. Pages 311-337.

White, H. 1988. Economic Prediction Using Neural Networks: The Case of IBM Daily Stock Returns. *Proceedings of the IEEE International Conference on Neural Networks*. II. Pages 451-458.

Willey, T. 1992. Testing for Nonlinear Dependence in Daily Stock Indices. *Journal of Economics and Business*. Vol. 44, No. 1. Pages 63-74.

Bargaining by Artificial Agents in Two Coalition Games: A Study in Genetic Programming for Electronic Commerce[1]

Garett Dworman
dworman@opim. ...

Steven O. Kimbrough
kimbrough@...
wharton.upenn.edu

James D. Laing
laing@...

The Wharton School, University of Pennsylvania
1300 Steinberg Hall – Dietrich Hall
3620 Locust Walk
Philadelphia, PA 19104-6366
(215)898-5872

ABSTRACT

Artificial agents, coevolving under a machine learning regime, offer a promising basis for modeling adaptive behavior in multilateral negotiations and developing practical applications in electronic commerce. Results from simulations of bargaining in two coalition games demonstrate that simple artificial agents, adapting to one another under a genetic programming protocol (Koza, 1992), formulate effective strategies for negotiating agreements that both approximate those prescribed by the theory of cooperative games and rival the performance of humans negotiating in a laboratory situation.

1. Introduction

Despite substantial progress in getting artificial agents (AAs) to negotiate effectively in stylized settings, they are not yet ready for deployment in real-world applications. In such situations, there are both common and conflicting interests among the relevant players (Schelling, 1960: 5):

> ...the ability of one participant to gain his [or her] ends is dependent to an important degree on the choices or decisions that the other participant[s] will make.

The research we describe here is part of our ongoing research, aimed at advancing our fundamental understanding of artificial agents in these mixed-motive bargaining contexts. We organize the paper as follows. Section 2 outlines our motivation for the type of computational investigations we report in the sequel. Sections 3-6 then describe the experimental design and results from simulations of negotiations among artificial agents in two coalition games. These studies involve three coevolving agents whose strategies are encoded as genetic programs.

2. Motivation

There are two (complementary) kinds of reasons for developing the requisite theory and technology for building effective artificial agents for mixed-motive bargaining. These are: (1) reasons of electronic commerce (broadly construed to include activities by organizations, such as governmental bodies, whose purposes are not mainly commercial) and (2) theoretical reasons pertaining to the understanding of negotiation in mixed-motive contexts.

First, for purposes of electronic commerce, important efficiencies would likely often result if negotiations could be entrusted to artificial agents. We have in mind, for example, the incremental replacement, or augmentation, of human negotiators with artificial agents acting in their stead for fairly routine, frequently-encountered bargaining situations, such as the sort of buying and selling that is often supported by EDI (electronic data interchange) protocols. Here, one might hope that artificial agents could negotiate prices, quantities, delivery terms, and so forth in a manner sufficiently effective, speedy, reliable, and economic that human negotiators could be substantially replaced.

Second, there are ample theoretical reasons for this research. As we shall argue immediately below, however briefly, existing theory is inadequate for the general task of instructing boundedly-rational agents (including all artificial agents) on how to bargain in

---

[1] File: GP96_k08.doc. Revised: 960409.

mixed-motive, information-poor contexts. These are just the contexts encountered in electronic commerce. We are extending work, by others and by ourselves, that presents artificial agents with computationally-effective bargaining strategies in such information-poor contexts. Examination and evaluation of these strategies will have, we believe, both theoretical import and practical significance.

We shall now elaborate upon these points by offering in broad outline the general argument and motivation for this work. This will serve to motivate and frame the context of the specific results we present below. For lack of space, we have severely limited the explanatory remarks.

(1) Mixed-motive situations of conflict, calling for negotiation, are both ubiquitous and important, both within and without the realm of electronic commerce. *Remark*: In fact, the alternatives (pure conflict situations and pure coordination situations) are comparatively rare in actual applications.

(2) It is reasonable to hope that artificial agents (AAs) could be built that would perform credibly for many bargaining situations of import in electronic commerce. *Remark*: A substantial and growing series of experimental implementations amply attests to the plausibility of this hope (cf., Abreu and Rubinstein, 1988; Andreoni and Miller, 1990; Axelrod, 1984; Dworman, 1994; Dworman, Kimbrough and Laing, 1995a; Marimon, McGratton and Sargent, 1990; Miller, 1989; Oliver, 1996a,b).

(3) There are basically two ways for AAs to acquire the knowledge or information they need in order to negotiate effectively: (a) the knowledge can be encoded explicitly in them by humans, or (b) the AAs can undergo a machine-based learning regime and thereby obtain a functional equivalent to the requisite knowledge.

(4) Current theory is insufficient for task (3a). Although existing theory is rich in insights and provides useful benchmarks (more on this in the sequel), it cannot tell us how to program AAs for most mixed-motive bargaining contexts of interest. *Remarks*: (a) Relevant game-theoretic solution techniques almost invariably make assumptions (e.g., of shared prior probability estimates, common knowledge of agents' preferences and other aspects of the game, and perfect rationality) that do not apply to real agents. Further, until quite recently, game theory has focused almost exclusively on outcomes that, under perfect rationality, constitute equilibria of the game, paying scant attention to processes off the equilibrium path and the many vagaries of nonrational behavior. Consequently, previous game theory has had little to offer by way of instruction for programming AAs to negotiate effectively in complex real-world contexts. (b) Recent work in AI (e.g., Rosenschein and Zlotkin, 1994) has aimed at circumventing this problem by seeking to design mechanisms (*aka*: protocols) under which AAs may negotiate using pre-specified strategies, known *a priori* to be appropriate. While this is certainly a useful approach, and should be explored, its range of potential applicability is far from covering all the requirements for electronic commerce. In short, for very many attractive potential applications of AAs in negotiations, we simply are not yet up to the task of instructing the AAs how to bargain.

(5) Hence, if AAs are to serve effectively as negotiators in practical contexts, they will have to be instructed through a machine-learning regime. *Remark:* There remains the possibility of extracting practical bargaining knowledge from humans and implementing the AAs as expert systems, but this is not likely to be sufficiently general or cost effective.

(6) For present purposes, evolutionary computation is the most, or at least a very, promising approach for machine-based learning of effective bargaining strategies in mixed-motive contexts. *Remarks*: (a) Evolutionary computation has the tremendous advantage of yielding inspectable, interpretable bargaining strategies, which may, once discovered, be implemented in other contexts. (b) To date, the successful implementations of AAs that learn to negotiate have (as far as we know) all used some form of evolutionary computation (see references under point (2), above).

This line of reasoning leads easily to the suggestion that an extensive and well-designed series of computational experiments is required in order to produce artificial agents that can negotiate effectively in mixed-motive bargaining contexts. In what follows, we describe results from a series of experiments having exactly this aim.

3. Two Games

We report in this paper results from simulations, using genetic programming, of negotiations by AAs in two games. In both games, three agents (A, B, and C) negotiate competitively to decide which two-agent coalition forms and to agree on a division between the coalition partners of the points constituting this coalition's value. The agent who is excluded from the coalition wins zero. Neither game is symmetric, in that each coalition (*ij*) of two players has a unique value. In Game 1: $v(AB)=18$, $v(AC)=24$, $v(BC)=30$. In Game 2: $v(AB)=180$, $v(AC)=240$, $v(BC)=300$. This is a dynamic environment in which three agents are coevolving throughout while they negotiate a considerably more difficult decision problem than has been addressed in most previous research.

Each game is played as follows: To start the game, we make a null offer to the agent selected to be the first initiator. If the agent accepts, the game ends and all agents receive zero points. Otherwise, this agent makes an offer to another agent. If this other agent accepts, the game ends, and the coalition partners win the payoffs as specified in the offer, while the third player wins zero.

Table 0. Tableau for strategies of each agent

Fitness	Average number of points attained per game in which agent attained the floor.
Parameters	Generations = 4,000 Population size = 50 Creation procedure = Ramped half-and-half (Initial population: random) Selection = Fitness proportionate
Syntactic Structure	strategy :: if-then-else-stmt if-then-else-stmt :: (IF-THEN-ELSE condition-stmt action-stmt action-stmt) condition-stmt :: (condition-fn condition-stmt condition-stmt) | condition action-stmt :: if-then-else-stmt | action condition-fn :: AND | OR | NOT action :: ACCEPT | offer condition :: {player-id amount-lower-bound amount-upper-bound} offer :: {player-id amount}

If the responder does not accept, then it becomes the initiator and makes an offer, tacitly rejecting the previous offer. The game continues in this fashion until an agreement is reached, or until a specified maximum number (here, 6) of offers has been rejected, in which case the game ends by default, and all three agents receive a payoff of zero.[2]

4. Experimental Design

4.1 Strategy Trees

Table 0 summarizes our simulation.[3] Each agent's alternative strategies for playing the game are represented by a population of 50 strongly-typed GP trees. Each strategy is a nested IF-THEN-ELSE statement. The condition statement applies only to the most recent offer made to the agent. If the condition statement evaluates to TRUE then the first of the two action statements is evaluated. Otherwise, the second action statement is evaluated. The condition statement may be a condition or a Boolean combination of conditions. Action statements are either ACCEPT, or an offer (which implies a rejection of the previous proposal), or another IF-THEN-ELSE statement. Therefore, the strategies can be of almost arbitrary complexity by nesting Boolean condition statements and IF-THEN-ELSE action statements.[4] The output of a strategy is either the ACCEPT symbol or an offer. The condition and offer terminals are represented by a triple of values and a pair of values, respectively. These complex data structures (CDSs) were introduced because a coalition structure consists of two pieces of information: a set of players in the proposed coalition, and a distribution of the coalition's value over those players. Therefore, combinations of player names and payoff amounts are essential building blocks for constructing strategies.

Conditions specify contingencies for the strategy and evaluate to TRUE or FALSE, whereas actions evaluate to ACCEPT or an offer, where an offer terminal is represented in the form {Player, Amount}. An offer implies a rejection of the previous offer to which the rule is responding. A condition CDS – written as {Player Lower-Bound Upper-Bound} – evaluates to TRUE if, in the proposed agreement, the specified player would receive a payoff between Lower Bound and Upper Bound (inclusive). For example, suppose B makes the offer (A 5) and that A is responding to this offer with the rule (IF-THEN-ELSE {A 2 8} ACCEPT {C 18}). Taking $v(AB)=18$ and $v(C)=0$ into account, this implies both the proposed coalition agreement ((B 13)(A 5)) and the payoff (C 0) to the player omitted from the coalition. Given this offer and rule, the condition {A 2 8} is satisfied, and therefore A accepts the proposal.

4.2 Genetic Operators

We employ four genetic operators to create new strategies, and constrain these operators to ensure that the syntactic structure is maintained. *Crossover* swaps subtrees from two parents. *Mutation* changes the symbol at a randomly picked function node or a terminal node in the tree. If a terminal node is chosen then a new CDS may be created for that node. We provide two additional operators – *CDS crossover* and *CDS mutation* – to evolve the genetic material inside the CDS terminals. Crossover and mutation operate on entire subtrees or nodes of a program tree; they do not operate on the values inside a CDS. Therefore, without CDS-specific operators, the values inside a CDS would never

[2] The restriction to 6 offers per game is not very limiting. In our simulations, almost all agreements are reached within 3 offers. Initially, we limited games to 10 offers, but discovered that any game that went beyond 5 offers continued until the game ended by default. Therefore, we lowered the limit to 6 to speed up the simulations.

[3] For greater detail, see (Dworman, Kimbrough, and Laing; 1995b).

[4] Actually, we limit the complexity of the strategies to a maximum tree depth of 8. See Dworman, Kimbrough, and Laing (1995b).

change, and the agent would be prevented from evolving new offers or conditions. Both of the CDS operators work on a randomly chosen field within a CDS. CDS mutation toggles a random bit in the chosen field, thereby creating a new value. CDS crossover selects two CDSs, one from each parent, then picks a random point in the chosen field's bit string and swaps the right-hand sides of that field's bit string in the two CDSs. Again, both CDS operations are constrained to ensure legal syntactic structures. When a child is created without crossover (i.e., reproduction) it is probabilistically subjected to mutation and CDS mutation. When children are created via crossover the post-crossover children are probabilistically subjected to mutation, CDS crossover, and CDS mutation.

4.3 Strategy Evaluation

We simulated negotiations in the three-player cooperative games defined earlier. Each agent is represented by a population of 50 strategies. A tournament is used to evaluate all the strategies. Ideally, every strategy in an agent's population of rules would play each combination of opponent agents' strategies and do so three times, so that each of the players in the combination is given a turn to begin the negotiations. Unfortunately, that would require $3(50^3)$ = 375,000 games per generation, and thus is not feasible. Instead, selecting randomly without replacement, we match each strategy of a player with exactly 250 combinations of the other two players' strategies. Each selected combination then plays three games, each initiated by a different player. Thus, each strategy plays 750 games in the tournament. A strategy's strength (fitness) is based on its score in the tournament, which we calculate as the average payoff (points) won by the strategy in the outcomes of games in which, sometime during the negotiations, it obtained the floor.

4.4 Simulation Termination

We terminate the simulations when a specified number of generations (here 4000) has been completed, rather than using a stopping criterion based on the extent to which optimum behavior is approximated. We are interested in the behaviour of this system over time, and the extent to which these simple adaptive agents, which do not calculate optimal strategies, can achieve optimal or near-optimal performance in dynamic, complex, and multilateral bargaining situations.

5. Results

This section analyzes data from simulations through 4000 generations of these artificial agents negotiating coalition agreements in two games. The simulations are based on a twelve-cell experimental design consisting of a Cartesian product of parameter values for the genetic programming representation.[5] In both games, the values of the coalitions are specified by a characteristic function, *v(ij)*, which indicates the total number of points to be divided between agents *i* and *j* if they can reach an agreement.

We first address Game 1, in which the values of the two-agent coalitions are *v*(AB)=12, *v*(BC)=24, *v*(BC)=30, and all other coalitions have a value of zero. For Game 1, in each of the twelve cells of the design we conducted separate runs for three different seeds of the pseudo-random number generator. We recorded data at the random start (generation 0) and then every five generations, for a total of 801 reports per run. Thus the results for Game 1 are based on 12(cells) × 3(seeds) = 36 simulations , as described in N= 36×801 = 28,836 reports, each of which reports variables that summarize results from up to 3 (agents) × 50 (rules per agent) × 750 (games per rule) = 112,500 distinct bargaining sequences played in the tournament during that generation. Later, we compare results with nine simulations of negotiations by these artificial agents playing Game 2, in which the coalition values are ten times those of Game 1. Also we compare the performance of our artificial agents with that achieved by humans in a laboratory study conducted by Kahan and Rapoport (1974).[6]

5.1 Frameworks for Analysis: Game-Theoretic Equilibrium and System Stability

Our primary focus in this paper is to determine the extent to which these artificial agents, so cognitively limited by design, can, through coevolution, develop effective strategies for negotiating in this rather complex environment. As benchmarks, we compare the performance of the artificial agents and human subjects with that prescribed for rational play by the theories of cooperative and noncooperative games.

The Quota Solution. Various solution concepts in the theory of cooperative games prescribe for the games in this paper a *quota solution* (e.g., see Shubik, 1982, p. 177n), akin to a vector of equilibrium prices, which specifies the equilibrium values of the three agents' positions in the game when each agent seeks to maximize its total payoff under perfect competition for coalition agreements. For Game 1, the quota solution is given by the vector $q = (q_A, q_B, q_C) = (6, 12, 18)$, indicating that if coalition *(ij)* forms, then the coalition members should divide their coalition's value *v*(*ij*) (=

---

[5] For all runs reported in this paper, mutation probability = 0.5. The 12-cell design is based on the full Cartesian product (2×2×3) of the following parameter sets: crossover fractions = {0.2, 0.5}, leaf crossover probabilities = {0.2, 0.5}, leaf mutation probabilities = {0.2, 0.5, 0.8}.

[6] Their Game IV, Experiments 1-3 combined, *n* = 48 triads. The characteristic function of their Game IV is *v*"(AB)=66, *v*"(AC)=86, *v*"(BC)=106, else *v*"(S)=0. To compare payoffs achieved by human subjects in Game IV with artificial agents in our Game 1, we use the following transformation for positive-valued coalitions: *v*(S) = 0.3[*v*"(S)-6] if *v*"(S)>0; thus, *v*(AB)=18, *v*(AC)=24, and *v*(BC)=30.

Table 1. Best Rule's Deviation from Quota Solution, by Settledness, Game 1

	Number of Agents With Best Rule Settled			
	three	two	one	zero
Mean Abs. Deviation from Quota, Best Rule:	2.520	3.074	3.579	4.159
(n)	(3,158)	(7,439)	(9,951)	(8,288)
(n/N)	(.110)	(.258)	(.345)	(.287)

Table 2. Deviations from Quota and Coalition Payoffs in Settled Systems, by Generation, Game 1

	Generation (thousands)			
	0–1	1^+–2	2^+–3	3^+–4
(n)	(845)	(803)	(791)	(719)
Fraction of Cases Settled	(.110)	(.258)	(.345)	(.287)
Mean Abs. Deviation from Quota:				
best rule	2.716	2.945	2.290	2.066
median rule	2.692	2.861	2.242	2.037
Mean Payoff as Coalition Member:				
Agent A				
best rule	8.919	8.697	8.413	8.478
median rule	8.400	8.138	7.919	7.885
Agent B				
best rule	11.303	13.381	11.088	11.969
median rule	10.897	12.861	10.610	11.347
Agent C				
best rule	19.093	18.359	18.730	17.817
median rule	18.669	17.779	18.414	17.387

$q_i + q_j$) into the respective shares, q_i and q_j. It would be quite remarkable if the artifical agents coevolve to reach agreements approximating the quota solution, because we measure the agents' fitness by the *average payoff* they win in the tournament, and *not by how close these agreements are to the quota solution.*

Nash Equilibrium. Whereas cooperative game theory takes a macroscopic view of the situation by considering dominance relations among the alternative agreements that are possible, a more microscropic analysis considers the strategies agents use to guide their negotiations. The well-defined rules governing negotiations in this game enable us to model the bargaining situation as a noncooperative game in extensive (tree) form. A vector of strategies, $s = (s_A, s_B, s_C)$, constitutes a *Nash equilibrium* if each agent's strategy in s is a best response to the other two agents' strategies in s (Nash, 1951). Although multiple Nash equilibria can be expected in cooperative games, and do occur in ours, special types of Nash equilibria support the quota solution (Selten, 1988, 247-267; Laing, 1991). In this sense the two solution concepts are complementary, and the quota solution is consistent with perfectly rational play. But the Nash equilibrium concept makes a quite stringent requirement: each agent should exploit fully the opportunities presented by the other agents' strategies.

Settledness: A Measure of System Stability. In particular, we wish to focus on the behaviour of the system of artificial agents when, in a well-specified sense, it reaches a state of "equilibrium." In each of our runs, "equilibrium" is only transitory: the system of three agents does reach a period of relative stability only to be driven by random shocks into rather chaotic episodes, before once more settling into a relatively stable condition.

To measure the extent to which the system is presently stable, we employ the following operational

Table 3. Mean Absolute Deviation from Quota and Coalition Payoffs of Best Rules in First Settled Systems, Game 1 (n=36)

Mean Abs. Deviation from Quota:	Payoff as Coalition Member Agent A	Agent B	Agent C	Average Generation First Settled
2.156	8.453	10.423	18.897	413

Table 4. Best Rule's Deviation from Quota Solution, by Settledness and Game, in Three Cells of the Experimental Design (N= 9×801 = 7,209)

	Number of Agents With Best Rule Settled			
Mean Abs. Deviation from Quota, Best Rule:	three	two	one	zero
Game 1:	1.901	2.665	3.412	4.035
(n/N)	(.128)	(.287)	(.324)	(.260)
Game 2:	23.222	32.335	37.685	41.407
(n/N)	(.102)	(.274)	(.348)	(.275)

Table 5. Mean Payoff as Coalition Member of Best Rule in Settled Cases, by Game, in Three Cells of the Experimental Design

	Payoff as Coalition Member			
	Agent A	Agent B	Agent C	(n)
Game 1:	7.457	12.208	18.634	(923)
Game 2:	79.366	127.268	188.832	(737)

construct. Let *x(t)* denote any variable *x* at a point *t* in the sequence of generation reports. In this paper, we instantiate *x(t)* to be the average payoff won as a member of a coalition agreement by an agent's best rule, that is, the agent's rule that wins overall the greatest average payoff in the tournament during that generation. We define *x* to be *settled prima facie* at *t* if, for appropriate choices of the parameters m^* and d^*, both of these conditions obtain: [7]

(1) the m^* immediate predecessors of *x(t)*, with mean $x(m^*,t)$, have a standard deviation $s(m^*,t) \leq d^*$, and

(2) *x(t)* deviates no more than d^* from $x(m^*,t)$.

Further, if *x(t)* is settled prima facie, then describe as *settled retrospectively* any of its m^* immediate predecessors that deviates no more than d^* from $x(m^*,t)$. Finally, define *x(t)* to be *settled* if it is settled *prima facie* or retrospectively at *t*. We measure the extent of stability reached by the system during the generation at issue by the number of agents whose best rules are settled, in this sense, and say that the *system* is *settled* if the best rules of all three agents are settled. We now present results demonstrating that this operational construct provides a useful filter for examining the behaviour of these systems.

5.2 Coalition Payoffs vs the Quota in Game 1

Deviations from the Quota. The results displayed in Table 1 demonstrate that the more settled are these systems, the more they approach the cooperative game-theoretic equilibrium. When the best rules of all three agents' are settled, as they are in 11% of the cases, the coalition agreements they reach are roughly 2.5 points from the quota solution. The top portion of Table 2 indicates for all settled cases that the mean absolution deviation of the coalition payoffs achieved by both the best rules and the median rules of these artificial agents decreases somewhat over generations to within 2.0 points of the quota in the combined data for generations 3005–4000. This compares quite favorably with the agreements formed by humans: Kahan and Rapoport's (1974) subjects reached agreements (transformed to our payoff scale – see previous footnote) that *also* average 2.0 points from the quota solution.

Payoffs As Coalition Member. The lower portion of Table 2 displays the average payoffs achieved by each agent's best and median rules across the generations. There seems to be a weak trend across generations toward the quota payoffs of $q = (q_A, q_B, q_C) = (6, 12, 18)$. During the last thousand generations, the best rules

[7] We use m*=5 for both games, d*=0.5 for Game 1, and d*=5.0 for Game 2.

of the three agents, respectively, win payoffs as coalition members averaging roughly 8.5, 12 and 17.8. In contrast, human subjects in Kahan and Rapoport's experiments average 4.8, 12.7, and 17.7 (transformed – see prior footnote), respectively. Thus B and C approximate their quotas in both the simulation and human data. On the other hand, A averages 2.5 over quota in the simulation data, but 1.2 under quota in the human data. We speculate that a reason for this difference may lie in the fact that coalition BC formed overall in only 15.8% of the settled systems in our simulations, whereas the human subjects formed coalition BC in 72.9 % (35/48) of the time. Thus, A seems to enjoy a bargaining advantage in the simulation data, but to be disadvantaged by the human proclivity to form the BC coalition.

5.3 The First Settled System: A Predictor of Simulation Results

Of much importance in practical applications of artifical agents is the length of simulations needed to produce good results. We show in this section that the first-settled systems provide a reasonably good prediction about the overall simulation results. The first settled case can be reached early in the run. In the 36 runs of Game 1 in the 12-cell design the first settled system formed, on average, in generation 413. Two-thirds of the runs produced the first settled system within 500 generations, 11/12 within 1000 generations, and all within 2000 generations. In the 12-cell design, with 3 runs per cell, the median number of generations (per cell) to the first settled case ranged from 110 to 750. Table 3 presents results for the agents' best rules averaged over the 36 first-settled systems. Comparing these results with those for generations 3005-4000 displayed in the last column of Table 2, we see that the average deviation from the quota and the mean payoff won as a coalition member by Agent A is well predicted by the first-settled-case results, while the mean payoffs for Agent B (respectively, C) deviate about one and a half points (respectively, one point) from the average of all settled points in generations 3005–4000.

5.4 Payoffs and the Quota in a Scaled-Up Game

By multiplying the characteristic function of Game 1 by 10, we formed Game 2 with the characteristic function $v(AB)=120$, $v(BC)=240$, $v(BC)=300$, else $v(S)=0$, and the associated quota solution $q = (q_A, q_B, q_C) = (60, 120, 180)$. In all other respects, the two games are identical. While the overall search space for Game 1 is at least on the order of 10^{12} (Dworman, Kimbrough and Laing, 1995), this transformation to Game 2 increases the size of the search space by a factor of 10^6. Yet, in a run of 4000 generations with 3 players and 50 rules per player each generation, the maximum number of alternative rules created in the simulation is 6×10^5. Thus, given the size of the problem, it is noteworthy that the artificial agents do so well in Game 1. It would be even more remarkable if they cope successfully with Game 2.

We conducted simulations of Game 2 using three cells of the twelve-cell design and the three seeds of the pseudo-random number generator, for a total of nine runs.[8] To compare the results of these two games, we now limit our attention to the results obtained from the nine runs of Game 1 in these same three cells.

The results displayed in Tables 4 and 5 indicate that, in the scale of Game 1, the mean deviation from the quota solution and average payoffs won by coalition members in the settled cases of Game 2 are within a half point of those obtained in Game 1. We think it remarkable that these artificial agents cope with the upscaling of the original game with such aplomb.

5.5 Opportunity Costs and Nash Equilibrium

Thus, these artificial agents succeed in reaching agreements in the neighborhood of the cooperative game-theoretic solution, but do they satisfy the more stringent requirements for a Nash equilibrium? To do so, they must employ strategies that are optimal responses to the bargaining environment, and therefore must incur no opportunity cost.

We operationalize opportunity cost in this situation as follows. Consider any game in the tournament in which the rule of, say, Agent B obtains the floor to respond to an offer. At this point B can accept the offer or make a proposal to one of the other agents. We compute B's *opportunity cost* for this game as the maximum payoff the rule could win by making an offer that would be accepted immediately, minus the actual payoff achieved by B in the final outcome of negotiations in this game of the tournament.[9] At a Nash equilibrium, each agent's opportunity cost, thus measured, would be zero.

Table 6 displays the mean opportunity costs incurred in the settled systems for all runs of Game 1 in the 12-cell design by each agent's best and median rules, controlling for generations. Although Agent A employs strategies that are excellent responses to the opponents' strategies, Agents B and C do not. Moreover, B and C do not improve in this regard over generations – if anything, they get worse.

Although these simple adaptive agents coevolve to negotiate agreements that approximate the cooperative game-theoretic solution, even the best strategies of Agents B and C are quite suboptimal reponses to the

[8] Recall that in all runs reported in this paper, mutation probability = 0.5. The parameter values for each of the three cells is given by its respective element in the following vectors: crossover fractions = (0.5, 0.2, 0.2), leaf crossover probabilities = (0.2, 0.2, 0.2), leaf mutation probabilities = (0.5, 0.8, 0.5).

[9] Note that, by this measure, an agent's opportunity cost in a particular game could be negative: the agent might accept an offer of a payoff that is greater than it could achieve by making an offer that would be accepted.

Table 6. Mean Opportunity Costs of Best and Median Rules in All Settled Cases in Game 1, Twelve-Cell Design, by Agent and Generation

	Generation (thousands)			
	0–1	1^+–2	2^+–3	3^+–4
(n)	(845)	(803)	(791)	(719)
Mean Opportunity Cost:				
Agent A				
best rule	0.673	0.818	-0.810	-1.412
median rule	1.194	1.395	-0.345	-0.876
Agent B				
best rule	2.706	3.037	2.636	3.611
median rule	3.355	3.942	3.435	4.745
Agent C				
best rule	3.872	3.706	4.133	4.325
median rule	4.960	4.929	5.170	5.359

strategies of their opponents. Thus, the strategies of these artificial agents do not even approximate a Nash equilibrium. On the basis of experience in other laboratory investigations (e.g., Albers and Laing, 1991), we have good reason to suspect that Kahan and Rapoport's human subjects also used bargaining strategies that were suboptimal. Yet this suboptimality might not be apparent to an observer, because both human subjects and our artificial agents do reach negotiated outcomes that approximate a prominent cooperative solution. Understanding the processes underlying such behavior is an important goal for future research.

6. Conclusion

We are heartened by these results. The artificial agents, so simple in structure and handicapped by ignorance about each other and the game, and lacking any memory of prior negotiations, nonetheless are able to coevolve within the situation to negotiate coalition agreements that not only approximate those prescribed by the theory of cooperative games but also rival the performance of humans. Moreover, the agents cope admirably when we scale up the original game, thus increasing the size of the strategy space, already on the order of at least 10^{12}, by a factor of 10^6. We think these are remarkable achievements that augur well for future applications in electronic commerce. On the other hand, these agents do not take full advantage of opportunities presented in the strategies of their opponents. Providing the agents with the wherewithal to respond more fully to opportunities in the bargaining environment is an important task for future research.

7. Acknowledgments

We gratefully acknowledge the comments and suggestions of four anonymous referees for GP-96.

8. Bibliography

Abreu, D. and A. Rubinstein. 1988. The structure of Nash equilibrium in repeated games with finite automata. *Econometrica* 56: 1259-81.

Albers, W., and J.D. Laing. 1991. Prominence, competition, learning, and the generation of offers in computer-aided experimental spatial games. In R. Selten, ed., *Game Equilibrium Models, III: Strategic Bargaining*. Berlin: Springer-Verlag, 141-185.

Andreoni, J. and J.H. Miller. 1990. *Auctions with Adaptive Artificially Intelligent Agents*. University of Wisconsin, Social Systems Research Institute Working Paper 9032.

Axelrod, R. 1984. *The Evolution of Cooperation.* New York: Basic Books.

Dworman, G. 1994. Games computers play: simulating characteristic function game playing agents with classifier systems. In *Proceedings of the First IEEE Conference on Evolutionary Computation IEEE World Congress on Computational Intelligence,* Volume 2. Los Alamitos, CA: IEEE Press, 696-701.

Dworman, G., S.O. Kimbrough, and J.D. Laing. 1995a. On automated discovery of models using genetic programming: bargaining in a three-agent coalition game. *Journal of Management Information Systems*, 12, 3, Winter 1995-1996, in press.

Dworman, G., S.O. Kimbrough, and J.D. Laing. 1995b. *Implementation of a Genetic Programming System in a Game-Theoretic Context.* The Wharton School, University of Pennsylvania, OPIM working paper 95-01-02.

Kahan, J.P., and Amnon Rapoport. 1974. Test of the bargaining set and kernel models in three-person games. In Anatol Rapoport (editor), *Game Theory as a Theory of Conflict Resolution.* Dordrecht, Holland: D. Reidel, 119-160.

Koza, J.R. 1992. *Genetic Programming: On the Programming of Computers by Means of Natural Selection.* Cambridge, MA: The MIT Press.

Laing, J.D. 1991. A noncooperative model of bargaining in simple spatial games. In R. Selten (editor), *Game Equilibrium Models, III: Strategic Bargaining*. Berlin: Springer-Verlag, 80-117.

Marimon, R., E. McGrattan, and T.J. Sargent. 1990. Money as a medium of exchange in an economy with artificially intelligent agents. *Journal of Economic Dynamics and Control* 14: 329-373.

Miller, J.H. 1989. *The Coevolution of Automata in the Repeated Prisoner's Dilemma.* Santa Fe Institute, Economics Research Program working paper 89-003.

Nash, J. 1951. Non-cooperative games. *Annals of Mathematics* 54: 286-95.

Oliver, J.R. 1996a. On artificial agents for negotiation in electronic commerce. In J.F. Nunamaker, Jr., and R.H. Sprague, Jr. (editors), *Proceedings of the 29th Hawaii Conference on System Sciences* 4. Washington, D.C.: IEEE Computer Society Press, 337-346.

Oliver, J.R. 1996b. *On Artificial Agents for Negotiation in Electronic Commerce.* Ph.D. dissertation, University of Pennsylvania, The Wharton School, Philadelphia, PA.

Rosenschein, J.S. and G. Zlotkin. 1994. *Rules of Encounter: Designing Conventions for Automated Negotiation among Computers.* Cambridge, MA: The MIT Press.

Schelling, T.C. 1960. *The Strategy of Conflict.* Cambridge, MA: Harvard University Press. Selten, R. 1988. *Models of Strategic Rationality.* Dordrecht, Holland: Kluwer.

Selten, R. 1988. *Models of Strategic Rationality.* Dordrecht, Holland: Kluwer.

Shubik, M. 1982. *Game Theory in the Social Sciences: Concepts and Solsutions.* Cambridge, MA: The MIT Press.

Waveform Recognition Using Genetic Programming: The Myoelectric Signal Recognition Problem

Jaime J. Fernandez
Rice University
4110 St. Michael's Ct.
Sugar Land, TX 77479-2913
jjf@jjf.com
http://www.jjf.com/
http://tommy.jsc.nasa.gov/~jjf

Kristin A. Farry
NASA/Johnson Space Center
ER
Houston, TX 77058-3696
farry@mickey.jsc.nasa.gov
http://tommy.jsc.nasa.gov/~farry/kristin_farry.html

John B. Cheatham
Dept. of Mechanical Engineering and Materials Science
Rice University
Houston, TX 77005-1892
jbcjr@caesar.rice.edu
http://www.ruf.rice.edu/~jbcjr/cheatham.html

ABSTRACT

This paper presents a new strategy for multifunction myoelectric control of an upper arm prosthesis. We apply genetic programming to the classification of several anisometric myoelectric patterns (e.g., thumb flexion, extension and abduction), using features extracted from the myoelectric signal. This approach increases the number of possible control functions that can be extracted from myoelectric signals. This control scheme, combined with a multifunction myoelectric hand design, will provide the prosthesis user with more functions than now available. We present three different approaches each capable of recognizing 100% of the three motions. Finally, we present our ideas for the future of this project.

1. Introduction

The American Heritage Dictionary of the English Language defines myoelectric as "relating to the electrical properties of muscle tissue from which impulses may be amplified, and used especially in the control or operation of prosthetic devices."

There are two important uses for myoelectric signals in control systems. The first use and the one of primary interest to us is control of myoelectric prostheses for amputees. Present prostheses allow the amputee only a single motion choice from the myoelectric signal. Present day prosthesis control algorithms are not capable of recognizing different myoelectric patterns; instead, they simply look for some level of myoelectric activity which triggers a simple open or close command to the pincher-like prosthetic hand. The aim of this research is to develop an algorithm capable of recognizing three different patterns in a single myoelectric signal in order to control several different motions on a prosthetic hand.

Another important research interest is application of myoelectric signals for teleoperation purposes. NASA needs improvements in teleoperation of dextrous manipulators such as the Utah/MIT Dextrous Hand [Jacobsen 1986]. The three degree of freedom joysticks presently used for control of the space shuttle robotic arm are inadequate for intuitive and non-fatiguing teleoperation of dextrous hand manipulators. Mechanical exoskeletons are a possible joystick replacement; myoelectric control is also an emerging option [Farry 1994]. We mention myoelectric teleoperation because it provides us with a broad proving ground using normal subjects for myoelectric algorithms, some of which will be applicable to prosthesis control. In this paper we use myoelectric data from a normal subject, so our results are immediately applicable to teleoperation, although our ultimate goal is prosthesis control.

Figure 1 shows the context of our work. The myoelectric control processes are:

1. Measure the electrical activity associated with muscle activity. (the myoelectric signal).
2. Preprocesses the signal and extract distinguishing features.
3. Classify the myoelectric patterns into the specific motions or (in the amputee's case) motion intentions to generate them. This is the process that this paper primarily addresses.
4. Convert the human motions or intentions into robot or prosthesis kinematics and dynamics.

5. Convert the commands into the robot or prosthesis specific actuator commands.

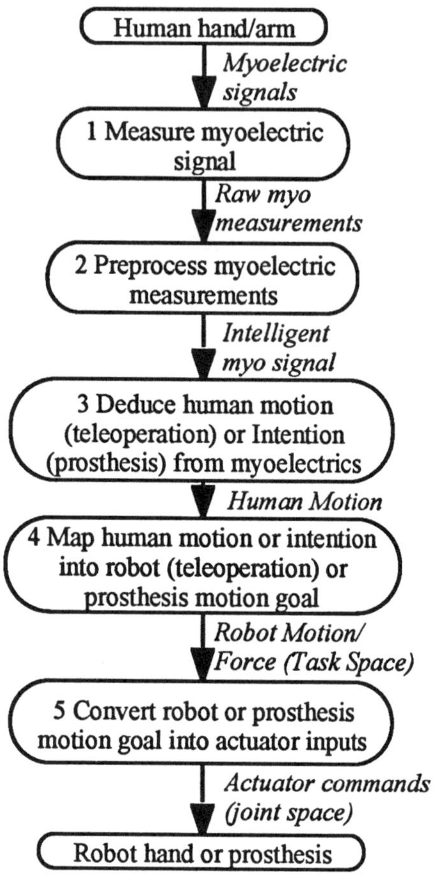

Figure 1 Real-time processes in intuitive myoelectric control.

2. Theory And Approach

We encourage the reader to study a physiology text [Berne et al, 1988] for a more detailed description of myoelectric signals. Briefly, an electrode placed on the skin surface can measure the passing of the action potentials propagating along muscle fibers underneath the electrode as the fibers contract. Skeletal muscle fibers are innervated (and hence stimulated to contract) in small groups known as single motor units (SMUs), the smallest voluntary contractible group. A single action potential in an SMU is associated with a twitch; sustained contractions require repeated firings (resulting in a string of action potentials) of an SMU. As more force is needed, more SMUs are fired or recruited. The electrode records the sum of all these action potentials after they propagate through the tissue to the skin surface [Hogan 1976].

This paper follows the feature extraction method from Hudgins [Hudgins et al 1993] and Saridis [Saridis 1982]. Hudgins in particular developed a simple approach for the classification of four different motions or muscle contractions. His classification scheme used five different features from several windows or time segments of each of the signals. His five features were Mean Absolute Value, Mean Absolute Value Slope, Zero Crossings, Slope Sign Changes and Waveform Length. He used a multilayer perceptron with back-propagation to classify myoelectric patterns from these features.

Our work differs from that of Hudgins by using additional features and using genetic programming for classification. We encourage the reader to study [Koza 1992] for a detailed explanation on genetic programming. Briefly, genetic programming creates computer programs in the Lisp or Scheme computer languages as the solution. Furthermore, it uses terminal and function sets to create the computer programs.

The first and most complicated task was the selection of the terminals for the genetic program. We began with Hudgins' features and defined five additional features to compose our terminal set. Furthermore, our features are calculated for a single 150 ms segment of the entire signal, while Hudgins divided each signal into six 40 ms segments. Our features are:

The *mean absolute value* (MAV) of the length-N signal $\overline{X}$ is given by $\overline{X} = \frac{1}{N}\sum_{k=1}^{N}|x_k|$, where x_k is the kth sample in the signal.

The *mean absolute value slope* (MAVS) is the difference between sums in adjacent samples *i* and *i*+1, as defined by $\Delta\overline{X}_i = \overline{X}_{i+1} - \overline{X}_i$.

The *number of zero crossings* (ZC) yields a simple frequency distribution measure. We included a threshold in the zero crossing calculation to reduce the noise induced zero crossings. Given two consecutive samples x_k and x_{k+1}, increment the zero crossing count (ZC) if $sign(x_k) * sign(x_{k+1}) = -1$ and $|x_k - x_{k+1}| \geq Threshold$.

The *number of slope sign changes* (SC) provides another simple measure of the frequency. Once again, a suitable threshold must be chosen to reduce noise induced slope sign changes. Given three consecutive samples, x_{k-1}, x_k and x_{k+1}, we increment the slope sign change count if $x_k > x_{k-1}$ and $x_k > x_{k+1}$, or $x_k < x_{k-1}$ and $x_k < x_{k+1}$, and $|x_k - x_{k+1}| \geq Threshold$ or $|x_k - x_{k-1}| \geq Threshold$.

Waveform length (LEN) provides information on the waveform complexity in each signal. This is the cumulative length of the waveform over the time segment defined as $length = \sum_{k=1}^{N}|\Delta x_k|$ where $\Delta x_k = x_k - x_{k-1}$ (difference in consecutive sample voltage values). The resultant value gives a measure of waveform amplitude, frequency, and duration all within a single parameter.

Average value (AVG) is a feature similar to the mean absolute value, except that the real value of the sample is used as opposed to the absolute value. We define it as $\overline{X} = \frac{1}{N}\sum_{k=1}^{N} x_k$.

Up slope (UP) is a feature which counts the number of individual samples which have a positive slope. It is defined by $x_k - x_{k-1} > 0$. This feature and the next one (Down Slopes) differentiate between a gradual increase in the waveform as opposed to a steep increase.

Down slope (DOWN) is a feature which counts the number of individual samples which have a negative slope, as defined by $x_k - x_{k-1} < 0$.

The *ratios of mean absolute values* compound feature which uses information from the previously described features. The features are the ratio of the MAV for channel one to the MAV for channel two, and the ratio of the MAV for channel two to the MAV for channel one. This ratio gives moderately good results in the differentiation of the signals. In some of the solutions the genetic program uses the first ratio, while in others it uses the second one, even though one is the reciprocal of the other.

The terminal set for the genetic program consists of the eight features previously described for each of the two myoelectric channels and the two ratios of the MAV, for a total of eighteen features. The terminal set also includes several constants. This gives the genetic program the ability to manipulate the data with some consistency. The majority of the constants are a series of numbers which start at 0.7 and change by an increment of 0.1 up to 1.3 (i.e., 0.7, 0.8, 0.9, 1.0, 1.1,1.2, 1.3).

The next task was defining the function set for the genetic program (Table 1). We selected several types of functions for their diverse properties. For example, we included trigonometric functions because the waveforms have cyclical features.

The fitness function is the most important constituent of the genetic programming recognition method. The fitness function must yield the same value for the same type of myoelectric signal and a different value for a different type of myoelectric signal. The values for the same type of myoelectric signal must fall in a small range of numbers while the number ranges for the other myoelectric signals must not overlap. An overlap occurs when the minimum range of one signal is less than the maximum range of another signal. Figure 2 illustrates an ideal classification scheme's separation.

Table 1 Functions for the genetic program.

Name	Symbol
Addition	+
Subtraction	-
Multiplication	*
Division	/
Square Root	sqrt(abs(x))
Sine	sin
Cosine	cos
Tangent	tan
Natural Logarithm	ln(abs(x))
Common Logarithm	log(abs(x))
Exponential	exp
Power Function	x^y
Reciprocal	1/
Absolute Value	abs
Integer or Truncate	int
Sign	sign

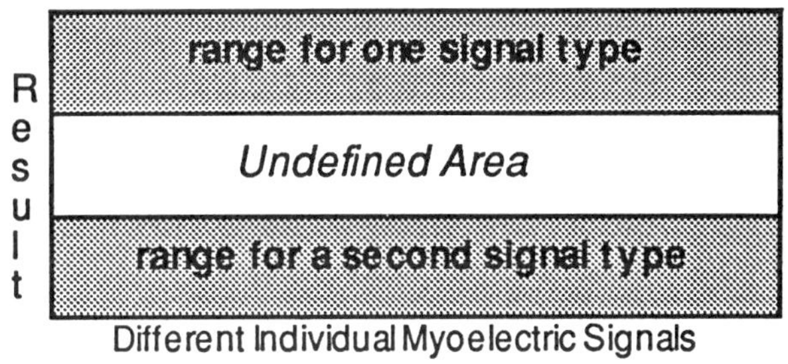

Figure 2 A graphical representation of the classification scheme, where the signal type could be extension, abduction or flexion.

The specific form of the fitness function was

$$\textit{Fitness Function} = \frac{\textit{Standard Deviation(Signal 1)} + \textit{Standard Deviation(Signal 2)} + \textit{Weight } (=10)}{\left|\textit{Average(Signal 1)} - \textit{Average(Signal 2)}\right|}$$

for the pair-wise solutions. The weight is a constant used to increase the importance of the difference between the average values of the two signals. This magnifies the difference between the two signal values.

3. Results

3.1 Data Set.

We tested our genetic programming approach with sixty different thumb abduction signals, fifty-three thumb extension signals, and fifty-four thumb flexion signals (Figure 3). During each motion, we recorded myoelectric signals from the Abductor Pollicis Longus and Extensor

Pollicis Brevis (APL/EPB) pair (Figure 4, Electrode 1) and from the Flexor Pollicis Longus (FPL) (Figure 4, Electrode 2). We used 6 mm recessed wet silver-silver chloride electrodes and band-pass filtered (3-300 Hz) the myoelectric signal before sampling it at 1000 Hz. Thus, we have two myoelectric traces for each motion. We simultaneously recorded thumb trajectory. This trajectory allowed us to find the precise starting locations of the motion. See [Farry 1994] for details of our data collection system.

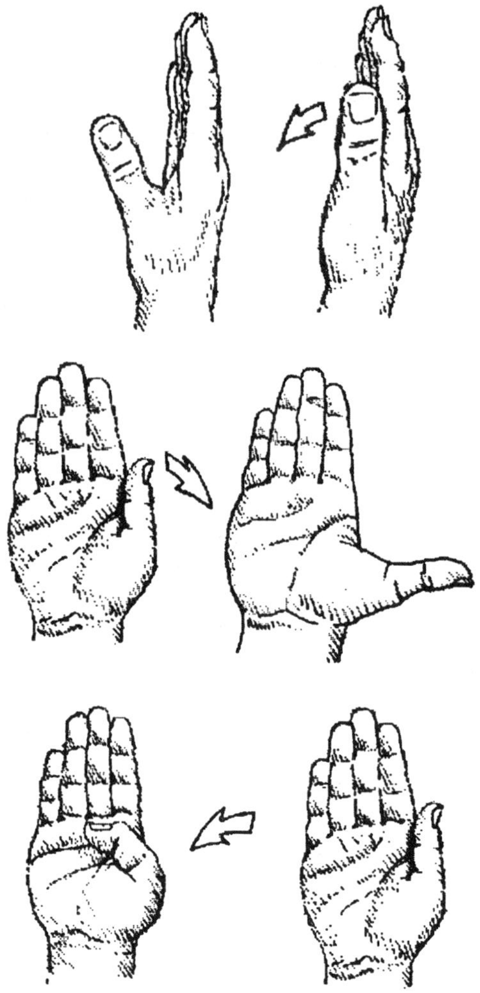

Figure 3 Thumb motions analyzed. Top: Abduction. Middle: Extension. Bottom: Flexion.

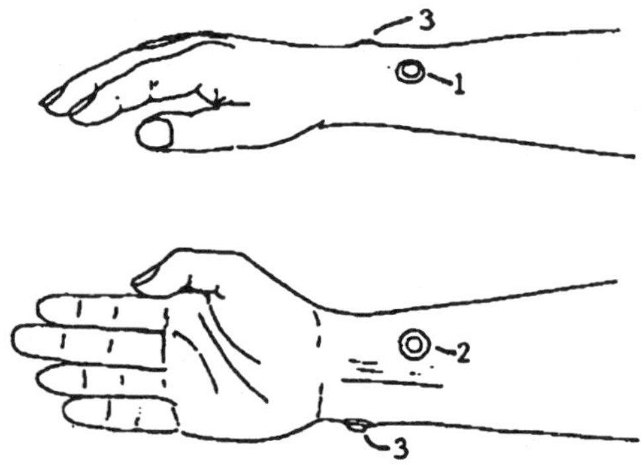

Figure 4 Electrode locations. Electrode 3 is ground. [Farry 1994]

Figures 5 through 8 illustrate several examples of our myoelectric signals. Figures 5 and 6 illustrate two different signals for the abduction motion; note that similar motions produce substantially different waveforms. Figures 7 and 8 illustrate sample extension and flexion signals respectively.

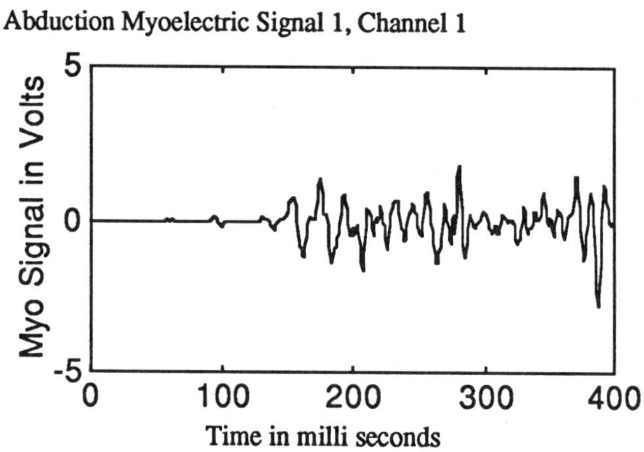

Figure 5 An abduction signal example.

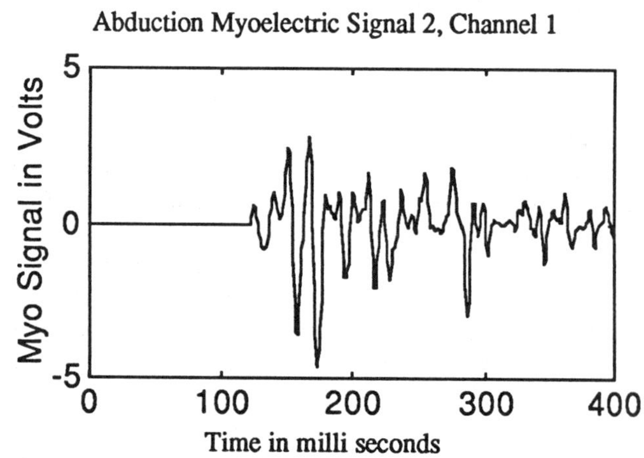

Figure 6 A second abduction signal example.

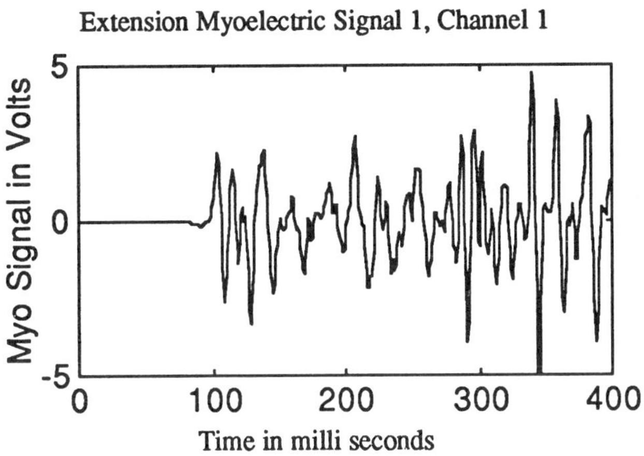

Figure 7 **An extension signal example.**

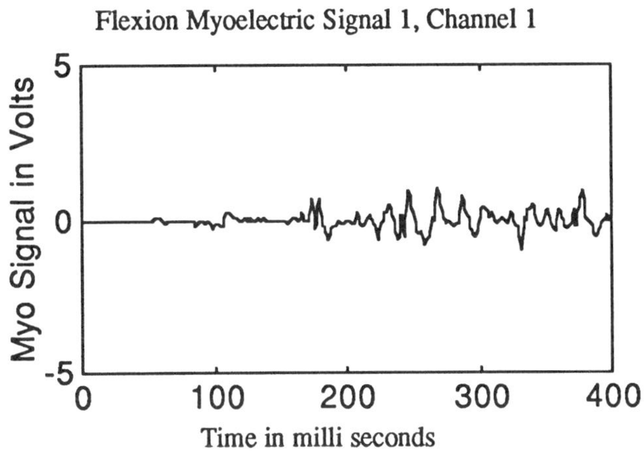

Figure 8 **A flexion signal example.**

3.2 Pair-Wise Genetic Programming Myoelectric Signal Classification

We implemented three different approaches to find a suitable solution to the problem of recognizing several myoelectric signals. For the first approach, we divided the three different myoelectric signal sets available into three pairs with three different combination possibilities. The pairs were: extension and flexion, abduction and extension, and abduction and flexion. Each of these pairs was then implemented in the genetic program to obtain a pair-wise equation. We computed the classification solution by substituting each individual thumb signal's motion features into each of the three pair-wise equations. The signal's classification was determined when the signal's computed solution falls within a sample range for two of three equations. For example, if a signal satisfied both the extension-flexion pair and the abduction-extension pair it was classified as extension.

Table 2 Tableau for the genetic programming solution to the recognition of myoelectric signal pairs.

Objective:	Find an Excel function that is able to recognize one of the three pairs of myoelectric signals.
Terminal Set:	`MAV1, MAVS1, ZC1, SC1, LEN1, AVG1, UP1, DOWN1, MAV2, MAVS2, ZC2, SC2, LEN2, AVG2, UP2, DOWN2, MAV1/MAV2, MAV2/MAV1, 0.7, 0.8, 0.9, 1.0, 1.1, 1.2, 1.3, 0.01, -1.0`
Function Set:	`+, -, * ,/ ,sqrt , sin, cos, tan, ln, log, exp, power, reciprocal, abs, int, sign`
Number of training myoelectric signals of each type:	20
Number of testing myoelectric signals:	40 (Total of 60 signals - 20 training signals = 40)
Fitness cases:	Three fitness cases
Raw fitness:	Fitness = First Contraction Standard Deviation + Second Contraction Standard Deviation + 10/Reciprocal of the Difference of the Averages.
Standardized fitness:	Same as raw fitness.
Parameters:	Population = 80. Generations = 100.
Success predicate:	The program finds the minimum fitness, and there are not overlaps in the solution.

Table 3 contains the three equations that the genetic program found for the three different pairs of motions. We gave twenty motion signal sets to the genetic program for it to find these equations. Figures 9,10, and 11 illustrate the solutions for the three possible pairs. Each of these three solutions perfectly distinguishes its pair of signal motions. Thus, they give a 100% accurate solution. The graphs illustrate the results using all of the signals (not just the twenty training signals) available for that particular thumb motion.

Table 3 The three different solutions for each pair of motions.

Extension and Flexion pair	exp(exp(exp(sign(ln(abs((abs(sin((tan(LEN1)/sign(sign((tan(log10(abs((LEN1*0.7))))/(log10(abs(LEN1))+sign(log10(abs((LEN1*0.7))))))))))-(MAV1/MAV2))))))))
Abduction and Flexion pair	exp(int(sqrt(abs((UP1+DOWN1)))))*int((int(sqrt(abs((MAV1/MAV2))))*sqrt(abs((UP1+DOWN1)))))
Abduction and Extension pair	exp((((exp(1.2)/sign((((1.1*sign(ln(abs((AVG2-((DOWN2/DOWN2)+((AVG2-((LEN1*LEN1)/log10(abs(abs(CHS2))))))/(AVG2+1.3))))))))*sqrt(abs((1.1/(AVG2+1.2)))))-((abs(((((DOWN2-1.1)*(1.2*1.1))*abs(sin((log10(abs((CHS2*(DOWN2*1.1))))-(LEN1*(CHS2*1.1))))))+ln(abs(ZC1))))*sign(ln(abs((AVG2-(tan((log10(abs((LEN1*(MAV1/MAV2))))+log10(abs((CHS2*1.1)))))/abs(sin(((DOWN2/DOWN2)-(AVG2+ln(abs(ZC1))))))))))))*(DOWN2+(exp((log10(abs((CHS2*(CHS2*1.1))))-(LEN1*1.1)))/DOWN2))))))-exp(1.3))/tan((sqrt(abs(abs(abs(cos(1.2)))))-sign((((abs(abs(cos(DOWN2)))*(MAV1/MAV2))*sign(ln(abs((AVG2-(tan((log10(abs((LEN1*(MAV1/MAV2))))+log10(abs((CHS2*1.1)))))/abs(sin(((DOWN2/DOWN2)-(AVG2+ln(abs(ZC1))))))))))))*sqrt(abs((((LEN1*(CHS2*ZC1))*1.1)/1.1))))))))

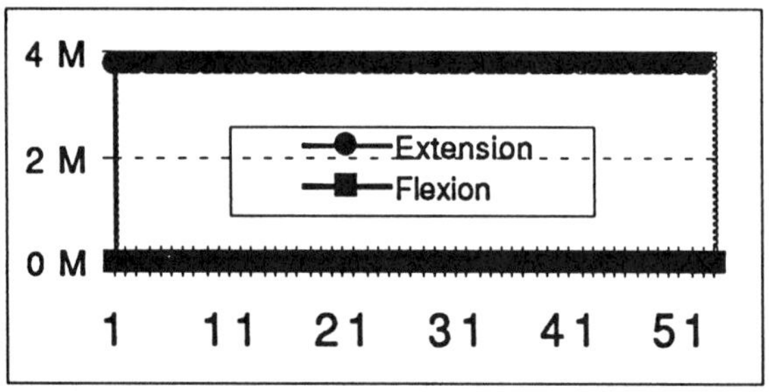

Figure 9 The graphical representation of the classification scheme for the extension-flexion pair.

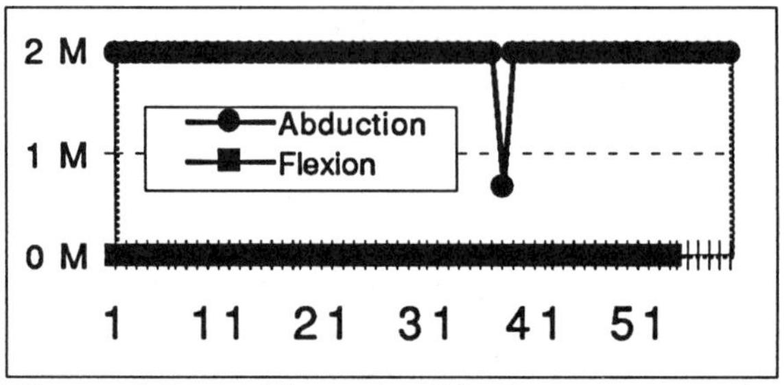

Figure 10 The graphical representation of the classification scheme for the abduction-flexion pair.

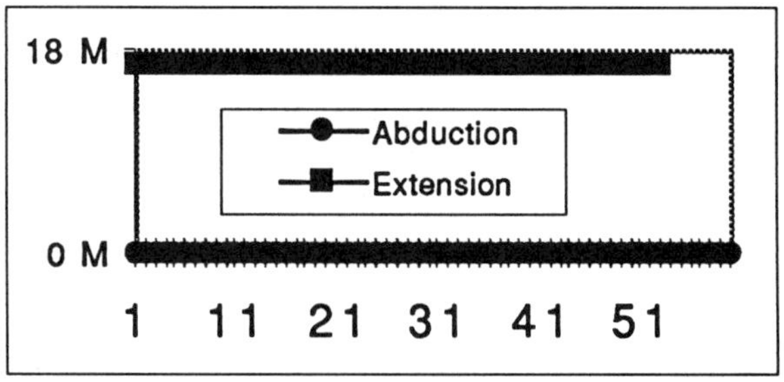

Figure 11 The graphical representation of the classification scheme for the abduction-extension pair.

3.3 Genetic Programming Applied To The Three Myoelectric Signals

Next, we used genetic programming to create a solution capable of distinguishing three different myoelectric signals simultaneously. The terminal and function sets were the same as the approach for the myoelectric signal pairs; however, we modified the fitness function to:

$$\text{Fitness Function} = STD(\text{Signal } 1) + STD(\text{Signal } 2) + STD(\text{Signal } 3) + \frac{100}{MIN\begin{bmatrix} \lvert \text{AVG of Signal 1} - \text{AVG of Signal 2} \rvert \\ \lvert \text{AVG of Signal 1} - \text{AVG of Signal 3} \rvert \\ \lvert \text{AVG of Signal 2} - \text{AVG of Signal 3} \rvert \end{bmatrix}}$$

Note: *STD* is standard deviation of the signals and *AVG* is the overage.

Table 4 Tableau for the genetic programming solution to the recognition of myoelectric signal pairs.

Objective:	Find an Excel function that is able to recognize the three different myoelectric signals.
Terminal Set:	MAV1, MAVS1, ZC1, SC1, LEN1, AVG1, UP1, DOWN1, MAV2, MAVS2, ZC2, SC2, LEN2, AVG2, UP2, DOWN2, MAV1/MAV2, MAV2/MAV1, 0.7, 0.8, 0.9, 1.0, 1.1, 1.2, 1.3, 0.01, -1.0
Function Set:	+, -, * ,/ ,sqrt , sin, cos, tan, ln, log, exp, power, reciprocal, abs, int, sign
Number of training myoelectric signals of each type:	22
Number of testing myoelectric signals:	38 (Total of 60 signals - 22 training signals = 38)
Fitness cases:	Four fitness cases
Raw fitness:	Fitness = Standard Deviation of Abduction + Standard Deviation Extension + Standard Deviation Flexion + 100/Maximum of the difference of the three reciprocals
Standardized fitness:	Same as raw fitness.
Parameters:	Population = 256. Generations = 4,000.
Success predicate:	The program finds the minimum fitness, and there are not overlaps in the solution.

Unlike the previous pair-wise equations, which were found on the first run, it took twenty three runs of the genetic program to find a 100% effective solution with no overlapping regions and no signals misclassified. We increased the number of training motions from twenty to twenty-two. The solution was a long string of functions and terminals containing most of the features. It gave a different resulting number for each of the three myoelectric signals without overlap in the results. The three-way solution is much larger than the pair-wise solutions in table 1 and because of space is not included in this paper. Table 5 displays the different features used by this solution. Figure 12 illustrates the classification scheme for the solution.

Table 5 The features used in the solution.

Features Available and Used		
Channel 1 Feature (APB/EPL)	Abbreviation	Used
Zero Crossings	ZC1	yes
Up Slopes	UP1	no
Down Slopes	DOWN1	yes
Slopes Changes	SC1	no
Average	AVG1	yes
Mean Absolute Value	MAV1	yes
Mean Absolute Value Slope	MAVS1	no
Length	LEN1	no
Channel 2 Feature (EPL)		
Zero Crossings	ZC2	yes
Up Slopes	UP2	yes
Down Slopes	DOWN2	yes
Slopes Changes	SC2	yes
Average	AVG2	no
Mean Absolute Value	MAV2	yes
Mean Absolute Value Slope	MAVS2	yes
Length	LEN2	yes

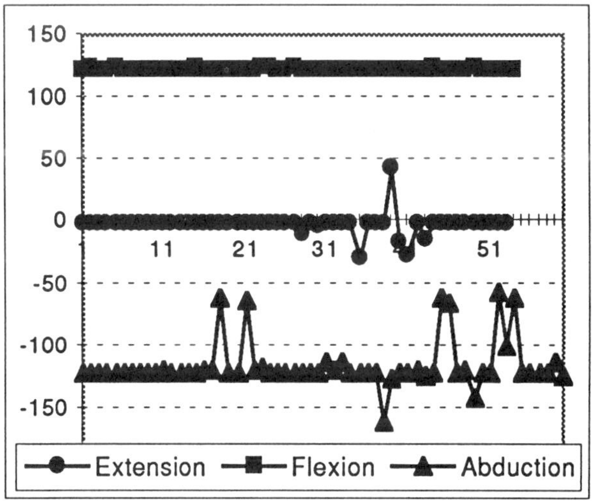

Figure 12 The graphical representation of the classification scheme for the solution of the three motions.

3.4 A Hybrid GP Solution To The Problem Of Myoelectric Signal Recognition

Our third approach to the problem of myoelectric signal recognition was a hybrid solution that used both solutions from the two previous approaches. Our analysis showed that the pair-wise solutions contained very simple relations among the features that gave 100% accuracy in distinguishing the flexion signal and up to 89% accuracy in distinguishing the other two signals. The best two relationships are in Table 6, which also displays the results for the ratio of the MAV of channel two over the MAV for channel one for comparison.

Table 6 Three very simple relationships that have a high accuracy in distinguishing the three myoelectric signals.

Relationship	Total Accuracy
LEN2 - LEN1	89 %
MAV2 - MAV1	85 %
MAV2 / MAV1	78 %

The hybrid solution consisted of first using one of the two simple relationships between the features (LEN2 - LEN1 or MAV2 - MAV1) to find a quick answer to about ninety percent of the signals. Afterward, the program applied the solution of the pair of abduction and extension to the remaining ten percent to complete the classification of all the signals.

4. Future Work

Tests are needed to determine the usefulness of this approach for prosthesis control. We performed all of our tests on pre-recorded data from a single normal subject. We need to tests our algorithm on many limb-deficient subjects and explore real-time implementation. Data collection from three amputee volunteers will begin this summer.

A better, yet more complicated improvement, is the modification of the genetic program in order to find its own features. Laird [Laird 1994] and Tackett [Tackett 1994] demonstrated the use of genetic programming for feature identification. Their work could be applied to this problem as well. The ability of the genetic program to find its own features will greatly reduce the learning curve necessary for the operator to apply the program to each individual user.

5. Conclusions

This project's main contribution is the successful application of genetic programming to recognize myoelectric signals. This is the first implementation of a genetic method to learn to recognize myoelectric signals. This implementation has many advantages over other implementations that solve similar problems. The most important advantage of this method is that it determines the most relevant characteristics of a waveform. In essence, it automates factor analysis. The genetic program found some simple mathematical relationships that in some cases are 100% effective in distinguishing one of the signals. For example, the simple relationship of the Length of signal from channel two minus the length of the signal from channel one (LEN2 - LEN1) provides 100% correct discrimination results for one signal type and 89% correct for the other two signals.

The genetic program learned the most important characteristics from a relatively small number of training signals. A maximum of twenty two signals of each of the three motions were used for training. Other solutions implemented with neural networks for this problem use a much larger number of training signals. This ability demonstrates one important advantage of genetic programming over neural networks. The small number of training signals would allow, an amputee to train the genetic program for prosthesis control much faster than the neural networks.

Acknowledgments

We wish to express our appreciation to Dr. Ian Walker of Rice University and to Mr. Rodrigo Fernandez for their help in proofreading the paper. We gratefully acknowledge partial support of this research by NASA grant# NAG 9-740. We also thank NASA's Johnson Space Center for use of myoelectric data collection equipment in their Dextrous Robotics Laboratory.

Bibliography

Berne, 1988. *Physiology*. St. Louis: The C.V. Mosby Co., 2nd. Ed. Pages 315-348.

Farry, Kristin A. 1994. *Issues in myoelectric Teleoperation of Complex Artificial Hands.* Ph.D. Dissertation, Electrical Engineering Dept., Rice University, Houston, Texas.

Fernandez Jaime J. 1995 *Myoelectric Signal Recognition Using Genetic Programming.* MS Thesis, MEMS Dept., Rice University, Houston, Texas.

Hogan N., 1976. A Review of the Methods of Processing EMG for Use as a Proportional Control Signal. *Biomedical Eng.*, Mar 1976, pages 81-86.

Hudgins B. 1993. A New Strategy for Multifunction Myoelectric Control. *IEEE Trans. on Biomedical Eng.* Volume 40, Number 1., January 1993., pages 82-94.

Koza, John R. 1992. *Genetic Programming: On the Programming of Computers by Means of Natural Selection.* Cambridge, MA: The MIT Press.

Koza John R. 1994 *Genetic Programming II: Automatic Discovery of Reusable Programs.* Cambridge, MA: The MIT Press.

Laird P. and Saul R. 1994. Automated Feature Extraction for Supervised Learning. *Proc. of the 1994 IEEE World Congress on Computational Intelligence*. IEEE Press.

Saridis G. and Gootee T. 1982. EMG Pattern Analysis and Classification for a Prosthetic Arm. *IEEE Trans. on Biomedical Eng.* BME-29, Number 6., pages 403-412.

Tackett Walter. 1994. *Recombination, Selection, And The Genetic Construction Of Computer Programs.* Ph.D. Dissertation, University Of Southern California. Available by public ftp to alife.SantaFe.edu.

Benchmarking The Generalization Capabilities Of A Compiling Genetic Programming System Using Sparse Data Sets

Frank D. Francone
Law Office of Frank D. Francone
4806 Fountain Ave. #77
LA, California 90027
ytns65a@prodigy.com
213-953-8584

Peter Nordin
University of Dortmund,
Dept. of Computer Science
LS11 D-44221 Dortmund Germany
nordin@ease.informatik.uni-dortmund.de. +49 231 9700-956

Wolfgang Banzhaf
University of Dortmund,
Dept. of Computer Science
LS11 D-44221 Dortmund Germany
banzhaf@LS11.informatik.uni-dortmund.de. +49 231 9700-953

ABSTRACT

Compiling Genetic Programming Systems ('CPGS') are advanced evolutionary algorithms that directly evolve RISC machine code. In this paper we compare the ability of CGPS to generalize with that of other machine learning ('ML') paradigms.

This study presents our results on three classification problems. Our study involved 720 complete CGPS runs of population 3000 each, over 500 billion fitness evaluations and 480 neural network runs as benchmarks. Our results were as follows:

1. When CGPS was trained on data sets that were not too sparse, CGPS performed very well, equaling the generalization capability of other ML systems quickly and consistently.

2. When CGPS was trained on very sparse data sets, CGPS produced individuals that generalized almost as well other ML systems trained on much larger data sets.

3. As between CGPS and multilayer feedforward neural networks trained on the same sparse data sets, CGPS generalized as well (and often better) than the neural network.

1. Introduction

Researchers in Machine Learning ('ML') frequently encounter problems of high dimensionality and too little data. An ideal ML system would, of course, be able to generalize well, even though it was trained on sparse data. This paper compares the effect of using sparse data sets on the generalization capability of our Compiling Genetic Programming System ('CGPS') with the generalization ability of a K-Nearest Neighbor classifier systems ('KNN') and Multilayer Feedforward Neural Networks ('MFLN').

We used three classification problems from the ELENA database—two real world problems and one difficult but artificially generated data set. The data sets that we used were the IRIS, the PHONEME and the GAUSSIAN 3D data sets [ELENA 1995].

ELENA provides a set of KNN benchmarks for each of these problems. The ELENA partners derived the KNN benchmarks using the entire ELENA data sets—usually in excess of 5000 training points—and optimized the KNN performance by varying the value of *K* until the best performance was had on the testing set [*Id*].

Our purpose in this paper was to test generalization performance where the number of training instances was severely limited. So the ELENA KNN benchmark figures are useful primarily to set the boundaries of what performance might be expected where we have the luxury of training on an abundance of data. Thus, the real question we address is in this paper is, where the training set is sparse, how close can CGPS and MLFN come to the KNN benchmarks that were derived on much larger data sets?

We performed 720 CGPS runs (240 on each problem) and 480 neural network runs (240 on each of two of the problems). Both the CGPS and MLFN runs were performed using a variety of parameters.

The results were gratifying. CGPS generated high quality solutions without using any experimenter knowledge of the problem domain. CGPS often produced solutions that generalized almost as well as ELENA benchmarks that had been trained on *much* larger data sets. Finally, CGPS' generalization performance was, by several different measures, superior to the performance of our 480 neural network benchmark runs that were trained on the same sparse data sets as were the CGPS runs.

2. An Introduction to CGPS

CGPS is the direct evolution of binary machine code through the use of evolutionary operators such as crossover and mutation. It would be impossible to describe the full CGPS system in the space allotted. CGPS has, however, been described in detail elsewhere [Nordin 1994, Nordin & Banzhaf 1995, and Nordin, Francone & Banzhaf, 1996].

In our description of CGPS below and in the works referenced, the reader will note that although CGPS is, in a very general sense, a 'Genetic Programming' system, it is also very different from 'Genetic Programming' as described in [Koza 1992] or as that term is used in the literature. Among other things, CGPS does not evolve trees, has no local memory, the CGPS genome is linear, crossover looks more like genetic algorithm string based crossover than like GP crossover, CGPS operates directly on the CPU's registers and CGPS uses only the CPU's instruction set. Thus, despite superficial similarities, CGPS should be regarded as an evolutionary algorithm that is quite distinct from canonical Genetic Programming.

2.1 Description of a CGPS Program

An evolved CGPS program is a sequence of binary machine instructions. Thus, an evolved CGPS program might be comprised of a sequence of three, 32 bit machine instructions. When executed, those three instruction would cause the CPU to perform three operations on the CPU's hardware registers. Here is an example of a simple, three instruction CGPS program that uses three hardware registers:

register 2 = register 1 + register 2 (1)
register 3 = register 1 * 128 (2)
register 3 = register 2 Div register 3. (3)

One of the three hardware registers in this sample CGPS program is selected as the output register. Once the output register is selected, a fitness evaluation for this sample CGPS program would consist of the following steps:

1. Initialize the hardware registers with the input values for the fitness instance;
2. Execute the above three instruction program {(1)-(3)} on the hardware registers as initialized; and
3. Evaluate the value in the selected output register for fitness against the fitness function.

While CGPS programs are apparently very simple, it is actually possible to evolve functions of great complexity using only simple arithmetic functions on a register machine [Nordin & Banzhaf 1995a].

2.2 Protocol for Reporting CPU Time

While there are many advantages to CGPS, the most pronounced advantage is probably speed [Nordin 1994]. Simply put, evolving machine code directly with no compilation is *much* faster than interpreted languages like LISP or compiled languages like C.

We conducted 720 runs with large populations (3000) on each run. We completed the runs in well under 3 weeks on a single Sun 20 Workstation. But this number greatly understates the actual speed of CGPS. On this project, at least 98% of the CPU time was occupied with a research measurement we call 'intron counting'. At the end of each generation, each individual in the population is evaluated for introns by replacing each instruction, one-by-one, with a No-OP instruction and then evaluating the fitness of the individual with the No-OP instruction included. If this replacement of an instruction has no effect on the individual's fitness, we count the instruction as an intron [Nordin, Francone, & Banzhaf 1996].

Intron counting has a *huge* effect on the time it takes to evolve solutions. For example, an individual that is 100 instructions long requires 100 separate fitness calculations that are devoted solely to intron counting. Where the average size of an individual is 100 instructions, intron counting increases the amount of time to evolve a population by a factor of 100. Applying this analysis to our 720 runs, we have calculated that intron checking in the first 20 generations of our runs occupied approximately 98% of the total CPU time in the run. Because the average size of individuals grew after generation 20, the impact of intron counting grew after generation 20—so a 98% figure actually understates the effect of intron counting substantially.

While intron counting yields useful research data [*id*], it also has nothing to do with the immediate job of evolving high quality individuals—intron counting is purely a research tool. Had we used a 'production' version of CGPS stripped of intron counting, this entire project—all 720 runs of 3000 individuals each—would have been completed in about eight hours on our Sun 20 workstation.

Henceforth, we will follow the practice of reporting CPU time figures for the production portion of the CGPS system, without including the 'research' overhead. This approach has two advantages. To begin with, it gives consistent performance figures from problem to problem and from research project to research project—research overhead varies on different projects. Further, although this approach to CPU time reporting is approximate, it is more representative of the time needed to evolve individuals than the gross figures that include research overhead.

A brief note here on the CPU time occupied by the MLFN runs is appropriate. On the GAUSSIAN 3D and the PHONEME problems, the MLFN runs took about 320 hours of CPU time on a 66 Megahertz 486 IBM compatible PC with 20 megabytes of RAM running in 32 bit DOS extended mode. Those same two problems occupied about 4.5 hours of CPU time in the CGPS runs. Of course, the Sun 20 Workstation is a much more powerful machine than the 486, even running in DOS extended mode. Any further comparisons of time are effectively meaningless because the platforms are so radically different.

3. Generalization Issues

3.1 Measuring Generalization

We evaluated the generalization of our CGPS and neural network runs as follows. To begin with, we separated our data into training and testing sets.

For CGPS runs, the best individual of each generation on the training set was evaluated on the testing set. The results of that test were preserved. After completing the CGPS run, we used the data preserved during the run, as described above, to select the best generalizing individual from the run. (We selected the best generalizing individual by selecting the individual with the highest number of 'hits' on the test data.) We then used that best generalizing individual to represent the generalization of the entire run.

Because the outputs available in the neural network package used were different than the outputs on our CGPS program, we had to choose the best generalizing network in a slightly different manner. For each neural network run, the network with the highest R^2 on the testing set was considered to be the best generalizer of the run. For comparison purposes with the CGPS system, we then calculated the number of 'hits' of this best generalizing network as a percentage of the total testing set.

The foregoing is a decent (though not optimal) way to test generalization [Masters 1995, Katz 1996]. We did not implement more optimal techniques such as the Jacknife and Bootstrap methods [Masters 1995] because the CPU time to perform them on the neural network runs made such techniques utterly impractical.

3.2 Improving The Measure of Generalization With a Third and Off Training Sample Data Set.

We acknowledge one respect in which the above methodology could have been improved. We choose which individual (or network) from a run is the 'best' generalizer by assessing its performance on the testing data. For example, if the best CGPS individual correctly classified 85% of the test cases in generation 10 but in generation 11, the best individual correctly classified only 80% of the test cases, we used the 85% figure to represent the generalization performance of that run.

But once we have chosen the best generalizer in the above manner, the testing set is, strictly speaking, no longer a testing set—it has become part of the training set. The reason? The testing set was used to select among different individuals.

The method that we used does provide useful information about the generalization capabilities of an ML system. In our prior experience, we have found that there is a decent correlation between generalization on the testing set and generalization on entirely new data. But to make our assessment of generalization more rigorous, a *third* data set ought to be added. So while our results here are meaningful, the ability of the best generalizer on the testing set to generalize on yet a third data set would measure generalization more accurately. We plan to implement this capability in our CGPS system in the near future.

3.3 No Free Lunches?

We acknowledge the ongoing controversy over the 'No Free Lunches' theorem and whether it is possible to find *any* algorithm that has *any* meaningful generalization capabilities [*Compare* Wolpert 1994 *with* Rao 1995]. It is not our purpose to add to that discussion here. We simply note that, at this point, the theoretical guidance offered by papers on this subject about how to judge problem 'domains' is sufficiently vague that ML yet remains an experimental science. It is in that spirit that we present these findings. Should it turn out that learning is effectively impossible in this universe, our work here will be for naught and our research group shall have to learn to occupy its time with pursuits other than experimental machine learning.

We do, however, agree with Rao's emphasis on the importance of off-training-set error, as we discussed above. And we repeat here Rao's admonition that "[R]esearchers should be careful not to say that Algorithm A is better than Algorithm B without mentioning that this holds with respect to the particular problem distribution." [*Id*]. That was good advice both before the 'No Free Lunches' Theorem and remains so today.

4. Description of the CGPS Portion of the Experiment

Our CGPS experimental setup stayed mostly constant from problem to problem. Table 1, contains the general experimental specification [Koza 1992].

Table 1. Experimental Specification for CGPS Runs

Objective	Classification of data sets from ELENA Database
Parsimony Factor	0, 0.1, or 1
Explicitly Defined Introns	Enabled or Not Enabled
Crossover/Mutation Mix	95/5, 80/20, 50/50, or 20/80
Terminal Set	Integer constants initialized from 0 - 10. May be mutated to the range, 0 - 255.
Function Set	Addition, Multiplication, Subtraction, Division
Number of Hardware Registers	One more than the number of classes in the data.
Number of Fitness Cases	75-100. Varies with problem.
Number of Testing Cases	75-100. Varies with problem.
Fitness Function	Sum of Absolute Deviations.
Hits	See discussion above.
Wrapper	None
Population Size	3000
Selection for Genetic Operators	Tournament. 4/2. Children replace the losing members of the tournament.
Termination Criteria.	200 Generations or Destructive Crossover falls to less than 10% of the total crossover events, whichever comes first.
Maximum Individual Size	256 instructions.
Total Number of Runs	720

Some of the parameters mentioned in Table 1 merit separate discussion.

4.1 Parameters That We Varied From Run To Run

We varied three of the parameters from run to run. These parameters were: Parsimony Factor, Explicitly Defined Introns, and Crossover/Mutation Mix.

There were twenty-four different combinations of values used for the above parameters (Table 1). Each of those twenty-four combinations of parameters was run for each of our three problem sets on ten different random seeds. As a result, we conducted a total of 240 runs per problem set. On the three problems reported here, we conducted a total of 720 runs.

4.2 Fitness Function

We had to define an appropriate fitness function for classification problems where the number of classes might vary from problem to problem. We did that as follows. We selected one hardware register as the output register. The value in the output register was then evaluated for fitness on a particular fitness case as follows:

If a particular fitness case should have been classified as being a member of *Class 0*, then the fitness of the individual for that fitness case is the absolute value of the difference between the value in the output register and *100*;

If a particular fitness case should have been classified as being a member of *Class 1*, then the fitness of the individual for that fitness case is the absolute value of the difference between the value in the output register and *200*; and so forth where there are more classes.

4.3 Measuring Testing Set Hits

Table 2 is the lookup table that we used to determine the meaning of a value contained in the Output Register of an individual for a particular fitness case.

If the Predicted Class Membership determined by Table 2 for a particular fitness case is correct, then that prediction is counted as a '*Hit*' for that fitness case. The total number of hits for an individual is then divided by the total number of test cases. Where the testing set is 100 in size, and where an individual gets 75 hits on the test set, that individual's Hit score would, therefore, be 75%.

Table 2. Class Membership Corresponding to Output Register Values

Output Register Value of . . .	Predicted Class Membership
50 - 149	Class 0
150 - 249	Class 1
250 - 350	Class 2

Table 2 was obviously designed arbitrarily. The effect of that design is that researcher knowledge of the domain space could not bias the results.

4.4 Function Set

We deliberately used a very simple function set—the arithmetic operators. The purpose of using the same function set that we used for previous symbolic regression work was to maintain as much independence as possible from experimenter knowledge of the problem domain.

4.5 Parsimony Factor

We used three different values for the parsimony factor—0, 0.1, and 1. A value of *n* for the parsimony factor means that the *n* times the length of the individual (measured by the number of instructions that comprise the individual) is subtracted, in the fitness function, from the fitness of the individual. A value of 0.1 for the parsimony factor, therefore, means that 1/10 of the length of an individual is subtracted from the fitness of the individual.

4.6 Explicitly Defined Introns

Explicitly Defined Introns ('EDI's') were developed in our previous work where we found that using EDI's to allow GP to vary the probability of crossover between instructions without effecting the fitness calculation of the individual improved fitness, generalization and speed. [Nordin, Banzhaf & Francone 1996.] We enabled EDI's in the manner described in the aforementioned work for one half of all runs.

4.7 Termination Criterion.

The maximum number of generations that we allowed the system on any CGPS run was 200 generations. However, we have previously reported a strong correlation between the exponential growth of the size of programs and a decline in the portion of total crossover events that can be characterized as destructive crossover [Nordin, Francone & Banzhaf 1996]. In fact, our earlier findings suggest that, at the point where such exponential growth occurs, all effective training is over [*Id*]. Accordingly, we monitored the rate of destructive crossover during training. When destructive crossover fell to 10% of the total crossover events in any given generation, we terminated the run even if it had not reached generation 200.

This early run termination approach saved a substantial amount of CPU time. Over 52% of our 720 runs were terminated by generation 80 because of the early termination criterion. Had these runs not terminated early, they would have continued the full 200 generations. Had we not used our early termination criterion, the total number of generations that our system would have needed to complete all 720 runs would have been 144,000 generations (720 runs times 200 generations per run). Instead, our system needed only 75,134 generations to complete the entire 720 runs. This savings was entirely due to our early termination criterion based on monitoring the destructive crossover rate relative to the total crossover rate.

5. Description of the Multilayer Feedforward Neural Network Portion of the Experiment

5.1 Selection of the MLFN

We decided to run the same sparse data that we used in our CGPS runs on at least one other established machine learning system. We spent some time deciding whether to use a Multilayer Feedforward Neural Network or a Probabilistic Neural Network ('PNN'). Generally speaking, a PNN is better for classification problems and is much faster than a MLFN. We had access to MLFN and PNN software from the same vendor so we discussed with the vendor which of the modules would work better on our sparse data set problem. It was suggested that the MLFN module would be more effective where the number of training samples were limited [Katz 1996]. As a result, we ran the same sparse data used for our CGPS runs on the N-TRAIN MLFN module from version 1.02 of the N-TRAIN Neural Network Development System [McCormick and Katz 1992]. N-TRAIN uses the backpropogation algorithm to train multilayer feedforward neural networks. N-TRAIN runs in 32 bit DOS extended protected mode. Thus, it is very fast for a PC based system.

5.2 Neural Network Architectures

We used a fully connected three layer MLFN with linear neurons in the input layer and sigmoid neurons in the other layers. We varied the number of hidden neurons according to the following protocol, which we set in advance of training and testing. For a network that has n independent variables as inputs, we did 30 training runs using 30 different random seeds for each network architecture that could be constructed consistent with the following rules:

1. The network has an input layer consisting of n linear neurons,
2. The network has one hidden layer consisting of k sigmoid neurons where $n <= k <= n*2$; and
3. The network has an output layer consisting of one sigmoid neuron.

So where there were three inputs for a problem, we did a set of 30 runs using 30 random seeds each for three layer architecture consisting of 3, 4, 5 and 6 hidden neurons, respectively.

Based on our past experience, this range of hidden neurons usually gives a good coverage of the better network structures. When the above runs were finished, we examined whether generalization performance appeared to be improving at either the high or low end of the hidden neuron architectures. If so, we performed runs for the next two numbers of hidden neurons in the direction that the network performance generalization was improving. By way of example, if the generalization performance in the above example appeared to be improving between 5 and 6 hidden neurons, we would do 30 runs using 30 random seeds for 7 and 8 hidden neurons architectures. If the performance continued to improve, would do two more hidden neurons and so forth. {We note, in retrospect, that this procedure of expanding the number of hidden neurons based on existing results probably biased the experiment in favor of the MLFN architecture. No such interative procedure was followed on the CGPS portion of the experiment. Were we to do this again, we would eliminate that portion of the experiment.}

After we finished the above procedure, we performed enough extra runs on that problem using new random seeds distributed evenly over all of the hidden neuron configurations used in the above procedure so that the total number of runs on that problem with the MLFN equaled 240 runs. Completing 240 runs with the MLFN—the same number of runs that we performed with the CGPS system—greatly improves the usefulness of the difference in means statistical test between the MLFN and the CGPS, which test we will use later [Crow, Davis & Maxfield 1960].

Based on this protocol, we performed 60 runs each with 3-6 hidden neurons on the GAUSSIAN 3D problem, which had 3 inputs. We also performed 30 runs each with 3-10 hidden neurons for the PHONEME problem set, which had 5 inputs. Therefore, the total number of runs we performed on the GAUSSIAN 3D problem set was 240 and the number on the PHONEME problem set was 240.

5.3 Data Preprocessing

Our CGPS runs required integer inputs. Neural networks require that inputs be scaled in the range 0 to 1 or -1 to 1 range. We scaled the inputs to the 0 to 1 range.

For reasons that are discussed below, we ran the MLFN only on the GAUSSIAN 3D and PHONEME problem sets. Each of those sets had only two output classes. Therefore, we gave the output data to the network as follows: All instances that were Class 0 were assigned an output value of 0—all those of Class 1 were assigned an output value of 1.

5.4 MLFN Parameter Settings

For the most part, we used the default parameter settings that come with the program—having found them in our previous work to be an effective parameter set. There were, however, three parameters that we changed from the default setting, each of which is detailed below.

The reader will note that we made our changes in the MLFN parameters using the suggestions of the proprietor of the MLFN software. His suggestions were excellent and greatly improved the performance of the MLFN system.

5.5 Learning Rate and Neuron Type Parameters

In our initial MLFN runs, we used the N-TRAIN default learning rate and neuron type settings. The results from those runs were unexpectedly poor. The vendor suggested that a much lower learning rate of 0.25 would be closer to optimum for our problems. In addition, he suggested that we use linear instead of sigmoid neurons on the input layer [Katz 1996]. We made those two adjustments to the parameters with greatly improved training results.

5.6 Error Tolerance Parameter Used to Determine Testing Set Hits

In order to compare the CGPS hits with MLFN hits, we had to measure the hit rate of the MLFN. In order to measure the MLFN hit rate, it is necessary to set a parameter called 'error tolerance' [McCormick and Katz 1992]. This proved to be the most difficult parameter to set in a principled manner.

Simply put, the error tolerance parameter determines how close to 0 or 1 the network output must be in order to be counted as a 'hit.' The program default setting of 0.10 meant that the network must output a value of less than

0.10 or greater than 0.90 to be counted as a 'hit.' All outputs between 0.10 and 0.90 were counted as wrong.

This default setting caused very poor performance by the MLFN because, the networks that had the best R^2 on the testing set were generating a lot of 'correct' answers that were between 0.10 and 0.90. (By 'correct', we mean that the network output was closer to the correct answer than it was to the wrong answer.)

The solution on this might initially seem to be to set the error tolerance to 0.4999. This, however, forces the MLFN to 'guess'. That is, where there are only two classes, the MLFN cannot do worse than 50% hits with an error tolerance of 0.4999. But the reader will note that the CGPS system was set up so that it was not forced to guess. So this solution was also not acceptable.

We finally elected to use an error tolerance of 0.25. This seemed to deliver good performance but to prevent the MLFN from using a range of outputs (between 0.25 and 0.75) where there was little or no relation between the network output and the correct answers. This is not an entirely satisfactory solution but it was the best we could devise given the differences between the paradigms.

5.7 Run Termination and Testing Granulation

Run termination on the MLFN problem was determined by number of passes of the backpropogtion algorithm through the training set. We started each run with 200 runs through the training set, tested and stored the results. Thereafter, we tested every 50 runs through the training set. Each run comprised 10,200 runs through the training set and was then ended.

We examined the course of training under this protocol on about 50 of the runs and satisfied ourselves in each case that: (1) the granularity of testing was sufficient to catch the best or very close to the best network in generalizing capabilities; and (2) that we had performed enough runs through the training set to assure that the run had indeed located the best maximum generalizer for that random seed and network configuration.

6. The IRIS Dataset.

6.1 The Full Dataset.

The IRIS dataset contains three classes and the total number of data instances is 150. Each of the classes is a type of IRIS plant [ELENA 1995, page 35]. This is a simple problem domain and, given a large enough training set, good results are to be expected [*Id*].

6.2 The KNN Benchmark.

The ELENA benchmark classifier for this problem is a KNN classifier run using the Leave-One-Out method with $K = 7$. A KNN (K-Nearest-Neighbors) classifier determines the class of a fitness case by examining the class membership of its K nearest neighbors. Performance is optimized by varying K for any particular problem domain Because of the use of the leave-one-out method, the dataset for deriving the KNN benchmark was 149 instances. The resulting error rate for the KNN classifier was between 0% and 7.3% [ELENA 1995 at 35-6]. We assume that the wide range of the error rate was due to a wide 95% confidence interval on this relatively small sample size (149). In short, a hit rate for CGPS of between 92.7% and 100% on the IRIS data would mean the CGPS result is statistically equivalent to the KNN benchmark.

7. IRIS Generalization Results

7.1 Sparse IRIS CGPS Training Set

Our training set size was 75, which is about half the size of the training set used in deriving the KNN benchmark. Despite the relatively sparse training set, CGPS did very well.

7.2 IRIS CGPS Results

Every IRIS CGPS run yielded at least one individual that generalized at a hit rate that was statistically indistinguishable from the ELENA KNN benchmark.

92% of the IRIS runs resulted in individuals that generalized better than or equal to 95% hits and 17% of the IRIS runs resulted in individuals that generalized better than 96% hits. Furthermore, CGPS found its benchmark-quality solutions quickly. In 78% of IRIS runs, our *GP system had evolved an individual that generalized as well as the KNN benchmark by the end of generation one of training.* In fact, our CGPS system generated at least one new benchmark quality solution approximately once every one half second.

7.3 IRIS MLFN Results.

We did not run the MLFN system on the IRIS data because, in discussions with the MLFN software vendor, he indicated that he had run the IRIS data and that his results were in the range that made them likely to be statistically indistinguishable from our CGPS runs. Given our time constraints and what turned out to be the simplicity of the IRIS problem, we believe it safe to conclude that the MLFN generalization would have been statistically indistinguishable from CGPS generalization on this problem. Accordingly, we did not run the IRIS problem on MLFN.

8. PHONEME Recognition Dataset.

8.1 The Dataset.

The PHONEME recognition database contains two classes of data—nasal vowels (Class 0) and oral vowels (Class 1) from isolated syllables spoken by different speakers. "This database is composed of two classes in 5 dimensions. There are 5404 patterns; 3919 for class zero and 1586 for class one. . . . [T]his number of samples is just sufficient for the database dimension." [ELENA 1995 at 30]. Note that, while 150 data points was sufficient for the IRIS problem, 5404 data points is regarded as 'just sufficient' for the PHONEME Recognition set because of its higher dimensionality and greater difficulty [*Id*].

8.2 The KNN Benchmark

The benchmark on this problem is a KNN classifier with *K* set to 20. The benchmark was derived from 5404 training instances. The mean misclassification rate calculated by the KNN Benchmark is 14.2% [ELENA 1995. Page 31]. Therefore, any CGPS or MLFN run that has Hits that equal or exceed 85.8% has generalized as well as the Benchmark.

9. Generalization Results on the PHONEME Dataset.

9.1 Sparse Training Set for CGPS and MLFN on the PHONEME Dataset.

We gave our system only 100 data points for training and another 100 for testing rather than the 5404 data points used to derive the KNN Benchmark. Thus, we withheld from the CGPS, over 98% of the training instances that were available to the Benchmark. And we did so on a data set where ordinary statistical analysis suggests that our system ought to need those extra training samples to do a good job [ELENA 1995. Page 30-1].

9.2 Relative Generalization Results of CGPS and MLFN on Sparse PHONEME Recognition Dataset.

Both CGPS and the MLFN did respectably at finding individuals that generalized well—many runs produced individuals that were of near benchmark quality despite the sparse training set. Table 3 contain the mean hit rates results for all, the best 25% and the best 10% of PHONEME Recognition runs for both MLFN and CGPS.

Table 3. PHONEME Recognition Results. Mean Hits by ELENA KNN Benchmark, by CGPS and by MLFN systems. All Runs, Best 25% Of Runs, Best 10% Of Runs, and Best Run.

System	All Runs	Best 25%	Best 10%	Best Run
KNN	85.8%	----	----	---
CGPS	77.6%	80.5%	81.2%	85%
MLFN	71.6%	79%	80.9%	82%

By any measure on this problem, CGPS outperformed the MLFN in generalization. The differences in performance are, for the most part statistically significant. To test statistical significance, we performed the test for a hypothesized difference between two means assuming unequal variances. Table 4 sets forth our results. P is the probability that the actual difference between the CGPS and MLFN means is less than the hypothesized difference.

Table 4. PHONEME Recognition Results. Statistical Significance Level of Hypothesized Difference Between the CGPS Mean Hits and The MLFN Mean Hits.

Differences	All Runs	Best 25%	Best 10%
Actual Diff.	6%	1.5%	0.3%
Hypothesized Diff.	5%	0.9%	0.05%
P	0.0141	0.0341	0.0884

Measured in CPU time, CGPS generated a solution that generalized at least 90% as well as the Benchmark figure about once every 1 minute. And, finally, CGPS generated a solution that was almost identical to the Benchmark figure about once every 55 minutes.

10. The GAUSSIAN 3D Dataset

10.1 The Full Dataset.

GAUSSIAN 3D is a three input, two class database that is generated artificially. Class 0 is a the set of points with a normal distribution across the three input axes with zero mean and standard deviation of 1. Class 1 is a similar series of points except that the standard deviation is 2. [ELENA 1995. Page 14.] This is an difficult classification problem because the classes are linearly non-separable and because of overlap between the two classes [Figure 1].

10.2 The Benchmarks.

There are several benchmarks available on this problem:

1. The theoretical Bayes confusion boundary is 21.4%. It is not possible to generalize better than this on this problem [ELENA 1995. Page 16]. Therefore a Hit rate of 78.6% should be the best possible performance of any classifier on these data.
2. ELENA's KNN classifier (K = 35) using the leave-one-out method yields an error on 5000 training samples of 22.2%. Therefore, a hit rate of 77.8% would be competitive with the KNN benchmark [ELENA 1995]
3. A Binary Boltzman Machine and A Learning Vector Quantization system attained results virtually the same as the Bayes minimum but had to use 4,500,000 and 100,000 training samples respectively to attain that result [ELENA 1995].

10.3 Generalization Results On The GAUSSIAN 3D Dataset.

The task we assigned to CGPS and the MFLN on the GAUSSIAN Data set seemed daunting. We gave them only 100 training samples. In short, we withheld fully 98% of the 5000 training samples that were available to the KNN benchmark measure. Figure 1 is a projection of 100 points from the GAUSSIAN 3D data set into two dimensions. CGPS' and the MLFN's jobs were to distinguish the darker diamonds (class 0) from the lighter dashes (class 1).

Again, both CGPS and the MLFN did respectably at finding individuals that generalized well. Table 5 contains the mean hit rates results for all, the best 25% and the best 10% of the GAUSSIAN 3D runs for MLFN and CGPS.

Table 5. GAUSSIAN 3D Results. Mean Hits by ELENA KNN Benchmark, by CGPS and by MLFN systems. All Runs, Best 25% of Runs, Best 10% of Runs and Best Run

System	All Runs	Best 25%	Best 10%	Best Run
KNN	77.8%	----	----	----
CGPS	57.4%	63.9%	67%	72%
MLFN	56.8%	62.4%	64.9%	68%

Once again CGPS outperformed the MLFN in the overall mean and on the upper tail of the generalization distribution. Again, the differences in are statistically significant. Table 6 presents the same types of statistical results for the GAUSSIAN 3D data set as did Table 4 for the PHONEME recognition data.

Table 6. GAUSSIAN 3D Results. Statistical Significance Level of Hypothesized Difference Between the CGPS Mean Hits and The MLFN Mean Hits.

Differences	All Runs	Best 25%	Best 10%
Actual Diff.	0.6%	1.5%	2.14%
Hypothesized Diff.	0.2%	0.5%	1%
P	.0601	.0264%	.0484

Measured in CPU time, CGPS generated an individual 90% as good as the KNN benchmark every 18 minutes.

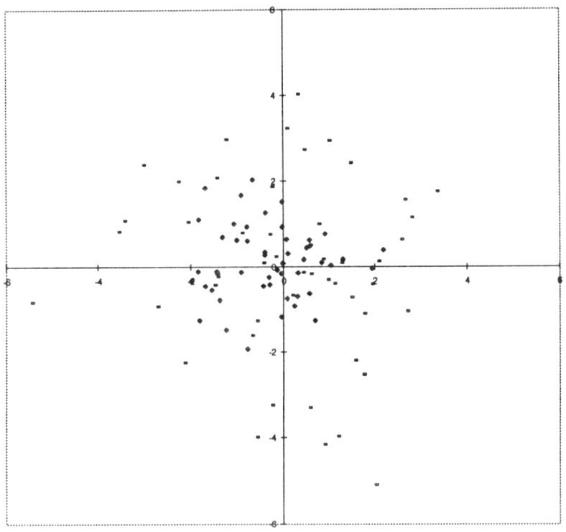

Figure 1. Sparse GAUSSIAN 3D Data Set Projected into Two Dimensions. 100 Data Points.

11. Discussion.

11.1 The Future of the IRIS Problem

The easiest conclusion to draw from our data is that the IRIS data set is probably not a very good test of any machine learning system—it is too easy. CGPS evolved programs by the end of Generation 1 that were only a few instructions long that did extremely well in classifying the IRIS data. The ease of the problem is also indicated by reported MLFN results [Katz 1996]. Of course, a finding that a machine learning system could *not* do well on the IRIS data would be quite a significant finding. Other than that, researchers should probably stop using the IRIS data as being indicative of very much of anything.

11.2 CGPS and Generalization on Sparse Data Sets.

For the most part, our other results speak for themselves. CGPS generated solutions that generalized on the other two problem sets well and quickly. It did so somewhat more effectively than the MLFN used here. CGPS did well even though we used simple functions and made no adjustments to help the system solve the particular problems presented. In short, our solutions were as free from experimenter knowledge of the solution space as we could make them.

CGPS' ability to generalize from sparse data sets was quite consistent. From the time figures noted above, with enough CPU time, CGPS was able to extract good benchmark or near benchmark quality individuals using sparse data. Of course, on the GAUSSIAN 3D and PHONEME problems, additional data was available—so in a sense, it was not necessary to utilize the extra CPU time to generate solutions that generalized well. We could just as well have used larger training sets. But in the real world, there is frequently too little data for the dimensionality of the problem. This suggests that CGPS may be a good choice for classification problems where the available data is sparse.

11.3 Relative Distribution of Generalization Results: MLFN vs. CGPS.

One of the most interesting comparisons between the CGPS and MLFN paradigms comes from examining the distribution of their relative generalization scores over all runs. Figures 2 and 3 show those distributions for the PHONEME and the GAUSSIAN 3D data sets. What jumps out is that MLFN results are spread fairly uniformly over a wide range—a normal or flattened normal distribution. The CGPS results are concentrated very heavily around the mean, have a very short low end tail, and have a long tail in the upper end of the distribution. This long tail is the reason CGPS outperformed MLFN in the high quality solution regions [Tables 3-6]. Given CGPS greater freedom of representation than MLFN, this tightly bunched distribution is, intuitively, the opposite of what we expected.

More fundamentally, the relative uniformity in CGPS results from run to run occurred despite the rather radical set of parameter changes that we made among runs. This suggests that mere repetition of runs with different parameters and random seeds may not a particularly good approach to using CGPS. The same search space appears to keep being explored over and over [*See* Mahoud 1995]. This implies that recent research where information gleaned from one run is used to improve the performance on subsequent runs may well be a *very* important future direction for research [Whigham 1995, Zannoni & Reynolds 1996].

Acknowledgments

We would like to thank the ELENA partners for assembling the ELENA databases and the benchmark data.

We also thank the reviewers and William Langdon, whose questions and comments were very helpful.

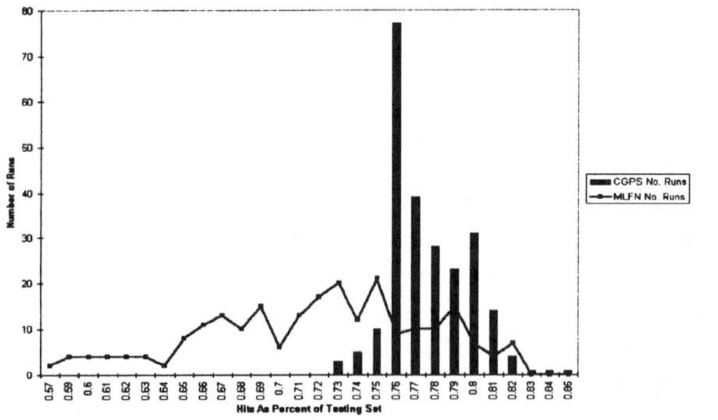

Figure 2. Phoneme Recognition Problem. Comparative Histograms of all CGPS Runs (bars) and MLFN Runs (lines) Sorted by Generalization Performance.

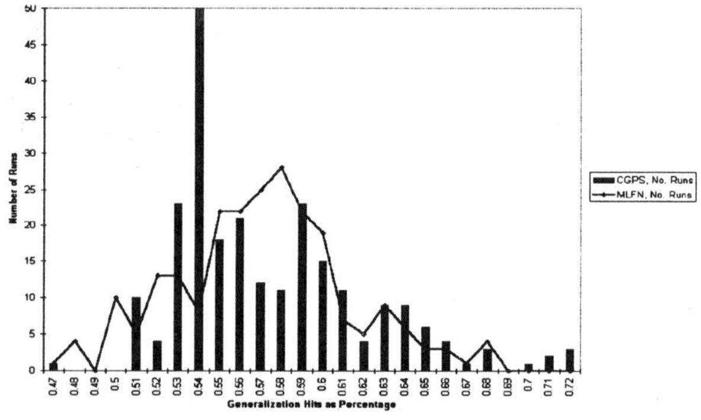

Figure 3. Gaussian 3D Problem. Comparative Histograms of all CGPS Runs (bars) and MLFN Runs (lines) Sorted by Generalization Performance.

Jeffrey Katz of Scientific Consulting, Inc. helped us configure the parameters on his N-Train Neural Network system for benchmarking purposes.

We would also like to acknowledge support from the Ministerium fur Wissenschaft und Forschung des Landes Nordrhein-Westfalen, under grant I-A-4-6037.I.

Bibliography

Crow, E., Davis, F. and Maxfield, M. 1960. *Statistics Manual*, NY, NY. Dover Publications, Inc.

ELENA Partners, The. Jutten, C., Project Coordinator 1995. Esprit Basic Research Project Number 6891, Document Number R3-B1-P.

Katz, J. 1996. Designer and owner of Scientific Consultant Services, Inc., proprietor of N-Train neural network software. Personal communication.

Koza, J. 1992. *Genetic Programming,* Cambridge, MA: MIT Press.

Mahoud, S. 1995, *Niching Methods for Genetic Algorithms*, IlliGAL Technical Report No 95001. University of Illinois. Dept. of General Engineering.

Masters, T. 1995, *Advanced Algorithms for Neural Networks.* NY, NY, John Wiley and Sons Inc. Pages 335-376.

McCormick, D. and Katz, J. 1992. *N-TRAIN Neural Network Development System, Users Manual, V 1.02.*

Nordin, J.P. 1994. A Compiling Genetic Programming System that Directly Manipulates the Machine Code. In *Advances in Genetic Programming,* K. Kinnear, Jr. (ed.), Cambridge MA: MIT Press.

Nordin, J.P., Francone, F. and Banzhaf, W. 1996. Explicitly Defined Introns and Destructive Crossover in Genetic Programming. *Advances in Genetic Programming 2,* K. Kinnear, Jr. (Editor), Cambridge MA: MIT Press.

Nordin, J.P., Banzhaf W. 1995, Complexity Compression and Evolution. In *Proceedings of Sixth International Conference of Genetic Algorithms,* Morgan Kaufmann Publishers, Inc.

Nordin, J.P., Banzhaf, W. 1995a. Evolving Turing Complete Programs for a Register Machine with Self Modifying Code. In, *Proceedings of Sixth International Conference of Genetic Algorithms,* Morgan Kaufmann Publishers, Inc.

Rao, B, Gordon, D. & Spears, W. 1995. For Every Generalization Action, Is There Really an Equal and Opposite Reaction. Analysis of the Conservation Law for Generalization Performance. In Prieditis, A. and Russell, S. (editors). *Proceedings of the Twelfth International Conference on Machine Learning.* San Francisco, CA: Morgan Kaufmann Publishers, Inc. Pages 471-9.

Whigham, P. 1995. Gramatically Based GP. In Rosca, J. (editor). *Proceedings of the Genetic Programming Workshop of the 12th Annual Machine Learning Conference.*

Wolpert, D. H. 1994. *Off-Training Set Error And A Priori Distinctions Between Learning Algorithms*, Technical Report, Santa Fe Institute, Santa Fe, N.M.

Zannoni, E. and Reynolds, R., Extracting Design Information from Genetic Program Using Cultural Algorithms. In *Proceedings for the Fifth Annual Conference on Evolutionary Programming*, 1996.

A Comparison between Cellular Encoding and Direct Encoding for Genetic Neural Networks

Frédéric Gruau
CWI, Kruislaan 413
1098 SJ Amsterdam
The Netherlands
gruau@cwi.nl

Darrell Whitley
Computer Science Dept.
Colorado State University
Fort Collins, CO 80523 USA
whitley@cs.colostate.edu

Larry Pyeatt
Computer Science Dept.
Colorado State University
Fort Collins, CO 80523 USA
pyeatt@cs.colostate.edu

ABSTRACT

This paper compares the efficiency of two encoding schemes for Artificial Neural Networks optimized by evolutionary algorithms. Direct Encoding encodes the weights for an a priori fixed neural network architecture. Cellular Encoding encodes both weights and the architecture of the neural network. In previous studies, Direct Encoding and Cellular Encoding have been used to create neural networks for balancing 1 and 2 poles attached to a cart on a fixed track. The poles are balanced by a controller that pushes the cart to the left or the right. In some cases velocity information about the pole and cart is provided as an input; in other cases the network must learn to balance a single pole without velocity information. A careful study of the behavior of these systems suggests that it is possible to balance a single pole with velocity information as an input and without learning to compute the velocity. A new fitness function is introduced that forces the neural network to compute the velocity. By using this new fitness function and tuning the syntactic constraints used with cellular encoding, we achieve a tenfold speedup over our previous study and solve a more difficult problem: balancing two poles when no information about the velocity is provided as input.

1 Introduction

For reinforcement learning problems a training set of input-output pairs does not exist. Instead, the behavior of a particular instantiation of an artificial neural network (ANN) is used to supply a reinforcement signal that evaluates performance by providing either positive or negative feedback. Particularly difficult reinforcement learning problems are those that require a recurrent network. Genetic algorithms would appear to be particularly well suited for training neural networks for this class of reinforcement learning problems, since they require only feedback based on performance. On the other hand, there are well known problems when attempting to train neural networks using evolutionary algorithms that emphasize recombination.

Wieland [9] introduced several interesting test problems for neurocontrol that involve balancing one or more poles fixed to a cart moving on a finite track. Possible actions on the system include applying a force to the cart to push it to the left or right. This can be a bang-bang controller, such that there is one fixed force that is applied to the cart, or the force may be variable depending on the output of the controller. Note that this corresponds to having a discrete (bang-bang) output action or a continuous (variable force) output action. Traditionally, there are 4 inputs to such a system: pole angle and pole velocity as well as cart position and cart velocity. Wieland offered a solution to the pole balancing problem using only the pole angle and cart position as inputs. Wieland's solution involves the use of a fully recurrent network which can potentially learn to compute velocity information.

For the classic pole balancing problem, velocity information is supplied as an input and the pole angle is restricted to ± 12 degrees; in this case a good linear solution is known to exist. Our recent experimental studies also suggest that good linear solutions exist for some problems that involve balancing two poles when velocity information is provided. We have produced results similar to Wieland's for 1 and 2 poles and have also evolved cellular encodings for neural networks to solve these same problems. While a recurrent network should be required when velocity information is not provided, we have found that neural networks with no recurrent connections are able to balance poles from multiple initial states for 100,000 time steps. These networks are able to do pole balancing by implementing cyclic strategies that keep the system in a state of oscillation. A new evaluation function is introduced that penalizes oscillation and forces the networks to learn to compute velocity information.

Networks were developed in two ways for the current study. First, networks were evolved using cellular encoding. This method generates both the weights and the architectures of the neural networks. Second, neural networks with fixed architectures were trained using a genetic algorithm to find the weights. Following the approach originally used by Wieland, each weight was encoded using 8 bits with values distributed among the values $\{-1, -\frac{253}{255}, -\frac{251}{255}, ..., \frac{253}{255}, 1\}$ with no representation for zero [9]. Several networks of varying sizes were tried and the best network for solving each problem was selected. The fixed architecture networks included a bias input, which was not used by Wieland. Also, the number of neurons used was different than those reported by Wieland. Relatively small populations were used (100 strings) with a high mutation rate (30%).

The goal of the paper is to continue the comparison between direct encoding and cellular encoding started in [8]. The use of the new evaluation function and new syntactic constraints has allowed us to solve the simpler version of the pole balancing problem more than ten times faster with cellular encoding (CE). This speedup is both in terms number of evaluations and wall clock time. We also solve a more difficult version of the pole balancing problem where two poles are balanced and no information about velocity is supplied. We had not previous attempted this problem (e.g. [8]) and this problem was not addressed by Wieland.

In general our results indicate that cellular encoding is able to optimize both the architecture and the weights at the same time. In contrast, direct encodings assumes an a priori architecture which is suboptimal in the general case. Achieving good performance with direct encoding requires testing many architectures individually. Due to the number of architectures and parameters available, this can take considerable effort. As a result, better overall performance can be obtained with cellular encoding.

One benefit of using cellular encoding is that the structure of the resulting neural networks fits the problem at hand and can be analyzed. In particular, when velocity information is supplied as an input, good linear solutions to both the one-pole and the two-pole problems were produced. The linear solutions are not obvious and represent a somewhat unexpected result for the problem of simultaneously balancing two poles. For the case where the velocity is not known solutions with one or two hidden units were produced.

2 Reproducing Wieland's Results

This section describes the experiments done using a direct encoding of the weights, following [9]. The balancing of one pole is a four input problem. These are the pole angle and angular velocity, as well as the cart position and velocity. One variant of the pole balancing problem involves balancing one pole with out using the velocity information; in this case only pole angle and cart position are used as inputs. Another variant involves balancing two poles of different length and mass, where the length and mass of the small pole are ten percent of that of the big pole. This problem has 6 inputs: angle and velocity for each pole, plus the cart position and velocity. The most difficult problem is a combination of the last two: balancing two poles without velocity information. This problem has three inputs: cart position and two inputs for the angles of the large and small pole. The big pole length and mass are 1 meter and 1 kilogram. The length of the cart track is 4.8 meters. The inputs are scaled before being input to the neural network. Cart position is scaled from ± 2.4 meters to an input range of ± 1. Pole angles ranging from ± 36 degrees are scaled to ± 1. Pole angles outside this range produce a failure signal. Cart velocity ranging from ± 1.5 meters per second and pole angular velocity ranging from ± 115 degrees per second are scaled to ± 1. Velocities outside of these ranges are allowed, but the resulting inputs will also be scaled accordingly and produce inputs that are outside the ± 1 range. The simplest measure of performance is given by testing how long a neural controller can balance the pole(s) without allowing the poles angles to go outside the range of ± 36 degrees and without crashing into the ends of the tracks.

We attempted to reproduce Wieland's results for 1 and 2 poles. A population size of 100 was used in all experiments, and the mutation rate was 30%. Instead of using exactly the same genetic algorithm as Wieland, we used the Genitor software package; this is a steady state genetic algorithm using linear ranking for selection purposes. We used both a fixed mutation rate as well as an adaptive mutation strategy. The results reported in this paper used the adaptive mutation strategy: the rate of mutation is increased for parents that are more similar. Genitor uses a selection bias on the linear ranked gene pool to select genes for crossover. The selection bias was 2.0, indicating that the best individual in the population was twice as likely to be selected for crossover as the mean individual in the population.

The networks used by Wieland had a fixed number of fully connected hidden units and as many inputs as specified by the specific problem. In addition to

Wieland's inputs, one more input unit with a fixed value of 1.0 was added as a bias. One of the hidden units is arbitrarily chosen to be the output unit. Since the network is fully connected, it does not matter which unit is selected to be the output, as long as the same unit is used in each trial. The classic 1-pole 4 input problem was solved using a single hidden unit and 5 input units, including the bias unit. The 1-pole 2 input problem was solved with 3 hidden units and 3 input units, including the bias input. The 2 pole, six input problem was solved using 5 hidden units and 7 input units, including the bias unit. We were unable to find an architecture to solve the 2-pole problem without velocity information. The networks we used are slightly different from those used by Wieland, who did not use a bias input. Wieland used 6 neurons to solve the 1-pole 2 input problem and 10 neurons to solve the 2 pole, six input problem. Wieland does not report a solution to the 2-pole problem without velocity information. The network sizes and genetic algorithm parameters that we used were chosen after many trials and represent the best performance that we could obtain. The results of the experiments are reported in Table 1, and will be compared to those obtained with cellular encoding.

3 Cellular Encoding With Real Weights

Cellular encoding is a language for local graph transformations that controls the division of cells which grow into an artificial neural network [3] [2]. Each cell has an input site and an output site and can be linked to other cells with directed and ordered links. A cell also possesses a list of internal registers that represent local memory. The registers are initialized with a default value, and are duplicated when a cell division occurs. The registers contain neuron attributes such as weights a threshold value. The graph transformations can be classified into cell divisions and modifications of cell registers.

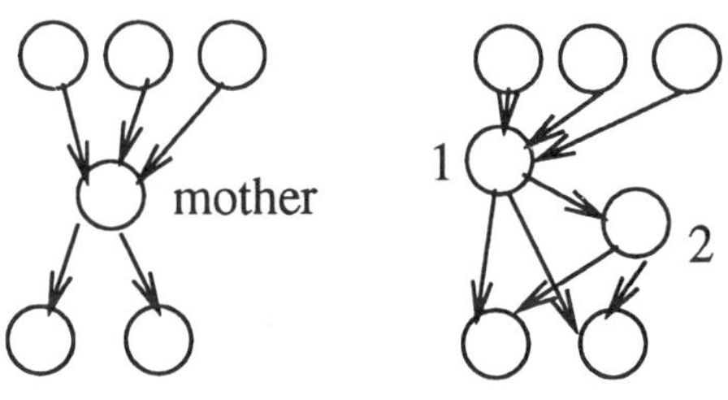

Figure 1: Illustration of the CPO division (a) Before the mother cell divides (b) The two child cells.

A cell division replaces one cell called the parent cell by two cells called child cells. A cell division must specify how the two child cells will be linked. For practical purposes, we give a name to each graph transformation; these names in turn are manipulated by the genetic algorithm. In the *sequential* division denoted `SEQ` the first child cell inherits the input links, the second child cell inherits the output links and the first child cell is connected to the second child cell. In the *parallel* division denoted `PAR` both child cells inherit both the input and output links from the parent cell. Hence, each link is duplicated. The child cells are not connected. Examples of these types of cell division are given by Gruau and Whitley [3]. In general, a particular cell division is specified by indicating for each child cell which link is inherited from the mother cell. In this paper, we used another division called `CPO` illustrated in figure 1. It is similar to a sequential division, except that the output links are duplicated in both child cells.

Before describing the instructions used to modify cell registers it is useful to to describe how a neural network unit performs a computation. The default value of the weights is 1, and the bias is 0. The default transfer function is the identity. Each neuron computes the weighted sum of its inputs, applies the transfer function and then multiplies the result by a coefficient which is 1 by default. See the figures 3 and 4 for examples of neural networks. The instruction `MULT` x sets that coefficient to x. It is worth mentioning that in [8], we did not used a multiplicative coefficient, and imposed weights to scale between -1 and $+1$. As a result, the genetic program could not directly produce ANNs with no hidden units. It produced one network with 19 units; however, this network could be simplified to an ANN with no hidden units, since all the neurons were computing identity function (see section on the analysis of the ANN structure) or they add a constant output. In fact, the encoding generated excess neurons only to obtain weights greater than 1. By using a multiplicative coefficient, the encoding can produce directly ANNs without hidden units (assuming no hidden units are required.) The ANN's computation is performed with integers; the activity is coded using 12 bits so that 4096 corresponds to activity 1. The instruction `SBIAS` x sets the bias to $x/4096$. The instruction `STEP` sets the transfer function to the clipped linear function between -1 and $+1$. The `WEIGHT` instruction is used to modify cell registers. It has k integer parameters, each one specifying a real number in floating point notation: the real is equal to the integer between -255 and 256 divided by 256. (The resulting weight distributions are similar, but not identical to those used by Wieland.) The parameters are used to set the k weights of the first input links. If a neuron happens to have more than k input links, the weights of the supernumerary input links will be set by default to the value 256 (i.e., $\frac{256}{256} = 1$).

The cellular code is a *grammar-tree* with nodes labeled by names of graph transformations. Each cell carries a duplicate copy of the grammar tree and has an internal register called a reading head that points to a particular position of the grammar tree. At each step of development, each cell executes the graph transformation pointed to by its reading head and then advances the reading head to the left or to the right subtree. After cells terminate development they lose their reading-heads and become neurons.

The order in which cells execute graph transformations is determined as follows: once a cell has executed its graph transformation, it enters a First In First Out (FIFO) queue. The next cell to execute is the head of the FIFO queue. If the cell divides, the child which reads the left subtree enters the FIFO queue first. This order of execution tries to model what would happen if cells were active in parallel. It ensures that a cell cannot be active twice while another cell has not been active at all.

We also used the control program symbol `PROGN`. The program symbol `PROGN` has an arbitrary number of subtrees, and all the subtrees are executed one after the other, starting from the subtree number one.

Consider a control problem where the number of control variables is n and the number of sensors is p. We want to solve this control problem using an ANN with p input units and n output units. At the beginning of the development the initial graph of cells consists of p input units connected to a reading cell which is connected to n output units. The input and output units do not read any code, they are fixed during all the development. In effective these cells are pointers or place-holders for the inputs and outputs. The initial reading cell reads at the root of the grammar tree. It will divide according to what it reads and generate all the cells that will eventually generate the final decoded ANN.

3.1 Syntactic Constraints and Genetic Operators

We used a BNF grammar as a general technique to specify both a subset of syntactically correct grammar-trees and the underlying data structure. The default data structure is a tree. When the data structure is not a tree, it can be a `list`, `set` or `integer`. By using syntactic constraints on the trees produced by the BNF grammar, a recursive nonterminal of the type `tree` can be associated with a range that specifies a lower and upper bound on the number of recursive rewritings. In our experiments, this is used to set a lower bound m and an upper bound M on the number of neurons in the final neural network architecture. For the `list` and `set` data structure we set a range for the number of elements in these structures. For the `integer` data structure we set a lower bound and an upper bound of a random integer value. The `list` and `set` data structures are described by a set of subtrees called the "`elements`." The `list` data structure is used to store a vector of subtrees. Each of the subtrees is derived using one of the elements. Two subtrees may be derived using the same element. The `set` data structure is like the `list` data structure, except that each of the subtrees must be derived using a different element. So for example, the rule

$$< A >::= (list[2..2]of(0)(1))$$

generates the trees ((0)(0)), ((0)(1)), ((1)(0)), ((1)(1). The rule

$$< A >::= (set[2..2]of(0)(1))$$

generates only the trees ((0)(1)) and ((1)(0)).

```
<axiom> ::= <nn>

<nn>[0..8]  ::=  PAR(<nn>)(<nn>)
               | CPO(<nn>)(<nn>)
               | SEQ (<nn>)(<nn>)
               | <attribute>

<attribute> ::=
          (PROGN : set[0..4] of
              (WEIGHT: list[8..8] of
                  (integer[-255..+255]))
              (SETMULT(integer[-8..+8]))
              (SBIAS(integer[-4096..+4096]))
              (STEP) )
```

Figure 2: Syntactic constraints used to restrict the space of possible solutions

Figure 3.1 shows the syntactic constraints. The nonterminal <nn> is recursive. It can be rewritten recursively between 0 and 8 times. Each time is it rewritten recursively, it generates a division and adds a new ANN unit. Thus the final number of ANN units will be between 1 and 9. Note that in this particular case, the size of the phenome is proportional to the size of the genome, therefore constraints of the grammar in fig. 3.1 which controls the size of genome result directly in constraints on phenome growth which controls the phenome size.

The nonterminal <attribute> is used to implement a subset of four possible specializations of the ANN units. The first 8 weights can be set to values between -1 and $+1$. The multiplicative coefficient can be set to a value that ranges from -8 to $+8$, and the bias is set to a value between -1 and $+1$. The transfer function can be set to a clipped linear function instead of the default transfer function. Since the lower bound on the `set` range is 0, there can be 0 specializations generated, in which case the ANN unit will compute the sum of its inputs and apply the identity function. Because the upper bound on the `set` is 4, all the 4 specializations can be generated. In this case, the neuron will make a weighted sum, subtract the bias, apply the clipped linear function, and multiply by a coefficient.

Crossover. Crossover must be implemented such that two cellular codes that are syntactically correct produce an offspring that is also syntactically correct (i.e. that can be parsed by the BNF grammar). Each terminal of a grammar tree has a primary label. The primary label of a terminal is the name of the nonterminal that generated it. Crossover with another tree may occur only if the two root symbols of the subtrees being exchanged have the same primary label. This simple mechanism ensures the closure of the crossover operator with respect to the syntactic constraints.

Crossover between two `trees` is the classic crossover used in Genetic Programming as advocated by Koza [5], where two subtrees are exchanged. Crossover between two `integers` is disabled. Crossover between two `lists`, or two `sets` is implemented like crossover between bit strings, since the underlying arrangement of all these data structures is a string.

Mutation. To mutate one node of a tree labeled by a terminal t, we replace the subtree beginning at this node by a single node labeled with the nonterminal parent of t. Then we rewrite the tree using the BNF grammar. To mutate a `list, set` or `array` data structure, we randomly add or suppress an element. To mutate an integer, we add a random value uniformly distributed between ± 32.

Each time an offspring is created, all the nodes are mutated with a small probability. For `tree`, `list` and `set` nodes the mutation rate is 0.05, while for the `integer` node it is 0.5.

4 Comparative Experiments for Cellular Encoding and Direct Encoding

We used a parallel Genetic Programming Algorithm with 32 subpopulations. A more complete description of this algorithm is given by Gruau [4]. The parallel genetic program combines the advantages of the massively parallel model [1] and the island model [6]. Individuals are distributed on islands where each island is a grid forming a 2-D torus. Islands are also arranged as a grid and individuals can migrate only to the four neighbor islands. Not all the sites of a 2-D grid are occupied. The density of population is kept around 0.5. During mating, a site s is randomly chosen on the grid. Two successive random walks are performed and the best individuals found on the 2 random walks are mated. The offspring is placed on site s.

The migration rate is 1%. An individual is exchanged between two adjacent processors after 100 individuals have been created with crossover. The MIMD parallel machine used is a paragon with 400 processors. Each of the 32 subpopulations is 128 individuals and resides on one processor, so the total population is 4096.

4.1 The Evaluation Function

Defining the fitness function can have a critical impact on the quality of results since it is mainly the fitness which shapes the solution. Our initial experiments used a fitness based on the number of time steps the pole could stay balanced [8]. Our early experiments indicated that learning to control the system for a single random start state was easy to achieve, but that the resulting generalization was poor. To alleviate this problem the neural network had to balance the pole starting from 11 initial positions, for 1000 time steps in each of those positions. Those initial positions were chosen at the frontier of the domain where it seemed impossible to recover. This enabled us to generate recurrent ANNs to balance one pole on a cart, using only the cart position and the pole angle as input. Our assumption was that the recurrent connections makes it possible to remember the previous position of the cart and the pole, and to compute the speed. More recently, we discovered that this hypothesis does not appear to be true. Using the same setting as the one described in [8] we evolved nonrecurrent networks to see whether the recurrent links were important. The genetic program could generate an ANN without hidden units (no hidden units) that could balance the pole for the 11 initial positions and for over 100,000 time steps. Obviously this network does not compute the speed! It is able to balance the pole and the cart by keeping them swinging between left and right. The amplitude of the oscillation is subjected to very small increase or decrease over the time. This is due to round off errors. Starting from the resting position, 100,000 time step is not sufficient for the amplitude of the oscillation to reach the fatal value where the system crashes, although the system will crash sooner or later. To force the ANN to compute the speed, we included in the fitness a term which penalizes oscillations.

We now present the new fitness function. We used a single initial position where all the variables are zero except the big pole angle whose value is θ. Since the fitness is more complicated, a single fitness case is sufficient to provide enough precision to allow for good generalization. We used $\theta = 18$ degrees for the one pole problem and $\theta = 4.5$ degrees for the two pole problem. The cart and poles where simulated for 1000 time steps. The fitness is the weighted sum of two fitnesses. The first fitness is the number of time steps the system could stay balanced, divided by 1000. The second fitness is 0 if the pole was balanced less than 100 time steps, else it is a constant K multiplied by the inverse of the sum of the absolute values of all the four normalized variables { cart position, cart velocity, big pole angle, big pole velocity }, over the last hundred time steps. The constant K was chosen to be 0.75. Obviously, if the pole is swinging, then the angle and velocities are going to be high in absolute value, and the inverse of the sum is going to be very low. Thus this second fitness directly penalizes swinging. By penalizing oscillation the fitness function is also biased toward finding neural networks that can return the pole to a centered upright position. If the second fitness is greater that one, then it is set to one. The total fitness is 10% of the first fitness plus 90% of the second fitness. On all the runs we did two generalization tests:

- We balanced the pole for 100,000 time steps.
- We checked to make sure that the ANN was able to bring the cart and the pole back to a centered upright position when started from the initial position in the training set.

Note that the result on the fixed architecture are reported with the old fitness function; this weakens the comparison.

4.2 The Single Pole Problem

We experimented on two variants of the classic problem of balancing a single pole attached to a cart on a finite track. The most common variant of this problem uses 4 inputs (pole angle and velocity, cart position and velocity) and a bang-bang control strategy. Following Wieland we tested a variant with 4 inputs and a continuous control strategy where the force applied to the cart is the output unit's activity multiplied by 10. Thus, the magnitude of the force that is applied smoothly varies between -1 and +1, as opposed to the bang-bang control. Again following Wieland, we experimented with a variant of the problem where only 2 inputs are given, (pole angle and cart position) and an ANN with a parallel dynamic must be used to compute the relevant velocity information. This problem was also solved using a continuous control strategy.

For the parallel dynamic case, the initial activation of all neurons is set to 0. For each computation the network is relaxed three times in parallel before its output is considered.

	CE code		Direct code	
Problem	Learning	Gen.	Learning	Gen
1-pole 4 inputs	1400	250	142	466
1-pole 2 inputs	21,000	225	1276	206
2-poles 6 inputs	34,000	569	410	510
2-poles 3 inputs	840,000	300	—	—

Table 1: Results of the experiments. The learning column reports the average number of evaluations needed to find the solution, the Gen. column report the performance in generalization

We ran a generalization test based on that used by Whitley et al.[7] where 625 initial settings of the cart and of the pole are generated. Each of the normalized 4 input variables representing cart position, cart velocity, pole position, and pole velocity take the following 5 values: 0.05, 0.25, 0.5, 0.75, 0.95. (Note that these values actually scale to positions -0.9, -0.5, 0, 0.5 and 0.9 over the ± 1 input range.) This results in $5^4 = 625$ test cases. Generalization is tested by counting the number of test cases for which the ANN can balance the pole for 1000 time steps. Whitley et al. reported an average of 406 successes over 625 test cases using all 4 inputs and a bang bang control strategy.

The results obtained using cellular encoded networks as well as fixed architectures (Direct Encoding) using our implementation of Wieland's methods are reported in Table 1. Each result represents an average over 30 experiments.

Learning time increases and generalization decreases when the neural network must learn to compute the velocity information on its own. This is due to the lack of precision with which the speed is computed by the neural network.

Cellular Encoding needs more evaluations than the fixed architecture coding to solve these problems. However, it should be noted that the numbers reported for the fixed architecture are the best that we could obtain after many trials with different architectures. Overall, it took a great deal of human intervention to get the fixed architecture to solve the problems rapidly and reliably. At the beginning, fixed encoding needed more than 200,000 evaluations to solve the one pole problem with 2 inputs and no velocity, or the 2 pole problem with 6 inputs (including velocity). The time spent finding parameters for quick and reliable convergence is not shown in the table. Also, we were unable to find a fixed architecture solution to the 2 pole, 3 input problem without velocity information. Cellular Encoding did find a solution to this problem.

4.3 Two Poles

Table 1 also reports results for the experiments involving two poles. The ANNs have 6 input units which are respectively the cart position, the cart velocity, the big pole position, the big pole velocity, the small pole position and the small pole velocity. The failure angle for the two pole problem is ±36 degrees.

Sampling over the full ranges of the input variables is impractical for testing generalization because the two pole system can be controlled within only a narrow range of values. We defined new ranges of the variables for training and testing purposes. The actual input ranges did not change; only the values used for training and testing. The intervals for the cart variable are the same as the ones used in the learning set: the position varies between ±2.16 meters, the velocity varies between ±1.35 meters per second. The interval for the big pole is the same as the one used in [8]. The angle varies between ±3.6 degrees, and the pole velocity between ±8.6 degrees per second. We then used 625 test cases using settings at 0.05, 0.25, 0.5, 0.75 and 0.95 intervals within these reduced ranges. The small pole was always set to zero, because it moves so quickly. The poles could be balanced on average for 569 of the 625 initial settings using these reduced ranges.

Finally, we ran an experiment with two poles, but without using information about pole or cart velocities. The ANN had three inputs: the cart position, the big pole angle, and the small pole angle. We run a single experiment using a larger number of processors (64 processors) and a larger population per processor (256) for a total population of 16,384. A solution was found in 51 generations. For this solution, if the big pole is set to 4.8 degrees, the ANN is able to bring the two poles and the cart back to the null position after 200 time steps. We were not able to solve this problem with the fixed architecture encoding.

5 Analysis of the ANN Structures

One Pole For the four input problem where the velocity is given the genetic program found solutions with no input units. It thus automatically generated minimal networks.

Figure 3 (a) shows an example of neural networks generated by the genetic program for the one pole problem without information about velocity. All these ANNs show the same features which enable them to compute the speed. First, the minimum path between the input and the output unit is smaller than three. The reason for this is that the ANN is relaxed three times before its output is considered. If the minimum path was greater than three, the current state of the cart and the pole could not affect the control decision taken by the ANN until the next time step. Second, some intermediate neurons are used to retain information about the previous state. As a result there are other possible paths between the input and the output units with length strictly greater than three. When the information flows through these neurons, it takes four or more time steps to go from the input layer to the output layer. These neurons are used to register the previous cart position and pole position, and to compute the speed. In the ANN represented in figure 3, there are 6 hidden units; three of them use the clipped linear transfer function. Their transfer functions are piecewise linear, and are equal to the identity around the origin. Those three neurons are never saturated. That is, the net input always stays in the linear part of the transfer function. Therefore we can replace the transfer function by the identity function. The ANN was simplified by hand by merging together the ANN units that compute the identity. The network is mathematically equivalent to the ANN shown in figure 3 (b), which has to be relaxed only 2 times and has only two hidden units. We tested this solution to verify that it was working in the same way. This ANN is simple enough that its behavior can be fully understood. If the pole is far from equilibrium, the variables (cart position and pole angle) are high, the two hidden units are saturated, and no information can be used about velocities. The resulting control is not smooth and one can see the pole swinging very violently. When the variables are smaller, then the two hidden units are not saturated, they can compute the velocities, and the pole is very quickly brought to rest. Thus there are two modes of computation which can clearly be distinguished. For a test case where the pole angle is set at 18 degrees, this is far from equilibrium and the pole swings violently. After hundred time steps it changes its mode and suddenly the pole is smoothly brought back to rest.

Two Poles with velocity. About 80% of the neural networks generated by the genetic program to control the two poles had no hidden units. The remaining 20% had one or two hidden units that could be simplified. Wieland suggested that the 2 pole problem was very difficult; Wieland used 10 neurons to solve this problem. Moreover, we had designed a hand coded solution to this problem before running the GP that had 16 hidden units. The threshold of the output unit is 0, and the weights of one solution (after scaling by 1/256) are −0.16, −0.23, −1.36, −2.9, 1.47, 0.87. The solution found by the genetic algorithm is not obvious. It begins with four negative weights. The third and fourth negative weights indicate that if the big pole is to the right (resp. left) then push the car to the left (resp.

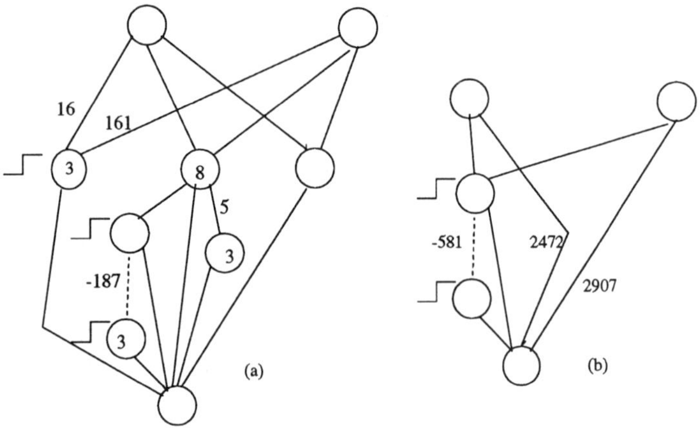

Figure 3: (a) An ANN generated for the one pole without velocity information. Inputs are shown at the top of the network diagram. We specify the ANN unit clipped linear transfer function using a schematization of the graph of this function. Inside the circle we represent the multiplicative coefficient when it is different from 1. The bias are not represented, because they are all 0. We also represent the weights when they are different from 256. (b) This is the equivalent ANN obtained after simplification. It has now only 3 hidden units.

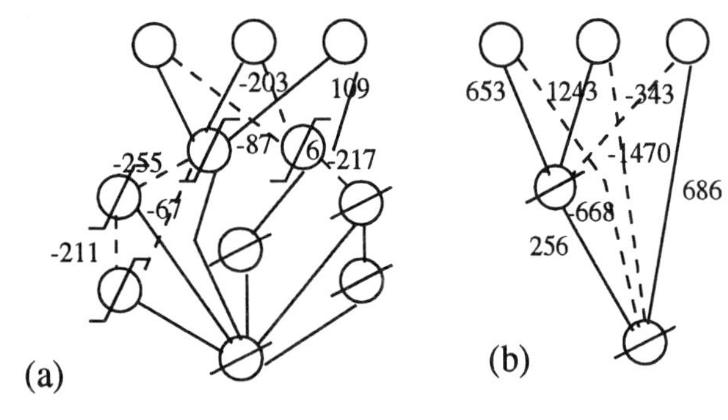

Figure 4: (a) ANN found by the GP for the balancing of two poles. It has seven hidden units. (b) The ANN (b) is mathematically equivalent to the ANN (a) it has only one hidden units.

right). This at first seems counterintuitive, because it is going to make the big pole bend down even worse. But then, weight 5 and weight 6 are positive; when the small pole makes an angle bigger that the big pole in absolute value, then the cart is pushed right (resp. left) and the two poles are brought back to the null position. The first two small negative weights tend to push the cart back to the center.

Two Poles without velocity. Figure 4 shows a mathematical equivalent ANN that has only one hidden unit, and must be relaxed only one time instead of three. We picked a solution found by the genetic algorithm that has seven hidden units, which are all computing the identity. Therefore they can be all simplified. We need to keep one hidden unit however to maintain the delay that makes it possible to compute the speed.

6 Conclusions

This paper compares a new version of Cellular Encoding with direct encoding on the problem of balancing one or more poles on a cart moving on a fixed track. The new version of Cellular Encoding allows for data structures encoding the weights of the neural network; it also uses syntactic constraints to bound certain properties of the neural networks, such as the number of hidden units. The use of real-valued weights in conjunction with Cellular Encoding addresses a criticism of previous work with cellular encoding that was restricted to evolving Boolean networks. We compared Cellular Encoding with a Direct Encoding method that produces architectures with a fixed size.

Cellular Encoding required more evaluations to solve the problems than the Direct Encoding method. However, the large amount of human effort, intervention, and experience required to find the proper architecture for Direct Encoding makes it less attractive than Cellular Encoding. The solutions found by Cellular Encoding are smaller than those obtained using methods and architectures developed by Wieland on these same problems. The advantage of cellular encoding is that it could automatically find small architectures whose structure and complexity fit the specificity of the problem. In the cases studies here these are often architectures without hidden units or architectures that can simplify to one or two hidden units. This provides new insight about the complexity of the problem we are solving. In particular, the results indicate that competitive linear solutions exist not only for the problem of balancing one pole between ±36 degrees, but also the problem of balancing two poles.

The problem of balancing two poles without velocity information represents a more complex control task.

This problem could be solved using cellular encoding, which evolved networks using only one or two hidden units. This problem could not be solved however with the fixed architecture coding.

Acknowledgements

This work was supported in part by NSF grant IRI-9312748 and by a TMR european fellowship to Frédéric Gruau. We thank San Diego Super Computer Center for providing an access to their Paragon.

Bibliography

[1] Collins, R. and Jefferson, D. (1991) Selection in Massively Parallel Genetic Algorithms. *Proc. 4th International Conf. on Genetic Algorithms*, Morgan-Kaufmann, pp 249-256.

[2] Gruau, F. (1994) Neural Network Synthesis using Cellular Encoding and the Genetic Algorithm PhD Thesis, Ecole Normale Supérieure de Lyon. anonymous ftp: lip.ens-lyon.fr (140.77.1.11) pub/Rapports/PhD PhD94-01-E.ps.Z (English) PhD94-01-F.ps.Z (French)

[3] Gruau, F. and Whitley, D. (1993) Adding learning to the the cellular developmental process: a comparative study *Evolutionary Computation* 1(3):213–233.

[4] Gruau, F. (1994 Automatic Definition of Modular Neural Networks, *Adaptive Behavior* 3(2) 151–184

[5] Koza, J. (1992) Genetic Programming: On the Programming of Computers by Means of Natural Selection, Cambrige, MA *the MIT Press.*

[6] Muehlenbein, H., Schomish, M. and Born, J. (1991) The parallel genetic algorithm as function optimizer, *Parallel Computing* 17:619–632.

[7] Whitley, D., Dominic, S., Das, R. and Anderson, C. (1993) Genetic reinforcement learning for neurocontrol problems, *Maching Learning* 13:259–284.

[8] Whitley, D., Gruau, F., and Pyeatt, L. (1995) Cellular Encoding Applied to Neurocontrol *Proc. 6th International Conf. on Genetic Algorithms*, Morgan-Kaufmann, pp 460–467.

[9] Wieland, A. (1990) Evolving Controls for Unstable Systems. *Connectionist Models: Proc. 1990 Summer School*, Morgan Kaufmann, pp 91-102.

Entailment for Specification Refinement

Thomas Haynes, Rose Gamble, Leslie Knight and Roger Wainwright
Department of Mathematical & Computer Sciences
600 South College Avenue
The University of Tulsa
Tulsa, OK 74104–3189
e–mail: [haynes,gamble,knight,rogerw]@euler.mcs.utulsa.edu

Abstract

Specification refinement is part of formal program derivation, a method by which software is directly constructed from a provably correct specification. Because program derivation is an intensive manual exercise used for critical software systems, an automated approach would allow it to be viable for many other types of software systems. The goal of this research is to determine if genetic programming (GP) can be used to automate the specification refinement process. The initial steps toward this goal are to show that a well–known proof logic for program derivation can be encoded such that a GP–based system can infer sentences in the logic for proof of a particular sentence. The results are promising and indicate that GP can be useful in aiding program derivation.

1 Introduction

Specification refinement, a part of formal program derivation, is the process of transforming an abstract specification of a software system into a corresponding implementation that is provably correct [Chandy and Misra, 1988, Gamble *et al.*, 1994, Roman *et al.*, 1993]. The process involves the stepwise refinement of the specification in which each step preserves correctness. Thus, the final specification and implementation are provably correct by construction. Formal program derivation is largely a manual process requiring expert knowledge and intuition to determine which refinements to perform and then to prove that the refinement steps do not violate the original specification. However, there exist methodologies and techniques that can be formulated as expert rules of thumb. These methodologies and techniques can be difficult to formally state and thus put in an automated system. In addition, some refinements come by an inductive leap as the designer determines the details for the next specification and finds that those details preserve correctness. Such an inductive leap is also difficult to automate.

Currently, there are few systems that perform program derivation using rules of thumb. Theorem provers are one way to transform one specification to another [Lowry, 1991], but they require full domain knowledge of the possible transformations. Because of this lack of automation in program derivation, it is rarely used in large software engineering projects. With increased use, program derivation would lead to higher quality software because functional reliability could be guaranteed. Thus, there is need for a system that incorporates both theorem proving, expert knowledge, and inductive search to make program derivation a viable part of the software development process.

We investigate the current technology of genetic programming (GP) [Koza, 1992] to aid in the automated process of program derivation. We begin by using GP to search for the proof of entailment of a sentence in a knowledge base of theories from the UNITY [Chandy and Misra, 1988], which is a popular language for expressing the formal specification of concurrent programs and for using program derivation to develop concurrent implementations that are provably correct. These initial steps, of encoding the theories for use by the GP and showing proof of entailment, present a framework to establish the efficacy of GP technology for use in program derivation and for creating an automated system utilizing GP to correctly refine specifications. Because each refinement step must preserve the correctness of the original specification, automating the process requires a theorem prover to verify correct refinements.

The paper is divided into the following sections. The next section provides a brief overview of UNITY and program derivation. Section 3 presents the design and implementation details for the GP system. Section 4 presents experimental results. In Section 5 we discuss the system design with respect to the results. Section 6 provides a discussion relevant to our initial goal. Section 7 concludes the paper.

2 The UNITY Language

The UNITY language [Chandy and Misra, 1988, Kaltenbach, 1994] includes specification constructs in which *safety* and *progress* assertions can be made about a program. The UNITY proof logic allows one to perform formal program derivation. A program developer specifies the problem with an abstract initial specification. The proof logic guides the programmer to refining that specification and proving that the refinement at each step is correct. When the specification is sufficiently detailed, a program is directly constructed that satisfies the specification. The program is then refined to execute on specific architectures in specific languages with its correctness guaranteed. Using UNITY, a designer can concentrate on creating the best program to solve a problem without worrying about the final architecture on which the program will be executed. Though there are many tricks to specification and program refinement, experience and intuition account for a large majority of the effort involved.

In UNITY, there is a progress property called leads-to, which is written as $p \mapsto q$. This assertion means that if p is true in some program state, then eventually the program will reach a state in which q will become true, regardless of the value of p when q becomes true. Typically, an initial specification will include a leads-to property that states that the initial state must lead to a goal state, as stated below in (P1).

$$Initial\_State \mapsto Goal\_State \qquad \text{(P1)}$$

Because this specification may be abstract, refinement steps are made to add detail while preserving correctness. For example, we may divide the property into intermediate goals that correspond to the completion of a program task. In the properties below, we see that from an initial state the program should reach a state in which Task 1 will begin. From that state, eventually Task 2 will begin, which eventually leads to the goal state. Due to the transitivity of leads-to, (P2), (P3), and (P4) together form a refinement of (P1).

$$Initial\_State \mapsto Task\_1 \qquad \text{(P2)}$$
$$Task\_1 \mapsto Task\_2 \qquad \text{(P3)}$$
$$Task\_2 \mapsto Goal\_State \qquad \text{(P4)}$$

Overall, the process of specification refinement includes a cycle of the following major steps: (1) create an abstract specification, (2) suggest refinements to the specification, and (3) prove the refinements preserve correctness. In the process of refining properties, the developer has to make an inductive leap to suggest possible refinements that specify a valid direction. Then the proof logic must be brought to bear on the new properties to show that they are a correct refinement. Though the process guarantees that correct programs will be developed, a must in safety-critical applications, it is a very difficult and tedious process. In this respect, automation is the best chance to incorporate formal program derivations into large software development.

In this paper, we determine the viability of using GP for the task of automating specification refinement. The initial portion of the research is to determine if step (3) above can be performed by the GP. Therefore, we prove that the entailment of a sentence α from a knowledge base of theories can be inferred automatically by a GP-based system [Koza, 1992]. In our example above, the sentence is (P1) and the knowledge base consists of (P2)-(P4). The inference rule representing the transitivity of leads-to shows that (P2)-(P4) entails (P1). The results of our initial research suggests that the possibility exists, with extensions to the current GP technology, that step (2) can be automated as well.

2.1 Introduction to GP

Genetic programming is an extension of genetic algorithms for generating computer programs [Koza, 1992]. By applying the rules of natural selection, the GP will evolve a program to solve a problem. Given a set of functions and terminals, the GP will generate a random population of programs. The members of the population, chromosomes, are represented by parse trees, where the leaves are terminals (constants or state data) and the branches are functions. New and possibly better programs are created by combining or modifying the subtrees through crossover and mutation. The "goodness" of a program, also referred to as its fitness, is determined by how well the program solves the problem.

Two properties restrict or govern the GP method. The first is closure – every function must handle all terminals and output produced by other functions. Closure ensures the creation of valid programs across genetic operations. The second is sufficiency – the designer must supply enough functions and terminals to solve the problem.

In traditional GP, all of the terminal and function set members must be of the same type. Closure ensures that all inputs to functions can be handled. Montana [Montana, 1995] introduced strongly typed genetic programming (STGP), in which the variables, constants, arguments, and returned values can be of any type. The only restriction is that the data type for each element be specified beforehand.

3 Proving Entailment of UNITY Theorems

Our intent is to encode sentences using UNITY operators to form a knowledge base KB, and a target sentence α, and, then, to use propositional inference rules from the UNITY proof logic to show that the knowledge base entails that sentence, $KB \models_i \alpha$. We employ a GP package, which utilizes strong typing [Montana, 1995], to apply the inference rules. The main domain specific tasks that a GP researcher must tackle are the determination of what comprises both the terminal and function sets and what is the fitness criteria for a chromosome. The terminal set shown in Table 1 is comprised of enough predicate variables to solve the problems. Both a subset of the theorems presented in Chandy and Misra [Chandy and Misra, 1988] and some additional simple logic rules were chosen as the propositional inference rules used to prove entailment. The subset and the simple rules combine to form the function set as shown in Table 2. Both the terminal and function set provide sufficient functionality and representation to prove the entailment of sentences from the provided KB.

Terminal	Type	Purpose
P	Boolean	The predicate P.
Q	Boolean	The predicate Q.
R	Boolean	The predicate R.
S	Boolean	The predicate S.
T	Boolean	The predicate T.

Table 1: Terminal Types

The members of the function set map trees from GP space into trees in propositional logic space. A simple heuristic in the logical inference process is to only apply inferences when there is a match between the "arguments" of the hypothesis and sentences in the KB [Chandy and Misra, 1988, Russell and Norvig, 1995]. Strong typing enables the GP system to guide the pattern matching rule. The restriction of only considering valid child nodes has a side effect of reducing the size of the possible search space [Haynes *et al.*, 1995, Montana, 1995]. A corollary of this is that STGP chromosomes are more representational than vanilla GP chromosomes, and hence tend to be smaller.

3.1 Fitness Evaluation

Each chromosome represents a propositional logical inference from the KB and is evaluated to determine its fitness. The typical GP evaluation function evaluates the parse tree in a preorder traversal, since short–circuiting can occur inside macro functions. The evaluation function for the inference engine has a postorder traversal, which is due to the mapping of GP space into propositional logic space. Since each child subtree must be known before a function can be applied, there is no short–circuiting of arguments.

The fitness function evaluates the soundness of an inference from the KB. To aid this process, two additional knowledge bases are maintained: a newly derived sentence, KB_n, and a derived sentence, KB_d. We also define KB_o as the given sentences and $KB = KB_o + KB_d$. The KB_n is comprised of the newly derived sentences for the current generation. The KB_d begins as an empty list. After each generation, the sentences from the KB_n are moved to the KB_d. The KB_d allows us to track sentences derived by the inference engine.

The fitness function maps the chromosome from a parse tree to a series of operations of inference rules in the propositional logic space. The resultant series of rules can be represented as a tree structure, which we will call a proof tree. The fitness evaluation has different categories to allow for the evolution of better answers. The fitness evaluation is as follows:

- N points for true sentences in the proof tree that are from the KB_o.
- M points for true sentences in the proof tree which are in the KB_d.
- $4 * (M + N)$ points if the proof of $KB \models_i \alpha$ is at the root of the proof tree.
- $2 * (M + N)$ if the proof tree contains the proof of $KB \models_i \alpha$.
- $M + N$ points for the final proof tree being true.

The KB_o is separated from the KB_d to bias the inference engine to build a proof from the sentences in the KB_o. The KB_n is maintained separately from the KB_d to allow a reward for the discovery of new sentences (Note that KB_n does not contribute anything during fitness evaluation.). We are concerned with both fairness of derivation of a sentence and in detecting random generation of a sentence versus derivation from the KB_o. We enforce fairness in that a chromosome which derives a sentence without proving its validity will not be rewarded either before or after another chromosome in the current generation has derived the sentence while proving its validity. Chromosomes in later generations may treat any appearance of a sentence derived in an earlier generation as valid.

These separations of sentences are symptomatic of a basic difficulty of composing fitness functions for genetic programs: how are good sub–structures identified? For example, if the fitness function simply determines if the mapping of the chromosome to an inference procedure results in the target sentence α, then with a randomly generated initial population, it is very unlikely that any chromosome derives α. Thus additional terms are needed in the fitness function to differentially evaluate these "bad" derivations. The key aspect of the family of genetic algorithms is that even though a particular structure is not effective, it may contain useful sub–parts which when combined with other useful sub–parts, will produce a highly effective structure [Koza, 1992]. The fitness function should be designed such that useful sub–structures are assigned due credit.

Function	Return	Arguments	Equivalent UNITY Statement
And	And	Generic A and Generic B	$p \wedge q$
AntiReflexivity	Unless	Generic A	$\frac{p}{p \ unless \ \neg p}$
FDisjunction	LeadsTo	LeadsTo A and LeadsTo B	$\frac{p \mapsto q, \ p' \mapsto q'}{p \vee p' \mapsto q \vee q'}$
Implication	Implies	LeadsTo A	$\frac{p \Rightarrow q}{p \mapsto q}$
Implies	Implies	Generic A and Generic B	$p \Rightarrow q$
LawOfExpansion1	Generic	Or A	$\frac{p \wedge q \ \vee \ p \wedge \neg q}{p}$
LawOfExpansion2	Or	Generic A and Generic B	$\frac{p, \ q}{p \wedge q \ \vee \ p \wedge \neg q}$
LeadsTo	LeadsTo	Generic A and Generic B	$p \mapsto q$
Not	Not	Generic A	$\neg p$
Or	Or	Generic A and Generic B	$p \vee q$
PSP	LeadsTo	LeadsTo A and Unless B	$\frac{p \mapsto q \ , \ r \ unless \ b}{p \wedge r \mapsto (q \wedge r) \vee b}$
Reflexivity	Unless	Generic A	$\frac{p}{p \ unless \ p}$
SimpleLogic1	Implies	Generic A and Generic B	$\frac{p \ , \ q}{p \wedge q \ \Rightarrow \ p}$
Transitivity	LeadsTo	LeadsTo A and LeadsTo B	$\frac{p \mapsto q \ , \ q \mapsto q}{p \mapsto r}$
Unless	Unless	Generic A and Generic B	$p \ unless \ q$

Table 2: Function Types

A potential negative consequence of assigning credit for sub–structures, *in this domain*, is that a good sub–structure might appear many times in a chromosome, and thus, falsely inflate the total fitness of that chromosome. Note that in typical GP research, this is seen as a benefit as it is the foundation of the schema theorem [Haynes, 1996]. However, we want the inference engine to derive new valid sentences, not to re–derive the same sentence over and over again. The fitness function for the inference engine has to carefully balance these two concerns: allow multiple occurrences of a sub–structure and spend time re–deriving the same sentence or not allow multiple occurrences and have the population prematurely converge.

Finally, it should be pointed out that by collecting valid sentences in KB_d, we are forming a collective memory of previous successful entailments. Given two populations composed of exactly the same individual chromosomes, the fitness evaluation can be different depending on the chromosomes of all previous generations.

3.2 Mapping from Chromosome to Proof Tree

The terminal set represents five predicates, which are sufficient in number to solve the problems we will consider. The function set does not verify the validity of the predicates, but rather it manipulates proof trees in the predicate space to determine the validity of sentences composed of the atoms the predicates represent. These trees are constructed from the leaves up to the root because the chromosomes, which are trees, are evaluated from the leaves to the root. The task of the fitness function is to apply functions in the GP space to the corresponding proof trees represented by the child nodes of the function node.

As a terminal is encountered, it is added as a single node tree to the tree manager, which is a list containing all the proof trees for a chromosome. If the terminal is a valid sentence by itself, i.e. it is in either KB_o or KB_d, then its validity is marked as *TRUE*, else its validity is marked as *DONT_CARE*. After all of the child nodes of a function node have been processed, the proof trees of the children are then combined into the proof tree of the current node by applying the associated UNITY operator for that function node. Its validity is determined by the validity of the proof trees associated with the child nodes, and in some cases on a characteristic property of the UNITY operator itself.

As an example, we will build a proof tree for the problem of showing

$$\frac{p \ \mapsto \ q}{p \wedge \neg q \ \mapsto \ q} \quad or \quad \frac{p \ \mapsto \ q}{p \wedge \neg q \ \mapsto \ q \wedge \neg q \ \vee \ q}.$$

Since we do not provide a function to rewrite $q \wedge \neg q \ \vee \ q$ as q, we have logically rewritten the problem to prove that $p \wedge \neg q \ \mapsto \ q \wedge \neg q \ \vee \ q$. While adding an additional UNITY operator will help, simple editing of the output will not: the problem has to be restated in order for the sentence to be resolved. An evaluation of the optimally encoded chromosome, **PSP(LeadsTo**(P,Q)**, AntiReflexivity**(Q), is shown in Figure 1.

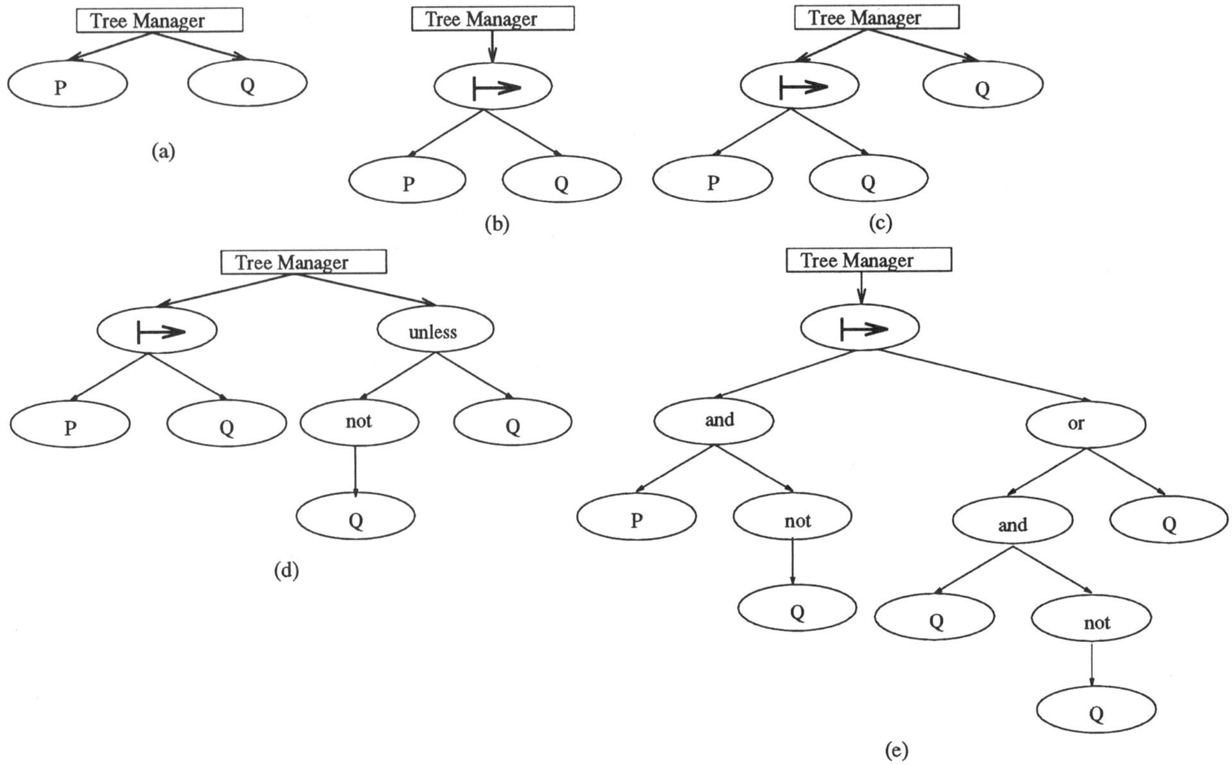

Figure 1: Effect of transformation of S–expression to proof tree. (a) Addition of terminals to Proof Tree Space. (b) Application of **LeadsTo** to Terminals. (c) Another terminal is added to the tree manager. (d) Application of **AntiReflexivity**, an unary operator. (e) The final result in Proof Tree Space.

4 Experimental Results

We present one problem that is illustrative of the workings of the inference engine:

$$\frac{p \mapsto q \ , \ q \mapsto r \ , \ r \mapsto s}{p \mapsto s}.$$

This problem can be solved using two applications of transitivity. With population sizes of 500 and 1000, the solution appears after the randomly generated population performs crossover and mutation. With a population size of 100, the solution chromosome does not appear without some work by the STGP system. We expected that a larger population size would be needed. However, we are using an STGP system, not a vanilla GP system. The majority of the work is being done by the STGP algorithm, which generates more representational chromosomes, i.e. it reduces the search space [Haynes *et al.*, 1995, Montana, 1995]. If we add another level of transitivity, $S \mapsto T$, then even with a population size of 2000, the solution does not appear in the early generations. Since strong typing is needed to ensure legitimate proof trees are generated, see Section 5, we are not able to test an untyped system. However, we can generate an initial population, and we find that even for population sizes of 16,000 useful sub–trees do not appear in that population. For a typed system, building blocks do appear immediately. For the original problem, with two applications of transitivity, the solution appears within these 16,000 chromosomes, but it does not for the modified problem, i.e. an additional level of transitivity.

A difficulty is shown in Figure 2, which graphs the best chromosome's fitness and average fitness per generation. The average fitness is tracking the fitness of the best chromosome. This is a linear curve, which is not typical of the curves one gets from most GP problems. (This linearity is similar to that artificially induced in the research of McPhee and Miller [McPhee and Miller, 1995] into introns.) The system is exhibiting symptoms caused by an ever growing maximum fitness. As new true sentences are derived, they are added to the KB_d to be matched against later. For example, $\mathbf{AR}(Q)$ will slowly progress to $\mathbf{AR}(\mathbf{AR}(\mathbf{AR}(\ldots(\mathbf{AR}(Q))\ldots)))$. This growth of true, yet meaningless, sentences will continue until restricted by the solution space as a function of the tree size. As the KB_d grows, the chromosomes can exploit it, resulting in ever increasing fitness scores. However, these scores are not indicative of the chromosomes converging to a good solution. As shown in Figure 2, the fitness dramatically jumps once the solution appears within a subtree. Also, within 5 generations, the solution rises to the top of the tree. This supports our hypothesis that finding the sentence α inside the chromosome is a good building block which will allow the inference engine to find the final solution.

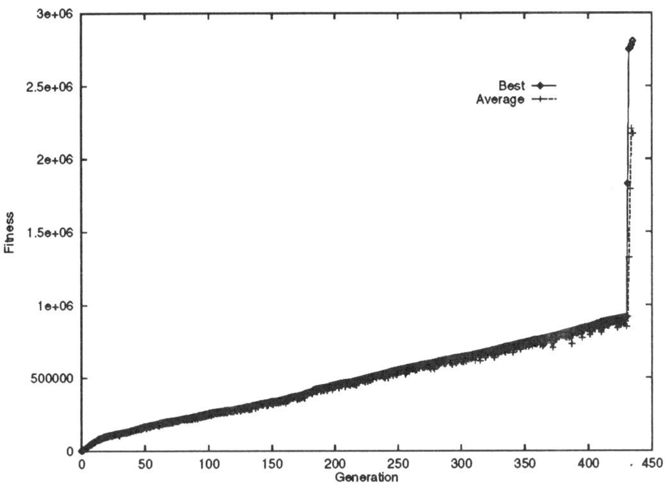

Figure 2: Best and Average Fitness for double transitivity problem.

The S–expression in Figure 3 corresponds to the chromosome which scored the best fitness in generation 431 (the first generation to derive the proof). The line marked ⋆ denotes a **Transitivity** which is applied on two **LeadsTo**: **TR**(**LeadsTo**(P, Q),**LeadsTo**(Q,S)). From the form of this derivation, it can be seen that the sentence: **TR**(**LeadsTo**(Q,R),**LeadsTo**(R,S)) must have been arrived at earlier, and the sentence **LeadsTo**(Q,S) must have been added to KB_d. Thus the chromosome can back up its claim that the sentence **LeadsTo**(Q,S) is true by referring to the proof which placed this sentence in KB_d. Finally, in generation 436, a chromosome corresponding to the ⋆ line in Figure 3 is generated. Our ultimate goal of a chromosome succinctly proving the entailment of the sentence has been achieved.

```
FD( FD( PSP(
⋆               Trans( LeadsTo( P, Q ), LeadsTo( Q, S ) ),
                Reflex( Unless( Q, R ) ) ),
        PSP( PSP(  LeadsTo( R, P ), Unless( S, R ) ),
                AR( P ) ) ),
    FD( PSP( PSP(  LeadsTo( P, P ), Unless( P, P ) ) ,
                AR(   Implies( P, Q ) ) ),
        PSP( PSP(  LeadsTo( P, P ), Unless( P, P )),
                AR(   Implies( S, R ) ) ) ) ) )
```

Figure 3: Generation 431's best S–expression.

5 Design Issues

Our method of credit apportionment presents several unresolved difficulties. It is difficult to differentiate between derived and randomly generated true sentences. We would like to be able to reward discovery of a true sentence (actual learning) rather than (or more than) random generation of a true sentence. Also, as the KB_d grows, the maximum fitness also grows, which makes it difficult to compare the members across generations.

It also became apparent that while the system learned that **AR**(Q) was true, it could not learn that **Unless**(Q,$\neg Q$) was just as true. We altered our code to detect when the functions **Unless**, **LeadsTo**, **Implies**, and **Implication** were being applied to the same subtree. If they were, then these derived sentences were added to the KB_n. In addition, **Unless** can detect if it is being applied to a subtree, and the negation of that subtree.

This inference engine also raises some interesting concerns about the standard manner in which GP systems are constructed. The traditional GP system is designed with only closure properties to ensure that the function set can handle any input. This simple restriction will not work with our system. For example, if the interior grandchild nodes of a **Transitivity** node are not the roots of identical subtrees, then the evaluation and the side effects of the node are meaningless and undefined. What proof tree should be passed up the GP S-expression? Our system handles this by carrying out the transitivity with the two exterior grandchild subtrees, and by marking the validity of the resultant proof tree as *FALSE*. The subtler problem is what to do if the two proof trees manipulated by the **Transitivity** node do not have **LeadsTo** nodes in their roots? This problem also points out that there must be function nodes underneath the **Transitivity** node.

Our system handles this class of problems by enforcing strong typing [Montana, 1995] for each function and its children. Tables 1 and 2 show the types used for the terminal and function sets. By strictly enforcing type instantiations across both the random generation of chromosomes and crossover, the whole issue of correct proof trees can be avoided. Unfortunately the type system does not allow for context sensitive typing. All of the supported functions use "single" levels of typing. The law–of–expansion, which was used in several problems from Chandy and Misra [Chandy and Misra, 1988], presents a problem in which the grandchildren of the **LawOfExpansion1** node must honor a specific function, i.e. **Or**, when in the context of this node. This requires the GP to be context sensitive when applying crossover or mutation to a parse tree. For example, the child nodes of **Transitivity** are of type `leads-to` and could conceivably be recombined with other nodes with type `leads-to` children. The law-of-expansion has a type `Or` with two children of type `And`, which under any other circumstance could be worked on independently, but in this instance must be manipulated together. This problem restricts the context in which functions of type `And` and `Or` can be used. The type hierarchy presented in

Haynes *et. al.* [Haynes *et al.*, 1996] solves this problem.

5.1 *An Alternative Approach*

The proof of entailment of a sentence in a knowledge base is achieved by the GP derived chromosomes beginning to sense patterns in the given sentences. The current fitness evaluation "encourages" the derivation of linear subtrees: find a valid sentence, and make it longer. An alternative approach is to generate a KB_s of both simple true sentences of the form **Unless**(Q,Q) and the KB_o to form a fixed size table which can be enumerated. The terminal set would become instances of this enumerated KB_s. The function set would then manipulate these sentences to form the desired solution. The KB_d and KB_n would not be utilized.

We want to reward the discovery of a valid sentence, and its subsequent usage in future generations. It should both be combined with other functions and valid sentences to derive the desired final sentence, and start appearing in as many chromosomes as possible. The current implementation does not reward the discovery of how to prove sentences. For example, if one chromosome discovers that the subtree **TR(LeadsTo(**P,Q**),LeadsTo(**Q,R**))** is *TRUE* because either the two **LeadsTo** subtrees are given or have been proved earlier, then the entire subtree is not highly valued. Instead, the subtree **LeadsTo(**P,R**)** is much more compact and can appear more often in a chromosome which leads to a higher fitness value. Smaller building blocks are less likely to be torn apart.

By not adding to the KB_d, and by forcing the chromosomes to operate on a fixed set of true sentences, we can free ourselves from this problem. Also, we will get rid of the ever increasing maximum fitness, and the inability to examine fitness graphs to determine the viability and progress of the generations.

6 Discussion

As mentioned earlier, we want to take the lessons learned from the inference engine to create a GP system that will refine a UNITY program. The difficulties we foresee in using GP to refine a UNITY program are:

- Encoding the UNITY program in a manner that will permit the GP to work with it effectively.
- Translating the presented heuristics for refining a program into information the GP can use.
- Evaluating the fitness of the new program with respect to its functionality as compared to the old program.
- Evaluating the fitness of the new program with respect to the amount of parallelism introduced.

These difficulties are all related. Since a UNITY program must first be developed by hand, we know that a functional program exists. We also know that heuristic strategies exist for refining the initial program that can be executed by a human in a short period of time.

Olsson has conducted research into utilizing evolutionary techniques for synthesizing programs from specifications [Olsson, 1995]. His work differs significantly from ours in that he uses input/output pairs to evaluate the fitness of programs and his programs are in Standard ML. The input/output pairs are sample input with desired output. Our fitness cases are based solely on the application of UNITY operators which preserve correctness. Thus our transformation operators are bound up in the fitness set and not in the evaluation function. Both techniques are similar in that they only handle small programs.

Olsson also does not believe that GP is powerful enough to do program refinement. Since the appearance of his paper, many of the drawbacks he has highlighted have been or are being addressed and evaluated. Specifically GP systems can handle recursion [Brave, 1996], iteration [Spector and Alpern, 1995], and the automatic creation of functions not foreseen by the programmer [Koza, 1995].

7 Conclusion

This paper demonstrates that genetic programming is a viable method for proving the entailment of a sentence from a knowledge base of UNITY theorems. While the current implementation spends an inordinate amount of time proving trivially true sentences, it can find the correct solution. This shortcoming also applies to more traditional theorem provers. It is also shown, in Section 5, that strong typing greatly aids in the implementation of the GP Theorem Prover. Some further work in this area is highlighted by the **Law of Expansion** function. Note that DeMorgan's Laws will suffer this same problem. Finally, we are at work on an alternative evaluation function which will not suffer from the problem of ever increasing maximum fitness.

Acknowledgments

We wish to thank the anonymous referees for their insightful comments.

References

[Brave, 1996] Scott Brave. Using genetic programming to evolve recursive programs for tree search. In P. Angeline and K. E. Kinnear, Jr., editors, *Advances in Genetic Programming 2*. MIT Press, 1996.

[Chandy and Misra, 1988]
K. Mani Chandy and Jayadev Misra. *Parallel Pro-*

gram Design: A Foundation. Addison-Wesley, 1988.

[Gamble *et al.*, 1994] R. F. Gamble, G.-C. Roman, W. E. Ball, and H. C. Cunningham. Applying formal verification techniques to rule-based programs. *International Journal of Expert Systems*, 7(3):203–238, 1994.

[Haynes *et al.*, 1995] Thomas Haynes, Roger Wainwright, Sandip Sen, and Dale Schoenefeld. Strongly typed genetic programming in evolving cooperation strategies. In L. Eshelman, editor, *Proceedings of the Sixth International Conference on Genetic Algorithms*, pages 271–278. Morgan Kaufmann, 1995.

[Haynes *et al.*, 1996] Thomas Haynes, Dale Schoenefeld, and Roger Wainwright. Type inheritance in strongly typed genetic programming. In Kenneth E. Kinnear, Jr. and Peter J. Angeline, editors, *Advances in Genetic Programming 2*, chapter 18. MIT Press, 1996.

[Haynes, 1996] Thomas Haynes. Duplication of coding segments in genetic programming. In *Proceedings of the Thirteenth National Conference on Artificial Intelligence*, Portland, OR, August 1996.

[Kaltenbach, 1994] Markus Kaltenbach. Model checking for UNITY. Technical Report TR94-31, Department of Computer Sciences, The University of Texas at Austin, May 9, 1994.

[Koza, 1992] John R. Koza. *Genetic Programming: On the Programming of Computers by Means of Natural Selection.* MIT Press, 1992.

[Koza, 1995] John R. Koza. Two ways of discovering the size and shape of a computer program to solve a problem. In L. Eshelman, editor, *Proceedings of the Sixth International Conference on Genetic Algorithms*, pages 287–294. Morgan Kaufmann, 1995.

[Lowry, 1991] Michael R. Lowry. Automating the design of local search algorithms. In Michael R. Lowry and Robert D. McCartney, editors, *Automating Software Design*, pages 515–546. AAAI Press, 1991.

[McPhee and Miller, 1995] Nicholas Freitag McPhee and Justin Darwin Miller. Accurate replication in genetic programming. In L. Eshelman, editor, *Proceedings of the Sixth International Conference on Genetic Algorithms*, pages 303–309. Morgan Kaufmann, 1995.

[Montana, 1995] David J. Montana. Strongly typed genetic programming. *Evolutionary Computation*, 3(2):199–230, 1995.

[Olsson, 1995] Roland Olsson. Inductive functional programming using incremental program transformation. *Artificial Intelligence*, 74(1), March 1995.

[Roman *et al.*, 1993] G.-C. Roman, R. F. Gamble, and W. E. Ball. Formal derivation of rule-based programs. *IEEE Transactions on Software Engineering*, 19(3):277–296, March 1993.

[Russell and Norvig, 1995] Stuart Russell and Peter Norvig. *Artificial Intelligence: A Modern Approach.* Prentice Hall, 1995.

[Spector and Alpern, 1995] Lee Spector and Adam Alpern. Induction and recapitulation of deep musical structure. In *Proceedings of Internatinal Joint Conference on Artificial Intelligence, IJCAI'95 Workshop on Music and AI*, 1995.

Genetic Programming of Near-Minimum-Time Spacecraft Attitude Maneuvers

Brian Howley
Org: 73-15/Bldg: 150
Lockheed Martin Missiles and Space Co
1111 Lockheed Way
Sunnyvale, CA 94089
howley_brian@mm.ssd.lmsc.lockheed.com

ABSTRACT

Genetic programming is used to discover near optimal control laws for the minimum time reorientation of a spacecraft. Two types of acceleration limited maneuvers are considered: maneuvers with zero initial and final angular velocities, and rate limited maneuvers with non-zero initial and terminal velocities. An indexed terminal set was devised to enforce symmetry over the three degrees of freedom. Results for the zero initial and final rate maneuvers are within 2% of published numerical solutions to the problem. The non-zero initial and final rate solution outperforms a manual solution but does not generalize to arbitrary maneuvers.

1. Introduction

Time optimal satellite attitude maneuvers are an important class of problem for both practical and theoretical reasons. Many spacecraft missions require agile spacecraft that can quickly and accurately change orientation. Examples include earth observing spacecraft which must precisely point large optics to perform remote sensing. The faster these maneuvers can be performed, the more images that can be collected.

Figure 1 shows a spacecraft, represented as an orthogonal triad of unit vectors in orbit around the Earth. The spacecraft's yaw axis is pointed to the Earth's center. In order to remain pointed at the Earth's center, the spacecraft must rotate about its pitch axis during the course of the orbit. An agile spacecraft will point to many different locations on the Earth's surface and the maneuvers require rotations about all three spacecraft axes.

Minimum time reorientation is a problem in optimal control. Scrivener and Thompson, 1994, give a recent survey of the literature. The general solution is a two point boundary value problem which is difficult to evaluate [Bryson and Ho, 75]. Practical applications require approximate solutions which can be easily implemented in a

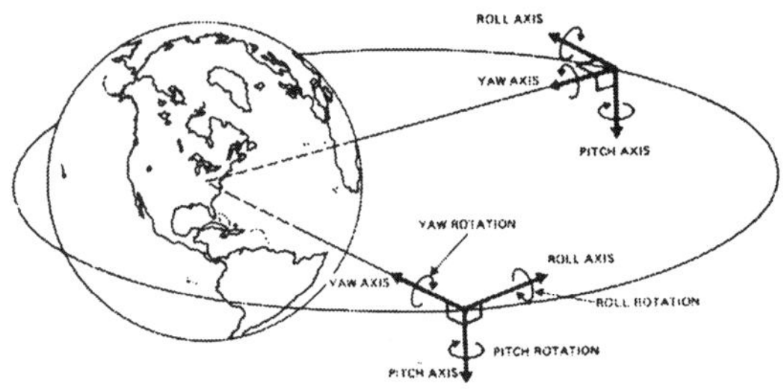

Figure 1: Satellite rotations about three axes.

real time environment. Optimal control theory does not support such a frame work. Genetic programming (GP) performs an efficient search for near optimal solutions over a domain that is inherently easy to implement under computer control.

This paper is concerned with two classes of maneuver: rest-to-rest maneuvers and rate limited, non-zero terminal velocity maneuvers. The rest-to-rest maneuvers start and end with zero angular rates. This is an important class of problems that has been solved numerically [Bilimoria and Wie, 93] and therefore can serve as a benchmark for genetic programming performance. The rate limited, non-zero terminal velocity maneuvers (RLNZ maneuvers) are a more general class. Non zero initial and final maneuver rates are typically encountered when the spacecraft is pointing a scanning sensor or tracking an object. The spacecraft actuators limit maximum angular rates during a maneuver.

2. Background

Problems involving satellite orientation require a set of parameters that quantitatively describe attitude with respect to a known frame of reference. The roll, pitch, and yaw angles shown in Figure 1 are an intuitive description of attitude for small rotations about a known frame of reference. However, when describing large rotations (greater than 90 degrees) these angles are difficult to use in a numerical simulation. For this reason, an alternative set of parameters, known as Euler parameters [Wertz, 78], is used in this study.

Euler parameters are derived from Euler's rotation theorem, illustrated in Figure 2. Euler's theorem states that the rotation of a rigid body can always be expressed as a rotation through some angle about a fixed axis [Wertz, 78]. Thus, a roll, pitch, and yaw rotation sequence can be

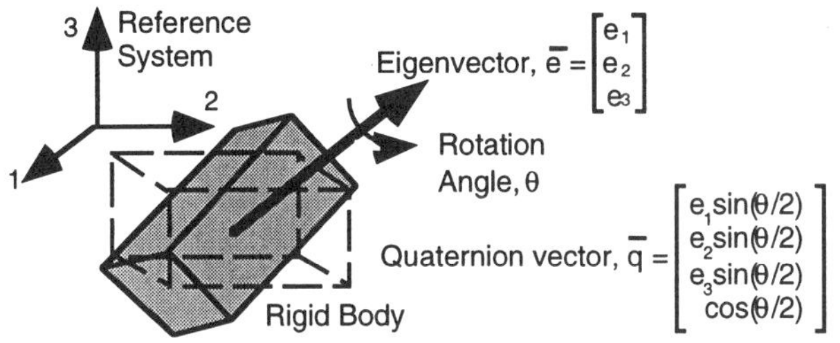

Figure 2: Euler Parameters in the quaternion vector describe the rotation of a rigid body about the eigenvector.

replaced by a single rotation. The Euler parameter set consists of four elements arranged in the quaternion vector in Figure 2.

Spacecraft can use thrusters, magnetic devices, and momentum exchange devices to control attitude. This paper is concerned with a particular kind of momentum exchange device known as a reaction wheel [Wertz, 78]. Reaction wheels are used for applications that require smooth and precise pointing at moderate to low torques.

Figure 3 shows an arrangement of three reaction wheels mounted along the X, Y, and Z axes of the spacecraft reference frame. The three wheels are operated in combination to provide a net torque in any direction. The torque produced by each wheel is limited by capabilities of the reaction wheel motor. An envelope defining the maximum values for the vector sum of the wheel torques is shown in the right hand portion of Figure 3. For example, if reaction wheel 1 exerts the maximum negative torque on the spacecraft and reaction wheels 2 and 3 exert maximum positive torques, then the summed torque is the point marked (-1, 1, 1). Reaction wheel speeds are limited by vibration and bearing wear. These limits impose constraints on the spacecraft angular rate which are also characterized by the cubic surface in Figure 3.

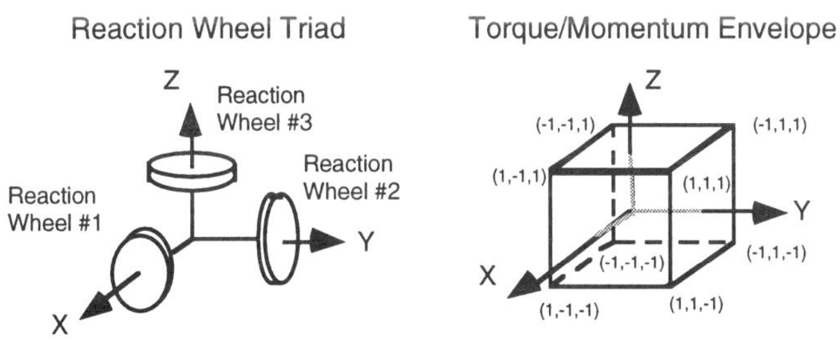

Figure 3: Reaction wheels and the torque/ momentum envelope.

A set of nonlinear differential equations describe spacecraft motion subject to reaction wheel torques. The equations of motion simplify under the following conditions: 1) the spacecraft inertia is equal in all directions, and 2), no external torques act on the spacecraft. The first condition is true for homogeneous spheres or cubes. This is a rough, but often used approximation to true spacecraft mass properties. The second condition is true when the spacecraft does not use thrusters to control attitude and when disturbing torques arising from atmospheric drag or solar pressure can be neglected. If these two conditions are true and if the spacecraft starts with an initial total angular momentum (angular momentum the wheels and the spacecraft body), the spacecraft equations of motion are given in equations 1a through 1g [derived from Wertz, 1978]. These equations constitute the dynamic model of the spacecraft.

$$\begin{aligned} \dot{\omega}_x &= u_x \\ \dot{\omega}_y &= u_y \\ \dot{\omega}_z &= u_z \\ \dot{q}_1 &= \tfrac{1}{2}(\omega_x * q_4 + \omega_z * q_2 - \omega_y * q_3) \\ \dot{q}_2 &= \tfrac{1}{2}(\omega_y * q_4 + \omega_z * q_3 - \omega_z * q_1) \\ \dot{q}_3 &= \tfrac{1}{2}(\omega_z * q_4 - \omega_z * q_2 + \omega_y * q_1) \\ \dot{q}_4 &= -\tfrac{1}{2}(\omega_x * q_1 + \omega_y * q_2 - \omega_z * q_3) \end{aligned} \tag{1a-1g}$$

ω_x, ω_y, and ω_z are the X, Y, and Z components of the spacecraft angular rate vector. u_x, u_y, and u_z are the control accelerations (wheel torque divided by spacecraft inertia) about the X, Y, and Z axes. Finally, q_1, q_2, q_3, and q_4 are elements of the quaternion vector of Euler parameters.

Minimum time attitude maneuvers are a problem of optimal control. The mathematical solution to the minimum time optimal control problem is a vector of nonlinear differential equations with unknown boundary conditions. Although the numerical solution to the problem is difficult to solve, the optimal control formulation provides some important insights. The optimal controls without spacecraft rate constraints are always saturated (either maximum positive or negative acceleration) [Bilimoria and Wie, 93]. Furthermore, the optimal control law can be expressed as a switching function which itself is a function of the spacecraft attitude and angular rate. Once the switching function is known, the optimal control can be determined from knowledge of rate and attitude.

For attitude maneuvers restricted to a single axis of rotation, the minimum time control is the bang-bang control law shown in equation 2 [Bryson and Ho, 75]:

$$u = \begin{cases} \mathrm{u}_{\max} \text{ if } -\theta > \dfrac{\omega|\omega|}{2\mathrm{u}_{\max}} \\ -\mathrm{u}_{\max} \text{ if } -\theta < \dfrac{\omega|\omega|}{2\mathrm{u}_{\max}} \end{cases} \tag{2}$$

θ is the angle of rotation about the eigenaxis and the inequality inside the bracket is the switching function. Switching function control is illustrated in Figure 4. For θ greater than the switching curve, a negative control acceleration forces the spacecraft angle and angular rate along a trajectory which intersects the switching curve. Upon reaching the curve, the control acceleration changes sign, modifying the angle/angular rate trajectory. The switching process repeats until angle and angular rate are forced to zero. Koza uses GP to solve the linear analog to the minimum time eigenaxis rotation problem [Koza, 92].

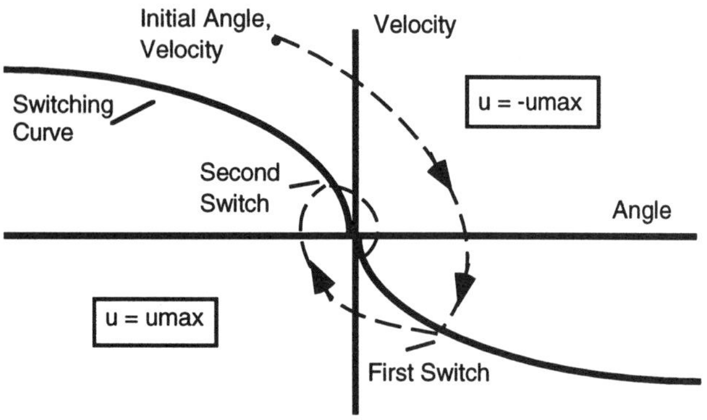

Figure 4: The bang-bang control switching function.

3. Problem Definition

The relationship between the control law and spacecraft motion is shown in the block diagram in Figure 5. The switching function control law compares the current attitude and rate against desired values and issues acceleration commands. The reaction wheels convert the acceleration commands into torques which affect spacecraft motion per equations 1a through 1g.

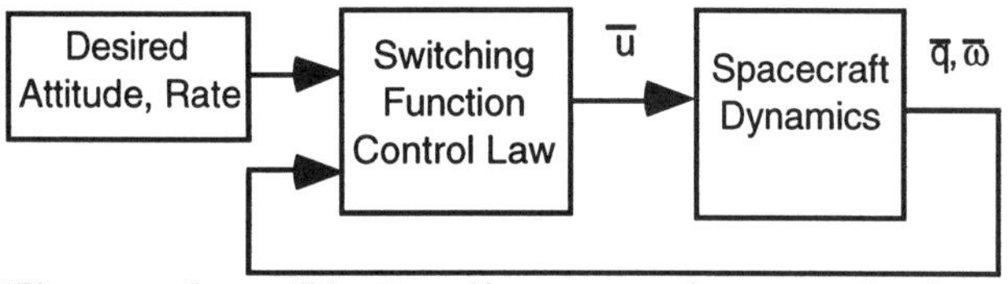

Figure 5: Block diagram of control law implementation.

Table 1 summarizes the physical parameters of the system. The maximum spacecraft acceleration and velocities are a function of the reaction wheel torque and momentum capabilities and spacecraft inertia. The values in the table are representative of a small (800 Kg) spacecraft.

Table 1: Spacecraft Simulation Parameters.

Parameter	Values	Comments
Max Acceleration	0.004 rad/sec^2	1 N-m reaction wheel torques and 250 Kg-m^2 per axis spacecraft inertia.
Max Velocity	0.06 rad/sec (unlimited in rest-to-rest maneuvers)	15 Kg-m^2/sec reaction wheel momentum capacity.
Simulation Termination Condition	Attitude and velocity error magnitudes within limits	Error limits varied over course of run to reduce simulation time. See discussion in Section 4.
Integration Time Step	0.1 to 0.5 sec. Modified Euler integration	Varied with termination condition. See discussion in Section 4

The genetic program operates on a population of control laws. Each individual control law is a character string or S-expression which can be interpreted to calculate a value for the acceleration command. The S-expression is a structured listing which relates a group of problem specific function and terminal operations. An example S-expression of the bang-bang control law is represented as a parse tree in Figure 6. In the figure, the problem specific functions are circled.

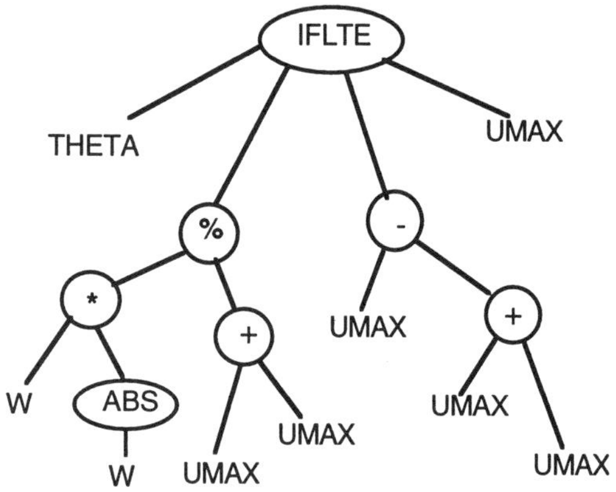

Figure 6: Parse tree for bang-bang control law

IFLTE is the "if less than equal" function. If the first argument is less than the second argument IFLTE returns the third argument, otherwise it returns the fourth argument. Other problem specific functions are add, subtract, multiply, protected divide (%), and absolute value (ABS). The parse tree terminals THETA, W, and UMAX return numerical values.

Table 2 summarizes the tailored genetic program for the minimum time attitude maneuver problem. The terminal set includes problem specific parameters and variables. THETA is the eigenaxis angle of rotation required to take the vehicle from its current orientation to the final desired orientation. UMAX is the maximum acceleration capability of each reaction wheel. WMAX is the maximum spacecraft angular rate about each axis (unused in the rest-to-rest maneuver run). WMAG is the magnitude of the current angular velocity. E1I, E2I, E3I are the X, Y, and Z components of the eigenvector. Similarly, W1I, W2I, W3I, and WF1I, WF2I, WF3I are the X, Y, and Z components of the current and desired final angular velocity. The E*I, W*I, and WF*I terminals are indexed with the evaluation call as described in Section 4. The function set in Table 2 includes basic arithmetic functions the ABS, the IFLTE conditional operator and the automatically defined functions ADF0, ADF1, and ADF2. The automatically defined functions [Koza, 92] used only arithmetic functions. "--" is a negation operator that returns the value of the input argument multiplied by -1. Although the negation function is not required for closure, it was convenient for crafting "expert" individuals against which genetically derived individuals are compared.

The wrapper transforms the vector of acceleration commands output by an S-expression to one of the eight vertices of the torque envelope shown in Figure 3. The

Table 2: Tableau for Near-Minimum-Time Attitude Maneuver Problem.

Objective:	Find a switching function control law to slew a spacecraft from an arbitrary orientation and angular rate to a desired orientation and angular rate in minimum time.
Terminal Set:	THETA, UMAX, WMAX, WMAG, E1I, E2I, E3I, W1I, W2I, W3I, WF1I, WF2I, WF3I
Function Set:	+, -, *, %, --, ABS, IFLTE, ADF0, ADF1, ADF2
Fitness Cases:	7 (RLNZ) or 8 (rest-to-rest) maneuvers with different initial attitude and angular rates and different final rates. The final attitude is always the {0, 0, 0, 1} quaternion. See Table 3.
Raw Fitness:	The sum of the time required to complete the maneuver (or to time out) plus a variable penalty for terminal attitude and velocity errors. See discussion in Section 4.
Standardized Fitness:	Same as raw fitness.
Hits:	Attitude and velocity error magnitudes within limits before time out.
Wrapper:	Converts a positive return value to UMAX and a negative return value to -UMAX. Control is set to zero if the acceleration violates velocity constraints.
Parameters:	Rest-to-Rest : 5000 Individuals (one seeded individual) for 51 generations. RLNZ: 10000 Individuals (two seeded individuals) for 74 generations. Genetic Operations: Node crossover 80%, Leaf crossover 10%, Copy 10%
Success Predicate:	None

wrapper thus ensures that each S-expression complies with the acceleration constraints in Table 1. For the rate limited maneuvers, the wrapper performs an additional check for violation of the Table 1 rate limits. If the angular velocity about a particular axis is at saturation and the commanded acceleration would cause the vehicle to exceed rate limits, the wrapper sets the command to zero.

Fitness evaluation for the minimum time control problem requires numerical integration of the equations of motion (equations 1a to 1g) over a number of fitness cases. The S-expression under evaluation provides three acceleration commands which are integrated to compute the spacecraft's attitude and rate. At each time step, the eigenvector direction, eigenaxis angle of rotation, and angular velocity magnitude are calculated. The integration continues until either the spacecraft attitude and rate are within an error bound of their final values or the time out period has expired.

The fitness cases for both the rest-to-rest and RLNZ maneuvers are given in Table 3. Each maneuver is described by the spacecraft's initial and desired final attitude and attitude rate. The initial and final attitude rates for the rest-to-rest maneuvers are zero and are not included in the table. Rate dimensions in the table are in radians per second. The fitness cases also include a time limit (in seconds) to perform the maneuver. Fitness cases were chosen over a range of initial attitude and rates with the hope that the genetic programming solution will generalize to work for any maneuver.

The GP software is a modified version of "DGPC" written in C by David Andre and provided by the Stanford Department of Computer Science. GP runs were performed on an IBM RS6000 workstation.

4. Time Saving Techniques

Solutions for the rest-to-rest maneuver were not found with population of sizes less than 5000 individuals. Solutions for the RLNZ problem were found using 10,000 and 15,000 individuals. Because of the large population sizes and the numerical integration required for each fitness evaluation, a great deal of effort went into limiting computer processing time.

The first time saving technique used was the inclusion of Automatically Defined Functions (ADFs) in the problem function set. Automatically Defined Functions can greatly reduce the number of fitness evaluations required to find a solution [Koza, 92]. The Rest-to-Rest maneuver runs used two ADFs and the RLNZ maneuver runs used three ADFs.

The fitness evaluation requires a vector of three control accelerations at each time step. One way to get three different values from a single S-expression is use a structure with multiple Result Producing Branches (RPBs) [Koza, 92]. A difficulty with multiple RPBs is that the genetic operations are spread out over three branches, so the GP typically runs for many generations before a solution is found. Furthermore, multiple RPBs do not ensure symmetry. For instance, the physics of this problem suggest that the controls required to maneuver the spacecraft 90 degrees about the X axis would also apply to a 90 degree rotation about the Z axis.

To overcome difficulties with multiple RPBs, a vector indexing scheme was devised. A single individual is evaluated once for each control axis. After each evaluation, an index is incremented so that in the indexed terminals, E*I, W*I, and WF*I return different values. For instance, W1I returns the X, Y, and Z components of angular velocity in the first, second, and third evaluation of the individual. Likewise, W2I returns the Y, Z, and X components of angular velocity in the first, second and third evaluation of

Table 3: Fitness Cases.

Fitness Case	Rest-to-rest Maneuvers		Rate Limited Non-Zero Terminal Velocity Maneuvers			
	Euler Parameters	Time	Euler Parameters	Initial Rates	Final Rates	Time
1	{0.0,0.0,1.0,0.0}	120	{0.408,0.408,0.408,0.707}	{0.04,0.03,0.02}	{-.02,0.03,-.04}	150
2	{0.707,0.0,0.0,0.707}	60	{0.408,0.408,0.408,0.707}	{0.02,0.04,0.03}	{-.02,-.04,-.03}	150
3	{0.0,0.707,0.0,0.707}	60	{-.707,0.0 ,0.0 ,0.707}	{0.00.0.06,0.00}	{-.06,0.00,0.00}	75
4	{-.707,0.0,0.0,0.707}	60	{0.0 ,0.6 ,0.0 ,0.8 }	{0.00,-.06,0.00}	{0.00,0.06,0.00}	60
5	{0.0,-.707,0.0,0.707}	60	{0.0 ,0.0 ,1.0 ,0.0 }	{0.06,0.00,0.00}	{0.00,0.00,-.06}	60
6	{0.408,0.408,0.408,0.707}	40	{0.306, 0.230,0.0 ,0.924}	{0.00,0.00,0.00}	{0.00,0.06,0.00}	60
7	{0.408,-.408,0.408,0.707}	40	{0.0 ,0.0 ,0.0 ,1.0 }	{0.06,0.00,0.00}	{0.00,0.06,0.00}	45
8	{-.408,0.408,-.408,0.707}	40	No RLNZ Case 8			

the individual. This technique enforces symmetry and reduces the processing required to find a solution.

Another method for reducing computer time was to adjust the fidelity of the numerical computation as the population evolves. The integration step size controls numerical accuracy and the granularity of the position and velocity terminal errors. The rest-to-rest attitude maneuvers used an integration step size of 0.25 seconds with a termination error bound of 5 degrees and 1 degree per second until an individual scored eight hits. The time step was then reduced to 0.1 second and the error bounds reduced to 2 degrees and 0.2 degrees/second. The rate limited maneuvers are slower and used a larger integration time step. The RLNZ time step varied from 0.5 to 0.2 seconds and the error bounds varied from 5 degrees to 1 degree and from 1 to 0.5 degrees per second.

The fitness measure was also adjusted during the course of the run. During the early portion of the runs, fitness included both the maneuver elapsed time and a penalty for errors in the final attitude and rate. The penalty function promoted individuals which were close to the target attitude and rate after the maneuver time limit. After an individual scored the maximum number of hits, the fitness measure was changed to the maneuver elapsed time only.

Yet another method of reducing simulation times was to evaluate only a subset of the fitness cases during the early portion of the run. For both the rest-to-rest and the non-zero terminal velocity maneuvers, the most difficult maneuvers were identified and solved first (case 1 in the rest-to-rest and cases 1 and 2 in the RLNZ). Once an individual scored a hit on these maneuvers, the populations were exposed to the remaining fitness cases.

5. Results

Expert individuals were created for both types of maneuvers to provide a reference for evaluating the GP results. The expert individuals are shown in Figure 7. Both are related to the single axis bang-bang control law. The expert individual for the rest-to-rest maneuver is similar to the bang-bang switching function (Figure 6) but the angle of rotation is multiplied by the eigenvector component about the axis of control (`E1I`). The expert individual for the RLNZ maneuvers adds a correction to the switching function angle (θ in equation 2) to compensate for the non-zero final velocity.

Genetic programming results for the rest-to-rest maneuvers are summarized in Table 4. The table compares the maneuver times of each fitness case for the optimal solution, the expert individual, and the GP best individual. The optimal solution was calculated from results in Bilimoria and Wie but modified by the acceleration constraint in Table 1 and simulation termination when with in 2 degrees and 0.2 degree/second of the final orientation and rate. The expert individual times are within 4.2% of the optimal solution. Initial GP runs found individuals that scored eight out of eight possible hits, but could not outperform the expert individual. To see if GP could improve the expert solution, a GP was run with an initial population (5000 individuals) seeded with a single copy of the expert individual. The GP solution improves on expert the individual's performance to within 2% of the optimal solution. The improvement is greatest for case 1 which is a 180 degree rotation about the Z axis (see Table 3).

Rest-to-rest Expert Individual:

```
RPB_Num 0
   (ADF0(THETA)(E1I)(W1I)(UMAX))
ADF_Num 0
   (IFLTE(*(ARG0)(ARG1))(%((ARG2)
   (ABS(ARG2)))(+(ARG3)(ARG3)))
   ((ARG3)(+(ARG3)(ARG3)))(ARG3))
ADF_Num 1
   (-(ARG0)(+(ARG0)(ARG0)))
```

RLNZ Expert Individual:

```
RPB_Num 0
   (IFLTE(-(*(THETA)(E1I))(IFLTE(WF1I)
   (-(UMAX)(UMAX))(--(ADF1(WF1I)(W1I)
   (UMAX)))(ADF1(WF1I)(W1I)(UMAX))))
   (ADF0(W1I)(WF1I)(UMAX)(UMAX))
   (--(UMAX))(UMAX))
ADF_Num 0
   (%(*(-(ARG0)(ARG1))(ABS(-(ARG0)
   (ARG1))))(+(ARG2)(ARG2)))
ADF_Num 1
   (ABS(%(*(ARG0)(-(ARG0)(ARG1)))
   (ARG2)))
ADF_Num 2
   (-(ARG0)(+(ARG0)(ARG1)))
```

Figure 7: Rest-to-rest and RLNZ expert individuals.

Table 4: Rest-to-Rest Maneuver Times (sec).

Fitness Case	*Optimal Solution	Expert Individual	Best Individual
1	50.43	55.2	51.5
2	37.43	39.0	37.9
3	37.43	39.0	37.9
4	37.43	39.0	37.9
5	37.43	39.0	37.9
6	29.26	29.7	29.9
7	29.26	29.7	29.9
8	29.26	29.7	29.9
Total (Hits)	**287.93**	**300.3 (8/8)**	**292.8 (8/8)**

***Bilimoria and Wie results modified by acceleration limits and simulation termination criteria**

Figure 8 shows the rest-to-rest maneuver average and best individual fitness over the course of the seeded GP run. Despite the expert individual seed, best individual performance does not approach 300 seconds until generation 10 because the time varying fitness measure discussed in Section 4. The solution begins to stagnate after generation 23 The entire 50 generation run required about 83 hours of CPU time on an RS 6000 computer.

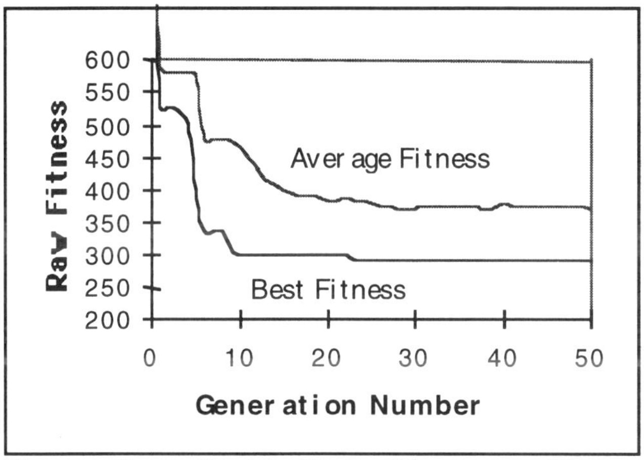

Figure 8: Rest-to-Rest Maneuver Fitness Evolution.

Results for the RLNZ maneuver GP are summarized in Table 5. The difficulty of this problem is apparent in the poor performance by the RLNZ "expert" individual (Figure 7), which could solve only four of the seven fitness cases. The so called expert was derived through trial and since a suitable example could not be found in the surveyed literature. The best of run individual scored a perfect 7 out of 7 hits. The best individual's lower fitness is achieved by solving the maneuvers that the expert individual could not solve. Against maneuvers that the expert does solve, the best individual provides only minor improvement. Optimal solutions to the RLNZ maneuvers are unknown and therefore not included in Table 5.

Table 5: RLNZ Maneuver Times (sec).

Fitness Case	Expert Individual	Best Individual
1	**150.	70.25
2	**150.	76.75
3	61.0	60.75
4	51.0	51.0
5	**60.	50.75
6	45.5	45.25
7	37.5	36.5
Totals (Hits)	****555.0 (4/7)**	**391.25 (7/7)**

**** Run timed out - no solution.**

The GP solution to the RLNZ problem required a two step procedure. After designing the expert individual in Table 5, the relative difficulty of the first two fitness cases was recognized. The first GP run used a population of 15,000 individuals to solve only the first fitness case. After twelve generations at four CPU hours, a single solution to the first fitness case was found. The next step repeated the GP but used an initial population of 10,000 including two seeded individuals: the best individual from the initial run, and the expert individual in Table 5. The population was exposed to successive fitness cases after achieving a threshold number of hits as described in Section 4. After ten generations an individual with seven out of seven hits was discovered. The run was terminated after 74 generations.

Figure 9 shows the RLNZ maneuver average and best individual fitness over the course of the run. The solution appears stagnated after generation 25 until a 60 second drop in fitness at generation 68. The 74 generation run required about 340 CPU hours on an RS 6000 computer.

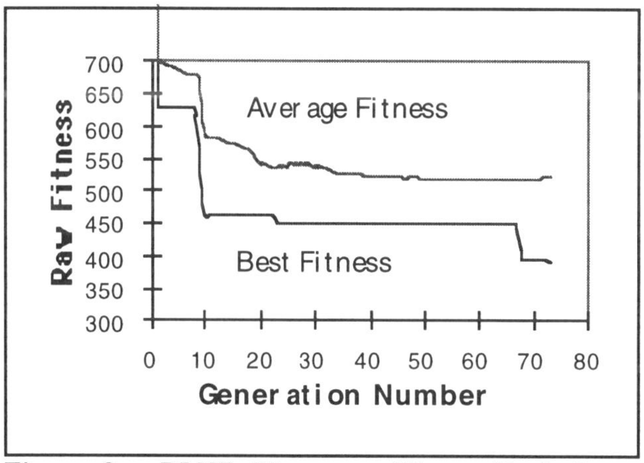

Figure 9: RLNZ Maneuver Fitness Evolution.

Figure 10 shows time histories of the acceleration command, the angular velocities, and the Euler parameters for the case 1 rest-to-rest maneuver (Table 3) . The maneuver is a 180 degree rotation about the spacecraft Z axis. At initiation, the Z axis is aligned as desired but the spacecraft is facing in the opposite direction. The obvious solution is a simple eigenaxis rotation: spin the spacecraft 180 degrees about the Z axis and follow the bang-bang

Optimal Solution Results
(Time 50.43 seconds)

Best Individual Results
(Time 51.5 seconds)

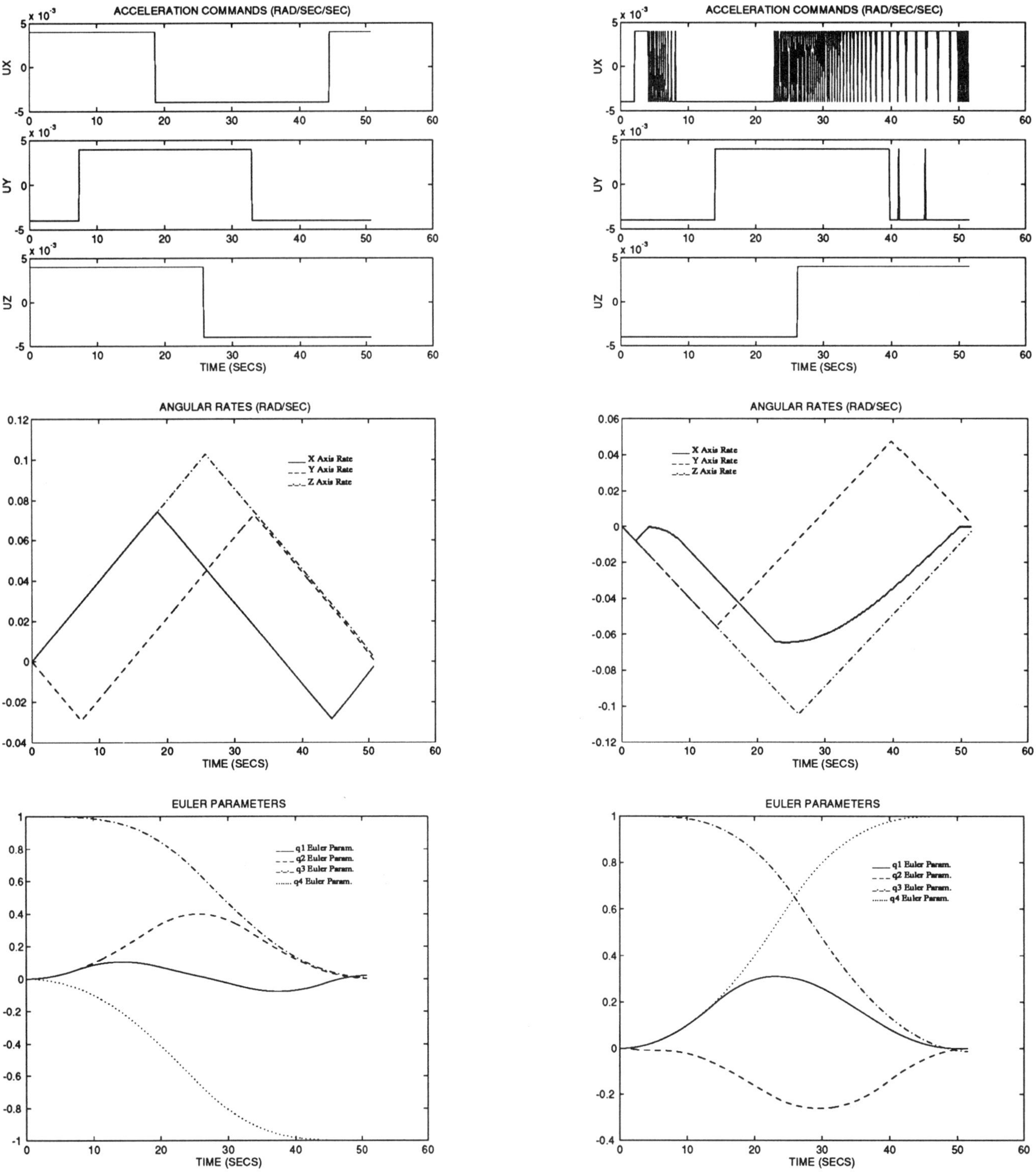

Figure 10: Optimal and best individual solutions to the 180 degree Z-axis rest-to-rest maneuver.

control switching logic (equation 2) to determine when to accelerate and decelerate vehicle. However, Bilimoria and Wie show that the a simple eigenaxis rotation is not the optimal solution. Instead, as shown at the top left of Figure 9, torques are applied to accelerate all three axes. The X and Y axis torques tilt the spacecraft Z axis away from the original desired direction up to 45 degrees during the course of the rotation. The tilt is manifest in the X and Y axis angular rates and the q1 and q2 Euler parameters in the middle and bottom left plots in Figure 9. The Z axis tilt allows the spacecraft to increase the total acceleration (vector sum of X Y and Z accelerations) in the desired direction of motion and hence reduce maneuver time. Note that at the end of the maneuver, the angular rates and the q1 and q2 Euler parameters return to zero. Note also that the desired orientation, q1 through q3 zero and q4 at +/- 1 is also achieved. The acceleration command is always saturated at +/- 0.004 rad/sec/sec, which is consistent with the GP wrapper described in Section 3.

The rest-to-rest maneuver best individual results are shown to the right of Figure 10. The results lack the symmetry of the optimal solution but does accelerate all three axes and tilt the Z axis by 45 degrees during the course of the maneuver. The GP was able to evolve complex dynamic behavior. The X axis acceleration command dithers at high frequency over parts of the maneuver. This reduces average acceleration and increases maneuver time. However, the maneuver achieves the desired final rates and orientation. The difference in the final quaternion between optimal and best individual solutions ({0,0,0,-1} vs. {0,0,0,1}), is the result of clock wise versus counter-clock wise rotations to the desired orientation.

Figure 11 shows the RLNZ best individual acceleration and rates for the 180 degree Z axis rest-to-rest maneuver. Maneuver time is necessarily longer than in the previous solutions because of the rate limits imposed by reaction wheel constraints. The rate limit holds the Z axis acceleration command to zero for over half of the maneuver. The maximum angular velocity is thus kept to 0.06 radians per second (versus over 0.1 radians per second in Figure 10).

The generality of the GP derived solutions was investigated by testing the best of run individuals against a randomly generated set of 100 fitness cases. Results are summarized in Table 6. The initial quaternion vectors were normalized to ensure a unity magnitude. Rest-to-rest maneuver results are at the top of the table. Both the expert and GP best individuals scored a perfect 100 out of 100 hits. Total time to complete the 100 maneuvers was 547 seconds less for the GP best individual, a 10% improvement over the expert solution.

RLNZ results are shown in the bottom half of the table. Simulation runs were repeated with different limits on initial and terminal velocities. In all cases, maximum vehicle rates over the course of the maneuver were limited to +/- .06 rad/sec. GP solution performance degrades with increasing terminal velocities. Furthermore, the RLNZ best individual does not perform as well as the RLNZ expert (Figure 7). This indicates that the RLNZ best individual was over specialized and that seven fitness cases are too few for a general solution. A general RLNZ solution will require more fitness cases and a commensurably greater investment of computer CPU. The fact that a general solution for the rest-to-rest maneuver was found with only eight fitness cases demonstrates the difficulty of the RLNZ maneuver problem.

Table 6: GP results with random fitness cases.

Maneuver	Expert Indv. Results ([1]Fitness/Hits)	Best Indv. Results (Fitness/Hits)
[2]Rest-to-Rest	5232 / 100	4685 / 100
[3]RLNZ:		
+/- 0 rad/sec	5249 / 100	5249 / 100
+/- .02 rad/sec	8716 / 91	9011 / 90
+/- .04 rad/sec	14132 / 52	14820 / 48
+/- .06 rad/sec	17063 / 30	17051 / 39

[1]Fitness is total time in seconds to complete 100 maneuvers. Hits are total scored out of a max of 100.
[2]Rest-to-Rest maneuvers are not rate limited.
[3]Values are uniform distribution of terminal velocity values.

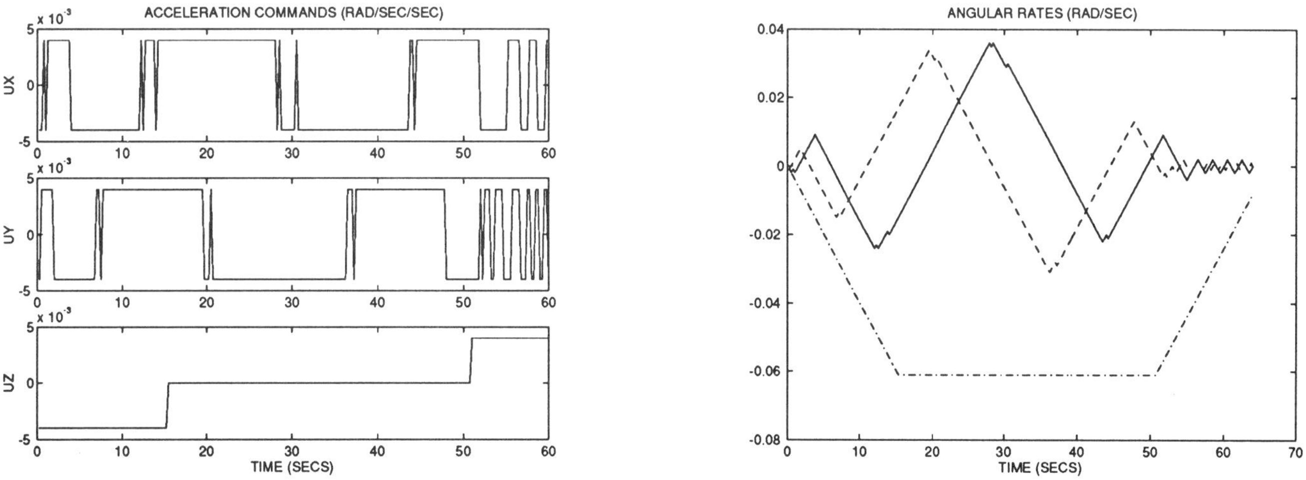

Figure 11: RLNZ best individual solution to 180 degree Z-axis rest-to-rest maneuver (64.0 sec).

6. Conclusions

Genetic Programming (GP) was used to find near-minimum-time control laws for two classes of spacecraft attitude maneuvers: rest-to-rest and rate-limited, non-zero terminal velocity (RLNZ) maneuvers. Hand crafted expert solutions to the control problem were created to provide a basis for evaluating GP results. In addition, optimal solutions to the rest-to-rest maneuvers were derived from the previously published results [Bilimoria and Wie, 1993].

The GP solution for the rest-to-rest maneuvers is within 2% of the optimal result and generalizes to solve 100 randomly generated maneuvers.

RLNZ maneuvers permit non-zero initial and terminal rates and constrain maximum vehicle angular rates to comply with actuator limitations. A GP solution was found that successfully solves all evaluated fitness cases. However, the solution did not generalize to solve randomly generated fitness cases.

A number of techniques were used to reduce processing time requirements. An indexed terminal set was created to enforce symmetry over three axes of control. This technique should be applicable to a number of engineering problems.

Acknowledgments

The author wishes to thank Dr. John Koza for his guidance, support, and instruction, and David Andre for providing the GP code that made this work possible. The author would also like to thank Sandy Graves for her valued help in editing and clarifying this document.

Bibliography

Bilimoria, Karl D. and Wie, Bong. 1993. Time-Optimal Three-Axis Reorientation of a Rigid Spacecraft. *Journal of Guidance, Control, and Dynamics*. Volume 16. Pages 446-452.

Bryson, Arthur E. and Ho, Yu-Chi. 1975. *Applied Optimal Control*. Washington DC: Hemisphere Publishing.

Koza, John R. 1992. *Genetic Programming: On the Programming of Computers by Means of Natural Selection*. Cambridge, MA: The MIT Press.

Scrivener, Sandra L. and Thompson, Roger C. 1994. Survey of Time-Optimal Attitude Maneuvers. *Journal of Guidance, Control, and Dynamics.* Volume 17. Pages 225-233.

Wertz, James R. (editor). 1978. *Spacecraft Attitude Determination and Control.* Dordrecht, The Netherlands: Kluwer Academic Publishers.

Evolving Evolution Programs: Genetic Programming and L-Systems

Christian Jacob

Lehrstuhl für Programmiersprachen
Martensstr. 3, Universität Erlangen-Nürnberg
D-91058 Erlangen, Germany
jacob@informatik.uni-erlangen.de
http://www2.informatik.uni-erlangen.de/IMMD-II/Persons/jacob

ABSTRACT

Parallel rewrite systems in the form of string based L-systems are used for modeling and visualizing growth processes of artificial plants. It is demonstrated how to use evolutionary algorithms for inferring L-systems encoding structures with characteristic properties. We describe our *Mathematica* based genetic programming system *Evolvica*, present an L-system encoding via expressions, and explain how to generate, modify and breed L-systems through simulated evolution techniques. Extensions of genetic programming operators and expression generation methods strongly relying on templates and pattern matching are shown by example.

1. Modeling growth processes by L-systems

The formation of structures in nature is always the result of complex growth processes. One idea of evolutionary biology is that structure formation can be interpreted as the execution of »programs« or rewrite rules. These programs are highly parametrized where the parameters are set by the environment in which development and interaction processes take place. Generally, rewriting has proved to be a useful technique for defining complex objects by successively replacing parts of simple initial objects using a set of rewrite rules or productions. In the scope of this article we will focus on a special type of character based rewrite systems, commonly termed L-systems (Lindenmayer systems), which are used in theoretical biology for describing and simulating natural growth processes, Lindenmayer (1968). All letters in a given word are replaced in parallel and simultaneously. This feature makes L-systems especially suitable for describing fractal structures, cell divisions in multicellular organisms (Jacob (1994), Jacob (1995)), or flowering stages of herbaceous plants (Prusinkiewicz and Lindenmayer (1990)), as we will demonstrate in this paper.

D0L-systems (D0 means deterministic with no context) are the simplest type of L-systems. Formally, a D0L-system can be defined as a triple $G = (\Sigma, P, \alpha)$ where $\Sigma = \{\sigma_1, \ldots, \sigma_n\}$ is an alphabet, α, the initial start word commonly referred to as the axiom, is an element of Σ^*, the set of all finite words over the alphabet Σ. The structure preserving D0L-mapping $P : \Sigma \to \Sigma^*$ is defined by a set of contextfree productions or rewrite rules $\sigma \to P(\sigma)$ for each $\sigma \in \Sigma$. Let us consider the following L-system which models a simple growth process:

Σ = { f, sprout, stalk, pd, leaf, bloom}
α : sprout (4)
p_1 : sprout (4) $\to$ f stalk(2) [pd(60) leaf(0) sprout(3)] bloom(0)
p_2 : sprout(t<4) $\to$ sprout(t+1)
p_3 : stalk(t>0) $\to$ f f stalk(t-1)

Applying the rules p_1, p_2, and p_3 first to the start word α, and then repeatedly to the resulting expressions generates the following sequence of words:

α : sprout (4)
$P(\alpha)$: f stalk(2) [pd(60) leaf(0) sprout(3)] bloom(0)
$P^2(\alpha)$: f f f stalk(1) [pd(60) leaf(0) sprout(4)] bloom(0)
$P^3(\alpha)$: f f f f f stalk(0) [pd(60) leaf(0)
f stalk(2) [pd(60) leaf(0) sprout(3)] bloom(0)] bloom(0)
...

In the example L-system given above production p_3 encodes elongation of a stalk segment, and p_1 describes part of a stalk and a bloom, together with another sprout segment enclosing a leaf and branching from the main axis. In order to translate these expressions into spatial structures resembling three-dimensional plant architectures we have to define a geometrical interpretation of the strings. More specifically, we will use a variant of the so-called turtle geometry, the basic idea of which is to interpret some of the alphabet symbols Σ as commands to maneuver a virtual drawing tool, known as a turtle, in 3D space. The turtle can thus change its position and orientation and draw lines or graphical elements while being moved around in three dimensions. Figure 1 shows an animation sequence for the turtle interpretation of a more complex parametrized L-

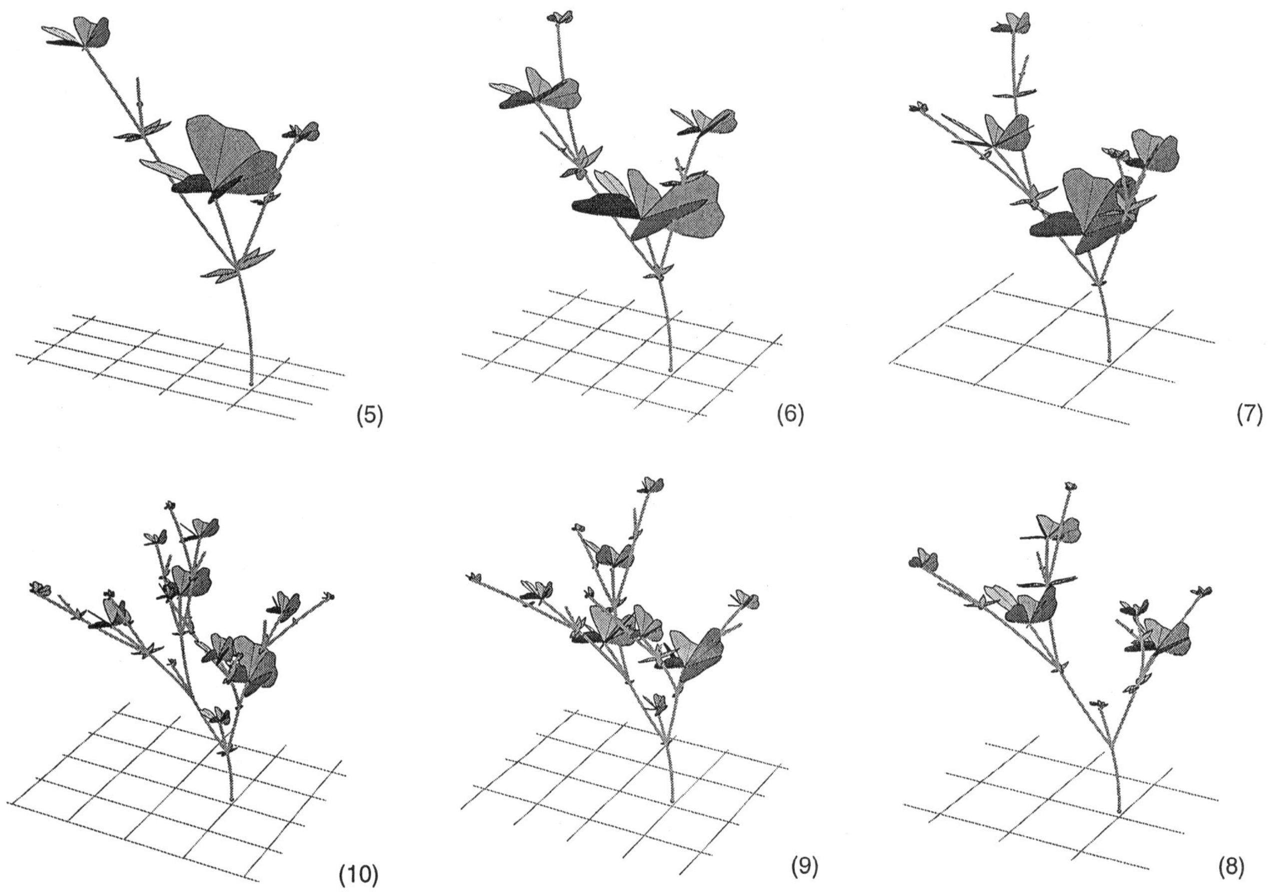

Initial start word: *sprout*(4)

Sprout developing leaves and flower:

p_1 : *sprout*(4) → f *stalk*(2) [pd(60) *leaf*(0)]
pu(20) [pu(25) *sprout*(0)]
[pd(60) *leaf*(0)] pd(20)
[pu(25) *sprout*(2)]
f stalk(1) *bloom*(0)

Riping sprout:

p_2 : *sprout*(t < 4) → *sprout*(t+1)

Stalk elongation:

p_3 : *stalk*(t > 0) → f f *stalk*(t-1)

Changing leaf sizes:

p_4 : *leaf*(t) → *leaf*(t + 1.5)
p_5 : *leaf*(t > 7) → *Leaf*(7)
p_6 : *Leaf*(t) → *Leaf*(t - 1.5)
p_7 : *Leaf*(t < 2) → *leaf*(0)

Growing bloom:

p_8 : *bloom*(t) → *bloom*(t + 1)
p_9 : *bloom*(7) → *bloom*(1)

Figure 1 L-system modeling of growth processes for plantlike geometrical structures

system describing growth sequences of sprouts, leaves and blooms of an artificial flower. Besides the elementary turtle commands this D0L-system encodes macros for generating graphical representations of the leaves, blooms and stalks. All the non-italic terms (f, pu, pd) represent commands to move the turtle (f: forward, b: backward) and change the drawing tool´s orientation by rotation around its longitudinal, lateral, and vertical axes (rl/rr: roll left/right, pu/pd: pitch up/down, yl/yr: yaw left/right), thus translating a one-dimensional string into a 3D-object resembling a plant. In order to be able to generate branching structures a kind of stacking mechanism for the turtle´s position and orientation is necessary. For each string of the form s_1 [s_2] s_3 the strings s_1, s_2, and s_3 are interpreted in sequence, however, before starting the interpretation of s_2 the current turtle position and orientation are pushed on a stack, so that, having finished interpreting s_2, the turtle is reset to its prior coordinates and orientation.

```
1.   LSystem[□ AXIOM,□ LRULES],                                              (1)
2.   AXIOM[sprout[4]],                                                       (1)
3.   LRULES[
       LRule[LEFT[], PRED[ sprout[4] ], RIGHT[],
             SUCC[SEQ[SEQ[f],SEQ[stalk[2]], STACK[PD[60],leaf[0]], ...,
             □□ SEQ,
             SEQ[f],SEQ[stalk[1]], bloom[0]]]],
       LRule[LEFT[], PRED[ sprout[t<4] ], RIGHT[], SUCC[sprout[t+1]]],
       ...
       LRule[LEFT[], PRED[ bloom[6] ], RIGHT[], SUCC[ bloom[1]]],
      □□□ LRule
     ],                                                                      (1)
4.   LRule[LEFT[],□ PRED,RIGHT[],□ SUCC],                                    (1)
5.   PRED[sprout[aIndex]],                                                   (1)
6.   SUCC[□ SEQ | □ STACK],                                                  (1)
7.   SEQ[□□ [□ sprout | □ stalk | □ leaf | □ bloom | f |
        □ YL | □ YR | □ PU | □ PD | □ RL | □ RR | □ SEQ]],                  (1)
8.   SEQ[□□ [□ sprout | □ stalk | □ leaf | □ bloom | f |
        □ YL | □ YR | □ PU | □ PD | □ RL | □ RR ]],                          (4)
     ...
     sprout[sproutIndex], ..., bloom[bloomIndex]                         (1,...,3)
```

Figure 2 Templates for generating L-system encodings

2. Genetic L-system programming

The inference problem for L-systems involves finding a proper axiom and rewrite rules in order to model a given structure formation or growth process. It turns out that the manual synthesis of L-systems modeling morphogenesis of a particular (biological) species is an overall difficult and sometimes tedious process. But what methods do we have at hand for generating L-systems automatically, or at least semi-automatically? As P. Prusinkiewicz points out, random modification of production rules generally gives little insight into the relationship between L-systems and the geometrical structures they encode (Prusinkiewicz and Lindenmayer (1990)). Algorithmic approaches reported in the literature up to now are still too limited to be of practical value for complex structure formation tasks (Prusinkiewicz and Lindenmayer (1990), p. 39 and 62). A partial remedy might be to use evolutionary methods for L-system inference. In the following sections we will discuss our ideas about the use of genetic programming techniques for breeding populations of L-systems that describe growth processes of plant structures which are evaluated with respect to their ability to represent growth processes with special characteristics.

Genetic Programming (GP) has been introduced as a method to automatically develop populations of computer programs through simulated evolution (Koza (1992), Koza (1994)). Considering L-systems as rulebased development programs it is easy to define program evolution. Each program encoded as a symbolic expression has to be interpreted and is assigned a fitness value dependent on the optimization task to be solved. On the basis of these fitnesses the individual programs struggle for „survival of the fittest" and for the chance to become members of the next generation. In order to introduce variations into the program encoding structures genetic operators like mutation or crossover – gleaned from nature's mutating and recombining operators on cell genomes – are applied. The evolution process develops new populations of programs from generation to generation, the interplay of modifying operators and selection hopefully leading to ever better programs.

2.1 Encoding L-systems

In order to use expression evolution for L-systems an encoding scheme has to be defined which will take into account even the more general case of context dependent IL-systems (with I referring to the number of context symbols, Koza (1993) encodes for context-free L-systems only). In context-sensitive IL-systems the rewriting of a letter depends on m of its left and n of its right neighbors, where m and n are fixed integers. These systems are denoted as (m,n)L-systems which resemble context-sensitive Chomsky grammars, but – as L-system rewriting is parallel in nature – every symbol is rewritten in each derivation step; this is especially important whenever there is an overlap of context strings.

Each IL-system rule has the form $l < p > r \rightarrow s$ with l, p, r and s denoting the *left context*, *predecessor*, *right context* and *successor*, respectively. The symbols "<" and ">" separate context and predecessor strings. Thus each rule can be represented by an expression of the form

```
LRule[ LEFT[ l ], PRED[ p ],
            RIGHT[ r ], SUCC[ s ] ].
```

Accordingly, an L-system with its axiom and rule set is encoded by an expression of the form

```
LSystem[ □ AXIOM, LRULES[ □□ LRule] ],
```

where we use a pattern notation with □ F denoting a term of the form F[...], and □□ F representing a non-empty sequence of □ F expressions. So our example L-system of section 1 would be represented as follows:

```
LSystem[AXIOM[sprout[4]],
  LRULES[
    LRule[LEFT[],PRED[sprout[4]],RIGHT[],
      SUCC[f,stalk[3],
           STACK[pd[60],...],bloom[0]]],
    LRule[LEFT[],PRED[sprout[t<4]],RIGHT[],
      SUCC[sprout[t+1]]],
    LRule[LEFT[],PRED[stalk[t>0]],RIGHT[],
      SUCC[f,f,stalk[t-1]]],
```

where L-system bracketing of the form [s] is now represented as `STACK[s]`.

2.2 Stochastic generation of L-system codings

One of the main differences of our GP approach to the GP paradigm introduced by Koza (1992) is the use of high-level building blocks for expression generation as well as modification. Instead of just defining a set of function symbols together with their arities, each expression from a template pool, as depicted in figure 2, serves as a possibly partial description of a genotype, encoding an L-system in our case. Each of these templates (marked with 1., 2., 3., ...) is associated with a set of attributes like, e.g., predicates constraining the set of subexpressions that can be plugged in. Thus the encoded L-system productions are restricted to context-free forms (4., LEFT[], RIGHT[]) with their left-hand sides (4., □ PRED) constrained to *sprout[i]*, with *i* replaced by an integer number from the interval, say, *[0,4]* (5.), and their right-hand sides (4., □ SUCC) defined either as a sequence (SEQ) of expressions (6., 7. and 8.) or a bracketed expression sequence (6., □ STACK).

Each expression is constructed from a start pattern (here: LSystem[□ AXIOM,□ LRULES]) by recursively inserting matching expressions from the expression pool until all pattern blanks – marked by □ – have been replaced by according subexpressions. Of course, one has to take care that this construction loop eventually ends.

Some further remarks should be made about the encoding scheme of figure 2. The templates basically describe the D0L-system of figure 1. However, the additional □□ SEQ pattern within the first LRule expression (3.) enables the generation system to create rule variations by inserting new expression sequences, i.e. one or more expressions of the form SEQ[...]. Accordingly, the L-system encoding might be enhanced by additional rules by replacing the □□□ LRule pattern with a (possibly empty) sequence of LRule[...] expressions. Each sequence of expressions is constructed via alternative templates (7. and 8.), where SEQ[□□ [a|b|c]] denotes an arbitrary sequence of expressions taken from the set {a,b,c} as, e.g., SEQ[a,a,c,b] or SEQ[c,c,c,a,b,a] with the sequence length restricted to, say, between 4 and 6. Whenever there are several templates matching one pattern the expressions are selected with probabilities proportional to their weights (see the bracketed numbers right from the expressions in figure 2, serving as fitness values for the templates). The pattern matching is not unique for the □ SEQ pattern (6.) for which there are two matching templates (7. and 8.), a recursive (SEQ[.... □ SEQ]) and non-recursive version which will be selected with a probability of 0.8, thus implicitly restricting the nesting of expressions.

2.3 Variations on L-system expressions

The following is a small collection of operators used for generating variations on arbitrary expressions (figure 3). As with the expression generating operators the basic idea is to use again templates for controling which subexpressions should be in the scope of each operator:

- *Mutation* replaces a randomly selected subexpression by an expression – with the same head expression – generated from the template pool.
- *Crossover* is a recombination operator between two or more expressions. Subexpressions with the same head are selected within the expressions and interchanged.
- *Deletion* erases expression arguments whenever this is possible according to restrictions of the number of arguments for the selected expression.
- *Duplication* inserts a copy of a randomly selected argument as an additional subexpression.
- *Permutation* interchanges the sequence of arguements of an expression.

These operators are defined for general expressions and are not especially tailored to L-system encodings. Subexpressions are chosen according to operator specific selection schemes based on pattern matching mechanisms. Thus it is, e.g., possible to restrict recombinations to subexpressions with head SUCC, i.e. expressions of the form SUCC[...], which allows to interchange right-hand sides of L-system productions, or permute only LRule expressions thus changing the ordering of rules. With a set of templates defined for each operator the effects of genetic operators can be constrained in a problem specific way which turns out to be important especially in the case of interactive »breeding« of expressi-

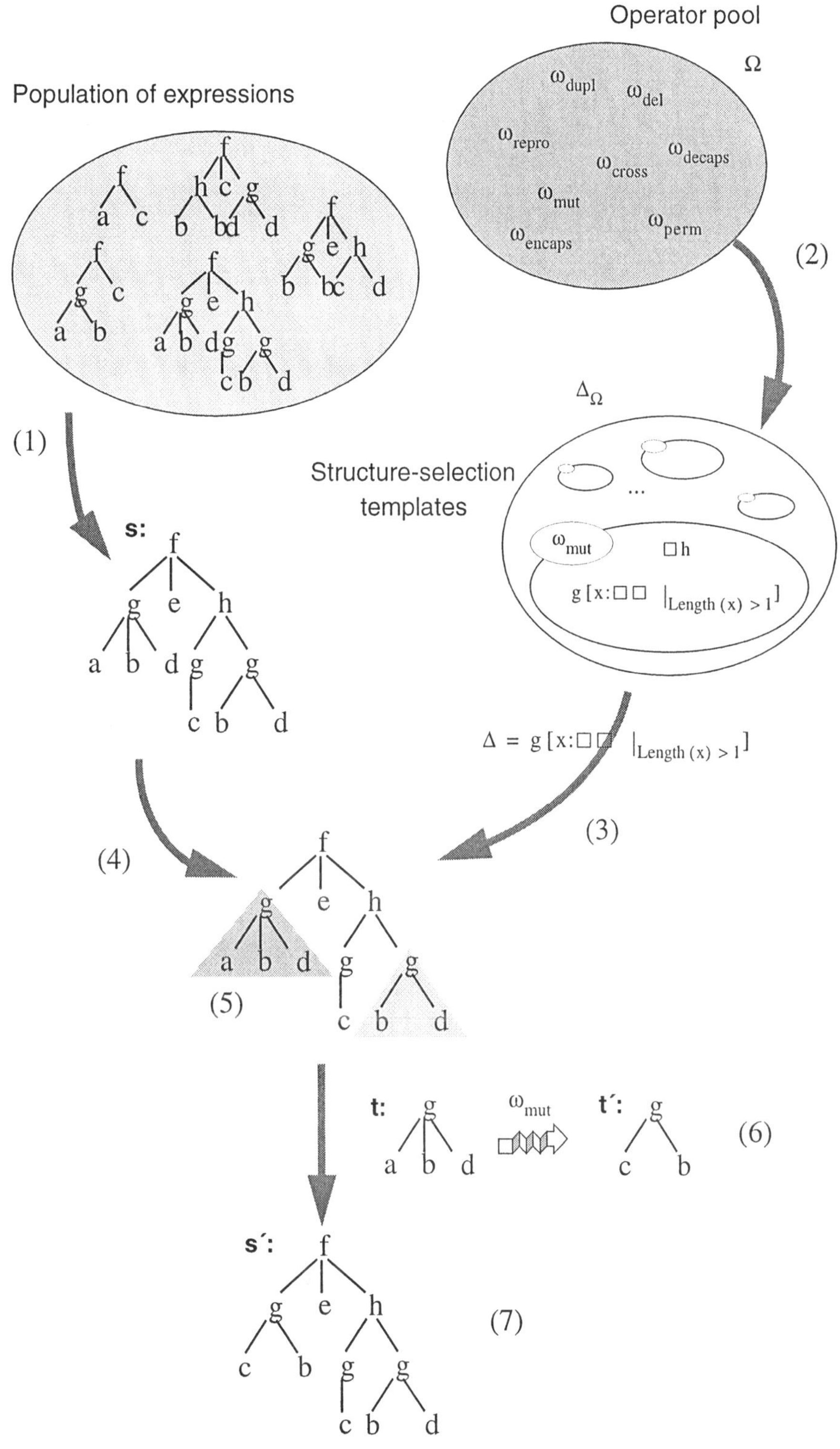

Figure 3 Example of expression variations by using genetic operators with pattern matching capabilities

ons as in the case of evolution-based L-system inference.

Figure 3 shows the general scheme for the application of genetic operators, exemplified by the mutation operator, and how these operators extensively use pattern matching techniques.

(1) An expression s to be modified by a genetic operator is selected from the population of expressions.
(2) An operator – attributed each with a weighting selection probability – is chosen from an operator pool Ω.
(3) The operator to be applied, the mutation operator for this example, then selects an expression template Δ from its operator specific pattern pool Δ_Ω.
(4) According to the Δ-template matching subexpressions of s are identified.
(5) Among the set of those matching subexpressions one is chosen at random.
(6) Finally, the operator specific modifications are applied.
(7) In the case of the mutation operator this means that the subexpression is replaced by a newly generated expression with the same head.

3. Breeding artificial flowers

With all the ingredients described above we are now able to demonstrate how plant like structures with specific characteristics can be developed with the help of L-system evolution. Suppose we want to breed flowers similar to those introduced in section 1. However, suppose we want to breed species which spread out far in x-, y- and z-direction and carry as many leaves and blooms as possible. The fitness function used for the following example evolution takes just this criterion into account and is defined as follows:

$$\text{fitness (plant)} = \sum_{i=0}^{M} \Delta x_i \cdot \Delta y_i \cdot \Delta z_i + 2 \cdot B_i + L_i$$

Δx, Δy, Δz: extension of the geometric plant structure in x-, y-, and z-dimension

B: the number of blooms blossoming

L: the number of leaves the plant carries

M: the maximum number of L-system iterations

The total fitness is computed as the sum of the plant's extension, blooms, and leaves for each L-system iteration. This means that fitness does not only depend on how the plant flourishes at a certain growth stage but on the whole structure generation process from an initial sprout to a full-grown plant.

In the following simple example we start with a small population of six L-system genomes each generated with the templates of figure 2. A maximum number of $M = 6$ L-system iterations is used. Figure 4 shows three stages of a typical evolution sequence over eight generations. The individuals are selected proportional to their fitness. Table 1 lists the set of operators and operator-specific templates that were used for evolutionary L-system synthesis.

Table 1 Genetic operators, their application probabilities, and according sets of templates with template selection probabilities

Operator	Prob.	Selection templates	TProb.
Mutation	0.35	□ LRule	0.2
		□ SEQ	0.4
		□ STACK	0.4
Crossover	0.35	□ LRule	0.2
		□ SEQ	0.4
		□ STACK	0.4
Deletion	0.1	LRULES[□□ |l> 1]	0.2
		SEQ[□□ |l> 1]	0.4
		STACK[□□ |l> 1]	0.4
Duplication	0.1	- ditto	- ditto
Permutation	0.1	- ditto	- ditto

Even with a very limited number of individuals, as in our example, it turns out that the evolution system is able to create rather complex plantlike structures even within only a few generations. Figure 4 shows such a typical evolution run with an exponential increase of the best fitness values ranging from below 40 at the beginning to above 2500 in generation 8. From initially rather small individuals widespread plants evolve carrying bunches of blooms and leaves. On the phenotypical level it is striking that even in the very first generations (Gen. 2) the expressed plantlike structures exhibit a large range of divesity although all L-system encodings are just variants of the basic genotype template of figure 2. Those templates basically ensure, to a certain degree, that plantlike geometrical structures arise whereas the constrained composition of alternative patterns results in different morphogenetic characteristics of the same species or of a totally different species emerging (see, e.g., individuals 4 and 6 of generation 8).

Some of the generated individuals have a very low fitness (gen. 2, indiv. 5,6) carrying no or only a few leaves. However, there are also individuals possessing the characteristic branching structure of the basic plant (figure 1) while trying to meet the fitness criterion by growing an increasing number of blossoming branches. While at the beginning those branches are rather short carrying only a few blooms, finally long branches prevail in generation 5 where bunches of blooms and leaves appear on the bough segments. Fitness is increased by an order of magnitude up to values around 700. Further fitness improvement is achieved by elongating

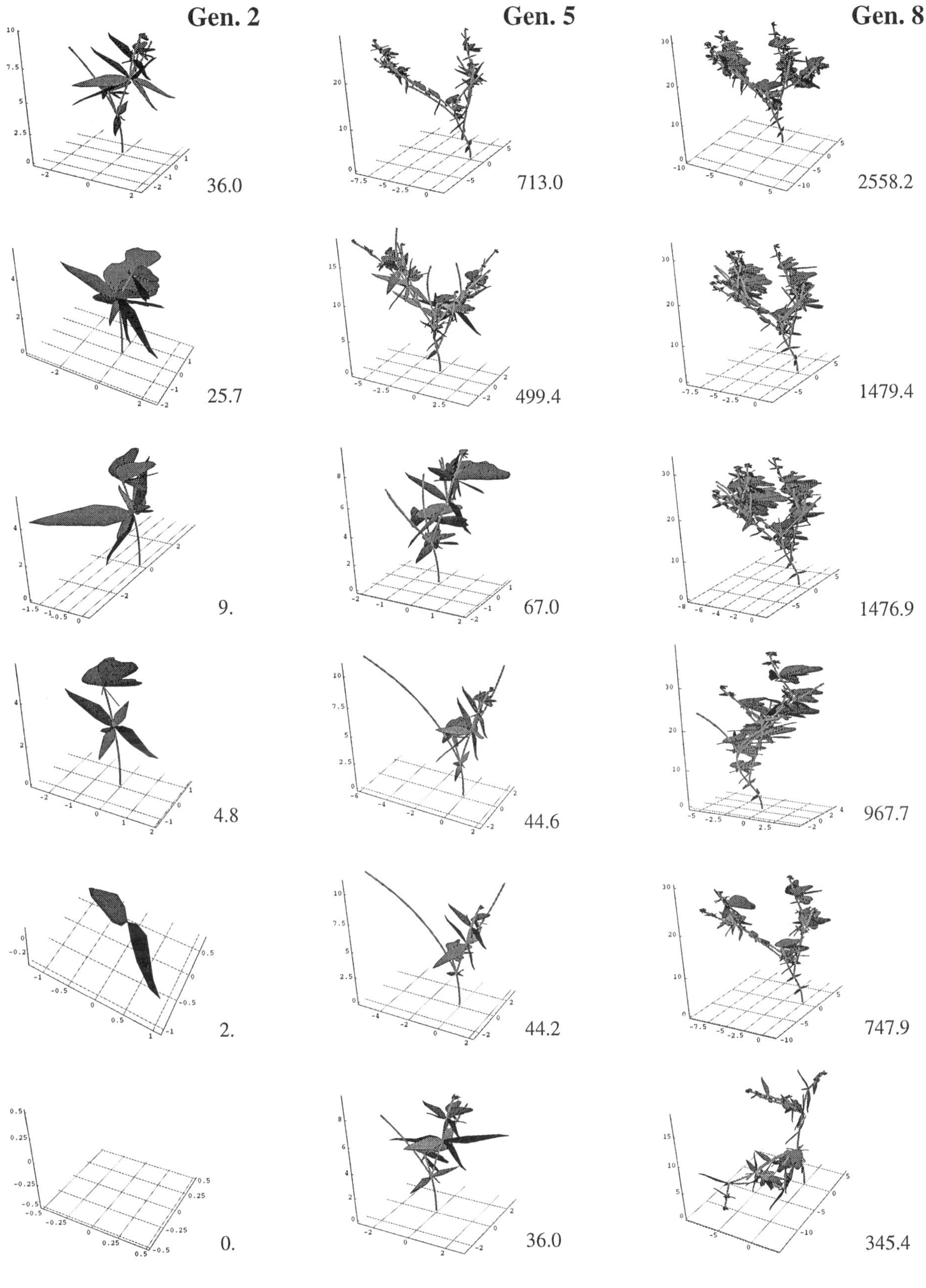

Figure 4 **Evolved plantlike geometrical structures encoded by L-systems**

the stalks as well as shortening the ripening periods. Thus it turns out that even with this very simple example we can observe effects found in natural evolution periods of plant species as well. The rudimentary phylogenetic tree of figure 5 might serve as a demonstration. Over seven generations we can show the phenotypical stages of evolution for the best plant up to the final (8th) generation. Initially, there is a plant with three main branches with the leftmost carrying no leaves or blooms. Its descendants in generations 3 and 4 have grown in width and height and exhibit larger blooms. A decisive evolution step arises with generation 5; now all three main branches (with a smaller branch in the middle) carry a great number of blooms and leaves. In generation 6 even more branches have been created and the species is still enhancing its overall dimension which finally leads to the widespread, flourishing plant structure of generation 8. Figure 5 compares the evolved L-system encodings of generations 2 and 8, where it can be seen that the explained phenotypical effects are, of course, due to an increase of complexity within the genotypes.

Generation 2:

```
Axiom: sprout(4)

sprout(4)     →   f stalk(2) [pd(60) leaf(0)] rr(90)
                  [pd(20) sprout(3)] rr(90) stalk(0)
                  rr(17) leaf(2) [pd(60) leaf(0)]
                  rr(90) [pd(20) sprout(2)] f stalk(1)
                  bloom(0)
sprout(t<4)   →   sprout(t+1)
...
bloom(6)      →   bloom(1)
sprout(2)     →   [stalk(4) yr(34)]
sprout(0)     →   [yr(18) yr(18)]
```

Generation 8:

```
Axiom: sprout(4)

sprout(4)     →   f stalk(2) [pd(60) leaf(0)] f stalk(2)
                  [pd(60) leaf(0)] rr(90) [pd(20) sprout(3)]
                  rr(90) stalk(3) bloom(4) rl(30)
                  [pd(60) leaf(0)] rr(90) [pd(20) sprout(3)]
                  f stalk(1) stalk(1) bloom(0)
                  [pd(20) sprout(3)] rr(90) stalk(0) rr(17)
                  leaf(2) [pd(60) leaf(0)] rr(90)
                  [pd(20) sprout(2)] f stalk(1) bloom(0)
sprout(t<4)   →   sprout(t+1)
...
bloom(6)      →   bloom(1)
sprout(2)     →   [pd(20) sprout(3)]
sprout(0)         [yr(18) yr(18)]
```

Figure 5 Examples of evolved L-systems

4. Conclusion

For all our implementations and experiments described in this article we have used *Mathematica*, a computer algebra system and programming environment based on symbolic expressions (Wolfram (1991)). What makes *Mathematica* especially useful for evolutionary algorithm applications as described above are its capabilities of easy expression manipulation which are important for defining genetic operators and L-system interpretations in a comfortable and flexible way. Finally, without *Mathematica's* graphics and animation tools we would not have enjoyed »playing evolution« and gained so much insight into the fascinating area of L-system design.

Currently we extend our evolutionary L-system inference software to include functional structures of natural genomes (operons, promoters, structure genes) into the L-system encoding thus trying to model influences of intragene control mechanisms on morphogenetic processes. Furthermore, our system will soon be able to simulate co-evolution among plant structures in 3D space which should give more insight into growth and structure formation in competitive environments. The L-system evolution package is part of *Evolvica*, a tutorial and programming environment for evolutionary algorithms, which is accessible via the World Wide Web (see my WWW home page for further details).

Bibliography

Jacob, C. 1994, *Genetic L-System Programming*. Parallel Problem Solving from Nature - PPSN III, Lecture Notes in Computer Science 866, Berlin, Springer.

Jacob, C. 1995, *Modeling Growth with L-Systems & Mathematica*. Mathematica in Education and Research, Vol. 4, No. 3, TELOS Springer. Pages 12-19.

Koza, J. R. 1992. *Genetic Programming: On the Programming of Computers by Means of Natural Selection.* Cambridge, MA: The MIT Press.

Koza, J. R. 1993. *Discovery of rewrite rules in Lindenmayer systems and state transition rules in cellular automata via genetic programming*. Symposium on Pattern Formation – SPF-93. Claremont, CA.

Koza, J. R. 1994, *Genetic Programming II: Automatic Discovery of Reusable Programs*. Cambridge, MA: The MIT-Press.

Lindenmayer, A. 1968, *Mathematical models for cellular interaction in development, Parts I and II*, Journal of Theoretical Biology, 18:280-315.

Prusinkiewicz, P., and Lindenmayer, A. 1990. *The Algorithmic Beauty of Plants*. New York, Springer

Wolfram, S. 1991, *Mathematica, A System for Doing Mathematics by Computer*, New York, Addison Wesley.

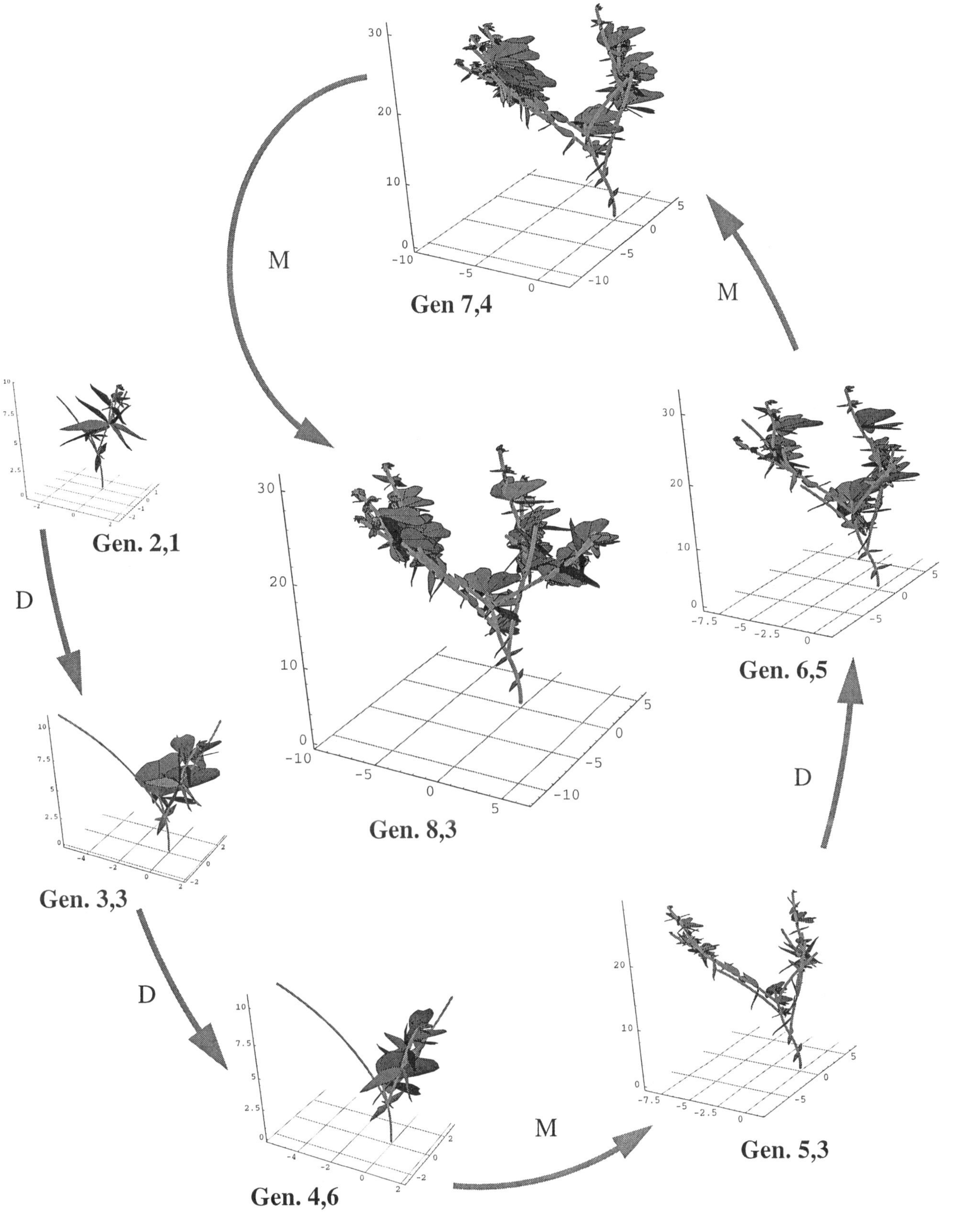

Figure 6 Demonstrating the evolution of phenotypic characteristics for the evolution experiment of figure 5. The numbers indicate the generation and individual index. D and M denote duplication and mutation, resp.

Genetic Programming using Genotype-Phenotype Mapping from Linear Genomes into Linear Phenotypes

Robert E. Keller Wolfgang Banzhaf
University of Dortmund
Department of Computer Science, LS11
D-44221 Dortmund, Germany
{keller,banzhaf}@ls11.informatik.uni-dortmund.de

ABSTRACT

In common genetic programming approaches, the space of genotypes, that is the search space, is identical to the space of phenotypes, that is the solution space. Facts and theories from molecular biology suggest the introduction of non-identical genospaces and phenospaces, and a generic genotype-phenotype mapping which maps unconstrained genotypes into syntactically correct phenotypes. Neutral variants come into effect due to this mapping. They enhance genetic diversity and allow for escaping local optima in phenospace via high-dimensional saddle surfaces in genospace. We propose a concrete mapping that maps linear binary genotypes into linear phenotypes of an arbitrary context-free programming language. Empirical results are presented which show that the mapping improves the performance of GP under mutation and reproduction.

1 Introduction

Common GP approaches (e.g. "Koza (1992)") do not distinguish between a genotype, i.e. a point in search space, and its phenotype, i.e. a point in solution space, that is search space and solution space are identical.

Recently, however, a strict distinction between the search space and the solution space as well as a genotype-phenotype mapping (GPM) from the search space into the solution space have been suggested in "Banzhaf (1994)". In that contribution, the author extends the common GP approach by a GPM and demonstrates the practicability of this *GP/GPM approach.*

Note that a GPM in itself is not a special GP variant. The principle behind a GPM is the distinction between the search space and the solution space of an underlying search process. The process locates points in the search space as results of the search. However, a potential solution must be a point from the solution space. Thus, a GPM is needed which maps an arbitrary search point onto a solution point. This principle can be used with every search process, like an arbitrary evolutionary algorithm.

Finally, it is hypothesized in "Banzhaf (1994)" that the GP/GPM approach enhances the performance of systems using evolutionary algorithms (EA).

The present paper focuses on a basic empirical investigation of this hypothesis with respect to the area of GP. To that end, we compare the performance of both approaches by applying them to the same non-toy test problem. In order to do so, we use our Binary-Genetic-Programming system, which incorporates both approaches.

One starting point towards the GP/GPM approach given in "Banzhaf (1994)" is Kimura's neutral theory of molecular evolution "Kimura (1968)" "Kimura (1983)" which postulates that molecular evolution is essentially driven by mutations almost neutral considering natural selection. This includes different genotypes (*neutral variants*) coding for precisely the same phenotype. The phenomenon of neutral mutations, according to Kimura, is a major reason for the high genetic diversity in natural populations. Note that this theory has been supported by empirical results "Mukai (1985)".

The motivation for applying this theory to the field of evolutionary algorithms is that an EA solving a problem often faces solving a *constrained* optimization problem. Not only should the EA evolve one or more phenotypes with highest possible quality (fitness), it is also forced to evolve only such phenotypes which obey given restrictions.

Common GP approaches, for instance, face a hard constraint given by the syntax of the underlying programming language: all generated individuals must be legal, that is syntactically correct. These approaches handle this constraint like GAs handle hard constraints: all operators modifying individuals, e.g. crossover, are constrained such that they only generate legal individuals.

Thus, large regions of the search space - which equals the solution space - are infeasible, thereby presenting a twofold problem to common GP approaches. First, the large potential of genetic diversity existing in the infeasible regions is not accessible. Second, on the search for high-quality individuals, sequences of individuals get generated. The larger the infeasible regions are, the higher is the probability they include such individuals which build short individual-sequences leading from the individuals of the actual generation to high-quality individuals of the next generation. To put it vividly: GP tries to wander from good to better individuals, and infeasible regions force it to detour.

This train of thought motivates the extension of a common GP approach by a distinction of search space and solution space, by unconstrained operators performing on the search space only, and by a GPM from search space into solution space. The search space equals the set of all genotypes, legal and illegal. A GPM maps each genotype into a legal phenotype, thereby matching the syntactical constraint.

By this extension, the problem of infeasible regions does not exist any more: the complete search space has become evolvable along short individual-sequences. Under mutation as one of the search operators, this leads to a random genetic drift in the population. In the ideal case, this drift leads to a genotype distribution such that for *each* legal phenotype p there exists a genotype g *in the actual generation* such that g is very close to some genotype h of p "Eigen (1992)". "Close" means there is a short individual-sequence leading from g to h. This enlarges the probability of finding p.

Especially, this enables evolution to "escape local optima on saddle surfaces", as it is put in "Banzhaf (1994)". If a population gets trapped in a suboptimum, there are many neutral variants mapping into the trapped phenotypes. Some of these variants will be close to neutral variants of phenotypes representing better (sub)optima. Thus, there is a high probability that evolution will find these phenotypes. On the other hand, there is a high probability that evolution will find worse phenotypes. Due to the fitness-based performance of reproduction and search operators, however, the corresponding worse genotypes will have few chances to proliferate.

In order to keep the comparison of both approaches transparent, certain restrictions will be imposed on the evolutionary process. Since mutation is the major search operator in Kimura's theory, we will use mutation as only search operator. Furthermore, the genotypes of the GP/GPM approach will all feature the same fixed length.

Note that it is not intended to solve the problem. The restrictions just mentioned will probably prevent the finding of a perfect solution anyway.

2 Representations, GPM, operators

Concerning the GP/GPM approach, some thought must be given to the representation of genotypes. In "Banzhaf (1994)", each genotype is identified with a binary string. This provides for the simplest and most universal representation of information, and it allows the use of simple search operators, which could even be standard GA operators. Additionally, it enables a sort of mapping such that the generation of neutral variants is enhanced, as will be shown subsequently. Finally, it provides exactly one type of search space, a space of binary strings, which is independent from the representation of phenotypes in the actual solution space.

As GPM to be actually used, an instantiation of a generic GPM "Banzhaf (1994)" motivated by molecular biology "Watson et al.(1987)", is taken. The generic GPM is a metaphor of the biological process of *protein synthesis*. During this process, DNA (genotype) gets transcribed into messenger RNA (mRNA), which is corrected with respect to the functionality of the protein it is coding for. The corrected mRNA then gets translated into amino acid sequences, which fold into a protein (phenotype) that features a certain functionality (behavior) with respect to the cell metabolism.

Protein synthesis is a very complex process. The GPM to be actually used is a simplified metaphor of this process. A binary string (genotype) gets transcribed into a *raw sequence* of symbols (*transcription*). Each such symbol is an element of either the function or the terminal set which both underlie a genetic-programming approach.

The raw sequence gets corrected, if necessary, according to the syntax of the used programming language, thus yielding a *legal symbol sequence* (*repairing*).

Then, *editing* turns this sequence into an *edited symbol sequence* by adding standard information, e.g. a main program frame enclosing the repaired sequence.

Finally, the last phase of the mapping, which is *compilation* of the edited symbol sequence, transforms this sequence into a machine-language program that can be executed in order to evaluate the fitness of the corresponding individual. Alternatively, *interpretation* of the edited symbol sequence can be used for fitness evaluation.

There remains the question of the representation of phenotypes in the GP/GPM approach. A legal symbol sequence s is the abstract expression of the behavior defined by that genotype that got mapped onto s. Since editing and compilation or interpretation do not add behavior-changing information to such a sequence, it is logical to consider this sequence as a phenotype. Thus, in the context of the GP/GPM approach, the solution space can be defined as the set of all legal symbol sequences.

Note that, with respect to the employed GPM/protein synthesis metaphor, a legal symbol sequence corresponds to a protein. Since amino acids are the essential components of a protein, an amino acid corresponds to a symbol.

We now detail the processes of transcription and repairing used during the GP/GPM approach. To that end, we introduce the notion of a *codon*. In molecular biology, this is a triplet of nucleic acids which uniquely encodes one amino acid, at most. Since an amino acid corresponds to a symbol, a codon corresponds to the encoding of a symbol.

This puts forward the question for the nature of the encoding to be used. At the beginning of this section, an argumentation in favor of a binary representation of a genotype has been given. This argumentation is also valid with respect to a codon. Thus, for the GP/GPM approach, a *codon* shall term a bit sequence of b bits length which encodes a symbol. In order to provide for the unique encoding of all symbols, b must be chosen such that for each symbol there is at least one codon which encodes this and only this symbol. For instance, if there are 20 symbols, $b \geq 5$ is a proper value, since 2^b is greater than 20.

The idea of codons implies that a genotype, which is a binary string by definition, consists of a sequence of n codons. Let the codons of a genotype be identified by positional numbers, starting with 0 and ending with $n-1$.

An arbitrary mapping M from the set of codons into the set of symbols can now be defined. Thus, M defines the encoding of symbols by codons. Transcription scans a genotype, starting at codon 0, ending at codon $n-1$. Using M, transcription maps each encountered codon onto the corresponding symbol, thereby creating a raw symbol sequence from the genotype.

In molecular biology, the analogy to our mapping M is called *genetic code*. The mapping M shall be termed likewise.

The natural genetic code is the mapping from codons into amino acids. It is important to note that this code is highly redundant: there are many different codons which encode the same amino acid. We will return to this point later with respect to the design of an artificial genetic code represented by some mapping M.

For later use, each codon of a certain artificial genetic code shall be identified with an index. That could be, for instance, the integer number represented by the codon. For instance, the codon 0110 could be identified with 6.

In order to define repairing, we only consider context-free LALR(1) (look-ahead-left-recursive, look ahead one symbol) grammars "Aho et al. (1986)". These grammars have a nice property with respect to compiler construction. An essential phase of the compiling of a program, that is a symbol sequence, is the parsing of this sequence in order to verify the syntactical correctness of the program. When parsing a symbol sequence according to a LALR(1) grammar, it is always and efficiently possible to compute the current *legal symbol set* with respect to the actually scanned symbol s. Each symbol of this set is syntactically correct in place of s. Especially, this set can be computed by solely considering the symbol preceding s.

Note that, since we only consider LALR(1) grammars for the GP/GPM approach, a GP system using the GP/GPM approach as described here can only evolve programs in languages that are defined by such a grammar. This, however, is no restriction with respect to the practicability of the approach, since many practically relevant languages like C are LALR(1) languages.

A symbol that represents a syntax error shall be called *illegal*. When parsing a raw sequence, it is highly probable that an illegal symbol s will be scanned. The legal symbol set can be computed with respect to s. In such a situation, a *minimal-distance set* can be defined. It is a subset of the actual legal symbol set, and it contains all those symbols whose codons are Hamming-closest to the codon of the illegal symbol s. With respect to the illegal symbol, such a symbol is called *closest*. Depending on the genetic code, there can be more than one such symbol.

We can now define the repairing of a raw sequence. The sequence gets parsed, and, in case an illegal symbol is encountered, it will be replaced by its closest symbol. If there are more than one closest symbol, that one will be taken which is encoded by a codon with the lowest index among all codons of closest symbols.

A common special case must be handled. Often, the described type of repairing will produce a symbol sequence that terminates unfinished. For instance, this sequence could be an unfinished arithmetical expression like $sin(a)*cos(b)+$. In order to handle that case, we assign a *termination number* to each symbol. This number indicates how appropriate the symbol is for shortest possible termination of an unfinished sequence. In the above expression, for instance, a variable would terminate the sequence, while an operator symbol like *sin* would call for at least three more symbols ('(', variable, ')'). By definition, the termination number of a symbol shall equal the number of these additionally needed symbols. Thus, in the example, a variable-symbol like 'a' has termination number 0, while a symbol like *sin* has termination number 3.

Note that the number of symbols *actually* needed for termination when using a certain symbol can be larger than its termination number. For instance, there could be open parantheses prior to the end of the unfinished sequence, which have to be closed. Such context-sensitive circumstances shall not be reflected in a termination number.

In case there are several symbols with equal minimal

termination numbers, that one shall be taken which is encoded by a codon with the lowest index among all codons of the symbols in question.

We give a simple example for the functioning of the described GPM. The underlying syntax shall be that of simple arithmetic expressions. Consider the redundant genetic code $000 \rightarrow a$; $001 \rightarrow b$; $010 \rightarrow +$; $011 \rightarrow *$; $100 \rightarrow a$; $101 \rightarrow b$; $110 \rightarrow +$; $111 \rightarrow *$.

The genotype 000 001 011 gets transcribed into the raw sequence $ab*$. Repairing scans a as first and legal symbol. It then scans b as an illegal symbol with 001 as its codon in the genotype. $\{+, *\}$ is the legal symbol set. The symbol closest to b is $*$. Thus, b gets replaced by $*$, thereby terminating the repair of the syntax error. The partially repaired raw sequence now equals $a**$. Repairing continues, replacing the last symbol $*$ with b. The repaired sequence $a*b$ is passed on to editing, which adds, for instance, a function frame. The edited sequence might look like

```
double fnc(double a, double b) {return a*b;}.
```

This sequence could now, for example, be integrated, together with other edited sequences, into a main program which could be given to any ANSI-C compiler.

Some more examples illustrate how the existence of many neutral variants for certain phenotypes can be explained out of the presented GPM. For instance, as shown above, 000 001 011 gets mapped into the phenotype $a*b$. However, 010 011 101, 000 011 001 and 100 111 101 get all mapped into $a*b$ as well. When analyzing this phenomenon, it becomes immediately clear that both the redundancy of the genetic code and the repair mechanism are responsible for this effect.

Note that the presented mapping mechanism indeed defines a *mapping* from genospace into phenospace: each genotype gets mapped always into exactly one phenotype. Otherwise, a genotype very likely would get associated with different phenotypes and thus with different fitness values over time. This would not allow for a proper fitness-related selection on genospace.

We now present the creation and mutation operators separately for the GP/GPM approach and the common GP approach. These operators perform selections on sets of codons, of symbols and of bits. All of these selections shall operate randomly and equally distributed.

For the GP/GPM approach, creation generates a random individual as random binary string which consists of n codons.

For the common GP approach, creation builds up a syntactically legal random symbol sequence which consists of at least n symbols. This is done by starting with an empty symbol string. The first symbol is selected from the set of symbols which are legal first symbols. The following symbol is selected from the set of symbols which are legal as follow-up symbols to the first symbol. This method iterates until the n-th symbol has been selected. The result is a legal sequence of n symbols. However, the sequence may be unfinished. In that case, it is terminated in the same way that has been described for the case of repairing. Note that this leads to individuals with varying length.

For the GP/GPM approach, two mutation operators shall be devised. The *coupled mutation*, applied to a genotype, selects a codon. It then flips two randomly selected different bits in that codon. This operator reflects the fact that some mutations tend to change nucleic acids in a coupled sort of way.

The *unrestricted mutation* selects a codon. In that codon, it then flips one randomly selected bit with 0.5 probability, or two randomly selected different bits with 0.35 probability, or three randomly selected different bits with 0.1 probability, or four randomly selected different bits with 0.05 probability. The probability distribution reflects the natural *principle of variation*: small changes occur more often than big changes.

For the common GP approach, mutation selects a symbol in the sequence. It then replaces this symbol by another randomly selected symbol which is legal in the respective position.

Note that this mutation operator does not produce vast changes in the genotype by replacing complete syntactical units as it is the case, for instance, with the mutation operator presented in "Koza (1992)". We feel that the presented operator shows a behavior closer to natural mutation and thus better matches the spirit of artificial evolutionary processes.

Note that an individual of the common GP approach is encoded as a symbol sequence, that is a string. It is not represented in a tree-like form like it is often done in work related to genetic programming. However, both the string encoding and the tree encoding are equivalent in the sense that each arithmetic expression represented as a string can be encoded as expression tree and vice versa.

Moreover, given corresponding mutation operators, each mutation done on an expression tree can be performed equivalently on the corresponding string and vice versa. It is interesting to note that this is also true with respect to crossover, although crossover is not used in the current context.

We finally sketch our binary genetic-programming system used for the empirical comparison of both approaches. The system incorporates the standard GP algorithm, as it is presented in "Koza (1992)", as kernel. Tournament selection on two individuals is used in order to select an individual for subsequent reproduction only or, alternatively, for reproduction followed by mutation. Both alternatives have 0.5 probability of getting applied as next operation on a selected individual.

Adjusted fitness is used as fitness measure. The square-error sum produced by an individual i over all fitness cases shall define i's standardized-fitness value, which is needed for the computation of i's adjusted-fitness value "Koza (1992)". Thus, all possible fitness values exist in $[0, 1]$. A perfect individual has fitness value 1.

In case the system uses the GP/GPM approach, the kernel calls the GPM between creation and fitness evaluation of the initial generation, and between the completed evolution of the actual generation and its fitness evaluation, respectively. Dependent on the actually used approach, the corresponding representations and operators are employed.

3 Method

The programming language to be used for the evolution shall be ANSI-C. Thus, a phenotype will be a symbol sequence obeying ANSI-C syntax. The domain of the test problem will suggest variables, unary and binary arithmetic functions, and paranthesis operators as elements of the terminal and function sets.

The test problem is a symbolic function regression on a four-dimensional parameter space. The function in question is

$$\sin(m) \cdot \cos(v) \cdot \frac{1}{\sqrt{e^q}} + \tan(a).$$

All parameter values shall be real-valued.

In order to allow for protection against division-by-zero, we devise the division function `D(x)` which returns the reciprocal value of its single argument. If the argument is zero, the result equals 1. We supply a protected square root function `sqrt(x)` which returns the square root of the absolute value of its argument. Furthermore, an overflow-protected exponential function **`exp(x)`** is provided which returns e^x. In case the value of x causes an overflow, the returned value is 1.

As codons, we supply the numbers 0 up to 15 in their binary representation, i.e. $0000, 0001, .., 1110, 1111$. In this order, the codons get mapped into a set of 16 symbols, featuring 14 different symbols:

$$+ \; * \; * \; D \; m \; v \; q \; a \; (\;) \; sin \; cos \; tan \; sqrt \; exp \;).$$

Note there is only little redundancy in this genetic code, so that in case the theory of neutral variants works at all, the GP/GPM approach will not have vastly more power than the common GP approach.

For the GP/GPM approach, the genotype length shall be 25, that is each genotype consists of 25 codons. Since there are 16 different codons in the genetic code, the search space contains 16^{25} or approximately 1.3E30 genotypes.

When using the unrestricted mutation operator which can mutate each single bit independently the GP/GPM approach faces $25 \cdot 4 = 100$ degrees of freedom, since each codon consists of 4 bit. In other words, in addition to the relatively large size of the search space, it is high-dimensional, featuring 100 dimensions.

For the common GP approach, the individual's length shall be at least 25, too. Thus, each individual is a symbol sequence of at least 25 symbols. Due to the above described phenomenon of unfinished sequences, the actual length of individuals after repairing can surpass 25. Imagine, for instance, the improbable case of an individual looking like

```
D(D(D(D(D(D(D(D(D(D(D(D  (a))))))))))))).
```

Prior to the spaces within the individual, there are 25 symbols. Shortest possible termination of the individual still requires the appending of 15 more symbols, most of them being ')', in order to close still open paranthesis levels.

However, the search space in the common GP approach has a size of much less than 14^{25} or approximately 4.5E28 individuals. That is because the symbols appended have not been evolved but non-stochastically computed, and because many symbol sequences are illegal. By using the mutation operator, which can mutate each of the 25 evolved symbols in a sequence independently, the common GP approach faces 25 degrees of freedom. Thus, the search space has only 25 dimensions.

Note that, for the GP/GPM approach, the maximal length of a phenotype can surpass 25, too, for the same reason that has been given above.

Due to the real-valued four-dimensional parameter space, a fitness case consists of four real input values and one real output value. We only supply 10 fitness cases in order to further increase the difficulty of the problem by lack of information. All experimental runs use a population size of 500 individuals and run for 50 generations.

In order to compare both approaches with respect to their performance, three series of experiments are run using the binary-genetic-programming system. Each series consists of 19 runs. All runs of the same series shall be indexed from 1 up to 19. All runs of the same series run with different randomizer seeds. Runs with identical indices, but from different series, run with the same randomizer seed.

A run from series 1 is done with the common GP approach. A run from series 2 features the GP/GPM approach, using the coupled mutation operator. A run from series 3 is done with the GP/GPM approach, using the unrestricted mutation operator. In all other respects, the runs perform under identical conditions, which have been described above.

4 Results and discussion

Fig. 1 shows the mean average fitness over time for all 3 series of runs. The graphs titled "GPM-coupled" and "GPM-unrestricted" belong to the series featuring the GP/GPM approaches with coupled and unrestricted mutation. The graph titled "Common" belongs to the series featuring the common GP approach.

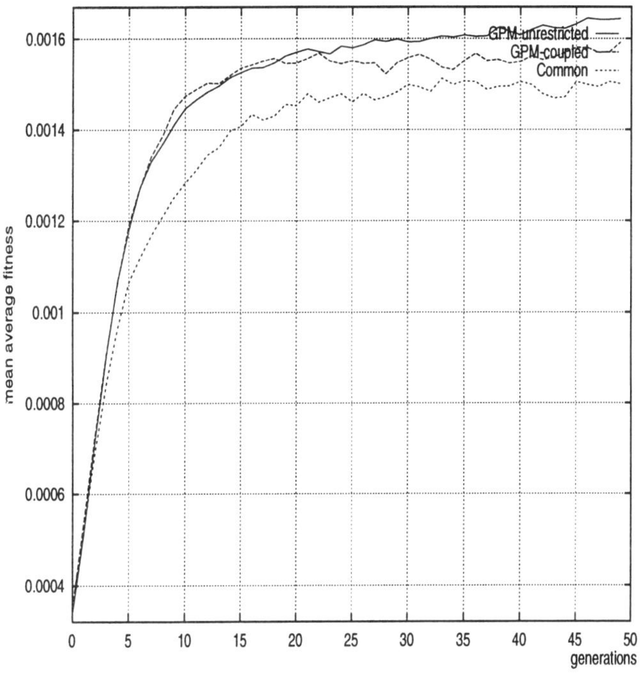

Figure 1 Mean average fitness of unrestricted, coupled and common GP approach

Both the coupled and the common GP approach start at generation 0 with practically identical fitness values, indicating that there is no bias in the GPM-independent initial conditions which might favor the GP/GPM approach.

Both graphs rise steeply until generation 5, with the GPM-coupled graph showing a clearly higher increase rate. At generation 5, the Common graph suffers a significant decrease in its inclination, while the GPM-coupled graph still rises strongly with only a light decrease in its inclination. This indicates the effectiveness of neutral variants helping to escape from local optima.

It is not until generation 9 that the inclination of the GPM-coupled graph decreases significantly. The inclination of both graphs dwindles down to more or less 0 over the next generations.

In the final generation, the coupled approach features a 6 percent higher fitness relative to the common GP approach. After generation 0, the fitness values of the common GP approach never reached or surpassed those of the coupled approach.

Both the coupled and the common GP approach get significantly outperformed by the unrestricted approach. As can be seen from the graphs, this approach behaves very similar to the coupled approach, at first. Especially, it starts in generation 0 with approximately the same fitness value as the other approaches. From generation 18 onward, however, the unrestricted approach clearly leaves the other approaches behind. In the final generation, the unrestricted approach features a 10 percent higher fitness relative to the common GP approach. After generation 1, the fitness values of the common GP approach never reached or surpassed those of the unrestricted approach.

We offer a hypothesis for the superiority of the unrestricted approach over the coupled approach. Given some codon, coupled mutation cannot mutate this codon into each other arbitrary codon, since it *always* flips 2 bits. For instance, no sequence of coupled mutations can change 0011 into 0001.

This implies that, given some genotype g, the search process cannot reach all genotypes. Thus, if the search process gets stuck in a local optimum, it can only reach certain other genotypes from there. If none of these genotypes has a better fitness than those representing the local optimum, the process cannot escape. Unrestricted mutation, however, can change each codon into each other codon, since it can flip one bit only. Thus, the search process can potentially escape.

In the final generation, the mean *best* fitness of the unrestricted and coupled approaches equal approx. 0.97 and 0.96 compared to approx. 0.93 for the common GP approach. The mean best-fitness values of the common GP approach never reached or surpassed those of the GP/GPM approaches.

Note also that the mean best-fitness values of all three approaches get rather close to the perfect fitness value 1.0, although no crossover operator has been used. This indicates that mutation should deserve more attention in GP instead of merely being considered a secondary operator.

Note that the superior performance in mean and best average fitness of both GP/GPM approaches compared to the common GP approach occur though search space size and dimensionality are significantly higher compared to the common GP approach, and despite the fact the used genetic code features not very much redundancy, which restricts the chance for the generation of neutral variants. This stresses the advantage of the GP/GPM approach.

Note that the genetic code used for the runs is redundant with respect to the multiplication and the closed-paranthesis operator. This might have positive and negative effects on the performance.

For instance, the redundancy in the multiplication operator could result in more such operators as parts of a phenotype. This can be advantageous since the problem function features two such operators.

On the other hand, the redundancy in the closed-paranthesis operator could pose a handicap to evolution, since it enlarges the probability that a needed long subexpression never emerges.

5 Conclusion

A first comparison of the common GP approach vs. the GP/GPM approach has been the main focus of this contribution. The presented empirical results argue in favor of the use of GPM, at least in GP systems where mutation is employed as a search operator. Furthermore, mutation appears to be a useful operator when applied by both the common or the GP/GPM approach.

6 Further research

Clearly, the genetic code is a key issue in GPM. It is a non-trivial task to determine the "optimal" code dependent on the actual problem. To put it more formally, the challenge is to define a GPM approach such that the resulting search space topology allows for an efficient evolution with respect to the underlying fitness landscape. Real-world problems, however, usually present "terra-incognita" landscapes.

Taking a closer look at *Evolutionsstrategien* "Schwefel (1995)" and the important subject of self-adaptation of parameters that control the search process suggests that an individual genetic code as part of a genotype might be beneficial for the search process as a whole. The individual codes could evolve along with the other features of the genotypes. This would be especially valuable in cases when a GP system gets confronted with dynamic fitness landscapes.

Furthermore, the results and their discussion suggest the use of a truly unrestricted mutation operator which flips between 1 to n bits under normal distribution, n being the number of all bits in the genotype. This operator should lead to even better performance of a GP/GPM approach than the presented unrestricted mutation operator.

The role of crossover in a GP/GPM approach and the relationship between crossover and mutation in that context call for intense research.

Obviously, the phase of repairing an illegal raw sequence is essential to a GP/GPM approach. We have presented a repairing type that replaces an illegal symbol by a legal one.

However, repairing could also be done by inserting or deleting symbols. For instance, the illegal raw sequence `a++b` can be turned into, for example, `a+a+b` by inserting `a` between the addition operators. On the other hand, the sequence can be transformed into `a+b` by deleting an addition operator.

Further research must go into potential connections between convergence properties of a GP/GPM approach and the type of employed repairing.

Finally, the GP/GPM approach and the common GP approach must be compared on a large set of further hard and diverse problems, using different genetic codes.

Bibliography

A.V. Aho et. al.: Compilers. Addison-Wesley. 1986.

W. Banzhaf: Genotype-Phenotype-Mapping and Neutral Variation – A case study in Genetic Programming. In: Y. Davidor et al. (eds.): Parallel Problem Solving from Nature III. Berlin: Springer. 1994.

M. Eigen: Steps toward Life: a perspective on evolution. Oxford: Oxford University Press. 1992.

M. Kimura: Evolutionary rate at the molecular level. Nature 217, 624-626. 1968.

M. Kimura: The Neutral Theory of Molecular Evolution. Cambridge: Cambridge University Press. 1983.

J.R. Koza: Genetic Programming. Cambridge (USA): MIT Press. 1992

T. Mukai: Experimental Verification of the Neutral Theory. In: T. Ohta et. al. (eds.): Population Genetics and Molecular Evolution. Berlin: Springer. 1985.

H.-P. Schwefel: Evolution And Optimum Seeking. New York: Wiley. 1995.

J.D. Watson et al.: Molecular Biology of the Gene. Amsterdam: Benjamin / Cummings Publishing Company. 1987.

Automated WYWIWYG Design of Both the Topology and Component Values of Electrical Circuits Using Genetic Programming

John R. Koza
Computer Science Dept.
258 Gates Building
Stanford University
Stanford, California 94305
koza@cs.stanford.edu
http://www-cs-faculty.stanford.edu/~koza/

Forrest H Bennett III
Visiting Scholar
Computer Science Dept.
Stanford University
Stanford, California 94305
fhb3@slip.net

David Andre
Visiting Scholar
Computer Science Dept.
Stanford University
Stanford, California 94305
andre@flamingo.stanford.edu

Martin A. Keane
Econometrics Inc.
5733 West Grover
Chicago, Illinois 60630
makeane@ix.netcom.com

ABSTRACT

This paper describes an automated process for designing electrical circuits in which "What You Want Is What You Get" ("WYWIWYG" – pronounced "wow-eee-wig"). The design process uses genetic programming to produce both the topology of the desired circuit and the sizing (numerical values) for all the components of a circuit. Genetic programming successfully evolves both the topology and the sizing for an asymmetric bandpass filter that was described as being difficult-to-design in a leading electrical engineering journal. This evolved circuit is another instance in which a genetically evolved solution to a non-trivial problem is competitive with human performance.

1. Introduction

The problem of circuit synthesis involves designing an electrical circuit that satisfies user-specified design goals.

A complete specification of an electrical circuit includes both its topology and the sizing of all its components. The *topology* of a circuit consists of the number of components in the circuit, the type of each component, and a list of the connections between the components. The *sizing* of a circuit consists of the component value(s) (typically numerical) associated with each component.

Considerable progress has been made in automating the design of certain categories of purely digital circuits; however, analog circuits and mixed analog-digital circuits are not as amenable to automation (Rutenbar 1993).

Once the user has specified the design goals for an electronic circuit, it would be ideal if an automated design process could create *both* the topology and the sizing of a circuit that satisfies the design goals. That is, it would be ideal to have an automated "What You Want Is What You Get" ("WYWIWYG" – pronounced "wow-eee-wig") process for circuit synthesis.

This paper shows how genetic programming was used to design an asymmetric bandpass filter that was described as difficult-to-design in an article in the *Analog Integrated Circuits and Signal Processing* journal (Nielsen 1995).

2. Automated Analog Design Tools

Existing techniques to automate the design process for analog and mixed analog-digital circuits typically require the user to supply a suitable topology; require the user to supply a reasonably good prototype circuit as a starting point; require repeated interactive intervention by the user; limit the user to certain specialized types of circuits; or are otherwise highly constrained. Examples include IDAC (Degrauwe 1987); OASYS (Harjani, Rutenbar, and Carley 1989); OPASYN (Koh, Sequin, and Gray 1990); SEAS (Ning, Kole, Mouthaan, and Wallings 1992); the expert design knowledge system of Maulik, Carley, and Rutenbar (1992); the works of Hemmi, Mizoguchi, and Shimohara (1994); Mizoguchi, Hemmi, and Shimohara 1994; and Higuchi et al. (1993); the "C-T filter design tool" of Nielsen (1995); and DARWIN (Kruiskamp and Leenaerts 1995). In DARWIN, circuit topologies are limited to a preestablished hand-designed set of 24 suitable opamp topologies while component values are adjusted by a genetic algorithm operating on a binary character string.

3. The Mapping between Electrical Circuits and Program Trees

Genetic programming is an extension of John Holland's genetic algorithm (1975) in which the population consists of computer programs of varying sizes and shapes (Koza 1992, 1994a, 1994b; Koza and Rice 1992).

Genetic programming breeds a population of rooted, point-labeled trees (i.e., graphs without cycles) with ordered branches. There is a considerable difference between the kind of trees bred by genetic programming and the labeled cyclic graphs encountered in the world of electrical circuits. Electrical circuits are cyclic graphs in which every line belongs to a cycle (i.e., there are no loose wires or dangling components). The lines of a graph that represent a circuit are each labeled. The primary label on each line gives the type of an electrical component (e.g., resistor). The secondary label(s), if any, give the component value(s).

Genetic programming can be applied to circuit synthesis if a mapping is established between the kind of point-labeled trees found in the world of genetic programming and the line-labeled cyclic graphs employed in the world of circuits. Developmental biology provides the motivation for this mapping. In *Cellular Encoding of Genetic Neural Networks*, Frederic Gruau (1992) described a clever technique, called *cellular encoding*, in which genetic programming is used to evolve the architecture of a neural network. Each individual program tree in the population is a specification for developing a complete neural network from a very simple embryonic neural network (consisting of a single neuron). Genetic programming is then applied to populations of network-constructing program trees in order to evolve a neural network capable of solving a problem. The growth process used herein for electrical circuits similarly begins with a very simple embryonic electrical circuit and builds a more complex circuit by progressively executing the functions in a circuit-constructing program tree. The result of this process is the topology of the circuit, the choice of the types of components that are situated at each location within the topology, and the sizing of the components.

Each program tree can contain (1) connection-modifying functions that modify the topology of the circuit (starting with the embryonic circuit), (2) component-creating functions that insert particular components into locations within the topology of the circuit in lieu of wires (and other components) and whose arithmetic-performing subtrees specify the value (sizing) for each such component, and possibly (3) automatically defined functions.

Program trees conform to a constrained syntactic structure. Each component-creating function in a program tree has zero, one, or more arithmetic-performing subtrees and one or more construction-continuing subtrees. Each connection-modifying function has one or more construction-continuing subtrees. The arithmetic-performing subtree(s) of each component-creating function consists of a composition of arithmetic functions and numerical constant terminals that together yield the numerical value for the component. The construction-continuing subtree specifies how the construction of the circuit is to be continued.

Both the random program trees in the initial population (generation 0) and all random subtrees created by the mutation operation in later generations are created so as to conform to this constrained syntactic structure. This constrained syntactic structure is then preserved by structure-preserving crossover with point typing (Koza 1994a).

4. The Problem

A *filter* is a one-input, one-output circuit that receives a signal as its input and passes the frequency components of the incoming signal that lie in a certain specified range (the *passband*) while stopping the frequency components of the signal that lie in other frequency ranges (the *stopband*).

In the *Analog Integrated Circuits and Signal Processing* journal, Ivan Riis Nielsen (1995) described stringent and asymmetric specifications for a difficult-to-design bandpass filter. Nielsen's bandpass filter (1995) is targeted for a modem application where one band of frequencies (31.2 to 45.6 kilohertz) must be isolated from another (69.6 to 84.0 KHz). Using a standard bandpass filter on this problem would require a tenth-order elliptic function.

A *decibel* (dB) is a unitless measure of relative voltage that is defined as 20 times the common logarithm of the ratio between the voltage at a particular probe point and a reference voltage (which, herein, is always 1.0 volts).

Nielsen specified that it would be *ideal* if the relative voltage within the passband were in the narrow region between –0.6 dB and 0.6 dB (i.e., the *passband ripple* around 0 dB is less than 0.6 decibels) and all the relative voltages outside the passband were below –120 dB (i.e., the *stopband attenuation* were at least 120 dB). These ideal characteristics are depicted by the dark region in figures 11, 12, and 13 (where the height of the ripple band is exaggerated).

Nielsen defined a set of *acceptable* characteristics (depicted by the light shading in figures 11, 12, and 13). Because of the importance of isolating the band of frequencies between 69.6 and 84.0 KHz, the attenuation there should be at least 73 dB (i.e., the relative voltage is below –73 dB). Less stringency is demanded elsewhere. The attenuation for frequencies below 20 KHz should be at least 38 dB (i.e., the relative voltage is below –38 dB). The relative voltages in the frequency band between 20 KHz and 31.2 KHz and in the band between 45.6 KHz and 69.6 KHz should be below 0 dB. The relative voltages in the band above 84.0 KHz should be below –55 dB. Nielsen's filter has a fixed 1 Ohm source resistor **RSOURCE** and a fixed 0.1 Ohm load resistor **RLOAD**.

5. Embryonic Electrical Circuit

The embryonic circuit used on a particular problem depends on the number of input signals and the number of output signals (probe points).

The embryonic circuit used herein contains one input signal, one output (probe point), Nielsen's fixed source resistor, and Nielsen's fixed load resistor, and two modifiable wires. The two modifiable wires (**Z0** and **Z1**) each initially possess a writing head (i.e., are highlighted with a circle in figure 1). A circuit is progressively developed by modifying the component to which a writing head is pointing in accordance with the functions in the circuit-constructing program tree. Each connection-modifying and component-creating function in the program tree modifies the developing circuit in a particular way and each also specifies the future disposition of the writing head(s).

The bottom part of figure 1 shows the embryonic circuit used for a one-input, one-output circuit. The energy source is a 22 volt sinusoidal voltage source **VSOURCE** whose negative (–) end is connected to node 0 (ground) and whose positive (+) end is connected to node 1. There is a fixed source resistor **RSOURCE** between nodes 1 and 2. There is a modifiable wire **Z1** between nodes 2 and 3 and another modifiable wire **Z0** between nodes 3 and 4. There is a fixed isolating wire **ZOUT** between nodes 3 and 5, a voltage probe labeled **VOUT** at node 5, and a fixed load resistor **RLOAD** between nodes 5 and ground. There is an isolating wire **ZGND** between nodes 4 and 0 (ground). All of the above elements of this embryonic circuit (except **Z0** and **Z1**) are fixed forever; they are not subject to modification during the process of developing the circuit. All subsequent development of the circuit originates from writing heads. Note that the output of the embryonic circuit is a constant zero volt signal **VOUT** at node 5.

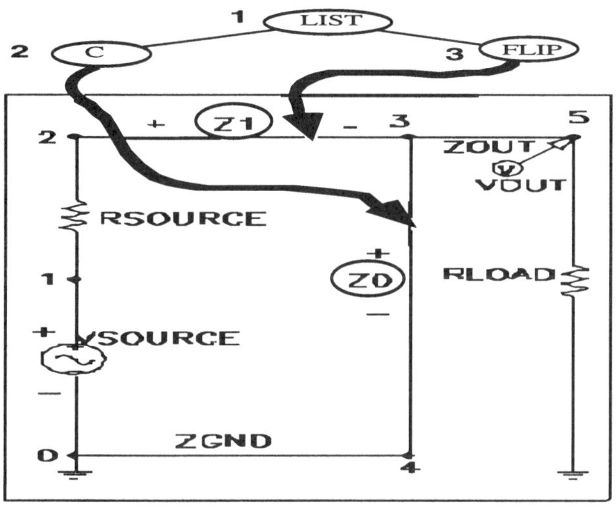

Figure 1 One-input, one-output embryonic electrical circuit.

The domain knowledge that went into this embryonic circuit is minimal. It consisted of the facts that (1) the embryo is a circuit, (2) the embryo has one input and one output, and (3) there are modifiable connections between the output and the source and between the output and ground.

A circuit is developed by modifying the component to which a writing head is pointing in accordance with the associated function in the circuit-constructing program tree. The figure shows a capacitor-creating `C` function and a polarity-reversing `FLIP` function just below the connective `LIST` function at the root of the program tree. The figure also shows one writing head pointing from the `C` function to modifiable wire **Z0** and another writing head pointing from the `FLIP` function to modifiable wire **Z1**. As will be described below, this `C` function will cause **Z0** to be changed into a capacitor and the `FLIP` function will cause the polarity of modifiable wire **Z1** to be reversed.

6. Component-Creating Functions

Each individual circuit-constructing program tree in the population generally contains component-creating functions and connection-modifying functions.

Each component-creating function inserts a component into the developing circuit and assigns component value(s) to the inserted component. Each component-creating function in a program tree points to an associated highlighted component (i.e., a component with a writing head) in the developing circuit and modifies the highlighted component in some way. Each component-creating function spawns one or more writing heads (through its construction-continuing subtrees). The construction-continuing subtree of each component-creating function points to a successor function or terminal in the circuit-constructing program tree.

The arithmetic-performing subtree of a component-creating function consists of a composition of arithmetic functions (addition and subtraction) and random constants (in the range –1.000 to +1.000). The arithmetic-performing subtree specifies the numerical value of the component by returning a floating-point value that is, in turn, interpreted as the value for the component in a range of 10 orders of magnitude (using a unit of measure that is appropriate for the particular type of component involved). The floating-point value is interpreted as the value of the component in the following way: If the return value is between –5.0 and +5.0, U is equated to the value returned by the subtree. If the return value is less than –100 or greater than +100, U is set to zero. If the return value is between –100 and –5.0, U is found from the straight line connecting the points (–100, 0) and (–5, -5). If the return value is between +5.0 and +100, U is found from the straight line connecting (5, 5) and (100, 0). The value of the component is 10^U in a unit that is appropriate for the type of component. This mapping gives the component a value within a range of 10 orders of magnitude centered on a certain value.

The two-argument capacitor-creating `C` function causes the highlighted component to be changed into a capacitor. The value of the capacitor is the antilogarithm of the intermediate value U (previously described) in nano-Farads. This mapping gives the capacitor a value within a range of plus or minus 5 orders of magnitude centered on 1 nF.

The two-argument inductor-creating `L` function causes the highlighted component to be changed into an inductor. The value of the inductor is in micro-Henrys within a range of plus or minus 5 orders of magnitude centered on 1 µH.

We describe one other function (not used in the main problem herein) to illustrate that numerous other component-creating functions may be used in this process.

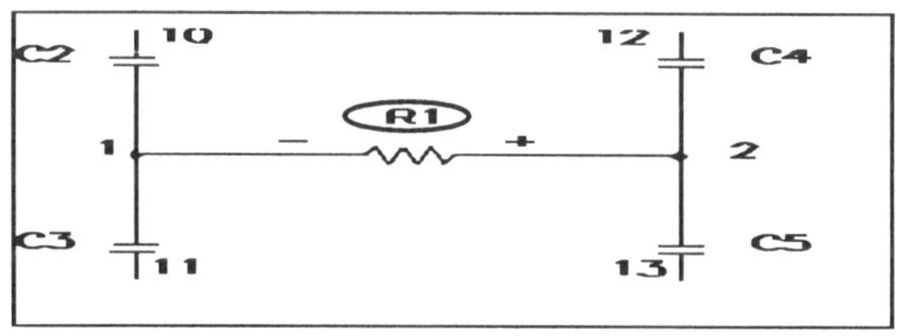

Figure 2 Circuit with resistor R1.

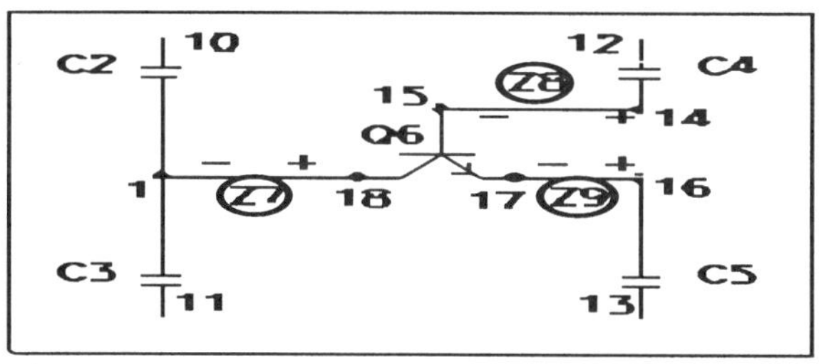

Figure 3 Result of applying `QT0` function.

Figure 2 shows a resistor **R1** (with a writing head) connecting nodes 1 and 2 of a partial circuit. Each of the 12 context-sensitive functions in the group of three-argument QT ("transistor") functions (called QT0, ..., QT11) causes a transistor to be inserted in place of one of the nodes to which the highlighted component is currently connected (while also deleting the highlighted component). Each QT function also creates five new nodes and three new modifiable wires. After execution of a QT function, there are three writing heads that point to three new modifiable wires. Figure 3 shows the result of applying the QT0 function to resistor **R1** of figure 2, thereby creating transistor **Q6**.

7. Connection-Modifying Functions

The topology of the circuit is determined by the connection-modifying functions. Each connection-modifying function in a program tree points to an associated highlighted component and modifies the topology of the developing circuit in some way. Each connection-modifying function spawns zero, one, or more writing heads.

The one-argument polarity-reversing FLIP function attaches the positive end of the highlighted component to the node to which its negative end is currently attached and vice versa. After execution of the FLIP function, one writing head points to the now-flipped original component.

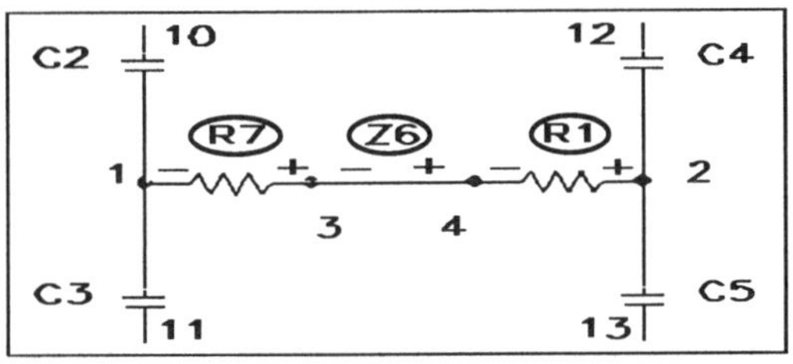

Figure 4 Result of applying SERIES.

The three-argument SERIES division function operates on one highlighted component and creates a series composition consisting of the highlighted component, a copy of the highlighted component, one new modifiable wire, and two new nodes. After execution of the SERIES function, there are three writing heads pointing to the original component, the new modifiable wire, and the copy of the original component. Figure 4 shows the result of applying the SERIES division function to resistor **R1** from figure 2. First, the SERIES function creates two new nodes, 3 and 4. Second, SERIES disconnects the negative end of the original component (**R1**) from node 1 and connects this negative end to the first new node, 4 (while leaving its positive end connected to the node 2). Third, SERIES creates a new wire (called **Z6** in the figure) between new nodes 3 and 4. The negative end of the new wire is connected to the first new node 3 and the positive end is connected to the second new node 4. Fourth, SERIES inserts a duplicate (called **R7** in the figure) of the original component (including all its component values) between new node 3 and original node 1. The positive end of the duplicate is connected to the original node 1 and its negative end is connected to new node 3.

The four-argument parallel division function PSS operates on one highlighted component to create a parallel composition consisting of the original highlighted component, a duplicate of the highlighted component, two new wires, and two new nodes. After execution of PSS, there are four writing heads. They point to the original component, the two new modifiable wires, and the copy of the original component. First, the parallel division function PSS creates two new nodes, 3 and 4. Second, PSS inserts a duplicate of the highlighted component (including all of its component values) between the new nodes 3 and 4 (with the negative end of the duplicate connected to node 4 and the positive end of the duplicate connected to 3. Third, PSS creates a first new wire **Z6** between the positive (+) end of **R1** (which is at original node 2) and first new node, 3. Fourth, PSS creates a second new wire **Z8** between the negative (-) end of **R1** (which is at original node 1) and second new node, 4. Figure 5 shows the results of applying the PSS function to resistor **R1** from figure 2. The negative end of the new component is connected to the smaller numbered component of the two components that were originally connected to the negative end of the highlighted component. Since **C4** bears a smaller number than **C5**, new node 3 and new wire **Z6** are located between original node 2 and **C4**. Since **C2** bears a smaller number than **C3**, new node 4 and new wire **Z8** are located between original node 1 and **C2**.

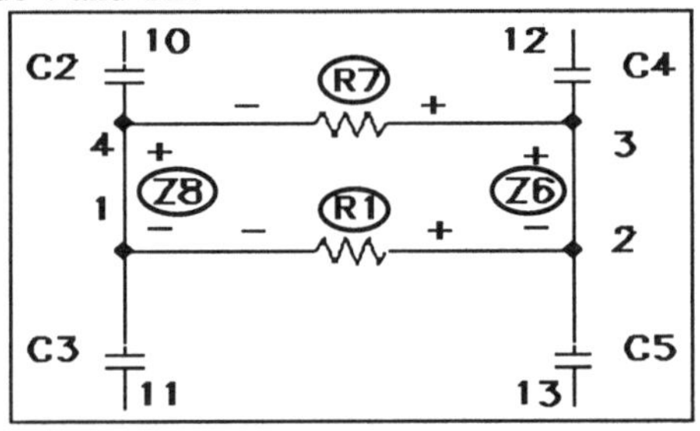

Figure 5 Result of applying PSS.

Eight two-argument functions (called VIA0, ..., VIA7) and the two-argument GND ("ground") function enable distant parts of a circuit to be connected together. After execution, writing heads point to two modifiable wires. The VIA functions create a series composition consisting of two wires that each possesses a successor writing head and a numbered port (called a *via*) that possesses no writing head. The port is connected to a designated one of eight imaginary layers (numbered from 0 to 7) of an imaginary silicon wafer. If more than one part of a circuit connects to a particular numbered layer, all of these parts become electrically connected as if wires were running between them. The two-argument GND function is a special "via" function that establishes a connection directly to ground.

The one-argument NOP function has no effect on the highlighted component; however, it delays activity on the developmental path on which it appears in relation to other developmental paths in the overall program tree. After execution of NOP, one writing head points to the original highlighted component.

The zero-argument END function causes the highlighted component to lose its writing head.

We describe two other functions (not used herein) to illustrate that numerous other connection-modifying functions can be employed in this process. The functions in the group of three-argument Y division functions operate on one highlighted component (and one adjacent node) and create a Y-shaped composition consisting of the highlighted component, two copies of the highlighted component (and all of its values), and two new nodes. The Y functions insert the two copies at the "active" node of the highlighted component. For the Y1 function, the active node is the node to which the negative end of the highlighted component is connected. Figure 6 shows the result of applying Y1 to resistor **R1** of figure 2.

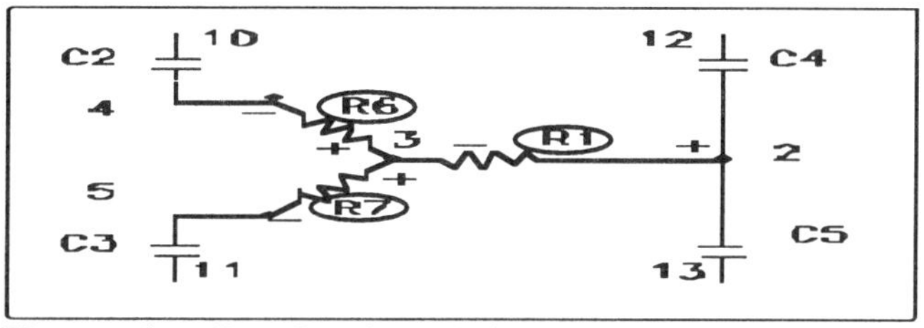

Figure 6 Result of applying the Y1 function.

The functions in the group of six-argument DELTA functions operate on one highlighted component by eliminating it (and one adjacent node) and creating a triangular Δ–shaped composition consisting of three copies of the original highlighted component (and all of its component values), three new modifiable wires, and five new nodes. Figure 7 illustrates the result of applying the DELTA1 division function to resistor **R1** of figure 2 when the active node (node 1) is of degree 3.

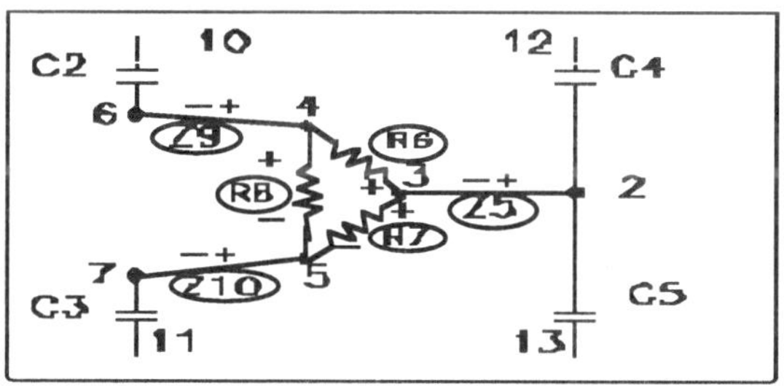

Figure 7 Result of applying DELTA1 function.

8. Preparatory Steps for the Run

8.1. Architecture and Embryonic Circuit

Since this LC filter design problem calls for a one-input, one-output circuit with a source resistor and a load resistor, the embryonic circuit in figure 1 is suitable. Since no automatically defined functions are used in this problem, the architecture of the overall program tree consists of two result-producing branches joined by the connective LIST function. That is, the embryonic circuit initially has two writing heads – one associated with each of the result-producing branches. The terminal set and function sets are the same for both result-producing branches.

8.2. Function and Terminal Sets

The function set, $\mathcal{F}_{aps}$ for an arithmetic-performing subtree associated with each component-creating function contains two two-argument arithmetic functions. That is,

$\mathcal{F}_{aps} = \{+, -\}$.

The terminal set, $\mathcal{T}_{aps}$, for an arithmetic-performing subtree consists of

$\mathcal{T}_{aps} = \{\mathfrak{R}\}$,

where $\mathfrak{R}$ represents floating-point random constants between –1.000 and +1.000.

The function set, $\mathcal{F}_{ccs}$, for the construction-continuing subtree of each component-creating function is

$\mathcal{F}_{ccs}$ = {C, L, SERIES, PSS, FLIP, NOP, GND, VIA0, VIA1, VIA2, VIA3, VIA4, VIA5, VIA6, VIA7}.

The terminal set, $\mathcal{T}_{ccs}$, for the construction-continuing subtree consists of

$\mathcal{T}_{ccs}$ = {END}.

8.3. Fitness Measure

Note that all of the above could be applied to *any* one-input, one-output LC circuit. It is the fitness measure that directs the evolutionary process (in the search space of circuit-constructing program trees) to a program tree that constructs the desired asymmetric bandpass filter. In other words, circuit structure arises from fitness.

The evaluation of fitness for each individual circuit-constructing program tree in the population begins with its execution. This execution applies the functions in the program tree to the very simple embryonic circuit, thereby developing the embryonic circuit into a fully developed circuit. A netlist describing the circuit is then created. The netlist identifies each component of the circuit, the nodes to which that component is connected, and the value of that component. Each circuit is then simulated to determine its behavior. The 217,000-line SPICE simulator was modified to run as a submodule within the genetic programming system. SPICE (an acronym for Simulation Program with Integrated Circuit Emphasis) is a large family of programs written over several decades at the University of California at Berkeley for the simulation of analog, digital, and mixed analog/digital electrical circuits (Quarles et al. 1994). The input to a SPICE simulation consists of a netlist describing the circuit to be analyzed and certain commands that instruct SPICE as to the type of analysis to be performed and the nature of the output to be produced.

In general, the fitness measure may incorporate any calculable characteristic or combination of characteristics of the circuit, including the circuit's behavior in the time domain, its behavior in the frequency domain, its power consumption, or the number, cost, or surface area of its components.

Since we are designing a filter, the focus is on the behavior of the circuit in the frequency domain. SPICE is requested to perform an AC small signal analysis and to report the circuit's behavior for each of 101 frequency values chosen from the range between 10,000 Hz to 200,000 Hz (in equal increments on a logarithmic scale).

Fitness is measured in terms of the sum, over these 101 fitness cases, of the absolute weighted deviation between the actual value of the relative voltage (in decibels) that is produced by the circuit at the probe point **VOUT** and the target value for relative voltage. The smaller the value of fitness, the better. A fitness of zero represents an ideal filter. Specifically, the standardized fitness is

$$F(t) = \sum_{i=0}^{100} [W(d(f_i), f)d(f_i)]$$

where f_i is the frequency of fitness case i; $d(x)$ is the difference between the target and the observed values at frequency x; and $W(y, x)$ is the weighting for a difference of y at frequency x.

The fitness measure does not penalize ideal values; it slightly penalizes every acceptable deviation; and it heavily penalizes every unacceptable deviation.

First consider the 87 sample points outside the desired passband. If the relative voltage is below –120 dB for a particular fitness case outside the desired passband, then the deviation is deemed to be zero for that fitness case. If the relative voltage is in the acceptable region (light shading in figures 11, 12, and 13) for a particular fitness case outside the desired passband, the absolute value of the deviation from –120 dB is weighted by 1 for that fitness case. If the relative voltage is in the unacceptable region (white in figures 11, 12, and 13) for a particular fitness case outside the desired passband, the absolute value of the deviation from –120 dB is weighted by 10 for that fitness case. This arrangement reflects the fact that the ideal relative voltage outside the desired passband is –120 dB or less, the fact that the previously described deviations are acceptable, and the fact that other deviations are unacceptable.

Next consider the 14 points inside the desired passband. If the relative voltage is between –0.6 dB and 0.6 dB, the absolute value of the deviation from 0 dB is weighted by a factor of 10.0; however, if the relative voltage lies outside that narrow region, the absolute value of the deviation from 0 dB is weighted by a factor of 100. This arrangement reflects the fact that the ideal relative voltage in the passband is 0 dB, the fact that a ripple of up to 0.6 dB centered at 0 dB is acceptable, and the fact that a relative voltage more than 0.6 dB away from 0 dB in either direction is unacceptable. The factors of 10 and 100 (rather than 1 and 10) were selected to approximately equalize the weight given to the relatively few sample points (i.e., 14) within the desired passband as compared to the relatively large number of sample points (i.e., 87) outside the desired passband.

Hits are defined as the number (0 to 101) of fitness cases for which the relative voltage is acceptable or ideal.

Some of the bizarre circuits that are created by genetic programming cannot be simulated by SPICE. Such circuits are assigned a high penalty value of fitness (10^8).

8.4. Control Parameters

The population size, M, is 640,000. Each subsequent generation of the run is created from the population of the preceding generation by performing 569,600 one-offspring crossover operations (89% of 640,000), 64,000 reproduction operations (10% of 640,000), and 6,400 mutation operations (1% of 640,000). A maximum size of 200 points was established for each of the two result-producing branches. The other minor parameters for controlling the runs were the default values specified in Koza 1994a (appendix D).

8.5. Parallel Implementation

This problem was run on a medium-grained parallel Parsytec computer system consisting of 64 80 MHz PowerPC 601 processors arranged in a toroidal mesh with a host PC Pentium type computer (running Windows). The so-called *distributed genetic algorithm* for parallelization was used with a subpopulation (deme) size of Q = 10,000 at each of D = 64 demes. On each generation, four boatloads of emigrants, each consisting of B = 2% (the migration rate) of the node's subpopulation (selected on the basis of fitness) were dispatched to each of the four toroidally adjacent processing nodes. Details of the parallel implementation of genetic programming can be found in Andre and Koza 1996.

9. Results

We made three runs of this problem. Two came close to solving the problem. We now discuss in detail the one run that produced an individual scoring 101 hits.

The best circuit of the 640,000 circuits from generation 0 has a fitness of 59,191, scores 72 hits, and is shown in figure 8. Figure 11 shows its behavior in the frequency domain. The horizontal scale is linear and ranges from 10,000 Hz to 200,000 Hz. The vertical scale is relative voltage in decibels. The curve represents the behavior of this circuit in the frequency domain. As can be seen, the curve in this figure peaks at –14.7 dB when the frequency is 38.40 KHz. This behavior does not even remotely satisfy Nielsen's design goals. However, the narrow peak at 38.40 KHz (which falls far short of the desired height) does, in fact, lie within the desired passband. That is, this inadequate circuit differentially passes frequencies near 38.40 KHz.

About 78% of the programs of generation 0 for this problem produce circuits that cannot be simulated by SPICE. However, the percentage of unsimulatable circuits drops rapidly as new offspring are created by genetic programming using Darwinian selection, crossover, and mutation. The percentage of unsimulatable programs drops to 37% by generation 1. Between generations 2 and 202, the average percentage of unsimulatable circuits is 13.7%. That is, starting as early as the third generation, the vast majority of the circuits are simulatable.

This observation supports the general principle that the individuals in the population in intermediate generations of a run of genetic programming (and random subtrees picked from them by the crossover operation) differ markedly from the individuals (and their randomly picked subtrees) in the randomly created population of generation 0 of the same run.

Crossover fragments from intermediate generations of a run of genetic programming are very different from the randomly grown subtrees supplied by the mutation operation. This fact is experimental evidence, for this non-trivial problem, that the population serves the vital role in the genetic algorithm of providing a reservoir of useful fragments to rapidly advance the search.

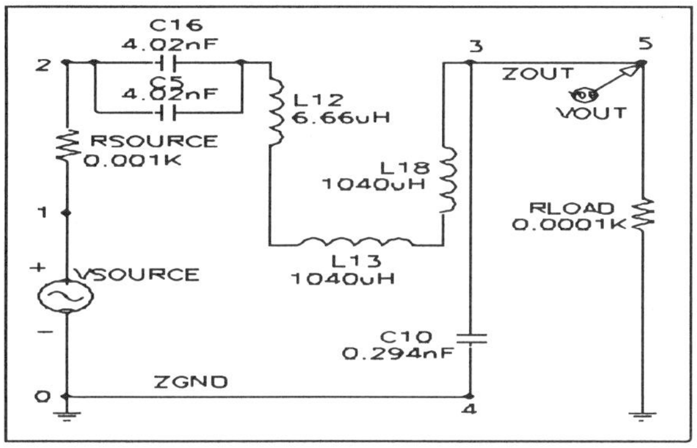

Figure 8 Best circuit of generation 0.

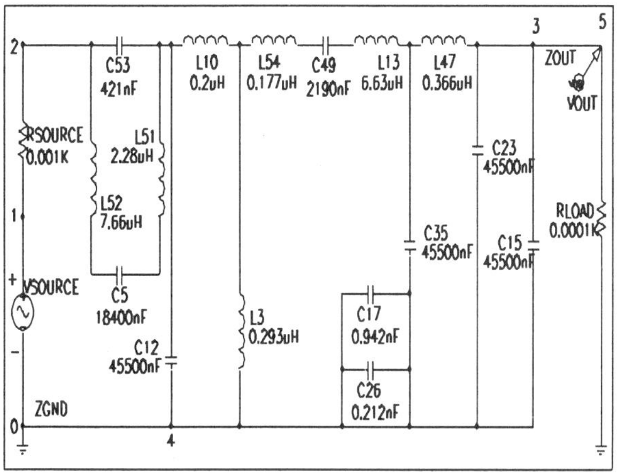

Figure 9 Best circuit of generation 32.

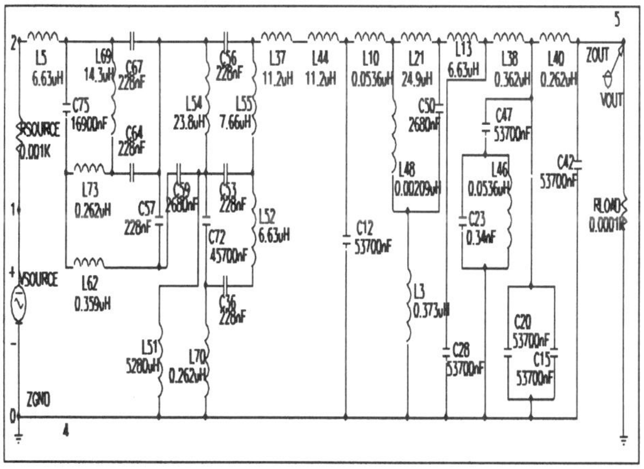

Figure 10 Best-of-run circuit of generation 199.

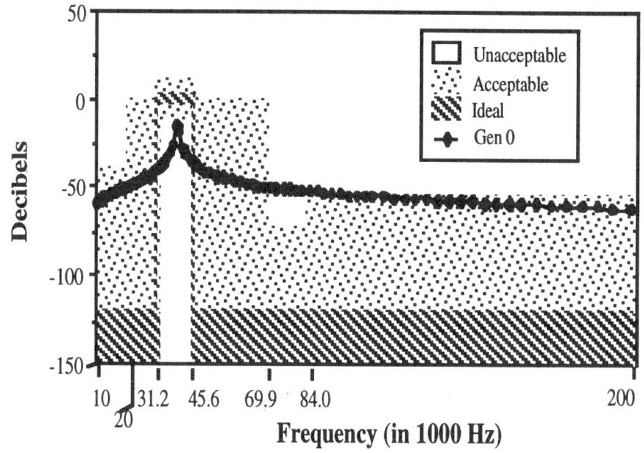

Figure 11 Frequency domain behavior of best circuit of generation 0.

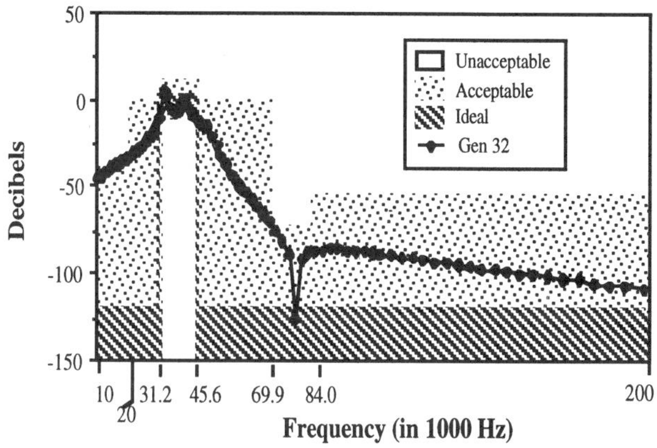

Figure 12 Frequency domain behavior of best circuit of generation 32.

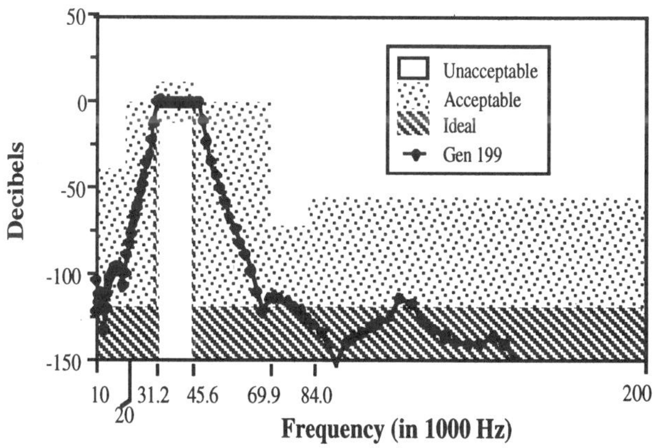

Figure 13 Frequency domain behavior of best-of-run circuit of generation 199.

Point-to-point search techniques (e.g., simulated annealing, hillclimbing) do not have this reservoir from which to draw.

In embarking on this project of trying to evolve electronic circuits using genetic programming, one of our major threshold concerns was whether any significant percentage of the randomly created circuits of generation 0 in this highly epistatic search space would be simulatable at all by SPICE. A second concern was whether the crossover operation would create any significant percentage of simulatable circuits. Neither of these concerns materialized with genetic programming on this problem. Darwinian selection apparently is very effective in quickly steering the population on successive generations into the portion of the search space where parents can successfully sire simulatable offspring by means of the crossover operation.

The best individual program in generation 32 has a fitness of 10,343.0, scores 87 hits, and is shown in figure 9. Figure 12 shows its frequency domain behavior.

The first 100%-compliant program achieving 101 hits appeared in generation 132 and has a fitness of 2252.5. However, we allowed the run to continue. The best-of-run program emerged in generation 199, has a fitness of 2024.0, scores 101 hits, and is shown in figure 10. Figure 13 shows its frequency domain behavior. As can be seen, 100% of the points are in compliance with the design specifications. Moreover, all the points in the passband are within 0.39 dB of 0 dB whereas the specifications merely require a passband ripple of less than 0.6 dB.

10. Related Work

The above techniques have recently been successfully applied to a variety of different problems of circuit synthesis.

10.1. Lowpass "Brick Wall" Filter

Consider the problem of designing a lowpass filter with passband below 1,000 Hz and a stopband above 2,000 Hz (as described in Koza, Bennett, Andre, and Keane 1996a). The voltages in the passband are to be between 970 millivolts and 1 volt (i.e., the passband ripple is 30 millivolts) and the voltages in the stopband between 0 volts and 1 millivolt.

In some runs of this problem, the genetically evolved lowpass filter has a topology that is similar to that employed by human engineers. For example, in one run, the 100% compliant evolved circuit (figure 14) had the recognizable ladder topology of a Butterworth or Chebychev filter (i.e., a composition of inductors horizontally with capacitors as vertical shunts).

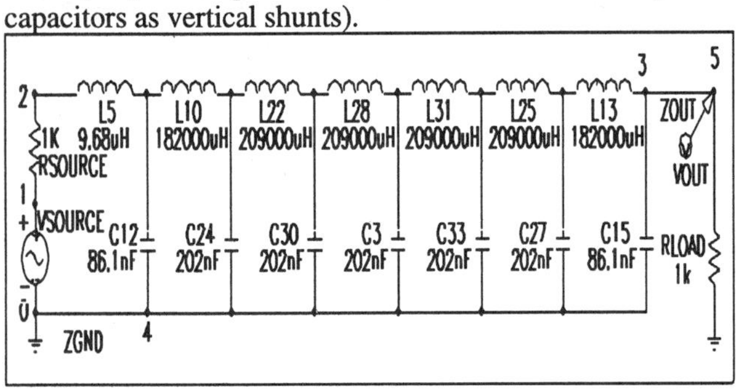

Figure 14 Seven-rung ladder circuit from generation 32.

In another run, a 100%-compliant recognizable "bridged T" arrangement (figure 15) was evolved (involving capacitors **C3** and **C15** and inductor **L11** in conjunction with **L14**).

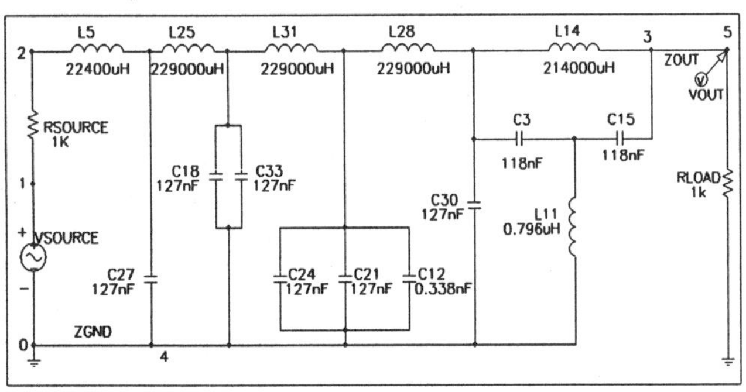

Figure 15 "Bridged T" from generation 64.

Figure 16 shows a 100%-compliant circuit from generation 212 of yet another run. It has a novel topology that no electrical engineer would be likely to create.

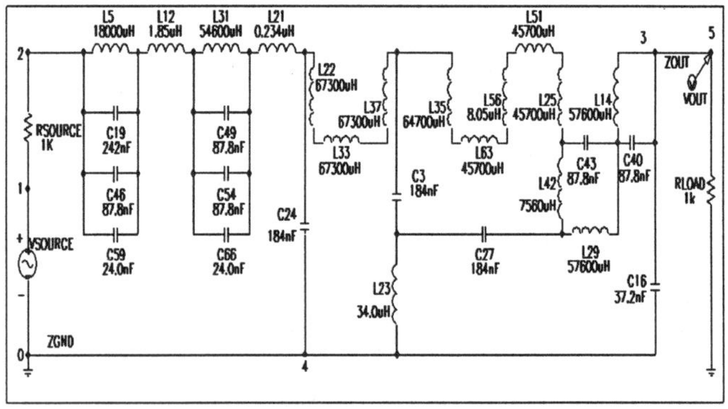

Figure 16 Best circuit from generation 212.

10.2. A 5 dB Amplifier using Transistors

Now consider the problem of designing an amplifier with an amplification factor of 3.5 over the frequency range of 20 Hz to 20,000 Hz. Amplifiers are active circuits; they require active components such as transistors. The major change in the function set needed to evolve the amplifier is to add the resistor-creating `R` function and the transistor-creating `QT` functions.

Figure 17 shows a genetically evolved 5 dB amplifier. The boxes highlight a recognizable voltage gain stage and a quasi-Darlington emitter-follower stage.

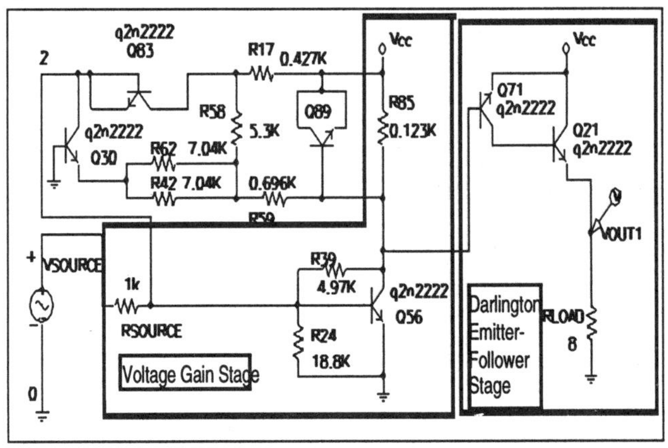

Figure 17 Genetically evolved 5 dB amplifier from generation 45.

10.3. A Crossover Filter

Now consider the problem of designing a crossover (woofer and tweeter) filter with a crossover frequency of 2,512 Hz (as described more fully in Koza, Bennett, Andre, and Keane 1996b). The voltages in the passband are to be between 970 millivolts and 1 volt (i.e., the passband ripple is 30 millivolts) and the voltages in the stopband between 0 volts and 1 millivolt. This problem requires a one-input, two-output embryonic circuit and requires that the fitness be measured at two probe points, but is otherwise similar to the previously discussed circuits. In one run, the best-of-run circuit (figure 18) from generation 137 scores 192 hits. The lowpass part of the best-of-run circuit has the Butterworth topology (but not the Butterworth component values). The highpass part of the best-of-run circuit also has the Butterworth topology (but not the Butterworth component values), except for additional capacitor **C36**. Note that highpass Butterworth filters have capacitors in series horizontally and inductors as vertical shunts. An analysis of this genetically evolved circuit in the frequency domain indicates that it delivers a response that is slightly better than the combination of lowpass and highpass Butterworth filters of order 7.

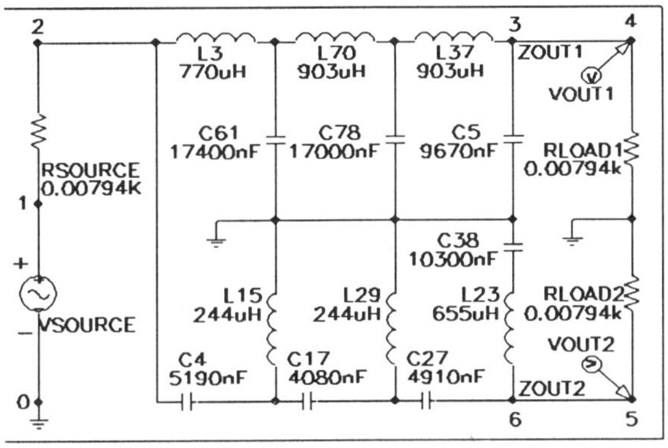

Figure 18 Best circuit from generation 137.

10.4. Related Paper in this Volume

Another paper (appearing next in this volume) on the subject of evolving circuits was presented at this conference.

Acknowledgments

Tom L. Quarles provided helpful advice concerning SPICE. Simon Handley and Scott Brave commented on earlier drafts.

Bibliography

Andre, David and Koza, John R. 1996. Parallel genetic programming: A scalable implementation using the transputer architecture. In Angeline, P. J. and Kinnear, K. E. Jr. (editors). *Advances in Genetic Programming 2.* Cambridge: MIT Press.

Degrauwe, M. 1987. IDAC: An interactive design tool for analog integrated circuits. *IEEE Journal of Solid State Circuits.* 22:1106–1116.

Gruau, Frederic. 1992. *Cellular Encoding of Genetic Neural Networks*. Technical report 92-21. Laboratoire de l'Informatique du Parallélisme. Ecole Normale Supérieure de Lyon. May 1992.

Harjani, R., Rutenbar, R. A., and Carley, L. R. 1989. OASYS: A framework for analog circuit synthesis. *IEEE Transactions on Computer Aided Design.* 8:1247–1266.

Hemmi, Hitoshi, Mizoguchi, Jun'ichi, and Shimohara, Katsunori. 1994. Development and evolution of hardware behaviors. In Brooks, R. and Maes, P. (editors). *Artificial Life IV: Proceedings of the Fourth International Workshop on the Synthesis and Simulation of Living Systems.* Cambridge, MA: MIT Press. Pages 371–376.

Higuchi, T., Niwa, T., Tanaka, H., Iba, H., de Garis, H. and Furuya, T. 1993. Evolvable hardware – Genetic-based generation of electric circuitry at gate and hardware description language (HDL) levels. Electrotechnical Laboratory technical report 93-4, Tsukuba, Ibaraki, Japan.

Holland, John H. 1975. *Adaptation in Natural and Artificial System.* Ann Arbor, MI: University of Michigan Press.

Koh, H. Y., Sequin, C. H. and Gray, P. R. 1990. OPASYN: A compiler for MOS operational amplifiers. *IEEE Trans. on Computer Aided Design.* 9:113–125.

Koza, John R. 1992. *Genetic Programming: On the Programming of Computers by Means of Natural Selection.* Cambridge, MA: MIT Press.

Koza, John R. 1994a. *Genetic Programming II: Automatic Discovery of Reusable Programs.* Cambridge, MA: MIT Press.

Koza, John R. 1994b. *Genetic Programming II Videotape: The Next Generation.* Cambridge, MA: MIT Press.

Koza, John R., Bennett III, Forrest H, Andre, David, and Keane, Martin A. 1996a. Automated design of both the topology and sizing of analog electrical circuits using genetic programming. In Gero, John S. and Sudweeks, Fay (editors). *Artificial Intelligence in Design '96.* Dordrecht: Kluwer Academic Publishers. 151-170.

Koza, John R., Bennett III, Forrest H, Andre, David, and Keane, Martin A. 1996b. Four problems for which a computer program evolved by genetic programming is competitive with human performance. *Proceedings of the 1996 IEEE International Conference on Evolutionary Computation.* IEEE Press.

Koza, John R., and Rice, James P. 1992. *Genetic Programming: The Movie.* Cambridge, MA: MIT Press.

Kruiskamp, Wim and Leenaerts, Domine. 1995. DARWIN: CMOS opamp synthesis by means of a genetic algorithm. *Proceedings of the 32nd Design Automation Conference.* New York, NY: Association for Computing Machinery. Pages 433–438.

Maulik, P. C. Carley, L. R., and Rutenbar, R. A. 1992. A mixed-integer nonlinear programming approach to analog circuit synthesis. *Proceedings of the 29th Design Automation Conference.* Los Alamitos, CA: IEEE Press. Pages 698–703.

Mizoguchi, Junichi, Hemmi, Hitoshi, and Shimohara, Katsunori. 1994. Production genetic algorithms for automated hardware design through an evolutionary process. *Proceedings of the First IEEE Conference on Evolutionary Computation.* IEEE Press. Volume I. Pages 661-664.

Nielsen, Ivan Riis. 1995. A C-T filter compiler - From specification to layout. *Analog Integrated Circuits and Signal Processing.* 7(1):21–33.

Ning, Z., Kole, M., Mouthaan, T., and Wallings, H. 1992. Analog circuit design automation for performance. *Proceedings of the 14th IEEE CICC.* New York: IEEE Press. Pages 8.2.1–8.2.4.

Quarles, Thomas, Newton, A. R., Pederson, D. O., and Sangiovanni-Vincentelli, A. 1994. *SPICE 3 Version 3F5 User's Manual.* Department of Electrical Engineering and Computer Science, University of California, Berkeley, California. March 1994.

Rutenbar, R. A. 1993. Analog design automation: Where are we? Where are we going? *Proceedings of the l5th IEEE CICC.* New York: IEEE Press. 13.1.1-13.1.8.

Use of Automatically Defined Functions and Architecture-Altering Operations in Automated Circuit Synthesis with Genetic Programming

John R. Koza
Computer Science Dept.
258 Gates Building
Stanford University
Stanford, California 94305
koza@cs.stanford.edu

David Andre
Visiting Scholar
Computer Science Dept.
Stanford University
Stanford, California 94305
andre@flamingo.stanford.edu

Forrest H Bennett III
Visiting Scholar
Computer Science Dept.
Stanford University
Stanford, California 94305
fhb3@slip.net

Martin A. Keane
Econometrics Inc.
5733 West Grover
Chicago, Illinois 60630
makeane@ix.netcom.com

ABSTRACT

This paper demonstrates the usefulness of automatically defined functions and architecture-altering operations in designing analog electrical circuits using genetic programming.

A design for a lowpass filter is genetically evolved in which an automatically defined function is profitably reused in the 100% compliant circuit. The symmetric reuse of an evolved substructure directly enhances the performance of the circuit. Genetic programming rediscovered the classical ladder topology used in Butterworth and Chebychev filters as well as the more complex topology used in Cauer (elliptic) filters.

A design for a double-passband filter is genetically evolved in which the architecture-altering operations discover a suitable program architecture dynamically during the run. Two automatically defined functions are profitably reused in the genetically evolved 100% complaint circuit.

1. Introduction

Human-designed electrical circuits are replete with instances of the hierarchical reuse of previously designed subcircuits. Indeed, design of a large electronic circuit would be impractical if the human designer had to start from first principles and rethink the design of each subcircuit each occasion time it is needed.

The previous paper in this volume – "Automated WYWIWYG Design of Both the Topology and Component Values of Electrical Circuits Using Genetic Programming" – demonstrated that complex structures such as electrical circuits can be evolved by means of natural selection (Koza, Bennett, Andre, and Keane 1996).

The regularity, symmetry, and reuse of substructures found in human-designed electrical circuits suggests that automatically defined functions might be useful in the problem domain of circuit synthesis. Accordingly, this paper explores the usefulness of automatically defined functions and architecture-altering operations for the problem of automated circuit synthesis.

Section 2 provides background on automatically defined functions and the architecture-altering operations. Section 3 describes the design of a lowpass filter in which an automatically defined function is profitably reused in the genetically evolved electrical circuit. Section 4 describes the design of a double-passband filter in which the architecture-altering operations successfully evolve the architecture of the solution and in which automatically defined functions are profitably reused in the genetically evolved solution.

2. Background

Genetic programming is an extension of John Holland's genetic algorithm (1975) in which the population consists of computer programs of varying sizes and shapes (Koza 1992, Koza and Rice 1992).

The book *Genetic Programming II: Automatic Discovery of Reusable Programs* (Koza 1994a, 1994b) describes how to evolve multi-part programs consisting of a main program and one or more reusable, parameterized, hierarchically-called subprograms. This book is based on the premise that no approach to automated programming is likely to be successful on non-trivial problems unless it provides some hierarchical mechanism to exploit, by reuse and parameterization, the regularities, symmetries, homogeneities, similarities, patterns, and modularities inherent in problem environments.

An *automatically defined function* (*ADF*) is a function (subroutine, subprogram, DEFUN, procedure, or module) that is dynamically evolved during a run of genetic programming and that may be called by a calling program (or subprogram) that is concurrently being evolved. When automatically defined functions are being used, a program in the population consists of a hierarchy of one (or more) reusable function-defining branches (i.e., automatically defined functions) along with a main result-producing branch. Typically, the automatically defined functions possess one or more dummy arguments (formal parameters) and are reused with different instantiations of these dummy arguments. During a run, genetic programming creates different subprograms in the function-defining branches of the overall program, different main programs in the result-producing branch, different instantiations of the dummy arguments of the automatically defined functions (function-defining branches), and different hierarchical references between the branches.

When automatically defined functions are used, it is necessary to determine the architecture of the yet-to-be-evolved programs. The specification of the architecture consists of (a) the number of function-defining branches in the overall program, (b) the number of arguments (if any) possessed by each function-defining branch, and (c) if there is more than one function-defining branch, the nature of the hierarchical references (if any) allowed between them.

The architecture-altering operations (Koza 1994c, 1995a) provide one way to automate this architectural choice so that it can be made dynamically during a run of genetic programming. The six architecture-altering operations are motivated by the naturally occurring mechanisms of gene duplication and gene deletion in chromosome strings as described in Susumu Ohno's book *Evolution by Gene Duplication* (1970). In that book, Ohno advanced the thesis that the creation of new proteins (and hence new structures and new behaviors in living things) begins with a gene duplication.

The potential usefulness of automatically defined functions and architecture-altering operations have been previously demonstrated by both proof-of-principle ("toy") problems and by non-trivial problems. For example, on the transmembrane segment identification problem (Koza and Andre 1996), the results produced by using genetic programming with automatically defined functions and the architecture-altering operations were competitive with the human-written algorithm for that problem.

3. Evolving a Lowpass Filter Using Automatically Defined Functions

A *filter* is a one-input, one-output circuit that receives a signal as its input and passes the frequency components of the incoming signal that lie in a certain specified range (the *passband*) while stopping the frequency components of the signal that lie in other frequency ranges (the *stopband*).

Consider a circuit synthesis problem in which the design goal is to design a lowpass filter using inductors and capacitors with a passband below 1,000 Hertz and a stopband above 2,000 Hz. The voltages in the filter's passband are to lie in the narrow range between 970 millivolts and 1 volt (i.e., the permissible passband ripple is 30 millivolts) and the voltages in the stopband are to lie in the narrow range between 0 volts and 1 millivolt. The circuit is to be driven from an alternating-current input signal with a 2 volt amplitude with a source resistance of 1,000 Ohms and a load resistance of 1,000 Ohms.

3.1. Preparatory Steps

This paper assumes that the reader is familiar with the use of genetic programming in electrical circuit design described in the previous paper in this volume (Koza, Bennett, Andre, and Keane 1996).

3.1.1 Embryonic Circuit and Program Architecture

A one-input, one-output embryonic circuit with two writing heads is suitable for this problem. Thus, there are two result-producing branches in each program tree in the population.

In this section of this paper, we pre-specified that the common architecture of each individual in the population would consist of four zero-argument automatically defined functions and that there would be no hierarchical references among the automatically defined functions. Consequently, the architecture of every program tree in the population throughout the entire run consists of a total of six branches joined by a `LIST` function, as shown in figure 1.

Figure 1 Program architecture.

3.1.2 Function and Terminal Sets

The function set, $\mathcal{F}_{ccs}$, for each construction-continuing subtree is

$\mathcal{F}_{ccs}$ = {`C`, `L`, `SERIES`, `PSS`, `FLIP`, `NOP`, `THGND`, `THVIA0`, `THVIA1`, `THVIA2`, `THVIA3`, `THVIA4`, `THVIA5`, `THVIA6`, `THVIA7`, `ADF0`, `ADF1`, `ADF2`, `ADF3`}.

The `C`, `L`, `SERIES`, `PSS`, `FLIP`, and `NOP` functions are defined in Koza, Bennett, Andre, and Keane (1996).

The eight three-argument "via" functions (called `THVIA0`, ..., `THVIA7`) and the three-argument "ground" function (called `THGND`) enable distant parts of a circuit to be connected together. These functions are similar to the `VIA` and the `GND` functions, except that they each create three (instead of two) modifiable wires.

`ADF0`, `ADF1`, `ADF2`, and `ADF3` are automatically defined functions.

The terminal set, $\mathcal{T}_{ccs}$, for each construction-continuing subtree is

$\mathcal{T}_{ccs}$ = {`END`, `CUT`}.

The zero-argument `CUT` function causes the highlighted component to be removed from the circuit. The `CUT` function also causes the writing head to be lost.

The function set, $\mathcal{F}_{aps}$, for each arithmetic-performing subtree is

$\mathcal{F}_{aps}$ = {+, -}.

The terminal set, $\mathcal{T}_{aps}$, for each arithmetic-performing subtree is

$\mathcal{T}_{aps} = \{\Re\}$,

where $\Re$ represents floating-point random constants between −1.000 and +1.000.

3.1.3 Fitness Measure

The evaluation of fitness for each individual circuit-constructing program tree in the population begins with its execution. This execution applies the functions in the program tree to the embryonic circuit thereby developing the embryonic circuit into a fully developed circuit. A netlist describing the circuit is then created. Each circuit is then simulated to determine its behavior using a version of the SPICE3 (Quarles et al. 1994) simulator that we modified to run as a submodule within the genetic programming system.

The SPICE simulator is requested to perform an AC small signal analysis and to report the circuit's behavior for each of 101 frequency values chosen over five decades of frequency (from 1 Hz to 100,000 Hz). Each decade is divided into 20 parts (on a logarithmic scale).

Fitness is measured in terms of the sum, over these 101 fitness cases, of the absolute weighted deviation between the actual value of the voltage in the frequency domain that is produced by the circuit at its output probe point and the desired voltage at that point. The smaller the fitness, the better (with a fitness of zero being best).

The fitness measure is constructed so that it does not penalize ideal values; it slightly penalizes acceptable deviations; and it heavily penalizes unacceptable deviations.

Specifically, the procedure for each of the 61 points in the three-decade interval from 1 Hz to 1,000 Hz (i.e., the desired passband) is as follows: If the voltage at the output probe point equals the ideal value of 1.0 volts in this interval, the deviation is 0.0. If the voltage is between 970 millivolts and 1,000 millivolts, the absolute value of the deviation from 1,000 millivolts is weighted by a factor of 1.0. If the voltage is less than 970 millivolts, the absolute value of the deviation from 1,000 millivolts is weighted by a factor of 10.0. This arrangement reflects the fact that the ideal voltage in the passband is 1.0 volt, the fact that a 30 millivolt shortfall is acceptable, and the fact that a voltage below 970 millivolts in the passband is not acceptable.

The procedure for each of the 35 points between 2,000 Hz and 100,000 Hz (i.e., the desired stopband) is as follows: If the voltage is between 0 millivolts (the ideal value) and 1 millivolt, the absolute value of the deviation from 0 millivolts is weighted by a factor of 1.0. If the voltage is more than 1 millvolt, the absolute value of the deviation from 0 millivolts is weighted by a factor of 10.0. This arrangement reflects the fact that the ideal voltage in the stopband is 0 millivolts, the fact that a 1 millivolt ripple above 0 millivolts is acceptable, and the fact that a voltage above 1 millivolt in the stopband is not acceptable.

The deviation is deemed to be zero for each of the 5 points in the interval between 1,000 Hz and 2,000 Hz (i.e., the "don't care" band of frequencies).

Hits are defined as the number (from 5 to 101) of fitness cases for which the voltage is acceptable or ideal or which lie in the "don't care" band.

3.1.4 Control Parameters

The population size, M, is 640,000. The crossover percentage is 89% (producing 569,600 offspring), the reproduction percentage was 10%, and the mutation percentage was 1%. A maximum size of 300 points was established for each of the branches in each overall program. The other minor parameters were the default values in Koza 1994a (appendix D). The problem was run on a parallel computer system with a migration rate of B = 2% as described in Koza and Andre 1996.

3.2. Results for the Lowpass Filter

3.2.1 Initial Random Generation

The best circuit from generation 0 has a fitness of 58.6 and scores 52 hits (out of 101). Figure 3 shows that this best-of-generation circuit from generation 0 consists of a single inductor and a single capacitor. The topological arrangement of these two components is that of the first rung of a classical ladder. Figure 9 shows the frequency domain behavior of this best-of-generation circuit from generation 0.

3.2.2 Emergence of Reuse – A Two-Rung Ladder

The best circuit from generation 9 (figure 4) has the two-rung ladder topology. The ladder topology for a lowpass filter consists of repeated instances of various *series* inductors (so named because they run "in series" horizontally across the top of the figure) and repeated instances of various vertical *shunt* capacitors. The classical Butterworth or Chebychev filters are based on the ladder topology (Van Valkenburg 1982). Automatically defined function ADF0 supplies a group of three inductors (equivalent to one 154,400 μH inductor). Figure 2 shows this twice-called two-ported substructure developed by ADF0.

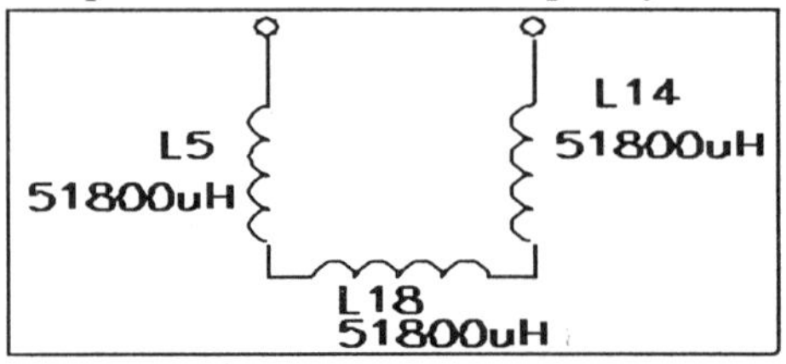

Figure 2 Twice-called two-ported automatically defined function ADF0 from generation 9.

Figure 10 shows the behavior in the frequency domain of the best-of-generation circuit from generation 9. Figures 10 – 14 were made with 200 points per decade for greater detail.

3.2.3 A Thrice-Called Substructure Further Enhances Performance

The best circuit from generation 16 (figure 5) has a fitness of 4.1 and scores 90 hits. This circuit has a three-rung ladder topology. When electrical engineers design Butterworth or Chebychev filters, additional rungs on the ladder (in conjunction with properly chosen numerical values for the components) generally improve the level of performance of the filter (at the expense of additional power consumption, space, and cost). Figure 11 shows the behavior in the frequency domain of the best circuit from generation 16. ADF0 is used three times in creating this best circuit of generation 16. Figure 15 shows this thrice-called automatically defined function.

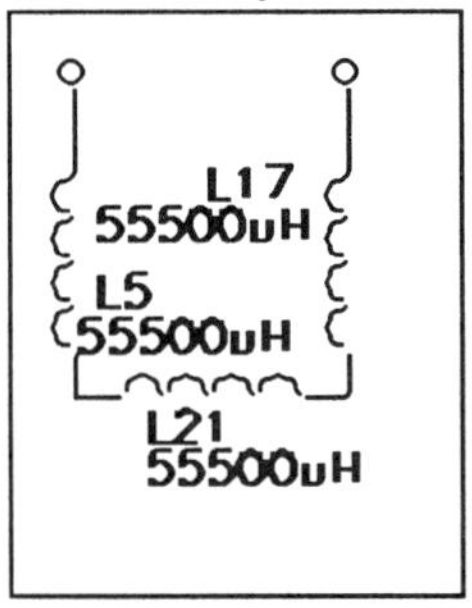

Figure 15 Thrice-called two-ported automatically defined function ADF0 from generation 16.

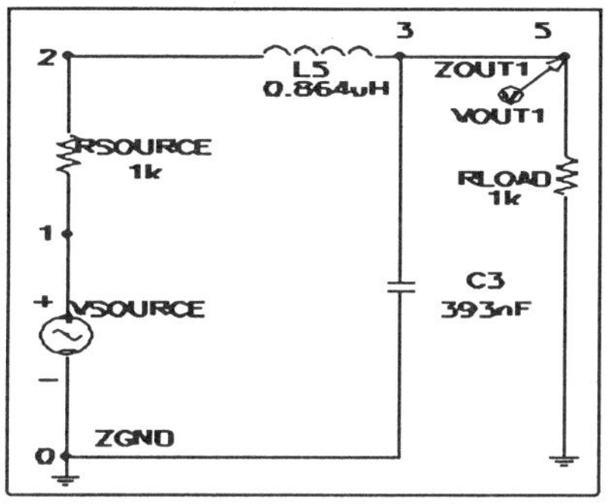

Figure 3 The best circuit from generation 0 is a one-rung ladder.

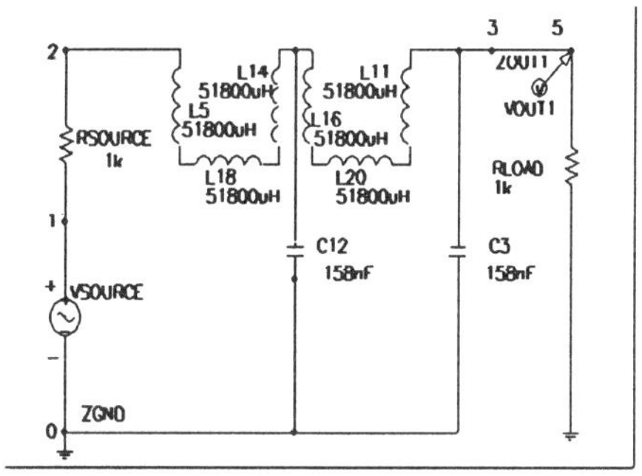

Figure 4 The best circuit from generation 9 is a two-rung ladder.

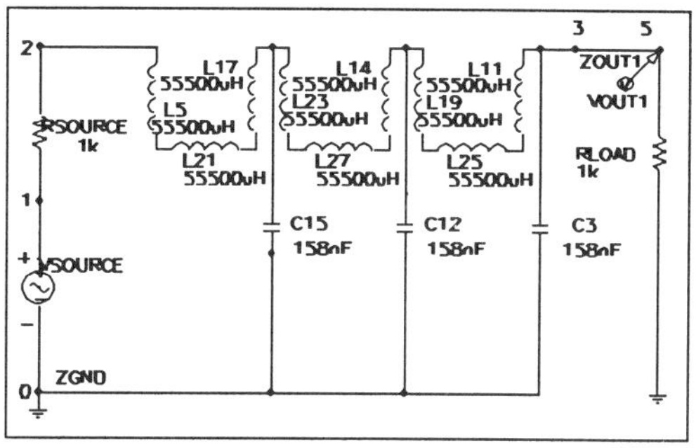

Figure 5 The best circuit of generation 16 is a three-rung ladder.

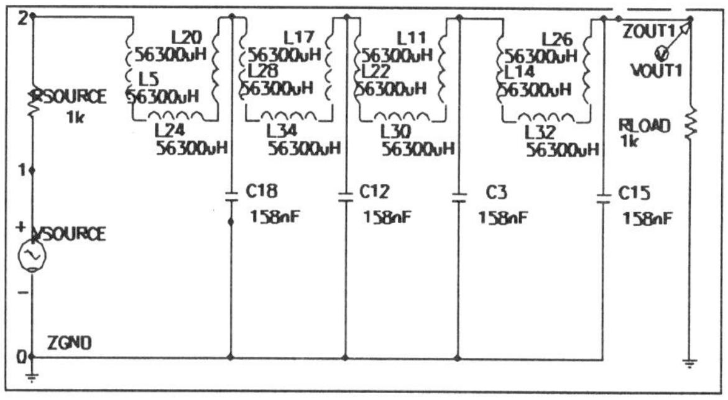

Figure 6 The best circuit of generation 20 is a four-rung ladder.

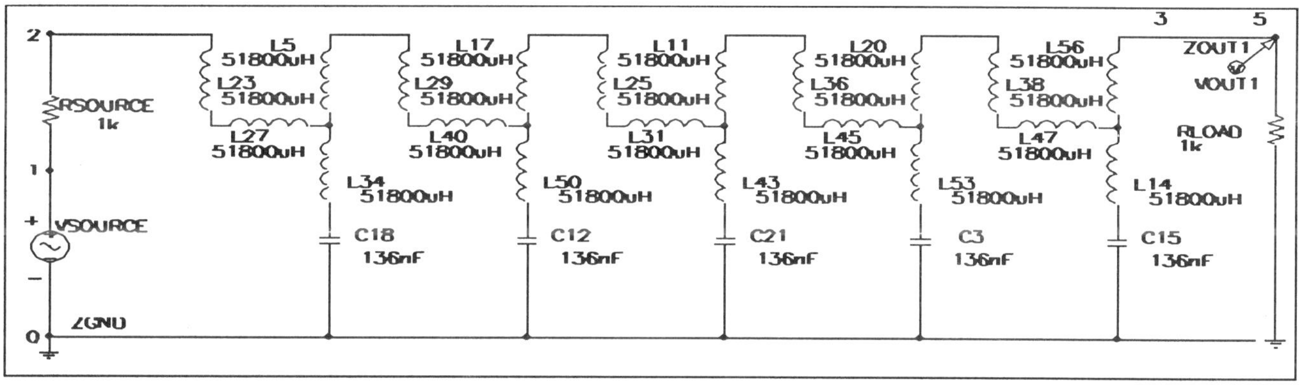

Figure 7 The 100% compliant best circuit of generation 31 has the Cauer (elliptic) topology.

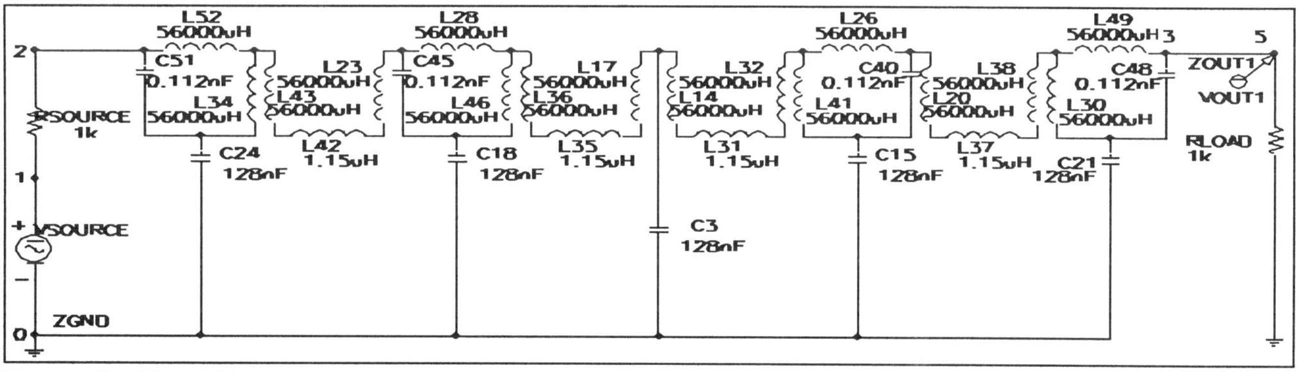

Figure 8 The 100% compliant best-of-run circuit from generation 35.

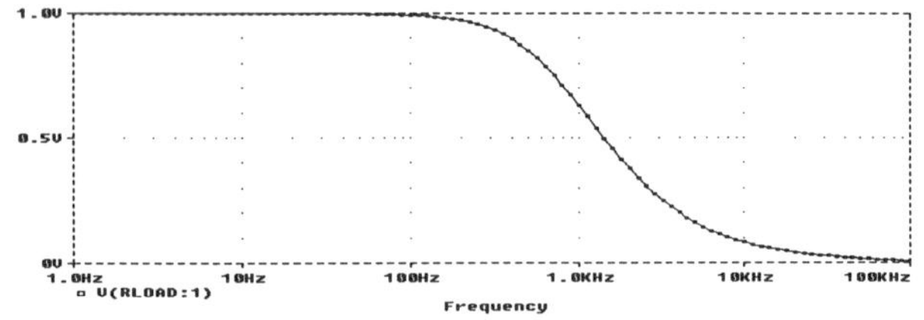

Figure 9 Frequency domain behavior of best circuit of generation 0.

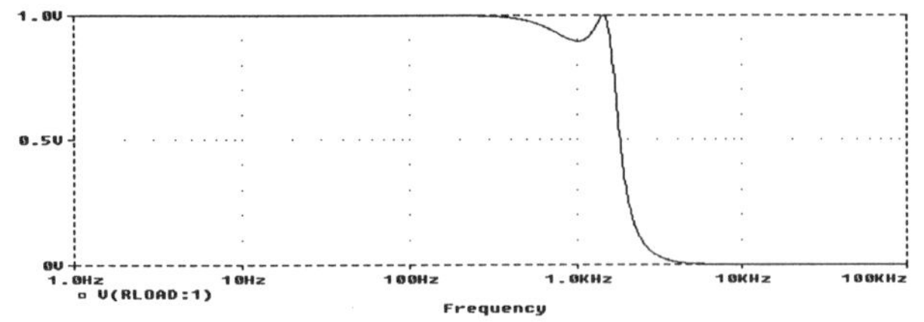

Figure 10 Frequency domain behavior of best circuit and generation 9.

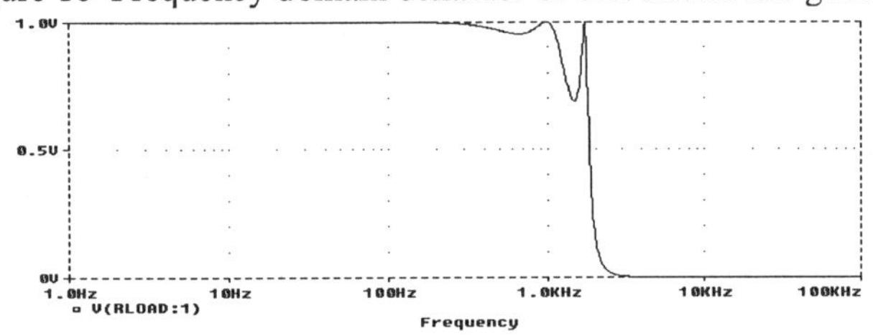

Figure 11 Frequency domain behavior of best circuit from generation 16.

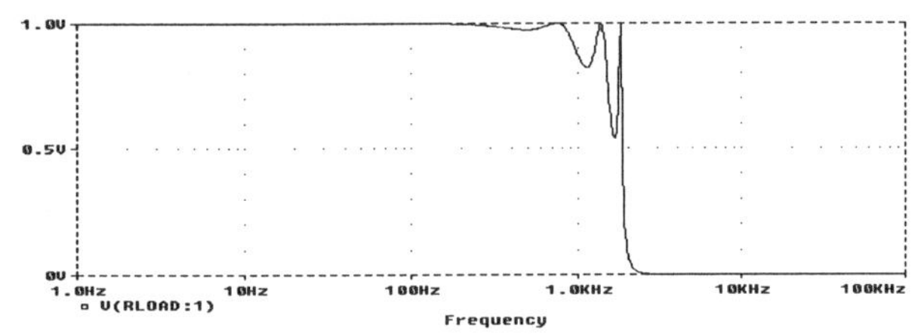

Figure 12 Frequency domain behavior of best circuit from gener

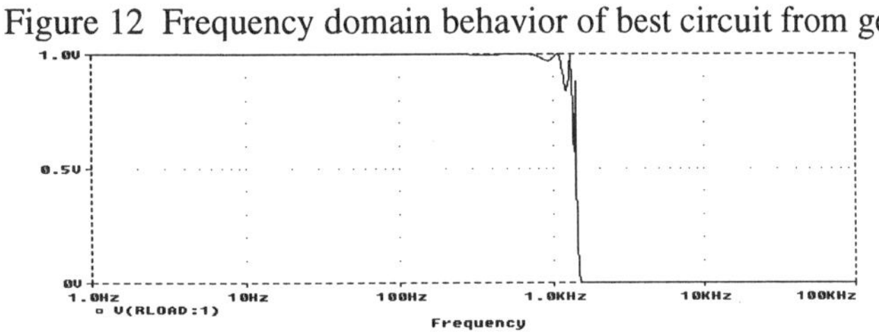

Figure 13 Frequency domain behavior of best circuit from generation 31.

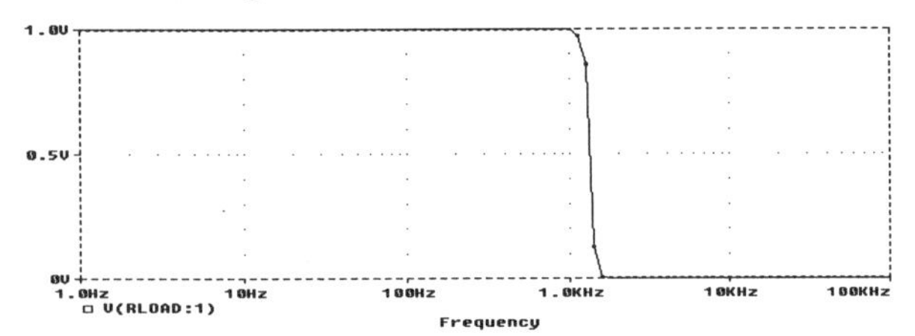

Figure 14 Frequency domain behavior of best-of-run circuit from generation 35.

3.2.4 A Quadruply-Called Substructure Further Enhances Performance

The best circuit from generation 20 (figure 6) has a fitness of 2.8 and scores 96 hits. It is a four-rung ladder.

ADF0 (figure 16) is used four times.

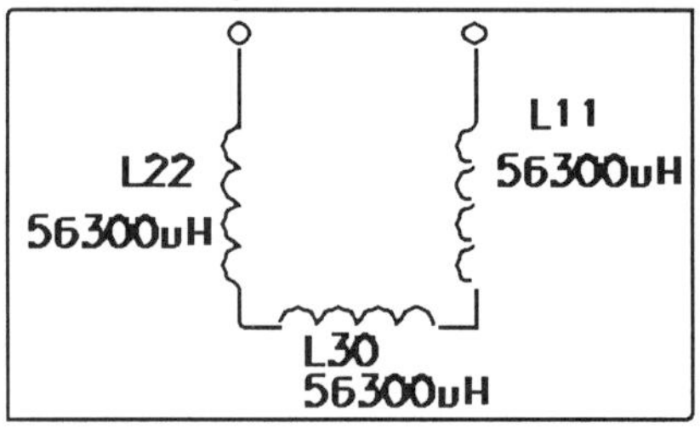

Figure 16 Quadruply-called two-ported automatically defined function ADF0 from generation 20.

Figure 12 shows the behavior in the frequency domain of the best circuit from generation 20. The enhanced performance of this circuit is the consequence of the additional repeated calls to ADF0. Note that in figures 10 – 12, there is ripple in the passband, but none in the stopband (i.e., these circuits exhibit a Chebychev-like response).

3.2.5 Emergence of the Elliptic Topology Further Enhances Performance

The best circuit in generation 31 (figure 7) has a fitness of 0.0850 and scores 101 hits (i.e., is 100% compliant). This individual has 41 points in its RPB0 and 7 points in its RPB1; it has 52, 297, 22, and 2 points in ADF0, ADF1, ADF2, and ADF3, respectively. Figure 13 shows its behavior in the frequency domain.

Automatically defined function ADF0 is interesting because it constructs a three-ported substructure. This ADF0 (figure 17) is used five times in the best circuit from generation 31.

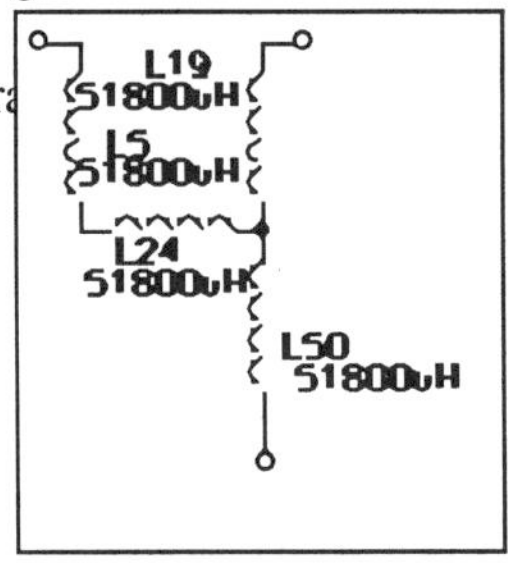

Figure 17 Quintuply-called three-ported automatically defined function ADF0 from generation 31.

This genetically evolved 100% compliant circuit is especially interesting because it has the topology of an elliptic (Cauer) filter.

After all of the pairs and triplets of inductors are consolidated, it can be seen that the circuit has the equivalent of six inductors horizontally across the top of the circuit and five vertical shunts. Each shunt consists of an inductor and a capacitor (e.g., **L34** and **C18** appear in the first shunt). This is the topology of the elliptic invented by Cauer.

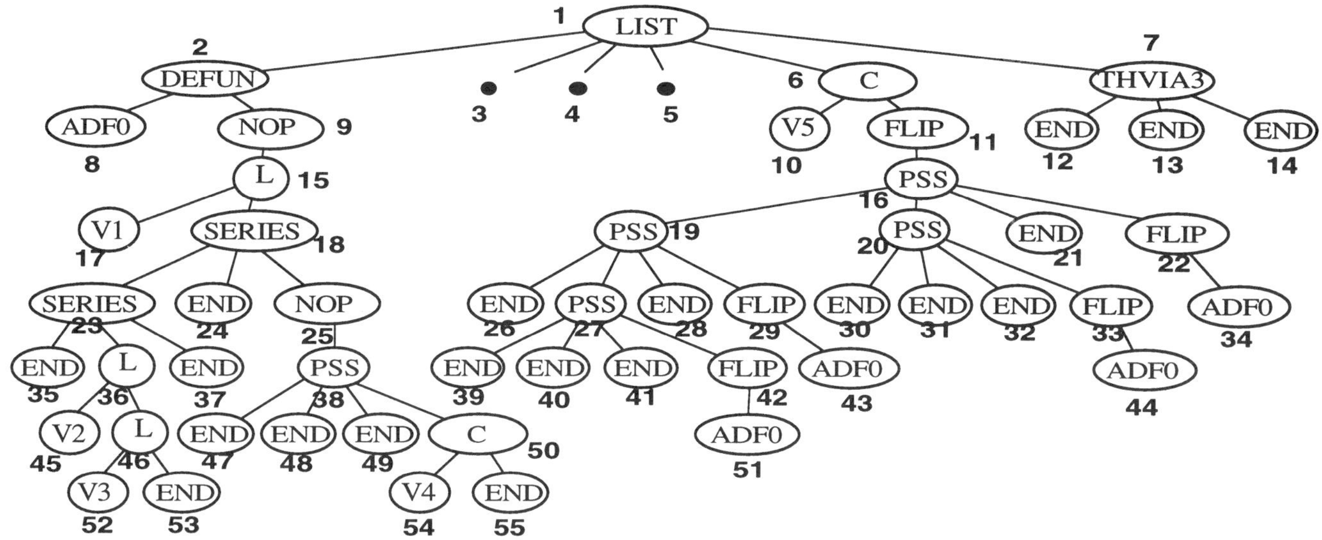

Figure 18 Best-of-run individual from generation 35.

The Cauer filter was a significant advance (both theoretically and commercially) over the Butterworth and Chebychev filters. For example, for one illustrative set of specifications, a fifth-order elliptic filter could equal the performance of 17th order Butterworth filter or an eighth order Chebychev filter. The eighth order Chebychev filter has one more component than the fifth order elliptic filter. As Van Valkenburg (1982, page 379) explains:

> "Cauer first used his new theory in solving a filter problem for the German telephone industry. His new design achieved specifications with one less inductor than had ever been done before. The world first learned of the Cauer method not through scholarly publication but through a patent disclosure, which eventually reached the Bell Laboratories. Legend has it that the entire Mathematics Department of Bell Laboratories spent the next two weeks at the New York Public library studying elliptic functions. Cauer had studied mathematics under Hilbert at Goettingen, and so elliptic functions and their applications were familiar to him."

Thus, this run of genetic programming illustrates the rediscovery of the ladder topology used in Butterworth and Chebychev filters as well as the more complex topology used in elliptic (Cauer) filters.

3.2.6 An Even Better Circuit

The best circuit in generation 35 (figure 8) has a fitness of 0.00752 (i.e., about an order of magnitude better than that of the best-of-generation individual from generation 31) and is also 100% compliant (i.e., scores 101 hits).

Only ADF0 is referenced by the result-producing branches.

Figure 14 shows the "brick wall" behavior in the frequency domain of this best-of-run best circuit from generation 35.

Notice this circuit's two-fold symmetry involving the repetition of a total of four modular substructures. The symmetry of the best-of-run circuit from generation 35 is a consequence of its quadruply-called three-ported automatically defined function, ADF0 (figure 19). Two inductors (**L52** and **L34**) and one capacitor (**C51**) form a triangle. There is an additional induction element (specifically, a series composition of three inductors, **L43**, **L42**, and **L23**) branching away from the triangle at the node where inductors **L52** and **L34** meet.

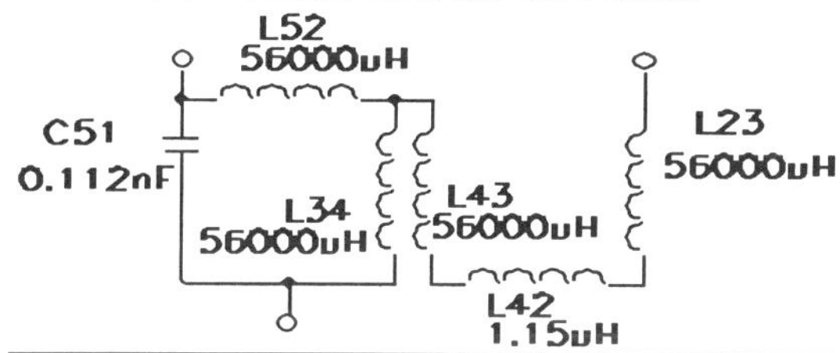

Figure 19 Quadruply-called three-ported automatically defined function ADF0 of best-of-run circuit from generation 35.

Figure 18 presents the best-of-run individual from generation 35 as a rooted, point-labeled tree with ordered branches. Note that ADF0 is invoked at points 34, 43, 44, and 51 of the figure. Since ADF0 is the only function-defining branch that is actually referenced by either result-producing branch, the three unreferenced function-defining branches (ADF1, ADF2, and ADF3) are represented by the filled circles labeled 3, 4, and 5. The complete arithmetic-performing subtrees are not shown, but, instead, are abbreviated as V1 through V5.

4. Evolving a Double-Bandpass Filter Using Architecture-Altering Operations

In the previous section, the user pre-specified the number of automatically defined functions as one of the preparatory steps for the run of genetic programming. In this section, the architecture-altering operations will be used to evolve the program architecture dynamically during the run.

The goal is to design a double-bandpass filter using inductors and capacitors as primitive components. Specifically, the goal is to design a double-bandpass filter whose first passband starts at 1,000 Hz and ends at 2,000 Hz, whose second passband starts at 1,000,000 Hz and ends at 2,000,000 Hz, and whose passband and stopband ripple are that of an elliptic filter (as detailed below).

The regularity inherent in a filter with two passbands suggests that automatically defined functions may potentially be useful for this design problem.

4.1. Preparatory Steps

4.1.1 Embryonic Circuit and Program Architecture

A one-input, one-output embryonic circuit with one writing head is suitable for this problem. Since the embryonic circuit has one writing head, each program in the initial population of programs has one result-producing branch. Since we did not use the operation of branch creation on this problem, each program tree in the initial random population at generation 0 also has one one-argument automatically defined function. Thus, each program tree at generation 0 starts with a uniform architecture consisting of a `LIST` function joining `ADF0` and the result-producing branch, `RPB`. The number of automatically defined functions in the eventual solution will be determined dynamically, during the run, using the architecture-altering operations of branch duplication and branch deletion. Thus, starting with generation 1 of each run, the population becomes architecturally diverse.

4.1.2 Function and Terminal Sets

For any branch, the function set, $\mathcal{F}_{ccs}$, for each construction-continuing subtree is

$\mathcal{F}_{ccs}$ = {`ADF0`, `C`, `L`, `SERIES`, `PSS`, `FLIP`, `NOP`, `THGND`, `THVIA0`, `THVIA1`, `THVIA2`, `THVIA3`, `THVIA4`, `THVIA5`, `THVIA6`, `THVIA7`}.

Here `ADF0` is the first automatically defined function.

The terminal set, $\mathcal{T}_{ccs\text{-}rpb}$, for each construction-continuing subtree in the result-producing branch is

$\mathcal{T}_{ccs\text{-}rpb}$ = {`END`, `CUT`}.

The terminal set, $\mathcal{T}_{ccs\text{-}adf}$, for each construction-continuing subtree in any function-defining branch is

$\mathcal{T}_{ccs\text{-}adf}$ = {`ARG0`, `END`, `CUT`}.

Since architecture-altering operations can create automatically defined functions, the set of potential new functions, $\mathcal{F}_{potential}$, is

$\mathcal{F}_{potential}$ = {`ADF1`, `ADF2`, `ADF3`, `ADF4`}.

The function set, $\mathcal{F}_{aps}$, for an arithmetic-performing subtree in any branch and the terminal set, $\mathcal{T}_{aps}$, for an arithmetic-performing subtree in any branch are the same as in the previous section of this paper.

4.1.3 Fitness Measure

The SPICE simulator is requested to perform an AC small signal analysis and to report the circuit's behavior for each of 176 frequency values chosen over seven decades of frequency (from 10 Hz to 100,000,000 Hertz). Each decade is divided into 25 parts (using a logarithmic scale).

Fitness is measured in terms of the sum, over these 176 fitness cases, of the absolute weighted deviation between the actual value of the voltage that is produced by the circuit at the output probe point and the target value for voltage.

The frequency range is divided into two passbands, three stopbands, and four "don't care" regions.

The procedure for each of the 8 points in each of the two passbands (a total of 16 points) is as follows: If the voltage is between 960 millivolts and 1,000 millivolts (the ideal value), the absolute value of the deviation from 1,000 millivolts is weighted by a factor of 1.0. If the voltage is less than 960 millivolts, the absolute value of the deviation from 1,000 millivolts is weighted by a factor of 10.0.

The procedure for each of the 120 points in the three stopbands is as follows: If the voltage is between 0 millivolts (the ideal value) and 40 millivolts, the absolute value of the deviation from 0 millivolts is weighted by a factor of 1.0. If the voltage is more than 40 millivolts, the absolute value of the deviation from 0 millivolts is weighted by a factor of 10.0.

There are 10 "don't care" points just before each passband and 10 "don't care" points just after each passband. The deviation is deemed to be zero for each of the 40 points in these four "don't care" bands.

Hits are defined as the number (between 40 and 176) of fitness cases for which the voltage is acceptable or ideal or that lie in one of the "don't care" bands.

4.1.4 Parameters

The control parameters were the same as in the previous section, except for the following:

(1) The architecture-altering operations are intended to be used sparingly on each generation. The percentage of operations on each generation after generation 5 was 87.5% one-offspring crossovers; 10% reproductions; 1% mutations; 1% branch duplications; 0% argument duplications; 0.5% branch deletions; 0% argument deletions; 0% branch creations; and 0% argument creations. Since we do not want to waste large amounts of computer time in early generations where only a few programs have any automatically functions, the percentage of operations on each generation before generation 6 was 82.0% one-offspring crossovers; 11% reproductions; 1% mutations; 5.0% branch duplications; 0% argument duplications; 1% branch deletions; 0% argument deletions; 0.0% branch creations; and 0% argument creations.

(2) The maximum number of automatically defined functions is five.

(3) The number of arguments for each automatically defined function is one.

(4) In randomly choosing functions during the creation of the initial random population, the `C`, `L`, `SERIES`, `PSS`, `FLIP`, `NOP`, `THGND`, and `ADF0` functions were each assigned a relative weight of 8, while each of the eight `THVIA` functions were assigned a relative weight of 1.

4.2. Results for Double-Bandpass Filter Using Architecture-Altering Operations

This run illustrates the progressive addition of function-defining branches and references to them.

The best circuit from generation 0 has a fitness of 720.3 and scores 54 hits; however, the result-producing branch of the program tree does not reference its 45-point automatically defined function `ADF0`.

In generation 2, the first best-of-generation individual with a referenced automatically defined function appears (with a fitness of 697.3 and 58 hits).

The first best-of-generation individual with two automatically defined functions appears in generation 22 (with a fitness of 184.7 and 110 hits); however, there are no references to either automatically defined function.

In generation 29, the first best-of-generation individual with two *referenced* automatically defined functions appears (with a fitness of 105.4 and 128 hits).

The first circuit scoring a full 176 hits is produced in generation 82 (with a fitness of 0.731).

A 100% complaint best-of-run individual emerged at generation 89 with four automatically defined functions. This individual has a fitness of 0.549 and scores 176 hits. The result-producing branch has 296 points and the function-defining branches have 82, 71, 64, and 82 points, respectively. ADF0 is called four times; ADF1 is called once; ADF2 is ignored; and ADF3 is called twice.

Figure 23 shows the best-of-run circuit from generation 89. Note that **L79**, **C78**, and **L66** are superfluous. Boxes are used to indicate the parts of the overall circuit that are developed by the automatically defined functions. The letters, A, B, C, and D indicate the interface points at which the substructure specified by each automatically defined function connects to the remainder of the circuit.

Figure 20 shows the three-ported substructure developed from ADF0 at each of the four places in the result-producing branch where ADF0 is invoked. The figure shows ADF0's three ports (A, B, and C) indicating the interface points at which this substructure developed from ADF0 connects to the remainder of the circuit.

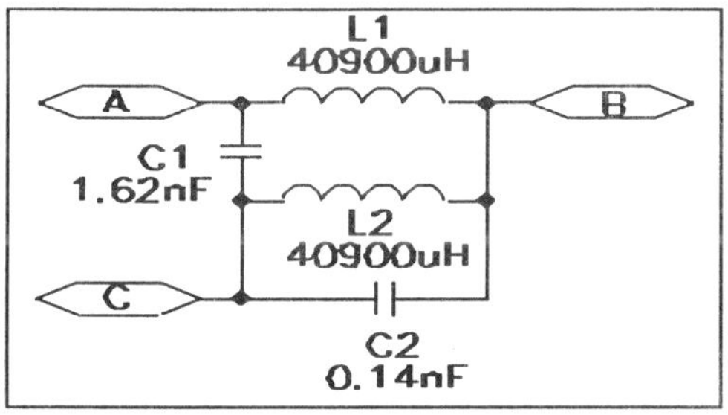

Figure 20 Three-ported quadruply-called ADF0 from generation 89.

Figure 21 shows the three-ported substructure developed from ADF1. ADF1 is invoked once.

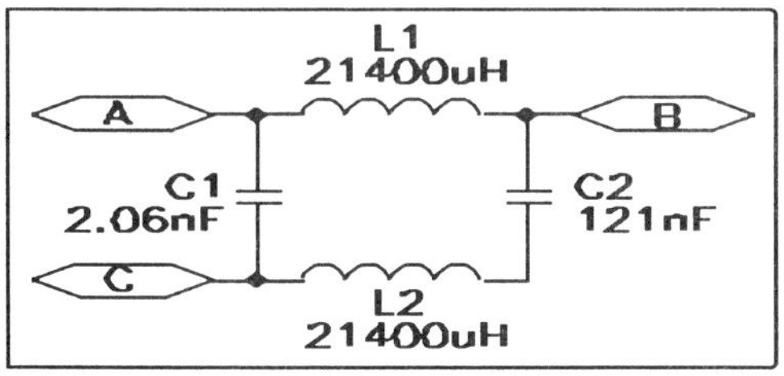

Figure 21 Three-ported ADF1 from generation 89.

Figure 22 shows the four-ported substructure developed from ADF3 at each of the two places in the result-producing branch where ADF3 is invoked. The figure shows ADF3's four ports (A, B, C, and D) indicating the interface points at which this substructure developed from ADF3 connects to the remainder of the circuit.

The insertion process for automatically defined functions (even those with no arguments) is context-sensitive. For example, in figure 23, the substructures developed by the two invocations of ADF3 are different. Specifically, one invocation employs ports A, B, and C (of figure 22) while the other employs ports A, B, and D. This difference was caused by parts of the overall program tree outside ADF3.

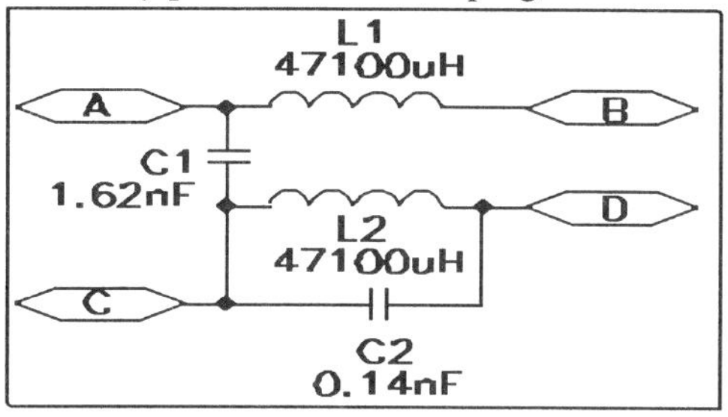

Figure 22 Four-ported twice-called ADF3 from generation 89.

Figure 24 shows the behavior in the frequency domain of the best-of-run circuit from generation 89. All 136 fitness cases that are not part of the "don't care" bands satisfy the design goals of this problem. Notice that the two small non-monotonic spikes lie in the "don't care" bands.

5. Conclusions

This paper demonstrates the usefulness of automatically defined functions and the architecture-altering operations in runs of genetic programming in which electrical circuits are being designed.

Acknowledgments

Simon Handley made helpful comments on drafts of this paper.

Bibliography

Andre, David and Koza, John R. 1996. Parallel genetic programming: A scalable implementation using the transputer architecture. In Angeline, Peter J. and Kinnear, Kenneth E. Jr. (editors). *Advances in Genetic Programming 2*. Cambridge, MA: MIT Press.

Holland, John H. 1975. *Adaptation in Natural and Artificial System.* Ann Arbor, MI: University of Michigan Press.

Koza, John R. 1992. *Genetic Programming: On the Programming of Computers by Means of Natural Selection.* Cambridge, MA: MIT Press.

Koza, John R. 1994a. *Genetic Programming II: Automatic Discovery of Reusable Programs.* Cambridge, MA: MIT Press.

Koza, John R. 1994b. *Genetic Programming II Videotape: The Next Generation.* Cambridge, MA: MIT Press.

Koza, John R. 1994c. *Architecture-Altering Operations for Evolving the Architecture of a Multi-Part Program in Genetic Programming.* Stanford University Computer Science Department technical report STAN-CS-TR-94-1528. October 21, 1994.

Koza, John R. 1995. Evolving the architecture of a multi-part program in genetic programming using architecture-altering operations. In McDonnell, John R., Reynolds,

Robert G., and Fogel, David B. (editors). 1995. *Evolutionary Programming IV: Proceedings of the Fourth Annual Conference on Evolutionary Programming.* Cambridge, MA: MIT Press. Pages 695–717.

Koza, John R. and Andre, David. 1996. Classifying protein segments as transmembrane domains using architecture-altering operations in genetic programming. In Angeline, Peter J. and Kinnear, Kenneth E. Jr. (editors). 1996. *Advances in Genetic Programming 2.* Cambridge, MA: MIT Press. In Press.

Koza, John R., Bennett III, Forrest H, Andre, David, and Keane, Martin A. 1996. Automated WYWIWYG design of both the topology and component values of analog electrical circuits using genetic programming. In Koza, John R., Goldberg, David E., Fogel, David B., and Riolo, Rick L. (editors). *Genetic Programming 1996: Proceedings of the First Annual Conference, July 28-31, 1996, Stanford University.* Cambridge, MA: MIT Press. In this volume.

Koza, John R., and Rice, James P. 1992. *Genetic Programming: The Movie.* Cambridge, MA: MIT Press.

Ohno, Susumu. 1970. *Evolution by Gene Duplication.* New York: Springer-Verlag.

Quarles, Thomas, Newton, A. R., Pederson, D. O., and Sangiovanni-Vincentelli, A. 1994. *SPICE 3 Version 3F5 User's Manual.* Department of Electrical Engineering and Computer Science, University of California, Berkeley, California. March 1994.

Van Valkenburg, M. E. 1982. *Analog Filter Design.* Fort Worth, TX: Harcourt Brace Jovanovich.

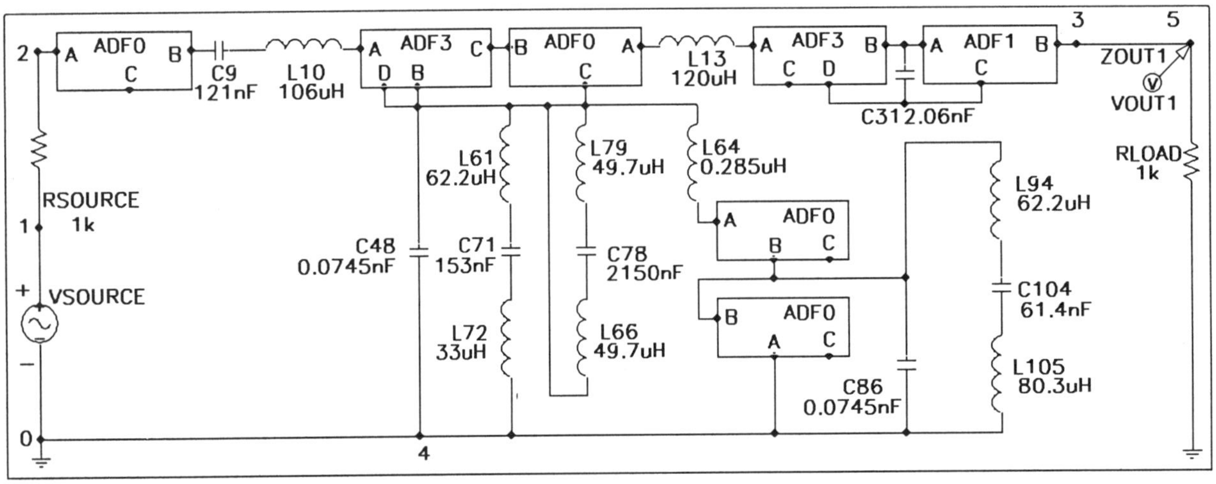

Figure 23 Best-of-run circuit from generation 89.

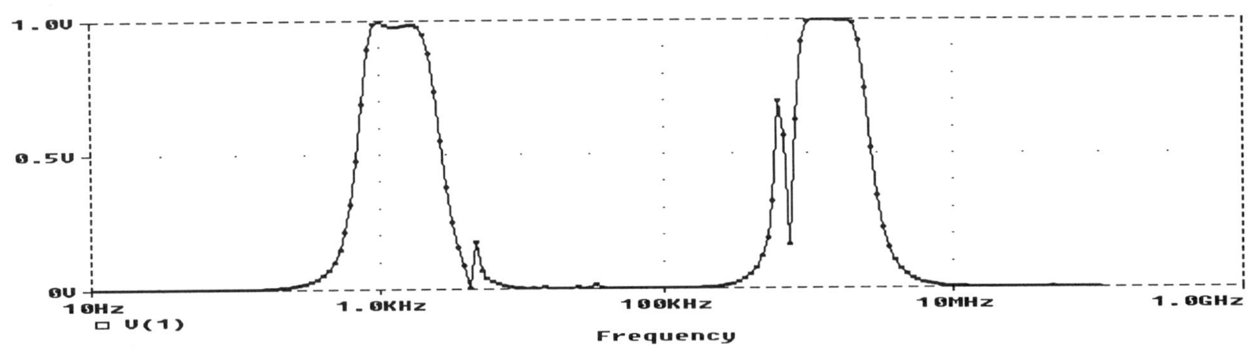

Figure 24 Frequency domain behavior of the best-of-run circuit from generation 89.

Using Data Structures within Genetic Programming

W. B. Langdon
Genetic Programming Group,
Dept. of Computer Science,
University College, London,
Gower Street, London WC1E 6BT, UK
Email: W.Langdon@cs.ucl.ac.uk
www: http://www.cs.ucl.ac.uk/staff/W.Langdon/
Tel: +44 (0) 171 380 7214, Fax: +44 (0) 171 387 1397

Abstract

Provision of appropriately structured memory is shown, in some cases, to be advantageous to genetic programming (GP) in comparison with directly addressable indexed memory.

Three "classic" problems are solved. The first two require the GP to distinguish between sentences that are in a context free language and those that are not given positive and negative training examples of the language. The two languages are, correctly nested brackets and a Dyck language (correctly nested brackets of different types). The third problem is to evaluate integer Reverse Polish (postfix) expressions.

Comparisons are made between GP attempting to solve these problems when provided with indexed memory or with stack data structures.

1 Introduction

Human programmers have long recognised that program production, maintenance and testing are eased if the software is written so that it be independent of memory access implementation details. This is achieved by using abstract data structures, such as stacks, queues and lists, to provide an interface between programs and memory.

We seek to show that data abstraction can be beneficially used within genetic programming (GP). So far work by Teller [Teller, 1994], Andre [Andre, 1994] and others shows GP can automatically create programs which explicitly use directly addressable (indexed) memory to solve problems which cannot be solved without memory and we have demonstrated that GP can automatically generate abstract data structures such as stacks, queues [Langdon, 1995] and lists [Langdon, 1996]. In this paper we show that GP can evolve programs which solve problems using such data structures. In two cases we show better GP performance when using data structures compared to directly addressable memory. In the remaining case (which is the first problem presented) the evolved solution uses an unexpected data structure which is appropriate to the problem rather than indexed memory when both are available.

The first two examples (Sections 2 and 3) show GP automatically producing programs which recognize whether a sample of a language is within a specific context free language. The third (Section 4) shows GP can evolve programs which evaluate Reverse Polish (postfix) expressions. These were chosen in the expectation that data structures at least as powerful as stacks would be required to solve them.

2 Balanced Bracket Problem

Other work on GP evolving language recognizers has concentrated upon using GP to evolve tree based specifications for abstract machines, such as finite state machines [Dunay *et al.*, 1994], deterministic push down automata [Zomorodian, 1995], machines composed of simple Turing machines [Dunay and Petry, 1995] or special memory nodes within the tree [Iba *et al.*, 1995]. In contrast we wish to investigate the impact of providing data structures v. indexed memory and so our GP executes the GP tree directly.

The balanced bracket problem is to recognise sentences composed of sequences of two symbols, `(` and `)`, which are correctly nested. E.g. `(())` is correctly nested but `())` is not. A limit of ten symbols per sentence was assumed.

2.1 Architecture

Two automatically defined functions (ADFs) [Koza, 1994] are available to assist the main result producing branch (or tree). The first, adf1, has no arguments and has the same terminal and function sets (except arg1) as the main tree. The second, adf2, has one argument but cannot contain terminals and functions with side effects. This allows a cache of previous values returned by it to be maintained, thus reducing run time.

Table 1 shows the parameters used and the terminals and functions provided, n.b. they include indexed memory but not stacks. Parameters not shown are as [Koza, 1994, page 655]. Table 9 gives the actions of most terminals and functions.

Earlier work with indexed memory had shown even large population had a great tendency to converge to partial solutions which effectively trapped the whole population preventing further progress. Partitioning the population so crossover is restricted to near neighbours has been used to reduced this. In this experiment the population is treated as a 100×100 torus with each member of the population occupying a different square on its surface. Each time a new individual is created a 3×3 square neigbourhood on the torus (known as a deme) is selected at random. Parents and the individual their offspring will replace are selected from this deme rather than from the whole population [Langdon, 1995; Tackett, 1994; Collins, 1992].

2.2 Fitness Function

The fitness of each trial program was evaluated on a fixed set of 175 example sentences containing both correctly nested (positive) and incorrectly nested brackets (negative). The test case includes all the positive cases up to a length of ten symbols and all the negative examples up to a length of four. The number of negative examples grows rapidly with sentence length and so above a length of four a limited number negative examples were chosen at random (see Table 2). The program is run once for each symbol in the sentence. Thus each program is run 1403 times (674 for (and 729 with an argument of)). (This and the two other test cases, Sections 3.2 and 4.2, are available via anonymous ftp node ftp.io.com directory pub/genetic-programming/code in file GPdata_gp96.test.tar.Z).

All stored data within the program initialised to zero at the start of each test sentence. The value returned by the program on the last symbol of the sentence gives its verdict as to whether the sequence is correctly nested.

2.3 Results

In the first run a general solution was produced by generation 19, which contained 88 primitives. A simplified version of it is shown in Figure 1.

In contrast to earlier work [Zomorodian, 1995] where GP was obliged to evolved push down automata the evolved solution is effectively a register machine (n.b. less powerful that a push down automata). The evolved solution (cf. Figure 1) only uses a single integer memory

Table 2: Nested Bracket Test Cases

Length	Positive		Negative	
1			all	2
2	all	1	all	3
3			all	8
4	all	2	all	14
5			random	4
6	all	5	random	5
7			random	5
8	all	14	random	14
9			random	14
10	all	42	random	42
Totals		64		111

cell (aux1), in which it counts the depth of nesting. At the end of a legal sentence this count must be zero, further should the brackets be unbalanced before the end is reached, this is recognised and aux1 is also used as a flag indicating this. This solution not only passes all the fitness tests and is a general solution to the problem but (given suitable redefinition of max) is a solution for sequences of any length.

Given the readily found general solution did not exhibit stack like behavior it was decided not to repeat this problem with a GP that had stack primitives.

3 Dyck Problem

Once again the problem is to recognise which sentences are correctly bracketed, however there are now four types of bracket pairs, (,), [,], {, }, ‘, ’. E.g. {}[] is correctly bracketed but [} is not. As with the nested brackets problem, a limit of ten symbols per sentence was assumed.

Two sets of experiments have been conducted, the first provided the GP with primitives which implement a stack for it and the second provided primitives like those from which it has been shown GP can evolve stack data structures [Langdon, 1995].

3.1 Architecture

In the first experiments (stack given) no ADFs were used, whilst in the second there are three ADFs, having 0, 1 and 0 arguments. It was hoped that these could evolve to operate like pop, push and top. Each could be called from the main tree, additionally the third (top?) could be called by the first.

The terminals, functions and control parameters used in these two experiments are as Section 2 except where given in Table 3. The differences between the two experiments in this section are shown in the middle and right hand columns of Table 3.

Table 1: Balanced Bracket Problem

Objective:	Find a program that classifies sequences of ((represented by 1) and) (-1) as being correctly nested or not.
Primitives: (any tree)	ADD, SUB, PROG2, IFLTE, Ifeq, 0, 1, -1, max, forwhile, i0
(rpb, adf1)	adf2, aux1, read, write, swap, Set_Aux1
(rpb, adf2)	arg1
(rpb only)	adf1
Max prog size	$4 \times 50 = 200$. In initial population each tree is limited to 50 primitives.
Fitness cases:	175 fixed test cases.
Fitness:	Number of test cases passed (scalar).
Selection:	Tournament group size of 4 used for both parent selection and selecting programs to be removed from the population. Steady state population (elitist).
Wrapper:	Zero represents False (i.e. not in language) otherwise True.
Parameters:	Pop = 10,000, G = 50, 3×3 demes, no CPU penalty, no aborts.
Success predicate:	A program passes all tests.

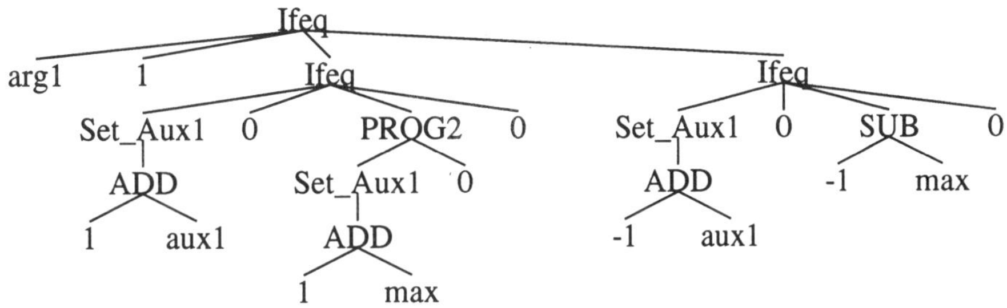

Figure 1: Solution to Bracket Problem (Simplified)

Table 3: Dyck Language Problem

Objective:	Find a program that classifies sequences of four types of bracket (((represented as 5),) (71), [(13),] (103), { (31), } (137), ‘ (43) and ’ (167)) as being correctly nested or not.		
Primitives:	Common	Stack Given	Index Memory
All trees:	ADD, SUB, PROG2, IFLTE, Ifeq, 0, 1, max, aux1	Makenull, Empty, Top, Pop, Push	read, write, inc_aux1, dec_aux1
rpb: as all plus	ifopen, ifmatch, ARG1, Set_Aux1		adf1, adf2, adf3
adf1: as all plus			adf3
adf2: as all plus			arg1, arg2
Max prog size	Initial tree limit 50	50	$4 \times 50 = 200$
Fitness cases:	286 fixed test cases.		
Fitness:	Number of correct answers returned.		
Selection:	Tournament size 4 (After 1st solution CPU penalty used giving a two dimensional fitness value, fitness niching used with a sample of up to 81 (9×9) nearest neighbours.)		
Wrapper:	Zero represents True (i.e. in language) and all other values False.		
Parameters:	Pop = 10,000, G = 50, Pareto, 3×3 demes, CPU penalty only after 1st solution found, Abort on first error in sentence.		
Success predicate:	A program passes all tests.		

3.2 Fitness Function

The fitness of every trial program is determined by presenting it with a series of symbols from test sentences and counting how many times it correctly classifies each as to whether it is the last of a correctly balanced sequence.

This problem is more complex than that in Section 2 and so the test cases are longer. To constrain the increase in runtime, forwhile loops were not used.

The number of possible test cases is much larger than in Section 2 and so it was not practical to include sentences longer than eight symbols and even for lengths of six and eight symbols it was necessary to select (at random) positive test cases to include.

In a correctly matched sentence there will be equal numbers of opening and closing brackets of each type but this is unlikely to be true in a random sequence of brackets. To penalise partial solutions based on a counting mechanisms, like that evolved in Section 2, the test cases include examples where there are equal numbers but which are not correctly nested (referred to as "balanced" in Table 4).

As before it was not practical to include all cases and so longer negative examples (both balanced and not balanced) were selected at random. Even so the fitness tests are much longer than that in Section 2 and so to keep runtime manageable the number of times each program must be run was reduced by:

- Only using the first half of the test cases (i.e. tests up to length six) unless a program passes them all, were upon test sentences of length seven and eight were also used.
- Using the value returned for each symbol, rather than just the last in the sentence, and ignoring the rest of the sentence after the first error. As some shorter sentences are identical to the start of longer ones they need not be tested explicitly and so could be removed from the test cases. The test cases after removing such duplicates are summarised in the right hand side of Table 4.

Initially brackets were coded as $\pm1, \pm2, \pm3, \pm4$ but general solutions proved difficult to find. Instead, despite the use of "balanced" negative examples, partial solutions based upon summing up symbol values dominated. Since the purpose of the experiment was to investigate learning correct nesting of symbols rather than learning which symbols match each other the problem was simplified by providing the GP with two new primitives (ifmatch and ifopen, cf. Table 9) which say which symbols match each other. To further discourage partial solutions based on summing symbol values the symbols were recoded as prime values with no simple relationships between them (cf. Table 3).

After an individual which passes all the tests is found the GP run is allowed to continue using a modified fitness function which includes a CPU penalty. Each program's fitness now contains two orthogonal terms, the original score and the ⌊average⌋ number of instructions run per program execution. Tournament selection is still used for reproduction and deletion but now uses Pareto comparison [Goldberg, 1989], so passing tests and using little CPU are equally important.

In the event of no member of the tournament being better than all the others the best members of the tournament are compared with a sample of the adjacent population. The one that is dominated the least by the rest of the population is selected. This introduces a selection pressure to be different from the rest the population so allowing high scoring and low CPU programs to co-exist. This is similar to fitness niching [Horn *et al.*, 1993].

3.3 Results

In three runs given the stack primitives general solutions were evolved by generation seven to 23 (in three identical runs but using simple non-demic (normal) populations, two runs produced solutions in generations 30 and 39). Evolution was allowed to continue after the first individual to pass all the tests was found. Under the influence of the CPU penalty faster but still general solutions were found. Figure 2 shows one of the fastest solutions produced in a run using demes after 50 generations. As in Section 2 the solutions are not only general solutions to the given problem, but given a deep enough stack would work with any sentences of any length.

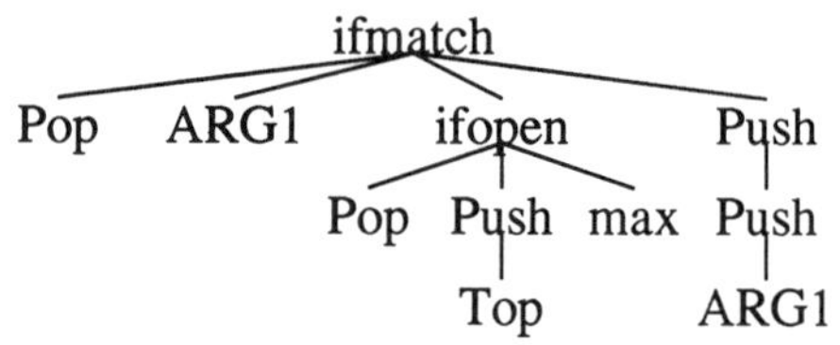

Figure 2: One of the Fastest Solutions to the Dyck Problem evolved after 50 Generation

In contrast none of 15 runs using the indexed memory primitives passed all the tests. Some of the more promising runs were extended beyond 50 generations up to 140 generations without finding a solution. The best (produced after 84 generations) still failed 3 tests (on sequences of up to six symbols). It showed some stack like behaviour which enables it to pass 13 of the tests of length seven and eight but also showed some signs of over fitting to the specific the test cases used rather than having learnt to solve the general problem.

A program which always returns zero (i.e. True) has a fitness of zero because it will always fail on the first symbol of each test sentence (a sentence of odd length must be unbalanced). In contrast a program which never returns zero will always be correct on the first symbol of each sentence and so will get the opportunity to be tested

Table 4: Dyck Language Test Cases

Length	Positive	Negative		After Removing Duplicates			
		Balanced	Random	Positive	Balanced	Random	Score
1		all 8					0
2	all 4	all 60				9	18
3			16			10	30
4	all 32	all 24	16	27	16		172
5			16			16	80
6	random 32	random 32	32	32	32	32	576
7			16			16	112
8	random 32	random 32	32	32	32	32	768
			Totals	91	112	83	1756

on the second symbol which it may also pass. For the actual test cases used a program which never returns zero has a fitness of 714. While aborting a test sentence on the first error reduces the number of times programs are run, it may also make it more difficult to evolve a solution. In both experiments the GP population quickly learns not to return zero, but when using indexed memory it appears to be more difficult than when given a stack to escape this local optima and learn to return zero at some points.

4 Reverse Polish Calculator

In this section the GP evolves a four function (+, −, / and ×) calculator, i.e. evaluates integer arithmetic expression. The problem is simplified by presenting the expression in Reverse Polish Notation (postfix), which avoids consideration of operator precedence, and by avoiding expressions which include divide by zero. No limit on the length of expressions was assumed, however the expressions tested were between three and fifteen symbols long (see Table 6).

Two sets of experiments were made, the first provided the GP with primitives which implement a stack for it and the second provided primitives like those from which it has been shown GP can evolve stack data structures. Since all storage including the supplied stack is initialised before the evolved programs can use it the Makenull operation is not needed. Therefore the terminal set was simplified by not including Makenull and Empty (which is also not needed) in these experiments.

4.1 Architecture

Each individual within the population consists of five separate trees (num, plus, minus, times and div) plus zero or two ADFs. As in sections 2 and 3 each test sentence is presented a symbol at a time to the GP, however in this case the appropriate tree is selected. E.g. if the symbol is an integer, then the num tree is executed with the integer as its argument. Each tree returns a value as the current value of the expression (num's answer is ignored).

In the first experiments (stack given) no ADFs were used, whilst in the second there are two ADFs, having 0 and 1 arguments respectively. It was hoped that these could evolve to operate like pop and push. Each could be called from the five main trees.

The terminals, functions and control parameters are as Section 3 except where given in Table 5. Nb. the long run times encountered with these experiments led to the decision to include a CPU penalty of $\lfloor \text{average} \rfloor$ number of primitives executed per program run. However initially only programs with long run times are penalised (by setting the penalty to zero for fast programs, i.e. where it was ≤ 50). Should a program be evolved which passes all the fitness test then the CPU penalty is increased by applying it to all programs.

4.2 Fitness Function

In each individual in the population a separate score is maintained for its five operations (num, plus, minus, times and div) plus the CPU penalty. Each time the individual returns the correct answer (and this is checked) the score for each of its operations that has been used since the last time its result was checked is incremented. As in Section 3, these scores are not combined and each contributes as a separate objective in multi-objective Pareto selection tournaments.

The fixed test cases were created before the GP was run. A number of test cases were devised by hand and the remainder were selected at random. However the randomly selected data values (from the range −99 . . . +99) proved to be unsatisfactory for expressions containing "/" because division of two randomly selected integers has a high chance of yielding zero or an integer near it and therefore were changed by hand.

Table 5: Reverse Polish Notation (RPN) Expression Evaluation Problem

Objective:	Find a program that evaluates Reverse Polish (postfix) arithmetic expressions.		
Primitives:	Common	Stack Given	Index Memory
$+ - \times /$ trees:	ADD, SUB, MUL, DIV, PROG2, 0, 1, aux1, Set_Aux1	Top, Pop, Push	read, write, inc_aux1, dec_aux1, adf1, adf2
num: as ops plus	arg1		
adf1: as ops but			no adfs
adf2: as ops but			no adfs and add arg1
adf3: as ops but			no adfs and add arg1, arg2
Max prog size	Initial tree limit 50	$5 \times 50 = 250$	$7 \times 50 = 350$
Fitness cases:	127 fixed test cases.		
Fitness:	Number of correct answers returned.		
Selection:	Pareto tournament size 4, CPU penalty (initial threshold 50 per operation), fitness niching used with a sample of up to 81 other members of the population.		
Wrapper:	Value on num ignored. No wrapper on other trees.		
Parameters:	Pop = 10,000, G = 50, Pareto, no demes, no threshold on CPU penalty only after 1st solution found, abort on first wrong answer given in expression.		
Success predicate:	A program passes all tests.		

Table 6: Length of Reverse Polish Expressions at each Point where Answers are Checked in the Fitness Test Cases

length	1	2	3	4	5	6	7	8	9	10	11	12	13	14	15	Total
No. of cases			10	3	55	27	44	2	36	1	5		8		3	194

The following rules were used to create the test cases:

- include a higher proportion of minus and divide (cf. Table 7).
- be deeply nested (Table 8).
- divide by zero was avoided.
- Where necessary randomly generated data values were manually changed so that only a few divisions yield values in the range $-3 \ldots +3$.
- To avoid problems with overflow, randomly generated expressions did not allow: arguments to addition or subtraction outside the range $-10^8 \ldots +10^8$ or arguments to multiplication or division outside the range $-65535 \ldots +65535$.
- Data values set by hand were chosen so neither: the product of two arguments of divide nor the square of the second argument exceeded 2,147,483,647.
- In all but one test expression there is one less arithmetic operation than data values. So only the final result should be left after evaluating the expression.

  One case with fewer arithmetic operation was included. In this case there should be multiple data values left after evaluating the expression.

As before it was necessary to constrain run time. This was done by checking answers during the evaluation of each expression and aborting evaluation following the first error detected and removing test cases which essentially duplicated others. This left 127 test expressions which include 194 points where the trial program's answer is checked.

Table 7: Reverse Polish Expression Test Cases

Operation	No.	Max Score
num	550	163
plus	67	58
minus	103	85
times	85	64
divide	156	127
	420	
Totals	970	497

Table 8: Depth of Expression Nesting in RPN Test Cases (depth after each symbol)

depth	1	2	3	4	5	6	Total
number	387	390	149	31	12	1	970

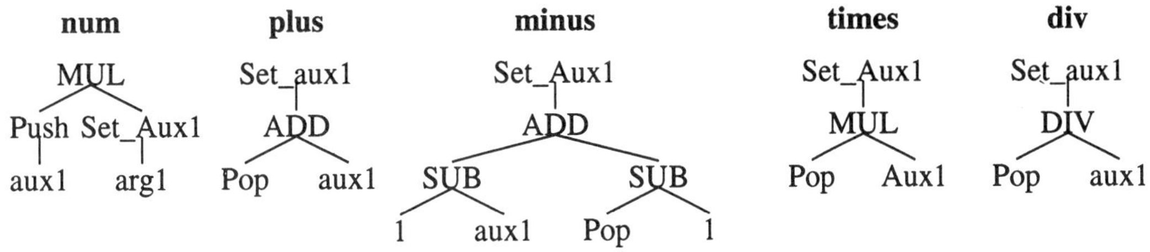

Figure 3: One of the Fastest Solutions to the RPN Problem evolved after 50 generations

4.3 Results

In eleven runs using stack primitives, six produced solutions which passed all the tests, these were found between generations 11 and 23. In four cases the first programs to pass all the tests were also general solutions to the problem. In the other two the first solutions failed special cases such as $1 - 1$ and $x/y = 0$, however in both runs general solutions were evolved before 50 generations.

Under the action of the increased CPU penalty, solutions which took about one third of the CPU time of the first solution found were evolved (see Figure 3).

In 59 runs with stack primitives replaced by indexed memory (see right hand side of Table 5) no program passed all the tests. The highest no. tests passed (148 of 194) was achieved by a program which used the first ADF to implement DIVR (i.e. standard divide but with the arguments in reversed order, see Table 9) and the second to approximate both push and pop on a three level stack. Other unsuccessful trials included adding a third ADF (with two arguments) in the hope that this might evolve the DIVR functionality (best 102 in 33 runs, of which 16 ran out of time before 50 gens) and supplying SUBR and DIVR functions (in place of SUB and DIV) where the best score was 116, in 38 runs.

The non-commutative functions ($-$ and $/$) appear to be more difficult to evolve than commutative ones because the arguments on the stack are in the opposite order to that used by the SUB and DIV functions. (The problem can be readily solved, when given the stack primitives, by replacing SUB and DIV by SUBR and DIVR which process their arguments in the opposite order, i.e. the order on the stack.) The div tree has to use some storage (i.e. aux1) to reverse the order of the stack arguments (sub can simply negate the result of performing the operation in the wrong order to get the right answer). The need to use aux1 makes the problem deceptive, in that many programs can obtain high scores using aux1 as the top of the stack and only fail tests which require deeper use of the stack.

Some of the difficulty the GP with indexed memory found may have been due to trying to use aux1 both as stack pointer (for which inc_aux1 and dec_aux1 make it suitable) and as the top of stack (as evolved in many cases where the stack primitives were given).

5 Conclusions

These experiments have shown GP can solve two new problems (inducing a Dyck language (Section 3) and evaluating Reverse Polish Notation (RPN) expressions, Section 4) and GP can solve the nested bracket problem (Section 2) without requiring an intermediate step generating an abstract machine.

All three examples were solved by GP using the appropriate data structure for the problem. The two more complex examples (Dyck and RPN) proved to be more difficult (no solution has been found to date) for GP when provided with indexed memory than when provided with a stack, despite indexed memory being more powerful.

It has often been argued, e.g. [Kinnear, Jr., 1994, page 12], that functional primitives used with GP should be as powerful as possible, in these examples we have shown appropriate data structures are advantageous, that is GP can benefit from data abstraction.

The failure of GP to solve the problems when provided with the more general (i.e. more powerful) directly addressable memory data structure, is worrying as it shows (at present) data structures should be chosen with care and it may not be sufficient to simply over provide, with more powerful structures than are needed. It was hoped that GP could evolve appropriate data structures as it solved the problem but this has yet to be demonstrated.

A Pareto CPU penalty term to fitness has again been used to curb growth in program size and CPU time.

Acknowledgments

W. B. Langdon is funded by the EPSRC and the National Grid Company plc. I would like to thank my supervisors (M. Levene and P. C. Treleaven), and Andy Singleton for the original version of GP-QUICK on which my code was based.

Table 9: Actions Performed by Terminals and Functions

Primitive	Purpose
`DIV`(x,y)	**if** $y \neq 0$ **return** x/y **else return** 1
`SUBR`(x,y) `DIVR`(x,y)	As `SUB` and `DIV` except yield $y-x$ and y/x, i.e. operands reversed.
`max`	constant 10 ($\geq$ max input size).
`PROG2`(t,u)	evaluate t; **return** u
`ARG1, arg1, arg2`	arguments of current operation or ADF
`aux1`	an auxiliary variable (i.e. in addition to indexed memory).
`Set_Aux1`(x)	`aux1` = x; **return** `aux1`

`forwhile`(s,e,l)	**for** `i0` = s; `i0` $\leq e$; `i0++`
	    if timeout (128) **exit loop**
	    if l returns zero **exit loop**
	return `i0`

`i0`	Yields value of loop control variable of most deeply nested loop or zero if not in a loop in current tree. N.b. loop control variable in one tree cannot be accessed in another (e.g. an ADF).
`IFLTE`(x,y,$t1$,$t2$)	**if** $x \leq y$ **return** $t1$ **else return** $t2$
`Ifeq`(x,y,$t1$,$t2$)	**if** $x = y$ **return** $t1$ **else return** $t2$
`ifopen`(x,$t1$,$t2$)	**if** x = 5, 13, 31 or 43 **return** $t1$ //i.e. opening symbol **else return** $t2$

`ifmatch`(x,y,$t1$,$t2$)	**if** x = 5, 13, 31 or 43 evaluate y //i.e. opening symbol
	    if (x,y) = (5,71), (13,103), (31,137) or (43,167) **return** $t1$ //x and y match
	    else return $t2$
	else return $t2$

`Makenull`	clear stack; **return** 0
`Empty`	**if** stack is empty **return** 0; **else return** 1
`Top`	**if** stack is empty **return** 0; **else return** top of stack
`Pop`	**if** stack is empty **return** 0; **else** pop stack and **return** popped value
`Push`(x)	Evaluate x; **if** $<$ 99 items on stack push x; **return** x **else return** 0
Indexed memory is held in store[$-l$ ... $+l$], where $l = 63$, i.e. a total of 127 cells.	
`read`(x)	**if** $\lvert x\rvert \leq l$ **return** store[x] **else return** 0
`write`(x,d)	**if** $\lvert x\rvert \leq l$ store[x] = d; **return** original contents of store[x] **else** evaluate d; **return** 0
`swap`(x,y)	**if** $\lvert x\rvert \leq l$ **and** $\lvert y\rvert \leq l$ exchange contents of store[x] and store[y] **if** $\lvert x\rvert > l$ **and** $\lvert y\rvert \leq l$ store[y] = 0 **if** $\lvert x\rvert \leq l$ **and** $\lvert y\rvert > l$ store[x] = 0 **return** 1

References

[Andre, 1994] David Andre. Evolution of mapmaking ability: Strategies for the evolution of learning, planning, and memory using genetic programming. In *Proceedings of the 1994 IEEE World Congress on Computational Intelligence*, volume 1, pages 250–255, Orlando, Florida, USA, 27-29 June 1994. IEEE Press.

[Collins, 1992] Robert J. Collins. *Studies in Artificial Evolution.* PhD thesis, UCLA, Artificial Life Laboratory, Department of Computer Science, University of California, Los Angeles, LA CA 90024, USA, 1992.

[Dunay and Petry, 1995] Bertrand Daniel Dunay and Frederic E. Petry. Solving complex problems with genetic algorithms. In L. Eshelman, editor, *Genetic Algorithms: Proceedings of the Sixth International Conference (ICGA95)*, pages 264-270, Pittsburgh, PA, USA, 15-19 July 1995. Morgan Kaufmann.

[Dunay *et al.*, 1994] B. D. Dunay, F. E. Petry, and W. P Buckles. Regular language induction with genetic programming. In *Proceedings of the 1994 IEEE World Congress on Computational Intelligence*, pages 396–400, Orlando, Florida, USA, 27-29 June 1994. IEEE Press.

[Goldberg, 1989] David E. Goldberg. *Genetic Algorithms in Search Optimization and Machine Learning.* Addison-Wesley, 1989.

[Horn *et al.*, 1993] Jeffrey Horn, Nicholas Nafpliotis, and David E. Goldberg. Multiobjective optimization using the niched pareto genetic algorithm. IlliGAL Report no. 93005, Illinois Genetic Algorithm Laboratory, University of Illinois at Urbana-Champaign, 117 Transportation Building, 104 South Mathews Avenue, Urbana, IL 61801-2296, July 1993.

[Iba *et al.*, 1995] Hitoshi Iba, Hugo de Garis, and Taisuke Sato. Temporal data processing using genetic programming. In L. Eshelman, editor, *Genetic Algorithms: Proceedings of the Sixth International Conference (ICGA95)*, pages 279–286, Pittsburgh, PA, USA, 15-19 July 1995. Morgan Kaufmann.

[Kinnear, Jr., 1994] Kenneth E. Kinnear, Jr. A perspective on the work in this book. In Kenneth E. Kinnear, Jr., editor, *Advances in Genetic Programming*, chapter 1, pages 3–19. MIT Press, 1994.

[Koza, 1994] John R. Koza. *Genetic Programming II: Automatic Discovery of Reusable Programs.* MIT Press, 1994.

[Langdon, 1995] W. B. Langdon. Evolving data structures using genetic programming. In L. Eshelman, editor, *Genetic Algorithms: Proceedings of the Sixth International Conference (ICGA95)*, pages 295–302, Pittsburgh, PA, USA, 15-19 July 1995. Morgan Kaufmann.

[Langdon, 1996] William B. Langdon. Data structures and genetic programming. In Peter J. Angeline and K. E. Kinnear, Jr., editors, *Advances in Genetic Programming 2*, chapter 20. MIT Press, 1996.

[Tackett, 1994] Walter Alden Tackett. *Recombination, Selection, and the Genetic Construction of Computer Programs.* PhD thesis, University of Southern California, Department of Electrical Engineering Systems, 1994.

[Teller, 1994] Astro Teller. The evolution of mental models. In Kenneth E. Kinnear, Jr., editor, *Advances in Genetic Programming*, chapter 9, pages 199–219. MIT Press, 1994.

[Zomorodian, 1995] Afra Zomorodian. Context-free language induction by evolution of deterministic push-down automata using genetic programming. In E. S. Siegel and J. R. Koza, editors, *Working Notes for the AAAI Symposium on Genetic Programming*, pages 127–133, MIT, Cambridge, MA, USA, 10–12 November 1995. AAAI.

Evolving Teamwork and Coordination with Genetic Programming

Sean Luke*
seanl@cs.umd.edu
http://www.cs.umd.edu/~seanl/

Lee Spector* †
spector@cs.umd.edu
http://hamp.hampshire.edu/~lasCCS/Home.html

*Department of Computer Science
University of Maryland
College Park, MD 20742

†School of Cognitive Science and Cultural Studies
Hampshire College
Amherst, MA 01002

ABSTRACT

Some problems can be solved only by multi–agent teams. In using genetic programming to produce such teams, one faces several design decisions. First, there are questions of team diversity and of breeding strategy. In one commonly used scheme, teams consist of clones of single individuals; these individuals breed in the normal way and are cloned to form teams during fitness evaluation. In contrast, teams could also consist of distinct individuals. In this case one can either allow free interbreeding between members of different teams, or one can restrict interbreeding in various ways. A second design decision concerns the types of coordination–facilitating mechanisms provided to individual team members; these range from sensors of various sorts to complex communication systems. This paper examines three breeding strategies (clones, free, and restricted) and three coordination mechanisms (none, deictic sensing, and name–based sensing) for evolving teams of agents in the Serengeti world, a simple predator/prey environment. Among the conclusions are the fact that a simple form of restricted interbreeding outperforms free interbreeding in all teams with distinct individuals, and the fact that name–based sensing consistently outperforms deictic sensing.

1 Teamwork and Specialization

The most common kind of task found in multi–agent problems is one in which the task can be done *faster* or *more easily* by dividing it up among many simultaneous agents. The more agents working on the problem, the larger the speedup. But some tasks may not only be solved better by using multiple agents, but can *only* be solved, or only effectively solved, by using teams of agents working together. For example, consider the problem faced by Pac Man monsters, which are slower than Pac Man on all but long straightaways. If there was only one monster then the monster would rarely catch even poor players, and the game would be trivial.

Many such problems are solvable by multiple agents each of which uses essentially the same algorithm. These *homogeneous* tasks are common candidates for the study of emergent behavior because one (often simple) set of rules can give rise to complex patterns of behavior. An example of a homogeneous task is self–replication of cellular automata. Lohn and Reggia [1995] used genetic algorithms to breed cellular automata that spontaneously emit identical copies of themselves. Each cell in the automata grid can be thought of as an individual helping to create these duplicate patterns: the problem is homogenous because all cells follow the same cellular automata rule set. An example in the genetic programming (GP) literature is the evolution of herding behaviors [Reynolds 1993]. Reynolds used GP to evolve "critters" which reacted with a herd instinct to outside predators. He evolved a single controller which moved each critter based on its position and information about its neighbors and predators. Each critter used this same controller algorithm to make movement decisions.

In contrast, *heterogeneous* tasks can only be solved (or solved easily or effectively) with multiple individuals each of which uses a distinct specialized algorithm. For example, Haynes *et al* chose a simple grid–based predator/prey game to test a variety of breeding strategies on a team of (typically) distinct GP individuals [Haynes *et al* 1995]. And Andre's MAPMAKER system required two specialized agents (one with the sensors, the other able to pick up objects) to gather gold strewn in a 2-dimensional grid [Andre 1995]. Without the interaction of two unique algorithms, this task would be much more difficult, perhaps impossible.

The distinction between homogeneous and heterogeneous approaches can be recast as an issue of breeding policy. In the homogeneous approach, individuals breed as in ordinary genetic programming, and the individuals are then cloned for fitness evaluation. In the heterogeneous approach, individual team members breed separately, which makes possible additional strategies for interbreeding within and between teams.

Table 1: Basic Lion Operators

Name	# Args	Description
`last`	0	One unit in the direction the lion went last. On the first move, return a random normal vector.
`rand-dir`	0	A random normal vector.
`gazelle`	0	The smallest vector from the lion to the gazelle.
`+`	2	Add two vectors.
`-`	1	Negate a vector.
`*2`	1	Multiply the magnitude of a vector by 2.
`/2`	1	Divide the magnitude of a vector by 2.
`->90`	1	Rotate a vector clockwise 90 degrees.
`rand`	1	Given a vector, return a vector in the same direction, whose magnitude varies randomly between 0 and the magnitude of the original vector.
`inv`	1	"Invert" a vector's magnitude, so that small vectors are "more significant" than large vectors. This is done similarly to the method used in the gazelle algorithm: Let v be the vector, and $\|max\| = \sqrt{(w/2)^2 + (h/2)^2}$, where w is the width of the world (15 units), and h is the height of the world (also 15 units). Then inv returns $\frac{v}{\|v\|}(\|max\| - \|v\|)$.
`ifdot`	4	Evaluate the first and second arguments. If their dot product is greater than or equal to 0, then evaluate and return the third argument, else evaluate and return the fourth argument.
`if>=`	4	Evaluate the first and second arguments. If the magnitude of the first argument is greater than or equal to the magnitude of the second argument, then evaluate and return the third argument, else evaluate and return the fourth argument.

In many multi–agent domains, agents must coordinate their actions with one another. Various mechanisms may facilitate the evolution of such coordinated behavior. In some cases it is sufficient for agents to be able to sense one another in some way, though in other cases it may be necessary for agents to have more elaborate communication mechanisms. Even in the case of simple sensing, there are many possibilities. For example, one can allow each agent direct access to the positions of each distinct fellow agent, referenced by agent name; we call this *name–based sensing*. Another approach, used for example in evolving herding behaviors, allows only sensing relative to the agent— for example, "nearest neighbor agent," or "front of the pack." We call this *deictic sensing*. It may also be possible that team members can evolve to coordinate with one another in a hard–coded way with *no sensing* of neighboring agents.

This paper examines three breeding strategies (clones, free, and restricted) and three coordination mechanisms (none, deictic sensing, and name–based sensing) for evolving teams of agents in the Serengeti world, a simple predator/prey environment. We show that a simple form of restricted interbreeding outperforms free interbreeding in all teams with distinct individuals, that name–based sensing consistently outperforms deictic sensing, and that as the sensing becomes increasingly direct (more name–based), heterogeneous approaches work better than homogeneous approaches. In the paper we describe the breeding strategies, followed by descriptions of Serengeti world, of our runs, and of our results. We conclude with a brief discussion of directions for further research.

2 Breeding Distributed Agents with Genetic Programming

Heterogeneous agents may be difficult to evolve with GP because existing GP methods are designed primarily to evolve one single algorithm. There are two common ways to tackle this problem. The first is to use *co–evolution* [Koza 1992] to select individuals from a population to form a team at trial–time. Co–evolving specialized cooperation might be made more effective if the population is divided into subpopulation *demes* [Tackett and Carmi 1994]; each subpopulation could provide a specialized member to join each team. However, selecting team members from the population complicates the evolution process with the *credit assignment problem:* when a trial is complete, which individuals get more credit for its success or failure [Haynes *et al* 1995]?

Haynes *et al* note that another way to generate a team is to consider the whole team as one GP individual. This eliminates the credit assignment problem, since there is only one individual per trial. An easy way do this is to take advantage of GP facilities for *Automatically Defined Functions* (ADFs) [Koza 1994] or *Automatically Defined Macros* (ADMs) [Spector 1996]. For teamwork purposes, ADFs and ADMs can be thought of as subindividuals within a main individual, each of which may serve a distinct specialized purpose. By modifying existing ADF systems, it's possible to treat each ADF as a separate agent; GP systems like *lil-gp* [Zongker and Punch 1995] allow the evaluation of each ADF in an individual as if the ADF was an individual entity. We implemented our heterogeneous breeding strategies this way,

Table 2: Deictic Sensing Lion Operators

Name	# Args	Description
`nearest`	0	The vector from the lion nearest the gazelle, to the gazelle.
`lion`	0	A vector from the lion to its nearest neighbor lion.
`rlion`	0	A vector from the lion to the first neighbor lion encountered in a clockwise sweep. The sweep begins in the direction the lion moved last. All the minimal vectors to each lion are gathered, and the first one found in the sweep is returned.
`llion`	0	A vector from the lion to the first neighbor lion encountered in a counterclockwise sweep. The sweep begins in the direction the lion moved last. All the minimal vectors to each lion are gathered, and the first one found in the sweep is returned.

Table 3: Name–Based Sensing Lion Operators

Name	# Args	Description
`lion-1`	0	A vector from the lion to lion #1.
`lion-2`	0	A vector from the lion to lion #2.
`lion-3`	0	A vector from the lion to lion #3.
`lion-4`	0	A vector from the lion to lion #4.

since it is functionally very similar to the clear method of producing homogeneous individuals. In both cases we grow standard GP individuals, and only one "individual" is tested at a given time. For homogenous teams, individuals are tested by cloning them to form teams, and the resulting teams are tested in the environment. Heterogeneous teams, however, are formed from the individual's collection of ADFs.

We allowed for two different kinds of breeding strategies for heterogeneous teams. *Free Breeding* allows any member of a team to freely breed with any other member of another team. *Restricted Breeding* allows team member 1 to breed only with other team member 1's, and team member 2 to breed only with other team member 2's, etc. This restriction further promotes specialization since it breaks team members up into separate breeding pools.

3 The Serengeti World

To test the effectiveness of homogeneous and heterogeneous team algorithms under a variety of sensing capabilities, we chose a predator/prey domain nicknamed the Serengeti. The Serengeti world is a toroidal, continuous 2-dimensional landscape, 15 units on a side. In this world roam a gazelle and a group of lions, initially placed in random positions. The gazelle moves 3 units in one bound, but each lion may only move one unit at a time. The gazelle and lions take turns moving until the maximum number of turns is used up, or a lion has moved to within 1 unit of the gazelle (the gazelle is "killed" and the simulation stops). The programs for the animals return 2-dimensional vectors which determine the direction they move. Animals may move in *any* direction; they are not limited to, say, the cardinal directions. The goal is to generate a lion "pack" that regularly gets as close as possible to the gazelle.

Predator/prey pursuit domains like the Serengeti have been used often to test both evolved and non–evolved agent coordination with genetic programming or distributed AI [Korf 1992, Haynes *et al* 1995, Benda *et al* 1986]. Commonly, predator/prey domains exist in a simple grid world with limited movement abilities (up, down, right, and left), and predators often move faster than the prey (though sometimes the object is not just to "catch" the prey but to surround it with all four predators). In contrast, the Serengeti allows for continuous positions and a much more difficult–to–catch prey. We feel the Serengeti is a useful domain to test heterogeneous versus homogeneous team strategies for two important reasons. First, it presents a problem that is *impossible* for a single agent (one lion) to solve. Since the gazelle moves faster than a lion, the only way to effectively bring down the gazelle is for a group of lions to hunt together. Second, the Serengeti does not clearly lend itself to heterogeneous teams of specialized members: the Serengeti is symmetric and has no obstacles or hill slopes, etc. that give clear niche opportunities for team members. On the other hand, Serengeti world does not clearly promote homogeneous teams either: group members start in random positions, and obvious homogeneous algorithms like "head directly at the gazelle" or "circle the gazelle" turn out to be suboptimal solutions [Korf 1992]. This is especially true for the Serengeti because, unlike earlier domains, in this domain the prey moves much faster than the predators.

Table 4: Trial Results

Sensing	Restricted Breeding Average	Restricted Breeding Best	Free Breeding Average	Free Breeding Best	Clones Average	Clones Best
Deictic	1.65	0.13	2.03	0.23	1.52	0.20
Name–Based	1.33	0.03	1.79	0.07	1.93	0.22
None	2.20	0.49	2.23	0.50	2.18	0.45

Table 5: Control Results

	One Lion Average	One Lion Best	One Random Average	One Random Best	Four Random Average	Four Random Best
Controls	7.39	5.43	7.87	6.74	4.41	2.57

3.1 The Gazelle

The difficulty of the domain is directly affected by the intelligence of the gazelle. We chose an avoidance algorithm that runs quickly, but still makes it surprisingly difficult for the lions to corner the gazelle.

The gazelle's heuristic is as follows: Let $\|max\| = \sqrt{(w/2)^2 + (h/2)^2}$, where w is the width of the world (15 units), and h is the height of the world (also 15 units). This is the largest possible distance from one object to another in the Serengeti. Then the gazelle moves 3 units in the direction of the vector given by

$$-\sum_{v \in V} \frac{v}{\|v\|}(\|max\| - \|v\|)$$

where V is the set of smallest vectors from the gazelle to the lions. In other words, the gazelle modifies the magnitude of vectors to lions so that closer lions are more significant than far-away lions, and then moves away from the sum of the vectors.

3.2 Lions

For genetic programming breeding purposes, each lion group was treated as a single individual, since the group stays together and is graded together based on its combined success. We tested groups with either one lion or four lions (labeled lion numbers 1, 2, 3, or 4); within a group, a lion was a single GP function tree using 2-element vectors as its data type. All lions used at *least* the basic function set shown in Table 1.

The lions were tested with either deictic sensing ability, name–based sensing, or no sensing ability. Lion groups with deictic sensing were given the four additional operators `nearest`, `lion`, `rlion`, `llion` (Table 2). Lion groups with name–based sensing were given the four additional operators `lion-1`, `lion-2`, `lion-3`, `lion-4` (Table 3). Lion groups with no sensing were given no additional operators. We implemented the three breeding strategies as follows: *Clones* really consisted of a single function tree, alternately given four different sets of environment variables to simulate four identical lions. *Restricted* and *Free Breeding* groups had four separate function trees, one per lion. In addition to combinations of breeding strategy and sensing operators, we performed control runs for one lion using the basic operators, one lion moving completely randomly, and four lions moving completely randomly.

4 Test Runs

We performed 1200 random runs under the *lil-gp* GP system with the Serengeti problem: 100 runs for each of the nine combinations of sensing and breeding strategies, plus 100 runs for each of the three control configurations. We assessed the fitness of groups based on the average of 5 simulation trials. At the beginning of each simulation, the lions and gazelle started in random positions. During each turn, the gazelle moved first, then the lions. This was repeated 15 times, or until the gazelle was "killed" as explained before. On each turn, lions moved one unit in the direction of the final vector returned. If a lion returned a 0-length vector, it moved randomly 1 unit. At the end of each simulation, the lion group's fitness was based on the distance from the gazelle to the lion nearest the gazelle. If this lion finished within 1 unit of the gazelle, the lion group grade was 0. Otherwise, the group's grade was the distance of the nearest lion to the gazelle, minus 1. Lower scores were better.

Each run lasted for 51 generations, with a population of 500, a maximum tree size of 70 and a maximum tree depth of 17. We used crossover (90% of the time) and reproduction (10% of the time) to generate new individuals.

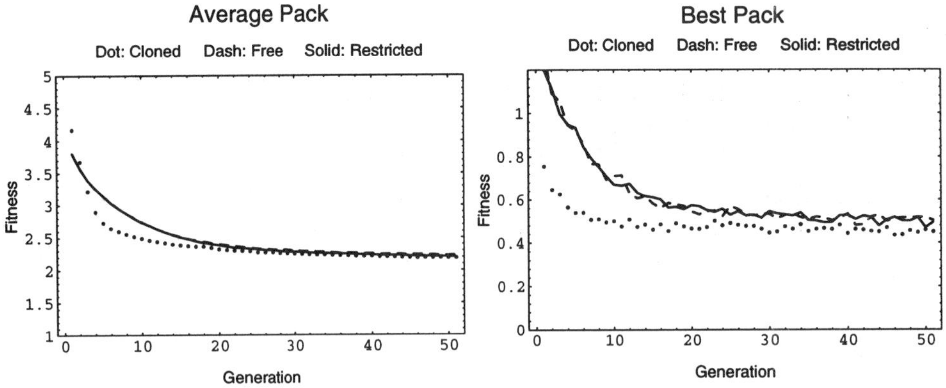

Figure 1: No–Sensing Results

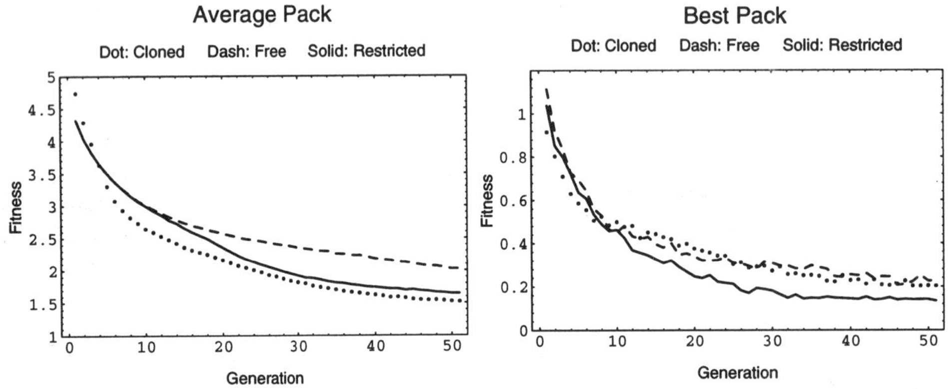

Figure 2: Deictic Sensing Results

5 Results

The results are summarized in Tables 4 and 5. *Average* gives the fitness of the average group in the population at the 51st generation, and *Best* gives the fitness of the best group.

When there was no sensing at all, the best algorithm was typically "go to the gazelle" (in other words, `gazelle` or minor variants). This simple algorithm seemed to predominate in the best lion groups. Given that the better algorithms were often very simple, the groups of cloned lions (which only had to evolve one tree) evolved faster and ultimately slightly better than the heterogeneous groups. Overall, the no–sensing results were not as good as those obtained with sensing. No–sensing results are summarized in Figure 1.

This pattern held more or less for deictic sensing; the clone groups evolved faster and after 51 generations the population as a whole came out ahead, though restricted breeding usually generated the best lion groups. This might be because restricted breeding really was producing the best individuals even if its average might be worse than the average groups under other breeding strategies. Or perhaps this was because restricted–breeding's diversity helped create unusual groups that happened to be well–suited for the initial lion and gazelle positions dealt them that generation. Deictic sensing results are summarized in Figure 2.

In the case of name–based sensing, however, the heterogeneous groups clearly had the upper hand. For both the best of the population and for the population as a whole, the heterogeneous groups produced significantly better results than the clone groups, which fared only slightly better than when they had no sensing at all. Name–based sensing results are summarized in Figure 3.

In all cases, restricted breeding produced as good or better results than free breeding. Furthermore, name–based sensing produced better results than deictic sensing, which produced much better results than no sensing at all. This suggests that even in this highly symmetric domain, creating and enabling specialized team members still proved more profitable than using a uniform multi–agent method.

As expected, the controls did miserably. One lone evolved lion fared slightly better than one randomly-moving lion. Four randomly-moving lions predictably did much better still. But all three did much worse than any evolved four–lion scheme. It's clear that the evolved four–lion schemes are succeeding on their own merit and not just because there are four lions filling space rather than one. Control results are summarized in Figure 4.

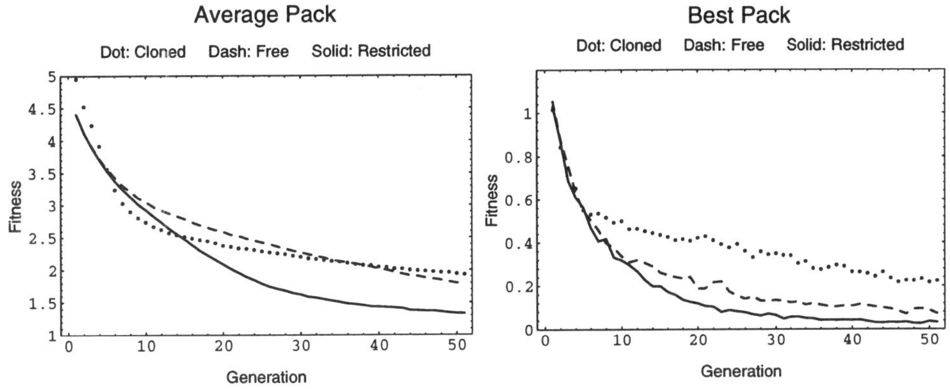

Figure 3: Name–Based Sensing Results

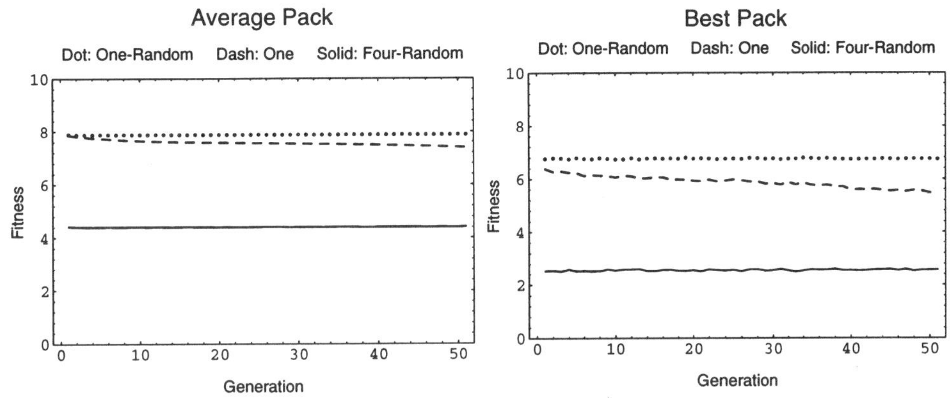

Figure 4: Control Results

6 Conclusions and Future Work

It appears that when given direct information about each other's whereabouts, distinct agents fare best in the Serengeti domain. Uniform agents only do well when coordinating with deictic sensing. And no agents did particularly well when denied the ability to sense teammates. It seems that, although it might require slightly more time for evolution, using a heterogeneous multi–agent strategy may be worthwhile even in a domain where it's unclear that it can help.

Further, restricted breeding consistently evolved faster than free breeding. While there is no reason that free–bred agent teams couldn't eventually develop as well as restricted teams, restricted breeding evolved so much more rapidly that it demonstrated a clear advantage. It seems better to try restricted breeding when developing heterogeneous agents.

In this experiment, a more direct form of sensing proved to be the best method for solving a surprisingly difficult problem. This suggests that even more sophisticated interactive strategies might do better still. One natural extension of this work would be to allow agents to *communicate* with one another through more sophisticated methods than simple directional cues. For example, agents might evolve to communicate instructions to specialized follower–agents through simple memory storage to which both have access, such as the kind used in MAPMAKER [Andre 1995]. Another open issue is whether advantages conferred from developing multiple independent agents are really worth the computational expense, even if they do perform better than uniform agents.

More domains must be studied to determine if the results for the Serengeti generalize to other interesting multi–agent domains. In addition, the conditions that were studied in this paper represented only a few of the many possible breeding strategies and coordination mechanisms. Nature provides other illustrative examples; for example, bee colonies consist of many hundreds of identical individuals from perhaps a half–dozen specialized classes (queens, drones, workers, etc.). This sort of hybrid breeding strategy could also be tested in Serengeti world, and may have more widespread applicability.

Acknowledgements

This research was supported in part by grants from ONR (N00014-J-91-1451), AFOSR (F49620-93-1-0065), and ARPA contract DAST-95-C0037.

We are particularly grateful to Glen Henshaw and the General Motors Truck Division for his assistance in the development and preparation of this paper.

Thanks also to Mark Feinstein for helping to correct some of our more glaring errors concerning animal behavior.

Bibliography

Andre, D. 1995. The Automatic Programming of Agents that Learn Mental Models and Create Simple Plans of Action. In *Proceedings of the Fourteenth International Joint Conference on Artificial Intelligence*. Chris S. Mellish, editor. 741–747. San Mateo, CA: Morgan Kaufmann.

Benda, M., B. Jagannathan, and R. Dodhiawala. 1986. On Optimal Cooperation of Knowledge Sources. Technical Report BCS-G2010-28, Boeing AI Center, Boeing Computer Services, Bellevue, WA.

Haynes, T., S. Sen, D. Schoenefeld and R. Wainwright. 1995. Evolving a Team. In *Working Notes of the AAAI-95 Fall Symposium on Genetic Programming*. Eric V. Siegel and John R. Koza, editors. 23–30. AAAI Press.

Korf, R. 1992. A Simple Solution to Pursuit Games. In *Working Papers of the 11th International Workshop on Distributed Artificial Intelligence*. 183–194.

Koza, J.R. 1992. *Genetic Programming: On the Programming of Computers by Means of Natural Selection*. Cambridge, MA: The MIT Press.

Koza, J.R. 1994. *Genetic Programming II: Automatic Discovery of Reusable Programs*. Cambridge, MA: The MIT Press.

Lohn, J. and J. Reggia. 1995. Discovery of Self-Replicating Structures using a Genetic Algorithm. In *1995 IEEE International Conference on Evolutionary Computing*. 678–683. Perth: IEEE.

Reynolds, C.W. 1993. An Evolved, Vision–Based Behavioral Model of Coordinated Group Motion. In *Proceedings of the Second International Conference on Simulation of Adaptive Behavior*, Jean-Arcady Meyer, *et al,* editors. 384–392. Cambridge, MA: The MIT Press.

Spector, L. 1996. Simultaneous Evolution of Programs and their Control Structures. In *Advances in Genetic Programming 2*, edited by P. Angeline and K. Kinnear. Cambridge, MA: MIT Press, in press.

Tacket, W.A., and A. Carmi. 1994. The Donut Problem: Scalability, Generalization and Breeding Policies in Genetic Programming. In *Advances in Genetic Programming*, Kenneth E. Kinnear, Jr., editor. 143–176. Cambridge, MA: The MIT Press.

Zongker, D., and B. Punch. 1995. *lil-gp 1.0 User's Manual.* Available through the World-Wide Web at http://isl.cps.msu.edu/GA/software/lil-gp, or via anonymous FTP at isl.cps.msu.edu in the /pub/GA/lilgp directory.

USING GENETIC PROGRAMMING TO DEVELOP INFERENTIAL ESTIMATION ALGORITHMS

BEN McKAY	MARK WILLIS	GARY MONTAGUE	GEOFFREY BARTON
Dept. of Chemical Engineering	Dept. of Chemical & Process Engineering	Dept. of Chemical & Process Engineering	Dept. of Chemical Engineering
University of Sydney	University of Newcastle	University of Newcastle	University of Sydney
NSW 2006	Newcastle upon Tyne	Newcastle upon Tyne	NSW 2006
Australia	NE1 7RU, UK	NE1 7RU, UK	Australia
ben@chem.eng.usyd.edu.au	Mark.Willis@ncl.ac.uk	Gary.Montague@ncl.ac.uk	barton@chem.eng.usyd.edu.au

ABSTRACT

Genetic Programming (GP) is used to develop inferential estimation algorithms for two industrial chemical processes. Within this context, dynamic modelling procedures (as opposed to static or steady-state modelling) are often required if accurate inferential models are to be developed. Thus, a simple procedure is suggested so that the GP technique may be used for the development of dynamic process models. Using measurements from a vacuum distillation column and an industrial plasticating extrusion process, it is then demonstrated how the GP methodology can be used to develop reliable 'cost' effective process models. A statistical analysis procedure is used to aid in the assessment of GP algorithm settings and to guide in the selection of the final model structure.

1. Introduction

The measurement of key industrial process variables at a rate suitable for on-line control is a problem common to many industrial processes. Due to the high installation and maintenance costs of equipment, measurements are often unavailable. When appropriate instrumentation is available, on-line measurement is often restricted by long analysis cycle times. This often leads to the monitoring of key process variables through the use of off-line laboratory assays. This can result in the process deviating from desired operating conditions with long disturbance recovery times, leading to unsatisfactory product variability and a reduction in profit. Often, these adverse effects cannot be acceptably overcome by the use of existing advanced control algorithms. Efforts towards alleviating this problem have included the development of inferential estimators. Inferential estimators are reliant on the desired primary output variable being related to other more easily measured secondary variables. Where this is the case, it is possible to infer the 'difficult to measure' variable at the rate at which the secondary measurements become available. Consequently, when the inferential model is used on-line as a soft-sensor (an analyser based on software) improved control performance can result simply because of the higher sampling rate.

There are two general methodologies that may be adopted when building inferential models. Firstly, a "bottom up", mechanistic approach can be taken. This is generally the preferred approach for engineering modelling tasks however, it requires an understanding of the basic physical and chemical properties of the system. In the chemical and allied industries an incomplete understanding of the process often precludes this mechanistic approach. A common alternative is to approach the problem from the "top down" and generate an input-output model of the process from plant data. Of the many input-output modelling methodologies available two procedures predominate, artificial neural networks and linear regression techniques.

Linear regression models represent a simple classical approach to the development of static process models. They can be extended for the modelling of dynamic systems by the inclusion of a time series of past inputs and outputs. Such models are known as auto-regressive models with exogenous inputs (ARX models) and are often used as the basis for the design of control algorithms. In most applications these models are sufficiently accurate to be used successfully for controller design. However, as industrial processes are often non-linear, it has been found that the use of a non-linear model can be beneficial in some situations. This is particularly the case for inferential estimation where an accurate process description is required.

Applications within the process industries indicate that neural networks can adequately represent non-linear process system behaviour (e.g. see Qin and McAvoy, 1992; Willis et al, 1992, 1995; McKay et al, 1996). While neural networks can be directly applied to determine non-linear static mappings, a number of modifications to the basic network structure must be made to extend neural networks for dynamic modelling. For instance, one approach is to adopt the philosophy of the ARX and NARX (non-linear ARX) models and supply the network with a time series of system inputs and outputs. While a second approach to modelling dynamics is to incorporate first order filters (Terzuolo, 1969) into the neuron processing elements in what is called the locally recurrent network. The values of the filter time constants are not known a priori and must therefore be determined, in addition to the network weights, when the neural network is trained.

While neural networks can provide an extremely effective "black box" modelling tool for both steady-state and dynamic systems, the technique has some inherent disadvantages and limitations. For instance, the need for relatively large amounts of training and verification data, which may be restrictive in industrial situations if experiments have to be performed to obtain the data. Another significant drawback of neural networks is that they are not designed to provide any insight into the underlying nature of the process. In an industrial context, it is normally highly desirable to gain as much information as possible about the underlying process, not merely to make accurate numeric predictions. It is for this reason we turned to Genetic Programming (GP) to develop inferential estimation algorithms.

2. GP algorithm details

Our algorithm has been implemented in MATLAB an environment specifically designed to facilitate control system modelling, design and analysis. This environment provides an excellent development platform, with graphics visualisation tools, optimisation and symbolic maths capabilities, as well as convenient routines for string manipulation. Throughout this work, the heuristically selected GP algorithm parameters shown in Table 1 were used.

A feature of the algorithm is that it will explore a wide variety of novel model structures. Generally as the complexity of a model increases, so its ability to generalise (i.e. to fit unseen data as well as the training data) is compromised. Consequently, large tree structures can cause over-fitting of the data. In order to prevent the formation of overly large trees, the fitness function was penalised according to the size of a tree.

Table 1 Default Parameters used by the Genetic Programming Algorithm.

Functional set:	+, -, /, *, ^, sqrt, square, log, exp	Mutation Probability	$P_m = 0.2$
Relative abundance of operators:	equal	Crossover Probability	$P_c = 0.8$
Pop. Size:	$N = 25$	Fitness function	Correlation
No. of Generations:	$G = 50$	Generation Gap	$P_{gap} = 90\%$
Total No. of runs:	$J = 50$	Constant-Input Probability	$P_{mvc} = 0.1$

2.1 Optimisation Of Individual Expressions

Each member of the population may contain a number of constants. Rather than randomly generating the values of these constants, each expression was optimised to obtain the best possible values. The optimised individuals then replaced the unoptimised expressions in the population. Utilising the 'optimal' constant values for each suggested solution will:

- Ensure the best possible performance from each model structure.
- Provide a mechanism by which to recover disrupted building blocks (i.e. improve the portability of successful subtrees).

The last two points can be demonstrated by the following example. Suppose the objective is to fit a model between the following input data sets $a = \{0.91, 0.76, 0.26, 0.05, 0.74\}$, $b = \{1.71, 1.16\ 0.26, -0.35, -0.06\}$ and the corresponding output set, $y = \{5.00, 1.67, 1.00, 0.71, 0.56\}$. Two candidate models may be,

$$y = 5.6*(a-b)^2 \quad (1)$$

$$y = 1/(a - 0.16) \quad (2)$$

which have correlation fitness values of 0.5 and 0.1, respectively. If the crossover operator was applied to these two expressions, one possible offspring would be,

$$y = 1/((a-b)-0.16) \quad (3)$$

which has a correlation fitness value of only 0.2. However, if the value of the constant in this new expression is optimised, the following model is obtained,

$$y = 1/((a-b)-1.0) \quad (4)$$

which yields a fitness value of 1.0. Thus, without performing an optimisation of the model constants, it is very difficult to successfully transfer "building blocks" from one mathematical expression to another. This means that models with an inherently good structure could be discarded due to sub-optimal constants.

In our algorithm, each time a new mathematical expression is generated, regression constants are determined using the Levenberg-Marquardt method of non-linear least squares optimisation (Edgar and Himmelblau, 1989). The default settings for MATLAB's optimisation toolbox were used for all runs. Thus, the global search in the structural (i.e. topological) space is combined with local parameter optimisation. It may be anticipated that this procedure would be computationally intensive, however in our work to date, solution times have been tractable.

3. Dynamic modelling

GP may be applied in a straightforward manner for the development of non-linear models that describe the static (or steady state) behaviour of a process. However, within the domain of estimation and control the process model is often required to capture the dynamics of the system. A number of techniques have been suggested in the literature for dynamic modelling using GP. For instance, GP has been used to approximate the impulse response of a linear time invariant system (Koza, 1993). Unfortunately many process engineering system exibit non-linear gains and as such this approach would not be appropriate. In another application, Iba et al (1993) used GP to perform time series prediction for a chaotic system. However, their algorithm was restricted to the use of polynomial functions, thus unnecessarily limiting the type of solutions that could be obtained. Bettenhausen et al (1995) and Sharman et al (1995) developed GP algorithms that could build dynamic models. However, their techniques required recursive nodes which can cause the fitness evaluation and optimisation steps to become extremely time consuming. In this contribution, we propose an alternative approach, that incorporates the use of process engineering knowledge and allows reasonable solution times to be achieved.

An approach to dynamic modelling commonly adopted by control engineers, is to characterise the dynamic response of a system in the form of simple Laplace transfer functions. Generally, a first order plus dead-time model is postulated (a first order differential equation with delayed input). While the process dead-time and dynamics can be approximated in this fashion, many process systems exhibit non-linear gains, the functional form of which can be very difficult to determine. Nevertheless, if non-linear gains can be identified along with approximate time constants and dead-times, studies have demonstrated that good control performance can be achieved (Hong and Lee, 1985). Therefore, in using GP as a dynamic modelling tool, the methodology adopted is to make use of the control engineers knowledge by initially specifying process transfer functions and delays. Assuming that the process dynamics can be represented by a first order dynamic response, a simple estimate of the dynamic response of the output is given by:

$$y(s) = \sum_{i=1}^{N} G_{pi}(s)u_i(s) \qquad (5)$$

In other words the output is a linear combination of the 'N' process inputs being passed through the first order filters $G_{pi}(s)$. Re-defining:

$$u_i'(s) = G_{pi}(s)u_i(s) \quad i=1,\ldots.n \qquad (6)$$

creates a new terminal set $u'_i(s)$ with the approximate process dynamics removed. Once the data has been processed, the static GP algorithm can then be used to combine the filtered inputs together in such a way as to provide an accurate dynamic model. In this manner, the algorithm will determine the non-linear process gains, process variable interactions and, if required, compensate for residual dynamics.

4. Analysis procedure

GP is probabilistic by nature, and thus for non-trivial symbolic regression problems each time the algorithm is used it is expected that it will arrive at different approximate solutions. In order to capitalise on this inherent variation, it is useful to perform multiple runs and to employ a statistical analysis of the results. In particular, this provides the following information:

- The frequency of runs that produce successful results. This allows the adequacy of the program settings to be assessed. If the probability of obtaining a successful result (P_{suc}) is low, then it may be necessary to adjust some of the parameters of the algorithm. In particular, it may be necessary to increase either the population size, the number of generations or both of these parameters.
- An analysis of the probability of occurrence of each input variable ($P_{inp,i}$ $i=1,..n$) in the best model from each run. This will allow the user to obtain a 'feel' for the relevance of a particular input to the modelling task. This is particularly important in the context of chemical process systems, as the determination of the process inputs that affect an output can be as time consuming as the actual model development.

A run is considered a success if the root mean square (RMS) error on the verification set is below a pre-specified error tolerance. This error tolerance must be determined in a heuristic fashion for each application. As a run can either be successful or not, the probability of success can be defined as follows:

$$\Pr(success) = \lim_{J\to\infty} \frac{X}{J} = P_{suc} \qquad (7)$$

where X is the number of acceptable runs and J is the total number of runs. The success of each run is a binomial variable, as it can take only one of two values (successful or unsuccessful). Moreover, it should be noted that for any specified number of GP runs ($J < \infty$), the value obtained is merely an estimate of the probability of success, $\hat{P}_{suc}$. For the results to be meaningful, there must be some measure of the accuracy of the estimate $\hat{P}_{suc}$. Confidence intervals relate $\hat{P}_{suc}$ to P_{suc} by bounding the latter unknown true value with a specified certainty. A confidence statement for the true probability P_{suc} is:

$$\Pr(L < P_{suc} < U) = 1-\alpha \qquad (8)$$

where L and U are the lower and upper bounds, α is the confidence coefficient, and P_{suc} lies within (L,U) with $100(1-\alpha)\%$ confidence. Based upon the binomial distribution standard statistical tables can be used to determine the values of L and U for given values of $\hat{P}_{suc}$ and J.

In a similar fashion, the probability that a given input variable will be present in a successful model may also be described by a binomial distribution. Thus, following the procedures detailed above, it is possible to calculate the bounded probability that a given input is required to obtain a successful model.

4.1 Selection Of The Best Set Of Models

Having reduced the number of models to a set that conforms to the user's specification on the verification data, the choice of final model structure can be further refined by performing an F-test. This identifies the structures that do not perform significantly worse than the "best" model. To do this, it is necessary to check that the error residuals on the verification data of the selected models are of zero mean and can be adequately approximated by a normal distribution. This allows the comparison of the standard deviation of the error residuals between various models. The F-test is then used to determine whether the standard deviation of the residuals of one model is significantly greater than another (or the F-statistic is used to test the null hypothesis that $\sigma_1 = \sigma_2$). Consequently, the final model set represents equally acceptable solutions to the original modelling problem. The choice of which model to use if multiple models are still present is then a matter of preference.

5. Application results

In this section, two industrial processes are used to highlight the utility of the GP methodology in Process Engineering. First, a steady-state modelling example is presented: the algorithm is used to develop an inferential estimator for product composition in a distillation column. Next dynamic extruder data is used to develop a dynamic inferential estimation algorithm for viscosity.

5.1 Development Of An Inferential Estimation Algorithm For A Vacuum Distillation Column

The vacuum column considered is equipped with 48 trays, a steam-heated reboiler and a total condenser (see Figure 1). The feed F (which consists primarily of two components) is split into two product streams the distillate (D) and bottoms (B).

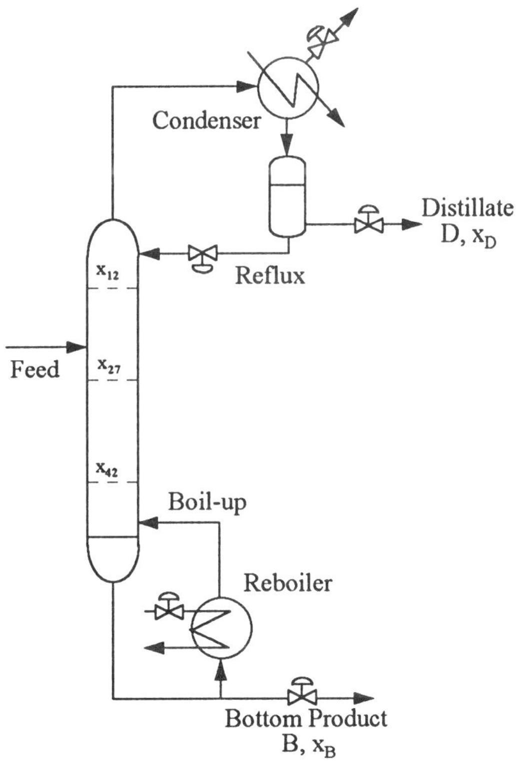

Figure 1: Vacuum Distillation Column

The overhead vapour stream is cooled to form a liquid (in the condenser) and it subsequently flows into a reflux drum. The liquid from the reflux drum is partly pumped back into the column with the remainder being removed as the distillate product. The distillate product is rich in the more-volatile components. At the base of the column, the liquid product stream is rich in the less-volatile components. A liquid stream is also taken from the bottom of the column, boiled with steam (in a reboiler) and returned to the base of the column as a vapour. The control objective is to maintain the quality of either the top, bottom or both product streams. While composition analysers may be used to measure product quality, investment and maintenance costs often restrict their use. Thus, a reliable inferential estimator would provide a cheap alternative to direct measurement and allow tighter control of product composition.

5.1.1 Process Description

The column design specifications and operating data are summarised in Table 2. Since the system is essentially binary, compositions can be estimated based on temperature and pressure measurements. However, the necessary sensors are only installed on trays 12, 27 and 42. As a result, reliable composition estimates can only be obtained for these three trays.

Table 2 Binary vacuum distillation column data.

Number of trays	48	Top composition x_D (%)	99.8
Column diameter (m)	0.9	Bottom composition x_B (%)	2.0
Feed tray	24	Feed composition x_F (%)	80-95
Tray efficiency(%)	44	Feed flowrate F (l/h)	50-400
Relative volatility	1.6	Top pressure (mbar)	30

The objective of this exercise was to develop a model that could be used to infer the bottom product composition given the available measurements. The control system on this column ensures that it operates close to steady-state, thus allowing the dynamics between the tray and the product compositions to be neglected.

5.1.2 *Results*

Input-output data was obtained from a phenomenological model (a structured model describing the underlying physics and chemistry of the system) of the column. The model was comprised of several hundred differential and algebraic equations describing the column dynamics (Gani et al, 1986). To develop an estimator, 150 records of steady-state composition estimates from trays 12 (x_{12}), 27 (x_{27}) and 42 (x_{42}) together with the corresponding value of x_B were used. A second set of 50 data records was used for model validation.

When our GP algorithm was applied, 16 out of the 50 runs were successful. Success was defined as achieving a root mean square (RMS) error of less than 0.02 on the verification data set. Using the binomial distribution, we are thus 95% confident that 20-47% of all runs would meet the required tolerance on the validation data set. This success rate was considered somewhat low. In an attempt to improve the performance of the algorithm, the population size was doubled and another 50 runs were carried out. The number of successful runs increased to 24. Therefore, the proportion of successful runs was increased to 31-60% (at the 95% confidence level), a significant improvement.

Based upon the complete set of successful models, we are 95% confident that 54-85% would contain input variable x_{12}, while 89-100% of all models would contain x_{27} and x_{42}. Therefore, it would appear that the composition on tray 12 is less important than the other two. This result is intuitive, as it would be expected that the compositions near the top of the column would be more weakly related to the bottom product composition.

An F-test was then used to find the structures that were not significantly worse than the best model (as determined by the minimum RMS error on the verification data set). The set of successful models was reduced to five, four of which were identical after simplification. The final two models were:

$$x_B = \left(-2.25 + x_{12} + \frac{-3.166 + 2.574\left(-0.6688 + x_{27}\right)^2}{-2.323 + x_{42}}\right)^2 \quad (9)$$

$$x_B = x_{42}^{\left(3.431 x_{12} + \left(27.0576 - 27.07 x_{12}\right) x_{42} + 1.724 x_{27} - 2.63\right)} \quad (10)$$

Both models performed equally well in terms of their performance as an estimator, yet neither appeared to have captured any significant phenomenological process knowledge. Thus, Equation (9) was chosen as the estimator because of its predominance in the final set of successful models. This model had an RMS on the training data set of 0.011 and a RMS on the verification data set of 0.015. Figure (2) shows the actual and predicted bottom composition on the verification data set, and the corresponding residual error.

5.2 Development Of An Inferential Estimation Algorithm For A Plasticising Extruder

A typical extruder consists of a barrel, inside which one or more helical screws rotate to convey the feed material towards a die at one end, as illustrated in Figure 3. The section nearest the feed point is referred to as the solids conveying zone, where the screw channels are only partially filled and there is no pressure build-up. At some point along the extruder, the channels become completely filled and the temperature and pressure increase considerably as a result of viscous heat dissipation and material transport.

This section is referred to as the melt zone. Additional temperature control may be provided by heating or cooling sections along the barrel.

5.2.1 Process Description

The extrusion process under investigation forms part of a wire coating operation. A base polymer is continually supplied to the extruder barrel together with a cocktail of chemical additives, known as monosil. These two materials are then heated to approximately 220°C and forced down the length of the extruder barrel by an archimedal screw. The increase in temperature causes the decomposition of the peroxide contained in the monosil creating peroxide radicals. These radicals then remove hydrogen radicals from the polymer chains, which results in the formation of chain radicals. The chain radicals subsequently attack the silane molecules, also present in the monosil, to form grafted chain radicals. A die applies a cylinder of polymer to a wire cable which is continually drawn through the die from a pay out drum. The coated wire is then cooled by passing it through baths of cold water, prior to being wound onto a powered take up drum.

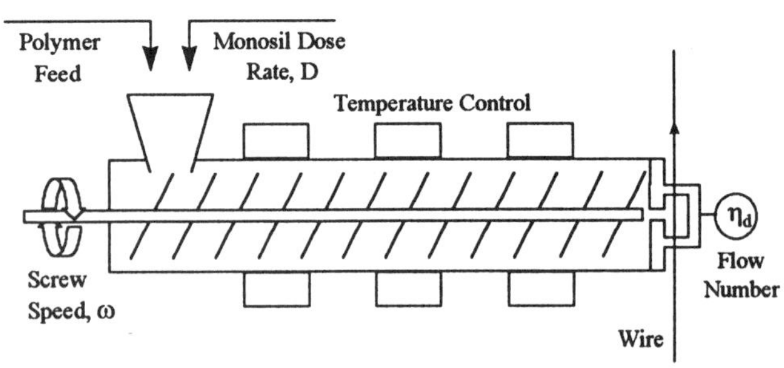

Figure 3: The Pilot Scale Extruder

The grafted chain radicals react with water in the hot water baths, over a period of five hours, to form crosslinks within the polymer. The ultimate tensile strength of the polymer is dependent upon the degree to which this crosslinking occurs, which is in turn governed by the extent to which the grafting reaction proceeds within the extruder.

The tensile strength of the polymer is currently measured infrequently by destructive testing. On-line monitoring of the polymer strength would greatly improve product quality. Previous investigations have shown that there is a high correlation between the ultimate tensile strength of the polymer and the viscosity of the polymer (measured in terms of the flow number, which is proportional to the inverse of viscosity) as it leaves the extruder barrel. Fortunately, this viscosity can be measured relatively easily with the use of a rheometer, which BICC Cables have now installed on one of their wire coating process lines. However, it is not economically viable for BICC Cables to install a rheometer on every wire coating process that they operate world-wide. Therefore the aim of this work was to develop an inferential model of the polymer viscosity from the readily available on-line measurements.

5.2.2 *Results*

A fully instrumented extruder at BICC Cables was used for data generation. Multi-level step tests were performed, implementing changes in monosil and screw speed over a wide range of operating conditions. Here the objective was to obtain information over the entire range of operation in order to capture the inherent non-linearities. All process conditions were logged every minute and around 48 hours of data was gathered in total. Data screening to remove obvious outliers was carried out prior to model generation. Due to the flow nature of the process, long and variable delays arise between system inputs and the output. In this application, an external time delay model was used to time shift data before it entered the process model.

As the extruder is an experimental rig, many more measurements are available than would typically be measured on a production plant. A subset of this information was used, limiting the potential performance of the model but maintaining the practicality of the solution. Therefore, the inputs made available for modelling purposes were screw speed (w) and the monosil dose rate (D), while the output was the polymer flow number at the die (h_d). A section of data 900 minutes long was used for model building and a further 400 minutes was used for validation. In accordance with the dynamic modelling methodology outlined above, linear first order plus dead time transfer functions were used to model the dynamic relationship between D, w and h_f The parameters of the model were first approximated by hand and then fine tuned using a Levenburg-Marquardt non-linear least squares optimisation algorithm. The transfer functions were then used to pre-filter the input variables to provide the terminal set for the GP algorithm. The GP algorithm was then used to find a non-linear function of the filtered inputs that captured the non-linear gains of the process, and compensated for any residual dynamics.

When our GP algorithm was applied, 20 out of the 50 runs were successful. Success was defined as achieving a RMS error of less than 0.135 on the verification data set. Using the binomial distribution, we are thus 95% confident that 26 - 55% of all runs would meet the required tolerance on the validation data set. This success rate was considered acceptable. Based upon the complete set of successful models, we are 95% confident that 89-100% of all models would contain D, w . Therefore, as expected both variables are equally important to our model.

An F-test was then used to find the structures that were not significantly worse than the best model (as determined by the minimum RMS error on the verification data set). The set of successful models was reduced to 12. Based upon the latter set of 12, all models performed equally well in terms of their performance as an estimator, yet none appeared to have captured any significant phenomenological process knowledge. Therefore, adopting the principle of "Occam's razor" the model depicted in Figure (4) was selected to infer polymer viscosity, where 's' is the Laplace operator and D' and w' represent the filtered inputs that are the terminal set for the GP. Figure (5) shows the actual and predicted viscosity on both the test and verification data set. It should be recalled that in developing this model, only a subset of process information was used, limiting the potential performance of the model but maintaining the practicality of the solution. Nevertheless, on the training data set the RMS error was only 0.1 and this only increased to a value of 0.129 on the verification data This RMS compares favourably with that of a locally recurrent neural network (McKay et al, 1996) however, the model has a more parsimonious structure. For this reason,

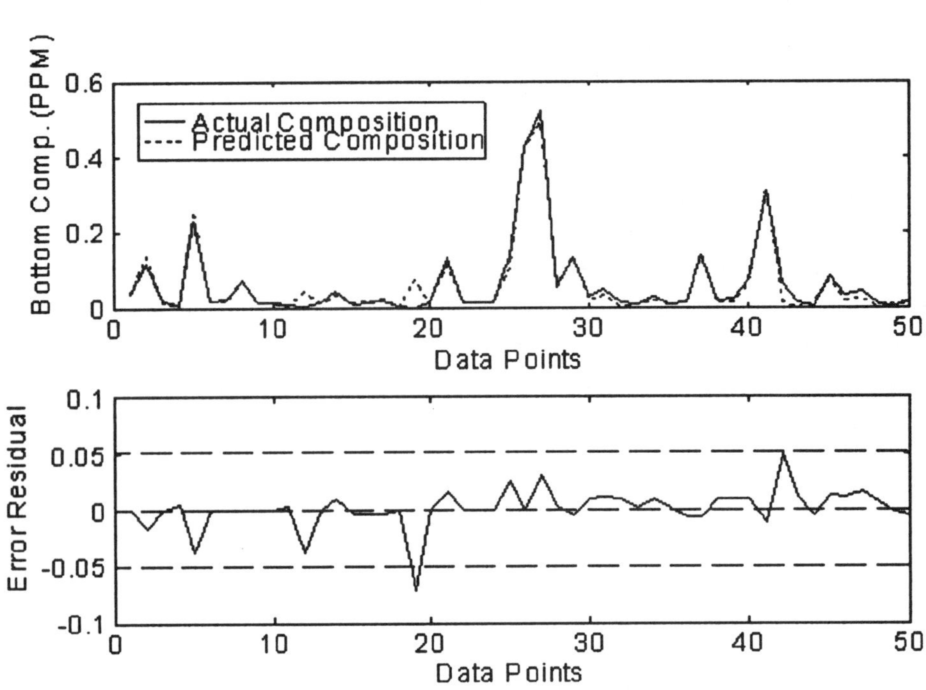

Figure 2: Prediction of Composition in the Distillation Column

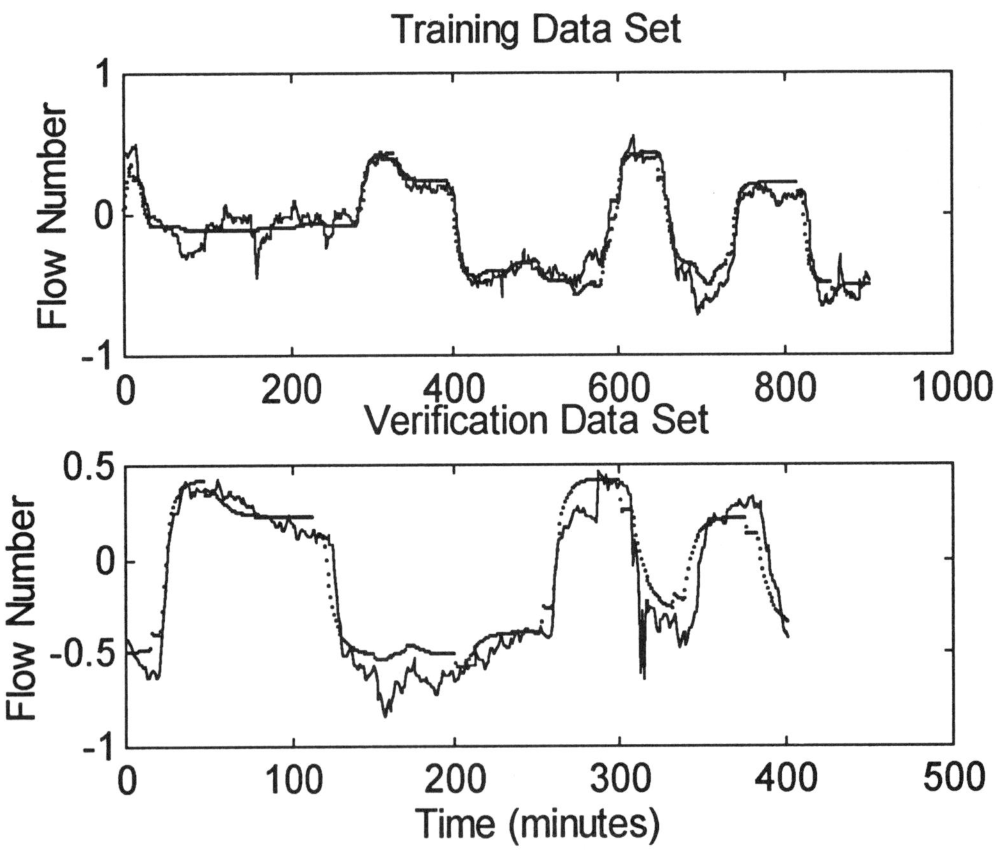

Figure 5: Prediction of Viscosity in the Pilot Scale Extruder

current work is aimed at the development and application of a closed loop control strategy based upon this model. set.

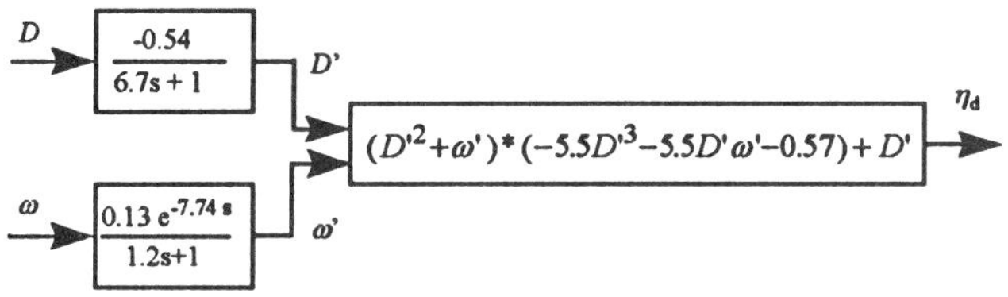

Figure (4) Estimator for Viscosity.

6. Conclusions

The application of GP to the development of inferential models of chemical process systems has been considered. Two industrial examples were used to highlight the utility of this approach. The results revealed that in each case the GP algorithm could generate an accurate input-output model based solely on observed data. One possible perceived disadvantage of the input-output models obtained using GP is that the identified structure does not provide detailed phenomenological information about the system being modelled (a characteristic that this approach shares with other black-box modelling techniques, including neural networks). However, in each case the final model form clearly indicates the relative contribution of each input to the output. This allows the sensitivity of the model to changes in various inputs to be judged with relative ease.

Acknowledgements

The authors would like to acknowledge Pat Duffy and his colleagues at BICC Cables, Wrexham for providing the process data and access to the pilot plant. In order to carry out this work Ben McKay was supported by the British Council.

Bibliography

Edger, T.F. and Himmelblau, D.M., (1989) 'Optimisation of chemical processes', McGraw-Hill.

Hong, J.C. and Lee, H.H. (1985) 'Fixed Bed with Catalyst Deactivation: On-line Estimation of Deactivation, Control and Design', AICHE Journal 31, 302.

Iba, H., Kurita,T., Garis,H. and Sato,T., 'System Identification Using Structured Genetic Algorithms, Proc. of the 5th Int. Conf. on Genetic Algorithms, Urbana-Champaign, USA, pp 276-286, (1993).

Koza, J.R., 'Finding an Impulse Response Function Using Genetic Programming', Proc. of the ACC, pp 2345-2350, San Francisco, USA, (1993).

McKay, B., Lennox, B., Willis, M.J., Barton, G.W. and Montague, G.A, (1996) 'Extruder modelling: A comparison of two paradigms' Submitted to Control '96 Exeter, U.K.

Terzuolo, C. A. (Ed), (1969), 'Systems Analysis to Neurophysiological Problems', University of Minnesota, USA, pp86-91

Qin, S.J. and McAvoy, T., 'A data based modelling approach and applications', Dycord' 92, 321- 326, Maryland, USA, (1992).

Willis, M.J., Di Massimo, C., Montague, G.A., Tham, M.T. and Morris, A.J., 'Artificial neural networks in process estimation and control', Automatica, 28(6), pp 1181-1187, (1992).

Willis, M.J., Montague, G.A. and Peel, C., (1995) 'On the application of artificial neural networks to process control', Applications of Neural Networks', A. Murray (Ed), Kluwer Academic Publishers, pp 191-219.

Bettebhausen, K.D., Marenbach, P, Freyer, S., Rettenmaier, H., Nieken, U., (1995) 'Self-organising structured modelling of a biotechnological fed-batch fermentation by means of genetic programming', First IEE/IEE Int. Conf. on Genetic Algorithms in Engineering Systems: Innovations and Applications, Sheffield, UK, No. 414, pp481-486.

Sharman, K.C., Alcazar, A.I.E, Li, Y., (1995) 'Evolving signal processing algorithms by genetic programming', First IEE/IEE Int. Conf. on Genetic Algorithms in Engineering Systems: Innovations and Applications, Sheffield, UK, No. 414, pp473-480.

Dynamics of Genetic Programming and Chaotic Time Series Prediction

Brian S. Mulloy
Rick L. Riolo
Robert S. Savit
Program for Study of Complex Systems (PSCS)
University of Michigan, Ann Arbor MI USA 48109-1120
bmulloy@umich.edu, rlriolo@umich.edu, savit@umich.edu
(313) 763-3233

Abstract

An investigation into the dynamics of Genetic Programming applied to chaotic time series prediction is reported. An interesting characteristic of adaptive search techniques is their ability to perform well in many problem domains while failing in others. Because of Genetic Programming's flexible tree structure, any particular problem can be represented in myriad forms. These representations have variegated effects on search performance. Therefore, an aspect of fundamental engineering significance is to find a representation which, when acted upon by Genetic Programming operators, optimizes search performance. We discover, in the case of chaotic time series prediction, that the representation commonly used in this domain does not yield optimal solutions. Instead, we find that the population converges onto one "accurately replicating" tree before other trees can be explored. To correct for this premature convergence we make a simple modification to the crossover operator. In this paper we review previous work with GP time series prediction, pointing out an anomalous result related to overlearning, and report the improvement effected by our modified crossover operator.

1 Introduction

Applications of Genetic Programming (GP) to the task of chaotic time series prediction have had mixed results. The following is a brief review of previous studies.

The first study was done by [Koza, 1992, pp. 507-513] who applied GP to the task of predicting next values of the logistic equation. The results of this short-term[1] prediction were expectably good. As noted by [Koza, 1992, pp. 508], the logistic equation is one the simplest equations which displays chaotic behavior. More challenging problems would need to be studied to prove GP as a paradigm capable of performing chaotic time-series prediction.

A more difficult task was attempted by [Oakley, 1994], who studied a more complicated time series and a far more challenging prediction task. The time series studied was produced by the Mackey-Glass equation [Mackey and Glass, 1977]. The prediction task considered was a non-standard long term prediction akin to a squared-error regression[2] . This problem proved to be too difficult for the GP methods used. Many of the solutions found were simply constants in time and the other solutions did not even qualitatively resemble the original time series.

However, [Iba *et al*, 1993] had introduced "a new GA (Genetic Algorithm)-based approach to the solution of system identification" and applied it to another time series prediction problem. This new system, called STROGANOFF (STructured Representation On Genetic Algorithms for NOn-linear Function Fitting), was a hybrid of both GP and a "method of heuristic self-organization" due to [Ivakhnenko, 1971].

STROGANOFF was applied to the tasks of time series prediction and pattern recognition. The time series considered was also generated with the Mackey-Glass equation. However, the version used by [Iba *et al*, 1993] had a lower attractor dimension than that used by [Oakley, 1994] and the prediction task was short-term

[1] Here short-term prediction is taken to mean prediction of only the next value, x(t+1), of a time series. Whereas long-term prediction is taken to mean prediction of $\tau = 2$ or more steps into the future, $x(t+\tau)$. Generally short-term prediction is an easier problem than long-term.

[2] The methods used by [Oakley, 1994] and the others are described in more detail in section 2.3 of this paper.

instead of long-term. Later, [Iba, 1995] compared the performance of STROGANOFF to the performance of a traditional GP. He showed that STROGANOFF performed better than traditional GP on the Mackey-Glass time series, and also on the Ikeda map [Ikeda, 1979] and the Lorenz equation [Lorenz, 1963].

In summary, although the above mentioned studies of traditional GP on the logistic equation and of a hybridized GP on the Mackey-Glass equation have shown encouraging results, traditional GP has not performed well on the Mackey-Glass equation.

Nonetheless, traditional GP seems to be well suited for the task of time-series prediction. All the necessary data can be easily embedded in the terminal set and a potpourri of nonlinear functions can be created from an elementary function set. Furthermore, one can easily vary the GP fitness metric, which will facilitate the application of novel time-series prediction methods with GP.

Therefore, we investigate some of the problems encountered when GP is applied to the task of chaotic, time-series prediction. We find two causes for GP's reportedly poor performance and present a solution which gives qualitatively better results than those previously reported for traditional GP. First, we show that the over-fitting reported by [Iba, 1995] is not a typical result. Rather, Iba's result may have been a pathological anomaly. Second, we show that GP converges prematurely when performing a short-term time series prediction. We present a solution to this problem of convergence and report the resulting increase in GP performance. The rest of the paper is structured as follows. Section 2 explains the problem of time series prediction, and reviews in more detail the work of previous authors who have applied GP to the time series prediction task. Section 3 describes in detail the methods used in this study, and section 4 presents our results. Section 5 summarizes our work and section 6 is a discussion of possible future studies.

2 Background

2.1 Time Series Prediction

The problem of time series prediction is this: given values of the past, x, one must find a function, $\tilde{f}$, which predicts values of the future. Past values, $x(t-\phi)$, can be considered as a vector,

$$\mathbf{x}(t) = (x(t), x(t-1), ..., x(t-\phi)).$$

Future values, $x(t+\tau)$, are estimated by a function of previous values, $\tilde{f}(\mathbf{x}(t))$. In this paper we consider the problem of searching for $\tilde{f}(\mathbf{x}(t))$, given the vector

$$\mathbf{x}(t) = (x(t), x(t-1), ..., x(t-10)),$$

which will predict the actual value $x(t+1)$.

To evaluate how close a particular function is to predicting the actual value of a time series one must impose a metric. A simple method, and the one considered in this paper, is the sum of squared errors between the actual time series values and the predicted values,

$$S = \sum_{t}^{T} \left(x(t+1) - \tilde{f}(\mathbf{x}(\mathbf{t}))\right)^2.$$

The time series considered in this study is the Mackey-Glass equation [Mackey and Glass, 1977] with the parameters given below. The equation for the discretized map is

$$x(t+1) = x(t) + \frac{bx(t-\Delta)}{1+(x(t-\Delta))^c} - ax(t), \quad (1)$$

with $a = 0.1$, $b = 0.2$, $c = 10.0$ and $\Delta = 17$.[3] This is a commonly studied map for time series prediction research [Farmer and Sidorowich, 1987]. Our method of generating the map is described in section 3.1.

2.2 Comparing Performance of Predictors

In order to evaluate the performance of different predictors we use the method also used by [Iba, 1995]. This method is due to [Casdagli, 1989]. Once a predictor function, $\tilde{f}$, is found, we can measure its performance on an evaluation set by estimating the following:

$$\sigma^2\left(\tilde{f}\right) = \frac{1}{N} \times \sum_{t}^{N} \left|x(t) - \tilde{f}(\mathbf{x}(t))\right|^2 / var, \quad (2)$$

where

$$var = \frac{1}{N} \times \sum_{t}^{N} \left|x(t) - \frac{1}{N} \times \sum_{t}^{N} x(t)\right|^2,$$

and N is the number of points evaluated. $N = 10^3$ in [Iba, 1995] and we use the same number of evaluation points in this study. In the results section we use this measure to compare the performances among different solutions.

2.3 Previous Work

The following is a more detailed review of the work of previous authors.

The earliest work with GP time series prediction was that of Koza [Koza, 1992, pp. 507-513] who considered the logistic map. Koza performed short-term prediction,

[3] With these parameters the attractor dimension was estimated to be 2.1 [Farmer and Sidorowich, 1987].

estimating the current value, $x(t)$, with a GP generated function of the previous map value, $\tilde{f}(x(t-1))$. He used as a fitness function the sum of the absolute error between the actual map value and the predicted value over $t = \{1, 2, \ldots, 10\}$ fitness cases, $\sum_t^{10} \left| x(t) - \tilde{f}(t) \right|$. Koza's results were encouraging, showing qualitatively good performance on the training set.

Another study was that of [Oakley, 1994], who considered the discretized version of the delay differential equation due to Mackey and Glass discussed in the previous section. He performed a challenging long term prediction task, and estimated the current value, $x(t)$, with a GP generated function,

$$\tilde{f}\left(\tilde{f}(t-1), \tilde{f}(t-2), ..., \tilde{f}(0), \mathbf{x}\right).$$

The dependent variables were values previously generated by the GP's $\tilde{f}$ and 31 of the map values, $\mathbf{x}$. The map values used were $\mathbf{x} = \{x(-1), x(-2), x(-3), \ldots, x(-31)\}$. Notice that these input values, $\mathbf{x}$, were fixed and did not vary with t. Oakley used as his fitness function the sum of the squared errors between the predicted and actual values over T fitness cases, $\sum_t^T \left(x(t) - \tilde{f}(t)\right)^2$. This was a challenging problem for the following reason. Each fitness case, t, depended upon $t-1$ values which were previously generated by the GP's predictor function. For example, on the $20th$ fitness case 19 of the possible input values were generated by the GP function itself. Because of this reflexive relationship the inaccuracy of initial predictions was magnified with each fitness case. Therefore the accuracy of predictions rapidly decreased with each additional fitness case. In total,

$$\sum_t^T (t-1) \approx T^2$$

self-generated data points were used. Therefore as the number of fitness cases, T, increased the difficulty of the problem increased as T^2.

A more efficient way of performing long-term prediction may be to generate

$$\{x(t), x(t+1), ..., x(t+\tau)\}$$

using the previous map values

$$\{x(t-1), x(t-2), ..., x(t-\phi)\}.$$

Here τ is the distance into the future the GP will be trained to predict and ϕ is the maximum past value used from the map (both are constants). Notice, instead of T^2 self-generated values being used, only $\tau \times T$ self-generated values are used. Therefore, as one uses more

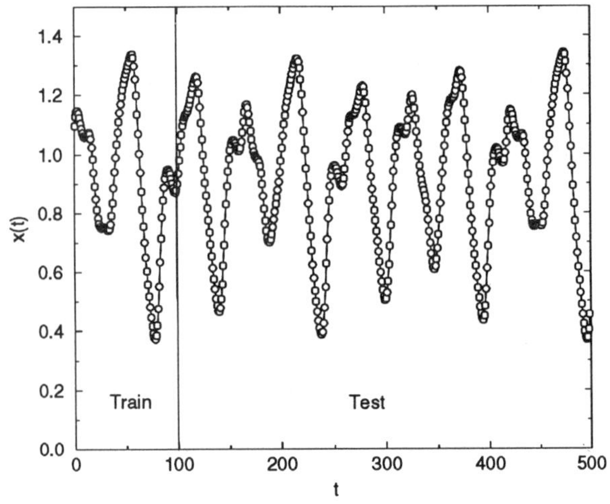

Figure 1: 500 point Mackey-Glass time series.

training data the problem does not, in general, become more difficult; instead the GP may more easily evolve a τ-steps ahead predictor.

A way to think of the difference between these two methods is to picture a window which shows previous values of the time series. The GP functions can only use values which appear in this window. In the first method the window is locked on some initial values, $x(-1), x(-2)$, etc. As the fitness case, t, increases the predictor function moves further and further away from this window. In the second method, however, the window moves freely with each fitness case so the function always uses the same number of direct map values, $x(t-1), x(t-2)$, etc. The second method presented here may prove to be a better use of the data.

The most recent GP study was that of [Iba, 1995], which was primarily a comparison between the STROGANOFF system and traditional GP[4]. STROGANOFF and GP were both applied to the task of short term prediction on several chaotic maps [Iba, 1995]. The methods and results of [Iba, 1995] will be studied in detail in sections 3 and 4 of this paper.

3 Methods

3.1 *Time Series Generation*

To generate the Mackey-Glass time series with $\Delta = 17$, 17 double precision pseudo-random numbers are generated. These values are used to seed the Mackey-Glass map. (This is necessary because the map value at time t depends upon the past value $t-17$.) Then 1500 points are generated, using the Mackey-Glass map (Eqn. 1).

[4]See section 1.

The first 1000 points are discarded. The remaining 500 points are used as the time series. The first 100 points are used as the training set and the next 400 points are used as the test set. In total only 117 values are needed to train the GP.

In this paper the same time series is used for all GP runs[5]. The 500 point time series is plotted in Figure 1.

3.2 GP Parameters

As explained previously, the task of time series prediction is accomplished by using previous values of a time series to predict future values. Therefore, the GP terminal set will consist of previous values of the time series. Choosing which past values to employ is difficult. Ideally a very large number of past values would be given in the terminal set and the GP would find those values appropriate for prediction. However, as the size of the terminal set increases so too does the size of the search space and hence the difficulty of the search task. The automatic discovery of useful terminals is an interesting and perhaps fundamental issue for both time series prediction research and GP dynamics research[6]. As a first step toward understanding this issue we investigate the terminal set used by [Iba, 1995], which employs the ten previous map values and an ephemeral random constant,

$$\mathbf{T} = \{x(t), x(t-1), ..., x(t-10), \mathcal{R}\}.$$

The ephemeral random constant ranges from $[-1.0, 1.0]$.

The function set used is due to [Oakley, 1994]. The only requirement of the function set is its ability to produce nonlinear expressions. The function set[7] is

$$\mathbf{F} = \{+, -, *, \%, sin, cos, k^{10}\},$$

where % represents protected division[8] and k^{10} represents the raising of a number to the tenth power. Fitness is determined by comparing the values returned by a GP tree to each value of the 100 fitness cases (i.e. the 100 point training set). To be clear, the GP evolution does not depend in any way on the 400 point test set. The test set is used to evaluate performance only after the GP has found its best tree. The raw fitness, r, is the sum over the fitness cases (training set) of the squared error between the actual Mackey-Glass values and the GP predicted values, $r = \sum_{t}^{N_{train}} \left(x(t+1) - \tilde{f}(\mathbf{x}(t))\right)^2$. The standardized fitness, s, in this case is the same as the raw fitness, r. The adjusted fitness, a, of each individual is the traditional GP value $a = 1.0/(1.0+s)$.

The parameters we use are exactly those reported in [Iba, 1995]: The population size is 5000, the number of generations is 101 (including the first random population). The initial population is generated using the ramped-half-and-half method. The maximum depth of new individuals is 6. The maximum depth of individual after crossover is 17. Fitness proportionate selection is used. 90% of the selected individuals undergo crossover, with 70% internal and 20% external crossover points being selected. The remaining 10% of selected individuals undergo reproduction. The results shown are based on 20 runs for each experimental condition. Each of the 20 runs will have a different initial population. Across the different groups of 20 runs we use the same 20 initial populations.

3.3 Crossover

We use two types of the Crossover operator in the experiments reported in this paper. The first type is just "normal Crossover" (NX), as described in [Koza, 1992]. The second type differs from the first only in that we forbid one-node tree crossovers (FONTX). These two crossover operators are described in more detail below.

For normal crossover, upon selection of two parent trees a node in each tree is chosen at random. In GP there are two node types, internal (function) and external (terminal). Each node type has its own probability of being selected during crossover. Traditionally, the internal points are given a much higher probability of being selected than the external points[9]. Once the nodes have been selected the subtrees at these nodes are swapped (crossed-over) between the two parents, producing two offspring.

Crossover is intended to search for new, better solutions. However, inherent in the process of GP crossover is a probability for the offspring to be exact duplicates of their parents, i.e. each crossover operation has a probability of degenerating into a reproduction operation. The simplest circumstance where crossover results in a reproduction is if the root node of each parent is selected.

---

[5] The 17 seed values for this time series are 1.097410, 1.128060, 1.145130, 1.147970, 1.139460, 1.123850, 1.105310, 1.087330, 1.072550, 1.062700, 1.058500, 1.059560, 1.064340, 1.070190, 1.073410, 1.069520, 1.054130; the rest of the time series can be generated from these data.

[6] See also the paper of [Green and Savit, 1991, pp. 534-540] in which the question of which variables contain the most information relevant to a specific criterion was discussed.

[7] This function set may not be the optimal function set. The sin, cos, and k^{10} functions are not necessary to express a robust nonlinear function. Furthermore, unlike the arithmetic functions $\{+, -, *, \%\}$, each of which has an "inverse function" (i.e. adding b to a is undone by subtracting b from $a+b$), the functions sine, cosine, and k^{10} have no inverse functions in this function set. This inability to produce trees which display essentially null behavior (introns) may inhibit the fitness of the GP (see [Nordin and Banzhaf, 1995]).

[8] Protected division returns 1.0 if the denominator is 0 and returns the result of division otherwise [Koza, 1992].

[9] The justification for this distribution is to promote the "recombining of small substructures or building blocks" [Koza, 1992]. However, the building block hypothesis has not been confirmed [O'Reilly and Oppacher, 1994, Altenberg, 1994].

Because there is only one root node in any tree, the probability of selecting the root node is just the inverse of the internal node count multiplied by the probability of internal crossover. Therefore the probability of crossover degenerating into reproduction due to root node selection is the product of each parent's probability for root point selection,

$$P_{r\triangleleft c} = P_{ip}^2 * \frac{1}{N_{ip}^A N_{ip}^B}.$$

Thus as a tree grows its total probability of reproduction from root-node crossover decreases.

A case which is especially problematic for time-series prediction is the selection of two trees that consist of single nodes, e.g., $x(t-1)$. This situation can cause difficulty for time series prediction because the value $x(t-1)$ is often a much better than random predictor for $x(t)$. Such trees can spread rapidly early in GP runs, before more sophisticated predictors can be found. When both trees are one-node-trees the probability of reproduction from crossover is unity. The reproductive crossover of these small, fairly fit trees can cause the one-node trees to spread very rapidly in the population, leading to a kind of premature convergence which prevents GP from ever finding better, more complex solutions[10].

The second type of crossover we use is designed to correct for this tendency toward reproduction by crossover of one-node trees. This crossover operator, "forbid one-node tree crossover" (FONTX) does not allow two one-node parent trees to cross with each other. Instead if two one-node trees are selected for crossover, they are both discarded and new trees are selected from the population.

4 Results

This section begins with the investigation of overlearning as reported by [Iba, 1995]. Then the results of our two experimental runs are given. Please note that the tree considered in section 4.1 is found in a run whose results are described in section 4.2. This order of presentation was chosen to keep two strands of thought separate. The first strand is that a given GP tree, once found, does not inherently suffer from overlearning of the training data. The second strand relates to premature convergence and depends upon contrasting our two experimental runs.

The first set of runs are a replication of the previous work of [Iba, 1995] and the results of this run are referred to as "Normal Crossover" (NX). The second method differs from the first only in the way crossover is accomplished (see Section 3.3) and is referred to as "forbid one-node tree crossover" (FONTX). Furthermore, for purposes of comparison the results reported by [Iba, 1995] are referred to as "Previously Reported Results" (PRR).

Table 1: Comparison of the previously reported results (PRR) with results found in this study (NX) for $T1$.

Study	Mean Squared Train Error	Mean Squared Test Error	$log_{10}\sigma\left(\tilde{f}\right)$
NX	6.69×10^{-5}	6.52×10^{-5}	-1.44
PRR	6.50×10^{-6}	1.5×10^{-4}	-1.43

4.1 Overlearning

In the results reported by [Iba, 1995] the best tree found was

{$T1$} (- (+ X[t-1] X[t-1]) X[t-2]) ,

which is equivalent to

$$\tilde{f}\left(\mathbf{x}\left(t\right)\right) = 2x\left(t-1\right) - x\left(t-2\right).$$

This tree, as will be reported in section 4.2, is found in 13 of 20 runs performed in this study.

Table 1 shows the mean squared error observed in the training and test sets, as well as the measure in equation 2 calculated over 10^3 evaluation points, for tree $T1$. The results as reported by [Iba, 1995] (PRR) and as observed in this study (NX) are shown in Table 1. As can be seen, the comparative measure values, $log_{10}\sigma\left(\tilde{f}\right)$, are very nearly identical between the two studies. However, there is a large discrepancy between the two studies in how training data relates to test data. In this study we find the difference between training and test data to be essentially negligible. However, the difference reported by [Iba, 1995] is more than an order of magnitude. What is the origin of this discrepancy?

Although the time series used in each study is generated from the same Mackey-Glass mapping, the actual values in a particular series may differ. Therefore by testing a particular tree on many other time series generated by the same map, with different seeds, one can create a statistical picture of the relationship between training performance and test performance. To accomplish this, we generate 10^3, 500 point time series according to the methods given above (Section 3.1), changing the random seed in each case. We then measure the

[10] [McPhee and Miller, 1995] have studied, both experimentally and theoretically, the causes and effects of crossover degenerating into reproduction (called "accurate reproduction"). They point out that accurate reproduction, tree growth, and the discovery of solution trees may be related phenomena. Our results strengthen this hypothesis. The accurate replicator we find is a one node tree, and it inhibits the growth of larger, possibly fitter replicators.

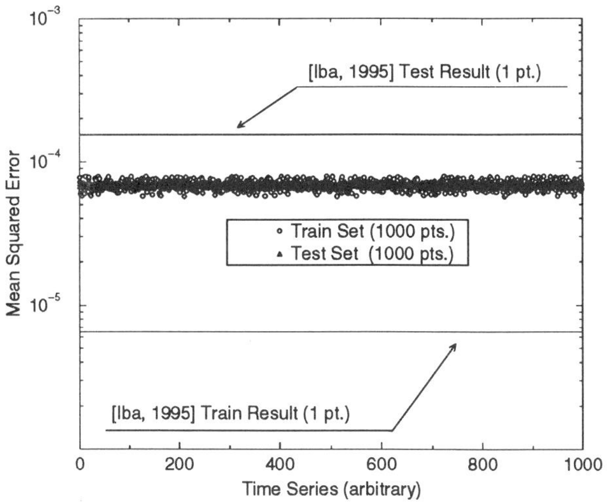

Figure 2: A scatter plot of mean squared error for 1000 coupled train and test time series. Only the seed values differ between each of the 1000 time series. For comparison, the results reported by Iba are plotted as lines.

Table 2: Statistical summary of Figure 2.

	Minimum Mean Squared Error	Maximum Mean Squared Error	Average Mean Squared Error
Training	5.7×10^{-5}	7.9×10^{-5}	6.8×10^{-5}
Testing	6.3×10^{-5}	7.2×10^{-5}	6.8×10^{-5}

mean squared error over the first 100 points (the training fitness) and the mean squared error over the next 400 points (test fitness). As can be seen from Figure 2, the distributions of both training performance and test performance are centered around the same mean value. A statistical summary of Figure 2 is given in Table 2.

Clearly, a performance decrease of two orders of magnitude between training data and test data is not expected for tree $T1$ on this prediction problem. Furthermore, the values reported by [Iba, 1995] are out of the range of the maximum and minimum values over the 1000 data points (i.e $6.50 \times 10^{-6} < 5.7 \times 10^{-5}$ and $1.5 \times 10^{-4} > 7.9 \times 10^{-5}$). Therefore it seems likely that the results reported by [Iba, 1995] are an anomaly. In the remainder of the results section we will not refer to these seemingly anomalous data. However we will consider other results reported by [Iba, 1995].

Table 3: Comparison of performance of best trees from Normal Crossover ($T0$,$T1$) and FONTX ($T2$) runs.

Tree	Mean Squared Train Error	Mean Squared Test Error	$\log_{10} \sigma(f)$
T0 (NX)	5.24×10^{-5}	5.08×10^{-5}	-1.53
T1 (NX)	6.69×10^{-5}	6.52×10^{-5}	-1.44
T2 (FONTX)	1.49×10^{-5}	1.19×10^{-5}	-1.82
STROG.	4.70×10^{-6}	5.06×10^{-6}	-2.00

4.2 Normal Crossover (NX)

Now we consider the problem of premature convergence. We perform 20 runs using normal crossover (NX). The two overall best trees GP finds using NX are the following:

$\{T0\}$

```
( % (+ X[t-1]
(* 3.270900e-02
(* (cos (sin (- X[t-2] -3.483200e-03)))
(sin (- X[t-8]
(+ X[t-1]
(* 3.270900e-02 X[t-1] )))))))
( % X[t-2] X[t-1] ))
```

$\{T1\}$ (- (+ X[t-1] X[t-1]) X[t-2]).

The tree $T1$ is interesting, because in the previous study by [Iba, 1995] it was the best tree overall found by traditional GP. The mean squared error of each tree over the training and testing sets is given in Table 3. One can see that the difference between training and test performance is negligible for each tree.

In finding these trees, using normal crossover, GP converges prematurely in eighteen of the twenty runs performed. In each of the 18 cases where GP converges the entire population is comprised of only the tree $x(t-1)$. Figure 3 shows the average, worst and best fitness for a typical run which converges to the tree $x(t-1)$. The convergence of these fitness measures typically is indicative of a converged population. More directly, Figure 4 shows the count of the number of instances of the one-node tree $x(t-1)$ in the population over the generations; clearly this tree has come to dominate the population, which explains why these GP runs cannot find better solutions.

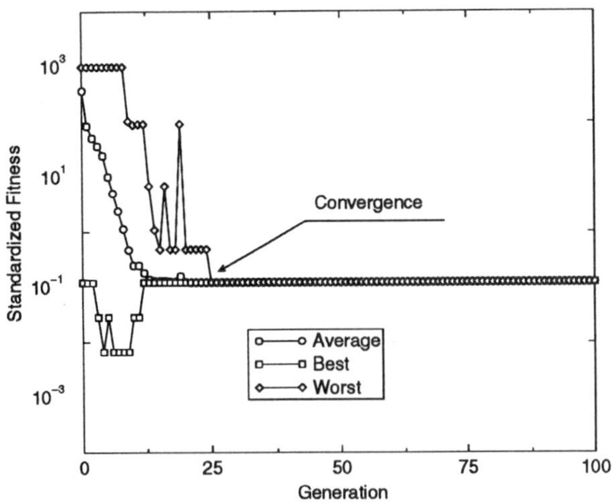

Figure 3: Best, average and worst fitness vs. generation for the run producing tree $T1$. The entire population converges onto the tree $x(t-1)$ at generation 25.

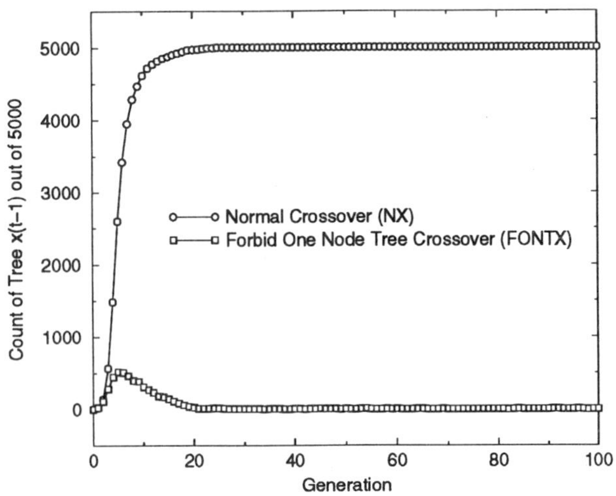

Figure 4: Count of the tree $x(t-1)$ vs. generation for one run using Normal Crossover (NX) and one run using Forbid One-Node Tree Crossover (FONTX). By generation 30, the NX population is composed entirely of the tree $x(t-1)$. Whereas the FONTX population is essentially bereft of the tree $x(t-1)$ by generation 30.

4.3 Forbid One-Node Tree Crossover (FONTX)

Now we consider a method of crossover which corrects for the observed premature convergence. We perform 20 runs by forbidding one-node tree crossover (FONTX).

These FONTX runs no longer exhibit premature convergence, as seen by the dynamics of the fitness measures (Figure 5). The average fitness is gradually sloped and new best trees are found throughout the evolution. However, the best tree of each generation usually becomes extinct the generation after it is found. Aside from this, the evolution appears robust. Figure 4 shows that one-node trees are, for the most part, a negligible part of the population. These runs also find trees with better fitness than were found with normal crossover. The best overall individual found using this technique is the following:

{$T2$}

```
(% X[t-1]
(% (% X[t-2]
(% (* (% X[t-3] X[t-2]) X[t-1])
(% X[t-2]
(% (- X[t-3] X[t-3])
(* (- X[t-3] X[t-3]) X[t-7]))))) X[t-1])).
```

This tree is equivalent[11] to,

$$\tilde{f}(\mathbf{x}(t)) = \frac{x^3(t-1)}{x^3(t-2)} \times x(t-3),$$

which is a nonlinear expression, unlike tree $T1$.

Table 3 shows the results of previously found trees including the tree found by STROGANOFF. The best tree we find by FONTX, $T2$, is better than those we find with normal crossover. However, the tree found by FONTX in this case is still not as good as that found by STROGANOFF. Nonetheless, the improvement from this one simple change to the crossover operator is considerable, and makes it seem plausible that traditional GP can produce competitive performance on time series prediction problems. It also suggests that the operator-representation relationship[12] may be crucial to GP success.

4.4 Comparisons

For brevity we include in Table 4 the average results of both experimental methods, NX and FONTX. Given are the averages over twenty runs of both the generation in which the best of run tree was found and the mean squared errors of the training and test sets.

[11]The tree is equivalent because division by zero with the protected division operator, %, returns 1.0.

[12]This terminology was used by [Altenberg, 1994].

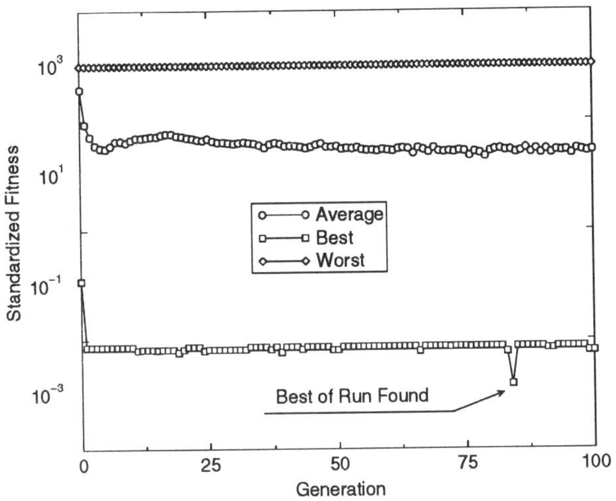

Figure 5: Best, average and worst fitness vs. generation for the FONTX run producing tree $T2$. The population does not converge.

Table 4: Comparison of average statistics from Normal Crossover (NX) and Modified Crossover (FONTX) for 20 runs each.

Method	Gen. Best Found	Mean Squared Train Error	Mean Squared Test Error
NX	4.8	1.0×10^{-4}	9.8×10^{-5}
FONTX	31.3	6.0×10^{-5}	5.8×10^{-5}

5 Discussion and Conclusions

We now summarize our investigation. First, [Iba, 1995] reported that GP had over-learned its training data; while the performance of a particular tree on a set of training data had been acceptable, when the same tree was applied to other data its performance diminished by two orders of magnitude. We reproduced the simulations performed by [Iba, 1995], finding the same results except for the phenomenon of overlearning. To verify our results we applied the found solution tree to 10^3 different data sets. We showed that the GP solution did not exhibit the reported overlearning on any of the 10^3 trials.

Second, one does not usually think of premature convergence of tree composition as being a problem for GP. However, we found in the case of time series prediction that premature convergence may be an inevitable consequence of the traditional GP crossover operator acting on the time-series terminal set.

Because the tree $x(t-1)$ was a better than random predictor it was selected for breeding at the early stages of GP evolution. Due to its small size it had a high probability of being accurately replicated by the crossover operator. As more clones of $x(t-1)$ were created the probability of reproducing still more clones increased. Therefore, the one-node tree expanded exponentially, quickly dominating the entire population.

To correct this problem of the population converging onto one tree we modified the crossover operator so that it could not, in this special case, produce a clone. Applying this method of crossover, we no longer observed the problem of convergence and the trees found by GP were better than those found without using this method.

6 Future Work

The method used here yielded better results than previously realized with traditional GP applied to short term chaotic time series prediction. However, the performance was not as good as other non-GP methods available. Nonetheless, plenty of room for improvement exists within the GP paradigm.

The most obvious direction for future work is to investigate other cases in which crossover degenerates into reproduction. For example if one parent is a one-node tree and the other parent is a two-node-tree there is still a 50% chance of reproduction from crossover. Possibly this and other "higher order" effects decrease diversity in the crucial early stages of GP evolution. By forbidding these other higher order cases one may effect further improvements of the kind reported here.

We pointed out that GP generally lost its best trees the generation after they were found. This could be remedied by enforcing a rule which would keep the best of the population from generation to generation (i.e. an elitist policy [Koza, 1992].

From the point of view of time series analysis, one can investigate the use of more sophisticated fitness measures. We have done preliminary runs with a metric due to [Manuca and Savit, 1996]. Our results have been encouraging.

[Prugel-Bennett and Shapiro, 1994] have studied the effects of increasing selection pressure during the evolution of a GA. We have applied this concept very roughly to GP. Surprisingly, in the time series prediction task considered in this paper we found better results than any presented here. The best tree found was:

$\{T3\}$

```
(+ (+ X[t-1]
(- (+ (- X[t-3] X[t-2]) X[t-1]) X[t-2]))
(* (- X[t-1] X[t-2])
(cos (- (- (+ X[t-1] X[t-1]) X[t-3]) X[t-3]))))
```

with a training fitness of 9.15e-06 and a testing fitness of 8.16e-06. This area of GP research shows considerable promise.

Acknowledgements

We thank Dr. Hitoshi Iba for providing the technical reports referenced throughout this work. We thank the reviewers, including Dr. Nicholas F. McPhee who made us aware of his work with Justin Miller, for their helpful comments. B.S. Mulloy thanks T.P.B. for thoughtful comments and suggestions. Also we thank Douglas Zongker and Dr. Bill Punch for providing *lil-gp 1.0*. This work was supported by The Program for Study of Complex Systems at the University of Michigan. R.L.Riolo was also partially supported by NSF Grant IRI-9224912.

References

[Altenberg, 1994] Altenberg, Lee. The Evolution of Evolvability in Genetic Programming. In Kinnear, Kim (editor). *Advances in Genetic Programming.* Cambrige, MA: The MIT Press. Pages 47-74.

[Casdagli, 1989] Casdagli, Martin. 1989. Nonlinear Prediction of Chaotic Time Series. *Physica D.* Volume 35. Pages 335-356.

[Casdagli, 1992] Casdagli, Martin *et al.* 1992. Nonlinear Modeling of Chaotic Time Series: Theory and Applications. In Hyun, Jong and John Stringer (editors). *Applied Chaos.* John Wiley and Sons, Inc. Pages 335-380.

[Farmer and Sidorowich, 1987] Farmer, Doyne and John J. Sidorowich. 1987. Predicting Chaotic Time Series. *Physical Review Letters.* Volume 59, Number 8. Pages 845-848.

[Green and Savit, 1991] Green, Matthew L. and Robert Savit. 1991. Dependent variables in broad band continuous time series. *Physica D.* Volume 50. Pages 521-544.

[Iba *et al*, 1993] Iba, Hitoshi, Takio Kurita , Hugo de Garis and Taisuke Sato. 1993. System Identification using Structured Genetic Algorithms. In Forrest, Stephanie. (editor). *Proceedings of the Fifth International Conference on Genetic Algorithms.* San Mateo, CA: Morgan Kaufmann. Pages 279-286.

[Iba, 1995] Iba, Hitoshi. 1995. *A Numerical Approach to Genetic Programming for System Identification.* Electrotechnical Laboratory technical report ETL-TR-95-23. Tsukuba Science City, Japan. 1995. (contact iba@etl.go.jp).

[Ikeda, 1979] Ikeda, K. 1979. Multiple-valued Stationary Sate and its Instability of the Transmitted Light by a Ring Cavity System. *Opt. Commun.* Volume 30. Pages 257-261.

[Ivakhnenko, 1971] Ivakhnenko A. G. 1971. Polynomial Thoery of Complex Systems. *IEEE Transactions on Systems, Man, and Cybernetics* Volume 1, Number 4. Pages 364-378.

[Koza, 1992] Koza, John. 1992. *Genetic Programming: On the Programming of Computers by means of Natural Selection.* Cambridge, MA: The MIT Press.

[Lorenz, 1963] Lorenz, E. N. 1963. Deterministic Non-Periodic Flow. *J. Atmos. Sci.* Volume 20. Pages 130-141.

[Mackey and Glass, 1977] Mackey, M.C. and Glass L. 1977. Oscillation and Chaos in Physiological Control Systems. *Science.* Volume 197. Pages 287-289.

[Manuca and Savit, 1996] Manuca, Radu and Robert Savit. 1996. Model Misspecification Tests, Model Building and Predictability in Complex Systems. *Physica D.* In press.

[McPhee and Miller, 1995] McPhee, Nicholas Freitag and Justin Darwin Miller. 1995. Accurate Replication in Genetic Programming. In Eshelman, Larry J. (editor). *Proceedings of the Sixth International Conference on Genetic Algorithms.* San Francisco, CA: Morgan Kaufmann. Pages 303-309.

[Nordin and Banzhaf, 1995] Nordin, Peter and Wolfgang Banzhaf. 1995. Complexity Compression and Evolution. In Eshelman, Larry J. (editor). *Proceedings of the Sixth International Conference on Genetic Algorithms.* San Francisco, CA: Morgan Kaufmann. Pages 310-317.

[Oakley, 1994] Oakley, Howard. 1994. Two Scientific Applications of Genetic Programming: Stack Filters and Non-Linear Equation Fitting to Chaotic Data. In Kinnear, Kim (editor). *Advances in Genetic Programming.* Cambrige, MA: The MIT Press. Pages 369-389.

[O'Reilly and Oppacher, 1994] O'Reilly, Una-May and Franz Oppacher. 1994. *The Troubling Aspects of a Building Block Hypothesis for Genetic Programming.* Santa Fe Institute Working Paper 94-02-001. 1994.

[Prugel-Bennett and Shapiro, 1994] Prugel-Bennett A. and J.L. Shapiro. 1994. Analysis of Genetic Algorithms using Statistical Mechanics. *Physical Review Letters.* Volume 72, Number 9. Pages 1305-1309.

Genetic Programming, the Reflection of Chaos, and the Bootstrap: Towards a Useful Test for Chaos

E. Howard N. Oakley
EHN & DIJ Oakley
Brooklands Lodge, Park View Close
Wroxall, Ventnor
Isle of Wight PO38 3EQ, UK
howard@quercus.demon.co.uk

ABSTRACT

This study assessed the use of genetic programming (GP) to diagnose chaos. Fifty GP runs were performed on chaotic data, generated from the Mackey-Glass delay differential equation, on one surrogate with the same Fourier power spectrum and statistics but without chaotic dynamics, and on a random walk series. Single runs were performed on 50 different surrogates of the chaotic series. Fitness was measured across 5 separate forecast periods of 65 points each, each based upon 10 prior input data points.

Fittest program fragments for the chaotic series evolved later and were more complicated than those for the surrogates. Relative to fitnesses achieved by constant linear predictions, fitnesses from the chaotic series were also better. Random walk data resulted in an impoverished GP process, with quick evolution of simple program fragments but no later evolutionary improvement. This comparative test merits assessment on other datasets, and its implications with respect to the statistical bootstrap and GP estimation are discussed.

1. Introduction

Genetic programming (GP) is a technique which has been developed for the purpose of solving problems. To date, it has almost exclusively been used to generate program fragments which are potential solutions to a problem to which it is being directly applied. Thus, interest has either been focussed on the effectiveness of the fittest program fragments in solving particular problems, or on ways of improving the technique of GP to increase that effectiveness or to reduce the effort required. A classic illustration of this is the work of Koza (1994) on Boolean parity problems.

In the course of previous work which used GP to predict chaotic time series (Oakley (1994b); Oakley (1994a)), it was claimed that the course of GP differed between chaotic and non-chaotic input data series, thus that the internal process occurring reflected the nature of the data (Oakley (1995)). If that is the case, it has important consequences for both the theory and application of GP. It suggests that more attention should be devoted to understanding the dynamics of the internal process and how it may reflect characteristics of the data.

Chaos is notoriously easy to postulate (Glass and Kaplan (1994)) but hard to diagnose and investigate (Ruelle (1990)). If long noise-free data series are available, there is a range of tests which can be used to determine if the underlying system is chaotic or stochastic, and to estimate the dimension of the chaos (Casdagli *et al.* (1992)). Even given suitable data series, though, existing tests tend to be complicated, computationally intensive, and usually involve subjective interpretation. The original motivation behind this series of studies was to investigate the suitability of GP for system identification and prediction of short noisy time series derived from biological systems; such datasets are not amenable to existing diagnostic tests for chaos, leading to speculation but little evidence that many biological systems are chaotic (Glass and Kaplan (1994)). Any test which was capable of use on short noisy datasets would therefore be of immediate practical value.

Considerable effort has been devoted to improving the efficiency of GP using automatically defined functions (ADFs) (Koza (1994)). Although this has lacked a thorough theoretical foundation, the value of ADFs in reducing the number of individuals which have to be evaluated before 'perfect' solutions are found, indeed in some problems in allowing solution at all, has been demonstrated. However, as with most variations of the basic technique, there is little knowledge as to when it should be employed. A better understanding of the processes at work within populations of program fragments during genetic programming should enable the development of a more sound theoretical basis for the technique, and thus for the selection of variants. This is particularly true for variations such as ADFs which have direct effects on the program fragments being evolved. If it is true that the internal process during GP reflects the nature of the data being fitted, and that this is an important feature of GP, any intervention which alters the internal process could disrupt this reflection.

Another area of both theoretical and practical importance is the relationship between genetic programming and other methods for solving problems; there has been interest in the hybridisation of GP with other machine learning (ML) methods such as neural networks. Given the overwhelming influence of chance within GP, it is perhaps surprising that there has been little combination of statistical methods with GP.

One relatively modern statistical technique which shares some features with, but which also contrasts with, GP is the bootstrap, devised and developed by Efron and Tibshirani (1993). Although derived from the same phrase as the term used in computing[1], its statistical meaning is quite different, applying to a resampling scheme based on Monte-Carlo methods. Under the bootstrap method, a large number of random samples are drawn with replacement from the available data, and a parameter (or antecedent 'plug-in' estimator) estimated on each bootstrap sample. The distribution of results from the bootstrap samples can then give a best estimate and confidence intervals of a statistic for the population as a whole. This contrasts with conventional statistical methodology in using probability directly in the derivation of estimators, and is particularly suited to implementation on fast computers.

For example, a bootstrap estimate of the standard error of an arbitrary statistic $s(\mathbf{x})$ could be made thus. A large number B of bootstrap samples are generated by random sampling (with replacement) from the original data set. The statistic $s(\mathbf{x})$ is then calculated for each of the bootstrap samples, and the standard deviation of those B values is then the bootstrap estimate of the standard error of $s(\mathbf{x})$.

The bootstrap has already been the source of inspiration for the use of surrogate data series in the investigation of chaos (Theiler *et al.* (1992)), and the basis of the claim for the value of GP in testing for chaos (Oakley (1995)). However, only one surrogate series was studied previously, conclusions being based upon multiple GP runs on that and truly chaotic datasets. This did not yield any information on the likelihood of those results occurring by chance, nor their dependence on that particular surrogate series.

When many GP runs are performed on a single problem, they can be viewed as applying the bootstrap using the fitness parameter. The initial population is a bootstrap sample of the vast space of program fragments which is then used to evolve a succession of non-random populations with increasing fitness, eventually yielding the fittest individual arising from that bootstrap sample. Choosing the fittest individual found from many GP runs can then be seen as a bootstrap best estimator.

This study therefore set out to compare the outcome of performing many repeated GP runs on: a known chaotic dataset, the Mackey-Glass delay differential equation (Mackey and Glass (1977))

$$\frac{dx_t}{dt} = \frac{bx_{t-\Delta}}{1 + x_{t-\Delta}{}^c} - ax_t$$

(where a is 0.1, b is 0.2, c is 10.0, and Δ is 30.0); surrogates to it which lack its chaotic qualities; and on random walk data which is purely stochastically determined. Rather than concentrating on the predictive performance of the fittest program fragments evolved, it adopts the bootstrap approach to understanding how the GP process might be influenced by the nature of the data series. Whilst its particular objective was to assess more thoroughly the practicality of using GP combined with surrogate data as a test for chaos, it was also intended to allow excursive insights into these other areas.

[1] *to pull oneself up by one's bootstrap*, attributed to the legendary liar Baron von Munchausen, was the technique which he purported to use to regain the surface when he had fallen to the bottom of a deep lake.

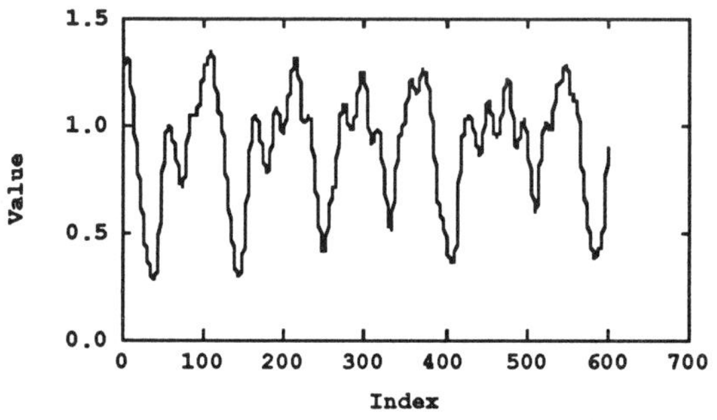

Figure 1: The Mackey-Glass delay differential series.

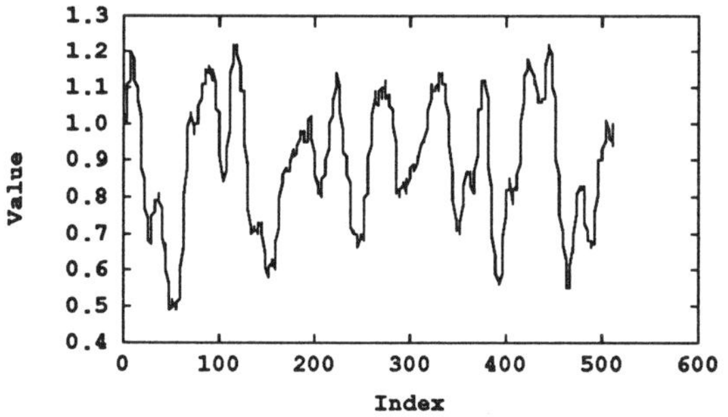

Figure 2: The first surrogate to the Mackey-Glass series.

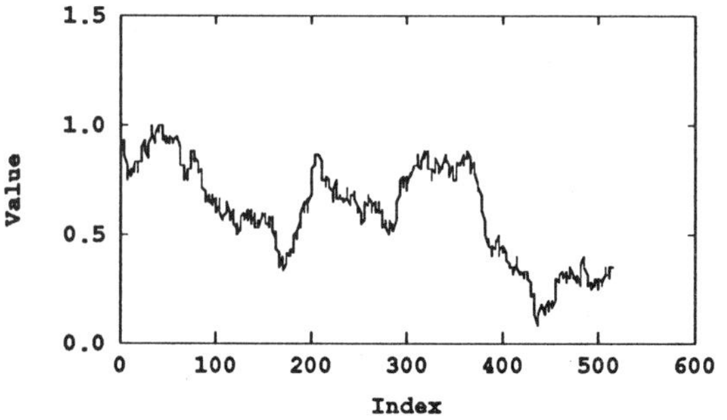

Figure 3: The random walk series.

Table 1: Tableau for fitting data series

Objective	Predict next 65 points at 5 places in series
Terminal set	Embedded data at $t = 1, 2, 3, 4, 5, 6, 11, 16, 21, 31$; $\Re$
Function set	`+, -, %, *`
Fitness cases	Actual members of the data series
Raw fitness	Sum over the 325 fitness cases of squared error between predicted and actual points
Standardised fitness	Same as raw fitness
Hits	Predicted and actual points are within 0.001 of each other
Wrapper	None
Parameters	$M = 500$. $G = 51$
Success predicate	None
Max. depth of new individuals	6
Max. depth of new subtrees for mutants	4
Max. depth of individuals after crossover	17
Fitness-proportionate reproduction fraction	0.1
Crossover at any point fraction	0.2
Crossover at function points fraction	0.7
Selection method	Fitness-proportionate (by normalised fitness)
Generation method	Ramped half-and-half

NOTE: `%` is the protected divide operation (Koza (1992)).

2. Methods

2.1. Data

Two primary data series were generated for these experiments. A 600-long exemplar of the Mackey-Glass delay differential equation was approximated using established Runge-Kutta techniques (Oakley (1995)), as shown in Figure 1. A random walk series was generated by randomly allocating a sign to each member of a suitably scaled series of randomised floating point numbers which were in turn used as successive increments to a seed, as shown in Figure 3. In both cases, the first portion of the original generated series was discarded to ensure that the data used were free of influence from startup processes.

Surrogate data series were generated using the first method of Theiler *et al.* (1992), which ensures that they possess the same Fourier spectra and statistics as the original data. In essence, this requires the application of the discrete Fourier transform to the data, randomization of the resulting phases, symmetrizing the phases (to ensure that the eventual output is real), and then inverting the discrete Fourier transform to generate the surrogate series. Fifty surrogate series were generated for the exemplar of the Mackey-Glass delay differential equation, of which the first is shown in Figure 2.

2.2. Genetic Programming

GP runs were performed using Koza's Simple Lisp code (Koza (1992)), incorporating implementation-specific optimisations to accelerate the evaluation of S-expressions. Code was compiled and run under Macintosh Common Lisp 2.0.1 (Apple Computer, Inc.) and Macintosh Common Lisp 3.0p2 (Digitool, Inc.), respectively on a Macintosh IIfx and a Power Macintosh 6100 computer.

No modifications were employed to incorporate ADFs, and the settings for all GP runs are shown in Table 1. Fifty complete runs were performed using the original Mackey-Glass data, 50 using the random walk data, 50 using the first surrogate of the Mackey-Glass data, and one run each using the 50 different surrogates of the Mackey-Glass data. The same 50 different random number seeds were used for the first three sets of runs, whilst the same random number seed was used for each of the 50 runs on different surrogates.

It is emphasised that fitnesses were not derived from the assessment of single forecast series, but from the sum of five complete forecast series, each of which used input data immediately prior to the forecast period, and each forecast extended 65 steps into the future. Intervals between forecasts were chosen to ensure that there was no overlap between input and forecast data, so that input data and forecast series varied over the 5 applications to each input dataset, and so that the maximum possible number of forecast points were evaluated over each dataset.

In contrast to previous work (Oakley (1995)), the function set was restricted to the four arithmetical operators.

2.3. Analysis

For each GP run, the fittest S-expression evaluated over the 51 generations was recorded, together with the generation at which it first appeared, its fitness, and the number of left parentheses contained within that S-expression. These data were pooled for each set of 50 runs and analysed using Systat and Statistica statistical analysis software. Some of the fittest S-expressions resulted in constant linear fits i.e., predicted constant values for all 325 fitness cases; one example of the fitness of such an S-expression was recorded for each set of runs, as an indication of the fitness of such a minimal forecast.

3. Results

Characteristics of the different input datasets are given in Table 2. The other 49 surrogates were statistically indistinguishable from surrogate series 1.

Fittest S-expressions evolved during runs using the original Mackey-Glass series displayed the relative richness seen previously, in terms of the range of generations at which they appeared, the complexity of the expressions themselves, and the range of different expressions and fitnesses. Particularly impressive was the apparent bi-modal frequency distribution (Figure 4) of the generation at which the fittest was found; thus it is clear that continuing the GP beyond the 51 generations employed would have still stood a worthwhile chance of finding fitter individuals.

In contrast, the fittest S-expressions generated from the surrogate data were achieved in relatively few generations (Figure 5), were simpler, and occurred more commonly than those of the chaotic data. This was still more true of S-expressions from the random walk data. These results are compared in Table 3. It is also important to note that the overall fittest S-expressions found for surrogate series were considerably fitter than those for the original series, and that the fittest S-expression for the random walk series was fitter still.

In spite of these observations of the fittest individuals found in each run, a few populations evolved progressively more complicated – but less fit – S-expressions when trying to fit surrogate and random walk data. But more commonly, the fittest of even late generations were simple and less fit variations of the fittest of earlier generations. Thus the GP process was impoverished and dominated by simple S-expressions of second-rank fitness, and there was little advantage in running more generations.

4. Discussion

4.1. Chaotic versus Stochastic Systems

The random walk series used in this study was not matched to the original Mackey-Glass series for Fourier power spectra or statistics, and the effects of this are seen in the wide differences in fitnesses achieved, although the means and variances of the series (Table 2) did not differ by much. Examination of the data series suggests an explanation, in that the Mackey-Glass series exhibited nearly 6 'pseudo-cycles' over the 500 individuals used in GP, whilst the random walk data had less than 3 (compare Figures 1 and 3). Thus, constant linear predictions were considerably fitter when applied to the random walk data (see the entries for typical linear fitness in Table 3). This calls into question the practice of using normalised error measures to compare forecasting techniques applied to different datasets

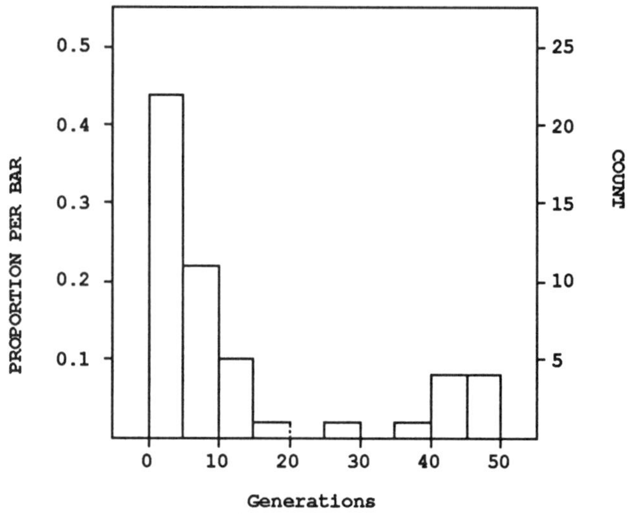

Figure 4: Frequency distribution of generations at which fittest S-expression was found, Mackey-Glass series.

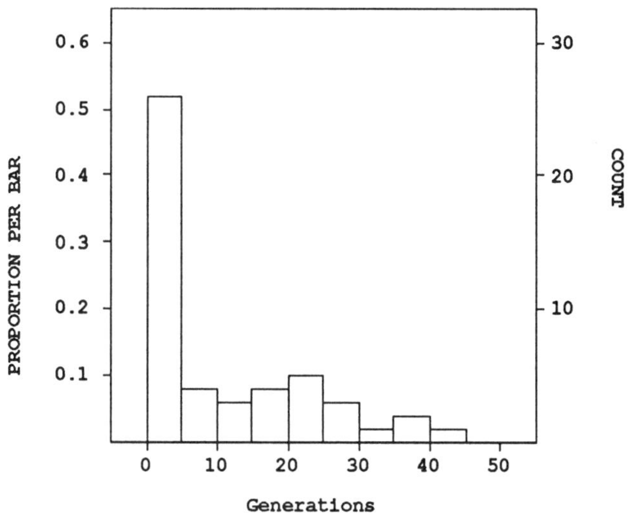

Figure 5: Frequency distribution of generations at which fittest S-expression was found, surrogate series 1.

Table 2: Characteristics of three input data series

Series	Mean	Number	Variance	Std. dev.
Mackey-Glass	0.8987	600	0.0696	0.2639
Surrogate 1	0.8975	512	0.0315	0.1774
Random walk	0.6059	512	0.0482	0.2196

Table 3: Summary of fittest S-expressions resulting

Series	Mackey-Glass	Surrogate 1	Surrogates	Random walk
Number	50	50	50	50
Mean generations to fittest	13.38	11.18	8.660	7.200
Std. dev. of generations	16.46	12.95	11.97	9.010
Median generations	5	4	3	3
No. of generations $\geq$ 25	10	8	7	4
No. of generations $\geq$ 40	9	2	2	1
Mean best fitness	10.22	4.305	4.230	0.4916
Std. dev. of best fitness	3.371	1.073	1.543	0.0571
Median best fitness	10.51	4.079	3.997	0.5105
No. of duplicate fitnesses	3	6	0	27
Overall best fitness	3.851	2.833	1.799	0.3507
Typical linear fitness	11.44	5.110	5.334	0.5246
Mean left parentheses	9.120	4.600	6.880	4.480
Std. dev. of left parens.	17.64	7.420	10.05	4.917
Median left parens.	3	2	3	2
No. left parens $\geq$ 10	8	4	8	7
No. left parens $\geq$ 20	6	2	4	1

(Gershenfeld and Weigend (1994)); instead, it is proposed that any comparisons of fitness of prediction achieved between datasets should be standardised against the fitness resulting from a constant linear fit. When expressed against the typical linear fitness value for its dataset, the overall best fitness for the Mackey-Glass data is considerably better than that for the random walk series, at 0.3366 and 0.6685 respectively, with a smaller difference between mean best fitnesses at 0.8934 and 0.9371. Because the use of fitness measures during GP runs is not normally susceptible to these effects, this issue should not affect the choice of fitness measure for GP, only comparisons between best fitnesses achieved over different datasets.

Otherwise, the original Mackey-Glass series and the random walk data were easily distinguishable in all respects, with the latter resulting in early convergence to simple fittest S-expressions. The remarkably high number of duplicate fitnesses among the runs on random walk data reflects the general poverty of the populations and the GP process. This agrees with previous claims that the 'richness' of the GP process mirrors the variety of the fitness landscape (Oakley (1995)). There is no evidence that running GP for more generations on the random walk data would result in any improvements in best fitness. This reflects the lack of any 'deeper structure' (attractors or similar) within the data, in contrast to the richer structures within the chaotic data.

4.2. Chaotic versus Surrogate Systems

When fitnesses are expressed relative to the typical constant linear fitness value for that series of runs, the mean best fitness values across the three series derived from Mackey-Glass data are very similar: 0.8934, 0.8425 and 0.7930 for the original chaotic series, the multiple runs on the first surrogate, and single runs on multiple surrogates. However, a different picture emerges when the overall best fitnesses are so compared: 0.3366, 0.5544 and 0.3373 respectively. Again, use of the typical linear fitness value to make comparisons between fitnesses derived from different datasets appears valid, although because of the similarity in Fourier power spectra, normalisation against the standard deviation of each series could also be appropriate in this instance.

The contrasts between chaotic and surrogate series are considerably less than those between chaotic and stochastic series. However, there are still some differences which are substantial enough to merit discussion. Among the Surrogate 1 runs, only 4% evolved their fittest S-expressions after 40 generations had been evaluated, whilst 18% of the original data runs did so. The visual contrast between Figures 4 and 5 is impressive, even if their statistical distinction is more difficult.

Chaotic data also resulted in the fittest S-expressions being more complicated, as shown by the frequencies of S-expressions with 20 or more left parentheses: 4% for Surrogate 1, against 12% for the chaotic series. In general, the more complex S-expressions evolved during runs on surrogate data did not achieve the best fitnesses, again suggesting that continuing runs for more than 51 generations would not be likely to result in much improvement in fitness. However there is good evidence (Figure 4) that additional generations could have led to further improvements in fitness for the prediction of chaotic data. The GP process is not as impoverished with surrogate data as when used on purely stochastic data, but it is not as rich as when used on truly chaotic data, with its 'deeper structure' (attractors).

4.3. The Bootstrap

Two different variations on the bootstrap were used in this study. The first used a single surrogate series, on which multiple GP runs were performed, whilst the other performed single GP runs on each of multiple surrogates. In the first, resampling was only performed from the vast space of possible S-expressions, and they constituted the bootstrap samples. In the latter, the bootstrap samples were drawn from the vast set of possible surrogate data series, on which resampling of S-expressions was not performed. Examination of the measures shown in Table 3 shows how remarkably similar the outcomes were. The only differences which appear to be of significance are a tendency towards rather more complex fittest S-expressions, as shown by measures based on the count of left parentheses, although in general these do not reach the values shown by true chaotic data.

This is an important result with respect to the value of performing multiple GP runs on a single representative dataset. Although in this case there is little difference in programming or computational effort between that and running GP on multiple datasets, in other problems it is much easier to obtain a single dataset. This does not mean that the actual S-expressions are necessarily equivalent – applying the fittest S-expression derived from one surrogate to another surrogate series would of course be expected to result in a reduction in observed fitness – but that the estimators of the GP process would be equivalent. This result also suggests that more formal exploration of the bootstrap approach to GP would be worthwhile. One important factor which is likely to have blurred any initial distinction between the results from the two variations of the bootstrap is the process of evolution. Assuming that this is driven by selection pressure towards a common goal, the eventual outcome should be qualitatively if not quantitatively similar.

4.4. Two Views on Estimators

Conventional ML practice uses training datasets to generate estimators, perhaps to classify data, and then applies the estimators to unseen data in order to assess their efficacy. If the accuracies of training and assessment are dependent on the quantity of data (for instance the length of time series) to which they are applied, then striking the optimal balance between the quantities devoted to each can be difficult. If only very short datasets are available, then this approach can fail altogether.

By contrast, statistical estimators are normally gained from the largest samples possible, unless a complete census is available. The bootstrap is a variation on this, although the technique is still applied to the whole sample available, no data being withheld for the purposes of assessing the accuracy of the estimate. Indeed, the bootstrap approach to measurement of confidence intervals can be seen as opposite in intention to the two phases of ML.

In many real-world situations, notably in the study of short time series, keeping half the data unseen will seriously weaken the prediction estimator generated from the other half. Consideration of the graphs shown in Figures 1 to 3 demonstrates that halving the length of each dataset would permit only 2 prediction series in the random walk and surrogate data, as opposed to the 5 used here, in turn restricting the general predictive power of the fittest S-expressions resulting from GP. Furthermore, there would be a wide disparity between fitnesses measured during ML training and assessment phases if the first half of each dataset was used for training and the second for assessment. Real data series are often non-stationary too, further frustrating efforts towards the two-phase ML approach.

There is a need for more work in this area to clarify the costs and benefits of the ML two-phase approach, against that of the statistical best estimator, before an informed choice can be made in the prediction of time series.

4.5. Proposed Test

Given this more extensive study, the previous proposal (Oakley (1995)) can now be refined into a formal test which should be applied to a wider range of chaotic and non-chaotic datasets. The data series to be studied should be subjected to a large number (50 or more) of predictive runs using a GP technique similar to that employed here, with the fittest program fragments from each run being recorded with respect to the generation in which they first occurred, their fitness, and size or complexity. If one of these program fragments does not produce a constant linear prediction, then the fitness of a typical constant linear prediction should be measured separately, to enable standardisation of fitnesses between series.

One or more surrogate datasets should be generated according to the first method of Theiler *et al.* (1992), and the same GP should be performed on these surrogates. The user has the option of running the same large number of GP runs on a single representative surrogate, or of running GP once on many individual surrogates, as has been performed in this study. If there is any doubt as to how representative a single surrogate might be, then the latter would be preferable, and should in general be encouraged until GP is better understood in terms of the bootstrap. Unfortunately, rigorous statistical tests of significance are not immediately available, and further work needs to be done to determine the most appropriate form of comparison.

It is not possible to say at this stage whether the GP method used should exclude ADFs or other variations of the basic technique. Intuitively, it might appear that the effects of ADFs in manipulating the content of program fragments, and in encouraging early convergence, would be disadvantageous. Indeed, the results of this study could be used to argue against the whole principle of ADFs. However, formal assessment of the effects of ADFs is required before any conclusions should be drawn.

5. Conclusions

Standard GP, without the use of ADFs, has been shown to be a useful tool in trying to distinguish one chaotic series from its surrogates and purely stochastic data. The process occurring within the GP, as shown by the fittest program fragments across many runs, appears to relate to the degree of complexity in the data being predicted. The less stochastic, and the greater the influence of chaotic attractors, the richer will be the GP process. This is marked by convergence in later rather than earlier generations, fitnesses (relative to 'simple linear prediction' fitness) which tend to be better, and more complex program fragments. In contrast when trying to predict stochastic data the GP process is impoverished. These differences can be used as a diagnostic test for chaos, on the short data series commonly encountered in biological sciences, for example.

This study has also raised interesting and potentially important issues with respect to the relationship between GP and the bootstrap, even viewing the application of multiple GP runs as a form of bootstrapping. A statistical approach may also be warranted with respect to the most efficient use of limited amounts of data for best estimate prediction – whether a bootstrap view can compete with ML-based two-phase estimation. Finally, although ADFs might appear to be disadvantageous in this work, they merit formal assessment.

Acknowledgements

This work has been influenced by numerous discussions with participants at the Genetic Programming section of the AAAI Fall Symposium series, November 1995, at MIT, Cambridge MA.

Bibliography

[Casdagli *et al.*, 1992] M. Casdagli, D. des Jardins, S. Eubank, J. D. Farmer, J. Gibson, and J. Theiler. 1992. Nonlinear modeling of chaotic time series: theory and applications. In J. H. Kim and J. Stringer (editors). *Applied Chaos.* New York, NY: Wiley-Interscience. Pages 335–380.

[Efron and Tibshirani, 1993] B. Efron and R. J. Tibshirani. 1993. *An Introduction to the Bootstrap, Monographs on Statistics and Applied Probability.* Volume 57. New York, NY: Chapmân and Hall.

[Gershenfeld and Weigend, 1994] N. A. Gershenfeld and A. S. Weigend. 1994. The future of time series: learning and understanding. In A. S. Weigend and N. A. Gershenfeld (editors). *Time Series Prediction: Forecasting the future and understanding the past.* Reading, MA: Addison-Wesley. Pages 1–70.

[Glass and Kaplan, 1994] L. Glass and D. Kaplan. 1994. Complex dynamics in physiology and medicine. In A. S. Weigend and N. A. Gershenfeld (editors). *Time Series Prediction: Forecasting the future and understanding the past.* Reading, MA: Addison-Wesley. Pages 513–528.

[Koza, 1992] J. R. Koza. 1992. *Genetic Programming: On the Programming of Computers by Means of Natural Selection.* Cambridge, MA: The MIT Press.

[Koza, 1994] J. R. Koza. 1992. *Genetic Programming II: Automatic Discovery of Reusable Programs.* Cambridge, MA: The MIT Press.

[Mackey and Glass, 1977] M. C. Mackey and L. Glass. 1977. Oscillation and chaos in physiological control systems. *Science.* Volume 197. Pages 287–289.

[Oakley, 1994a] E. H. N. Oakley. 1994. The application of genetic programming to the investigation of short, noisy, chaotic data series. In T. C. Fogarty (editor). *Evolutionary Programming, Lecture Notes in Computer Science.* Number 865. Berlin, Germany: Springer-Verlag. Pages 320–332.

[Oakley, 1994b] E. H. N. Oakley. 1994. Two scientific applications of genetic programming: stack filters and non-linear equation fitting to chaotic data. In K. E. Kinnear Jr. (editor). *Advances in Genetic Programming.* Cambridge, MA: The MIT Press. Pages 369–389.

[Oakley, 1995] E. H. N. Oakley. 1995. Genetic programming as a means of assessing and reflecting chaos. In *Genetic Programming, AAAI-95 Fall Symposium Series.* American Association for Artificial Intelligence. November 1995. Pages 68–72.

[Ruelle, 1990] D. Ruelle. 1990. Deterministic chaos: the science and the fiction. *Proceedings of the Royal Society of London A.* Volume 427. Pages 241–248.

[Theiler *et al.*, 1992] J. Theiler, B. Galdrikian, A. Longtin, S. Eubank, and J. D. Farmer. 1992. Using surrogate data to detect nonlinearity in time series. In M. Casdagli and S. Eubank (editors). *Nonlinear Modeling and Forecasting.* Redwood City, CA: Addison-Wesley. Pages 163–188.

Solving Facility Layout Problems Using Genetic Programming

Jaime Garces-Perez, Dale A. Schoenefeld and Roger L. Wainwright

Department of Mathematical and Computer Sciences
The University of Tulsa
600 South College Avenue
Tulsa, OK 74104-3189
email: [schoend,rogerw]@utulsa.edu
url: http://euler.utulsa.edu/~schoend

Abstract

This research applies techniques and tools from Genetic Programming (GP) to the facility layout problem. The facility layout problem (FLP) is an NP-complete combinatorial optimization problem that has applications to efficient facility design for manufacturing and service industries. A facility layout is represented as a collection of rectangular blocks using a slicing tree structure (STS). We use a multiple purpose genetic programming kernel to generate slicing trees that are converted into candidate solutions for an FLP. The utility of our techniques is established using eight previously published benchmark problems. Our genetic programming techniques that evolve STSs are more natural and more flexible than all of the previously published genetic algorithm and simulated annealing techniques. Previous genetic algorithm techniques use a two-phase optimization strategy. The first phase uses clustering techniques to determine a near optimal fixed tree structure that is represented as a chromosome in a genetic algorithm. Within the constraints implied by the fixed tree structure, genetic algorithm techniques are applied during the second phase to optimize the placement of facilities in relation to each other. Our genetic programming technique is a single phase global optimization strategy using an unconstrained tree structure. This yields superior results.

1 Introduction

The problem of facility layout is to decide the proper positioning of a collection of facilities on a planar site. Each facility has a required area and there is an interconnection cost between each pair of facilities. The interconnection cost between any pair of facilities is a quantitative flow of "materials" cost that respects both the amount of material that must be moved and the distance between the two facilities. The facility layout problem is an NP-complete combinatorial optimization problem that has applications in many contexts including architectural space planning, manufacturing cell layout and VLSI design [Tam and Li, 1991].

Historically, the Facility Layout Problem (FLP) has been modeled as a quadratic assignment problem, a quadratic set covering problem, a linear integer programming problem, a mixed integer programming problem, and a graph theoretic problem. Kusiak and Heragu provide a comprehensive survey of the facility layout problem [Kusiak and Heragu, 1987].

More recently, FLP research has progressed in various ways. More sophisticated problem solving frameworks, including genetic algorithms (GA) and simulated annealing, have been applied [Tam, 1992a, Tam, 1992b, Smith and Tate, 1993, Kado, 1995]. A variety of layout representations have also been developed. One promising layout representation technique uses a Slicing Tree Structure (STS) due to Otten [Otten, 1982]. The STS is often a useful layout representation; however, there are some physical layouts that cannot be represented by an STS, and there are other layouts for which the STS is not suitable. An STS specifies a general facility layout topology that, after application of some procedure, determines a specific facility layout.

This research applies techniques and tools from the Genetic Programming (GP) problem solving framework to the facility layout problem. We represent a facility layout as a collection of rectangular blocks using the Slicing Tree Structure. We use a multiple purpose genetic programming kernel written by Fraser [Fraser, 1994] to evolve slicing tree structures that correspond to optimal or near optimal solutions to the FLP. Our solutions are competitive with previously published facility layouts for eight benchmark problems that are commonly found in the literature, and that were recently used by Kado [Kado, 1995] and by Tam and Li [Tam and Li, 1991]. Our solutions are superior to Kado's results for the two largest problems containing, respectively, 20 and 30 facilities. Our genetic programming techniques applied to STSs are more natural and more flexible than previously published genetic algorithm and simulated annealing techniques. Previously published genetic algorithm techniques use a two-phase optimization strategy. The first phase uses a clustering strategy to determine a near optimal fixed tree structure. The second phase represents the fixed tree structure resulting from phase one as a fixed length chromosome and, using this fixed tree structure, attempts to optimize the placement of facilities in relation to each other [Kado, 1995]. Our genetic programming is a single phase global optimization strategy that does not constrain the tree structure. We determine the facility layout corresponding to an STS by using a variation of the "bottom-up" interpretation used by Kado [Kado, 1995]. To the best of our knowledge, our research is the first to apply GP to the facility layout problem.

2 The FLP Problem

We use the following formulation of the Facility Layout Problem (FLP). Other variations of the FLP are reviewed by Tam and Li [Tam and Li, 1991].

There are n two-dimensional rectangular facilities that must be positioned in a site area. The site area can be any subset of the Euclidean plane (i.e., we consider an FLP without room limitations.) The site area contains no obstructions (i.e., we can use any part of the Euclidean plane for any facility). Each facility, F_i, is modeled as a rectangular block having area, A_i, determined by the facility requirement. Only a horizontal or a vertical orientation of each rectangular block is acceptable. If a block is allowed to be placed either horizontally or vertically, it is a *free orientation* block; otherwise it is a *fixed orientation* block. The *aspect ratio*, r_i of a block is defined as the ratio of the width of the block to its height. A block with a variable aspect ratio defined over a range of values is a *flexible* block; otherwise it is a *rigid* block. Each block can be placed in any location within the site area but blocks cannot overlap one another. Let $w_{i,j}$ represent the weight of the material flow and let $d_{i,j}$ represents the Euclidean distance from the center of *block* i to *block* j. Our objective is to minimize

$$flow = 0.50\Sigma_{i<j}(w_{i,j}d_{i,j}^2).$$

Our objective function is the same objective function used by Tam and Li [Tam and Li, 1991] and by Kado [Kado, 1995]. Our objective function is the function resulting by insisting that there are no aspect ratio violations and by letting $a = 1$, $b = 2$, $P_a = 0.5$, $P_b = 10^6$, and $P_c = 0$ in the following generalized flow function presented by Kado [Kado, 1995]:

$$flow = P_a\Sigma_{i<j}(w_{i,j}^a d_{i,j}^b) + P_b\Sigma_i(asp\_break)_i + P_c(total\_area).$$

In this more general formulation, the term $(asp\_break)_i$ is used to indicate the extent that the i-th facility violates the given aspect ratio limitation. The term $(total\_area)$ is the minimum rectangular area that encloses all facilities.

Another objective function, resulting from alternate parameter settings and frequently used in other research, is given by

$$flow = 2.00\Sigma_{i<j}(w_{i,j}d_{i,j}).$$

3 The STS Representation Technique

A Slicing Tree Structure (STS) for n facilities is a binary expression tree where each operand (terminal) node is a label given by a facility index (0, 1, 2, ..., or n-1), and each operator (internal) node expresses the relation between its children substructures. The operator is U if the facility substructure represented by the left child is just Up from the facility substructure represented by the right child. Similarly, the operator is D, L, or R, respectively, if the facility substructure represented by the left child is just Down, just Left, or just Right, of the facility substructure represented by the right child. An STS is constrained to have exactly n terminal nodes and each index must occur exactly once as a terminal node. It has been noted that different STSs may represent the same relationship among facilities [Wong and Liu, 1986, Kado, 1995]. For example, using the interpretation that we describe later, the expression tree illustrated in Figure 1(a) with symbolic expression given by

```
(D ( D 2 ( R 0 1 ) ) ( R 3 4 ) )
```

and the expression tree illustrated in Figure 1(b) with symbolic expression given by

```
( D 2 ( D ( R 0 1 ) ( R 3 4 ) ) )
```

will each determine the same initial facility layout shown in Figure 2. The redundancy is also increased by using four operators rather than two since, e.g., (L 1 0) and (R 0 1) represent the same substructure. Hence, the initial facility layout determined by the expression tree

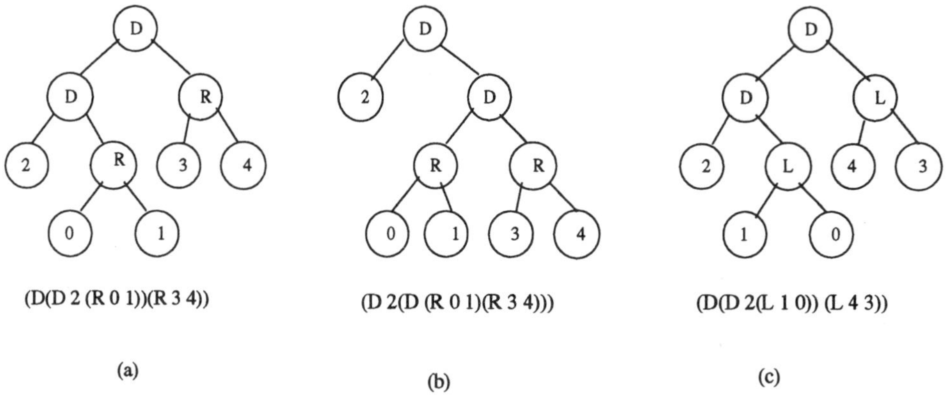

Figure 1: Equivalent Expressions

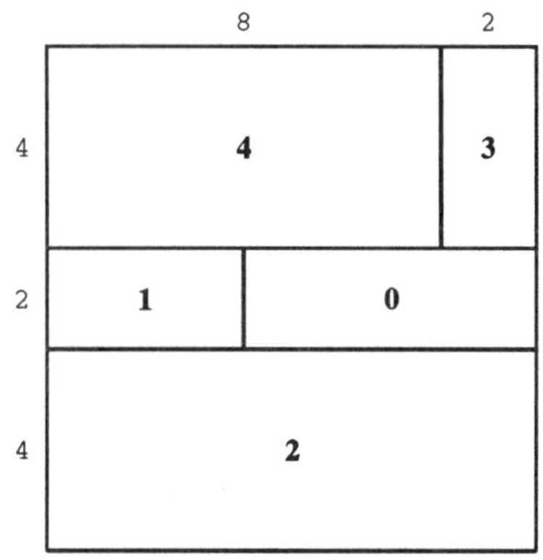

Figure 2: Top-Down Conversion of (D(D 2(R(0 1))(R 3 4)) for Dataset Simple-05

in Figure 1(c) will also be the same as the initial layout determined by Figure 1(a). Other researchers have argued that these redundancies lead to a richer search and better performance in an evolutionary GA problem solving framework [Kado, 1995]. We believe the same is true in an evolutionary GP framework. It has also been shown that the four operators are not redundant in other FLP variations that have obstructions within the site area [Kado, 1995].

There are two primary approaches for converting an STS-expression into a layout. Given a rectangular area equal to the total area required for all facilities, the "top-down" approach starts at the root node of the STS and recursively divides the available region into two subregions. The placement of a division, which is horizontal if the operator is U or B and vertical if the operator is L or R, is calculated by computing the area subtotals for the facilities on either side of the root node. The approach is broadly illustrated in Figure 2 using the elementary data set named Simple-05 from Table 1 with the expression from Figure 1(a). The top-down approach is useful for other FLP variations that have room limitations and obstructions [Tam, 1992a]. Without post-processing, the top-down strategy does not necessarily produce rectangles that respect the aspect ratio constraints.

Facility Number	Area	Aspect Ratio	Width	Height	Orientation
0	12	0.75	3	4	Fixed
1	8	0.50	2	4	Fixed
2	40	1.60	8	5	Fixed
3	8	2.00	4	2	Fixed
4	32	0.50	4	8	Fixed

Table 1: Simple-05 Data

A "bottom-up" approach for converting an STS expression into a layout begins by assembling any two sibling rectangles represented by terminal nodes in the manner specified by their parent operator. We align the centers of the two rectangles horizontally if the parent operator is L or R, and vertically if the parent operator is U or D. As we recursively ascend the STS expression, we use the centers of rectangles that bound groups of rectangles to determine alignment. This approach is illustrated in Figure 3, again using the data set Simple-05 from Table 1 with the expression from Figure 1(a). The resulting facility layout is not necessarily optimal. However, if each initial rectangle satisfies its required aspect ratio and if the assembly does not change the shape of any rectangle, the resulting facility layout will, of course, be in compliance with all of the aspect ratio constraints. The bottom-up approach is useful for our FLP variation since we assume that there are no room limitations [Cohoon *et al.*, 1991].

Our approach for converting an STS expression into a layout is similar to the approach used by Kado. We believe there are two reasons for obtaining superior results with our techniques. The first is due to an improved technique for converting the STS expression into a facility layout. The second reason is due to using a Genetic Programming framework that does not constrain

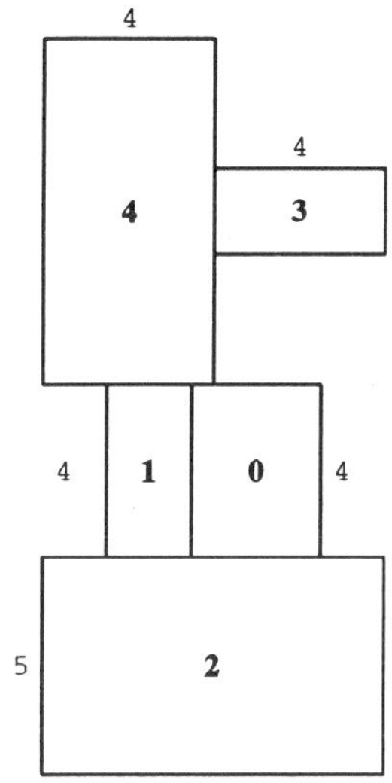

Figure 3: Bottom-Up Conversion of (D(D 2(R(0 1))(R 3 4)) for Dataset Simple-05

the structure of a facility layout. We refer to our approach as the Top-Down Bottom-Up (TDBU) conversion technique. The major steps of the TDBU technique for converting an STS expression to a facility layout are as follows:

1. Begin with a square region where the area of the square is equal to the total area required by all of the n facilities. Partition the square into a collection of rectangles representing a temporary facility layout by using the STS and a top down approach.

2. Each rectangle resulting from Step 1 will have the correct area but will likely violate its aspect ratio constraint. We scale the entire square region resulting from Step 1 by $k * scaleW$ in the direction of the the width and $k * scaleH$ in the direction of the height. We choose *scaleW* so that the width of each rectangle in the resulting scaled region is greater than or equal to the minimum width that is required for its corresponding facility, with either orientation, to be in compliance with its aspect ratio range. We select *scaleH* in a similar manner. It is important to note that even if the new width of each scaled rectangle exceeds the product of its old width and *scaleW* and even if the new height of each scaled rectangle exceeds the product of its old height and *scaleH*, it is not guaranteed that each scaled rectangle contains, as a subset, a subrectangle of the required area with aspect ratio in the required range. We choose k large enough so that the rectangles resulting from the next step, Step 3, have the required area, the required aspect ratio, and do not overlap. Two different values of k, as long as each is large enough, will determine the same facility layout.

3. Each rectangle resulting from Step 2 that is out of compliance with its aspect ratio range is contracted in one dimension as it is expanded in the other dimension to bring its aspect ratio into compliance, while leaving its center location and its area fixed. The direction for expansion and the amount of the expansion is determined in order to modify the rectangle as little as possible. Each rectangle, now in compliance with its required aspect ratio range, is then contracted in both dimensions in a way that exactly preserves the aspect ratio while reducing its area to the area required by the associated facility.

4. The non-overlapping rectangles resulting from Step 3 are packed to remove empty space using the STS and the bottom-up conversion technique.

4 Genetic Algorithms and Genetic Programming

Using the conventional genetic algorithm, individuals in a population are usually fixed-length character strings. Specification of the representation scheme starts with the selection of the string length and the alphabet size. The representation scheme maps each possible character string to a candidate solution in the search space of the problem. The evolutionary process starts with an initial population of randomly generated character strings and is driven by a fitness measure which assigns a fitness value to each possible character string. The GA works by selection, recombination, and mutation on the fixed length character strings [Koza, 1992].

The genetic program is an extension of the conventional genetic algorithm in which individuals in a population are expression trees (programs). Specification of the representation scheme starts with identifying a set of terminals and a set of functions. The representation scheme "evaluates" each possible expression tree to a candidate solution in the search space of the problem. The evolutionary process starts with an initial population of randomly generated expression trees and is driven by a fitness measure which assigns a fitness value to each possible expression tree. The GP works by selection, recombination, and mutation on expression trees [Kinnear, 1994, Koza, 1992, Koza, 1994].

5 Data Sets, Techniques, and Results

Eight different data sets with 5, 6, 7, 8, 12, 15, 20, and 30 facilities, respectively, were used to establish the validity of our procedure. The data sets are based on a test bank provided in Nugent et al. [Nugent *et al.*, 1969] as modified by Tam and Li [Tam and Li, 1991] to add aspect ratio constraints. Although the modified problems have aspect ratio constraints, they have no room limitations and no obstructions. Tam and Li's approach employs a divide and conquer strategy which first partitions the facilities into different clusters. The layout of each cluster is generated, and the overall layout is derived by treating each cluster as a large facility. Tam and Li's procedure uses an expensive sequential Lagragian method that is supplemented with a quasi-Newton procedure [Kado *et al.*, 1995]. Their procedure could not practically be used on the two largest problems in the test bank. The best flow obtained by Tam and Li for the other six problems is recorded in the TL91 column of Table 2.

Kado et al. [Kado *et al.*, 1995] provide a comprehensive performance evaluation of numerous variations of a two-phase optimization technique that uses cluster analysis and genetic algorithms. In each variation of their technique, clustering is used to fix the structure of a slicing tree, at least for the generation of an initial population of linear, fixed-length chromosomes that are used in their GA search. Their GA, with one or more of a variety of crossover and mutation operators is used to evolve their best reported flows. Their best flow for each of the eight test problems is indicated in the KA95 column of Table 2. Although partially due to a desire to narrow the search space, another motivation for using clustering techniques to create a fixed STS is to force an expression tree into a linear, fixed length chromosome so that standard crossover and mutation operators make sense. As an example of the problem, one can easily show that the uniform GA crossover applied to two postfix expressions, each corresponding to a different STS, does not necessarily produce valid postfix expressions. A variety of strategies are required to create and manipulate reasonable chromosomes representations using standard GA search strategies.

Our research also uses a Slicing Tree Structure to represent a candidate solution to the Facility Layout Problem. Our STSs are manipulated by a GP. The elegance of our technique is simply that it is more natural to use a GP that represents and manipulates STSs as trees than it is to use a GA that represents and manipulates STSs as linear chromosomes. The set of terminal nodes, corresponding to the set of facility labels, is the set {0, 1, ..., n-1}. The set of function nodes, corresponding to placement of the facilities (or bounding rectangles of collections of facilities) relative to each other, is the set {U, D, L, R}. Each expression trees is constrained to have exactly n terminal nodes and each element of the set {0, 1, ..., n-1} must be represented in exactly one terminal node. The GP kernel used for our research is Gpc++, Version 0.40 [Fraser, 1994]. Our results are shown in the GP96 column of Table 2.

Our results are significantly superior to Kado's for the two largest 20 and 30 facility problems. Our results are equal or better that Kado's for four of the other six benchmark problems. After experimentation, each of our flow values listed in the GP96 column of Table 2 was obtained using the Gpc++ parameters indicated in Table 2. The *Pop* column indicates the size of a population in the genetic program. The *Gen* column indicates the number of generations that the genetic program was allowed to run. The *First* column indicates the generation when the reported result first occurred. The crossover rate is indicated as a percentage in the *Cr* column and the mutation rate is indicated as a percentage in the *Mu* column. If mutation was indicated, either a single internal (operator) node was changed, two subtrees were swapped, or two internal nodes were swapped, with probability indicated, respectively, by the percentages in column *M1*, *M2*, and *M3*. If crossover was indicated, a tournament selection was used to select parents. Since any valid STS has exactly n terminal nodes, crossover points in the two parents were selected by repeated attempts, if required, to insure that the number of terminal nodes in each tree remains the same. Finally, since each element of the set {0, 1, ..., n-1} must occur exactly once in a valid STS tree, a deterministic procedure is applied to alter some terminal nodes in each tree resulting from crossover.

The data set for the TL91-05 benchmark problem is presented in Table 3. Our facility layout for TL91-05 is presented in Table 4. Table 4 gives the coordinates of the center, the width, the height, the area, and the two aspect ratios for each of the five facilities. Figure 4 graphically displays the facility layout for TL91-05. Our value of flow for TL91-05 is 226 (see Table 2). The data set for the TL91-20 and other benchmark problems is in the literature [Kado, 1995]. Our facility layout for TL91-20 is presented in Table 5. Figure 6 graphically displays the facility layout for TL91-20. Our value of flow for TL91-20 is 14,033 (see Table 2), vastly superior to Kado's result of 16,393. In fact, the results shown in Table 2 clearly show the superior results of our techniques in six (6) of the eight (8) benchmark problems, including the two larger data sets. Figure 5 shows the best and the average fitness values for each generation of our GP when applied to TL91-20. Our facility layout for TL91-30 is presented in Table 6. Figure 7 graphically displays the facility layout for TL91-30. Our value of flow for TL91-30 is 39,018 (see Table 2), also superior to Kado's result of 41,095. Corresponding to the GP96 flow values shown in Table 2, the generating slicing tree programs are shown in Table 7.

Facility Number	Flow (Traffic)	Area	Aspect Ratio Low	Aspect Ratio High	Orientation
0	0 5 2 4 1	24	0.80	1.00	Free
1	5 0 3 0 2	16	0.75	1.15	Free
2	2 3 0 0 0	36	0.60	1.85	Free
3	4 0 0 0 5	8	0.30	1.10	Free
4	1 2 0 5 0	21	0.90	1.18	Free

Table 3: TL91-05 Benchmark Data

Problem	TL91	KA95	GPK	GP96	Pop	Gen	First	Cr	Mu	M1	M2	M3
TL91-5	247	228	228	226	600	1,000	24	50	50	15	70	15
TL91-6	514	361	361	384	600	1,000	33	50	50	15	70	15
TL91-7	559	596	615	568	600	1,000	175	50	50	15	70	15
TL91-7	559	596	596	595	1,500	4,000	979	50	50	15	70	15
TL91-8	839	878	878	878	1,500	4,000	1,657	50	50	15	70	15
TL91-12	3,162	3,283	3,576	3,220	1,500	3,000	168	70	40	25	50	25
TL91-15	5,862	7,384	7,612	7,510	1,500	3,000	1,383	70	40	25	50	25
TL91-20	—	16,393	14,026	14,033	1,500	3,000	2,521	50	50	15	70	15
TL91-30	—	41,095	39,176	39,018	1,500	3,000	2,933	75	40	25	50	25

Table 2: Flows for TL91 Facility Flow Benchmark Problems

Facility	XC	YC	Width	Height	Area	Ht/Wt	Wt/Ht
0	5.8171	6.7685	4.7060	5.0998	24.0	1.0837	0.9228
1	1.7321	6.7685	3.4641	4.6188	16.0	1.3333	0.7500
2	4.0851	11.6134	7.8434	4.5899	36.0	0.5852	1.7088
3	6.5740	2.1093	2.1637	3.6974	8.0	1.7088	0.5852
4	3.0032	2.1093	4.9780	4.2186	21.0	0.8475	1.1800

Table 4: GP96 Facility Layout for the TL91-05 Benchmark Problem

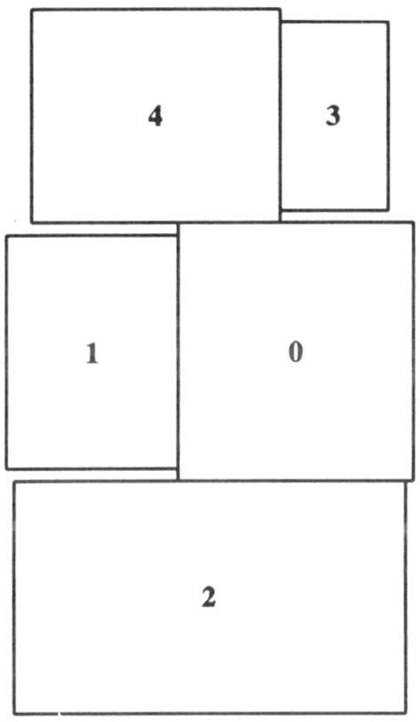

Figure 4: GP96 Facility Layout for the TL91-05 Benchmark Problem

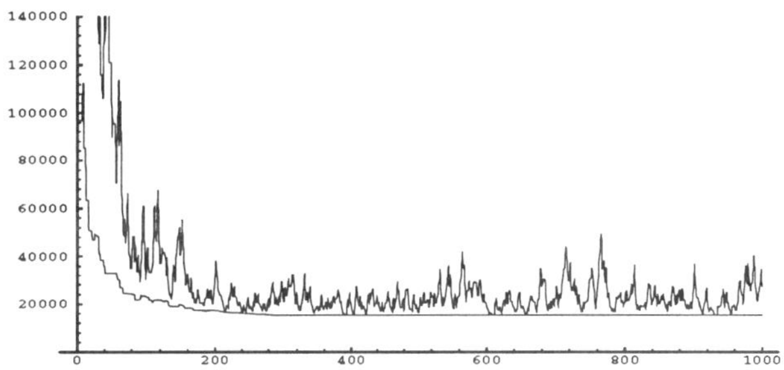

Figure 5: GP96 Number of Generations Versus Fitness Values for the TL91-20 Benchmark Problem

For each data set, the result reported in the GP96 column of Table2 was obtained by applying our variation of Kado's technique (for converting an STS expression to a facility layout as indicated in step 1 through step 4 above) to the best chromosome generated by our GP evolution. The corresponding result reported in the GPK column of Table 2 was obtained by applying Kado's STS conversion technique (using, in fact, Kado's own conversion code) to the best chromosome generated by our GP evolution. Examination of the GPK column and the GP96 column in the TL91-6 row and the first of the two TL91-7 rows in Table 2 indicate that the STS conversion technique used by Kado and the technique used by this research are not the same. The inclusion of the two TL91-7 rows in Table 2, corresponding to different GP parameter settings, is intentional since there is a significant contrast between the results in the GP96 and the GPK columns.

Although the STS trees generated by Kado's GA technique are not reported in our tables, we also note that Kado's evolved STS tree is identical to the STS tree generated by our GP technique for four of the rows in Table 2: TL91-5, TL91-6, the second TL91-7 row, and TL91-8.

Facility	XC	YC	Width	Height	Area	Ht/Wt	Wt/Ht
0	11.0826	14.2293	5.4443	4.4083	24.0	0.8097	1.2350
1	6.5279	2.4190	3.4641	4.6188	16.0	1.3333	0.7500
2	2.3979	10.8775	4.4113	8.1609	36.0	1.8500	0.5405
3	9.1856	2.6868	1.6803	4.7610	8.0	2.8334	0.3529
4	16.0265	9.8863	4.2186	4.9779	21.0	1.1800	0.8475
5	16.0265	14.4587	4.1998	4.1669	17.5	0.9922	1.0079
6	9.2599	8.9128	1.7990	2.0011	3.6	1.1123	0.8990
7	12.0826	9.9128	3.6453	4.2247	15.4	1.1589	0.8629
8	16.0265	2.3570	4.2426	4.7140	20.0	1.1111	0.9000
9	6.5279	13.5035	3.3288	5.8581	19.5	1.7598	0.5682
10	11.9700	2.2341	3.7639	4.2509	16.0	1.1294	0.8854
11	6.5279	9.1514	3.1623	2.8460	9.0	0.9000	1.1111
12	16.0266	6.0557	3.3541	2.6833	9.0	0.8000	1.2500
13	2.3979	4.1906	4.7958	5.2129	25.0	1.0870	0.9200
14	11.1484	5.2894	2.1508	1.8598	4.0	0.8647	1.1565
15	13.0454	5.2894	1.6432	1.8257	3.0	1.1111	0.9000
16	9.2599	10.9133	2.0000	2.0000	4.0	1.0000	1.0000
17	6.5279	6.2284	3.0000	3.0000	9.0	1.0000	1.0000
18	9.1856	6.3350	1.7748	2.5354	4.5	1.4286	0.7000
19	11.9700	7.0099	3.1623	1.5811	5.0	0.5000	2.0000

Table 5: GP96 Facility Layout for the TL91-20 Benchmark Problem

6 Future Work

We are currently articulating the precise distinction between Kado's STS conversion technique and the one used by our research. We are also developing alternate crossover and mutation techniques for our genetic programming evolution of STS trees. The objective of alternate crossover and mutation techniques is to duplicate or enhance other genetic algorithm strategies used by Kado. Further, we are evaluating our techniques with alternate flow functions including, in particular, a flow function where the distance is not squared. Finally, we are applying our techniques to other benchmark data sets that are in the literature. Some of the data sets and reported results use Manhattan distance and others have room limitations or obstructions in the site area.

7 Conclusions

This research finds optimal or near optimal solutions to the Facility Layout Problem by applying standard GP techniques to a Slicing Tree Structure that determines a facility layout. As noted before, the elegance of our technique is that it is quite natural to use genetic programming techniques to manipulate slicing trees. The superior nature of our results is due, in part, to our single phase global optimization strategy that does not constrain the tree structure in comparison to, say, a two phase optimization technique that first optimizes the structure of a slicing tree, and, then, optimizes the placement of facilities by using the fixed structure. All of our current research is in the context of eight FLPs where each facility has an associated aspect ratio but there are no room limitations and no obstructions in the site area. Clearly, the facility layout problem appears to be more conducive to a genetic programming strategy than a genetic algorithm strategy and this is indicated by our results.

References

[Cohoon *et al.*, 1991] J. P. Cohoon, S. U. Hegde, W. N. Martin, and D. S. Richards. Distributed genetic algorithms for the floorplan design problem. *IEEE Transactions on Computer-Aided Design of Integrated Circuits and Systems*, 10(4):483–492, April 1991.

[Fraser, 1994] Adam Fraser. Genetic programming in C++. Technical Report 040, Cybernetic Research Institute, 1994.

[Kado *et al.*, 1995] K. Kado, P. Ross, and D. Corne. A study of genetic algorithm hybrids for facility layout problems. In *Proceedings of the Sixth International Conference on Genetic Algorithms*, pages 499–505. Morgan Kaufmann, 1995.

[Kado, 1995] K. Kado. An investigation of genetic algorithms for facility layout problems. Master's thesis, University of Edinburgh, 1995.

FACILITY	XC	YC	WIDTH	HEIGHT	AREA	HT/WT	WT/HT
0	6.7144	3.4736	4.3818	5.4772	24.0	1.2500	0.8000
1	6.6301	17.4374	4.6188	3.4641	16.0	0.7500	1.3333
2	2.3178	5.3531	4.4113	8.1609	36.0	1.8500	0.5405
3	5.7157	14.3383	2.9261	2.7341	8.0	0.9344	1.0702
4	2.1213	18.7712	4.2186	4.9780	21.0	1.1800	0.8475
5	15.5903	14.0473	5.9160	2.9580	17.5	0.5000	2.0000
6	9.8926	8.7108	1.7392	2.0699	3.6	1.1901	0.8402
7	10.7142	2.4149	3.1886	4.8298	15.4	1.5147	0.6602
8	2.1213	13.9252	4.2426	4.7140	20.0	1.1111	0.9000
9	14.0511	7.7601	3.2914	5.9246	19.5	1.8000	0.5555
10	6.7144	8.0917	4.2564	3.7590	16.0	0.8831	1.1323
11	17.1198	6.7601	2.8460	3.1622	9.0	1.1111	0.9000
12	13.9608	17.1613	2.7523	3.2700	9.0	1.1881	0.8417
13	13.7829	21.1942	5.2128	4.7958	25.0	0.9200	1.0870
14	16.5587	11.6444	2.1693	1.8439	4.0	0.8500	1.1765
15	11.8107	14.0473	1.6432	1.8258	3.0	1.1111	0.9000
16	17.1198	9.3412	2.0000	2.0000	4.0	1.0000	1.0000
17	5.7427	11.4712	3.0000	3.0000	9.0	1.0000	1.0000
18	8.1301	11.4712	1.7748	2.5354	4.5	1.4286	0.7000
19	11.7941	17.1614	1.5811	3.1623	5.0	2.0001	0.5000
20	6.6301	21.0133	4.3386	3.6878	16.0	0.8500	1.1765
21	10.7141	6.2528	3.1623	2.8460	9.0	0.9000	1.1111
22	10.7142	11.1570	3.1886	2.8226	9.0	0.8852	1.1297
23	15.4741	2.3999	5.2128	4.7958	25.0	0.9200	1.0870
24	14.3894	11.6444	2.1693	1.8439	4.0	0.8500	1.1765
25	11.5838	8.7108	1.6432	1.8257	3.0	1.1111	0.9000
26	10.0175	14.0473	2.0000	2.0000	4.0	1.0000	1.0000
27	16.8370	17.1613	3.0000	3.0000	9.0	1.0000	1.0000
28	10.1162	17.1614	1.7748	2.5355	4.5	1.4286	0.7000
29	8.0931	14.3383	1.8288	2.7341	5.0	1.4950	0.6689

Table 6: GP96 Facility Layout for the TL91-30 Benchmark Problem

[Kinnear, 1994] Kenneth E. Kinnear, Jr., editor. *Advances in Genetic Programming.* MIT Press, Cambridge, MA, 1994.

[Koza, 1992] John R. Koza. *Genetic Programming: On the Programming of Computers by Natural Selection.* MIT Press, Cambridge, MA, USA, 1992.

[Koza, 1994] John R. Koza. *Genetic Programming II: Automatic Discovery of Reusable Programs.* MIT Press, Cambridge Massachusetts, May 1994.

[Kusiak and Heragu, 1987] A. Kusiak and S. Heragu. The facility layout problem. *European Journal of Operational Research*, 29:229–251, 1987.

[Nugent *et al.*, 1969] C. E. Nugent, T. E. Vollmann, and J. Ruml. An experimental comparison of techniques for the assignment of facilities to locations. *Operations Research*, 16:150–173, 1969.

[Otten, 1982] R. H. J. M. Otten. Automatic floorplan design. In *Proceedings of the 19th ACM-IEEE Design Automation Conference*, pages 261–267, 1982.

[Smith and Tate, 1993] A. Smith and D. Tate. Genetic optimization using a penalty function. In *Proceedings of the Fifth International Conference on Genetic Algorithms*, pages 499–505. Morgan Kaufmann, 1993.

[Tam and Li, 1991] K. Y. Tam and S. G. Li. A hierarchial approach to the facility layout problem. *International Journal of Production Research*, 29(1):165–184, 1991.

[Tam, 1992a] K. Y. Tam. Genetic algorithms, function optimization, and facility layout design. *European Journal of Operational Research*, 63(2):322–346, December 1992.

[Tam, 1992b] K. Y. Tam. A simulated annealing algorithm for allocating space to manufacturing cells. *International Journal of Production Research*, 30(1):63–87, 1992.

Problem	GP96 Generated Program
TL91-5	((U(U(L 4 3)(L 1 0)) 2))
TL91-6	((U(U(L 5(U 4 3))(L 1 0)) 2))
TL91-7	((U(U 2(L 1(L 6 5)))(L(L 0 3) 4)))
TL91-7	((U 2(L(L(U 1 0)(U 6 3))(U 5 4))))
TL91-8	((U(U 2(L(L 1 7) 0))(U(L 6 3)(L 5 4))))
TL91-12	((U(U(U 4(L 5 9))(L(L 10(L 3(L 6 7))) 1))(U(L 11(L 8 0)) 2)))
TL91-15	((L(L(U 8 0)(U(L(U(U 10(L 7 6))(L 11 12))(U 4 13))(U(L(L 1 3) 14)2)))(U 5 9)))
TL91-20	((L(U 13 2)(L(U 1(U 17(U 11 9)))(L(U(L(U 3 18)(U 10(U(L 14 15) 19))) (U(L(U 6 16) 7) 0))(U(U(U 8 12) 4) 5)))))
TL91-30	((L(U(L 2(U 0 10))(L(U 8 4)(U(L 17 18)(U(L 3 29)(U 1 20))))) (U(U(U(L(U(U 7 21)(U(L 6 25) 22))(U 23(U(L 9(U 11 16))(L 24 14)))) (L(R 15 26) 5))(L(L 28 19)(L 12 27)))13)))

Table 7: GP96 Generated Facility Layout Programs for TL91 Benchmark Problems

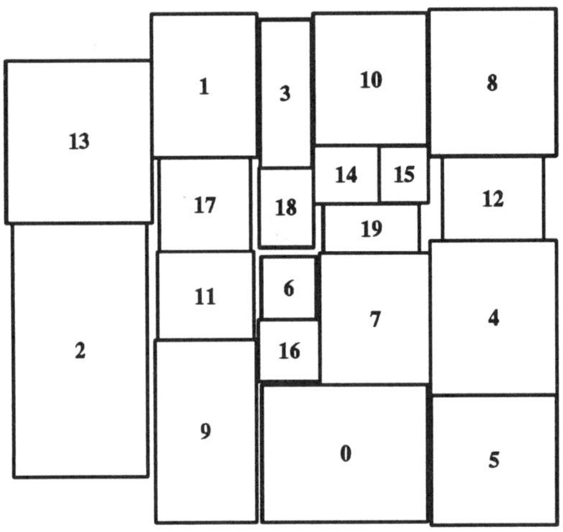

Figure 6: GP96 Facility Layout for the TL91-20 Benchmark Problem

[Wong and Liu, 1986] D. F. Wong and C. L. Liu. A new algorithm for floorplan design. In *Proceedings of the 23rd ACM/IEEE Design Automation Conference*, pages 101–107, 1986.

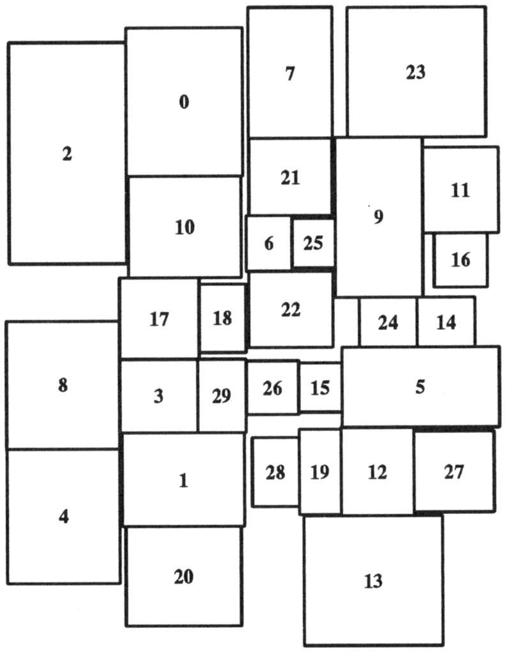

Figure 7: GP96 Facility Layout for the TL91-30 Benchmark Problem

Variations in Evolution of Subsumption Architectures Using Genetic Programming: The Wall Following Robot Revisited

Steven J. Ross, Jason M. Daida, Chau M. Doan, Tommaso F. Bersano-Begey, and Jeffrey J. McClain

The University of Michigan
Artificial Intelligence Laboratory and Space Physics Research Laboratory
2455 Hayward Avenue Ann Arbor, Michigan 48109-2143
stevross@engin.umich.edu, daida@eecs.umich.edu, doan@engin.umich.edu, tombb@engin.umich.edu, jeffjm@engin.umich.edu

ABSTRACT

The wall following robot is examined as a potential benchmark problem for applications of genetic programming (GP) to emergent robotic behavior. This paper describes experiments that were performed to characterize the performance, solution space, search space, and robustness using GP with and without automatically defined functions (ADFs). GP with ADFs was unable to significantly outperform GP without ADFs on this problem. A sub-optimal modality was discovered across all four program architectures. Many of the optimal solutions that were discovered tended to limit the number of sensors used for wall-following behavior; some used as few as three sensors. Tests for robustness indicate a "handedness" to the evolved solutions, which does seem to contribute to solution brittleness.

1. Introduction

1.1 Background

Several key works have demonstrated that GP shows promise as a means for behavior-based robot programming. The wall following robot, box pushing robot, lawn mower problem, and artificial ant are a few of the several examples presented by [Koza 1992, 1994b]. [Reynolds 1994a, 1994b] demonstrated that GP can evolve solutions for obstacle avoidance and corridor following behavior. [Handley 1994] demonstrated that an ADF GP can evolve a robotic planner that operates on predicates. Recent work by [Nordin 1995] demonstrated that a compiling GP system can evolve robust real-time obstacle avoidance behavior for real robots in noisy physical environments.

However, because many of these previous efforts have focused on demonstrating the solutions GP is capable of evolving, little is known about the overall nature of many of these problems. Subsequently, only a few suitable benchmarks exist as a basis for measuring gains by a particular GP methodology for behavior based or animat programming. Consequently, the goal of this work is to further characterize the problem of the wall following robot. Our eventual goal is to craft a suitable benchmark for measuring performance of new GP methodologies that may apply to problems involving emergent robotic behavior.

This work is divided into several sections. Sections 1.2 to 1.3 discuss considerations of the wall following robot as a benchmark. Section 2 presents a brief summary of the wall following robot problem and describes the experiments performed in this work. Section 3 presents and discusses the results in terms of overall performance, possible modalities in the solution space, observations on the search, and robustness of evolved solutions. Section 4 highlights a few conclusions from this paper.

1.2 Benchmark considerations

The wall following robot is particularly attractive as a benchmark because it is a well-known problem in GP [Koza 1992a], as well as in robotics [Brooks 1992]. With respect to GP, the problem is rich, as a number of different solutions are possible with only a few general functions and terminals. The problem also has solutions that are easily identifiable and has behaviors that are understandable to a lay audience. Furthermore, the problem is attractive because of previous work. [Koza 1992b] has demonstrated that GP can evolve a successful solution to this problem, provided code for the successful individual, and described behaviors seen in the evolution of the successful run. [Koza 1992a] adds to this work by demonstrating that the problem can be solved with an even more general implementation of the terminal set. [Koza 1994b] provides a performance curve for an experiment using the [Koza 1992b] terminal set.

With respect to robotics, this problem is also attractive because it has been examined as a case study in subsumption architectures [Brooks 1986, 1992, Mataric 1990]. The problem contains great potential for evolutionary robotic programming since the problem's functions and terminals have a close analog to realizable hardware components. For this reason, the problem avails itself to pragmatic concerns such as the effect of damage to any of the sensors or motors on an evolved robot program. Other similar problems like the lawn mower problem and its variants [Koza 1994a] have been well characterized, but have used component abstractions that are not easily realized in hardware. The lack of sensors and types of motor functions used in these problems link them tightly to the realm of animat problems like the artificial ant [Jefferson et al. 1992]. These types of idealizations aid in the discovery of fundamental principles but constrain them to the realm of computational life rather than as a means of evolving controls for physically embodied robots [Brooks 1992].

Consequently, the wall following robot is a reasonable marriage between problems in genetic programming and robotics and is suitable for studying aspects of evolutionary robotic programming from the perspectives of both domains. Furthermore, the physical representation of the robot and the introduction of ADFs into GP provide a means to study hypothesis regarding the use of GP techniques to build behavior-based programs to run real robots. [Brooks 1992] proposes that for GP to be an effective tool for programming real robots, the language that GP is operating on should include a macro capability. Brooks hypothesizes that the macro capability of the language should be effective if the macros can capture symmetries inherent to programming the robot and should greatly accelerate the production of programs and traversal of the search space. From the GP perspective, the wall following robot is a difficult problem and studying it may provide further insight into fundamental techniques needed for scaling GP and identifying break points in this problem.

For the wall following robot to be considered as a benchmark, several elements must be added to the knowledge of the problem. An independent verification of the results is necessary. Additional characterization is needed regarding the solution space, search space, average case performance of GP, and robustness of solutions. ADFs should be applied to test their effectiveness.

1.3 Contributions

This paper:

- Provides an independent verification of a well-known problem in GP.
- Characterizes several modalities of this problem's solution space to help qualify the problem for future use as a GP benchmark.
- Provides a limited test of Brooks' macro-language hypotheses in behavioral-based robot programming by applying ADFs.
- Gives a first assessment on robustness of evolved solutions for this problem.

2. Overview of Experiments

This section starts with a brief description of the problem and details of our problem specific code. The choice of ADF architectures and tests of robustness used in the experiments are also discussed. It includes a description of the parameters used and implementation notes.

2.1 Code description

Our implementation of the robot follows that which is described in [Koza 1994b, 1992a, 1992b]. The TOTO robot has 12 sensors, each covering an area of 30°. The robot can move forward 1.0 ft, backward 1.33 ft, turn left 30°, and turn right 30°. Koza's room has an irregular shape with protrusions on the south and east walls. The objective of the wall following robot problem is to evolve a computer program that allows a simulated TOTO robot to move along the perimeter of Koza's room.

Table 1 presents the key elements of our problem specific code used in the experiments. Note that the derived

Table 1: Wall following robot problem specific code

Terminal Set	Twelve sonar measurements [S00, S01..S11]; Derived minimum of measurements [SS]; minimum safe distance and preferred edging distance from wall [MSD, EDG]; Primitive motor functions [MF, MB, TR, TL]
Function Set	If-Less-Than-Or-Equal-To macro [IFLTE(arg1,arg2,arg3,arg4)] (i.e., IF(arg1 <= arg2) then arg3, else arg4); Connective function [PROGN2(arg1, arg2)] (i.e., eval arg1 return eval arg2)
Fitness Function	Fitness cases: one fitness case Koza's irregular room with an initial starting location near the middle of the room(13.8, 13.8)[1] and an initial facing direction of south(270°) Hit: robot touches a 2.3 square foot tile along the wall of the room Raw fitness: number hits in 400 time steps Standardized fitness: total number of wall tiles minus the number of hits Success predicate: 56 out of 56 hits

Table 2: Program Architectures

Experiment I	Non-ADF: Terminal set includes all 19 terminals described in Table 1 Function set includes both functions described in Table 1
Experiment II	One result producing branch, and two function-defining branches(ADF0, ADF1) Function-defining branches require no arguments and can only be called by the result-producing branch ADF0, ADF1, and RPB include all functions and terminals described in Table 1
Experiment III	One result-producing branch, and two function-defining branches(ADF0, ADF1) Function-defining branches require no arguments and can only be called by the result-producing branch ADF0, ADF1, and RPB include all functions and terminals described in Table 1 with these notable exceptions: • MF, MB are omitted from terminal set of ADF0 • TR, TL are omitted from the terminal set of ADF1
Experiment IV	One result-producing branch, and four function-defining branches(ADF0, ADF1, ADF2, ADF3) Function-defining branches require no arguments and can only be called by the result-producing branch ADF0, ADF1, ADF2, ADF3, and RPB include all functions and terminals described in Table 1

value `SS` was included in the terminal set and movement and turning functions return the minimum of the two forward looking sensors as described in [Koza 1992a] Example I.

2.2 Experiments

Experiment I was performed without ADFs to verify our results against Koza's. Experiments II, III, and IV used ADFs for two reasons. First, ADFs are a logical extension of Koza's work. Second, they provide a test of Brooks' hypothesis [Brooks 1992] that a macro capability in the language of GP would greatly accelerate the production of good programs on problems in emergent robotics. Choices regarding ADF architectures were made using the method of prospective analysis [Koza 1994a] since no previous work suggests an optimal architecture for this problem.

A choice common to all ADF architectures was that all function-defining branches would take no arguments and could only be called by the result-producing branch. We chose this type of hierarchy for the ADF architectures because Mataric's subsumption architecture used mutually exclusive behaviors [Mataric 1990]. Experiment II used an ADF architecture with two function-defining branches because it was the maximum default supported by the ADF kernel found in [Koza 1994a]. Experiment III used an ADF architecture with two function-defining branches whose terminal sets were restricted to try to coax GP into using ADFs more effectively. (Preliminary results indicated that many of the evolved individuals were not taking advantage of ADFs in the search.) Experiment IV used an architecture with four function-defining branches to see if GP would evolve a solution similar to Mataric's that relied on four behaviors [Mataric 1990]. Table 2 summarizes the program architectures used in each experiment.

2.3 Robustness

The robustness of each successful individual was tested by evaluating the code produced over additional fitness cases not used in the evolution of the solution. One additional fitness case consisted of moving the initial starting position of the robot eight feet to the west. The other case consisted of changing the initial direction of the robot from facing south to facing north.

2.4 Run parameters

All four experiments used the same parameters as a means of control. The parameters used were identical to those found in [Koza 1992a, 1994b]: 10% probability of reproduction, 90% probability of crossover, maximum depth of six for new individuals, ramped-half-and-half method of generation for individuals in the initial population, a maximum depth of 17 for individuals after crossover, and fitness-proportionate-with-overselection selection method. Each experiment consisted of 61 runs[2] with a population size of 1000 for a maximum of 57 generations.

2.5 Implementation notes

Runs were performed concurrently in a distributed computing environment, consisting mostly of Sun SparcStation 20 and HP 715 workstations using Allegro Common Lisp. Runs averaged from about 48 - 100 hours[3]. We used a lookup table for sensor values and terminated the execution of programs when the state of the robot stabilized as recommended in [Koza 1992a]. The "on-the-fly" redimensioning technique was not used.[4] Experiment I was performed with the canonical GP kernel found in [Koza 1992a]. Experiments II, III, and IV were performed with the ADF GP kernel found in [Koza 1994a] and a variant to support the use of architectures with four function-defining branches.[5] *Mathematica* was used in creating the lookup table of sensor values and in visualizing each of the solutions.

3. Results and discussion

This section discusses the results of our experiments in regards to performance of GP, modalities in the solution space, observations on the search space, and robustness of solutions.

3.1 Performance

Table 3 presents results that show the overall performance of the program architectures used in the four experiments. Table 4 presents a detailed breakdown of the successful runs in Table 3.

No experiment produced enough successful runs to generate meaningful performance curves. However, the success rate for Experiment I was consistent with the performance curve shown in [Koza 1994b].[6] Furthermore, the successful solutions were comparable to the best individuals presented in [Koza 1994b, 1992a, 1992b] as they contained a similar number of program nodes, used a similar number of time steps, and exhibited the same types of behavior. Consequently, the independently produced problem specific code consistently reproduced Koza's earlier results.

GP with ADFs was able to evolve at least one successful solution for each program architecture used. Solutions with ADF architectures contained a similar number of program

Table 3: Overall Performance

	Trials	Successes	Success Rate (%)
Experiment I	61	6	9.8%
Experiment II	61	5	8.1%
Experiment III	61	1	1.6%
Experiment IV	61	2	3.2%

Table 4: Characteristics of successful runs

Experiment	Generation Found	Nodes	Time Steps
I	9	15	375
	16	23	310
	19	87	397
	25	57	353
	36	89	380
	52	193	393
II	10	123	327
	11	161	286
	34	93	399
	45	152	374
	47	121	390
III	9	55	388
IV	22	37	365
	39	155	363

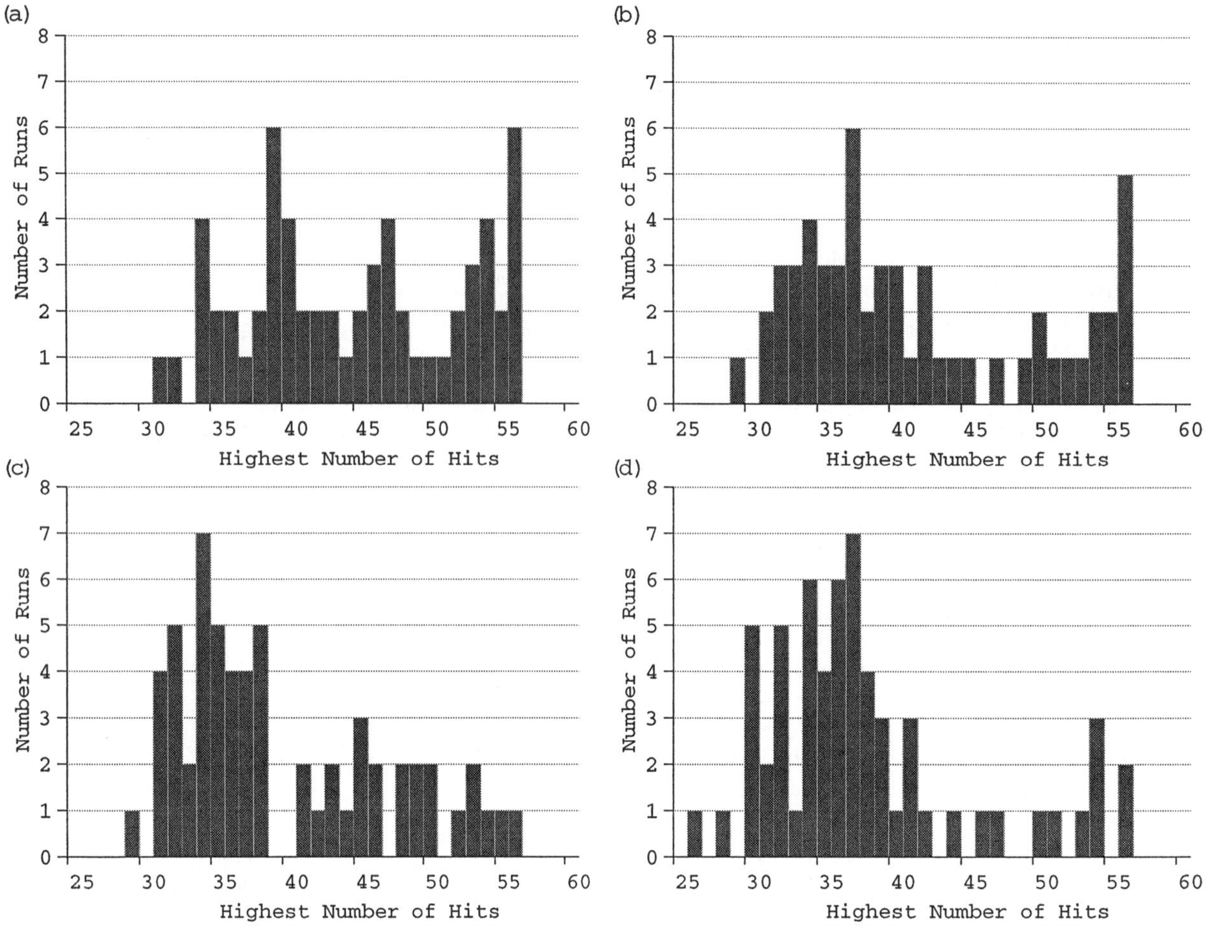

Figure 1: Hits histograms. Experiment I(a), Experiment II(b), Experiment III(c), Experiment IV(d)

nodes and used a similar number of time steps as those with non-ADF architectures. In spite of these successes, GP with ADFs did not out-perform GP without ADFs. Experiment II had a success rate comparable to that of Experiment I. The success rates of Experiments III and IV were slightly lower. Further experiments are needed to determine whether ADFs will improve the performance of GP on this problem. Once enough successful solutions are evolved to generate meaningful performance curves, the computational effort with ADFs (E_{with}) can be compared with that without ($E_{without}$) to determine if they are indeed more effective for this problem.

3.2 Fitness Modalities

Figure 1 presents histograms that provide an overall picture of the number of hits the best-of-run individuals converged upon. The horizontal axis of each histogram refers to the number of hits. The vertical axis is the frequency of runs whose best-of-run individual scored the specified number of hits.

The histograms for all four experiments display a prominent grouping near the 35 to 40 hit range suggesting a sub-optimal modality in the problem. Histograms for Experiments II, III, and IV also suggest something other than a unimodal distribution. Examination of best-of-run individuals from runs at the common sub-optimal modality revealed that the sub-optimal modality is not correlated to any feature inherent to the room or a specific type of behavior exhibited by the individuals. Further experiments need to be performed to determine if the modality is inherent to other aspects of the solution space.

3.3 Search Space

Figure 2 presents spider graphs that provide an overall picture of the manner in which the available primitives were used in optimal and sub-optimal[7] programs evolved in the experiments. Each spider graph consists of 21 axes, one for each of the available primitives. The spider graphs are normalized and present the ratio of occurrences of a primitive to occurrences of all primitives used in active branches of each program.[8]

The sub-optimal solutions found in each of the experiments used `IFLTE` and `PROGN2` frequently, movement and turning sparingly, and a wide range of sensors. The optimal solutions used `IFLTE` and `PROGN2` frequently, and movement and turning sparingly. However, the optimal solutions needed only a subset of the available sensors. Consequently, GP appears to be narrowing the search space by focusing on a few sensors. [Brooks 1992] hypothesized GP may search the solution space more effectively if the robot is started with only some of its sensors and actuators, adding more as fundamental

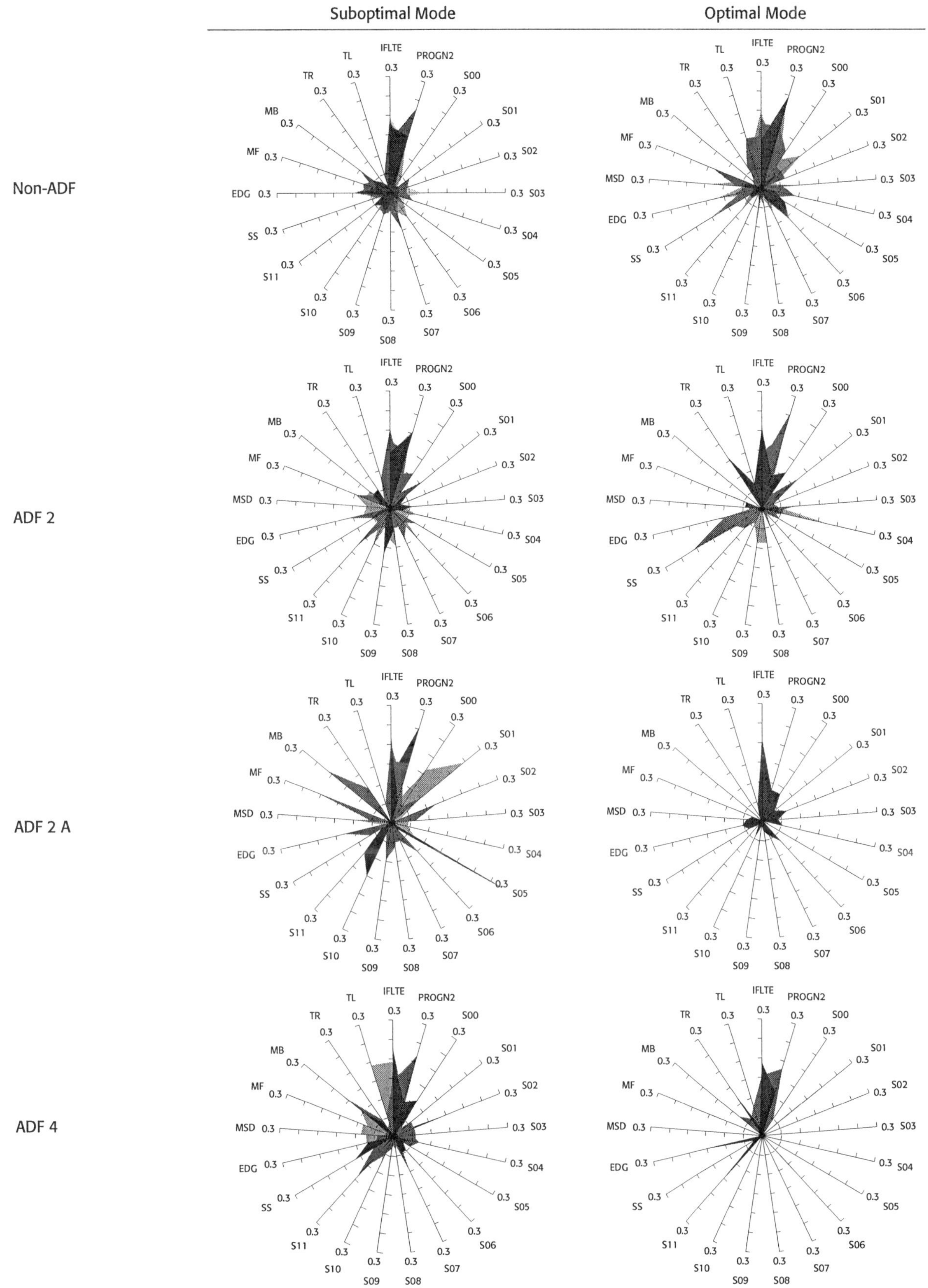

Figure 2: Spider graphs of functions and terminals used.

behaviors are found. Further experiments need to be performed to test the effectiveness of this method.

[Brooks 1990] also hypothesized that a macro capability in the language of GP, especially one that took advantage of symmetry and regularity in a robot, would accelerate the production of good programs. ADFs can be considered as a macro capability in the language of GP because they provide a means for GP to repeatedly use a function. A qualitative observation of suboptimal and optimal solutions across ADF architectures revealed that result-producing branches did not use function-defining branches effectively. Many of the result-producing branches consisted of a statement that called just one function-defining branch (e.g., `(progn (defun ADF 1...) (defun ADF2...) (defun ADF3...) (defun ADF4...) (values (ADF1))`. Consequently, either this suggests that the overhead associated with ADFs may outweigh their benefit in GP's search of the solution space for this problem or that ADFs are not being used in a manner to take advantage of symmetry in the robot.

3.4 Robustness

Figure 3 presents the range of successful solutions found in the experiments. Figures 4 and 5 illustrate the successful individuals executed on fitness cases not used in their evolution.

These figures display a diverse range of solutions and robustness but reveal a commonality in the problem. The figures suggest that GP is parsing the problem into three tasks: wall finding, wall following, and cornering. Strategies for finding the wall range from looping until the wall is hit to walking directly towards the wall once a particular room feature is identified. Many of the solutions exhibited tight looping and oscillating snake-like motion solving the tasks of wall following and maneuvering around corners with one behavior. However, some solutions handled wall following and cornering separately walking straight along the wall and performing tight cornering.

An ideal test of robustness would have consisted of executing the solutions in one to several additional rooms. However, executing the solutions from different initial positions in the same room was sufficient to show the failure modes for the evolved solutions. The solutions evolved were general in the sense that most of them still exhibit wall-following behavior when executed on fitness cases not used to evolve the solution. However, most of the evolved solutions were relatively brittle. Only one of the 14 solutions was able to hit all 56 wall tiles for both tests of robustness. The majority of the individuals timed out, got stuck, or missed a significant number of wall tiles. Some individuals performed well on one of the additional fitness tests but poorly on the other. This may be attributed to an evolved "handedness" of the solutions. The solutions performed better on the robustness tests when the robots, after finding a wall, traveled along that wall around the room in the same direction (i.e., clockwise or counterclockwise) that the robots traveled along the wall in the fitness case used in their evolution. This result suggests a modification in the fitness case used. In particular, we hypothesize that a room that would inherently demand a solution requiring a robot to switch directions during a run may yield a higher percentage of robust solutions. The results suggest that a single fitness case (consisting of one initial facing direction, initial position, and room shape) was

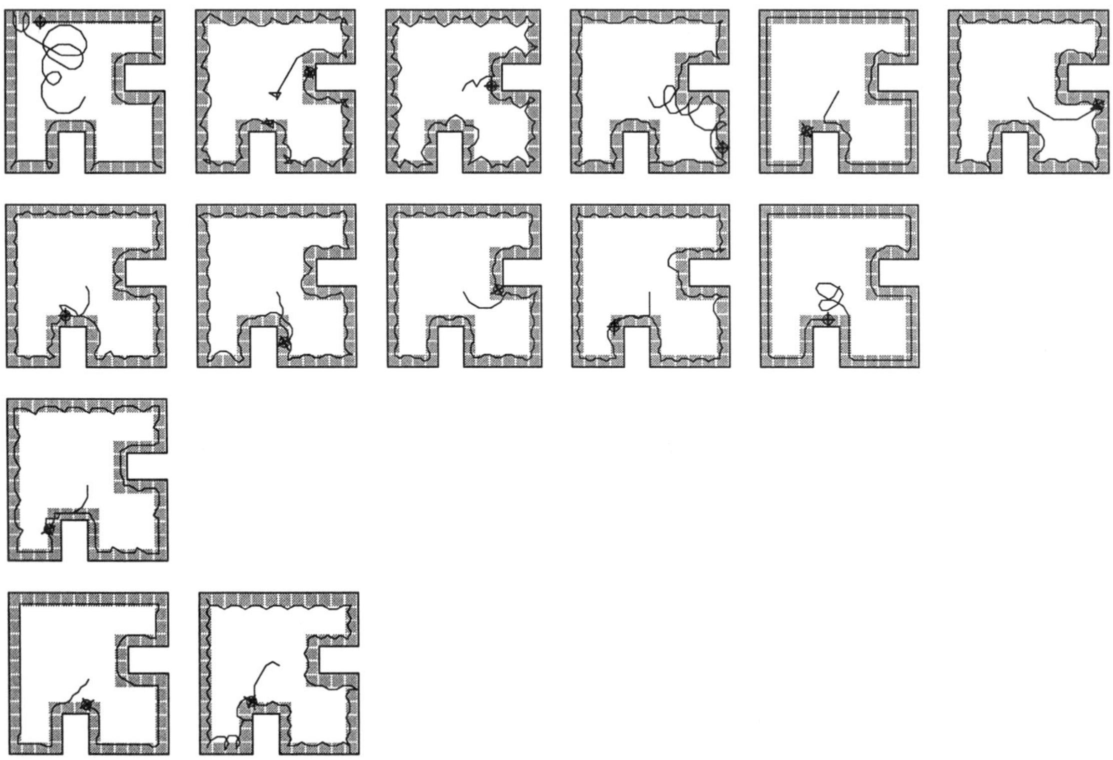

Figure 3: Successful Solutions. Experiment I(row 1), II(row 2), III(row 3), IV(row 4)

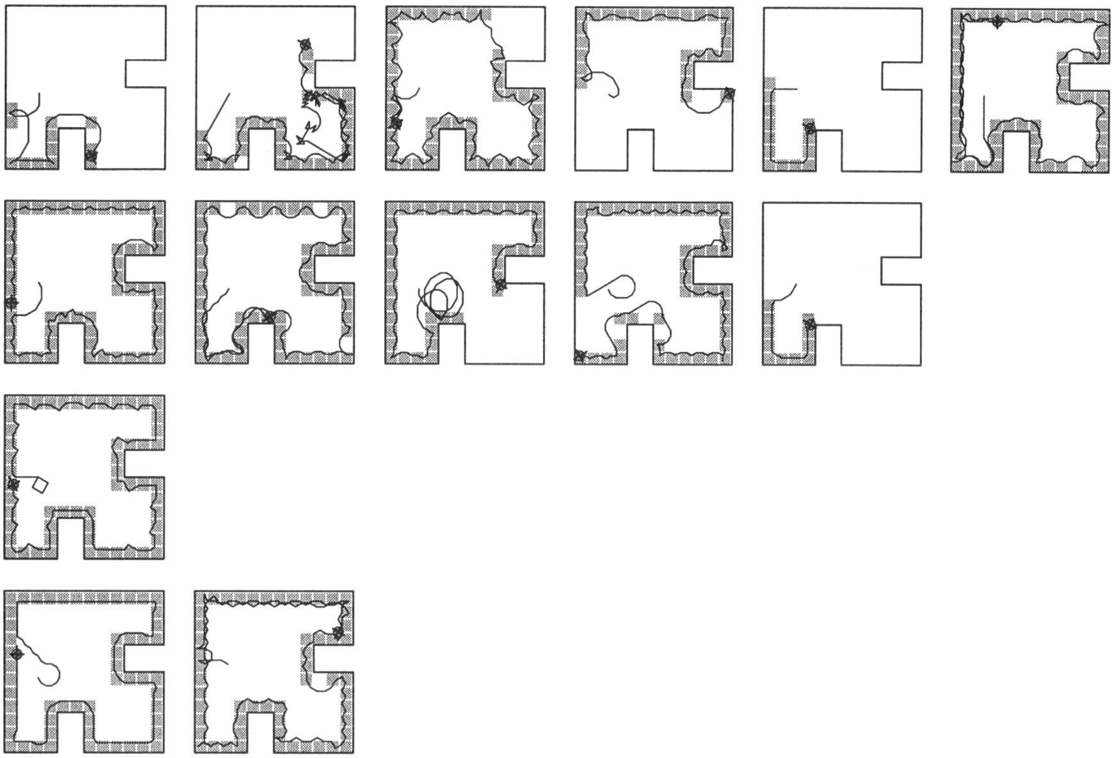

Figure 4: Robustness: initial location changed. Experiment I(row 1), II(row 2), III(row 3), IV(row 4)

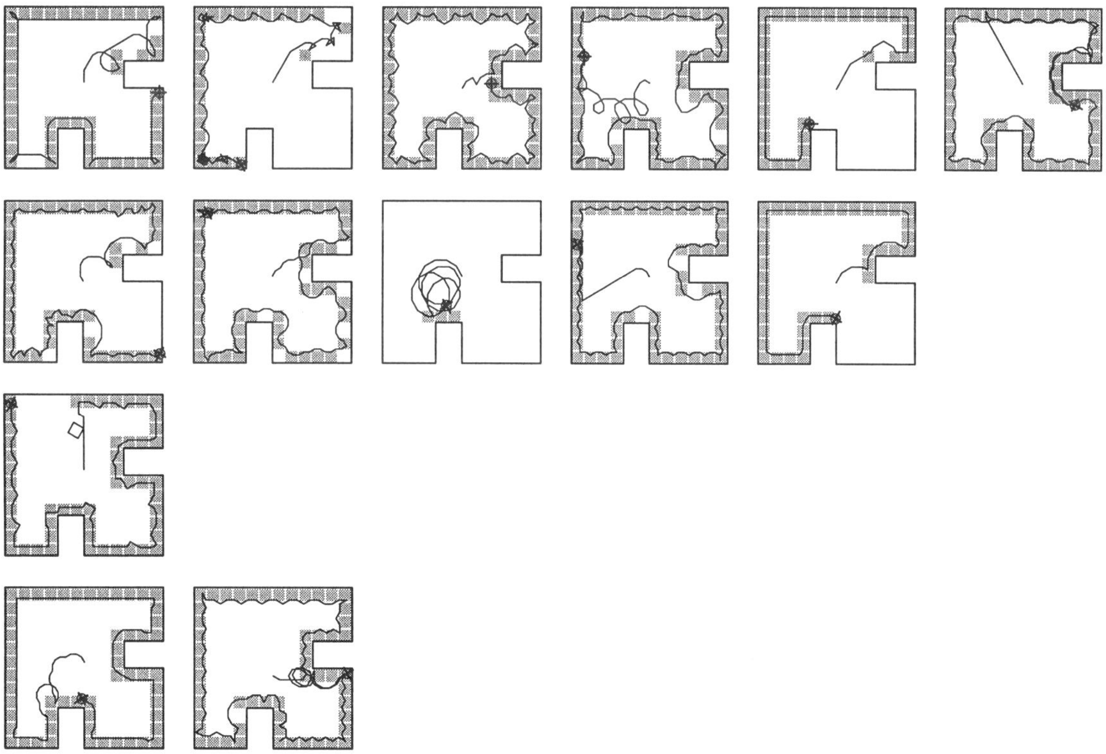

Figure 5: Robustness: initial direction changed. Experiment I(row 1), II(row 2), III(row 3), IV(row 4)

not sufficient to evolve a high percentage of robust solutions to the wall-following problem. Further experiments are needed to determine whether multiple starting positions in the same room, different rooms, or the suggested modification of the room fitness case would help in evolving a higher percentage of robust solutions.

4. Conclusions

This paper contributes several key elements to the benchmark for the wall following robot problem including:

- An independent implementation of the wall following robot problem produced results similar to those presented by Koza (Section 3.1)
- Identification of a modality in this problem that exists across architectures and cannot be attribute to a feature of the room or the robots' behavior (Section 3.2)
- Optimal solutions seem to need only a subset of available sensors(Section 3.3)
- A test of ADFs that suggests the introduction of ADF architectures to this problem does not improve GP's ability to search the solution space as it is unable to use them effectively (Section 3.3)
- The solutions evolved using GP on this problem are relatively brittle (Section 3.4)

Future work includes:

- Further tests with ADFs
- Identification of possible causes for the sub-optimal modality
- Use of another fitness set, consisting of either a different room or the same room but with several different starting locations for the robot
- Tests of Brooks' incremental functionality (sensors/actuators) hypothesis

(The software used to create this benchmark is available at http://www-personal.engin.umich.edu/~daida/)

Acknowledgments

This research has been partially funded with internal grants within U-M. We gratefully acknowledge the following people: J. Koza, J. Rice, for their original work with GP and evolutionary robotics; R. Riolo, for discussions on GP; S. Daida, for logistical support; the reviewers for their insightful feedback; E. Linas, for help in preparing figures for the camera-ready paper; J. Ross, M. A. Ross, and A. Ross, for their support.

Bibliography

Brooks, R. A. 1986. A Robust Layered Control System for a Mobile Robot. *IEEE Journal of Robotics and Automation*. March. pp. 14-23.

Brooks, R. A. 1992. Artificial Life and Real Robots. In Varela, F.J., and Bourgine, P.(editors). *Towards a Practice of Autonomous Systems: Proceeding of the First European Conference on Artificial Life.* The MIT Press.

Handley, Simon. G. 1994. The Automatic Generation of Plans for a Mobile Robot via Genetic Programming with Automatically Defined Functions. In Kinnear, K. E. Jr. (editor). *Advances in Genetic Programming*. The MIT Press.

Jefferson, D., R. Collins, C. Cooper, M. Dyer, M. Flowers, R. Korf, C. Taylor, and A. Wang, 1992. Evolution as a Theme in Artificial Life: The Genesys/Tracker System. *Artificial Life II*, SFI Studies in the Sciences of Complexity. Vol X. Addison-Wesley.

Koza, J.R. 1994a. *Genetic Programming II: Automatic Discovery of Reusable Programs.* Cambridge, MA: MIT Press.

Koza, J.R. 1994b. Evolution of a Subsumption Architecture that Performs a Wall Following Task for an Autonomous Mobile Robot. In Hanson S. J., et al.(editors) *Computational Learning Theory and Natural Learning Systems.* The MIT Press.

Koza, J.R. 1992a. *Genetic Programming: On the Programming of Computers by Means of Natural Selection.* Cambridge, MA: MIT Press.

Koza, J.R. 1992b. Evolution of subsumption using genetic programming. In Varela, F.J., and Bourgine, P.(editors). *Towards a Practice of Autonomous Systems: Proceeding of the First European Conference on Artifical Life.* The MIT Press.

Mataric, Maja J. 1990. *A Distributed Model for Mobile Robot Environment-Learning and Navigation.* MIT Artificial Intelligence Laboratory technical report. AI-TR-1228.

Nordin, P., W. Banzhaf 1995. Genetic Programming Controlling a Miniature Robot. *In Working Notes of the AAAI-95 Fall Symposium Series, Symposium on Genetic Programming.* MIT

Reynolds, C.W. 1994a. An evolved vision-based behavioral model of obstacle avoidance behavior. In Langton, C.G. (editor) *Artificial Life III*, SFI Studies in the Sciences of Complexity. Volume XVII. Addison-Wesley.

Reynolds, C.W. 1994b. Evolution of obstacle avoidance behavior. Using noise to promote robust solutions. In Kinnear, K. E. Jr. (editor). *Advances in Genetic Programming*. The MIT Press.

---

[1] Koza's original work starts the robot at a position near the center of the room(12, 16). This work is a slight variation starting it in the center of the room at position(13.8, 13.8).

[2] The initial goal was 60 runs. The hostile nature of the computing environment caused many of our runs to be killed so multiple runs were started for redundancy. Experiment II completed one additional run so for consistency, we decided to perform an extra run for each of the other experiments.

[3]Our runs were considerably slower than that reported in [Koza 1992a, 1994b]. In retrospect, we believe that three factors contributed to this slowdown. First, we may have unintentionally penalized runtime performance with our code implementation. Second, our tests ran in the background in a computing environment that is shared extensively by students, staff, and faculty. Third, we used a generic workstation rather than a Texas Instrument Explorer II Plus workstation (a LISP-specific machine) used in Koza and Rice's original work.

[4]To control all experiments as much as possible, we decided that at no time during a run would we intervene. This non-intervention policy consequently precludes "on-the-fly" redimensioning and extensioning.

[5] The GP kernel found in [Koza 1994a] only supports architectures with up to two function-defining branches.

[6] The performance curve found in [Koza 1994b] suggests that "processing a total of 1,176,000 individuals (i.e. 2,000 times 28 generations times 21 runs) is sufficient to yield a solution to this problem with 99% probability."

[7] The sample of sub-optimal programs was taken from the modality common across all architectures.

[8] A distinction is made between active branches and non-active branches as a basis for comparison between the non-ADF and ADF architectures. The non-ADF programs consisted of only one branch that must be activated. The ADF architectures consisted of one result-producing branch that must be activated and several function-defining branches which may be activated. The primitives in a function-defining branch were counted each time the designated function-defining branch was called.

MASSON: Discovering Commonalities in Collection of Objects using Genetic Programming

Tae-Wan Ryu
Department of Computer Science
University of Houston
Houston, Texas 77204-3475
twryu@cs.uh.edu
URL: http://www.cs.uh.edu/~twryu

Christoph F. Eick
Department of Computer Science
University of Houston
Houston, Texas 77204-3475
ceick@cs.uh.edu
URL: http://www.cs.uh.edu/~ceick

ABSTRACT

For the current flood of data, automatic tools for searching or analyzing data are necessary. Accordingly, knowledge discovery in databases is getting more and more attention. This paper centers on the problem of discovering the common characteristics that are shared by a set of objects belonging to an object-oriented database. In our approach, commonalities within a set of objects are described by object-oriented queries that compute this set of objects. The paper discusses the architecture of a knowledge discovery system, called MASSON, which employs genetic programming to find such queries, and presents an example run of the system to illustrate how the system works; we will show how interesting queries that describe commonalities within a set of objects are automatically generated, modified, evaluated, and selected; we will also discuss how the search for the "best" query is conducted by the MASSON system. Specific problems such as the generation of constants in queries, how to cope with type violations and other constraints when creating object-oriented queries, and query evaluation are discussed in some detail.

1. Introduction

Currently, in corporate, governmental, and scientific organizations scanners and other electronic devices generate large volumes of data that is routinely stored in on-line databases. The size of such databases is growing rapidly every year. This makes it very hard for human being to analyze these databases for interesting relationships or information; for example, cash registers in super markets or satellites in space produce volumes of data that are usually so large that it is no longer possible for a human data analyst to inspect the produced data collections manually. On the other hand, the generated data might contain valuable patterns, such as what products are usually bought together by customers of the super market, whose discovery would give the super market a strategic advantage over its competitors. Consequently, computer assistance and powerful analytical tools for extracting useful and interesting patterns from databases are desirable.

This paper centers on the following specific knowledge discovery problem. Suppose we have a database which stores properties of persons and relationships between persons and other classes, we have a given set of persons, and we are interested in knowing:

"What are the common characteristics that the objects that belong to this set share."

For example, we have a set of drug-dealers or a set of customers that have changed phone company frequently, and we would like to know what the objects belonging to this set have in common; that is, we would like to find the typical profile of drug dealers or customers. To answer such a question will certainly be a non-trivial problem in conventional database environment; for example, a lot of queries have to be written by the data analyst to manually perform this task, which would be very time consuming. In this paper we will describe the features of a computerized knowledge discovery system, MASSON, which employs application independent algorithm, genetic programming

(GP) "Koza (1990), Koza (1994), Kinnear (1994)" that facilitates this task for a data analyst. In GP, many dynamic program structures are randomly generated and those program structures are genetically evolved based on their fitness values. GP starts searching from a broad problem space to possible solution states. It searches for multiple solutions in parallel, which is important because we are interested in finding as many commonalities as possible. Moreover, MASSON performs knowledge discovery in the context of object-oriented databases. In our approach, commonalities between a set of objects are specified using object-oriented queries that extract the members belonging to this set from an object-oriented database.

Knowledge discovery in databases (KDD) has become an active research area, and many techniques and systems have been already developed "Piatetsky-Shapiro and Frawley (1991), Matheus et al. (1993)." Some of these techniques include database summary "Peter and Willi (1991), Lee and Kim (1993)" or abstract extraction "Dhar and Tuzhilin (1993)," and database classification "Agrawal et al. (1993)." "Yoon and Kerschberg (1993), Agrawal et al. (1993), Zhong and Ohsuga (1995)" use machine learning techniques to discover general rules for a database. Most techniques focus on extracting general patterns or rules by analyzing the contents of the whole database. The proposed techniques are not directly applicable to our problem, since they propose to extract useful knowledge from a whole relation or database, and do not focus on the commonalities of a given set of objects. There are other areas related to machine learning such as inductive logic programming (ILP), AQ family "Michalski et al. (1986)", and ID3-based decision tree approach "Quinlan (1993)." ILP is a formal framework for inductively learning logic programs (i.e., Prolog) from examples "Muggleton (1992), Lavrac and Dzeroski (1994)". According to "Lavrac and Dzeroski (1994)", ILP can be used as a knowledge discovery tool. However, ILP assumes that positive and negative examples need to be given to discover concepts that describe the given examples. AQ family and ID3-based decision tree approach learn if-then rules and attributes from also positive and negative examples. Similarly, "Chen and She (1994)" applies machine learning techniques to help user writing appropriate queries for information retrieval system based on positive and negative examples.

The rest of this paper is organized as follows: Section 2 compares the conventional approach with our approach. In Sections 3 and 4, KDD-system MASSON and the GP approach it employs is discussed. Section 5 shows our system at work. Section 6 discusses further research areas and section 7 summarizes the paper.

2. Retrieving Results by Query and Deriving Queries from Results

In the conventional approach, searching useful information in databases starts with a query devised by a user to get a set of objects, while our approach starts with a given set of objects, and the target of the search process is to find a query or a set of queries that computes this set of objects. In this paper, we call the conventional approach *retrieving results by query* and our approach *deriving queries from results.*

2.1. Retrieving Result by Query

Retrieving results by query is the process of extracting information based on the user's information requirements, background knowledge, and other useful information about the database. Based on those inputs, a user derives a query, submits it to database management system (DBMS), and the DBMS returns the results of the query to the user. *Retrieving results by query* is the common approach to extracting information from commercial databases. However, in order to be able to write queries that extract meaningful information from the database, it is necessary to understand the schema and the relationships of data in the database. For many databases it is not always easy to write useful queries unless the schema and relationships are explicitly specified. If this is not the case, we may not even know what queries we have to write, especially in databases whose interrelationships of data are quite complex and/or hidden. Retrieving useful and interesting information from databases depends heavily on the programmer's capability and background knowledge about the database.

2.2. Deriving Queries from Results

Deriving queries from results is the process of finding a desired query or a set of queries from a set of objects. In this setting, unlike the *retrieving results by query* approach, the user's role is not to derive a query but the KDD-system will derive the query by accessing schema information, data in a database, and domain knowledge given by the user. The user of the KDD-system, on the other hand, does not need in-depth knowledge about the database schema since the KDD-system works fairly autonomously. We claim that this approach actually discovers useful or interesting information implicitly stored in a database, since a user is not directly involved in generating queries and does not need the database information that was necessary in the *retrieving results by query* approach.

To illustrate *deriving queries from results* approach, suppose we have a personal database which contains information about persons and their other activities, and a police officer has two suspects {Joe, Mary} who are

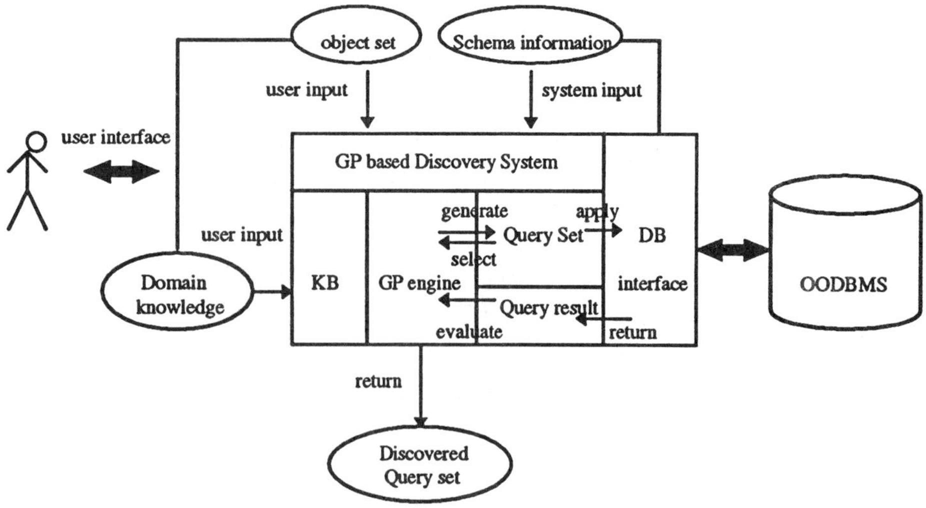

Figure 1: Architecture of MASSON

involved in a particular case and wants to find out what Joe and Mary have in common. How can he get this information from a database, given only a set of instances? A simple solution to this problem is to conduct a time-consuming search process in which we have to find a query or set of queries that returns the exactly same instances, {Joe, Mary} in our case, as its result. During the process many queries have to be manually written and executed. Suppose the following SQL form of query has been found,

"(SELECT ssn name address
FROM person purchase
WHERE (amount-spent > 1000) and
(payment-type = 'cash')and
(store-name = 'flea-market'))"

which returns {Joe, Mary} as its result and expresses that they both have spent more than $1,000 cash for shopping in a 'flea-market'. This information might lead the police officer to investigate suspicious activities in flea-markets. Of course, there may be several other activities and characteristics that Mary and Joe have in common.

In the remainder of the paper, we present an approach to finding such commonalities automatically employing GP to conduct the search process.

3. The Architecture of MASSON that Discovers a Query Set

We are developing a prototype system called MASSON that interfaces with an object-oriented database management system (OODBMS). Figure 1 depicts the architecture of the system. MASSON takes a set of objects, domain knowledge, and schema information as its input. The GP engine will generate many random queries based on those given inputs. The system will apply all of these queries to OODBMS and then OODBMS will return those results. The system will then evaluate those returned results with the given input, according to the fitness values of individual queries. A query, or a set of queries, that matches the given object set is selected, or a best query is selected if no other queries match the given object set.

3.1. An Object-Oriented Database

We assume a database model based on object-oriented and functional data model (OOFDM) "Shipman (1981), Gray and Kulkarni (1992)." Basically, OOFDM combines both features of object-oriented and functional data model. We are developing this model with a general KDD-system in mind, because in this model we can represent knowledge more appropriately and easily than the conventional relational database model "Sibley (1976), Gray and Kulkarni (1992)."

The fundamental concepts underlying the object-oriented data model are to perceive the real world as a variety of inter-related objects, and to view these objects at different levels of detail "Gray and Kulkarni (1992)". The functional data model that is similar to Dayal's model "Shipman (1981)", represents relationships among data simply by means of the mathematical notion of a function "Prabhakar and Navathe (1991), Sibley (1976)". The functional database schema diagram shown in Figure 2 represents our experimental object-oriented database based on OOFDM that contains information of persons and their related activities. Each class (or entity set) has slots and their values. The relationship between a class and an attribute is a function that maps the objects in that class into the domain of the attribute. For example, for a class *person* with attributes *name* and *address*, rather than defining the *person* object with attributes *name* and

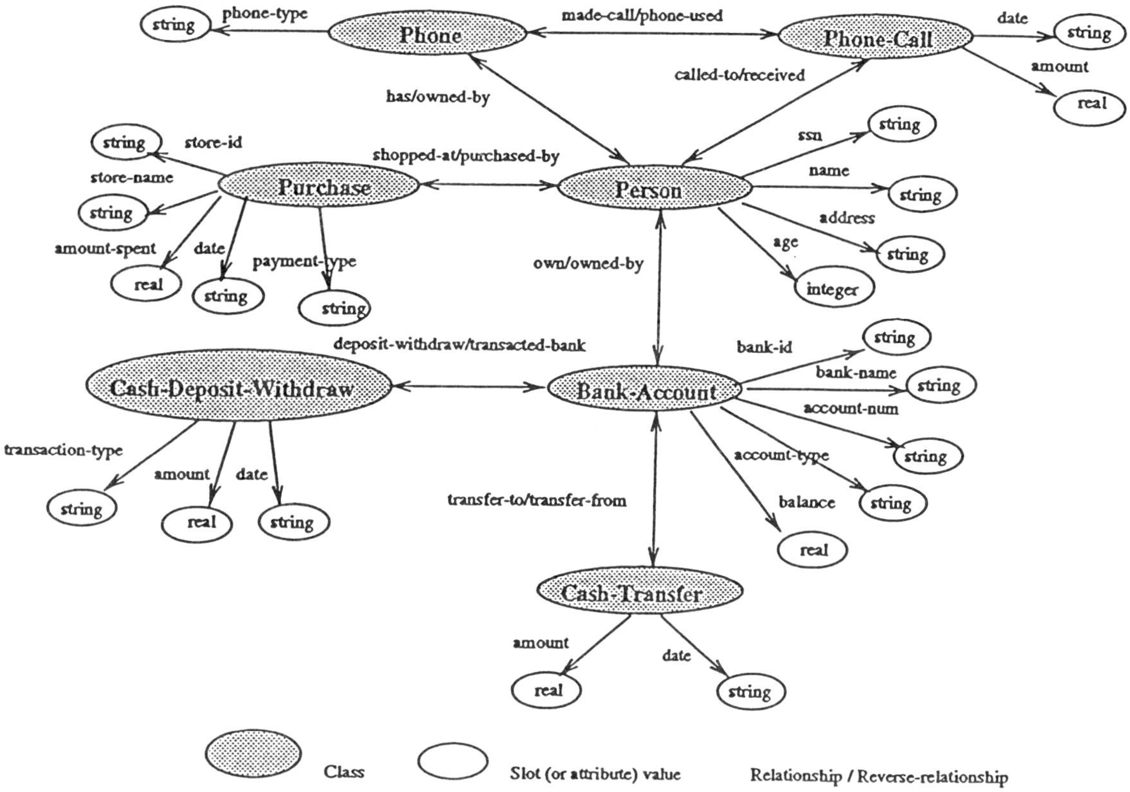

Figure 2: Schema Diagram for Personal Database

address which are strings, *name* and *address* may be regarded as functions which map *person* into string. A relationship or reverse relationship links a class to another class. The MASSON system was implemented by the PCL (Portable Common Loops) version of the CLOS (Common Lisp Object System) implementation.

3.2. Basic Navigational Query Operators

We used the following basic navigational query operators defined by "Czejdo and Eick (1996)" as a query language for an object-oriented database.

(SELECT Class-1 [(Predicate)])

This operator selects all the objects belonging to *Class-1* that satisfy *Predicate*. *Class-1* can be either a class name or a query that belongs to the class *Class-1*. *Predicate* is a conditional predicate and optional. If there is no predicate in this operator, it selects all the objects of the class.

(RESTRICTED Class-1 Predicate)

This operator restricts the objects in the given set to those that are related to another class, according to the given *Predicate*. *Class-1* and *Predicate* are the same as the ones used for *SELECT* operator, except that slots for the predicate of *RESTRICTED* must be only relationship slots, while slots for the predicate of *SELECT* must be non-relationship slots.

(RELATED Class-1 Relationship-r Class-2)

This operator selects all the objects from *Class-1* that are related to objects in *Class-2*. This is a relationship operator that relates one class *Class-1* and the other class *Class-2* through *Relationship-r*. *Class-1* can be either a class name or a query that belongs to the class *Class-1*. *Relationship-r* can be one of the relationship slots among the relationship set defined in the class *Class-1*. *Class-2* must be a class name or a query that belongs to the class *Class-2* and related to another class *Class-1* through *Relationship-r*.

(GET-RELATED Class-1 Relationship-r Class-2)

This is the inverse operator of *RELATED*. This operator selects all the objects from *Class-2* that are related to the objects in *Class-1*. *Class-1*, *Class-2*, and *Relationship-r* have the same format as the arguments in *RELATED*.

In addition, there are set operators *UNION*, *INTERSECTION*, *DIFFERENCE*. These are binary operators which take only compatible objects (same class or subclass) as arguments.

4. Using Genetic Programming to Search for Interesting Queries

To search interesting queries, we need a tool to intelligently generate many queries and evaluate them to see whether those queries compute the same set of objects given by the user. The GP can be applied to this kind of problem very well. "Kraft et al. (1994)" showed that GP can be a viable method to generate Boolean queries for information retrieval systems, in order to search certain documents effectively. However, in contrast to Boolean queries used in information retrieval systems, in our application, object-oriented queries are typed, support inheritance, return objects of a particular type and non-matching types result in syntactically incorrect queries and involve constants, slots, and operators with significant syntactical restrictions and type constraints similar to the previous works by "Montana (1994), Koza (1994), Haynes et al. (1995)."

4.1. Function Set and Terminal Set

Basically, the set of possible program structures in GP is the set of all possible compositions of functions and terminals "Koza (1990)." Table 1 shows the function set and terminal set in this application. The function set consists of query operators defined in section 3.2 and conditional operators such as Boolean operators and comparison operators.

Table 1: Function Set and Terminal Set

Function Set	{SELECT, RELATED, GET-RELATED, RESTRICTED}, {UNION, INTERSECTION, DIFFERENCE}, {AND, OR, NOT}, {>, >=, =, <, <=}
Terminal Set	CLASS-SET, SLOT-SET, VALUE-SET

However, in our application there are many query operators with different syntax. Each operator has a different syntax and an assuming database is object-oriented. So it has a specific function set (or operator set) and terminal set for its own use to create an individual. Accordingly, all the individuals consisting of a population must be created under the constraints of each query operator "Montana (1994), Haynes et al. (1995)." Constraints can be the operator's syntax and type of a query, or domain knowledge for choosing an appropriate slot value for creating the query. Comparison operators (i.g., <, =, >, etc.) and Boolean operators are only used for predicate of a query. When comparing the symbolic operands, only '=' is used. The terminal set consists of CLASS-SET, SLOT-SET, and VALUE-SET. The CLASS-SET consists of the names of classes defined in the section 3.1, for example, {*Person*, *Phone*, *Phone-call*, *Purchase*, *Bank-account*, *Cash-deposit-withdraw*, *Cash-transfer*}. The SLOT-SET consists of all the slots for each class, and the VALUE-SET consists of numeric values and symbolic values which are the slot values. The numeric values consist of integer/real numbers in the range of $[\Im_{min}, \Im_{max}]$ where $\Im_{min}$ and $\Im_{max}$ is minimum and maximum integer/real number defined in the database schema. The symbolic values consist of symbolic slot values such as 'bank-id' or 'address'.

4.2. Creating Initial Population and Choosing Constants for a Query

When creating an individual query, we have first to determine the type of objects the particular query has to return. After the result type has been selected, an operator is randomly chosen from the operator set that return instances of the selected type. Then the process is repeated recursively for each parameter of the query. Initially, the pre-defined number of queries that are syntactically legal, are generated randomly, forming the initial population.

Choosing a slot value (numeric or symbolic constants) from the value set for a query is a difficult problem, because of the large number of possible values to choose from. For example, suppose we need to select a slot value for the slot *amount-spent* of the class *Purchase* as $1,000. We can choose this value randomly from the integer number set in the range of $[\Im_{min}, \Im_{max}]$. However, if the range is too large, say [1, 5000], then the probability to get the value of 1000 from the range is 1/5000, which is very low. The problem even gets worse with symbolic constants because it usually requires domain-specific knowledge to create good symbolic constants, e.g. constants that describe street names. We cope with this problem by only picking constants that occur in the underlying database. For example, when generating a slot value for a person's *amount-spent* on a purchase, first a *Purchase* instance is randomly selected and its *amount-spent* slot value is selected as the constant to be generated. Note that, using this approach, constants that occur more frequently in the database are more likely to be used in queries generated in the knowledge discovery process.

4.3. Breeding New Population

Each individual is evaluated based on a pre-defined fitness function. Fitter queries are then selected with higher probability to breed a new population, using three genetic operators: selection, crossover, and mutation. The selection operator is used to choose certain individuals based on their fitness values for generating the next generation. The crossover operation creates two new offsprings by exchanging subtrees between the two parents if we represent a query as a tree. The mutation operation produces a new offspring by replacing one parent subtree with a newly created subtree of query. The selection-crossover-mutation cycle is repeated until user defined termination criteria are satisfied.

When performing crossover, type compatibility of operator arguments and operator results has to be guaranteed. To illustrate how our crossover operation works, suppose we have following pair of queries (numbers represent crossover points marked in the depth first order):

query-1:

(SELECT (RELATED person has phone) (= name "John"))
❶ ❷ ❸ ❹ ❺

query-2:

(GET-RELATED (SELECT purchase (> amount-spent $1000))
❶ ❷ ❸ ❹
shopped-by
(SELECT person (= address "Dallas")))
❺ ❻ ❼

First, the crossover point in *query-1* is selected randomly. However, choosing a crossover point for the *query-2* needs careful consideration because of type compatibility. Both of the crossover points must have the same type or subtypes of each other. Suppose point 2 which belongs to the class of *Person* is chosen as the crossover point in *query-1*. Then the crossover point 5 or 6 which also belongs to the class of *Person* must be chosen in *query-2* for type compatibility. If there is no matching crossover point, then the crossover operation for the parents are ignored. On the other hand, if there are more than one matching points, one point is randomly selected. Once both of the crossover points are selected, subtrees of the two parents (as shown in boldface above) are switched. The newly generated queries by choosing the point 2 in *query-1* and the point 6 in *query-2* look as follows:

(SELECT person (= name "John"))

(GET-RELATED (SELECT purchase (> amount-spent $1000))
shopped-at
(SELECT (RELATED person has phone)
(= address "Dallas")))

For the mutation operation, a subtree or conditional predicate of a query will be replaced by a newly created subtree or conditional predicate. If a newly selected class name replaces an old subtree, then this gives the same effect as simplification of a query.

4.5. Fitness Measure

We use the following fitness function f to evaluate each individual query i in a population:

$$f\,(n_i, h_i) = T - (h_i * h_i)/n_i\,,$$

where $n_i > 0$, $T \geq h_i$, and $i =$ 1, 2, ... population size (T is the cardinality of the set of objects whose commonalities have to be determined, h_i is the number of hits for an individual query i, n_i is the cardinality of query i's result).

This function is our standardized fitness function, which means that the smaller the fitness value, the fitter the individual is "Koza (1990)." The above fitness function depends on the number of hits and the cardinality of the query result. If a query does not make any hit ($h_i = 0$), then it has the value T, which is the worst fitness value. If, on the other hand the result of query i perfectly matches the set of objects given to the system, then its fitness value $f_i(n_i, h_i)$ is 0 ($h_i = n_i = T$ in this case). We call this case *perfect hit*. However, if a query is too general, it may contain superfluous instances (false positives) even if it made 100% of hits. If, on the other hand, a query is too specific then the number of hits of the query is less than the number of target instances - the difference is the number of false negatives. To cope with false positives and false negatives, we put the number of instances for an individual query, n_i as denominator. Moreover, because we take the square of the number of hits, false positives are punished less severely than false negatives by the fitness function: we are more interested in queries that return the target set or at large portions of the target set, even if they return objects that do not belong to the target set.

5. MASSON at Work

In order to illustrate how MASSON works, we assume that a personal database is given, which contains personal activities or information that are organized according to the schema diagram depicted in Figure 2. An object set {"Kenneth Shifrin", "William Sharkey", ...} with 22 instances was obtained by the following query:

"Select persons who have transferred cash to anyone more than 4 times."

(GET-RELATED (RESTRICTED bank-account (> transfer-to 4))
owned-by
person)

This object set was given to the MASSON system. The query discovery process starts with 100 randomly generated queries. Table 2 shows the summary for the best query at each generation. f_i, h_i, and n_i is a fitness value, the number of hits, and the cardinality respectively for the best query in a generation. f_a is the average standardized fitness value which is computed as follows: $f_a = (\Sigma f_i)/P$, where P is the population size and (Σf_i) is the total fitness value of a population.

Table 2: Summary for the Best Individual Query i Discovered at Each Generation.

(T = 22, P = 100, the number of generations = 200)

Generation	f_a	f_i	h_i	n_i	Perfect hit ?
0	21.68	15.25	9	12	N
2	21.57	17.16	22	100	N
11	20.70	16.46	12	26	N
27	19.39	8.00	14	14	N
33	15.69	0.00	22	22	Y
47	2.83	0.00	22	22	Y
81	1.87	0.00	22	22	Y
199	2.54	0.00	22	22	Y

The average standardized fitness (average fitness for short) for the initial generation is about 21.68. The best individual query for this generation has a fitness measure of 15.25 with 9 hits. The total number of instances for the result of this query are 12 as shown in the Table 2 and the query and its natural language interpretation was:

"Select persons who have transferred cash more than 3 times, received phone calls less than 3 times, and shopped less than 2 times."

```
(GET-RELATED
 (RELATED
  (RESTRICTED BANK-ACCOUNT(> TRANSFER-TO 3))
   OWNED-BY
  (RESTRICTED PERSON (<= CALLED-FROM 3)))
  OWNED-BY
 (GET-RELATED
  (RELATED PURCHASE PURCHASED-BY PERSON)
   PURCHASED-BY
  (RESTRICTED PERSON (< SHOPPED-AT 2))))
```

The best individual of the initial generation computes many superfluous instances, which do not belong to the object set. Therefore, we can not say this query reflects the common characteristics of the object set. Many individuals in this population return an empty set. This means that either many queries are semantically meaningless or they do not reflect those activities for the given object set in this database.

New population is created for second generation through breeding the previous population by applying genetic operators selection, crossover, and mutation operations to parents selected with probabilities. The best individual query with the fitness of 17.16 and 22 hits was:

"Select persons who have telephone."

```
(GET-RELATED PHONE OWNED-BY PERSON)
```

In this generation, even though the query hits 100% to the input object set, it also returns a large number of superfluous objects, because 50% of the persons in the database have a phone.

At generation 11, the fitness value for the best individual query is 16.46 with 12 hits, and with 26 objects returned, and the query was:

"Select persons who own two bank accounts."

```
(RESTRICTED PERSON (= OWN 2))
```

Even though the number of hits was decreased at this generation, the fitness of the best individual and the average fitness of a population are getting smaller.

At generation 27, we can see an overall improvement. The average fitness is decreased to about 19.39, and the fitness value of the best query is 8.00 with 14 hits. The query was:

"Select persons who have transferred cash through checking account more than 4 times."

```
(GET-RELATED
 (SELECT
   (RESTRICTED BANK-ACCOUNT (> TRANSFER-TO 4))
   (>= ACCOUNT-TYPE 1))
  OWNED-BY
 (RELATED PERSON OWN BANK-ACCOUNT))
```

As seen in Table 2, the number of hits are the same as the number of instances that this query computed. This means that this query does not have any superfluous instances.

At generation 33, the original query,

"Select persons who have transferred cash to anyone more than 4 times."

```
(GET-RELATED
 (RESTRICTED BANK-ACCOUNT (> TRANSFER-TO 4))
 OWNED-BY PERSON)
```

as the one used to get the object set was finally found. This is the first query that made a *perfect hit,* which is the case of ***$f_{33}=0$ (since $T=h_{33}=n_{33}=22$).*** Note that this is the query that was used to get the object set. However, this query is just one of many possible queries that may compute the given object set.

At generation 47, the average fitness value is significantly decreased to 2.83, and the best query that also made a perfect hit was:

"Select persons who have transferred cash to anyone more than 4 times and are age less than 30 or greater than 20."

```
(GET-RELATED
 (RESTRICTED BANK-ACCOUNT (> TRANSFER-TO 4))
 OWNED-BY
 (SELECT PERSON (OR (< AGE 30) (> AGE 20))))
```

At this point, we can observe an interesting fact. Even though this query made a perfect hit to the object set, this query is not the same query as the original query used to obtain our object set. On the other hand, it still computes the exact instance set while it has different semantics. The predicate in boldface is an additional predicate. In fact, the system may not discover the exact query that we expected at this generation; rather, it may discover different unexpected query like the one found at generation 47 that shows common activities for the provided object set. Therefore, if we have found many queries that compute the input instance set, then we may say we have actually discovered many knowledge implicitly stored in a database.

The query found at generation 81 is another query which is different from the one found at generation 33, but still made a perfect hit and the query was:

"Select persons who have transferred cash to anyone more than 4 times and their bank-id is not 'T171'."

```
(GET-RELATED
 (RESTRICTED
  (RESTRICTED
   (RESTRICTED
     SELECT BANK-ACCOUNT (NOT (= BANK-ID 'T171')))
     (> TRANSFER-TO 4))
   (> TRANSFER-TO 4))
  (> TRANSFER-TO 4))
 OWNED-BY PERSON)
```

This query is similar to the query found at generation 33 except that it has additional predicate, "`(NOT (= BANK-ID 'T171'))`".

In summary, MASSON found two different other queries that generate the target set as well as the original query. From generation 47 to the last generation, 199, the average fitness of each generation oscillates between the value 1.00 and 3.00. We experimented 4 other different queries based on complexity of those queries. For the two relatively simple queries, MASSON found syntactically same queries as the ones used to obtain the input object set and other syntactically and semantically different queries. However, for the other two relatively complex queries, MASSON could not find the exactly same queries for 200 generations but found syntactically and semantically different queries with perfect hit.

6. Future Works

There are many areas that are and have to be addressed by our current and future research. The first area is the empirical evaluation of the knowledge discovery power of the MASSON system. We are currently conducting experiments in which we try to measure what percentage of object-oriented queries can be found by the system at what cost. The second area is the development of more sophisticated fitness functions to drive the evolution of queries in MASSON. Our idea is to make the system more interactive, so that the user can drive the evolution process by redefining the fitness function interactively to reflect what he considers to be interesting. Moreover, we plan to develop an interpretation tool for the end-user of the system that facilitates understanding of the queries that were derived by the system. The third area is the use of domain knowledge more intensively to enhance system performance. Currently, our system uses the object-oriented database schema as its main source of knowledge, but other sources of knowledge are worthy of consideration, such as statistical knowledge, and domain specific constraints. The fourth area is the development of a simplifier for queries that removes redundant computations for a query and normalizes queries: one problem that the search process currently faces is that the same query can be written in too many different ways, which slows down the search process, because the system has a tendency to search for different forms of the same good query, instead of different good queries.

7. Summary

The paper focused on the problem of a data analyst who has a set of objects within an object-oriented database, and who wants to find the common interrelationships or characteristics that are shared by the objects belonging to the set. We dealt with this problem by introducing *deriving queries from results* approach in which we try to find an object-oriented query or a set of queries that returns an object set given by a user from a database. We briefly discussed the features of a KDD-system MASSON that employs GP to search for "good" queries. GP is appropriate for this kind of problem, since we can get many interesting diverse queries, the population of the search process changes dynamically, and the fitness value of each query imposes enough selective pressure to drive the search process to good solutions. Discovering a query or a set of queries that compute a given set of objects is important since the found queries describe the common characteristics of the set of objects given to the KDD-system. Therefore, we argue that the presented approach is a new approach for knowledge discovery in databases that employs GP in an object-oriented framework. Moreover, we briefly described the techniques employed in the MASSON system and its implementation, and we presented an example that demonstrated how MASSON works.

The major contribution of this paper is the presentation of a new approach to discovering commonalities within a given set of objects and showing that GP can be a viable method for the problem of knowledge discovery in databases.

Acknowledgments

We thank Ben S. Hadad, Susan Norman, Nic McPhee, Sid Maxwell, and other anonymous reviewers whose comments and suggestions helped improve the quality of this paper.

Bibliography

Kinnear, Kenneth E. Jr. 1994. *Advances in Genetic Programming*. Cambridge, MA: The MIT Press.

Koza, John R. 1990. *Genetic Programming: On the Programming of Computers by Means of Natural Selection*. Cambridge, MA: The MIT Press.

Koza, John R. 1994. *Genetic Programming: Automatic Discovery of Reusable Programs*. Cambridge, MA: The MIT Press.

Haynes, T.D., Schoenefeld, D.A., and Wainwright, R.L. 1995. Type Inheritance in Strongly Typed Genetic Programming. Kinnear, Kenneth E. Jr. and Angeline, Peter J. (editors). *Advances in Genetic Programming 2*. Cambridge, MA: The MIT Press. Chapter 19.

Montana, D.J. 1994. Strongly typed genetic programming. Technical Report 7866. Bolt Beranek and Newman, Inc.

Czejdo, B.D. and Eick, C.F. 1996. Knowledge Discovery in Object-Oriented Databases Using Genetic

Programming and Universal Symbolic Transformation System. submitted for publication.

Paepcke, Andreas 1993. *Object-Oriented Programming: the CLOS perspective*. Cambridge, MA: The MIT Press.

Lawless, J.A. and Miller, M.M. 1990. *Understanding CLOS*. Digital Press.

Gray, Peter M.D., Kulkarni, K.G., and Paton, N.W. 1992. *Object-Oriented Databases: A Semantic Data Model Approach*. Englewood Cliffs, NJ: Prentice Hall.

Prabhakar, S. and Navathe, S. 1991. A Knowledge Base Design and Application Prototyping Tool Based on an Enhanced Functional Data Model. *International Conference on Tools for AI*.

Shipman, D.W. 1981. The Functional Data Model and DAPLEX. *ACM Transactions on Database Systems*.

Sibley, E. 1976. The Development of Database Technology. *ACM Computing Surveys*, March.

Piatetsky-Shapiro, G. and Frawley, W.J. 1991. Knowledge Discovery in Databases: An Overview. *Knowledge Discovery in Databases*. AAAI/The MIT press. Pages 1-27.

Peter, H. and Willi, K. 1991. A support System for Interpreting Statistical Data. *Knowledge Discovery in Databases*. AAAI/The MIT press. Pages 325-345.

Matheus, C.J., Chan, P.K., and Piatetsky-Shapiro, G. 1993. Systems for knowledge discovery in databases. *IEEE Transactions on Knowledge and Data engineering*. Volume 5.

Yoon, J.P. and Kerschberg, L. 1993. A framework for knowledge discovery and evolution in databases. *IEEE Transactions on Knowledge and Data engineering*. Volume 5.

Agrawal, R., Imielinski, T., and Swami, A. 1993. Database mining: a performance perspective. *IEEE Transactions on Knowledge and Data engineering*. Volume 5.

Lee, D.H. and Kim, M.H. 1993. Discovering Database Summaries through Refinements of Fuzzy Hypotheses. *IEEE Transactions on Knowledge and Data engineering*. Volume 5.

Dhar, V. and Tuzhilin, A. 1993. Abstract-Driven Pattern Discovery in Databases. *IEEE Transactions on Knowledge and Data engineering*. Volume 5.

Zhong, N. and Ohsuga, S. 1995. Managing/refining structural characteristics discovered from databases. *Proc. of the 24th Hawaii International Conference on System Sciences.* Volume 3. Pages 283-92.

Kraft, D.H., Petry, F.E., Buckles, B.P., and Sadasivan, T. 1994. The use of genetic programming to build queries for information retrieval. *Proc. of the First IEEE Conference on Evolutionary Computation.* Volume 1. Pages 468-73.

Chen, H. and She, L. 1994. Inductive Query by Examples (IQBE): A Machine Learning Approach. *Proc. of the 27th Annual Hawaii International Conference on System Sciences.*

Muggleton, S. 1992. *Inductive Logic Programming*. Academic Press.

Lavrac, N., and Dzeroski, S. 1994. *Inductive Logic Programming Techniques and Applications*. England: Ellis Horwood.

Quinlan, J. 1993. *C4.5: Programs for Machine Learning*. San Mateo, CA: Morgan Kaufmann.

Michalski, R., Mozetic, I., Hong, J., and Lavrac, N. 1986. The multi-purpose incremental learning system AQ15 and its testing application on three medical domains. In Proc. *5th National Conference on Artificial Intelligence*, San Mateo, CA: Morgan Kaufmann. Pages 1041-1045.

Cultural Transmission of Information in Genetic Programming

Lee Spector* †
lspector@hampshire.edu

Sean Luke †
seanl@cs.umd.edu

*School of Cognitive Science and Cultural Studies
Hampshire College
Amherst, MA 01002

†Department of Computer Science
University of Maryland
College Park, MD 20742

ABSTRACT

This paper shows how the performance of a genetic programming system can be improved through the addition of mechanisms for non-genetic transmission of information between individuals (culture). Teller has previously shown how genetic programming systems can be enhanced through the addition of memory mechanisms for individual programs [Teller 1994]; in this paper we show how Teller's memory mechanism can be changed to allow for communication between individuals within and across generations. We show the effects of indexed memory and culture on the performance of a genetic programming system on a symbolic regression problem, on Koza's Lawnmower problem, and on Wumpus world agent problems. We show that culture can reduce the computational effort required to solve all of these problems. We conclude with a discussion of possible improvements to the technique.

1 Culture and Evolution

In most evolutionary computation systems, individuals are assessed for fitness independently. Each individual is placed into an initialized environment, a fitness test is performed, and the environment is then re-initialized for the next individual. By contrast, in biological populations (and in many Artificial Life systems; see, e.g., [Brooks and Maes 1994]) each individual acts in a shared environment that may bear the marks of contemporaries and of predecessors. Some animals use these marks to their advantage, and some intentionally modify the environment to communicate with others. Modifications to the environment may be ephemeral (e.g., calls, spoken language) or long-lasting (e.g., nests, written language). Through these mechanisms fitness-enhancing information may be transmitted to others, including offspring, by non-genetic means.

We use the term "culture" to refer to any information transmitted between individuals by non-genetic means. Our definition is similar to that of Bonner, who describes his use of the term as follows:

> By culture I mean the transfer of information by behavioral means, most particularly by the process of teaching and learning. It is used in a sense that contrasts with the transmission of genetic information passed by the direct inheritance of genes from one generation to the next. [Bonner 1980, p. 9]

Culture, in this sense, is employed by many animals other than humans. Higher primates provide many examples, but one can also argue that much simpler animals, even insects, make significant use of culture [Bonner 1980].

One can view the evolution of a culture-using species as a complex interaction between two processes, one genetic and one cultural. Dawkins coined the term "meme" to serve the function for cultural transmission that "gene" serves for genetic transmission; that is, a meme is a unit of information transmitted by behavioral means [Dawkins 1976]. The interactions between cultural and genetic transmission, and their combined effects on fitness and adaptation, have been the subject of much discussion in evolutionary biology. For our purposes we may simply note that the difference in mode of transmission leads to a significant difference in the speed with which a new piece of information may spread through a population. Again, quoting Bonner:

> If, for instance, a favorable genetic mutation appears in one individual, even with reasonable positive selection pressures, it will take many life cycles and perhaps hundreds of years for the new gene to be present in an appreciable number of individuals in the population. It is only in some lethal circumstances that the genetic structure of a population can change rapidly.
>
> ... On the other hand a cultural change can be exceedingly rapid. A new fad or dress style may take over a whole nation in a matter of days or weeks. [p. 18]

The difference in speed of transmission leads to a complementary difference in stability; the gene pool is less vulnerable to sudden dangerous fluctuations than is the meme pool. As a result one can expect that a combination of the two transmission modes will be most beneficial for the evolution of certain complex systems.

Several researchers have previously experimented with cultural elements in evolutionary computation systems. For example, Bankes has explored "meme-based" adaptive systems in multiplayer games, although the bulk of this work is concerned with the evolution of memes *within* individuals [Bankes 1995]. Reynolds has developed a framework of dual inheritance "cultural algorithms" in which a "belief space," containing generalizations of the representations of individuals, helps to guide the evolution of a population [Reynolds 1994]. Others have explored related ideas in a range of evolutionary computation contexts (e.g., [Hutchins and Hazlehurst 1993]).

In contrast to previous research, this paper describes a cultural mechanism that is a straightforward extension of the memory mechanisms used by individuals. The same mechanisms that are used by individuals to build and maintain state information are extended to allow for the communication of information between individuals within and across generations. The virtues of this scheme include simplicity, a better match to biological notions of culture, and a demonstrable positive impact on the speed of evolution for several test problems.

The remainder of this paper is organized as follows: after a brief digression on performance in genetic programming we describe a simple technique for adding cultural transmission to genetic programming systems. We then show how the addition of culture can reduce the computational effort required to produce solutions for symbolic regression, Lawnmower, and Wumpus world problems. We conclude with a discussion of possible improvements to the technique.

2 Genetic Programming and Performance

Genetic programming is a technique for the automatic generation of computer programs by means of natural selection [Koza 1992]. The genetic programming process starts by creating a large initial population of programs that are random combinations of elements from problem-specific function and terminal sets. Each program in the initial population is then assessed for fitness, and the fitness values are used in producing the next generation of programs via a variety of genetic operations including reproduction, crossover, and mutation. After a preestablished number of generations, or after the fitness improves to some preestablished level, the best-of-run individual is designated as the result and is produced as the output from the genetic programming system.

The impact of alternative approaches to genetic programming can only be assessed by measuring performance over a large number of runs. This is because the algorithm includes random choices at several steps; in any particular run the effects of the random choices may obscure the effects of the alternative approaches.

To analyze the performance of a genetic programming system over a large number of runs one can first calculate $P(M,i)$, the cumulative probability of success by generation i using a population of size M. For each generation i this is simply the total number of runs that succeeded on or before the ith generation, divided by the total number of runs conducted. Given $P(M,i)$ one can calculate $I(M,i,z)$, the number of individuals that must be processed to produce a solution by generation i with probability greater than z.[1] $I(M,i,z)$ can be calculated using the following formula:

$$I(M,i,z) = M * (i+1) * \left\lceil \frac{\log(1-z)}{\log(1-P(M,i))} \right\rceil$$

Koza defines the minimum of $I(M,i,z)$ as the "computational effort" required to solve the problem with the given system.[2]

3 Memory and Culture in Genetic Programming

Teller has developed a simple mechanism that allows programs produced by genetic programming to make use of memory. The mechanism, called *indexed memory*, consists of a linear array of memory locations and two functions, `READ` and `WRITE`, that are added to the function set. The memory is initialized (e.g., to 0) at the start of each fitness case. `READ` takes a single argument and returns the contents of the memory location indexed by that argument. `WRITE` takes two arguments, a memory index and a data item, and stores the data item in the memory at the specified index. `WRITE` returns the *previous* value of the specified memory location. Teller showed that indexed memory can help to evolve correct programs for certain problems, and that the combination of indexed memory and iteration allows genetic programming to produce any Turing-computable function [Teller 1994]. Others have further examined the utility of indexed memory; for example, Andre has experimented with problems that *require* the use of memory, and has explored the ways in which evolved programs use indexed memory in solving these problems [Andre 1995].

We implement "culture" through a simple modification to Teller's indexed memory: all individuals share the same memory, and it is initialized only at the start of a genetic programming run. This means that the contents of memory at the end of one fitness evaluation are available for use in the

[1] For the analyses in this paper a value of z=99% is always used.

[2] The $P(M,i)$ and $I(M,i,z)$ measures were developed by Koza and are discussed on pages 99 through 103 of [Koza 1994].

next fitness evaluation. If individual programs are subjected to multiple fitness cases, an individual program can use this mechanism to pass information to itself from one fitness case to the next. In any event, programs that are evaluated later may use information left in memory by those that were evaluated earlier. A program may pass information to itself (in later fitness cases), to its contemporaries, to its offspring, and to unrelated members of future generations. The order in which the population is evaluated for fitness can clearly have an effect; we randomize evaluation order within each generation to prevent systematic exploitation of this effect.

Note that the culture is the collective product of all individuals throughout evolutionary time. On the positive side, this means that a "good idea" developed by any individual in any population may be preserved for use by all other individuals. On the negative side, this means that one destructive individual can destroy a great deal of valuable information. For most of the problems that we have studied so far, the positives outweigh the negatives—the availability of a culture speeds evolution. One avenue for future work, however, is to minimize the effects that destructive individuals can have on the culture (see below).

A program evolved with culture may function correctly only when run with a particular initial memory state. The initial memory state is therefore an intrinsic part of the program, and the genetic programming system should report the appropriate initial memory state along with the best-of-run program. If one later wishes to run such a program outside of the context of the evolutionary process, one should first re-initialize the cultural memory with the reported memory state.

4 Symbolic Regression

To assess the utility of cultural transmission we ran the *lil-gp* genetic programming system [Zongker 1995] on a symbolic regression problem. We compared the performance of the system on versions of the problem in which the programs had access to no memory, to ordinary indexed memory, and to culture.

The goal of the symbolic regression problem, as described in [Koza 1992], is to produce a function that fits a provided set of data points. For each element of a set of (*x*, *y*) points, the program should return the correct *y* value when provided with the *x* value. For our experiments we obtained our data points from the equation $y = x^4 + x^3 + x^2 + x$. One can view the task of the genetic programming system as that of "rediscovering" this formula from the data points used as fitness cases. We used 20 fitness cases, with randomly selected *x* values between -1 and 1. A program was considered to be successful if it returned a value within 0.01 of the target value for all 20 cases.

For our experiments we used a function set consisting of the two-argument addition function +, the two-argument subtraction function -, the two-argument multiplication function *, the two-argument protected division function %, two one-argument trigonometric functions SIN and COS, the one-argument exponential function EXP, and the one-argument protected logarithm function RLOG (as in [Koza 1992]). We used a single terminal for the independent variable X, and ephemeral random constants between -1 and 1 (also as in [Koza 1992]). We used tournament selection (tournament size = 5), a 90% crossover rate, a 10% reproduction rate, no mutation, a population size of 1000, and a maximum of 51 generations per run.

For the runs with indexed-memory and with culture we added a 100-element memory array and READ and WRITE functions that behave as described above. We mapped the "index" arguments of these functions to the proper range ([0–99]) by multiplying them by 100 and by then taking them modulo 100.

We performed 100 runs with no memory, 100 runs with indexed memory, and 100 runs with culture. The results can be summarized as follows:

Condition	Computational Effort
No memory	899,000
Memory	1,131,000
Culture	551,000

These results show that the addition of an ordinary indexed memory *increases* the computational effort required to solve the regression problem. Presumably this is because the search space includes many programs that compute the desired function ($y = x^4 + x^3 + x^2 + x$) without using memory. One might additionally conjecture that the unhelpful functions READ and WRITE act as noise in the function set, contributing little to success while filling positions that could instead be filled with useful functions. Nevertheless, when the memory is preserved between fitness evaluations as culture, the system is able to take advantage of this, and the computational effort is reduced to 61% of that required when no memory is available.

5 The Lawnmower Problem

Similar results were obtained for Koza's 64-square Lawnmower problem [Koza 1994]. The goal in the Lawnmower problem is to find a program for controlling the movement of an autonomous lawnmower on an 8-by-8 torroidal world. All values manipulated by functions for this problem are vectors of integers modulo 8 of the form (i, j), where $0 \leq i, j \leq 7$. The terminal set consists of the 0-argument function (MOW), the 0-argument function (LEFT), and random vector constants modulo 8 ($\Re_{v8}$). (MOW) moves the lawnmower in the direction it is currently facing, mows the lawn at the new location, and returns the vector value (0,0).

(LEFT) turns the lawnmower 90° to the left and returns the vector value (0,0). The function set consists of the operators V8A, FROG, and PROGN. V8A is a 2-argument vector addition function that adds vector components modulo 8. FROG is a 1-argument movement operator that jumps the lawnmower to the coordinate produced by adding (modulo 8) its vector argument to the lawnmower's current location. FROG acts as the identity operator on its argument. PROGN is a 2-argument sequencing operator that returns the value of its second argument.

For the runs with indexed-memory and with culture we added a 2-dimensional, vector-indexed, 64-element (8-by-8) memory array along with appropriate READ and WRITE functions.

We performed 100 runs with no memory, 100 runs with indexed memory, and 100 runs with culture. In all cases we used a population size of 500, two automatically-defined functions (as in [Koza 1994]), and a maximum of 51 generations. The results can be summarized as follows:

Condition	Computational Effort
No memory	12,000
Memory	16,000
Culture	10,000

These results show the same pattern as the results for symbolic regression, although the effect is less extreme. Again, memory by itself slows evolution, but evolution proceeds most quickly with culture.

6 Wumpus World

Wumpus world [Russell and Norvig 1995] is a problem environment that is more complex than those described above in several interesting ways. The use of genetic programming for the evolution of Wumpus world agents has been described elsewhere [Spector 1996]. In this section we present only an informal description of the problem and the results of our culture experiments. See [Spector 1996] for more information on the Wumpus world simulator, the function and terminal sets, and other parameters.

Wumpus world is cave represented as a grid of squares surrounded by walls. The agent's task is to start in a particular square, to move through the world to find and to pick up the piece of gold, to return to the start square, and to climb out of the cave. The cave is also inhabited by a "wumpus"— a beast that will eat anyone who enters its square. The wumpus produces a stench that can be perceived by the agent from adjacent (but not diagonal) squares. The agent has a single arrow that can be used to kill the wumpus. When hit by the arrow the wumpus screams; this can be heard anywhere in the cave. The wumpus still produces a stench when dead, but it is harmless. The cave also contains bottomless pits that will trap unwary agents. Pits produce breezes that can be felt in adjacent (but not diagonal) squares. The agent perceives a bump when it walks into a wall, and a glitter when it is in the same square as the gold.

The wumpus world agent can perform only the following actions in the world: go forward one square; turn left 90°; turn right 90°; grab an object (e.g., the gold) if it is in the same square as the agent; release a grabbed object; shoot the arrow in the direction in which the agent is facing; climb out of the cave if the agent is in the start square.

The agent's program is invoked to produce a single action for each time-step of the simulation. The program returns one of the valid actions and the simulator then causes that action, and any secondary effects, to happen in the world. The agent can maintain information between actions by use of a persistent memory system. The agent's program has a single parameter, a "percept" that encodes all of the sensory information available to the agent. The agent's program can refer to the components of the percept arbitrarily many times during its execution; that is, sensing is free.

Agents are assessed on the basis of performance in four randomly generated worlds. Four new random worlds are generated for each fitness evaluation; this helps to prevent the production of agents that are overfitted to a particular set of worlds. In each world the agent is allowed to perform a maximum of 50 actions, and the agent's score is determined as follows: 100 points are awarded for obtaining the gold, there is a 1-point penalty for each action taken, and there is a 100-point penalty for each unit of distance between the agent and the gold at the end of the run. "Standardized fitness" values (for which lower values are better [Koza 1992]) are the average of the scores from the four random worlds, subtracted from 100. Agents are not explicitly rewarded for climbing out of the cave, although less action penalties are accumulated if an agent climbs out and thereby ends the simulation. An agent is considered to have solved the problem if its average score in four random worlds is greater than zero. To have obtained such a score an agent must have grabbed the gold in at least one and usually two or more of the four random worlds. This may be difficult; in many cases it is necessary to risk death in order to navigate to the gold, and in some cases the gold may be unobtainable because it is in a pit or in a square surrounded by pits.

1709 runs of the lil-gp genetic programming system were performed on the wumpus world problem, 400 without memory, 509 with ordinary indexed memory, and 800 with culture.[3] A C-language re-implementation of Russell and Norvig's wumpus world simulator was used for fitness evaluation. The population size was 1,000 and the maximum number of generations per run was 21. Tournament selec-

[3] Note that computational effort comparisons are indeed valid between sets of different numbers of runs. For these computationally expensive runs we started processes on several machines and stopped them when it was clear that we had obtained sufficient data; the exact numbers of runs are therefore arbitrary.

tion was used with a tournament size of 7. The results can be summarized as follows:

Condition	Computational Effort
No memory	1,710,000
Memory	2,100,000
Culture	1,386,000

These results show that when culture is used the computational effort required to produce a successful agent for this version of Wumpus world is reduced to 66% of that required when ordinary indexed memory is used. They also show, surprisingly, that indexed memory once again increases computational effort over the "no memory" condition. It appears that the reactive strategies are at least as useful as knowledge-based strategies for this domain.

The version of Wumpus world used in this experiment differs from that used in previous work [Spector 1996]; while re-implementing the system in C we made other changes as well. In particular, the new version operates on a true 6-by-6 world. In the old version the world's "walls" occupied space so the actual playing area in a "6-by-6" world was only 4-by-4. In addition, the old version charged an agent an explicit 100-point penalty for dying. In the new version the only penalty for death is implicit—after one dies one can no longer get closer to the gold or pick it up.

Wumpus world can be varied in many other ways as well, and the relative utility of memory and culture may differ in each of these variants. We are currently conducting experiments with several variants and it is clear from the preliminary results that culture is *not always* beneficial in Wumpus world. In particular, for a configuration similar to that of [Spector 1996] it appears that ordinary indexed memory is best, followed by culture. Further experiments and analysis will be required to discover the reasons for culture's utility in various circumstances.

7 Discussion and Future Work

It is clear from the reported experiments that culture can have a significant beneficial impact on computational effort for certain problems. Fortunately, it is easy to add culture to other genetic programming applications: one simply includes an indexed memory that is shared between all fitness evaluations, and arranges for the appropriate memory state to be returned along with the best-of-run program. If the target application already includes an indexed memory, then simply by sharing this memory between all fitness evaluations the computational effort required to produce a correct program may be decreased. If the application does not already include an indexed memory, then a shared indexed memory can usually be added with little difficulty.

Although we have examined memory and culture as alternatives in this paper, there is no reason why they can't be combined. By providing two versions of the `READ` and `WRITE` functions, one each for ordinary indexed memory and one each for culture, one can give evolving programs access to both mechanisms; for some problems, this may provide the maximum benefit.

Another avenue for future work is to minimize the effects that destructive individuals can have on the culture. One way to do this would be to have each fitness test occur using a copy of the culture in a local memory. The changes to the memory could then be transferred to the culture in a fitness-proportionate way — the greater the fitness of an individual, the greater its impacts on the culture. Unfortunately, it may be the case that low-fitness individuals nonetheless contribute to the culture in important ways; fitness-proportionate impact on the culture would prevent the population from benefiting from these contributions. An alternative scheme for minimizing the effects of destructive individuals would be to use many sub-populations, each of which has its own sub-culture, and to periodically merge sub-populations and sub-cultures. Many variations on this theme exist, and many may be worthy of further exploration.

The possibility also exists that some destructive individuals may be highly fit. Some such individuals may in fact be using destruction to maintain an advantage over others in the population; they could make use of information in the culture and then destroy it so that it would not be available to others.[4] In order to detect this sort of activity one could collect information about access to the culture by individuals of various fitness levels throughout a run. Such information, combined with genealogical audit trails [Koza 1992], would also permit one to explore broader questions about how and why individuals make use of culture.

While culture has been useful in most of the domains to which we have applied it, further work must be conducted to characterize the relationship between domain characteristics and the expected impact of the technique. Here again we may gain insights from biological examples; not *all* animals use culture, and there are presumably many niches in which culture confers no adaptive advantage. On the other hand, culture appears to be extremely useful in certain situations (e.g., for humans), and we may expect correspondingly large impacts in certain genetic programming domains.

8 Conclusions

Culture, the transmission of information between individuals by non-genetic means, can in some cases improve the performance of a genetic programming system. This paper showed how culture can be implemented as a simple modification of Teller's indexed memory mechanism. The paper demonstrated that culture can improve the performance of a genetic programming system for three problems: symbolic regression of $y = x^4 + x^3 + x^2 + x$, the 64-square Lawnmower

[4] Frank D. Francone drew this possibility to our attention.

problem, and a version of Wumpus world. Further research should examine the effects that destructive individuals may have on a shared culture, delineate the cases in which culture is helpful, and examine the ways in which culture is actually employed by successful individuals.

Acknowledgments

The authors wish to thank the reviewers for several helpful comments; in particular, Frank D. Francone raised more interesting issues than we were able to explore in the context of this paper — but they will help to guide our ongoing research on the project. This research was supported in part by grants from ONR (N00014-J-91-1451), AFOSR (F49620-93-1-0065), and ARPA contract DAST-95-C0037.

Bibliography

Andre, D. 1995. The Evolution of Agents that Build Mental Models and Create Simple Plans Using Genetic Programming. In *Proceedings of the Sixth International Conference on Genetic Algorithms*, edited by L.J. Eshelman. pp. 248–255. San Francisco, CA: Morgan Kaufmann Publishers, Inc.

Bankes, S. 1995. Evolving Social Structure in Communities of Agents Through Meme Evolution. In *Proceedings of The Fourth Annual Conference On Evolutionary Programming*. Cambridge, MA: The MIT Press.

Bonner, J.T. 1980. *The Evolution of Culture in Animals.* Princeton, NJ: Princeton University Press.

Brooks, R.A., and P. Maes, eds. 1994. *Artificial Life IV: Proceedings of the Fourth International Workshop on the Synthesis and Simulation of Living Systems.* Cambridge, MA: The MIT Press.

Dawkins, R. 1976. *The Selfish Gene.* Oxford: Oxford University Press.

Hutchins, E. and B. Hazlehurst. 1993. Learning in the Cultural Process. In *Artificial Life II*, edited by C.G. Langton, C. Taylor, J.D. Farmer, and S. Rasmussen. Addison-Wesley Publishing Company.

Koza, J.R. 1992. *Genetic Programming: On the Programming of Computers by Means of Natural Selection.* Cambridge, MA: The MIT Press.

Koza, J.R. 1994. *Genetic Programming II: Automatic Discovery of Reusable Programs.* Cambridge, MA: The MIT Press.

Reynolds, R.G. 1994. An Introduction to Cultural Algorithms. In *Proceedings of the Third Annual Conference on Evolutionary Programming.* pp. 131–139. River Edge, NJ: World Scientific.

Russell, S.J., and P. Norvig. 1995. *Artificial Intelligence, A Modern Approach.* Englewood Cliffs, NJ: Prentice Hall.

Spector, L. 1996. Simultaneous Evolution of Programs and their Control Structures. In *Advances in Genetic Programming 2*, edited by P. Angeline and K. Kinnear. Cambridge, MA: MIT Press, in press.

Teller, A. 1994. The Evolution of Mental Models. In *Advances in Genetic Programming*, edited by K.E. Kinnear Jr. pp. 199–219. Cambridge, MA: The MIT Press.

Zongker, D., and B. Punch. 1995. *lil-gp 1.0 User's Manual.* WWW: http://isl.cps.msu.edu/GA/software/lil-gp, or via anonymous FTP to isl.cps.msu.edu, in the directory "/pub/GA/lilgp".

Code Growth in Genetic Programming *

Terence Soule
Laboratory for Applied Logic
Computer Science Dept.
University of Idaho
Moscow, Idaho 83844-1010
tsoule@cs.uidaho.edu

James A. Foster
Laboratory for Applied Logic
Computer Science Dept.
University of Idaho
Moscow, Idaho 83844-1010
foster@cs.uidaho.edu

John Dickinson
Computer Science Dept.
University of Idaho
Moscow, ID 83844-1010
johnd@cs.uidaho.edu

Abstract

In this paper we examine how quickly the programs generated using genetic programming grow in size. We found that without a constraint mechanism the programs will grow indefinitely regardless of whether or not the growth acts to improve the programs' solutions. This growth is dominated by non-functional code. If the non-functional code is removed the growth is dominated by functional, but non-executed code. Two methods of controlling the growth: removing non-functional code, and selective pressure applied by penalizing longer programs, are compared. Only the later method appears to be effective in bounding the programs' size.

1 Introduction

Considerable study has gone into how close programs generated using genetic programming (GP) come to the optimal solution: how effective they are [Koza, 1992, Koza, 1994]. The experiments in this paper study how efficiently genetic programming solves problems: how much code is generated compared to the amount of code actually needed. In addition, we measure how much of the generated code is actually non-functional.

*This work supported by funding from the URO FY96 Seed Grant program, and through funding received from the NSF-Idaho EPSCoR project under NSF Cooperative Agreement number OSR-9350539.

One of the interesting features of the programs generated using GP is that they almost invariably grow in size and incorporate large amounts of non-functional code which has no net effect, but does increase the programs' lengths. Unfortunately, there are several reasons for generally preferring shorter programs. Shorter programs typically require less time and less space to run. This is particularly important during the GP process which may need to store and evaluate populations of hundreds or thousands of programs. In addition, shorter programs tend to show better generalization performance than do longer programs. This means that efforts to limit code size will be important in all but the simplest GP applications. Reducing the amount of non-functional code is an obvious place to start.

However, several researchers have hypothesized that the growth of code during GP and the creation of non-functional code are both important contributors to the production of effective programs. Iba et al. have suggested that the creation of large programs is a necessary part of the exploration of the solution space [Iba et al., 1994]. Nordin and Banzhaf have argued that non-functional code is also important in shielding useful code from the harmful effects of crossover [Nordin and Banzhaf, 1995]. They attempted to use external pressures to reduce the amount of non-functional information in variable length genetic algorithm individuals. Although the amount of non-functional information was reduced almost to zero the system failed to find the optimal solution, unlike in the case without external pressure.

In contrast Zhang and Muhlenbein used an implementation of Occam's Razor to limit code size in a successful GP to design neural nets [Zhang and Muhlenbein, 1995]. Their model attempted to balance parsimony and performance. This resulted in populations which cycled between growth and reduction and produced reasonably effective individuals.

We have used selective pressure to control code size in a relatively successful genetic program for approximating the maximum clique problem [Soule et al., 1996].

Overall, the relative importance of code growth and non-functional code is still unclear. Although there are definite benefits to shorter, more parsimonious programs some studies suggest that code growth, particularly of non-functional code, is important to the evolution of effective programs. This makes attempts to restrict code growth and improve efficiency difficult.

Looking at biological evolutionary processes is also not very enlightening. In humans roughly 80 – 90% of DNA does not code for functional proteins, although some of this DNA does have a structural function. Similar percentages hold for most other higher organisms. On the other hand, prokaryotes have almost no non-functional genetic material. It has been hypothesized that organisms which compete via rapid cell division experience strong selective pressure for streamlined genomes – an efficiency consideration which may be relevant for GP.

Comparing code growth to performance improvements will help indicate how important the growth is in producing effective solutions. The ability to use selective pressure to control code growth and non-functional code without interfering with the production of effective solutions places a limit on the importance of these factors to the GP process. It also leads to more efficient and generalizable solutions.

We examined the rate of code growth in a simple GP, separating the code into functional and non-functional categories. Then we measured the effect of removing the non-functional code or of applying selective pressure. We found that the size of the genetically generated programs grows indefinitely, regardless of whether the additional code acts to improve the programs' fitness. Additionally, most of this growth is caused by increases in the amount of non-functional code. Despite the prevalence of non-functional code, removing this code at every generation does not halt the programs' growth. Instead the programs generate code which, while functional, is never actually executed. These results imply that much of the programs' growth is not caused by pressure to improve the solution, but, instead, is an innate part of the genetic programming process. However, selective pressure was found to dramatically reduce the overall growth and the percentage of non-functional code while still producing effective programs. This implies that most of the growth seen in unrestrained GP and most of the non-functional code is not necessary for the evolution of effective programs.

2 The Experiment

2.1 The Test Arena

Our test case was a GP for a robot guidance problem. The task is to generate a program which will guide a robot through an obstacle-filled room. The room, a simple maze, is shown in Figure 1. This problem was chosen to make automated detection of useless code relatively simple. Branches of statements which fail to change the robots position or direction are clearly useless.

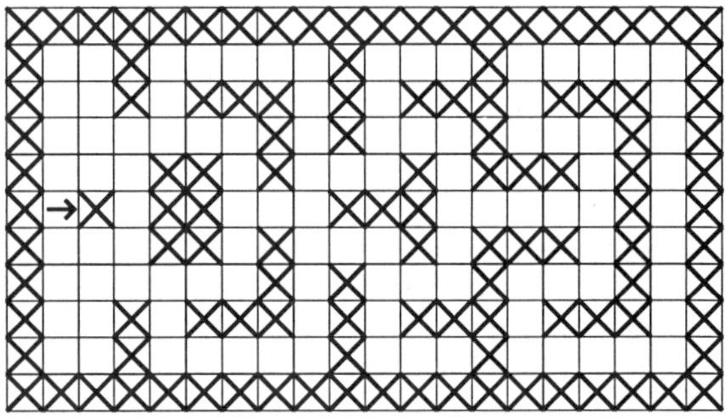

Figure 1: The Room

The initial position and direction of the robot is shown by the arrow at the left of the figure. The fitness of a program is based on the horizontal distance the robot has traveled when the program terminates. Hence, the fitness varies from one to eighteen (the length of the room). No penalties are assessed for running into walls (although of course it does not advance the robot) or for the number of steps taken. To avoid infinite loops the programs are only allowed 3000 statement executions. This number is more than sufficient to cross the room, but should avoid incidentally penalizing programs which might cross the room using a very long path. Of course, some programs will terminate in less than 3000 steps.

We developed a simple, interpreted language for the genetic program (which was written in C++). The language has five terminal statements (which have no arguments and no return values): `left`, `right`, `forward`, `back`, and `no_op`. These statements change the robot's position or direction in the obvious manner. The `no_op` statement has no effect. It was included so that branches of non-functional code could be removed and replaced by a `no_op` without interfering with the syntax of the language, which requires binary branching.

The language also contains three control statements which take two other statements as arguments: **progn**, **if**, and **while**. The **progn** statement simply executes its first argument followed by its second argument. Both the **if** and **while** statements take a predicate, either **wall_ahead** or **no_wall_ahead**. The **if** statement executes its first argument if the predicate is true and its second argument otherwise. As long as its predicate is true the **while** statement executes its first argument and then its second argument. This language produces programs which have the structure of a binary tree. This means that the crossover operation is relatively simple to accomplish through the exchange of branches.

For example, a typical program is

```
progn(left,
while(no_wall_ahead,forward,right))
```

The corresponding program tree and the path along which the program would move the robot are shown in Figure 2. The robot ends at location 2,4 facing the barrier to the South. The program itself would be awarded a fitness of two.

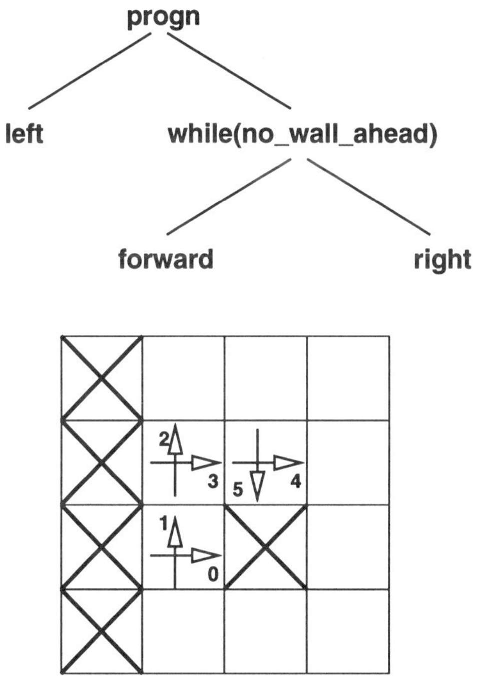

Figure 2: A Sample Program Tree and Path

For each trial the genetic program was run for fifty generations with a population of 500 individual programs. The initial population was generated randomly, but each program always began with a control statement. Thus, the initial programs consisted of at least three statements. The average size of the programs in the initial population was approximately ten statements.

The crossover rate was 0.6667 and the mutation rate was 0.001. Mutation was limited such that statements could only mutate into other statements of the same type: terminals to terminals, controls to controls, and predicates to predicates. At each generation a new population was chosen from the previous one using the stochastic remainder technique which is based on the the fitness ranking of the individuals.

Program fitness was determined by the number of columns between the robot's position at the start and termination of the program. More exacting fitness functions are possible. For example it would be reasonable to consider the number of steps the program used in determining its fitness. However, such fitness functions are likely to exert an additional influence on the code size possibly obscuring the relationships between code growth and performance that we were measuring.

2.2 The Test Cases

We used four test cases: 1) a simple genetic program, 2) a genetic program in which the test programs were reduced at each generation by editing out some of the non-functional code, 3) a genetic program in which only the initial, random population had non-functional code removed, and 4) a genetic program in which the fitness function was modified to penalize longer programs. The use of a penalizing function applies selective pressure in favor of shorter programs.

The non-functional code fell into two basic categories: code which did nothing and code which could never be executed. The first category includes any of the control statements with **no_op**'s for both arguments, **progn**'s with a **no_op** for one of the arguments, and **left-right** combinations. The second category is a result of nested conditionals. For example, consider the code:

```
if(wall_ahead,while(no_wall_ahead,X,Y),Z)
```

where X,Y, and Z are any additional statements, or lists of statements. The statements denoted by X and Y will never be executed. The **while** is only reached if there is a wall ahead in which case the **while**'s conditional is false. Hence, the **while** statement and its attendant branches can all be replaced by a single **no_op** without affecting the program's performance. The `if` statement must have a statement in its first branch making the `no_op` necessary. Thus, the edited program would become

```
if(wall_ahead,no_op,Z)
```

It is also possible for code to be functional, but to never be executed. For example, the statement Z in the above example will never be executed if the robot is always facing a wall when the `if` statement is executed. However, in such a case the non-execution of the statement Z is dependent upon the room and is not an inherent feature of the code, so the statement is not considered non-functional and is not removed.

Although these tests remove much of the non-functional code they do not guarantee that it will all be removed. The detection of all non-functional code is an unsolvable problem, since it is reducible to the program equivalence problem, which is non-recursive [Hartley Rogers, 1967]. Therefore all non-functional code cannot be removed. A more exhaustive search could find and remove more of the non-functional code, but the differences between the edited and unedited test cases in our results are definitive enough that more rigorous editing was not warranted.

Interestingly, in a few cases the removal of non-functional code did change the fitness of the edited program. This could occur if, for example, the non-functional code was also an infinite loop. By removing the loop, sections of code which followed the loop were executed where previously they had not been. Similarly, because of the 3000 step maximum, removing an early section of non-functional code might allow a later section of code to be executed. Overall, changes in fitness due to editing were very rare and lowered the fitness roughly as often as they improved it. Hence, the effects were negligible.

The fitness function used to penalize longer programs subtracted one point of fitness for every fifteen statements beyond the first fifty. We felt that the limit should be close to the length of the shortest program which could solve the task. Otherwise there is a significant risk that programs will evolve primarily for size rather than ability. Preliminary trials bore this out. When the cutoff was set too low, program performance suffered. The programs worked to avoid the penalties of a longer program rather than to produce effective solutions. Sample trials found a program of fifty-one statements which successfully crossed the room, so we rounded the the cut-off value to fifty. Further work has shown that shorter programs can successfully complete the task.

3 The Results

All of the data were obtained by averaging the results of fifty separate trials. Figure 3 shows the program length of the best programs of each generation for the four test cases: the standard genetic program (control), the genetic program with individuals edited at every generation (edited programs), the genetic program with an edited initial population (edited initially), and the genetic program with individuals subjected to selective pressure against length (selective pressure). Figure 4 shows the average program lengths averaged across all fifty trials.

For the earlier generations (roughly the first twenty) the length of the best programs are generally greater than that of the average program. However, after that the length of the best and average programs are very similar. This suggests that once the programs reach a certain minimum size a longer program is not any more likely to be effective than a shorter program. The one exception occurs when selective pressure is applied. Here the best programs are actually slightly shorter than the average because the shorter programs are not penalized for length.

Comparison of the edited and unedited programs shows that for the initial random programs almost fifty percent of their length is non-functional code. Simply removing this code in the initial generation has a lasting effect on the code size, although it does not appear to effect the basic growth curve.

We found that non-functional code is responsible for most of the code growth in the control case. The amount of functional code grows much more slowly and, unlike the total code size, its growth begins to level off. This suggests that removing the non-functional code at each generation would restrict the overall program growth. However, the edited programs show that this does not happen. Even when the programs are edited at each generation their growth curve is still very dramatic. We found that when the non-functional code is removed the GP responds by producing large programs consisting entirely of potentially functional code. However, most of this code is never executed, since it is stored in branches of the program which are never reached during execution. This non-executed code accounts for most of the edited programs' growth, while the executed code remains relatively small. More significantly, the growth in the amount of executed code appears to level off just as the growth in the functional code did in the unedited case.

In contrast, penalizing longer code through the fitness function appears to be considerably more effective at reducing code length. More importantly,

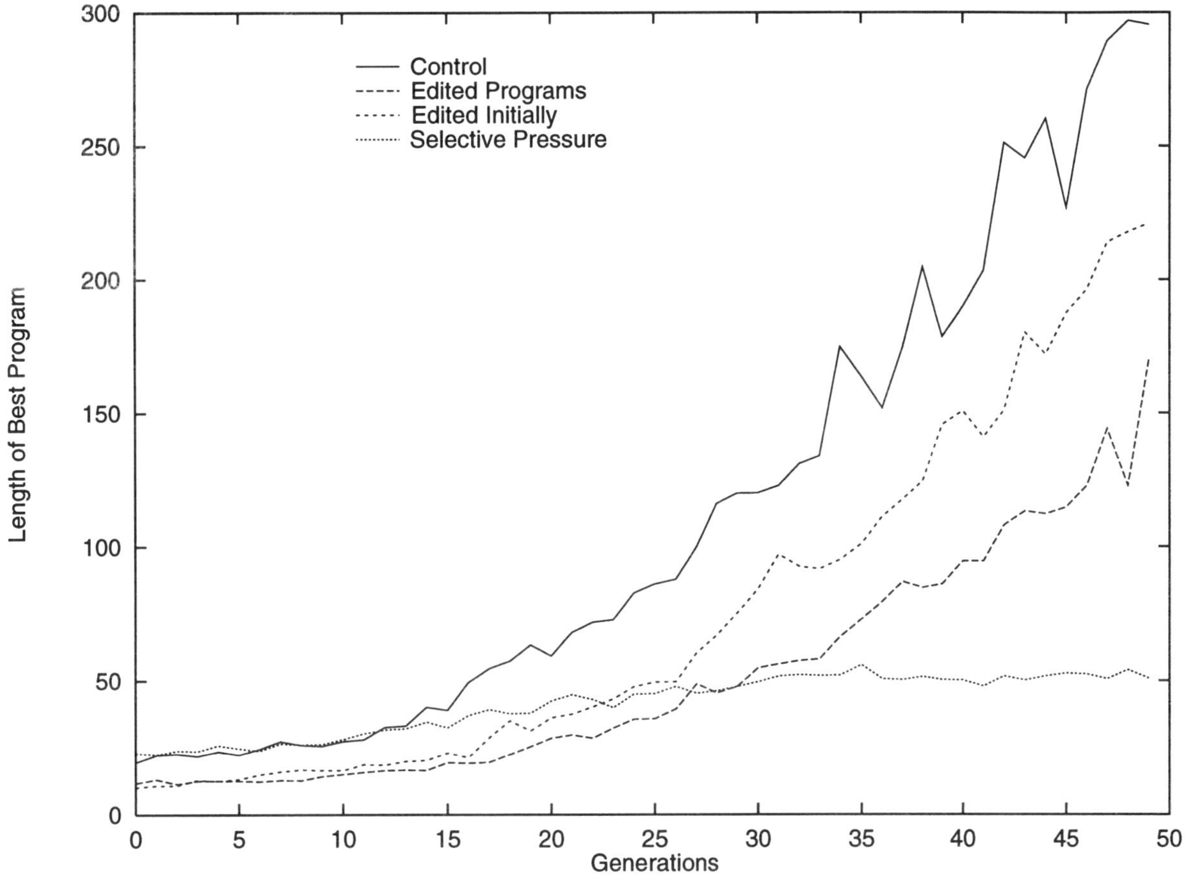

Figure 3: Best Program Lengths

such selective pressure limits the overall amount of growth.

Figure 5 shows the fitness for the best program of each generation for each of the four test cases. The test cases are labeled in the same manner as in the previous figures. All of the cases show roughly the same performance results, including the programs subjected to selective pressure. This strongly indicates that the rapid growth seen in the other test cases is not necessary for the production of effective programs.

Figure 5 also shows that the success of the programs tends to level off towards the later generations. This plateau corresponds to the convergence of the population towards a particular solution and implies that further rapid improvement is unlikely. In some cases the achieved solution is optimal, and in others it is sub-optimal, thus the average fitness over all fifty trials is slightly lower than the the maximum possible (eighteen). Figure 6 shows the average fitness for all of the programs of each generation for the four test cases. The actual fitness values are shown, so the programs subjected to selective pressure are slightly more successful in crossing the room than the data indicate. They have lost points due to their length rather than to poor performance. The data in Figure 4 show that towards the later generations (beginning around generation thirty-five) the average program length with selective pressure was between fifty and sixty. Thus, the fitnesses in Figure 6 are approximately one point less than the distance traveled.

Figure 6 also shows a leveling of the fitnesses towards the end of the run. The plateau lags slightly behind that of the best programs (Figure 5) in a fashion typical of genetic programs. In contrast, Figure 3 and Figure 4 show that no similar effect appears in the code size for the first three test cases. Even though the genetic program is no longer improving significantly (indeed in many of the trials an optimal solution has been found and no further improvement is possible) it continues to make larger and larger programs.

We hypothesize that this phenomenon is caused by crossover. Consider two identical and highly successful programs (program 1 and program 2) which

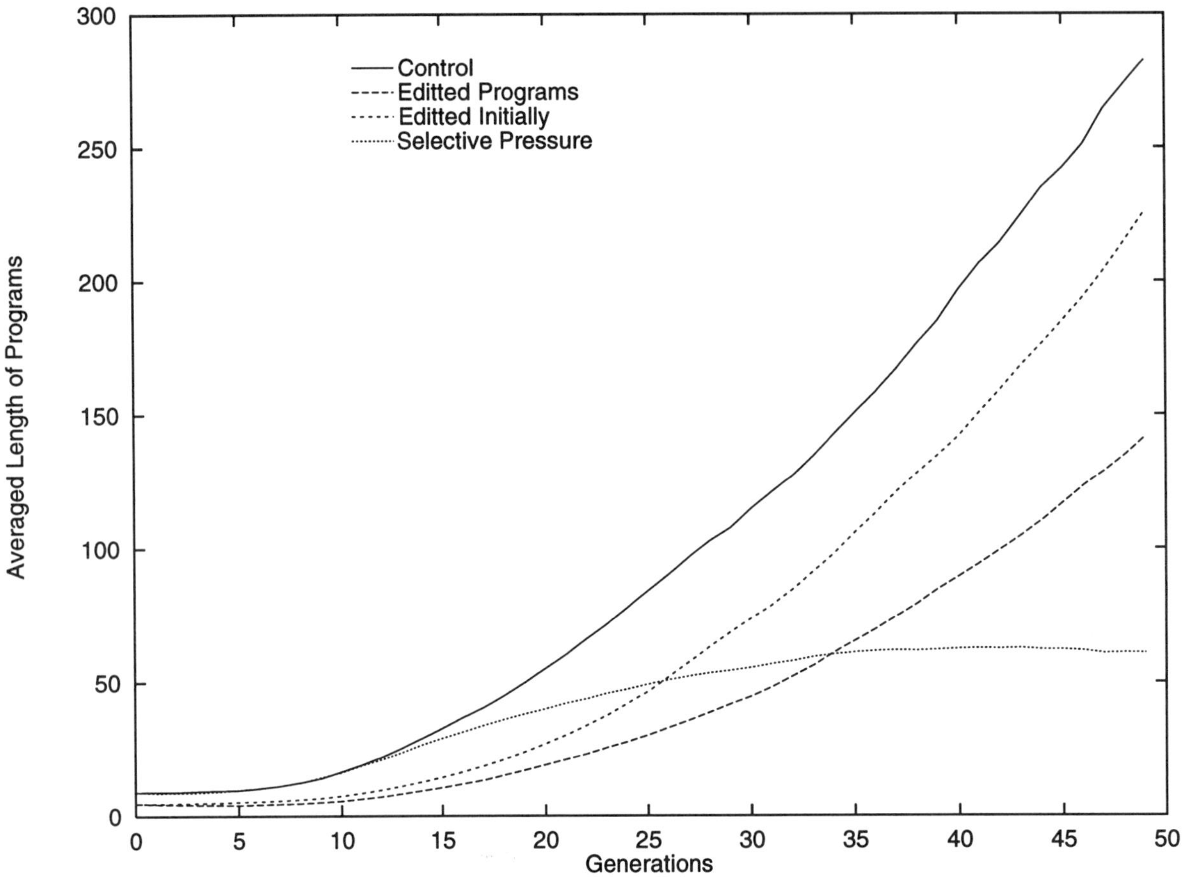

Figure 4: Average Program Lengths

are chosen to be crossed with each other. Program 1 has a relatively small branch removed which is replaced by a much larger branch from program 2. The opposite occurs in program 2; it loses the large branch and gains a smaller branch. It seems likely that after the crossover, program 1, which lost less of its original code and gained more of program 2's code, is going to have a better chance of remaining a fairly fit individual. Whereas program 2, which lost more code and only gained a little to replace it, is more likely to lose fitness. Hence, there is a slight, but distinct, statistical advantage for the longer of two programs created by crossover. This advantage would act to keep the code size growing even after an optimal program had been found, as is occurring in these trials.

This point deserves emphasizing. Even though the programs' performance has begun to level off, they are growing in size just as rapidly as ever. Removing the non-functional code is not sufficient to halt this growth. Only the use of a penalty function and selective pressures are effective, and their success in limiting code size is extremely dramatic.

Given that the penalty function is so successful, it is reasonable to consider what portion of the non-functional code it is removing. Figure 7 shows the average program lengths at each generation for the programs subjected to selective pressure and the average program lengths after these programs have been edited. Clearly, the programs still contain a considerable amount of non-functional code. However, the selective pressure is sufficient to keep the amount of this code from growing indefinitely as in the previous cases.

While we have not considered programs which are limited to less than some fixed size, we feel that some predictions may be made concerning this case. First, to avoid the possibility that even programs of the maximum size cannot complete the task it is necessary to fix the limit at a large value. Given the rate of program growth which we observed, and the fact that most of this growth is either non-functional or non-executed, it is reasonable to expect that the programs will grow rapidly towards the size limit. Additionally, most of this growth will consist of non-functional code.

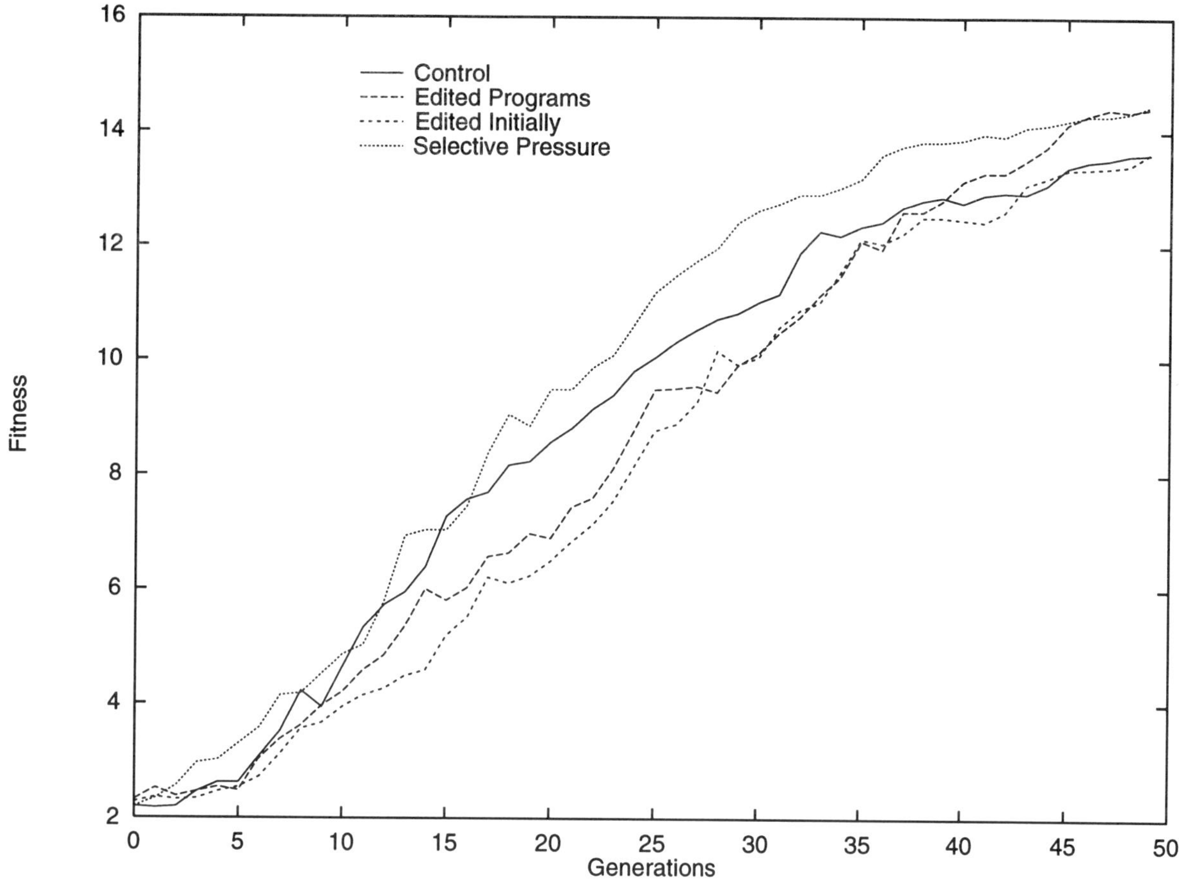

Figure 5: Best Program Fitness

The bound on program size is usually defined as a maximum depth. This restricts the program's size, but also artificially influences its shape. For programs to incorporate the maximum number of functions they must be relatively symmetrical. An additional problem stems from the fact that the enforcement of the size limit normally occurs during crossover. If a program created during crossover exceeds the accepted limit it is discarded and one of its parents is used in the next generation instead. This requires additional computation to detect oversized offspring and, perhaps more importantly, also artificially influences the programs' development. Effectively, there is a decline in the crossover rate as the average program size grows and more products of crossover are discarded.

Using a fitness function which penalizes larger programs avoids most of these problems. It is true that programs appear to rapidly approach the size at which penalties are introduced. However, because this limit is not completely restrictive, it can be set much lower than otherwise. This means that the usual program size will be much lower without running the risk of making the task unsolvable.

4 Conclusions

Code growth in GP is an issue which certainly deserves attention. In these experiments code growth was extremely rapid and gave every indication of continuing, without corresponding increases in fitness. We hypothesize that this growth is due to a slight preference for the larger of the two programs created during crossover. This would cause an overall increase in the program sizes. However, this doesn't appear to be an important factor in evolving effective programs. Incorporating a balancing preference for shorter programs should restrict this growth without harming the overall results.

Most of the code growth is attributable to increases in the amount of non-functional code. However, removing this code does not halt the growth–instead functional, but non-executed, code is substituted for non-executable code. Removing non-functional code does not appear to harm the evolution of effective programs and does lead to program

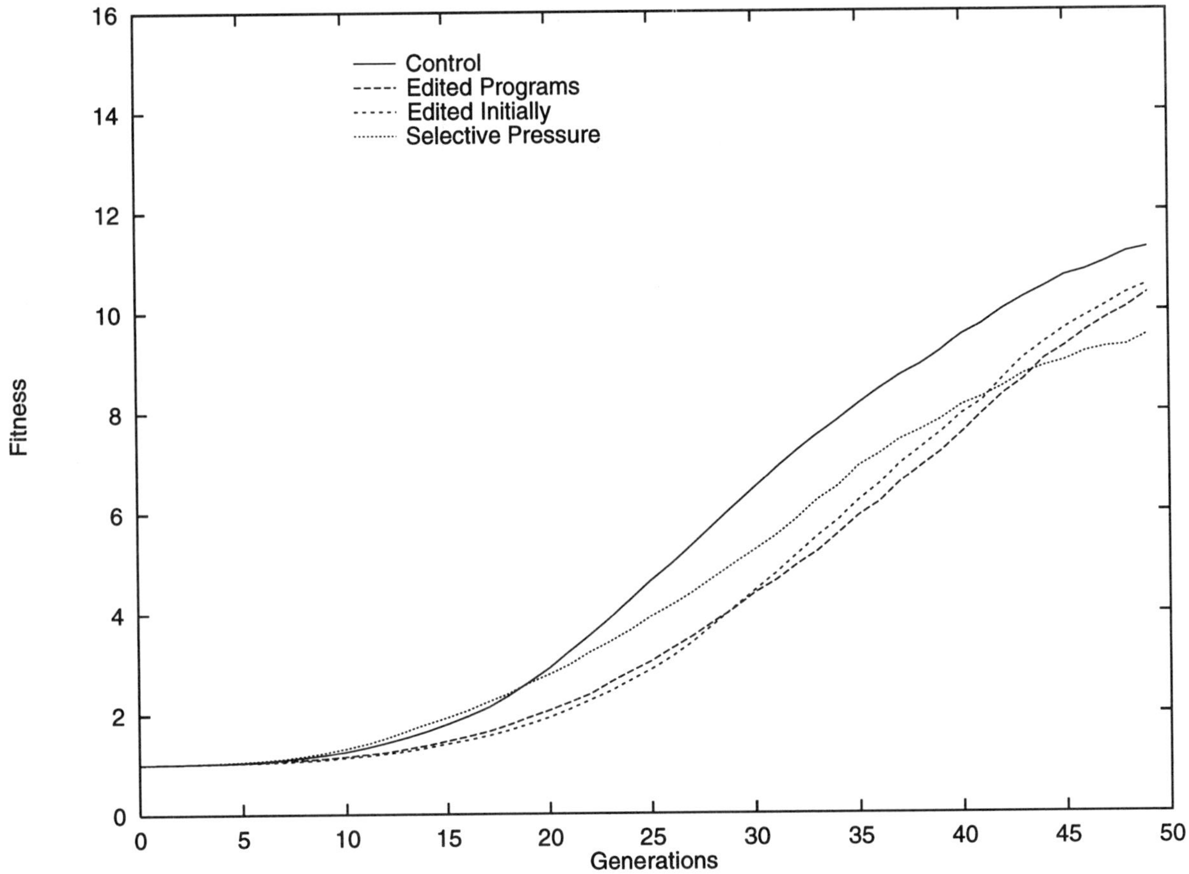

Figure 6: Average Program Fitnesses

which are considerably shorter than when editing is not used.

Using a fitness function which penalizes large programs generates selective pressure which does act to limit program growth. Furthermore, in our experiments selective pressure does not hurt the production of effective programs. Thus, it appears that selective pressure can be used to evolve programs which are both effective and efficient. It is evident that all of the code growth and non-functional code seen in our control and edited cases, as well as many other GP applications, is not necessary to the evolution of effective programs.

The one difficulty in using a penalty function is that it may require estimating the minimum size of programs which can solve the problem. However, the growth of unrestricted programs is so rapid and the effect of the penalty function is so dramatic that even an extremely liberal guess at the minimal program size should produce beneficial results. The effects of underestimating the required program size, and thereby penalizing programs which are too small to optimally solve the problem still needs to be studied, as does effectiveness of other types of penalty functions.

Other factors being equal, shorter code has the inherent benefits of requiring less resources and generalizing better than longer code. However, it is also possible that shorter code is also more amenable to the GP process. If, as Nordin and Banzhaf suggest, non-functional code shields functional code from the harmful effects of crossover, it also decreases the chances of helpful crossovers occurring. Thus, too much code growth would lead to a stagnation of the GP process.

References

[Hartley Rogers, 1967] Hartley Rogers, J. (1967). *Theory of Recursive Functions and Effective Computability.* McGraw–Hill.

[Iba et al., 1994] Iba, H., de Garis, H., and Sato, T. (1994). Genetic programming using a minimum description length principle. In Kenneth E. Kinn-

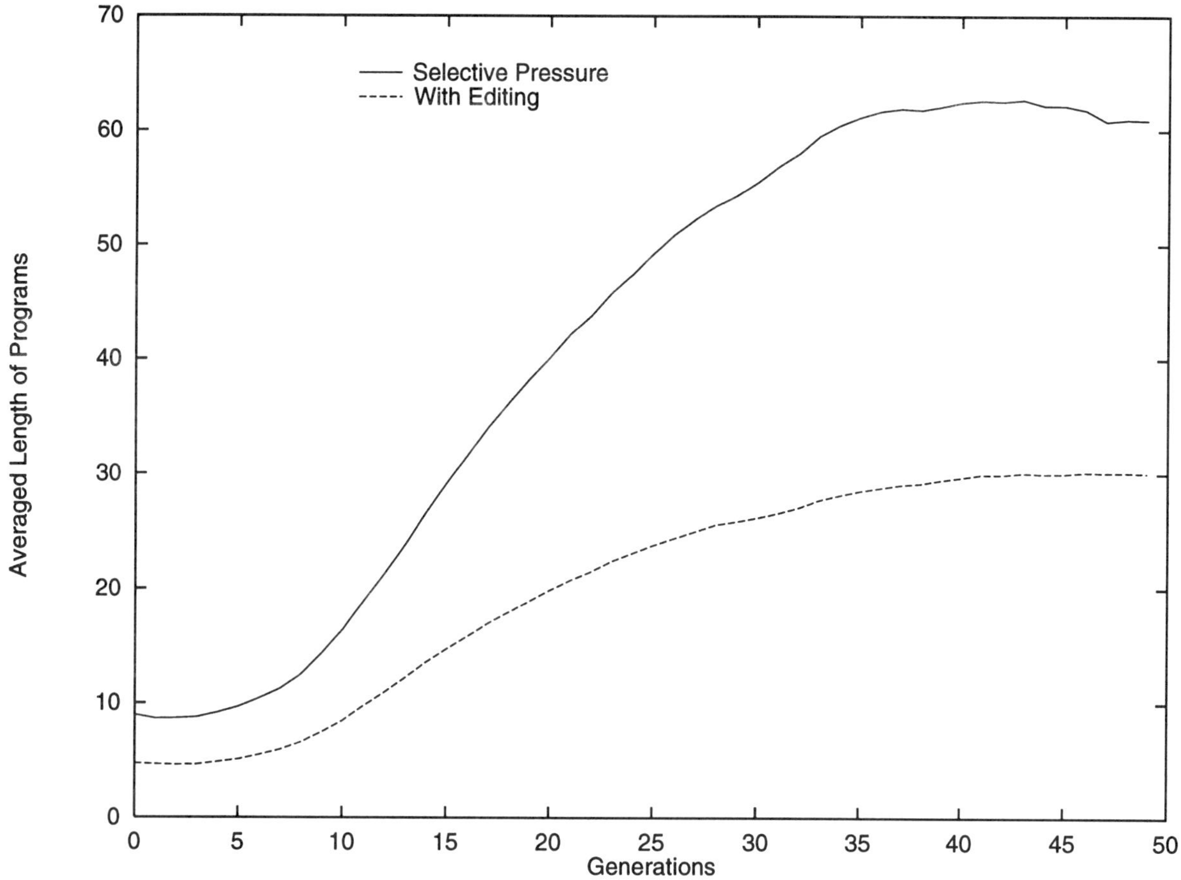

Figure 7: Non-functional Code in Penalized Programs

ear, J., editor, *Advances in Genetic Programming*, pages 265–284. Cambridge, MA: The MIT Press.

[Koza, 1992] Koza, J. R. (1992). *Genetic Programming: On the Programming of Computers by Means of Natural Selection.* Cambridge, MA: The MIT Press.

[Koza, 1994] Koza, J. R. (1994). *Genetic Programming II: Automatic Discovery of Reusable Programs.* Cambridge, MA: The MIT Press.

[Nordin and Banzhaf, 1995] Nordin, P. and Banzhaf, W. (1995). Complexity compression and evolution. In Eshelman, L. J., editor, *Proceedings of the Sixth International Conference on Genetic Algorithms*, pages 310–317. San Francisco, CA: Morgan Kaufmann.

[Soule et al., 1996] Soule, T., Foster, J. A., and Dickinson, J. (1996). Using genetic programming to approximate maximum clique. In Koza, J. R., Goldberg, D. E., Fogel, D. B., and Riolo, R. R., editors, *Genetic Programming 1996 (this conference).*

[Zhang and Muhlenbein, 1995] Zhang, B. and Muhlenbein, H. (1995). Balancing accuracy and parsimony in genetic programming. *Evolutionary Computation*, 3(1):17–38.

High-Performance, Parallel, Stack-Based Genetic Programming

Kilian Stoffel*
stoffel@cs.umd.edu

Lee Spector* †
lspector@hampshire.edu

*Department of Computer Science
University of Maryland
College Park, MD 20742

†School of Cognitive Science and Cultural Studies
Hampshire College
Amherst, MA 01002

ABSTRACT

HiGP is a new high-performance genetic programming system. This system combines techniques from string-based genetic algorithms, S-expression-based genetic programming systems, and high-performance parallel computing. The result is a fast, flexible, and easily portable genetic programming engine with a clear and efficient parallel implementation. HiGP manipulates and produces linear programs for a stack-based virtual machine, rather than the tree-structured S-expressions used in traditional genetic programming. In this paper we describe the HiGP virtual machine and genetic programming algorithms. We demonstrate the system's performance on a symbolic regression problem and show that HiGP can solve this problem with substantially less computational effort than can a traditional genetic programming system. We also show that HiGP's time performance is significantly better than that of a well-written S-expression-based system, also written in C. We further show that our parallel version of HiGP achieves a speedup that is nearly linear in the number of processors, without mandating the use of localized breeding strategies.

1 Performance of Genetic Programming Systems

Genetic programming is a technique for the automatic generation of computer programs by means of natural selection [Koza 1992]. The genetic programming process starts by creating a large initial population of programs that are random combinations of elements from problem-specific function and terminal sets. Each program in the initial population is then assessed for fitness, and the fitness values are used in producing the next generation of programs via a variety of genetic operations including reproduction, crossover, and mutation. After a preestablished number of generations, or after the fitness improves to some preestablished level, the best-of-run individual is designated as the result and is produced as the output from the genetic programming system.

The performance impact of alternative approaches to genetic programming can only be assessed by measuring performance over a large number of runs. This is because the algorithm includes random choices at several steps; in any particular run the effects of the random choices may easily obscure the effects of the alternative approaches.

To analyze the performance of a genetic programming system over a large number of runs one can first calculate *P(M,i)*, the cumulative probability of success by generation *i* using a population of size *M*. For each generation *i* this is simply the total number of runs that succeeded on or before the *i*th generation, divided by the total number of runs conducted. Given *P(M,i)* one can calculate *I(M,i,z)*, the number of individuals that must be processed to produce a solution by generation *i* with probability greater than *z*.[1] *I(M,i,z)* can be calculated using the following formula:

$$I(M,i,z) = M * (i+1) * \left\lceil \frac{\log(1-z)}{\log(1-P(M,i))} \right\rceil$$

Koza defines the minimum of *I(M,i,z)* as the "computational effort" required to solve the problem with the given system.[2]

2 High Performance Genetic Programming

Genetic programming is a computationally intensive enterprise, and recent work has focused on ways in which high performance computation techniques can increase the set of problems to which it can be applied. Since the underlying genetic algorithm appears to be parallelizable, and since several parallel implementations of string-based genetic algorithms have been described in the literature, there has been

[1] For the analyses in this paper a value of z=99% is always used.

[2] The *P(M,i)* and *I(M,i,z)* measures were developed by Koza and are discussed on pages 99 through 103 of [Koza 1994].

particularly strong interest in parallelizing genetic programming systems.

Most previous high performance genetic programming systems have either been wedded to particular computer architectures, or have been hampered by high overheads in distributing fitness evaluation and in coordinating reproduction across parallel machines. For example, Koza and Andre have implemented an S-expression-based genetic programming system on a network of transputers and have achieved considerable speed improvements over serial implementations [Koza and Andre 1995]. Their approach relies on a hardware base that is less expensive than parallel supercomputers but is nonetheless somewhat exotic. They used local breeding strategies with migration; they report that this reduces computational effort as compared with a panmictic (globally interbreeding) population on the problems they tried. Others have reported similar improvements from the use of localized breeding strategies, both in genetic programming and in string-based genetic algorithms (e.g., [Collins and Jefferson 1991]). It is not yet clear, however, that particular localized breeding strategies are always beneficial, and it is important that high-performance genetic programming systems not be swamped with communication overheads when less localized or global (panmictic) strategies are used.

Nordin and Banzhaf have developed a "compiling" genetic programming system that directly manipulates SPARC machine code [Nordin and Banzhaf 1995]. Through a combination of compact fixed-length representations, simplified memory management, and non-interpreted program execution they have achieved dramatic speed improvements over traditional genetic programming systems. Their system depends on details of the SPARC architecture and would require significant redesign for other machines.

Juillé and Pollack have developed a SIMD (single instruction multiple data) parallel genetic programming system that represents programs as S-expressions but "precompiles" them into programs for a virtual stack machine prior to execution [Juillé and Pollack 1995]. They note that the variation in S-expression sizes across the population can introduce overhead, and they use local breeding strategies to reduce interprocessor communication costs.

S-expression-based program representations are responsible for several of the performance limitations of previous systems. These representations make strong demands on a system's memory allocation subsystems and they are expensive to manipulate and to move between processors. In addition, less prior research has been conducted on the optimization and parallelization of S-expression interpreters than on techniques for more common models of computation.

These limitations of previous systems lead us to consider the idea of genetic programming with small, linear programs that can be executed on stack-based virtual machines for which portable, high performance interpretation techniques have already been developed. We were encouraged by previous work that explored the performance benefits of linear program representations (e.g., [Keith and Martin 1994]). We were further encouraged by Perkis's work on stack-based genetic programming [Perkis 1994], in which he used linear programs that were executed on a stack-based virtual machine. In contrast to the work of Keith and Martin and of Juillé and Pollack, Perkis performed string-based genetic operations (e.g. string-based crossover) directly on the linear programs. The safety of the resulting programs was guaranteed by specifying that all functions take their arguments from the stack, and that function calls that occur with too few items on the stack simply do nothing. Using this scheme, Perkis reported lower computational efforts than were required using traditional S-expression-based genetic programming.

In HiGP we have combined elements of the stack-based program frameworks of Perkis and of Juillé and Pollack, the linear chromosomes of string-based genetic algorithms (also used in Nordin and Banzhaf's and Perkis's systems), the efficient, low level execution model of Nordin and Banzhaf, a machine independent virtual stack machine, and high-performance parallel programming techniques.

3 The HiGP Virtual Stack Machine

HiGP programs are executed on a virtual machine that is similar to a pushdown automaton. The virtual machine consists of three components: an input tape containing a linear program, a pushdown stack, and a finite-state control unit. The contents of the input tape are restricted to a small set of words that have been defined as HiGP operators. The contents of pushdown stack are restricted to double precision floating point numbers. The finite-state control unit reads the input tape and executes, for each word, the function call for the corresponding operator. The operators may perform arbitrary computations and manipulate the values on the stack. They may also reposition the read head on the input tape; this allows for the implementation of conditionals and loop structures. Return values are generally read from the top of the stack at the end of program execution.

The system includes two basic stack operators, `pop` and `dup`. The `pop` operator removes the topmost element from the stack, while the `dup` operator pushes a duplicate of the top element onto the stack. The system also includes a family of `push` operators that correspond to the terminal set in a traditional genetic programming system; each `push` operator pushes a single pre-determined value onto the stack.

The system also includes a `noop` operator that does nothing. This is necessary because all programs in the system have the same length, and because we do not wish to predetermine the number of actual problem-solving operators that should appear in solution programs. With the inclusion of the `noop` operator the fixed program size becomes a size

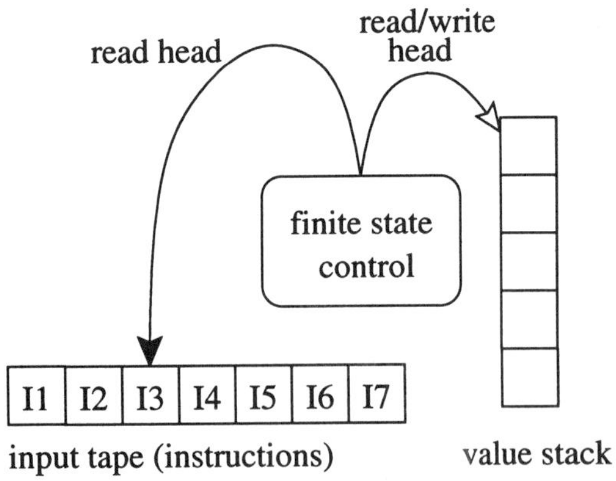

Figure 1: The virtual stack machine

limit, analogous to the depth limits used in S-expression-based genetic programming systems. "Shorter" programs are encoded by filling in extra program steps with noops. Note that this provides the same flexibility with respect to program size as do S-expression-based genetic programming systems. If one suspects that a large program may be required, then one can set the program size to an arbitrarily large number. In such a case, however, the system may still produce efficient, parsimonious programs; it may do so by producing programs that consist mostly of noops.

Any additional, problem-specific operators must take their input values from the stack and must push their results back onto the stack. When there are not enough values on the stack for an operator it is skipped by the finite-control unit and the stack remains untouched (as in [Perkis 1994]).

A simple example may help to clarify the operation of the virtual stack machine. Consider the following program:

```
push-x noop push-y * push-x push-z noop
- + noop noop
```

The noops in this program have no effect and the remainder is equivalent to the Lisp expression:

```
(+ (* x y) (- x z))
```

and to the C expression:

```
(x * y) + (x - z)
```

For the test examples presented in this paper we used only the four basic arithmetic operators: + (addition), - (subtraction), * (multiplication) and % (protected division [Koza 1992]). Although conditionals, loops, and other control structures were not used for the examples in this paper, several have been implemented for HiGP. For example, one implemented control structure reads a value from the stack and executes the next instruction if the value, when truncated, is zero; otherwise it jumps over the next instruction. Another control structure jumps over a number of instructions; the number is obtained by truncating the value on the top of the stack. These control structures, along with several other features of our virtual stack machine, are similar to those of the FORTH programming language [Brodie 1981]. The extension of the virtual stack machine with additional operators and control structures is trivial; the FORTH language provides examples of what such extensions might look like.

4 Genetic Programming in HiGP

One nice feature of the HiGP program representation is that one can use the same genetic operators that are used in string-based genetic algorithms. Each gene in the chromosome represents an operator for the virtual machine. Since the operators communicate with one another only by means of the stack, and since stack-underflow is handled gracefully (by ignoring offending operators), every possible ordering of operators represents a syntactically correct program. Arbitrary string-based manipulations of the chromosomes are therefore acceptable. In fact, it would be possible to use many existing, off-the-shelf genetic algorithm packages to perform the genetic operations on HiGP programs. In the current implementation we use three standard genetic operators: reproduction, single-point crossover, and point mutation. The population is initialized by setting each gene in each chromosome to a random member of the operator set.

An additional consideration for genetic programming is the implementation of ephemeral random constants [Koza 1992]. In HiGP constants are not included directly in programs (chromosomes) because a small, fixed gene size is important for optimized, high-performance execution and manipulation of HiGP programs. Because some applications call for constants with representations that are larger than the optimal gene size (for example, double precision floating-point constants in an application with a small number of operators and constants) HiGP maintains a table in which constants are stored. Inside the programs the constants are represented by their positions in the table. The "execution" of a constant causes the corresponding value from the table to be pushed onto the stack. The ephemeral random constants in the table are initialized at the beginning of a run and remain unchanged during the entire run; this differs from Koza's implementation, in which new constants may be generated by the mutation operator at any point in a run. Our point mutation operator can transform one constant into another, or into an operator, but the set of available constants does not change during a run. The size of the constant table is limited only by the size of the alphabet for the virtual stack machine, which is in turn determined by the operator word-size. Because the word-size is a variable parameter in the current implementation, any number of ephemeral random constants can be supported.

5 Advantages of the HiGP approach

HiGP has several advantages over traditional S-expression-based genetic programming systems. HiGP programs are of fixed length and our experience is that they can be quite small. We have solved several problems with programs only 32 operators in length, with less than a byte required for each operator. This makes it easy to optimize memory allocation, to create populations with millions of individuals, and to efficiently pass programs between processors in multiprocessor architectures. Load-balancing during fitness evaluation is also eased by the fixed program length. These factors combine to allow for nearly linear speedups with added processors, even when a global breeding strategy is used. Of course many problems will require longer programs, but all components of the system should scale at worst linearly with program length.

In addition, the evolutionary dynamics of systems of linear programs appear in some cases to be superior to those of S-expression-based genetic programming (e.g., in [Perkis 1994] and below). We do not yet have a theoretical explanation of why this should be so.

6 Example: Symbolic Regression

The goal of the symbolic regression problem, as described in [Koza 1992], is to produce a function that fits a provided set of data points. For each element of a set of (x, y) points, the program should return the correct y value when provided only with the x value. For our experiments we obtained our data points from the equation $y = x^9$. One can view the task of the genetic programming system as that of "rediscovering" this formula from the data points used as fitness cases. We used 20 fitness cases, with randomly selected x values between -1 and 1.

We used a function set consisting of the two-argument addition function +, the two-argument subtraction function -, the two-argument multiplication function `*`, and the two-argument protected division function `%`. We used a single "terminal" operator `push-x` that pushes the x value onto the stack. We did not use ephemeral random constants or any other constants. We ran HiGP with tournament selection (tournament size = 5), a 90% crossover rate, a 10% reproduction rate, and no mutation. We conducted 100 runs, each with a population size of 500, and each for a maximum of 30 generations. The length of each individual program was set to 32 operators.

We also conducted 100 runs of the lil-gp S-expression-based genetic programming system [Zongker 1995] on this problem. We used parameters that were as similar as possible to those used for HiGP: function set={`+, -, *, %`}, terminal set={`X`}, tournament selection (size=5), 90% crossover, 10% reproduction, no mutation, 100 runs, population=500, generations=30. The depth limit presented a more difficult problem since the HiGP concept of *length* limit and the lil-gp concept of *depth* limit are quite different. Lowering the depth limit in lil-gp improves its execution speed considerably, but in some cases this may make it more difficult or impossible to find correct programs. It is not clear, in general, how the depth limit influences computational effort. For this reason we ran lil-gp with a range of depth limits. We used a depth limit of 4 for the low end of the range; this is the lowest limit for which lil-gp was able to find any correct programs at all. We used 17 for the high end of the range; this is the default for lil-gp and, according to our data, much larger than optimal.

We believe that lil-gp, which is implemented in C, is a well-written program that provides a reasonable comparative benchmark for S-expression-based genetic programming. One way to establish the relative time performance of the two systems would be to compare the times that each system takes for a complete set of 100 runs. However, the number of generations computed by the two systems are different, as indicated by the computational effort results reported below. To remove this factor from the time performance comparisons we computed the time used by each system to evaluate one generation. For HiGP the execution time per generation is nearly constant, because of the fixed length of the programs and because of the absence of conditionals or looping operators in the test problem. For lil-gp the time for the execution of one generation can change significantly as the structures of the S-expressions evolve. We therefore used the mean value over 100 runs for all evaluated generations.

Our results, obtained on a Sun SPARC 5, are summarized in Table 1. HiGP was clearly superior both with respect to

Table 1: Results for HiGP and for lil-gp using different depth limits

		time/gen.	*comp. effort*
HiGP		0.08	32,729
	depth		
lil-gp	4	*0.16*	212,521
	5	0.18	135,933
	6	0.21	71,481
	7	0.23	93,013
	8	0.25	82,210
	9	0.28	66,323
	10	0.30	*65,216*
	11	0.30	70,116
	12	0.34	77,030
	13	0.36	76,404
	14	0.35	71,481
	15	0.38	92,436
	16	0.39	90,379
	17	0.41	79,737

computational effort and with respect to time performance. The best computational effort for lil-gp was obtained with a depth limit of 10, but even this effort (65,216) was nearly double that required by HiGP (32,729). In addition, the time performance of lil-gp with a depth limit of 10 was quite poor; the time required per generation was 0.30 seconds, while HiGP required only 0.08 seconds per generation. The best time performance for lil-gp was obtained with a depth limit of 4, but even this time (0.16 seconds) was double that of HiGP (0.08 seconds). In addition, the computational effort required by lil-gp with a depth limit of 4 was quite poor; it was 212,521, which is over six times greater than the effort required by HiGP (32,729).

7 Parallel HiGP

The parallel version of HiGP was developed to run on MIMD (multiple instruction multiple data) supercomputers, but it was not tailored for any one particular system. We used the MPI (message passing interface) communication library, which is available for many different parallel computer systems. This greatly simplifies the process of porting HiGP to various different computer systems. The system was written within the SPMD (single program multiple data) paradigm; all nodes of the parallel system execute identical programs but on different data.

The population is distributed evenly across the processors, with each processor maintaining its own local copy of a fitness table for the entire population. This allows much of the breeding process (including selection) to occur locally, along with fitness evaluation.

During the fitness evaluation phase, all processors evaluate the fitness of their individuals in parallel. Because each processor hosts the same number of programs, and because all of the evolving programs are the same length, all of the processors complete fitness evaluation at nearly the same time. Load balancing may be more complicated for applications with function sets that include functions of widely varying runtimes, and for applications that include conditional and looping operators in their function sets. At the end of the fitness evaluation phase all computed fitness values are exchanged between processors and all fitness tables are updated.

During the breeding phase each processor independently produces its own segment of the next generation. The choice of which genetic operators to apply is determined locally, using the percentages specified for the whole population. For example, if it has been specified that 60% of the next generation should be produced by crossover, 30% by reproduction, and 10% by mutation, each processor can choose operators according to these percentages locally and the correct global percentages will result. Selection is also performed locally, using the local fitness tables to perform either fitness-proportionate or tournament selection. Interprocessor communication is only required to exchange the actual individuals that have been selected for use in genetic operations (and then only when they are not local).

Notice that the breeding strategy is global (panmictic) even though selection occurs locally on individual processors. The need for interprocessor communication is minimized, and there are low communication overheads both for the distribution of fitness values and for the transportation of the small linear programs. Notice also that it would be trivial to modify HiGP to make use of localized breeding strategies; although little would be saved with respect to communication costs, improvements in computational effort might result for certain problems.

nodes	time	Efficiency
1	91.41	—
2	47.12	96.9
4	27.32	83.6
8	16.82	67.9
16	10.89	53.4

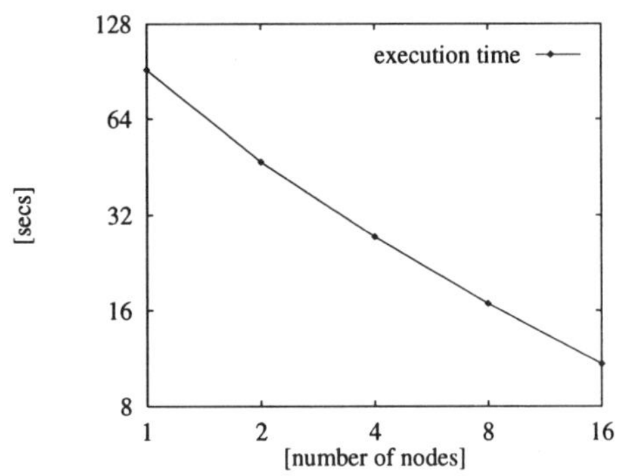

Figure 2: Parallel execution times for one run on up to 16 processors on an IBM SP2.

The HiGP strategy leads to a very efficient implementation, as Figure 2 shows. We were able to execute a program which took 91.41 seconds on one single node of the IBM SP2 nearly 9 times faster on 16 nodes in 10.89 seconds. This leads to an efficiency (number of times faster divided by number of nodes) of more than 50%, which is good for a communication-intensive program. Note also that the graph in Figure 2 is nearly linear.

8 Conclusions

HiGP is a new high-performance genetic programming system that combines techniques from string-based genetic algorithms, S-expression-based genetic programming systems, and high-performance parallel computing. HiGP is fast, flexible, and easily portable. It also has a clear and efficient parallel implementation that is not tied to any particular parallel computer architecture.

HiGP manipulates and produces linear programs for a stack-based virtual machine, rather than the tree-structured S-expressions used in traditional genetic programming. This appears to have several benefits; along with the optimizations that it allows for memory management and parallelization, it appears to decrease the computational effort required to produce a correct program, at least for the regression problem that we presented.

We showed that HiGP's time performance is significantly better than that of lil-gp, a well-written S-expression-based system which is also written in C. We further showed that our parallel version of HiGP achieves a speedup that is nearly linear in the number of processors, without mandating the use of localized breeding strategies.

Acknowledgments

The research described in this paper was supported in part by grants from ONR (N00014-J-91-1451), AFOSR (F49620-93-1-0065), and ARPA contract DAST-95-C0037.

Bibliography

Brodie, L. 1981. *Starting FORTH.* Prentice Hall.

Collins, R.J. and D.R. Jefferson. 1991. Selection in Massively Parallel Genetic Algorithms. In *Proceedings of the Fourth International Conference on Genetic Algorithms,* edited by R.K. Belew and L.B. Booker. pp. 249–256. San Mateo, CA: Morgan Kaufmann Publishers.

Juillé, H., and J.B. Pollack. 1995. Parallel Genetic Programming on Fine-Grained SIMD Architectures. In *Working Notes of the AAAI-95 Fall Symposium on Genetic Programming.* Eric V. Siegel and John R. Koza, editors. AAAI Press.

Keith, M.J., and M.C. Martin. 1994. Genetic Programming in C++: Implementation Issues. In *Advances in Genetic Programming,* edited by Kenneth E. Kinnear, Jr. pp. 285–310. Cambridge, MA: The MIT Press.

Koza, J.R. 1992. *Genetic Programming: On the Programming of Computers by Means of Natural Selection.* Cambridge, MA: The MIT Press.

Koza, J.R. 1994. *Genetic Programming II: Automatic Discovery of Reusable Programs.* Cambridge, MA: The MIT Press.

Koza, J.R., and D. Andre. 1995. Parallel Genetic Programming on a Network of Transputers. Stanford University, Department of Computer Science Technical Report CS-TR-95-1542.

Nordin, P., and W. Banzhaf. 1995. Evolving Turing-Complete Programs for a Register Machine with Self-modifying Code. In *Proceedings of the Sixth International Conference on Genetic Algorithms (ICGA-95).* Larry J. Eshelman, editor. San Francisco, CA: Morgan Kaufmann Publishers, Inc.

Perkis, T. 1994. Stack-Based Genetic Programming. In *Proceedings of the 1994 IEEE World Congress on Computational Intelligence,* pp. 148–153. IEEE Press.

Zongker, D., and B. Punch. 1995. *lil-gp 1.0 User's Manual.* WWW: http://isl.cps.msu.edu/GA/software/lil-gp, or via anonymous FTP to isl.cps.msu.edu, in the directory "/pub/GA/lilgp".

Search Bias, Language Bias and Genetic Programming

P. A. Whigham

Department of Computer Science
University College, University of New South Wales
Australian Defence Force Academy
Canberra ACT 2600 AUSTRALIA
Email: paw@csadfa.cs.adfa.oz.au

Abstract

The use of bias with automated learning systems has become an important area of research. The use of bias with evolutionary techniques of learning has been shown to have advantages when complex structures are evolved. This is especially true when the semantics of the evolving population of structures is not explicitly represented or analysed during the evolution. This paper describes a framework which brings together two types of bias, namely search bias (the way new structures are created) and language bias (the form of possible structures that may be created). The resulting system extends genetic programming (GP) by allowing declarative bias with both the form of possible solutions that are created and the method by which they are transformed.

1 Introduction

"All our experiences in AI research have led us to believe that for automatic programming, the answer lies in *knowledge*, in adding a collection of expert rules which will guide code synthesis and transformation."[Lenat, 1984]

Douglas Lenat, 1983

The use of bias to assist learning has pervaded the field of computer science for many years[Michalski, 1984, Pazzani and Kibler, 1992, Mitchell et al., 1984]. Typically the bias represented a preference over selecting one hypothesis from another given many possible situations. This form of bias was often called *heuristic knowledge*. These heuristics were used in two ways; to guide a system when no solution was known to exist and to restrict the search space when the solution was known to be computationally intractable.

1.1 Forms of Bias

Bias is the set of all factors that influence the selection of a particular hypothesis. There are three major kinds of bias, as follows.

1. **Selection Bias:** The method used to select one hypothesis over another is termed the selection bias. In terms of an evolutionary learning framework this bias is the result of some type of probabilistic fitness selection scheme. Note that any measure affecting the fitness of an individual, such as a parsimony metric, is also a selection bias.

2. **Language Bias:** The hypotheses that may be created are usually restricted in the forms that they represent. This usually occurs because of a restricted description language. In terms of GP this bias is reflected in the function and terminal sets.

3. **Search Bias:** The method by which a learning system transforms one hypothesis into another may be viewed as a search through the space of possible hypotheses. Search bias refers to the factors that control this transformation. In terms of GP this bias is represented by the crossover and mutation operators and the maximum depth of any created program tree.

1.2 Theoretical Arguments for Bias

The prospect of building inductive systems which could create any reasonable generalisation, with limited knowledge, was set back in 1984 by the theoretical results of Valiant[Valiant, 1984]. He introduced the notion of example complexity, which defined the number of observations required to reach a specified level of certainty (confidence) that an induced hypothesis is likely to be correct. He concluded that the example complexity was proportional to the logarithm of the size of the hypothesis space. The implication of this work was that, to ensure a problem remained tractable, the hypothesis space should be restricted.

The No Free Lunch Theorem has demonstrated that all learning systems will perform poorly over some problems. This theorem implies that to solve a function optimisation problem a search algorithm should be tailored to the salient features of the problem [Wolpert and Macready, 1996]. This may be viewed as an argument for allowing bias in the language and search methods, thus allowing a learning system to be shaped towards the important features of a problem. Further, if the bias can be represented explicitly (i.e. declaratively) then the user may alter the bias as more information about the solution becomes available, without altering the underlying learning system.

1.3 Grammatically Biased Learning

The concept learning system LEX [Mitchell et al., 1984] used a grammar to represent a generalised description of possible forms of a solution. This system was designed to use heuristic knowledge to aid learning about the domain of symbolic integration. The grammar described the form of legal problem states that could occur during learning. The *sentential forms* derivable from the grammar represented possible generalisations that were used to modify the heuristic knowledge.

The concepts used in LEX were extended to allow the concept description language to be modified during the learning phase [Utgoff, 1986]. This program (named STABB) had the ability to *learn its own inductive bias.* Utgoff assumed that bias could be represented by restricting the hypothesis space defined by an incomplete concept description language.

From the domain of Inductive Logic Programming (ILP) the system GRENDEL has been defined that uses an *antecedent description language*[Cohen, 1993]. This language is used to generate a set of literals as a bias when learning first-order descriptions. GRENDEL used an extension of FOIL [Quinlan, 1990], a first order inductive learning system that used horn clauses as the description language. GRENDEL used an extension of context-free grammars to represent the possible literals that could be expressed. In addition *sentential forms* of the language could be used to represent the body of a partially generated clause. This work brought together many forms of knowledge that could be incorporated under the one representation.

Formal grammars have been used to represent the evolution of a hardware design [Mizoguchi et al., 1994]. This approach used a hardware description language, SFL(Structured Function description Language), which was defined as a set of production rules in a context-free grammar. This work closely parallels the system described here, since the derivation trees were maintained as the representation of the population. However, there was no method for representing an explicit search bias.

The author has presented work using grammars to represent the language bias with GP [Whigham, 1995a, Whigham, 1995b]. However, this work did not use the framework of language and search bias developed in this paper.

Recently there has been increasing interest in using formal grammars to represent bias in an evolutionary framework. Wong et al.[Wong and Leung, 1995] in his LOGENPRO system demonstrated that a logic grammar could be used to combine GP and ILP. The emphasis of this work was to show that a grammar could be used to represent many different forms of learning. This work was similar to the work presented here, in that the crossover and mutation operators applied to derivation trees, rather than the symbols explicitly representing the programs (i.e. GP). However, the derivation trees were not maintained in the population. This led to issues of ambiguity when attempting to parse a program to create the derivation tree representing this program. Additionally, there was no way to represent an explicit search bias with this work. Similar work has been described [Gruau, 1996] which suffered from the same difficulty arising from reconstructing the derivation trees when applying crossover and mutation.

This paper defines a program induction system based upon the genetic programming paradigm[Koza, 1992]. The system allows explicit language bias to be described using a context-free grammar. The derivation trees associated with generated programs are maintained and used to control the subsequent structure of programs when they are transformed by genetic operators. Hence, the problems of ambiguity with a grammar are removed both in terms of parsing a program to create a derivation tree and in forcing the user to define an unambiguous grammar. The genetic operators are defined in such a way as to allow an explicit search bias (in terms of which parts of a program are most likely to be explored) and to allow directed modification of a program, based on the current structure of the derivation tree that forms the program. A context-free grammar is used to generate an initial population. The declarative structure offered by a grammar allows a clear and easily modifiable definition of language bias when searching for program solutions. The defined genetic operators allow a variety of search bias strategies to be imposed when evolving programs.

2 The Structure of a Biased GP System

The structure of the program induction system is shown in Figure 1. Program derivation trees are selected based on their fitness. These trees are probabilistically modified by the specified genetic operators, thereby creating the next generation of programs. The remainder of this paper will describe how bias may be introduced in both the initial creation of programs and during the search for new programs using various genetic operators.

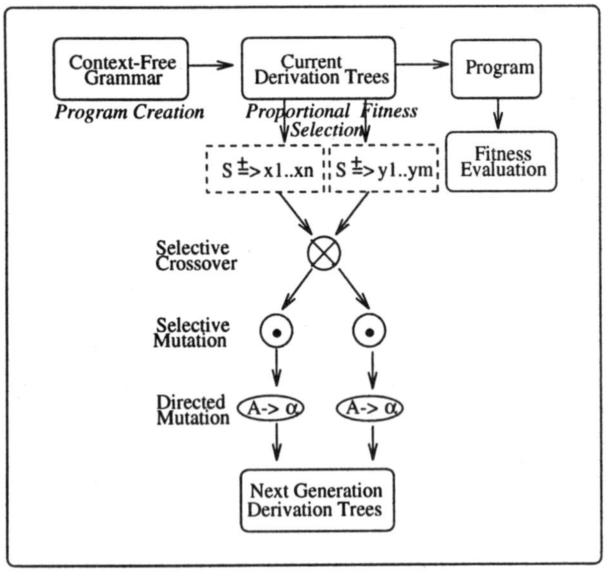

Figure 1: The Structure of a Biased GP System.

3 The Initial Population Bias

A grammar may be used as a generator of sentences to represent the initial population of programs. The goal of this process is to create a mixture of different programs in terms of overall size, shape (referring to the program structure viewed as a parse tree) and functionality. The notion is to cover a large area of the possible program search space and to include all of the basic program components that will contribute to a solution. Each member of the population is forced to be structurally unique. This ensures that a measure of syntactic diversity (although semantic diversity is not guaranteed) is initially defined.

To create a population of programs using a context-free grammar the following steps are required.

1. Label each production to indicate the *minimum depth of derivation tree* to create a string composed only of terminals. This concept will be referred to as min-depth-tree.
2. Specify the number of programs to be created. Associated with this statement is a maximum depth of derivation tree allowed during the creation of the programs. Normally there will be a range of these statements to allow a variety of differently shaped programs to be created.
3. Generate the programs by randomly selecting productions, commencing with the *start symbol* S and guided by min-depth-tree.

3.1 Labelling the Productions

Each production of the form $A \rightarrow \alpha$, where $A \in N, \alpha \in \{N \cup \sum\}^*$, is assigned a value for min-depth-tree. This represents the smallest depth of derivation tree, from A, that creates a string which is composed only of terminals. For example, from the nonterminal, A, min-depth-tree is determined by the maximum depth of the derivation tree, rooted in A, created using the following derivation steps.

$$A \stackrel{+}{\Rightarrow} x_1 x_2 \ldots x_n \text{ where } x_1 x_2 \ldots x_n \in \sum^*$$

The following algorithm is used to label min-depth-tree for each production, P, from a grammar, G.

1. For each production of the form $A \rightarrow x_1 x_2 \ldots x_n$, where $x_1 x_2 \ldots x_n \in \sum^*$, min-depth-tree = 1.
2. **WHILE** all productions from P have not been assigned **DO**
   (a) For each unassigned production, $A \rightarrow \alpha$, where $\alpha \in \{N \cup \sum\}^*$,
   For each nonterminal $B \in \alpha$, if a production $B \rightarrow \beta$ has min-depth-tree set, assign the depth for $A \rightarrow \alpha$ as the maximum of min-depth-tree$\{B\} + 1$.

For example, the following grammar, G_{6m}, has productions labelled with min-depth-tree, shown as $< depth >$ before each production.

$$
\begin{array}{l}
G_{6m} = \\
\quad \{S, \\
\quad\quad N = \{B, T\}, \\
\quad\quad \sum = \{and, or, not, if, a0, a1, d0, d1, d2, d3\}, \\
\quad\quad P = \\
\quad\quad\quad \{< 4 > S \rightarrow B \\
\quad\quad\quad\quad < 3 > B \rightarrow and\ B\ B \mid or\ B\ B \\
\quad\quad\quad\quad < 3 > B \rightarrow not\ B \mid if\ B\ B\ B \\
\quad\quad\quad\quad < 2 > B \rightarrow T \\
\quad\quad\quad\quad < 1 > T \rightarrow a0 \mid a1 \mid d0 \mid d1 \mid d2 \mid d3 \\
\quad\quad\quad \} \\
\quad \}
\end{array}
$$

3.2 Creating the Initial Population of Programs

The initial program population is defined by a set of parameters which specify the number of programs to be created with some upper bound on the depth of the generated derivation trees.

Initial Population Creation	
Parameters	Specifications
POPULATION SIZE	500
CREATE MAXIMUM DEPTH 4	100
CREATE MAXIMUM DEPTH 5	200
CREATE MAXIMUM DEPTH 6	200

Table 1: An Example of Defining the Initial Population.

Table 1 shows an example of specifying an initial population. The **CREATE MAXIMUM DEPTH** parameter is used to specify the number of programs to be created with various maximum depths of derivation tree. The following algorithm is used to create the population, based on the parameter **CREATE MAXIMUM DEPTH** depth.

1. Select the start symbol S. Label this as the current nonterminal A
2. Select, at random, a production $P_1 \in P$ of the form $A \rightarrow \alpha$ with min-depth-tree $\leq$ depth.
3. Select each nonterminal $B \in \alpha$ and label B as the current nonterminal. $depth = depth - 1$. Repeat steps 2 and 3.

3.3 Selecting Productions with Further Bias

The previous algorithm assumes that there is equal probability of selecting any valid production from a particular nonterminal. There may be a desire to express a bias towards the generation of certain program strings in the initial population. This bias is explicitly represented by associating a *merit weighting* with each production. The probability of a production being selected during the creation of the initial population is now directly proportional to the merit weighting of each production. If we represent the merit weighting for a production from some nonterminal A as w_A, then the probability of selecting this production can be defined as

$$\frac{w_A}{\sum_A w_A}$$

where $\sum_A w_A$ represents the sum of the probabilities of all productions from A which satisfy the min-depth-tree requirements. This proportionate selection of productions allows a preference to be given in terms of the composition of the initial population in a declarative manner.

4 Bias and Search using Genetic Operators

Genetic operators may be defined that modify programs during the evolution of a solution. Often the form of the language defined by the grammar is structured so that some parts of a program should have more emphasis on search than other parts. This section describes three genetic operators that allow explicit emphasis to be defined on how the search for new programs is to proceed.

4.1 Selective Crossover $\otimes$

The crossover search operator creates new programs by mixing the contents of two parent programs. In the context of a population, represented by derivation trees, crossover operates by swapping the derivations associated with two occurrences of the same nonterminal. By ensuring that the rooted derivation trees that are swapped have matching nonterminals, the new programs must also be members of the language defined by the grammar. The crossover is described as *selective* because the operation may be specified to have legal crossover sites from a subset of the total nonterminals, represented by the set $\otimes$. There may be more than one definition for $\otimes$, in which case each selective crossover is performed in the defined order. The crossover operator may change the depth of the derivation trees associated with a program. To ensure that the created programs do not grow in an unbounded fashion, a limit is placed on the maximum depth of any program created by crossover. This parameter is referred to as **MAX-DEPTH-PROGRAMS** and reflects the maximum depth of any program that is expected to be required to produce a solution. Given two derivation trees, ρ_1 and ρ_2, selective crossover is defined as follows:

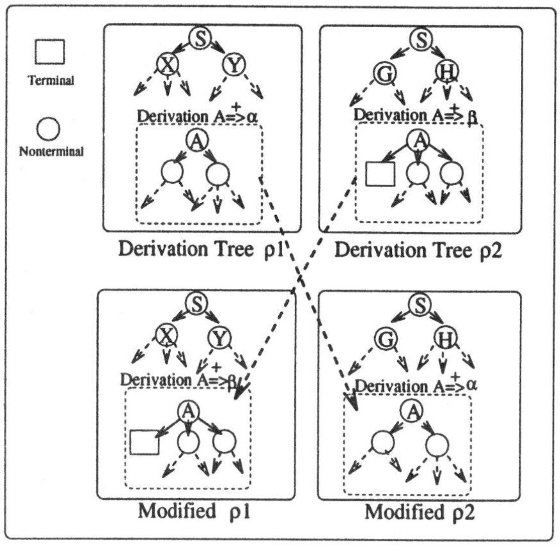

Figure 2: Crossover Based on Derivation Trees.

1. Randomly select a nonterminal $A \in \otimes$ from ρ_1. If no nonterminal from the set $\otimes$ exists in ρ_1, crossover is complete.
2. Randomly select a nonterminal $A \in \otimes$ from ρ_2. If no nonterminal from the set $\otimes$ exists in ρ_2, crossover is complete.
3. Swap the derivation trees below these matching nonterminals, thereby creating 2 new programs.
4. If either new program exceeds **MAX-DEPTH-PROGRAMS** the crossover is ignored. The parent programs ρ_1 and ρ_2 then remain unchanged.

Figure 2 shows crossover viewed as an operation on derivation trees. Crossover may also be described in

Selective Crossover	
Parameters	Specifications
MAX-DEPTH-PROGRAMS	15
$\otimes = \{A, B, C\}$	40%
$\otimes = \{E, F\}$	50%

Table 2: Specification of Selective Crossover.

terms of how the final strings of a program are mixed. If we have two programs, with derivations

$S \stackrel{+}{\Rightarrow} x_1 \ldots x_n \; A \; z_1 \ldots z_i \stackrel{+}{\Rightarrow} x_1 \ldots x_n y_1 \ldots y_m z_1 \ldots z_i$

$S \stackrel{+}{\Rightarrow} a_1 \ldots a_j \; A \; c_1 \ldots c_k \stackrel{+}{\Rightarrow} a_1 \ldots a_j b_1 \ldots b_r c_1 \ldots c_k$

crossover on the nonterminal A will produce two new strings of the form

$S \stackrel{+}{\Rightarrow} x_1 \ldots x_n \; A \; z_1 \ldots z_i \stackrel{+}{\Rightarrow} x_1 \ldots x_n b_1 \ldots b_r z_1 \ldots z_i$

$S \stackrel{+}{\Rightarrow} a_1 \ldots a_j \; A \; c_1 \ldots c_k \stackrel{+}{\Rightarrow} a_1 \ldots a_j y_1 \ldots y_m c_1 \ldots c_k$

4.1.1 Search Bias and Selective Crossover

Table 2 shows an example of the specification of selective crossover. The maximum allowable depth of a derivation tree that can be created has been set to 15. There are 2 crossover operations; one using the nonterminals $\{A, B, C\}$, occuring with a probability of 40%, and one using the nonterminals $\{E, F\}$, occuring with a probability of 50%. These operations represent a declarative bias in the search for new programs.

4.2 Selective Mutation $\odot$

Mutation has traditionally been associated with a *small change* in the structure of an object to which it applies. The mutation operator introduced here follows Koza's GP mutation [Koza, 1992]. However, as a selective operator it allows the user to be more discriminating about how a program is modified.

Mutation applies to a single derivation tree (and the program that it represents). The mutation is referred to as selective because the operation may be specified to have legal mutation sites from a subset of the total nonterminals, represented by the set $\odot$. There may be more than one definition for $\odot$, in which case each selective mutation is performed in the defined order. Mutation is portrayed in Figure 3 as an operation on a derivation tree using the language defined by the grammar, G. Given a derivation tree ρ_1 the selective mutation algorithm is defined as follows.

1. Randomly select some nonterminal $A \in \odot$ for mutation. If no nonterminal A exists in ρ_1, mutation is complete.

2. Delete the derivation tree rooted in the nonterminal A.

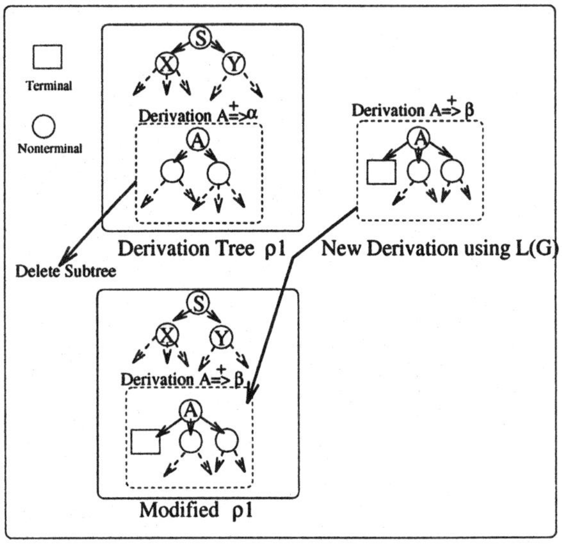

Figure 3: Mutation Based on Derivation Trees.

Selective Mutation	
Parameters	Specifications
MAX-DEPTH-PROGRAMS	15
$\odot = \{A, B\}$	10%
$\odot = \{C\}$	5%

Table 3: Specification of Selective Mutation.

3. Create a new derivation tree associated with A using the grammar, G. The depth of each complete derivation tree is limited by the parameter **MAX-DEPTH-PROGRAMS**. This operation applies the same technique as that of generating the initial population. However, the generation commences with the nonterminal A rather then the start symbol S.

4.2.1 Search Bias and Selective Mutation

Specifying selective mutation is similar to that of selective crossover. The nonterminals indicating the selective sites for mutation are specified along with the probability of mutation occuring. These parameters are shown in Table 3, where two mutation operations are defined. One is defined over the nonterminals A and B with a probability of 10%, the other over the nonterminal C with a probability of 5%.

4.3 Directed Mutation $A \rightarrow \alpha$

The selective mutation operator is potentially very destructive, in contrast with the normal view of mutation which implies *a small change in structure*. Hence, a second mutation operation is defined that allows small changes to be controlled in the structure of a program.

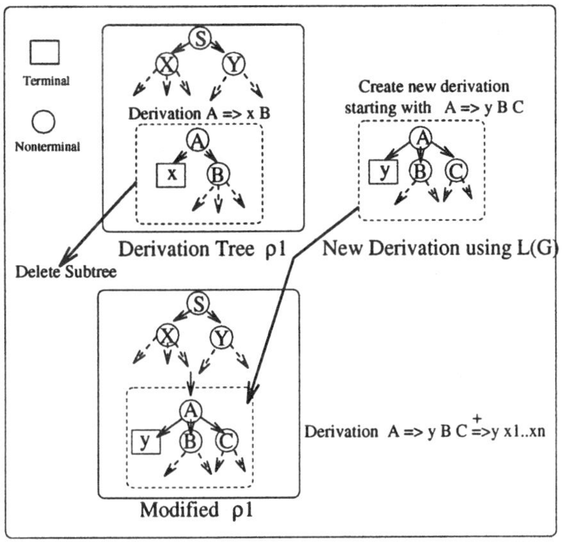

Figure 4: Directed Mutation $A \rightarrow x\ B \rhd y\ B\ C$.

This operation is described as *directed mutation.*

A directed mutation specifies that one particular production used in a program derivation should be replaced by a second production. The productions may contain both terminal and nonterminal symbols. If the replaced production contains nonterminal symbols they are randomly generated, using the grammar G, in the same manner as selective mutation. The general form of a directed mutation is

$$A \rightarrow \alpha \rhd \beta$$

where $A \in N$ and $\alpha, \beta \in \{N \cup \sum\}^{\star}$.

Note that $A \rightarrow \alpha$ and $A \rightarrow \beta$ are legal productions defined in the grammar. Directed mutation may be considered as a specialisation of selective mutation, where the selection site for mutation is specified and the form of mutation is defined explicitly. An example of directed mutation is shown in Figure 4.

Given a derivation tree ρ_1, the directed mutation algorithm is defined as follows:

1. Randomly select some derivation representing the production $A \rightarrow \alpha$ from ρ_1. If no derivation of this form exists in ρ_1, directed mutation is complete.

2. Delete the derivation tree associated with this production.

3. Create a new derivation tree associated with A using the production $A \rightarrow \beta$.

4. For each nonterminal in β, create a new derivation tree using the grammar, G. The depth of each complete derivation tree is limited by the parameter **MAX-DEPTH-PROGRAMS**.

Directed Mutation	
Parameters	Specification
MAX-DEPTH-PROGRAMS	15
$A \rightarrow B\ x \rhd y\ C$	10%
$B \rightarrow x_1 \rhd y_1$	5%

Table 4: Specification of Directed Mutation

4.3.1 Search Bias and Directed Mutation

As shown in Table 4, the specification for directed mutation is similar to previous genetic operators. The parameter **MAX-DEPTH-PROGRAMS** is required as new derivations may have to be randomly generated when the substitution of one production for another is performed. The most important role of directed mutation is to allow small changes in the represented program. For example, the directed mutation $B \rightarrow x_1 \rhd y_1$ allows a program to mutate the string x_1 to the string y_1. This is reminiscent of a standard genetic algorithm, single-bit mutation where one bit in a string is flipped, based on some probability.

4.3.2 A Generalisation of Directed Mutation

The *directed mutation* operator selects one derivation and replaces it with a second derivation. This concept can be extended to allow larger derivation tree structures to be specified. For example a derivation from A such as

$$A \overset{A \rightarrow xyC}{\Rightarrow} x\ y\ C \overset{C \rightarrow bD}{\Rightarrow} x\ y\ b\ D$$

may be replaced (mutated) by the derivation steps

$$A \overset{A \rightarrow Ec}{\Rightarrow} E\ c \overset{E \rightarrow xF}{\Rightarrow} x\ F\ c \overset{F \rightarrow Ga}{\Rightarrow} x\ G\ a\ c$$

This generalisation allows derivation trees of an arbitrary complexity to be selected for mutation. The derivation tree that replaces the specified mutation site may also be specified by any number of derivation steps. This allows general structures to be distinguished in the population.

5 Some Examples

A language bias for the 6-multiplexer problem, based on G_{6m}, could be represented by forcing all programs to initially start with an if function, using the address lines of $a0$ or $a1$. This bias is represented in the grammar, $G_{6m-if-address}$.

$G_{6m-if-address} =$
$\{S,$
$N = \{B, T, ADDRESS\},$
$\sum = \{and, or, not, if, a0, a1, d0, d1, d2, d3\},$
$P =$
$\{S \rightarrow if\ ADDRESS\ B\ B$

$$ADDRESS \rightarrow a0 \mid a1$$
$$B \rightarrow and\ B\ B \mid or\ B\ B \mid not\ B \mid if\ B\ B\ B \mid T$$
$$T \rightarrow a0 \mid a1 \mid d0 \mid d1 \mid d2 \mid d3$$
$$\}$$
$$\}$$

This represents a language bias. Note that to create this bias, based on the grammar, G_{6m}, no changes to the learning system were required (other than creating the new grammar).

A search bias for the following grammar, $G_{lisp-member}$, could be stated using directed mutation to force repeated patterns to be mutated into a recursive call. $G_{lisp-member}$ defines a lisp-like language representing a definition of the *member* function. The terminal string *member* is a recursive call to the program (*member x y*) which has been composed using $G_{lisp-member}$.

$$\begin{array}{l} G_{lisp-member} = \\ \quad \{S, \\ \quad N = \{M, TEST, RESULT, AL\}, \\ \quad \sum = \{cond, atom, eq, car, cdr, member, \\ \qquad x, y, true, nil\}, \\ \quad P = \\ \qquad \{S \rightarrow M \\ \qquad M \rightarrow cond\ TEST\ RESULT\ M \mid \epsilon \\ \qquad TEST \rightarrow atom\ AL \mid eq\ AL\ AL \\ \qquad TEST \rightarrow member\ x\ AL \mid true \mid nil \\ \qquad RESULT \rightarrow eq\ AL\ AL \mid member\ x\ AL \\ \qquad RESULT \rightarrow true \mid nil \\ \qquad AL \rightarrow car\ AL \mid cdr\ AL \mid x \mid y \\ \qquad \} \\ \quad \} \end{array}$$

The directed mutation may state that a repeated application of the *cdr* function, on the list argument *y*, is an attempt to move along the list. This implies that a recursive call may be a useful function to replace the repeated operation *y*. The directed mutation for this condition is represented as follows.

$$\begin{array}{l} RESULT \Rightarrow eq\ AL\ AL \Rightarrow eq\ x\ AL \Rightarrow eq\ x\ cdr\ AL \\ \qquad \Rightarrow eq\ x\ cdr\ cdr\ AL \Rightarrow eq\ x\ cdr\ cdr\ y \\ \qquad\qquad \rhd \\ \qquad \Rightarrow member\ x\ AL \Rightarrow member\ x\ cdr\ AL \\ \qquad \Rightarrow member\ x\ cdr\ y \end{array}$$

6 Conclusion

This paper has described a evolutionary computation technique which learns computer programs from example. The use of a context-free grammar to represent declarative bias and structure allows a domain expert to narrow the possible search space and therefore have a greater possibility of discovering solutions. The grammar allows both the typing to be automatically maintained with program constructs and also the structure (i.e. how functions are combined) to be explicitly stated and controlled.

The genetic operators of *selective crossover, selective mutation* and *directed mutation* are used to transform a program from one generation to the next. These operators allow a search bias to be explicitly defined. Hence the system brings together a search and language bias in an evolutionary context within a declarative framework.

Acknowledgments

The author would like to thank Bob McKay and Xin Yao for useful discussions and support. William Cohen must also be thanked for encouraging the use of grammars with genetic programming. The suggestions of the reviewers are kindly acknowledged.

References

[Cohen, 1993] Cohen, W. W. (1993). Grammatically biased learning: Learning logic programs using an explicit antecedent description language. Technical report, AT and T Bell Laboratories, Murray Hill, NJ.

[Gruau, 1996] Gruau, F. (1996). On using syntactic constraints with genetic programming. In Angeline, P. and Kinnear, Jr., K., editors, *Advances in Genetic Programming 2*, chapter 19, pages 402–417. USA:MIT Press.

[Koza, 1992] Koza, J. (1992). *Genetic Programming:on the programming of computers by means of natural selection.* A Bradford Book, The MIT Press.

[Lenat, 1984] Lenat, D. (1984). The role of heuristics in learning by discovery: Three case studies. In Michalski, R., Carbonell, J., and Mitchell, T., editors, *Machine Learning: An Artificial Intelligence Approach*, chapter 9, pages 243–306. Springer-Verlag.

[Michalski, 1984] Michalski, R. S. (1984). A theory and methodology of inductive learning. In Michalski, R., Carbonell, J., and Mitchell, T., editors, *Machine Learning: An Artificial Intelligence Approach*, chapter 4, pages 83–134. Springer-Verlag.

[Mitchell et al., 1984] Mitchell, T., Utgoff, P., and Banerji, R. (1984). Learning by experimentation: Acquiring and refining problem-solving heuristics. In *Machine Learning: An Artificial Intelligence Approach*, chapter 6, pages 163–190. Springer-Verlag.

[Mizoguchi et al., 1994] Mizoguchi, J., Hemmi, H., and Shimohara, K. (1994). Production genetic algorithms for automated hardware design through an evolutionary process. In *Proceedings of the First IEEE Conference on Evolutionary Computation.* Piscatawa, NJ, USA:IEEE Press.

[Pazzani and Kibler, 1992] Pazzani, M. and Kibler, D. (1992). The utility of knowledge in inductive learning. *Machine Learning*, 9(1).

[Quinlan, 1990] Quinlan, J. (1990). Learning logical definitions from relations. *Machine Learning*, 5(3):239–266.

[Utgoff, 1986] Utgoff, P. (1986). *Machine Learning of Inductive Bias*. Kluwer Academic Publishers.

[Valiant, 1984] Valiant, L. (1984). A theory of the learnable. *ACM*, 27(11):1134–1142.

[Whigham, 1995a] Whigham, P. (1995a). Grammatically-based genetic programming. In J.Rosca, editor, *Proceedings of the Workshop on Genetic Programming:From Theory to Real-World Applications*, pages 33–41. Morgan Kaufmann Pub.

[Whigham, 1995b] Whigham, P. (1995b). Inductive bias and genetic programming. In *First International Conference on Genetic Algorithms in Engineering Systems:Innovations and Applications*, pages 461–466. UK:IEE.

[Wolpert and Macready, 1996] Wolpert, D. and Macready, W. (1996). No free lunch theorems for search. Technical Report SFI-TR-95-02-010, Santa Fe, NM, 87501.

[Wong and Leung, 1995] Wong, M. and Leung, K. (1995). Applying logic grammars to induce subfunctions in genetic programming. In *Proceedings of the 1995 IEEE Conference on Evolutionary Computation*, pages 737–740. USA:IEEE Press.

Learning Recursive Functions from Noisy Examples using Generic Genetic Programming

Man Leung Wong
Department of Systems Engineering and Engineering Management
The Chinese University of Hong Kong
mlwong@se.cuhk.hk

Kwong Sak Leung
Department of Computer Science and Computer Engineering
The Chinese University of Hong Kong
ksleung@cs.cuhk.hk

ABSTRACT

One of the most important and challenging areas of research in evolutionary algorithms is the investigation of ways to successfully apply evolutionary algorithms to larger and more complicated problems. In this paper, we apply GGP (Generic Genetic Programming) to evolve general recursive functions for the even-n-parity problem from noisy training examples. GGP is very flexible and programs in various programming languages can be acquired. Moreover, it is powerful enough to handle context-sensitive information and domain-dependent knowledge. A number of experiments have been performed to determine the impact of noise in training examples on the speed of learning.

1. Introduction

One of the most important and challenging areas of research in evolutionary algorithms is the investigation of ways to apply them to larger and more complicated problems. One approach to make a large problem more tractable is to discover problem representations automatically. Koza (1994) uses the even-n-parity problem to demonstrate extensively that his approach of hierarchical Automatically Defined Functions (ADFs) can facilitate the solving of the problem.

Koza shows that the even-7-parity problem can be solved using GP with hierarchical ADFs. He finds that about 1440000 functions, I(M, i, z), should be evaluated to obtain at least one solution with 99% probability. In each fitness calculation, 128 fitness cases must be evaluated. Thus, 1440000*128 = 184320000 fitness cases should be processed. Unfortunately, the solutions found can only solve the even-n-parity problem with a particular value of n. If a different value of n is used, GP with hierarchical ADFs must be applied again to find other programs that can solve the new even-n-parity problem.

Clearly, the solution found is not general enough to solve all instances of the even-n-parity problem for all $n \geq 0$. In this paper, we apply a novel Generic Genetic Programming (GGP) approach to evolve general recursive functions for the even-n-parity problem from noisy training examples. The next section describes the GGP approach which is powerful enough to handle context-sensitive information and domain-dependent knowledge. The even-n-parity problem is discussed in Section 3. The subsequent section describes a number of experiments that evaluate the impact of noise in training examples on the rate of inducing general recursive functions. Section 5 is the conclusion.

2. Generic Genetic Programming (GGP)

Program induction generates a computer program that can produce the desired behavior for a given set of situations. Two of the approaches in program induction are Inductive Logic Programming (Dzeroski and Lavrac 1993; Muggletion 1992) and Genetic Programming (Koza 1992, 1994; Kinnear 1994). These approaches are restrictive because Inductive Logic Programming (ILP) can only learn logic programs and Genetic Programming (GP) usually evolves S-expressions in Lisp.

Moreover, since their formalisms are so different, these two approaches cannot be integrated easily although their properties and goals are similar. If they can be combined in a common framework, then many of the techniques and theories obtained in one approach can be applied in the other one. The combination can greatly enhance the information exchange between these fields.

Generic Genetic Programming (GGP) is a novel approach that combines GP and ILP (Wong and Leung 1995b). GGP is a generalization and extension of GLPS (Wong and Leung 1995a). GLPS (the Genetic Logic Programming System) can induce logic programs from noisy examples using Genetic Algorithms (Goldberg 1989; Holland 1975). Using GGP, programs in various programming languages such as Lisp, Prolog, and Fuzzy Prolog can be evolved (Wong and Leung 1995b). The approach is also powerful enough to handle context-sensitive information and domain-dependent knowledge which can be used to accelerate the learning speed and/or improve the quality of the programs (Wong and Leung 1995c).

GGP can induce programs in various programming languages. This is achieved by accepting or choosing grammars of different languages to produce programs in these languages. Most modern programming languages are specified in the notation of context-free grammar (CFG). However, logic grammars are used in GGP because they are more powerful than that of CFG, but equally amenable to efficient execution. In this paper, the notation of definite clause grammars (DCG) is used (Pereira and Warren 1980). The details of logic grammars and the representation method are described in the Appendixes. On the other hand, the genetic operators are detailed in (Wong and Leung 1996).

In GGP, populations of programs are genetically bred using the Darwinian principle of survival and reproduction of the fittest along with genetic operations appropriate for processing programs. GGP starts with an initial population of programs generated randomly, induced by other learning systems, or provided by the user. Logic grammars provide declarative descriptions of the valid programs that can appear in the initial population. A high-level algorithm of GGP is presented in table 1.

Table 1 The high level algorithm of GGP

1. Generate an initial population of programs.
2. Execute each program in the current population and assign it a fitness value according to the fitness function
3. If the termination criterion is satisfied, terminate the algorithm. The best program found in the run of the algorithm is designated as the result.
4. Create a new population of programs from the current population by applying the reproduction, crossover, and mutation operations. These operations are applied to programs selected by fitness proportionate or tournament selections.
5. Rename the new population to the current population.
6. Proceed to the next generation by branching back to the step 2.

2.1 Learning logic programs

To illustrate that GGP can learn logic programs, we perform an experiment and compare the performance of GGP and FOIL in inducing logic programs in the domain of the chess endgame problem (Quinlan 1990).

In this domain, the target predicate `illegal(WKf, WKr, WRf, WRr, BKf, BKr)` states whether the position where the white king is at (WKf, WKr), the white rook at (WRf, WRr) and the black king at (BKf, BKr) is not a legal white-to-move position. The background knowledge is represented by two predicates, `adjacent(X, Y)` and `less_than(W, Z)`, indicating that rank/file `X` is adjacent to rank/file `Y` and rank/file `W` is less than rank/file `Z` respectively. GGP uses the logic grammar in table 2 for this problem. In this grammar, `[adjacent(?X, ?Y)]` and `[less_than(?X, ?Y)]` are terminal symbols. The logic goal `member(?X, [WKf, WKr, WRf, WRr, BKf, BKr])` will instantiate `?X` to either `WKf`, `WKr`, `WRf`, `WRr`, `BKf`, or `BKr`. The logic variables of the target logic program are `WKf`, `WKr`, `WRf`, `WRr`, `BKf`, and `BKr`.

Table 2 The logic grammar for the chess endgame problem

```
start        ->     clauses.
clauses      ->     clauses, clauses.
clauses      ->     clause.
clause       ->
  consq, [:-], antes, [.].
consq        ->
  [illegal(WKf,WKr,WRf,WRf,BKf,BKr)].
antes        ->     antes, [,], antes.
antes        ->     ante.
ante         ->
  {member(?X,
   [WKf,WKr,WRf,WRf,BKf,BKr])},
  {member(?Y,
   [WKf, WKr, WRf, WRf, BKf, BKr])},
  literal(?X, ?Y).
literal(?X, ?Y)    ->     [?X = ?Y].
literal(?X, ?Y)    ->     [ ~ ?X = ?Y].
literal(?X, ?Y)    ->
  [ adjacent(?X, ?Y) ].
literal(?X, ?Y)    ->
  [ ~ adjacent(?X, ?Y) ].
literal(?X, ?Y)    ->
  [ less_than(?X, ?Y) ].
literal(?X, ?Y)    ->
  [ ~ less_than(?X, ?Y) ].
```

The training set contains 1000 examples (336 positive and 664 negative examples). The testing set has 10000 examples (3240 positive and 6760 negative examples). Different amount of noise is introduced into the training examples in order to study the performance of both systems in learning programs from noisy environment. To introduce n% of noise into argument X of the examples, the value of argument X is replaced by a random value of the same type from a uniform distribution, independent to noise in other arguments. For the class variable, n% positive examples are labeled as negative ones while n% negatives examples are labeled as positive ones. In this experiment, the percentages of noise introduced are 5%, 10%, 15%, 20%, 30% and 40%. Thus, the probabilities for an example to be noisy are respectively 27.36%, 48.04%, 63.46%, 74.78%, 88.74% and 95.47%. Background knowledge and testing examples are not corrupted with noise.

A chosen level of noise is first introduced in the training set. Logic programs are then induced from the training set using GGP and FOIL. Finally, the classification accuracy of the learned programs is estimated on the testing set. For GGP, the initial population of programs are induced by a variation of FOIL using a portion of the training examples. The population size is 10 and the maximum number of generations for each experiment is 50. Since GGP is a non-deterministic learning system, the process is repeated for five times on the same training set and the average of the five results is reported as the classification accuracy of GGP.

Many runs of the above experiment are performed on different training examples. The average results of ten runs are summarized in figure 1.

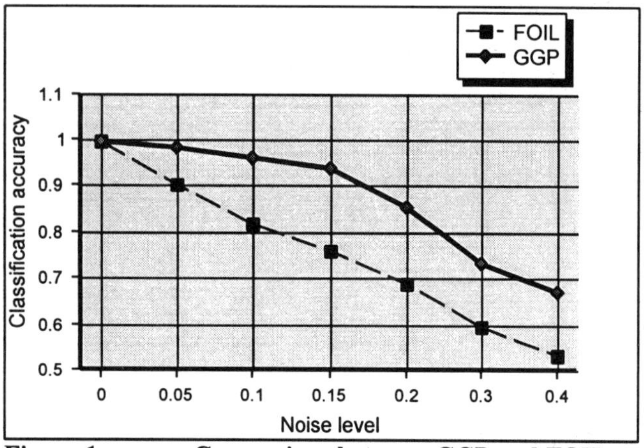

Figure 1 **Comparison between GGP and FOIL**

The performance of both systems are compared using paired t-test. From this experiment, the classification accuracy of both systems degrades seriously as the noise level increases. Nevertheless, the classification accuracy of GGP is better than that of FOIL by at least 5% at the 99.995% confidence interval at all noise levels (except the noise level of 0%). The largest difference reaches 24% at the 20% noise level. One possible explanation of the better performance of GGP is that the Darwinian principle of survival and reproduction of the fittest is a good noise handling method. It avoids overfitting noisy examples, but at the same time, it can finds interesting and useful patterns from these noisy examples. The experiment demonstrate that GGP is a promising alternative to other inductive logic programming systems and sometimes is superior for handling noisy data.

2.2 Learning programs in Fuzzy Prolog

The goal of this experiment is to induce a Fuzzy Prolog program that describes the fuzzy relation `can-reach` intensionally. Li and Liu (Li and Liu 1980) described the detailed definitions of the syntax and semantics of Fuzzy Prolog. Consider the fuzzy network in figure 2, this network represents the fuzzy relation `linked-to(X, Y)` that denotes node X is directly linked to node Y with truth value f, where f ∈ (0, 1]. In this network, the edges represent the instances of the `linked-to` relation and the number on an edge is the truth value of the corresponding instance. For example, the truth value of the instance `linked-to(0, 1)` is 0.9.

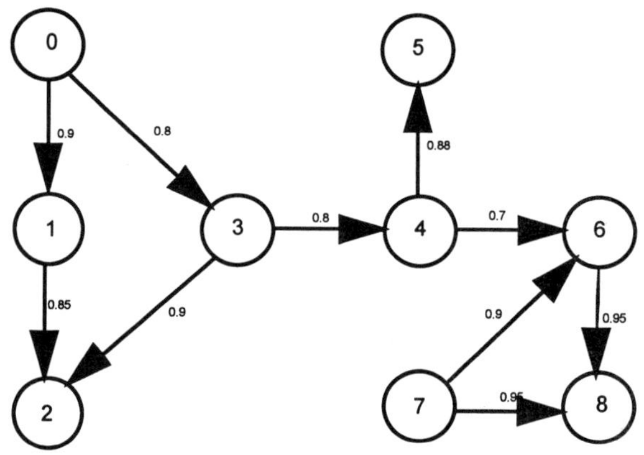

Figure 2 **A fuzzy network**

A fuzzy relation is associated with a n-ary fuzzy predicate and can be described extensionally as a set of ordered pairs. Thus the relation `linked-to(X, Y)` can be represented as:

```
linked-to(X, Y) =
    {(<0,1>, 0.9),   (<0,3>, 0.8),
     (<1,2>, 0.85),  (<3,2>, 0.9),
     (<3,4>, 0.8),   (<4,5>, 0.88),
     (<4,6>, 0.7),   (<6,8>, 0.95),
     (<7,6>, 0.9),   (<7,8>, 0.95)}
```

The first element of an ordered pair is a n-tuple of constants that satisfies the associated fuzzy predicate. The second element of the ordered pair is the corresponding truth value. Other fuzzy relations can be obtained from the fuzzy network. One of them is `can-reach(X, Y)` which is represented explicitly as:

```
can-reach(X, Y)=
    {(<0,1>, 0.9),    (<0,2>, 0.765),
     (<0,3>, 0.85),   (<0,4>, 0.72),
     (<0,5>, 0.648),  (<0,6>, 0.567),
     (<0,8>, 0.510),  (<1,2>, 0.85),
     (<3,2>, 0.9)     (<3,4>, 0.8),
     (<3,5>, 0.72),   (<3,6>, 0.63),
```

```
(<3,8>, 0.567), (<4,5>, 0.88),
((4,6>, 0.7),    (<4,8>, 0.63),
(<6,8>, 0.95),   (<7,6>, 0.9),
(<7,8>, 0.95) }
```

The negative instances of this relation can be found using the close world assumption (Li and Liu 1990). Thus, the set of negative instances is:

```
{(<0,0>, 0), (<0,7>, 0), (<1,0>, 0),
 (<1,1>, 0), (<1,3>, 0), (<1,4>, 0),
 (<1,5>, 0), (<1,6>, 0), (<1,7>, 0),
 (<1,8>, 0), (<2,0>, 0), (<2,1>, 0),
 (<2,2>, 0), (<2,3>, 0), (<2,4>, 0),
 (<2,5>, 0), (<2,6>, 0), (<2,7>, 0),
 (<2,8>, 0), (<3,0>, 0), (<3,1>, 0),
 (<3,3>, 0), (<3,7>, 0), (<4,0>, 0),
 (<4,1>, 0), (<4,2>, 0), (<4,3>, 0),
 (<4,4>, 0), (<4,7>, 0), (<5,0>, 0),
 (<5,1>, 0), (<5,2>, 0), (<5,3>, 0),
 (<5,4>, 0), (<5,5>, 0), (<5,6>, 0),
 (<5,7>, 0), (<5,8>, 0), (<6,0>, 0),
 (<6,1>, 0), (<6,2>, 0), (<6,3>, 0),
 (<6,4>, 0), (<6,5>, 0), (<6,6>, 0),
 (<6,7>, 0), (<7,0>, 0), (<7,1>, 0),
 (<7,2>, 0), (<7,3>, 0), (<7,4>, 0),
 (<7,5>, 0), (<7,7>, 0), (<8,0>, 0),
 (<8,1>, 0), (<8,2>, 0), (<8,3>, 0),
 (<8,4>, 0), (<8,5>, 0), (<8,6>, 0),
 (<8,7>, 0), (<8,8>, 0)}
```

The 19 positive and 62 negative instances are used as the fitness cases. The fitness function finds the sum, taken over all 81 fitness cases, of the absolute values of the difference between the desired truth value and the truth value returned by the generated program. A fitness case is said to be covered by a program if the truth value returned is within 0.05 of the desired value. GGP terminates if the maximum number of generations of 25, is reached or a Fuzzy Prolog program that covers all fitness cases is found. The logic grammar for this problem is shown in table 3. The background knowledge is represented by the fuzzy relation `linked-to(X, Y)` and the predicate `random(0, 1, ?A)` is a logic goal.

Table 3 The logic grammar for the can-reach problem

```
start          ->   clauses.
clauses        ->   clauses, clauses.
clauses        ->   clause.
clause         ->   {random(0, 1, ?A)},
                    consq,[<-(?A)], antes,
                    [.].
consq          ->   [can-reach(X, Y)].
antes          ->   antes, [,], antes.
antes          ->   ante.
ante           ->
                    {member(?A,[W, X, Y, Z])},
                    {member(?B,[W, X, Y, Z])},
                    literal(?A, ?B).
literal(?A, ?B)     ->
                    [ linked-to(?A, ?B) ].
literal(?A, ?B)     ->
                    [ can-reach(?A, ?B) ].
```

A number of trials have been performed using a population size of 100. Correct programs can be found in all trials. The following correct simplified program is found in one trial:

```
can-reach(X, Y)<- (1)
                 linked-to(X, Y).
can-reach(X, Y)<- (0.9)
                 linked-to(X, Z),
                 can-reach(Z, Y).
```

3. The Even-n-parity Problem

The boolean even-n-parity function of n boolean input arguments returns true (`T`) if an even number of the arguments are true, otherwise it returns false (`nil`). Koza (1994) uses GP with ADFs to induce hierarchical functions from training examples (fitness cases) to solve the problem. The training set contains all 2^n combinations of the n boolean input arguments. The standardized fitness of an S-expression is the sum of the error between the value returned by the S-expression and the correct value of the even-n-parity function.

Since all 2^n fitness cases, for a particular value of n, are used as the training examples, it is unclear whether GP can discover the regularities of the even-n-parity problem and induce a general function. Moreover, GP can only solve an instance of the even-n-parity problem for a particular value of n. If a different value n is required, GP must be used again to induce another function for the new instance of the problem. A better solution is a recursive function that solves all instances of the problem for all $n \geq 0$.

A general recursive function is given below:

```
(defun parity (L)
 (if (null L)
   T
   (AND
     (OR (first L)
         (parity (rest L)))
     (NAND (first L)
           (parity (rest L))))))
```

To evolve recursive functions using GGP, we have to determine the terminals, the primitive functions, the fitness cases, the fitness function, and the termination criterion. The terminal set is {`L`, `T`, `nil`} where `L` is the input argument of the recursive function to be learned, `T` and `nil` are boolean truth values. The argument `L` is a list of boolean values and any number of boolean values can exist in the list `L`. The set of primitive functions is {AND, OR, NAND, NOR, `ifnil`, `first`, `rest`, `parity`}

The boolean functions AND, OR, NAND, and NOR take two boolean input arguments and return one boolean value. The function `ifnil` takes three arguments. The first argument must be an S-expression that returns a list of boolean values. The last two arguments must be S-expressions that produce boolean output values. The function `ifnil` checks whether the first input argument returns an empty list. If the list is empty, `ifnil` returns the boolean value of the second S-expression as the output value of the function, otherwise it returns the boolean value of the third S-expression.

The primitive function `first` takes a list of boolean values as its argument and returns the first boolean value of the list if the list is not empty. Otherwise, the function generates an exception signal to indicate that an illegal operation has been attempted to get the first element from an empty list. The primitive function `rest` must take a list of n boolean values, for any value of $n \geq 0$, as its argument. If the input list is not empty, it returns a list containing the last n-1 elements of the input list, otherwise, the function generates an exception signal to indicate an illegal action has been tried. The primitive function `parity` takes a list of boolean values as its input and returns a boolean value. This function recursively calls the recursive function being evolved by GGP.

There are three data types: BOOLEAN, LIST, and SMALLER-LIST, that can be used to specify the primitive functions. The data type LIST contains lists of n boolean values, for any value of $n \geq 0$. The data type SMALLER-LIST contains lists of m boolean values for any value of m < n. The logic grammar for this problem is given in table 4.

In this grammar, we employ the argument of the non-terminal grammar symbol `s-expr` to designate the data type of the result returned by the S-expression generated from the grammar symbol. For example, the S-expression (AND (OR T nil) (NAND nil T)) returning a boolean value can be generated from the non-terminal grammar symbol `s-expr(BOOLEAN)`. Similarly, the non-terminal grammar symbol s-expr(LIST) can produce the S-expression (rest (rest L)) that returns a list of boolean values.

The terminal grammar symbols **[T]**, **[nil]**, and **[L]** in rules 12, 13, and 17 of the grammar form the terminal set of the problem. The terminal grammar symbols **[AND]**, **[OR]**, **[NAND]**, and **[NOR]** in rules 20, 21, 22, and 23 represent primitive functions that perform ordinary boolean operations. Rule 14 of the grammar specifies that these primitive functions take two boolean input arguments and return one boolean value.

The terminal symbol **[first]** in rule 16 represents the primitive function that returns the first element of a list of boolean values. Rule 16 also specifies that the primitive function takes a list of boolean values as its argument and returns a boolean value. The terminal symbol **[rest]** in rule 19 represents the primitive function that takes a list of n boolean values as its argument and returns a list containing the last n - 1 elements of the input list. Rule 19 also declares the arity of the function and the data types of its argument and output value.

Since an item of the SMALLER-LIST data type is also a list of boolean values, it should belong to the LIST data type. This fact is specified in rule 18. This example shows that a hierarchy of data types can be declared easily using grammar rules.

The primitive function represented by the terminal symbol **[parity]** in rule 15 must take a list of the SMALLER-LIST data type. This rule avoids a non-terminating recursive function such as: (defun parity (L) (ifnil L T (parity L))) to be evolved by GGP.

Table 4 The logic grammar for the even-n-parity problem

```
11:start                    ->
      [ (defun parity (L) ],
      [ (ifnil L T ],
      s-expr(boolean), [ )) ].
12:s-expr(BOOLEAN)          ->     [ T ].
13:s-expr(BOOLEAN)          ->     [ nil ].
14:s-expr(BOOLEAN)          ->
      [ ( ],op, s-expr(BOOLEAN),
      s-expr(BOOLEAN), [ ) ].
15:s-expr(BOOLEAN)          ->
      [ ( ], [ parity ],
      s-expr(SMALLER-LIST), [ ) ].
16:s-expr(BOOLEAN)          ->
      [ ( ], [ first ], s-expr(LIST),
      [ ) ].
17:s-expr(LIST)             ->     [ L ].
18:s-expr(LIST)             ->
      s-expr(SMALLER-LIST).
19:s-expr(SMALLER-LIST) ->
      [ ( ],[ rest ],
      s-expr(LIST),[ ) ].
20:op                       ->     [ AND ].
21:op                       ->     [ OR ].
22:op                       ->     [ NAND ].
23:op                       ->     [ NOR ].
```

The idea of applying knowledge of data type to accelerate learning has been investigated independently by Montana (1994) in the Strongly Typed Genetic Programming (STGP). The previous paragraphs show that knowledge of data types can be represented easily using logic grammars and thus GGP can emulate the effect of STGP easily as a type of knowledge among many others.

The first rule of the grammar represents the domain-specific knowledge that the recursive function to be evolved must return `T` if the input argument `L` is an empty list, and the outermost statement of the function must be the correct base statement. The results of the experiments are discussed in the next section.

4. Experiments

Three experiments have been conducted. They differ in the fitness cases used. In each experiment, the population size is 500 and the maximum number of generations is 50. The probabilities of performing crossover and mutation are respectively 0.7 and 0.1. The maximum depth of derivation trees generated in the initial population is 12. The maximum depth of trees produced by crossover and mutation is also 12. All experiments are repeated for 60 times.

The first experiment evaluates the ability of GGP in inducing recursive functions for the even-n-parity problem. The even-0-, 2-, and 3- parity problems are used in the training process. The training set contains all 13 fitness cases from these even-parity problems. The standardized fitness value of an evolved function is the total number of misclassifications on the 13 fitness cases. The evolution terminates if the maximum number of generations of 50 is reached or a function that classifies all fitness cases correctly is found. In order to avoid the problem caused by inefficient functions, an execution time limit is enforced. After executing 100 primitive functions, if the evolved function fails to find a result for a fitness case, it will be terminated. In this case, it is assumed that the function will misclassify the corresponding fitness case.

It is possible that an evolved function will generate exceptions during its execution for some fitness cases, because it is illegal to perform the first/rest operation on an empty list. If the function produces an exception, it is assumed that it will misclassify the corresponding fitness cases.

In the 60 trials, GGP successfully evolves 16 functions that classify all fitness cases correctly. The generated functions are then tested on the even-i-parity problems, where i ∈ {0, 1, 2, 4, 5, 6, 7, 8, 9, 10}. All of them can successfully solve all the problems. They are further analyzed manually and it is found that these 16 functions are correct recursive functions for the general even-n-parity problem.

The I(M, i, z) for z of 99% reaches a minimum value of 335000 at generation 9 (Koza 1992). Since there are 13 fitness cases in the training set, 335000*13 = 4355000 fitness cases should be processed. Thus, GGP can find a general, recursive function for the even-n-parity problem very efficiently. On the other hand, GP with hierarchical ADFs evaluates 184320000 fitness cases to find a function that solves the even-7-parity problem only. In other words, GGP can solve the even-7-parity problem about 42 times faster than GP with hierarchical ADFs.

The second experiment evaluates the ability of GGP in inducing recursive functions for the even-n-parity problem from noisy training examples. The even-0-, 2-, and 3- parity problems are used in the training process. The training set contains all 13 fitness cases from these even-parity problems. To introduce noise into the training examples, one of them is randomly selected and the result of the selected example is modified from T to nil or from nil to T. The fitness function and the termination criterion are the same as those of the first experiment.

GGP successfully evolves 12 correct recursive functions for the general even-n-parity problem. The I(M, i, z) for z of 99% reaches a minimum value of 450000 at generation 9. Since there are 13 fitness cases in the training set, 450000*13 = 5850000 fitness cases should be processed. It is found that 1.34 times effort must be used if a noisy training set is employed compared to the first experiment. On the other hand, GGP can solve the even-7-parity problem about 32 times faster than GP with hierarchical ADFs.

The third experiment is similar to the second experiment. The only difference between them is that 2 of the 13 training examples are changed to introduce more noise into the training examples. The selected examples are modified from T to nil or from nil to T.

GGP successfully evolves 5 correct recursive functions for the general even-n-parity problem. The I(M, i, z) for z of 99% reaches a minimum value of 765000 at generation 16. Since there are 13 fitness cases in the training set, 765000*13 = 9945000 fitness cases should be processed. By comparing with the result of the second experiment, it is found that 1.7 times effort must be used if there is more noise in the training set. On the other hand, GGP can solve the even-7-parity problem about 18.53 times faster than GP with hierarchical ADFs.

5. Conclusion

In this paper, we described a method of evolving recursive functions for the even-n-parity problem from noisy training examples. Three experiments have been performed to study the impact of noise in training examples on the speed of learning recursive functions. The numbers of fitness cases processed to induce general recursive functions with 99% probability are summarized in table 5. These experiments demonstrated that GGP can evolve recursive functions from noisy training examples. However, more computation effort is required if more noise exists in training examples.

Table 5 The numbers of fitness cases processed to induce general recursive functions with 99% probability

The first experiment (0% noise)	4355000
The second experiment (7.7% noise)	5850000
The third experiment (15.4% noise)	9945000

For future work, we will use the techniques of the inductive learning systems such as THESYS (Summers 1977) and ADATE (Olsson 1995) so that GGP can induce recursive functional programs more efficiently.

Acknowledgments

The research is sponsored by the RGC Earmarked research grant of UGC reference number CUHK 486/95E.

Appendix A: Logic Grammars

Logic grammars are the generalizations of CFGs. Their expressiveness is much more powerful than those of CFGs, but equally amenable to efficient execution (Pereira and Warren 1980). The logic grammar for some simple S-expressions is given in table 6.

A logic grammar differs from a CFG in that the logic grammar symbols, whether terminal or non-terminal, may include arguments. The arguments can be any term in the grammar. A term is either a logic variable, a function or a constant. A variable is represented by a question mark '?' followed by a string of letters and/or digits. A function is a grammar symbol followed by a bracketed n-tuple of terms and a constant is simply a 0-arity function. Arguments can be used in a logic grammar to enforce context-dependency.

Thus, the permissible forms for a constituent may depend on the context in which that constituent occurs in the program. Another application of arguments is to construct tree structures in the course of parsing, such tree structures can provide a representation of the semantics of the program.

Table 6 A logic grammar

```
1:start      ->
       [(*], exp(W), exp(W), exp(W),
       [)].
2:start      ->
       {member(?x,[W, Z])}, [(*],
       exp-1(?x), exp-1(?x), exp-1(?x),
       [)].
3:start      ->
       {member(?x,[W, Z])}, [(+],
       exp-1(?x), exp-1(?x), exp-1(?x),
       [)].
4:exp(?x)    ->      [(/ ?x 1.5)].
5:exp-1(?x)  ->      {random(1,2,?y)},
                     [(/ ?x ?y)].
6:exp-1(?x)  ->      {random(3,4,?y)},
                     [(- ?x ?y)].
7:exp-1(W)   ->      [(+ (- W 11) 12)].
```

The terminal symbols, which are enclosed in square brackets, correspond to the set of words of the language specified. For example, the terminal `[(- ?x ?y)]` creates the constituent `(- 1.0 2.0)` of a program if `?x` and `?y` are instantiated respectively to 1.0 and 2.0. Non-terminal symbols are similar to literals in Prolog; `exp-1(?x)` in table 6 is an example of non-terminal symbol. Commas denote concatenation and each grammar rule ends with a full stop.

The right-hand side of a grammar rule may contain logic goals and grammar symbols. The goals are pure logical predicates for which logical definitions have been given. They specify the conditions that must be satisfied before the rule can be applied. For example, if the variable `?y` has not been bound, the goal `random(1, 2, ?y)` in table 6 instantiates `?y` to a random floating point number between 1 and 2. Otherwise, the goal checks whether the value of `?y` is between 1 and 2.

Domain-dependent knowledge can be represented in logic goals. For example, consider the following grammar rule:

```
a-useful-program      ->
    first-component(?X),
    {is-useful(?X, ?Y)},
    second-component(?Y).
```

This rule states that a useful program is composed of two components. The first component is generated from the non-terminal `first-component(?X)`. The logic variable `?X` is used to store semantic information about the first component produced. The logic goal then determines whether the first component is useful according to the semantic information stored in `?X`. Domain-dependent knowledge about which program fragments are useful is represented in the logical definition of this predicate. If the first component is useful, the logic goal `is-useful(?X, ?Y)` is satisfied and some semantic information is stored into the logic variable `?Y`. This information will be used in the non-terminal `second-component(?Y)` to guide the search for a good program fragment as the second component of a useful program.

The special non-terminal `start` corresponds to a program of the language. In table 6, some grammar symbols are shown in bold-face to identify the constituents that cannot be manipulated by genetic operators. For example, the last terminal symbol **`[)]`** of the second rule is revealed in bold-face because every S-expression must end with a ')', and thus it is not necessary to modify the ')' symbol. The number before each rule is a label for later discussions and is not part of the grammar.

Appendix B: Representations of programs

One of the fundamental contributions of GGP is in the representations of programs in different programming languages appropriately so that initial population can be generated easily and the genetic operators such as reproduction, mutation, and crossover can be performed effectively. A program can be represented as a derivation tree that shows how the program has been derived from the logic grammar. GGP applies deduction to randomly generate programs and their derivation trees in the language declared by the given grammar. These programs form the initial population. For example, the program `(* (/ W 1.5) (/ W 1.5) (/ W 1.5))` can be generated by GGP given the logic grammar in table 6. It is derived from the following sequence of derivations:

```
start => [(*] exp(W) exp(W) exp(W) [)]
      => [(*] [(/ W 1.5)] exp(W) exp(W)
         [)]
      => [(*] [(/ W 1.5)] [(/ W 1.5)]
         exp(W) [)]
      => [(*] [(/ W 1.5)] [(/ W 1.5)]
         [(/ W 1.5)] [)]
      => (* (/ W 1.5)(/ W 1.5)(/ W 1.5))
```

This sequence of derivations can be represented as the derivation tree depicted in figure 3.

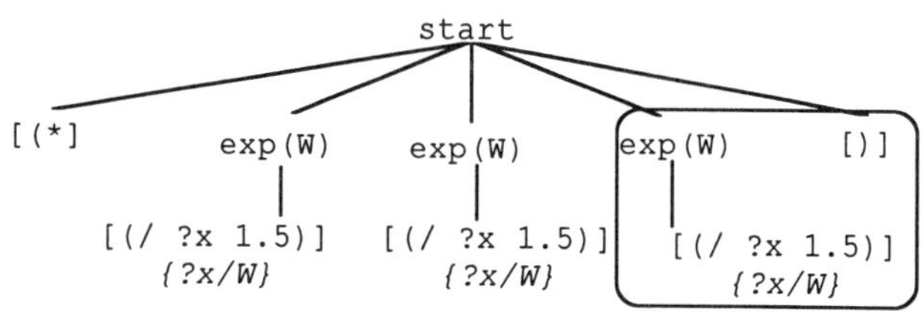

Figure 3 A derivation tree of the S-expression in Lisp `(* (/ W 1.5) (/ W 1.5) (/ W 1.5))`

The bindings of logic variables are shown in italic font and enclosed in a pair of braces. The sub-trees enclosed in a rectangular are frozen. In other words, they are generated by bold-faced grammar symbols and they cannot be modified by genetic operators.

One advantage of logic grammars is that they specify what is a legal program without any explicit reference to the process of program generation and parsing. Furthermore, a logic grammar can be translated into an efficient logic program that can generate and parse the programs in the language declared by the logic grammar (Pereira and Warren 1980). In other words, the process of program generation and parsing can be achieved by performing deduction using the produced logic program. Consequently, the program generation and analysis mechanisms of GGP can be implemented using a deduction mechanism based on the logic programs translated from the grammars. In the following paragraphs, we discuss the method of implementing GGP using a Prolog-like logic programming language.

The differences between the logic programming language used and Prolog are listed as follows:

- A variable is represented by a question mark ? followed by a string of letters and/or digits.
- The elements of a list can be separated by either commas or spaces. For example, `[a b c]` and `[a, b, c]` are equivalent.
- A pair of `'|'` is used to represent a frozen terminal symbol. For example, the symbol **[)]** in the second rule of the grammar in table 6 is translated into |)|.
- A pair of braces encloses a sequence of logic goals appearing in a logic grammar.
- If there are a number of clauses C_1, C_2, ..., C_n that match a goal G, the ordering of evaluating these clauses is determined randomly.

Using the difference list approach (Sterling and Shapiro 1986), a grammar rule of the form:

A_0 -> A_1, A_2, ..., A_n.

is translated into a logic program clause of the form:

A_0' :- A_1', A_2', ..., A_n'.

in the logic programming language. Here, if A_i , for some i between 0 and n, is a non-terminal with M arguments, then A_i' is a literal with M+3 arguments. The predicate symbols of A_i and A_i' are the same. For example, A_i is translated into exp(?X, ?Tree, ?S_j, ?S_{j+1}), for some j, if A_i is exp(?X). The literal exp(?X, ?Tree, ?S_j, ?S_{j+1}) states that the sequence of symbols between ?S_j and ?S_{j+1} is a sentence of the category represented by the non-terminal symbol `exp(?X)`. The derivation tree of the sentence is stored in the logic variable `?Tree`.

A terminal symbol such as `[a b c]` is translated to a literal with 3 arguments: connect([a b c], ?S_j, ?S_{j+1}), for some j. The predicate `connect` is defined as:

```
connect(?A, ?S0,?S1):-
    append(?A, ?S1, ?S0).
```

This predicate declares that the list of symbols stored in the logic variable `?A` can be found in the sequence of symbols between `?S0` and `?S1`.

If A_k, for some k between 1 and n, is a pair of braces enclosing a sequence of pure logic goals, i.e., A_k has the form of {G_0, G_1, ..., G_m}, then A_k' is obtained from A_k by removing the pair of braces.

For example, the grammar depicted in table 6 can be translated into the logic program presented in table 7. In the clause 1' of the logic program shown in table 7, the compound term `tree(start, [(*], ?E1, ?E2, frozen(?E3), |)|)` indicates that it is a tree with a root labeled as `start`. The children of the root include the terminal symbol `[(*]`, a tree created from the non-terminal `exp(W)`, another tree created from the non-terminal `exp(W)`, a frozen tree generated from the non-terminal **exp(W)**, and the frozen terminal |)|.

Thus, a derivation tree can be generated randomly by issuing the following query:

```
?- start(?T, ?S, []).
```

This goal can be satisfied by deducing a sentence that is in the language specified by the grammar. One solution is:

```
?S = [(* (/ W 1.5) (/ W 1.5)
     (/ W 1.5))]
```

and the corresponding derivation tree is:

```
?T = tree(start, [(*],
          tree(exp(W), [(/ W 1.5)]),
          tree(exp(W), [(/ W 1.5)]),
          frozen(tree(exp(W),
          [(/ W 1.5)])), |)|)
```

This is exactly a representation of the derivation tree shown in figure 3. In fact, the bindings of all logic variables and other information are also maintained in the derivation trees to facilitate the genetic operations that will be performed on the derivation trees.

Alternatively, initial programs can be induced by other learning systems such as FOIL (Quinlan 1990) or given by the user. GGP analyzes each program and creates the corresponding derivation tree. If the language is ambiguous, multiple derivation trees can be generated. GGP produces only one tree randomly.

Using the logic program in table 7, a given program such as `(* (/ W 1.5) (/ W 1.5) (/ W 1.5))` can be analyzed using the following query:

```
?-start(?T,
[(* (/ W 1.5) (/ W 1.5) (/ W 1.5))],
[]).
```

The given program is correct if the above goal can be satisfied and the corresponding derivation tree will be bound to the logic variable `?T`. Since the logic grammar in table 6 is ambiguous, the corresponding logic program may produce multiple derivation trees for a given program. Since the search strategy of the underlying deduction mechanism selects randomly one clause to explore with backtracking from all unifiable clauses, the sequence of generating the derivation trees of a particular program is also random. Consequently, GGP takes the first tree returned from the query to represent the given program.

Table 7 A logic program obtained from translating the logic grammar presented in table 6

```
1':start(tree(start,[(*],?E1,?E2,
        frozen(?E3), |)|), ?S0, ?S5)
    :- connect([(*], ?S0, ?S1),
        exp(W, ?E1, ?S1, ?S2),
        exp(W, ?E2, ?S2, ?S3),
        exp(W, ?E3, ?S3, ?S4),
        connect([)], ?S4, ?S5).
2':start(tree(start,
        {member(?x, [W, Z])},
        [(*], ?E1, ?E2, frozen(?E3),
        |)|),?S0, ?S5)
    :- member(?x, [W, Z]),
        connect([(*], ?S0, ?S1),
        exp-1(?x, ?E1, ?S1, ?S2),
        exp-1(?x, ?E2, ?S2, ?S3),
        exp-1(?x, ?E3, ?S3, ?S4),
        connect([)], ?S4, ?S5).
3':start(tree(start,
        {member(?x, [W, Z])},
        [(+], ?E1, ?E2, frozen(?E3),
        |)|),?S0, ?S5)
    :- member(?x, [W, Z]),
        connect([(+], ?S0, ?S1),
        exp-1(?x, ?E1, ?S1, ?S2),
        exp-1(?x, ?E2, ?S2, ?S3),
        exp-1(?x, ?E3, ?S3, ?S4),
        connect([)], ?S4, ?S5).
4':exp(?x, tree(exp(?x),
     [(/ ?x 1.5)]),?S0, ?S1)
    :- connect([(/ ?x 1.5)],?S0,?S1).
5':exp-1(?x, tree(exp-1(?x),
        {random(1,2,?y)},
        [(/ ?x ?y)]),?S0, ?S1)
    :- random(1, 2, ?y),
        connect([(/ ?x ?y)], ?S0, ?S1).
6':exp-1(?x, tree(exp-1(?x),
        {random(3,4,?y)},
        [(- ?x ?y)]),?S0, ?S1)
    :- random(3, 4, ?y),
        connect([(- ?x ?y)], ?S0, ?S1).
7':exp-1(W,
        tree(exp-1(W),
        [(+ (- W 11) 12)]),?S0, ?S1)
    :- connect([(+ (- W 11) 12)],
        ?S0, ?S1).
```

Bibliography

Dzeroski, S. and Lavrac, N. (1993). Inductive Learning in Deductive Databases. *IEEE Transactions on Knowledge and Data Engineering*, **5**, pp. 939-949.

Goldberg, D. E. (1989). *Genetic Algorithms in Search, Optimization, and Machine Learning*. MA: Addison-Wesley.

Holland, J. H. (1975). *Adaptation in natural and artificial systems*. Ann Arbor: The University of Michigan Press.

Kinnear, K. E. Jr., editor (1994) *Advances in Genetic Programming*. Cambridge MA: MIT Press.

Koza, J. R. (1992). *Genetic Programming: on the Programming of Computers by Means of Natural Selection*. Cambridge, MA: MIT Press.

Koza, J. R. (1994). *Genetic Programming II: Automatic Discovery of Reusable Programs*. Cambridge, MA: MIT Press.

Li, D. and Liu, D (1990). *A Fuzzy Prolog Database system*. Great Britain: Research Studies Press Ltd.

Montana, D. J. (1994) Strongly Typed Genetic Programming. Technical Report 7866, Bolt, Beranek, and Newman.

Muggleton, S. (1992) Inductive Logic Programming. In S. Muggletion (ed.), *Inductive Logic Programming*, pp. 3-27. London: Academic Press.

Olsson, R. (1995) Inductive Functional Programming using Incremental Program Transformation. *Artificial Intelligence*, **74**, pp. 55-81.

Pereira, F. C. N. and Warren, D. H. D. (1980). Definite Clause Grammars for Language Analysis - A Survey of the Formalism and a Comparison with Augmented Transition Networks. *Artificial Intelligence*, **13**, pp. 231-278.

Quinlan, J. R. (1990). Learning logical definitions from relations. *Machine Learning*, **5**, 239-266.

Sterling, L. and Shapiro, E. (1986). *The Art of Prolog*. MA: MIT Press.

Summers, P. D. (1977) A Methodology for LISP Program Construction from Examples. *JACM*, **24**, pp. 161-175.

Wong, M. L. and Leung, K. S.(1995a) Inducing Logic Programs with Genetic Algorithms: The Genetic Logic Programming System, *IEEE Expert*, **9**, no. 5. pp. 68-76.

Wong, M. L. and Leung, K. S. (1995b). An Induction System that Learns Programs in different Programming Languages using Genetic Programming and Logic Grammars. In *Proceedings of the 7th IEEE International Conference on Tools with Artificial Intelligence*. pp. 380-387. CA: IEEE Computer Society Press.

Wong, M. L. and Leung, K. S. (1995c). Applying Logic Grammars to Induce sub-functions in Genetic Programming. In *Proceedings of the 1995 IEEE International Conference on Evolutionary Computing*.

Wong, M. L. and Leung, K. S. (1996). Evolving recursive functions for the even-parity problem using genetic programming. In P. J. Angeline and K. E. Kinnear, Jr. (Eds.) *Advances in Genetic Programming 2*. Cambridge MA: MIT Press.

SHORT GENETIC PROGRAMMING PAPERS

Classification using Cultural Co-evolution and Genetic Programming

Myriam Z. Abramson
Management Systems Designers, Inc.
131 Park Street, NE
Vienna, VA 22180
mabramso@nlm.nih.gov

Lawrence Hunter
National Library of Medicine
Bldg. 38A, MS-54
Bethesda, MD 20894 USA
hunter@nlm.nih.gov

Abstract

Genetic programming has been used to evolve decision trees for classification [Koza, 1992] but its role as an inductive learner remains elusive. This paper examines a possible role of genetic programming as a learner in a cultural co-evolution model of learners developed by Hunter. We demonstrate that this cooperation allows classification to evolve much faster than their individual components and report encouraging results on a class of problems.

1. Introduction

1.1 Cultural co-evolution Learning

Cultural co-evolution learning [Hunter, 1995] interleaves multistrategy inductive inference with a search through the space of possible representations of the training data. The co-evolution in this scheme arises through the evolution of the parameters of learners (genes) and the evolution of their data representations (culture).

All learning methods are profoundly effected by the representations they use. In addition, the results of the application of a learning method identifies significant combinations of features, which can be used to re-represent the training data for future learning. Cultural co-evolution learning involves both the evolution of a set of diverse learning agents, and changes of the initial problem representation based on the results found by the evolving learners. Re-representations are extracted from each learner, and then compete for adaptation by other learners in the population. Since the fitness of learners depends on the representations they use, and since those learners generate new representations themselves, the learners and representations can be said to "co-evolve."

If there were only a single type of learner in the population, this approach would reduce to constructive induction. However, since a learner using a particular induction method (say, neural networks) can use a representation generated by a different method (say, decision tree induction), cultural co-evolution can discover and exploit synergies among induction methods. The cultural co-evolution learning approach has been implemented at the National Library of Medicine in a system called COEV. A block diagram of the system is shown in Figure 1.

1.2 Cultural co-evolution and Genetic Programming

A central assumption of cultural co-evolution learning is that induction that fails to completely solve a classification problem may nonetheless have discovered interactions among its input features which are significant. An "unsuccessful" classification may instead be an "incomplete" or "partial" classification; partial results might compose with other partial results to ultimately produce a better classification. For this reason, individual learning agents may contribute to the performance of the COEV system as a whole by producing good partial results, even they are not as good as other methods at solving the entire problem. It is a property of genetic programming to approximately solve a problem if a solution is not found [Koza, 1992]. Figure 2 demonstrates improving average fitness of a population of genetic programs (GPs), indicating that partial solutions are being found and learning

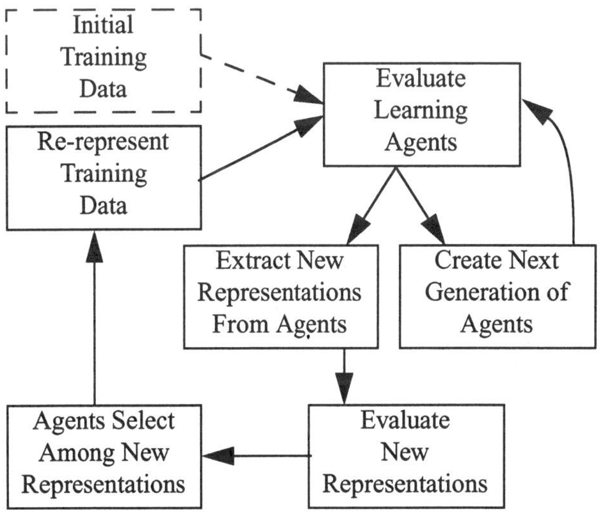

Figure 1. Cultural co-evolution Process Block Diagram

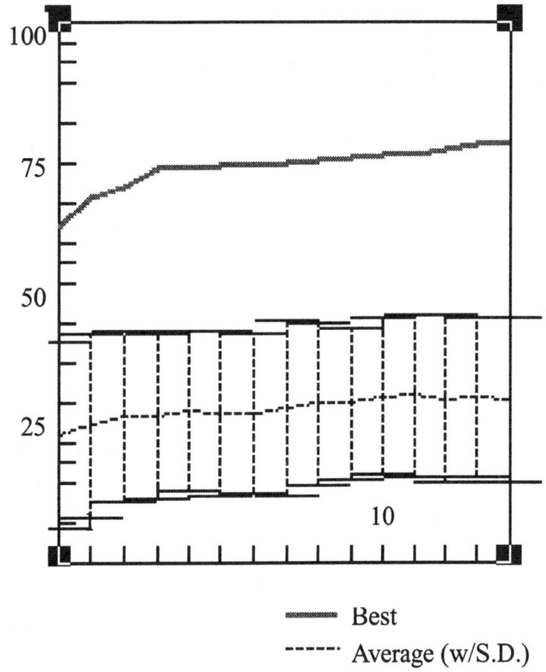

Figure 2. First MONK Problem - 100 GP Learners

is occurring. Use of genetic programming has been effective in the COEV system primarily for this reason.

Cultural co-evolution can be viewed as a way to partition the search space like a "divide and conquer" method to explore disparate regions of the space [D'Hasseleer, 1994]. The synthesis happens in COEV when partial solutions from the different learners are exchanged, using a fitness proportionate selection scheme. As a result of assimilating new representations, the biases of the various learning agents lead to new ways of searching the hypothesis space. Both the representations used and the type of search through hypothesis space are important in determining whether a learner can solve a problem. As will be seen below, for classification problems, the strength of genetic programming seems to be in the exploration of the representation space; when a "good" representation has been found, other induction methods, especially "greedy" ones using the entropy measure, appear to be faster and more effective at finding the correct hypothesis.

The search space of a GP population is defined by its terminal set and its search operators by its function set. Each GP individual is an sexpression born from this initial search space that in turn participate in the search space. The determination of when and how to modify the search space comes for free, so to speak. An area where genetic programming can help in classification problems is to search a different space of concepts than empirical inductive learners. The combination of those two kinds of learners in a co-evolution model can solve efficiently problems with such hybrid representation space. It is worth pointing out that the relevance of a hybrid representation space also occurs when the representation language is not completely adequate for the representation of the problem leading to the enumeration of exceptions. This paper shows that genetic programming makes this co-evolution model equivalent to a multistrategy constructive induction learner [Bloedorn, et al., 1993].

1.3 Details of the COEV implementation

Each individual in a COEV population has both a vector of parameters that control the inference method (which we call "genes") and a set of feature definitions, constructed as compounds of the primitive features originally used to describe the problem (which we call "memes"). An example gene might specify the number of hidden nodes for a neural network, or the pruning severity for an inductive decision tree builder. An example meme might specify that an input feature to the learner should be constructed by taking the ratio of two of the primitive features, or some boolean combination of primitive features.

Subpopulations of individuals using the same learning are defined within the overall population. We enforce locality at the reproductive (gene) level but uses a panmictic mating scheme at the cultural (meme) level since meme exchanges cross subpopulation boundaries. New memes are introduced into the GP population as individual sexpressions, which then participate in crossover and mutation, and are thus incorporated into GP individuals. In the current model, meme exchange alternates with reproduction. Extraction of new memes from an evaluated population of learners generally involves parsing the output of a learner into a form that can be used as a feature definition (e.g. the left hand sides of rules generated by a rule induction method each become memes). For GP, each sexpression in a population is treated as a potential meme.

1.4 Details of the GP implementation

A few changes to the standard genetic programming code provided with Koza's book were made to implement a population of GP learners. The maximum depth of the lisp expression has been set to 5. Each lisp expression is created with a random depth. Changes made by the mutation and crossover operators have to obey this maximum depth. In the case of mutation, the size of the subtree generated cannot produce an invalid tree when added to the mutation point. In the case of crossover, the tree is "repaired" by replacing out-of-bounds nodes by a terminal. A reproduction "loop" has been added to the genetic programming mutation and crossover operators so as not to produce an invalid sexpression, that is, a tautology or a contradiction. In this domain, this is equivalent to removing introns and testing for an effective sexpression [Nordin, 1995]. Keep in mind that the full sexpression is used in the GP population and not the simplified one. Future work could include a "soft" brood selection [Altenberg, 1994]. We also augmented the simplification code of sexpressions from Koza's book and included algebraic rules. Finally, the initial population is not guaranteed to be unique. The Koza tableau is presented in table 1.

For classification to be possible, there must be a closure between the classes to predict and the output of a sexpression. Target identification problems and boolean expressions have this closure property. Much more work is required of GP for multiple-class problems. There must also be a closure between the output of the different learner populations and the GP population terminal set since memes are exchanged

1.5 RELIEF for feature fitness calculations

The fitness of the concepts generated by the inductive learners and the GP learners are measured by RELIEF [Kira & Rendell, 1992] which is the concept evaluation function in COEV. It is this RELIEF score that is used as the fitness function in the GP population.

The RELIEF algorithm computes the Euclidean distance of each example against all the other examples. Scores of the difference between the distance of the nearest example in the same class and the nearest example in a different class is computed for each feature in each example and the average of those scores constitutes the fitness (or relevance of the feature). This algorithm therefore pits each example against each other to compute the "closest" example in the different classes and indirectly tests each feature cooperation with each other. An example will illustrate the results of this algorithm. Suppose that the target concept is $F0 \oplus F1$ and F2, F3, F4 and F5 are irrelevant features. The training set contains all possible values of those variables or 64 examples. A fitness function that would test how well a feature predicted the class of the example would give each feature an equal probability to succeed since the number of times that the feature value matches the outcome is the same in all cases. In contrast, Table 2 shows the fitness scores produced by the RELIEF algorithm. Features with a negative score are deemed irrelevant and ignored from the meme pool in the meme exchange phase.

Table 1: Tableau for GP

Objective	Find the target concept
Terminal Set	Features: head-shape, body-shape, jacket-color, is-smiling, holding, has-tie, (eql head-shape round), (eql head-shape square), (eql head-shape octagon), (eql body-shape round), (eql body-shape square), (eql body-shape octagon), (eql holding sword), (eql holding balloon), (eql holding flag), (eql jacket-color red), (eql jacket-color yellow), (eql jacket-color green), (eql jacket-color blue)
Function Set	AND, OR, NOT, and EQ
Fitness Cases	The test set provided
Raw Fitness	The relevance score computed by RELIEF [-1,1]
Standardized Fitness	Raw Fitness * 100
Hit	A sexpression scores 100.0
Parameters	Evaluation Ticks = 15 Population Size=100 Runs = 10 Crossover Rate = 75% Mutation Rate = 33% Reproduction selection = Fitness Proportionate Parsimony = None

Table 1: Tableau for GP

Table 2: Relief Scores for $F0 \oplus F1$ problem

F0	F1	F2	F3	F4	F5
0.45	0.55	-0.20	-0.18	-0.35	-0.25

Some of RELIEF's limitations are worth noting however. Measuring the distance between examples is sensitive to irrelevant features. Our experiments found that the ratio of examples to features must be 2 to 1 to get some accurate results. Another limitation is that spurious relationships can affect accuracy. A canonicalization of features into conjunctive normal form is used in COEV to prevent unnecessary duplication of features.

1.6 The Empirical Inductive Learner

We used an algorithm called LFC (lookahead feature construction) [Ragavan, et. al., 1993] as our empirical inductive learner in this experiment. LFC builds decision trees with the ID3 information gain utility measure combined with a lookahead search for combination of operators. The depth of the lookahead search is one of the parameters being evolved. This learner therefore performs extremely well on complex concepts, that is concepts involving a boolean combination of attributes and is well suited to MONK-type problems. Any other inductive category formation program such as C4.5 [Quinlan, 1993] could have been used as well, with similar results.

2. Experimental Design

The three Monks problems [Thrun, et. al., 1991] have been devised to allow comparison between different learning algorithms. More importantly, they have been devised to compare the degree of generality in the solutions achieved by different learning algorithms. The domain consists of 6 discrete-value attributes and all possible examples have been divided into a training and test datasets.

Our testbed consists of the first MONK problem ((head-shape = body-shape) *or* (jacket-color = red)) and a variation on this problem ((head-shape = body-shape) *and* (jacket-color = red)) which we will call the fourth MONK problem. This problem is significant because it involves the induction of a relationship, head-shape = body-shape, rather than just the identification of appropriate values for features. Discovering such relationships is difficult for most machine learning systems.

For the MONK problems, 124 of the 432 possible examples were randomly selected to constitute the training set and the remaining kept for the test set. We compare the effectiveness of both GP and LFC alone with their effectiveness in the cultural co-evolution framework. We tried two such cultural co-evolutions: one that divided the representation space between the two population of learners--GP searching only the more abstract space--and one that did not restrict the search space. The comparative results are done on the average over 10 runs. The code is written in CMU Common Lisp and the experiments were conducted on an SGI R3000.

The evolution is measured in "ticks" since in co-evolution mode, meme exchange alternates with reproduction. A tick represents a synchronization point where the accuracy of the learning agents is evaluated (Figure 1). After a meme exchange, a tick is equivalent to 100 evals since meme exchange affects all learners. After reproduction, a tick can be less than that. Roughly, in those experiments, a tick is equivalent to 100 evals. Future work will include an interpolation to evals.

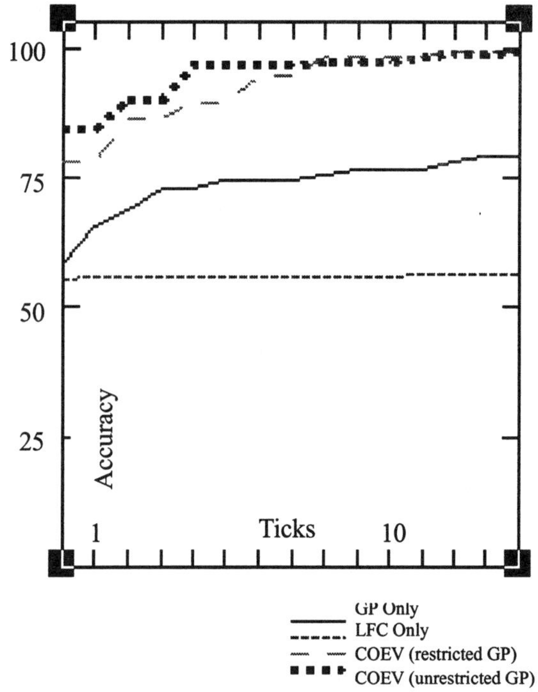

Figure 3. First MONK Problem - Average Best Offline

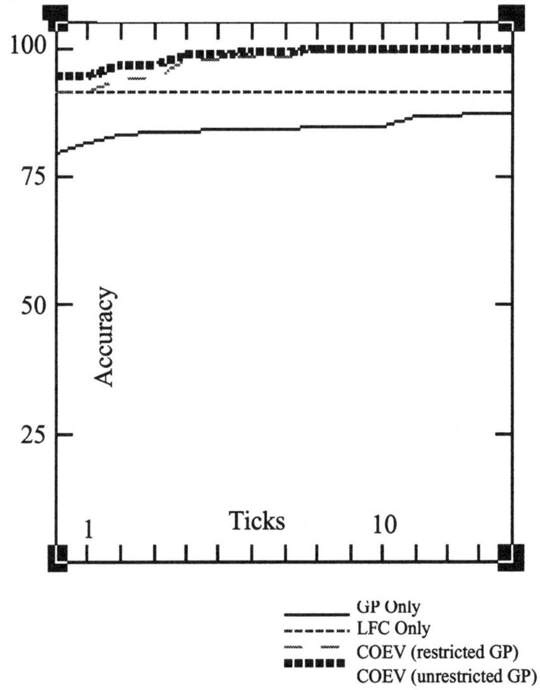

Figure 4. Fourth MONK Problem - Average Best Offline

3. Results

We ran 10 trials on a population of 100 LFC learners, a population of 100 GP learners. In these trials, the "genes" (learning parameters) were allowed to vary, thereby searching for optimal parameter settings, but no "memes" were exchanged. Two other sets of 10 trials were run on a co-evolving population (with "meme" exchange): one with a cultural co-evolution of 50 LFC and 50 restricted GP learners (the GP's representation space was restricted to head-shape, body-shape, holding, jacket-color, has-tie and is-smiling) and a cultural co-evolution of 25 LFC and 75 unrestricted GP learners. All trials were run for 15 ticks. No hits (i.e. populations with one or more individuals with 100% accuracy) were obtained on the evolution of the individual learners for the two problems. In contrast, cultural co-evolution of GP and LFC solved these two problems quite quickly. The Koza tableaux below in tables 3-6 show the computational resources required for obtaining a hit with a 99% probability [Koza 1992, Chapter 8]. Figures 3 and 4 show the learning curves for the four different attempts to solve the two problems.

Table 3: First MONK problem, 50 GPs 50 LFCs

Tick	First Time Hits	Cumulative Probability P(M, i)	Number of Independent Runs	Number of Individuals to Process
3	7	70%	4	1200
5	1	80%	3	1500
7	9	170%		

Table 4: First MONK problem, 75 GPs, 25 LFCs

Tick	First Time Hits	Cumulative Probability P(M,i)	Number of Independent Runs	Number of Individuals to Process
1	3	30%	13	1300
3	5	80%	3	900
5	8	160%		

Table 5: Fourth MONK problem, 50 GPs, 50 LFCs

Tick	First Time Hits	Cumulative Probability P(M,i)	Number of Independent Runs	Number of Individuals to Process
3	10	100%	1	100

Table 6: Fourth MONK problem: 75 GPs, 25 LFCs

Tick	First Time Hits	Cumulative Probability P(M,i)	Number of Independent Runs	Number of Individuals to Process
1	7	70%	4	400
3	6	130%		

4. Discussion

Cultural co-evolution learning generally demonstrates significant synergies between different learning methods, and the experience reported here bears that out. The biases associated with each learning method each have their strengths and weaknesses, and none is universally appropriate. Cultural co-evolution allows diverse learning systems to interact in mutually advantageous ways, without having to specify in advance how those interactions should take place.

Genetic programming exhibits quite different biases in searching hypothesis space than other induction methods, and it therefore seems to contribute significantly to the effectiveness of cultural co-evolution learning systems. COEV has been extended to add GP to its collection of learning methods (full COEV uses many different learning methods--the implementation used here was simplified to just two for illustrative purposes).

The exact form of the correct solutions arrived at by COEV are not always optimal. Like other GP systems, they sometimes contain redundant or unreachable code ("introns"). Figure 5 shows an example of some solutions found in the COEV system that included introns

```
(and (eql jacket-color 'red)
     (eq has-tie has-tie)
     (eq head-shape body-shape))

 (and (eql jacket-color 'red)
      (or (eq body-shape head-shape)
         (eq head-shape is-smiling)))
```

Figure 5. Cultural Co-evolution solutions involving Introns.

Because of the relatively small number of values for head and body shape, it is possible for an inductive learner to solve the problem by enumerating all of the head and body shapes, rather than by identifying the equality relationship between them. Some inductive learners (such as C4.5) can find the correct solution without cultural co-evolution in this way (Table 7). Additional data (not shown) demonstrates that this exhaustive enumeration approach quickly breaks down when the number of possible values is increased only moderately, and in such situations, cultural co-evolution can still solve the problem.

Table 7: Fourth MONK Problem C4.5

Tick	First Time Hits	Cumulative Probability P(M,i)	Number of Independent Runs	Number of Individuals to Process
4	1	10%	44	17600
8	1	20%	21	16800
12	1	30%	13	15600

Given a large enough population and enough clock ticks, GP would probably arrive at the correct solution, since its representation space does contain the solution. However, the cultural co-evolution method is at least an order of magnitude faster at this problem (and quite likely many orders of magnitude faster). The representation space is just too large for GP to search alone (Figure 2).

The effectiveness of GP in cultural co-evolution (as elsewhere) depends crucially on the function set. In this case, the function set included not only boolean combinations, but also EQ, which is appropriate for datasets with nominal terminals. In significant problems, that is, in ones where many different types of terminals and possible interactions among them are possible, selection of the function set can be difficult. A large function set makes the GP search space intractably large, while too small a function set means that the correct combination may be overlooked. Since the one of the main goals of cultural co-evolution learning is to remove much of the onus of "learning engineering" (i.e. selecting input representations, parameter settings, the function set, etc.) from the human setting up the problem, and place it inside the program/learning theory, this question cannot be evaded by hoping that the researcher will select the "right" function set.

What is required is a general set of operators that can be applied in all situations. Typed-GPs must be used, since not all operators can be applied to all types of data. As the number of applicable operators in the function set grows, the problem of failing to find appropriate combinations due to the size of the search space grows worse. It is not clear if it is possible to develop such a set of operators for a truly general purpose learning system. Working out the appropriate type system and set of operators, and determining possible fundamental limits is an important aspect of our future development efforts.

5. Conclusion

This paper presented GP as significant contributor to a cultural co-evolution learning system. The unique biases of GP as an inductive learner complemented a more traditional inductive learning system, demonstrating significant synergy. In our opinion, GP should be seen as an important contributor to multistrategy learning generally,.and to cultural co-evolution learning in particular. GP can provide the necessary exploration of the representation space that is missing in empirical inductive learners.

Future work will generally involve extending and applying the cultural co-evolution model as a whole and also improving the specifically genetic programming aspects of the system. For example, other transformations, such as counting arguments [Michalski, 1983], might profitably be added to the GP function set.

Acknowledgment

M. Abramson wishes to acknowledge the invaluable teaching of K. DeJong obtained at George Mason University and his encouragements and comments in this project.

Bibliography

[Altenberg, 1994] Altenberg, L., The Evolution of Evolvability in Genetic Programming, Advances in Genetic Programming, 1994, MIT Press.

[Bloedorn, et. al., 1993] Bloedorn E., Michalski R.S., and Wnek J., Multistrategy Constructive Induction: AQ17-MCI, Proceedings of the Second International Workshop on Multistrategy Learning (MSL-93).

[D'Hasseleer, 1994] D'Haeseleer P., Effects of Locality in Individual and Population Evolution, Advances in Genetic Programming, 1994, MIT Press.

[Hunter, 1995] Hunter, L., Cultural co-evolution Learning: Synergistic Evolution of Learning Agents and Problem Representations, Technical Report, available by anonymous FTP at lhc.nlm.nih.gov:/pub/hunter.

[Kira & Rendell, 1992] Kira, K. & Rendell, L. A., The Feature Selection Problem: Traditional Methods and a New Algorithm, Proceedings of the Tenth National Conference on Artificial Intelligence.

[Koza, 1992] Koza, J,. Genetic Programming: On the Programming of Computers by Means of Natural Selection, Cambridge, MA, MIT Press.

[Michalski, 1983] Michalski, R.S. "A theory and Methodology of Inductive Learning," Machine Learning: An Artificial Intelligence Approach, Vol. I, R.S. Michalski, J.G. Carbonell and T.M. Mitchell (Eds.), Palo Alto, CA: Morgan Kaufmann, 1983.

[Nordin, 1995] Nordin, P. & Banzhaf W., Complexity, Compression and Evolution, Proceedings of the Fifth International Conference on Genetic Algorithms, Morgan Kaufmann, 1995.

[Quinlan, 1993] Quinlan J. Ross, C4.5: Programs for Machine Learning, 1993, Morgan Kaufman Publishers.

[Ragavan, et. al., 1993] Ragavan, H., Rendell, L., Shaw M., Tessmer A., Learning Complex Concept Acquisition through Feature Construction, Proceedings of the 13th International Joint Conference on Artificial Intelligence, 1993.

[Thrun, et., al., 1991] Thrun, S.B., Bala, J., Bloedorn, E., Bratko, I., Cestnik, B., Cheng, J., De Jong, K., Dzerowski, S., Fahlman, S.E., Hamann, R., Kaufman, K., Keller, S., Kononenko, I., Kreuziger, J., Michalski, R.s., Mitchell, T., Pachowicz, P., Vafaie, H., Van De Velde, W., Wenzel, W., Wnek, J., Zhang, J., "The MONK'S Problems: A Performance Comparison of Different Learning Algorithms," Carnegie-Mellon University, Pittsburgh, PA, CMU-CS-91-197, 1991.

Type-Constrained Genetic Programming for Rule-Base Definition in Fuzzy Logic Controllers

Enrique Alba
Dpto. de Lenguajes y CC.CC.
Universidad de Málaga
Campus de Teatinos (2.2.A.6)
29071-Málaga (ESPAÑA)
alba@tecma1.ctima.uma.es

Carlos Cotta
Dpto. de Lenguajes y CC.CC.
Universidad de Málaga
Campus de Teatinos (2.2.A.6)
29071-Málaga (ESPAÑA)
cotta@tecma1.ctima.uma.es

José M. Troya
Dpto. de Lenguajes y CC.CC.
Universidad de Málaga
Campus de Teatinos (2.2.A.13)
29071-Málaga (ESPAÑA)
troya@tecma1.ctima.uma.es

ABSTRACT

A genetic programming approach to fuzzy logic controller (FLC) design is presented in this paper. We propose an encoding that represents fuzzy rule-bases as type-constrained syntactic trees implemented as variable-length strings. This encoding is applied to the cart centering problem and compared with other approaches: an intuitive FLC done by hand, an FLC obtained by a traditional genetic algorithm operating on fixed length strings and the analytical optimal solution (*bang-bang* rule). The results of the GP model are better (158 time steps) than the intuitive solution (249 time steps) and comparable with the GA solution (149 time steps). The analytical optimal solution docks the cart in 129 time steps.

1. Introduction

The genetic programming (GP) paradigm (Koza, 1992) is a relatively new field of research focused on the automatic breeding of computer programs by means of genetic algorithms (GAs) (Holland, 1975). The population of the GA is composed of syntactically correct trees each one representing a computer program that solves a given task. This task is presented to the GA as a set of examples that express the behaviour the solution tree must exhibit. The tree is composed of a set of functions (internal nodes of the tree) and a set of terminal symbols (leaves) that together conform a suboptimal solution to the problem.

In this work we try a new approach in which individuals (trees) are not computer programs in a traditional sense. Every tree will represent a set of fuzzy rules intended to solve a control problem (the cart centering problem -Fig. 1-, described and analytically solved in (Koza, 1992)). The underlying target programming language is neither C, PASCAL nor LISP. We use a fuzzy interpreter in order to compile and run the resulting fuzzy rules on our specific problem. We are interested in studying the new problems this field presents just as the power of the GP paradigm in solving a new field of application.

In this paper a brief review of fuzzy logic controllers and a description of the cart centering problem is first outlined, followed by the GP model used for the experiments. Finally, we present a performance analysis of our approach and compare it with a tailored fuzzy solution in order to conclude some interesting results in terms of efficacy, efficiency and new operators.

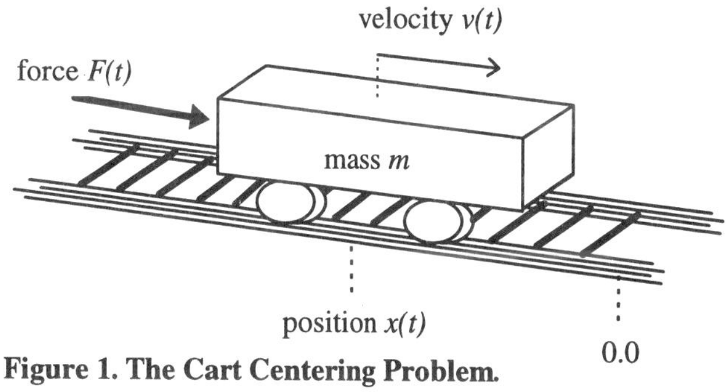

Figure 1. The Cart Centering Problem.

2. Fuzzy Logic Controllers

A Fuzzy Logic Controller (FLC) is a rule-based system that tries to incorporate the flexibility of human-decision making by means of the use of fuzzy set theory (Lee, 1990). The rules of an FLC incorporate fuzzy linguistic terms (e.g. temperature is warm) described by membership functions. These functions are intended to represent human expert's conception of the linguistic terms, thus giving an approximation of the confidence with which a precise numeric value is described by a linguistic label.

Rules take the form IF [conditions] THEN [actions], where conditions and actions are linguistic labels applied to input and output variables respectively (e.g. IF temperature IS hot THEN refrigeration IS high). A set of such fuzzy rules constitutes the fuzzy rule-base of the FLC. The system uses this rule-base to produce precise output values according to actual input values. This control process is divided into three stages:

- fuzzification: calculate fuzzy input, i.e. evaluate input variables with respect to the corresponding linguistic terms in the condition side.
- fuzzy inference: calculate fuzzy output, i.e. evaluate activation strength of every rule and combine their action sides.
- defuzzification: calculate actual output, i.e. convert fuzzy output into a precise numerical value.

The fuzzy interpreter used in this work performs fuzzification via triangular membership functions, uses the *min* intersection operator, the *maxmin* method for fuzzy inference and the *centroid* procedure for defuzzification (for comparation purposes with (Thrift, 1991)).

3. The Cart Centering Problem

In this section a description of the cart centering problem will be made. Next, it will be shown how the principles of fuzzy control can be applied to this problem, outlining an ad-hoc solution.

3.1. Problem Description

The cart centering problem definition involves a cart with mass m that can move along an infinite one-dimensional frictionless track. The problem consists in centering the cart, in minimal time, by applying an external time-variant force (of maximal magnitude F) so as to accelerate the cart in either the positive or the negative direction of the track. Given the values for the position $x(t)$ and velocity $v(t)$ of the cart at time t, this problem can be further formulated as calculating a force $F(t) = F[x(t), v(t)]$ in order to place the cart in position $x(t_f)=0$ with velocity $v(t_f)=0$ in minimal t_f.

Though simple, this is a typical optimal control problem and has a well-known solution (the *bang-bang* rule referenced in (Koza, 1992)). This exact time-optimal solution consists in applying a force $F(t)$=F when

$$-x(t) > \frac{v(t)^2 \, sgn[v(t)]}{2\frac{F}{m}}$$

and -F otherwise, where $x(t)$ and $v(t)$ are the position and velocity of the cart at time t.

This system can be simulated using Euler's method, i.e. considering time steps of duration τ and applying Newton's Laws:

$$x(t+\tau) = x(t) + \tau \cdot v(t)$$

$$v(t+\tau) = v(t) + \tau \frac{F(t)}{m}$$

As in (Thrift, 1991) and (Feldman, 1993), we will consider $\tau = 0.02$ sec. and $m = 2.0$ kg.

3.2. Fuzzy Approach

As an optimal control problem, the cart centering problem is well-suited for fuzzy control. In order to develop a FLC for this system, the first step consists in determining condition (input) variables. It is straightforward to see that position $x(t)$ and velocity $v(t)$ of the cart are the input variables in this problem. The membership functions defining the fuzzy sets over these inputs must be also specified in this step (Karr, 1991). We will consider the functions shown in Figure 2, i.e. five partitions representing the linguistic terms negative large (NL), negative small (NS), zero (ZE), positive small (PS) and positive large (PL).

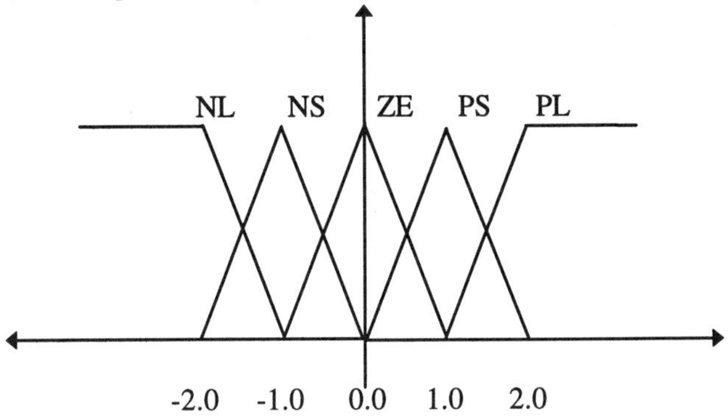

Figure 2. Fuzzy sets for position, velocity and force.

The second step is to identify action (output) variables. In this problem, there is just one output variable, the force $F(t)$ applied to the cart. For simplicity, the same fuzzy sets above will be used to describe this output variable.

The third step consists in the production of the rule-base. This rule set may be designed by a human expert. The use of fuzzy linguistic terms allows an easier incorporation of human knowledge. For example, it seems intuitive in the cart centering problem that the applied force must almost always be directed towards the origin (since we want to place the cart in the 0 position) and, just when we are close to that point, we must reverse that force to stop the cart. The following fuzzy rules express these rules-of-thumb:

IF *pos* IS PL THEN *for* IS NL
IF *pos* IS PS THEN *for* IS NL
IF *pos* IS ZE AND *vel* IS PL THEN *for* IS NL
IF *pos* IS ZE AND *vel* IS PS THEN *for* IS NL
IF *pos* IS ZE AND *vel* IS NS THEN *for* IS PL
IF *pos* IS ZE AND *vel* IS NL THEN *for* IS PL
IF *pos* IS NS THEN *for* IS PL
IF *pos* IS NL THEN *for* IS PL

where *pos*, *vel* and *for* stand for position, velocity and force respectively.

The FLC defined by the membership functions and the rule-base described above is not the optimal one. It can be improved by means of a learning technique that modifies the rule set or the interpretation of the linguistic terms. Genetic algorithms have been used for that purpose in both ways (see (Thrift, 1991), (Karr, 1991), (Lee and Takagi, 1993), (Feldman, 1993)). We propose the use of GP to produce rule bases that can be used as a starting point for further refinements by a human expert.

4. Genetic Programming of Fuzzy Logic Controllers

The genetic programming model will be described in this section. The basic encoding is first shown. Next, the type system is exposed. Finally, some general details about the whole model are outlined.

4.1. Encoding a Rule-Base

As stated above, the rule-base of the FLC is a list of `IF-THEN` rules. Each of them can be easily represented as a binary tree: the root being an `IF` node and left and right branches representing the condition and the action sides respectively. Likewise, both conditions and actions can be expressed as trees. On the one hand, a variable paired with a fuzzy set can be represented as a tree with an `IS` root-node, the variable name in the left branch and the fuzzy set in the right branch. On the other hand, a conjunction of such terms can be expressed as a tree with an AND root-node and two branches representing nested conjunctions or pairs (variable, fuzzy set). Figure 3 shows an example.

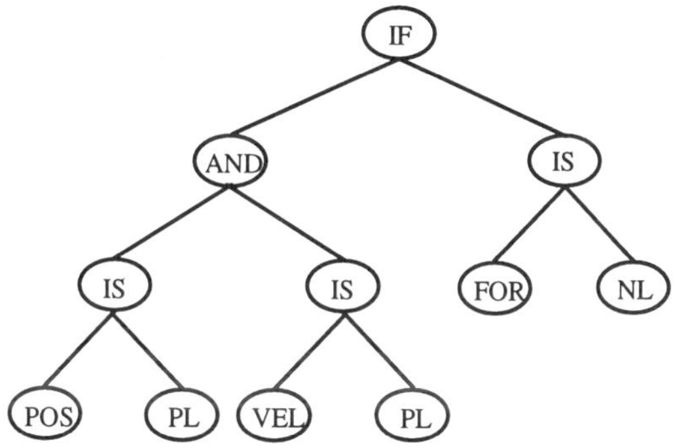

`if` *`pos`* `is PL and` *`vel`* `is PL then` *`for`* `is NL`

Figure 3. A syntactic tree and the rule it represents.

A list of rules might be represented as a list of trees. However, we think it is more appropriate to use a single tree representation. It becomes then necessary to incorporate an additional symbol to join many rules. This extra symbol (`RLIST`) combines rules in the same fashion that the AND symbol joins terms, i.e. a list of rules is a tree with an `RLIST` root-node and two branches representing single rules or analogous lists of rules.

This is a very flexible and powerful representation of FLC rule-bases. It deals well with the fact that an FLC may have rules with very different structures among them as well as a variable number of rules. This aspect contrasts with other GA-based approaches in which the structure (Thrift, 1991) (Lee and Takagi, 1993) or the number of rules in the FLC (Feldman, 1993) is arbitrarily fixed.

4.2. Adding Types to Nodes

The flexibility of this tree representation makes crossover a very powerful tool since it allows interchange at several levels: rules, terms or even variable names. However, the traditional crossover operator may produce nonsense rules. For example, consider the case in which a variable name is substituted by a whole rule. There are three possible solutions for that problem:

- Define closed functions, so those ill-defined rules are reinterpreted in some predefined way. For example, the `IS` function might consider anything different from a variable name in the left branch as a dummy variable with a fixed value.
- Repair the tree, deleting incorrect subtrees and adding new subtrees if necessary.
- Perform a restricted crossover, i.e. select crossover points in such a way that the resulting trees be correct (Montana, 1993).

Since the first option will produce a high number of useless trees composed mostly of dummy rules, and the second one is computationally expensive and will produce offspring very different from each parent (due to the repairing mechanism), we have chosen the third option.

In our system, every symbol has a type. The crossover operator first selects a random node in one parent; subsequently, a node of the same type is randomly selected in the other parent and the corresponding subtrees are swapped, thus producing two valid offspring. The type of each symbol is determined by means of the partitions defined by the equivalency relationship "is interchangeable with". For example, since an `EL` node (empty list) may be swapped with an `RLIST` node, they must have the same type. Table 1 shows the partitions.

Table 1. Four basic type partitions and the symbols belonging to each one.

TYPE	SYMBOLS
Type I	{ `IF, RLIST, EL` }
Type II	{ `AND, IS` }
Type III	{ ... variable names ... }
Type IV	{ ... fuzzy sets ... }

However, considering just the types above will lead to the creation of syntactically correct but semantically incorrect trees. For instance, a rule could be spawned in which output variables appear in the condition side and/or input variables appear in the action side (e.g. IF *for* IS NL THEN *pos* IS ZE). If this kind of situations were allowed, the GP system would waste valuable resources in evaluating and storing such useless rules. To avoid this problem, we define two subtypes for both type II and III.

Every subtype can be seen as the variety of the corresponding type that appears in either the condition side or the action side. For example, for type VBLE (type III), we define the subtypes VBLE_IN and VBLE_OUT, representing input and output variables respectively. A complete list of types and subtypes is shown in Table 2. Types are listed in the same order as in Table 1.

A list of all symbols (including those specific for the cart centering problem) is shown in Table 3.

Table 2. The meaning of the types and subtypes of the nodes in the tree.

TYPE	MEANING	SUBTYPES	MEANING
LIST	List of rules	-	-
COND_ACT	Condition or Action	CONDITION ACTION	To separate branches of an IF structure
VBLE	Variable	VBLE_IN VBLE_OUT	Input variable Output variable
SET	Fuzzy Set	-	-

Table 3. Symbols, their types, argument description and a brief explanation of their role.

SYMBOL	MEANING	ARGUMENTS	TYPE - SUBTYPE
POS	POSITION of the cart	Terminal	vble - vble_in
VEL	VELOCITY of the cart	Terminal	vble - vble_in
FOR	FORCE to be applied	Terminal	vble - vble_out
PL	POSITIVE LARGE fuzzy set	Terminal	set
PS	POSITIVE SMALL fuzzy set	Terminal	set
ZE	ZERO fuzzy set	Terminal	set
NS	NEGATIVE SMALL fuzzy set	Terminal	set
NL	NEGATIVE LARGE fuzzy set	Terminal	set
EL	EMPTY LIST	Terminal	list
IF	IF_THEN conditional rule	(condition,action)	list
LEFT_AND	AND in left hand side of IF	(condition,condition)	cond_act - action
LEFT_IS	IS in left hand side of IF	(vble_in,set)	cond_act - action
RIGHT_AND	AND in right hand side of IF	(action,action)	cond_act - condition
RIGHT_IS	IS in right hand side of IF	(vble_out,set)	cond_act - condition
RLIST	A list of rules	(list,list)	list

4.3. GP details

We have used an elitist generational GA to evolve trees. These trees are bounded to have a maximum depth of 6 levels in the initial generation and 15 in subsequent generations. Initially, the population is generated using a half-ramping mechanism (Koza, 1992). We follow the BNF grammar shown in Figure 4 to produce correct trees. Notice that only the last two rules (describing input and output variables of the problem) are problem-dependent. The rest of rules can be used in any other fuzzy problem.

Figure 4. BNF grammar of correct trees. Trees are described and implemented as postordered strings.

```
<TREE>     ::= EL | <IF> | <RLIST>
<RLIST>    ::= <TREE> <TREE>  RLIST
<IF>       ::= <COND> <ACT> IF
<COND>     ::= <L_IS> | <L_AND>
<L_IS>     ::= <VBLEIN> <SET> LEFT_IS
<L_AND>    ::= <COND> <COND> LEFT_AND
<ACT>      ::= <R_IS> | <R_AND>
<R_IS>     ::= <VBLEOUT> <SET> RIGHT_IS
<R_AND>    ::= <ACT> <ACT> RIGHT_AND
<SET>      ::= NL | NS | ZE | PS | PL
<VBLEIN>   ::= VEL | POS
<VBLEOUT>  ::= FOR
```

Every tree is generated in a top-down fashion, i.e. choosing a root-node and then producing appropriate subtrees in a similar way. To perform this procedure, a node-type relationship list is maintained. This list indicates which types are valid for a specific node according to its parent node (see Table 3). The possible relationships of a node with a type are shown in Table 4. For the very first node (the root of the whole tree) only type I (LIST) is allowed. This implies that the root of a tree representing a rule-base may be only `EL`, `IF` or `RLIST`.

Table 4. Relationships between a node and its type for the initial generation of trees.

Node-Type Relationship	Meaning
NONE	Type not allowed
ANY	Type allowed
LEFT_FLAVOR	Only *left* subtype allowed
RIGHT_FLAVOR	Only *right* subtype allowed

When generating a node, a random type is chosen and subsequently a random symbol belonging to that type is assigned to that node. The selection of this random symbol depends on the level of the node. For example, an `IF` node cannot appear in levels below 3 because its branches must hang two levels down (an `IS` subtree) at least. The same is true for AND nodes. Moreover, when generating a complete tree, `IS` nodes only appear at level 2 (appearing in a higher level will produce an incomplete tree) and `RLIST` nodes must reside at levels higher than 3 because the `EL` symbol is not allowed to appear in complete trees.

In every generation, the two best individuals are copied to the next generation. The rest of the population is obtained through crossover operations between selected individuals. The selection mechanism selects 50% of the population according to scaled fitness and 50% randomly. This selection procedure has revealed itself as appropriate in other domains and represents a good tradeoff between exploitation and exploration.

To calculate one individual's fitness, 16 simulations of the problem are performed, starting from equally spaced points in the input surface. These simulations are executed for a maximum number of 500 time steps. The cart is considered to be centered if both position and velocity are close enough to 0 (if *max*(|*pos*|,|*vel*|)<.5 as in (Thrift, 1991)). If the cart could not be centered, the maximum number of steps is taken. The mean time of all simulations is subtracted from 500 to obtain a fitness value (to be maximized).

Table 5. Tableau for the cart centering problem.

Fitness cases	16 equaly spaced points (*x*,*v*) taken from the interval [(-2.5, -2.5), (2.5, 2.5)].
Raw fitness	Mean time, over the 16 fitness cases, needed to center the cart. Timeout contribution is 500 time steps.
Selection	Elitist (2 best), 50% fitness proportional and 50% randomly.
Crossover	restricted for crossing only compatible types. Pc=1.0.
Mutation	none.
Replacement	Generational.
Parameters	M=500 individuals, G=51 generations.

5. Results

In this section we show the results the GP model has produced. For comparison purposes, we also include the results of the intuitive solution described in section 3.2. and some results from related works.

5.1. Optimal and intuitive solutions

To measure the performance of a given FLC, we use the mean time to center the cart throughout 100 simulations of the problem, starting from random initial points in the domain [-2.5, 2.5]. We exclude those situations in which the initial state is also a solution state. As stated in section 3.1., the parameters of the simulations were $\tau = .02$ sec, $m = 2.0$ kg. Force has a cut-off value of 2.5 N. Simulations are executed for a maximum number of 500 time steps.

Whilst the optimal solution averaged 129 time steps to center the cart, our naive solution took 249 time steps, timing-out 14 times. Excluding those cases, the mean time is 205 time steps, still far from the optimal solution. In Figure 6, the control surfaces of both solutions are shown.

We tested the 18 rule-FLC described in (Thrift, 1991) too. Its performance is better than our *naive* solution. The mean time to center the cart is 149 time steps (for the parameters of our simulator).

5.2. The GP solution

We performed 10 runs of our GP system. Following (Koza, 1992) we chose a population size of 500, for a maximum number of 51 generations (the initial one plus 50 additional generations). Figure 5 shows the best and the mean fitness in every generation, averaged for the 10 runs.

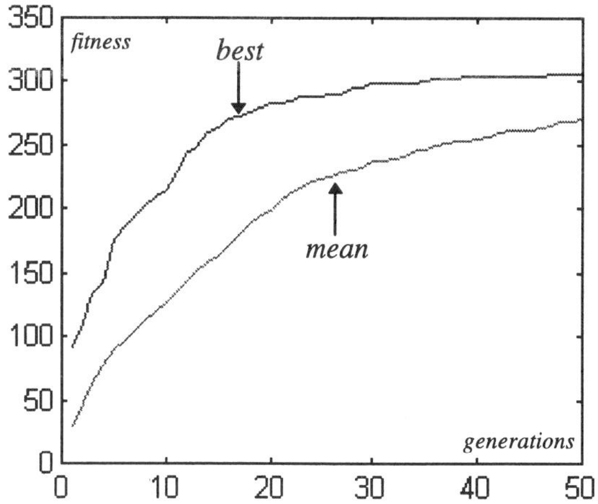

Figure 5. Best and mean fitness versus generations averaged for 10 runs of the GP system.

The best generated individual obtained a fitness value of 321 (averaging 158 time steps to dock the cart), and corresponded to the following 9 rule-FLC:

1- IF *pos* IS PL THEN *for* IS NL
2- IF *pos* IS PS THEN *for* IS NL
3- IF *pos* IS NL THEN *for* IS PL
4- IF *vel* IS PL THEN *for* IS NL
5- IF *vel* IS NS THEN *for* IS PL
6- IF *vel* IS NL THEN *for* IS PL
7- IF *pos* IS ZE AND *vel* IS PS THEN *for* IS NL
8- IF *pos* IS NS AND *vel* IS ZE THEN *for* IS PL
9- IF *pos* IS PS AND *pos* IS PL AND *vel* IS ZE AND *vel* IS NS THEN *for* IS NL AND *for* IS NS AND *for* IS PL

Actually, the rule-base above was obtained editing this best individual. In fact, rules #2, #3 and #6 were duplicated and rules #1 and #4 were repeated three times. That indicates they are important for the FLC to perform well. The best individual also contained two dummy rules whose condition side is equivalent to an empty fuzzy set, so they carry no influence on the output.

Notice the structure of rule #9. According to the grammar shown in Figure 4, this rule is syntactically correct. It is also semantically correct since input variables only appear in the condition side and output variables are located just in the action side. Fuzzy inference and defuzzification will give an interpretation to this rule. As stated in section 3.2., the GP system was intended to generate rule-bases for an FLC with fixed predefined fuzzy sets over inputs and outputs. However, new fuzzy sets can be expressed by means of the union and intersection of those predefined fuzzy sets, allowing their evolution in some sense.

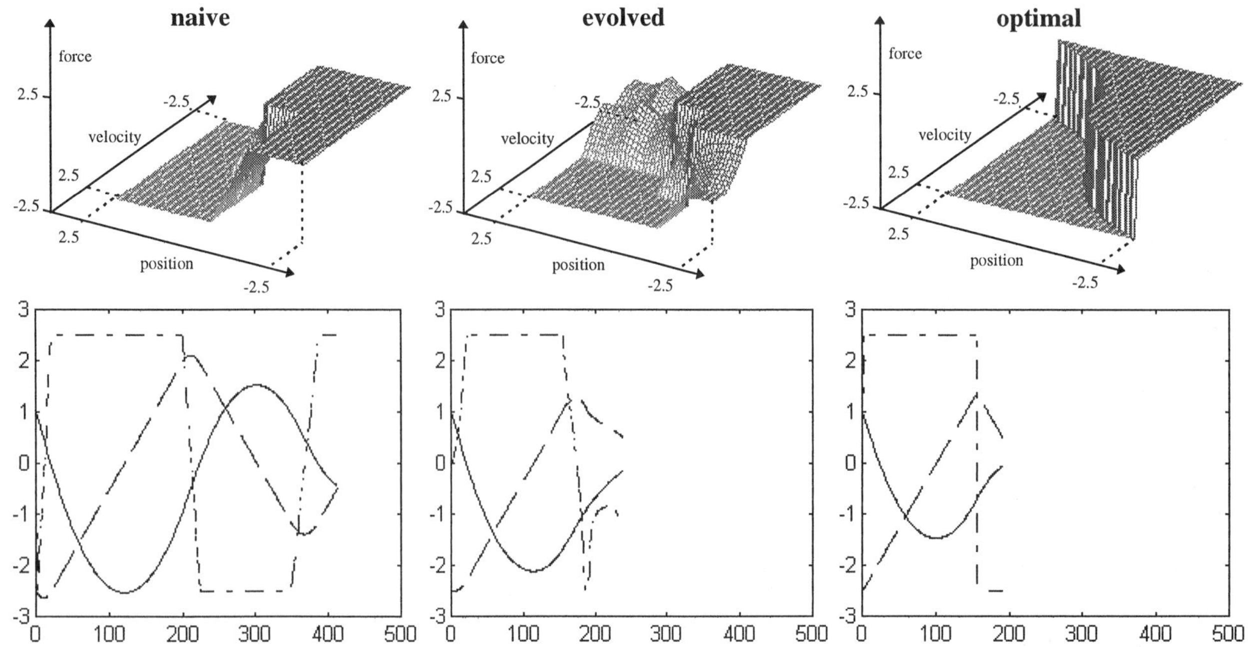

Figure 6. Control surfaces and responses for pos = 1.0 m, vel = -2.5 m/s of the naive (left), evolved (middle) and optimal solution (right). The solid, dashed and dashdotted lines represent position, velocity and force respectively.

6. Conclusions

A GP approach to FLC design has been presented. Genetic programming seems to be well suited to solve this task as it allows the evolution of structures working at a higher conceptual level than simple genetic algorithms: GP encoding of an FLC rule-base is much more flexible and natural than binary strings and since nothing is supposed on whatever the structure of a good rule or even the number of them should be (excluding of course syntactically or semantically incorrect rules), any underlying (known or unknown) structure can evolve to improve the solution.

The results obtained are encouraging. A 9-rule FLC has been generated clearly outperforming an intuitive solution. Its performance is comparable with an 18-rule FLC described in (Thrift, 1991) with just the half number of rules. The extensibility of this approach is also remarkable. In other GA-based systems in which the structure or the number of rules are fixed, increasing the number of variables involved in the test problem implies a growth in chromosome length, needing large populations to keep diversity and therefore leading to very long runs of the GA. In the GP system, it would be necessary to increase the maximum depth of trees to allow more complex rules and FLCs to evolve, but, since loss of diversity is not a main issue in GP (Koza, 1992), population size could be kept constant (or just slightly increased).

Future work could be directed to evolve more sophisticated FLCs including modifiers and credibility factors (as in fuzzy networks (Feldman, 1993)). More difficult test problems (e.g. the cart-pole balancing) and novel GP techniques as ADF (Kinnear, 1994) for composed-set detection will be tried too.

Bibliography

Feldman, D.S. 1993. Fuzzy Network Synthesis with Genetic Algorithms. In Forrest S. (Ed.), *Procs. of the Fifth International Conference on Genetic Algorithms,* Morgan Kaufmann Pub., San Mateo CA. Pages 312-317.

Holland, J.H. 1975. *Adaptation in Natural and Artificial Systems*, University of Michigan Press, Ann Arbor.

Karr, C.L. 1991. Design of an Adaptive Fuzzy Logic Controller using a Genetic Algorithm. In Belew R.K., Booker L.B. (Eds.), *Procs. of the Fourth International Conference on Genetic Algorithms*, Morgan Kaufmann Pub., San Mateo CA. Pages 450-457.

Kinnear, K.E. 1994. *Advances in Genetic Programming*, Chapters 5-6. MIT Press, Cambridge MA.

Koza, J.R. 1992. *Genetic Programming*, MIT Press, Cambridge MA.

Lee, C.C. 1990. Fuzzy Logic in Control Systems: Fuzzy Logic Controller - Parts I & II, *IEEE Transacts. on Systems, Man and Cybernetics*. 20:2. Pages 404-435.

Lee, M.A. Takagi, H. 1993. Dynamic Control of Genetic Algorithms using Fuzzy Logic Techniques. In Forrest S. (Ed.), *Procs. of the Fifth International Conference on Genetic Algorithms*, Morgan Kaufmann Publ., San Mateo CA. Pages 76-83.

Montana, D.J. 1993. *Strongly Typed Genetic Programming*, Bolt Beranek & Newman, Inc. Technical Report #7866.

Thrift, P. 1991. Fuzzy Logic Synthesis with Genetic Algorithms. In Belew R.K., Booker L.B. (Eds.), *Procs. of the Fourth International Conference on Genetic Algorithms*, Morgan Kaufmann Pub., San Mateo CA. Pages 509-513.

The Evolution of Memory and Mental Models Using Genetic Programming

Scott Brave
Computer Science Dept.
Stanford University
Stanford, California 94305
brave@cs.stanford.edu

ABSTRACT

This paper applies genetic programming to the evolution of intelligent agents that gradually build internal representations of their surroundings for later use in planning. The method used allows for the creation of dynamically determined representations that are not pre-designed by the human creator of the system. In an illustrative path-planning problem, evolved programs learn a model of their world and use this internal representation to plan their successive actions. The results show that the proposed method is successful in evolving programs that solve the planning problem and is thus a worthy basis for further investigation.

1. Introduction

Intelligent behavior depends on the ability to learn and retain information about the world for later use. Though many tasks are solvable by purely reactive systems, more complex tasks require the creation and utilization of mental models.

If genetic programming is to be applied to produce agents capable of building and using stored representations, mechanisms for the manipulation of memory must be introduced. The most basic form of memory, as described by Koza (1992), involves functions which read and write values to a small fixed number of individually named storage locations. For example, there might be two settable variables `M0` and `M1`, and two functions `SET_M0` and `SET_M1` to set their values (figure 1a).

Teller (1994a) extended this basic notion of memory by providing the evolving programs with functions to write and access a linear array indexed memory (as shown in figure 1b). In his work, Teller demonstrates the evolution of indexed memory on a simple problem which requires the use of state for success. In this problem, the agent's task is to push several boxes to the walls of a grid world. Memory is required for success because the evolved program is unable to sense whether a box is at the wall. Teller (1994b) proves that his memory paradigm is Turing complete. However, there is no evidence in Teller's work of anything resembling a mental model or internal representation of the world. In fact, the simplest solution to the box pushing problem is to use the indexed memory as a counter.

Andre (1994) extended Teller's indexed memory to a mental model that was isomorphic to the world (figure 1c). The task here is for an agent to collect all of the gold scattered in a five by five toroidal grid. To require the use of a memory representation, the evaluation of an individual occurs in two stages. In the first stage, the agent is allowed to look around the world and write to a five by five memory, but not pick up any gold. In the second stage, the agent can access its memory, but is unable to see the gold.

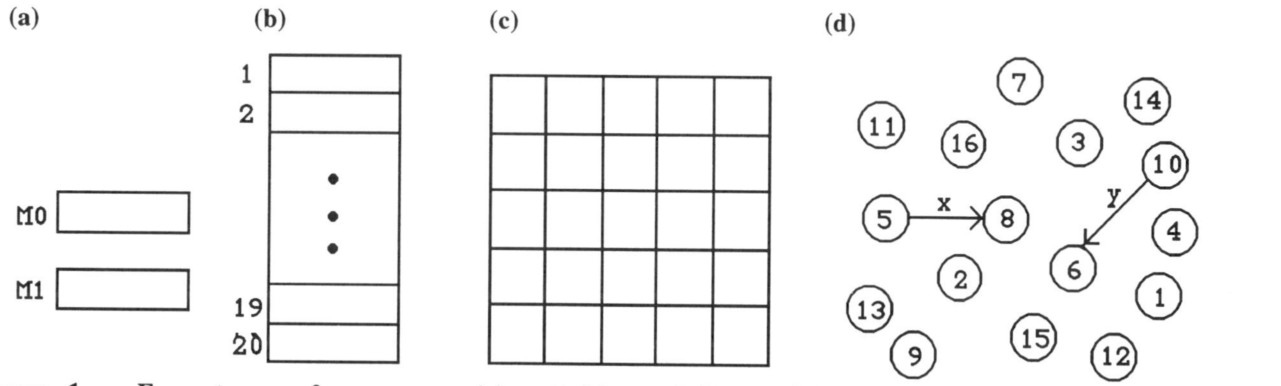

Figure 1 Four types of memory (a) settable variables, (b) linear indexed memory, (c) 5x5 memory isomorphic to a 5x5 world, and (d) pallet of isolated nodes with dynamically labeled connections.

Because there is a penalty for attempting to pick up non-existent gold, an evolved solution must make use of its memory to represent the location of the gold. Although this system has the potential to create a plan (a mental representation of the path to follow to reach all of the gold, for example), such planning is not needed to solve the problem. Moreover, such behavior does not evolve. What evolves instead is a mental representation that the agent, in the second phase, searches much in the way it would search for the gold itself. Because the provided memory maps directly to the world, the agent can move both in the world and in memory to a specific location and then check to see if the memory indicates gold before attempting to pick up. We would prefer that the structure of memory be defined by the agent and not the human researcher.

This paper presents a problem and a method which both allows for the dynamic structuring of memory and requires the creation of plans. While evolutionary methods have often been used to produce specific representations or plans, the possibility of evolving the ability to represent and plan has not been fully explored.

2. Path-Planning Problem

The to-be-evolved agent acts in a world represented by an N by N toroidal grid. After exploring the world, the agent is required to produce a plan for reaching every arbitrary location in the world from every arbitrary starting point. We use a dual-phase fitness evaluation similar to that used in Andre's gold collection problem (Andre 1994). In the exploration phase, the agent may move around the world and write to memory, while in the planning phase the agent may only refer to its memory to produce the specified navigation plans. As seen in figure 1d, the form of memory used is a set of isolated nodes that the agent may structure freely to create a directed graph. The number of nodes is specified, but the connections between the nodes and the labels written to these arcs are dynamically created by the agent while exploring in phase one. This memory is intended to serve as the basis for creation of the mental model which will be used in the second phase for planning. In the exploration phase, the agent is also supplied with a small buffer for use as short-term memory (to potentially remember a small number of past moves). Figure 2 illustrates the interaction between the agent, the world, short-term memory, and the mental model in the exploration phase.

In the planning phase, the agent must make use of the model created in the exploration phase to produce a navigational plan for each fitness case. The agent writes this plan to an external buffer as shown in figure 3.

The dual-phase fitness is accomplished by evolving individuals with two main branches. The first branch is evaluated once for each location from which the agent is allowed to explore. This number varies from one (if the agent is only allowed to explore starting from a single location) to the total number of locations in the world. The memory is preserved between evaluations of the branch. Next, the second branch is evaluated once for each fitness case (a specified starting and ending location) and writes

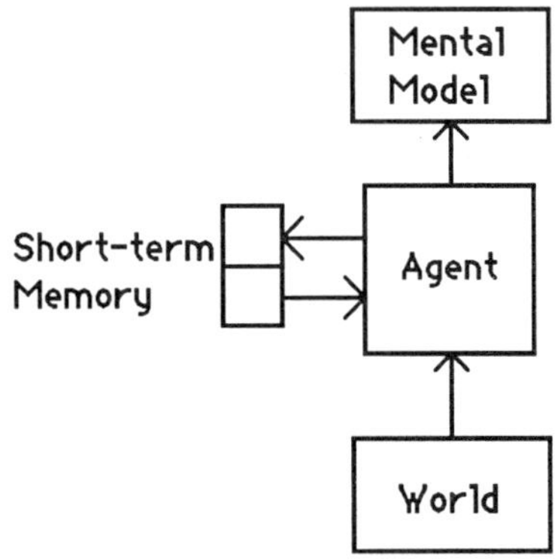

Figure 2 Interaction between the agent, the world, short-term memory, and the mental model in the exploration phase.

directions to an external buffer. This branch is given access to memory through functions that manipulate a node pointer. When the second branch is finished evaluating, the plan in the buffer is executed from the start location. The distance away from the end location is the agent's fitness for that case and the sum of its fitness over all cases is the individual's fitness.

The experiments in this paper use world sizes of 3x3 and 4x4 with the buffer for short-term memory of sizes 1 and 2 respectively. The first phase is evaluated either once or NUMBER_OF_LOCATIONS times depending on the run. All possible combinations of distinct start and end locations comprise the set of fitness cases. As a simplification for these first experiments, the number of available nodes is equal to the number of locations in the world and the direct mapping of location numbers to node numbers is implicitly required. For example, if a specific fitness case has location 1 as its start and location 5 as its end, the memory pointer of the second phase will start at node 1 and the agent will be able to determine when the pointer is at node 5. Although this does force some direction to the structure of memory, the number of nodes used and their connections are still dynamically determined by the actions of the first branch.

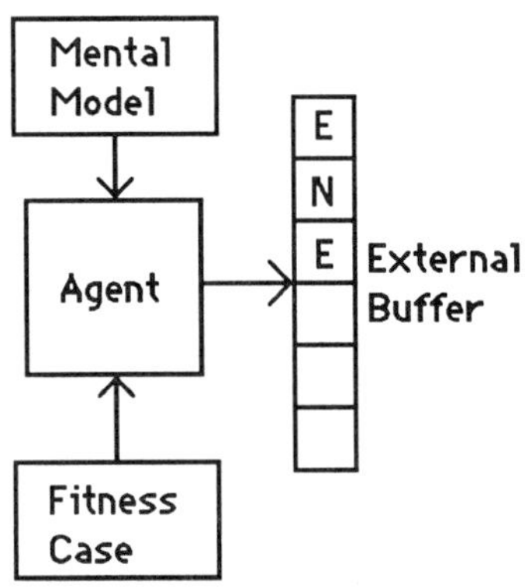

Figure 3 Interaction between the agent, the world, and the mental model in the planning phase.

Additionally, this simplification is one which can easily be removed with the addition of arithmetic functions for mapping in later experiments.

3. Genetic Programming Preparatory Steps

A tableau summarizing the genetic programming parameters is given as table 1. Four of the functions available to branch 1, evaluated during the exploration phase, allow for movement within the world. NORTH, SOUTH, EAST, and WEST move the agent one grid space in the corresponding direction. Because the world is toroidal, movement is always successful.

Four PUT functions are made available for writing to the short-term memory buffer. Each function writes a different value to the next available space in the buffer. If the buffer is full, space is made by shifting the values so that the oldest value is removed. Although the four values that can be written to the buffer are necessarily interpreted by the second branch as a one-to-one correspondence to directions, the described method does not require this. The addition of conditionals and arithmetic to both branches would allow for arbitrary interpretations to evolve. The CLEAR function simply erases the buffer. The worth of this function was not investigated.

CONNECT1 creates a directed arc from the previous node (the node that corresponds to the previous location visited) to the current node (the node that corresponds to the current location). The value written to this arc is the current value of the short-term memory buffer (possibly NULL). After creating the arc, the buffer is cleared. CONNECT1 is used in both the experiments that use a simple 3x3 world and those that use a 4x4 world. CONNECT2 is identical to CONNECT1 except that the arc originates from the node corresponding to the location visited two steps ago instead of the location visited previously. This function is used only with 4x4 worlds. The CONNECT functions will not connect a node to itself nor will they create an arc from one node to another if an arc already exists between the two nodes. Additionally, the number of arcs leaving any one node was limited to 5. Attempting to create an arc originating from a node with 5 arcs leaving it already does nothing.

The model created by the explorer, however, accomplishes nothing unless it is used by the second branch to aid in planning. To access the model, the second branch is given a node pointer which is placed at the node corresponding to the starting location before each fitness case evaluation. Movement within the model is achieved through the GO_CHILD function. This function moves the node pointer to the next unvisited child, marks all of this child's children as unvisited, and then evaluates its argument. After the argument is evaluated, the node pointer is reset to point to the parent node. In addition the GO_CHILD function takes care of writing to the external buffer which holds the plan to be executed. When the node pointer is moved along an arc to a child, the value on the arc is written to the next available space in the buffer. When the node pointer is moved backward along an arc (when resetting the pointer to the parent), the value on the arc is removed from the buffer.

IF_EQU executes its first argument if the node pointer is pointing to the node corresponding to the goal location and executes its second argument otherwise. The plan currently specified in the buffer is executed when the EXECUTE function is called. The fitness of the individual depends only on the final location of the agent after execution of the specified plan from the starting location.

Table 1 Tableau for the path-planning problem.

Objective:	Find a program to control an agent that, after exploring the world, can produce a correct plan for reaching every arbitrary location in the world from every arbitrary starting point.
Function set for the exploration phase main branch (branch 1):	NORTH, SOUTH, EAST, WEST, PUT_N, PUT_S, PUT_E, PUT_W, CLEAR, PROGN(arg1, arg2), CONNECT1, CONNECT2 - used only in 4x4 worlds, ADF0, ADF1
Function set for ADF0:	Same as exploration main branch without ADF0.
Function set for ADF1:	Same as exploration main branch without ADF0 or ADF1.
Function set for the planning phase main branch (branch 2):	IF_EQU(arg1, arg2), GO_CHILD(arg1), EXECUTE
Function set for ADF2:	Same as planning main branch without ADF2.
Function set for ADF3:	Same as planning main branch without ADF2 or ADF3.
Fitness cases:	All combinations of distinct starting and goal locations in the world.
Fitness:	Sum over all fitness cases of the Manhattan distance away from the goal after execution of the plan.
Parameters:	Population size (*M*): 5,000 Maximum number of generations (*G*): 150
Maximum number of nodes per individual:	Each main branch: 200 Each ADF: 100
Success predicate:	A fitness of 0 indicating that the program is able to produce a correct plan for all fitness cases.

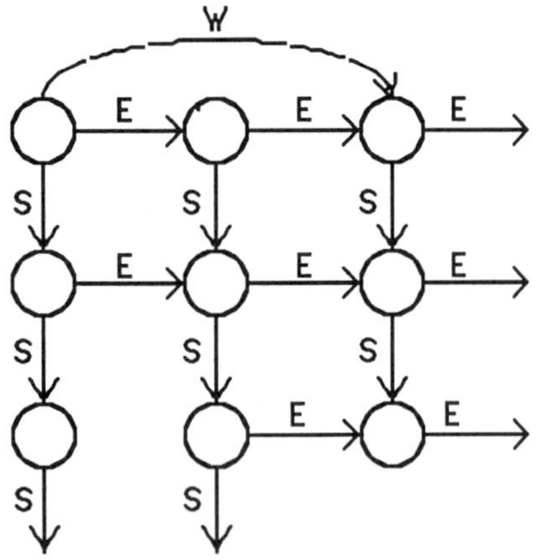

Figure 4 Example of model created by solution to 3x3 world. Solution appeared on generation 39.

All experiments in this paper use a population size (M) of 5,000 and 150 as the maximum number of generations (G). Each main branch (maximum of 200 nodes) is given access to two automatically defined functions (maximum of 100 nodes each). Automatically defined functions (ADFs) are reusable subroutines which are defined once but may be called repeatedly (Koza 1994). As mentioned previously, the fitness function used is the sum of the distance away from the goal over all possible combinations of distinct start and goal locations.

4. Results

The first runs were made using a 3x3 world with a single evaluation of the first branch starting from location 0 (i.e. the explorer was only allowed to explore the world from a single location). The size of the short-term memory buffer is one for 3x3 worlds. Figure 4 shows the model created by the solution which appeared on generation 39 of one run, and figure 5 shows the model of a solution found on generation 30 of another run (location 0 corresponds to the top left node). For illustration, the nodes of the model are

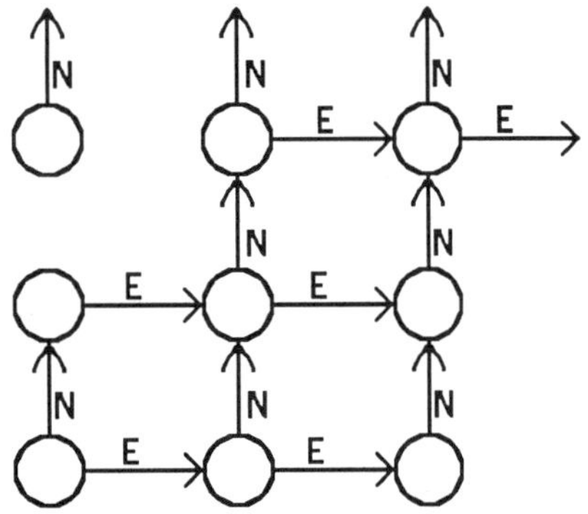

Figure 5 Example of model created by solution to 3x3 world. Solution appeared on generation 30.

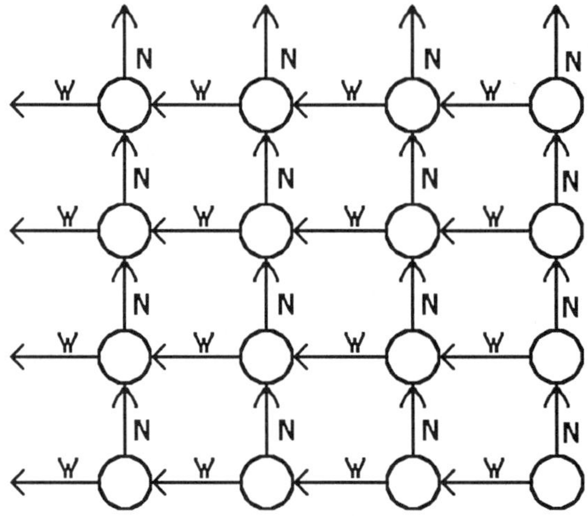

Figure 6 Example of model created by solution to 4x4 world. Solution appeared on generation 64 without iteration of the first branch and on generation 36 with iteration.

drawn with the same spatial orientation as the corresponding world locations; however, we must remember that there is no inherent spatial orientation of the nodes.

Recall that the fitness of the individual depends only on the final distance away from the goal location and not the efficiency of the plan. Models which contain all information about the world are thus not necessary. Four of the ten runs executed using the 3x3 world found solutions (on generations 32, 39, 27, and 37).

The evolution of model building was next attempted on 4x4 worlds using a short-term memory buffer size of two. Only one of five runs was able to find a solution by generation 150. The solution, which appeared on generation 64, built the model shown in figure 6.

The symmetry of this and the other models arises form the inclusion of ADFs in the function set. Upon inspection of this run, it was clear that a large part of the evolution was

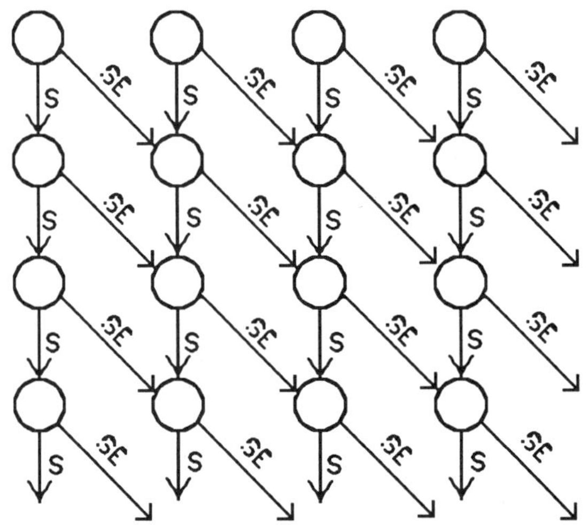

Figure 7 Example of model created by solution to 4x4 world. Solution appeared on generation 82 with iteration of the first branch.

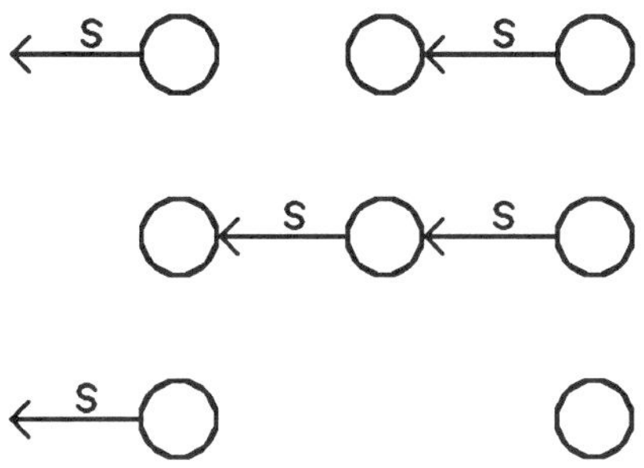

Figure 8 Generation 12, fitness = 98.

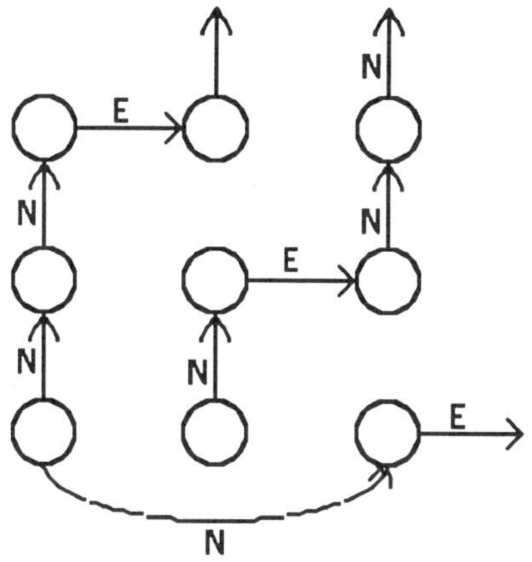

Figure 10 Generation 24, fitness = 54.

spent developing the iteration of one or two of the ADFs. To address this issue, iteration was made explicit in the fitness evaluation. The first branch of the program was executed `NUMBER_OF_LOCATIONS` times with the agentstarting at each location in the world. As expected, the runs using this method were more successful (nine of ten runs solved on generation 61, 131, 36, 88, 93, 82, 78, 127, and 43). A solution which built the identical model to that shown in figure 6 appeared on generation 36. The model in figure 7 was created by a solution found on generation 82 of another run using the iterated first branch.

5. Discussion

To better understand the evolution of mental model creation we will look at the run whose solution created the model shown in figure 5. Because the fitness cases do not include combinations of locations where the start and the goal are the same, individuals are rewarded for any non-empty plan. While the worst possible fitness on the 3x3 world is 108 (assigned to an individual which doesn't move), a fitness as good as 90 is achievable simply by random plans. This fact encourages model building and use in early generations regardless of relation to the world. Figure 8 shows one such individual which appeared on generation 12 and received a fitness of 98 (location 0, where the explorer starts, corresponds to the bottom left node).

Passing the threshold fitness of 90 requires some intelligent creation and use of a mental representation. The model in figure 9 was created by the best individual at generation 16 (location 0 corresponds to the bottom left node). Although the agent mislabels the arcs to western nodes as east, it correctly labels some of the northern nodes and is thus able to achieve a fitness of 88.

By generation 24 an individual which correctly associates directions in memory with movement in the world has appeared with a fitness of 54. This individual builds the model shown in figure 10 (location 0 corresponds to the bottom left node). The model created by the solution is repeated as figure 11 (location 0 corresponds to the top left node).

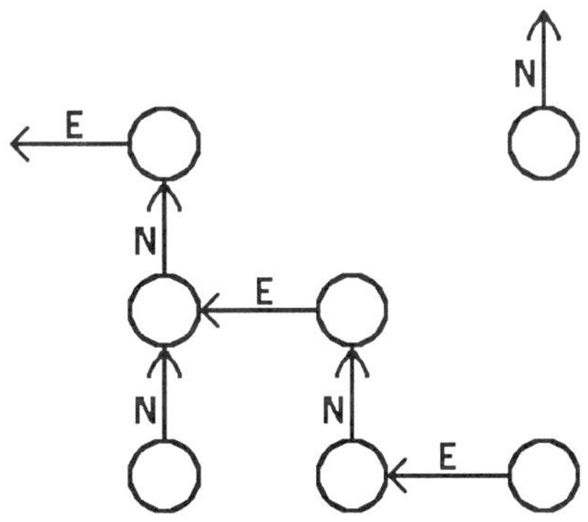

Figure 9 Generation 16, fitness = 88.

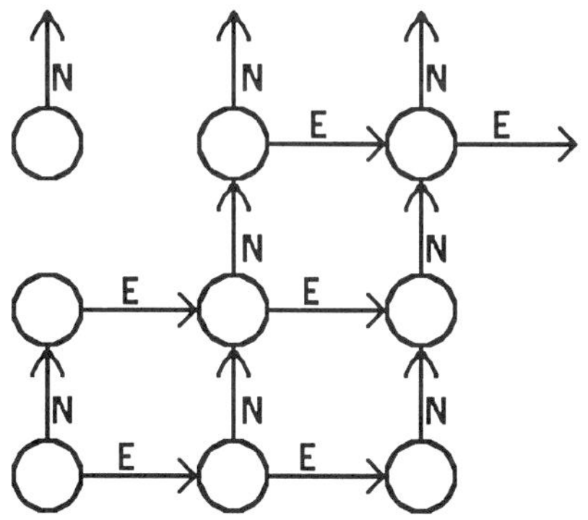

Figure 11 Generation 32, fitness = 0.

6. Conclusion

This paper has presented an approach for investigating the evolution of agents that build and utilize mental models. The method used has the advantage that the provided memory is not pre-designed by the human designer of the system, but rather dynamically structured by the evolving agent. The results show that genetic programming is capable of producing intelligent agents that use such a memory to build a mental model and solve the path-planning problem. Although several simplifications were used in the experiments in this paper, the described method proves a promising basis for further exploration of this area.

7. Future Work

Future work will proceed in two directions. The first will address simplifications used in this paper. Instead of implicitly forcing the direct mapping of node numbers to locations, we will allow arbitrary mappings and interpretations to evolve. This can be accomplished by the introduction of arithmetic and conditional functions into the function set. Additionally, with such functions in place, the number of nodes will no longer need to be equal to the number of locations in the world. This step will allow mental models to evolve without any direction from the researcher. A second avenue for future work will be the application of the described method to more complicated worlds. The toroidal graph world, for example, could be extended to include obstacles.

Acknowledgments

The author is grateful for the time and support offered by John Koza and Nils Nilsson. All runs were executed using DGPC (Dave's Genetic Programming Code) written by David Andre.

Bibliography

Andre, David. 1994. Evolution of mapmaking: learning, planning, and memory using genetic programming. *Proceedings of the First IEEE Conference on Evolutionary Computation.* IEEE Press. Volume I.

Koza, John. 1992. *Genetic Programming: On the Programming of Computers by the Means of Natural Selection.* Cambridge, MA: MIT Press.

Koza, John. 1994. *Genetic Programming II: Automatic Discovery of Reusable Programs.* Cambridge, MA: MIT Press.

Teller, Astro. 1994a. The evolution of mental models. *Advances in Genetic Programming.* (Kenneth Kinnear, Ed.). Cambridge, MA: MIT Press.

Teller, Astro. 1994b. Turing completeness in the language of genetic programming with indexed memory. *Proceedings of the First IEEE Conference on Evolutionary Computation.* IEEE Press. Volume I.

Automatic Generation of Object-Oriented Programs Using Genetic Programming

Wilker Shane Bruce
Nova Southeastern University
School of Computer and Information Sciences
3100 SW 9th Avenue
Fort Lauderdale, Florida 33315
brucews@alpha.acast.nova.edu
http://scis.acast.nova.edu/~brucews/

April 2, 1996

Abstract

This research addresses the application of genetic programming to the generation of object-oriented programs. An extended chromosome data structure is presented where the set of methods associated with an object is stored as an array of program trees. Modified genetic operators are defined to manipulate this structure. Indexed memory is used to allow the programs generated by the system to access and modify object memory. These extensions to the standard genetic programming environment result in a system that can simultaneously generate all of the methods associated with an object.

Experiments were performed to compare sequential generation of individual object methods with two variants of simultaneous generation. The first variant used information about both method return values and object internal memory state in its fitness function. The second variant used information about method return values only. It was found that simultaneous generation of methods is possible in the domain of simple collection objects both with and without the availability of internal memory state in the fitness function. It was also found that this technique when successful is several orders of magnitude more computationally expensive in terms of number of individuals generated in the search than the sequential generation of individual methods.

1 Introduction

Genetic programming [Cramer, 1985; Koza, 1989; Koza, 1992; Koza, 1994; Kinnear, Jr., 1994] has recently been proposed as a method for the automatic generation of programs from specifications of their desired input/output behavior. Early research in genetic programming concentrated on the generation of functional programs where no memory is manipulated during program execution. In more recent research the induction of procedural programs has been attempted through the extension of the language element set used by genetic programming to construct programs. Both ad hoc and formal extensions have been proposed to add looping/recursion constructs and memory manipulation constructs to the program construction language [Teller, 1993; Teller, 1994a; Teller, 1994b].

Object-oriented programming is a method for organizing programs where the fundamental unit of programming structure, known as an object, consists of a persistent private memory and a well defined set of procedures that may exclusively access and modify this private memory [Meyer, 1988]. Object-oriented analysis, design and programming is rapidly becoming the predominant software engineering methodology used for new product development in the 1990's [Booch, 1994; Meyer, 1988; Rumbaugh, 1991; Shlaer and Mellor, 1988].

Previous research has not addressed the issue of applying genetic programming to the task of producing object-oriented programs. Several practical questions come to mind when dealing with this issue. These include questions like (1) how can features unique to the object-oriented paradigm such as object state and inheritance be represented in the genetic programming process; (2) is it possible using automatically defined functions to simultaneously induce all of the methods associated with an object in a single genetic programming run; and (3) is is more efficient to use simultaneous induction of all of the methods or sequential induction of individual methods?

This paper presents research reported in full detail in [Bruce, 1995] in which genetic programming is applied to the automatic generation of object oriented programs.

2 Literature Review

Genetic programming was invented by Cramer [Cramer, 1985] and independently rediscovered and popularized by Koza [Koza, 1989; Koza, 1992; Koza, 1994]. It is a variation of Holland's genetic algorithm [Holland, 1975] in which the simulated genetic material consists of tree representations of computer programs instead of binary character strings. The genetic operators are suitably modified to work with the program tree representation. Crossover (sexual reproduction) is performed in genetic programming by swapping subtrees between copies of the two parent trees. Reproduction (cloning) is performed by making an exact copy of the parent tree. Mutation (asexual reproduction) is performed in one of two ways. In the first form, a single subtree from a copy of the parent tree is replaced by a randomly generated subtree. In the second form, a single node from a copy of the parent tree is replaced by a randomly selected node that satisfies the requirement that the resulting tree still represent a syntactically correct program.

Fitness computation in genetic programming consists of the execution of the program tree against a number of cases in the training set. The program tree may be directly executed by interpretation or may be compiled and executed. The fitness of each individual training case is generated by measuring the distance between the result obtained by running the program against the inputs associated with the training case and the desired result of that computation. This differs from Holland's approach where the simulated genetic material must be converted from a binary string representation into a representation that has meaning in the problem environment and is typically treated as unexecutable static data in that environment.

Several extensions to genetic programming are used in this work. Montana [Montana, 1994] has extended genetic programming to limit the manner in which programs can be built so that all generated programs are syntactically correct with regards to data typing information associated with each language element. This extension has the effect of reducing the size of the set of programs that can be constructed from the language element set.

Teller [Teller, 1993; Teller, 1994a; Teller, 1994b] has extended genetic programming through the use of indexed memory, a general mechanism by which the programs that are constructed by genetic programming can manipulate state information. Language primitives are defined that allow a constructed program to read from and write to a memory that is organized as an indexable array. Depending on how the fitness function initializes and maintains the state data, indexed memory can be used to manipulate both global and local state information.

Koza [Koza, 1994] has extended genetic programming to allow the simultaneous induction of a main program and its callable subroutines through a method called automatically defined functions (ADFs). This method takes advantage of the LISP language feature that allows a program to dynamically define a subroutine and then call that subroutine within the same program. The user of a genetic programming system that includes automatically defined functions must specify the number of subroutines to be induced and the number of arguments accepted by each subroutine.

A fixed program tree structure is created that sequentially defines each of the subroutines and then calls the main program. The main program definition and each of the subroutine definitions consist of one subtree in the full program tree. Each subtree has assigned its own set of language elements that may be used in program construction. The crossover operator is modified to only swap subtrees between corresponding routine definitions in the two parent trees. The fixed portion of the trees may not be modified. The language element set associated with the main program includes the names of the subroutines that are dynamically generated.

The program trees produced by genetic programming with ADFs allow for the generation of hierarchical programs where subroutines may be used to specify code that is used in more than one place in the problem solution. It has been shown that the use of ADFs allows genetic programming to solve more difficult programming problems.

Finally, Langdon [Langdon, 1995] has extended genetic programming to allow the simultaneous induction of each of the programs associated with an abstract data type. A new representation for the simulated genetic material is defined where each program associated with the abstract data type is stored as a single program tree in an array of trees. A new crossover operator is defined that generates copies of the two parent's genetic material, randomly selects a tree from each of the copies at the same array position, and swaps randomly selected subtrees between the chosen trees. Experiments are reported where the programs for the stack and queue abstract data types are simultaneously induced using a fitness function that only views the return value information from sequences of program calls. The key emphasis of this work is on whether or not cooperative self-organization of the layout of memory can be produced during the simultaneous induction process. Induction of the stack object was straightforward. A significant amount of engineering of the fitness function was required to induce the queue object.

3 Methodology

The induction of an object-oriented program by genetic programming requires that those features that differentiate object-oriented programming from functional or procedural programming be representable in the language used to build methods. Two specific features make object-oriented programming different from procedural programming. They are the representation of object state as a global memory that persists beyond the execution of an individual program and the use of inheritance to share definition between objects.

Indexed memory is used in this research to represent the memory associated with an object. Standard indexed memory read and write functions are provided in the language used to construct methods by genetic programming. When strongly typed genetic programming is used to associate data type information with language elements, a separate indexed memory (along with separate read and write functions) is used to store values of each of the individual data types. Unlike the approach used in [Langdon, 1995] where more elements are available in the indexed memory than are required to solve the problem, indexed memory is limited in this research to only the number of elements required to solve the problem.

Inheritance can take two forms in object-oriented programming. Both memory structure and methods may be inherited from another object. Both forms of inheritance can easily be represented in the language used by genetic programming to construct methods. A terminal symbol representing the object from which inheritance occurs can be made available in the language. Method inheritance can be represented by making available functions that invoke methods on the inherited object. Memory structure inheritance can represented by making available terminal symbols or functions that allow read and/or write access to the inherited memory. The persistence characteristic of objects can be represented by having the fitness function only initialize object memory before the first in a sequence of method invocations representing a case. By allowing the state to be maintained between subsequent method invocations, the global nature of object state is simulated during the fitness calculation.

In order to allow the simultaneous induction of the set of methods associated with an object, the simulated genetic material structure was modified so that each individual consists of an array of program trees. During fitness calculation each tree is used to represent one method of the object. This is the same simulated genetic material structure used by Langdon in [Langdon, 1995]. Our use of the structure was developed independently and without knowledge of Langdon's work. The new structure represents a generalization of Koza's ADF representation [Bruce, 1995]. In particular, the representation has the flexibility to also allow induction of a single method by setting the size of the array to one.

Table 1: Experimental Design Parameters

Parameter	Experiment					
	1a	1b	2a	2b	3a	3b
Induction Type						
Individual Induction	X	X				
Simultaneous Induction			X	X	X	X
Fitness Function						
Uses Desired Object State	X	X	X	X		
Uses No Object State					X	X
Strongly Typed Genetic Programming						
No	X		X		X	
Yes		X		X		X

Genetic operators were defined to work with the new representation. The reproduction operator creates a copy of the array of program trees. The crossover operator creates a full copy of each of the two parents. An index position in the array is randomly chosen in order to identify one tree from each array upon which crossover will occur. Both of the trees to be used for crossover are taken from the same position in each of the two parent arrays. Subtrees are then swapped between these two trees to generate the new individuals. These operations are similar to those defined by Langdon. The experiments reported in this paper also use a point mutation operator which was not used by Langdon. The point mutation operator creates a copy of its parent. A position in the array is then randomly chosen in order to identify the tree which will be modified. Within that tree a single node is selected and randomly mutated to generate a new syntactically correct program.

A set of six experiments were conducted to explore the issues of the feasibility of simultaneous object induction, the feasibility of cooperative self-organization of object memory, and the efficiency of the simultaneous induction approach in comparison to individual induction of each method. Experiments were performed both with and without the use of strongly typed genetic programming to determine the effect of this extension on efficiency. The specific design parameters of each of the six experiments are presented in Table 1.

In experiments 1a and 1b each of the methods associated with the object were individually induced. The fitness function in these experiments used information on both the desired return value and the desired state of object memory after method invocation. In experiments 2a and 2b all of the methods associated with the object were simultaneously induced. The fitness function in these experiments also used information on both the desired return value and the desired state of object memory after method invocation. In experiments 3a and 3b all of the methods associated with the object were simultaneously induced. However, the fitness function used in these experiments only applied information on the desired return value from each of the method invocations in a sequence of invocations. Experiments 3a and 3b required the genetic programming system to generate a set of programs that cooperatively self-organize their memory. Experiments 1a, 2a and 3a were run without the use of strongly typed genetic programming. Experiments 1b, 2b and 3b were run with the use of strongly typed genetic programming.

Two fitness functions were used in these experiments. In experiments 1a, 1b, 2a and 2b the function used information on both the desired method return value and on the desired pre-invocation and post-invocation state of the object memory to compute fitness. A fitness case consisted of one method invocation. To evaluate a case, the object memory would first be set to the specified pre-invocation state. The method would then be invoked using the specified parameters. The method return value and the post-invocation object memory values were used to compute a distance between the actual state and the desired state. The distance measure to evaluate a case was designed to give equal weight to object memory locations that should change and object memory locations that should not change. Space limitations preclude a full description of the algorithm here. It is described in full detail in [Bruce, 1995].

Experiments 3a and 3b only used information on the desired method return value. In this fitness function a single fitness case consists of a sequence of object method invocations where each method invocation has a desired return value. The first method in the sequence is always the object constructor. Methods in the sequence are invoked until the sequence is complete or a run-time error is encountered. Fitness is computed as the distance between the actual and desired method return values. Details of the algorithm are available in [Bruce, 1995].

In each experiment twenty runs of the genetic programming system were performed using the statistical cross-validation method. In this method a different subset of the available training data is set aside on each run for use as generalization test data. The best individual generated in each run is stored along with its training and generalization fitness. The number of individuals created before this best individual is found is also retained for each run. This information is used to compute the average training and generalization fitness of the best individual found over all of the runs in the experiment and to produce a prediction of how many individuals would have to be generated within the genetic programming search process in order to obtain a correct individual with a ninety-nine percent probability of success.

In each of the experimental runs a population size of one thousand is used. The system is run until an individual is generated that achieves a perfect score on the training data set or until one hundred generations have passed. A steady state control strategy is used along with a 2 player tournament strategy for selection and a 2 player inverse tournament strategy for replacement. The settings of the remaining genetic programming system parameters used in the experiments are listed in [Bruce, 1995].

Table 2: Stack Experiment Results

Stack Search Comparison

Experiment	Predicted Search Size	Average Training Fitness	Average Test Fitness
1a	3,917	0.000000	0.0000000
1b	962	0.000000	0.0000000
2a	8,876,970	17304.779689	648.8931485
2b	6,561,630	9261.644809	457.8397025
3a	§	627.950000	33.0500000
3b	2,859,032	278.500000	15.2500000

Finally, the space of constructable programs was limited by only making available for the construction of each method those language elements that were required to produce a correct solution. This differs from most genetic programming research where the exact language elements involved in a problem solution are not known in advance of the execution of the algorithm and one of the data points achieved by a problem solution is the definition of such a set. Since the correct method definitions were known in advance in this research such a limitation is warranted to improve efficiency of the algorithm. A description of the language elements used in the experiments is presented in [Bruce, 1995].

4 Results

Experiments were performed in three problem domains: stack, queue, and priority queue. Five methods were induced in each domain: `Init`, `Empty?`, `Full?`, `AddDataItem`, and `RemoveDataItem`. The semantics of each method are those expected when working with these simple data structures. The numerical results of the experiments are presented respectively for each domain in Tables 2, 3 and 4. The average fitness values presented in these tables are distance measures with a value of zero representing the best possible average fitness.

The feasibility of the solution language extension approach for supporting the features of object-oriented programming was shown in experiments 1a and 1b where the individual induction of a correct definition of each object method was achieved in all domains except the priority queue domain in experiment 1a. In this domain no correct solution for `RemoveDataItem` was generated under the experiment 1a constraints. Generating correct definitions using strongly typed genetic programming in

Table 3: Queue Experiment Results

Queue Search Comparison			
Experiment	Predicted Search Size	Average Training Fitness	Average Test Fitness
1a	198,242	0.8442746	0.1172600
1b	194,432	0.8642746	0.1232600
2a	§	13929.0333080	1052.8361500
2b	§	9854.2132315	651.5527255
3a	§	193.3500000	7.7000000
3b	8,676,180	803.5000000	47.5000000

Table 4: Priority Queue Experiment Results

Priority Queue Search Comparison			
Experiment	Predicted Search Size	Average Training Fitness	Average Test Fitness
1a	§	1254.133871	66.0076243
1b	530	0.000000	0.0000000
2a	§	19554.786017	1120.9302835
2b	§	447.743999	8.3005480
3a	§	137068.850000	7214.1500000
3b	§	7912.200000	493.0000000

these experiments proved to be very easy given the type constraints of the language elements and the limitation that the only elements available for building a method were those that contributed to the correct solution of the problem. As was expected, the stack domain proved to be the easiest domain to solve, the queue domain proved to be more difficult to solve, and the priority queue domain proved to be very difficult to solve.

The simultaneous induction experiments were somewhat less successful. In the stack domain at least one correct definition for the full set of methods was found in all of the experiments except experiment 3a where simultaneous induction was attempted using no object state information and untyped genetic programming. In this experiment no best program correctly induced the `Full?` method.

In the queue domain the only successful simultaneous induction experiment was experiment 3b where strongly typed genetic programming was used along with an objective function that did not consider object state information in computing fitness. Similar to the failure in the stack domain, `Full?` proved to be difficult to learn in the best generated programs for these experiments.

In the priority queue domain none of the simultaneous induction experiments was successful in generating a correct set of methods. The `RemoveDataItem` method was incorrect in all of the best generated programs for these experiments. This was not unexpected given the difficulty of inducing this method in the individual induction experiments.

Predictions for the number of individuals that would have to be created in the search to get a correct set of method definitions with 99% probability of success were computed using the method in [Koza, 1992]. No computation was possible for the failed simultaneous induction experiments. These are noted in Tables 2, 3 and 4 by the use of the § character. In all of the experiments the use of individual induction proved to be more efficient than simultaneous induction using a number of individuals searched measure. Strongly typed genetic programming generally outperformed untyped genetic programming as was expected.

The simultaneous induction approach had difficulty in converging to a correct solution once several correct method definitions were generated in the best generated program. A combination of the general weakness of the genetic algorithm in local search situations and the fact that the crossover and mutation operators had no knowledge of which methods needed to be adjusted to move towards convergence is the most probable cause of this difficulty. The use of a Pareto fitness measure where each individual method has its own distance measure along with a fitness proportional tree selection strategy within the individual could help solve this problem. A brood selection approach [Tackett and Carmi, 1994] in which one child is generated by applying the operator at array position one, a second child is generated by applying the operator at array position two, and so on could also help to focus the search and to alleviate this problem.

5 Conclusions

The experiments performed in this research have confirmed Langdon's result that simultaneous induction of programs operating on a single data structure is possible. They have also shown that it is possible to cooperatively self-organize memory in the stack and queue domains under the constraints of the experiments performed.

The software engineering implications of the research are four-fold. First, indexed memory proved to be an effective method for representing the state of an object. Second, the use of solution language extensions to support features of the object-oriented programming paradigm was also effective. Next, strongly typed genetic programming proved to be more effective than untyped genetic programming. Finally and most importantly, the individual induction of object methods proved to be several orders of magnitude more efficient than the simultaneous induction of object methods. Given the current state of the art, individual induction of methods should be used when information on the state of object memory after method invocation is available for use by the fitness function.

For those situations where the fitness function does

not have access to information on the desired state of object memory after each method invocation, simultaneous method induction should be performed as a two pass process in which those methods that modify memory are generated in the first pass and those methods that query memory are generated in the second pass. This will allow the genetic programming system to solely deal with the cooperative self-organization of memory in the first pass. Since induction of those methods that only query memory can not be performed in an enlightened manner until the structure of the object memory is fixed, this should simplify the simultaneous induction process. Additional research is required to determine if changing the genetic operators to give more focus to the process of selecting which method to modify will improve efficiency to the point that simultaneous induction is competitive with individual induction.

Acknowledgments

I would like to thank Dr. Phil Adams, Dr. Edward Lieblein, and Nova Southeastern University for their contributions in making this research possible. I would like to thank the anonymous reviewers for their insightful comments on the paper.

References

[Booch, 1994] Grady Booch. *Object-Oriented Analysis and Design with Applications.* The Benjamin/Cummings Publishing Company, Inc., Redwood City, CA, second edition, 1994.

[Bruce, 1995] Wilker Shane Bruce. *The Application of Genetic Programming to the Automatic Generation of Object-Oriented Programs.* PhD thesis, School of Computer and Information Sciences, Nova Southeastern University, Fort Lauderdale, FL, December 1995.

[Cramer, 1985] Nichael Lynn Cramer. A representation for the adaptive generation of simple sequential programs. In *Proceedings, International Conference on Genetic Algorithms and their Applications*, pages 183–187, 1985.

[Holland, 1975] John H. Holland. *Adaptation in Natural and Artificial Systems.* University of Michigan Press, Ann Arbor, Michigan, 1975.

[Kinnear, Jr., 1994] Kenneth E. Kinnear, Jr., editor. *Advances in Genetic Programming.* MIT Press, 1994.

[Koza, 1989] John R. Koza. Hierarchical genetic algorithms operating on populations of computer programs. In *Proceedings of the Eleventh International Joint Conference on Artificial Intelligence (IJCAI-89)*, pages 768–774, San Mateo, CA, 1989. Morgan Kaufman.

[Koza, 1992] John R. Koza. *Genetic Programming. On the Programming of Computers by Means of Natural Selection.* Complex Adaptive Systems. MIT Press, Cambridge, MA, 1992.

[Koza, 1994] John R. Koza. *Genetic Programming II. Automatic Discovery of Reusable Subprograms.* MIT Press, Cambridge, MA, 1994.

[Langdon, 1995] William B. Langdon. Evolving data structures using genetic programming. In L. Eshelman, editor, *Proceedings of the Sixth International Conference on Genetic Algorithms (ICGA95)*, pages 295–302, San Francisco, CA, July 1995. Morgan Kaufmann Publishers Inc.

[Meyer, 1988] Bertrand Meyer. *Object-Oriented Software Construction.* Prentice-Hall, 1988.

[Montana, 1994] David J. Montana. Strongly typed genetic programming. BBN Technical Report 7866, Bolt Beranek and Newman, Inc., Cambridge, MA, March 1994.

[Rumbaugh, 1991] James Rumbaugh. *Object-Oriented Modeling and Design.* Prentice-Hall, Englewood Cliffs, NJ, 1991.

[Shlaer and Mellor, 1988] Sally Shlaer and Stephen J. Mellor. *Object-Oriented Systems Analysis: Modeling the World in Data.* Yourdon Press Computing Series. Yourdon Press, Englewood Cliffs, NJ, 1988.

[Tackett and Carmi, 1994] Walter Alden Tackett and Aviram Carmi. The unique implications of brood selection for genetic programming. In Z. Michaelewicz, editor, *Proceedings of the First IEEE World Congress on Evolutionary Computation. IEEE World Congress on Computational Intelligence. Volume 1*, New York, NY, 1994. IEEE Press.

[Teller, 1993] Astro Teller. Learning mental models. In *Proceedings of the Fifth Workshop on Neural Networks: An International Conference on Computational Intelligence: Neural Networks, Fuzzy Systems, Evolutionary Programming, and Virtual Reality*, pages 147–152. The Society for Computer Simulation, 1993.

[Teller, 1994a] Astro Teller. The evolution of mental models. In Kenneth E. Kinnear, Jr., editor, *Advances in Genetic Programming*, chapter 9. MIT Press, Cambridge, MA, 1994.

[Teller, 1994b] Astro Teller. Genetic programming, indexed memory, the halting problem, and other curiosities. In *Proceedings of the Seventh Florida Artificial Intelligence Research Symposium*, 1994.

Evolving Event-Driven Programs

Mark Crosbie, Eugene H. Spafford
COAST Laboratory
Dept. of Computer Sciences
Purdue University
West Lafayette IN 47907-1398
{mcrosbie, spaf}@cs.purdue.edu
(317) 494-9313

April 10, 1996

Abstract

This paper examines how Genetic Programming has shortcomings in an event-driven environment. The need for event-driven programming is motivated by some examples. We then describe the difficulty in handling these examples using the traditional genetic programming approach. A potential solution that uses colored Petri nets is outlined. We present an experimental setup to test our theory.

1 Introduction

In the event-driven paradigm, a program responds to asynchronous events that occur in a system. The program performs no processing of its own, other then to respond to events. These events could correspond to user input (e.g. keyboard or mouse input) or system events (interrupts, alarms or warnings). We view the occurrence of these asynchronous events over time as an *event stream*. This stream is handled by a *dispatch loop* that calls an appropriate function to handle each event in turn (see Figure 1). Once the event has been handled, control is returned to the dispatch loop. Such a program is event-driven because the sequence of events in the event-stream completely governs what processing takes place.

Event-driven programs are well suited to handling real-world situations. It is often more natural to view systems as event-driven, such as in *real-time* systems. Such a system has to handle many different types of asynchronous event, and correctly handle each event within strict time limits. Failure to do so could result in catastrophic loss or death. On a less dramatic note, a graphical user interface is another example of an event-driven system. The myriad of possible user inputs (both keyboard and mouse events) are best represented as an event stream and handled by an event dispatcher (as an example, see [Petzold C. (1992)]).

Genetic Programming (GP) [Koza (1992)] is a powerful solution discovery paradigm that attempts to solve problems by "evolving" solution programs. These solution programs are evaluated relative to some predetermined test cases, and are ranked based on their solutions to these test cases. Typically, a genetic program is represented as a parse tree, or LISP s-expression. Such representations facilitate manipulation of the program by genetic operators (mutation, crossover, reproduction etc.) Moreover, the underlying assumption behind these representations is that a program will be executed as a single thread of execution. It will run, carry out computations, and produce a result.

This "traditional" approach to GP is not well suited to handling event-driven systems. The solution programs tend to query their environment by calling procedures, rather than be driven by events in the environment. For example, a solution program to reverse a truck up to a loading dock (as in [Koza (1992)]) may need input from a sensor indicating that the truck has reached the loading dock. In the event-driven model, the sensor would generate an event in the event stream that indicates that the rear of the truck is within a certain distance of the loading dock. An event handler would then take the appropriate action (such as applying the brakes). In the traditional GP model, the solution program would have to continuously poll the value of this sensor before it could take the correct action.

This paper will outline an approach to handling asynchronous events that uses a state machine graph representation. The representation of the graph makes it amenable to evolution using genetic programming style operators. We will present an application of this to computer security.

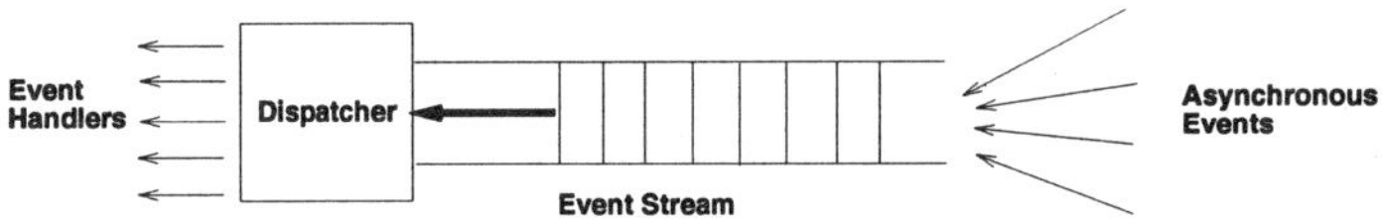

Figure 1: An event-stream and dispatcher

2 Motivation — Problems with GP in an event-driven system

To motivate this work, we will present an application of genetic programming in an event-driven framework. We consider a system that will detect intruders on a computer system. There are three aspects we will examine in relation to this problem. These are

1. Determining inter-event correlations. The occurrence of two or more events in a certain order is often more interesting than the occurrence of each individual event. Our solution must be able to represent such partial orders.

2. Handling previously unforeseen event interactions. Events may occur in previously unforeseen ways, and our solution should handle this case.

3. Handling asynchronous events. When a GP is being trained, asynchronous occurrences can be minimized, but in a real-world application, interrupts and other events can occur at any time. A solution must be able to handle events as they occur.

2.1 Intrusions and Intrusion Detection

Computer security is concerned with *confidentiality*, *integrity* and *availability* [Russell and Gangeni (1991)] of resources on a system. Users must be able to trust that the system will keep their data secure from prying eyes, operate as they expect it to, and be available when they need to use it.

An *intrusion* can be defined as [Heady et al. (1990)]

> any set of actions that attempt to compromise the integrity, confidentiality or availability of a resource.

An intrusion attempts to subvert the security policy in force on a system to gain access to information, to alter the system's operation, or to deny the system to other users. Intrusions are enacted by users — a user runs a program that executes actions in excess of that user's authorization, for example. This would be classified as an intrusion. Similarly, a user who attempted to overload a system and prevent other users from working would also have initiated an intrusion (a denial of service).

To detect and report intrusions, a system must have some form of *intrusion detection system* (IDS) installed. An IDS must [Mukherjee et al. (1994)]

> identify, preferably in real time, unauthorized use, misuse, and abuse of computer systems.

An IDS will often use *audit data* from a system to detect intrusions. Audit data is generated on a running system as a process executes system calls. Each system call is logged to a file, along with a time and date stamp and pertinent information about the call. This is an *audit file*. It provides a snapshot of the activity on the system over a given period of time. As most intrusions involve either executing a program or modifying a file, the record of system calls should be sufficient to detect an intrusion attempt.

An IDS will examine relationships between discrete events on a system to see if certain abnormal patterns are occurring. By analysing these patterns, an IDS attempts to detect misuse of a computer system. This is termed *misuse intrusion detection*. There are potentially hundreds of types of audit events on a large system. Each audit event is notable in and of itself, but it is the combination of these events that is far more interesting. We may not care if a file is created, but we may be suspicious if a file is created and is immediately changed to have a new creation date. The occurrence of both these events, in a specific sequence is far more important than the occurrence of each individual event.

An Intrusion Detection System must be able to handle unknown event interactions. Intruders are continually changing the way they break into computer systems, and an IDS must be able to detect these new intrusion mechanisms. Luckily, however, most intrusions are based on a common set of exploitations which are combined in a variety of ways to subvert a computer system. For example, overwriting executable files is a common to many viral attacks, independent of the actions of the virus itself. These common sequences of actions must be detected by our proposed solution, even if they form part of an overall more complex intrusion attempt.

Our goal is to build a *real-time* intrusion detection system that can respond to intrusion attempts as they occur. In our prototype described in [Crosbie and Spafford (1995b)] and [Crosbie and Spafford (1995a)], audit data is fed directly from the system kernel into the IDS. The audit data arrives asynchronously — there is no way of predicting when the next event will occur. The genetic program that we evolve must be able to handle this asynchronous data. We anticipate that this event-driven GP model will be incorporated into our Autonomous Agents architecture as outlined in our previous work.

Despite its strengths in discovering solutions, traditional genetic programming is not well suited to handling the "real-world" conditions we outlined above. Our proposal is that we can adapt the traditional model of genetic programming to allow handling related asynchronous events. Instead of using a parse tree or s-expression to represent code, we propose using a directed graph. The graph is manipulated by genetic operators to change both its topology and contents.

3 Related Work

In [Ryan (1995)] an approach to evolving event driven genetic programs is presented. He evolves fighting tanks in a closed world. He splits his population in two: *main programs* (MPs), which register events to be handled, and *callback functions* (CBs) each of which handle an event. Events are prioritized so higher priority events can interrupt the handlers for lower priority ones.

His early results were unsatisfactory as he states that the

> more successful members of the MP population [used] only one or two callback functions in a sensible manner

He gives reasons for this poor performance, and alters the setup of his experiments to evolve better performing individuals. However, the approach he outlines will not work well as the problem size increases. In the IDS example we gave above, the number of events will be in the hundreds. Associating a population of solutions per event is too costly in terms of speed of evolution and storage space.

He found that in some cases, event handlers for one event queried the the value of a sensor for another event. This is a form of rudimentary event correlation. However, it is not a powerful enough mechanism to express partial orderings of events, which would be necessary in the IDS problem.

Despite the success of his work, we felt it was not a powerful enough model to handle all the complexities of a real-world event-driven system. In particular, we need to handle a large number of events and determine inter-event correlations. His model was not able to handle this efficiently.

The approach we outline below is based closely on the colored Petri net approach in [Kumar and Spafford (1994)]. More information on colored Petri nets can be found in [Jensen (1992)] and [Lakos (1994)]. Kumar's work shows that potential intrusions on a system can be encoded as *patterns* that can be represented by colored Petri nets. These patterns can represent partial orders, and are efficiently evaluated. Furthermore, they are ideal for handling event-driven systems. We propose taking this model as a basis from which GP can be extended to handle event-driven systems.

4 Directed Graph Approach

In this model, a directed graph represents a state machine. *Tokens* move through the state machine and multiple tokens are allowed. A token can contain data items, termed the token's *color*. These local variables are bound to the token once, and usually do not change.

Transitions between states are determined by *guards*. A guard is a boolean expression that must be satisfied before a token can move from one state to another. Evaluating the guard will generally use some values from the token's color. If a guard is true, the token is moved into the next state in the graph. If a guard is not true, the token may not move out of the state it is in.

Tokens encapsulate information about events. For example, if a mouse event occurred, a token could be created which contained the position of the mouse, it's velocity and its button status as colors. These could be considered "fields" in the token "record". Figure 2 shows a graph with a token in the start state. This is the *initial marking* of the graph.

This simple graph illustrates how a graph can naturally express some important constructs. If the token is in state 2, we know that guard A has been satisfied. If a token has moved into state 3, we know guards A and B must be satisfied. Thus, we can naturally represent **conjunction** by the position of the token in the state machine. Furthermore, if tokens exist in states 3 and 4, then we can conclude that guards A *and* (B *or* C) must hold. Thus we can represent a **disjunction** of any number of clauses by placing guard expressions on the out-edges of a vertex. If a token satisfies two or more guards, it will arbitrarily choose one of the transitions to take — there is no concurrency in the path of a single token through the state machine. There can, however, be multiple tokens in the state machine at any one time.

Each state can have an *action* associated with it. When a token enters the state, the action for that state is fired. The actions associated with a state are executed in sequence. An action routine may use values bound to the token in the state as arguments.

Iteration can also be expressed — the back edge from state 4 to state 1 (the initial state). If a token satisfies guard D, then it may move back into state 1. It will continue doing this for as long as it satisfies guard D while in state 4. This is a natural expression of a looping construct.

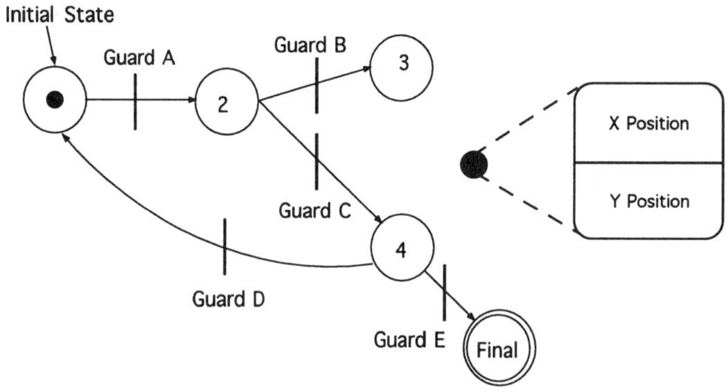

Figure 2: Initial Marking of a state graph

Figure 2 also shows the bindings associated with a token. Using the previous example, the bindings associate the X and Y position of a mouse with the token. A mouse event will cause a token to be created in the initial state, and it will bind the current X and Y position of the mouse to the token (the token's color). These two fields can be used while evaluating guard expressions.

This model is better at handling event-driven systems than traditional GP. Inter-event correlations can be represented by the position of a token in the state machine. As noted above, if a token has transited through a certain number of states, then we can conclude that the conjunction of the guards to those states must hold. If those guards can test for the occurrence of an event, then the truth of a particular boolean guard expression will express a relationship between events.

For every event that occurs, we can introduce a new token into the graph. This token will move through the state machine for as long as it satisfies guards. The state machine can handle asynchronous events, because it itself is event-driven. If an event occurs, and a token is waiting to transit an edge where the guard references that event, then the token will move to the next state immediately.

In summary, we can use the state machine to not only represent logical constructs in a program, but also to detect potential correlations between events. It is flexible enough that it can handle the occurrence of asynchronous events.

5 Evolving state machine graphs

Our representation for a state graph must handle three components:

1. *Transitions* are directed edges between state vertices.

2. *Guards* are boolean expressions which must be satisfied before a transition is taken.

3. *Actions* are commands executed when a token enters a state.

When we are evolving a state graph, each of these data items must be considered separately. Different operators will be applied to evolve boolean guards and transitions in the graph.

Traditionally, GP operates on a homogeneous population of potential solution programs. By combining pieces of these individuals, GP attempts to evolve an optimal solution program to a given problem. The entire population must be *evolution compatible* in that we can take any two components of any two solutions and interchange them using the crossover operator, and still have a viable solution as a result. We are attempting to evolve an optimal graph to solve a particular problem, but we have three distinct components being evolved — the actual topological layout of the graph itself, the boolean expressions on each edge and the action primitives per state. We cannot attempt to co-evolve a population of graphs and of boolean expressions: they are just not evolution compatible.

Our solution is to consider a fully connected graph — every vertex has an edge to every other state in the graph. As we evolve guards on these edges, some guards will never be satisfiable — they always evaluate to false. No token can ever transit that edge, and so the state it leads to is unreachable. It is as if the edge never existed in the graph. This removes the heterogeneity in our graph representation. We can maintain a population of fully connected graphs, and only evolve the boolean expressions on the edges.

Some expressions will evaluate to false, and those edges can now be considered "removed" from the graph. Every few generations, the graph can be simplified and "dead" vertices (those which are only reached by transiting a false guard) can be removed. Figure 3 shows how vertex 2 can be removed from the graph as it is no longer reachable by a token from any state as the guards to state 2 will always evaluate to false. In this figure, ~

is boolean negation, **and** and **or** are boolean conjunction and disjunction. The arguments to the boolean guards are colors from the token transiting that guard. Thus, as `A and ~A` will always be false, we know no token will ever transit that edge, and we can consider the destination vertex of the edge for removal.

Another optimization is to remove guards that always evaluate to false. This reduces the diversity of the population — we are removing potentially useful subexpressions which could be evolved further in later generations. However, this reduction may become necessary as our graphs may have hundreds or thousands of states. As evolution progresses, we expect our population members to be converging on an optimal solution. We can remove guards that evaluate to false and simplify our population somewhat, and still converge to a solution. This process is shown in Figure 4.

Petri nets are *composable* — a set of vertices and edges can be reduced to a *meta-vertex* — the meta-vertex is a pointer to a subgraph which we have extracted from the main graph. This is analogous to automatically defined functions (ADFs) [Koza (1994)]. It allows the evolutionary process to build useful subgraphs and then evolve a main graph around them, much as ADFs allow genetic programming to discover useful building blocks. As graphs grow in size, this ability to abstract common features into meta-vertices is a useful evolutionary tool.

6 Experimental Setup

We are currently designing an experimental setup to test our hypothesis. In this setup, we are using the IDIOT system [Kumar (1995)] to perform the intrusion detection. It specifies a Petri net and state machine using a simple language that specifies transitions and the guards on those transitions. We will evolve GP programs that specify the guards on edges which are translated into this language by an intermediate parser.

6.1 Training and fitness

Genetic programming assigns each potential solution program a *fitness* score based on how well it solves the problem at hand. In our case, the fitness score will be determined by how well the state graph can detect certain well-known intrusion attempts. A similar approach to evolution was taken in [Crosbie and Spafford (1995b)] where *scenarios* were used to present training situations to the solution programs. A scenario is a collection of commands that may or may not comprise an intrusion attempt. Programs were assigned a fitness score depending on how they classified each scenario set. This approach can also be used here, where a training scenario encodes an attack pattern and we attempt to evolve a state machine to detect the attack.

We will initially attempt to evolve patterns that will detect some simple intrusions — executing particular programs, using `lpr` to copy non-owned files, writing to executable files or writing to non-owned files, for example. We already have hand-coded solutions to these problems, and so we expect the evolved solutions will match the efficiency of these hand coded solutions.

We will then attempt to evolve patterns to detect attacks for which we have no efficient hand-coded solutions. If the evolved solution can detect these attacks, we feel that this will demonstrate the ability of genetic programming to evolve event-driven programs to solve problems in specific domains.

7 Conclusions

We have shown how event-driven systems have significant value when dealing with real-world situations. We have motivated this using an example from computer security in which we attempt to detect intruders on a computer.

We have identified some weaknesses in the traditional genetic programming approach to problems. These weaknesses arise from GP's inability to handle asynchronous events. We have presented what we feel will be a solution to some of these shortcomings based on a directed graph approach, and outlined an experimental setup to test our hypothesis.

References

[Koza (1992)] Koza, John, 1992. *Genetic Programming.* MIT Press.

[Koza (1994)] Koza, John, 1994. *Genetic Programming II.* MIT Press.

[Denning (1987)] Denning, Dorothy E, 1987. An Intrusion Detection Model. *IEEE Transactions on Software Engineering* February: 222

[Garfinkel and Spafford (1991)] Garfinkel S., and Spafford E. 1991. *Practical UNIX Security.* O'Reilly and Associates Inc.

[Russell and Gangeni (1991)] Russell D., and Gangeni Sr. G. 1991. *Computer Security Basics.* O'Reilly and Associates Inc.

[Heady et al. (1990)] Heady R., Luger G., Maccabe A., and Servilla M. 1990. *The Architecture of a Network-level Intrusion Detection System*, Technical Report, CS90-20. Dept. of Computer Science, University of New Mexico.

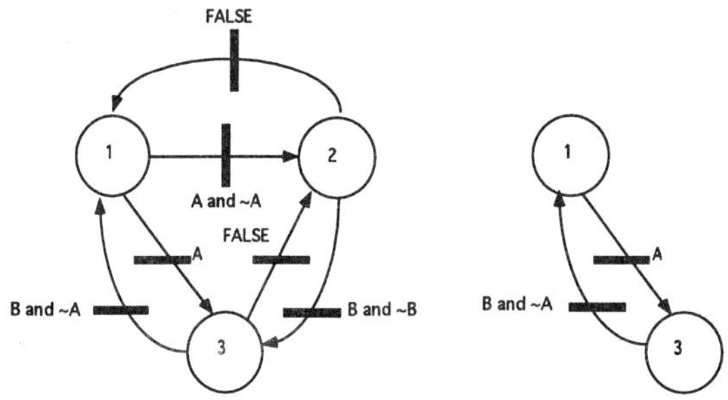

Figure 3: A complete graph with guards, and after dead node elimination

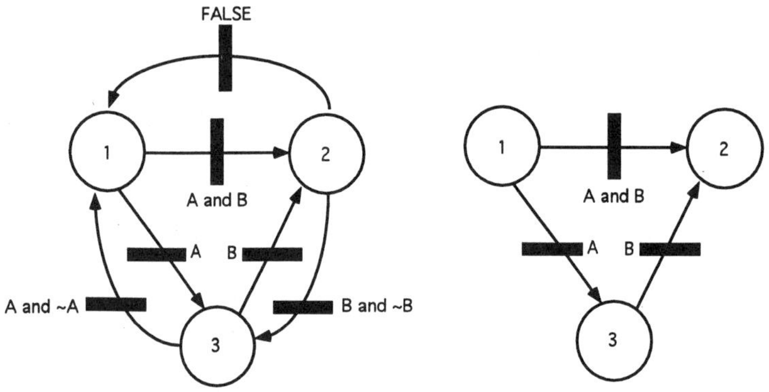

Figure 4: Simplification of a graph by removing useless guards

[Mukherjee et al. (1994)] Mukherjee B., Heberline L. T., and Levitt K. 1994. Network Intrusion Detection. *IEEE Network* May/June 1994.

[Crosbie and Spafford (1995a)] Crosbie M., Spafford E. Defending a Computer System using Autonomous Agents. *In Proceedings of the 18th National Information Systems Security Conference*, October 1995.

[Petzold C. (1992)] Petzold, Charles. 1992. *Programming Windows 3.1.* Microsoft Press, Redmond, Washington.

[Crosbie and Spafford (1995b)] Crosbie M., Spafford E. Applying Genetic Programming to Intrusion Detection. In *Proceedings of AAAI Fall Symposium on Genetic Programming.* AAAI, November 1995.

[Jensen (1992)] Jensen, Kurt. 1992. *Coloured Petri Nets - Basic Concepts.* Springer-Verlag.

[Lakos (1994)] Lakos, Charles. *Object petri nets - definition and relationship to coloured nets.* Technical Report, University of Tasmania. April 1994.

[Kumar and Spafford (1994)] Kumar, S. and Spafford E. 1994. A pattern matching model for misuse intrusion detection. In *Proceedings of the 17th National Computer Security Conference.* October 1994.

[Kumar (1995)] Kumar, Sandeep. 1995. *Classification and Detection of Computer Intrusions.* PhD thesis, Purdue University. August 1995.

[Ryan (1995)] Ryan, Conor. GPRobots and GPTeams - competition, co-evolution and cooperation in genetic programming. In *Working Notes for AAAI Symposium on Genetic Programming.* November 1995.

Computer-Assisted Design of Image Classification Algorithms: Dynamic and Static Fitness Evaluations in a Scaffolded Genetic Programming Environment

Jason M. Daida, Tommaso F. Bersano-Begey, Steven J. Ross, and John F. Vesecky
The University of Michigan
Artificial Intelligence Laboratory and Space Physics Research Laboratory
2455 Hayward Avenue, Ann Arbor, Michigan 48109-2143
daida@umich.edu, tombb@umich.edu, stevross@umich.edu, jfv@umich.edu

ABSTRACT

This paper discusses several issues in applying genetic programming to image classification problems in geoscience and remote sensing. In particular, this paper examines the role in using dynamic and static fitness evaluation functions. This paper also examines a few of the aspects in human-computer interactions that facilitate computer-assisted learning and problem solving (i.e., scaffolding) for our system. We describe a possible means for visualizing and summarizing a solution space without having to resort to an exhaustive search of individuals.

1. Introduction

1.1 Background

Classification (alternately, *pattern classification*, or *pattern identification*) denotes a process in which a decision rule is applied to categorize a set of image data (Haralick and Shapiro (1993)). This process often represents a key step in transforming image data into information. Francone et al. (1996) suggest that genetic programming (GP) can be used to evolve robust classification algorithms. However, there are certain trade-offs that are involved in using GP for classification tasks with image data, particularly with satellite image data used in the fields of geoscience and remote sensing. Of these, one key trade-off involves computational effort and human supervision.

In an "ideal" world, image classification tasks would require minimal, if any, human supervision and most of the programming would be done by a computer. Unfortunately, the computational effort associated with classifying satellite image data is nontrivial. A typical satellite image may consist of one to hundreds of megabytes, which often involves a fair amount of CPU time for processing. Common geoscience problems can also often involve the classification of many images. When one considers that GP involves the evaluation of many candidate solutions, each of which could require seconds to hours of computation per image, the prospect of using just one image for evaluation becomes daunting; the cost of using just one image as a fitness set becomes prohibitive. Of course, one could reduce the amount of computation involved by reducing the size of the fitness set from whole images to subimage samples.

Although the idea of using subimage samples may seem straightforward, in practice, it is not. One often encounters situations where there exists incomplete knowledge about the variations in the features to be classified. In such situations, what exactly constitutes the minimal set that completely describes the categories of interest is unknown. It becomes entirely possible that a subimage sample may in fact represent an ill-posed problem. Moreover, the available computational resources might further constrain the size of a subimage sample to such an extent that the only practical sample corresponds to an ill-posed problem. In either circumstance, a workaround involves human supervision.

In Daida et al. (1996), we introduced a GP system for the computer-assisted design of image classification algorithms. The system has been used to solve a difficult pattern recognition problem that has had no prior satisfactory solution. The system has been designed to classify megabyte-sized grayscale images, even though the corresponding fitness set may consist of only a handful of manually classified pixels (test points). The system has featured a canonical GP with a dynamic evaluation function (i.e., a function that increases the number of the fitness cases that are evaluated during the course of a run).

In designing this system, we have presumed the following. First, there exists incomplete knowledge on the user's part about the classification problem at hand. Second, there exists a high probability of a user formulating an imperfect fitness set (at least initially). Third, that GP provides "logical" outcomes to a presented fitness set. Fourth, that learning on the user's part about the classification task at hand would result in the crafting of better fitness sets, which would in turn generate better solutions. In this sense, GP is used as a kind of agent that handles the task of programming while the user is left to concentrate on the task of program specification. In other words, GP "scaffolds" a user to learn about the problem's essentials by illuminating consequences of specifications made with incomplete knowledge, while hiding the potentially distracting details associated with programming.

Although this system has performed well, we left for later the task of qualifying the necessity of using GP in a scaffolded fashion. In particular, this paper shows just how ill-posed fitness sets can be, if only because large fitness sets are not computationally tractable. We also left for later the task of qualifying the reasoning for using dynamic, as opposed to static, fitness evaluation functions. This paper subsequently addresses these issues.

1.2 Previous Work

We have observed that most work in GP has not required scaffolding, since fitness, as defined by an evaluation function, has usually been sufficient in determining program success. We note that although we have not yet found other works using GP as a scaffolding technology for image analysis, we have found a similar use of GP in the area of computer-produced art (e.g., Sims (1991)).

The idea for using a dynamic evaluation function—one in which either the fitness cases or the evaluation function itself changes during the course a run—is not new. Koza (1992) used such functions in many examples. However, the particular strategy on which we based our work stems from Goldberg's (1989) work in genetic algorithm classifiers and Holland's (1986) work in default hierarchies and learning. Of interest to us has been how their strategies applies to image analysis problems that involve using GP to extract information from image data. Although Tackett (1993) has used GP with this type of image analysis problem, he used a different evaluation strategy. To date, we have encountered few works that evaluate the effectiveness of Holland's and Goldberg's type of dynamic evaluation function with GP systems for image analysis. We do note, however, that Schoenauer (1996) has used this type of evaluation function to increase the numerical precision of computational models. (See also Siegel (1994) and Angeline and Pollack (1993).)

2. Highlighted Methods

This section details two notable distinctives about our system described in Daida et al. (1996) and briefly mentions how these distinctives relate to the experiment discussed in this paper.

2.1 Dynamic Fitness Evaluation

For our system, dynamic fitness specifically refers to evaluating a set of program benchmarks that changes in membership during the course of a GP run. The following paragraphs supplement the qualitative description given in Daida et al. (1996).

Let P_0 represent the set of initial test cases, i.e.,

$$P_0 = \{a_0, a_1, \ldots, a_M\},$$

where a_i represents a test case (an image test point in our case), and M is a positive integer.

Furthermore, let P' denote the set of additional test-case arrays, i.e.,

$$P' = \{\{b_{00}, b_{01}, \ldots, b_{0I(0)}\}, \{b_{10}, b_{11}, \ldots, b_{1I(1)}\}, \ldots, \{b_{K0}, b_{K1}, \ldots, b_{KI(K)}\}\},$$

where P' is a K-element array of arrays (which do not necessarily have a uniform length), b_{kl} is a test case, and $I(k)$ is (integer length - 1) of the kth test-case array.

Given an evaluation function f such that

$$f(c_{individual}, P, t, trigger): \quad c_{individual}(p(t, trigger)) \rightarrow raw\ fitness\ score,$$

where $c_{individual}$ denotes an individual from a population of programs, $P = p(t, trigger)$ denotes the set of test cases to be evaluated by $c_{individual}$ at a time t, let p be a mapping such that

$$p(t, trigger) = \begin{cases} P_0, & t = 0. \\ P_0 \cup P'(1), & t_1 < t < t_2, \text{ where } t_1 \text{ represents the first time } trigger = TRUE \text{ and } t_2 \text{ represents the second time } trigger = TRUE. \\ P_0 \cup \bigcup_{k=1}^{K'} P'(k), & t_{K'} < t < t_{K'+1}, \text{ where } t_{K'} \text{ represents the } K'\text{th time } trigger = TRUE \text{ and } t_{K'+1} \text{ represents the next time } trigger = TRUE. \\ P_0 \cup \bigcup_{k=1}^{K} P'(k), & t_K < t \le t_{Max}, \text{ where } t_{Max} \text{ represents the total amount of allowable time steps.} \end{cases}$$

For our purposes, we let t denote generational time, e.g., $t = 0$ denotes generation zero. (Note: function p can also be expressed in terms of individual time, which would be relevant for some cases of adaptation.) The Boolean trigger in function p represents the conditional in which the test-case set is incremented with new test cases. Although there are several different ways to trigger an increment, for both this paper and in Daida et al. (1996), the trigger used was strictly performance based. In particular, an increment was triggered when enough hits were scored for a particular set P, i.e.,

$$trigger = \begin{cases} TRUE, & (Number\ of\ Hits) \ge (\text{Length}(P)\ -\ Margin). \\ FALSE, & \text{Otherwise.} \end{cases}$$

We denote *Margin* as a parameter that a user defines, where *Margin* is an nonnegative integer such that $Margin \le Length(P)$ for all sets P.

2.2 Scaffolding

Figure 1 shows a flowchart of how we have used GP in a system for the computer-assisted design of image classification algorithms. For this paper, we deliberately chose not to use the full configuration in Daida et al. (1996), but instead allowed GP to execute without human supervision for two series of runs (i.e., feedback loops indicated by the dashed gray lines were not used). (This scenario could happen in actuality: a user would execute a series of GP runs to get a probabilistic assessment of her hypothesis before she would amend any inputs to the GP kernel). This partial configuration was used for this paper to test scaffolding and fitness evaluation functions.

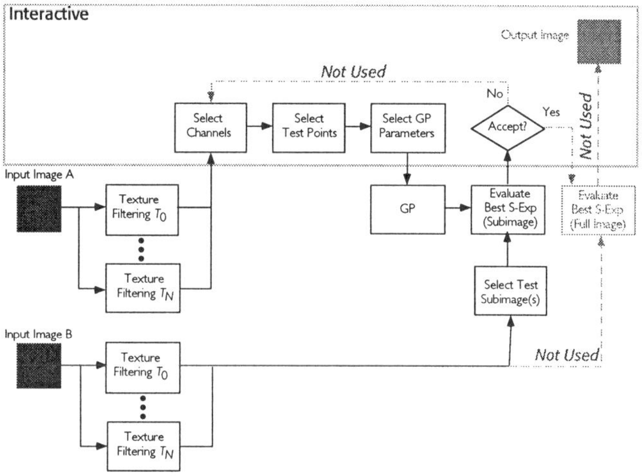

Figure 1. Described Scaffolded GP System.

3. Procedure

3.1 Domain Problem

The domain problem featured in this paper is identical to that described in Daida et al. (1995, 1996). The problem involves extraction of light-gray curvilinear and blob features from a light-gray background. The images are synthetic aperture radar (SAR) low-resolution data products that depict scenes of the Arctic Ocean and surrounding seas. The particular SAR of interest is mounted on the European Space Agency ERS-series of satellites. The curvilinear and blob-like features in these scenes of the ice cover correspond to pressure-ridges and rubble fields, respectively. This particular problem has been difficult to solve because of the low signal-to-noise ratio of the desired features to background. There has been no record of an algorithm that satisfactorily solves this problem for ERS SAR imagery prior to Daida et al. (1996).

3.2 Experiment Design

The dynamic case portion of this experiment consisted of executing 20 runs of system configuration shown in Figure 1 with the dynamic evaluation function described in Section 2.2. Each run used a different random number seed. The static case portion of this experiment consisted of executing 20 runs of GP with a static evaluation function. Identical random number seeds were used for both dynamic- and static-case runs. All other common parameters, terminals, functions, and an out-of-sample evaluation image remained constant. Table 1 summarizes some of those items.

Note that the following items have been presented in greater detail in Daida et al. (1996). Information in parentheses refers to where in that work a description for a particular item is given, namely: terminal set (Figure 21.4), function set (Tables 21.1–21.3), and image data (in particular, the April 23, 1992 data shown in Figure 21.3b). We further note that implementing the function set for closure under eight-bit image data does involve more than what can be conveyed in this paper. For example, addition can be implemented in at least two ways: saturation 255 or modulo 255. (Under saturation 255 arithmetic, the sum of any number greater than 255 is 255.) We included both implementations in the function set as P and +, for saturation 255 and modulo 255 addition, respectively.

The following parameters were used for both dynamic- and static-case portions: maximum number of generations were equal to 50; size of population, 300; maximum depth of new individuals, 6; maximum depth of new subtrees for mutants, 4; maximum depth of individuals after crossover, 17; fitness proportionate reproduction fraction, 0.1; crossover at any point fraction, 0.2; crossover at function points fraction, 0.7; selection method, fitness-proportionate; generation method, ramped-half-and-half. Note that the population size and number of generations used were modest and chosen in part so that we could complete a run on an HP 715 within a few CPU hours.

The following parameters were used for the dynamic evaluation function: 15 initial test points, subsequent increments of nine test points, *Margin* of six hits. Note that both dynamic and static cases used the same set of test points. The most significant difference was that the dynamic evaluation function made the test points gradually available by small increments, while the static evaluation function made all 53 test points available as a single chunk.

Figure 2a shows the out-of-sample test image that was used. This 128×128 pixel image is a subset from the original non-geocoded April 23, 1992 ERS SAR data product that has been calibrated in gray scale. (Note: image attributes of geocoding and calibration are mostly for domain science purposes. Geocoding refers to image data that has been geometrically corrected according to a specified latitude-longitude grid. Calibration allows for direct comparison between images taken from different times and different satellites.) The subimage shown in Figure 2a has been contrast enhanced for publication. For comparison purposes, Figure 2b shows the classification results using the best known algorithm (from the switch filter, described in Daida et al. (1996)).

Table 1. Settings and Parameters

Terminal Set:	Consists of an array of size N of manually selected test points, which contains data from an image and its corresponding filtered versions (i.e., 5x5 Laplacian of a 3x3 Mean, 5x5 Laplacian, 3x3 Mean, 5x5 Mean); a random floating-point variable.
Function Set:	Arithmetic operators defined to operate on matrices and constants in any combination (+, −, ×, ÷, P, M), and the threshold operator If-Less-Than-or-Equal (≤).
Fitness Cases:	List of N manually selected control points from benchmark. Pixels that are part of pressure ridge features are given a gray-scale value of 255 and all other non-feature pixels take a value of zero. The number of current fitness cases depends on the type of evaluation function used.
Raw Fitness:	The number of hits.
Standardized Fitness:	(Number of fitness cases, N) - (Number of hits).
Hits:	The number of fitness cases for which an individual program's output is less than 9 away from the target gray-scale value of 255 or 0.
Success Predicate:	A run ends when the maximum number of generations is reached or when an S-expression scores the maximum number of hits.

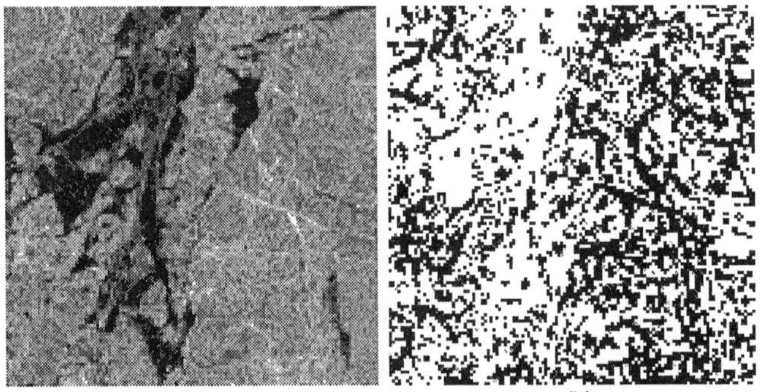

(a) (b)

Figure 2. (a) Subimage from April 23, 1992. Image of sea ice near Beaufort Sea gyre. ERS-1 © ESA 1992. (b) Reference classification.

3.3 Experiment Rationale

The experiment design of Section 3.2 allowed us to:

- isolate the role of the evaluation function by completely removing human supervision from fitness evaluation. In particular, it allowed us to ask whether supplying test cases in toto or by increments would make for significant differences in classification output. We have hypothesized that a dynamic fitness function would likely yield better solutions than a static fitness function, given identical sets of fitness test cases.
- examine the case where no scaffolding is allowed. Although our use of scaffolding does indicate that a human is involved in some aspect of fitness evaluation, we have designed the system to limit that involvement mostly to actions taken between GP runs (i.e., evaluations of out-of-sample results, as well as changes made to fitness cases and GP parameters). Still, we have hypothesized that even this amount of restricted access is (for all practical purposes) necessary because the problem is likely to be ill-posed. For this case where no scaffolding is allowed, one can test for this hypothesis by showing that the out-of-sample results are not well correlated to fitness scores.

3.4 Implementation Notes

Most of the implementation did not change from that described in Daida et al. (1996). All of the runs were concurrently executed on a network of HP 715 UNIX workstations using Allegro Common LISP.

Each run takes about four hours to complete. Although the workarounds described in Daida et al. (1996) can and have allowed for the processing of 1024×1024 pixel images, we chose to use 128×128 pixel subimages for out-of-sample inspection. We have found that this size of subimage is sufficient to inspect all of this experiment's images on a monitor without having to sacrifice the detail necessary for a qualitative assessment. (At least this is true for our domain problem. Other problems may require larger subimage sizes than what we used.)

4. Results

Figure 3 shows the performance curves associated with the experiment. All of the curves are displayed as box plots, which summarize the statistical variability measured for each generation.

Figures 3a and 3b show the box plots associated with the average standardized fitness measured for each generation for static and dynamic evaluation functions, respectively. A perfect standardized fitness score is zero hits and the worst case score is 53 hits, which is indicated by dotted lines. Figures 3c and 3d show the box plots associated with the raw fitness score for the best individual of each generation for static and dynamic evaluation functions, respectively. A perfect score is 53 hits, which is also indicated by dotted lines.

Figures 4 shows an excerpt from the out-of-sample test image for static and dynamic evaluation functions. For each row, each image corresponds to an output from a best-performing individual from a single run. Only the first individual that the kernel identified as having the highest raw fitness score has been visualized. (A run can generate many individuals with the same score.) Black (gray value 255) represents regions that supposedly have pressure-ridge or rubble-field features. The images in each row are organized by trial so that the main difference between images in a column is the type of evaluation function (and not a random-number seed). Visualized are the results corresponding to the first ten random-number seeds.

5. Discussion

5.1 Interpreting Performance Curves

The shapes of the performance curves are the most different at the outset, from zero to about seven generations. According to Figure 4, the static evaluation function allowed for GP to create an individual that solves about 75% of the presented problem (approximately 38 hits in raw fitness score) in generation zero, which indicates that most of the solution has been generated by a pure random process. In contrast, the dynamic evaluation function allowed for the GP to solve at most 30% of the presented problem (15 hits, raw fitness score) also by a pure random process. The short rise of the average standardized fitness curve for the dynamic case suggests that the presented problem has increased in difficulty for several generations. Figure 4 suggests that one effect, then, of using dynamic evaluation was to shift the predominant means of obtaining a solution from a pure random process to one of crossover and reproduction.

The shapes of the performance curves are the most similar towards 50 generations. The particular fitness set used in this experiment did yield an individual with either a perfect or near-perfect score for all of the runs, whether static or dynamic.

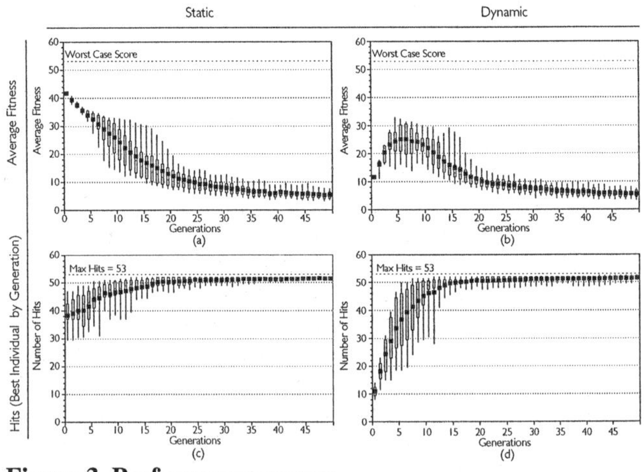

Figure 3. Performance curves.

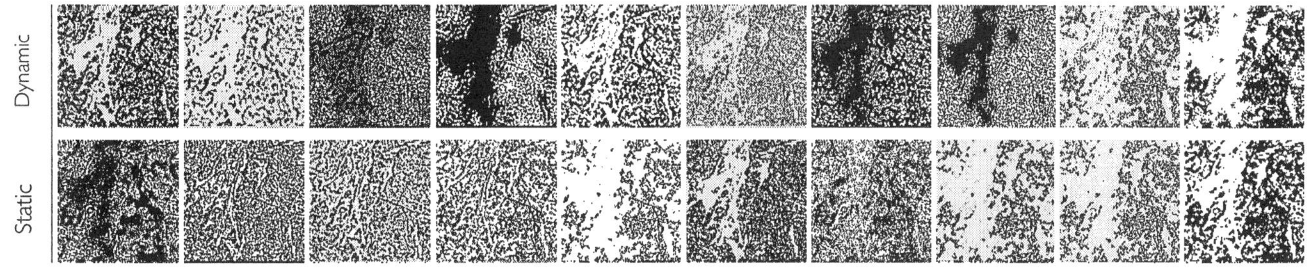

Figure 4. Excerpt of Out-of-Sample Results.

5.2 Classification Output and Scaffolding

The results presented in Figures 3 and 4 indicate that the out-of-sample classification results were not well correlated with high fitness scores (which is an expected finding for this experiment). At face value, the performance curves shown in Figure 3 would have suggested that all algorithms obtained from runs under either static or dynamic fitness evaluation functions correspond to reasonable classification solutions. However, when these algorithms classified imagery that contained data exclusive of that used in fitness calculations, the quality of classification output has varied greatly. Even perfectly scoring algorithms yielded questionable results that looked nothing like the solution shown in Figure 2b. These results did strongly suggest that the provided fitness set was ill-posed, which is what we expected for the case of no scaffolding, and which is why we have advocated the use of scaffolding.

The out-of-sample processing requirement for image analysis tasks is often many times larger than the number of in-sample test cases with which GP can feasibly process. Considering the out-of-sample requirement for just *one* ERS low-resolution (1024×1024 pixel) data product, the ratio of the number of out-of-sample cases to the number of in-sample cases for our fitness set is almost 20,000:1. (In an operational environment, a 2,000,000,000:1 ratio would not be unreasonable.) Even if one could accurately and precisely quantify every relevant aspect of a feature for processing (which is unlikely at the start of problem solving, *cf.* Congalton (1991)), the possibility of undersampling would still remain. For that reason, using GP as a means for scaffolding (which would include a human in an evaluation loop with GP) in image analysis tasks makes more sense than using GP without human supervision.

5.3 Likelihood of Success

Are there any predictors that might indicate whether a reasonable solution is likely? Fortunately, the answer may be yes. To assess whether a set of parameters, terminals, and function sets might yield a solution, a user would likely end up executing a large number of runs, if only because GP is a probabilistic method. Furthermore, for ill-posed cases, a single run can generate upwards of several thousand individuals that have scores that would merit human inspection. In any case, without a predictor, a user could be faced with a fairly laborious task of inspecting each candidate without necessarily being assured of whether any solution would be possible for that particular set of GP parameters or fitness test cases.

Figures 5a and 5b show the result of doing a matrix addition and normalization over the twenty images for dynamic and static cases, respectively. White represents low average values; and black, high values. What has surprised us was how close the resulting averaged images were to the solution shown in Figure 2b. Part of this surprise came from noting that none of the images in Figure 4 looked like the solution. Indeed, some of those images looked quite different. Another part of this surprise came from also noting that the number of images averaged was fairly modest, given a considerable number of individuals (e.g., over 5000 in the dynamic case) having similar fitness scores.

We have hypothesized that each image represents the outcome of one or more genotypic building blocks that have evolved over the course of a run. The expression of those building blocks would depend on what building blocks exist and what blocks become expressed, which depend upon the structures that link these blocks together. Because components in the function set were Cartesian operators, an average image would represent a projection in image space of those building blocks that commonly occur. An average image would subsequently provide a rough indication of allowable and disallowable solutions.

If this hypothesis were the case, thresholding an averaged image would then yield a likelihood map of classified image data for some probability (i.e., given M runs, the probability that all the components necessary for a solution are attained). Assuming that the values associated with the probability of assembling all necessary building blocks remains relatively constant, the most likely map of classified image data should be comparable between static and dynamic cases for some fixed threshold.

Figure 6 shows the result of thresholding the averaged images. Figure 6a shows a subimage of the original data shown in Figure 2a. Figure 6b shows the corresponding classified solution for reference purposes. Figure 6c shows the thresholded averaged image for the dynamic case, while Figure 6d shows the thresholded averaged image for the static case. The thresholded values corresponding to Figures 6c and 6d were kept identical.

Figure 6 indicates that the dynamic evaluation function has a

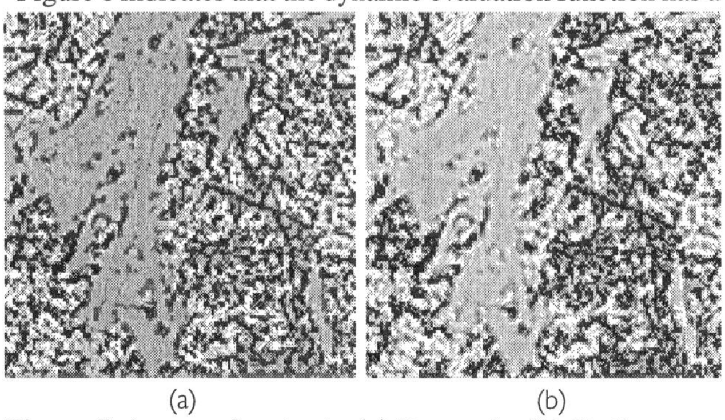

Figure 5. Averaged outputs. (a) Dynamic. (b) Static.

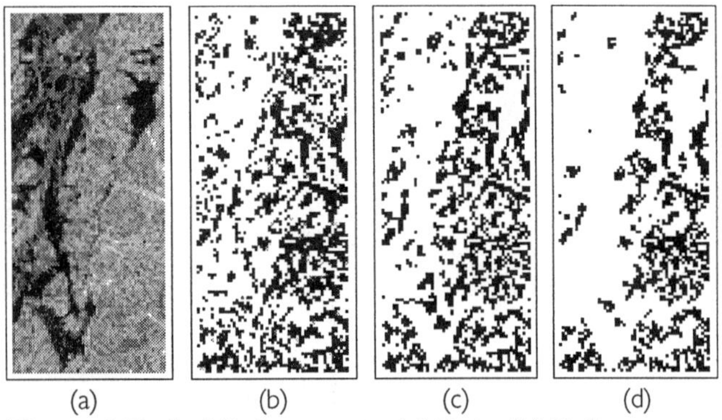

Figure 6. Probablistic outputs. (a) Data. (b) Reference solution. (c) Dynamic. (d) Static.

greater likelihood of obtaining a solution in comparison with the static evaluation function (i.e., Figure 6c is more similar to Figure 6b than is Figure 6d). What is missing in the static case is the building block(s) associated with classifying ridge and rubble features at the edge of multiyear ice. In other words, the probability of obtaining code that detects just the presence of bright curvilinear features at the boundaries between light gray and dark gray regions appears to be fairly low for the static case. Note that the required code would not be an edge detector. (If it were, all the edges between the light and dark gray regions would be incorrectly classified as ridge features.)

At the time of this experiment, we did not have the means to track building blocks for entire populations of algorithms and consequently did not have the means for testing this hypothesis. We have subsequently left this assignment for future work.

6. Conclusions

This paper has examined some of the issues involved in using GP in the computer-assisted design of image classification algorithms. Although it would be desirable to have the computer to do much, if not all of the work in deriving an appropriate classification algorithm, for practical reasons, we have indicated that a human should be involved in at least part of fitness evaluation. In particular, we have indicated that there is a trade-off in the number of fitness cases considered during evaluation versus the available computational resources. We have also pointed out that there may exist uncertainty about the quality of fitness cases used. In any case, we have indicated than an image classification problem could be ill-posed, which was certainly the case for this paper's experiment.

This paper has also has described the role of using either dynamic and static evaluation functions in deriving a solution. Results have suggested that use of a dynamic evaluation function would be more likely to evolve a classification solution than would a static evaluation function using the same set of fitness test cases. A novel means of visualizing and predicting the solution space for a given set of parameters, functions, and terminal sets was also described.

Acknowledgments

This research has been partially funded by grants from the Office of Naval Research, the Naval Research Laboratory (Stennis) and the Space Physics Research Laboratory (U-M). We appreciate the thoughtful critiques given to us by the reviewers of this paper. We gratefully acknowledge I. Kristo, P. Angeline, M. Bassin, S. Daida, J. Hommes, J. Koza, R. Onstott, R. Riolo, E. Soloway, and A. Wu for their invaluable comments and assistance.

Bibliography

Angeline, P.J., & J.B. Pollack, 1993. Competitive environment evolve better solutions for complex tasks. In S. Forrest (ed.), *Proceedings Fifth International Conference on Genetic Algorithms.* San Mateo: Morgan Kaufmann Publishers, Inc., pp. 264–270.

Congalton, R.G., 1991. A review of assessing the accuracy of classifications of remotely sensed data. *Remote Sensing of the Environment*, **37**:1, pp. 35-36.

Daida, J.M., J.D. Hommes, S.J. Ross, & J.F. Vesecky, 1995. Extracting curvilinear features from synthetic aperture radar images of arctic ice: Algorithm discovery using the genetic programming paradigm. *Proceedings 1995 International Geoscience and Remote Sensing Symposium: Quantitative Remote Sensing for Science and Applications*, IEEE Press., pp. 1415–1417.

Daida, J.M., J.D. Hommes, T.F. Bersano-Begey, S.J. Ross, & J.F. Vesecky, 1996. Algorithm discovery using the genetic programming paradigm: Extracting curvilinear features from synthetic aperture radar images of arctic ice. In P. Angeline and K. Kinnear, Jr. (eds.), *Advances in Genetic Programming II.* Cambridge: MIT Press.

Francone, F.D., P. Nordin, & B. Banzhaf, 1996. Benchmarking the real-world generalization capabilities of an advanced, compiling genetic programming system. *Proceedings Genetic Programming '96.* In press.

Goldberg, D.E., *1989. Genetic Algorithms in Search, Optimization, and Machine Learning*, Addison-Wesley, Reading.

Guzdial, M., 1995. Software-realized scaffolding to facilitate programming for science learning. *Interactive Learning Environments*, **4**:1, pp. 1-44.

Haralick, R.M., & L.G. Shapiro, 1993. *Computer and Robot Vision: Volume II.* Reading: Addison-Wesley Publishing Company.

Holland, J.H., K.J. Holyoak, R.E. Nisbett, & P.R. Thagard, *1986. Induction: Processes of Inference, Learning, and Discovery.* Cambridge: MIT Press.

Koza, J.R., 1992. *Genetic Programming: On the Programming of Computers by Means of Natural Selection*, Cambridge: MIT Press.

Schoenauer, M., M. Sebag. F. Jouve, B. Lamy, & H. Maitournam, 1996. Identification of mechanical behavior by genetic programming: Part I rheological formation. In P. Angeline and K. Kinnear, Jr. (eds.), *Advances in Genetic Programming II.* Cambridge: MIT Press.

Siegel, E.V., 1994. Competitively evolving decision trees. In K. Kinnear, Jr. (ed.), *Advances in Genetic Programming.* Cambridge: MIT Press.

Sims, K., 1991. Artificial evolution for computer graphics. *Computer Graphics*, **24**:4, pp. 319-328, July 1991.

Tackett, W.A., 1993. Genetic programming for feature discovery and image discrimination. *Proceedings 2nd International Conference on AI Planning Systems*, pp. 303-309.

Improved Directed Acyclic Graph Evaluation and the Combine Operator in Genetic Programming

Herman Ehrenburg
CWI, Kruislaan 413
NL-1098 SJ Amsterdam
ehrenbur@cwi.nl

ABSTRACT

The use of a directed acyclic graph (DAG) to represent a population in genetic programming offers several advantages, only one of which is the efficient use of space. We improve on existing methods to evaluate a DAG and offer two new ways of evaluating a population. The first method uses a linked list and a negligible amount of space. In the second method, each node is evaluated only once on all fitness cases and the results are cached. We also introduce two genetic operators in connection to the use of a DAG. The first is a simpler alternative to crossover. The second is a context-preserving genetic operator based on the building block hypothesis, which accurately combines two similar trees.

1 Introduction

Natural evolution can be viewed as a process that preserves genetic information. In order to preserve information, DNA is constantly copied resulting in many identical pieces of genetic information. Methods like genetic programming Koza (1992) imitate the process of evolution by multiplying genetic information. Unlike natural evolution however, such methods do not have to copy information. Instead it suffices to have multiple references to the same genetic information. In genetic programming the structures undergoing evolution are trees. Applying the idea of multiple referencing instead of copying to genetic programming, we represent the population by a DAG, rather than the standard forest of trees, where each individual (program) in the population is a tree rooted at a node in the DAG (Figure 1).

The idea of using a DAG was first published in Handley (1994). Handley showed that space usage is greatly reduced (for a variety of problems) by using a DAG. Under the restriction that functions have no side effects, Handley also discussed two methods to evaluate the population more efficiently than in conventional genetic programming. However, the most time efficient of these methods still relies on recursion (i.e. depth first traversal) like conventional genetic programming.

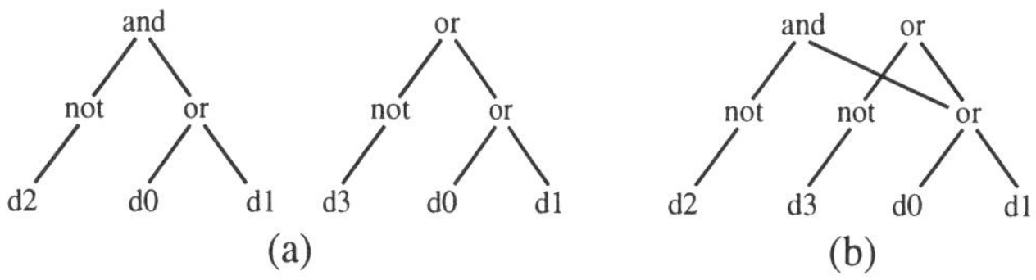

Figure 1: A population of size 2 represented as (a) a forest and (b) a DAG

Here we introduce two improved and faster methods of evaluating a population using a DAG representation *without* using recursion. Again under the restriction, that functions have no side effects. The two evaluation methods are distinguished by their use of space. The first method uses little space in the evaluation process. The second method is more memory intensive and based on the availability of a sufficiently large amount of space to cache the results of all nodes over all fitness cases.

Moreover, we introduce two genetic operators in connection to the use of a DAG. The first is a simpler alternative to the crossover operator, instead of two offspring it only produces one offspring. The second is a context-preserving genetic operator, based on the building block hypothesis, which becomes feasible by using the DAG representation. This operator accurately combines two similar individuals, which both have improved fitness over a (possibly imaginary) common parent.

In the remainder of this paper, Section 2 describes methods to evaluate a DAG. Section 3 describes operations to update a DAG. Section 4 describes two new genetic operators. Section 5 describes how experiments were performed and some results of measurements which did not fit into the preceding sections.

2 Evaluation Methods

Time performance of genetic programming for many, if not most, problems is determined by the efficiency of the evaluation of the fitness of the trees in the population. So it is imperative to have fitness evaluation of the population executed as efficiently as possible. If we choose to evaluate each tree in the population separately, we evaluate each tree on a particular fitness case by starting at the leaves of the tree and working our way up to the root

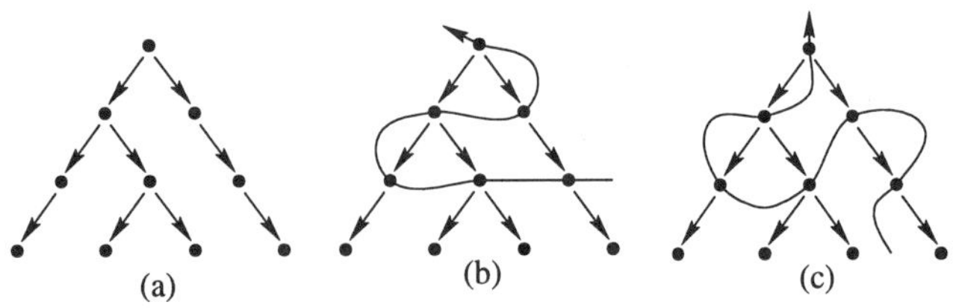

Figure 2: Evaluation order of a tree (a) the direction of each arc is contrary to the order in which the nodes are evaluated, (b) and (c) show two possible evaluation orders.

of the tree. The evaluation order of nodes is restricted only in the sense that children have to be evaluated before parents. To illustrate this, in Figure 2 (a) connected nodes should be evaluated in the order contrary to the direction of the arc. Thus there is a partial ordering on the nodes which restricts the evaluation order. In Figure 2 (b) and (c), we show two examples of the order in which a tree can be evaluated. Note that we consider leaves not to be evaluated because they have no children, they simply are assigned a value before evaluation.

We assume that all individuals in the population are evaluated on the same fixed and bounded set of fitness cases during each generation and that functions have no side effects. In this case one can more efficiently evaluate a DAG corresponding to the entire population as a whole, instead of all individual trees separately. This has the advantage that one avoids reevaluating a node which has been evaluated already. To do so, we proceed as follows. The DAG has a set of minimal nodes and a set of maximal nodes induced by the direction of the arcs. We work our way up from the minimal nodes (corresponding to the leaves in the forest representation of the population) to the maximal nodes of the DAG (corresponding to the roots of the trees in the forest representation) in the DAG representation of the entire population. Evaluation order of nodes is again restricted in the sense that children have to be evaluated before their parents. In Figure 3 (a) connected nodes should be evaluated in the order contrary to the direction of the arc. The partial ordering on the nodes indicated by the arcs again restricts the evaluation order. Figure 3 (b) and (c) show two examples of the order in which a DAG can be evaluated.

2.1 Evaluation Methods: Previous Work

Traditional evaluation is the evaluation method used in conventional genetic programming as described in Koza (1992). In traditional evaluation the trees in the population are each recursively evaluated (i.e. a depth first traversal) on every fitness case during a generation.
Fitness case-indexed caching is due to Handley (1994). Here the population is recursively evaluated and the results of the evaluation of each node over all fitness cases are cached on the first evaluation of a node on a fitness case. Using this method each node in the population has to be evaluated only once on each particular fitness case. If the DAG representation is used, this method is optimal in the sense that it never does a duplicate evaluation, either on identical nodes or on the same fitness case. Fitness case-indexed caching has the drawback of recursively descending the trees of the population, in addition to requiring a large amount of space. The amount of space required is of the order of number of fitness cases times number of nodes.
Evaluation context-indexed caching is also due to Handley. The method is based on the fact that the results of evaluation of some nodes only depend on a subset of the terminal set. As a consequence, these nodes do not have to be evaluated over all fitness cases, but only on a subset induced by their dependency. Evaluation context-indexed caching can be combined with the other evaluation methods discussed in this paper, and gives the appropriate improvement. Therefore, we do not compare its advantage with the others methods.

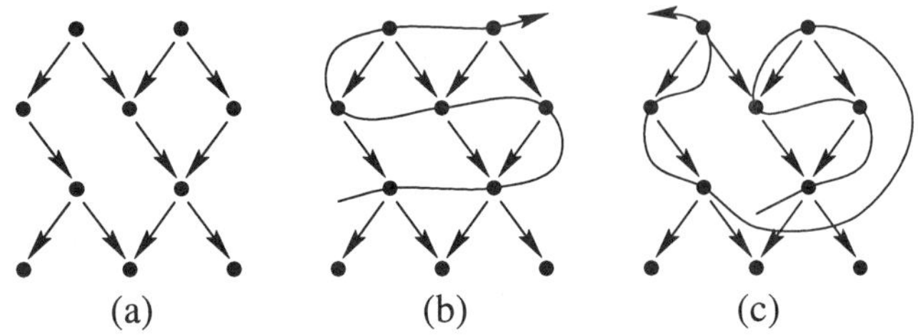

Figure 3: Evaluation order of a DAG (a) the direction of each arc is contrary to the order in which the nodes are evaluated, (b) and (c) show two possible evaluation orders.

2.2 New Evaluation Methods

Creation evaluation improves on fitness case-indexed caching, by not recursively evaluating the nodes in the DAG, but instead evaluating each node over all fitness cases the moment it is added to the population. In the case of using a DAG, this corresponds to the moment a node is newly inserted to the DAG (see Section 3.1). Creation evaluation has the best time performance in case the DAG representation is used. The amount of space required is of the order of number of fitness cases times number of nodes (like required for fitness case-indexed caching, see Section 2.1).

In Figure 4 results of experiments are shown, in which evaluation times where measured for fitness case-indexed caching evaluation and for creation evaluation. (See Section 5 on details of the experiments reported in this paper). The measurements confirm our expectation that creation evaluation is faster than fitness case-indexed caching. Creation evaluation had by the 30th generation only 29% of the running time of fitness case-indexed caching. The actual evaluation of nodes took 41% of the running time of creation evaluation, less time than the overhead to evaluate the individuals at each generation. The initial population takes more time than the generations immediately following it, because no results are

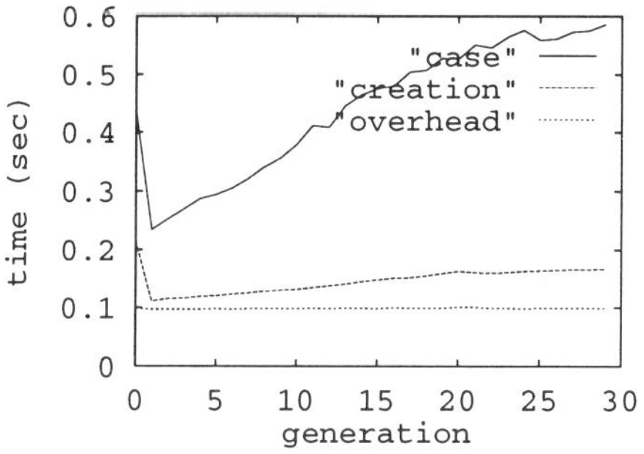

Figure 4: Evaluation time of evaluation methods, based on caching averaged over 100 runs fitness case-indexed caching and creation evaluation (evaluation time overhead of creation evaluation is also shown)

cached yet. The upward slope of both graphs is caused by the increasing number of nodes in the DAG and, for fitness case-indexed caching, also by the increasing average depth of the trees in the population.

List evaluation improves on traditional evaluation. Here a list of all nodes in the population is maintained. Nodes are added to the end of the list, at the time they are added to the population. Evaluation of the nodes in the current list, is done in the order they were added. This ensures that children are evaluated before their parents. When the DAG representation is used, this method is optimal in the sense that no duplicate evaluation of an identical node is done during the evaluation of the population for a particular generation. Note that using this method for each node only the result of the evaluation on one fitness case has to be temporarily stored, which takes a negligible amount of memory, comparable to the memory required for evaluation by traditional evaluation. List evaluation is both time and space efficient.

In Figure 5 results of experiments are shown in which evaluation times were measured for list evaluation and conventional evaluation. The measurements confirm our expectation that list evaluation is faster. List evaluation had by the 30th generation only 50% of the running time of traditional evaluation. The upward slope of both graphs is caused by the increasing number of nodes in the DAG and, for traditional evaluation, by the increasing average depth of the trees in the population. The initial population takes more time than the generations immediately following it, because the number of nodes in the DAG at generation 0 is greater than in generation 1 (The diversity of nodes in the randomly created population of generation 0 is somewhat greater).

Implementation of the list in C or C++ is best done as a (pointer) linked list. By directly addressing the next node in the list to be evaluated, no time is wasted in the evaluation process. Removing nodes (Section 3.4) is also easy using a linked list. Evaluation of each node is most efficiently done as described in Keith and Martin (1994).

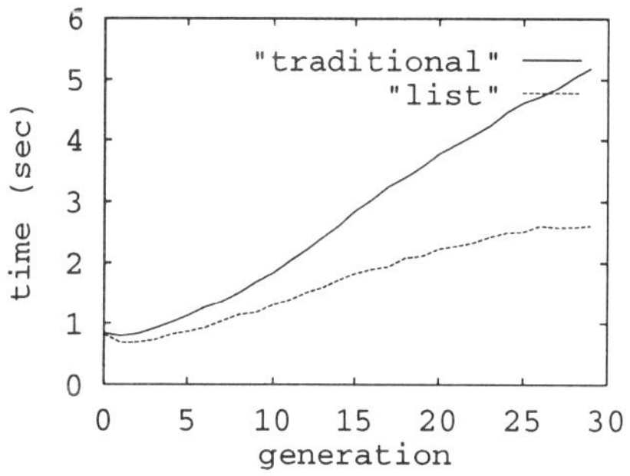

Figure 5: Evaluation time of evaluation methods not based on caching, averaged over 100 runs

3 Updating the DAG

From now on, we will identify a node with its associated subtree. We call two nodes *equivalent*, if their associated (sub)trees are equivalent. Two (sub)trees are equivalent, if their root nodes are labeled with the same function and if the children of the root nodes are pairwise equivalent (possibly without respecting order, see Section 5.1). Two leaves are equivalent, if they are labeled with the same terminal.

3.1 Insertion of Nodes

To ensure the correct operation of the DAG, we make the restriction that parent nodes can only be inserted after their children have been inserted. Thus, the insertion of the nodes of a tree is restricted by the same partial ordering which also restricts evaluation, as explained in Section 2. So when creating trees to fill the initial population, we insert the nodes of a tree in the order restricted by the partial ordering, we start at the leaves and end with the root. Modification of a constituent tree in the DAG is somewhat complicated. In Section 3.3 we show how genetic operators modify a tree in the DAG and how nodes are inserted in these operations.

Using a DAG, it is desirable to be able to locate a node in the DAG, given an equivalent node. If this can not be done, it is not possible to store equivalent, but separately evolved nodes uniquely, which rapidly leads to a lot of duplicate entries of small trees. To prevent duplicate entries, we need to check if an equivalent node is already present in the DAG, when inserting a node. To achieve this, we can store the nodes in a data structure on which an efficient search operation is defined, such as a hash table (as in Handley (1994)) or a red black tree (Cormen, Leiserson, and Rivest (1990)).

A red black tree is less time efficient, using $O(\log n)$ time per insertion, in comparison to a hash table, which is $O(1)$, but this is of secondary importance, since time performance in genetic programming is, in most cases, dominated by evaluation time. The red black tree does lack disadvantages of a hash table, which consist in finding a suitable hash function, determining table size, and handling collision. A red black tree requires a relational

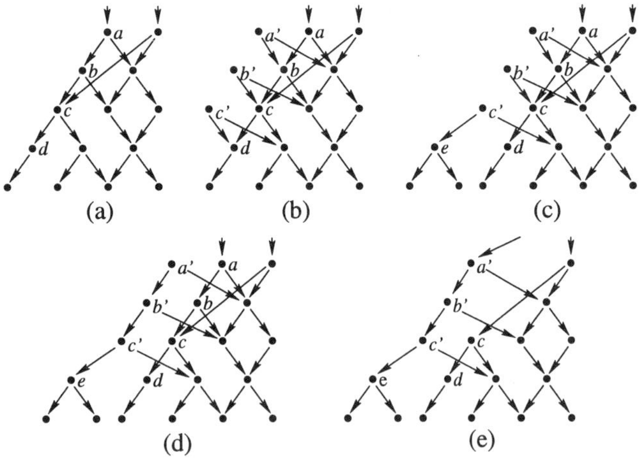

Figure 6: Modification of a constituent tree.

operator which gives a complete ordering on nodes.

Section 3.2 discusses how to actually compare nodes.

3.2 Comparing Trees

Comparing (sub)trees (i.e. nodes) for equality in a DAG is simple, since equivalent nodes are stored as one (see Section 3.1). Thus in a language like C or C++, one can simply compare the addresses of the root nodes. On insertion of a newly created node into the DAG, it is necessary to be able to compare it to the nodes in the DAG. Using a hash table, we need to compare nodes on equality in case of a collision. Using a red black tree, we need a relational operator to descend the red black tree. In both cases nodes are compared by comparing the type of the function and if equal comparing the children of the node. Since the children are part of the DAG, comparing their addresses suffices as explained. Since the nodes in a red black tree need to be strictly ordered, a red black tree requires the relational operator to give a strict ordering.

3.3 Tree Modification

Suppose we want to modify a tree in a DAG by replacing a subtree with an other (sub)tree. How this is done, is best explained by an example. In Figure 6 (a) a picture of a DAG is shown, containing two trees. The two arcs at the top indicate the two references by members of the population. In the tree with root a, we are going to replace the subtree with root d. We start the modification operation at the root a and we descend to the parent c of the root node d of the subtree to be replace. While descending, we make a copy of all the nodes we encounter on the path from root to parent node, Figure 6 (b). We replace the root d of the old subtree by the root e of the new subtree in the copy c' of the parent node. The new parent node c' is then inserted into the DAG, Figure 6 (c). This process is repeated for the copy of the grandparent node b', in which the parent node c is replace by the copy of the parent node c'. The process is repeated for the grand-grandparent a' etc., until the root

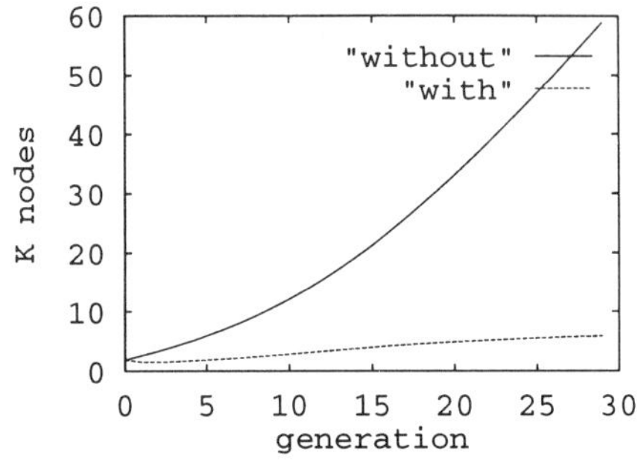

Figure 7: Space usage "with" and "without" removal of nodes, averaged over 100 runs.

is reached (in this example a is the root, so "etc." does not apply), Figure 6 (d). Finally we delete all the nodes in the DAG, that are no longer referenced, first a and then b, Figure 6 (e). This ends the tree modification.

3.4 Removal of Nodes

In order to save space, one can do garbage collection, consisting of deleting nodes from the DAG, that are no longer referenced. Deletion of a node from the DAG can only be performed, if one is sure the node is not used anymore. To know that a node is not used anymore, one has to maintain a reference counter for each node, indicating the number of nodes and members of the population referring to this node. So on insertion of a node into the DAG, the reference count of each of its children has to be increased by one. The reference count of a node is decreased by one, every time a member of the population or node referring to it is removed. As soon as the reference count of a node becomes zero, we can delete it. When deleting a node, we decrease the reference count of its children. In Figure 7 the effect on space usage with or without removal of nodes is shown. Removal of nodes reduces space usage up to about 10-fold in generation 29.

4 Genetic Operators

4.1 The Substitute Operator

The crossover operator in traditional genetic programming produces two offspring from two parents. There is, however, no need to produce two offspring, instead of one. Intuitively it seems obvious, that producing one or two offspring should make no difference in the search performance of genetic programming (all else being equal). This intuition is confirmed by experiments we did, see Figure 8. To distinguish the two operators, we call the single offspring producing operator *substitute*.

Because substitute is conceptually simpler we prefer it over crossover regardless of how the population is represented. There is however also a practical reason for using substitute instead of crossover in the case of using the DAG representation. Using a forest representation, crossover is a simple operation to perform, it

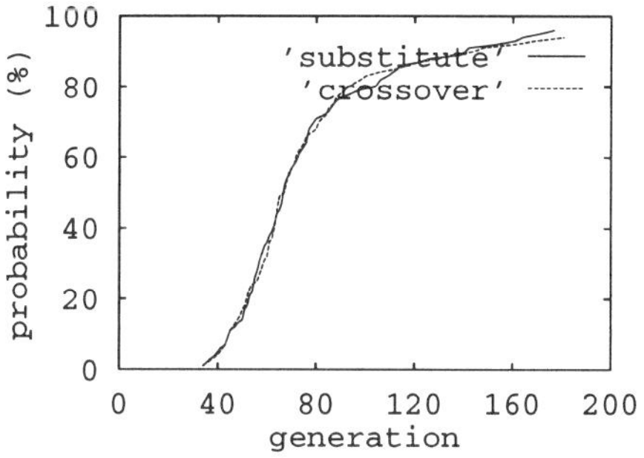

Figure 8: Cumulative probability of success averaged over 100 runs of 200 generations using substitute and crossover.

literally amounts to exchanging subtrees or redirecting pointers to subtrees. Using a DAG representation performing crossover is unnecessarily complicated and inefficient. Substitute on the other hand is comparatively simple to perform, it amounts to replacing a subtree as explained in Section 3.3.

4.2 The Combine Operator

Inspired by the so-called building block hypothesis, we implemented a context-preserving genetic operator, we call *combine*. The combine operator starts by selecting two individuals (i.e. trees) from the population, that are similar. Two individuals are called *similar*, if both individuals can be viewed as the result of a substitute (see Section 4.1) on the same parent of which a subtree was replaced (so not the whole tree). This parent may or may not have existed, it is an imaginary parent. In fact there are always two imaginary parents. Figure 9 shows two similar individuals i_1 and i_2 and their two imaginary parents p_1 and p_2 and the subtrees in which they differ s_1, s_2, s_3, and s_4. Using a DAG, comparison for similarity is easily done by simply pairwise comparing the children of the root nodes of the two individuals (see Section 3.2). If at least one pair of corresponding children compares equal and one or two pairs compare unequal, the children are similar.

Next the combine operator creates two children. The first by replacing s_2 with s_4 in i_1 producing a child identical to p_1. The second by replacing s_1 with s_3 in i_1 producing a child identical to p_2. Figure 9 illustrates this process. In each individual i_1 and i_2, the two paths from the root of the individual to the subtrees are shown. The two paths are identical till they reach the common ancestor node, which is shown. Descending simultaneously in i_1 and i_2, we pairwise compare the children of each node. As long as there is a single differing pair of children in each node, we descend to those children. We have reached the common ancestor, the moment two (or more) pairs of children differ (we will ignore the case in which more than two pairs differ). This is the node on which a substitute is performed twice, producing p_1 and p_2. Experiments with combine are on the way.

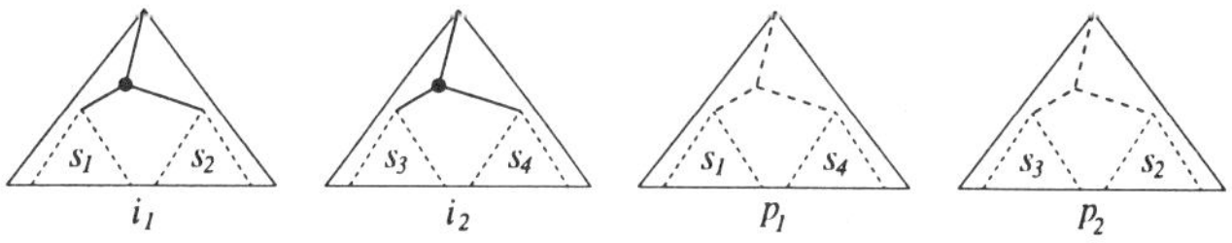

Figure 9: Two similar individuals i_1 and i_2 and their imaginary parents p_1 and p_2. Each with a different pair of subtrees s_1, s_2, s_3, and s_4 and the paths leading to them. The common ancestor node on the paths is shown in i_1 and i_2.

5 Experiments

The experiments in this paper where all done on the 6-multiplexer problem with a population size of 500 and function set {NAND}. The selection method used is (two) tournament selection, because this gives better results on this problem than fitness proportionate selection. Genetic operations are done in the conventional way as described in Koza (1992), where crossover was replaced by substitute (see Section 4.1). Times were measured on a "SGI Indy" with a "IP22" processor running at 100 MHz. We used a 32-bit integer to represent a boolean, so we purposely did not make use of the 32-fold speedup by evaluating 32 fitness cases in parallel, in order to obtain results which are also indicative for non boolean problems. Results of experiments are shown over the first 30 generations, because the shortest successful run over 100 runs had only 30 generations. The choice of the 6-multiplexer was made to make caching of evaluation results possible on our hardware. The choice of the function set {NAND} was made because the fast execution on our hardware of the NAND function minimizes the time needed for the actual node evaluation.

5.1 Commutative Functions

A small reduction in space usage by the DAG can be achieved by ordering the children of nodes labeled with *commutative* functions, such as "NAND" or "+". By definition, trees are (functionally) equivalent, if they only differ in the order of subtrees of nodes labeled with commutative functions. So, if we reorder the arguments of commutative functions, this should have no influence on the performance of genetic programming. By (re)ordering we are able to recognize functionally equivalent trees, that become equivalent because of ordering. This ability can be exploited by genetic operators, like combine (see Section 4.2). Using a DAG, we (re)order nodes labeled with commutative functions according to a strict relational operator on nodes before insertion. In Figure 10, we show an example of space reduction by rearranging the order of subtrees.

5.2 Ancestors

Using a DAG maintaining all individuals of one or more previous generations, the *ancestors*, becomes more feasible. In Figure 11, we show the results of an experiment to measure the additional space requirements, if all an-

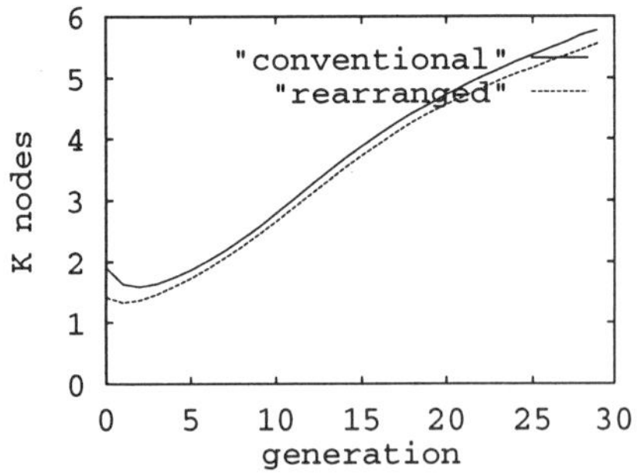

Figure 10: Space usage with ("rearranged") and without ("conventional") ordering arguments of functions, averaged over 100 runs.

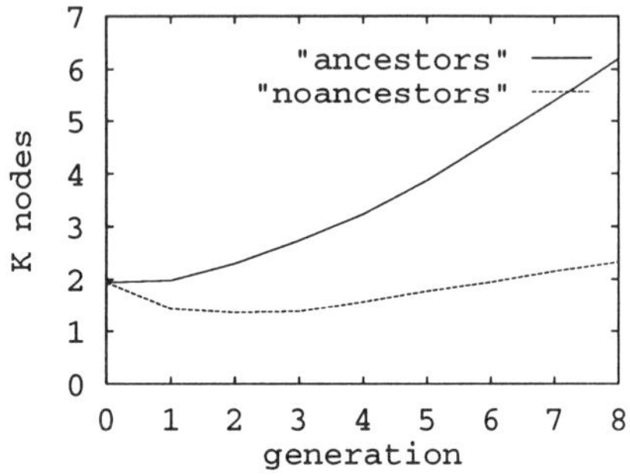

Figure 11: Space usage with ("ancestors") and without ("noancestors") ancestors.

cestors of the individuals in the current population are kept. Although space is saved by using a DAG the number of nodes seems to grow linear.

6 Conclusions

We have shown how to make better use of the DAG representation in genetic programming. Two new evaluation methods were introduced which do not rely on recursively descending trees of the population. As a result these methods are faster than previous methods, as was confirmed by experiments we did. We also gave an exposition of the operations needed to update the DAG. Two genetic operators related to the DAG were also discussed. One called substitute is an equivalent but simpler alternative to crossover. The other called combine was inspired by the "building block hypothesis", it realizes an accurate context-preserving genetic operator.

7 Future work

The discussed methods to evaluate a population represented by a DAG are restricted to genetic programming without the use of functions that have side effects. It might be possible to combine the use of a DAG with efficient evaluation and functions that have side effects. For example, the result of side effects can be expressed as states. These states could be made part of the input. If the number of states is very large however, this might not be a feasible option, since the number of fitness cases to be evaluated may become too large.

The nodes in the DAG can be evaluated in parallel as long as the evaluation order described in Section 2 is respected. It would be interesting to compare this to the parallel evaluation of a forest representation.

The DAG representation reduces redundancy in the population. No information is lost, if reference counts are maintained (see Section 3.4). In a sense statistical information is extracted in the form of reference counts to nodes. It might be interesting to see if the information contained in the DAG in the form of reference counts can be used in the evolutionary process, for instance by biasing the choice of subtrees in genetic operations.

The size of the function set obviously influences the number of possible trees. By reducing the number of functions in the function set, the number of possible trees is reduced. Although Handley showed that for various sizes of function sets the DAG representation greatly reduces space usage, it would be interesting to see how exactly the space and time saving techniques are related to function set size. Based on our experiment with commutative functions (see Section 5.1), we expect the effect of function set size on space reduction to be small.

Acknowledgments

I thank Paul Vitányi my adviser at CWI for advice and commenting and checking the manuscript, John R. Koza of Stanford University for inviting me to Stanford and arranging talks for me on the subject of this paper, James P. Rice of Stanford University for advice and locating an omission in my implementation, Martin C. Martin of Carnegie Mellon University for reminding me of the "32-parallelization trick", Jeroen van Maanen of CWI and Maarten Keizer of University of Utrecht for discussions.

Bibliography

Cormen, T.H., Leiserson, C.E. and Rivest, R.L. 1990. *Introduction to Algorithms.* Cambridge, MA: The MIT Press. Pages 263-280.

Handley, S. 1994. On The Use of a Directed Acyclic Graph to Represent a Population of Computer Programs. *Proceedings of the 1994 IEEE World Congress on Computational Intelligence.* Los Alamitos, CA: IEEE Press. Pages 154-159.

Keith, M.J. and Martin, M.C. 1994. Genetic Programming in C++; Implementational Issues. In Kinnear, Kenneth E. Jr. (editor). *Advances in Genetic Programming.* Cambridge, MA: The MIT Press. Page 287.

Koza, J.R. 1992. *Genetic Programming: On the Programming of Computers by Means of Natural Selection*, Cambridge, MA: The MIT Press.

An Adverse Interaction between Crossover and Restricted Tree Depth in Genetic Programming

Chris Gathercole and Peter Ross
Department of Artificial Intelligence
University of Edinburgh
80 South Bridge
Edinburgh EH1 1HN
U.K.
{chrisg,peter}@aisb.ed.ac.uk

Abstract

The Crossover operator is common to most implementations of Genetic Programming (GP). Often, there is some form of restriction on the size of trees in the GP population. This paper concentrates on the interaction between the standard crossover operator and a restriction on tree depth demonstrated by the MAX problem, which involves returning the largest possible value for given function and terminal sets and maximum tree depth.

Some characteristics and inadequacies of crossover in normal use are highlighted and discussed. Subtree discovery and movement takes place mostly near the leaf nodes, with nodes near the root left untouched, where diversity drops quickly to zero in the tree population. GP is then unable to create fitter trees via the crossover operator, leaving a mutation operator as the only common, but ineffective, route to discovery of fitter trees.

1 Introduction

The Crossover operator is common to most implementations of Genetic Programming (GP), providing a simple but powerful method for recombining genetic material in a population. Crossover seems to be in widespread use in its simplest form, described in Section 3, mixing two parent trees through the exchange of randomly selected subtrees to produce one or two child trees. Mutation operators are often used in combination with crossover. Also common to most GP is some form of upper limit on tree size, necessary to prevent the population expanding to exceed available computer resources.

This paper introduces the MAX problem for GP, in Section 2, a convenient mechanism for looking at the machinations of crossover. The task is to produce the largest possible value for a given function and terminal set and maximum tree depth. Ostensibly an easy problem for GP to solve, results for several variations of the MAX problem, given in Section 6, confirm some inadequacies of the crossover operator in normal use. These are highlighted in an analysis in Section 7. Even with the mitigating effects of some mutation operators, described in Section 4, a loss of diversity in the upper levels of trees in the population due to crossover, discussed in Sections 6 and 8, leads to premature convergence onto sub-optimal solutions. This is made irreversible through the interaction of crossover and the restriction on tree depth.

The tendency of crossover to ignore the upper tree levels should be well known, but the extent of its negative impact on population diversity and premature convergence are made more apparent here through the use of the MAX problem.

2 The MAX Problem for GP

The various MAX problems can be expressed as MAX-depth-D-{Function Set}-{Terminal Set}. The task is to find a tree which returns the largest possible value for a given terminal and function set and (in this paper) with a depth limit, D, where the root node counts as depth 0. GP is given a maximum number of generations (the cutoff) in which to find an optimal tree, after which it is considered to have failed. The cutoff is made sufficiently large so as to give GP a good chance of finding the optimal tree before reaching the generation limit. In every run in this paper the number of generations needed to find the optimal tree (if it was found) was much less than the cutoff.

There are several advantages of the MAX problem for looking at GP as stated, including: results are known quickly; the solution space is easy to visualise; the optimal trees are known in advance; the problem can be varied and made more difficult in small steps.

The MAX problem for GP is analogous to the Ones-Max problem for GAs, [J.D.Schaffer *et al.*, 1991], where an individual consists of a fixed length binary string, and its fitness is simply the sum of its bit values. Thus the fittest individual has all of its bits set to on. Although essentially a simple problem, in practice a GA's population often converges to a state where every individual has some bits set to off in the same positions as every other individual, if the chromosome is sufficiently long. Thus GA Crossover, one of the usual methods (called operators) of combining individuals together, will result in new individuals with the same bits set to off. GA mutation is then the only operator which can change the off bits to on, resulting in only a slow progression of the population towards containing the best possible individuals.

For GP, **MAX-depth-D-{+}{1}** is a simple problem, where the optimal tree simply consists of a full tree of '+'s and '1's. For a given depth D, the maximum possible tree value is 2^D, and there is only one such distinct tree. The depths tested were from 3 to 8.

In **MAX-depth-D-{*,+}{n}**, the '*' (times) function is of no use in terms of returning large values unless both its arguments are greater than 1. This requires the use of the '+' function to add the small terminals together, creating values greater than 1. So the '+' function is most effective at the base of the tree, and the '*' function is most effective at the top of the tree.

For **MAX-depth-D-{*,+}{1}**, the optimal tree consists of '1's in the bottom layer, '+'s in the penultimate layer, '+'s or '*'s in the next-to-penultimate layer, and '*'s in any layers above that. For a given depth D, the maximum possible tree value is $4^{2^{D-2}}$, where $D >= 2$, and there are 2^{D-2} distinct optimal trees. The depths tested were from 3 to 6.

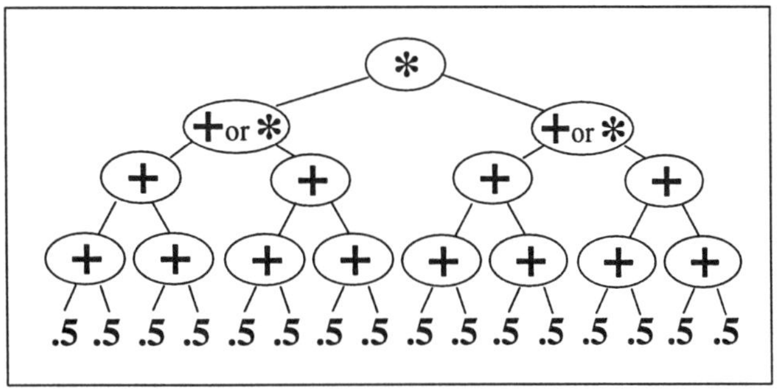

Figure 1: An optimal tree for MAX-depth-4-{*,+}{0.5}

The MAX problem can be made progressively more difficult for GP by decreasing the size of the constant terminal (in this paper, to 0.5 and 0.25). With **MAX-depth-D-{*,+}{0.5}**, the constant has been reduced from 1 to 0.5, and now an optimal tree consists of one more layer of '+'s and one less layer of '*'s. The generic optimal tree is shown in Figure 1. For a given depth D, the maximum possible tree value is $4^{2^{D-3}}$, where $D >= 3$, and there are 2^{D-3} distinct optimal trees. The depths tested were from 3 to 7.

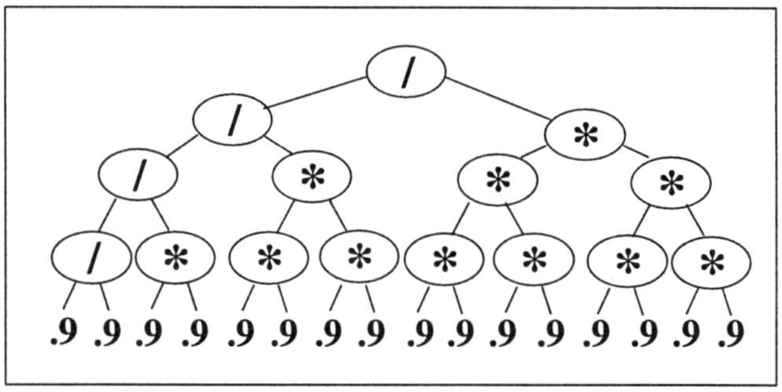

Figure 2: An optimal tree for MAX-depth-4-{*,/}{0.9}

The variation **MAX-depth-D-{*,/}{0.9}** involves the '/' (divide) function instead of the '+' function. The two functions must be combined together asymmetrically to create an optimal tree. The '*' function's role becomes that of providing as small a value as possible which, as the second argument of the '/' function, is turned into a large value. The optimal tree for MAX-depth-D-{*,/}{0.9}, shown in Figure 2, has a different structure from the optimal tree for the MAX-depth-D-{*,+}{n} problems. Most of the tree layers consist of an unbalanced mix of the two functions.

3 Crossover in GP

The standard crossover operator in GP [Koza, 1992], given two parent trees, can in theory bring a subtree from anywhere in one parent tree and swap it with a subtree anywhere in the other parent tree, to produce two new child trees containing a mixture of genetic information from both parent trees. There are no restrictions on the selection of subtrees. Any node in either tree can be chosen as a crossover point. There is no direct equivalent of a static location as with genes in a chromosome in GAs. Thus, if a particular node, e.g. a '+'-node, appears anywhere within any tree in the GP population, then it is possible for GP crossover (assuming such a tree is ever selected to be a parent) to spread that particular node to anywhere in a tree in the next generation. Only when that particular node disappears from all trees in the population, will GP crossover be unable to spread it into the next generation.

In a typical GP setup, operators such as crossover work with some idea of a maximum allowed depth or maximum number of nodes allowed for each tree in the population. This restriction is imposed since it is impractical from the point of view of CPU and memory usage to allow a GP's trees to grow unbounded, as they would otherwise do [Blickle and Thiele, 1994]. A consequence of this restriction on tree size is that the crossover operator may no longer be able to swap all possible pairs of subtrees between two parents and still produce 'legal' trees. This effect is minimal when the population consists only of trees much smaller than the size limit, but the effect becomes more marked as GP trees usually tend to increase in size in later generations.

There are many simple ways of dealing with crossover's propensity for producing illegal trees. The simplest is to select just one of the two child trees, at least one of which is guaranteed to be legal, and is the method used in this paper. Other methods include reselecting the crossover points in some way until both children are legal, pruning over-large trees, or default to making the children identical copies of their parents. Thus, in practice, crossover in GP is not free to move just any subtree to any part of another tree. The tree size restrictions limit the effectiveness of crossover and, as discussed later, can hinder GP's discovery of better trees.

4 Mutation Operators

At this point it is worth mentioning the more common situation where crossover is used along with another operator, usually some form of mutation, creating one offspring from one parent. One often used variant randomly generates a new subtree in a parent tree to produce a child. This method is useful for introducing small effective subtrees to a population but is not likely to introduce large effective subtrees. Another form of mutation operator replaces an individual node in a parent tree with another node of the same arity to produce a child. This method is useful for reintroducing nodes back into a population but only if the population can make immediate use of them. In the case of MAX-depth-D-{+,*}{0.5}, this would speedily bring about the discovery of the optimal tree since replacing a high-level '+'-node with a '*'-node creates a fitter tree. Looking at MAX-depth-D-{/,*}{0.9}, the left-hand side of the trees often requires the introduction of large effective subtrees which are unlikely to be created by mutating a subtree and would not be created by mutating an individual node. Thus incorporating either or both of the mutation operators as described above would not help GP find the optimal tree.

5 Experiments

The MAX problem variations discussed in this paper are MAX-depth-D-{+}{1}, {+,*}{1}, {+,*}{0.5}, {+,*}{0.25}, and {*,/}{0.9}. The depths were between 3 and 8, where the upper depth limit depended on the size of the optimal tree value or complete failure by GP at smaller depths. Fifty runs were made for each variation, with the only operator being crossover. These runs were then divided according to whether or not they were successful, as detailed in Section 6. For comparison, MAX-depth-D-{+,*}{0.5} and {*,/}{0.9} were repeated with the addition of the mutation operators mentioned in Section 4.

An effort was made to keep the GP used in this paper simple since the aim was to demonstrate some qualitative aspects of the crossover operator, and not to strive for optimal GP performance. The implementation used generational replacement with elitism on a population size of 200, with tournament selection of size 6. A tree's fitness was the maximum possible return value minus the tree's return value. Thus a fitness score of zero was optimal, with worse fitnesses being large and positive. Standard crossover was used, with two parents producing one child, where all trees were limited to a maximum depth D. For the purposes of this paper, a crossover point was chosen by randomly selecting one node from all of the nodes in a tree. Crossover could have been made more likely to produce a child different from its parents by requiring that at least one of the crossover points was a function (i.e. non-leaf) node, since all terminal nodes were identical. Runs were stopped after 500 generations as failures, or earlier as successes if the optimal tree was produced.

6 Results

The average number of generations for the successful runs for each of the variations of the MAX problem, in Figure 3, shows the MAX problems increasing in difficulty for GP with increasing depth, unsurprisingly.

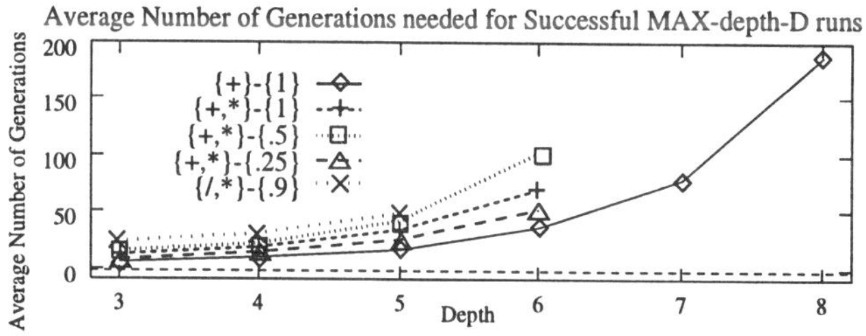

Figure 3: Average number of generations needed by the successful runs, showing difficulty increasing with depth

The percentage of runs which ended in failure, shown in Figure 4, increases rapidly with depth for all but MAX-depth-D-{*,+}{1} and {+}{1}. The populations in these failed runs reached a state where it was impossible or extremely unlikely to produce the optimal tree. The populations in successful runs were, in effect, 'lucky' to discover the optimal trees before the disappearance of crucial subtrees from the populations.

Adding the mutation operators, shown in Figure 5 has

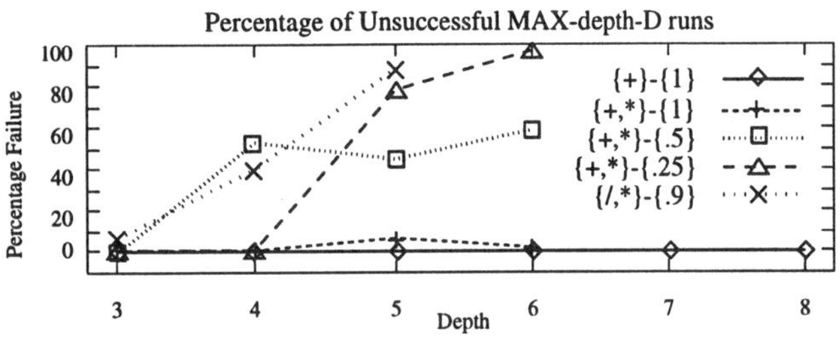

Figure 4: Percentage of runs ending in failure, using crossover

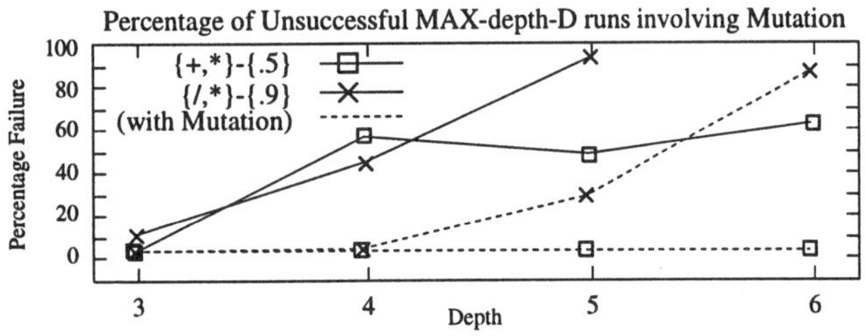

Figure 5: Percentage of runs ending in failure, using crossover only (solid lines) and mutation (dashed lines)

helped MAX-depth-D-{+,*}{0.5} to reach 100% success at all the depths tested. The MAX-depth-D-{/,*}{0.9} runs, on the other hand, still have difficulty in finding the optimal trees. Looking at the runs which failed, the entire population had usually converged to be a duplicate (or very close copy) of the 'best of run' tree.

The following trees exemplify the sub-optimal trees discovered by GP in the variations of the MAX problem (where depth 0 refers to the root node):

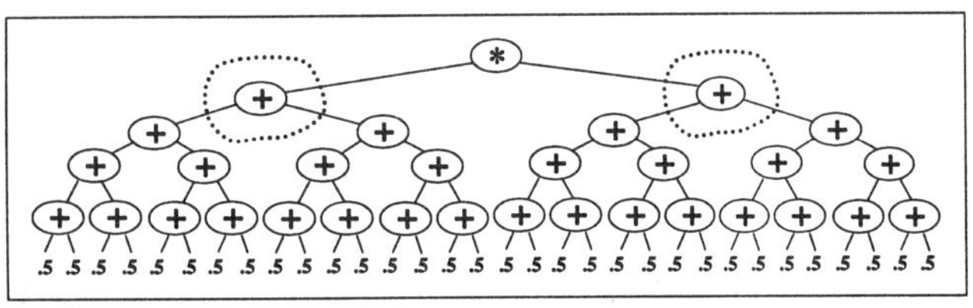

Figure 6: MAX-depth-5-{*,+}{0.5}, sub-optimal tree

For **MAX-depth-5-{*,+}{0.5}**, the 'best of run' tree shown in Figure 6 is not very different (in terms of node changes) from the optimal tree. Each node at depth 1 is a '+' instead of a '*'. However all the fit trees in the population also had no '*'-nodes in this layer, and were all the same size and shape (i.e. full to depth 5). The only '*'-nodes in the population were at depth 0. Given this situation, crossover was no longer able to improve on 'best of run' tree. An operator which mutated individual nodes could easily construct a fitter tree from this one.

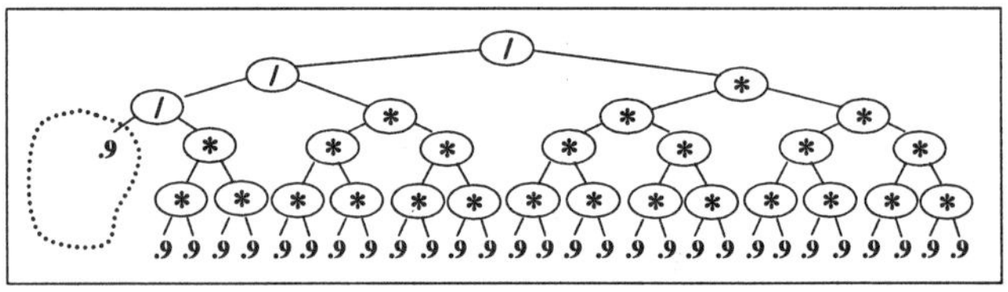

Figure 7: MAX-depth-5-{*,/}{0.9}, sub-optimal tree

For **MAX-depth-5-{*,/}{0.9}**, the 'best of run' tree shown in Figure 7 needs two more '/'-nodes down the left-most side to become the optimal tree. Given that all the fit trees in the population came to be duplicates of this tree, crossover could not improve upon it. Any subtree not containing a '/'-node would have a smaller value than the '.9'-node at depth 3, and thus would result in the whole tree having a smaller fitness. The only '/'-nodes are to be found at depths 0, 1, and 2, and crossover cannot bring those subtrees down to start at depth 3 since that would create an illegal tree. An operator which mutated subtrees might be able to construct a fitter tree from this one.

7 Analysis of the crossover operator

This section presents a simple analysis of the crossover operator in GP, showing its likely impact on GP trees where there is a restriction on tree depth, and the population consists of 'full' trees, i.e. trees which have filled out to the maximum allowed depth. This situation was frequently reached whilst doing the MAX problems. The following calculations apply to GPs with function sets involving functions of arity of two, e.g. like the MAX problems. If the largest function arity is greater than 2, the problem is exacerbated.

D - the maximum tree depth

l - tree layer, from 0 (the root node) to D

$\mathbf{L}_l$ - the number of nodes in layer l
$= 2^l, \quad where \quad 0 \leq l \leq D$

$\mathbf{N}_l$ - the number of nodes in all layers from 0 to layer l inclusive
$= \sum_{j=0}^{l} L_j = \sum_{j=0}^{l} 2^j = 2^{l+1} - 1$

$\mathbf{N}_D$ - the total number of nodes in a tree of depth D
$= 2^{D+1} - 1$

$\mathbf{C(any)}_l$ - the likelihood of any node(s) in a tree, in layer l, experiencing some crossover activity, i.e. when a crossover point occurs in any layer from 0 to l

$$= \frac{N_l}{N_D} = \frac{2^{l+1}-1}{2^{D+1}-1} \simeq \frac{2^{l+1}}{2^{D+1}} = \frac{1}{2^{D-l}}$$

C(layer)$_l$ - the likelihood of crossover occurring within a layer l, i.e. when the two crossover points are in the same layer

$$= \frac{L_l^2}{N_D^2} = \frac{2^{2l}}{(2^{D+1}-1)^2} \simeq \frac{2^{2l}}{2^{2D+2}} = \frac{1}{2^{2(D-l+1)}}$$

C(2 legal offspring) - the likelihood of crossover, based on a random choice of crossover points, producing two legal offspring

$$= \frac{legal}{total} = \frac{\sum_{j=0}^{D} L_j^2}{N_D^2} = \frac{\sum_{j=0}^{D} (2^j)^2}{(2^{D+1}-1)^2} \simeq \frac{\sum_{j=0}^{D} 2^{2j}}{2^{2(D+1)}}$$

$$\simeq \frac{2^{2D}}{2^{2(D+1)}} = \frac{1}{4}, \qquad for\ large\ D$$

Looking at **C(any)$_l$** shows that the upper layers receive relatively little attention from the crossover operator, especially as the trees grow large. In the MAX problem as used in this paper, the tree sizes in question are fairly small (e.g. **D**=6), even so the upper tree layers are mostly unaffected by crossover.

Looking at **C(layer)$_l$** shows how an exchange of subtrees within the upper layers is much less likely again. This implies that there is little or no spread of subtrees within the upper layers in later generations.

Looking at **C(2 legal offspring)** shows that crossover will frequently produce illegal offspring once the trees in the population fill out. For each pair of offspring produced by crossover in this way, there will always be at least one legal offspring, but there is a high probability that the other will be illegal. Either this illegal offspring has to be modified in some way or simply take just one offspring from each crossover operation.

Given two full trees, selecting just one offspring produced by crossover will mean either that subtrees were swapped between the same layer in each tree, or that a subtree from a lower level has been raised to a higher level. It is impossible for a subtree from a higher level to be brought down to a lower level and still produce a legal tree, since it would exceed the depth limit. Thus crossover, producing a single offspring from each pair of parent trees, operating on a population of full trees, will produce the following distribution of offspring in the next generation:

- mostly through the exchange of low level subtrees
- some through the raising of a low level subtree to a higher level
- very few through the exchange of high level subtrees
- none through the lowering of a high level subtree to a lower level

If subtree discovery takes place in the lower levels, crossover would be very effective at spreading these new subtrees through the population. If, on the other hand, improvements in subtrees solely or largely take place in the higher levels, crossover will be very slow to spread these new subtrees through the population.

8 Discussion

The analysis in Section 7 highlights one of the main biases that crossover brings to GP, even without considering populations of full trees. Subtree discovery and the spread of subtrees takes place at lower levels, mostly involving the leaf nodes. The effectiveness of this is highly dependent on the problem in hand. Immediately beneficial subtrees are quickly spread within the trees and through the population, at the expense of other subtrees of less immediate benefit. For MAX-depth-D-{+}{1}, this results in a speedy discovery of the optimal tree.

The situation mentioned above for crossover in general is made worse when a tree depth restriction is incorporated. Whilst the GP population consists of trees which are shallower than the depth restriction, it is still possible, though unlikely for crossover to move subtrees around freely. When the trees have expanded to reach the depth restriction, the situation changes. The only movement of subtrees possible via crossover is from lower levels to higher levels or within the same level. In the case of the other MAX problems, such as MAX-depth-D-{+,*}{0.5}, this can result in the purging the '*'-nodes from the population, perhaps leaving only a few of these function nodes in the higher levels of some of the trees, where they will remain relatively untouched. The benefit of these nodes only emerges after the trees reach an appreciable size, by which time it is unlikely or impossible that they can be spread by crossover. Once the trees have reached the depth limit, the only way the higher levels are affected is through the promotion of lower level subtrees, which contain no '*'-nodes, and the movement of subtrees within the same level. If there are no '*'-nodes in any tree at a particular high level, it now impossible for crossover to introduce a '*'-node to this level; the population has converged to being duplicates of a sub-optimal tree, and no further improvement is possible.

Some early experiments indicate that effects similar to those seen with a tree depth restriction are also seen with a restriction on the number of tree nodes. A shallow scan of published papers has indicated that this method is used more often than a maximum tree depth, since it lends itself well to various memory and cpu-efficient GP implementations. The trees in a GP population still expand with the generations, quickly reaching their size limit. At this stage, crossover affects mainly the peripheral (i.e. leaf) nodes, and becomes unable to modify high levels of the trees effectively, although perhaps not as often as with the depth restriction.

9 Modifications to Crossover

Several other investigations have focused in some way on crossover in GP. These include: [Rosca, 1995] which looks at causality in GP, relating the changes in the structure of GP trees with changes in the properties of the trees, [D'haeseleer, 1994] which looks at context preserving crossover in GP, attempting to ensure that swapped subtrees will still be effective, and [O'Reilly and Oppacher, 1995] which looks at hybrids of operators such as crossover and Hill Climbing.

There is no doubt that it should be possible to modify crossover or its use so that it is more likely to result in the discovery of optimal (or at least better) solutions, even if this is at the expense of speed on those problems where it does well already. In practice, where the focus of a paper is not on the operators themselves, 'standard' crossover still seems to be used as one of the main GP operators, usually in combination with other operators such as mutation. More attention should be paid to ensure that the other operators are capable of overcoming crossover's shortcomings. Simple mutation operators have been shown here to be insufficient. Using very large population sizes to boost mutation's chance of constructing large useful subtrees only obscure the problem.

The method of selecting crossover points could be modified to ensure that the tree layers get a more even spread of crossover activity. [Koza, 1992] uses crossover where internal nodes have a 90% chance of being selected as a crossover point compared with 10% for leaf nodes. However, the higher tree levels will still experience much lower crossover activity than the lower levels. It would be better to scale the selection probability according to depth or the number of nodes in each level, perhaps targeting sections of tree which have a low node variance across the population.

For problems where tree size is part of the solution, allowing unlimited tree sizes is obviously not applicable. But for others, where the optimal tree sizes are unknown, parsimony (penalising large trees) appears to work well, especially when it is used only to discriminate between trees which would otherwise have the same fitness. Thus trees can grow as large as is needed to do better on a problem, but there is continual selection pressure for smaller trees which do just as well.

10 Conclusions

The MAX problems for GP show how the crossover operator can be directly responsible for loss of diversity within a population, leading to premature convergence to a sub-optimal solution. When used in combination with a restriction on tree depth, the premature convergence becomes irreversible. Subtree discovery and movement takes place mostly near the leaf nodes, with nodes near the root left mostly untouched. Diversity drops quickly to zero near the root node in the tree population, resulting in GP being unable to create 'fitter' trees via the crossover operator. The addition of simple mutation operators is not sufficient to overcome these problems. Care should be taken to ensure that the spread of subtrees throughout the GP population is not stifled by crossover.

This should not be taken as an attack on the crossover operator and GP, or a claim that the MAX problem epitomises all GP problems. Instead it gives a better understanding of how crossover works in practice, often in combination with tree size restrictions, and enables the user to be aware of its potential failings.

Acknowledgments

Many thanks to EPSRC for PhD grant #93314680.

Bibliography

[Blickle and Thiele, 1994] Tobias Blickle and Lothar Thiele. Genetic programming and redundancy. In J. Hopf, editor, *Genetic Algorithms within the Framework of Evolutionary Computation (Workshop at KI-94, Saarbrücken)*, pages 33–38. Max-Planck-Institut für Informatik (MPI-I-94-241), 1994.

[D'haeseleer, 1994] P. D'haeseleer. Context preserving crossover in genetic programming. In *Proceedings of the 1994 IEEE World Congress on Computational Intelligence*, volume 1, pages 256–261. IEEE Press, 1994.

[J.D.Schaffer *et al.*, 1991] J.D.Schaffer, L.J. Eshelman, and D. Offuttt. Spurious correlations and premature convergence in genetic algorithms. In G.J.E. Rawlins, editor, *Foundations of Genetic Algorithms*, pages 102–112. Morgan Kauffman, 1991.

[Koza, 1992] J.R. Koza. *Genetic Programming: on the Programming of Computers by means of Natural Selection.* MIT Press, Cambridge, MA, 1992.

[O'Reilly and Oppacher, 1995] Una-May O'Reilly and Franz Oppacher. Hybridized crossover-based search techniques for program discovery. Technical Report 95-02-007, Santa Fe Institute, 1995.

[Rosca, 1995] Justinian P. Rosca. An Analysis Of Hierarchical Genetic Programming. Technical Report 566, University of Rochester, 1995.

The Prediction of the Degree of Exposure to Solvent of Amino Acid Residues via Genetic Programming

Simon Handley

Computer Science Department
Stanford University
Stanford, CA 94305
http://www-leland.stanford.edu/~shandley
(415) 723-4096
shandley@cs.stanford.edu

ABSTRACT

This paper applies a new extension to genetic programming (Handley 1996, this volume) to a problem on which genetic programming has previously been applied (Handley 1994). The problem is to predict the degree of exposure to solvent (the *buriedness*) of a protein's amino acid residues given only the protein's primary structure. The best programs evolved are better than the programs evolved in an earlier set of experiments that used a problem-specific ad-hoc function set (Handley 1994).

1. Introduction

A fundamental problem in molecular biology is the prediction of tertiary structure from primary structure: the *protein folding problem*. The goal of protein folding is the prediction of one feature of a folded protein (the 3 dimensional coordinates of its backbone atoms) from another feature (the sequence of amino acid residues that make up the protein). The protein folding problem is of enormous practical importance because the latter feature (the primary structure) is much easier to experimentally establish than is the former (the tertiary structure).

A related problem is the *buriedness problem*: the prediction of the degree of exposure to the solvent (the *buriedness*) of each amino acid residue in a folded protein. Some amino acid residues will have a buriedness of 0%: these are in the core of the protein and are likely hydrophobic. Other residues will have a buriedness of 100%: these are on the surface of the protein and are probably hydrophilic.

The buriedness problem is interesting in so far as it is closely related to the protein folding problem The information in the patterns of exposure to solvent of the amino acids in a compact globular protein is likely to be very similar to the information required to determine the protein's fold.

A previous paper (Handley 1994) showed that genetic programming could produce programs that predicted the degree of exposure with a high degree of accuracy. The programs evolved in that paper implemented convolutions. Each program was effectively a pattern that was iterated over each protein by the wrapper.

This paper demonstrates that a new function set—called "statistic-computing zones"—can be applied to this problem and that results comparable to the more problem-specific function set can be achieved.

This paper applies a new technique ("statistic-computing zones") to an existing problem ("the buriedness problem"). The technique is described in detail in Handley 1996 (this volume).

2. Previous Work

Holbrook, Muskal and Kim (1990, 1993) trained neural networks to classify windows of contiguous residues (a central residue plus a fixed number of flanking residues on either side). This classification was either two-state ([0%...20%) and [20%...100%]) or three-state ([0%...5%), [5%...40%) and [40%...100%]). They trained by backpropagation both perceptrons and one-hidden-layer nets on a training set of 20 proteins from the Brookhaven Protein Data Bank (PDB) (Bernstein 1977) and cross-validated on a testing set of 5 proteins, also from the PDB. They achieved a prediction accuracy of $Q_3 = 72.0\%$ on the two-state prediction problem and $Q_3 = 52.0\%$ for three-state prediction[1].

3. Statistic-Computing Zones

The functions used in this paper are based on a problem-independent set of functions called the "statistic-computing zone" function set. The motivation behind these functions and their intended problem domains are described in (Handley 1996); the remainder of this section will simply describe the functions.

The statistic-computing zones function set contains two groups of functions: First, those that compute convolutions (by sliding patterns down the sequence that are independently evaluated at each position). Second, those that compute statistics (a more general type of computation in which temporary results can be accumulated).

[1] Q_3 is the number of correct predictions over the total number of predictions (also called accuracy).

3.1. Functions That Compute Convolutions

The following functions together can represent patterns that are applied independently to each element of a sequence:

- Leaves of the program tree that represent constants and problem inputs. Constants can be scalars such as `3.14` or sequences of scalars, such as `[1,2,3]`.
- Arithmetic functions. For example,

$$(+X\,Y)=\begin{cases} X+Y & X,\ Y \text{ both numbers} \\ [(+X_1\,Y),\ldots,(+X_n\,Y)] & X \text{ sequence, } Y \text{ number} \\ [(+X\,Y_1),\ldots,(+X\,Y_n)] & X \text{ number, } Y \text{ sequence} \\ [(+X_1\,Y_1),\ldots,(+X_n\,Y_n)] & X,\ Y \text{ both sequences} \end{cases}$$

Some examples: `(+ 1 2)` → `3`; `(+ [1,2] 3)` → `[(+ 1 3),(+ 2 3)]` → `[4,5]`; and `(+ [1,2] [3,4])` → `[(+ 1 3), (+ 2 4)]` → `[4,6]`. If both arguments are sequences and the sequences are of different lengths then the shorter sequence is padded with zeros. e.g.: `(+ [1,2] [3])` → `(+ [1,2] [3,0])` → `[4,2]`.

- The conditional function,

$$(\mathit{if}\ P\,T\,F)=\begin{cases} T & P \text{ is a number, } P\neq 0 \\ F & P \text{ is a number, } P=0. \\ [(\mathit{if}\ P_1\,T_1\,F_1),\ldots, (\mathit{if}\ P_n\,T_n\,F_n)] & P,T,F \text{ sequences} \end{cases}$$

If P is a sequence and T and/or F are numbers then T and/or F are converted into sequences: `(if [0, 1] 2 3)` → `(if [0, 1] [2, 2] [3, 3])` → `[(if 0 2 3), (if 1 2 3)]` → `[3, 2]`. P, T and/or F are padded with zeros if they are of different lengths.

- Functions to shift sequences left and right: `(shiftl [1,2,3])` → `[2,3,0]` and `(shiftr [1,2,3])` → `[0,1,2]`.

The shift functions are the key to computing convolutions. For example, to convolve the pattern $X_i'=(X_{i-1}+X_i+X_{i+1})/3$ over the sequence X to create the sequence X' we write,

```
(/ (+ (shiftr X) X (shiftl X)) 3),
```

where for convenience `(+)` is extended to take three arguments. Here's a trace of the evaluation of this program on the list X = `[1,2,3]`:

```
X →                                   [1,2,3],
(shiftr X) →                          [0,1,2],
(shiftl X) →                          [2,3,0],
(+ (shiftr X) X (shiftl X)) →         [3,6,5],
(/ (+ (shiftr X) X (shiftl X)) 3) →
                                      [1,2,5/3].
```

3.2. Functions That Compute Statistics

Statistics differ from convolutions in that the computation at each element has access to the results of the computations at other elements. For example, to compute the mean of a sequence X we could apply the pattern $X_i'=X_{i-1}'+X_i$ to X resulting in X'. The computation of X_i' has access to X_{i-1}' (and X_{i-1}' has access to X_{i-2}', etc.). Statistics are computed with the functions from the previous section plus the `scz` function which delineates a part of the program on which the X_i' values are accessible to patterns. The implementation of `scz` is discussed in (Handley 1996). Here's an example of how the `scz` function is called:

```
(scz (+ (shiftr sczX) sczX) [1,2,3]).
```

The first argument is the pattern, the second argument is the input sequence (called X above). The zero-arity function `sczX` returns the sequence being computed (called X' above). The pattern is applied from left to right, when the pattern is applied to the ith position, the X' sequence contains the computed values from 1 to i-1.

4. Applying the Statistic-Computing Zones Function Set To This Problem

The functions in the previous section are problem-independent: they are intended to be augmented with problem-specific functions. To recap, the statistic-computing zones function set contains the following functions:

{ `X`, $\Re$, `+`, `if`, `shiftl`, `shiftr` } ∪ { `scz`, `scz-t`, `sczX` }.

The function `X` is replaced in this problem by the function `res` that returns a sequence of numbers corresponding to the primary sequence of the protein. The function $\Re$ is replaced at program creation time by random values (either numbers or sequences of numbers). The function + is joined by `-`, `*`, `/`, `<`, `=`, `max`, and `min`. The functions < and = return the value 1 for true and 0 for false.

The function `scz` delineates a statistic-computing zone. The function `scz-t` is similar but only returns the last ("tail") element of X': `(scz-t P X) ≡ (car (last (scz P X)))`. The function `sczX` contains the value computed by the second argument of `scz` while executing the first argument of `scz`, and some default value (such as 0) elsewhere. In the current implementation, calls to `scz` cannot be nested (calls to `scz` inside of the first argument of another `scz` just return the second argument, `X`).

In addition, the following functions were added:

- **Unary**. The `unary` function is designed to create frequency matrices. (Frequency matrices are useful for counting how many times elements appear in a sequence. For example one frequency matrix for the sequence `[4,1,3,4,4]` is `[1,0,1,3]`.) `Unary` takes a positive integer as argument and returns a sequence of `0`s and `1`s that sort of represents that number in base 1:

```
(unary 0) → [1];
(unary 1) → [0,1];
(unary 2) → [0,0,1]; etc.
```

Table 1 The Brookhaven identifiers of the 122 proteins used in this paper.

1CSE_E, 1CSE_I, 1ECO, 1FKF, 1FNR, 1FXD, 1GCR, 1GD1_R, 1GP1_B, 1GPB, 1HIP, 1HOE, 1HRH_B, 1HSA_D, 1HSA_E, 1HYP, 1IFB, 1LPE, 1LZ1, 1MBD, 1MSB_B, 1NN2, 1NXB, 1OVA_D, 1PAZ, 1PGX, 1PHH, 1PRC_C, 1PRC_M, 1PRC_H, 1R69, 1RBP, 1RHD, 1RNB_A, 1RNH, 1ROP_A, 1SN3, 1TFD, 1TIE, 1TPK_C, 1UBQ, 1UTG, 1WSY_A, 1WSY_B, 1YCC, 256B_B, 2AZA_B, 2CA2, 2CCY_B, 2CDV, 2CPP, 2CSC, 2ER7_E, 2FCR, 2GN5, 2HAD, 2HBG, 2HMZ_D, 2LH7, 2LIV, 2LTN_D, 2MCM, 2OVO, 2PAB_B, 2RHE, 2RSP_B, 2SAR_B, 2SDH_B, 2SGA, 2SIC_I, 2SNS, 2SOD_Y, 2STV, 2TRX_B, 2TS1, 2TSC_B, 2WRP_R, 2YHX, 3ADK, 3B5C, 3BCL, 3CHY, 3CLA, 3GAP_B, 3GRS, 3IL8, 3PGK, 3RUB_S, 3SDP_B, 451C, 4BP2, 4CPV, 4ENL, 4FXN, 4I1B, 4LZM, 4MDH_B, 4PFK, 4PTP, 4SGB_I, 4XIA_B, 5ABP, 5ACN, 5CPA, 5HVP_B, 5P21, 5PTI, 5RUB_B, 5RXN, 5TIM_B, 6LDH, 6TMN_E, 7RSA, 8ADH, 8ATC_C, 8ATC_D, 8CAT_B, 8DFR, 9ICD, 9PAP, 9RNT, 9WGA_B.

Non-integers are rounded; negative numbers have their sign flipped. Sequences are recursed-on:

```
(unary [X1,...,Xn]) →
    [unary(X1),...,unary(Xn)].
```

`Unary` is the only function in this paper whose output can be larger than its input. Thus its inputs must be limited. This is done two ways: First, the depth to which `unary` can call itself is limited to 2 (in this paper). Second, integer arguments greater than 7 (in this paper) return 0. So, for example,

```
(unary [[[1]]]) → 0; (unary 999) → 0.
```

- **Even**. The `even` function takes an integer and returns `1` if it is even, `0` otherwise:

```
(even 3) → 0; (even 4) → 1; etc.
```

Real numbers and sequences are handled as in `unary`.

Finally, the output of each program is scaled to be in the range 0% to 100%. That is, to predict the buriedness of a protein with a given program, the program is first run on each amino acid residue in the primary structure. This results in a list of floating point numbers, one per residue. I then scale these numbers so that the largest is 100.0 and the smallest is 0.0. Each number is then interpreted as the predicted percentage buriedness of the corresponding residue.

5. Parameter Settings

The fitness measure should reward programs that perform well at the desired task. Here I want to evolve programs that are good predictors of buriedness. So the fitness of a program is defined as the similarity between its predictions and the known buriednesses (as computed by Kabsch and Sander (1983) from the 3 dimensional coordinates in the PDB). I quantified similarity as the RMS error between the two real-valued curves (predicted and observed buriednesses as a function of residue number). RMS error was used for this experiment (as opposed to those in Handley 1994, that used correlation) because RMS error is easier to interpret.

The population size should be large enough that there is enough genetic diversity to solve the problem. It should be small enough that useful results can be obtained in a reasonable amount of time on whatever computers are available. For the results shown here I used a population size, M, of 19,200.

Similarly, the maximum number of generations should be chosen to optimize computer time. The runs reported here used $G = 51$ generations (1 initial random population + 50 subsequent evolved populations).

6. The Data

I used 122 canonical proteins from the PDB that Tod Klingler of Stanford University selected for me (personal communication). These 122 proteins are canonical in the sense that there is minimal sequence identity and evolutionary homology between them. Table 1 shows these proteins' identifiers.

Thirty two proteins were held out as an evaluation set. The remaining 90 proteins were 3-way jackknifed. The evolutionary process was driven by the proteins in the training set (60 proteins). The difference between the predictive performance on the training and testing sets gives a measure of how much each program is overfitting the training data; that is, the degree to which a program does better on the training set than the testing set shows to what degree the program is looking at irrelevant aspects of the problem (i.e., aspects that aren't shared by the proteins in the testing set).

7. Results

I did 1 run on a 64-node parallel machine (a two-dimensional toroidal mesh) with demes of size $Q = 300$ (for a total population of $M = 19{,}200$), per-generation emigration of 5% of each deme in each direction, and $G = 51$.

The best-of-generation (on the training set) individual that was fittest on the testing set appeared in generation 18 and was

```
   (if (shiftr (= res (max res res)))
(scz_t (shiftr (- (max (- sczX res)
sczX) [-9.6585,-8.66606,-5.06984,
7.57551,-8.91838,-1.68983,-9.87357,-
4.03295,4.66446])) [-3.8593,5.52721,
7.89694,-2.36325,-5.25199,9.75733,
5.81191,7.05549])(+ (scz_t (shiftr (-
(max (- sczX res)(- (max (- sczX res)
sczX) [-9.6585,-8.66606,-5.06984,
7.57551,-8.91838,-1.68983,-9.87357,-
4.03295,4.66446])) [-9.6585,-8.66606,-
5.06984,7.57551,-8.91838,-1.68983,-
9.87357,-4.03295,4.66446]))(if (shiftr
(= res 9.37449))(scz_t res
```

Table 2 The best individual's performance. RMS error was used as fitness, sum-difference is the sum of the differences between observed and predicted. Note that the correlations are much lower than those reported in Handley 1994. Correlation is not driving the evolution in these experiments, RMS error is.

	RMS	Sum-differences	Correlation	Q_3 (binary)	Q_3 (tertiary)
training set	28.4	24.4	0.10	92.0%	90.2%
testing set	28.6	24.7	0.08	88.9%	86.7%
evaluation set	28.0	23.8	0.11	91.4%	90.0%

```
[4.69383,8.87702,-8.98234])(- (max
[7.90397,-1.74158,-1.58522,6.16291,-
8.75552,-7.99935,-9.19641] sczX)(if (min
(+ [-8.51402,-0.0421677,-3.62687,-
1.38171,4.35159] (scz_t sczX (+
[3.40444,3.80769,-7.79385,-1.0323,
3.95082,-7.00434,2.41654,-2.46169,
3.48103] (if res (* [9.06141,-4.2153,-
6.84452,-3.12939] (= -1.1646 res))
(shiftl (- (* [9.06141,-4.2153,-6.84452,
-3.12939] [-7.95689,-6.44818,-7.07708,-
4.42886,-2.87947,2.58291])(if res (*
[9.06141,-4.2153,-6.84452,-3.12939] [-
7.95689,-6.44818,-7.07708,-4.42886,-
2.87947,2.58291]) (- -9.05959 (+ (min [-
2.95232,7.64877,-7.53649,5.94216,
3.4814,7.40892] (min [-2.95232,7.64877,-
7.53649,5.94216,3.4814,7.40892] (scz_t
res (scz_t res sczX)))) [9.06141,-
4.2153,-6.84452,-3.12939]))))))))
(shiftl res)) [-0.515766,-2.82477,
7.69331,-8.41727,0.2566,-2.70493,
1.46101,-6.88955,5.51359,-4.91248]
res)))) [6.63336,3.54814,1.58864,-
2.40145])).
```

Table 2 shows how this individual scored on various fitness measures.

8. Conclusions

This paper has shown that a new function set, statistic-computing zones, can evolve programs that are both better than the programs evolved by a more ad-hoc function set and also better than other researcher's results on a real problem. The statistic-computing zones function set is a general-purpose function set in the sense that it was not designed to solve just this problem. Even though it is not specifically tailored to predicting exposure to solvent it has proven to be capable of evolving programs that approximately solve this problem.

Acknowledgments

David Andre, Forrest Bennett and John Koza made many useful comments.

References

Anfinsen, C. B. 1973. Principles that govern the folding of protein chains. *Science* 81: 223–30.

Bernstein, F. C.; Koetzle, T. F.; Williams, G. J. B.; Meyer, J., E. J.; Brice, M. D.; Rodgers, J. R.; Kennard, O.; Shimamouchi, T.; and Tasumi, M. 1977. The protein data bank: A computer based archival file for macromolecular structures. *Journal of Molecular Biology* 112: 535–42.

Handley, Simon G. 1994. The prediction of the degree of exposure to solvent of amino acid residues via genetic programming. *Proceedings of the Second International Conference on Intelligent Systems for Molecular Biology.* Pages 156–60. Altman, R. et al. (Eds). August 14–17, Stanford, CA. AAAI Press: Menlo Park, CA.

Handley, Simon G. 1996. A New Class of Function Sets For Solving Sequence Problems. In Koza, John R., Goldberg, David E., Fogel, David B., and Riolo, Rick L. (editors). Genetic Programming 1996: Proceedings of the First Annual Conference, July 28–31, 1996, Stanford University. Cambridge, MA: MIT Press.

Holbrook, S. R.; Muskal, S. M.; and Kim, S.-H. 1990. Predicting surface exposure of amino acids from protein sequence. *Protein Engineering* 3 (8): 659–65.

Holbrook, S. R.; Muskal, S. M.; and Kim, S.-H. 1993. "Predicting protein structural features with artificial neural networks." In *Artificial Intelligence and Molecular Biology*, ed. Hunter, L. 161–94. AAAI Press.

Kabsch, W. and Sander, C. 1983. Dictionary of protein secondary structure: Pattern recognition of hydrogen-bonded and geometrical features. *Biopolymers* 22 (1983): 2577–637.

Koza, John. R. 1992. *Genetic Programming: On the Programming of Computers by Means of Natural Selection.* Cambridge, MA: MIT Press.

Koza, John. R. 1994. *Genetic Programming II: Automatic Discovery of Reusable Programs.* Cambridge, MA: MIT Press.

A New Class of Function Sets For Solving Sequence Problems

Simon Handley

Computer Science Department
Stanford University
Stanford, CA 94305
http://www-leland.stanford.edu/~shandley
(415) 723-4096
shandley@cs.stanford.edu

ABSTRACT

This paper presents an extension of genetic programming that facilitates the solution of problems whose instances are most naturally represented as arbitrary-length sequences (*sequence problems*). This new technique is simple, easy to implement, expressive, efficient, and doesn't require arbitrary time-outs. The technique is demonstrated on a concrete problem: classifying pinochle poker hands.

1. Introduction

This paper presents an extension of genetic programming that facilitates the solution of problems whose instances are most naturally represented as arbitrary-length sequences (*sequence problems*). This new technique is simple, easy to implement, expressive, efficient, and doesn't require arbitrary time-outs.

The remainder of this paper is organized as follows. Section 2 defines what I call "sequence problems". Section 3 examines existing machine learning techniques and how they can be applied to sequence problems. Section 4 describes the new technique. Section 5 discusses the technique's merits . Finally, section 6 demonstrates the use of the technique on a concrete problem: classifying pinochle poker hands.

2. Sequence Problems

The following are examples of *sequence problems*:

- *Time series analysis*: given a sequence of values $Y_1,...,Y_t$; predict the value Y_{t+1}.
- *Image analysis*: given a 2-dimensional image, $I_{0,0},..., I_{m,n}$, compute summaries such as "Is there a cylinder in this image?".
- *Classifying poker hands*: Does a poker hand, such as {9♥, 10♦, J♠, Q♥, K♦}, contain a flush? A straight?
- *Parity*: Given an arbitrary-length sequence of Boolean values, such as [true, false, true] or [false], compute the parity of the sequence.
- The *intron/exon problem*: given a sequence of DNA bases, return a sequence of the same length of Boolean values, where contiguous true values denote introns (DNA sequences that are not expressed as protein) and contiguous false values denote exons (sequences that are expressed as protein) (Handley 1995).
- The *DNA motif recognition problem*: given a sequence of DNA bases, return a sequence of the same length of Boolean values, where a true at position i means that the protein of interest binds to the DNA sequence at that point.
- The *secondary-structure prediction problem*: given a sequence of amino acid residues, classify each residue as to the pattern of hydrogen-bonds it enters into with its neighbours in the folded protein.

These problems share two important characteristics: One, their problem instances are sequences (of DNA bases, numbers, etc.) or sequences of sequences (as in the image analysis problem). Two, the elements of these sequences are all of the same "data type", they all have similar "semantics" (all elements are DNA bases or all are readings from some experiment, or all are intensities, etc.). The term "semantics" is used loosely here. Another way to put this is to say that the elements must be (largely) translation-invariant or position-independent. Shifting a problem instance's sequence left or right a small amount should not greatly affect the result. For example, in the image analysis problem, if the given raster image is shifted left by one pixel then the resulting summary statistic will likely be unchanged.

An example of a problem that fails the second criterion is the Boolean 11-multiplexer: if its 11 arguments are put into an 11-element sequence then it satisfies the first criterion. The 11 elements, however, do not have the same semantics: the first 3 are interpreted as address bits and the remaining 8 are interpreted as data bits. Even if the 11-multiplexer's problem instances are represented by two sequences (address-bits and data-bits) it fails the translation-invariance test.

An example of a problem that passes the second criterion is the Boolean 11-parity problem. The 11-parity problem's instances are 11-element sequences whose elements are all of the same type and the 11-element sequences are roughly translation-invariant (ignoring boundary effects, shifting the sequence left or right does not affect the result).

3. Is a New Technique Necessary?

Most machine learning techniques (such as neural nets (Rumelhart 1986), decision trees (Quinlan 1986, 1987), the genetic algorithm (Goldberg 1989; Holland 1992), and "classic" genetic programming (as defined in Koza 1992)) require that a problem's instances be represented as fixed-length feature vectors. For example, "classic" genetic programming requires that each problem instance (i.e., each fitness case) be represented by a fixed set of terminals, each being a scalar.

Sequence problems have been solved with the above machine learning techniques by converting the problem of classifying a sequence into a collection of problems, each being to classify a fixed-length window of the sequence. This attempt to shoehorn sequence problems onto non-sequence-oriented techniques suffers from two main problems: (1) how to choose a window size, and why should the window size be fixed? and (2) how to represent solutions that need non-local information, i.e. solutions that are more than just simple convolutions?

The technique described in this paper is an elegant solution to both of these problems. It puts no restrictions on the window size, indeed it has no explicit notion of a window. The "windows" are adaptive (they are wider on some parts of the sequence than others). This technique also elegantly represents both simple convolutions (sliding patterns over sequences) and more complex computations that require non-local information.

It may at first seem that sequence problems can easily be solved with genetic programming by just adding a looping function. Unfortunately, extending genetic programming to solve sequence problems is not this easy. The various reasons why this is so are discussed in section 5.

4. The "Statistic-Computing Zones" Function Set

The function set described in this section is intended as a basis for the solution of a number of different problems. Problem-specific functions will need to be added to solve particular problems. Thus, the function set described here is more properly a class of function sets.

The functions below all operate on *values*; a value is either a floating point number or a sequence (list) of values. Examples of values are `3.14`, `[1,2]`, and `[1, [2,3]]`. Values could also be based on integers or Booleans for problems in non-floating point domains.

The functions below are divided into two groups: those that compute convolutions and those that compute statistics.

A *convolution* is a mapping from one sequence to another such that the computation done at each element is independent of the computations done at the other elements. An example is a smoothing filter: $X_i' = (X_{i-1} + X_i + X_{i+1})/3$. (The following convention is used hereafter: the sequence X (and its elements the X_is) is the input to a program (composition of functions) and the sequence X' (and its elements the X_i's) is the output.)

A *statistic* is a less restrictive form of convolution in which the computation done at an element can depend on the computations done at other elements. For example, the formula $X_i' = X_{i-1}' + X_i$ when applied to each element of X returns a sequence of partial sums of X ($X_j' = \sum_{i=1}^{j-1} X_i$). The existence of the term X_{i-1}' in the right hand side of the formula is the reason that it is not a convolution: it is not possible to compute X_i' without first computing X_{i-1}' etc.

Statistics are a generalization of convolutions: That is, the set of convolutions is a subset of the set of statistics. A convolution can be thought of as a degenerate statistic that just happens to not refer to any X_i's. Although the distinction may seem minor, it will prove useful when designing function sets.

4.1. Convolution-Computing Functions

The following functions compute convolutions:

- Leaves of the program tree that represent constants and problem inputs. For example the function `X` may return a sequence representing an input to a problem (a dependent variable, analogous to the use of `X` in the symbolic regression problems in (Koza 1992,1994)). Constants can be scalars such as `3.14` or sequences of scalars, such as `[1,2,3]`.
- Arithmetic functions (or, in Boolean domains, logical functions). For example,

$$(+XY) = \begin{cases} X+Y & X,\ Y \text{ both numbers} \\ [(+X_1\,Y),\ldots,(+X_n\,Y)] & X \text{ sequence, } Y \text{ number} \\ [(+X\,Y_1),\ldots,(+X\,Y_n)] & X \text{ number, } Y \text{ sequence} \\ [(+X_1\,Y_1),\ldots,(+X_n\,Y_n)] & X,\ Y \text{ both sequences} \end{cases}$$

Some examples: `(+ 1 2)` → `3`; `(+ [1,2] 3)` → `[(+ 1 3),(+ 2 3)]` → `[4,5]`; and `(+ [1,2] [3,4])` → `[(+ 1 3), (+ 2 4)]` → `[4,6]`. Unequal-length sequences are padded with zeros : `(+ [1,2] [3])` → `(+ [1,2] [3,0])` → `[4,2]`.

- The conditional function,

$$(\mathit{if}\ P\,T\,F) = \begin{cases} T & P \text{ is a number, } P \neq 0 \\ F & P \text{ is a number, } P = 0 \\ [(\mathit{if}\ P_1\,T_1\,F_1),\ldots, (\mathit{if}\ P_n\,T_n\,F_n)] & P,T,F \text{ sequences} \end{cases}$$

If P is a sequence and T and/or F are numbers then T and/or F are converted into sequences: `(if [0, 1] 2 3)` → `(if [0, 1] [2, 2] [3, 3])` → `[(if 0 2 3), (if 1 2 3)]` → `[3, 2]`. P, T and/or F are padded with zeros if they are unequal-length sequences.

- Functions to shift sequences left and right: `(shiftl [1,2,3])` → `[2,3,0]` and `(shiftr [1,2,3])` → `[0,1,2]`.

The shift functions are the key to computing convolutions. For example, to convolve the formula $X_i' = (X_{i-1} + X_i + X_{i+1})/3$ over the sequence X to create the sequence X' we write,

```
(/ (+ (shiftr X) X (shiftl X)) 3),
```

where for convenience (+) is extended to take three arguments. Here's a trace of the evaluation of this program on the list $X =$ [1,2,3]:

```
X →                                    [1,2,3],
(shiftr X) →                           [0,1,2],
(shiftl X) →                           [2,3,0],
(+ (shiftr X) X (shiftl X)) →          [3,6,5],
(/ (+ (shiftr X) X (shiftl X)) 3) →
                                       [1,2,5/3].
```

4.2. Statistic-Computing Functions

The second group of functions compute statistics. Recall that statistics are computed from sequences by applying to each element a formula whose inputs may contain "prior" outputs. Consider, for example, the formula that sums up X: $X'_i = X'_{i-1} + X_i$. Statistics cannot be computed by the convolution-computing functions presented in the previous section due to the order of evaluation. Consider, for example, this ill-fated attempt to compute a mean:

```
(+ (shiftr X) X).
```

The left half of figure 1 shows how this program is evaluated when $X =$ [1,2,3]. The final result is X' = [X_1, X_1+X_2, X_2+X_3]. What we want, however, is X' = [(+ 0 X_1), (+ X_1' X_2), (+ X_2' X_3)]=[X_1, X_1+X_2, X_1+X_2+X_3]. There is no way that the convolution-computing functions described above can compute this X' because X'_1 and X'_2 aren't known until steps 4.1 and 4.2 respectively but they are needed in steps 1.2 and 1.3 respectively.

Clearly the order of evaluation must change. Normally genetic programming uses an *evaluator* that walks over the program tree and evaluates nodes in postfix order:

function evaluate(program, fitness case):
  loop over the nodes in the program in postfix order:
    compute the result of each node
end

For example, in the symbolic regression problem (Koza 1992) the nodes making up the program (+ (* 3 X) (% 5 1)) are evaluated in the following order: 3, X, *, 5, 1, %, +. When the fitness case is a sequence this becomes:

function evaluate(program, sequence):
  loop over the nodes in the program in postfix order:
    loop over the elements in the sequences that represent the arguments:
      compute this element of this node
end

For example, to evaluate the program (+ [1,2] [3,4]) we first evaluate [1,2] (first evaluating 1, then 2); second we evaluate [3,4] (first evaluating 3, then 4); third we evaluate (+ [1,2] [3,4]) (first evaluating 1+3, then 2+4).

Consider what happens if we swap the two loops:

function evaluate(program, sequence):
  loop over the elements in the sequence:
    loop over the nodes in the program in postfix order:
      .....
end

The right half of figure 1 shows this new order of evaluation for the program (+ (shiftr X) X). Note that it is now possible to compute statistics.

The two orders of evaluation are functionally identical: both evaluations compute X'= [X_1, X_1+X_2, X_2+X_3]. Now, however, it is possible to use the previously computed elements of X' instead of X. That is, at step 2.2' of figure 1, the value of X'_1 is available when computing X'_2, etc. So, we can compute statistics with the same functions that we use to compute convolutions, we just change the order of evaluation and use the X'_i values in place of the X_i values.

X'_1: evaluate(`(+ (shiftr sczX) sczX)`, `[1,2,3]`, 1)
→ evaluate(`(shiftr sczX)`, `[1,2,3]`, 1) + evaluate(`sczX`, `[1,2,3]`, 1)
→ evaluate(`sczX`, `[1,2,3]`, 0) + 1
→ 0 + 1 → 1

X'_2: evaluate(`(+ (shiftr sczX) sczX)`, `[1,2,3]`, 2)
→ evaluate(`(shiftr sczX)`, `[1,2,3]`, 2) + evaluate(`sczX`, `[1,2,3]`, 2)
→ evaluate(`sczX`, `[1,2,3]`, 1) + 2
→ 1 + 2 → 3

...

Resulting in X'= `[1,3,6]`.

Figure 2 Trace of the evaluation of `(scz (+ (shiftr sczX) sczX) [1,2,3])`.

The `scz` function delineates a part of the program on which to use this "inverted" order of evaluation. The sequence X' is referred to by the zero-arity function `sczX` when evaluating `scz`. Here's how to compute the partial sums of the sequence `[1,2,3]`:

`(scz (+ (shiftr sczX) sczX) [1,2,3])`.

The first argument to `scz`—"`(+ (shiftr sczX) sczX)`", above—is the program to be evaluated under the new semantics. The second argument to `scz`—"`[1,2,3]`", above—is the sequence from which the statistics are going to be computed. Here is a simplified account of what `scz` does:

```
function scz( program P, sequence X )
    for i ← 1 to |X| do
        Xi ← evaluate( P, X, i)
    return X
end
```

where + and `if` are as defined in section 4.1, and leaves (`X`, `sczX` and ephemeral random constants) evaluate to their *i*th elements. The evaluator gets two extra parameters, *i*, and *X*, that are ignored when computing convolutions and that are passed down to the leaves when computing statistics. The *i* parameter is incremented/decremented by shift-left/-right functions. Figure 2 traces the evaluation of the program `(scz (+ (shiftr sczX) sczX) [1,2,3])` that computes partial sums.

4.3. Summary Of The "Statistic-Computing Zones" Function Set

To summarize, the statistic-computing zones function set contains the following functions:

{ `X`, $\Re$, +, `if`, `shiftl`, `shiftr` } ∪
{`scz`, `scz-t`, `sczX` }.

The function `X` is replaced in a particular problem by one or more zero-arity functions that return the sequence(s) that define each fitness case (problem instance). The function $\Re$ is replaced at program creation time by random values (either scalars or sequences). The function + will be joined by other similar functions such as `-`, `*`, `/`, `and`, and `or`. These functions constitute the convolution-generating functions. The function `scz` delineates a statistic-computing zone. The function `scz-t` is similar but only returns the last ("tail") element of X': `(scz-t P X) ≡ (car (last (scz P X)))`. The function `sczX` returns (an

Step	Action		Step	Action	
1.1	X_1	→ 1	1.1'	X_0	→ 0
1.2	X_2	→ 2	1.2'	(shiftr X)$_1$	→ 0
1.3	X_3	→ 3	1.3'	X_1	→ 1
1	**X**	**→ [1,2,3]**	**1'**	**(+ (shiftr X) X)$_1$**	**→ 1**
2.1	(shiftr X)$_1$	→ 0	2.1'	X_1	→ 1
2.2	(shiftr X)$_2$	→ 1	2.2'	(shiftr X)$_2$	→ 1
2.3	(shiftr X)$_3$	→ 2	2.3'	X_2	→ 2
2	**(shiftr X)**	**→ [0,1,2]**	**2'**	**(+ (shiftr X) X)$_2$**	**→ 3**
3.1	X_1	→ 1	3.1'	X_2	→ 2
3.2	X_2	→ 2	3.2'	(shiftr X)$_3$	→ 2
3.3	X_3	→ 3	3.3'	X_3	→ 3
3	**X**	**→ [1,2,3]**	**3'**	**(+ (shiftr X) X)$_3$**	**→ 5**
4.1	(+ (shiftr X) X)$_1$	→ 1			
4.2	(+ (shiftr X) X)$_2$	→ 3			
4.3	(+ (shiftr X) X)$_3$	→ 5			
4	**(+ (shiftr X) X)**	**→ [1,3,5]**			

Figure 1 The normal order of evaluation (left) and the inverted order of evaluation (right), demonstrated on the program `(+ (shiftr X) X)` and the data `X=[1,2,3]`.

element of) the value computed by the second argument of `scz` while executing the first argument of `scz`, and some default value (such as 0) elsewhere. In the current implementation, calls to `scz` cannot be nested (calls to `scz` inside of the first argument of another `scz` just return the second argument, `X`).

5. Evaluation Of The "Statistic-Computing Zones" Function Set

It remains to evaluate the statistic-computing zones function set. I define the utility of a function set as its efficiency at solving problems. By "efficiency" I mean the expected number of individuals that have to be processed to find a solution. An example of a measure of efficiency is Koza's computational effort (E) (Koza 1992, 1994).

Ultimately, function set comparisons must be based on data: which works better for which problems, which fitness cases, which parameter settings, etc. Sequence problems, however, are too computationally expensive (each problem instance (fitness case) in the intron/exon problem, for example, can contain tens of thousands of bases) so we must rely mostly on more abstract measures:

Elegance. Elegance is a slippery concept but it is clearly closely related to measures such as size, complexity, number of special cases, and generality. The notion of computing statistics by declaring a part of the program to be a "statistic-computing zone" and using a slightly different semantics is elegant because it requires only three new functions (`scz`, `scz-t` and `sczX`), no special cases and a minor extension of the evaluator (adding the i and X parameters).

Expressiveness. The statistic-computing zone functions can represent the types of computation that appear in human-written solutions to sequence problems such as the intron/exon problem, the motif recognition problem and poker hands classification (see the next section).

Ease of coding. The above function set is easy to implement. The only complexity is dealing with sequences of different lengths. A complete discussion is beyond the scope of this paper but it involves tagging each value with an integer ("distance of this element from the end of the sequence") and a Boolean ("this value is a scalar") and propagating these tags up out of the evaluator's recursion.

Worst-case efficiency. All the functions proposed are worst case linear in the length of the sequences. This is very important when trying to reduce the amount of computation wasted on random low-fitness programs. For example, functions that do explicit loops (loops in which the initialization, updating and termination criteria are subject to evolution) are often worst case quadratic in the length of their input sequences and always require time-outs. The statistic-computing zones function set doesn't contain any explicit loops.

Efficient use of information. Functions such as `(first [1,2,3])` $\rightarrow$ `1` are inefficient in the sense that they throw away information. One of the design goals of the statistic-computing zones function set was to ensure that as many of a program's nodes as possible contribute to the resultant fitness. The `first` function, for example, throws away information and wastes computation: only one of the computed elements has any effect on the fitness.

No time-outs are required. The term *time-out* refers to the practice of halting the execution of a program when some resource (time or space) is exhausted. Time-outs must be used in genetic programming if one or more of the primitive function's resource usage is not tightly bounded a priori. Unfortunately, time-outs butcher the semantics of programs and hence make the resulting fitnesses more arbitrary and less representative of the program's behaviour.

All the functions in the statistic-computing zones function set require time and space that is linear in the size of their inputs[1]. Therefore, time-outs are not needed.

6. An Example Application: Classifying Pinochle Poker Hands

In the remainder of this paper I will show how the statistic-computing zones functions can be implemented on a real problem: classifying poker hands. The purpose of this section is to demonstrate the steps involved in applying this new technique to a concrete problem.

This technique has also been applied to a problem in molecular biology (Handley 1996).

6.1. Terminology

A *pinochle deck* is a set of 24 cards. Each *card* has two characteristics or features: a *suit* (one of: hearts, clubs, diamonds and spades; or ♥, ♣, ♦, and ♠ for short) and a *denomination* (one of: "9", "10", "Jack", "Queen", "King", "Ace"; or 9, 10, J, Q, K and A for short). An example of a card is 9♦ ("nine of diamonds"). A *hand* is a set of 5 cards chosen without replacement from the deck of 24 cards. For example, {9♦, 10♣, A♦, J♣, K♣} is a hand, but {9♦} is not a hand and neither is {9♦, 9♦, A♦, J♣, K♣}. For the purposes of this paper, five different types of hands are recognized: flushes, straights, royal flushes, 3-of-a-kinds, pairs (2-of-a-kinds) and two-pairs. The appendix defines these classes of hands.

6.2. Five Two-State Classification Problems

The problem of classifying a poker hand was represented as five two-state classifications: flush vs. non-flush, straight vs. non-straight, royal-flush vs. non-royal-flush, 3-of-a-kind vs. non-3-of-a-kind, pair vs. non-pair, and two-pairs vs. non-two-pairs.

[1]The `unary` function appears to be an exception but it actually isn't: it requires time and space proportional to the product of the input sequence's length and the size of the input alphabet (`unary` is only intended for discrete domains).

Table 1 The number of fitness cases used in each problem. There are 42,504 possible hands. The number of negative fitness cases was chosen arbitrarily to be either the same as the number of positive fitness cases or some multiple thereof. The "number used" column gives the subset of the "total number" of fitness cases that were actually used.

Problem	positive	negative	total number	number used
flush vs. non-flush	24	2,400	2,424	2,424
straight	9432	9,432	18,864	18,864
royal-flush	8	800	808	808
3-of-a-kind	4,560	4,560	9,120	9,120
pair	32,400	9,720	42,120	10,000
two-pair	13,200	13,200	26,400	10,000

6.3. Applying Genetic Programming: Choosing Values For Parameters

Before doing a run of genetic programming there are four major parameters to set: the fitness measure, the class of programs, the population size and the termination criterion.

6.3.1. The Fitness Measure

The programs being evolved assign hands into one of two classes, such as flush vs. not-flush. Each problem (such as flush vs. non-flush) is defined by a set of examples: both positive (such as {9♥, 10♥, J♥, Q♥, K♥}) and negative (such as {9♥, 10♦, J♠, Q♥, K♦}). The performance of the two-state classifiers was defined as their accuracy, Q_3, the proportion of examples correctly classified:

$$Q_3 = \frac{\text{true positives} + \text{true negatives}}{\text{number of examples}}.$$

The choice of parents to participate in the genetic operations was based on this fitness measure.

6.3.2. The Class of Programs

The programs evolved return values but the problem requires two-state classifications so the wrapper first converts the returned value into a number (by repeatedly summing the sequence) and then thresholds the number into a Boolean. The function set used was the statistic-computing zone functions of section 4.3 plus the problem-specific functions in the following sections.

6.3.2.1. Suit and Denom

The `(suit)` and `(denom)` zero-argument functions return sequences corresponding to the suits and the denominations of a particular example, respectively. If, for example, a program is being tested on the example {9♥, 10♦, J♠, Q♥, K♦}, then the function `(suit) = [2,4,1,2,4]` (where ♠ → 1, ♥ → 2, ♣ → 3 and ♦ → 4) and `(denom) = [1,2,3,4,5]` (where 9 → 1, 10 → 2, J → 3, Q → 4, K → 5 and A → 6).

*6.3.2.2. +, -, *, /, =, <, Max, Min*

These functions are all patterned after the + function in section 4.1. The = and < functions return 0 for false and 1 for true (consistent with the `if` function).

6.3.2.3. Unary

The `unary` function creates frequency matrices. (Frequency matrices are useful for counting how many times elements appear in a sequence. For example one frequency matrix for the sequence `[4,1,3,4,4]` is `[1,0,1,3]`.) `Unary` takes a positive integer as argument and returns a sequence of `0`s and `1`s that sort of represents that number in base 1:

```
(unary 0) → [1];
(unary 1) → [0,1];
(unary 2) → [0,0,1]; etc.
```

Non-integers are rounded; negative numbers have their sign flipped. Sequences are recursed-on:

```
(unary [X1,...,Xn]) →
        [unary(X1),...,unary(Xn)].
```

`Unary` is the only function in this paper whose output can be larger than its input. Thus its inputs must be limited. This is done two ways: First, the depth to which `unary` can call itself is limited to 2 (in this paper). Second, integer arguments greater than 7 (in this paper) return 0. So, for example, `(unary [[[1]]])` → `0` and `(unary 999)` → `0`.

6.3.2.4. Even

The `even` function takes an integer and returns `1` if it is even, `0` otherwise:

```
(even 3) → 0; (even 4) → 1; etc.
```

Real numbers and sequences are handled a la `unary`.

6.3.3. The Population Size

The populations each contained 2,000 individuals.

6.3.4. The Termination Criterion

Each run was terminated when either (a) a total of 51 generations had been created (i.e., 1 randomly created and 50 subsequently evolved), or (b) a 100% correct individual appeared.

6.4. Results

Table 1 shows the number of fitness cases for each problem.

The following individual correctly classifies 100% of the 2,424 examples of flushes and non-flushes:

```
(= (shiftr (+ (min (max denom suit) (shiftr
suit)) (shiftl (if sczX sczX suit)))) (= suit (/
denom denom))).
```

The following individual correctly classifies 83.5% of the 18,864 examples of straights and non-straights :

```
(= (unary (scz-t (if [3.22303,9.46608,
-3.58727,8.81175,-0.949967,-6.09327] [3.21362,
-8.66399,4.34161,9.50997,-5.96245,9.08361,
8.15578] [-9.14842,2.48139,4.75838,-5.86998,
3.27015,1.46147]) (unary suit))) (+ (< (even
denom) (= denom suit)) (if (+ 0.634709 sczX) (=
denom denom) (/ denom suit)))).
```

The following individual correctly classifies 100% of the 808 examples of royal-flushes and non-royal-flushes:

```
(scz (unary (+ (shiftl (if sczX [-7.99712,-
7.6549,4.01414,5.7299,2.34682,2.97555,9.99954]
denom)) (- (max sczX suit) (shiftr suit)))) (min
(< (min (+ sczX sczX) (- sczX sczX)) (if (max
(shiftl sczX) (= suit denom)) (min 0.755303
[3.07053,6.37961,2.11467,1.21851,-0.486063,-
9.25682,0.644129]) (even suit))) (even (max
(shiftl sczX) (even denom))))).
```

The following individual correctly classifies 99.77% of the 9,120 examples of 3-of-a-kinds and non-3-of-a-kinds:

```
(= (if (shiftl (unary denom)) (even sczX) (+
(scz -9.32915 (* (shiftl denom) (shiftl
2.61339))) (scz-t denom denom))) (shiftr (-
(unary denom) (shiftl (= (scz -8.94314 4.59043)
suit))))).
```

The following individual correctly classifies 94.74% of the 10,000 examples of pairs and non-pairs:

```
(unary (< (= (min denom denom) (shiftr
denom)) (= (unary (shiftr suit)) (min (min sczX
denom) (shiftl sczX))))).
```

The following individual correctly classifies 94.79% of pairs:

```
(/ (shiftl (shiftr (/ sczX suit))) (if (unary
(scz-t denom denom)) (* (shiftl (max (even suit)
(= denom denom))) (= (even (even sczX)) (*
(shiftr suit) (shiftl sczX)))) (< (shiftr denom)
(+ sczX denom)))).
```

The following individual correctly classifies 94.7% of the 10,000 examples of two-pairs and non-two-pairs:

```
(* (shiftl (+ (if (even [2.317,1.84999,-
7.19038,-8.73673,1.79512,-9.48448,-5.7193])
(unary (scz-t (= [-4.52698,-4.98831,1.45474,
9.86205,-8.44215] 5.7647) (even [2.21378,
6.97966,7.09646,-9.83506,2.08966,0.930429,
-2.27754]))) suit) (unary (shiftl (< (+
-0.129668 denom) (shiftr denom)))))) (* (shiftl
(< (+ -0.129668 denom) (shiftr denom))) (scz-t
suit -1.58118))).
```

7. Conclusions

This paper has described an extension of genetic programming that facilitates the solution of problems whose instances are most naturally represented as arbitrary-length sequences. This new technique is simple, easy to implement, expressive, efficient, and doesn't require arbitrary time-outs. The technique was further demonstrated by applying it to a concrete problem, the classification of pinochle poker hands.

Acknowledgments

David Andre, Forrest Bennett, John Koza and the reviewers made many useful comments.

Appendix: Definitions of The Classifications

For the purposes of this paper, there are five different types of hands:

flushes. A hand is a *flush* if all five cards have the same suit:

$$\text{flush}(\{d_1 s_1, \ldots, d_5 s_5\}) \equiv s_1 = s_2 = \ldots = s_5.$$

straights. A hand is a *straight* if all five cards' denominations form a contiguous stretch:

$$\text{straight}(\{d_1 s_1, \ldots, d_5 s_5\}) \equiv \text{rank}(d_{i+1}) = \text{rank}(d_i)+1,\ 1 \le i < 5,$$
$$\text{where rank}(9) = 1,\ \text{rank}(10) = 2, \text{etc.}$$

royal flushes. A hand is a *royal flush* if it is both a flush and it contains the highest-value (highest rank) straight:

$$\text{royal-flush}(H) \equiv \text{flush}(H) \wedge d_1=10, d_2=J, d_3=Q, d_4=K, d_5=A,$$
$$\text{where } H = \{d_1 s_1, \ldots, d_5 s_5\}.$$

3-of-a-kinds. A hand is a 3-of-a-kind if it contains 3 or more cards of the same denomination. More generally,

$$k\text{-of-a-kind}(\{d_1 s_1, \ldots, d_5 s_5\}) \equiv k \text{ or more of } \{d_1, \ldots, d_5\} \text{ are the same.}$$

pairs. A *pair* is another name for a 2-of-a-kind.

two-pairs. A hand is a *two-pairs* it if contains two pairs that involve at most one card in common.

References

Goldberg, David E. 1989. *Genetic algorithms in search, optimization, and machine learning*. Reading, MA: Addison-Wesley.

Handley, Simon G. 1995. Classifying Nucleic Acid Sub-Sequences as Introns or Exons Using Genetic Programming. in *Third International Conference on Intelligent Systems for Molecular Biology*. Robinson College, Cambridge, England: AAAI Press.

Handley, Simon G. 1996. The Prediction of the Degree of Exposure to Solvent of Amino Acid Residues via Genetic Programming. In Koza, John R., Goldberg, David E., Fogel, David B., and Riolo, Rick L. (editors). Genetic Programming 1996: Proceedings of the First Annual Conference, July 28–31, 1996, Stanford University. Cambridge, MA: MIT Press.

Holland, John H. 1992. *Adaptation in natural and artificial systems: An introductory analysis with applications to biology, control, and artificial intelligence*. Cambridge, MA: MIT Press.

Koza, John R. 1992. *Genetic Programming: On the Programming of Computers by Means of Natural Selection.*, Cambridge, MA: MIT Press.

Koza, John R. 1994. *Genetic Programming II: Automatic Discovery of Reusable Programs*. Cambridge, MA: MIT Press.

Quinlan, J. R. 1986. Induction of decision trees. *Machine Learning*, 1(1), p. 81–106.

Quinlan, J. R. 1987. Simplifying decision trees. *International Journal of Man-Machine Studies*, 27, p. 221–34.

Rumelhart, David E., McClelland, James L., and the PDP Research Group. 1986. *Parallel distributed processing: explorations in the microstructure of cognition*, Cambridge, MA: MIT Press.

Evolving Edge Detectors with Genetic Programming

Christopher Harris and Bernard Buxton

Department of Computer Science
University College, London
Gower Street
London
WC1E 6BT
United Kingdom
Email: C.Harris@cs.ucl.ac.uk, B.Buxton@cs.ucl.ac.uk
Telephone: Chris Harris +44 171 419 3679, Bernard Buxton +44 171 380 7294

ABSTRACT

Edge detection is the process of detecting discontinuities in signals and images. We apply genetic programming techniques to the production of high-performance edge detectors for 1-D signals and image profiles. The method, which it is intended to extend to the development of practical edge detectors for use in image processing and machine vision, uses theoretical performance measures as criteria for the experimental design.

1. Introduction

Edge detection is a classic problem in machine vision applications. It forms the first part of the process of converting a raster image to a symbolic description of the scene and for many other high-level tasks, such as stereo vision, segmentation, and object recognition and location. A large body of work exists devoted to the production of high quality detectors, and to what constitutes a good detector. We use genetic programming (GP) to produce functions that perform well on a set of synthetic and real-world test signals, using theoretical criteria to judge performance.

The purpose of this work is twofold: (i) to use existing theory as an inspiration for experimental design in genetic programming and (ii) to use GP as a technique for developing detectors that can give optimal (or near optimal) performance on real signals and images for which no analytic or simple numerical solution of the problem is possible. In this work we use edge detection theory to provide a fitness function for our experiments and utilise GP as a means of obtaining optimal detectors automatically, without user intervention. The development of such a facility would have important implications in machine vision applications where much human effort is often devoted to developing and tuning algorithms and techniques to particular applications.

2. The Edge Detection Task

Edge detection involves locating and highlighting local discontinuities in signals. This is usually achieved by convolving the signal with an operator (specified by some functional form over a given interval) to produce a response signal. For discretely sampled signals the operator is sampled to produce a filter. Most edge detectors are designed to respond strongly to areas of the signal with large first derivative, edges being marked by peaks in the response signal.

[Canny 1986] was the first to produce a set of theoretical criteria for determining the performance of an edge detector. He proposed that a detector should have a good signal-to-noise ratio (SNR) so that it produces a strong response to an edge in the signal, and that it should have good localisation properties so that a detected edge is as close as possible to the position of the 'true' edge. Finally, it should only give a single response to an edge, i.e. it should not respond to maxima in the signal due to noise.

For the first two of these criteria, Canny derived an expression in terms of integrals of the operator function and signal that would measure the performance of the operator with respect to each criterion. An optimal operator would maximise the product of the SNR and localisation criteria. Canny calculated the optimal operator for idealised step edge signals and presented the first derivative of a Gaussian function as an efficient approximation to this optimum (see Figure 10).

[Spacek 1986] took Canny's work and extended the theoretical performance measure to incorporate the third criterion of noise suppression. An expression for an optimal operator was derived as an equation depending on six parameters. Optimising four of these, Spacek found an operator that performed better than Canny's under slightly different boundary conditions. A cubic spline function was found to be a convenient approximation to the optimum.

[Petrou] presented further improvements by optimising Spacek's expression in all six parameters. She then compared Canny's approximation, Spacek's approximation, Spacek's optimum and the new optimum on some real world images. Although theoretical performance was maximal for the optimal operator, in practice this did not translate to a statistically significant improvement when applied to a real world image.

Details of the criteria used by Canny in deriving an optimal detector are presented in section 4.3.

3. Related Work

Other automatic techniques have been applied to edge detection with promising results. These techniques were applied directly to images rather than the mathematics of the underlying detectors, and in two dimensions.

[Chao 1994] used a Hopfield neural network to perform edge detection on a grey-level image. The whole image was mapped to a 2-D Hopfield net, one pixel to each neuron. Results were found to be comparable to a Sobel [Gonzalez 1992] operator on grey-level noisy images.

[Srinivasan 1994] used a 2-stage network to detect edges. The first stage encoded information from a receptive field (an area of image around the target pixel). The second stage took this encoding and produced edge vector components as outputs. Performance matched that of Canny's operator on a single training image.

[Bhandarkar 1994] used genetic algorithm optimisation techniques to detect edges in a grey-level image. Using 2-D chromosomes to represent 'edge-images', optimisation was performed on a whole image using a complicated cost function based on region dissimilarity, and edge curvature, thickness, and fragmentation.

4. Experimental Design

The aim of this experiment is to use GP to evolve a good edge detector, as specified by the criteria mentioned earlier. The selection pressure comes from maximising the performance of candidate edge detectors on a mixture of synthetic and artificial signals.

4.1 Anatomy of an Edge Detector

An edge detector consists of a function in a variable x over a specified interval. For this experiment, candidate functions are generated and sampled over the interval [-1,1] These samples are normalised to produce an operator. The operator is convolved with a training signal to produce a response signal. The candidate trees are produced from the primitives in tables 1 and 2.

Table 1. Terminal Set

`x`	The x position of the candidate function. This will range from -1 to +1.
`0.0`	A constant.
`1.0`	A constant.
`2.0`	A constant.

Table 2. Function Set

`+ -` `* /`	Arithmetic operators with protected division. Dividing by 0 returns 0.
`rlog`	Protected ln function. Takes absolute value of parameter.
`exp`	Exponent.
`pow`	Protected power function. Raises first parameter to the power of the second parameter. Only allows integer powers of negative numbers, but any power of positive numbers.
`sin`	Sine function.
`cos`	Cosine function.

4.2 Training data

Forty 1-dimensional signals of varying lengths were used for training data. Six of these were synthetic and contained idealised 'step' edges of infinite gradient, as shown in Figure 2.

In applications, edges can occur at a variety of scales and vary in width. To incorporate this property into the training set, each of the idealised signals was smoothed with a Gaussian smoothing operator to produce a further six signals. The original signals were also corrupted with normally distributed noise, to produce a total of 18 synthetic signals. A further 22 signals were extracted from real-world images taken from a video camera (Figure 1). Edges in these images were much less obvious and the signals were often noisy. The data thus consists of roughly equal portions of synthetic and real signals.

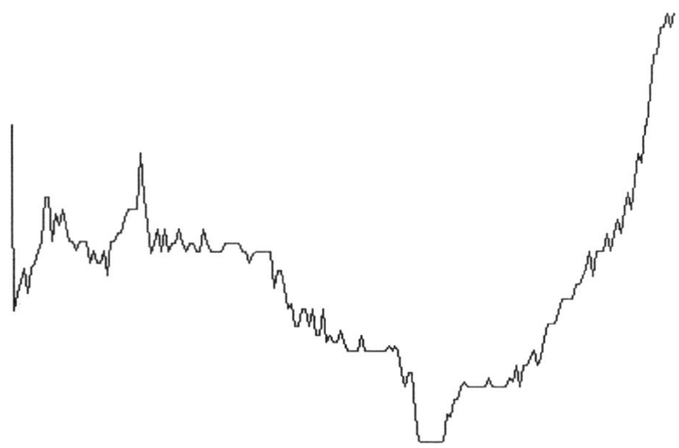

Figure 1. A real-world noisy signal

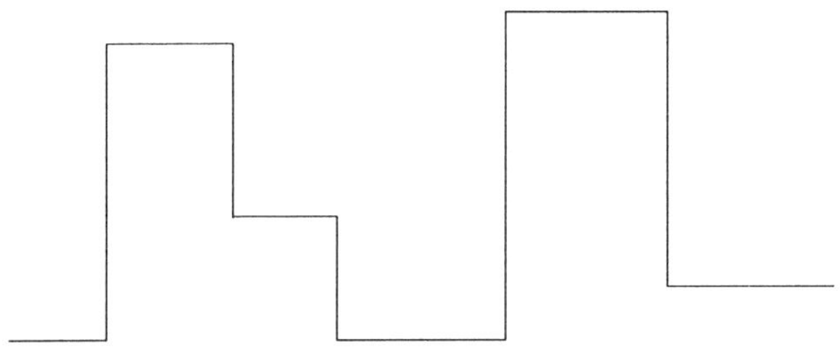

Figure 2. A signal with step edges

Each signal contained several edges, the locations of which were marked by hand and recorded. A total of 102 edges were marked over all signals. To reduce computational load, each generation was tested on a random sample of 10 of the 40 signals, which changed after each generation.

4.3 The Fitness Function

Candidate functions were assessed on three criteria for each signal, corresponding to the criteria in [Canny 1986].

Signal-to-noise ratio. A measure of the response of the operator to the edge as opposed to the rest of the signal. This is defined as the area under the nearest peak to the edge location. It is assumed here that some maximum in the response exists. The area under the peak is found by first locating a maximum in the response of a single or two pixels width. Then, the nearest trough or zero-derivative point of inflexion to this peak is found and is used to define the base level of the response, above which the response to the edge is measured. A point to the other side of the peak is then found at which the signal crosses the base level, giving the total extent of the peak. The size of response to the edge is the sum of responses in this area, over and above the base level, as illustrated in Figure 3. A good detector will have a large response.

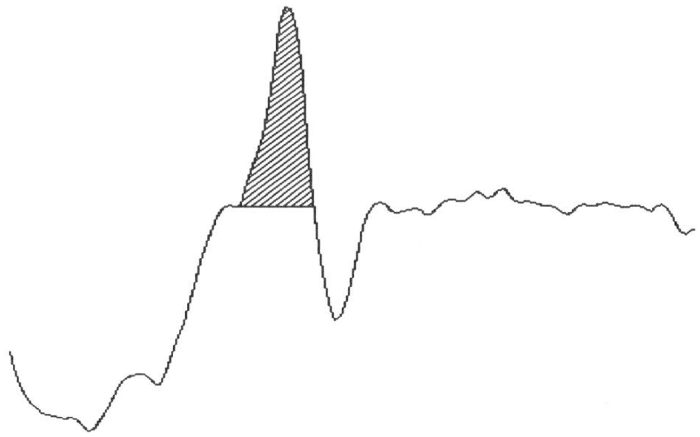

Figure 3. Area captured by a peak in signal response

Localisation. This is a measure of the accuracy in positioning of the response. A good detector will produce a peak in response, close to the position of the edge in the original signal. Performance is measured by the average displacement of the peak from the 'true' edge position, over all edges in the training set. The 'true' position is marked by hand by a human operator. A lower value here indicates a better performance.

Single response criterion. This criterion is restated to say that the operator should respond only to real edges and not respond to features near the edge caused by noise in the signal. This can be measured quantitatively by finding the average separation of peaks in the response of the operator to a signal consisting entirely of normally distributed noise. A larger peak separation means that the operator has better noise-suppression qualities.

Each candidate was sampled at resolutions of 9, 13, and 17 pixels to give operators that would respond to edges of different width. Each operator was applied to ten signals and the responses were analysed to give average response strength S(f), localisation error L(f), and noise suppression figures C(f).The overall fitness of each candidate is calculated as $\frac{S(f)C(f)}{L(f)}$.

For step-like edges, a good edge detector will normally be a combination of a smoothing operator (noise suppression measure) and a first-derivative operator (large response to change in intensity).

4.4 Tableau of edge detection problem

Table 3. Tableau of problem

Objective:	To respond strongly and accurately to edges in a signal, and not to respond to non-edge sections.
Terminal Set:	`x 0.0 1.0 2.0`
Function Set:	`+ - * / rlog exp sin cos pow`
Fitness Cases:	10 1-D signals drawn randomly from a pool of 40.
Raw Fitness:	as detailed elsewhere
Standard Fitness:	(Max raw fitness) - raw fitness
Population size:	200
Mutation rate:	10%
Crossover rate:	60%
Terminal Probability:	45%
Mutation Terminal Probability:	50%

"Terminal Probability" and "Mutation Terminal Probability" control the average size of program trees in the population. For a fuller explanation, see [Harris, 1996].

5. Results

Six runs were made using the GP-COM system [Harris, 1996]. Each run produced an operator that performed well when compared to Canny's Gaussian first-derivative approximation. Each run was terminated when an operator was found, that exceeded the performance of Canny's operator over the sample of 10 signals.. The operators' performances are detailed in table 4, along with scores for Canny's approximation. These are raw fitness scores, high values indicating better performance, and were obtained by testing the operators over the entire set of 40 signals. This allows a fair comparison of the operators.

The operators are detailed in Figures 4-10. Each figure contains a plot of the evolved operator over the interval [-1,1], the program tree that produced the function, and the boiled-down functional form of the operator which eliminates the superfluous parts of the program tree.

Table 4. Results of experimental runs

Operator (f)	S(f)	L(f)	C(f)	Overall Fitness
1.1	41.36	0.3583	3.938	119.9
1.2	38.37	0.4217	4.242	114.5
1.3	41.76	0.6367	3.865	98.62
1.4	35.54	0.4900	4.285	102.2
1.5	33.68	0.4733	4.786	109.4
1.6	30.66	0.5167	4.876	98.56
Canny	39.57	0.4467	3.430	93.81

$$\frac{1}{\cos(\cos(e^{x})) - e^{x}}$$

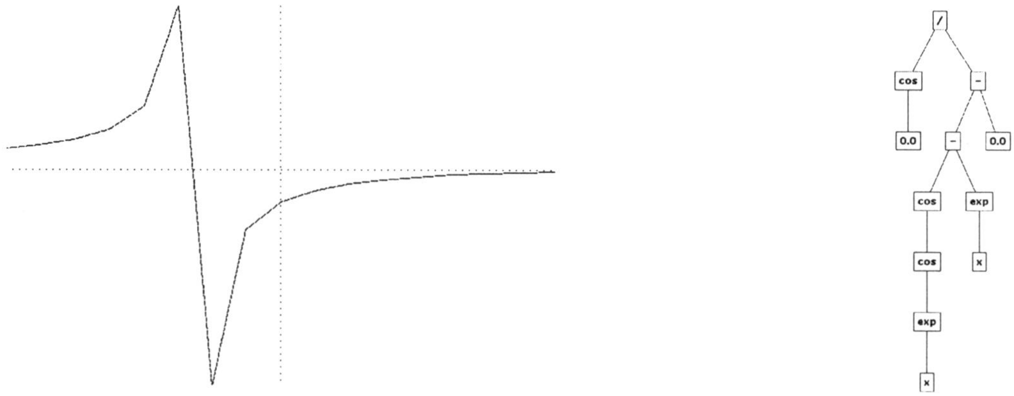

Figure 4. Operator 1.1

$$\frac{\sin(x)\,\mathrm{rlog}(x)}{x^{x}}$$

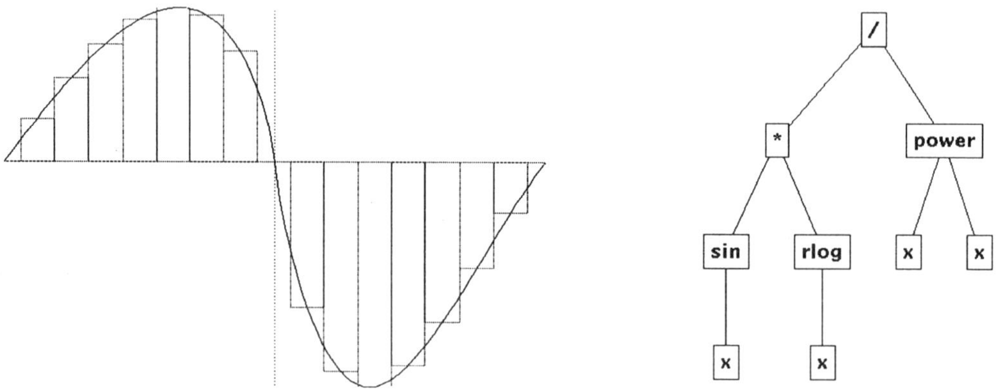

Figure 5. Operator 1.2

$$\frac{\sin(2)}{-x-\dfrac{\mathrm{rlog}(x+2)}{2-x}}$$

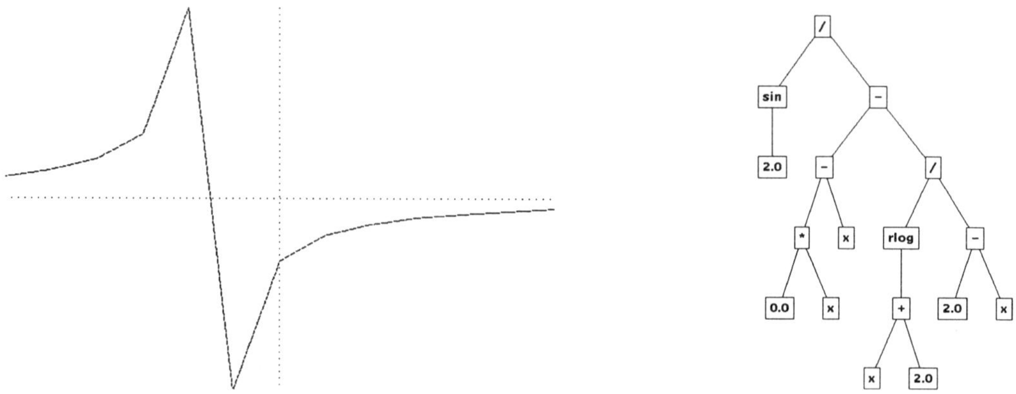

Figure 6. Operator 1.3

$x \, \mathrm{rlog}(x)$

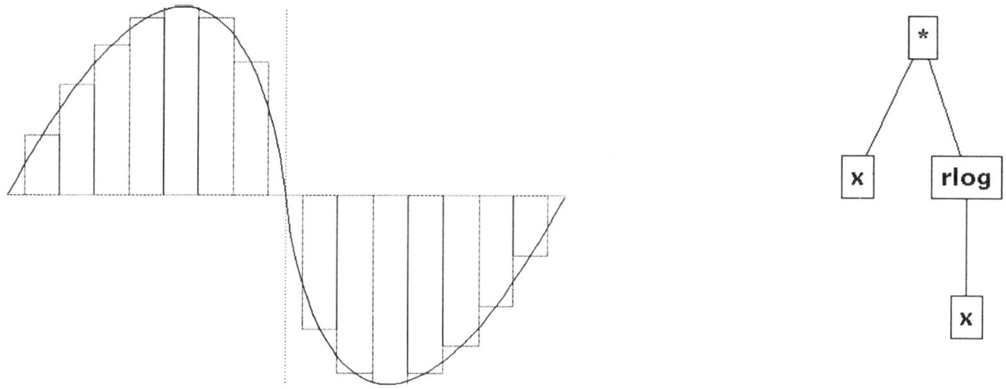

Figure 7. Operator 1.4

$x \sin(\mathrm{rlog}(x))$

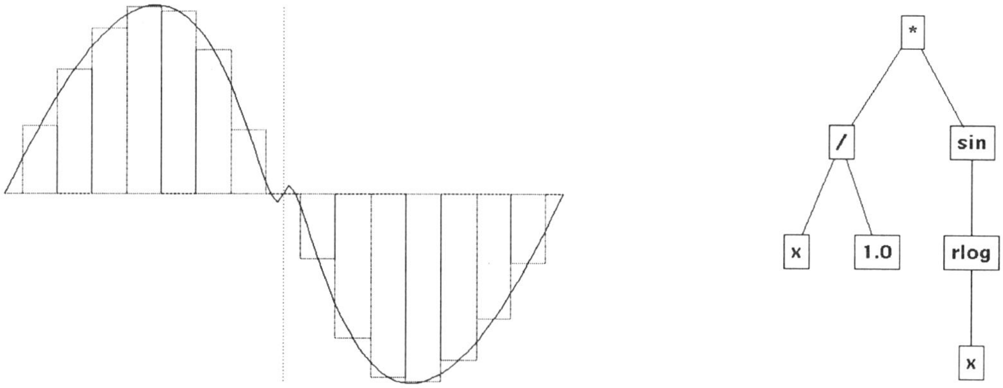

Figure 8. Operator 1.5

$x - 2x\cos(x)$

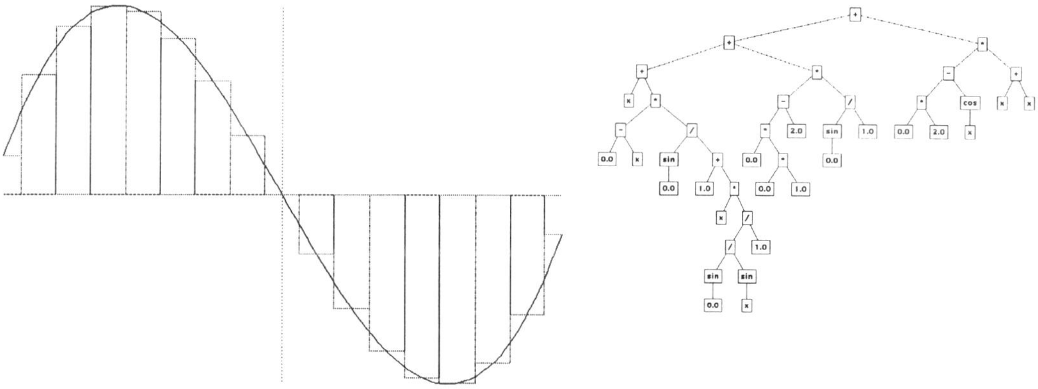

Figure 9. Operator 1.6

$$-\frac{x}{\sigma^2} e^{\frac{-x}{2\sigma^2}}$$

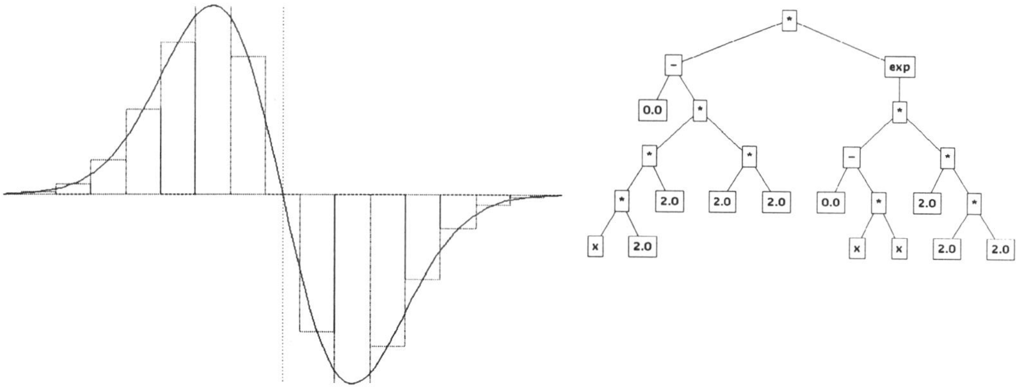

Figure 10. Approximation to Canny optimal operator with first derivative of Gaussian. Here, σ was set to 0.25.

Operators 1.1 and 1.3 do not look like good edge detectors (i.e. not like the derivatives of Gaussian smoothing functions) when sampled at high resolutions. These operators both have large discontinuities around their zero-crossings near -0.3, which are very evident at higher sampling rates. These discontinuities have the potential to dominate the make-up of the operator but the low sampling rate used in the experiments reduces this effect considerably. At these low sampling rates (as in Figures 4 and 6), they reasonably approximate the shape of a Canny-like edge detector. A high fitness should therefore be taken in context - genetic search found a function that makes a good detector *at the specified sample rates*.

The other four evolved operators (1.2, 1.4, 1.5, 1.6) look fairly similar and performed well with respect to noise suppression. A good response to first derivative involves an operator with high first derivative at its zero-crossing. This in turn implies a high-frequency aspect to the operator which is contrary to the requirements for noise suppression. This mutual exclusion principle was noted by Canny and means that maximising one property invariably involves reduced performance in the other. Hence the overall fitness is not based on maximising either value but their product.

The differences in response strength correspond quite well to the shape of the operators at their extrema. Those operators which are flatter at the extrema, such as 1.1, 1.2, 1.3, and Canny, have the best response strength. This is because more of the influence of the operator is located toward the centre of the edge, reducing the influence of the peripheries. A larger response to sharp edges is the result.

The Canny operator performs reasonably well on response strength but falls down when it comes to noise suppression. Increasing σ here would increase noise suppression performance, and reduce response strength, by dilating the operator. The reduced performance of the Canny operator here can also be attributed to the training data. Canny's approximation was to an optimal detector for step edges, yet step edges and smoothed step edges made up only half of the edges in the training set. It is thus to be expected that this function is not optimal for this data set.

7. Conclusions

We have shown an application of genetic programming to the evolution of operators designed to detect edges in 1-dimensional signals. The design of the experiment was based upon theoretical analysis of the ideal edge detector and its properties as, for example, first carried out in [Canny, 1986]. The fitness function was adapted from a set of criteria that a good edge detector would fulfil, and stated in terms of the responses of those detectors to a series of signals, in the digital domain. A number of detectors were found that out-performed Canny's approximation to an optimal detector, on a mixture of synthetic and real-world signals.

Acknowledgements

This work was carried out under the Postgraduate Training Partnership, between SIRA Ltd and University College London.

Bibliography

Andre, D. 1994. Automatically Defined Features: The Simultaneous Evolution of 2-Dimensional Feature Detectors and an Algorithm for Using Them. In *Advances in Genetic Programming,* Kenneth E Kinnear (ed). MIT Press.

Bhandarkar, S.M., Yiqing Zhang, Potter, W.D. 1994. An edge detection technique using Genetic Algorithm based optimisation. *Pattern Recognition.* 27(9). pp 1159-1180.

Canny, J. 1986. A Computational Approach to Edge Detection. *IEEE Transactions on Pattern Analysis and Machine Intelligence.* 8(6).

Chih-Ho Chao, Atam P. Dhawan. 1994. Edge detection using a Hopfield neural network. *Optical Engineering* 33(11).

Gonzalez, R., Woods, R. 1992. *Digital Image Processing.* Addison-Wesley, Reading, Mass.

Harris, C.P., Buxton, B.F. 1996. *GP-COM: A Distributed, Component-Based Genetic Programming System in C++.* Department of Computer Science Research Note RN/96/2, University College London.

Johnson, M.P., Maes, P., Darrell, T. 1994. Evolving Visual Routines. In *Artificial Life IV: Proceedings of the fourth International Workshop on the Synthesis and Simulation of Living Systems*, Rodney Brooks and Pattie Maes, ed. MIT Press.

Koza, J.R. 1992. *Genetic Programming: On the Programming of Computers by Natural Selection.* MIT Press.

Petrou, M., Kittler, J. *On the optimal edge detector.*

Srinivasan, V., Byatia, P., Ong, S.H. 1994. Edge detection using a neural network. *Pattern Recognition* 27(12). pp 1653-1662.

Spacek, L.A. 1986. Edge detection and motion detection. *Image Vision Computing* 4. pp 43-55.

Tackett, W.A. 1993. Genetic Programming for Feature Discovery and Image Discrimination. In *Proc 5th International Conference on Genetic Algorithms,* ICGA '93. Morgan Kauffman.

Teller, E., Veloso, M. 1995. PADO: A New Learning Architecture for Object Recognition. In *Symbolic Visual Learning*, Katsushi Ikeuchi and Manuela Veloso, ed. Oxford University Press.

Toward Simulated Evolution of Machine-Language Iteration

Lorenz Huelsbergen
Bell Laboratories
Lucent Technologies
600 Mountain Avenue
Murray Hill, NJ 07974
lorenz@research.bell-labs.com

Abstract

We use a simulated evolution search (genetic programming) for the automatic synthesis of small iterative machine-language programs. For an integer register machine with an addition instruction as its sole arithmetic operator, we show that genetic programming can produce exact and general multiplication routines by synthesizing the necessary iterative control structures from primitive machine-language instructions. Our program representation is a virtual register machine that admits arbitrary control flow. Our evolution strategy furthermore does not artificially restrict the synthesis of any control structure; we only place an upper bound on program evaluation time. A program's fitness is the distance between the output produced by a test case and the desired output (multiplication). The test cases exhaustively cover multiplication over a finite subset of the natural numbers ($\mathbb{N}_{10}$); yet the derived solutions constitute general multiplication for the positive integers. For this problem, simulated evolution with a two-point crossover operator examines significantly fewer individuals in finding a solution than random search. Introduction of a small rate of mutation further increases the number of solutions.

1 Introduction

Genetic programming (GP) [8] uses principles of evolution (populations, fitness measures, recombination) to perform directed non-linear searches in the space of computer programs. GP has successfully produced solution programs in many problem domains, *e.g.* [7, 9, 8]. For the most part however, GP solutions are restricted to non-iterative (non-looping) programs. To both broaden the domain of GP applications and to scale to larger problem sizes, it is necessary to extend GP to iterative programs. To this end, we study the evolution of *unrestricted iteration* in the context of machine-language programs.

Little GP research to date addresses the problem of general iteration. Early work modeled the synthesis of machine-language programs as evolutionary processes [5, 4], but did not outperform random search. Applications of genetic programming that do use iteration do so by restricting its form to a single type of loop and by bounding its extent to a fixed number of iterations. Cramer [3] describes simulated evolution of programs in a machine language with a high-level LOOP construct parameterized by an expression e and an integer variable denoting the number of iterative executions of e; the system externally limits a program's running time. Koza [8] similarly introduces the *Do-Until* (DU) iterative language construct; parameterized by an expression e and a predicate p, DU iteratively evaluates e until p is true. Each DU operator in the program is, however, restricted in that the number of its iterations is externally bounded from above. Kinnear also uses a specialized loop construct, *dobl*, to evolve sorting algorithms [6]. As with iteration, GP experiments with recursion restrict the scope and generality of the recursion. Recent work by Brave [2] investigates restricted recursion over trees to solve state-space exploration problems. The external system again restrictively curbs the maximum depth of a recursion. With constructs such as LOOP, DU, or *dobl* as its only means of iteration, GP can only evolve programs with structured control flow.

An alternate approach to the problem of GP iteration is taken here—the automatic synthesis of general (*i.e.* unrestricted) iteration from control-flow primitives. Our program representation is an instance of a virtual register machine (VRM). The particular VRM under investigation (denoted VRM-$\mathcal{M}$) is a small integer instruction set that includes add, move, increment/decrement, set/clear, compare, and branch (conditional and unconditional) instructions. To synthesize a *for* loop in

Address	Instruction	Comment
0:	`Cmp(R2,R3)`	compare counter R_2 to zero (R_3= 0)
1:	`Je(+4)`	if counter zero, then exit
2:	`Add(R0,R1)`	$R_0 \leftarrow R_0 + R_1$
3:	`Dec(R2)`	decrement counter
4:	`J(-4)`	and loop ...

Figure 1: A hand-coded VRM-$\mathcal{M}$ program that solves the multiplication problem. The input multiplier and multiplicand are initially in registers R_1 and R_2; other registers are initially zero. The product accumulates in R_0.

VRM-$\mathcal{M}$ for example, GP must construct a counter, an increment or decrement of this counter, a comparison of the counter to its terminal value, and appropriate branch instructions based on this comparison.[1] Since loops are not explicit in this representation, it is not feasible to place time bounds on individual loops (*cf. Do-Until* [8])—as in *e.g.* [3], we instead limit the total number of instructions a program may execute, *i.e.* its running time.

This paper's result is that GP can automatically synthesize multiplication routines[2] (*solutions*) by constructing a loop that has the VRM-$\mathcal{M}$'s `Add` instruction at its core. (A hand-coded solution to the *multiplication problem* is in Figure 1.) Informally, the fitness of an individual VRM-$\mathcal{M}$ program p is computed by exhaustively evaluating p with inputs (i, j), for all $i, j \in \mathbb{N}_{10}$, and computing the distance of p's result from the desired product ij. Although the test cases cover only a small finite subset of the natural numbers, they suffice to produce general multiplication (bounded only by a machine's register precision) for the positive integers. All solutions discovered to date are general.

Genetic programming finds solutions to the multiplication problem in the VRM-$\mathcal{M}$ language using a single genetic operator (two-point crossover) to recombine fit individuals during the construction of successive generations. Introduction of a mutation operator—applied with low probability—further increases the likelihood that a given GP run finds a solution. This problem is solved exactly by only a small percentage ($\approx 3\%$) of the GP runs. Even so, we find that GP solution of this problem (with and without mutation) processes significantly fewer individual programs than a random search.

The results presented here were obtained in a new system—called Finnegan and briefly described in the next section—for experimentation with simulated evolution of programs. Section 3 specifies the virtual machine language; Section 4 details the experimental setup for the multiplication problem. Results are in Section 5. We conclude with a discussion in Section 6.

[1] A high-level *if* control structure also requires such synthesis from primitive VRM-$\mathcal{M}$ instructions.

[2] Cramer [3] also studied simulated evolution of multiplication in machine languages with high-level loop constructs.

2 Finnegan System

Finnegan is a framework for experimenting with simulated evolution. It is written in the Standard ML (SML) programming language [12] and implemented using the Standard ML of New Jersey (SML/NJ) compiler [1]. The SML/NJ implementation supports parameterized modules [10] which are indispensable in an experimentation framework. SML's strong static typing ensures that module implementations match their declared interfaces (signatures). In Finnegan, one is therefore able to quickly interchange GP's variable components (*e.g.*, representations, fitness functions, selection mechanisms) by selecting modules from libraries of such components.

3 Virtual Register Machine $\mathcal{M}$

The representation used for the multiplication problem is an instance of a virtual register machine called VRM-$\mathcal{M}$. The notation VRM-$\mathcal{M}_{(m,n)}$ names a particular VRM-$\mathcal{M}$ that consists of external state (m integer registers), internal state (a program counter and flag) and a sequence of n immutable instructions.

3.1 External State: Registers

We define the *register state* as a vector

$$\vec{R} \equiv \langle R_0, \ldots, R_{m-1} \rangle$$

of m integers; the register state constitutes the machine's mutable memory. The precision of a register is inherited from the underlying implementation.[3] Most program instructions (*e.g.*, `Add`) modify registers directly. All program input and output is communicated through the register state; that is, program inputs are supplied in the initial state $\vec{R}_{initial}$ and outputs are taken from the final register state, $\vec{R}_{final}$. Registers may also be used to initially supply the program with constants. Alternately, the program can synthesize necessary constants in otherwise unused registers.

[3] Integers in the SML/NJ compiler used for Finnegan are signed and 31 bit.

3.2 Internal State: PC, Flag

In addition to the external register state, VRM-$\mathcal{M}$ maintains two pieces of internal state: a program counter (PC) and a comparison flag (*Flag*). The program counter is an integer that selects which instruction to fetch and execute. Branch instructions modify the PC to point to the branch's target; all other instructions always increment the PC to point to the next instruction. The *Flag* reflects the result of the last comparison instruction executed. It can assume the values *less*, *greater*, and *equal*. *Flag* is initially undefined which we denote as $\perp$ (bottom). Only the comparison instructions (see below) can modify the *Flag* state.

3.3 Instruction Set

A *program* is a vector of n instructions:

$$\vec{I} \equiv \langle I_0, \ldots, I_{n-1} \rangle$$

The program counter naturally corresponds to an index of $\vec{I}$. A program *terminates* when $PC = n$; that is, when evaluation steps off the end of the program. (Our evaluation strategy also limits the maximum number of instructions evaluated; see Section 5.) Figure 2 contains VRM-$\mathcal{M}$'s instructions and their operational semantics. The instruction set consists of a single nullary instruction `Nop` which does nothing, a single arithmetic instruction `Add`, a register move instruction `Mov`, a register comparison instruction `Cmp`, an unconditional branch `J`, conditional branches (`Jl`, `Jg`, `Je`), instructions to initialize registers (`Set` and `Clear`), and instructions to increment (`Inc`) and decrement (`Dec`) a given register.

Since any finite implementation of integer arithmetic may result in overflow or underflow, we define the arithmetic operators $\oplus$ and $\ominus$ (used in Figure 2) to be conventional integer addition and subtraction respectively, except that they produce zero when the result would exceed the implementation's precision.

Branches are always relative to the program counter. We chose to forego absolute branch addressing since relative addressing readily admits code relocation, which is in essence what the application of genetic operators (§4.4 below) to VRM-$\mathcal{M}$ code does. Note that the operational semantics rewrites a branch to an address < 0 as a branch to I_0 (*i.e.* $PC \leftarrow 0$) and a branch past the end of the program $(n-1)$ as termination (*i.e.* $PC \leftarrow n$). A jump instruction I_j can therefore branch to any one of $n+1$ distinct addresses.

The number of syntactically-distinct instructions for VRM-$\mathcal{M}_{(m,n)}$ is

$$S(m,n) \equiv 3m^2 + 4m + 4(n+1) + 1 \qquad (1)$$

because there are three binary register instructions, four unary register instructions, four relative branch instructions, and one nullary instruction (`Nop`). The number of possible (syntactic) programs in a such a machine is then:

$$S(m,n)^n \qquad (2)$$

3.4 Evaluation Function

Evaluation of an n-instruction VRM-$\mathcal{M}_{(m,n)}$ program $\vec{I}$ is with respect to an m-register input state $\vec{R}$ and an integer number ($k > 0$) of evaluation steps (instructions). The evaluation function $\mathcal{E}$ maps a triple to a pair:

$$\mathcal{E} : \left(\vec{I}, \vec{R}, k\right) \rightarrow \left(\vec{R}', PC\right) \qquad (3)$$

$\mathcal{E}$ produces a pair consisting of the final register state $\vec{R}'$ and final PC after evaluation of at most k instructions. A program that terminates within k instructions has $PC = n$. Since our VRM-$\mathcal{M}$ evaluator is an interpreter (essentially Figure 2), it can be easily halted after evaluation of k instructions.

4 Experimental Setup

In addition to a program representation, a GP experiment is parameterized by a number of variable components: a set of test cases, a fitness function, a population selection mechanism, and a set of genetic operators. This section fixes the components used in our experiments with the multiplication problem. We defer elaboration of the quantitative settings (*e.g.*, population size) to the next section (§5).

4.1 Test Cases

A test case t for the multiplication problem is a pair (i, j). The *answer* for t is the product ij. Over any infinite domain, the set of all test cases is also infinite; we therefore restrict ourselves to test cases over a finite subset of the natural numbers, $\mathbb{N}_{10}$.[4] Denote the set of all such test cases as T:

$$T \equiv \{(i,j)\,|\forall i, \forall j \in \mathbb{N}_{10}\} \qquad (4)$$

4.2 Fitness Function

We define registers R_1 and R_2 as input registers (multiplier and multiplicand) and register R_0 as the output register (product).[5] The raw fitness of an n-instruction program $\vec{I}$ for test-case $t \equiv (i,j)$ is given by

$$\mathcal{F}_{(i,j)}\left(\vec{I}\right) \equiv |R'_0 - ij| + |PC - n| \qquad (5)$$

where, using (3), $\mathcal{E}(\vec{I}, \vec{R}, k) = (\langle R'_0, \ldots \rangle, PC)$, input registers $\vec{R} \equiv \langle 0, i, j, 0, \ldots, 0_{m-1} \rangle$, and k is the maximum

[4] We note that GP driven by this set finds exact solutions to the multiplication problem; we have not ascertained whether a smaller test-case set would also suffice (see §5 and §6).

[5] A VRM-$\mathcal{M}$ that can solve the multiplication problem must therefore contain at least $m = 3$ registers.

$$\texttt{Nop} \equiv \left(PC \leftarrow PC + 1 \right.$$

$$\texttt{Mov}(R_{dst}, R_{src}) \equiv \left(\begin{array}{l} PC \leftarrow PC + 1 \\ R_{dst} \leftarrow R_{src} \end{array} \right.$$

$$\texttt{Add}(R_{dst}, R_{src}) \equiv \left(\begin{array}{l} PC \leftarrow PC + 1 \\ R_{dst} \leftarrow R_{src} \oplus R_{dst} \end{array} \right.$$

$$\texttt{Cmp}(R_a, R_b) \equiv \left(\begin{array}{l} PC \leftarrow PC + 1 \\ Flag \leftarrow \begin{cases} less & \text{if } R_a < R_b \\ greater & \text{if } R_a > R_b \\ equal & \text{otherwise} \end{cases} \end{array} \right.$$

$$\texttt{J}(offset) \equiv \left(PC \leftarrow \min(\max(0, PC + offset), n) \right.$$

$$\texttt{Jl}(offset) \equiv \left(PC \leftarrow \begin{cases} \min(\max(0, PC + offset), n) & \text{if } Flag = less \\ PC + 1 & \text{otherwise} \end{cases} \right.$$

$$\texttt{Jg}(offset) \equiv \left(PC \leftarrow \begin{cases} \min(\max(0, PC + offset), n) & \text{if } Flag = greater \\ PC + 1 & \text{otherwise} \end{cases} \right.$$

$$\texttt{Je}(offset) \equiv \left(PC \leftarrow \begin{cases} \min(\max(0, PC + offset), n) & \text{if } Flag = equal \\ PC + 1 & \text{otherwise} \end{cases} \right.$$

$$\texttt{Inc}(R_a) \equiv \left(\begin{array}{l} PC \leftarrow PC + 1 \\ R_a \leftarrow R_a \oplus 1 \end{array} \right.$$

$$\texttt{Dec}(R_a) \equiv \left(\begin{array}{l} PC \leftarrow PC + 1 \\ R_a \leftarrow R_a \ominus 1 \end{array} \right.$$

$$\texttt{Set}(R_a) \equiv \left(\begin{array}{l} PC \leftarrow PC + 1 \\ R_a \leftarrow 1 \end{array} \right.$$

$$\texttt{Clear}(R_a) \equiv \left(\begin{array}{l} PC \leftarrow PC + 1 \\ R_a \leftarrow 0 \end{array} \right.$$

Figure 2: Operational semantics for the virtual register machine VRM-$\mathcal{M}$ used to solve the multiplication problem. Evaluation commences with $PC = 0$ and $Flag = \perp$ (uninitialized). The arithmetic operators $\oplus$ and $\ominus$ are defined to be conventional integer addition and subtraction respectively, except that when their result exceeds the implementation's precision (overflow or underflow), they produce zero.

number of evaluation steps. The total *raw fitness* of a program $\vec{I}$ with respect to the test-case set T is the sum of the raw fitness values of the individual test cases:

$$\mathcal{F}\left(\vec{I}\right) \equiv \sum_{t \in T} \mathcal{F}_t\left(\vec{I}\right) \tag{6}$$

The fitness function has two parts: the distance of a computed result from the desired product and the distance of the program counter (after k steps) from the halting position n. The $|PC - n|$ term favors—in some sense—programs that after k steps are "close" to the proper termination point, n. In this manner, $\mathcal{F}$ selects for programs that resemble multiplication and that terminate.

4.3 Population Selection

Population selection, for the construction of successive generations, is performed via *proportional selection* (see, *e.g.*, [8, 11]). Let P be a population (set) of N programs. Our selection mechanism first normalizes an individual's raw fitness (6) to the unit interval:

$$\widehat{\mathcal{F}}\left(\vec{I}\right) \equiv \frac{1}{\mathcal{F}\left(\vec{I}\right) \sum_{p \in P} \mathcal{F}(p)^{-1}} \tag{7}$$

We then select a N random reals, $X \equiv \{x_1, \ldots, x_N\}$, in the unit interval. For each $x \in X$, we select program $\vec{I}$ for the next generation if $x < \widehat{\mathcal{F}}\left(\vec{I}\right)$ and there does not exist an $\vec{I'}$ such that $x < \widehat{\mathcal{F}}\left(\vec{I'}\right) < \widehat{\mathcal{F}}\left(\vec{I}\right)$.

4.4 Genetic Operators

We use a single recombination operator that performs two-point crossover. For some experiments, we augment this operator with mutation.

4.4.1 Two-Point Crossover

Crossover of two n-instruction programs $\vec{I_i}$ and $\vec{I_j}$ first selects a subsequence of instructions starting at a random point $0 \leq p_i < n$ in program $\vec{I_i}$. The length l of the subsequence is chosen randomly such that $p_i < p_i + l < n$. A random point p_j, $0 \leq p_j + l < n$, is then chosen in program $\vec{I_j}$. Finally, the l instructions in $\vec{I_i}$ starting at p_i are interchanged with the l instructions in $\vec{I_j}$ starting at p_j.

Crossover in a population P is performed by first selecting a subset $P' \subseteq P$ of programs from the population; an individual program is randomly selected for P' with probability $Prob_{xover}$. The programs in P' are then randomly paired and the crossover operator is applied to each pair. The pairs resulting from crossover replace the corresponding original pairs in the population.

4.4.2 Mutation

With probability $Prob_{mutate}$, a program instruction in the population is changed to a random instruction. A random VRM-$\mathcal{M}_{(m,n)}$ instruction (§3) is generated by first selecting an opcode at random and then, when necessary, randomly generating the opcode's operands (*i.e.* registers and relative addresses).

5 Results

We performed two sets of experiments for the multiplication problem using the GP configuration of the previous section (§4)—two-point crossover (Experiment I) and two-point crossover with mutation (Experiment II). We also compared the number of programs examined by GP solution of this problem to the number required for a successful random search. All experiments used Finnegan compiled with SML/NJ version 108.15 and the pseudorandom number generator from the SML/NJ library.

For experiments I and II, we studied VRM-$\mathcal{M}_{(4,8)}$; that is, VRM-$\mathcal{M}$ with four registers and eight instructions. Each experiment consisted of 2000 GP runs with $Prob_{xover} = 0.25$ on a population of 1024 (initially random) programs. Evaluation of an individual was arbitrarily[6] limited to $k = 100$ steps. A run was deemed complete when the fitness of the best individual remained constant for 50 consecutive generations (stasis). Experiment I did not use mutation ($Prob_{mutate} = 0.0$); experiment II had a slight mutation rate ($Prob_{mutate} = 0.001$).

Experiment I evaluated 6.2×10^7 programs overall with a yield of 28 exact solutions to the multiplication problem, or an exact solution approximately every 71 runs (2.2×10^6 evaluations per solution).[7] Experiment II evaluated 6.8×10^7 programs overall and gave 60 exact solutions with an exact solution appearing approximately every 33 runs (1.1×10^6 evaluations per solution).

Some solutions discovered by experiment I are in Figure 3. Solutions (a) and (b) closely resemble a human's coding of this problem (*cf.* Figure 1) with (b) exhibiting unstructured control flow. Solution (c) always executes at least one loop iteration and, in the case of a zero multiplier, then proceeds to undo this incorrect subcomputation. Solution (d) is interesting because it accumulates *two* separate products—in R_0 and R_2. In (d), R_2 is used as multiplier input, loop counter, and product accumulator. Unlike most solutions, (d) requires only

[6] A system might find suitable k values automatically by performing runs using $k = 2^i$ for successive integer values of i until a run yields a solution.

[7] Finnegan eliminates some redundant evaluations; a program that moves unchanged from generation g to generation $g+1$ is not reevaluated. Eliminated evaluations do *not* appear in reported totals.

Address	Instruction
0:	Cmp(R_1,R_3)
1:	Je(+7)
2:	Add(R_0,R_2)
3:	Dec(R_1)
4:	Jg(-4)
5:	~~Mov(R_2,R_3)~~
6:	~~Jg(+2)~~
7:	~~Add(R_1,R_1)~~

(a) Generation 30

Address	Instruction
0:	J(+6)
1:	~~Jl(+7)~~
2:	~~Je(-2)~~
3:	Inc(R_3)
4:	~~Nop~~
5:	Add(R_0,R_1)
6:	Cmp(R_2,R_3)
7:	Jg(-6)

(b) Generation 15

Address	Instruction
0:	Dec(R_1)
1:	Cmp(R_3,R_1)
2:	Add(R_0,R_2)
3:	Jl(-3)
4:	Je(+4)
5:	~~Jl(0)~~
6:	Mov(R_0,R_1)
7:	Inc(R_0)

(c) Generation 13

Address	Instruction
0:	Cmp(R_0,R_2)
1:	Dec(R_2)
2:	Je(+5)
3:	~~Je(+3)~~
4:	Cmp(R_0,R_2)
5:	Add(R_0,R_1)
6:	Add(R_2,R_1)
7:	Jl(-6)

(d) Generation 28

Figure 3: Solutions to the multiplication problem discovered by GP using the VRM-$\mathcal{M}_{(m,n)}$ representation and halted after $k = 100$ steps. Instructions that do not affect a program's result have a ~~line~~ through them.

three VRM-$\mathcal{M}$ registers (but seven instructions).

Examination of 10^9 random VRM-$\mathcal{M}_{(4,8)}$ programs found 104 general solutions to the multiplication problem.[8] On average, a solution via random search required examination of $\mu = 9.5 \times 10^6$ individuals per solution with sample variance $\sigma^2 = 7.2 \times 10^{13}$. In this regard, GP's directed search strategy is quite effective. However, examination of a program by random search is usually much faster than examination by GP. This is because a random search can discard the current program as soon as a single test case fails whereas GP must construct fitness values by evaluating programs against all test cases. With respect to the multiplication problem, it is yet unclear how the absolute efficiences of GP and random search relate.

6 Conclusion

We have shown that—using a machine-language representation with addition as its sole arithmetic operator—genetic programming can effectively synthesize unstructured loops from primitive instructions (comparisons and relative branches) to evolve general multiplication. Furthermore, the number of unique programs processed by GP solving this problem is significantly less than the number processed by a random search, on average.

Questions concerning the absolute efficiency of a GP solution relative to random search (as noted in the previous section) are of foremost importance. To this end, we are pursuing a two-pronged approach. First, we are designing optimizations for the Finnegan system that further remove redundant program evaluations and that reduce the costs of such evaluations. This can directly increase GP's efficiency. Second, we are turning our attention toward harder problems (synthesis of multiplication over $\mathbb{Z}$ instead of $\mathbb{N}$, for example), the assumption being that unrestricted-iterative GP may effectively scale to harder problems for which random search is intractable.

To establish unrestricted iteration as a valuable device for genetic programming, future work must successfully apply it to richer representations and solve problems of higher complexity.

[8] Combined with equation 2, these empirical data suggest that multiplication is surprisingly dense in the space of VRM-$\mathcal{M}_{(4,8)}$ programs.

Acknowledgment

Thanks to Chris Fraser for helpful comments on drafts of this paper.

References

[1] A. W. Appel and D. B. MacQueen. A Standard ML compiler. *Functional Programming Languages and Computer Architecture*, 274:301–324, 1987.

[2] S. Brave. Evolving recursive programs for tree search. In P. Angeline and K. E. Kinnear Jr., editors, *Advances in Genetic Programming II*, chapter 10. MIT Press, Cambridge, MA, 1996. (To appear).

[3] N. L. Cramer. A representation for the adaptive generation of simple sequential programs. In *Proceeding of the International Conference on Genetic Algorithms and their Applications*, pages 183–187. Texas Instruments, July 1985.

[4] R. M. Friedberg. A learning machine: Part I. *IBM Journal of Research and Development*, 2:2–13, 1958.

[5] R. M. Friedberg, B. Dunham, and J. H. North. A learning machine: Part II. *IBM Journal of Research and Development*, 3:282–287, 1959.

[6] K. E. Kinnear Jr. Evolving a sort: Lessons in genetic programming. In *International Conference on Neural Networks*, New York, NY, 1993. IEEE.

[7] K. E. Kinnear Jr., editor. *Advances in Genetic Programming*. MIT Press, Cambridge, MA, 1994.

[8] J. Koza. *Genetic Programming: On the Programming of Computers by the Means of Natural Selection.* MIT Press, Cambridge, MA, 1992.

[9] J. Koza. *Genetic Programming II.* MIT Press, Cambridge, MA, 1994.

[10] D. B. MacQueen. Modules for Standard ML (revised version). *Polymorphism*, II(2), October 1985.

[11] Z. Michalewicz. *Genetic Algorithms + Data Structures = Evolution Programs.* Springer Verlag, Berlin, 1992.

[12] R. Milner, M. Tofte, and R. Harper. *The Definition of Standard ML.* MIT Press, 1990.

Robustness of Robot Programs Generated by Genetic Programming

Takuya ITO
School of Information Science,
Japan Advanced Institute
of Science and Technology,
15 Asahidai, Nomi,
Ishikawa, 923-12 Japan
takuya@jaist.ac.jp

Hitoshi IBA
Machine Inference Section,
Electrotechnical Laboratory,
1-1-4 Umezono, Tsukuba,
Ibaragi, 305 Japan,
iba@etl.go.jp

Masayuki KIMURA
School of Information Science,
Japan Advanced Institute
of Science and Technology,
15 Asahidai, Nomi,
Ishikawa, 923-12 Japan
kimura@jaist.ac.jp

ABSTRACT

The goal of this research is to study whether robot programs which are generated by means of genetic programming (GP) are robust. The simulated robot task is a box moving problem. In order to test the robustness of generated programs, we conduct two experiments. In the first experiment, the initial conditions of the robot are fluxed. This is to prevent the overgeneralization of the specific situations. In the second one, the robot's sensors and actuators are assumed to be noisy. As a result of experiments, we have observed that the robot behaves robustly on the both experiments. GP has generated programs which evaluate almost effective commands for each sensor state. We suppose that the robot behaves robustly due to the redundancy of the programs, and we discuss how this redundancy is realized in GP framework for our robot programs.

1 Introduction

Recently, genetic programming (GP) is applied for various domains, not only for toy-world but also for real-world. The goal of this paper is to study whether generated robot programs by means of GP are robust.

We think that the robustness in GP has two meanings:–

1. The first one is the robustness of generated programs. In this case, the robustness means how much generated programs adapt to an environment.
2. The second one is the robustness of the whole species. In this case, the robustness means to select the best fitness from the GP genetic pool when an environment is changed in the middle of generation.

In this paper, we have studied the first robustness, i.e., the robustness of generated programs. The second robustness is mentioned in [Ito, 1996].

Even if a generated program successfully controls a robot for a given task on a specific environment (establishment), it does not necessarily show the robustness of the program. In general, the more adaptation (or the evolution) proceeds, the more genes adapt heavily upon the training environment [Reynolds, 1994]. Thus, the resultant genes (programs) are inefficient when the environment or initial conditions of the robot are changed. We do not focus on the generation of programs which control a robot so as to behave efficiently on the training data but focus on the generation of programs which control a robot so as to conduct a given task in any environment.

Some GP researchers described the evolution of an autonomous robot (agent). For instance, Koza experimented in the wall following [Koza, 1991] and the box moving [Koza and Rice, 1992]. Reynolds experimented in the evolution of obstacle avoidance behavior with noise of sensor and actuator [Reynolds, 1994]. Handley researched on the planning of a robot by "Genetic Planner" [Handley, 1994]. Haynes also studied the construction of a program, which allowed an agent to correctly sense and mark the presence of items in any environment [Haynes and Wainwright, 1995]. Nordin controlled a real robot (Khepera), which the task was the obstacle avoidance [Nordin and Banzhaf, 1995]. Although the robustness is assumed to be essential for these robot tasks, there has been no researchers to investigate the robustness, without some exceptions mentioned above.

We have constructed an autonomous robot simulator (Fig.1). The model of which is a behavior-based robot [Maes, 1993]. In this simulator, we assume that there are only five objects in the workspace (Fig. 2).

For constructing the simulated robot, we have referred to Mars. Mars is also an autonomous robot simulator, constructed by RWCP (Real World Computing Partnership) in Japan [Matsubara *et al.*, 1995], [Kuniyoshi *et al.*, 1995]. Our simulated robot has five types of sensors. They are a touch sensor, five radar sensors, an eye sensor, a target sensor and a bumped sensor :–

Touch Sensor reports the IDs of object that the robot is touching.

Eye Sensor reports the IDs of object which are in the eye-sensor area.

Radar Sensors return the distance between the nearest object and the robot with each five radar sensors (they are covered each 30° from +60° to −60°).

Bumped Sensor reports whether the robot bumps an object or not.

Target Sensor reports the distance between the robot and the goal when the goal is in robot's sensor range.

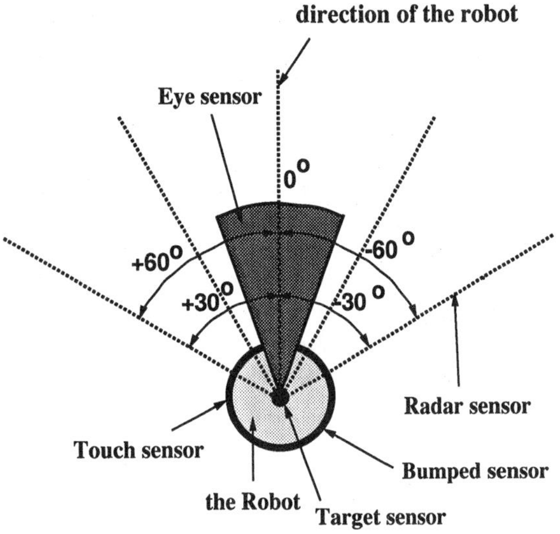

Fig. 1: Sensors of the Simulated Robot

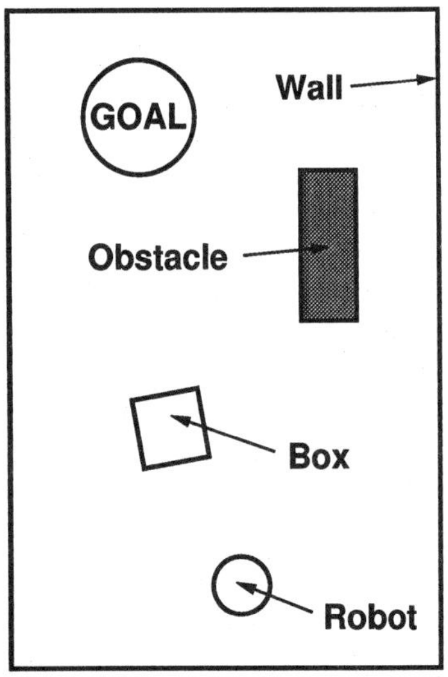

Fig. 2: the Workspace of the Robot

The motivation of constructing the above robot simulator is derived from our design principle. We constructed this robot simulator so as to close to a real world as much as possible. In this sense, we have chosen the actual experimental settings, i.e., (1) the robot's coordinates are arbitrary floating point numbers but not grid-based, and (2) the robot commands are taken from a real robot.

In order to examine robust robot programs generated by GP, we have conducted the following two experiments.

1. Changing the initial conditions, i.e., coordinates and direction of the robot which are generated randomly. This is to test whether robot programs which do not depend on the initial conditions of the robot are generated.
2. Introducing noisy sensors and actuators. Dealing with noise is significantly important for two reasons. First, as Brooks mentioned in the context of "embodiment learning" [Brooks, 1991], there seems to be no ideal sensors and actuators on real robots. Second, as Reynolds discussed in [Reynolds, 1994], assuming noises is helpful for preventing to evolve brittle control programs and encouraging the evolution of robust programs.

The rest of this paper is organized as follows. Section 2 introduces experimental set up. Section 3 describes the first experiment, in which the initial conditions of the robot are fluxed (i.e., to prevent an evolution which depends on specific situations). Section 4 describes the second experiment in a noisy environment (i.e., noisy sensors and actuators are assumed). Section 5 discusses the robustness of generated programs. Conclusions are given in Section 6.

2 Experimental Set Up

In our experiments, we used ADF GP [Koza, 1994]. Our ADF GP has two function definition branches. Population is 500, and max generation is 50.

2.1 Task Overview

The robot task we have chosen is a box moving. This task consists of (1) recognizing the difference between a box and an obstacle, (2) bringing the box to a given goal in a limited time.

2.2 Terminal Set

As commonly used in GP applications [Koza, 1991], we assigned the robot commands (i.e., 0-argument function) to the GP terminal set as follows

:Spd moves forward with the robot speed in its direction. The robot speed is assumed to be constant.

:Turn-Left turns 10° to the left (counterclockwise).

:Turn-Right turns 10° to the right (clockwise).

:Grasp grasps an object (box or obstacle) which is in its arm-reach area. When the object is a box, the box moves with the robot. When the robot grasps an obstacle, it cannot move.

:Release releases the object (box or obstacle) that the robot grasps.

Each command takes one time step to evaluate. Maximum of time step is 800.

Maes mentioned that if a system of robot does not maintain a history of what it did in the past, it falls into deadlock or loop situation [Maes, 1990]. One very simple solution of this problem is to introduce some randomness in the system. Therefore, in our simulated robot, when the robot's coordinates remain the same in past twenty time steps, it walks randomly in ten time steps in order to get away from its deadlock situation.

2.3 Non-Terminal Set

In order to make the robot respond its sensor inputs, we have used the following non-terminal set

(Prog2 p1 p2) evaluate sequentially all the argument forms and return the value of the second argument (p2).

(Case-Radar p1 ··· p5) evaluate the i-th argument (pi) if the value of the radar sensor i is the smallest among the five radar sensors.

(Case-Touch p1 ··· p5) evaluate an appropriate argument depending upon the value of the touch sensor, i.e., if the touched object is the box then evaluate p1; else if it is the goal then evaluate p2; else if it is the obstacle then evaluate p3; else if it is the wall then evaluate p4; else (i.e., if nothing is touched) evaluate p5.

(Case-Eye p1 ··· p5) evaluate an appropriate argument depending upon the value of the eye sensor. An evaluation of this function is the same as function Case-Touch.

(If-Bumped p1 p2) evaluate p1 if the robot bumps into something, and evaluate p2 otherwise.

(Radar-Target p1 ··· p3) evaluate p1 if the robot approaches the goal, evaluate p2 if the robot goes away from the goal, and evaluate p3 if the goal is not in the target sensor range.

(Case-Grasp p1 ··· p3) evaluate p1 if the robot grasps the box, evaluate p2 if the robot grasps the obstacle, and evaluate p3 if the robot does not grasp anything.

2.4 Fitness Function

Fitness is

$$\begin{cases} \dfrac{dist(G, B_t)}{dist(G, B_0)}, & \text{if the robot brought the box} \\[2ex] \dfrac{dist(G, B_t) + dist(B_t, R_t)}{dist(G, B_0)}, & \text{else if the robot did not bring the box once} \end{cases}$$

where B_0 is the initial position of the box, B_t is the position of the box at t time step, G is the position of the goal, R_t is the the position of the robot at t time step, $dist(x, y)$ is the Euclidean distance between x and y.

This definition reflects how the box approaches the goal as a result of moving. Although the definition does not necessarily represent the hardness of problem, we have chosen this definition because of its simplicity. Note that if the robot does not move the box at all, the distance between the box and the robot is added to the fitness for the sake of penalty.

It is too difficult to justify whether the generated programs are robust or not. In this paper, we discuss the robustness of acquired robot programs by GP through the fifty test cases which are not the same data as training one.

3 Experiment1 (fluxed initial robot's conditions)

In this section, we experimented in changing the initial conditions of the robot randomly at each generation. Fig. 3 shows the best and the average fitness of the training and test cases. This is a typical example of our result. Each fitness is as good as small value.

The fitness of the test cases is plotted for the best individual of the training data at each generation. As it can be seen in Fig. 3, the fitness of the test cases is almost steadily improved when generations proceed.

The average fitness seems to be gradually improved with the training cases. The fitness for the training cases have the keen peak at the 33th generation. This peak indicates that the robot has fallen into a deadlock.

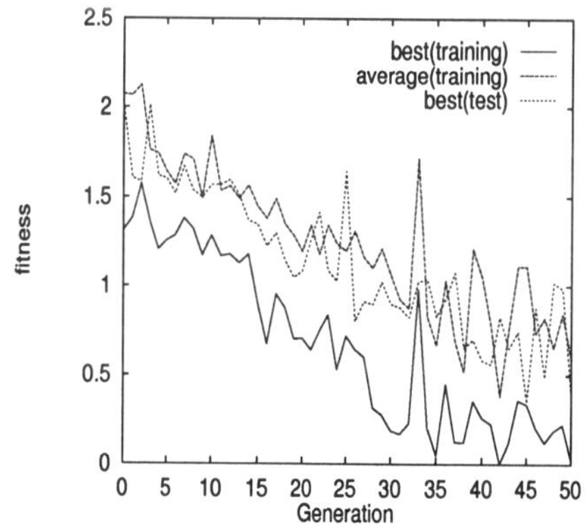

Fig. 3: Fitness of the Best and the Average on the Training and the Test Cases

The result of this experiment shows that the fitness of the test cases is improved for each generation. The robust program is generated by sensor fusion even for fluxed training cases. Thus, each robot command is evaluated to conduct a given task for five sensors suitably. For instance, Fig. 10 is a part of the generated program at the 50th generation. This fragment of the generated program shows that "if the robot bumps, turn to the right". It means that the robot can change its direction when it bumps something. Then, the robot can go forward without falling into deadlock. And also its program shows that "if the robot touches the box, grasp it and if the robot touches the obstacle, release it.". Thus, this generated program has a sort of robustness. However, it may be difficult to generate programs which fuse all sensors by means of GP because of it is hard to generate programs which all arguments of all functions are all effective combinations. We mention it in detail in discussions.

Figures 4–9 show how the robot behaves for certain test cases at every 10 generations. In these figures, ○ indicates the robot's start position and ● depicts the robot's end position.

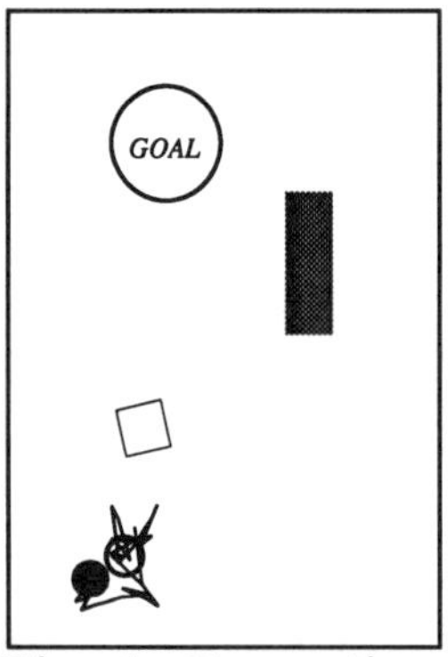

Fig. 4: Generation 0

Fig. 5: Generation 10

Fig. 6: Generation 20

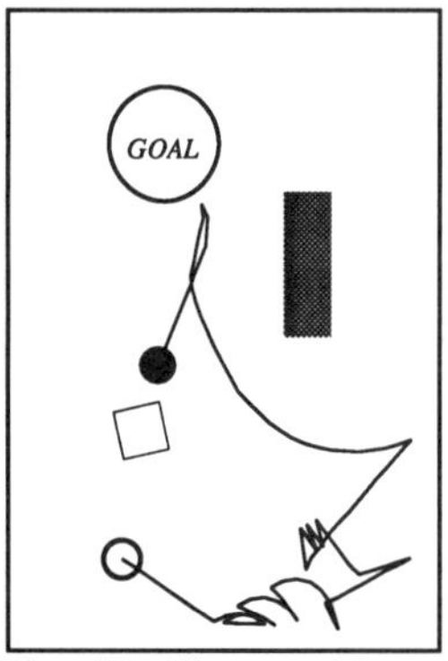

Fig. 7: Generation 30

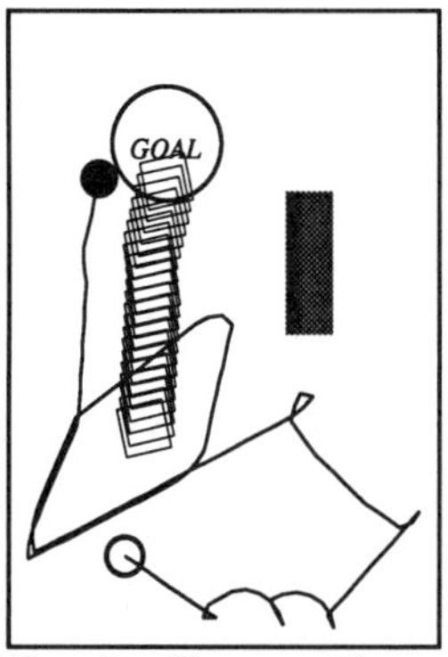

Fig. 8: Generation 40

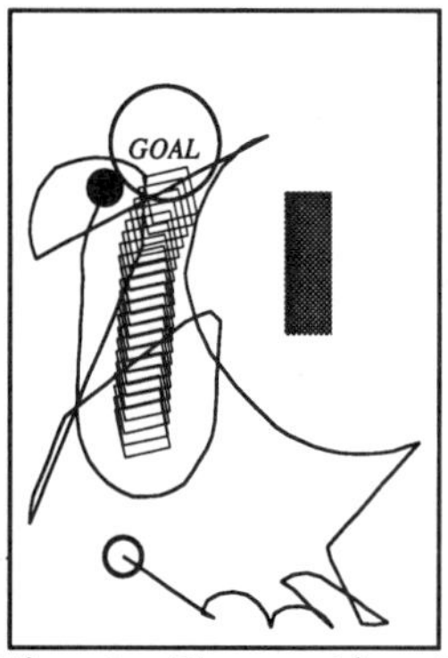

Fig. 9: Generation 50

```
.....(If-Bumped :Turn-Right (Case-Radar
:Turn-Right :Turn-Right :Turn-Right :Spd
:Spd)) (Case-Touch :Grasp :Turn-Right
:Release :Turn-Left :Turn-Right).....
```

Fig. 10: the Best Program of the 50th Generation

4 Experiment2 (noisy sensors and actuators)

In this section, we experimented in noisy situations. The noise is introduced as follows :–

Radar Sensor All five radar sensors report the distance between the nearest object and the robot with a maximum of 11% noise.

Eye Sensor The radius of an eye sensor is perturbed with a maximum of 5%. The error rate of its recognition of ID is set to be a maximum of 19%.

Touch Sensor The error rate of its recognition of ID is set to be a maximum of 10%.

Bumped Sensor The bumped sensor reports an error with a maximum of 5%.

:Spd The speed of the robot is changed with a maximum of 8% error.

:Turn-Left (Right) The error of steering rate is set to be a maximum of 10%

Fig. 11 shows the best and the average fitness of the training and the test cases. These values are almost improved for every generation except the 45th generation. The fitness of the test cases has a keen peak at the 45th generation. The peak indicates that the robot has fallen into a deadlock. The best program at the 45th generation is overgeneralization of the training cases. Thus, the generated program is performed well on the training cases, however it is not robust one.

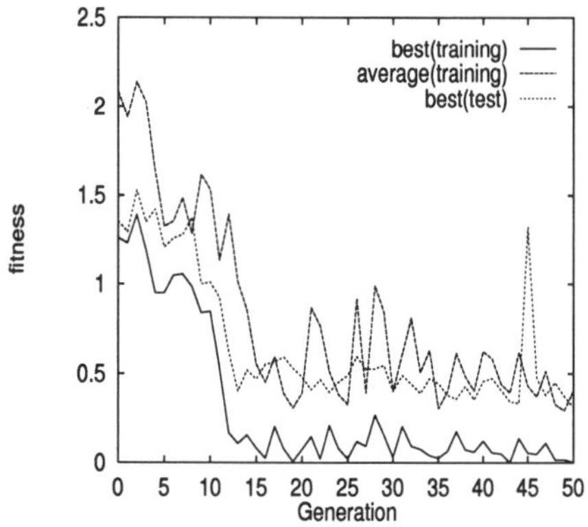

Fig. 11: Fitness of the Best and the Average on the Training and the Test Cases

5 Discussions

5.1 Redundancy of the programs

The previous experiments showed that generated programs by GP is capable of fusing sensor information so as to achieve a sort of robustness. However, the robot could not behave robustly by means of merely sensor fusion. We consider that redundancy is important for generating a robust program. We define the redundancy of the program as follows

"There are some fragments in the generated program which are not effective on some situations but effective on other situations."

In other words, the generated program can afford to handle many building blocks. These Building blocks mean fragments of programs which fit highly on each situation. The redundancy promotes that generated programs store not only a part of structure which depends on the training data but also a part of structure which is effective on an unknown test data. When the generated program stores very few building blocks, it becomes a "Tight Program" which is dependent heavily upon the training data.

For achieving the redundancy in programs, we designed carefully a non-terminal set. For instance, suppose that a non-terminal set consists only of "if" functions. The combination of arguments of each function must be almost effective for behaving robustly in any environment. We have conducted an experiment for investigating this hypothesis.

Fig. 12 depicts the fitness curves of the training and the test cases when a non-terminal set does not include prog2.

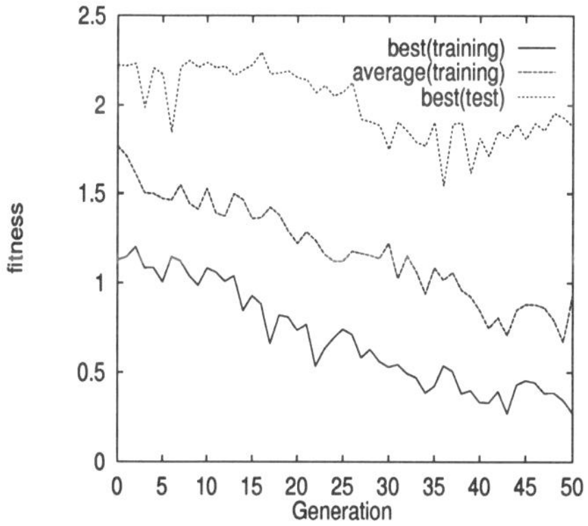

Fig. 12: Fitness of the Best and the Average on the Training and the Test Cases (without prog2)

The best fitness of the training cases is improved from generation to generation. However the fitness of the test cases is not improved. Even on the training cases, the curve of the best fitness is not decreased after the 35th generation. This shows the difficulty of generating an effective solution by this non-terminal set (i.e., "if" functions). Fig. 13 and Fig. 14 display the behaviors of the robot for the training and the test cases, respectively. For the training case, the robot walks complicatedly and brings the box to the given goal (Fig. 13). However the robot walks straightly on the test case (Fig. 14). Thus, the generated program is supposed to control the robot on specific training data.

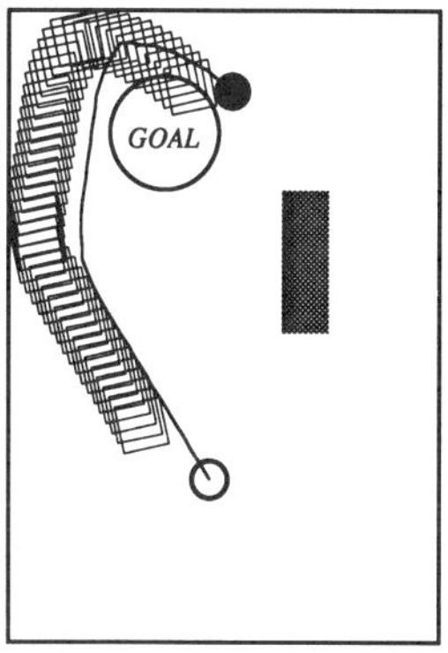

Fig. 13: Training Case (Generation 50)

Fig. 14: Test Case (Generation 50)

From this experimental result it seems to be necessary to include a function which introduces redundancy into the program. We adopted prog2 for this purpose. This is because the robot can behave robustly when first argument of prog2 is effective even if second argument is not effective (and vice versa). Consider the acquired program shown in Fig. 15. If the robot touches the box and it does not bump into something, the robot conducts the release command (**:Release**). This is not actually a suitable program for the sake of bringing the box to the goal. However the robot evaluates the grasp command (**:Grasp**) anyhow by means of prog2. Thus, the robot can grasp the box after all.

```
...(Case-Touch (If-Bumped :Grasp (Prog2
(Prog2 :Release :Turn-Right) (Radar-Target
:Grasp :Grasp :Grasp)))...
```

Fig. 15: the Best Program of the 50th Generation (Experiment 2)

The other realization of generating redundancy is assumption of introns. Nordin introduced Explicitly Defined Introns (EDI's) against the destructive effect of crossover [Nordin *et al.*, 1995]. This produced the better generalization, compared with a non-EDI's version.

5.2 Planning by prog2

The other effect of prog2 is that the generated program can plan behavior of the robot for conducting the given task (i.e., planning) because prog2 evaluates its arguments sequentially. In the above sentence, planning means to analyze the situation in which a robot finds itself and then to develop a strategy for achieving the robot's goal - presumably by finding a series of actions that can be expected to have a desirable outcome [Ginsberg, 1993]. On Handley's genetic planner, the non-terminal set consists of only prog2, prog3 and prog4. Thus, the acquired program which was generated by the genetic planner is planning of the robot motion entirely. In our experiments, the non-terminal set has "if" functions besides prog2. The effect of prog2 which is planning of behavior is reduced more than only prog2. For instance, on Fig. 15, if the robot touches the box and bumps into something, the robot conducts only **:Grasp**. There is no effects of prog2 in this case. It means that the robot does not plan its behavior. However, when the robot does not bump into something, the robot conducts **:Release**, **:Turn-Right**, **:Grasp**. In this case, the robot can plan its behavior to conduct the given task.

6 Conclusion

This paper investigated the robustness of acquired robot programs. We have realized the following points

- The robustness of generated programs by means of two experiments. In the first experiment, the initial conditions of the robot is fluxed each generation. In the second experiment, we assumed that there are noises of each sensor and actuator.
- The generated programs have allowed the robot to behave robustly by means of sensor fusion in these experiments.
- The redundancy of program was effective for generating robust robot program.

In these experiments, the workspace (the coordinates of the box, the goal and the obstacle) is fixed for each generation. We plan to conduct an experiment with a more difficult problem when the workspace is gradually changed on one generation (i.e., the problem becomes more difficult). We have experimented on a simulated robot in these experiments. We plan to study this problem on a real robot in the future.

Acknowledgments

We are thankful to Walter Tackett for the use of his Genetic Programming system. We would like to thank Satoshi Sato for his helpful comments, and TuBao Ho for useful discussions.

Bibliography

[Brooks, 1991] Brooks, R. A. Artificial Life and Robots. In Varela, F.J. and Bourgine, P., editors, *Proceedings of the First European Conference on Artificial Life (ECAL-91)*, pages 3–10, MIT Press, 1991

[Ginsberg, 1993] Ginsberg, M. *Essentials of Artificial Intelligence.* Morgan Kaufmann, 1993

[Handley, 1994] Handley, S. G. The Automatic Generation of Plans for a Mobile Robot via Genetic Programming with Automatically Defined Functions. In Kenneth E. K., Jr., editor, *Advances in Genetic Programming*, chapter 18, pages 391–407, MIT Press, 1994

[Haynes and Wainwright, 1995] Haynes, T. and Wainwright R. A Simulation of Adaptive Agents in a Hostile Environment. In *Proceedings of the 1995 ACM Symposium on Applied Computing*, pages 318–323, ACM Press, 1995

[Ito, 1996] Ito, T. Robustness of Robot Programs Generated by Genetic Programming. Master's thesis, Japan Advanced Institute of Science and Technology, 1996, in Japanese.

[Koza, 1991] Koza, J. R. Evolution of Subsumption Using Genetic Programming. In *Proceedings of the First European Conference on Artificial Life (ECAL-91)*, pages 110–119, MIT Press, 1991

[Koza and Rice, 1992] Koza, J. R. and Rice, J. P. Automatic Programming of Robots Using Genetic Programming. In *Proceedings of Tenth National Conference on Artificial Intelligence*, pages 194–201, AAAI Press / MIT Press, 1992

[Koza, 1994] Koza, J. R. *Genetic ProgrammingII: Automatic Discovery of Reusable Programs.* MIT Press, 1994

[Kuniyoshi *et al.*, 1995] Kuniyoshi, Y., Motomura, Y. and Hiraki, H., *et al.* Multiple Autonomous Robot Simulator and Information Integration Research. In *RWC Information Integration Workshop*, pages 304–313, 1995, in Japanese

[Maes, 1990] Maes, P. Situated Agents Can Have Goals. In Maes, P., editor, *Robotics and Autonomous Systems*, pages 49–79, MIT Press, 1990

[Maes, 1993] Maes, P. Behavior-Based Artificial Intelligence. In Meyer, J.-A., Roitblat, H. L. and Wilson, S. W., editors, *From Animals to Animats 2: Proceedings of the Second International Conference on Simulation of Adaptive Behavior (SAB-92)*, pages 2–10, MIT Press, 1993

[Matsubara *et al.*, 1995] Matsubara, H., Hiraki, H. and Motomura, Y. *et al.* Towards a Study of Cooperative Learning Using a Robot Simulator. In *Bulletin of the Electrotechnical Laboratory*, 59(3):61–68 Ohmsha, 1995, in Japanese

[Nordin and Banzhaf, 1995] Nordin, P. and Banzhaf, W. Genetic Programming Controlling a Miniature Robot. In *Working Notes for the AAAI Symposium on Genetic Programming*, pages 61–67, 1995

[Nordin *et al.*, 1995] Nordin, P., Francone, F. and Banzhaf, W. Explicitly Defined Introns and Destructive Crossover in Genetic Programming. In Rosca, J. P., editor, *Proceedings of the Workshop on Genetic Programming: From Theory to Real-World Applications*, pages 6–22, 1995

[Reynolds, 1994] Reynolds, C. W. An Evolved, Vision-Based Model of Obstacle Avoidance Behavior. In Langton, C. G., editor, *Artificial Life III*, pages 327–346. Addison-Wesley, 1994

[Reynolds, 1994] Reynolds, C. W. Evolution of Obstacle Avoidance Behavior: Using Noise to Promote Robust Solutions. In *Advances in Genetic Programming*, pages 221–241. MIT Press, 1994

Signal Path Oriented Approach for Generation of Dynamic Process Models

Peter Marenbach
Institute of Control Engineering
Darmstadt University of Technology
Landgraf-Georg-Strasse 4
D-64283 Darmstadt, Germany
mali@rt.e-technik.th-darmstadt.de

Kurt D. Bettenhausen
Institute of Control Engineering
Darmstadt University of Technology
Landgraf-Georg-Strasse 4
D-64283 Darmstadt, Germany
kdb@rt.e-technik.th-darmstadt.de

Stephan Freyer
Technical Development
BASF AG
D-67056 Ludwigshafen
Germany
stephan.freyer@ze.basf-ag.de

ABSTRACT

The article at hand discusses a tool for automatic generation of structured models for complex dynamic processes by means of genetic programming. In contrast to other techniques which use genetic programming to find an appropriate arithmetic expression in order to describe the input-output behaviour of a process, this tool is based on a block oriented approach with a transparent description of signal paths. A short survey on other techniques for computer based system identification is given and the basic concept of SMOG (Structured MOdel Generator) is described. Furthermore latest extensions of the system are presented in detail, including automatically defined sub-models and qualitative fitness criteria.

1 Introduction

Within the field of automatic control theory the term model stands for the description of a technical system or process which can be used as a starting point for tasks like diagnosis, simulation or control. Since a model is developed with respect to such a given task, it does not need to contain more details than those which are really relevant. Therefore the search for an appropriate model can be seen as an optimization problem with the aim to compromise between accuracy and parsimony of the process description.

This optimization problem in general is a difficult and time consuming task, where results depend very much on the experience and process knowledge of the engineer or scientist working on it. Therefore the question arises how model generation can be supported by the use of computers. Techniques such as parameter estimation or Neural Networks are often said to provide a way to automatic modelling. However, both of these techniques are based on the principle that a given structure – differential equations or network topology respectively – is adapted to fit the process' input-output behaviour by tuning parameters or coefficients within that structure. In contrast to techniques which are just optimizing parameters within a fixed structure, Genetic Programming (GP, for an introduction see Koza 1992) provides a way for the automatic modification of symbolic structure representations. In this paper a tool for the self-organizing development of nonlinear dynamic process models is presented, which has the capability to generate well structured mathematical and therefore transparent models in a block oriented form. This approach offers a number of possibilities to make use of a priori knowledge like fundamental physical relationships.

In the following section the mentioned ordinary techniques are described in more detail and other GP approaches for system identification are discussed. Section 3 describes basic principles and realization details of the Structured MOdel Generator *SMOG*. In sections 4 and 5 some major extensions of previous versions of SMOG are presented and a new conception for a qualitative formulation of fitness criteria is introduced. A survey on a number of successful applications of this tool, a short summary, and a preview concerning the future practical and theoretical work are given in the last section.

2 Related work

Most of the computer aided modelling techniques are based on the principle that a usually fixed structure is given and the parameters or coefficients included in this structure are modified in order to optimize the model's accuracy comparing the simulated model response with training data. Therefore these techniques are often described as data driven modelling approaches. In this section two very popular classes of these techniques as well as other investigations using a GP approach for modelling are discussed briefly.

2.1 Parameter estimation

The classical approach for developing a mathematical-physical process model starts with an analysis of the dominating physical aspects of the process. Even for a relative simple process this often leads to systems of differential equations that cannot be reduced to a concentrated and handy form without a significant number of simplifications. Due to these simplifications, the values of parameters which were derived from physical reflections soon loose their original meaning. Therefore data driven techniques are required to estimate these parameters which are usually described by the term system or parameter identification (Isermann 1981).

However, a priori knowledge of the mathematical structure of the model is essential for a successful application (and the model has to be linear in the unknown parameters, which is usually no serious restriction in practice). For complex processes – especially non-artificial systems like biological fermentation processes, where the complexity cannot be influenced by constructive changes – the required a priori knowledge is often not available. Therefore a suitable model structure is usually developed iteratively by combining partially available physical knowledge with structural ideas derived from examinations of the measured input-output behaviour. With each modification of the model structure, the set of parameters has to be re-estimated. This makes modelling a time consuming and therefore expensive task.

2.2 Neural Networks

In cases as described above the use of Neural Networks (NN, Rumelhart *et al.* 1986) or other associative memory systems (Tolle and Ersü 1992) as process models is suggested as a possibility to reduce modelling efforts. These techniques approximate nonlinear mappings by interpolating between information "learned" from a given training data set. Only a minimum of a priori knowledge about the structure of the process is needed for the configuration, the user has just to choose the network topology and which input variables to use[1].

From a mathematical point of view, the difference between a NN model and a physically motivated process description based on differential equations is that the NN uses a very general structure in which the process behaviour is approximated by a superposition of basis functions while differential equations are a well structured and concentrated form. This leads to some remarkable disadvantages of NNs: Since the NN as a black box model is just approximating the input-output behaviour, no direct insight into the process structure and its underlying relationships can be gained. This usually leads to problems concerning the acceptance of these approaches in industrial applications. Furthermore, long term predictions, which are essential for bioprocesses[2], based on NN models are difficult. This is due to the fact that NNs are usually trained to make a one step ahead prediction of the estimated process state. Therefore long term predictions are achieved by recursive predictions which lead to an error propagation.

[1] In most cases NNs are restricted to learn static manifolds and dynamic process behaviour can only be approximated by additionally introducing time delayed input values. Therefore the choice of external connections of a NN is a non-trivial, often underestimated task.

[2] Often the first few hours, the so-called lag phase, are very important for the further development of the fermentation process.

Another problem is that a process engineer wants to have a structured mathematical description of his process, even if an associative memory which perfectly reproduces the observed process behaviour would be available. This is due to the fact that information about what influences the different process variables is an important help for understanding and modification of the process.

2.3 Other GP approaches

From the different disadvantages of the techniques described above the question arises whether there could be a way for the automatic and data driven generation of transparent models by optimizing both parameter and structure. In the chapter on *symbolic regression* Koza (1992) presented a technique to generate mathematical expressions which were able to reproduce a functional relationship based on given training data. The tool STROGANOFF presented by Iba *et al.* (1993) was one of the first applications of the GP paradigm explicitly for system identification. Both of these approaches use GP to generate mathematical equations from input variables and elementary arithmetic operators. Due to the difficulties to evaluate the coefficients within the model with a pure GP approach[3] and the fact that there are a number of well known efficient techniques especially for this task, the GP part of STROGANOFF generates symbolic coefficients which are then evaluated with respect to training data. McKay *et al.* (1995) presented the application of a similar approach for the modelling of a chemical process. However both of these approaches were not able to generate models which include dynamic explicitly. Instead – similar to NNs – time delayed input values had to be included into the *terminal set.*

The extended system $\Re$-STROGANOFF (Iba *et al.* 1995) provides a mechanism to generate models with explicit dynamics. This is achieved by propagating the model output values as additional input values for the next time step. Combined with the generation of state variables in addition to the output values this enables $\Re$-STROGANOFF to produce time discrete systems of first-order difference equations similar to a state-space representation.

However, from the users point of view, who wants to have the opportunity to integrate as much of his process knowledge as possible in an easy and transparent way, the equation oriented representation based approaches described above have certain disadvantages. Therefore the approach presented in the following section is based on a block oriented representation which allows the concentration on statements according to the underlying signal paths without the need to make assumptions concerning the process dynamics.

3 The structured model generator SMOG

The aim of SMOG is to develop a methodology which provides data driven model generation, modification, and ver-

[3] Even for simple problems the solutions found by a pure GP approach for the *symbolic regression* examples in (Koza 1992) filled complete pages.

ification in close cooperation with process experts. To reduce modelling effort it has to be easy to integrate as much as possible available knowledge on sub-models and/or non-measurable or proven relationships into the automatic modelling process. Similar to the standard engineering approach, SMOG internally distinguishes between two tasks: One is the search for an appropriate – i.e. accurate and parsimonious – model structure and the other is the optimization or identification of the structure's inherent set of parameters.

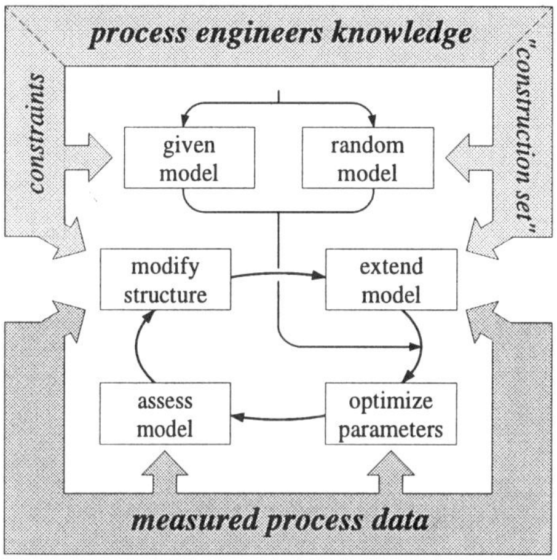

Figure 1 Influences on the (automatic) modelling procedure.

Therefore modelling of dynamic systems can be seen as a hybrid task which is influenced by existing process knowledge and by measured process data (fig. 1). There are a number of well known techniques which can be used to adjust the parameters – directed search methods, Genetic Algorithms or parameter estimation as discussed in section 2. However, for symbolic structure search GP is one of few computer based approaches.

3.1 Symbolic model representation and object oriented GP

When trying to apply the GP approach for symbolic model structure search at first a suitable representation scheme has to be evolved. As mentioned in section 2 a number of disadvantages arise from the (obvious) use of an equation oriented representation. Looking at the form a chemist expresses his process knowledge, it appears that it is basically oriented on signal paths and only a vague idea about dynamics exists. Therefore from the very beginning (Marenbach 1994) SMOG uses a block diagram representation – often used in control theory – which is based on a description of interactions between input and output variables in the sense of a directed graph. As an example, the model depicted as a block diagram in fig. 2 (left) corresponds to the following system of first-order differential equations:

$$\begin{aligned} \dot{x}_1(t) &= x_2(t) \\ \dot{x}_2(t) &= K \cdot u_2(t - T_d) - \omega^2 \cdot x_1(t) + 2d\omega \cdot x_2(t) \\ y(t) &= u_1(t) + x_2(t) \end{aligned} \tag{1}$$

As it can easily be seen from this example, the block diagram provides a much quicker overview about the influence of each input variable as equ. 1. With increasing dimension of the in-

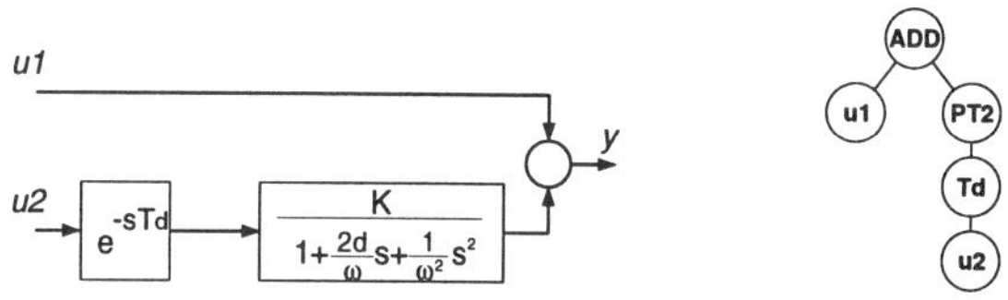

Figure 2 Simple process model: depicted as a block diagram (left) and as a tree structure (right).

put vector and growing complexity of the model this gets more and more important. Fig. 2 shows the model (left) and the corresponding tree structure (right) which is internally used by the GP algorithm. As it is shown in this example, nodes in the tree structure represent not only simple (arithmetic) operations, but also more or less complex transfer elements like

- dynamic elements with time-delay of different order,
- non-linear elements like switches or limiters,
- elements that let the system interpret subexpressions as feed-forward or feedback loops and
- domain specific elements like Monod kinetics or bell-shaped characteristics.

Most of these elements include certain parameters – e.g. gain and delay time of a first-order time-delay element –, that are initialized by random. During the evolutionary process, these parameters are not explicitly modified by the genetic operators but – as described above – adapted by a conventional search algorithm which is applied to the entire model. That means similar to the coefficients in STROGANOFF, the symbolic model includes symbolic parameters which are not represented explicitly by *terminals* but are implicitly included as parts of the transfer elements.

Compared with the operators usually used in GP, the model blocks in SMOG are of higher complexity. The fact that each node in the tree structure can have private variables – namely its parameters – may cause difficulties concerning an interpretation of these tree structures as computer programs. On the other hand, the concept presented here can be seen as a kind of Object Oriented GP (OOGP), where a model block is an object with private variables, a predefined public interface to other objects and the assignment of numerical values to symbolic parameters can be seen as an object's instantiation. Besides some academic aspects this interpretation can be useful in a case where the evolutionary generated models are not fed into a special interpreter program but are coded in a conventional programming language.

3.2 Fitness evaluation

As pointed out above, the aim of SMOG is to generate models by compromising between accuracy and parsimony. The most common way to force GP systems to produce small solutions is to introduce some penalty value for the model complexity which is considered for fitness evaluation. Problems arise from the fact that in general the values defined for accuracy and this penalty value are of different order of magnitude and that the parsimony pressure does not only influence the selection of a single individual but has a strong impact on the evolutionary development of the whole population (for more details see Nordin and Banzhaf 1995).

SMOG assesses the accuracy of a model i by an error value E_i which is defined as the root-mean-square error between given process behaviour and simulated model response. The models complexity C_i is evaluated as the sum of a heuristically defined complexity value of each block within the model. Based on these two objectives, two different formulas to calculate *raw fitness* F_i respectively *standardized fitness* f_i were implemented:

raw fitness F_i	*standardized fitness* f_i	
$F_i = qE_i + rC_i$	$f_i = \frac{1}{1+qE_i+rC_i}$	(2)
$F_i = E_iC_i$	$f_i = \frac{1}{1+E_iC_i}$	(3)

While a weighted sum (equ. 2) is very often used in all fields of optimization in applications a number of good results were achieved with formula 3. A detailed discussion how equ. 3 can be interpreted in the sense of qualitative fitness criteria is given in section 5.

3.3 Introducing existing knowledge

The most important advantage of SMOG compared to the usage of NNs is, that it provides a number of possibilities to introduce existing process knowledge in an easy and transparent way. This is achieved by the following mechanisms:

Configuring the model construction set. At the beginning of evolutionary model generation an initial population is – as long as no other option is chosen – composed by random. The creation of models by random as well as its manipulation by the genetic operators is based on a collection of elementary transfer elements placed in a so called *model construction set.* The user has the chance to configure the content of this construction set and to influence the selection probability of the blocks from the set.

Restricting the ranges for numerical parameters. When a block is taken from the model construction set, initial values are chosen for its private parameters by random. The range from which this is done is different for each type of parameter and can be modified by the user.

Defining super blocks. Similar to *modules* as defined by Angeline (1994) a *super block* means a special block, which itself consists of a sequence of blocks. A super block is treated as a normal block and can therefore not be divided by crossover or mutation.

Defining a fixed sub-model. For all models of the population a *fixed sub-model* can be defined. This means that each individual starts with the same subtree. The fixed sub-model must not be changed by genetic operators. In contrast to super blocks which may appear anywhere in a model several times a fixed sub-model appears in every model at the same place.

Starting with a given population. It is possible to start SMOG with a given initial population taken from earlier runs or created by hand. Furthermore, a mechanism was implemented to insert individuals – e.g. from a parallel population – into the actual population on the run.

4 Extensions of the original system

4.1 Automatic defined sub-models

The previous implementation of SMOG – which was tested at different example processes (section 6) – was only able to generate multiple-input single-output models in a tree structure with the root node as the output. Additionally, the limitation on only one single tree led to considerable restrictions concerning the usage of known partial models caused by the suppression of forks in the direction of the flow of information. The tree structure allows only forks from the output towards the input variables. Following the mechanism of *automatic defined functions* (ADF) proposed by Koza (1992) (for detailed examinations see Koza 1994), the new release of SMOG supports internal states and additional outputs.

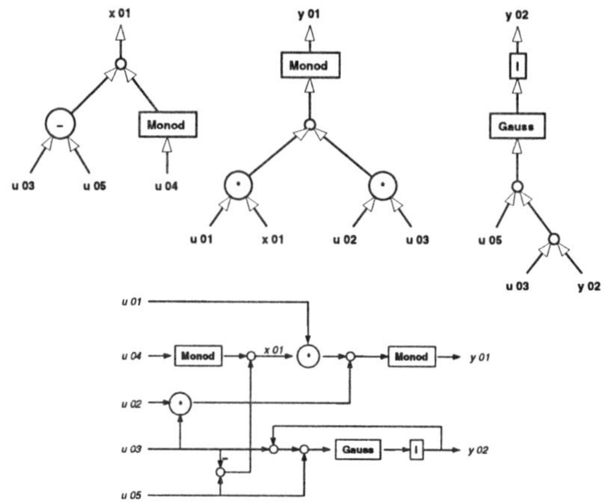

Figure 3 Example for a model with two outputs and one internal state: tree structure (top) and block diagram (bottom).

Internal states similar to ADFs are results of fixed predefined subtrees. They are very similar to the outputs in principle. The main difference is that trees representing an output return a value for error calculation. Internal states are used to allow information forks only. Fig. 3 shows an example for a model with two outputs and one internal state. In contrast to ADF the automatic defined sub-models in SMOG do not have parameters, because this would not be useful for model transparency.

4.2 Generalization test

As known from other learning approaches – e.g. NN or associative memories – the increasing adaptation on a set of training data can lead to a hyper specialization or overfitting on the training data by memorizing the data instead of finding underlying relationships. Similar effects can take place during a structured modelling procedure[4], especially, if the examined structure does not fit the real process and therefore can only be used for memorizing.

One possibility – which is also easy to implement – to avoid this phenomenon is to test the generalization of the model instead of its ability to adapt. This can be realized by using only one part of the available data for the parameter optimization but all data for the error calculation so that overfitting does not lead to high fitness values.

5 Qualitative fitness criteria

An alternative for the integration of two or more optimization objectives to achieve a single quality or fitness value is the search for the so-called Pareto optimality (Goldberg 1989, Fonseca and Fleming 1993). These and other techniques described in literature lead to a population of Pareto optimal individuals which are more or less equally distributed in the Pareto optimal subspace. But this is not the main aim for the application of SMOG. Instead of this, only one or up to a few models should be generated and selected automatically granting the operators compromise between accuracy and complexity.

The number of different approaches and the permanent discussion in the *gp-mailing-list* have shown, that the operator is in most cases able to define his ideas about a compromise between several objectives qualitatively, but – especially for more than two objectives – the development of an adequate precise mathematical formula is an iterative task.

Fuzzy set theory offers an approach for the imitation of qualitative formulations in a computer framework taking the lexical vagueness into account. The following section will show, how fuzzy methods enable an easy formulation of fitness criteria for multi-objective optimization. Such defined criteria will be called *qualitative fitness criteria* in the future.

Definition 5.1 (Qualitative fitness criteria) *Qualitative fitness criteria integrate – possibly competing – optimization objectives by using the logical operators* and, or *and* not *based on the fuzzy set theory and the given operators.*

Starting with a visual interpretation of this general definition taking the given task into account, the next steps will show, how easily higher dimensional aims can be formulated.

In the first step the introduced measures E_i for error and C_i for complexity have to be normalized fulfilling $0 \leq \tilde{E}_i \leq 1$ and $0 \leq \tilde{C}_i \leq 1$. Interpreting the values of $\tilde{E}_i$ and $\tilde{C}_i$ as grades of membership – e.g. $\tilde{E}_i \rightarrow 1$ means "error is high", $\tilde{E}_i \rightarrow 0$ equals "error is not high" or even "error is low" and appropriate interpretations for $\tilde{C}_i$ – qualitative fitness criteria can be formulated like

IF $\tilde{E}$ is not high *and* $\tilde{C}$ is not high **THEN** f is high.
IF $\tilde{E}$ is not high *or* $\tilde{C}$ is not high **THEN** f is high.

[4]The idea that size of an individual evolved by a GP system is related to how well it generalizes is discussed controversial. However, generalization has to be an important criterion for model assessment.

Using the algebraic operators as the realization of logical *and* and *or*, this corresponds to

$$f_i = (1 - \tilde{E}_i) \cdot (1 - \tilde{C}_i) \quad (4)$$

respectively for the logical *or*

$$f_i = (1 - \tilde{E}_i) + (1 - \tilde{C}_i) - (1 - \tilde{E}_i) \cdot (1 - \tilde{C}_i) \quad (5)$$

The interpretation of operator $\frac{1}{1+\mu}$ as logical *not* leads to the following qualitative interpretation of equ. 3:

IF *not*($\tilde{E}$ is high *and* $\tilde{C}$ is high) **THEN** f is high.

According to inversion theorems of the boolean algebra follows

IF $\tilde{E}$ is low *or* $\tilde{C}$ is low **THEN** f is high.

The question of a useful normalization method has not been discussed yet. General possibilities are simple scaling methods (Goldberg 1989), an approach for the optimization of a vector performance index proposed by Kreisselmeier and Steinhauser (1979) or separate rankings for each objective with subsequent scaling. However a separate normalization of each criterion does not only allow the application of fuzzy operators as shown above, it offers general advantages since the fitness criterion is decoupled from the specific modelling task. After normalization the achievable minimum error value is not influenced by effects like noise in measured data.

A first impression of the effect of fitness evaluation using equ. 4 and 5 can theoretically be gained from the visualization of the fitness function's contour lines shown in fig. 4 which also shows the distribution of the individuals – placed in a not

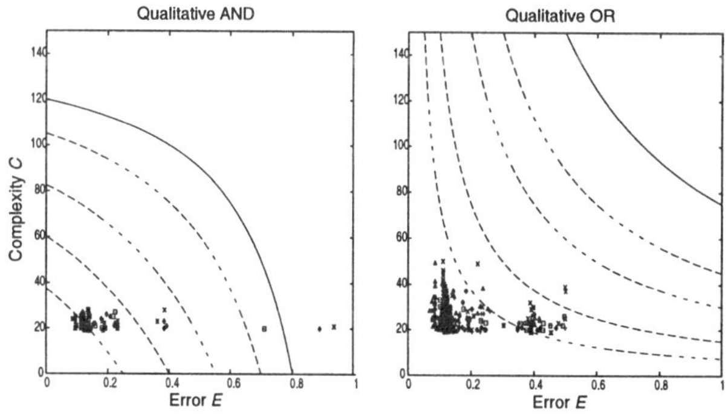

Figure 4 Contour lines of the fitness functions and distribution of individuals based on equ. 4 (left) and on equ. 5 (right).

normalized (E_i, C_i)-plane – in the 200th generation of several runs where normalization was done by separate ranking.

Since normalized quality measures are independent of e.g. the order of magnitude of the error value, preferences between objectives can easily be introduced. Fig. 5 shows the resulting contour lines after extending equ. 4 and 5 to

$$f_i = (1 - \tilde{E}_i^{\alpha_E}) \cdot (1 - \tilde{C}_i^{\alpha_C}) \quad \text{respectively} \quad (6)$$

$$f_i = (1-\tilde{E}_i^{\alpha_E})+(1-\tilde{C}_i^{\alpha_C})-(1-\tilde{E}_i^{\alpha_E})\cdot(1-\tilde{C}_i^{\alpha_C}) \quad (7)$$

with $\alpha_E = 6$ and $\alpha_C = 1$ in the first and $\alpha_E = 1$ and $\alpha_C = 6$ in the second case to achieve preference of solutions with smaller error in both cases.

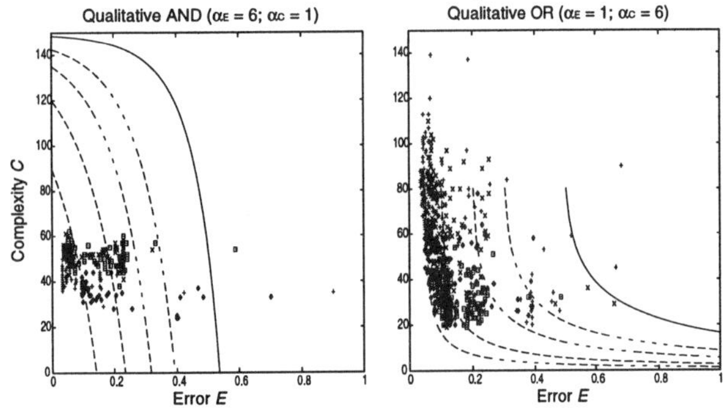

Figure 5 Contour lines of the fitness functions and distribution of individuals based on equ. 6 (left) and on equ. 7 (right).

6 Conclusion

In this paper the general framework of a tool for data driven generation of structured process models is presented. In the past two years its capabilities have been proven in different applications, e.g. the modelling of a river barrage system in cooperation with AEG Daimler Benz Industrie AG, Germany, (Marenbach *et al.* 1995) and an industrial biological fed-batch fermentation process within an cooperation with BASF AG, Germany (Bettenhausen *et al.* 1995). Furthermore SMOG was tested on several artificial problems and on data taken from an industrial steel rolling process at Daimler Benz AG, Germany (Pohlheim and Marenbach 1996).

Besides this description the underlying approach is compared with other work in the field of GP and – as an extension of the previous system – several new techniques are introduced.

References

Angeline, Peter John (1994). Genetic programming and emergent intelligence. In: *Advances in Genetic Programming* (Kenneth E. Kinnear, Jr., Ed.). Chap. 4. MIT Press.

Bettenhausen, Kurt Dirk, Peter Marenbach, Stephan Freyer, Hans Rettenmaier and Ulrich Nieken (1995). Self-organizing structured modelling of a biotechnological fed-batch fermentation by means of genetic programming. In: *Int. Conf. on Genetic Algorithms in Engineering Systems: Innovations and Applications* IEE. London, UK. pp. 481–486.

Fonseca, Carlos M. and Peter J. Fleming (1993). Genetic algorithms for multiobjective optimization: Formulation, discussion and generalization. In: *5th Int. Conf. Genetic Algorithms*. Morgan Kaufmann. San Mateo, CA. pp. 45–52.

Goldberg, David E. (1989). *Genetic algorithms in search, optimization, and machine learning*. Addison-Wesley Publishing.

Iba, Hitoshi, Hugo de Garis and Taisuke Sato (1995). Temporal data processing using genetic programming. In: *6th Int. Conf. Genetic Algorithms*. Morgan Kaufmann. San Francisco, CA. pp. 279–286.

Iba, Hitoshi, Takio Kurita, Hugo de Garis and Taisuke Sato (1993). System identification using structured genetic algorithms. *5th Int. Conf. Genetic Algorithms* pp. 297—285.

Isermann, Rolf (1981). *Digital Control Systems*. Springer.

Koza, John R. (1992). *Genetic Programming: On the Programming of Computers by Means of Natural Selection*. The MIT Press. Cambridge, MA.

Koza, John R. (1994). *Genetic Programming II: Automatic Discovery of Reusable Programs*. MIT Press. Cambridge, MA.

Kreisselmeier, G. and R. Steinhauser (1979). Systematische Auslegung von Reglern durch Optimierung eines vektoriellen Gütekriteriums. *Regelungstechnik* **27**(3), 76–79 (in German).

Marenbach, Peter (1994). Selbstorganisierende nichtlineare Modellbildung – Ansätze und Untersuchungen am Beispiel des dynamischen Verhaltens einer Fluß-Stauhaltung. Master's thesis. TH Darmstadt (in German).

Marenbach, Peter, Kurt Dirk Bettenhausen and Bernd Cuno (1995). Selbstorganisierende Generierung strukturierter Prozeßmodelle. *at* **43**(6), 277–288 (in German).

McKay, Ben, Mark J. Willis and Geoffrey W. Barton (1995). Using a tree structured genetic algorithm to perform symbolic regression. In: *1st Int. Conf. on Genetic Algorithms in Engineering Systems: Innovations and Applications*. IEE. London, UK. pp. 487–492.

Nordin, Peter and Wolfgang Banzhaf (1995). Complexity compression and evolution. In: *6th Int. Conf. on Genetic Algorithms*. Pittsburgh.

Pohlheim, Hartmut and Peter Marenbach (1996). Generation of structured process models using genetic algorithms. In: *AISB'96 Workshop on Evolutionary Computing*.

Rumelhart, D.E., G.E. Hinton and R.J. Williams (1986). Learning internal representation by error propagation. In: *Parallel Distributed Processing* (David E. Rumelhart and James L. McClelland, Eds.). Vol. 1. MIT Press. Cambridge, MA.

Tolle, Henning and Enis Ersü (1992). *Neurocontrol*. Vol. 172 of *Lecture Notes in Control and Information Sciences*. Springer-Verlag. Berlin.

Evolving Control Laws for a Network of Traffic Signals

David J. Montana and Steven Czerwinski
BBN Systems and Technologies
70 Fawcett Street, Cambridge, MA 02138
dmontana@bbn.com, czerwin@mit.edu

Abstract

Optimally controlling the timings of traffic signals within a network of intersections is a difficult but important problem. Because the traffic signals need to coordinate their behavior to achieve the common goal of optimizing traffic flow through the network, this is a problem in collective intelligence. We apply a hybrid of a genetic algorithm and strongly typed genetic programming (STGP) to the problem of learning control laws which optimize aggregate performance. STGP learns the single basic decision tree to be executed by all the intersections when deciding whether to change the phase of the traffic signal. The genetic algorithm learns different constants to be used in these decision trees for different intersections, hence allowing specialization based on differences in geometry and traffic flow. Preliminary experimental work shows that our approach yields good performance on a variety of network configurations and that it can evolve control laws which induce cooperation, communication, and specialization among the traffic signals.

1 Background

1.1 Networks of Traffic Signals

Most traffic signals in use today use preset timings. These preset timings are optimized off-line for the traffic patterns at a particular time of day. At predetermined times during the day, a new set of timings is downloaded to each traffic signal in the network. Two popular packages for doing the off-line optimization are MAXBAND [Little *et al.*, 1981] and TRANSYT-7F [FHWA, 1986].

For the purposes of this paper, we are interested in a different, newer type of control known as adaptive traffic signal control. This type of control takes into consideration the variations in traffic flow (both variations around the average and variations from the average) which naturally occur. Based on sensory inputs of the current traffic conditions, the control algorithm modifies the signal timings over time.

SCOOT [Robertson and Bretherton, 1991] starts with a set of preset timings as optimized by TRANSYT. At fixed intervals it modifies the splits, offsets, and cycle times of each signal so as to try to improve the performance evaluation of this signal and its immediate neighbors. OPACS [Gartner *et al.*, 1991] also has each signal modify its timing to optimize some local evaluation criterion; its major innovation is the use of dynamic programming as a way to efficiently look further ahead in its predictions of future performance. SCATS [Sims, 1979] and Chiu's fuzzy logic approch [Chiu, 1992] both use heuristic rules as a way to adjust signal timings; the major difference is that Chiu's rules are fuzzy rules.

A shortcoming of all these techniques is that a potential adaptation is evaluated based only on its effect on a small neighborhood of signals over a small interval of time. The global effects of this adaptation on the full network are ignored despite the fact that these effects are often significant. Hence, for our work we draw on an area of research that focuses on how to coordinate individual actions to produce desired aggregate behavior.

1.2 Collective Intelligence

The concept of *collective intelligence* has been developed in the fields of Robotics and Artificial Life. The basic idea is that many individual agents each executing simple, deterministic behaviors can interact in such a way that the aggregate behavior of the group is complex and not fully predictable from knowledge of the individual behaviors. One early example of artificial collective intelligence is the Boids simulation, which simulates flocking of birds [Reynolds, 1987]. A variety of work has followed, including some using real robots, e.g. [Mataric, 1994].

One important issue in collective intelligence and in our work is that of communication [Arkin and Hobbs, 1993]. Explicit communication can aid coordination among agents, but there generally is a cost (in the case of traffic signals a cost which is measurable as the price of the extra wires and extra interfaces required).

Another important issue in collective intelligence, and the key one for our work, is learning the control strategies which allow agents to cooperatively perform a task. For learning control laws for a group of robots, [Mataric, 1994] uses a type of reinforcement learning. For learning control of a colony of ants foraging for food, [Collins and Jefferson, 1992] use a genetic algorithm to learn the weights and connectivity of an artificial neural network. Most relevant to our work are those approaches using genetic programming (GP) as the learning paradigm. Koza's work on learning strategies for ants foraging for food was the first to use GP in this context [Koza, 1990]. Subsequent work includes Ryan's use of GP to evolve teams of callback functions in an event-driven system [Ryan, 1995]. Most similar to our work is that of [Haynes *et al.*, 1995], which uses strongly typed GP (STGP) to evolve strategies for packs of predators trapping a prey.

1.3 Genetic Programming Enhancements

Strong Typing: Strongly typed GP (STGP) [Montana, 1995] is an enhancement to standard GP [Koza, 1992] where (i) all the functions know what data types they take as arguments and what data type they return, and (ii) the routines for tree generation, mutation, and crossover all ensure that data types are consistent. The big advantage of STGP over GP when using typed data is that STGP eliminates the huge overhead of having to evaluate syntactically incorrect parse trees (which greatly outnumber the syntactically correct parse trees).

GA/GP Hybrids: An issue in GP is how to generate constants for use in the parse trees. The standard approach is to define a terminal $\Re$ which generates a random numerical constant in some range each time it is initially placed in a tree [Koza, 1992]. This method works well when the constants required are single, isolated numbers. However, especially when using STGP, the constants required are often not single numbers but rather more complex data structures such as the matrices in [Andre, 1994]. Extending the terminal $\Re$ to generate random instances of these data structures makes the search for correct values for these constants essentially a random search, which is known to be impractically inefficient for higher-dimensional search. A more practical alternative is to utilize a hybrid GA/GP search [Andre, 1994; Nguyen and Huang, 1994]. The GA performs the search to find the right values for the constants, while the GP searches the space of parse trees. The chromosome representation contains both the parse tree and the constant values, and there are different genetic operators for manipulating the parse trees and the constants.

2 Our Approach

2.1 The Simulation

The simulator we are using is a special version of the TRAF-NETSIM traffic simulator. TRAF-NETSIM has been developed by the U. S. Federal Highway Administration and has been one of the standards of the traffic engineering field. It is a microsimulator, which means that it simulates individual vehicles rather than average flows. It allows the user to specify via a file the geometry of the network, the traffic flows, and the signal timings (which for the standard TRAF-NETSIM are fixed).

A special version of TRAF-NETSIM has been developed for evaluating adaptive traffic signal control. It contains two major enhancements. First, sensors can be placed in the road in order to detect the passage of vehicles. Second, the fixed signal timings have been replaced with hooks for a user-implemented adaptive control law. This control law utilizes the sensor outputs in determining when to change the signal phase. The control law is executed for each traffic signal once every second.

We have linked our enhanced genetic programming code with the simulator to allow execution of a specified parse tree as the main part of an adaptive control law. The high-level logic of this control law is as follows. We require that there be exactly four phases for the signal, two green phases and two corresponding yellow phases. (This simplifying assumption excludes situations such as delayed-green phases and left-turn-only phases.) The yellow phases all last some fixed amount of time, which for the experiments described below was five seconds. To determine when to end a green phase, we execute the parse tree every second. The parse tree returns a Boolean value which is true if the signal should change at the current time and false if it should not.

We require that there be two sensors for each link of an intersection. One should be placed right next to the intersection to count the cars as they enter the intersection. The other should be placed further away to count the cars as they approach the intersection. The placement of this second sensor should be such that the queue for that link is considered "full" if cars are stopped back to that sensor. In the experiments described below, we always placed this second sensor 120 feet from the signal.

2.2 Phase-Dependent Constants

A phase-dependent constant is a constant in a parse tree which has a different value for each phase of each intersection. A simple example of how a phase-dependent constant can be used is the parse tree (> SINCE-CHANGE CONSTANT-1), where SINCE-CHANGE is the time since the current signal last changed color and CONSTANT-1 is a phase-dependent constant. This tree implements fixed-timings control where the length of each phase is given by the value of CONSTANT-1 for that phase. Phase-dependent constants are useful because they provide a mechanism to have each phase of each intersection execute a single basic parse tree while allowing for the fact that each phase is constrained by its own particular geometry and traffic patterns and re-

Function	Argument Types	Return Type
NUMBER-CARS	PHASE	INTEGER
WAITING	PHASE	BOOLEAN
APPROACHING	PHASE	BOOLEAN
QUEUE-FULL	PHASE	BOOLEAN
OR	BOOLEAN BOOLEAN	BOOLEAN
AND	BOOLEAN BOOLEAN	BOOLEAN
NOT	BOOLEAN	BOOLEAN
>	INTEGER INTEGER	BOOLEAN
SINCE-CHANGE		INTEGER
CONSTANT-i		INTEGER
MOVING		PHASE
STOPPED		PHASE

Figure 1: Baseline functions with their types.

quires customization of its parse tree for these factors.

Because we have assumed that each intersection has exactly two phases, each phase-dependent constant will have $2N$ values, where N is the number of intersections. These $2N$ values can be thought of as a $2N$-valued real string, which can be evolved using a standard string-based genetic algorithm. To evolve phase-dependent constants simultaneously with the parse tree, we use a GA/GP hybrid as described in Section 1.3.

2.3 The Functions

Figure 1 shows the functions utilized in all the experiments described below. Note that the PHASE data type enumerates the $2N$ different green phases discussed in Section 2.2. We now describe these functions.

NUMBER-CARS examines all incoming links for the specified phase and returns the maximum number of cars between the two sensors for any link (calculated as the difference of the vehicle counts of the two sensors).

WAITING tells whether there is a car currently stopped waiting for the specified phase to turn green (as detected by there being a car sitting directly over the sensor closest to the intersection for any incoming link).

APPROACHING tells whether there is a car which is approaching the specified phase (as detected by the vehicle count of the second sensor of an incoming link exceeding that of the first).

QUEUE-FULL tells whether there are cars backed up to the second sensor for any incoming link of the specified phase (as detected by there being a car sitting directly over this second sensor).

SINCE-CHANGE returns the time since the current intersection (i.e., that intersection for which the control law is currently being executed) last changed its phase.

CONSTANT-1 and CONSTANT-2 are two phase-

Function	Argument Types	Return Type
DS-NUM-SAME	PHASE	INTEGER
DS-NUM-CROSS	PHASE	INTEGER
DS-WAIT-SAME	PHASE	BOOLEAN
DS-WAIT-CROSS	PHASE	BOOLEAN
DS-GREEN-SAME	PHASE	BOOLEAN
DS-GREEN-CROSS	PHASE	BOOLEAN

Figure 2: Additional communications functions

dependent constants that are simultaneously evolved along with the parse tree (see Section 2.2). They return the entry in the real-valued string which corresponds to the current green phase of the current intersection.

MOVING (STOPPED) returns the current green (red) phase of the current intersection.

Figure 2 shows six additional functions used only in one of the experiments. We refer to these functions as "communications functions" because they require data from the sensors of neighboring intersections. In a real traffic network, implementing such functions would require extra expenses, e.g. additional wires being laid in the ground. Therefore, we consider these experimental functions whose potential contributations need to be determined first in simulation. The functions are:

DS-NUM-SAME returns the maximum number of cars between the two sensors for any link downstream from the specified phase.

DS-NUM-CROSS returns the maximum number of cars between the two sensors for any link crossing an intersection downstream from the specified phase.

DS-WAIT-SAME returns whether there is a car waiting at any link downstream from the specified phase.

DS-WAIT-CROSS returns whether there is a car waiting at any link crossing an intersection downstream from the specified phase.

DS-GREEN-SAME returns whether any link downstream from the current phase has a green signal at the downstream intersection.

DS-GREEN-CROSS returns whether any link crossing an intersection downstream from the current phase has a green signal at that intersection.

Note that all these communications functions address downstream links. This is because there is always implicit communications from the upstream links based on the number of cars approaching, while there is no such mechanism for implicit communication from the downstream links. Hence, communication from the downstream links is more likely to be useful.

2.4 The Evaluation Function

The basic measure of effectiveness we use for evaluating performance is delay. Delay is defined as the total amount of time lost due to the traffic signals. This in-

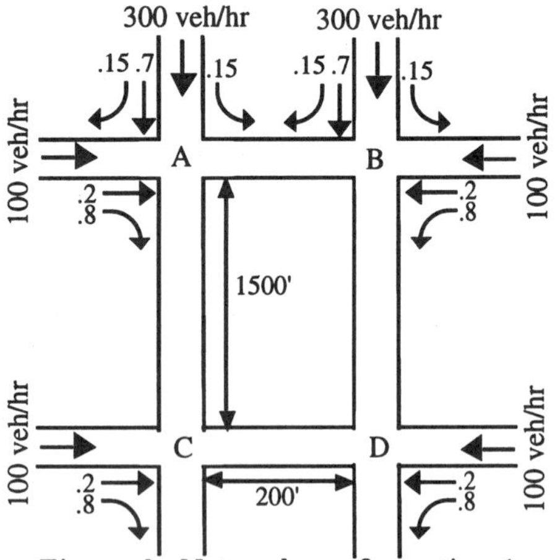

Figure 3: Network configuration 1.

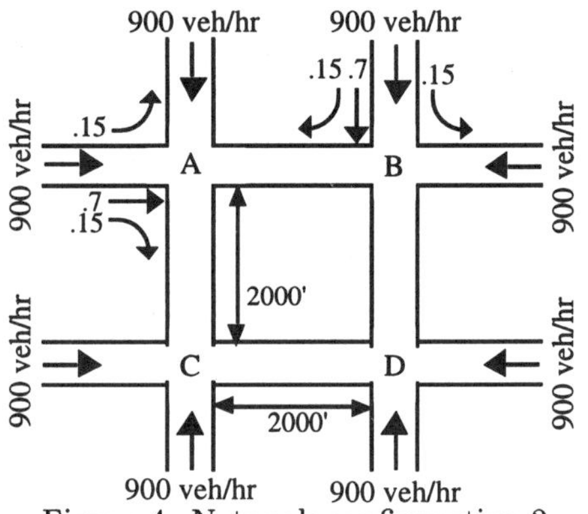

Figure 4: Network configuration 2.

cludes not just the time stopped at a traffic signal but also the time lost traveling at less than free-flow speed when decelerating to a stop and then accelerating again.

The precise criterion is the sum over each phase of each intersection of the number of cars passing this phase times the square of the average delay for this phase. This measure can be easily calculated from the statistics TRAF-NETSIM keeps and represents a reasonable tradeoff between efficiency and equity. By weighting the contribution by the number of cars, we make sure that we give more weight to the phases with larger number of cars, hence promoting efficiency. However, by using the square of the average delay rather than just the average delay, we ensure that the phases with smaller number of cars cannot be ignored, hence promoting equity.

3 Experimental Results

3.1 The Network Configurations

For the experiments we have performed so far, we have used three different network configurations. Figure 3 shows the first configuration. It contains two streets running north-south, both one-way in the south direction, and two two-way streets running east-west. Vehicles approaching an intersection from the east or west turn onto the one-way street with a probability of 0.8 and go straight with a probability of 0.2. Vehicles approaching an intersection from the north turn left with a probability of 0.15, turn right with a probability of 0.15, and go straight with a probability of 0.7.

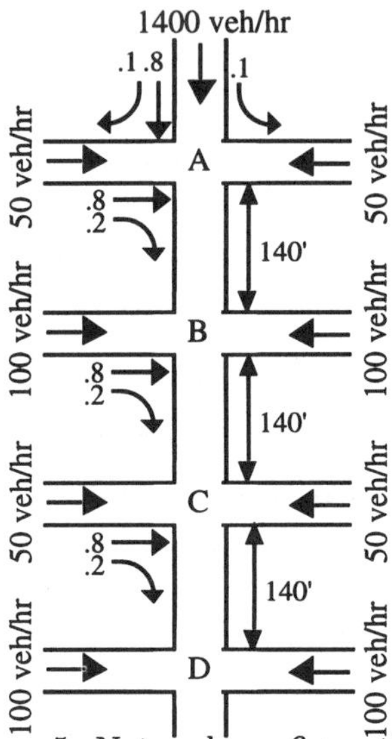

Figure 5: Network configuration 3.

There are six sources of vehicles entering the network. These sources randomly generate vehicles with an exponential distribution for the time between vehicles. The two sources on the north-south streets generate an average of 300 vehicles per hour, while the four on the east-west streets generate 100 vehicles per hour.

The intersections are arbitrarily labeled A-D. This ordering of the intersections is used when accessing the vector of values for a phase-dependent constant. The two phases of intersection A are the first two entries in this vector, then the two phases of intersection B, and so on. By convention, the north-south phase of an intersection precedes the east-west phase.

Figure 4 shows the second configuration. It contains two streets running north-south and two east-west, all two-way streets. The network is fully symmetric.

Figure 5 shows the third configuration. It contains one one-way street running north-south and four two-way streets running east-west.

Note that all three geometries contain four intersections. This was a good number for preliminary work, but to prove that our approach is practical requires that in future work we investigate larger networks and the scaling issues involved.

Also note that we have not yet considered a configuration where the average traffic flows vary with time. Since the ability to adjust for such variations is potentially the main benefit of adaptive control, experiments with time-varying flows are a high priority for future work.

Configuration	1	2	3
Population Size	1000	2000	5000
Evaluations	1000	2000	13000
Tree Score (Training)	22.62	540.7	20.13
Tree Score (Test)	21.81	558.0	20.99
Fixed Score (Training)	24.20	579.6	24.78
Fixed Score (Test)	24.83	640.2	28.76

Figure 6: Summary of experimental results.

3.2 The Results

For each configuration, we performed three experiments. The first experiment was to make a run with the approach described above to see what the best individual was and how well it performed. For all three configurations we used the functions in Figure 1, but only for the third configuration did we used the communications functions of Figure 2.

The second experiment was to use a genetic algorithm to optimize fixed-cycle timings to see how well a fixed-cycle approach could perform on the same problem. In this case, the individuals were real-valued strings of length twelve, three entries for each intersection. Two of the entries for each intersection were the phase splits (i.e., how long each phase stays green), while the third entry was the offset, which tells at what point in the cycle to start. The phase split entries could vary between 0 and 30, while the offset entries could vary between 0 and 99 (indicating the percentage of the cycle completed at the start). The population size was 1000 and the number of evaluations 10,000 for each run. The rationale for this experiment is to compare the adaptive control law we evolved with the best possible fixed-cycle control.

The third experiment was to make three runs with three different random seeds using the best individual from the STGP/GA run and then the best individual from the fixed cycles run. The rationale for this experiment is to see how well the two techniques generalize to test cases with the same geometry and flow statistics.

A summary of the runs is given in Figure 6. Note that these results clearly indicate that the adaptive control we evolve: (i) is superior to fixed-cycle control and (ii) generalizes well to new situations with the same statistics. Some other observations are the following. The good generalization of the trees was probably due to the simplicity of the trees. As more complex trees are required (for example, when we start looking at variations of the average flows), we expect generalization performance to drop unless we compensate with longer simulations. In two of the three cases, the best individual was found in the initial population, and evolution gave no benefit. This is often the case when using strong typing (which restricts the search space) and when the best tree is simple [Montana, 1995]. Finding more complex trees should require the full power of evolution rather than just a random search.

We now analyze the the best individual for each STGP/GA run. Note that we have reduced each of these individuals to their simplest form to simplify analysis.

Configuration 1: The best individual had tree

```
(OR (> CONSTANT-2 CONSTANT-1)
    (APPROACHING STOPPED))
```

and constants

```
Constant 1 = [ 27 10 11 16 14 12 20 13 ]
Constant 2 = [ 6 24 4 18 11 30 13 17 ]
```

Observe that the constant values for the north-south phases (the odd-numbered entries) are such that (> CONSTANT-2 CONSTANT-1) is false while for the east-west phases (even-numbered entries) (> CONSTANT-2 CONSTANT-1) is true. Hence, for the north-south phases the tree is equivalent to (APPROACHING STOPPED), while for the east-west phases the tree is always true. Therefore, the strategy is to always keep each signal green for the north-south street except when a car is detected approaching on the east-west street. This makes sense when we observe that (i) the traffic flows are light and (ii) the north-south streets are more heavily traveled than the east-west streets. Note that this is a good example of specialization using the phase-dependent constants.

Configuration 2: The best individual had tree

```
(AND (APPROACHING MOVING)
     (WAITING STOPPED))
```

This tree succeeds because it provides a good trade-off between allowing vehicles which are moving to maintain their momentum and not forcing vehicles which are stopped to wait too long. It may seem counterintuitive to wait until a vehicle is approaching a green light before changing to yellow until we remember that the sensor is only 120 feet from the intersection. Hence, not only the vehicle which triggered the sensor but also one or two vehicles behind it can make it through the intersection before the signal turns red. Because this configuration has a moderately heavy traffic load, if there is not currently a vehicle approaching the signal, there soon will be, often as part of a small group. Keeping the signal green long enough to allow a group through allows maintenance of momentum. There is coordination of effort (i.e., cooperation) because each traffic signal tends to hold vehicles to form them into small groups, which another traffic signal can exploit.

Configuration 3: The best individual had tree

```
(AND (APPROACHING STOPPED)
     (NOT (DS-WAIT-SAME MOVING)))
```

Observe that (DS-WAIT-SAME MOVING) is always false for all the east-west phases and the north-south phase of the southernmost intersection. Hence, for these five phases, the tree is equivalent to (APPROACHING STOPPED), which for the east-west phases is almost always true due to the large traffic flow on the north-south street. For the other three north-south phases, the term (NOT (DS-WAIT-SAME MOVING)) is for momentum preservation. The timing is such that keeping the phase green when the downstream intersection has cars waiting on the same street guarantees that the additional cars which are let through the intersection have no delay at the downstream intersection.

4 Conclusions and Future Work

The results are preliminary (due to the limited size and complexity of the problems) but encouraging. We were able to train our controller for a variety of geometric configurations and traffic conditions. The fact that our approach outperformed fixed timings shows that it is indeed adapting to variations around the average and hence fulfilling some of the promise of adaptive control. It was able to coordinate the individual traffic signals for aggregate goals, both with and without explicit communication. Finally, our approach allowed specialization, i.e. different phases of different intersections executing qualitatively and quantitatively different behaviors.

However, many key questions remain unanswered. These include: (i) how to make our approach handle a large number of intersections, (ii) how our approach handles variations in the average traffic flows, (iii) how our approach compares in performance with other adaptive control approaches, and (iv) how well a control strategy evolved for a simulation performs in a real network.

Acknowledgments

This work was supported under FHWA Grant DTFH61-94-X-00026. We would like to thank: Christina Andrews, who helped us get started with TRAF-NETSIM; and Alberto Castillo, who assisted with the experiments.

References

[Andre, 1994] D. Andre. Automatically defined features: The simultaneous evolution of 2-dimensional feature detectors and algorithm for using them. In K. Kinnear, editor, *Advances in Genetic Programming*, pages 477–494. MIT Press/Bradford Books, 1994.

[Arkin and Hobbs, 1993] R. C. Arkin and J. D. Hobbs. Dimensions of communications and social organization in multi-agent robotic systems. In *Proc. Second Int'l Conf. on Simulation of Adaptive Behavior*, pages 486–493, 1993.

[Chiu, 1992] S. Chiu. Adaptive traffic signal control using fuzzy logic. In *Proc. Intelligent Vehicles '92*, pages 98–107, 1992.

[Collins and Jefferson, 1992] R. J. Collins and D. R. Jefferson. AntFarm: Towards simulated evolution. In *Artificial Life III*, pages 579–601, 1992.

[FHWA, 1986] U. S. FHWA. TRANSYT-7F self-study guide, 1986.

[Gartner *et al.*, 1991] N. H. Gartner, P. J. Tarnoff, and C. M. Andrews. Evaluation of optimized policies for adaptive control strategy. *Transportation Research Record*, 1324:105–114, 1991.

[Haynes *et al.*, 1995] T. Haynes, R. Wainwright, and S. Sen. Strongly typed genetic programming in evolving cooperation strategies. In *Proc. Sixth Int'l Conf. on Genetic Algorithms*, pages 271–278, 1995.

[Koza, 1990] J. R. Koza. Genetic evolution and co-evolution of computer programs. In *Artificial Life II*, pages 603–629, 1990.

[Koza, 1992] J. R. Koza. *Genetic Programming*. MIT Press/Bradford Books, Cambridge, MA, 1992.

[Little *et al.*, 1981] J. Little, M. Kelson, and N. Gartner. MAXBAND: A program for setting signals on arteries and triangular networks. *Transportation Research Record*, 795:40–46, 1981.

[Mataric, 1994] M. J. Mataric. Learning to behave socially. In *Proc. Third Int'l Conf. on Simulation of Adaptive Behavior*, pages 453–462, 1994.

[Montana, 1995] D. J. Montana. Strongly typed genetic programming. *Evolutionary Computation*, 3(2):199–230, 1995.

[Nguyen and Huang, 1994] T. Nguyen and T. Huang. Evolvable 3D modeling for model-based object recognition systems. In K. Kinnear, editor, *Advances in Genetic Programming*, pages 459–475. MIT Press/Bradford Books, 1994.

[Reynolds, 1987] C. W. Reynolds. Flocks, herds, and schools: A distributed behavioral model. *Computer Graphics*, 21(4):25–34, 1987.

[Robertson and Bretherton, 1991] D. Robertson and R. D. Bretherton. Optimizing networks of traffic signals in real time - the SCOOT method. *IEEE Trans. on Vehicular Tech.*, 40(1):11–15, 1991.

[Ryan, 1995] C. Ryan. GPRobots and GPTeams - competition, co-evolution, and co-operation in genetic programming. In *AAAI Symposium on Genetic Programming*, pages 86–93, 1995.

[Sims, 1979] A. Sims. The Sydney Coordinated Adaptive Traffic system. In *Proc. ASCE Conf. on Computer Control of Urban Traffic Systems*, pages 12–27, 1979.

Distributed Genetic Programming — Empirical Study and Analysis —

Tatsuya NIWA niwa@etl.go.jp Distributed Systems Section

Hitoshi IBA iba@etl.go.jp Machine Inference Section

Electrotechnical Laboratory (ETL)
1-1-4 Umezono, Tsukuba city, Ibaraki, 305, Japan.

ABSTRACT

This paper introduces a distributed parallel implementation of GP (Genetic Programming), DGP. We implemented our DGP on the parallel system AP–1000+ of Fujitsu, which consisting of 32 processors. Several experimental results are shown comparing three architectures of DGP, i.e. the ring type, the one-way torus, and the two-way torus. We provide evidence for the effectiveness of our DGP by comparative experiments with the partitioned GP, i.e. a DGP without migration, and the canonical GP. In order to analyze the performance, we discuss the change of the structural diversity with generations for solving the deceptive problem.

1 Introduction

Tanese [1989] proposed the distributed genetic algorithm (DGA) as a way of efficiently parallelizing the canonical genetic algorithm (CGA). In the DGA, the global population is divided into several subpopulations, one per processor. Each processor runs the CGA independently on its subpopulation. Inter-processor communication occurs during the migration phase at regular intervals (i.e. migration interval). During migration, a fixed proportion of each subpopulation is selected and sent to another subpopulation. In return, the same number of migrants are received from some other subpopulation and replace individuals selected according to some criteria. This migration can occur either asynchronously or synchronously. Because the time-consuming measurement of fitness is performed independently at each separate processing node, this approach to parallelization delivers an overall increase in performance that is nearly linear with the number of independent processing nodes. Tanese found that the DGA achieved near-linear speedup on a 64-processor NCUBE/six hypercube computer. Belding [1995] compared the DGA's performance on the Royal Road class of fitness functions to that of the CGA. He found similar experimental results and drew a more general conclusion as to the DGA performance. Empirical studies and theoretical analyses were also reported in Manderick and Spiessens [1989] and Mühlenbein [1989].

The superiority of DGA over CGA can be summarized as follows:

1. Running time is expected to be shorter because of the smaller size of subpopulation.
2. New building blocks are available through migration.
3. Rapid evolution to a global fitness peak is possible because each subpopulation searches for a solution independently.
4. The DGA maintains diversity across the subpopulations avoiding premature convergence this way.

This paper introduces a distributed GP (Genetic Programming), i.e. a parallel implementation of GP. Koza [1992] proposed three levels of parallelizing GP; the fitness evaluation level, the individual level, and the running level. Since then, several parallel GP implementations have been proposed and studied. Koza and Andre [1995] presented a transputer-based system for parallelize GP. They implemented GP in the C programming language on a host PC 486 computer and a network of 64 transputers. They found that the GP parallelization delivered more than linear speed-up in solving the Boolean 5-parity function. Juillè and Pollack [1996] proposed a parallel implementation of GP for a SIMD architecture, in which each processor efficiently evaluates a different S-expression. They implemented this approach on a MasPar MP-2 computer, and showed experimentally the effectiveness.

This paper is motivated by the previous studies mentioned above and focuses on the following points:–

1. Confirming the superiority of DGA over CGA in various problems.
2. Comparing empirically several types of DGP.

3. Analytical study of DGP performance in terms of structural diversity.

The effectiveness of our DGP is shown by several comparative experiments.

2 Distributed GP

We implemented a parallel GP (DGP), on a parallel system AP–1000+, a MIMD parallel computer developed by Fujitsu (see Fig.1). The machine used consists of 32 cell processors, each of which has SuperSPARC (50MHz) chip with 16MB local memory. The AP–1000+ has several networks for cell/host communication. A torus-type network is used to communicate between cell processors, and a broadcasting network is used to communicate between the host processor and cell processors.

We have implemented the DGP algorithm described below:–

1. The global population is divided into several subpopulations. One processor is allocated to each subpopulation in which a canonical GP (CGP) is executed.
2. At every generation, the following communication process occurs for each subpopulation:
   Sending Phase The best individual of a subpopulation is sent to its adjacent subpopulations.
   Receiving Phase If the fitness of the individual sent to a subpopulation is better than the best individual of the subpopulation, receive the individual and replace the worst individual with it.

We have chosen to implement sgpc1.1, a simple GP in the C language, as CGP. Considering the physical connections of AP–1000+, we introduced the following adjacency of 32 subpopulations (see Fig.2):

Ring type All processors are linked in a circle. This type of connection is denoted as "ring" type.

One-way Torus All processors are positioned on a 8×4 torus. They are connected toroidally and send messages in two directions, i.e. rightward and downward. This type of connection is denoted as "one-way torus".

Two-way Torus All processors are positioned on a 8×4 torus. They are connected toroidally and send messages in four directions, i.e. rightward, leftward, upward and downward. This type is denoted as "two-way torus".

Timing of message exchange is essential for parallelizing GP. Since CGP is executed separately in each subpopulation, the processing time per generation naturally differs from subpopulation to subpopulation. By the synchronized method, i.e. by exchanging messages between subpopulations synchronously at every generation, a subpopulation which has finished processing one generation has to wait for the other subpopulations which have not yet finished. This causes delays for the entire parallel GP. Therefore, we have chosen to use the asynchronous method for our parallel implementation of GP. The OS of the AP–1000+ supports asynchronous message exchange, which makes it easier to implement the asynchronous parallelization of GP.

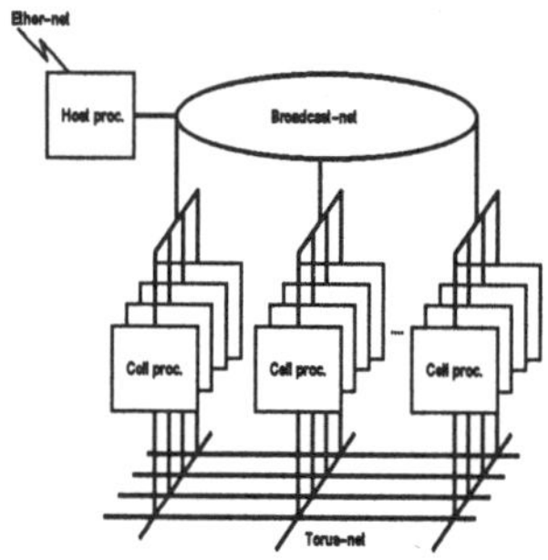

Figure 1: AP–1000+ Architecture

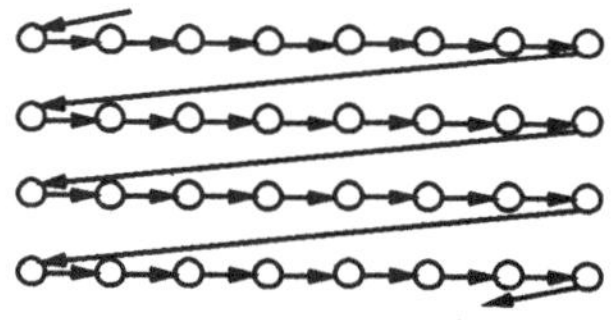

(a) Ring Type

(b) One-way Torus Type

(c) Two-way Torus Type

Figure 2: Connection Type

3 Experimental Results

This section studies empirically the performance of several types of parallel GP. The total population size is set to 3200, which is divided into 32 subpopulations of size 100. We conduct experiments in running three versions of DGP, i.e., the ring type, the one-way torus, and the two-way torus. For the sake of comparison, we also experiment in running the partitioned GP, i.e. a distributed GP without migration, and in running the canonical GP for the single population of size 3200 (i.e. panmictic GP). We have chosen sgpc1.1, a simple GP in the C language, as the canonical GP. The parameters used for this experiment are shown in Table 1.

GP is applied to the following problems:

Experiment 1 Discovery of trigonometric identities. Koza [1992; ch.10.1] used GP to find a new mathematical expression, in symbolic form, that equals a given mathematical expression, for all values of its independent variables. In the following experiment, our goal is to discover trigonometric identities; $\cos 2x = 1 - \sin^2 x$.

Experiment 2 Predicting a chaotic time series.
The Mackey-Glass time series is generated by integrating the following delay differential equation and is used as a standard benchmark for prediction algorithms:–

$$\frac{dx(t)}{dt} = \frac{ax(t-\tau)}{1+x^{10}(t-\tau)} - bx(t), \qquad (1)$$

Parameters of sgpc1.1 (for subpopulations)	Problem Name			
	cos(2x) (Exp.1)	Chaos (Exp.2)	4-parity (Exp.3)	Deceptive
Terminal Symbols	{X,1.0}	$\{X(t-1), X(t-2), \cdots, X(t-10), \Re\}$	{x0,x1,x2,x3}	{X,Y,Z}
Nonterminal Symbols	$\{+, -, *, \%, \sin\}$	$\{+, -, \times, \%, \sin, \cos, \exp_{10}\}$	{AND, OR, NAND, NOR}	{FOO,BAR,BAZ, NOP_1,NOP_2,NOP_4}
population_size	100	100	100	100
steady_state	0	0	0	0
max_depth_for_new_trees	6	6	6	3
max_depth_after_crossover	6	17	17	6
max_mutant_depth	4	4	4	3
grow_method	RAMPED	GROW	GROW	GROW
selection_method	FITNESSPROP	TOURNAMENT	FITNESSPROP	FITNESSPROP
tournament_K	6	6	6	6
deme_search_radius_sigma	1.0	1.0	1.0	1.0
crossover_func_pt_fraction	0.2	0.1	0.1	0.1
crossover_any_pt_fraction	0.2	0.7	0.7	0.7
fitness_prop_repro_fraction	0.1	0.1	0.1	0.1
parsimony_factor	0.00000	0.00000	0.00000	0.00000

Table 1: GP Parameters

with $a = 0.2$, $b = 0.1$ and $\tau = 17$. The trajectory is chaotic and lies on an approximately 2.1-dimensional strange attractor.

For the sake of comparison, all the parameters chosen were the same as those used in a previous study of Oakley [1994; p.380, Table17.3], except that the terminal set consisted of ten past data for the short-term prediction (see Table 1).

Experiment 3 Boolean concept formation (4-parity). To show the effectiveness as a Boolean concept learner, we conducted a simple experiment (4-parity), in which the goal function is the even parity function of 4 variables; $f(x_1, x_2, x_3, x_4) = x_1x_2x_3x_4 \vee \overline{x}_1\overline{x}_2x_3x_4 \vee \overline{x}_1x_2\overline{x}_3x_4 \vee \overline{x}_1x_2x_3\overline{x}_4 \vee x_1\overline{x}_2\overline{x}_3x_4 \vee x_1\overline{x}_2x_3\overline{x}_4 \vee x_1x_2\overline{x}_3\overline{x}_4 \vee \overline{x}_1\overline{x}_2\overline{x}_3\overline{x}_4$. Figs.3(a–1)(b–1)(c–1) plot the averaged error rates (i.e. standard fitnesses) versus generations. The average was taken over 20 runs. As can be seen in the figure, the performance difference is much more remarkable for symbolic regression (i.e. discovery of trigonometric identities) and the Boolean function generation (i.e. 4-parity). From these experimental results, the GP performance is ranked as follows:

$$\text{partitioned} < \text{panmictic} < \text{ring} \approx \text{one-way torus} \approx \text{two-way torus}$$

Note that $A < B$ means that B is superior to A. Figs.3(a–2)(b–2)(c–2) plot the averaged error rates versus the execution time. The curves for parallel GP are cut off in the middle of the graph because every processor reached the maximum generation (i.e. 100) at that execution time. The panmictic model, i.e. a single population of 3200 size, took quite a long time to reach a better solution so that the curve is omitted in this figure. There seems to be no obvious performance difference in three types of parallel GP. They all gave better results than the partitioned type. These experimental results show that it pays to parallelize GP at the cost of communication overhead between processors.

4 Structural Diversity of GP

Structural diversity describes the frequency with which unique building blocks are formed as evolution proceeds. In GP, building blocks (or schemata) are defined to be expressions or subexpressions in the form of tree structures. Let $TR(t)$ be the set of expressions or subexpressions at the generation t. $TR(t)$ is given by the enumeration method. Then the structural diversity $SD(t+1)$ is derived from the following equation introduced by Tackett [1995]:

$$SD(t+1) = \frac{|\, TR(t+1) \cap \overline{TR(t)} \,|}{pop\_size}. \quad (2)$$

Thus, the structural diversity is obtained by dividing the number of new schemata produced each generation by the population size. The population with large SD value is expected to maintain higher diversity and to prevent premature convergence.

In pursuit of structural diversity, we experiment in solving a constructional problem introduced by Tackett [1995]. The terminal and nonterminal sets are given as in Table 1, where FOO,BAR,BAZ,NOP_2 are 2-argument functions, NOP_1 is a 1-argument function, and NOP_4 is a 4-argument function. This problem uses tree templates T_m $(m = 1, \cdots, M)$, to which fitness values w_m are assigned. They include special symbols used for matching. The fitness of an individual is the maximum fitness of the tree templates which match the individual. If matching does not occur, the fitness is set to zero.

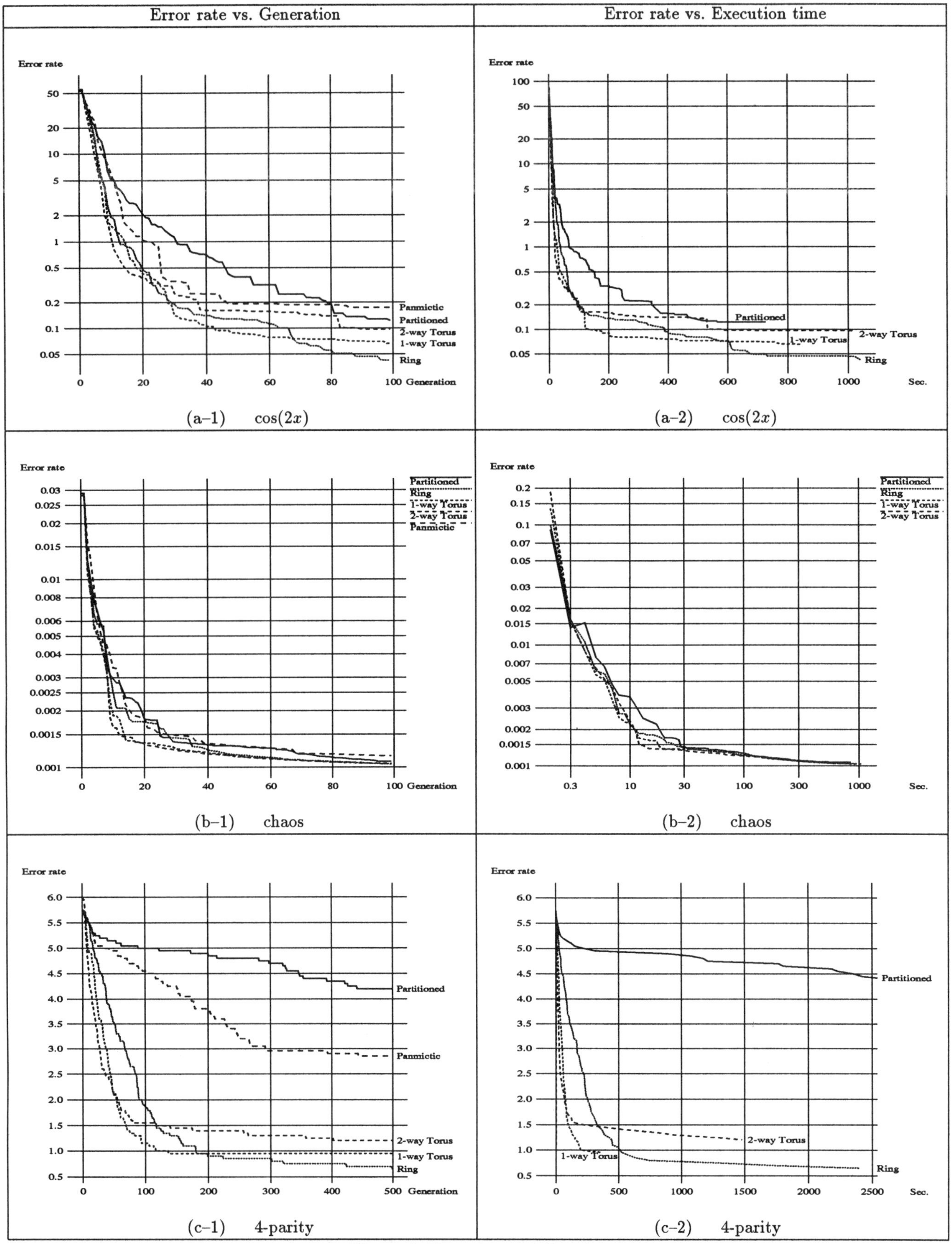

Figure 3: Experimental Results

Connection Type	Avr. of Suc. Gen.	No. of Suc.	No. of Runs	Struct. Div. Avr.	Struct. Div. Std. dev.
Partitioned	221.29	14	20	1.08	0.25
Ring	53.55	20	20	1.13	0.36
1-way T.	34.50	20	20	1.28	0.31
2-way T.	21.15	20	20	1.41	0.39
Panmictic	115.69	13	20	0.79	0.25

Table 2: Experimental Results (Deceptive Problem)

The goal of this problem is to construct the expression `(BAZ (BAR Z X)(FOO X Y))`. This expression is assigned the maximum fitness value, 1.0. A tree matching (i.e. including) `(BAR Z X)` and `(FOO X Y)` is assigned a fitness value of 0.125. The fitness of a tree matching `(BAZ $ANY (FOO X Y))` and `(BAZ (BAR Z X) $ANY)` is 0.25. `$ANY` is a wild card matching any expression. The two problems introduced by Tackett differ in the fitness assignment of `(BAZ (BAR Z X)(FOO X Y))`. For the Royal Road problem a value of 0.5 is assigned, while for the deceptive problem a value of 0.125 is assigned. In the former case intermediate solutions are assigned fitness value which increase monotonically, so as to provide stepping stones to the solution. In the latter case, there is a fitness valley between intermediate solutions and the final solution. Refer to Forrest and Mitchell [1993] Goldberg [1989] for more details of Royal Road and deceptive functions in GA.

We conducted an experiment in solving the deceptive function. The parameters used are given in Table 1. For this problem, we have used 9 processors (3×3 for the torus model) instead of 32, because of tracing the structural diversity. The result is shown in Table 2. The table clearly shows the superiority of DGP over CGP or the partitioned GP. Fig.4 illustrates the structural diversity for typical runs of the five types of GP, typical run means one whose terminal generation (i.e. the generation at which a solution is acquired) is close to the averaged generation given in Table 1). The horizontal axis represents the generation. The curves are terminated at the terminal generations, i.e. 363 for the partitioned GP, 54 for the ring, 34 for one-way torus, 28 for two-way torus, and 81 for the panmictic model. Fig.4(b) shows the enlarged part of the first 80 generations of Fig.4(a). The vertical axis represents the 9 processors respectively, except for the panmictic model (Fig.4(f)), in which the diversity of the unique processor is shown. In the figure, the gray level is proportional to the structural diversity. In other words, the whiter, the higher structural diversity, and the blacker, the lower structural diversity. As can be seen in the figure, compared to the partitioned GP, the white areas are relatively dense for three types of parallel GP, which means structural diversity is higher. On the contrary, the black areas are expanding (i.e. the diversity is becoming lower) with generations for the panmictic model so as to degrade the GP search, i.e. fall into the premature convergence. The averages and standard deviations of the structural diversity are listed in Table 2.

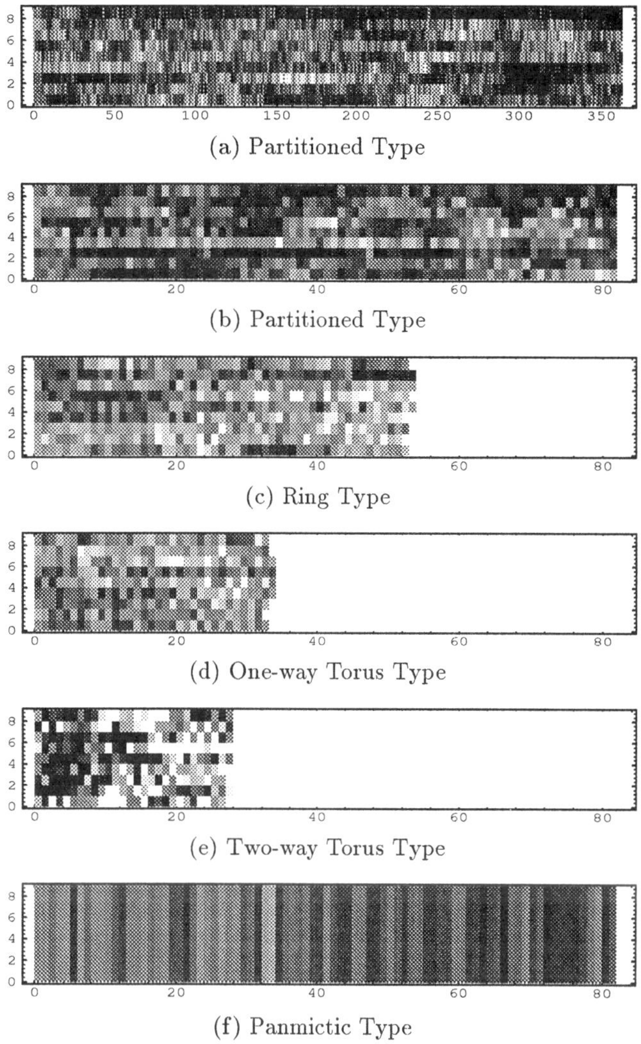

(a) Partitioned Type

(b) Partitioned Type

(c) Ring Type

(d) One-way Torus Type

(e) Two-way Torus Type

(f) Panmictic Type

Figure 4: Diversity

5 Discussion

The previous experiments provide evidence for the superiority of the distributed GP. With these results, we can attribute the success of the distributed GP to the following points:

1. A single population is liable to fall into a local solution and to degrade the search efficiency.
2. Dividing a population into smaller subpopulations makes it possible to prevent premature convergence to a local solution.
3. Exchanging and recombining different local solutions between subpopulations enable escape from local.

Koza and Andre [1995] also proposed a different version of DGP and implemented it on a transputer-based system. The difference between the two approaches lies in the following points:

1. In our DGP model, one individual is migrated, whereas various immigration rates were tested in Koza's DGP.
2. We accept an immigrant only if its fitness is better than the best fitness of the subpopulation to which it is sent. On the other hand, in Koza's DGP all immigrants are accepted.
3. Various iteration intervals were examined in Koza's DGP, whereas immigration occurs every generation in our model.

There are many ways to design a parallelized version of GP. In general, the hardware architecture constrains the implementation model of the parallelization. Our design principle is derived from the narrow communication aisle in AP–1000+. Koza empirically suggested 8% migration rate for solving Boolean function generation problems. Increasing the migration rate does not necessarily lead to performance improvement. We have conducted an experiment in running Koza's DGP for comparison. However, we have not observed any significant performance difference between the two approaches. This paper does not intend to express the superiority of our DGP over the other models. The comparative studies of different parallelized GP are our research concern in the near future.

The structural diversity is an essential index for analyzing the performance of GP. However, it is not a strict index for the following reasons:

1. The fitness of schemata is not included in the definition so that it does not necessarily reflect the number of useful building blocks.
2. Diversity across different subpopulations is not considered. For instance, a novel tree in a subpopulation may not be novel in another subpopulation.

A more precise description and theoretical analysis of schemata in GP are given by O'Reilly [1995]. We are now investigating more precisely indices for analyzing GP performance.

6 Conclusion

We implemented a DGP on a parallel system AP–1000+, which consists of 32 processors. The experimental results are summarized as follows:

1. Subpopulations even with a slight message exchange resulted in much better results than those without any exchange, i.e. the distributed GP is superior to the partitioned GP.
2. The distributed GP gave a better performance than the canonical GP, whose population size is equal to the total of the subpopulation sizes.
3. The appropriate amount of message exchange is problem specific.

We analyzed the performance of DGP in terms of structural diversity. This paper is a first step toward the analytical study of the parallelized GP performance.

Acknowledgments

This work was done as a part of Real World Computing Partnership (RWCP). We are grateful to Otsu, N. and Tsukamoto, M. for giving us the opportunity to use the AP–1000+ parallel computer. We thank Tackett, W.A. for providing his "sgpc1.1", which we used to implement our GPs, and also thank Manderick, B. and Connelly, J.C. for helpful comments on writing the paper.

References

[Belding,T.C., 1995] The Distributed Genetic Algorithm Revisited, in *Proc. of the 6th International Conference on Genetic Algorithms*, pp 114–121.

[Forrest,S. and Mitchell,M., 1993] Relative Building-Block Fitness and the Building-Block Hypothesis, in Whitley,L.D. (ed.), *Foundations of Genetic Algorithms 2*, Morgan Kaufmann, pp 109–126.

[Goldberg,D.E., 1989] Genetic Algorithms in Search, Optimization, and Machine Learning, Addison Wesley.

[Juillè,H. and Pollack,J.B., 1996] Massively Parallel Genetic Programming, in Angeline,P. and Kinnear,K.E.Jr. (ed.), *Advances in Genetic Programming II*, MIT Press.

[Koza,J.R., 1992] Genetic Programming, On the Programming of Computers by means of Natural Selection, MIT Press.

[Koza,J.R. and Andre,D., 1995] Parallel Genetic Programming on a Network of Transputers, Computer Science Dept. Stanford University, January 30.

[Manderick,B. and Spiessens,P., 1989] Fine-Grained Parallel Genetic Algorithms, in *Proc. of the 3rd International Conference on Genetic Algorithms*, pp 428–433.

[Mühlenbein,H., 1989] Parallel Genetic Algorithms, Population Genetics and Combinatorial Optimization, in *Proc. of the 3rd International Conference on Genetic Algorithms*, pp 416–421.

[Oakley,H., 1994] Two Scientific Applications of Genetic Programming: Stack Filters and Non-Linear Equation Fitting to Chaotic Data, in Kinnear,K.E.Jr. (ed.), *Advances in Genetic Programming*, MIT Press, pp 360–389.

[O'Reilly,U-M. and Oppacher,F., 1995] The Troubling Aspects of a Building Block Hypothesis for Genetic Programming, in Whitley,L.D. and Vose,M.D. (ed.), *Foundations of Genetic Algorithms 3*, Morgan Kaufmann, pp 73–88.

[Tanese,R., 1989] Distributed Genetic Algorithms, in *Proc. of the 3rd International Conference on Genetic Algorithms*, pp 434–439.

[Tackett,W.A., 1995] Greedy Recombination and Genetic Search on the Space of Computer Program, in Whitley,L.D. and Vose,M.D. (ed.), *Foundations of Genetic Algorithms 3*, Morgan Kaufmann, pp 271–297.

Programmatic Compression of Images and Sound

Peter Nordin and Wolfgang Banzhaf
Universität Dortmund LS11, Fachbereich Informatik, Dortmund, Germany
nordin,banzhaf@ls11.informatik.uni-dortmund.de

Abstract

The importance of digital data compression in the future media arena cannot be overestimated. A novel approach to data compression is built on Genetic Programming. This technique has been referred to as "programmatic compression". In this paper we apply a variant of programmatic compression to the compression of bitmap images and sampled digital sound. The work presented here constitutes the first successful result of genetic programming applied to compression of real full size images and sound. A compiling genetic programming system is used for efficiency reasons. Different related issues are discussed, such as the handling of very large fitness case sets.

1 Introduction

Programmatic compression is a very general form of compression. The basic idea behind this technique is that any system, which evolves programs or algorithms for generating data, can be viewed as a data compression system. The data that should be compressed are presented to the Genetic Programming system as fitness cases for symbolic regression. After choosing a function set that facilitates an accurate reproduction of the uncompressed data, the system then tries to evolve an individual program that, to a certain degree of precision, outputs the uncompressed data. If the evolved program solution can be expressed by fewer bits than the target data, then we have achieved a compression. In principle, this technique could address any compression problem with an appropriate choice of the function set and the method has a potential for both lossy and lossless compression. The initial studies performed by Koza [1]in the domain of very small bitmap images displays the difficulty of the problem and the huge computational effort needed for real world examples. We have applied a Compiling Genetic Programming System (CGPS) to the programmatic compression problem. A CGPS is up to 2000 times faster than a LISP Genetic Programming System (GPS) [5],[6], which enables the application of programmatic compression to real-world problems.

1.1 Compiling Genetic Programming

A CGPS is a GP system where all genetic manipulation of individuals is performed in binary machine code. Almost all computers today are of Von Neumann type, which means that programs and data reside in the same memory. Program information can thus be regarded as just another sort of data structure and it can be manipulated as any other data, in this case by the genetic operators. This arrangement enables the deletion of all interpreting steps, which results in a speed up of up to 2000 times compared to a LISP system, and 100 times compared to an interpreting C implementation. Some experiments described below consumed 10 CPU-days of execution. This kind of experiments would not be feasible with other types of GPSs. A LISP version of such an experiment could take more than 50 years while an interpreting C version still would need two years of execution time. These compression examples thus are real-life experiments and research, only feasible with a CGPS. For a more detailed description of the CGPS and comparison with other implementation methods, see [5],[6],[8].

2 Programmatic Compression (PC)

Programmatic image compression is briefly mentioned in [1]. Koza uses the 900 pixel values of a small 30x30 bitmap of a regular mathematical pattern as fitness cases for symbolic regression by a GPS. The system finds a good solution in less than 100 generations with a generation size of 2000 individuals and a function set consisting of arithmetic operators. This method has an important relation to the theoretically maximal compression achievable. If a Turing-complete function set is used then the shortest program that accurately produces the target data could be seen as one estimation of the Kolmogorov Complexity [3] which opens possibilities of a deeper theoretical analysis of the application. The GPS has at least the potential of finding this shortest possible compression of the data.

In our work we have used two real-world applications which both call for the use of very large fitness case sets. In the experiments described below we apply programmatic compression to bitmap images and sampled sound. The large potential of these two fields in the growing digital media arena cannot be overestimated.

We examine the compression in several different ways for instance with different function sets, fitness evaluation methods and parameter settings.

The majority of experiments were performed with a simple function set consisting of the most usual low level machine code operations such as arithmetic and logic operations. It is surprising that the most efficient search runs were performed using these machine instructions as the function set, regardless of application. These results strongly support the universal applicability of GP. Sometimes critique has been heard that GP only would manage to solve different problems because there was a specially adapted function set corresponding to the problem domain. In most of our runs we have used a function set that is not decided by us but by the manufacturer of a certain computer. The same set of operators was used for the completely different domains of sound and images, suggesting that it is *not* the choice of function set that is primarily responsible for the success of GP.

The advantage of programmatic compression is the large flexibility possible with a programming language as output of a compression process. This flexibility implies a possible larger compression factor sometimes approaching the theoretical limit. The use of machine code instructions makes decompression very fast. The decompression algorithm is simply a small and efficient binary machine code program. Decompression rates of 150 million bits/second per processor are sometimes possible on a SUN 20. This figure definitely exceeds the demand of, for instance, real-time full size video.

Another advantageous feature of the system is that each element of the target data (sample or pixel) can be accessed independently of the values of other elements. For instance, a sample with a certain index number can be obtained without the need to calculate the values of samples before and after that index number.

2.1 Chunking

The simplest and most obvious method to be applied when using PC is to present the target data as a continuous sequence of fitness cases. The evolved program then tries to reproduce the entire list of data when executed. While uncomplicated and elegant, this method has some disadvantages when used with very large fitness case sets. If the target data are very complex, there is a risk that an evolutionary search will not converge to a solution with acceptable quality. It is also harder to predict how many generations are needed before a solution is found. Therefore, we have used two different systems applied to each of the two domains sound and images. The first system treats all the fitness cases at the same time. The other system applies programmatic compression to equally sized sub-sets of the fitness cases and evolves a solution to each of them. Below we refer to this method as *chunking*. A time limit is imposed on each of the chunks which keeps the overall conversion time under control. The disadvantage of chunking is that the compression ratio could be lower since similarities between chunks will not be expressed in the individuals.

Chunking for sound is done by letting a CGPS evolve a solution for a fixed size sample chunk typically below the resolution time of the ear (70ms). In the Image domain the pictures were divided into small quadratic blocks, 8x8 or 16x16 pixels each.

The system was in this way presented with fitness case sets of sizes from 32 to 100000 integers. The largest fitness case set corresponded to 10 seconds of sound or a 256x256x8bit pixel image.

2.1.1 Two methods for long term GP-memory

In order to speed up the search in the chunked data approach two different "memory methods" were introduced. The first method *kept* the entire generation between chunks which speeded up the search when similar chunks were adjacent. The second method made a few individuals "read-only" in every generation. They could take part in reproduction and crossover but only as parents. They were never allowed to be over-written by any offspring. For every generation, less than 10 individual were frozen, to help maintain diversity and to save valuable genetic code for future chunks. The second method has the advantage that genetic information can be reused between non-adjacent chunks. In this way, the GPS is given an association capability that enables it to use knowledge obtained from a similar problem earlier in time. The results of this experiments show that these mechanisms can save considerable amounts of search time while the best solutions found tend to converge to a slightly worse fitness. The tradeoff between search time and quality makes the method worth further investigation.

2.1.2 Multi-level Compression

When this method is used in practice it is advisable to compress the resulting machine code segments further. In most experiments only half of 32 bits in the instructions were used in the programs, the rest of the instructions did not have any relevance in a CGPS. This suggests that the instructions in the result can be further compressed by a factor of two. In addition the resulting "16-bit instructions" can be ordered according to their frequency and then Huffman-coded. This method will have only a marginal effect on the overall decompression speed, because decompression of programs only has to be performed once per chunk.

Note that introns will be removed before packing the programes. The introns will be identified with the method described in [2].

3 Compression of Sound

Our first experiments with PC concerns sound generating programs. The difficulty in generating and predicting large and complex time series has previous been reported by Oakley [4]. In this case, the time series consists of sampled sound. The data are produced by the built in sampler in a SUN workstation, which produces sound samples with 8 bit resolution and 8kHz sampling frequency. A recording of a one second sound, is thus 8000 bytes long and corresponds to 8000 fitness cases for the CGPS.

3.1 Sound Generating Program architectures

There are several program architectures possible for individuals for time series problems. The first obvious individual structure is a function without inputs that has the ability to store values as side-effects. The individual is placed in a loop, the variables initialized to zero and then output of the program only depends on what is stored by the program in these variables during the cycles of the loop. The variables can be accessed as a stack, as fixed variables or as index memory. In a similar way the individual can be directly fed by its previous output. A different approach is to feed the individual programs with the index number of the fitness case and expect it to compute the output only from this index number. A combination of side effects and index number is also possible. Other architectures of the individuals are also possible. The individual could also be fed with recent fitness case output, but most of these methods are not suitable for efficient compression, because you have to supply output values. This method, however can be used for time series prediction into the near future. Table 1 summarizes the parameters used during training.
Of these approaches the index version without side-effects gave the best results during evolution of sound. An integer is thus given as input to the individual program and the output is taken to be a single sound sample. The integer index is subsequently incremented and fed to the same individual, which then produces the next sample. In its simplest case, the fitness is just the sum of the absolute values of the differences between the actual sound samples and generated samples. There are many other methods to establish fitness values and as we have experienced, they can have a considerable impact on the quality of the evolved results.

3.2 Fitness measurements

Oakley has previously reported the importance of choosing an appropriate fitness measurement when evolving a chaotic and oscillating target function [4]. He observed that when fitness was measured as the difference

Terminal set :	Integers 0-8192
Function set :	ADD, SUB, MUL, SHL, SHR, XOR, OR, AND desired value
Maximum population size :	100 - 100000
Crossover Prob :	90%
Mutation Prob :	5%
Selection :	Tournament Selection
Maximum number of nodes:	16-1024

Table 1: Summary of parameters used during training.

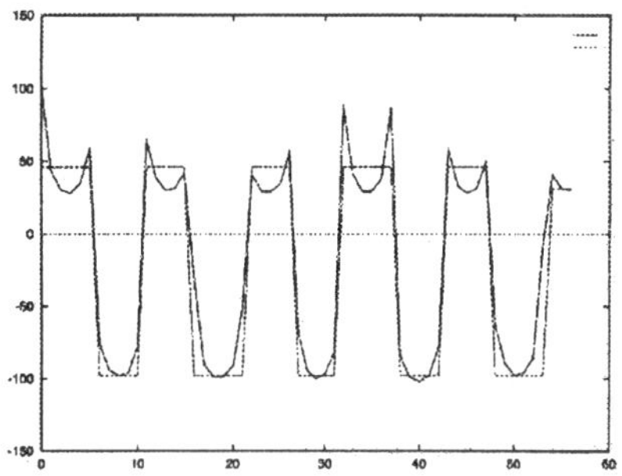

Figure 1: Adaption of generated sound to target sound (GP output is dotted)

between the target output and the actual output, the system would only converge to a constant output. The constant output represented the mean value of the target function. When we used the fitness measure of absolute value of differences, we noticed that for many initial generations, sometimes more than 100, the system displayed this behavior. However, a constant value is not of much use when the target is to evolve a sound. In the long run though the system always converged to a more accurate, oscillating output. Figure 3.2 shows the output of a program after the oscillation in the output has been adapted to a close fit of the target samples[1]. The original sound in this case is a sampled two-tone door bell. We can here see, how the output represented by the dotted line shows a very good adaption to the target function's frequency, amplitude and phase. A view of the entire sample also shows an adaption to the envelope, that is the long term changes in amplitude. However, the waveform in the output is simplified to a square wave. This behavior seemed hard to change by changing the parameters and varying the function set. When we changed the fitness function to the squared difference instead of the absolute value of the difference the result changed significantly. Not only did the convergence to an os-

[1]This figure, like most figures below, shows an excerpt of a longer sequence of samples. The details of the wave form of the full sequence is rather irregular, which might explain some irregularities in the output. It is impossible to show the full sequence in one diagram. If the training sequence was restricted to only the short sample spaces shown in the figures an even closer convergence would have been expected.

cillating output happen earlier, but the resulting curve also displayed much more details. We also tried fitness differences raised to three and raised to four without significantly better results. The crucial point could be that non-linearity is needed.

The results with the squared error both looked and sounded much better, but the details of the waveform in the generated output lacked the extreme peaks of the target sample. To increase the selection pressure towards accurately reproducing these pointed features we tried with a difference between two adjacent points as the fitness value. This fitness measure did not work well when adopted alone but when it was in a weighted sum with the normal absolute value fitness it displayed a more pointed behavior.

All of the fitness methods above have the disadvantage that they do not take into account the way the ear perceives sound. The output can look quite similar to the samples and there could still be a lot of audible distortion. The evolved sounds often had metallic overtones and sometimes noise added to it. To some extent this could probably be filtered out with a filter similar to an anti-aliasing filter, but the fact remains that there is a difference between looking similar and sounding similar.

The human inner ear could be considered to perform a transformation similar to a Fourier transform of the sound we hear. This is the reason why we perceive tones and not quickly changing sequences of pressure variations. This mechanic Fourier transform is necessary because the 'clock frequency' of the brain is not fast enough to process sound directly.

Another fitness measurement would thus be a comparison with the Fourier transformed spectra of the generated samples and the target functions. Oakley [4] used Fourier transforms as a means to force the system away from local minima of constant output. Our system showed oscillation without the Fourier transform but we instead applied it to improve sound quality by aligning fitness function judgment of the output with that of the human ear.

We thus implemented a fast Fourier transform (FFT) taking real-valued data and transforming them into the frequency domain represented by complex numbers. Fitness is measured as the squared difference between the transform of output and the transform of the target samples. Both the real and imaginary part of the transform were separately taken into account by the fitness function in order to assure that the phase of the output was treated appropriately.

The sound produced by the FFT fitness system was audibly better. The CPU time needed for this system was about twice that of the system without FFT, which was reasonable considering the quality improvement.

When FFT is applied in the fitness function then the exact amplitude of the curve is less important, instead

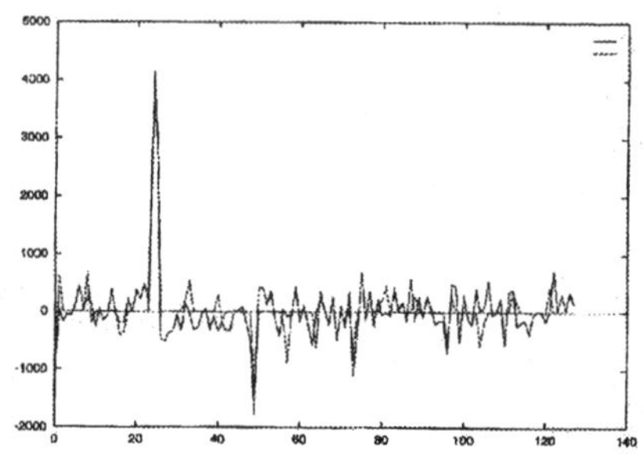

Figure 2: FFT spectrum of the sample(GP output is dotted)

features that influence the overtones of the system are dominating. Figure 3.2 shows the spectrum of the same sample. An external function in a CGPS is a function that is not part of the processor's instruction set and has to be used by a call instruction. Any C-function can be linked to the system and used as an external function.
The function set in the experiments presented above was eight low-level machine code instructions: addition, subtraction, multiplication, shift left, shift right, logical and, logical or and logical exclusive or.
Different experiments where also performed with the use of side effects in functions. The programs were given the possibility to store values in memory between the processing of the samples. In the extreme case the individual was not provided with an index number for the sample that should be generated instead the individual had to do its own book keeping with the use of memory variables. These programs took longer time to evolve and performed worse in fitness but had a softer sound with less overtones. Side effects were introduced as Save/Restore functions,the Swap function and indexed memory of different sizes.
The first reflection over sound generating programs is that it would be advantageous to use functions related to oscillation in the function set. We therefore tried to use the sinus function directly in the function set as well as functions that are components of a discrete differential equation for oscillation and thus together could produce a variety of different curves. To our surprise did none of these functions perform better than the individuals built with the simple machine code building blocks.
Automatically defined subfunctions (ADFs) [2] were also used without significantly improving results.

3.3 Evolving Large Samples Sequences

As mentioned above two methods were tried for evolution of large samples sequences. The simple linear method that tried to evolve the sample with one individual proved insufficient for large and complex sound. This method also had the disadvantage that its convergence behavior and evolution time requirements were

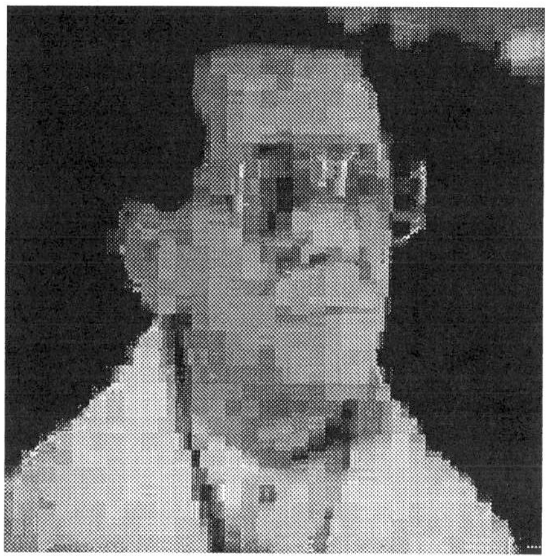

Figure 3: Target bitmap with fast evolution

Figure 4: An example of output from the programmatic compression system

hard to predict. It was possible to evolve the frequency, amplitude, phase, envelop and, to some extent, the wave form of a 16.000 samples sound lasting two seconds, if the sound was regular. However, to evolve changes in frequency in a sample, turned out to be harder. We succeeded to generate two different tones, that change at a specific point in the sample frequency, but we did not have any good results with more complex sounds. The more random output from earlier generations sometimes expressed complex melodies, but it did not evolve such melodies corresponding to a specific fitness case set. Even more complex sounds, like spoken words were impossible to mimic with this linear method and with sample length of audible size.

The solution we use instead for evolution of longer and more complex sounds is chunking. The sample sequence is divided into equally sized sequences and the CGPS in then applied to each chunk in turn. The chunk size is chosen so that the length of a chunk is below the resolution of the human ear. Pulses that come faster than 70ms are perceived as tones and not individual pulses. The fact that the evolved sound always is in phase assures that the sound pieces fit together well in the joint of two chunks. With this method it is possible to evolve complex sound lasting for a longer time. We have for instance evolved human speech with an understandable quality. Melodies of 100000 samples have also been evolved. In future work we would like to investigate the possibility of evolving the division points between chunks allowing for adaptive and varying chunk lengths.

3.4 Results

Compression ratios vary much depending on the regularity of the target sound. For regular sound the compression ratio can easily be 10 times or more, while the ratio for speech is around two times. The quality is still below telephone quality but we will in the future investigate several methods to improve it.

The CPU time for generating a ten second chunked sample is in the range of tens of hours on a SUN 20.

4 Compression of Pictures

Our system for programmatic compression of images bears many similarities with the sound compressing system. Instead of trying to evolve a one dimensional array of samples we here try to evolve a two dimensional array of pixels. In our experiments every pixel holds 8 bits of information. In the experiments we have used gray scale images because of the technically more complex handling of color in the display palette. The technique is however equally applicable to color pictures and to pictures with more than 8 bit quantification, i.e. 24 bit color. The output of the evolved individuals consists of eight bit numbers. If very fast decompression is required, it is possible to use the full 32 integer bits of the architecture. In this way more bits per second can be decompressed.

4.1 Program architecture

Basically the same method is used for images and sound. As a first method an individual is fed by the X and Y coordinates of the pixel and the output is interpreted as the value of the pixel. However a simpler method proved to give better results. Instead of supplying both the X and Y coordinates to the programs, we supplied a *one dimensional* index value of the pixel.

Chunking is a must when working with realistically sized images. In our experiments we have used pictures consisting of 256x256 pixel which were divided into squares of either 16x16 or 8x8 pixel. The definitely best results were achieved with 8x8 pixel blocks. This size corresponds to the size used by several other image compression techniques such as JPEG [9]. Different methods for side effect were also used, and unlike the sound example the pictures had a slightly better quality when the

Figure 5: The target bitmap

programs also had access to a small set of memory cells.

Figure 4.1 shows an example of an evolved image that used figure 4.1 as target data. Figure 3.2 shows a quickly (one hour) evolved picture with lower quality .

4.2 Fitness functions

Different fitness measurements were also applied, and similar to the sound domain, the squared error differences gave better results than the absolute value.

In future work we would like to use two dimensional FFT in the fitness function, as well as functions measuring differences between adjacent pixels in several directions.

The blocks of chunking are visible to a varying degree in the decompressed pictures. By incorporating the differences in color around the edges of blocks into the fitness function we hope to reduce this effect. We will also try different conventional smoothing techniques used i.e. with JPEG for the same purpose.

The CPU times for compressing a picture in our experiments on a SUN20 are in the range of 30 minutes to 10 days depending on the requested quality and compression ratio.

5 Summary and Conclusion

We have described experiments that have shown that programmatic compression can be used with other than toy problems in both the image and sound domain.

With these compression experiments we have demonstrated that GP can be used with very large fitness case sets if a compiling approach is applied. Some of our experimental runs have evaluated hundreds of billions of fitness cases, something that would take years on an interpreting GP system. The machine code approach provides for very fast decompression suitable even for the demand of real time full size video.

Acknowledgment

One of us (P.N.) acknowledges support by a grant from the Ministerium für Wissenschaft und Forschung des Landes Nordrhein-Westfalen under contract I-A-4-6037-I

Bibliography

[1] Koza, J. (1992) *Genetic Programming*, Cambridge, MA: MIT Press.

[2] Koza, J. (1993) *Genetic Programming II, Automatic discovery of reusable programs*, Cambridge, MA: MIT Press.

[3] Li M., Vitani P. (1990) Inductive Reasoning and Kolmorogov Complexity. In *Journal of Computer and System Sciences*, pp343-384

[4] Oakley H. (1994) Two Scientific Applications of Genetic Programming: Stack Filters and Non-Linear Equation Fitting to Chaotic Data In *Advances in Genetic Programming*, K. Kinnear, Jr. (ed.), Cambridge, MA , USA, MIT Press.

[5] Nordin, J.P. (1994) A Compiling Genetic Programming System that Directly Manipulates the Machine-Code. In *Advances in Genetic Programming*, K. Kinnear, Jr. (ed.), Cambridge, MA , USA, MIT Press.

[6] Nordin, J.P. , Banzhaf, W. (1995) Evolving Turing Complete Programs for a Register Machine with Self-Modifying Code. In: *Proceedings of the Sixth International Conference of Genetic Algorithms*, Pittsburgh, Penn., USA, Morgan Kaufmann Publishers

[7] Nordin, J.P. , Banzhaf W. (1995) Complexity Compression and Evolution. In: *Proceedings of the Sixth International Conference of Genetic Algorithms*, Pittsburgh, Penn. ,USA, Morgan Kaufmann Publishers

[8] Nordin, J.P. (1995) Compiling Genetic Programming. To appear in: *Handbook of Evolutionary Computation*, T. Bäck, D. Fogel (ed.), New York, NY ,USA , Oxford University Press.

[9] Pennebaker, W.B., Mitchell, J.L. (1993) JPEG still image data compression standard. Van Nostrand Reinhold, New York , USA

Investigating the Generality of Automatically Defined Functions

Una-May O'Reilly
M.I.T. Artificial Intelligence Lab,
Cambridge, MA, 02139
unamay@ai.mit.edu
http://www.ai.mit.edu/people/unamay/unamay.html

Abstract

This paper studies how well the combination of simulated annealing *and* ADFs solves genetic programming (GP) style program discovery problems. On a suite composed of the even-k-parity problems for k = 3,4,5, it analyses the performance of simulated annealing with ADFs as compared to not using ADFs. In contrast to GP results on this suite, when simulated annealing is run with ADFs, as problem size increases, the advantage to using them over a standard GP program representation is marginal. When the performance of simulated annealing is compared to GP with both algorithm using ADFs on the even-3-parity problem GP is advantageous, on the even-4-parity problem SA and GP are equal, and on the even-5-parity problem SA is advantageous.

1 Introduction

It is clear that genetic programming (GP) will be more valuable if its ability to scale to large program discovery problems of practical value is enhanced. A set of approaches with this motivation has focused on promoting modular solutions. Modular solutions allow reuse of procedures and efficient hierarchical design. Some systems work dynamically; they change the representation of candidate solutions while the algorithm proceeds. For example, in [Koza, 1992; Angeline and Pollack, 1991; Rosca and Ballard, 1996] various encapsulation operators have been added to GP to promote module creation and reuse.

In contrast, Automatically Defined Functions (ADFs) are a static representation approach to modularity. Basically, a candidate solution has a fixed template of main program and procedures and the definitions and invocations of the main program and procedures co-evolve in the process of the search. The ADF approach to representation is well described by its designer in [Koza, 1994] and summarized in Section 2 of this paper.

O'Reilly and Oppacher have shown [O'Reilly and Oppacher, 1996] that, with a hierarchical variable length mutation operator, simulated annealing (SA) is able to perform program discovery on problems expressed in exactly the style of GP. The "current" solution is a program is represented as a tree. It is executed to obtain its fitness and then a mutation operator is applied to the tree to produce a new candidate program for the SA acceptance criteria. In Section 2, we show how ADFs can be used in conjunction with SA on GP style problems.

ADFs change the fundamental structure of candidate solutions in terms of modularity and hierarchy plus they change the nature of the fitness landscape (i.e. the search space in terms of the syntactic distance between two solutions and the fitness relationships of neighbouring solutions that are one syntactic change apart). The experiments in this paper provide preliminary insight into two related issues:

First, does the performance advantage conferred by ADFs to GP search generalize to SA? That is, as problem size increases, is there a benefit in terms of computational effort, likelihood of success and program parsimony to using ADFs with SA versus not using them? Does the payoff point in problem size occurs at similar points with SA and GP?

Second, is there a greater advantage to using ADFs with evolution-based search, i.e. GP, than to using them with SA? In other words, do programs using ADFs interact more ideally with the crossover operator and selection strategy of GP than with the mutation operator and move acceptance criteria of SA?

The paper proceeds as follows. Section 2 provides background on ADFs and SA. Section 3 motivates and presents our goals. In Section 4 we describe the methodology and problem suite we have chosen. In Section 5 we present the results of our experiments and analyze them. The paper ends with a summary and suggestions for future work.

2 Background

ADFs pursue the general goal of promoting modularity. With ADFs, a candidate solution is composed of a result producing branch (rpb) and one or more procedures called ADFs. It is evaluated for fitness by execution of the rpb which could result, in turn, in the invocation of its ADFs. To maintain the *a priori* designated basic program structure (i.e. one rpb and the declarations of the ADFs), a program has a fixed template which does not undergo adaptation. All expressions *within* this template, however, do undergo adaptation. Distinct primitive sets are designated for the rbp and each ADF to facilitate hierarchical sequence and provide the benefits of parameterization. They impose a syntactic constraint that some primitives can not be parameters of others because they are not within the correct program scope. To address this, GP with ADFs uses a modified crossover operator which respects branch typing. The crossover operator first randomly selects any point within the program template of the first parent and determines from what primitive set it is drawn. It then restricts where it chooses a crossover point in the second parent to the program area that uses the same set. Based on this design, both main program and procedures co-evolve in GP with ADFs.

Simulating Annealing [Aarts and Korst, 1989] is a widely used optimization algorithm that contrasts with GP's population and crossover based approach, because it is a single point search which generates a new program from the current one via mutation and chooses to transfer the search to it depending on an acceptance criteria. This criteria has both a deterministic component: always accept the new program if it is as fit or fitter than the current one, and a probabilistic one: accept a less fit new program with a probability which depends on the fitness difference between current and new programs and the current temperature of the system. The temperature usually decreases exponentially over the course of the search.

SA can solve GP style problems competently by using a suitably designed mutation operator that mutates programs expressed as parse trees. O'Reilly and Oppacher [O'Reilly and Oppacher, 1994] introduced such a mutation operator, called HVL-mutate. HVL-mutate generates a candidate solution that is syntactically correct and which may differ in size and structure from its "parent". HVL-mutate works by performing a sub-operation of substitution, deletion or insertion (with equal probability) to the copy of the parent's program tree. It is described and depicted in detail in [O'Reilly, 1995].

ADFs are a static form of program representation (i.e. a template and separate primitive sets), so it is straight forward to modify SA to function with them. In SA with ADFs a candidate solution is, in fact, exactly like one in GP with ADFs. However, the HVL-mutate operator is changed in order to preserve syntactic correctness in the mutant because of the constraint imposed by different primitive sets among the rpb and ADFs. This modification is simple. Once a mutation sub-operation is determined, a mutation point within the template of the rpb and ADFs is chosen randomly. The deletion sub-operation requires no change: the tree rooted at the mutation point is replaced by its largest subtree. Both the substitute and insertion sub-operations are changed in a simple manner to use the primitive set of the mutation point when changing or adding new primitives to the program at the mutation point. Fitness evaluation of the mutant is exactly like GP: the rpb is executed and its results are measured against the desired ones.

3 Motivation and Goals

Progress has been made into understanding how and why ADFs work. Kinnear compared GP with ADFs to Angeline's Module Acquisition in [Kinnear, Jr., 1994]. For the even-4-parity problem he attributed the success of ADFs partially to a form of *structural regularity* that strongly increases the likelihood of evolving a correct solution. Structural regularity in a program with ADFs is implied by the large number of calls to ADFs from the rpb as well as the frequent use of parameters within each ADF. If the rpb of a program were expanded by replacing its ADF invocations with the expressions defining the appropriate ADF and with the formal parameters replaced by the actual parameters, one would observe many partially similar subtrees in the result. This regularity is much less likely to evolve without ADFs than with them. One change to an ADF has multiple and identical effects. The capacity of the ADF representation for concise repetition thus assures a form of structural regularity.

As the first two main points of Genetic Programming II [Koza, 1994], Koza, first, asserts that it is possible to interpret GP with ADFs as performing either a top-down process of problem decomposition or a bottom-up process of representational change to exploit identified regularities in the problem. Second, ADFs work by exploiting inherent patterns and modularities within a problem, though perhaps in ways that are very different from the style of programmers.

Koza also offers the "lens effect" in [Koza, 1994] as one explanation for why ADFs are advantageous over representations without them. For each problem he examined, the distribution of fitness values of a large random sample of candidate solutions had more extreme outliers and less variance than the distribution of a sample of candidate solutions not using an ADF representation. The difference, termed the "lens effect" appears to make usings ADFs advantageous since correct solutions are highly fit outliers.

Koza shows many examples in [Koza, 1994] which reinforce the assertion that, for some problems, ADFs pay off because they require less computational effort and produce smaller solutions than using GP without them. He reports that on three problems this payoff increased as problem size grows.

Existing research into ADFs has been done in the context of GP yet, obviously, ADFs are a style of program representation that can be used with a substitute for GP. This leads to our goals:

Goal 1: To find out whether ADFs work in the same manner with SA. We examine the ADF payoff as we increase the size of `even-parity` problems from 3 to 4 to 5 and use SA. We measure payoff in terms of the likelihood that an execution will succeed, the relative size of solutions with and without ADFs, and, the fitness of the best individual at the conclusion of an execution. Can SA with ADFs perform what is, effectively, the same top-down process of problem decomposition or bottom-up process of representational change with its sequential, point to point algorithm? Does the ADF potential for structural regularity confer advantages to SA? Will SA benefit, like GP, from conducting its search over a fitness landscape with the typical fitness distribution of the ADF representation?

Goal 2: In contrast to our first goal which evaluates ADFs in the context of SA, we wish to compare the performance of GP with ADFs to that of SA with ADFs. This will determine for this suite whether the fitness landscape of the problem represented with ADFs is ideally suited to one algorithm or the other.

4 Experimental Procedure

Problem Suite

To examine the scaling effects of ADFs with SA we chose one progression of a problem (i.e. versions that increase in size) which was used by Koza in Genetic Programming II [Koza, 1994]: even-k-parity in three versions – even-3-parity, even-4-parity, and even-5-parity. By choosing a problem Koza uses, we can compare GP with ADFs to SA with ADFs as well as make comparisons between GP, with and without ADFs, to SA, with and without ADFs. Instead of re-conducting the GP experiments we cite the results of Koza [Koza, 1994]. The same source should be consulted for details of the GP algorithm used. It is fairly standard in terms of minor and major parameters though the population size of 16000 is somewhat larger than normal.

The even-k-parity problem is to output `true` if an even number of the k boolean input variables are `true` and `nil` otherwise. The primitive set of the rpb includes the boolean input variables, ADF names and the boolean logic functions `and`, `or`, `nand` and `not`. Each ADF primitive set uses formal parameters and the boolean functions. No ADF primitive set references the boolean input variables directly, i.e. this is done via parameters. Without ADFs the primitive set is the same as that of the rpb except that the ADF names are excluded. Fitness cases are all combinations of the input variables and raw fitness is the number of test cases for which a program yields the incorrect answer. Tableaux and text which give the number of ADFs and formal parameters for each version of even-k-parity are available in sections 6.3 to 6.5 and sections 6.10 to 6.12 of [Koza, 1994].

SA Algorithm and Parameter Choices

Our SA algorithm is standard [Rich and Knight, 1991]. It always starts from a temperature of 1.5 and decreases it at regular intervals (parameter `stepsize`) towards a final temperature (parameter `T-final`) at an exponential rate. The parameter `T-final` is always set to 0.01% of the minimum fitness change. Both the `stepsize` and `T-final` of each run are noted in Table 1. The algorithm uses HVL-mutate either in its standard form or the form introduced to handle ADFs.

Problem	AD Fs?	`T-final`	`step size`	Ind Proc	N
`3-parity`	Y,N	0.01	1000	48,000	30
`4-parity`	Y,N	0.0625	4000	96,000	30
`5-parity`	Y	0.003125	2000	160,000	30
`5-parity`	N	0.003125	2000	160,000	16

Table 1: SA run parameters. N indicates number of executions of a run.

We conduct a run for each problem with ADFs and a run for each problem without ADFs. Except for `5-parity` without ADFs which had N=16 executions, each run consists of N=30 executions of the SA algorithm with the same parameter settings but different random seeds.

We chose the maximum fitness evaluations for our runs based on data from Koza's GP runs. Note that this data was obtained *without* optimized parameter settings. It was not feasible to process 816,000 individuals as these runs did. Instead, we referred to their plots and chose the generation either where the first solution occurred or where the minimum computational effort was spent. The decision depended on the amount of cpu resources the problem required and the amount we had available. We multiplied that value by the population size incremented by one (to account for processing generation 0 in GP) and used it as the maximum individuals an execution processed. These values are in the fifth column of Table 1.

Assessment

We did not base our comparisons on computational effort, E, defined by Koza [Koza, 1994]. E is determined *retrospectively* as the effort in the generation at which the total number of individuals, I(M,i+1,z), that must be processed in order to yield a solution for the problem with z=99% probability, for a population of size M, by generation i, is a minimum. The statistical significance of E comparison result can not be measured. Therefore, while theoretically E can provide insight, we adopt a practitioner's perspective. Foresight is not available to a practitioner. *A priori* a fixed maximum number of individuals for an execution to process must be designated. This value is chosen using experience and rule of thumb estimation. Where fitness evaluation dominates computation effort, the same maximum is usually given to both the algorithms or representations being compared. An execution terminates with success or when that maximum number of individuals are processed. All of our comparisons are based on runs that process the same number of individuals, i.e. maximally perform equivalent numbers of fitness evaluations. For GP data in comparisons between GP and SA, we use the value of Koza's P(M,i) where the product of M and (i+1) equals the maximum individuals processed in the SA run.

We compare two runs, whether they be with or without ADFs, or, GP and SA with the same representation, on the basis of the probability of success of an execution. We also compare the average percentage of individuals processed (Inds Proc'd) in each execution and just successful executions. These values are expressed as percentages of the maximum individuals processed. The average fitness values of the best individual at the end of each execution are also compared and expressed as a percent where a perfect solution has a fitness of 100%. The use of a *Student's t-test* (different means) measuring 95% confidence [Press, 1992] is implied when we indicate the difference between two averages is statistically significant. As a precondition of the t-test, an f-test is used to determine whether the compared samples have the same variance.

5 Results and Analysis

Goal 1

Tables 2 and 3 provide the run data for the 6 experiments run with SA: even-3-parity, even-4-parity, and even-5-parity, for SA using ADFs and SA not using ADFs. Table 3, columns 3 and 4 correspond in terms of maximum individuals processed. Column 3 lists the probability that an SA execution is successful. Column 4 lists the P(M,i) of GP runs from [Koza, 1994] where M multiplied by (i+1) equals the same maximum number of individuals processed by the SA run. For example, on even-3-parity each SA execution processed a maximum of 48,000 individuals. Because this corresponds to Generation 2 of a GP run with a population size of 16000, the value of P(M=16000, i=2) is in column 4.

Tables 2 and 3 indicate that SA with ADFs is capable of solving all three problems. What does this imply about the interpretation of "co-evolution" in conjunction with ADFs? Clearly, just as crossover and population based search are not essential to finding program discovery solutions, co-evolution can not be necessary for ADFs solutions to be found.

Problem	ADFs?	SA Success %-age	GP P(M,i) as %-age
`3-parity`	Y	63.3 (48.1)	94
`3-parity`	N	100.0	18
`4-parity`	Y	70.0 (45.8)	60
`4-parity`	N	66.7 (47.1)	0
`5-parity`	Y	33.3 (47.1)	5
`5-parity`	N	25.0	4

Table 2: Results of likelihood of success for SA with and without ADFs. Standard deviations are in parentheses

Problem	AD Fs?	Best Ind Fitness	Inds Proc'd	Inds Proc'd Succ Exec
`3-parity`	Y	94.8 (8.0)	89.1 (8.2)	82.8 (8.2)
`3-parity`	N	100.0	81.9 (8.2)	81.9 (8.2)
`4-parity`	Y	95.1 (9.4)	89.5 (14.9)	88.4 (9.8)
`4-parity`	N	97.6 (4.1)	95.2 (4.4)	92.8 (4.4)
`5-parity`	Y	97.0 (7.8)	97.2 (6.4)	90.1 (7.5)
`5-parity`	N	94.8 (4.5)	99.5 (0.9)	97.9 (0.3)

Table 3: Results of fitness and processing for SA with and without ADFs. Standard deviations are in parentheses

In the context of SA and GP, an extended interpretation of the series of changes which the rpb and ADFs of a program undergo should encompass, but not exclusively use, the notion of co-evolution. We suggest co-adaptation as an appropriate descriptor: ADFs facilitate the co-adaptation of the main program and procedures in a solution. The co-adaptation process may be simple, as in SA, where it proceeds one change at a time using random variation of the current candidate. Or, it may be complex, as in GP, where it proceeds with concurrent multiple changes of selected material each generation.

Problem	ADF effort Advantage for the run	ADF effort Advantage for succ. exec's	ADF Success Advantage
`3-parity`	-7.2 *	-0.9 *	-36.7
`4-parity`	+5.7	+3.4	+3.3
`5-parity`	+2.3 ?	+7.8 ?	+8.3 ?

Table 4: The advantage of ADFs when used with SA. Statistically significant values are marked with an asterisk, ? indicates statistical comparison could not be performed.

Next, Table 4 shows data which indicate the benefits

of using ADFs with SA versus not using them with SA. In the context of SA, we have measured three values. The first metric (columns 2 and 3) we call "ADF effort advantage". ADF effort advantage equals the difference (measured as a percentage) between the average number of individuals processed in a SA execution without ADFs and the average number of individual processed in a SA execution with ADFs. A positive value of ADF effort advantage indicates that using SA with ADFs requires less computational effort than using SA without them. For example, over the entire run of 30 executions, even-4-parity without ADFs processed 95.2% of the maximum individuals and with ADFs processed 89.5% of the maximum, therefore the ADF effort advantage of the entire run is +5.7%. Where a value is marked with an asterisk in the table, the ADF Effort advantage is statistically significant.

The table indicates that, under this measure, while even-3-parity is not a problem where it is effective to use ADFs with SA, for even-4-parity the decision is neutral. For even-5-parity using ADFs with SA appears advantageous over not using them, however, the small number of executions prevent us from testing the statistical significance of the advantage. The margin of advantage, were it significant, would still be small in absolute terms (between 2 and 8%). These results are true for a run and when just successful executions are examined (column 3).

The non-monotonicity of the values of column 2 of Table 4 is not an inconsistency. Although the maximum number of individuals processed is the same with and without ADFs (i.e. between columns), it changes for each version of the problem (i.e. between rows).

Turning to the likelihood of success, column 4 of Table 4 records a value we call "ADF Success advantage". It is the subtraction of the probability of success without ADFs from the probability of success with ADFs, expressed as a percentage. A positive value implies using ADFs with SA, with this maximum number of individuals being processed, is advantageous over not using ADFs with SA, on this problem. Unlike ADF effort advantage, because the probability of success is a consistent metric across problems, this progression is meaningful since direct comparisons for each problem are valid. This value is negative for even-3-parity and positive and increasing for even-4-parity and even-5-parity, however, only the value for even-3-parity is statistically significant. This implies that the payoff of ADFs on this scale, as problem size increases from 3 to 5, moves from disadvantageous to neutral with a hint that on larger sizes of the problem ADFs may become significantly advantageous. In future work we plan to try SA with even-6-parity.

The third measurement for evaluating ADFs with SA in Table 5 is Koza's R_s. R_s is the ratio of average structural complexity (i.e. size or the number of nodes in the

Problem	SA R_S	GP R_S
3-parity	0.73	0.92
4-parity	0.58	1.87
5-parity	0.53	1.91

Table 5: Parsimony comparison using ratio of average structural complexity of a perfect solution without ADFs to one with ADFs

program tree) of a perfect solution without ADFs to one with ADFs. A value below one indicates that not using ADFs yields more parsimonious solutions. We executed our runs with the maximum tree height parameter set to 15. In contrast to Koza's finding with GP (see column 3) using ADFs with SA on even-4-parity and even-5-parity does not yield more parsimonious solutions.

In summary, on these three problems, with the problem parameters and algorithm parameters chosen, as problem size increases, SA with ADFs versus SA without ADFs becomes computationally neutral and only marginally more likely to yield a correct solution. The correct solutions are not more parsimonious.

Interestingly, with GP, ADFs pay off immediately from even-3-parity onwards while, with SA, they first are detrimental and then are neutral. The break even point for this problem differs from GP to SA if we can believe that ADFs will pay off on larger versions of even-parity. The trend of column 4 in Table 3 hints at this possibility.

This suggests another interesting hypothesis for future work. That is, in general, independent of what reasonable adaptive heuristic search is chosen, when problem size increases, using ADFs improves likelihood of success and is advantageous in terms of computational effort, versus not using them. This hypothesis is accompanied by the qualification that the precise advantage of ADFs in these terms would differ from algorithm to algorithm (i.e. the break-even point in these terms might different on one problem's versions given a particular algorithm). For example, SA without ADFs found solutions to the even-4-parity problem well before GP could, yet with ADFs GP and SA had similar success likelihood on the even-4-parity problem. ADFs can be advantageous with both GP and SA but by different amounts.

Goal 2

Concerning the second goal of the paper, Table 6 assesses whether, on these three problems, GP with ADFs is better than SA with ADFs. We rely on a measurement we call GP success advantage. For a problem, it is the difference between the likelihood of success of an execution of GP with ADFs and a run of SA with ADFs. We calculate it by subtracting the corresponding values of columns 3 and 2 of Table 2 and express it as a percentage. A positive value indicates that, on this problem, with the parameter settings reported, GP with ADFs has a greater likelihood of success than SA with ADFs. Since the values are means of a sample and the sample size is known, we also check the t-scores of the comparison and note when the advantage (or disadvantage) is statistically significant. On even-3-parity GP with ADFs was positive in terms of GP success advantage over SA with ADFs. However, on even-4-parity there was not a statistically significant difference, and, on even-5-parity, GP success advantage is negative, indicating that SA with ADFs is superior in this aspect.

Problem	Search Success Advantage with ADFs	Search Success Advantage without ADFs
3-parity	+ 31 *	-82 *
4-parity	-10	-61 *
5-parity	-28 *	-21 ?

Table 6: Comparison Data of GP with ADFs and SA with ADFs. Statistically significant values are marked by an asterisk.

This is a simple, small comparison which did not use algorithms with optimized parameters but it suggests an informative question for future examination. Do ADFs confer a greater advantage to SA as problem size increases? More data (or at least one until one result that is contrary is found) needs to be investigated to answer this question.

6 Summary and Future Work

This paper made the following contributions: It showed how ADFs can be used with SA to solve program discovery problems in the style introduced by GP. In general, with either evolution-based search or simulated annealing search, using ADFs facilitates the co-adaptation of a main program and its procedures. We found that, the use of ADFs with SA rather than using SA alone, in terms of run success probability and computational effort, moved from disadvantageous to neutral as problem size increased from three to five in the even-parity problem. Contrary to GP, program parsimony was not in favour of ADFs with SA. We observed that ADF programs got larger as problem size increased. Finally, on the three problems we experimented with, we found that, when both algorithms used ADFs, GP was better than SA on the smallest, they were equal on the mid-sized problem and SA was superior on the largest. This suggests future work to examine whether the SA advantage over GP when they both use ADFs increases with problem size or whether it depends on what problem is solved.

Acknowledgements

The authour is grateful to the Santa Fe Institute and the Zoo at the AI Lab at MIT.

References

[Aarts and Korst, 1989] E. Aarts and J. Korst. *Simulated Annealing and Boltzmann Machines,*. Wiley, New York, NY, 1989.

[Angeline and Pollack, 1991] P. J. Angeline and J. B. Pollack. Coevolving high-level representations. In C. G. Langton, editor, *Artificial Life III, SFI Studies in the Sciences of Complexity*, volume XVII. Addison-Wesley, 1991.

[Kinnear, Jr., 1994] K. E. Kinnear, Jr. Alternatives in automatic function definition: A comparison of performance. In K. E. Kinnear, Jr., editor, *Advances in Genetic Programming*, chapter 6. MIT Press, 1994.

[Koza, 1990] J. R. Koza. Genetic programming: A paradigm for genetically breeding populations of computer programs to solve problems. Technical Report STAN-CS-90-1314, Department of Computer Science, Stanford University, Stanford, CA, 1990.

[Koza, 1992] J. R. Koza. *Genetic Programming: On the Programming of Computers by Means of Natural Selection.* MIT Press, Cambridge, MA, 1992.

[Koza, 1994] J. R. Koza. *Genetic Programming II: Automatic Discovery of Reusable Programs.* MIT Press, Cambridge, MA, 1994.

[O'Reilly and Oppacher, 1994] U.M. O'Reilly and F. Oppacher. Program search with a hierarchical variable length representation: Genetic programming, simulated annealing and hill climbing. In Y. Davidor, H.-P. Schwefel, and R. Männer, editors, *Parallel Problem Solving From Nature - PPSN III, Volume 866 of Lecture Notes in Computer Science*, pages 397–406, Berlin, 1994. Springer-Verlag.

[O'Reilly and Oppacher, 1996] U.M. O'Reilly and F. Oppacher. A comparative analysis of genetic programming. In P. Angeline and K. E. Kinnear, Jr., editors, *Advances in Genetic Programming 2*, chapter 9. MIT Press, Cambridge, MA, USA, 1996.

[O'Reilly, 1995] U.M. O'Reilly. *An Analysis of Genetic Programming.* PhD thesis, Carleton University, Ottawa, Ont, September 1995. Available at http://www.santafe.edu/~unamay.

[Press, 1992] W. H. Press. *Numerical Recipes in C.* Cambridge Press, Cambridge, MA, 1992.

[Rich and Knight, 1991] E. Rich and K. Knight. *Artificial Intelligence.* McGraw Hill, New York, 1991.

[Rosca and Ballard, 1996] J. P. Rosca and D. H. Ballard. Discovery of subroutines in genetic programming. In P. Angeline and K. E. Kinnear, Jr., editors, *Advances in Genetic Programming 2*, chapter 9. MIT Press, Cambridge, MA, USA, 1996.

Parallel Genetic Programming: an application to Trading Models Evolution

Mouloud Oussaidène
Bastien Chopard
CUI, Université de Genève
24 rue Général-Dufour
CH-1211 Genève 4
Switzerland
mouloud@cui.unige.ch
chopard@cui.unige.ch

Olivier V. Pictet
Olsen&Associates
Research Institute for Applied Economics, Seefeldstr. 233
CH-8008 Zurich
Switzerland
olivier@olsen.ch

Marco Tomassini
CSCS, Manno and
LSL, EPFL
CH-1015 Lausanne-Ecublens
Switzerland
marco.tomassini@di.epfl.ch

ABSTRACT

We present a parallel implementation of genetic programming on distributed memory machines. To overcome the time overhead due to uneven load associated with program evaluation, we propose and evaluate a non-preemptive dynamic scheduling algorithm for load balancing. The system is applied to the evolution of trading model strategies which is a compute-intensive application. Our results show that reasonable trading models can be inferred and that the system can produce a nearly linear speed-up for that application.

1. Introduction

Artificial evolutionary processes, such as genetic algorithms (GA) (Holland 1975), are based on reproduction, recombination and selection of the fittest members in an evolving population of candidate solutions.

(Koza 1992) extended this genetic model of learning into the space of programs and thus introduced the concept of genetic programming. Each solution in the search space is represented by a genetic program (GP), traditionally using the Lisp syntax. Genetic programming is now widely recognized as an effective search paradigm in artificial intelligence, databases, classification, robotics and many other areas .

We present here a parallel genetic programming system (PGPS) which speeds up the searching process. In complex applications, such as trading model optimization, the evaluation of each GP is very long and thus the time spent in the selection and reproduction phases is practically negligible compared with the population evaluation time. Parallelizing the evaluation phase on a parallel coarse-grain machine such as the IBM SP-2, can be done easily. After the reproduction phase, each GP is simply sent to a processing node for evaluation, independently of operations in other processing nodes (Oussaidène, Chopard and Tomassini 1995). However, the run time GP complexity with programs of widely differing sizes makes the design of a parallel algorithm and its implementation to obtain large speed-ups a nontrivial task. This irregularity causes some processing nodes to be idle while others are active.

We describe a parallel scheme in which the evaluation phase is separated from the rest of GA calculations. We first consider a static scheduling algorithm to distribute the genetic programs upon the processing nodes at the evaluation phase. Next, we improve the processor utilization by a dynamic load balancing algorithm based on run time GP complexity. Our implementation shows that, for problems of large enough size, the parallelization of genetic programming on distributed memory machines can deliver a nearly linear speed-up. The code was written in C++ using the PVM3 message passing library (Geist et al. 1994) and can be run on other parallel machines such as the Cray T3D. The sequential version of the code is based on (Fraser 1994).

The organization of this paper is as follows. The evolution of genetic programs is outlined in section 2. Section 3 discusses the PGPS architecture. Section 4 highlights the load balancing problem and analyzes two different techniques for solving the problem. The parallel GP system has been applied to a difficult problem of trading model search described in section 5. We discuss our experimental results in section 6 and concluding remarks are made in section 7.

2. Evolution of genetic programs

A GP can be regarded as a Lisp function. It is usually represented by a parse tree.

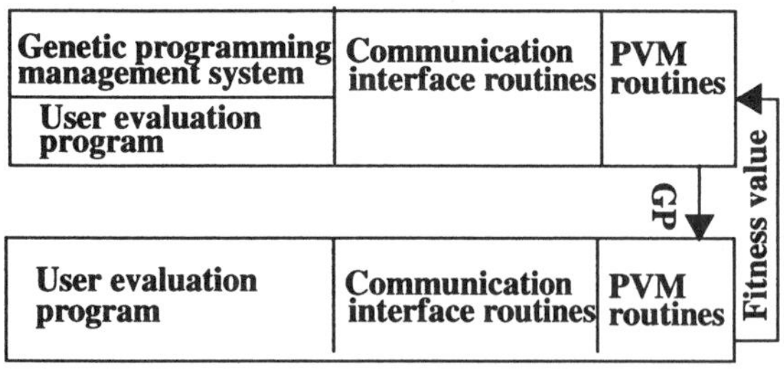

fig 1: PGPS scheme.

The number of nodes (including the terminal nodes) in the parse tree gives a measure of the complexity of that GP. Thus, the GP (* (+ x y) (- t b)) is of complexity 7. The terminal (non-terminal) nodes are randomly taken from some well defined *TerminalSet* (*FunctionSet*). In the creation phase, each parse tree is built in a recursive way, starting from the root node. To ensure the syntax validity and control the complexity expansion of GPs during this creation process, some rules must be observed. If the current node in the parse tree is a function taking r arguments then choose r nodes, from the *Terminal/FunctionSet*, to be child nodes. In the other case, the current node is a terminal so return to the parent node. A maximum depth, d, is fixed before the execution so that when a branch in the parse tree reaches this level then a terminal must be chosen. The depth level corresponding to the root node is $d = 0$.

Let r be the maximum function arity in the *FunctionSet*. At run time, the complexity, C , of a GP is therefore bounded by the formula (2.1).

$$\begin{pmatrix} 1 \le C \le \dfrac{r^{d+1}-1}{r-1} & \text{if} \quad (r \neq 1) \\ 1 \le C \le d+1 & \text{if} \quad (r = 1) \end{pmatrix} \qquad (2.1)$$

Two genetic operators are used in the reproduction phase; crossover and mutation. The crossover operator selects two GPs from the population and chooses one node (crosspoint) on each. Each node is, by definition, the root of some complete subtree. The two subtrees are extracted and each is swapped with the other. Note that the resulting GPs are syntactically valid and only the maximum depth constraint is checked for non-violation.

For illustration, if we choose the nodes 4 and 2 (numbering from the left) respectively on the GPs (+ x (* d (/ y z))) and (+ (* a b) 1) then the new GPs will be (+ x (* (* a b) (/ y z))) and (+ d 1) .

The mutation genetic operator is unary. A GP is modified according to some rules. Any terminal node can be replaced by any other terminal taken from the *TerminalSet*, a function node can only be replaced by another function (from the *FunctionSet*) which has the same number of arguments. This operator does not modify the complexity of the GP. The evaluation phase consists of assigning a fitness value to each GP in the population. This fitness calculation requires evaluating each GP as many times as there are fitness cases related to the problem to solve. Particularly, in learning trading models, each GP is evaluated on various price time series where each series element represents a fitness case. In this analysis, the problem size is given by $n = p \times g \times e$ where p, g, e indicate respectively the population size, number of generations and number of fitness cases.

3. PGPS scheme

Figure 1 shows the parallel scheme and the i/o communication interface between the genetic programming management system and the user defined problem. The architecture of PGPS follows the master-slave model. The master process is the genetic programming management system and the slave process is the user problem evaluation program. The genetic programming management system creates the initial population, applies the genetic operators (crossover and mutation) and performs the selection of genetic programs which will be the candidates for the reproduction phase. At the evaluation phase, the master process distributes the work load among all the processing nodes in the virtual machine, including the processor on which the master process runs. This computational load distribution can be accomplished by different strategies. In the next section we analyse a static scheduling algorithm and a dynamic one. The tasks assigned to the master process require no communication, they are executed locally. The communications between the master and slave processes are achieved using communication interface routines. The genetic programming management system packs each parse tree from its memory representation into a buffer and sends it to the slave processes using PVM routines. Each slave process receives the character string into a buffer and performs the unpack operation to build the equivalent parse tree in memory. After the fitness calculation of a genetic program, each slave process sends back the calculated fitness value to the master process. Note that GPs can be evaluated independently and therefore there is no communication between the slaves. The load distribution cost and the communication overhead are negligible compared with the computational cost associated with fitness calculation. Each message is sent individually and the messages belonging to the same destination are not clustered as a single message. This technique avoids delaying the fitness computations and allows the slave processes to be further kept busy. Thereby, fitness computations and work distribution overlap. The evaluation phase ends when the whole population is evaluated (each processing node has performed its assigned tasks). The termination signal of parallel execution is given when a preassigned maximum number of generations has been attained. The main steps of the parallel algorithm are shown in Figure 2.

Master process :

0: Load the slave processes.
1: Create the initial population.
2: Distribute the work load to the processing nodes.
3: Execute the tasks assigned to the Master process.
4: Receive the fitness values sent by the slave processes.
5: Perform the selection phase.
6: Perform the reproduction phase.
7: Repeat Steps 2 - 6 up to the maximum number of generations.
8: Terminate the slave processes.
9: End.

Slave process :

0: Receive a genetic program.
1: Calculate the fitness of the received program.
2: Send the fitness value to the Master process.
3: Repeat Steps 0 - 2 until reception of termination signal.
4: End.

fig 2: A parallel algorithm for Genetic Programming.

4. Load balancing

Load balancing refers to equal distribution of the computational load among the processing nodes. In the terminology used here, task is the fitness computation and task size refers to the GP complexity. A task is the smallest unit that can be scheduled on a processing node. Load balancing can be either static or dynamic.

4.1. A static scheduling algorithm

This algorithm distributes the work load in an ordered way defined at compilation time. The Round-Robin policy is used to assign tasks to the processing nodes. The criterion of load balancing upon the processors is the number of tasks assigned to each processor. Task i , in the task pool to be distributed over m processors, is assigned to the processing node given by $\lceil i/m \rceil$. This algorithm handles only static parameters, fixed at compilation time, and does not reflect dynamic factors which may change during the system evolution at run time. Even though the algorithm regulates the number of tasks assigned to each processor, irregularity in computational load may occur. This overhead is due to the fact that tasks are not of the same size and the work load on a processing node depends not only on the number of tasks to be performed, but also on the size of these tasks. This irregularity at the complexity level induces differences in the task evaluation time which may unbalance the work load on the processing nodes.

4.2. A dynamic scheduling algorithm

In the dynamic load balancing scheme, the decision of which processor should run a specified task is made at run time. The processor allocation is a function of population complexity at each generation of the evolutionary process. The work load on a processing node is considered as the sum of each task size assigned to that node. This problem can be formulated as follows.

Given m processing nodes M_i $(i = 1, ..., m)$ and p tasks T_j $(j = 1, ..., p)$, of size C_j , one wishes to find a schedule of minimum length. A schedule is optimal if the maximum task completion time is minimum. Note that this problem is known to be NP-hard. Let l_i be the work load of processor i . The algorithm is shown in figure 3. It is based on a "greedy" heuristic giving, on average, a well approximated solution to the problem.

Sort C_j *in Downward order*
$l_i \leftarrow 0 \quad (i \in [1...m])$
For $j \leftarrow 1$ to p *do*
  $i^* \leftarrow min\{i | l_i \leq l_k, 1 \leq k \leq m\}$
  Assign T_j to M_{i^*}
  $l_{i^*} \leftarrow l_{i^*} + C_j$
End

fig 3: A dynamic scheduling algorithm.

The tasks are sorted by descending size. The task distribution starts from the largest task. At each iteration, the algorithm assigns the current task to the least loaded processing node. When a processor is selected to perform a task its work load is increased by the size of that task.

5. Evolving trading models

Besides alleviating the restrictions of fixed-length representation of genetic structures, genetic programming provides a natural way to represent and evolve decision trees (Allen and Karjalainen 1993). The application presented here describes a way to use genetic programming to learn technical trading models for foreign exchange (FX) market. The recommendations given by a trading model are purely based on past prices of the exchange rate being analyzed. The price history is summarized in the form of variables called indicators.

A trading model is a system of rules catching the movement of the market and providing explicit trading recommendations for financial assets. A simple form of a trading rule could be

IF |I| > K THEN G := SIGN(I) (5.1)
ELSE G := 0

Where I is an indicator whose sign and value model the current trend and K is a threshold constant called *breaklevel*. The gearing, G, is the recommended position of the model. The value G = +1 corresponds to a 'buy signal', G = -1 corresponds to 'sell signal' and G = 0 corresponds to the neutral position (stay out of the market). A trading rule with a more complex strategy may use more than one indicator : typically, an indicator giving the current trend can be used in conjunction with an indicator reflecting the volatility of prices. Since no single indicator can ever be expected to signal all trend reversals, it is essential to combine a set of indicators so that an overall picture of the market can be built up. The trading models used in this study are decision trees seen as logical combinations of rules having the form (5.1). The decision trees are evolved using genetic programming.

An indicator is a function of time series and, in particular, all the indicators used here are functions of time and price history. The indicator construction is based on the concept of momentum. The price momentum is essentially the deviation between the current price and the mean of multiple exponential moving averages (EMA) used for smoothing the price time series *x(t)*. An EMA is characterized by its range (in days), *r*, and its order (the number of successive applications of EMA to the initial price time series). A full description of momentum computation is given in (Pictet et al. 1992). We use the notation $I_{x,r,j,n}$ to refer to the normalized price momentum based on *r* days EMAs of order in [*j* .. *n*].

The major problem in optimizing trading models is to avoid overfitting caused by the presence of noise. Overfitting means building trading models that fit a price history very well but generalize badly. In order to avoid this phenomena, the idea is to build genetic programs based on pre-optimized building blocks. These building blocks (robust indicators) were optimized using a niching genetic algorithm based on a fitness sharing scheme (Pictet et al. 1995). The fitness measure of a trading model quantifies not only the return but also the risk involved by the model. The total return of a model is obtained by continuously accumulating the return of each transaction (which is null for transactions starting from the neutral position). A transaction holds each time a new signal is generated.

If p_1 and p_2 are the prices of two consecutive transactions, then the return of the second transaction is given by the ratio $(p_2 - p_1)/p_1$.

The fitness function, termed X_{eff}, is defined as :

$$X_{eff} = \langle R \rangle - \frac{C}{2}\sigma^2$$

where $\langle R \rangle$ is the annualized average total return, $C = 0.1$ is a risk aversion constant and σ^2 is the variance of the total return over time. Since the variance is a measure of the stability of the return, then high effective return means also high stable return. The notion of robustness is directly related to the ability of generalizing the results beyond the training sample. For this purpose, each trading model is tested on more than one exchange rate time series. The fitness measure is then extended as follows:

$$X'_{eff} = \langle X_{eff} \rangle - \frac{\sigma_{eff}^2}{3}$$

where the first term is the average fitness value obtained for each exchange rate and the second term is proportional to the variance of these values.

The *FunctionSet* used in the construction of the trading strategies is composed of the logical operators {AND, OR , NOT , IF } and the *TerminalSet* is composed of the pre-optimized indicators and the values { +1 , -1}. Because of the presence of three possible values (-1, 0, +1) for the deal signal, we have redefined these logical operators. The OR operator returns the sign of the sum of its arguments, the NOT function returns the opposite decision of the argument, the AND function returns the neutral signal when one of its arguments is zero; otherwise returns the OR value. The IF function takes three arguments. It returns the second argument if the first one is true; otherwise it returns the third argument.

The indicators are revalued at each price event. Once updated, the normalized value $I_{x,r,j,n}(t)$ is used according to rule (5.1) to obtain the indicator signal $G_{x,r,j,n} = G(I_{x,r,j,n}(t))$. The different signals returned by the indicators are then embedded in the logical S-expression corresponding to the trading model. The value obtained by the evaluation of the S-expression represents the trading model recommendation.

6. Experimental results

This section discusses both parallel programming and trading model performances. Given the large size of our data set, the speed-up measurements were obtained by limiting the number of fitness cases to 1000 when fitting the logistic function (Oussaidène, Chopard and Tomassini 1995) and are consistent with the speed-ups obtained for the trading model application.

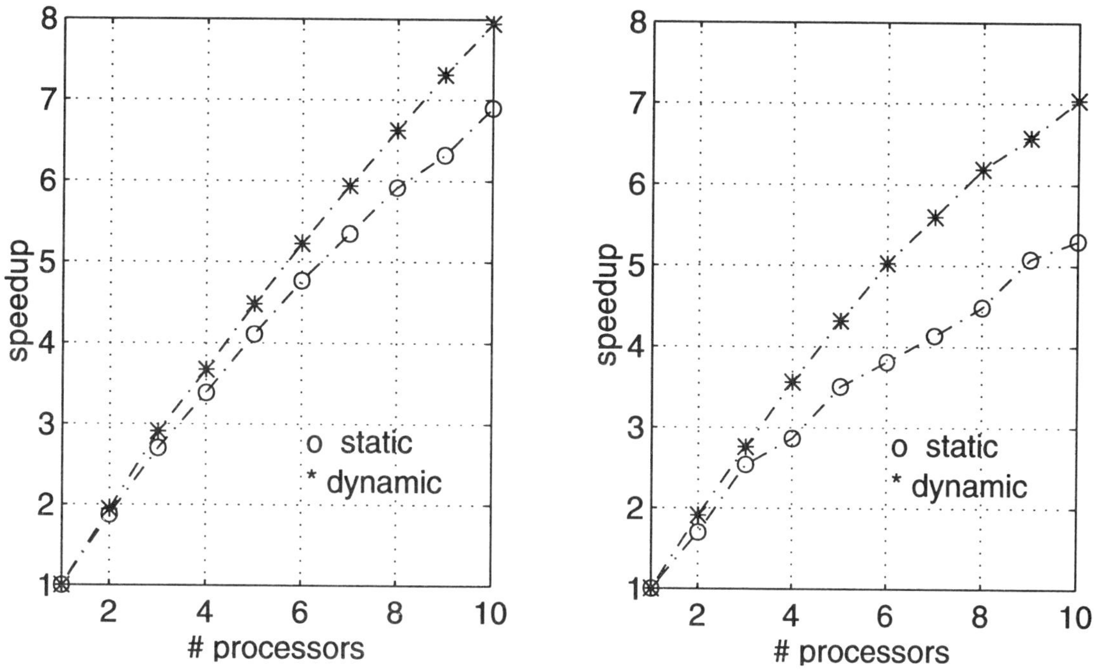

fig 4: speed-up of PGPS for $n = 100 \times 100 \times 1000$ *,* $d = 6$ *and* $n = 100 \times 10 \times 100$ *,* $d = 12$ *.*

6.1. Parallel programming performance

The parallel genetic programming system was implemented on an IBM SP-2 machine. SP-2 is a distributed memory machine using a network based computing model, providing two types of processing nodes: *Wide* and *Thin* nodes. The POWER2 processor in the *Wide* processing node runs at 66.7 MHz, giving a peak performance of 266 MFLOPS. The POWER processor in the *Thin* processing node runs at 62.5 MHz, giving a peak performance of 125 MFLOPS. The processing nodes share data via message passing over the *High Performance Switch* multistage packet switched Omega network (IBM 1995). The *switch* chip provides 40 MBytes/s peak channel bandwidth and 500 *nsec* hardware latency. We used from 1 to 10 *Thin* processing nodes. All the speed-up measurements reported here are calculated in the traditional form. For this, we have fixed not only the problem size but also the seed for the random number generator so that the output of a parallel execution is exactly the output of its corresponding sequential execution. Other authors suggested fixing only the problem size and to average over many runs (Koza and Andre 1995) but their model is based on evolving several loosely-connected subpopulations and is different from ours. The speed-ups of both static and dynamic scheduling algorithms are shown in Figure 4. The elapsed time was measured in dedicated mode. By setting the problem size as $n = 100 \times 100 \times 1000$ and the maximum depth as $d = 6$, the sequential execution takes 234.97 seconds on a single node of the SP-2. Using 10 SP-2 nodes, the static load balancing scheme completes in 34.05 seconds whereas the dynamic load balancing algorithm takes 29.57 seconds.

When increasing the run time maximum depth up to $d = 12$ and setting problem size as $n = 100 \times 10 \times 100$, the sequential execution time takes 743.77 seconds on a single node. On 10 nodes, the static scheduling algorithm completes in 140.2 seconds while the dynamic scheduling algorithm takes 105.68 seconds. We can remark that for relatively small depths, both static and dynamic load balancing schemes give approximately similar results. According to formula (2.1), the maximum task size grows exponentially with the maximum depth parameter. The crossover genetic operator introduces some variance in the task sizes within the task pool to be distributed and this variance becomes significant for high values of the maximum depth. The presence of this task size variability unbalances the work load distribution based on the static load balancing scheme which delivers a speed-up of 5.3 on 10 processors for a maximum depth of 12 while the dynamic load balancing scheme gives a speed-up of 7.

6.2. Trading model performance

The optimization of the trading models is performed on seven exchange rates (GBP/USD , USD/DEM , USD/ITL , USD/JPY , USD/CHF , USD/FRF , USD/

NLG) where each time series contains hourly data and is divided into alternating periods (in-sample / out-of-sample) of one and half years. The optimization period starts January 1, 1987 and ends June 30, 1994.
The GP evolutionary process was trained in the following manner. We performed ten in-sample runs evolving a population of 100 individuals over 100 generations (each run takes more than 30 hours on 4 SP-2 nodes). The best individuals issued from each in-sample run are then tested separately on the out-of-sample data. Hereafter, we report one of the best trading models which appears more systematically over the ten runs. The S-expression is given by

$$(\ OR (AND \ 1 \ G_{x,\,16,\,3,\,3}) \ (AND \ G_{x,\,16,\,2,\,4} \ G_{x,\,16,\,3,\,3}) \ )$$

For this trading model, the average yearly return $\langle R \rangle$ (in percent) and the fitness value X_{eff} in the training period are respectively (5.86% , 0.83). The values (5.77% , -5.63) were obtained in the out-of sample period.
We observe that all the indicator signals used in this S-expression are based on 16 days moving averages. Its underlying logical structure may be abbreviated as follows : if the indicator signal $G_{x,\,16,\,3,\,3} \neq -1$ then raise this position as the deal signal, otherwise validate the trend with a second indicator using the AND operation ($AND \ G_{x,\,16,\,2,\,4} \ G_{x,\,16,\,3,\,3}$) . It can be seen that the average yearly return seems to be stable in both learning and test periods. However, the return time series itself presents some fluctuations during the out-of-sample period. This observation raises the possibility that the price patterns captured by the model are no longer present in the test part of the data and also the risk sensitive return measure adopted is not necessarily the optimal one.

7. Concluding remarks

We have presented a parallel implementation of the genetic programming paradigm on distributed memory machines. The parallel scheme follows the Master-Slave model where the evaluation phase is performed by slave processes and the rest of the GA calculations is monitored by the master process. For problems of large enough size, the evaluation phase takes the most time to run and represents a potential source of parallelism. Good speed-up performances require an optimal load balancing algorithm. We presented and evaluated a dynamic scheduling algorithm based on run time task sizes. Our experimental results, on the IBM SP-2 machine, show that, for a problem requiring 1000 fitness cases, a nearly linear speed-up (8 over 10 processors) can be obtained. Reducing the number of fitness cases (to 100) reduces the speed-up performance curve (7 over 10 processors).
Genetic programming has been used to infer robust trading models. Although profitable, the average returns provided by the models present some variability, partly due to overfitting, during the test period. Our preliminary results show that genetic programming indicates a promising direction for trading model optimization which is not a well understood domain. Also, when addressing such highly complex problems it becomes necessary to decompose the problem representation so that an overall solution can be viewed as a combination of small modules.

Acknowledgments

The authors would like to acknowledge the Swiss National Science Foundation for its financial support. We would like to thank Olsen & associates research institute for providing the price data and making valuable contributions in understanding financial trading models.

Bibliography

Allen, F. and Karjalainen R. 1993. Using genetic algorithms to find technical trading rules. Working paper,Wharton school, university of Pennsylvania.
Fraser, A. P. 1994. Genetic programming in C++, public domain genetic programming system.
Geist, A., Beguelin A., Dongarra J., Jiang W., Manchek R. and Sunderam V. 1994. PVM3 User's Guide and reference manual, ORNL/TM-12187, September 1994.
Holland, J. 1975. Adaptation in Natural and Artificial Systems, University of Michigan Press, Ann Arbor.
IBM, 1995. Scalable POWER Parallel Systems, February 1995.
Koza, J. 1992. Genetic programming, MIT Press.
Koza, J. and Andre D. 1995. Parallel Genetic Programming on a Network of Transputers, Computer Science Department, Stanford University, Technical report CS-TR-95-1542.
Oussaidène, M., Chopard B. and Tomassini M. 1995. Programmation évolutionniste parallèle, RenPar'7 , PIP-FPMs Mons , Belgium. Libert G., Dekeyser J.L., Manneback P. (editors).
Pictet, O.V. , Dacorogna M.M. , Muller U. A. , Olsen R. B. and Ward J.R. 1992. Real-time trading models for foreign exchange rates , Neural Network World 6/92, 713-744.
Pictet, O.V., Dacorogna M.M., Chopard B., Oussaidene M., Schirru R. and Tomassini M. 1995. Using Genetic Algorithms for Robust Optimization in Financial Applications , Neural Network World 4/95, 573-587.

Genetic Programming for Image Analysis

Riccardo Poli
School of Computer Science, The University of Birmingham
Birmingham B15 2TT, UK
R.Poli@cs.bham.ac.uk

ABSTRACT

This paper describes an approach to using GP for image analysis based on the idea that image enhancement, feature detection and image segmentation can be re-framed as filtering problems. GP can discover efficient optimal filters which solve such problems but in order to make the search feasible and effective, terminal sets, function sets and fitness functions have to meet some requirements. We describe these requirements and we propose terminals, functions and fitness functions that satisfy them. Experiments are reported in which GP is applied to the segmentation of the brain in medical images and is compared with artificial neural nets.

1 Introduction

Genetic Programming (GP) has been applied successfully to a large number of difficult problems like automatic design, pattern recognition, robotic control, synthesis on neural architectures, symbolic regression, music and picture generation, etc [Koza, 1992, Koza, 1994, K. E. Kinnear, Jr., 1994]. However, a relatively small number of applications of GP in the domain of image processing and computer vision have been reported in the literature [Tackett, 1993, Koza, 1994, Andre, 1994, Teller and Veloso, 1995, Breunig, 1995, Riolo and Line, 1995], most of which are concerned with the problem of recognizing/classifying the structure represented in an image.

In this paper, more than with the high-level task of recognizing or classifying the structure(s) present in the image, the focus is on lower-level image-analysis problems such as image enhancement, feature detection and image segmentation. Our approach to solve these problems is based on the idea of using GP to discover effective problem-specific filters capable of highly and selectively emphasizing some characteristics of the image. On the ground of this behavior, these filters, coupled with simple thresholding wrappers, can be used to detect features of interest in the image (e.g. edges or texture) or to build pixel-classification-based segmentation algorithms.

In the next section an approach to low-level image analysis based on properly designed filters is described. Then, in Section 3, the characteristics that terminal sets, function sets and fitness functions should posses in order to enable GP to discover such filters are discussed, and one possible solution is proposed. Section 4 reports on some preliminary experiments in which GP is applied to the segmentation of the brain in magnetic resonance images and compared with artificial neural nets. We draw some final comments in Section 5.

2 Low-level Image Analysis as Filtering

Our approach to using GP for image analysis is based on the idea that most of the low level image analysis tasks, namely image enhancement, feature detection and image segmentation, can be re-framed as image filtering problems and that GP can be used to discover optimal filters which solve such problems.

2.1 Image Enhancement

In the case of *image enhancement* the objective can be to improve the actual quality of an image, for example by removing the noise present in it or by reducing the blurring due to the optics of the camera. Alternatively the objective can be to emphasize some structures of interest in the image, for example to help the visual perception of such structures. In both cases, image enhancement is normally performed with some kind of filter, which applied to the image transforms it into another image with the desired characteristics.

2.2 Feature Detection

In *feature detection* the objective is usually to locate the position (and possibly the shape) of given kinds of (usually small) structures present in the image. Examples of such structures can be edges, ridges/lines, bars, corners, textural elements of various kinds, but also any other problem-specific item. In most cases feature detection involves two steps: a pre-filtering of the image which enhances the features of interest, followed by a decision phase in which thresholding or maxima-detection techniques are used to locate the features of interest in the enhanced image.

2.3 Image Segmentation

In general, *image segmentation* consists of labelling the pixels in the image with a small number of labels so that pixels belonging to regions with (more or less) homogeneous gray level have the same label and pixels belonging to regions with significantly different gray level have different labels [Ballard and Brown, 1982]. Many techniques are available in the literature to solve this problem. A number of them consider the problem of segmenting an image as a pixel classification problem where each pixel is assigned to a class considering local information only (usually the gray levels of the pixels in a neighborhood of the pixel being classified, or information derived from them).

When the number of classes in known in advance and hand-segmented images are available neural networks and other statistical classification techniques can used for this purpose. In the case of neural networks, the approach is either to use a net with several outputs (one for each class) and assign each pixel to the class whose output neuron is maximally active, or to use as many different nets with a single output as the number of classes and select the class whose net provides the maximum output. In both cases there are as many input neurons as the pixels in the neighborhood and the net is used as a kind of *non-linear filter* (possibly with multiple outputs).

Quite often in practical applications the objective of segmentation is more specific and more difficult to achieve: it is to label the pixels belonging to one or more structures of interest as foreground and all the rest as background, even if there are conspicuous intra-class inhomogeneities and inter-class similarities. In these cases segmentation-as-pixel-classification, when at all possible, requires the exploitation of both statistical information about the gray levels of the structures of interest and (possibly more importantly) of other large scale regularities such as the typical shape of the structures of interest. These regularities can only be captured by giving the filter the information contained in large neighborhoods of pixels.

3 GP for Image Analysis

It should be noted that among the problems mentioned in the previous section, image segmentation is the most difficult one, and that standard filtering techniques (in particular linear FIR and IIR filters) usually provide insufficient discriminative power for a reliable classification. Only methods capable of considering all the available sources of information and of combining them in a strongly non-linear way, like artificial neural networks and some ad-hoc combinations of morphologic filters, seem to be able to give (relatively) good results.

However, both neural networks and morphologic operators offer only very peculiar kinds of non-linearities. If their relatively good performance derives from the use of large neighborhoods and the presence of non-linearities, it is arguable that a much larger space of methods exists with such features. Such a space can only be explored with an unbiased machine-learning technique, such genetic programming, which does not impose the use of a specific form of non-linearity, or a fixed shape for neighborhoods.

Although GP can be used to discover efficient optimal filters for image analysis, in order to make the search feasible and effective, terminal sets, function sets and fitness functions have to meet certain requirements.

3.1 Terminal set

A first idea to use GP for image filtering, and successively for pixel-classification, is to use a NN-like strategy, i.e. to use as many variables as the number of pixels in a neighborhood of fixed size, and then to apply the filter (i.e. run the program) to each pixel in the image (while appropriately shifting the neighborhood at each run). The variables would represent the gray level of the pixels in the neighborhood of the current pixel. The output of the program (i.e. of the root node) would be taken to be the output of the filter.

However, this approach presents some problems. The reason is that even neighborhoods of modest size can lead to a very large terminal set including tens to hundreds of variables. This has the consequence that, if large-scale statistical information is necessary to determine the correct output of the filter, GP will have to build very large (sub-)programs in order to extract such information. This can have a very serious impact on the computation load and memory requirements imposed by GP and ultimately on the feasibility of the approach.[1]

Good sets of terminals for image analysis should meet the following *requirements*:

- They should include a limited number of variables to maintain the size of the search space amenable.
- They should be capable of capturing all the information present in the image at different scales (maybe in conjunction with the function set).
- During a run of GP, the process of binding the variables for each evaluation of a program should not require complex calculations as fitness functions for image analysis (see below) include thousands of fitness cases. The ideal set would require only memory accesses. This is possible, for example, if all the operations necessary to evaluate the terminals can be performed once and for all in the initialization phase of GP and the results can be stored in separate arrays.
- If efficiency of the programs discovered via GP is sought, the computation load necessary to evaluate the aforementioned arrays should be as light as possible.

While the first two requirements can be met by many different sets of terminals, the third and fourth constrain considerably the possible choice.

[1] Actually, in some experiments (not reported) in which various combinations of functions and terminals were tried we had some difficulties with the afore-mentioned terminal set.

An example of a simple terminal set which satisfies these requirements is shown in Table 1, where $I(x, y)$ is the gray level of the pixel at coordinates (x, y), and (x_c, y_c) are the coordinates of the current pixel (i.e. of the center of the neighborhood). The terminals are the value taken by the convolution of the image with "box" operators (i.e. constant masks) of different sizes.

This set of terminals has the following features:

- It includes only a few terminals.
- It captures both fine scale and broader scale information. Also, in conjunction with the shifting macros described in the next section it provides full information (see below).
- The moving averages `I_3x3`, `I_7x7`, `I_15x15` etc. can be computed very efficiently, in the initialization phase, with an algorithm requiring only 4 additions and 1 division per pixel. Also, they can be stored in appropriate arrays, so that their evaluation at runtime requires only a memory access.

3.2 Function set

In addition to the usual closure and sufficiency properties, efficiency requirements similar to those of terminal sets apply to function sets for image analysis.

While closure is easy to achieve, sufficiency cannot be ensured, especially in problems for which no exact solution (i.e. filter) is known. A necessary step towards sufficiency is to use functions which can extract all the information contained in the image, when this is not already made available by the terminal set.

However, in order to meet the efficiency requirements, in most cases it is necessary not to include functions which perform real image processing operations. The reason is that the computation performed by such functions would depend on the values of their arguments. Those values may vary considerably between different programs, between different parts of a single program and more importantly from pixel to pixel. As a result it is impossible to pre-compute and store the values to be returned by such functions. Thus, the evaluation of the fitness of each individual may require a computation load orders of magnitude bigger than if image-processing functions are not used.

A function set that meets this requirements is shown in Table 2, where a, a_1 and a_2 are dummy arguments. It is worth noting that the macros `Xcp1`, `Xcm1`, `Ycp1` and `Ycm1` allow the construction of filters whose complexity is well beyond the one provided by the terminal set.[2] In addition, the presence of `Min`, `Max`, `Div` and `Mul` allows the construction of highly non-linear filters (e.g. mathematical-morphology-like operators). These operations in conjunction with `Add` and `Sub` allow the creation of any necessary constants, too.

[2] It is well known in image processing, that by combining properly shifted and scaled box operators, any filter (including Gaussian filters, derivatives of Gaussian, Laplacian filters, etc.) can be approximated to the desired degree of accuracy.

Terminals	
Name	*Definition*
`random`	Random constant generator
`I`	$I(x_c, y_c)$
`I_3x3`	$\frac{1}{9}\sum_{d_x=-1}^{1}\sum_{d_y=-1}^{1} I(x_c+d_x, y_c+d_y)$
`I_7x7`	$\frac{1}{49}\sum_{d_x=-3}^{3}\sum_{d_y=-3}^{3} I(x_c+d_x, y_c+d_y)$
`I_15x15`	$\frac{1}{225}\sum_{d_x=-7}^{7}\sum_{d_y=-7}^{7} I(x_c+d_x, y_c+d_y)$
⋮	Moving averages on larger neighborhoods

Table 1: A terminal set for image analysis.

Functions		
Name	*Arity*	*Definition*
`Add`	2	$a_1 + a_2$
`Sub`	2	$a_1 - a_2$
`Mul`	2	$a_1 \times a_2$
`Div`	2	$\begin{cases} a_1/a_2 & \text{if } \lvert a_2\rvert \geq 0.00001 \\ a_1 & \text{otherwise} \end{cases}$
`Max`	2	$\max(a_1, a_2)$
`Min`	2	$\min(a_1, a_2)$
Macros		
Name	*Arity*	*Definition*
`Xcp1`	1	Evaluates a with $x_c = x_c + 1$
`Xcm1`	1	Evaluates a with $x_c = x_c - 1$
`Ycp1`	1	Evaluates a with $y_c = y_c + 1$
`Ycm1`	1	Evaluates a with $y_c = y_c - 1$

Table 2: A function set for image analysis.

3.3 Fitness functions

If the objective is to develop filters for image enhancement, it is possible to use a symbolic-regression-like fitness function in which the output of the program at each location is compared with the desired output. The sum of the absolute errors made by the program for all the pixels of all the images of a training set can then be transformed into a fitness function with the usual scaling or mapping techniques. The availability of desired-output images and the huge number ($\approx 10^6 - 10^7$) of runs necessary to evaluate the fitness of each program hamper this approach. While the second problem can be reduced considerably by sampling the images of the training set at, say $10^3 - 10^4$, random points and running the program only in those locations, the first problem remains as hand retouching seems unacceptable for real-world images. The only solution is to use other, much more expensive filtering techniques (say a Kalman filter) or to use different image acquisition technologies to provide the desired output.

If the objective is to develop filters for feature detection or image segmentation, then using a symbolic-regression-like fitness function that forces the output of the program to exactly match the binary values representing the structures to be segmented/detected in a set of training images seems inappropriate. There are two reasons for this.

The first reason is that the output of the filter will have to be passed to a decision algorithms which in most cases will simply apply a threshold to it in order to make a decision as to the class to which to assign the pixel in (x_c, y_c). So all the computational effort spent by GP to build programs with binary outputs is unnecessary in this case. The search that GP has to do can be made considerably easier by using a wrapper for the filtering program, which simply sets the output to 1 if the filter's output is positive, to 0 otherwise.

The second reason has to do with the sensitivity/specificity dilemma [Poli and Valli, 1996] that any segmentation or detection algorithm has to face: with real-world images no algorithm can detect/segment the points belonging to the structures of interest (True Positives, TPs) without wrongly detecting also some points which belong to uninteresting structures (False Positives, FPs) and missing some points (False Negatives, FNs). Each algorithm will have both a *sensitivity* (number of TPs divided by the number of points to be detected) and a *specificity* (number of True Negatives, TNs, divided by the number of points not to be detected) smaller than 1. Changing the settings for the parameters of the algorithm can only increase the specificity and decrease the sensitivity or vice-versa. The optimum setting for an algorithm is usually one in which both the sensitivity and the specificity are "as big as possible".

The fitness function[3] used in symbolic regression is essentially the sum of the number of FPs and FNs: $f = FP + FN$. For a fixed training set, FP and FN determine the sensitivity and specificity of each program. However, optimizing their sum does not necessarily mean to achieve an optimum sensitivity/specificity trade-off. On the contrary, experiments with this fitness function have shown that, in most cases, GP tend to discover algorithms which are either very sensitive (and poorly specific) or vice versa.[4]

Unfortunately, optimising the sensitivity/specificity trade-off is quite difficult as, depending on the particular detection task, FPs and FNs can have a completely different impact on the quality of the results. For example, in some tasks it may be totally unacceptable to have false detections (FPs), while in others it is much better to have false detections than misses (FNs) (think for example of the detection of micro-calcifications in breast radiograms).

In our experience, the fitness function that has given the best results is:

$$f = FP + FN \exp\left(10\left(\frac{FN}{P} - \alpha\right)\right)$$

where P is the number of pixels belonging to the structures to be detected and α is a domain dependent parameter which represents the maximum percentage of misses above which the number of misses is considered unacceptable (typically $\alpha = 0.6 - 0.9$). Basically, if FN/P is sufficiently smaller than α, $f \approx FP$, i.e. GP tries to minimize the number of false detections. Instead, if FN/P is slightly bigger that α, $f \approx FN \exp(10\,(\frac{FN}{P} - \alpha))$, i.e. GP tries to minimize the number of misses.

[3]For the sake of simplicity in the following we will use the term "fitness" to mean "cost", i.e. a quantity to be minimised. A simple change of sign can transform the former into the latter.

[4]The fitness function $f = FP^2 + FN^2$ tends to behave slightly better but still does not give acceptable results when the number of pixels to be detected is much smaller than the pixels in the image.

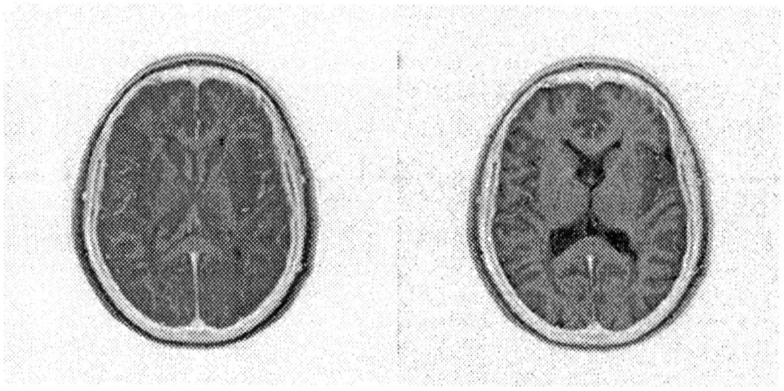

Figure 1: Two original MR images of the head.

4 Experimental Results

This section describes a small set of experiments performed with the functions and terminal described in the previous section. As one of our objectives is to obtain filters that can perform optimal segmentation and detection in complex real-world gray scale images, we did not even try to apply GP to simplified problems involving the filtering of binary images, synthetic images or small size images. On the contrary we decided to use real data from the medical domain and to face an extremely difficult problem: the segmentation of the brain in Magnetic Resonance (MR) images like the ones in Figure 1 (the images are maps of the intensity of two different physical properties of the same section of the head of a patient).

Pixel classification strategies based on neural networks or other statistical approaches have been used in the past for this task with limited success. The difficulty resides in the fact that some kinds of tissue in the brain present the same physical properties as some tissues in the skin. As a consequence, brain segmentation based on small neighborhoods is basically impossible [Coppini et al., 1992]. However, this does not mean that using large neighborhoods, the task becomes easy. To show this and also to provide a reference for comparison with GP, some experiments using neural networks in conjunction with medium- and large-size neighborhoods have been performed.

The architecture of the nets is shown in Figure 2. They are fully connected nets including: a) two squared "retinas" of

$N \times N$ input neurons (one for each input image) fed with the gray levels of the pixels in the neighborhood, b) ten hidden neurons, c) one output neuron whose activation determines whether the pixel at the center of the retinas belongs to the brain or not. The networks were trained with the backpropagation algorithm to segment the brain in the images in Figure 1.

Two different sizes for the retinas have been tried, 5×5 and 15×15, corresponding to medium and coarse scale (for 256×256 pixel wide images). Figures 3(a)–(b) show the output produced by the nets. Despite the very large number of degrees of freedom in the networks (more than 500 for the smaller net, more than 4500 for the larger one) and the presence of a *single image-pair in the training set*, the segmentation of the training set itself shown in Figures 3(c)–(d) looks very disappointing for both nets.[5] In particular, it can be observed that the small retinas of the first net provide insufficient information to avoid FPs in the skin region, while the large retinas of the second net lead to FNs due to the inability to correctly classify the brain edges. However, it should be noted that the poor quality of these results is subjective (i.e. it is the result of our demanding visual system) as the misclassification rates were relatively small: 3.90% for the 5×5 net and 1.84% for the 15×15 net. This is phenomenon is quite common in image analysis, where small variations in decimal figures can make big changes in the perceived quality of the results. For this reason image segmentation is in general such a difficult problem.

Figure 4 shows the results obtained in the same task by the following program with fitness $f = 16.6004$ discovered by GP after approximately 40,000 fitness evaluations:[6]

```
(Sub (Max  MR0_15x15 (Max (Max  (Mul (Max (Max (Mul (Max
MR0_15x15  MR0_15x15)       MR0_15x15)    (Sub    MR1_15x15
MR0_15x15))  MR0_15x15)     MR0_3x3) MR0_15x15)    (Mul (Add
(Mul   (Max MR0 (Max (Mul (Add  (Mul (Div  0.55 (Max (Max
(Mul MR0    MR0_7x7) (Max (Max  MR0_15x15 MR0_15x15) (Mul
(Mul   (Mul MR0_15x15  MR0_15x15) MR0_15x15) MR0_15x15)))
(Max (Mul (Add MR0_15x15 (Max  (Max (Mul (Max (Mul -0.75
MR0_15x15) (Max (Mul (Add MR0_15x15  (Mul (Mul (Add (Mul
MR1_15x15   MR0_7x7)  (Mul  (Min   MR0_15x15  MR0_15x15)
MR0_15x15)) MR1_15x15) MR0_15x15)) MR0_15x15) (Mul  (Div
(Add  (Mul     (Mul (Add  (Mul  0.55   MR1_3x3) (Mul (Min
MR1_15x15 (Mul (Max (Mul -0.75 MR0_15x15) (Max (Mul (Sub
MR0_15x15 MR0_15x15) MR0_15x15) (Mul (Div (Add (Mul (Add
(Mul  0.55   MR1_3x3)  (Mul  (Min   MR1_15x15 MR0_15x15)
MR0_15x15))  MR1)  MR0_15x15) 0.55)  (Mul (Max (Max (Add
MR1_3x3 MR0_15x15) (Sub MR0_15x15 MR0_15x15)) MR0_15x15)
MR0_15x15))))  MR0_15x15)) MR0_15x15))     MR1)       MR1)
MR0_15x15) 0.55) (Mul (Max  (Max (Add MR1_3x3 MR0_15x15)
(Sub MR0_15x15 MR0_15x15))   MR0_15x15)     MR0_15x15))))
MR0_15x15)  MR0_15x15)   (Mul  (Add  (Mul 0.55  MR0_3x3)
MR0_15x15) MR0_15x15))) MR0_15x15) MR0_15x15))) MR0_7x7)
MR0_15x15)   0.98)    MR0_15x15))         MR0_15x15)  0.55)
MR0_15x15))) 0.55)
```

Although the understandability of this program is minimum (comparable with that of a neural net), the segmentation it produces looks considerably better. GP performances are also objectively better as the misclassification rate is only 1.56%. To our knowledge this result is very difficult to achieve with

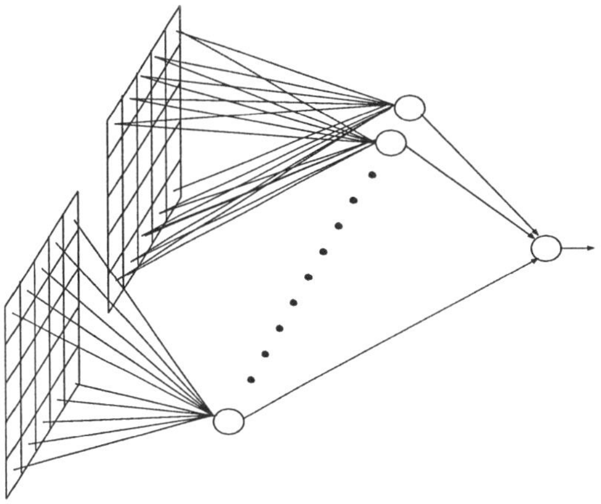

Figure 2: Architecture of NN used for MR image segmentation (for clarity some neurons and connections have not been drawn).

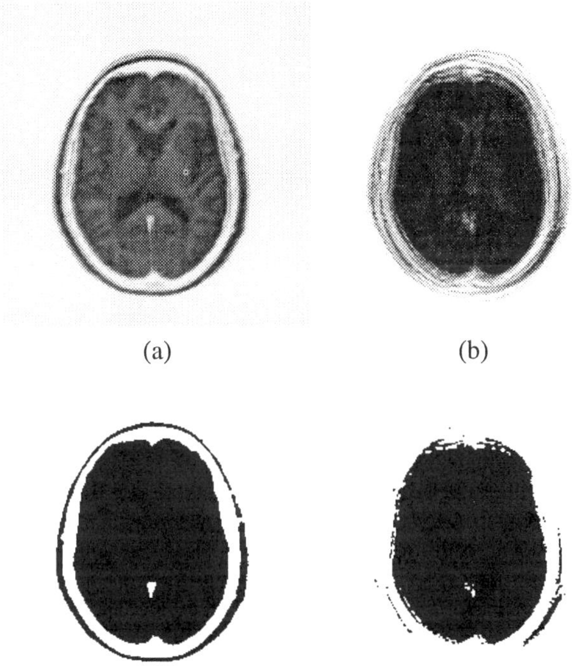

Figure 3: Segmentation of the images in Figure 1 produced by neural nets: (a) output of a net with 5×5 input retinas, (b) output of a net with 15×15 input retinas, (c) segmentation (via optimum thresholding) of the image in (a), (d) segmentation (via optimum thresholding) of the image in (b).

[5] Segmentation was obtained by thresholding the output of the nets. The threshold were chosen so as to minimise the difference between the segmented images and the training ones.

[6] The terminals with prefixes `MR0` and `MR1` represent the moving averages of the first and second image in Figure 1, respectively.

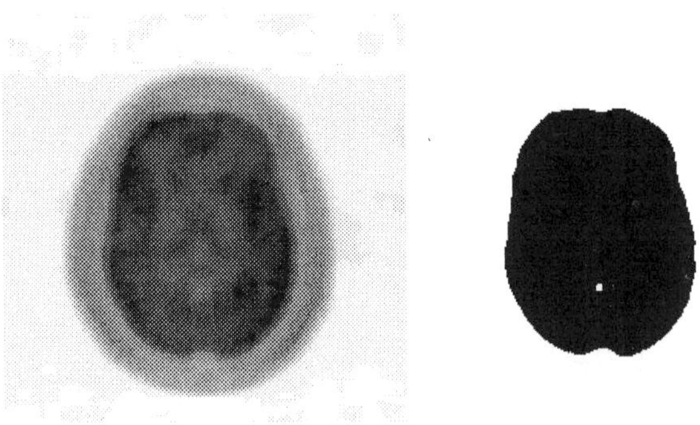

(a) (b)

Figure 4: Segmentation of the images in Figure 1 produced with GP: (a) output of the filtering program, (b) segmentation produced by the wrapper applied to the image in (a).

other techniques (even on a single image) unless anatomical knowledge is explicitly used.

5 Conclusions

In this paper we have presented an approach to image analysis based on the idea of using GP to discover optimal image filters. Although GP could be applied in a naive way to such a problem, we have outlined some criteria that terminal sets, function sets and fitness functions should satisfy in order to make the search feasible and produce efficient filters. On the ground of this criteria we have proposed some simple terminals, functions and fitness functions that in some preliminary experiments with medical images have enabled GP to outperform neural nets.

The research on genetic programming for image analysis is hampered by the tremendous demand of computational resources involved in fitness evaluation. This makes it very difficult if not impossible to produce the good result-documentation typical of genetic programming literature (e.g. plots of the probability of reaching a certain fitness value vs. generation number, the minimum computation effort required, generalisation capabilities of the filters discovered, etc.). However, the impressive performance shown by GP compared with NNs in the experimental results reported in this paper seems to suggest that there is a huge space of image analysis tools much more powerful than the ones used in image processing nowadays waiting to be discovered/designed. Whether genetic programming will be the ultimate tool to do that is certainly an open question to be addressed with future research.

Acknowledgements

The author wishes to thank Aaron Sloman and all the members of the EEBIC (Evolutionary and Emergent Behaviour Intelligence and Computation) group for useful discussions and comments. This research is partially supported by a grant under the British Council-MURST/CRUI agreement.

References

[Andre, 1994] Andre, D. (1994). Automatically defined features: the simulataneous evolution of 2-dimensional feature detectors and an algorithm for using them. In K. E. Kinnear, Jr., editor, *Advances in Genetic Programming*, chapter 23, pages 477–494. MIT Press.

[Ballard and Brown, 1982] Ballard, D. and Brown, C. (1982). *Computer Vision*. Prentice-Hall, Englewood Cliff, NJ.

[Breunig, 1995] Breunig, M. M. (1995). Location independent pattern recognition using genetic programming. In Koza, J. R., editor, *Genetic Algorithms and Genetic Programming at Stanford 1995*, pages 29–38. Stanford Bookstore, Stanford University.

[Coppini et al., 1992] Coppini, G., Poli, R., Rucci, M., and Valli, G. (1992). A neural network architecture for understanding 3D scenes in medical imaging. *Computer and Biomedical Research*, 25:569–585.

[K. E. Kinnear, Jr., 1994] K. E. Kinnear, Jr., editor (1994). *Advances in Genetic Programming*. MIT Press.

[Koza, 1992] Koza, J. R. (1992). *Genetic Programming: On the Programming of Computers by Means of Natural Selection*. MIT Press.

[Koza, 1994] Koza, J. R. (1994). *Genetic Programming II: Automatic Discovery of Reusable Programs*. MIT Pres, Cambridge, Massachusetts.

[Poli and Valli, 1996] Poli, R. and Valli, G. (1996). Hopfield neural nets for the optimum segmentation of medical images. In Fiesler, E. and Beale, R., editors, *Handbook of Neural Computation*, chapter G.5.5. Oxford University Press. (in press).

[Riolo and Line, 1995] Riolo, R. L. and Line, M. P. (1995). Automatic discovery of classification and estimation algorithms for earth-observation satellite imagery. In Siegel, E. S. and Koza, J. R., editors, *Working Notes for the AAAI Symposium on Genetic Programming*, pages 73–77, MIT, Cambridge, MA, USA. AAAI.

[Tackett, 1993] Tackett, W. A. (1993). Genetic programming for feature discovery and image discrimination. In *International Conference on Genetic Algorithms*.

[Teller and Veloso, 1995] Teller, A. and Veloso, M. (1995). A controlled experiment: Evolution for learning difficult image classification. In *Seventh Portuguese Conference On Artificial Intelligence*. Springer-Verlag.

Evolving Agents

Adil Qureshi
Genetic Programming Group,
Dept. of Computer Science,
University College London,
Gower Street,
London WC1E 6BT, UK
Email: A.Qureshi@cs.ucl.ac.uk
URL: http://www.cs.ucl.ac.uk/staff/A.Qureshi/

Abstract

The paradigm of *agent based computing* is becoming increasingly popular both in *distributed artificial intelligence* and as a general software engineering technique. The difficulty with agent based computing is that success depends not just on the correctness of any one agent, but on the *emergent behaviour* arising from the interaction of a *society* of agents. As a consequence, the problem of programming agents is non trivial and poorly understood.

In this paper we show that genetic programming can be used to automatically program agents which communicate and interact to solve problems. The programs evolved simultaneously define when and what to communicate, and how to use the communicated information to solve the given problem.

1 Introduction

In previous work GP approaches have concentrated on evolving *reactive agents*. The latter work on the principle that complex intelligent behaviour can emerge from the interaction of a collection of very simple agents. These agents obtain knowledge about actions and behaviours of other agents by sensing changes in the environment. Any communication is at a very low level, achieved implicitly by making changes to the environment rather than explicitly by exchanging messages with each other. Koza used GP to evolve reactive agents for the central food foraging problem [5]. Haynes evolved agents for the predator/prey (capture) problem [3], and others [2], [4]. Reynolds [8] co-evolved agents to simulate the game of tag. Manela [7] employed GAs to optimise architectural features of agents (including simple communication) for the pursuit problem. Andre [1] used GP to co-evolve program pairs, the first of which creates a map of the environment, and the second of which uses this map to control a mobile robot. The programs are run in sequence (writer first, then reader) and can be considered to be using a form of unidirectional communication.

In this paper we concentrate on using GP to evolve agents that communicate directly with each other, via message passing. These evolved agents must decide what information is to be communicated and how it is to be used to solve the global problem.

2 The Two Agents Problem

The goal of our experiments is to see if GP can be used to evolve agents that communicate to solve problems. For this purpose we have defined a simple problem for which it is necessary for communication to take place and furthermore, for which it is easy to measure the effectiveness of any communication. The problem involves two agents A and B that steer two vehicles placed on a two dimensional grid. The vehicles are constantly moving in the direction that they are currently pointing. All movement is discrete, so that the vehicles always have integer (x,y) coordinates. Initially the vehicles are pointing north. The steering mechanism allows the orientation of the vehicles to be changed to any one of the 8 compass directions. The goal is to evolve the code for agents A and B such that they steer the vehicles to enable them to meet. The vehicle model closely follows that used by Reynolds [8].

The vehicles are given energy proportional to the euclidian distance between them. The latter is an attempt to prevent the "meet you at (x,y)" strategy in which the agents obviate any need for communication having previously agreed to always move to the fixed position (x,y). A single energy unit is consumed each time a vehicle moves one step in any direction. The game ends when both vehicles run out of energy or when they meet.

Agents A and B know only their own positions, but both have access to two bi-directional unbuffered communication channels called ChannelA and ChannelB which can be used to send and receive messages to each other. Each communication channel is "wired" so that one end is connected to one vehicle and the other to the other vehicle. A message sent from one end can only be received at the other end, and vice versa.

The communication channels have been designed so they can only be used to send and receive messages in the form of integers. When an attempt is made to read from a channel the last message received is delivered (default value 0).

Clearly what we would like is for agents A and B to evolve behaviours that attempt to use the communication channels to exchange position information.

The success of the agents is measured using the following fitness function:-

$$Fitness(A) = Fitness(B) = \frac{D_{before}(A,B)}{D_{before}(A,B) + D_{after}(A,B)}$$

Where D_{before} and D_{after} are the distances between A and B before and after execution respectively. This function produces a range of fitness values varying between 0 and 1. A value of 0.5 indicates that the agents remained the same distance apart. Values below 0.5 indicate that the agents moved further apart. The value 0 is actually not possible since it implies that the agents were at an infinite distance apart when they ran out of energy. Values above 0.5 indicate that the agents moved closer together, and a value of 1 indicates that they met.

Note that for the agents to meet, we actually mean that they must be at the same coordinates at the same time. Since the vehicles are constantly moving, the latter implies synchronised movement which makes the problem more difficult.

2.1 Architecture

Tables 1 and 2 provide details of the architecture used to evolve programs to steer the agents. Note that all the primitives are integer based. The program makes use of a main branch and two ADFs [6]. ADF0 is a parameterised function and takes two arguments. Its function and terminal set do not contain any environment sensing or effecting facilities. All communication and most environment sensing facilities are available only to ADF1. The main branch has only ADF0 in its function set. The terminal set includes ADF1 and GetD, where the latter is the only environment sensing primitive not available to ADF1. The idea is that ADF0 is a computational unit and ADF1 is a communication unit.

A fixed set of 25 randomly generated initial positions (within the bounded region (-100,-100) to (100,100)) were used to evaluate each individual in the population. The raw fitness values for each of the 25 cases were averaged to arrive at an overall fitness value for the individual. The energy allocated to each individual was $1.5x$ the initial distance between the agents. This provides sufficient energy to allow solutions that don't use diagonals to evolve, but at the same time inhibits solutions that use the "meet you at (x,y)" strategy.

Each individual in the population was evaluated by using it to steer both A and B. This is achieved by alternatingly executing the GP once in the context of A, and then in the context of B. The result of each execution is used to steer the respective vehicle, which is then moved forward one unit. This process is repeated until either the two agents both run out of energy, or they meet. The pseudo code in table 3 illustrates this approach.

Table 3: Evaluation Pseudo Code

```
agent = A
while ((A.energy || B.energy) && (!met(A,B))) {
  if (agent.energy) {
    steering_delta = eval GP in context of agent
    agent.turn(steering_delta)
    agent.move()
    agent.energy--
  }
  if (agent == A) agent = B else agent = A
}
```

The turn() function adds the steering delta to the current direction. Negative delta values can be used to steer the vehicle anti-clockwise, and positive values clockwise. The turn() function guarantees that any resulting direction is always in the range 0-7.

Additional code was added to the move() function so that when enabled, it generates output for the LOGO turtle graphics language. This output provides a very simple means of visualising the performance of a GP for each of the test cases.

After the completion of a run, the best individual was re-evaluated using 1000 test cases instead of twenty five to check for the generality of the solutions. A further re-evaluation used the original 25 test cases, but with send operations disabled on both communication channels for both agents. The latter is used to provide a measure of the dependency of the solution on communication.

3 Results

The results for the 3 best out of 10 different runs are shown in table 4. Clearly the best individual is from run 7, which is also the most general solution. As expected, all three individuals score very low fitness values when both of the communications channels are disabled.

LOGO output tracing in "realtime" the movement of the agents, together with the UCB LOGO package can be obtained from the directory "/genetic/agents2" at the ftp site "cs.ucl.ac.uk" at UCL. The file named README explains how to view the results. Figure 1 shows snapshots of the output trace for the best individuals from the three best runs for sample test cases. The snapshots were taken at the point when the agents met or when

Table 1: Actions Performed by Terminals and Functions

Primitive	Purpose
`Add`(x,y)	**return** $x+y$
`Sub`(x,y)	**return** $x-y$
`Mlt`(x,y)	**return** $x*y$
`Div`(x,y)	**return** x/y
`IfGT`(x,y,$t1$,$t2$)	**if** $x > y$ **return** $t1$ **else return** $t2$
`RecvA`	**if** no message received yet at *ChannelA*, **return** 0; **else return** last message received
`RecvB`	**if** no message received yet at *ChannelB*, **return** 0; **else return** last message received
`SendA`(x)	send x down *ChannelA*; **return** x
`SendB`(x)	send x down *ChannelB*; **return** x
`GetX`	get the current X coordinate; **return** X
`GetY`	get the current Y coordinate; **return** Y
`GetD`	get the current direction D; **return** D
`Arg1, Arg2`	arguments of current ADF
`Zero`	constant value 0; **return** *0*

Table 2: The Two Agents Problem

Objective:	Find a program that steers the two agents so that they meet before they run out of energy.	
Primitives:	Functions	Terminals
Main	`Adf0`	`Adf1, GetD`
Adf0	`Add, Sub, Mlt, Div, IfGT`	`Arg0, Arg1, Zero`
Adf1	`SendA, SendB, Add, Sub, Mlt, Div, IfGT`	`RecvA, RecvB, GetX, GetY, Zero`
Fitness cases:	25 randomly generated fixed test cases.	
Energy:	$1.5 * Distance_{before}(A, B)$	
Raw Fitness:	$Fitness = \frac{Distance_{before}(A,B)}{Distance_{before}(A,B)+Distance_{after}(A,B)}$	
Fitness:	Average of the raw fitness scores of the 25 test cases times 10,000.	
Parameters:	Pop = 1000, G = 100, one dimensional demes (size 25), grid range (-100,-100) to (100,100).	
Termination Condition:	A program gets a fitness score of 10,000 or current generation exceeds G.	

they ran out of energy.

Table 4: Results

Run	Length	Fitness (25)	Fitness (25) No Comms	Fitness (1000)
0	74	8028	5081	6638
2	163	8122	4710	6300
7	211	8831	4996	8979

3.1 Run 0

The best individual from run 0 (table 5) achieves high fitness for the training test cases, but scores a considerably lower fitness value when faced with the more extensive unseen test cases. The strategy used by this individual is not very clearly defined.

X coordinates are exchanged using ChannelB, and the agent with the greater X value moves diagonally in the North-West direction until X coordinates are equalised. No clear strategy is apparent for equalising Y coordinates, however the agent which starts off with the greater X value never moves South, hence meetings are only be achieved when the other agent moves North.

Table 5: Best Individual from Run 0

Main: ((ADF0 ADF1 GetD))
ADF0: ((IfGT (Mlt Arg1 Arg1) (Add Arg0 Arg1) (Add (Div Arg1 Arg1) (Add Arg0 Arg1)) (Add (Mlt Arg1 Arg1) (Mlt (Mlt Arg1 Arg1) (IfGT Arg0 Arg0 Arg1 Arg0)))))
ADF1: ((Mlt (Mlt (IfGT GetX RecvB RecvA RecvB) (Mlt (Mlt (IfGT GetX RecvB RecvA RecvB) (IfGT GetX RecvB RecvA RecvB)) (SendB GetX))) (Mlt (Mlt (IfGT GetX RecvB RecvA RecvB) (Mlt (Mlt (IfGT GetX RecvB RecvA RecvB) (IfGT GetX RecvB RecvA RecvB)) (SendB GetX))) (SendB GetX))))
Fitness: 8028
Structural Complexity: 74

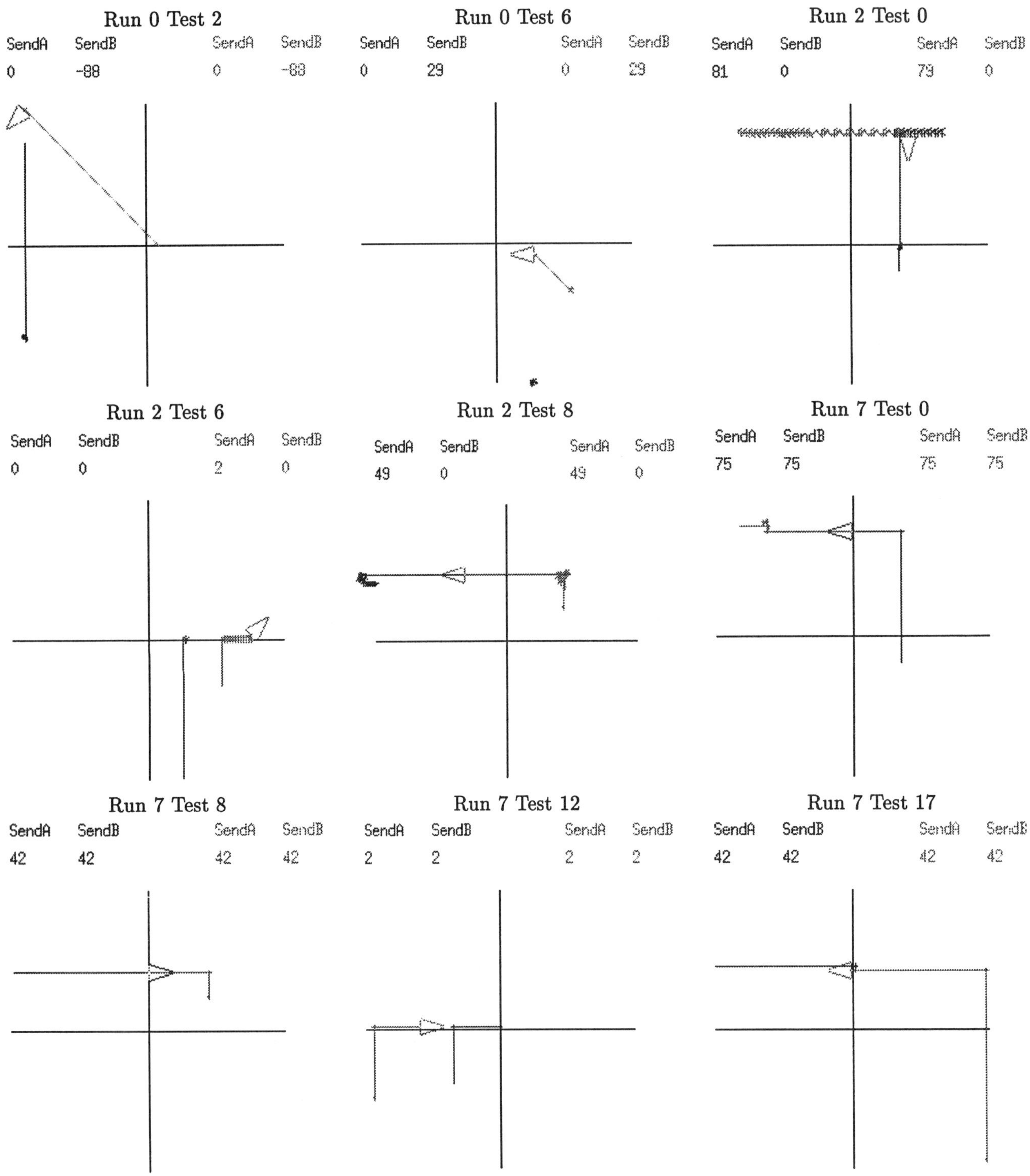

Figure 1: Sample Trace of Agent Movement

3.2 Run 2

The best individual from run 2 (table 6) again achieves high fitness for the training tests, but again is significantly less successful with the more extensive unseen test cases. The strategy used by this individual is also not very clear.

This time Y coordinates are exchanged using ChannelA, and the agent with the lower Y coordinate moves North until Y coordinates are matched. If both agents have negative Y coordinates, they both move North until they reach the line Y = 0. The agent with the higher Y coordinate (or the first to reach Y = 0 for the negative Y coordinates case) moves left or right in semi-circles. This makes it difficult for the other agent to match its Y coordinate exactly. The result is that when they are at approximately the same Y coordinates, they both end up following the cyclic movement. If during this movement, they synchronise Y values, they remove the vertical component from their movement and follow an East or West path. If they end up pointing towards each other, the latter facilitates a meeting.

Table 6: Best Individual from Run 2

Main: ((ADF0 ADF1 GetD))
ADF0: ((IfGT Zero Arg0 Arg1 (IfGT Zero Arg0 Arg1 Arg0)))
ADF1: ((IfGT (Div (Mlt RecvA Zero) (Mlt RecvA Zero)) (Div (Div RecvA RecvA) (Sub GetY RecvA)) (IfGT Zero Zero Zero RecvA) (IfGT (SendA GetY) (Sub GetY RecvA) (IfGT (Div (IfGT RecvB Zero GetY RecvA) (Sub GetY GetX)) (Sub (IfGT RecvB Zero GetY RecvA) (Sub GetY RecvA)) (Add (SendA GetY) (Sub (Sub GetY RecvA) RecvA)) (IfGT (Div RecvA RecvA) (Sub GetY RecvA) (IfGT (Div (IfGT RecvB Zero GetY RecvA) (Sub GetY GetX)) (Sub GetY RecvA) (IfGT (SendA GetY) (SendA (IfGT (SendA GetY) (Sub GetY RecvA) (IfGT (Div (IfGT RecvB Zero GetY RecvA) (SendA GetY)) (Div RecvA RecvA) (Sub (Div RecvA RecvA) (Sub GetY RecvA)) (IfGT (Div (SendA GetY) (Sub GetY GetX)) (SendA GetY) (Div (IfGT RecvB Zero GetY RecvA) (Sub GetY GetX)) (SendA GetY))) (SendA GetY))) (IfGT Zero Zero Zero RecvA) (Sub GetY RecvA)) (Div (IfGT RecvB Zero GetY RecvA) (Div (IfGT RecvB Zero GetY RecvA) (Sub GetY RecvA)))) (SendA GetY))) (SendA GetY))))
Fitness: 8122
Structural Complexity: 163

3.3 Run 7

The best individual for run 7 (table 7) not only achieves the highest fitness value, it also proves to be the most general solution when tested over 1000 different tests, scoring an even higher fitness value. The strategy used by this individual is very well defined.

The basic strategy is to equalise Y coordinates by making use of communication channel A. X-coordinates are not exchanged, so instead the agents have "agreed" to meet at the ordinate X = -1. This works best if the agents are on opposite sides of the line X = -1 or very close to this line. If however the agents are on the same side of the line X = -1, but distant from it, they waste energy by moving without getting any closer to each other, and are hence more likely to run out of energy before they meet.

The behaviour of an agent at any instant, can be approximated by the pseudo code in table 8.

Table 7: Best Individual from Run 7

Main: ((ADF0 GetD ADF1))
ADF0: ((Add (Sub Arg1 Zero) (Add (Mlt (Sub Arg1 Zero) (Sub Arg1 Zero)) (IfGT Arg0 Arg0 Arg0 Arg0))))
ADF1: ((IfGT (SendA GetY) (Div (SendA RecvA) (IfGT (Add GetX Zero) (SendB RecvA) (Add GetX Zero) (Div GetX Zero))) (Add (Div (IfGT RecvA RecvA Zero GetY) (IfGT (IfGT (SendB RecvA) (Sub (IfGT RecvA RecvA Zero GetY) (SendB RecvA)) (Div GetX Zero) (Add GetX Zero)) (Add GetX Zero) (Add (SendB RecvA) (Div GetX Zero)) (IfGT (SendB RecvA) (IfGT (IfGT (SendB RecvA) (IfGT RecvA RecvA Zero GetY) (Add GetX Zero) (IfGT RecvA RecvA Zero GetY)) (Sub (IfGT RecvA RecvA Zero GetY) (SendB RecvA)) (Div GetX Zero) (Add GetX Zero)) (IfGT (SendB RecvA) (Add (Div (IfGT RecvA RecvA Zero GetY) (IfGT (Div GetX Zero) (Add GetX Zero) (Add (SendB RecvA) (Div GetX Zero)) (IfGT (SendB RecvA) (IfGT (IfGT (SendB RecvA) (Div (IfGT RecvA RecvA Zero GetY) (SendB RecvA)) (SendB RecvA) (IfGT RecvA RecvA Zero GetY)) (Sub (IfGT RecvA RecvA Zero GetY) (SendB RecvA)) (Div GetX Zero) (Div (SendA GetY) (SendB RecvA)))) (IfGT (SendB RecvA) (Add (Div GetX Zero) (SendB RecvA)) (Div GetX Zero) (Add GetX Zero)) (SendB RecvA)))) (Div (IfGT RecvA RecvA Zero GetY) (IfGT RecvA RecvA Zero GetY))) (Div GetX Zero) (Add GetX Zero)) (SendB RecvA)))) (Div (SendA GetY) (SendB RecvA))) (Div GetX Zero)))
Fitness: 8831
Structural Complexity: 211

Table 8: Strategy of Best Individual from Run 7

```
if (my Y > 0) and (my Y >= remote Y)
    Send my Y
    if (my X > -1)
        move East
    else
        move West
else if (my Y > 0) and (my Y < remote Y)
    Send my Y
    move North
else if (my Y < 0)
    Send 2
    move North
```

4 Conclusion

We have demonstrated that it is possible to evolve agents that communicate with each other to solve a global problem. In doing so, the agents effectively decide what to communicate and when to communicate it. They then use the information communicated to solve the global problem. In summary behaviour directed by communication automatically defines who does what and when.

In the experiments performed the code generated only used one of the communication channels, for exchanging position information pertaining to only one of the axis. All the runs used heuristics to guide movement along the axis for which no information was exchanged. This problem could be due to any number of reasons, including the chosen architecture, loss of population diversity, or even the difficulty of the problem. Early analysis suggests that the problems are related to population diversity, but this will need to be investigated further. The results do however clearly show that GP programs can use communication to exchange information vital to solving a problem.

The long term goal of our project is to see if agents that use a very primitive language (for example, consisting of just verbs and nouns) to communicate and solve complex problems can be evolved.

5 Future Work

5.1 *Extending the Two Agent Problem*

Currently we do not take into account the total units of energy spent during the test cases. Hence agents that are more wasteful can potentially end up with the same fitness score as efficient ones. We will therefore use energy remaining as tie breaker when two individuals have the same fitness value.

It would be interesting to look at the effect of noisy communication channels, to see if noise tolerant communication protocols can be evolved. This will probably require the set of primitives to include indexed memory.

The communication channels are currently unbuffered, we could make them buffered, and furthermore add some cost so that each time a message is sent or received, energy is consumed. The amount of energy consumed would probably need to be much less than the cost of movement.

5.2 *The Three Agent Problem*

We are currently extending the two agent problem to a new variation of the pursuit problem. A third agent P is to be co-evolved, whose goal is to intercept A or B before they meet. P is equipped with communications facilities that allow messages to be received from and sent to both of the communication channels of A and B.

Experiments will be performed in passive and active phases. In the passive phase P will only be allowed listen to the communication channels connecting A and B. In the active phase, P will also be able to send messages to both A and B. The passive phase allows P to eavesdrop and hence obtain position information which should help it to intercept one of the agents. Agents A and B must hence use some form of encryption to prevent discovery. The active phase allows P to send decoy messages which would not be easy to distinguish between the real messages sent between A and B. This time A and B would need to develop methods of authentication. Clearly the set primitives would need to be extended to include indexed memory to facilitate the latter. In addition, a buffered communication model will probably be required.

Acknowledgements

A. Qureshi is funded by the EPSRC. I would like to thank my supervisors (Jon Crowcroft and Steve Hailes), and Adam Fraser for GPC++.

References

[1] David Andre. Evolution of mapmaking ability: Strategies for the evolution of learning, planning, and memory using genetic programming. In *Proceedings of the 1994 IEEE World Congress on Computational Intelligence*, volume 1, pages 250–255. IEEE Press, June 1994.

[2] Thomas Haynes, Sandip Sen, Dale Schoenefeld, and Roger Wainwright. Evolving a team. In E. S. Siegel and J. R. Koza, editors, *Working Notes for the AAAI Symposium on Genetic Programming*, pages 23–30, 445 Burgess Drive, Menlo Park, CA 94025, USA, 10–12 November 1995. AAAI.

[3] Thomas Haynes, Roger Wainwright, Sandip Sen, and Dale Schoenefeld. Strongly typed genetic programming in evolving cooperation strategies. In L. Eshelman, editor, *Genetic Algorithms: Proceedings of the Sixth International Conference (ICGA95)*, pages 271–278, San Francisco, CA, USA, July 1995. Morgan Kaufmann.

[4] Thomas D. Haynes and Roger L. Wainwright. A simulation of adaptive agents in hostile environment. In K. M. George, Janice H. Carroll, Ed Deaton, Dave Oppenheim, and Jim Hightower, editors, *Proceedings of the 1995 ACM Symposium on Applied Computing*, pages 318–323, Nashville, USA, 1995. ACM Press.

[5] John R. Koza. *Genetic Programming: On the Programming of Computers by Natural Selection.* MIT Press, Cambridge, MA, USA, 1992.

[6] John R. Koza. *Genetic Programming II: Automatic Discovery of Reusable Programs.* MIT Press, Cambridge Massachusetts, May 1994.

[7] Mauro Manela. *Contributions to the Theory and Applications of Genetic Algorithms.* PhD thesis, University College, London, December 1993.

[8] Craig W. Reynolds. Competition, coevolution and the game of tag. In Rodney A. Brooks and Pattie Maes, editors, *Proceedings of the Fourth International Workshop on the Synthesis and Simulation of Living Systems*, pages 59–69, Cambridge, MA, USA, 1994. MIT Press.

Genetic Programming for Improved Data Mining -- Application to the Biochemistry of Protein Interactions

M.L. Raymer[1,3] **W.F. Punch**[1] **E.D. Goodman**[2] **L.A. Kuhn**[3]

[1]Computer Science Department, A714 Wells Hall, {raymermi,punch}@cps.msu.edu
[2] Case Center for Computer Aided Engineering and Manufacturing, 112 Engineering Building, goodman@egr.msu.edu
[3] Department of Biochemistry, 502 Biochemistry Building, kuhn@sol.bch.msu.edu

Michigan State University, East Lansing MI, 48824

ABSTRACT

We have previously shown how a genetic algorithm (GA) can be used to perform "data mining," the discovery of particular/important data within large datasets, by finding optimal data classifications using known examples. However, these approaches, while successful, limited data relationships to those that were "fixed" before the GA run. We report here on an extension of our previous work, substituting a genetic program (GP) for a GA. The GP could optimize data classification, as did the GA, but could also determine the functional relationships among the features. This gave improved performance and new information on important relationships among features. We discuss the overall approach, and compare the effectiveness of the GA vs. GP on a biochemistry problem, the determination of the involvement of bound water molecules in protein interactions.

1. Introduction

A rigorous definition of the term *data mining* is difficult, but it has come to mean a search through a large database for "nuggets" of information that can be used for some particular purpose. We have previously addressed data mining problems using a genetic algorithm(GA). We view the problem of data mining as a problem of *feature extraction*, which we optimize with a GA. Two results are generated:

- a list of features which are important for distinguishing the particular data from the background of the large database, and a (typically much larger) list of features which are *not* important for distinguishing the data.
- of those important features, a relative weighting indicative of their importance for distinguishing the data.

The rest of this paper will review: feature extraction, our approach to optimizing feature extraction using a GA, our improved approach using a genetic program (GP), and an example of the effectiveness of both the GA and the GP on an important biochemistry problem, using the chemistry of proteins to predict their binding to other molecules.

2. The Basis for the Approach

2.1 Feature Extraction

Our data mining approach is based on feature extraction as described in classical pattern recognition literature. Broadly speaking, there are two steps: identification of the features and their form to be used in the recognition, thus defining the *feature space*, and the formation of the classification or *decision rules* used to separate the pattern classes in the feature space.

Definition of the feature space can be a difficult problem. Too few features (or non-representative features) may not provide enough information on which to base classification while too many features make the search process intractable. The rapid growth in the search space as additional features are added, which also increases the number of training samples required, is often referred to as the *curse of dimensionality*.

Feature selection methods attempt to find the best subset of size d of features of the original N features. "Best" is typically defined as that subset of cardinality d which gives the best classification (fewest misclassifications). Since an enumerative search of all size d subsets is computationally infeasible (essentially $\binom{N}{d}$), at least for any reasonably sized N, there are a number of heuristic approaches that have been used, including sequential forward selection[Jain82], branch and bound[Narendra77] and GA's[Siedlecki89, Kelly91, Punch93].

Feature extraction methods define a transformation to the feature space such that fewer features are required and the features used give better separation of the pattern classes. Thus feature extraction subsumes feature selection. Feature extraction is the approach we will focus on here.

Formally (following the work of Devivjer and Kittler[Devivjer82]), we can define the set of features used in classification as a vector y. The goal of feature extraction is to define a new feature vector $\hat{y}$ that meets the criterion for improved classification. We accomplish this transformation by defining a mapping function *W()* such that $\hat{y} = W(y)$.

The result of applying *W()* is both to create $\hat{y}$ such that $|\hat{y}| \leq |y|$ and to increase separation of pattern classes in the feature space as now defined by $\hat{y}$. Formally, we define *W()* by optimizing a criterion function *J(),* ideally the probability of correct classification. Selection of *W(y)* is defined as the optimal transformation among all *possible* transformations $\hat{W}(y)$ such that:

$$J\{W(y)\} = max(J\{\hat{W}(y)\})$$

While *W()* could consist of any transform, in the classification literature it typically is a linear mapping.

2.2 Using a GA for Feature Extraction

The essence of our earlier approach to feature extraction is twofold:

- Modify the original feature space based on a vector of weights generated by a genetic algorithm.
- Use a K-nearest-neighbor algorithm to evaluate the effectiveness of the weight vector in increasing separation between known pattern classes.

A Knn is a simple rule for classification, and its general approach is shown in Figure 1. Each feature of the test set is a dimension in the classification search space. For simplicity's sake, let us assume that there are only two features in the test we conduct, represented by the *x* and *y* axes in the figure. Known examples are then placed into this space as points based on their *known* feature values and labeled according to their known classification. An unknown can then be placed in the same space based on its feature values, and it can be classified based on its *K* nearest neighbors, where *K* is set to some integer value. Assume *K=3*; then in Figure 1, the unknown sample is labeled as type *X* based on the fact that 2 of its 3 nearest neighbors are of type *X*.

Our hybrid GA-Knn approach (see [Punch93] for more details) was inspired by work first reported by Siedlecki and Sklansky [Siedlecki89] but modifies and extends their approach in several ways. Siedlecki and Sklansky's goal was to find "the smallest or least costly subset of features for which the classifier's performance does not deteriorate below a certain specified level"[Siedlecki89]. This was done by constructing a GA chromosome which consisted of a binary encoding whose length (in bits) equaled the number of features. If a bit equaled one, that feature was preserved in the feature space and used by the Knn to classify exemplars into pattern classes. Each such string was penalized based on its performance (more penalty for worse performance in classifying the exemplars) and its length (more penalty for more 1's in the chromosome). Their approach was as good as many standard approaches, such as exhaustive search and branch and bound, and much better on large feature sets (20 features or more) where the standard approaches showed poor computational performance.

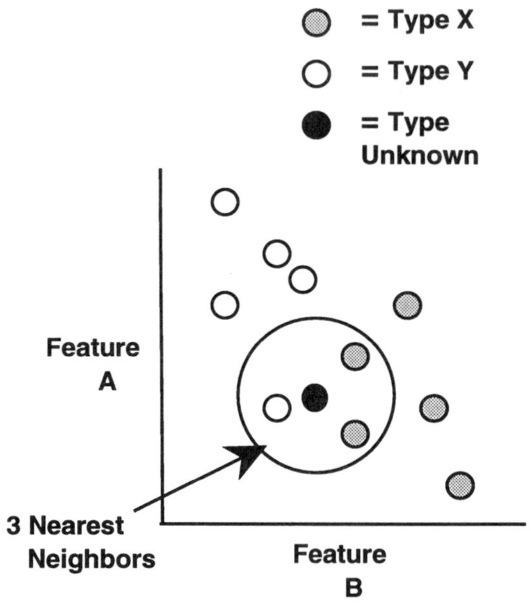

Figure 1 An example Knn classification. The unknown is classified as type X based on the majority of its K (in this case three) nearest neighbors.

The Siedlecki-Sklansky chromosome is an example of a 0/1 weighting of the importance of the features. That is, the feature space of the samples is uniformly modified by multiplying each exemplar's feature vector *y* by the GA weight vector *w* (in this case 0 or 1) and this modified Knn rule is used to classify the known samples via Knn. Thus the Siedlecki-Sklansky approach is one of feature selection.

We modified this approach to perform feature extraction by allowing the weights to be real values ranging over a broader scale, such as 0 to 10. Features are prenormalized to the range [1,10], giving the weights an interpretation of the relative importance of features to the classification task by scaling that feature's axis. Our feature extraction searches for a relative weighting of features that gives *optimal* performance on classification of the known samples. Those weights that move towards 0 indicate that their corresponding features are not important for discrimination, and any weight that moves towards the maximum indicates that the classification is sensitive to small changes in that feature. Thus each feature's dimension is elongated or shortened according to its importance in classification. Similar work was pursued by Kelly and Davis[Kelly91].

GA-Knn feature extraction is therefore a scaling of the feature space such that an optimal class separation can be achieved between the known classes. This weighting vector, once discovered, can then be used independently of the GA to classify unknown samples and can also be used to indicate which differences are important for class separation, providing focal points for further research in the application area.

2.3 Formal Definition of Feature Extraction for GA-Knn

From the definitions shown in Section 2.1, we formally describe the problem by defining the functions *W()* and *J()*.

Table 1: Results of Training and Unbiased Testing of Waters Displacement using the GA-Knn

Training Set	Water Status		Total %
	Conserved (Predicted/Observed)	Displaced (Predicted/Observed)	
1700 waters (including 157 active site waters used for testing), k=3, (biased test)	35/42, (83%)	86/115, (75%)	121/157, (77.1%)
1700 waters (with same 157 active waters), k=3, (test 67 unknown active site waters)	11/16, (69%)	39/51, (76%)	50/67, (74.6%)

- We define *w* to be the weight vector. It is a $1 \times |y|$ vector generated by the GA to be multiplied with each exemplar's feature vector *y*, yielding a new feature vector value $\hat{y}$. That is, *W(y)* is the transformation of all exemplar's *y* into $\hat{y}$ as in:

  $$\forall y \{\hat{y}_i = y_i \bullet w_i\}$$

  While $|y| = |\hat{y}|$, any $w[i]$ which is equal to 0 indicates that feature y[*i*] is effectively *not* used in the classification.

- The criterion function *J()* used is the performance of the Knn in properly classifying exemplars into pattern classes. The Knn is run on this new feature space and misclassifications noted. The *w* giving the fewest misclassifications is considered the "optimal" transformation.

2.4 Using the GA-Knn to Predict Protein Interactions

We tested the GA-Knn system on both artificially generated test data and on real-world application data. The approach proved particularly effective on "noisy" data, that is data where class exemplars had some random variance. We have applied it in areas such as: classification of soil samples based on microbial populations [Punch93], classification of characteristics of microbes with certain abilities (such as pesticide degradation and organic chemical degradation)[Pei95], and prediction of the role of water conservation or displacement in molecular binding [Raymer96].

The water displacement problem addresses predicting whether two molecules "bind" together, such as the binding of antibody to antigen or enzyme to substrate, and is a central problem of biochemistry with implications for understanding biological processes at the cellular level. The molecule which the protein binds is termed the *ligand* (typically another protein, carbohydrate, lipid or drug molecule) while the site of the ligand binding is called the *active-site* of the protein. One key consideration in such predictions is the role of water in protein-ligand binding. Water molecules (i.e., "waters") are observed in protein structures to be bound to the active site in the presence or absence of ligands [Kuhn92]. Various of these waters are important in mediating the binding process, by forming one or two water "bridges" between the protein and the ligand. Such water-mediated ligand interactions are essential to biological processes, yet the structural chemistry governing which active-site waters are "displaced" upon

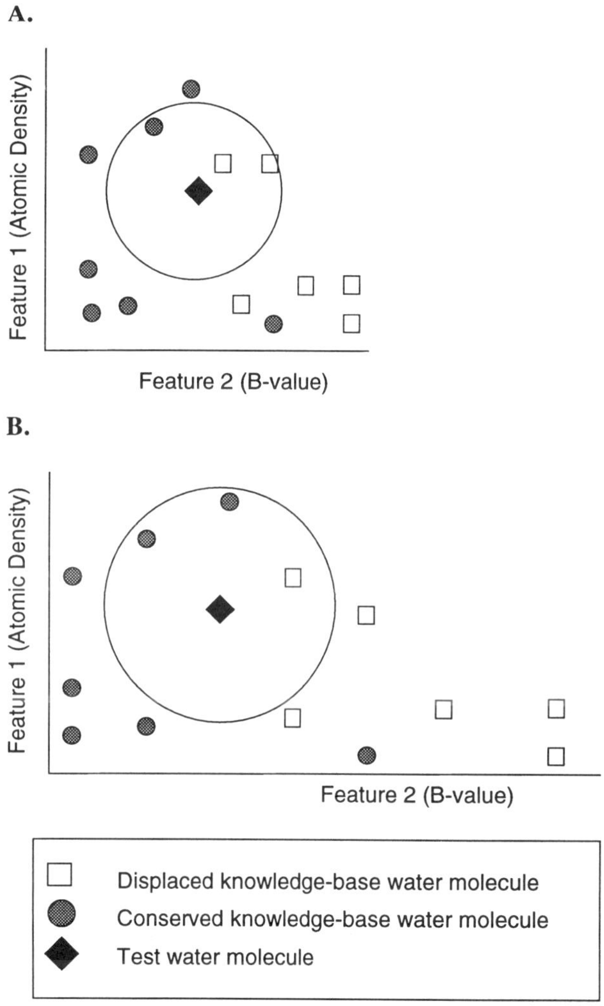

Figure 2 Scaling of the knn feature axis. Scaling of the x-axis (B value) changes the classification of the water in question. The GA searches for scalings in each dimension to improve classification of known waters.

ligand binding or "conserved" (participating in binding) has not been well-characterized. If the roles of the active-site waters could be predicted, this would have broad and important biotechnology applications in protein structural determinations and molecular simulations. The GA-Knn is used to predict which active-site waters participate in ligand binding. Training of the GA-Knn utilized information from the Consolv application. The Consolv is a database contains data on the environment of 1700 randomly-selected water molecules (850 conserved and 850 displaced) bound to 13 different protein structures as well as parameter settings to run the GA on the protein-waters problem. 157 these waters were from the active sites of these proteins, and 102 of these were used to train the GA-Knn. Four features were used to characterize the environment of each water molecule in the ligand-free structure: the water molecule's crystallographic temperature factor ("B-value", reflecting thermo-lability), the number of hydrogen bonds between water and protein, the number of protein atoms packed around the water molecule ("atomic density"), and the tendency of those protein atoms to attract or repel water ("hydrophilicity")[Kuhn95, Raymer96]. Each training water was labelled "conserved" or "displaced" based upon the known ligand-bound protein structure.

The resulting new Knn space weighted the four features according to importance, and this weighting was then used to predict the behavior of waters from the active site of 7 *new* proteins. This second test is unbiased and therefore used as predictor of the effectiveness of the GA-Knn. Results are indicated in Table 1. As shown, the modified Knn was able to predict with 75% accuracy which active-site water molecules were conserved or displaced in the 7 proteins on which it was *not* trained. An example of the action of the GA on this data set is shown in Figure 2.

Furthermore, where the algorithm mispredicted a water to be conserved (when it was in fact displaced), that water was often found to be displaced by a water-like polar atom in the bound ligand, indicating that the GA-Knn is correctly assessing the favorably of a protein's environment for binding water and similar polar atoms. The 75% predictive accuracy for waters involved in protein-ligand binding (90% accuracy if we include displacement by a polar ligand atom) shows the remarkable ability of the GA to address what was considered to be an intractable problem when the ligand-bound structure is unknown. This predictive accuracy exceeds all ab initio or knowledge-based methods for protein secondary structural prediction, of which the best algorithms approach 72% [Mehta95, Rost93].

3. Using a GP to Improve Performance

While the GA-Knn has proven very effective in data mining examples, it has limitations. The encoding employed above allows for only *linear* combinations of features. While a linear weighting is often sufficient, many natural processes have nonlinear dependencies. We addressed this problem by substituting a GP for the GA so we might derive both linear *and* nonlinear relationships. We previously addressed this problem [Punch93], by showing a GA could find weights for *predetermined* nonlinear functions. Here, we generalize to finding both the weights *and* the functions themselves. Further, for the protein-ligand waters problem, we show that such an approach leads to a "better" solution.

3.1 Modified Definition for GP Feature Extraction

We continue to employ the Knn approach, but improve how we modify the Knn space. Previously, our GA generated a vector w, which was $1 \times |y|$ ($|y|$ the number of features). We modify this as follows. Each element in the GP population will be a tree t, consisting of a root node n and $|y|$ subtrees. Each of these subtrees, denoted t_{y_i} where i ranges over all the subtrees, encodes a function based on the non-terminal/function set F and the terminal set T. For our runs, F = {+, -, *, $/_{\text{protected}}$} and T = {ERC, y_i} where ERC is an ephemeral random constant and y_i is the original value of the feature being tested (see [Koza92] for more details).

Each tree generates a new feature space modification for its associated feature in the following manner. Each training exemplar's feature element y_i is passed through its associated t_{y_i}, yielding a modified feature value $\hat{y}_i$. Testing of a new vector is then done as before (see Figure 2), setting the criterion function $J()$ to the performance of the Knn in properly classifying exemplars into pattern classes.

3.2 Implementation details for Application to the Protein Binding Problem

The GP part of the system was implemented in lilgp [lilgp95]. To maintain multiple subtrees for each member of the population, we hardcoded each individual's root note to consist of only calls to each of the required subtrees, then encoded each subtree as an Automatically Defined Function (ADF) using lilgp's ADF library. Typical runs were done with a population size of 100 (4 subtrees per individual for the protein problem, since there are 4 features), and run for 300 generations (or to convergence, whichever came first). Initialization was ramped-half-and-half with depths running between 2 and 6. Maximum tree depth was maintained at 17. Crossover was done at 90% (90% internal, 10% external) using fitness-proportionate selection, and reproduction at 10%, also using fitness-proportionate selection. No mutation was used in these runs. On a Sparc workstation (Model 502, 50MHz) using only the 157 active-site waters for training, it took 2 hours for 300 generations. Training on the 1700 waters on the same machine took ~80 hours for 300 generations.

4. GP Results

Table 2 shows that the GP approach provides an overall improvement in prediction accuracy of 79% as opposed to 75% for the GA runs. The form of the functions representing the best run of the 1700-training/157-target class are shown in Figure 3. Note that feature-3's function is a constant, and feature-0 is a linear function. Feature-1 approximates a delta function at 0, but reduces to a constant value of 0 since the features were normalized to the range [1,10]. Feature-2 has a complicated functional form, but resembles $-x^3$. Overall, the functions for feature-0 and feature-2 contribute the most to the classification process.

5. Discussion

While the GP performance on active-site waters was just slightly better overall than the GA, the GP results are more interesting for other reasons.

First, the GP typically found a better answer than the GA. Second, the GP was able to predict better using fewer fea-

Table 2: Results of Training and Unbiased Testing of Waters Displacement using the GP-Knn

Training Set	Water Status (Predicted/Observed)		Total %
training set (target set)	Conserved	Displaced	
1700 waters, including 157 active site waters used for testing, k=3, (biased test)	31/42, (74%)	93/115, (81%)	124/157 (79%)

Feature0:
```
(- x 47.70)
```

Feature3:
```
(* 52.73
52.73)
```

Feature1:
```
(/ 46.73
  (* x
    (* x
      (* x x))))
```

Feature2:
```
(* (- (+ x 6.19)
      (- (- 63.68 x)
         (- x x)))
   (* (+ (* (+ (* x x)
                (+ x
                   (+ 68.41 x))) 98.58)
         (* x x))
      (+ (+ x
            (* (+ (* (+ 24.90 71.11)
                     (* (* (+ (* x x)
                              (+ 24.90 71.11))
                           (+ (+ 68.41 x)
                              (/ (- 34.46 33.91) x)))
                        (- 89.92 x))) 71.11)
               (* (- 63.68 x)
                  (* (- (- (+ (/ (- 11.85 75.01)
                                 (* x x))
                              (* (+ 24.90 71.11)
                                 (* (* (- 89.92 x)
                                       (+ 95.45 x))
                                    (+ (* x x)
                                       (+ 24.90 71.11)))))
                           (* (* (* x x)
                                 (+ x x))
                              (- 31.87
                                 (/ (* (+ 68.41
                                          (- 79.11 59.11))
                                       (* (+ 95.45 x)
                                          (* x x)))
                                    (* x x)))))
                        (- (- 63.68 x)
                           (- x x)))
                     (* x
                        (+ (* x x)
                           (* (+ (* x x)
                                 (+ 24.90 71.11))
                              (+ (- (- x
                                       (+ (* x x)
                                          (* x x)))
                                    (- 79.11 59.11))
                                 (+ (* (+ 46.95
                                          (+ x
                                             (+ 68.41 x))) 98.58)
                                    (* x x))))))))))
         (- (* x x)
            (/ 46.95 40.26)))))
```

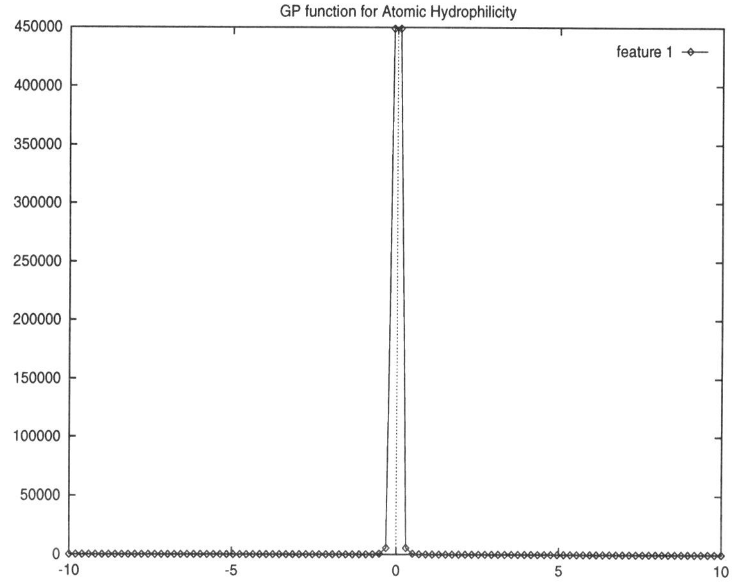

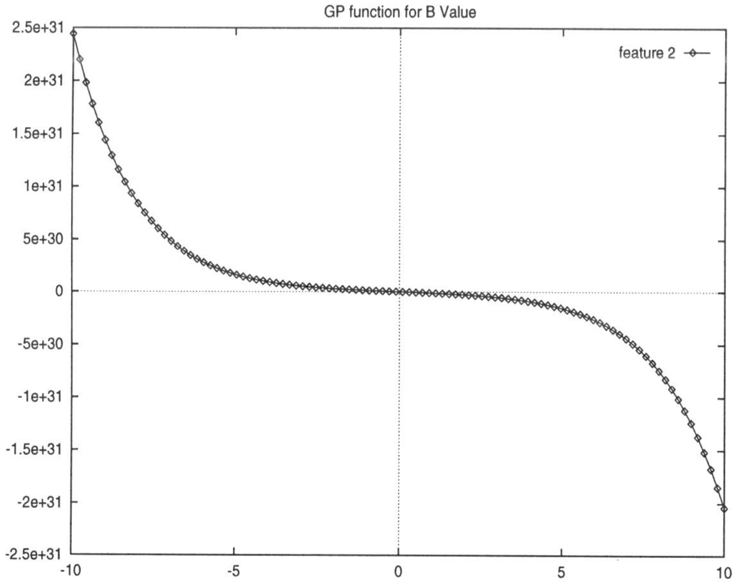

Figure 3 The 4 functions of the best Knn rule. Feature0's function is linear, and feature3's is constant. The plots of feature1 and feature2's function are shown. In the Gp-Knn, only values in the range [1,10] were used.

tures than the GA. In the GP run shown in Table 2, primarily two features were used to improve predictions relative to the GA results of Table 1 which employed four features. Third, the resulting functions were more scientifically interesting because they suggest functional dependencies for each of the four features. This could not be done in the GA since only a linear relationship was available. It is these relationships that make the GP-Knn approach more interesting for difficult problems like the protein-binding problem. Fourth, the population and individual best fitness for the GA and the GP (not shown) were quite different. The GA quickly reached a plateau for "best" fitness the GP for the population and the best individual were still improving, suggesting that longer run times would be useful.

Finally, we chose to develop a tree for each feature (as opposed to one tree incorporating all 4 features) to better indicate the kind of function each feature would require. As the number of features grows large, this could be circumvented by maintaining a single tree incorporating all features.

5.1 Future Directions

There are a number of variables we wish to explore in the GP runs. First, we are exploring the role of mutation in the generation of better answers in the GP. Previous work done on benchmark problems like the royal tree [Punch95] showed that mutation was important for maintaining diversity in the GP population. Second, experience has shown that overselection is an effective method even with populations as small as 100. We will explore overselection and other methods (such as tournament selection) as ways to improve GP performance on the data mining problem.

Finally, the results presented here show the power of the GA and GP, in combination with a Knn classifier, to solve what was considered an intractable problem in protein structural prediction, and to provide insight into the features governing protein binding.

Acknowledgments

We acknowledge support from both the GARAGe (Genetic Algorithms Research and Applications Group) and the Protein Structural Analysis and Design Lab. We also acknowledge the support of Procter & Gamble for the GARAGe, and the Research Excellence fund of Michigan for PSAL. Finally, we acknowledge the comments of our reviewers.

Bibliography

[Devijver82] P. A. Devijver, J. Kittler, *Pattern Recognition: A Statistical Approach*, Prentice Hall, London, 1982.

[Jain82] A. K. Jain and B. Chandrasekaran, "Dimensionality and Sample Size Considerations in Pattern Recognition Practice", *Handbook of Statistics*, Vol. 2 P. R. Krishnaiah and L. N. Kanal, eds. North-Holland, Amsterdam, 1982, 835-855

[Kelly91] J. D. Kelly and L. Davis, "Hybridizing the Genetic Algorithm and the K Nearest Neighbors Classification Algorithm", *Proc. Fourth Inter. Conf. Genetic Algorithms and their Applications (ICGA)*, 1991, 377-383.

[Koza92] J.R. Koza, *Genetic Programming, on the Programming of Computer by Means of Natural Selection*, MIT Press, 1992.

[Kuhn92] L.A. Kuhn, M.A. Siani, M.E. Pique, C.L. Fisher, E.D. Getzoff, and J.A. Tainer, "The Interdependence of Surface Topography and Bound Water Molecules Revealed by Surface Accessibility and Fractal Density Measures", *J. Mol. Biol.* 228, 13-22.

[Kuhn95] L.A. Kuhn, C.A. Swanson, M.E. Pique, J.A. Tainer, and E.D. Getzoff,"Atomic and Residue Hydrophilicity in the Context of Folded Protein Structures", *Proteins: Struct. Funct. Genet.*, 23, 536-547.

[lilgp95] D. Zongker and W.F. Punch, "lilgp 1.0 User's Manual," Technical Report, MSU Genetic Algorithms and Research Application Group (GARAGe). http://isl.msu.edu/GA

[Mehta95] P. K. Mehta, J. Heringa, and P. Argos, "A Simple and Fast Approach to Prediction of Protein Secondary Structure From Multiply Aligned Sequences with Accuracy Above 70%", *Protein Science* 4, 2517-2525.[

[Narendra77] P. M. Narendra and K. Fukunaga, "A Branch and Bound Algorithm for feature Subset Selection", *IEEE Trans. Computers*. 26 1977, 917-922.

[Pei95] M. Pei, E.D. Goodman, W.F. Punch and Y. Ding, "Genetic Algorithms for Classification and Feature Extraction," presented at *Classification Society Conference*, June 95.

[Punch93] W.F. Punch, E.D. Goodman, Min Pei, et al. "Further Research on Feature Selection and Classification Using Genetic Algorithms", *Proc. Fifth Inter. Conf. Genetic Algorithms and their Applications (ICGA)*, 1993, pg 557.

[Punch95] W.F. Punch, D. Zongker and E.D. Goodman, "The Royal Road Problem as a Benchmark for Genetic Programming: A Parallel Processing Example," to appear in *Advances in Genetic Programming II*, MIT Press, 1996.

[Raymer96] M.F. Raymer, W.F. Punch, S. Venkataraman, P.C. Sanschagrin, E.D. Goodman and L.A. Kuhn, "Predicting Conserved Water and Water-Mediated Ligand Interactions in Proteins Using a K-nearest-neighbor Genetic Algorithm". submitted to *J. Mol. Biol.*

[Rost93] B. Rost and C. Sander, "Prediction of Protein Secondary Structure at Better than 70% Accuracy", *J. Mol. Biol.* 232, 584-599.

[Siedlecki89] W. Siedlecki and J. Sklansky, "A Note on Genetic Algorithms for Large-Scale Feature Selection", *Pattern Recognition Letters*, 10 1989, 335-347.

Generality versus Size in Genetic Programming

Justinian P. Rosca
Computer Science Department
University of Rochester
Rochester, NY 14627
E-mail: rosca@cs.rochester.edu

Abstract

Genetic Programming (GP) uses variable size representations as programs. Size becomes an important and interesting emergent property of the structures evolved by GP. The size of programs can be both a controlling and a controlled factor in GP search. Size influences the efficiency of the search process and is related to the generality of solutions. This paper analyzes the size and generality issues in standard GP and GP using subroutines and addresses the question whether such an analysis can help control the search process. We relate the size, generalization and modularity issues for programs evolved to control an agent in a dynamic and non-deterministic environment, as exemplified by the Pac-Man game.

1 Introduction

Genetic Programming (Koza, 1992) has been applied to a variety of machine learning applications formulated mostly as classification or prediction problems. Some examples include the prediction of omega loops in proteins and the transmembrane problem, symbolic regression, character recognition, fitting to chaotic data, text classification, feature detection, image discrimination (Koza, 1992; Koza, 1994; Kinnear, Jr., 1994). Such applications have been formulated in the supervised learning framework. In this framework the goal of GP is to synthesize a program that maximizes some performance measure on a set of training examples, does not over-fit (over-specialize on) the training data, and most importantly, acquires a low error rate on unseen cases extracted from the underlying distribution of examples. Generality of learned solutions depends on how well they perform on such unseen examples.

Relatively few GP results have analyzed the generality of GP induced solutions. This led us to study the problematical generalization issue in GP from the machine learning perspective. Further this brought us close to an analysis of the complexity of representations evolved by GP. Many machine learning algorithms for learning or adapting a model from given data explicitly deal with complexity by taking a Bayesian approach. For instance, the minimum description length principle trades off the model code length (the complexity term) against the error code length (the data not explained by the model or error term). Complexity is naturally expressed as the size of code in bits of information.

Genetic encodings are adaptive encodings that are harder to analyze because "nature" packs in more code than would be apparently needed. Two distinguishing characteristics of the GP simulated evolution framework are its manipulation of procedural representations (programs) and the variable size of its representations. Programs are written in a user defined language and have variable complexity. Size is generally defined as the number of nodes in a tree representation of programs. Size plays an important role in the dynamics of GP. Size influences future evolution and simultaneously, it is also a result of evolution. For instance crossover and mutation changes are performed uniformly over the entire length of the structure evolved. It is obvious that atomic parts of the genome are less likely to be modified if the overall size of the genome increases over generations. This implicit evolvability of control parameters essentially distinguishes GP from genetic algorithms (GAs).

This paper reveals the program size issue by introducing the notions of effective size and size attractors. One problem often discussed is the increasing size of GP evolved structures or bloating. Big size programs cover up blocks of temporarily ineffective code that continue to be modified. GP processes small blocks of code, therefore modular GP approaches are suggested to fill a gap that exists between two conflicting theories about what can help GP effectiveness: introns, on one side, and direct use of parsimony pressure, on the other.

The outline of the paper is as follows. Section 2 presents the Pac-Man game as an application domain. Section 3 discusses the issue of generalization. Section 4 focuses on the issue of size in GP. Finally, we discuss the intricate relation between generality of solutions and their complexity and conclude.

2 Application Domain

2.1 Control Problems

General problem formulation of learning are agent-based. The agent (learning system) is not told what is the correct output for each input, but rather how it performs by means of occasional *rewards*. The agent receives rewards in response to actions taken. Often rewards are delayed. This is the reinforcement learning (RL) problem formulation. The task in RL is to learn a policy maximizing the expected future rewards. Formally, this corresponds to learning a function representing a good satisficing policy that maps perceptions and internal states to actions.

Several instances of RL-like problems in the GP literature are the obstacle avoiding robot (Reynolds, 1994), Pac-Man game (Koza, 1992; Rosca and Ballard, 1996), crawling and walking robots (Spencer, 1994). A reactive agent, embodied as a robot in a simulated world, pursues a task in the world. The goal of GP is to create a control program (specified in a domain-dependent language, as a preparatory step to applying GP) that controls the movement of the agent to accomplish the task. The fitness of a controller is based on how well the agent performs its goal in the environment or on rewards received from the environment. For example the obstacle avoiding agent or the crawling agent are "rewarded" proportional to the life of their collision-free walk.

2.2 The Pac-Man Game

We consider the problem of controlling an agent in a dynamic environment, similar to the well known Pac-Man game (see (Koza, 1992)). The problem is to learn a controller to drive the Pac-Man agent in order to acquire as many points as possible and to avoid being eaten by the monsters.

2.3 GP Representation

Programs that define agent controllers will be built based on perception, action and program control primitives. The agent perception primitives return the Manhattan distance to the closest food pellet, pill, fruit and closest or second closest monster. They are, respectively: SENSE-DIS-FOOD, SENSE-DIS-PILL, SENSE-DIS-FRUIT, SENSE-DIS-MON1, SENSE-DIS-MON2. The "sense blue" perception is combined with an if-then-else control primitive into the if-blue (IFB) lazy evaluation function which executes its first argument if monsters are blue, otherwise executes its second argument. The action primitives move the agent along the maze corridors and return the distance between the agent and the corresponding objects. For instance, ACT-A-PILL advances the agent on the shortest path to the nearest uneaten energizer (and if an energizer is found the monsters turn blue) while ACT-R-PILL retreats the agent from the nearest uneaten energizer. The other actions have analogous functions with respect to the closest monster, the second closest monster, fruit, and food: ACT-A-MON-1, ACT-R-MON-1, ACT-A-MON-2, ACT-R-MON-2, ACT-A-FRUIT, ACT-A-FOOD. If the shortest path/closest monster/food are not uniquely defined, then a random choice from the valid ones is returned by the corresponding function.

Table 1: GP setup for the Pac-Man game.

$\mathcal{T} \cup \mathcal{F}$	Representation A or B
Fitness cases	One or three runs
Fitness	Average number of points acquired by Pac-Man
Population size	500
Number of generations	100
Crossover rate	90% (20% on leaf nodes)
Reproduction rate	10%
Mutation rate	0

The above primitives were combined to form programs in two ways. In the first alternative (A) we used the if-less-than-or-equal (IFLTE) lazy function which compares its first argument to its second argument. For a "less-than" result the third argument is executed. For a "greater-or-equal" result the fourth argument is executed. In the second alternative (B) we used the if-then-else IFTE lazy evaluation function, relational operators ($<, =, \geq, \neq$), and logical operators (AND, OR, NOT). IFTE takes three arguments. If the result of its first argument is true then it evaluates its second argument, otherwise it evaluates its third argument. Representation (A) is taken from (Koza, 1992) with one change: all primitives return distances, as opposed to a mix of distances (senses) and directions (actions). In the latter alternative distances would be compared with actions without making much sense. In representation (B) primitives are *typed*. For example, all actions and control operators have type ACT, all logical expressions have type BOOL, distances have type DIST, etc. IFTE takes a BOOL first argument and two more ACT arguments. The programs that can be written in representation B are equivalent to decision trees. They are purely reactive and thus easier to understand and analyze.

3 Analysis of Generalization

Intuitively, more general solutions to our control problem should treat similar situations in similar ways. Situations are defined by agent perceptions and states. Formally, an agent policy is a mapping

$$p : \mathcal{P} \times \mathcal{S} \longrightarrow \mathcal{A}$$

where $\mathcal{P}$ is the set of perceptions, $\mathcal{A}$ the set of actions and $\mathcal{S}$ the agent state space. When searching for a policy, the size of the hypotheses space is the number of such mappings i.e. $||\mathcal{A}||^{||\mathcal{P}|| \times ||\mathcal{S}||}$. Explicitly representing the state space is undesirable from a learnability perspective. In contrast, an implicit representation such as a program has the potential of better generalization. This makes GP a candidate approach to learn policies. For the same reason, parameterized function approximators have been used to replace table lookup in reinforcement learning (Boyan and Moore, 1995; Sutton, 1996).

A program evolved by GP represents a compressed agent policy. It indicates how to control Pac-Man movements based on either the current sensor readings and internal state or memory for representation A, or just sensor readings for representation B. We are interested in the generalization performance of GP solutions.

3.1 Methodology

The performance of induced programs is often quantitatively measured by determining program performance on a set of *training cases*. This formulation of the induction problem is common to a variety of machine learning techniques. For real-world induction problems, we almost never have sufficient training cases, or we can not afford to consider too many training cases. The true performance (or true error rate in a classification problem) of the learning system is evaluated based on the performance of induced solutions on cases not used for training but kept aside for testing. Resampling techniques such as crossvalidation or bootstrapping (Weiss and Kulikowski, 1991) are used to better estimate performance for limited sample sets. For example (Johnson et al., 1994) used such techniques to evaluate the quality of GP evolved routines for detecting the left and right hand position in a person's bitmap silhouette.

In addition to the generalization analysis, further post processing can offer more confidence in the performance figures obtained. For example, if each fitness case is computationally expensive, a GP algorithm cannot afford to get a reliable average fitness estimate during evolution (by testing on many cases). Instead, final solutions can be tested more extensively. Apart from the average performance and deviation on unseen cases, one can compute a $P\%$ confidence interval in the true average performance (error) over a larger part of the final population. This would mean that with probability P the true performance rate is within that interval around the average value. We applied this method to analyze Pac-Man solutions. Note that the choice of the number of GP fitness cases is a balance in the accuracy-efficiency tradeoff. A larger number of fitness cases would offer better results but is very expensive to use (see also (Reynolds, 1994)).

3.2 Generalization in the Pac-Man Domain

Generality of the best solutions obtained with standard GP (SGP), GP with automatically defined functions (ADF) (Koza, 1994), the adaptive representation through learning (ARL) GP extension (Rosca and Ballard, 1996), and hand-coded solutions, is summarized in Table 2. The ADF approach uses two subroutines of two arguments each, while the ARL approach creates a library of subroutines on-the-fly by generalizing useful blocks of code. We tested all solutions on the same 100 random test cases. ARL achieved the best results. Hand solutions were carefully designed within a couple of hours of programming effort. Representation B, which is simpler, ranked better than A with each individual algorithm. But how confident can we be about these estimates? We determined the 95% confidence interval in the average number of points of a solution. For example, the ARL average for the solution evolved by training on three fitness cases using representation B has a confidence interval of ±90points, i.e. the true average is within this interval relative to the average with a confidence of $\frac{95}{100}$.

Table 2: Generalization performance of Pac-Man solutions: average fitness of solutions evolved through training on one or three cases. Averages for each approach are computed from the same 100 random test cases.

Rep.	SGP 1	SGP 3	ADF 1	ADF 3	ARL 1	ARL 3	Hand coded
A	3095	2906	1011	1569	3594	3611	1798
B	595	3370	2696	3875	3706	4439	1009

The smallest evolved programs were also the most general. More experimental evidence is needed to confirm this tendency. Among hand-coded programs, the smallest (made up of only 15 nodes) was also the most general, but its generality was worse than that of the best evolved programs.

4 The Dynamics of Size

Informally we refer to the *structural complexity* of a program by *size*. For standard GP, the size of a program is simply the number of nodes in the tree representation. For GP using subroutines, structural complexity is computed by adding up the sizes of all subroutines (or ADFs) (Rosca and Ballard, 1994). Next we present a new type of size analysis and show some experimental results.

4.1 How Size Varies

We present the "average path length" perspective on size change. This method focuses on the average path length variation of an individual after a genetic operation. The *average path length* of an individual i is defined as:

$$\overline{p(i)} = \frac{1}{L(i)} \sum_{j=1}^{L(i)} dist(root(i), j)$$

where $L(i)$ is the number of leaves of the tree representing i and $dist(\cdot, \cdot)$ is the distance between two nodes.

In current GP practice the nodes to act as "pivots" in

genetic operations are chosen from internal nodes with a probability of 90% and from leaf nodes with a probability of 10%. In turn, internal/leaf nodes are uniformly chosen. This results in an expected height of pivot nodes of approximately 2 (Rosca and Ballard, 1995). For example, if trees operated upon are complete binary tress, the expected height of pivot points $E[h]$ for a tree of height H can be showed to be:

$$E[h] = \frac{9}{10}(2 - \frac{H}{2^H - 1})$$

The number of paths affected by a crossover operation is small on the average. Equivalently, changes consist of small blocks of code. Moreover, the sum of the number of affected paths is an invariant of the operation, i.e the number of paths does not increase or decrease from the two parents to the resulting two children. Also, the number of nodes involved in the operation is conserved. It follows that the average path length over a two individual ensemble is crossover-invariant. Average path length of an offspring would be crossover invariant on the average if only one offspring is generated from every two selected parents.

Mutation (random replacement of a subtree with a new subtree) is path length invariant on *average* over generations only if newly generated subtrees have the same size as replaced subtrees. Mutation alone could determine a controlled drift towards trees of increased size if the substituted trees have bigger height.

The operation with a critical role in variations of size over time is fitness proportionate selection. At any given time search focuses on structures with sizes in a given range. Exploratory mutation or crossover will occasionally generate structures of size outside that range. If a more compact and fit program is discovered, then it will be preferentially selected, increasing the probability that the next generation population is more compact. Only if bigger structures have better fitness than smaller structures, the average path length of structures selected for reproduction, and thus size, can increase. There is no implicit tendency towards monotonically increasing or decreasing size over time.

On the average, genetic changes do not favor dramatic changes in size. Selection will determine the favorable changes in size, but those are generally small. A long-run tendency in size change is related to the probability of getting better solutions. If the better solutions are bigger or at least bigger structures will improve the probability of increasing fitness then we will notice a size increase.

4.2 Size Attractors

In the search space of programs we define the neighborhood of a program p relative to the population to be the set of programs obtained by highly probable changes of one small block to p. A *size attractor* is the average size of the best programs in a collection of neighborhoods of the population that have small size variations.

An analysis of size reveals a behavior resembling a walk through size attractors for different size values. If the average size of the current population is very different from the size of a potential solution then the search path is probably long and we may see a convergence in size to a local size attractor. The average size of the randomly generated initial population represents an important bias in this respect.

The next subsection introduces a definition of size that enables a finer analysis of the GP dynamics along the size attractor perspective.

4.3 Effective Size

When evaluating a program, not all code is used (evaluated). This is due to the fact that the function set may (and usually does) include *lazy* functions. A *lazy* function has its parameters evaluated only if they are needed (such as IFTE and IFLTE here). Sometimes, code that is used has no effect at all on the output, for any of the fitness cases but may be useful on new cases. This can not be known in advance, so when we talk about unused code our scope is the set of fitness cases.

Fortunately, for hierarchical genome representations as used in GP, useless code can be much more cheaply detected by analyzing code in a top-down manner. Moreover, the task of detecting useless code becomes straightforward if we partition the nodes of a program into *effective* and *ineffective* code. Effective code is code that is executed at least once, regardless of the overall result of the computation. The size of effective code from an individual is the individual's effective size. We disregard here information about which pieces of code contribute to useful computations, although differential behavior of code is a useful source of information that could be used for improving GP search too (Rosca and Ballard, 1996).

The effective size of a program is greater or equal than the difference between the size of the program and the total size of introns. In our experiments, instead of analyzing introns we track effective size. The focus on effective size reveals interesting adaptation phenomena.

4.4 Experimental Results

We have analyzed separately the effects of mutation and crossover on the dynamics of size and effective size.

Mutation was tested through a simulated-annealing procedure operating on discrete structures such as programs. As expected, mutating small blocks of code with blocks of approximately equal size determines relatively smooth changes in size. To contrast this behavior of the search procedure, we used a *hard* mutation operator that mutates a small block of code with a large one. Not surprisingly, in this case mutation and selection alone

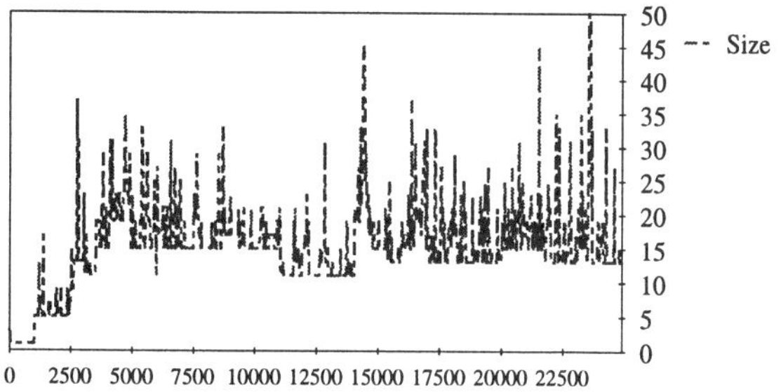

Figure 1: Size variation of the best of generation individual, over 25000 generations of changes (the equivalent of 50 GP generations with a population size of 500), in an iterative improvement test algorithm based on mutation run on representation A of the Pac-Man problem. The underlining effective size horizontals (not represented here for simplicity) corresponds to size attractors.

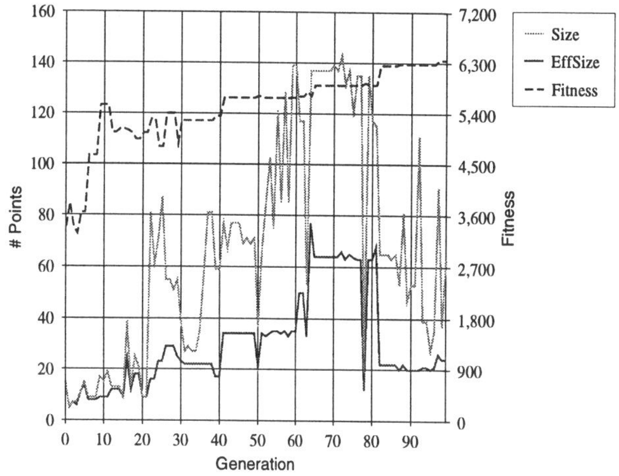

Figure 2: Size variation of the best of generation individual in an SGP run for the Pac-Man problem (representation A). Plateaus of the effective size plot are size attractors generators generated through crossover.

determine sharp oscillations over successive generations, but do not constantly increase the average size of the best individual in the long run. On the contrary, the behavior of the search procedure in this case outlines a dynamic adaptive process about local optima given by certain minimal patterns of trees (corresponding to the effective size or underlying horizontals in Figure 1). This suggests that for static fitness landscapes size is rather a characteristic of the difficulty of problem.

Figure 2 presents the evolution trace in a run of SGP with parameters from Table 1 for representation A of Pac-Man. Total size may vary quite dramatically, but the effective size change is much more moderate and closely corresponds to improvement in the best-of-generation individuals. The overall size tends to increase up to generation 60. Running SGP even longer shows that size drastically decreases after simpler but more effective solutions are discovered. This time one notices again clear plateaus in the effective size plot (size attractors), which correspond to the structure of the problem landscape. SGP manages to constantly avoid local optima. This is not the case for GP with ADFs. A typical run clearly showing the effect of a local optimum is shown in Figure 3. Effective size (as well as the size of the main branch) do not increase after generation 8. In this case, the increase in total size indicates the undesirable self protection against changes and the reach of a local optimum.

5 Putting It All Together

When learning a fixed size model that explains given data, a complexity fit is a method to make a good model choice with respect to generalization (Weiss and Kulikowski, 1991). Such a procedure is not relevant for GP, as GP simultaneously searches the space of all models given by the set of problem primitives using variable size structures.

Generality of solutions can be taken as a guiding principle for the search of representations using GP in inductive problems. Intuitively, we want small solutions that are also general, according to Ockham's razor principle. Biasing search towards smaller programs intuitively corresponds to biasing search towards an inductive model of lower dimensionality. A biased model has fewer degrees of freedom and lower variance (Geman et al., 1992). In statistics, this issue has been discussed as the bias-variance tradeoff. One aims for a low-bias, low-variance model but has to trade off one for another.

Parsimony pressure is often added into the GP fitness function to limit the size increase and improve efficiency and understandability of solutions. Adding the right size pressure component to fitness remains however, an art. One example is "disassortative mating" (Ryan, 1994). This GP algorithm selects parents for crossover from two different lists of individuals. One list of individuals is ranked based on fitness while the other is ranked based on the sum of size and weighted fitness. The goal is to evolve solutions of minimal size that solve the problem. However, by directly using the size constraint the GP algorithm is prevented from finding solutions. As reported, the disassortative mating algorithm improves convergence to a better optimum while maintaining speed.

A theoretically grounded method to balance the dimensionality of a model and its error is Rissanen's minimum description length principle (MDLP) (Li and Vitanyi, 1992). MDLP has been applied in GP to extend the fitness function of hybrid classification models (see (Iba et al., 1993) for a GP-regression tree hybrid and (Zhang and Muhlenbein, 1995) for a GP-neural net hybrid; in both cases GP manipulates tree structures, not programs). (Rosca and Ballard, 1994) has used MDLP to assess the suitability of an extension of GP with sub-

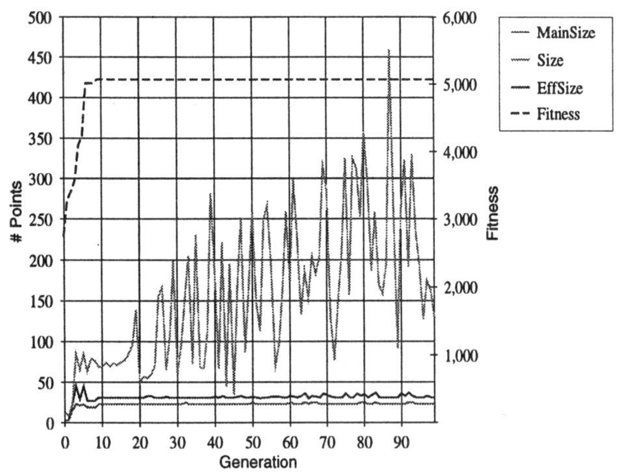

Figure 3: Size variation of the best of generation individual in a run of GP with ADF for the Pac-Man problem (representation A). The architecture of programs involves two subroutines of two arguments each, which can be invoked from the main program.

routines called adaptive representation.

In contrast to the two hybrid approaches above, there are two main problems with directly applying a parsimony pressure in standard GP. First, introns may be useful (see also (Angeline, 1994)). Second, the program space is much more rough and larger programs are not necessarily better. Iba outlined that the MDLP-based fitness measure can be applied problems satisfying the "size-based performance" criterion (Iba et al., 1994), where the more the tree structure grows the better its performance becomes.

Along the size perspective, GP research has focused on the intron issue and the bloating phenomenon. *Introns* are pieces of code with no effect on the output. An analysis of introns goes hand in hand with an analysis of *bloating*. Bloating is the phenomenon of size increase over time.

It has been conjectured that the increase in size is a "defense against crossover." This hypothesis was analyzed in detail in (Nordin et al., 1995) only for linear genome representations. The noticed increase in the size of programs was attributed to introns. The authors suggested that a representation which generates introns leads to better search effectiveness based on experiments with controlled crossover or mutation rate within intron fragments. Thus, introns may have a positive role in GP search protecting against destructive genetic operations.

However, the conclusion above is not warranted in the case of GP using tree representations. The overhead introduced by exponentially increasing tree sizes may offset the protective effects of introns. Moreover in (Nordin et al., 1995), the detection of introns is computationally expensive. In order to determine introns, each instruction has to be replaced with a NO-OP and the effect of this change is observed on the output of the program. Counting which instructions of a *sequential program* are introns takes time proportional to the product of the length of the program and the average time of an execution.

Tackett correctly pointed out that bloating cannot be selection-neutral (Tackett, 1994). He showed that the average growth in size is proportional to the selection pressure. We argue that this may not be the case in general and that the particular results obtained only analyzed a small number of generations with a small initial program size. If different biases were used, such as large random programs in the initial population, then size may have also decreased with an increase in the selection pressure. Tackett also suggested that the larger programs selected by GP contain expressions which are inert overall (introns), but contain useful subexpressions, thus correlating bloating with hitchhiking. The bloating-hitchhiking correlation is certainly valid for the Royal Road example he considered because bigger trees are more likely to have higher fitness by accumulating the rewarding subexpressions.

The code protection and hitchhiking functions of introns should be considered in conjunction with the capability of GP to modify unreachable blocks of code. Such blocks evolve and may become building blocks later. GP appears to rely on the function of small and fit blocks whose origin may be ineffective code. Thus, as an alternative to the use of introns to increase GP effectiveness we suggest to directly evolve modular programs. A direct way to evolve programs of smaller complexity is to discover subroutines as in ARL (Rosca and Ballard, 1996). As discussed before, evolved modular programs seem to have better generality indeed.

6 Conclusions

GP was used to learn a Pac-Man policy expressed as a program. Such an implicit representation of the agent policy satisfies two important properties discussed in this paper: compressibility and generalization. We presented various methods for assessing generalization for inductive problems. Generalization may be easier to achieve by means of small solutions but small solutions may be hard to come by in GP just by adding a size component in the fitness function. The paper also discusses the size problem in GP from the new perspective of size attractors and effective code. Effective size offers insights into the properties of the search space for a given problem and also indicates size attractors. More importantly, effective code can be used for modularization purposes (Rosca and Ballard, 1996) leading to a smaller complexity and increased generalization.

Monitoring size and effective size can be done almost without overhead in GP, as no additional program evaluations are needed. An increase in size can impose a memory overhead for GP with lazy evaluation domain primitives, decreasing the effectiveness of the GP procedure. An increase in effective size determines a time execution overhead, which can be serious for induction problems.

This work is one step for relating the size, generalization and modularity issues. Further work is needed to better understand how size can be used to more effectively control GP search. On the practical dimension, modular GP approaches offer an indirect but more promising path to trading off generalization and size than direct approaches based on parsimony pressure. On the theoretical dimension, a definition of a search landscape neighborhood based on *code size*, as presented in this paper, could be useful in further revealing the properties of GP search landscapes.

Acknowledgments

This research was sponsored by NIH/PHS research grant no. 1 R24 RR06853-02 and by the National Science Foundation under grant numbered IRI-9406481.

References

Angeline, P. J. (1994). *Evolutionary Algorithms and Emergent Intelligence.* PhD thesis, Computer Science Department, Ohio State University.

Boyan, J. A. and Moore, A. W. (1995). Generalization in reinforcement learning: Safely approximating the value function. In Tesauro, G., D.S.Touretzky, and T.K.Leen, editors, *NIPS 7.* MIT Press.

Geman, S., E.Bienenstock, and R.Doursat (1992). Neural networks and the bias/variance dilemma. *Neural Computation*, (4):1–58.

Iba, H., de Garis, H., and Sato, T. (1994). Genetic programming using a minimum description length principle. In Kinnear, K., editor, *Advances in Genetic Programming.* MIT Press.

Iba, H., Kurita, T., de Garis, H., and Sato, T. (1993). System identification using structured genetic algorithms. In *Proceedings of the Fifth International Conference on Genetic Algorithms.* Morgan Kaufmann Publishers, Inc.

Johnson, M. P., Maes, P., and Darrel, T. (1994). Evolving visual routines. *Artificial Life*, 1(4):373–389.

Kinnear, Jr., K. E. (1994). *Advances in Genetic Programming.* MIT Press.

Koza, J. R. (1992). *Genetic Programming: On the Programming of Computers by Means of Natural Selection.* MIT Press.

Koza, J. R. (1994). *Genetic Programming II.* MIT Press.

Li, M. and Vitanyi, P. M. B. (1992). Inductive reasonong and kolmogorov complexity. *Journal of Computer and Systems Sciences*, 44:343–384.

Nordin, P., Francone, F., and Banzhaf, W. (1995). Explicitly defined introns and destructive crossover in genetic programming. In Rosca, J. P., editor, *Proceeedings of the Workshop on Genetic Programming: From Theory to Real-World Applications (NRL TR 95.2, University of Rochester)*, pages 6–22.

Reynolds, C. W. (1994). Evolution of obstacle avoidance behaviour:using noise to promote robust solutions. In Kinnear, Jr., K. E., editor, *Advances in Genetic Programming*, chapter 10. MIT Press.

Rosca, J. P. and Ballard, D. H. (1994). Hierarchical self-organization in genetic programming. In *11th ICML*, pages 251–258. Morgan Kaufmann.

Rosca, J. P. and Ballard, D. H. (1995). Causality in genetic programming. In Eshelman, L., editor, *ICGA95*, pages 256–263, San Francisco, CA., USA. Morgan Kaufmann.

Rosca, J. P. and Ballard, D. H. (1996). Discovery of subroutines in genetic programming. In Angeline, P. and Kinnear, Jr., K. E., editors, *Advances in Genetic Programming 2*, chapter 9. MIT Press, Cambridge, MA, USA.

Ryan, C. O. (1994). Pygmies and civil servants. In Kinnear, Jr., K. E., editor, *Advances in Genetic Programming.* MIT Press.

Spencer, G. (1994). Automatic generation of programs for crawling and walking. In Kinnear, Jr., K. E., editor, *Advances in Genetic Programming*, chapter 15. MIT Press.

Sutton, R. S. (1996). Generalization in reinforcement learning: Successful examples using sparse coarse coding. In *NIPS 8.* MIT Press.

Tackett, W. A. (1994). *Recombination, Selection and the Genetic Construction of Computer Programs.* PhD thesis, University of Southern California.

Weiss, S. M. and Kulikowski, C. A. (1991). *Computer Systems that Learn.* Morgan Maufmann.

Zhang, B.-T. and Muhlenbein, H. (1995). Balancing accuracy and parsimony in genetic programming. *Evolutionary Computation*, 3(1):17–38.

Genetic Programming in Database Query Optimization

Michael Stillger *
Institut für Informatik
Humboldt-Universität zu Berlin
Unter den Linden 6, 10099 Berlin, Germany
stillger@dbis.informatik.hu-berlin.de
http://www.dbis.informatik.hu-berlin.de/~stillger

Myra Spiliopoulou
Institut für Wirtschaftsinformatik
Humboldt-Universität zu Berlin
Spandauer Str. 1, 10178 Berlin, Germany
myra@wiwi.hu-berlin.de
http://www.wiwi.hu-berlin.de/~myra

ABSTRACT

Database query optimization is a hard research problem. Exhaustive techniques are adequate for trivial instances only, while combinatorial optimization techniques are vulnerable to the peculiarities of specific instances. We propose a model based on genetic programming to address this problem, motivated by its robustness and efficiency in a wide area of search problems. We adapt the genetic programming paradigm to the requirements of the query optimization problem, showing that the nature of the problem makes genetic programming a particularly attractive approach to it.

1 Introduction

Genetic programming enjoys an increasing popularity in the mastering of difficult optimization problems. We propose a methodology applying genetic programming to the query optimization problem in relational databases.

Query optimization has been the subject of active research for more than 20 years. Recently, research has turned towards techniques based on combinatorial optimization. It has been observed, however, that their efficiency is affected by peculiarities of specific instances of the problem [Ioannidis and Kang, 1991]. The robustness of evolutional computation towards such problems has lead researchers to consider genetic algorithms for query optimization [Bennett *et al.*, 1991; Steinbrunn *et al.*, 1993], yielding encouraging results. However, the methods used in [Bennett *et al.*, 1991; Steinbrunn *et al.*, 1993] to represent the problem according to the genetic algorithms requirements are counterintuitive and prone to ambiguities. We show that a methodology based on genetic programming is closer to the nature of the problem, avoids much of the complexity and ambiguity of GA methods and better exploits the advantages of evolutional computation.

In the next section we briefly describe the query optimization problem and the techniques proposed to tackle it. In section 3, we present our genetic programming model for query optimization, specifying our GP representation and the GP operators applied on it. In section 4, we discuss the current status of our model. Section 5 concludes the study.

2 The Query Optimization Problem

The relational data model introduced by Codd [Codd, 1970] opened a new era in database technology. Relational database requests are formulated as declarative queries. A query can be represented as the conjunction of the operators in a "query graph": the graph nodes are database relations, while the edges represent operators applied on them, namely "selection", "join", "projection", "set operators" and "aggregates". Fig. 1 shows an example query graph.

Optimizing a query. The query graph is still a declarative representation of the query, from which several query execution plans (QEPs) can be produced, specifying different ways of executing the query. A QEP in relational databases is an "operator tree". Its non-leaf nodes are the query operators labelled with their execution algorithms, while the leafs are the database relations labelled with the access method (e.g. index) used to retrieve them. An edge represents the data stream output from a node and input to its parent. An edge denotes data flow; hence, the relations on which an operator is applied must appear in the leafs of the subtree below it.

Different QEPs can be produced by specifying different execution algorithms for the nodes and by changing the relative placement of the nodes while preserving the data flow semantics of the edges. All QEPs for a query

*The work of this author is supported by the German Research Council under contract DFG Fr 1142/1-1.

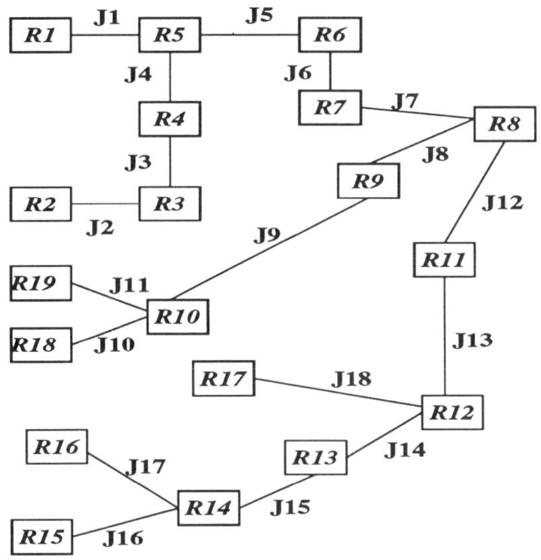

Figure 1: Join Graph for a 18 Join Query

constitute its "search space" S. Each database contains a module, the query optimizer, whose goal is the selection of the least expensive QEP in S. A cost model, usually measuring CPU and I/O time, attaches a value $cost(qep)$ to each $qep \in S$. Hence, the least expensive QEP is a $qep_0 \in S$ such that:

$$cost(qep_0) = \min_{qep \in S} cost(qep)$$

Optimizing a join query. The join is a binary relational operator associated with a constraint on its two input relations. It combines their tuples in a cartesian product fashion and uses the constraint to specify which tuples will be output. Joins appear in all but trivial queries. Their importance is such that much research is devoted to the optimization of queries comprised solely of joins. We also focus on join queries in our study.

The join optimization problem covers the development of join execution algorithms and the design of efficient join ordering techniques. For the execution of a join, several algorithms have been proposed, based on nested loops, on merging and sorting, and on hashing of the input relations [Graefe, 1993]. The join ordering problem concerns the relative placement of joins in a QEP.

Each repositioning of nodes and each change of a node's join algorithm produces a new QEP. Thus, the search space increases exponentially with the number of joins in the query. Optimizers scanning this space exhaustively are inappropriate for queries containing tenths of joins. Such queries occur in databases used for CAD/CAM, expert systems, data mining and decision support; their efficient processing is constantly gaining in importance.

For the optimization of large join queries, combinatorial optimization techniques with polynomially increasing optimization time have been studied in [Swami and Gupta, 1988; Ioannidis and Kang, 1990; Lanzelotte *et al.*, 1993; Spiliopoulou *et al.*, 1996] etc. However, their efficiency depends on the shape of the cost function, which can vary considerably [Bennett *et al.*, 1991]. Consequently, methodologies based on evolutional computation gain in interest, since they are more robust towards varying cost function shapes.

Genetic algorithms for join query optimization have been studied in [Bennett *et al.*, 1991; Steinbrunn *et al.*, 1993]. They transform QEPs into chromosomes, on which crossover, selection and mutation are applied. The fitness function is based on the query cost model, which is designed for QEPs. Therefore, the chromosomes must be transformed back to QEPs to compute their fitness. Those transformations are complicated and expensive. More gravely, the crossover operator may destroy the tree structure of a chromosome. So, the QEP must be "repaired" before computing its cost.

Despite these shortcomings, the results of [Bennett *et al.*, 1991; Steinbrunn *et al.*, 1993] are encouraging. This indicates that evolutional computation is appropriate for the query optimization problem, but the methodology has to be revised. This is the goal of our study. Rather than genetic algorithms, we use genetic programming, showing that the QEPs can be conveniently used as genetic programs.

3 Genetic Programming for Join Queries

We use genetic programming to optimize queries with up to hundreds of joins. The QEPs of a join query consist solely of join operators. We use the term "join tree" or "tree" to refer to the operator tree of the QEP of a join query, and the term "join-node" or "node" to refer to the non-leaf nodes denoting the operators.

3.1 The Representation Structure

A QEP can be observed as a program in an abstract tree representation, which is evaluated in a bottom-up way by the database system. The relations are the terminals and the joins are the functions of the genetic program. Hence, the QEP satisfies the structural requirements of the genetic programming method which applies the search paradigm of genetic algorithms developed in [Holland, 1975; Goldberg, 1989] on a tree representation of an abstract computer program.

The input and the output of all operators in a QEP are relations. So, the nodes of a QEP satisfy the closure property defined in [Koza, 1991] (chapter 6.1.1). However, the random placement of nodes in the QEP can introduce cross products and artificial joins.

"Cross products" materialize joins between relations that are not directly connected on the query graph; for the query graph of Fig. 1, the operator $R1 \bowtie R7$ in a QEP would be a cross product. "Artificial joins" are joins between two instances of the same relation: For the example query graph, consider a subtree of a QEP, in which joins $J1$ and $J5$ are performed simultaneously as in Fig. 2. Then, the fragments of $R5$ contained in the tuples output by the two joins satisfy different constraints. However, the fragments must satisfy the conjunction of the join constraints. Thus, an artificial join is introduced to enforce the conjunction constraint.

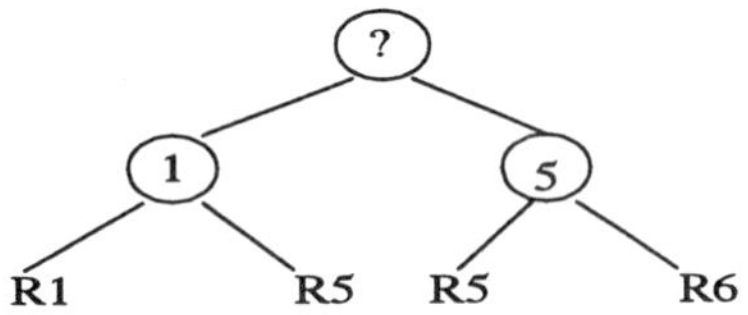

Figure 2: A QEP subtree with an artificial join

Cross products and artificial joins are expensive and increase the size of the search space. We therefore disallow them by introducing the notion of *valid* tree:

Definition 1: The join tree of a QEP is *valid* if:

1. it contains no cross products: the left, resp. right, relation of each join node is referenced in the left, resp. right, subtree below that node
2. it contains no artificial joins: each relation and each join-node appears only once in the tree

A QEP with a valid tree is a *valid* QEP.

Obviously, if a tree is valid, all its subtrees are valid. Hereafter, we consider only valid QEPs and use the terms (valid) "QEP" and "tree" interchangeably.

By this definition, the search space of a join query consists of all valid QEPs. To preserve the validity property, random placement of join nodes in a tree is not allowed. However, we define our crossover and mutation operators so that they always produce valid QEPs. Hence, the functions of the genetic program/QEP are closed in regard of our crossover and mutation.

Each join node has a label consisting of the join identifier, the algorithm identifier (e.g. Nested Loop 'l', Hash 'h', Sort merge 's') and the join constraint. It also contains pointers to its left and right subtree: the left and the right input streams are processed in a different way, so that the distinction between them is important for the node's execution time.

In Fig. 3, we show two valid QEPs T_1 and T_2 for the join graph in Fig. 1. Non-leaf nodes are surrounded by a circle. All nodes are identified by their number, being 12 for join $J12$ and **R12** for relation $R12$. In order to keep the figure simple, we omit the labels of the join nodes, as well as the leaf nodes and the edges pointing to them. We retain the join algorithm identifier and present the leaf nodes for the two marked subtrees S_1 and S_2 which are used in later examples.

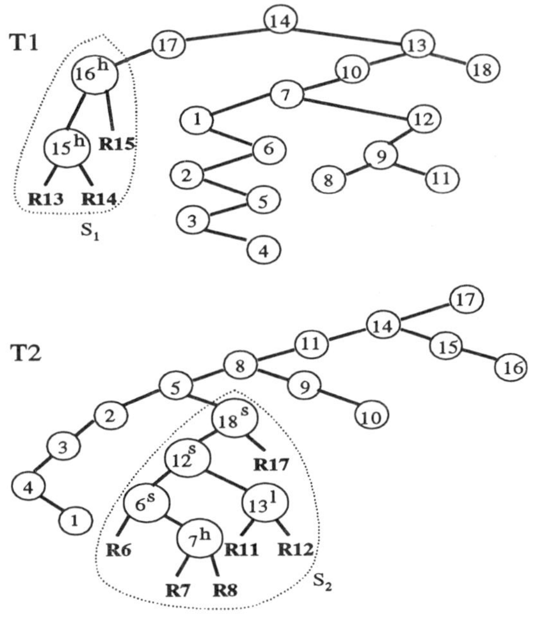

Figure 3: Join trees T_1 and T_2

3.2 Crossover, Selection and Mutation

The potential of genetic algorithms has not been sufficiently exploited in query optimization, because the string representation and the crossover operators used in [Bennett *et al.*, 1991] and [Steinbrunn *et al.*, 1993] were rather inappropriate: Small changes to a genom string produced large changes in the trees. Moreover, the Partially Matched Crossover operators adapted from [Goldberg, 1989] did not obey the semantics of building blocks for QEPs: an arbitrary string selected from a chromosome for crossover is often not a subtree. Hence, its insertion to another chromosome requires "repairing" the tree. The tree produced after repair has lost most of the structural information of its ancestor.

In our GP model, we use trees/QEPs as they are, so the structural information is properly inherited from ancestors to successors. Our crossover and mutation operators produce valid QEPs according to the GP principles.

Crossover. According to the GP paradigm, crossover is applied on two trees T_1, T_2, in which two subtrees S_1, S_2, have been selected, as marked in Fig. 3. Crossover produces two trees of the next generation, NG_1 and NG_2, which (a) must be *valid*, and each of them must (b) inherit most of the structure of its parent and (c) contain the selected subtree of the other parent.

Definition 2: *postorder(T_i)* is the list of join nodes produced by traversing tree T_i in postorder, i.e. left child–right child–root order.

Definition 3: *leaves_of(T_i)* is the set of relations in T_i.

Definition 4: $\gamma(nodelist, T\_set)$ is applied on an ordered list of join nodes, *nodelist*, and a set of trees or relations, *T_set*, and producing a valid tree as follows [1]:

```
foreach join J in nodelist
{
   A:= element of T_set that references
       the left relation of J;
   B:= element of T_set that references
       the right relation of J;
   build new tree T, where J is its root,
       A its left child and B its right child;
   remove A and B from T_set;
   insert T into T_set;
}
```

The γ-operator produces a valid tree if all trees in *T_set* are valid.

Definition 5: $crossover(T_1, S_1, T_2, S_2)$ is applied on two valid trees T_1 and T_2, from which two subtrees S_1 and S_2 are selected. It produces two valid trees NG_1 and NG_2 of the next generation using the γ-operator:

$$NG_1 := \gamma(\; postorder(T_1) - postorder(S_2), \; \{S_2\} \bigcup (leaves\_of(T_1) - leaves\_of(S_2)) \;)$$

$$NG_2 := \gamma(\; postorder(T_2) - postorder(S_1), \; \{S_1\} \bigcup (leaves\_of(T_2) - leaves\_of(S_1)) \;)$$

Example 1: We apply crossover to the trees T_1, T_2 of Fig. 3 using their marked subtrees S_1 and S_2. The next generation trees NG_1 and NG_2 are shown in Fig. 4.

To construct NG_1, we proceed as follows: We create the postorder list of T_1 and remove the nodes of S_2 from it. We then apply the γ-operator on the remaining postorder list and on the set of trees comprised of S_2 and the nodes of T_1 after eliminating those belonging to S_2. So, the *T_set* parameter to the γ-operator consists of the intact subtree S_2 and a number of nodes. By selecting the nodes from the postorder list and connecting them to form NG_1 we practically restore T_1: the only nodes missing are those belonging to S_2. This subtree however belongs to *T_set* and is attached to NG_1 intact. Hence, NG_1 inherits most of the structure of T_1 and contains S_2 in its original form. NG_2 is created in a similar way.

Our crossover operator ensures that most of the structural characteristics of the trees are inherited by their successors, while the two subtrees selected for crossover are attached intact to the new trees. Note that if S_1, S_2 have no nodes in common, crossover rebuilds S_1 in NG_1 exactly as it was in T_1 (same for S_2 in NG_2). Hence, *both* subtrees appear intact in *both* trees of the next generation. Thus, a probably good subtree of one ancestor is combined with the structure and the original join nodes of the other ancestor into a valid successor tree.

[1]A similar algorithm appears in [Bennett *et al.*, 1991] for the transformation between the tree and its chromosome representation.

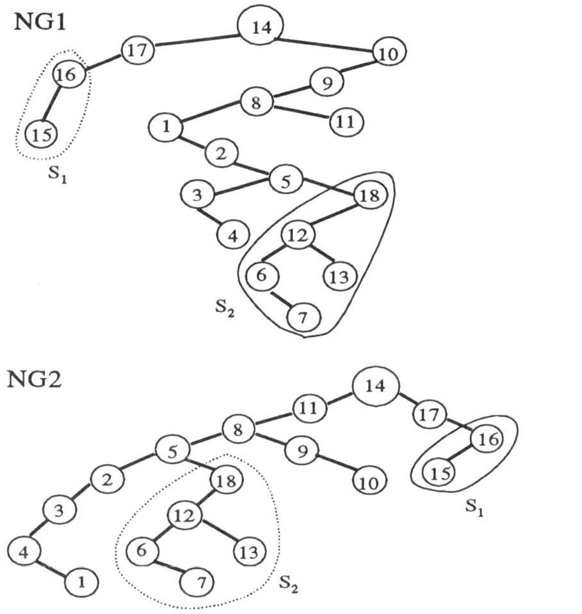

Figure 4: Next Generation: NG_1 **and** NG_2

Mutation. We consider two mutation operators:

- *mutate1*$(T, N, newAlg)$ changes the join algorithm into *newAlg* for a randomly selected node N in tree T. The structure of T is not affected, but its cost and hence its fitness value do change.
- *mutate2*$(T, N, parentOf(N))$ swaps the position of two neighbour nodes, the randomly selected node N and its parent. A repair action might be needed to produce a valid tree according to Def. 1.

Selection. Our selection operator is adopted from [Koza, 1991]. It uses a fitness-proportionate selection method. We formulate the next generation by recombining with crossover the 90% of the old population, mutating 5% and reproducing the remaining 5%.

3.3 Fitness Function and Cost Model

The fitness function of our GP model is based on the "cost" of a QEP, defined as the total execution time from the earliest retrieval of a database relation until the completion of the generation of the output by the root node of the QEP. Unlike other GP applications, the cost of a QEP is not computed by *executing* the QEP but *predicted* by a cost model that reflects the factors of execution time. Then, according to the fitness proportionate selection, the fitness of a QEP is its normalized execution cost, where the fittest QEP has fitness 1.

We consider a database in a multiprocessor environment. Parallelism is exploited by executing the join nodes simultaneously: nodes belonging to different subtrees can be executed concurrently, while neighbour nodes can be executed in "pipeline", where the parent node process the output of its children as it is produced.

In this environment, we consider two cost models. A cost model is a mathematical approximation of the real behaviour of a QEP. It must be as fine as possible to reliably reflect the actual QEP cost. The finer a cost model, though, the more complex and time-consuming is the evaluation of QEP cost. By considering two cost models with different search space shapes, we can study their impact on the efficiency of the search strategy.

The first model, "CM-1", reflects concurrent execution of nodes in different subtrees: once the input streams of a node are available, the node can start execution, possibly in parallel with other nodes. When a node completes execution, its output is written to disk, from which it is retrieved by the parent. The second cost model, "CM-2", also incorporates pipeline: the output of a node is not written to disk but sent immediately to its parent, which is executed simultaneously with its children. Those cost models are analyzed in [Stillger and Spiliopoulou, 1996].

4 Current Status

Our GP model is part of a larger model for parallel query optimization. For small join queries, the search space is scanned exhaustively [Spiliopoulou *et al.*, 1993]. For larger queries, two variations of iterative improvement have been implemented [Spiliopoulou *et al.*, 1996].

4.1 Architecture of the Model

The optimization module is parallel and has been implemented on several platforms; its most recent version runs on a network of Sun^{TM} workstations, organized according to the client-server paradigm. The clients scan different areas of the search space simultaneously. Parallel optimization is subject to one or more termination criteria, such as maximum optimization time or threshold value for QEP cost. When the termination criterion is met, the client sends the least expensive QEP to the server. The optimal QEP is the one with the lowest cost.

The exploitation of parallelism during optimization conforms nicely to the inherent parallelism of the *distributed genetic algorithm* approach [Koza, 1991] (chapter 22). The population is partitioned into subpopulations assigned to different clients/processors and processed concurrently, applying crossover, selection and mutation to generate the next population. The termination criterion may be extended to determine a maximum number of populations. When the termination criterion is met, the client sends the fittest QEP to the server, which selects the optimal QEP. We can enhance this scheme by migrating selected individuals from one subpopulation to another after a number of generations [Koza, 1991], thus establishing a cooperation among the clients for the generation of the optimal QEP.

4.2 Performance Experiments

In order to study the behaviour of our technique, we have compared it with iterative improvement, a combinatorial optimization technique frequently used for the optimization of large join queries, especially in parallel environments [Ioannidis *et al.*, 1992; Lin *et al.*, 1994; Spiliopoulou *et al.*, 1996; Swami and Gupta, 1988].

For our experiments, we have generated queries with 10 to 100 joins towards a database of relations with 10^3 to 10^6 tuples. The experimentation process is analyzed in [Stillger and Spiliopoulou, 1996], from which the following figures are obtained.

The genetic programming technique used a population of 100 to 1,000 QEPs, increasing with the query size. 50 generations were produced. To yield comparable results, iterative improvement produced an equal number of QEPs during its scan of the search space.

In Figures 5 and 6, we show the behaviour of our technique towards iterative improvement in terms of query cost improvement for each cost model. The cost values are normalized, using iterative improvement as the reference strategy.

For cost model CM-1 (Fig. 5), our technique is better than iterative improvement for small queries. For larger queries, it converges rapidly towards the optimal QEPs of iterative improvement.

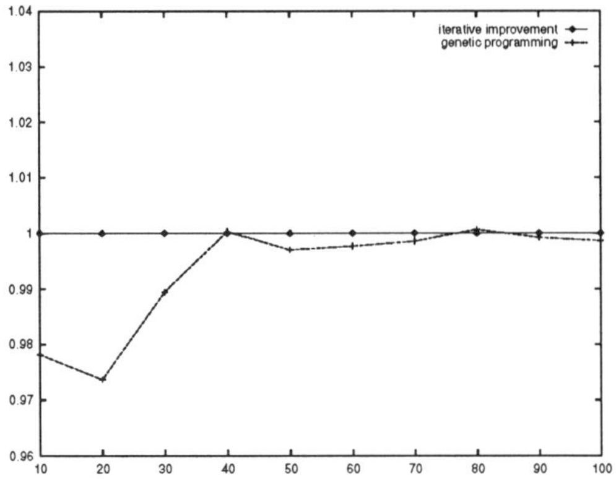

Figure 5: GP performance for cost model CM-1

For cost model CM-2 (Fig. 6), most optimal QEPs of GP are located within a distance of 10% from the optimal of iterative improvement. This distance becomes smaller as the query size increases.

5 Conclusions

In this study, we have proposed a genetic programming model for one of the hardest problems in databases, the query optimization problem. We have shown that the nature of the problem makes it particularly appropriate for GP, since the QEP of a query can be conveniently observed as a genetic program. We have specified the

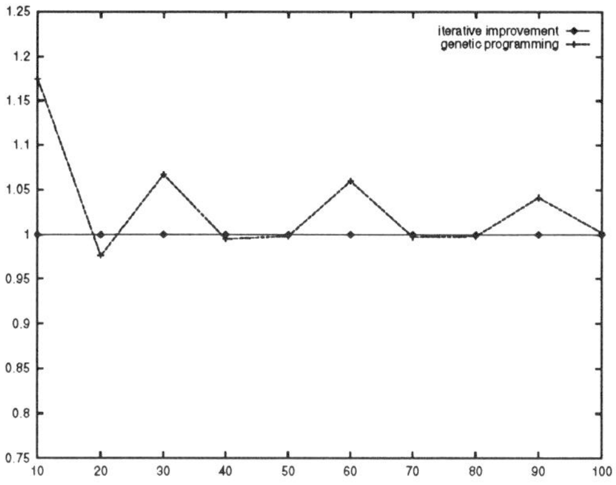

Figure 6: GP performance for cost model CM-2

search space of our model and presented the GP operators applied on the QEPs of this space to produce the successors of each population. Our experiments have demonstrated that our technique has comparable performance to a widely used classic technique.

We are currently performing a wide set of comparative experiments between our GP model and several other search techniques, including simulated annealing and tabu search. We consider various cost models with different search space shapes, and we study the impact of this difference on the behaviour of the search techniques. It is very important to test whether the GP model remains unaffected by the search space shape, since this would indicate that it is more appropriate for query optimization than the techniques currently used.

References

[Bennett *et al.*, 1991] Kristin Bennett, Michael C. Ferris, and Yannis Ioannidis. A genetic algorithm for database query optimization. Technical Report TR1004, University of Wisconsin, Madison, WI, 1991.

[Codd, 1970] E.F. Codd. A relational model of data for large shared data banks. *CACM*, 13(6):377–387, 1970.

[Goldberg, 1989] D.E. Goldberg. *Genetic Algorithms in Search, Optimization and Machine Learning.* Addison Wesley, Reading MA, 1989.

[Graefe, 1993] Goetz Graefe. Query evaluation techniques for large databases. *ACM Computing Surveys*, 25(2):73–170, 1993.

[Holland, 1975] J. Holland. *Adaption in Natural and Artificial Systems.* The University of Michigan Press, Ann Arbor, Michigan, 1975.

[Ioannidis and Kang, 1990] Yannis Ioannidis and Y.C. Kang. Randomized algorithms for optimizing large join queries. In *SIGMOD Int. Conf. on Management of Data*, pages 312–321, Atlantic City, NJ, 1990. ACM.

[Ioannidis and Kang, 1991] Yannis Ioannidis and Y.C. Kang. Left-deep vs. bushy trees: An analysis of strategy spaces and its implications on query optimization. In *SIGMOD Int. Conf. on Management of Data*, pages 168–177, Denver, CO, 1991. ACM.

[Ioannidis *et al.*, 1992] Yannis Ioannidis, Raymond T. Ng, Kyuseok Shim, and Timos K. Sellis. Parametric query optimisation. In *Int. Conf. on Very Large Databases*, pages 103–114, Vancouver, Canada, 1992.

[Koza, 1991] John R. Koza. *Genetic Programming.* The MIT Press, Cambridge, Massachusetts, 1991.

[Lanzelotte *et al.*, 1993] Rosana Lanzelotte, Patrick Valduriez, and Mohamed Zaït. On the effectiveness of optimization search strategies for parallel execution spaces. In *Int. Conf. on Very Large Databases*, pages 493–504, Dublin, Ireland, 1993.

[Lin *et al.*, 1994] E.T. Lin, E.R. Omiecinski, and S. Yalamanchili. Large join optimization on a hypercube multiprocessor. *IEEE Trans. on Knowledge and Data Engineering*, 6(2):304–315, 1994.

[Spiliopoulou *et al.*, 1993] Myra Spiliopoulou, Michalis Hatzopoulos, and Costas Vassilakis. Parallel optimization of join queries using a technique of exhaustive nature. *Computers & Artificial Intelligence*, 12(2):145–166, 1993.

[Spiliopoulou *et al.*, 1996] Myra Spiliopoulou, Michalis Hatzopoulos, and Yannis Cotronis. Parallel optimization of large join queries with set operators and aggregates in a parallel environment supporting pipeline. *IEEE Trans. on Knowledge and Data Engineering*, 1996. To appear.

[Steinbrunn *et al.*, 1993] Michael Steinbrunn, Guido Moerkotte, and Alfons Kemper. Optimizing join orders. Technical Report MIP9307, Faculty of Mathematic, University of Passau, Passau, Germany, 1993.

[Stillger and Spiliopoulou, 1996] Michael Stillger and Myra Spiliopoulou. Exploiting genetic programming in parallel query spaces. Technical report, Institut für Informatik, Humboldt-Universität zu Berlin, Berlin, Germany, 1996.

[Swami and Gupta, 1988] Arun Swami and Anoop Gupta. Optimization of large join queries. In *SIGMOD Int. Conf. on Management of Data*, pages 8–17, Chicago,IL, 1988. ACM.

Ontogenetic Programming

Lee Spector* †
lspector@hampshire.edu

Kilian Stoffel †
stoffel@cs.umd.edu

*School of Cognitive Science and Cultural Studies
Hampshire College
Amherst, MA 01002

†Department of Computer Science
University of Maryland
College Park, MD 20742

ABSTRACT

In nature, the structure and behavior of a mature organism is determined not only by its genetic endowment, but also by complex developmental processes that the organism undergoes while immersed in its environment (ontogeny). By contrast, an individual in a genetic programming system is generally expected to solve problems immediately, without the benefit of a developmental phase. Various morphological systems have been used in previous genetic programming systems to allow programs to "grow" into more complex forms prior to evaluation. Runtime memory mechanisms allow evolved programs to acquire information from their environments while they solve problems, and to change their future behavior on the basis of such information. ***Ontogenetic programming*** **combines these ideas to allow for runtime modification of program structure. In particular, an ontogenetic programming system includes program self-modification functions in the genetic programming function set, thereby allowing evolved programs to modify themselves during the course of a run. We demonstrate the ontogenetic programming methodology with the HiGP genetic programming system, and we show how ontogenetic programming can be used to solve problems that would not otherwise be solvable. We also discuss variants of the technique that could be used in other genetic programming systems.**

1 Phylogeny and Ontogeny

Biologists use the term *phylogeny* to refer to the developmental progression of a population or race through evolutionary time. Genetic programming systems use mechanisms inspired by those underlying biological phylogeny—in particular, genetic replication, mutation, and recombination combined with natural selection—to evolve programs that solve problems.

All but the simplest biological organisms also develop *as individuals*, both structurally and behaviorally, throughout their lives. Biologists refer to the developmental progression of an individual through its life span as *ontogeny*. In this paper we describe how rich ontogenetic components can be added to genetic programming systems, and we show how this can allow genetic programming to produce programs that solve more difficult problems.

Two sorts of ontogeny are evident in previously described genetic programming systems. The first consists of *morphological* components that allow programs to "grow" into different, perhaps more complex forms prior to execution. For example, in Gruau's technique of *cellular encoding* evolved programs are executed to produce neural networks; the networks that result from this developmental process are then assessed for fitness [Gruau 1994]. Similarly, Zomorodian has described a system in which evolved programs are executed to produce deterministic push-down automata [Zomorodian 1995]. Spector's automatically defined macros (ADMs) provide a morphological mechanism within the S-expression-based representational framework of traditional genetic programming; a program that includes ADMs is *macroexpanded* prior to execution, and the program that is executed may thereby be considerably larger and more complex than the evolved genotype [Spector 1995, 1996]. Morphological processes have also been used within other evolutionary computation paradigms; for example, Sims describes a system in which the morphological and behavioral components of virtual creatures are both described as directed graphs that evolve through the use of a graph-based genetic algorithm [Sims 1994]. Formal definitions of morphological processes in evolutionary computation, along with a good survey of past work, can be found in [Angeline 1995].

These sorts of morphological mechanisms may indeed add to the power of genetic programming, but they are simplifications of biological ontogeny in two important ways. First, they operate only prior to runtime; in contrast, biological ontogeny continues throughout the life span of an individual.

Second, they operate without environmental input; in contrast, a biological organism's environment may guide development in significant ways.

The second sort of ontogeny evident in previously described genetic programming systems involves the use of runtime memory mechanisms. Indexed memory [Teller 1994] and memory terminals [Iba et al. 1995] allow evolved programs to acquire and store information from their environments at runtime, and to use this information to guide future behavior. A program may thereby develop in certain ways throughout its "life span."

In theory, the ontogenetic mechanism provided by indexed memory is sufficient to support any developmental pathway for computer programs. One can imagine producing a program that includes a universal Turing machine; the program could periodically produce a new "more-developed" program in indexed memory and then begin to execute the new stored program. In practice, however, we cannot expect such remarkable programs to arise.

We use the term "ontogenetic programming" to describe an approach that combines the benefits of runtime memory with those of morphology. In ontogenetic programming, program self-modification operators are included in the set of functions that may be used by evolved programs. By use of these operators, an evolved program can dynamically change its own structure, and thereby change its future behavior, during the course of a run. A program's self-modification strategy is itself evolved; it may be arbitrarily complex and it may be conditionalized on runtime environmental inputs.

In our current implementation the self-modification and problem-solving components of programs are intermingled within a single program structure. But this is not essential; in other variants of the technique one might keep these components segregated. In either case, evolved programs can influence their own future behavior via direct manipulation of their own program structure.

Relations between evolution and ontogeny in evolutionary computation have previously been studied with genetic algorithms that produce neural networks (for example, [Parisi et al. 1992; Batali 1994]). Other work has explored the incorporation of runtime morphological development mechanisms into evolutionary computation frameworks (for example, [Dallaert and Beer 1994; Hemmi et al. 1994]). The work described in this paper applies related ideas to the production of computer programs that develop over time.

In the following sections we first discuss the adaptive utility of ontogeny. We then demonstrate ontogenetic programming using the HiGP genetic programming system. HiGP operates on linear programs for a stack-based virtual machine; program-modification mechanisms can be particularly simple when applied to these sorts of linear programs. The technique is not limited to such systems, however; at the end of the paper we discuss ontogenetic programming with more traditional S-expression-based representations.

2 Why Develop?

To some extent biological ontogeny is mandated by the physics of biological reproductive mechanisms. These produce, initially, very small structures; in many niches large size is adaptive (because, for example, it allows for faster movement or more effective hunting), and developmental processes are necessary to produce large structures. Ontogeny is therefore adaptive, if only because large size is adaptive, and because genes can produce large structures only by orchestrating developmental processes.

The situation is rather different for computer programs in a genetic programming system, at least with respect to size. Newly created programs may be of any size whatever, and it is not clear that large size is itself adaptive in any interesting niches. But developmental processes may nonetheless be adaptive for computer programs—developmental control of *size* may or may not be particularly useful, but developmental control of other phenotypic features is clearly useful in certain domains. More generally, developmental control of an arbitrary feature f of an individual is likely to be useful in two cases: 1) when there is an adaptive advantage to having an f value at maturity that differs significantly from the f values of new individuals, or 2) when there is an adaptive advantage to having different f values at different stages of an individual's life.

The second case applies to features of computer programs for many domains. Consider a program for an agent in an adventure game. It might be advantageous for such a program to change several times throughout the course of a run; it might be useful to have an initial, high-risk, exploratory phase, a more focused goal-seeking phase, and an analytical endgame phase. Each phase might be best served by rather different programs, and a program that could develop through several different forms during the course of its life span might have a significant advantage. This sort of development is enabled in ontogenetic programming by building the capability for self modification into the programs themselves—this allows them to evolve strategies for changing themselves as they run.

Consider the more abstract *sequential regression* problem, which shares features with many sequence prediction problems. As in ordinary symbolic regression [Koza 1992], the goal in sequential regression is to produce a program that returns the appropriate y value for each x value in a data set. Often the data set is generated by a known target function, and the regression problem can be viewed as one of rediscovering the target function from the data. In the sequential regression problem we stipulate also that the x values will be presented in a particular order. Programs have a "life span" insofar as they are run multiple times, one for each x value in the fitness-testing range, and insofar as they always encounter these x values in the same order. Development may be useful for sequential regression problems because different programs may be most appropriate for different ranges of the target function. It may also be useful because transitions from one value to the next may be better mediated by developmental operators than by domain operators.

Consider the particular sequential regression problem of rediscovering the target function x^2 over the non-negative integers. Suppose that we are using a function set that includes only addition and subtraction operators. Using ordinary genetic programming it is impossible to produce a correct program, because the provided function set is only capable of generating linear functions. But in an ontogenetic programming system with sufficiently powerful self-modification operators, a program can modify itself during the $x = n$ run to allow the subsequent run to produce a correct result for $x = n + 1$. We have in fact witnessed the evolution of such an ontogenetic program; upon each execution it inserts an additional "$+ x$" sequence into itself, thereby preparing itself for the next value. Although evolved with fitness cases that ranged only from 1 to 19, the program is general and works correctly for ranges from 1 to any positive integer (given unbounded space for program growth). In this case the ontogeny of the individual is properly matched to the dynamics of the individual's life course; as a result, the developmental process is able to extend the power of the available function set.

In some cases the benefits of development can be obtained more simply through the use of conditionals. For example, a program that develops in such a way that it replaces a body of code *a* with a body of code *b* at time *t* could in some environments be implemented with a conditional that executes either *a* or *b* depending on the result of a comparison of the current time to *t*.[1] Development is nonetheless more powerful; for example, a developing program may develop through an unbounded number of forms throughout its life span, while the number of execution paths for any conditionalized program is fixed.

Summarizing, there are a variety of situations in which development may be useful, for a variety of different reasons. Any problem for which programs have "life spans," in the sense used above, is a candidate for the application of ontogenetic programming. For such problems ontogeny may provide a capability that would otherwise be beyond the reach of the available function set.

In the next section we describe the HiGP implementation of ontogenetic programming, and in the subsequent section we demonstrate the utility of the approach for a binary sequence prediction problem.

3 Ontogenetic HiGP

HiGP is a new high-performance genetic programming system that combines techniques from string-based genetic algorithms, S-expression-based genetic programming systems, and high-performance parallel computing [Stoffel and Spector 1996]. The result is a fast, flexible, and easily portable genetic programming engine with a clear and efficient parallel implementation. The parallel version of HiGP scales nearly linearly with the number of available processors. HiGP manipulates and produces linear programs for a stack-based virtual machine (as in [Perkis 1994]), rather than the tree-structured S-expressions used in traditional genetic programming.

HiGP programs are executed on a virtual machine that is similar to a pushdown automaton. The virtual machine consists of three components: an input tape containing a linear program, a pushdown stack, and a finite-state control unit. The contents of the input tape are restricted to a small set of words that have been defined as HiGP operators. The contents of pushdown stack are restricted to double precision floating point numbers. The finite-state control unit reads the input tape and executes, for each word, the function call for the corresponding operator. The operators may perform arbitrary computations and manipulate the values on the stack. They may also reposition the read head on the input tape; this allows for the implementation of conditionals and loop structures. Return values are generally read from the top of the stack at the end of program execution.

The system includes two basic stack operators, `pop` and `dup`. The `pop` operator removes the topmost element from the stack, while the `dup` operator pushes a duplicate of the top element onto the stack. The system also includes a family of `push` operators that correspond to the terminal set in a traditional genetic programming system; each `push` operator pushes a single pre-determined value onto the stack.

The system also includes a `noop` operator that does nothing. This is necessary because all programs in the system have the same length, and because we do not wish to predetermine the number of actual problem-solving operators that should appear in solution programs. With the inclusion of the `noop` operator the fixed program size becomes a size *limit*, analogous to the depth limits used in S-expression-based genetic programming systems. "Shorter" programs are encoded by filling in extra program steps with `noop`s.

Any additional, problem-specific operators must take their input values from the stack and must push their results back onto the stack. When there are not enough values on the stack for a problem-specific operator it is skipped by the finite-control unit and the stack remains untouched (as in [Perkis 1994]). Additional information about HiGP may be found in [Stoffel and Spector 1996].

The ontogenetic version of HiGP results from including the following program self-modification operators in the function set:

segment-copy copies a part of the linear program over another part of the program. The function takes three arguments from the stack: the start position of the segment to copy, the length of the segment, and the position to which

[1] We thank an anonymous reviewer for drawing this possibility to our attention.

the segment should be copied. All positions are calculated relative to the current instruction. If there are not three values on the stack the instruction is skipped.

shift-left rotates the program to the left. The call takes one argument from the stack: the distance by which the program is to be rotated. If there is no value on the stack the program is rotated by one instruction to the left.

shift-right rotates the program to the right. The call takes one argument from the stack: the distance by which the program is to be rotated. If there is no value on the stack the program is rotated by one instruction to the right.

The position of the current instruction pointer is not changed by the execution of an ontogenetic operator. For example, if instruction #23 is a `segment-copy`, then after its execution instruction #24 will be executed, regardless of the fact that the *old* instruction #23 may now have been moved elsewhere.

4 Example: Binary Sequence Prediction

In this section we show how the ontogenetic version of HiGP can solve a problem that cannot be solved by the ordinary version of HiGP. We also show that the addition of indexed memory to the ordinary version of HiGP is not sufficient to produce solutions to this problem; the ontogenetic extensions provide benefits that indexed memory does not.

Consider a binary version of the sequential regression problem described above, with a target function that repeats the values [0 1 0 0 0 1] as x increases. The initial portion of the target function is {(0, 0) (1, 1) (2, 0) (3, 0) (4, 0) (5, 1) (6, 0) (7, 1) (8, 0) (9, 0) (10, 0) (11, 1) (12, 0)...}. The goal is to produce a program that returns the appropriate y value for each x value; the program will be run once for each x value, and the x values will be presented in order from 0 to some number n. In particular, we assessed fitness by testing each program on the range [0–19]. We use a "wrapper" function [Koza 1992] that maps positive y values to 1, and negative y values to 0. In other words, it is not necessary for an evolved program to produce the exact values 0 and 1; it is sufficient for it to produce either 0 or any negative number in place of each 0, and any positive number in place of each 1. This problem can be thought of as a binary sequence prediction problem, related to many other sequence prediction problems in the literature. In particular, Iba et al. describe a related binary oscillation task [Iba et al. 1995].

The function set for this problem consists of the 2-argument addition function `+`, the 2-argument subtraction function `-`, the 2-argument multiplication function `*`, the 2-argument protected division function `%` [Koza 1992], and three `push` functions: `push-x` (for the independent variable x), `push-0`, and `push-1`. The HiGP functions `dup` and `noop` were also included. Although this function set is simple it is nonetheless general and powerful; for example, one can obtain the effects of many conditionals through the clever use of arithmetic. For each fitness case the stack was initialized with a single copy of x, and the result was read from the top of the stack at the end of program execution.

We conducted 100 runs of the ordinary (non-ontogenetic) version of HiGP on this problem. We used a population size of 100, a maximum program size of 30, a crossover rate of 90%, and a reproduction rate of 10%. We allowed each run to continue for up to 20 generations. No correct solutions were produced by any of the 100 runs. We conclude that the problem cannot reliably be solved by ordinary HiGP and the given parameters.

We then conducted 100 runs of the ontogenetic version of HiGP on this problem (adding the `segment-copy`, `shift-right` and `shift-left` functions to the function set). We used the same population size and other parameters as in the non-ontogenetic case. 12 completely correct solutions were produced by the 100 runs. 10 of these solutions were general; although fitness was assessed only over the range [0–19], these programs produce correct results over the range [0–39], and they appear to be correct to any limit. The following is one of the correct evolved programs:

```
push-x % - shift-left push-x noop * *
dup % - + push-x % dup % shift-right
dup shift-left push-x shift-right * %
shift-right * + shift-right - - push-x
```

Because indexed memory also provides a limited ontogenetic capability, we also conducted 100 runs of the HiGP with indexed memory on this problem. We added a 30-element indexed memory, a 2-argument `write` function that stores the value of one argument in the location indexed by the other, and a 1-argument `read` function that pushes the indexed element of the memory onto the stack. We used the same population size and other parameters that were used for the runs described above. No correct solutions were produced by any of the 100 runs. We conclude that indexed memory does not provide sufficient power for HiGP to reliably solve this problem with the given parameters.

In summary, ontogenetic programming allows for the production of binary sequence prediction programs that change themselves in appropriate ways during the course of their "life spans" through the x range. This allows ontogenetic programming to reliably produce correct solutions to the binary sequence prediction problem, whereas the non-ontogenetic and indexed memory versions of our system were unable to produce any correct solutions.

5 Ontogenetic Programming with S-Expressions

Although the program self-modification functions of ontogenetic programming are easiest produce and to explain for linear programs, variants of the technique can be applied to the S-expression-based program representations of traditional genetic programming [Koza 1992].

One way to apply ontogenetic programming to S-expression-based programs is to include a `subtree-copy` function in the function set. The `subtree-copy` function takes two integer arguments, each of which indexes a subtree of the program through some standard (e.g., depth-first) traversal. The effect of a call to the `subtree-copy` function is to replace the subtree rooted at one of the indices with a copy of the subtree rooted at the other.

We have experimented with versions of the `subtree-copy` function, and although we have had some successes (e.g., for the x^2 sequential regression problem described above), several problems remain. One problem is that it is awkward to modify an S-expression during its execution. We avoided this problem by applying the `subtree-copy` operations to a copy of the program, and by replacing the program with its modified copy at the end of each execution. This means that a program could develop *between* the successive executions of its "life span," but not *during* the executions. In contrast, our HiGP implementation allows for program development during each single execution. This limitation of the `subtree-copy` model could be removed by using a more flexible S-expression evaluator.

A second problem with the `subtree-copy` model of ontogenetic programming is that the index arguments refer to specific subtrees of the entire program, indexed relative to the root of the S-expression. This means that it is unlikely that a particular call to `subtree-copy` will be transportable from one program to another; it will have a completely different effect in each program context into which it is placed. Crossover will therefore rarely produce high-fitness children from high-fitness parents — the fitness landscape will have sharp discontinuities. One way to mitigate this problem is to use a 3-argument `structured-subtree-copy` control structure. As with `subtree-copy`, the first two arguments to `structured-subtree-copy` produce integer indices. But the third argument is a *result-producing branch* that is executed to produce a value from the call to `structured-subtree-copy`. As a side-effect of each call to `structured-subtree-copy`, a subtree-copy operation is performed *within the result-producing branch.* Since the indices refer only to subtrees within the result-producing branch, and since they are indexed relative to the root of the result-producing branch, it is reasonable to assume that the entire call to `structured-subtree-copy` will be transportable to other program contexts.

A third problem with the `subtree-copy` model of ontogenetic programming is that it may produce arbitrarily large programs during execution. Whereas the HiGP program-modification functions never change the size of a program, the replacement of a small subtree by a large subtree in an S-expression program will indeed produce a larger program. Several techniques may be used to limit this growth. These include absolute depth limits that are enforced by converting offending `subtree-copy` calls into `noops`, and depth-based indexing schemes that ensure that replacement subtrees are selected only from those with acceptable sizes.

An alternative mechanism for ontogenetic programming with S-expression-based programs makes use of *dynamic ADFs* and *dynamic ADMs*. With ordinary automatically defined functions (ADFs [Koza 1994]) and automatically defined macros (ADMs [Spector 1995, 1996]) each individual has a fixed set of evolved sub-functions/macros that may be used by the main program and by each other. In contrast, dynamic ADFs and ADMs may be redefined by the individual at runtime, and the new definitions may be constructed on the basis of information in the individual's environment or memory. The implementation involves the inclusion function/macro definition and calling operators (e.g. variants of Lisp's DEFUN, FUNCALL, etc.) in the function set. One can refer to the dynamically defined functions/macros using integer labels; this allows one to use a mechanism much like indexed memory, with the memory containing callable and redefinable functions and macros rather than simple values. Special but straightforward measures must be taken to avoid problematic recursive definitions. Dynamic ADFs and ADMs allow programs to modify themselves at runtime through module redefinition, rather than through direct manipulation of program code. We are currently investigating the utility of this technique.

6 Conclusions

We have shown that it is possible to use genetic programming to produce programs that themselves develop in significant, structural ways over the course of a run. We use the term "ontogenetic programming" to describe our technique for achieving this effect, which involves the inclusion of program self-modification functions in the genetic programming function set. We described cases in which this may be useful, and we demonstrated the application of our HiGP implementation of ontogenetic programming to a binary sequence prediction problem. Although HiGP manipulates linearly-structured programs, and the program self-modification functions are particularly simple in this case, we also described several variants of the technique that are appropriate for more traditional S-expression-based program representations.

Acknowledgments

Mark Feinstein helped to develop our initial interest in ontogeny, and to refine our understanding of its role in biological systems. Discussions at the 1995 AAAI Fall Symposium on Genetic Programming [Siegel 1995] helped to further refine our approach to ontogenetic programming. We also acknowledge the helpful comments of several anonymous reviewers. This research was supported in part by grants from ONR (N00014-J-91-1451), AFOSR (F49620-93-1-0065), and ARPA contract DAST-95-C0037.

Bibliography

Angeline, P.J. 1995. Morphogenic Evolutionary Computations: Introduction, Issues, and Examples. In *Proceedings of the Fourth Annual Conference on Evolutionary Programming*, edited by J.R. McDonnell, R.G. Reynolds, and D.B. Fogel. Cambridge, MA: The MIT Press.

Batali, J. 1994. Innate Biases and Critical Periods: Combining Evolution and Learning in the Acquisition of Syntax. In *Artificial Life IV: Proceedings of the Fourth International Workshop on the Synthesis and Simulation of Living Systems,* edited by R.A. Brooks and P. Maes. pp. 160–171. Cambridge, MA: The MIT Press.

Dellaert, F., and R.D. Beer. 1994. Toward an Evolvable Model of Development for Autonomous Agent Synthesis. In *Artificial Life IV: Proceedings of the Fourth International Workshop on the Synthesis and Simulation of Living Systems,* edited by R.A. Brooks and P. Maes. pp. 246–257. Cambridge, MA: The MIT Press.

Gruau, F. 1994. Genetic Micro Programming of Neural Networks. In *Advances in Genetic Programming,* edited by Kenneth E. Kinnear, Jr. pp. 495–518. Cambridge, MA: The MIT Press.

Hemmi, H., J. Mizoguchi, and K. Shimohara. 1994. Development and Evolution of Hardware Behaviors. In *Artificial Life IV: Proceedings of the Fourth International Workshop on the Synthesis and Simulation of Living Systems,* edited by R.A. Brooks and P. Maes. pp. 371–376. Cambridge, MA: The MIT Press.

Iba, H., T. Sato, and H. de Garis. 1995. Temporal Data Processing Using Genetic Programming. In *Proceedings of the 6th International Conference on Genetic Algorithms, ICGA-95*, edited by Larry J. Eshelman, pp. 279–286. San Fransisco: Morgan Kaufmann Publishers, Inc.

Koza, J.R. 1992. *Genetic Programming: On the Programming of Computers by Means of Natural Selection.* Cambridge, MA: The MIT Press.

Koza, J.R. 1994. *Genetic Programming II: Automatic Discovery of Reusable Programs.* Cambridge, MA: The MIT Press.

Parisi, D., S. Nolfi, and F. Cecconi. 1992. Learning, Behavior, and Evolution. In *Toward a Practice of Autonomous Systems: Proceedings of the First European Conference on Artificial Life*, edited by F.J. Varela and P. Bourgine. pp. 207–216. Cambridge, MA: The MIT Press.

Perkis, T. 1994. Stack-Based Genetic Programming. In *Proceedings of the 1994 IEEE World Congress on Computational Intelligence.* pp. 148–153. IEEE Press.

Siegel, E.V., editor. 1995. *Collective Brainstorming at the AAAI Symposium on Genetic Programming*, http://www.cs.columbia.edu/ evs/gpsym95.html.

Sims, K. 1994. Evolving 3D Morphology and Behavior by Competition. In *Artificial Life IV: Proceedings of the Fourth International Workshop on the Synthesis and Simulation of Living Systems,* edited by R.A. Brooks and P. Maes. pp. 28–39. Cambridge, MA: The MIT Press.

Spector, L. 1995. Evolving Control Structures with Automatically Defined Macros. In *Working Notes of the AAAI Fall Symposium on Genetic Programming.* pp. 99–105. AAAI Press.

Spector, L. 1996. Simultaneous Evolution of Programs and their Control Structures. In *Advances in Genetic Programming 2*, edited by Peter J. Angeline and Kenneth E. Kinnear, Jr. Cambridge, MA: The MIT Press. In press.

Stoffel, K., and L. Spector. 1996. High-Performance, Parallel, Stack-Based Genetic Programming. In *Proceedings of The Genetic Programming 1996 Conference.* Cambridge, MA: The MIT Press.

Teller, A. 1994. The Evolution of Mental Models. In *Advances in Genetic Programming,* edited by Kenneth E. Kinnear, Jr. pp. 199–219. Cambridge, MA: The MIT Press.

Zomorodian, A. 1995. Context-Free Language Induction by Evolution of Deterministic Push-Down Automata Using Genetic Programming. In *Working Notes of the AAAI Fall Symposium on Genetic Programming.* pp. 127–133. AAAI Press.

Using Genetic Programming to Approximate Maximum Clique *

Terence Soule
Laboratory for Applied Logic
Computer Science Dept.
University of Idaho
Moscow, Idaho 83844-1010
tsoule@cs.uidaho.edu

James A. Foster
Laboratory for Applied Logic
Computer Science Dept.
University of Idaho
Moscow, Idaho 83844-1010
foster@cs.uidaho.edu

John Dickinson
Computer Science Dept.
University of Idaho
Moscow, ID 83844-1010
johnd@cs.uidaho.edu

Abstract

We have attempted to solve the Maximum Clique Problem using a simple genetic program. The program language consists of a *Union* operator and vertex numbers. Our results compare favorably with complex genetic algorithms. We hypothesize that our use of genetic programming is particularly effective at exploiting the mechanism embodied by the Building Block Hypothesis.

1 Introduction

Both genetic programming (GP) and genetic algorithms (GA) have proven themselves adept at solving a wide range of problems [Koza, 1992, Koza, 1994, Beasley *et al.*, 1993a, Beasley *et al.*, 1993b]. However, they each tend to be applied to problems requiring a particular type of solution. GAs are typically used where the answer is easily presented as a value or set of values whereas GP is more often used when the solution is a particular behavior or function. Although it is certainly possible to apply one technique to the other type of problem, it is rarely attempted.

The maximum clique problem (MC) is a natural problem for GAs, requiring only a set of vertex numbers for an answer. It has been the subject of several studies using GAs [Bui and Eppley, 1995, Carter and Park, 1993, Foster and Soule, 1995]. In these studies the basic GA was extended by features which improved the performance on MC. Despite these improvements to the basic GA our simple GP produced comparable results. We attribute this success to the GP's effective creation and use of building blocks.

We begin in Sections 1.1 and 1.2 by defining MC and reviewing some of the previous attempts to solve it. Section 2 introduces our Genetic Program. The results are presented in Section 3. We attempt to explain these results in terms of epistasis and the Building Block Hypothesis in Section 4.1. Finally, we present some conclusions in Section 4.2.

*This work supported by funding from the URO FY94 Seed Grant program, and through funding received from the NSF-Idaho EPSCoR project under NSF Cooperative Agreement number OSR-9350539.

1.1 The Maximum Clique Problem

We investigated the Maximum Clique Problem, defined as follows:

Given: A graph G

Find: The largest subgraph of G in which every vertex in the subgraph is directly connected to every other vertex in the subgraph (that is, the largest complete subgraph of G).

MC is a particularly difficult problem, since no polynomial time, deterministic algorithm exists for finding a satisfactory approximate solution unless P = NP (which is very unlikely):

Theorem 1 [Arora *et al.*, 1992] There is a constant $\epsilon > 0$ such that a polynomial time approximation of MC with accuracy at least $n^{-\epsilon}$ exists iff P = NP.

The accuracy of an approximation is the ratio of the size of the clique returned by the algorithm to the actual size of the maximum clique. MC is complete for hard to approximate NP optimization problems [Chang *et al.*, 1994]. So, any algorithm which does reasonably well at MC should make it possible to develop other algorithms which perform equally well on many other graph theory problems.

1.2 Previous Work

Carter and Park [Carter and Park, 1993] found that a simple GA performed poorly on MC, but that a scaled fitness improved the results. Their scaling raised their original, simple fitness values to a time dependent power. Thus, the fitness function changed in a time dependent manner similar to that used in determining the selection criteria in simulated annealing. With this scaling factor the preference for more fit individuals was much weaker at the beginning of a run than at the end when the scaling factor was larger. This change helped to minimize slow finishing in the GA. They also made and tested several other variations to the simple GA, including uniform crossover, greedy replacement, and diversity enhancement, resulting in slightly improved performances.

Bui and Eppley [Bui and Eppley, 1995] used a hybrid GA, the Genetic Maximum Clique Algorithm (GMCA), to study MC. GMCA incorporated a local optimization in each generation and had a preprocessing stage which ordered the vertices either by degree or in prefix order. The later variations were included with the expectation that they would act to place the vertices in the maximum clique close together on the chromosome, making it less likely that crossover would disrupt potential solutions. The localization guaranteed that the GA would at least find local optima and reduced the problem of slow finishing. The GMCA was applied to part of a test suite of graphs compiled by DIMACS, with generally positive results, although certain graph characteristics led to poorer performances.

DIMACS is the Center for Discrete Mathematics and Theoretical Computer Science, based at Rutgers University. It is an organization which regularly presents challenges to focus attention on open problems in Computer Science. In 1992 the DIMACS challenge centered on problems from graph theory, including MC. In addition to other work, they compiled a large suite of graphs which could be used for testing techniques for solving or approximating MC. For more information on past DIMACS challenges, including the full test suite of graphs from the 1992 challenge, see the DIMACS homepage at http://dimacs.rutgers.edu.

We [Foster and Soule, 1995] have used a Grouping Genetic Algorithm to approximate MC. In this variation of the basic GA an integer vector is used as the chromosome instead of a binary vector. The integer vector partitions the graph into subgraphs, which are evaluated as independent solutions. Individual fitnesses are based on the best of these solutions. Additionally, the fitnesses are scaled using a function which becomes more discriminating as better solutions are found. We are in the process of applying this algorithm to the entire DIMACS test suite. Our results to date show that the algorithm does perform better than the basic GA. However, as with the other algorithms, certain graphs have proven to be particularly difficult.

These studies are the only GA solutions to MC we have found in the literature. However, MC is an old, extensively studied problem which has proven consistently difficult to solve or approximate by any means. It's difficulty makes it an excellent problem for testing new approximation techniques. Any method which performs well on MC is likely to provide insight on how to approach many other, simpler problems. In addition, as a graph theory problem, MC is a representative of a very large class of important problems. In applying GP to MC we show that GP is an appropriate approach to graph theory problems, even though these have not been traditionally studied GP problems.

2 The Genetic Program

GP can be easily applied to MC using a very simple language. Our language uses statements consisting of the integers from 1 to M, where M is the number of vertices in the graph. The only non-terminal statement is `union` which takes two other statements as arguments. A typical program looks like

```
union(union(3,union(1,3))
   ,union(union(6,5),9))
```

Programs can also be presented as a tree as shown in Figure 1. Crossover is accomplished by swapping branches between two such trees. As might be expected this program guesses that the nodes $1, 3, 5, 6$, and 9 make up the maximum clique.

Our fitness function is designed to give higher fitness to chromosomes which select subgraphs with more edges, and to give an additional bonus when those subgraphs are cliques. Since there are a maximum of $\binom{v}{2}$ edges in a graph with v vertices, the density of a graph G is $e/\binom{v}{2}$, where e is the number of edges in G.

To formalize this idea, let σ be a chromosome, and $H \in \sigma$ denote the fact that some group in σ

Graph Name	Size	Max. Clique	Max. Found	Ave. Found	Standard Deviation
san400_0.9_1	400	100	60	49.8	5.1
brock200_1	200	21	19	15.2	3.3
hamming8-2	256	128	98	71.8	12.4
keller5	776	27	19	17.0	1.2
c-fat200-1	200	12	12	11.6	0.8

Table 1: GP results

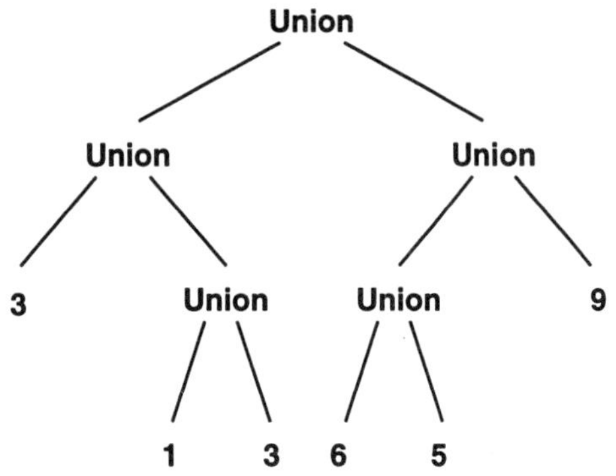

Figure 1: A Sample Program Tree

represents graph H. We assume that H has e edges and v vertexes. Furthermore, let $C(H)$ be a function such that $C(H) = 0$ if H is not a clique, and $C(H) = 1$ if H is a clique. Finally, let N be an upper bound on the size of the maximum clique.

The simple fitness function, then, is:

$$f(\sigma) = \left(\frac{1}{2}\left(e/\binom{v}{2}\right) + \frac{1}{2}\left(C(H)v/N\right)\right) \quad (1)$$

For non-cliques the first term in the fitness function is always positive and less than one half while the second term is zero. Thus, for non-cliques the fitness function returns a value between zero and one half. This value is higher for non-cliques with more edges, so there is selective pressure among non-cliques to add high degree vertices.

Since, a clique with v nodes has $\binom{v}{2}$ edges the first term of the fitness function is always one half for cliques. The second term is zero for non-cliques and less than or equal to one half for cliques; it is maximized when the maximum clique has been found. If N is the size of the maximum clique then the maximum value of the second term is one half and the fitness of cliques is between one half and one. Larger cliques will contain more vertices so they will have a higher fitness than smaller cliques.

In practice the size of the maximum clique will not be known and an upper bound on the maximum clique size is used for N. In this case the range of fitness values may be different, but the ordering of the subgraphs by fitness does not change.

To avoid exceeding memory limits and to reduce the amount of processing time it is necessary to limit the size of the generated programs. Our research has shown that selective pressure effectively controls the growth of programs [Soule *et al.*, 1996]. Selective pressure is applied by penalizing the longer programs. We tested several different penalty factors, eventually making the penalizing fitness function,

$$f_t(\sigma) = f(\sigma) - \left(\frac{1}{2M} * j\right) \quad (2)$$

where M is the input graph size and j is the number of nodes included by the program. If a program includes a vertex more than once each occurrence was counted separately. Therefore for the program shown in Figure 1, j would be six.

Based on Equation 1 the increase in fitness for increasing the size of an existing clique by one node is $1/(2N)$. Combining this with Equation 2 gives the net benefit of adding a new node to an existing clique as $1/(2N) - 1/(2M)$. If $N = M$ (the entire graph is a clique) then the net benefit is zero, otherwise adding a node to an existing clique is always beneficial. However, the more typical case involves a branch being added which contains several nodes only some of which may be new. In this case the net benefit is:

$$\frac{1}{2N} * n - \frac{1}{2M} * m \quad (3)$$

where m is the number of nodes in the branch and n is the number of new nodes in the branch. In all cases $n \leq m$. Thus, if N is relatively close to M (the maximum clique consists of a large portion of the entire graph) then it is likely that many crossovers will have a net harmful effect despite increasing the size of the clique encoded by the individual.

For one graph (hamming8-2, from the DIMACS test suite) the maximum clique consisted of half of the vertices in the graph, i.e. $M = 2N$. This was the largest clique, relative to the size of the entire graph, that we tested. For this graph performance was relatively poor until the fitness function was changed to:

$$f_t(\sigma) = f(\sigma) - \left(\frac{1}{20M} * j\right) \qquad (4)$$

Using a factor of 20 instead of 2 in the denominator creates a smaller penalty per node which acts to balance the smaller bonus per new node caused by the relatively large graph size and improves the results. However, even with this change the hamming8-2 graph proved to be the most difficult for our GP, suggesting that some of the original difficulty was caused by features other than the relative graph size.

We used a population size of 600, replacing the entire population at each generation. We used the stochastic remainder technique, which is based on ranked fitnesses, to select individuals for the next generation. The crossover rate was 0.9 and the mutation rate was 0.0007. Only the terminal nodes (vertex numbers) were subject to mutation. For this GP it is important to include mutation, otherwise it is possible that some vertices may permanently disappear from the population. Each trial was run ten times to reduce the variability in the results.

Graph Name	GP	Grouping GA	Bui and Eppley's GMCA
san400_0.9_1	60	53	50
brock200_1	19	11	20
hamming8-2	98	91	128
keller5	19	20	18
c-fat200-1	12	12	12

Table 2: Largest cliques found by three methods

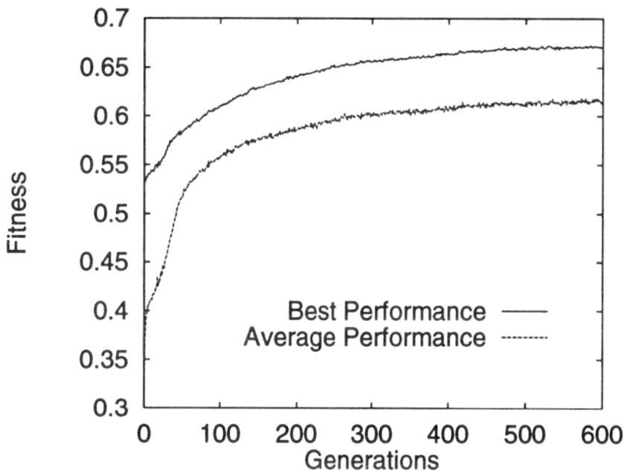

Figure 2: GP Performance on san400_0.9_1

3 Results

To date we have tested our GP on a number of the graphs from the DIMACS test challenge. The results are shown in Table 1.

Table 2 shows a direct comparison of the best results of the GP, our Grouping GA, and Bui and Eppley's GMCA. The results for both the GP and GMCA are taken from ten trials, while the Grouping GA results are from five trials. The data demonstrate that the GP's performance is comparable to that of the more complicated GAs. Additionally, these results were obtained from five very different types of graphs, which implies that GP is a robust approach to MC.

Figure 2 shows the best and average performances (averaged across all ten trials) of the GP on the graph san400_0.9_1.

Figure 3 shows the number of leaves in the average and best programs (again averaged across all ten trials) on the graph san400_0.9_1. Note that, each program contains one less `union` statements than it has leaves since it is a full binary tree. Thus, the total number of nodes in a program tree is $2N - 1$ where N is the number of leaves as shown in Figure 3.

Figure 3 show that at the end of ten trials the average length of the best programs in the ten trials is approximately 58. Table 1 shows that the average clique found by the same programs consists of 49.8 vertices. Thus, in this case the best programs contain an average of roughly 16 percent redundant code (code which includes vertices already included elsewhere).

4 Discussion and Conclusions

4.1 Discussion: The Building Block Hypothesis and MC

We believe the mechanisms described by the Schemata Theorem and the Building Block Hypothesis partially explain the positive results obtained using GP on MC. In particular, it appears that GP uses these mechanisms more effectively than the GA does. The Schema Theorem states that short pieces of the genetic code (identified by similarity templates known as schemata) with above average fitness are selected for reproduction and survive crossover with exponentially increasing frequency [Holland, 1975, O'Reilly and Oppacher, 1995]. The Building Block Hypothesis proposes that solutions

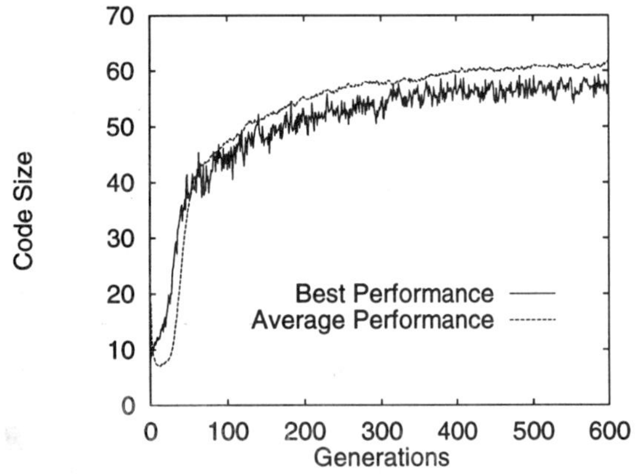

Figure 3: GP Code Size on san400_0.9_1

improve primarily by producing and testing various combinations of these highly fit schemata.

Unfortunately, the nature of MC makes a useful application of the Building Block Hypothesis difficult for GAs. Consider a large graph whose maximum clique consists of the vertices 3, 25, 31, 55, 56, 68, 81, and 90. Because these vertices are relatively evenly distributed there are no short, highly fit building blocks which will lead to the optimal solution. This is essentially the problem of epistasis. For GAs, epistasis is a measure of how much the fitness of one gene depends on the value of other genes and the separation of those genes on the chromosome. For MC the value of including one vertex depends on whether other vertices, which share a clique with the first vertex, are also included in the individual. Because the vertices of a clique are usually widely separated by the encoding, MC problems usually have a high epistasis. The GA hardness of a problem, how difficult it is to solving using a GA, has been linked to the degree of epistasis in the problem and its encoding [Davidor, 1991]. Thus, part of the difficulty of solving MC with GAs appears to be the epistasis inherent in the problem.

Bui and Eppley's heuristics for renumbering the nodes in an MC graph were attempts to limit the epistasis by placing important vertices close together in the encoding [Bui and Eppley, 1995]. Although their renumbering does not perfectly reorder the nodes it still appears to lead to better solutions than GAs which do not include renumbering. This is further evidence that epistasis contributes to the difficulty of approximating MC.

The GP approach avoids the problem of epistasis by not imposing a fixed ordering on the vertices. Although the vertices are numbered they may appear within programs in any order or grouping. Thus, a program can produce small, highly fit subtrees even though the maximum clique may consist of widely separated vertices. Because the GP avoids the problem of epistasis, and can produce these small, highly fit subtrees, it is reasonable to believe that the GP's performance is at least partially due to its more effective use of building blocks.

However, to make the claim that the GP is effectively exploiting the Building Block Hypothesis it is necessary to address several arguments raised by O'Reilly and Oppacher, who question whether the Building Block Hypothesis can be applied to GP at all [O'Reilly and Oppacher, 1995]. First, O'Reilly and Oppacher point out that the complex interactions between sections of code makes it difficult to assume that a highly fit solution can be created by combining sections of code which individually have relatively high fitnesses. However, in our case the maximum clique is found by a simple combination of subcliques, achieved by the `union` operator. There are no complicating interactions to interfere with the creation of a better solution from the joining of two lesser solutions. Of course, many such combinations will fail if the subtrees encode disjoint cliques. This does not mean that the Building Block Hypothesis fails for our GP, only that MC presents other difficulties in addition to epistasis.

O'Reilly and Oppacher also note that, unlike in GAs, GP subtrees are not easily represented as hyperplanes and the selection process cannot easily be viewed as competition between hyperplanes. Thus, the schema representation may not be the best route to understanding GP behavior. In general, this would appear to be a valid concern, but for our GP representation subtrees do represent hyperplanes of the solution space just as the GA schemata do.

Hence, despite O'Reilly and Oppacher's objections to a general Building Block Hypothesis for GP, it is reasonable that the Building Block Hypothesis does apply in this case. The ability of our GP to avoid the epistasis problem leads us to hypothesize that our GP approach to MC is relatively successful because it exploits the Building Block Hypothesis effectively.

4.2 Conclusions

We have presented an extremely simple genetic program for solving the maximum clique problem. The performance of the GP is comparable to that of relatively complex genetic algorithms on a variety of graph types. We attribute the effectiveness of the GP to its ability to avoid the problem of epistasis, which allows it to more effectively exploit the

Building Block Hypothesis. The objections raised by O'Reilly and Oppacher to a Building Block Hypothesis for GP do not apply to this application, because of the simplicity of the approach.

These results are particularly significant because the comparable performance of a simple GP and a complex GA suggests that a GP tuned for MC could produce superior results to those achievable with a GA. Additionally, MC is only one of many graph theory problems that are potentially difficult for GA because of the inherent amount of epistasis. The relative success of GP on one of the harder graph theory problems suggests that it may be applicable to many of the other problems from graph theory with similar success.

References

[Arora *et al.*, 1992] Sanjeev Arora, Carsten Lund, Rajeev Motwani, Madhu Sudan, and Mario Szegedy. Proof verification and the hardness of approximation problems. In *Proceedings 33rd IEEE Symp. on Foundations of Computer Science*, pages 14–23. IEEE, 1992.

[Beasley *et al.*, 1993a] David Beasley, David R. Bull, and Ralph R. Martin. An overview of genetic algorithms: Part I, fundamentals. *University Computing*, 15:58–69, 1993.

[Beasley *et al.*, 1993b] David Beasley, David R. Bull, and Ralph R. Martin. An overview of genetic algorithms: Part II, research topics. *University Computing*, 15, 1993.

[Bui and Eppley, 1995] T. Bui and P. Eppley. A hybrid genetic algorithm for the maximum clique problem. In Larry J. Eshelman, editor, *Proc. 6th Int. Conf. on Genetic Algorithms*, pages 478–483. San Francisco, CA: Morgan Kaufmann, 1995.

[Carter and Park, 1993] Bob Carter and Kihong Park. How good are genetic algorithms at finding large cliques: An experimental study. In *Proc. 2nd DIMACS Challenge: Cliques, Coloring and Satisfiability.* NSF, Center for Discrete Mathematics and Computer Science, October 1993.

[Chang *et al.*, 1994] W. I. Chang, Richard Gasarch, and C. Lund. On bounded queries and approximations. Technical Report TR-CS-94-05, University of Maryland, Baltimore County, Maryland, 1994.

[Davidor, 1991] Yuval Davidor. Epistasis variance: A viewpoint on GA-hardness. In Gregory J. E. Rawlins, editor, *Foundations of Genetic Algorithms*, pages 23–35. San Mateo, CA: Morgan Kaufmann, 1991.

[Foster and Soule, 1995] James A. Foster and Terence Soule. Using genetic algorithms to find maximum cliques. Technical Report LAL 95-12, University of Idaho, Moscow, Idaho, 1995.

[Holland, 1975] John H. Holland. *Adaptation in Natural and Artificial Systems.* Ann Arbor: University of Michigan Press, 1975.

[Koza, 1992] John R. Koza. *Genetic Programming: On the Programming of Computers by Means of Natural Selection.* Cambridge, MA: The MIT Press, 1992.

[Koza, 1994] John R. Koza. *Genetic Programming II: Automatic Discovery of Reusable Programs.* Cambridge, MA: The MIT Press, 1994.

[O'Reilly and Oppacher, 1995] Una–May O'Reilly and Franz Oppacher. The troubling aspects of a building block hypothesis for genetic programming. In L. Darrell Whitley and Michael D. Vose, editors, *Foundations of Genetic Programming 3*, pages 73–88. San Francisco, CA: Morgan Kaufmann, 1995.

[Soule *et al.*, 1996] Terence Soule, James A. Foster, and John Dickinson. Code growth in genetic programming. In John R. Koza, David E. Goldberg, David B. Fogel, and Rick R. Riolo, editors, *Genetic Programming 1996 (this conference)*, 1996.

Paragen - A novel technique for the autoparallelisation of sequential programs using GP

Paul Walsh
University College
Cork
paul@csvax1.ucc.ie

Conor Ryan
University College
Cork
conor@ravenloft.ucc.ie

Abstract

The Paragen system is a new technique for the automatic conversion of sequential programs into functionally equivalent parallel programs. This technique utilizes GP to generate a highly parallel program from an original sequential program, while preserving functionality. We show that a genetic search avoids the complexities introduced by standard data dependency techniques and can also introduce further efficiencies by automatically reordering program statements. This paper gives a brief introduction to the problem of autoparallelisation followed by a discussion of the design and implementation issues of the system. Results demonstrate the effectiveness of the Paragen system by the automatic conversion of a number of complex program segments.

1 Introduction

The performance of serial machines is limited by the "von Neumann" bottle neck, whereby at each instruction step, a single processor must write access data stored in memory. Therefore the computational requirements of a large number of applications must be met by parallel processing systems, which consist of a number of processors operating in parallel. Unlike the general purpose von Neumann architecture, there is no fixed parallel processing architecture, but they can be broadly classified as those machines that operate synchronously or asynchronously. Synchronous machines have a common clock where each processor executes the same instruction at each time step and are widely programmed in the Single Instruction Multiple Data (SIMD) paradigm. With asynchronous machines, however, each processor can operate independently and these machines are widely programmed in the Multiple Instruction, Multiple Data (MIMD) paradigm. The SIMD paradigm is generally less complex than the MIMD paradigm but is less flexible and can only exploit low levels of parallelism. Therefore in this paper we consider the more general purpose model of parallel computation, the MIMD paradigm.

Despite the power of parallel processing machines the software development cycle is considerably more complex and error prone than that for sequential software. The diversity and complexity of parallel processing architecture limits significantly the portability between different parallel machines, thus requiring the development of new software for each parallel machine configuration. Moreover there is already a vast amount of serial software for many applications, and this code would have to be totally rewritten to execute in parallel. Clearly there is a need for efficient software translators and generators to address these problems. Such parallel software development tools could perform automatic parallelisation (autoparallelisation) in full or offer the programmer the facility for exploiting low level parallelism in certain sections of code [Banarjee 93].

2 The Paragen System

The Paragen System was designed to be able to take a serial program, and automatically transform it into a functionally equivalent one. Individuals in the Paragen system are parse trees which map a program onto a parallel machine. Four functions, all of the form **XN**, were made available to the system, **P2, PN, S2** and **S3**. **X** denotes whether the subtrees are to be evaluated in **P**arallel or **S**erial. **N** denotes the number of subtrees attached to this node.

2.1 Terminals

Unlike virtually every other parallelisation technique, Paragen makes no attempt whatsoever to analyze the original program. In fact, Paragen completely disassembles a program into its constituent statements, and attempts to rebuild it in a parallel form using these statements and a combination of the parallel and serial functions available.

Each statement(line of code) in the original, serial program, is used as a terminal by Paragen. Automating the selection of terminals is a trivial task, as each can simply be read from the original program.

2.2 *Testing individuals*

The first step testing an individual in Paragen is to instantiate it on a (virtual) parallel machine. Individuals are then subjected to a number of test-cases, each of which involves different starting values for the variables used. The program is then executed on the parallel machine and the resulting values compared to the those values produced by the serial program. Each time the parallel version gets the same result as the serial version, its score is incremented.

2.3 *Selection in Paragen*

Most problems tackled by GP involve solving a set problem, such as sorting numbers, recognizing a shape, performing regression etc., but the problem of autoparallelisation is not quite so straightforward. The problem faced by Paragen is a two-pronged one.

To maintain a balanced population of individuals who are both correct and parallel, Paragen employs the *Pygmy Algorithm*[Ryan 94a]. The Pygmy Algorithm differs from normal GP selection in that there are *two* parent populations. Individuals in the first population are scored primarily on their correctness, but also take parallelism into consideration, while those in the second population are scored primarily for their parallelism, with some consideration being given to correctness. Each parent population is made up of a list of individuals, sorted on their score.

When an individual is tested, it is compared against the individuals in each of the parent populations. If the individual has scored high enough to enter the population, the lowest member of that population is removed. This approach of keeping a list of parents is known as a *Steady-State* approach and means that reproduction is no longer needed, as high performing individuals are always preserved.

3 Parallelisation Problems

Current techniques for autoparallelisation rely heavily on data dependency analysis techniques. This consists of analyzing the statements of the program to determine if there is any data dependency between them. If there is no chain of dependence between two statements then they can execute in parallel. For example, to analyse whether two statements S1 and S2 are independent then the set of used variables S1.U and S2.U, and the set of modified variables S1.M and S2.M must be determined. The statements S1 and S2 are independent if S1.M ∩ S2.U, S1.U ∩ S2.M and S1.M ∩ S2.M are empty [Lewis92]. These three data dependencies are known as flow, anti-dependency and output dependence respectively.

With current autoparallelisation techniques these dependencies must be determined before the correct transformation can be carried out. With the Paragen system, however, these dependencies are automatically determined by the fitness function. There are three major data dependencies:

3.1 *Flow Dependency*

In a sequential program a statement S1 is flow dependent on another statement S2 when the statement S1 writes to a variable that is subsequently read by statement S2.

3.2 *Anti-dependency*

A statement S2 is anti-dependent on statement S1 if S2 writes to a variable that was previously read by statement S1.

3.3 *Output-dependency*

A statement S2 is output-dependent on a statement S1 if S2 writes to a variable previously written to by statement S1.

Autoparallelisation techniques that rely on data dependency analysis have limitations due to the complexity of applying data dependency rules. Also the application of transformations to sequential code requires considerable processing [Blume 94].

The *Paragen* system avoids the complexities associated with data dependency analysis by utilising the power of Genetic Programming [Koza 92] to allow the automatic conversion of serial programs. Paragen does not employ any data dependency rules or analysis of any kind, either algorithmically or on the part of the programmer, thus considerably reducing the effort required for parallelizing code.

3.4 *Loops*

The parallelisation of loops is one of the most important aspects of autoparallelisation, as the bulk of computing workloads are controlled by loop type structures.In parallelizing loops, there may be data dependencies both within the loop itself and across the different loop iterations. Data dependencies that span loop iterations are known as cross iteration dependencies. This is especially significant as many parallelisation techniques attempt to execute the different iterations of a loop in parallel, but ignore any parallelism within the loop. This approach alleviates the data dependency constraint within the loop as often the original sequence of statement is preserved. However with the flexibility of the Paragen system we will also attempt to simultaneously extract parallelism both within the loop and across loop iterations. Another consideration in the parallelisation of loops is array access. Loop control structures are often used in the processing of arrays and this further complicates data dependency analysis techniques. The parallelisation of loop iterations in the Paragen system is specified by the **DoPar** function. **DoPar** takes one argument and evaluates it *n* times in parallel.

3.5 Arrays

Loops often access arrays, an activity which further complicates data dependency analysis techniques.

In the following example, which is an extended version of the problem described in [Bruanl 93], there is a data dependency across different loop iterations due to the statements S4 and S8. Here statement S4 of one iteration must be executed before statement S8 of the next iteration, due to the flow dependency caused by e[i-1]. This requires a synchronization mechanism to be used between different loop iterations. However, because the Paragen system also converts the statements within the loop iteration, by re-ordering and / or parallelisation, the effect of the synchronization mechanism on performance is minimized. In the case of a generated code being run on a machine that requires synchronization, a simple post-processing can be applied to the program to determine where synchronization mechanisms.

Our extensions to this problem are the introduction of a number of dependencies within the loop, between S1 and S2, between S6 and S7 and between S5 and S6.

```
for i := 1 to n do
   begin
     S1: a[i] := f[i] * 2;
     S2: f[i] := 57;
     S3: g[i] := d[i] + 8;
     S4: e[i] := e[i] + d[i];
     S5: b[i] := a[i] + d[i];
     S6: a[i] := b[i] + 2;
     S7: a[i] := d[i];
     S8: c[i] := e[i-1] * 2;
   end;

DoAcross i := 1 to n
   begin
      PAR-BEGIN
        S1: a[i] := f[i] * 2;
        S3: g[i] := d[i] + 8;
        S4: e[i] := e[i] + d[i];
      PAR-END
      PAR-BEGIN
        S8: c[i] := e[i-1] * 2;
        S5: b[i] := a[i] + d[i];
      PAR-END
      PAR-BEGIN
        S2: f[i] := 57;
        S6: a[i] := b[i] + 2;
      PAR-END
      S7: a[i] := d[i];
   end;
```

4 Paragen - The First Results

This paper serves very much as an introduction to the workings of Paragen. To illustrate how Paragen operates, it has been tested on a variety of common parallelisation problems from the literature[Bruanl 93], all of which tested Paragen's ability to deal with the dependancies described in Section 3.

Paragen successfully transformed all those programs, and was further tested using the example described in the previous section, the parallel version of the program is shown above. In this experiment, GP was run with a population of 100 individuals for 10 generations. The parallel code shown above is the code produced by Paragen.

4.1 Paragen and loopholes

As stated earlier, the fitness of a program is a combination of how correct it is and how parallel it is, but Paragen (and GP in general) proved to be quite adept at discovering loopholes. A common strategy, especially in the shorter examples, was for Paragen to execute all the instructions in parallel, regardless of dependencies. This gave the individual a high score in terms of parallelism - the program executing in only one time step - and a high score in terms of correctness was achieved by ensuring that each statement was executed so many times that the variables affected ended up with the correct values. This behavior was curbed by reducing an individual's fitness each time it repeated an instruction.

Paragen also noticed that in the case of output dependencies, where statement S2 writes to the same variable as S1, S1 could often be left out of the final program as it didn't affect any final values. Clearly, this is not correct, so it was necessary to reduce an individuals fitness each time it left out an instruction.

Another strategy that had to be discouraged was that of individuals who didn't do anything. These individuals exploited the fact that several variables used by many programs don't change, and by not doing anything these individuals would get a score by virtue of the fact that these variables will always contain the correct value at the end of the run. To prevent individuals such as these from having a chance of being selected, only those variables which appear on the left hand side of statements are considered when calculating an individual's score.

5 Conclusion and Future Directions

The Paragen system is a new auto-parallelisation tool which can be used to convert serial programs without any analysis for dependencies. Paragen has been successfully tested with code containing a number of data dependencies, and has successfully converted loops containing cross-iteration dependencies. As well as converting entire loops, Paragen can also determine sections of code within those loops that can be parallelized while still preserving any dependencies contained within.

Work is currently under way on using Paragen to help with data distribution. Data distribution involves dividing up data evenly so each processor stores some of it locally, which greatly reduces the amount of communication if all instructions that use the local data are executed on the same processor. The data is distributed by encoding the distribution in genetic form, so now both the distribution of the data and the instructions which operate on it are evolved.

The next step for Paragen is to produce code in the Occam language, using channels as necessary for synchronization, which would allow the generated code to be used on a wide variety of parallel machines.

The eventual goal for Paragen is to create parallel programs on the fly, with just a statement of the problem which is to be solved. This would eliminate the need to write the original, serial program, which would further automate the process of generating parallel code.

References

[Banarjee 93] Banerjee, U. et al, Automatic Program Parallelization, Proceedings of the IEEE, Vol. 81, No. 2, February 1993.

[Blume 94] Blume W., et al. Automatic Detection of Parallelism, IEEE Parallel and Distributed Technology, Fall 1994.

[Bruanl 93] Braunl, T. Automatic Parallelisation and Vectorization, Parallel Programming an Introduction, Prentice Hall, 1993, ISBN 0-13-336827-0.

[Burns 88] Burns, A. (1988) : Transforming Occam Programs. In Programming in Occam2 : Addison-Wesley, 1988, ISBN 0-201-17371-9.

[Koza 92] Koza, J. (1992) : *Genetic Programming.* Cambridge : M.I.T. Press.

[Lewis92] Lewis, G., El-Rewini, H., Introduction to Parallel Computing, Prentice Hall, 1992.

[Ryan 94a] Ryan, C. (1994) : Pygmies and Civil Servants. In Advances in Genetic Programming. Ed. K. Kinnear Jr. Cambridge : MIT Press

The Benefits of Computing with Introns

Mark Wineberg
Intelligent Systems Labratory
School of Computer Science
Carleton University
Ottawa, Ontario, Canada, K1S 5B6
wineberg@scs.carleton.ca

Franz Oppacher
Intelligent Systems Labratory
School of Computer Science
Carleton University
Ottawa, Ontario, Canada, K1S 5B6
oppacher@scs.carleton.ca

ABSTRACT

EPI (Evolutionary Program Induction) is an encoding scheme that allows a Genetic Algorithm with a linear, fixed size chromosome to perform the same tasks as does Genetic Programming. Our encoding scheme achieves this by using non-expressed genetic code, i.e. introns. The addition of introns, defined to more closely resemble actual biological usage than is usually done in the GP community, has the unexpected benefit of dynamically reducing the search space. This paper is our initial attempt to prove that this property holds.

1. Introduction

EPI (Evolutionary Program Induction with Introns), as described in Wineberg amd Oppacher (1994), is an encoding scheme that allows a Genetic Algorithm (GA) with a linear, fixed size chromosome to perform the same tasks with comparable speed as does Genetic Programming (GP) (see Koza (1992)). EPI is not merely a linear encoding scheme, but also maintains a fixed structure within the chromosome. This allows like alleles to compete against each other at a locus.

Our adherence to a linear, fixed structure necessitates the use of phenotypically[1] unexpressed chunks of genetic code, i.e. introns. These introns enable EPI to work comparably to GP while still allowing reliance and utilization of standard GA results and techniques. However, at first glance, we might expect introns to slow down convergence to a solution. In fact, we believe the opposite to be the case: introns dynamically reduce the search space and, thus, may reduce the search time. This paper is our initial attempt to establish this conclusion.

2. Evolutionary Program Induction with Introns: EPI

While GP is based on GA methods, there are differences. GA has fixed length chromosomes of fixed structure, and GP has non-linear, tree structured, variable length chromosomes. Accordingly GA and GP employ different operators, in particular, GA must use a mutation operator while GP does not have to, and the GP Crossover differs markedly from its GA counterpart.

The differences between GP, and a canonical GA, should be studied, to see which of the new, or modified, techniques, introduced by GP, are fundamental to the process of program induction through evolution, and which are merely expedient.

There are a few other advantages of couching program induction in a canonical GA framework. Two of these advantages are: the extant GA theory becomes applicable to the program induction domain; and techniques thoroughly studied in other GA domains become available to solve problems in program induction.

The reason that GP uses a non-homologous crossover is to accommodate the non-linear structures of the parse trees. The traditional crossovers in GA, such as the 1-point crossover, are defined only for linear chromosomes. A naive application of the traditional crossovers would render the resulting offspring programs completely meaningless. In order to be compatible with the genetic operators used in a canonical GA, a chromosome must, therefore, have the following two properties:

1) the chromosome must be linear
2) after any genetic operator is applied, the resulting chromosome must produce a syntactically correct computer program (syntactic closure).

Our approach to program induction, i.e., Evolutionary Program Induction with Introns, **EPI** for short, encodes parse trees in a way that conforms to both of the above requirements.

2.1 EPI Fixes the Tree Height

The property of syntactic closure requires the meaning of each locus to be fixed in the chromosome. This allows the crossover operator to exchange genes with their alleles, and not with incompatible genes. Satisfying syntactic closure presupposes the chromosome to have a fixed structure. If the structure were not fixed, the standard crossover operators would blindly cross nodes with incompatible data types. Were the chromosome to be decoded, nonsense would be produced.

As pointed out previously, GP uses a variable structured genotype. It should be noted, however, that the parse trees produced by the creation and reproduction routines

[1] Here the phenotype is a program that is evaluated after the decoding of the genome.

have a maximum tree height. If a parse tree is produced with a greater height than the maximum it is rejected.

This suggests a way of using parse trees as the genotype and yet still keep a fixed structure for the chromosome. If one views the genome as a full binary tree, expanded out to its maximum depth, with the original parse tree as a subtree, then the length of the genotype is actually unchanging. Therefore the length of the chromosome is now fixed.

2.2 Full Tree Embedding of the Parse Tree

While the above observation allows one to fix the size of the chromosome, it does not determine a fixed structure within the chromosome. A natural way to embed the original parse tree in the expanded tree is at the top, node for node, leaving empty gene loci below the parse tree's terminal nodes. Unfortunately, disconnected tree fragments result if these empty nodes mutate.

A simple observation leads to a slightly different embedding of the parse tree. While the terminals of the parse tree are also its leaves, this is not true for the above embedding of the parse tree in the expanded tree. It is, therefore, intuitive that the terminals should "sink" in the expanded tree, and become leaves there too. In this embedding of the parse tree, all internal nodes are operators, all leaves are terminators, and no empty gene loci are needed. Therefore when crossover is applied, each matched pair of loci will be drawn from the same set. If the pair consists of leaves of the expanded tree, both genes are from the terminal set, otherwise, both genes are from the operator set. The internal structure of the chromosome is now static.

The embedding of the parse tree into the expanded parse tree as defined so far is obviously incomplete. The expanded tree will be mismatched with the parse tree: there will usually be more operator nodes than operators, and more leaves than terminals in the original parse tree. To solve this problem, a small set of related operators, called selector operators, are added to the operator set.

A selector operator is a simple function that selects one, and only one, outgoing edge for traversal. In an n-ary tree we have n selectors; each selects one of the outgoing edges from a node. These operators can now be used to properly embed the parse tree into the expanded tree.

The basic idea for the embedding is to use the selectors to route the traversal of the expanded tree "around" areas that do not correspond to any part of the parse tree. The selectors form a chain hooking up the various code fragments. The elements that are not part of the code, nor in a selector chain, can hold any value, as long as the value is from the correct set: leaves must have genes from the terminal set, and internal nodes must have genes from the operator set (see Figure 1).

Once the internal structure of the chromosome is fixed, the canonical GA operators can be applied. Just as with GP's operators, any manipulation of the chromosome with any canonical genetic operators will lead to a syntactically correct program under EPI. Since we can now use any traditional GA operator, EPI opens the doors for using any operators established in other contexts for the purpose of program induction.

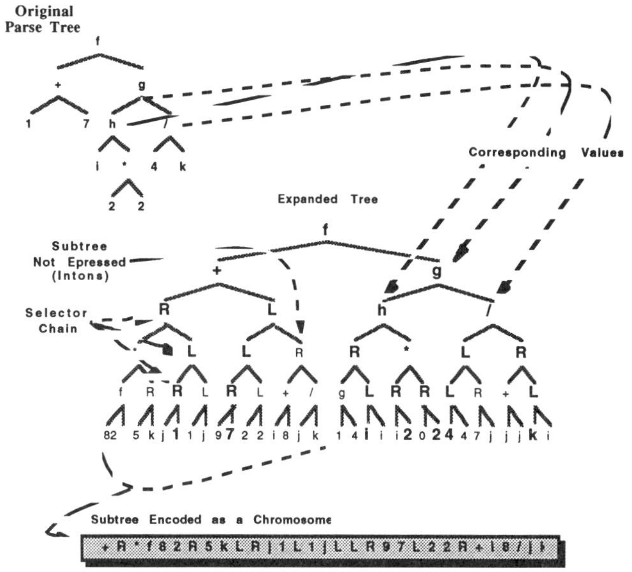

Operator Set, F = {+, -, *, /, f, g, h}
Terminal Set, T = {0, 1, 2, 3, 4, 5, 6, 7, 8, 9, i, j, k}

Figure 1: A Parse Tree Embedded in a Fully Expanded Tree

3 Introns

The terms 'intron' and 'exon' are borrowed from Genetics (see Hartl, Freifelder, and Snyder (1988) page 307ff., or Weaver and Hedrick (1995) page 219ff. for good descriptions of the concept of introns as used in genetics). We shall mean by 'intron' any genetic material that does not have a phenotypic expression; and by 'exon', similarly, we shall mean any coded genetic material that does have phenotypic expression. This concept of introns has been previously applied to genetic algorithms by Levenick (1991), Forrest and Mitchell (1993), and Wu and Lindsay (1995). This usage differs slightly from the current vogue of labeling as introns any "useless code", i.e.,code that does not affect the output of the program containing it (see Angeline (1994) for an example of this latter usage). The difference derives from the fact that we regard the program as the phenotype whereas many others in the GP community regard the behavior of the program as the genotype[2].

The notion of introns is naturally applicable to EPI. Like eukaryotic genomes, EPI has genetic material inside the chromosome that is not expressed when decoded. Subtrees under edges that are not selected by a selector operator, and subtrees under a node set to an operator that does not use all of its parameters, are not included in the decoded s-expression, and so do not have an effect when that s-expression is evaluated by the fitness function. It is natural, and in rather strict agreement with biological usage, to call these subtrees introns, and all the other nodes, which are expressed in the s-expression, exons. In other words,

[2] Of course, EPI really has both types of "introns"; it also frequently contains useless code!

introns can be thought of as non-selected subtrees in the expanded parse tree.

Introns hide genetic code from the selection process. This could be seen as slowing down convergence, since the selection pressures, that drive the system, are not applicable to the hidden code. Such code will not be bred out (except accidentally), even if it is useless or detrimental when expressed, nor recognized and promoted, even if it is beneficial. It is not even a good depository for safekeeping 'good' code; when hidden inside an intron, any schema can still be destroyed by the crossover operator (or the mutation operator if it is allowed to work within introns). The crossover operator cannot be prevented from selecting crossover points within an intron, since the other chromosome being crossed may have an exon at that locus.

In spite of the above concern, the introns may actually help the search process. An intron can be viewed as an analogue of a 'don't care' symbol explicitly expressed in the chromosome. In this respect it is comparable to the Classifier System's '#' symbol, except that introns affect more than one gene at a time.

Intron may dynamically block out segments of the genotype that are not needed for the ultimate solution. Since this reduces the amount of search space, the introns may actually speed up the convergence process. The positioning of the intron is influenced by selection pressure, and so introns can be seen as a large scale 'gene' that is expressed by blocking out unneeded segments of the genotype. The rest of the paper tries to establish this conclusion.

4 Designing the Experiment

How would one go about deciding whether introns reduce the search space? While introns by definition reduce phenotype length, the can only be said to reduce the search space if they cause the GA to consider fewer loci[3]. An obvious approach to show that introns reduce the search space would be to try the system with and without introns. Unfortunately the system doesn't work without the introns, so this type of controlled experiment is not feasible. Instead the experiment that shows that introns dynamically reduce the search space is designed to support the points below:

a) EPI's introns really behave as one would expect introns to behave
   i. the genetic code in an intron is not expressed in the phenotype
   ii. selection pressure does not act inside an intron
b) the expressed part of the chromosome shrinks
   i. EPI reduces the program size
   ii. there is a program length to which EPI converges regardless of the initial chromosome[4] length; this length may be different for different program induction tasks
   iii. the use of introns is the mechanism which causes the above reduction and convergence
c) loci, averaged across the population, become either introns or exons; in other words, the system decides which loci it needs
d) EPI performs comparable to a system whose chromosome size is fixed at the outset to the length to which EPI converges
   i. the performance criterion is probability of success
   ii. the time EPI takes to find the final chromosome size should be taken into account in the comparison
   iii. the time to converge to the final chromosome size should not be inordinately long
e) the performance of EPI after size convergence is comparable to other systems (e.g. Genetic Programming)

To date we have only run experiments in support of a), b), c) and d) iii.

5 Measures to Monitor EPI's Performance

In order to determine whether EPI's mode of processing reduces the search space using the methodology of the preceding section, we need to monitor how EPI progresses. For this purpose we need 3 measures: phenotype length, genic diversity, and intronal indeterminacy. The phenotype length is defined in terms of the number of exons or, equivalently, the number of nodes in the tree (program size). The other two measures are described below.

The most widely used measure of diversity in a population of binary string chromosomes is based on the Hamming distance between two binary strings. Unfortunately, since this measure presupposes a binary alphabet, and EPI has an n-ary, symbolic alphabet, we cannot use it. Therefore, we need another measure of diversity.

Diversity can be viewed as a synonym for randomness of the population. A well known measure of randomness is Shannon's entropy which is defined as follows:

$$H(X) = \sum_{i \ge 1} p_i \log\left(\frac{1}{p_i}\right)$$

where X is a random variable with a range R = $\{x_1\ x_2 \dots\}$, and p_i = prob(X=x_i). In each member of the population, each locus is set to some gene taken from the symbolic alphabet. We view each locus as a random variable, which takes on the values of the various genes. Formally, each locus L has a range R = $(g_1\ g_2 \dots)$ where g_i is a gene taken from the n-ary alphabet used by the system. Looking across the whole population, we determine the frequency of occurrence of each gene at that locus. This frequency can be viewed as a quasi-probability since it sums to 1. The prob(L=g_i) is estimated by these frequency counts (f_i estimates prob(L=g_i)). Therefore the randomness of a

[3] For example, if the smaller phenotypes occur anywhere in the genome throughout the population, then the search space itself is not reduced.

[4] This is true only if the chromosome length is not below that length initially.

locus in the population is the entropy of these frequency counts:

$$H(L) = \sum_{i \geq 1} f_i \log\left(\frac{1}{f_i}\right)$$

We call H(L) the **genic diversity** of a locus. The genic diversity of the chromosome is the average of the genic diversities across all loci.

This notion of the diversity at a given locus can be used for properties other than just the genic value at the locus. For example, each locus is either an intron or an exon. We can then use the diversity to measure the extent to which the system has determined whether the locus is an intron or an exon across the population. We call this new measure **intronal indeterminacy** and use it throughout our experiments.

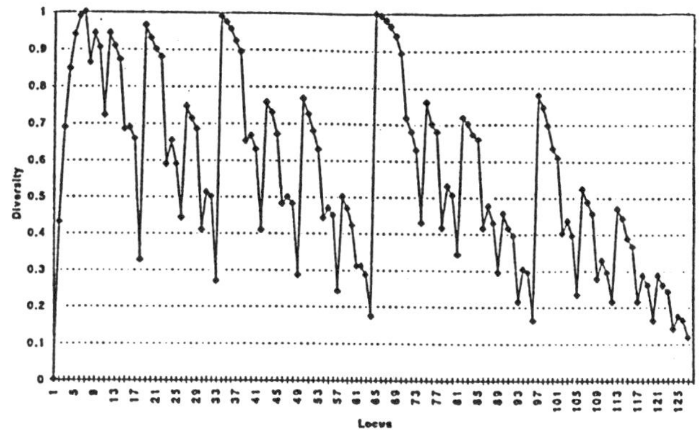

Figure 2i: Intronal Indeterminacy at the Start of a Run

6 Experiments

6.1 Experimental Setup

For our experiment we have chosen the task of symbolic regression of the function $f(x) = x^4 + x^3 + x^2 + x$ (as described in Koza (1992), p. 162ff). We use one-point crossover, rank selection, a population size of 500. Each generation is composed of 25% elite and 75% children from crossover reproduction where each locus in a new child has a mutation probability of 0.06. Each run of the experiment lasted a maximum of 51 generations. We used two different chromosome lengths, a short one of 15, and a long one of 127. Since in the implementation of a symbolic regression without constants all the leaves are the same (the free variable "x") we did not include this single terminal in the genome. Consequently, the maximum program size in the short chromosome experiment is 31, and is 255 in the long chromosome experiment. While keeping track of phenotype length, genic diversity, and intronal indeterminacy , we ran the experiment 50 times for both of the chromosome sizes.

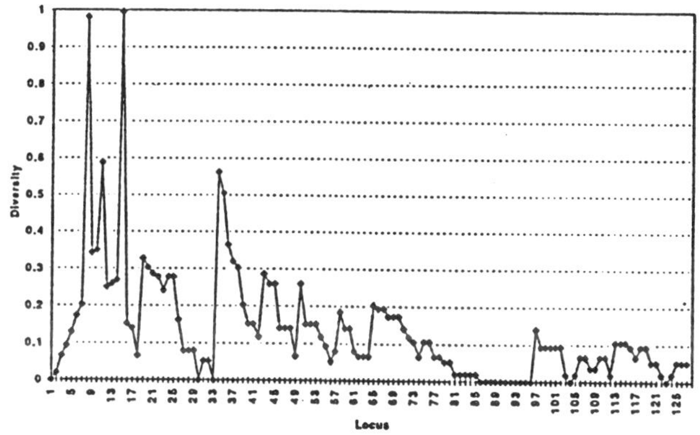

Figure 2ii: Intronal Indeterminacy at the End of a Run

6.2 Results and Observations

For reasons of space, we only show a few figures to make our points, but all figures represent typical results.

6.2.1 EPI's introns are true introns

The genetic code in an intron is not expressed in the phenotype

This is true by construction. EPI's very design ensures that our introns are not expressed in the final program that is tested by the fitness function.

Selection pressures do not act inside an intron

Inspecting figures 2i-2iii, one can see that the intronal indeterminacy drops dramatically from the beginning (2i) to the end (2ii) of the run. This means that the system has settled down into a state in which each locus is either an intron or an exon across the population. This is especially true of all but the leftmost section of the chromosome. When we examined the data closely, we noticed that in most of the population the loci on the right are introns. Looking at 2iii, we see that these very loci have a large genic

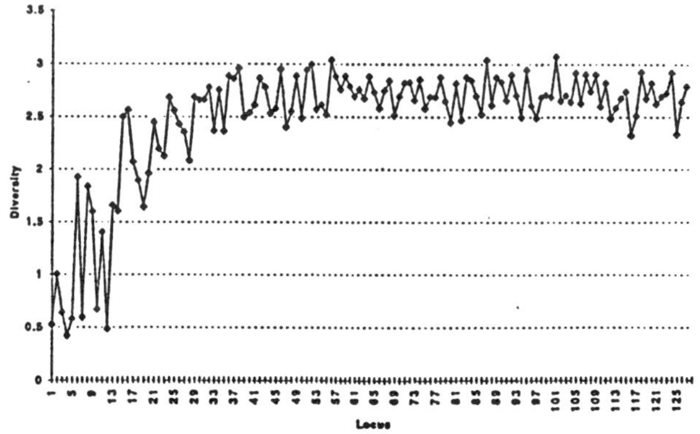

Figure 2iii: Genic Diversity at the End of a Run

diversity. In fact the diversity has not dropped much below the starting value of 3.4. This is just what one would expect if no selection pressure had been brought to bear.

6.2.2 EPI reduces the program size

On average at the beginning of a run the program size in the long chromosome experiment is 29.7 (out of a possible 155 symbols) with a standard deviation of 0.7. By the end of the run the program size has shrunk to an average of 16.9 with a standard deviation of 5.2. The drop in size is 13.2 which is almost 3 standard deviations of the average final program size.

Does EPI converge to a fixed program length?

Unfortunately we have not yet performed enough experiments to anwer this question. However we did try an experiment where the chromosome was much shorter than the program size obtained by the experiment above. This short chromosome experiment had results that were vastly different than those of the original long chromosome experiment.

Here the average program size at the beginning of a run is 9.5 (out of a possible 31 symbols) with a standard deviation of 0.2, but the average final size is 10.1 with a standard deviation of 1.7. The phenotype size did not shrink, but grew on average (although the two numbers are within each other's standard deviation). This is evidence that when there is enough space for the introns to do there work (as in the long chromosome experiment) they will indeed achieve length reduction, and conversly if the starting size is too small for introns to come into play, the phenotype size will not decrease, and may even grow slightly.

Introns cause the reduction of the phenotype size

There is another possible mechanism that could be responsible for the reduction of phenotype size. If the genic diversity drops, loci are frozen into having genes that are either introns or exons. Therefore, intron indeterminacy will be forced to drop too. However, our experiments show this not to be the case. Figure 3 shows the correlation between phenotype length and intronal indeterminacy as well as genic diversity. Each point on the graph represents a correlation taken over a window of 12 generations (the label on the x-axis is the midpoint of that window). As is apparent from the figure, intronal indeterminacy is consistently more strongly correlated with phenotype length than is genic diversity. This implies that it is the introns themselves, and not just a loss of diversity that fuels the reduction of the phenotype length.

Furthermore, the correlation between phenotype length and intronal indeterminacy is perfect up to generation 17. This correlation, together with the above fact (6.2a) that the introns are responsible for the size of a program, establishes our claim that the introns cause the reduction of the phenotype size.

6.2.3 EPI decides which loci it needs

Referring again to the drop of intronal indeterminacy from the beginning to the end of a run (Figures 2i-2iii), the resulting indeterminacy is almost zero. The system has determined for each locus across the population, whether or not it is to be an intron. It is not at all obvious that the system would do this. One might have guessed that while on average the phenotype length changes little from member to member in the population, the introns would be placed nonuniformly. But the above data shows that the system, on its own, and in no uncertain terms, has determined which parts of the chromosome it needs expressed to achieve a solution.

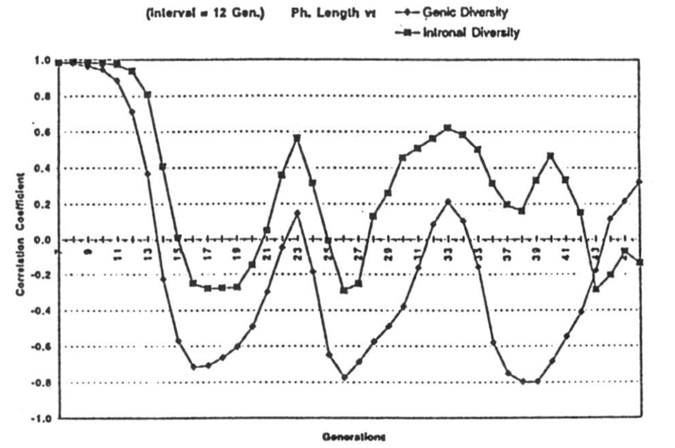

Figure 3: Correlations of Phenotype Length vs. Genic Diversity and Intronal Indeterminacy

Figure 4: Phenotype Length as the GA Progresses

6.2.4 EPI converges rapidly to the final chromosome size

Although we have not fully convinced ourselves yet, the evidence to date seems to indicate that EPI does indeed converge rapidly to the final chromosome size. In all the experiments that do settle to the average phenotype size, there are two phases: in the first phase the phenotype length drops rapidly, whereas in the second phase it remains fairly constant. Furthermore, the first phase exhibits a linear (or superlinear) drop-off that is always at least one exon per generation (see Figure 4).

7 Conclusion

In this paper we introduce a notion of introns the corresponds more closely to biological usage than is common in the GP community. These introns not only enable EPI to do program induction with results comparable to GP (see Wineberg and Oppacher (1994)) but they also reduce the search space and, thus, the search time.

Bibliography

Angeline, Peter J., 1994. Genetic Programming and Emergent Intelligence. In Kinnear, Kenneth E. Jr. (editor). *Advances in Genetic Programming.* Cambridge Massachusetts: MIT Press. Pages 75-98.

Forrest, Stephanie, Melanie Mitchell, 1993. Relative Building-Block Fitness and the Building Block Hypothesis. In Whitley, L. Darrell (editor). *Foundations of Genetic Algorithms 2.* San Mateo California: Morgan Kaufmann. Pages 109-126.

Hartl, Daniel L., David Freifelder, Leon A. Snyder 1988. *Basic Genetics.* Boston Massachusetts: Jones and Bartlett Publishers.

Koza, John R. 1992. *Genetic Programming: ·On the Programming of Computers by means of Natural Selection.* Cambridge Massachusetts: MIT Press.

Levenick, James R., 1991. Inserting Introns Improves Genetic Algorithm Success Rate: Taking a Cue from Biology. In Belew, Richard K., Lashon B. Booker (editors). *Proceedings of the Fourth International Conference on Genetic Algorithms.* San Mateo California: Morgan Kaufmann. Pages 123-127

Weaver, Robert F., and Philip W. Hedrick 1995. *Basic Genetics.* Dubuque, Iowa: Wm. C. Brown Publishers.

Wineberg, M., and F. Oppacher,. 1994. A Representation Scheme to Perform Program Induction in a Canonical Genetic Algorithm. In Davidor, Yuval, Hans-Paul Schwefel, Reinhard Männer (editors). *Parallel Problem Solving from Nature - PPSN III.* Berlin: Springer-Verlag. Pages 292-301.

Wu, Annie S., Robert K. Lindsay, 1995. Empirical Studies of the Genetic Algorithm with Noncoding Segments. In *Evolutionary Computation* 3(2). Pages 121-148.

GENETIC PROGRAMMING POSTER PAPERS

Co-Evolving Hierarchical Programs using Genetic Programming

Manu Ahluwalia
University of the West of England
Faculty of Computer Studies and Mathematics
Frenchay, Bristol, BS16 1QY, UK
m_ahluwa@btc.uwe.ac.uk

Terence C. Fogarty
University of the West of England
Faculty of Computer Studies and Mathematics
Frenchay, Bristol, BS16 1QY, UK
tcf@btc.uwe.ac.uk

The goal of this paper is to show that population size can be reduced by introducing different architectures (Automatic function Definition) in the initial population and the process of learning can be speeded up by having a separate sub-population for each automatic function. This provides the desired solutions of smaller sizes and also finds an effective way of scaling up the problem. A Standard Genetic programming, Genetic Programming with Automatic function defined functions (ADF) and Genetic Programming (with sub-population for each ADF) is implemented on 11-Multiplexer problem.

A Standard Genetic programming is implemented by selecting a function set {if, and} and a terminal set which contains all the inputs and outputs. Each new generation is created from the preceding generation by applying fitness proportionate reproduction to a small percentage of the population. The 100% correct solutions are found using a population size of 2000 (Table 1). This method would find difficulty in scaling up the problem because the fitness value associated with each program (hierarchical) is merely based on number of cases it solves. The fitness doesn't account for the constructional fitness (Kinnear 94). It is desirable to have additional copies of a useful code with in a program so when the tree size grows the probability of passing on the good genes is higher.

Automatic function definition (Koza 94) enables us to find sub-trees (ADF) with good fitness value which in turn can call other useful sub-trees. The main program will have same or different ADF's at various locations. As the population evolves it is possible to have multiple copies of useful codes (higher constructional fitness) in a program. The population is initialised by defining two automatic functions. and randomly generating a population size of 1200 (400 for each ADF and the main program). The programs which have less number of ADF's tend to disappear over the generations and the 100% correct solutions (smaller in size) are found in less number of evaluations.(Table 1).

Symbiosis is the phenomenon in which organisms of different species live together in close association, resulting in a raised level of fitness for one or more of the organisms. We simulate this model by having three sub-populations. Two automatic functions namely "ADF1" and "ADF2" are defined and they evolve in their own sub-populations, each receiving a separate measure of fitness. "ADF2" may call "ADF1" once or more. The third sub-population consists of main programs which may call "ADF1" or "ADF2" once or more. The main programs also evolves separately in the third sub-population. The population (population size 1200, 400 for each sub-population) is initialised by randomly creating individuals which are hierarchical computer programs. There are two ways in which fitness could be allocated to the individuals in their respective sub-populations.

(i) Each Automatic function is evaluated against all the possible combinations of input variables and the local fitness is given according to how many fitness cases each may solve. Therefore "ADF1", "ADF2" and the main program will have their own fitness (local fitness) value.

(ii) Assign a global fitness to individuals.The main program is evaluated against all the possible input variables to give the fitness value. This value is assigned to each ADF. Hence the "ADF1", "ADF2" and the main program will have the same fitness value. To implement this we draw (randomly) subroutines from the sub-population in such a way that a particular sub-routine is not used again by a different main program. But the unique "ADF1" and "ADF2" in each main program could be called once or more.

The co-evolving model is implemented with a small population size. It finds 100% correct solution (smaller in size) in fewer evaluations than the other two methods (Table 1).

Table 1 The total number of evaluations (different runs) is shown for each case. The number in the parenthesis shows the size of the smallest 100% correct program.

Simple GP	GP (ADF)	Co-Global	Co-Local
6,400 (106)	5,700 (73)	4,200 (65)	3,450 (56)
7,000 (120)	6,100 (80)	4,575 (69)	3,600 (50)
7,200 (132)	6,300 (84)	4,800 (70)	3,750 (60)

Bibliography

Koza, John R. 1994. *Genetic Programming II: Automatic Discovery of Reusable Programs.* Cambridge, MA: The MIT Press.

Kinnear, Kenneth E. Jr.(editor). 1994. *Advances in Genetic Programming*. Cambridge, MA: The MIT Press

GP Tools Available on the Web: A First Encounter

Anthony G. Deakin
Department of Computer Science
The University of Liverpool
Liverpool L69 3BX, England, UK
a.g.deakin@csc.liv.ac.uk
http://www.csc.liv.ac.uk/users/anthony

Derek F. Yates
Department of Computer Science
The University of Liverpool
Liverpool L69 3BX, England, UK
yatesdf@csc.liv.ac.uk
http://www.csc.liv.ac.uk/users/yatesdf

ABSTRACT

A number of genetic programming (GP) tools available on the Web were reviewed to assess the viability of GP freeware and to propose preliminary evaluation criteria for GP tools. Two of the example applications, symbolic regression and the Santa Fe ant, were compared across most of the tools. Three umbrella evaluation criteria were proposed: basic mechanics, additional features, operational issues. Modular design permits the basic GP tools to be applied generally. Factors such as run-seeding can influence the choice of tool. Operational issues are sensitive only to requirements for large memory and high speed. The available tools represent good quality working models and the Internet is increasingly becoming a key distribution medium for the tools.

1. Introduction

Some of the growing collection of GP tools available on the Internet were reviewed to generally establish the feasibility of using the Internet to obtain viable GP packages. A specific aim was to suggest preliminary review criteria for GP packages. These aims, which have not been previously addressed, are considered timely in view of the growth and development of the field, seen for example with the addition of automatically defined functions (ADFs) and co-evolution to the GP paradigm.

The approach taken is concerned less with the internal GP code than with the usability and utility of the GP packages available. A selection of tools located at the University College London mirror site for `ftp.io.com` were reviewed: Koza-book-gp-implementation; Geppetto; Gpc++; Gpquick; Gpdata. Prior to the start of the review, the UCL mirror site was last changed on August 14, 1995.

2. Methodology

The symbolic regression problems supplied with the tools were tested - Koza-book's `(0.5*x*x)`, Gpquick's `(w*(x-y)+2.5*z)`, Geppetto's and Gpc++'s $(x^4+x^3+x^2+x)$. It was straightforward to adjust Geppetto and Gpquick problems to `(0.5*x*x)` for comparison. The supplied ant problems (Santa Fe trail) were also tried on all packages (except Koza-book which comes with other examples). Additional example applications supplied with each package other than symbolic regression and the ant problem were also reviewed. Further issues that were considered include the quality of the documentation, the availability of support, expected future developments, the nature of the outputs, and additional features supplied with the packages such as co-evolution, run-seeding (useful for bootstrapping), and ADFs. Criteria for tool comparison were then identified and a tabular annotated comparison was made.

3. Results

One package, Lil-gp, that appears to be in common use, had not been discovered on the Web by the authors at the time of the review. In terms of basic mechanics, one other package, Sgpc, could not be compiled. Otherwise, most packages performed reasonably well on the tests. More detailed performance information is available. Considering extra features, one interesting criterion is repeatability of tests which are based on random numbers, which only Koza-book includes as standard. None of the packages contain an editor to simplify the generated programs (Lisp S-expressions which can contain many large introns) apart from Koza-book's editor for Boolean expressions. Operationally, Unix and PC versions were available for all packages as standard except for Geppetto and GPdata (Unix only).

4. Conclusions

The particular intended GP application is considered to be the main factor influencing the choice of which GP tool to use. Accordingly, the tools could need more or less adjustment to suit requirements. Any of the proposed evaluation criteria, basic mechanics, additional features and operational considerations can take precedence in this respect.

Speeding up Genetic Programming: A Parallel BSP implementation

Dimitris C. Dracopoulos
Brunel University
Department of Computer Science
and Information Systems
London, UK
Dimitris.Dracopoulos@brunel.ac.uk

Simon Kent
Brunel University
Department of Computer Science
and Information Systems
London, UK
Simon.Kent@brunel.ac.uk

1 Overview

For Genetic Programming to be accepted by the mainstream computer science community, an area which must be addressed is that of reducing the time taken to arrive at solutions. Fairly extensive work has been carried out on the parallelisation of genetic algorithms, Cantú-Paz (1995). However, parallelisation of Genetic Programming remains a relatively unimplemented area. This paper therefore discusses a parallel implementation of a Genetic Programming system which can be applied using very standard hardware, providing a faster GP system, for use by the genetic programming researchers and users.

The aim of the implementation was to parallelise the most computationally expensive part of the GP process; that is the fitness evaluation phase. The implementation makes use of the Bulk Synchronous Parallel (BSP) model of parallel computing, Leslie Valiant (1990) and the specific implementation used the Oxford BSP library. This was chosen for its simplicity; it has only six primitive operations. Tests were run on a subnet of SUN SPARCstation 5 machines with 70MHz microSPARC-II processors, and running SOLARIS 2.4, communicating using TCP/IP. The processors were organised in a Master-Slave paradigm in which a single master processor coordinates the run, with the slaves supporting the master during fitness evaluation.

The system was tested using the Artificial Ant problem. A reasonably complex version of the problem was chosen to most effectively demonstrate the speedup. This involves a 100×100 grid, of which the upper left 50×70 section contains a trail of 157 pieces of food. A population of 2000 individuals was used.

Taking all the overheads into consideration, calculations were made using the BSP Cost Model to predict the speedup which could be achieved with the BSP GP system. In practice it was found that the estimates were similar to the actual speedups achieved. The system was run with between one and eight processors. With two processors, a speedup of around 1.4 was achieved. With eight processors, this rose to a speedup factor of just under 3 times faster than a sequential run. A limitation of version 1.2 of the Oxford BSP library is its inability to access non-local dynamic data. This imposed the requirement of having to carry out additional processing to communicate population data. It is predicted that the speedup factor using eight processors could be increased to around 3.5, if direct dynamic data access were possible.

A clear speedup was achieved with the test problem. As problem size increases, communication overhead decreases, further improving the speedup achievable. Further work is being carried out using alternative parallel models. Results so far show that greater speedups can be achieved. Another improvement is that as more processors are added, the marginal improvement does not deteriorate so quickly.

2 Conclusions

The proposed parallelisation of an existing GP system can achieve very high speedups in large problems. In addition, the proposed BSP GP model is easy to implement and portable to many different platforms. The time required to parallelise an existing serial GP system with the BSP model is insignificant compared with the achieved speedups, so as to suggest that GP researchers should consider such an implementation when they deal with complex, time consuming problems.

Bibliography

[1] Erick Cantú-Paz. A summary of research on parallel genetic algorithms. Technical Report 95007, Illinois Genetic Algorithms Laboratory, July 1995.

[2] V. Scott Gordon and Darrell Whitley. Serial and parallel genetic algorithms as function optimizers. In Stephanie Forrest, editor, *Proceedings of the 5th International Conference on Genetic Algorithms*. Morgan Kaufman, 1993.

[3] Richard Miller and Joy Reed. The Oxford BSP Library users' guide. Technical report, University of Oxford, 1993.

[4] Leslie G. Valiant. A bridging model for parallel computation. *Communications of the Association for Computing Machinery*, 33(8):103–111, 1990.

Easy Inverse Kinematics Using Genetic Programming

Jonathan Gibbs
Computer Science Department, 225 Applied Sciences Building
University of California, Santa Cruz, 95064
jono@cse.ucsc.edu
http://www.cse.ucsc.edu/~jono/research

ABSTRACT

This paper describes a method for inverse kinematics which uses genetic programming. This method is able to automate low-level motion control, yet still give animators control over the visual qualities of their work. The main advantage of this method is its simplicity and its speed when dealing with complicated specifications.

1 Introduction

Current research in computer animation addresses the difficulties of traditional key-frame animation with automated methods. It is important that these methods relieve the animator of most low-level details, yet still give the animator control when wanted. One basic form of automatic control is goal generated motion, or inverse kinematics. Given a goal position in space and an articulated figure, an inverse kinematics algorithm must find the joint angles to place a certain part of the figure at the goal. The source of much complication is that the solution is often highly underconstrained. That is, many sets of joint angles can solve the problem. Further, the animator may have some visual attributes in mind which these solutions achieve to varying degrees. Other automatic control paradigms, such as controllers for animation and behavior-based animations, are similar goal-finding problems.

There are many methods available for solving the inverse kinematics problem. However, these methods are fairly complex even in simple situations, and the addition of joint limits, limited degrees of freedom, and energy constraints are either very complicated or not possible at all. Unlike other inverse kinematics algorithms, the method described here allows the animator to easily specify not only the goal position, but other constraints as well. This method can incorporate joint limits, limited degrees of freedom and relationships between joint angles through specification in the fitness function.

2 Solution and Results

To solve the problem using genetic programming, we need a fitness function to rate potential solutions based on the goal position and constraints on the figure's position. For our purposes complete numerical accuracy is not as important as visual appearance. We do not need a figure to move to the *exact* goal position any more than a human actor to would need to stand on an exact spot. This relaxation of the exact requirements suits genetic programming well, since the programs it produces are often approximations, and not exact solutions.

To compute the correct joint angles for a given figure and goal, we evolve one program for each degree of freedom. So, each joint has one to three programs associated with it. Evaluating these programs will tell us what angles to set the joints at. The fitness can be assigned to be proportional to the distance from the specified point on the figure to the goal position after all joint programs have been evaluated. This simple metric satisfies the basic inverse kinematics problem, but it doesn't allow for any control over the solution.

We can give the animator control by adding more parameters to the fitness function, restricting the functions to stay inside animator-defined joint limits, an adding energy constraints by penalizing programs based on a measure of how much energy they use to move the figure. Minimizing energy can work against movement towards the goal, so care must be taken to ensure the goal is met.

We can regularly find a solution with near perfect fitness for three segments arms inside 15 generations or only 3 seconds on a SGI Indigo2.

3 Conclusion

This method is able to produce results that are difficult to achieve by hand, and comparable to more complicated methods. While possibly not as robust as other methods, it is much simpler, and incorporates advanced features including joint limits, kinematic constraints, and joint relationships. This makes it easy for animators to change the behavior of the algorithm to suit their specific needs. This is also important for the larger goal of controlling more complex behaviors, where animators, not programmers, will be developing fitness functions. Genetic programming will be able to aid animators since writing fitness functions can be made simple enough for animators with little programming experience to write.

Noisy Wall Following and Maze Navigation through Genetic Programming

Andrew Goldish

Computer Science Dept., Brandeis University
415 South Street, Waltham, MA 02154
acg@cs.brandeis.edu

ABSTRACT

This paper discusses a project in which wall following and maze navigation algorithms were developed for a simulated robot in a noisy environment. The algorithms were developed using genetic programming and evaluated by determining how well they performed during a series of simulations.

Both the wall following and maze navigation algorithms were evolved from the following sets of terminals and functions. The terminals were instructions to move the robot ahead one unit, do nothing, or turn ninety degrees left or right. Branching and collision testing were implemented using two sets of functions. One set simulated bump sensors and caused the robot to execute one of two procedures depending on whether it sensed or did not sense an adjacent wall. The second set simulated longer-range sensors and caused the robot to execute one of two functions depending on whether it sensed or did not sense a wall within three units. Finally, instructions could be combined into blocks using a third series of functions. The maze navigation experiment included a pair of additional functions which allowed the robot to mark a grid cell and then execute one of two procedures depending on whether its current cell was marked. However, these procedures were rarely used and were generally ineffective.

Noise was implemented in the simulations as follows. Each instruction which involved sensing the presence of walls was assumed to malfunction and return the opposite of the intended result with a certain probability. Each turning instruction caused the robot to turn ninety degrees plus or minus an error term proportional to the probability. Finally, the instruction which caused the robot to proceed one unit was affected by noise in the same fashion as the turning instructions. Any collision with a wall increased the probability and magnitude of the errors in order to model damage the robot took from the collisions.

Several different kinds of noise were tested in the experiments. Some of the experiments included Gaussian noise, where the error in the turning and moving instructions had a standard deviation proportional to the probability. Other experiments included random noise in which the error term was selected randomly from a range of values. However, in most cases the heuristics were able to overcome the noise and produce a reasonable solution. The only times it failed were when there was too much noise for the learning heuristics to handle.

Each wall following algorithm's fitness was determined by seeing how closely the robot approached each wall segment, scoring each segment accordingly, and adding up the scores. The maze navigators, on the other hand, were evaluated by determining how much of the maze the robot was able to explore in the allotted time. Each candidate was tested four times and its overall fitness was defined to be the average of the four trials.

Experiments also were conducted in which overall fitness was defined as the minimum or maximum of the four trials. Although the fitnesses tended to be highest for the maximum operator and lowest for the minimum operator, the overall fitness performance graphs were similar to that of the average operator.

Both the maze navigation and wall following experiments were run for 25 generations under various initial noise rates. The wall following experiment, which used a mockup of an actual six-room apartment, produced the best results at a noise rate of about 3%. Higher noise rates caused the robot to stray too far from, or collide with, the walls, and lower noise rates caused the robot to retrace its steps too often to explore the entire room. The maze navigation experiment produced comparable results, with the peak fitness occurring with about 2% noise.

Most of the wall following strategies were based on the same procedure: "turn aside when you are about to collide with a wall." The maze navigators produced more complicated strategies, but the most important portions of the maze navigation procedures converged to algorithms similar to those in the wall following experiments.

CONCLUSION

Both experiments demonstrate that genetic programming can be used to evolve strategies for robot navigation in a noisy environment. More research must be conducted to determine if it is appropriate for other noisy applications.

Genetic Programming Classification of Magnetic Resonance Data

H. F. Gray
Computing Dept.
Aarhus University
8000 Aarhus C
Denmark
helen@daimi.aau.dk

R. J. Maxwell
MR Centre
Skejby University
Hospital
8200 Aarhus N
Denmark

I. Martínez-Perez, C. Arús
Universitat Autònoma
de Barcelona
Barcelona
Spain

S. Cerdán
Instituto de
Investigaciones
Biomédicas
CSIC
Madrid, Spain

ABSTRACT

Genetic programming (GP) is used to classify human brain tumours based on ^{1}H Magnetic Resonance spectra. Good classification was achieved by GP (compared to a neural network). GP classification used simple combinations of variables, corresponding to a small group of metabolites, facilitating biochemical interpretation.

1. Introduction

Pattern recognition methods have been used to classify tumours based on ^{1}H Magnetic Resonance (MR) spectra obtained *in vivo* or from biopsy samples [Howells 1992]. Although neural networks (NN) have been used for classification with this type of data [Howells 1992], it is usually difficult to make a biochemical interpretation of the classification scheme.

The task is to classify tumours from ^{1}H MR spectra which give information about biochemical compounds in a tissue. It would be useful to know what data in the spectra is being used to distinguish between different types of tumour. The specific data used was a database of ^{1}H MR spectra of human brain tumour biopsies which was collected as part of an EC Concerted Action (Biomed-1 PL920432).

2. Methods

The data from the spectra were pre-processed with principal component analysis and varimax rotation to give 20 variables.

The GP program used was in Lisp [Koza 1992]. The terminal set consisted of the 20 varimax scores. The functions used were arithmetic, logical, trigonometric and conditional. The selection method used was fitness-proportionate and the generation method was ramped half-and-half [Koza 1992]. The standardised fitness was taken to be the number of samples minus the number of hits.

In one experiment all 75 samples were used to train (with no testing set). A second experiment compared GP with NN. 10 samples were randomly removed from the original data set for use as a test set. All processing stages were then performed on the 65 remaining samples and the learned NN or best GP functions applied to the test set. The NN used back propagation with 20 input units and two hidden layers, 40 nodes per layer.

3. Results and Conclusions

In the first experiment the best individual classified 73/75 (97%) samples correctly. However, one of the most striking features of the analysis was that good classification (80-85% correct) could be obtained using very simple individuals with few function applications (Table 1).

Table 1 Some of the most successful Lisp functions

Function	% correct
`(-J B)`	80
`(-(+ R N) B)`	84
`(-R B)`	85
`(+(-(+(-(+(+(-J B)R)K)B)K)B)R)`	90

Examination of the varimax vectors corresponding to these functions showed that it was possible to classify most of the samples based on a very simple combination of metabolites e.g. glutamine, alanine and glutamate.

In the second experiment the best individual classified 90% correctly on the training set and 90% on the testing set. The test results were comparable with those from NN (80% correct, 10% uncertain, 10% wrong). The winning GP individual again included glutamine and alanine.

Good classification of human brain tumours based on ^{1}H MR spectra of biopsy extracts could be obtained using a GP approach. In addition, the most significant aspect of the analysis was that very simple functions gave classification results that were almost as good as the 'best-ever' functions.

Bibliography

Howells S.L. et al. 1992. An Investigation of Tumor ^{1}H Nuclear Magnetic Resonance Spectra by the Application of Chemometric Techniques, *Magnetic Resonance in Medicine*, Volume 28. Pages 214-236.

Koza, John R. 1992. *Genetic Programming: On the Programming of Computers by Means of Natural Selection.* MIT Press.

GP-COM: A Distributed, Component-Based Genetic Programming System in C++

Christopher Harris and Bernard Buxton

Department of Computer Science
University College, London
Gower Street
London
WC1E 6BT
United Kingdom
Email: C.Harris@cs.ucl.ac.uk, B.Buxton@cs.ucl.ac.uk

ABSTRACT

A genetic programming (GP) system is presented that mirrors the conceptual structure of the genetic programming cycle, maximising flexibility and re-use of code. This reduces the effort required to apply GP to a new problem domain.

1. Structure of a GP system

Genetic programming involves a cyclical process by which populations of candidate programs are manipulated based on their ability to solve a given problem. The GP cycle can conceptually be broken down into problem-dependent and problem-independent parts (see Figure 1.).

GP-COM is a genetic programming system based around this conceptual structure. Components in the system are loosely coupled and are defined only by their interfaces to other components. Components may be easily swapped to test potential improvements to the process, or to apply the system to a new problem domain. The system consists of a collection of components glued together by high-level scripts.

The central component is a population manager which performs all operations essential to the manipulation and management of populations of genetic programs. This component does not implement the cycle but is an 'engine' upon which are layered control scripts. The engine is flexible, and supports strongly typed primitives [Montana 1995], Automatically Defined Functions [Koza 1992] (also strongly typed), and demes. The execution of candidate programs is entirely independent of the software used to create and manipulate them. The system generates C code for each candidate, which is linked into a problem-specific program skeleton. This allows the system to be used with existing libraries of software as primitives.

GP-COM supports multiple concurrent runs through high-level UNIX processes. It can control multiple machines, of varying architecture, executing any number of simultaneous runs. A simple and flexible configuration process allows experimental parameters to be easily tuned for particular problems and runs.

The system has been successfully applied to symbolic regression problems and to edge detection in 1-dimensional signals [Harris 1996].

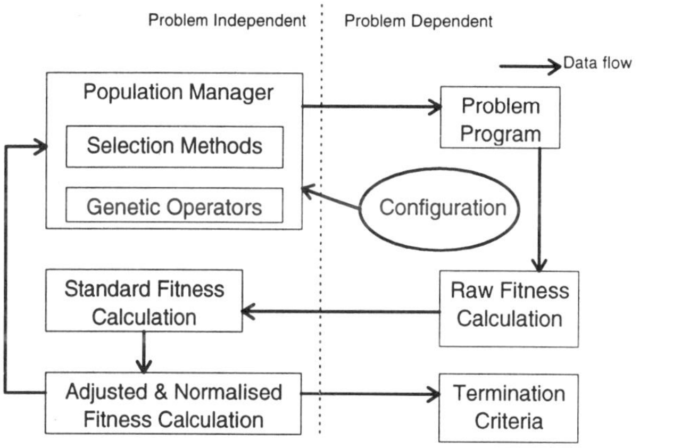

Figure 1. Conceptual structure of the GP cycle

Full details of GP-COM may be found in [Harris 1996a].

Bibliography

Harris, C.P, Buxton, B.F. 1996. *Evolving Edge Detectors.* Department of Computer Science Research Note RN/96/3, University College London.

Harris, C.P., Buxton, B.F. 1996a. *GP-COM: A Distributed, Component-Based Genetic Programming System in C++.* Department of Computer Science Research Note RN/96/2, University College London.

Koza, J.R. 1992. *Genetic Programming: On the Programming of Computers by Natural Selection.* MIT Press.

Montana, D.J. 1995. Strongly Typed Genetic Programming. *Evolutionary Computation* 3(2).

Clique Detection via Genetic Programming

Thomas Haynes & Dale Schoenefeld
Department of Mathematical & Computer Sciences
The University of Tulsa
600 South College Avenue
Tulsa, OK 74104–3189
e–mail: [haynes,dschoen]@euler.mcs.utulsa.edu

We present a genetic programming (GP) technique to find cliques in a graph [Haynes and Schoenefeld, 1996]. A collection of cliques in a graph can be represented as a list of a list of nodes which, in turn, can be represented by a tree structure. Given a graph $G = (V, E)$ a clique of G is a complete subgraph of G. We denote a clique by the set of vertices in the complete subgraph. Our goal is to find all cliques of G. Since the subgraph of G induced by any subset of the vertices of a complete subgraph of G is also complete, it is sufficient to find all maximal complete subgraphs of G. A maximal complete subgraph of G is a maximal clique. Each S–expression in a GP pool will represent sets of candidate maximal cliques. The function and terminal sets are F = {*ExtCon, IntCon*} and T = {*1,...,#nodes*}. *ExtCon* "separates" two candidate maximal cliques, while *IntCon* "joins" two candidate cliques to create a larger candidate.

The fitness evaluation rewards for clique size and rewards for the number of cliques in the tree. To gather the maximal complete subgraphs, the reward for size is greater than that for numbers. We also ensure that we do not reward for a clique either being in the tree twice or being subsumed by another clique. The first falsely inflates the fitness of the individual, while the second invalidates the goals of the problem. The algorithm for the fitness evaluation is: 1) Parse the S–expression into a sequence of candidate maximal cliques, each represented by an ordered list of vertex labels; 2) Throw away any duplicate candidate maximal cliques and any candidate maximal cliques that are subsumed by other candidate maximal cliques; and 3) Throw away any candidate maximal cliques that are not cliques.

$F = \alpha c + \sum_{i=1}^{c} \beta^{n_i}$ is the fitness formula, where $c = \#\ of\ valid\ candidate\ maximal\ cliques$ and $n_i = \#\ nodes\ in\ clique_i$. Both α and β are configurable by the user. β has to be large enough so that a large clique contributes more to the fitness of one S–expression than a collection of proper subcliques contributes to the fitness of a different S–expression.

Figure 1 is a ten node graph used to test the clique detection system. The best and average fitness per generation, Figure 2, steadily improves after generation 100, and the global optimum is found by generation 375. By generation 500, it is evident a plateau has been reached, and the average fitness is only increasing slightly.

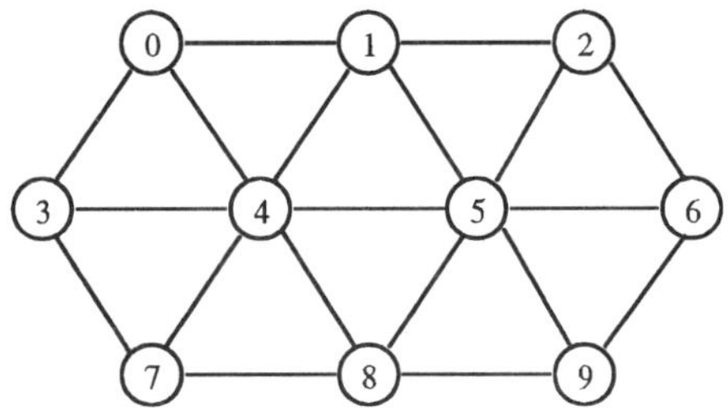

Figure 1: Example graph net10b.dat.

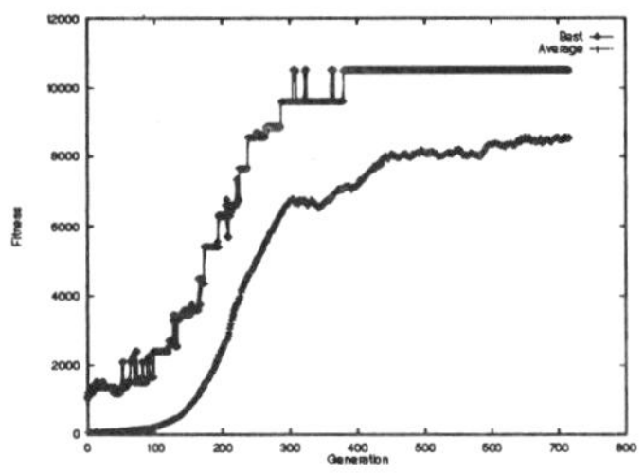

Figure 2: Best and Average fitnesses for graph.

The clique detector is better at discovering a large clique rather than all of the cliques. The largest clique is quickly found, while the list of all possible cliques takes much longer. Since the largest clique in a graph is not known beforehand, a large value for β must be supplied. This suggests a two–pass clique detector: First attempt to find the cardinality of the largest clique and then attempt to find all of the cliques in the graph.

References

[Haynes and Schoenefeld, 1996] Thomas Haynes and Dale Schoenefeld. Clique detection via genetic programming. Technical Report UTULSA-MCS-96-05, The University of Tulsa, March 15, 1996.

Functional Languages on Linear Chromosomes

Paul Holmes   Peter J. Barclay
Dept. of Computer Studies, Napier University, Edinburgh, EH14 1DJ, Scotland.
email: {p.holmes,pj.barclay}@dcs.napier.ac.uk
WWW: http://www.dcs.napier.ac.uk
tel: (+44) (0)131 – 444 2266

1 Introduction

Ayala & Kiger (1984) document a biological phenomenon in which co-adapted genes migrate and naturally cluster together into structures called *supergenes*. We have begun research into the implications of the supergene effect on the efficiency of evolutionary search. In particular, the applicability of the supergene effect to Genetic Programming (GP) is under investigation.

2 Supergenes

If the number of crossover points in a genotype is small compared to the number of genes, the likelihood of two adjacent genes being separated during crossover is correspondingly small. This means that most offspring will inherit both of the adjacent genes from one parent. If the genes are co-adapted (e.g. sex genes), most offspring will inherit matched, rather than mis-matched pairs, and thus be more viable.

If the set of genetic alterations includes translocations, the distance between two genes may change as generations pass, and individuals with co-adapted genes close together are favoured.

3 Implications for GP

In classical GP, evolving programs are stored and manipulated as parse trees. The *genotype* of a program is its parse tree, and the genotype is expressed by evaluating the parse tree.

The movement of genetic material from one part of a tree to another can alter its meaning, as shown in figure 1. One of us (Holmes 1995) presents an argument that this change in meaning could prevent migration and slow down evolutionary progress by wasting most of the opportunities of genetic change. A more economical mechanism for building on genetic adaptations is then suggested.

```
Parent 1     Parent 2     Child
    -            -          -
   / \          / \        / \
  *   1        *   2      2   *
 / \          / \            / \
9   7        9   8          9   7
```

Figure 1: Meaning changes as genes move.

4 Approach

A new programming system for GP has been designed and implemented, in which programs are expressed in a functional language and then transformed into linear chromosomes in such a way that meaning is independent of position in the genome.

The system allows chromosomes to undergo genetic alterations (crossover, mutation etc.) and then be executed as programs. The output of each evolved program is then assessed using a fitness function.

5 Further Work

Two threads of research are to be pursued:

- the theory of the supergene effect is to be explored to determine the conditions under which it can operate, and
- the potential usefulness of supergene clustering in genetic problem solving is to be investigated.

Bibliography

Ayala, F. J. & Kiger, J. A. (1984), *Modern Genetics*, second edn, Benjamin / Cummings, Menlo Park, CA. Translocations are described in chapter 21 (pp. 689–727). Chapter 25 has sections on genetic co-adaptation (pp. 842–846), and supergenes (pp. 851–854).

Holmes, P. (1995), The Odin genetic programming system, Technical Report RR-95-3, Department of Computer Studies, Napier University, Edinburgh, Scotland.

Improving the Accuracy and Robustness of Genetic Programming through Expression Simplification

Dale C. Hooper
Space Dynamics Laboratory
1695 N. Research Park Way
Logan, UT 84341
dale.hooper@sdl.usu.edu

Nicholas S. Flann
Department of Computer Science
Utah State University
Logan, UT 84322-4205
flann@nick.cs.usu.edu

Introduction Genetic Programming (GP) is clearly an inductive learning approach because the program discovered must reproduce the behavior of the desired program over all the input space, not just the space represented in the training examples [Koz92e], [Alt94]. When GP uses only consistency with the training examples as a guide, very large (bloated) programs can result that "overfit" the training examples and therefore perform poorly over the complete input space. To avoid this problem, the search process can be biased by applying Occam's razor to prefer simpler, smaller programs that are more likely to reproduce the desired program. This paper introduces a general way of doing this through expression simplification.

Approach Studies have shown that the size of programs in a GP search tend to increase monotonically, due mostly to the application of crossover operations; in particular, where large sub-expressions replace deep sub-expressions [Tac94b]. Much of this complexity is redundant in that sub-expressions are unreachable (introns), or the sub-expressions can be replaced by a much simpler, smaller expression which would compute the same result. This work applies an expression simplifier to identify and eliminate these introns and to rewrite the bloated expressions into simpler, but equivalent expressions. In this way the search process is biased towards producing smaller and simpler expressions.

Expression Simplification Expression simplification is a process of repeatly applying truth preserving transformations to an expression until a fixed point, which tends to reduce its overall size. For the experiments performed here, the simplification system employed over 200 rules. To control the application of simplification in the GP search process a new parameter is introduced, termed the *simplification rate*, that defines the probability that an individual in a generation will be simplified. With this parameter set to 0, we have standard GP.

Empirical Study The primary purpose of this study was to determine whether applying an expression simplifier improved the accuracy and robustness of a GP system, when tested on unseen data. A simple symbolic regression problem was used where a function was to be identified for characterizing limited experimental data. Runs were performed with different simplification rates, from 0 to 1.0, to determine both the effectiveness of the approach and any sensitivity to particular choices of the simplification rate parameter. All experiments were repeated 10 times and the average and standard deviation of results computed.

Results and Analysis The RMS error values measured over the unseen test set for the simplified runs were found to be significantly smaller than the RMS error values for standard GP runs, confirming our hypothesis. No significant increase in run times was noted with the introduction of simplification. Additionally, simplification was found to both prevent monotonic growth in program size and introduce new useful constants into the terminal mix. Furthermore, little sensitivity to the choice of simplification rate was noted, with values between 0.1 and 0.8 working equally well.

Conclusions We have shown that one effective way to implement a simplification bias into the GP search process is to apply an expression simplifier nondeterministically during search. Results demonstrate that simplification reduces program bloating, increases robustness and produces programs that are more accurate over the unseen data.

References

[Alt94] L. Altenberg. The evolution of evolvability in genetic programming. In Kenneth E. Kinnear, Jr., Editor, *Advances in Genetic Programming.* MIT Press, 1994.

[Koz92e] J. Koza. *Genetic Programming: On the Programming of Computers by Natural Selection.* MIT Press, Cambridge, MA, USA, 1992.

[Tac94b] W. A. Tackett. Recombination, Selection, and the Genetic Construction of Computer Programs. PhD thesis, USC, Department of Electrical Engineering Systems, 1994.

COAST: An Approach to Robustness and Reusability in Genetic Programming

Naohiro Hondo
Information Engineering,
Hokkaido Univ. N-13 W-8,
Sapporo, Hokkaido 060, Japan
hondo@complex.hokudai.ac.jp

Hitoshi Iba
Electrotechnical Lab.
1-1-4 Umezono, Tukuba
Science City, Ibaraki,305,Japan.
Iba@etl.go.jp

Yukinori Kakazu
Information Engineering,
Hokkaido Univ. N-13 W-8,
Sapporo, Hokkaido 060, Japan
kakazu@complex.hokudai.ac.jp

ABSTRACT

This paper presents a new approach to GP. Our goal is to realize reusability and robustness by means of the automatic discovery of functions. We propose an approach, named COAST, for reusability and robustness, which COAST has a library for storing certain subroutines for reuse them. We make use of the Wall Following Problem to illustrate the efficiency of this method.

1. Introduction

This paper introduces a new approach to the automatic discovery of functions for Genetic Programming (i.e., GP). This discovery of functions is aimed at improving GP in the reusability of GP programs, and in the robustness of the GP system. Reusability is defined here as reusing an acquired knowledge. Robustness means flexible behavior.

In order to realize the above, we propose a new system based on an ADF. GP system with ADF defines subroutines automatically. However, so far the reusability and robustness of GP have never been discussed using ADF. We aim at the acquisition of a generalized knowledge or subroutine and realizing the above improvements. A generalized knowledge refers to a rule which can be applied in various situations, giving the system flexible behavior.

In order to realize this facility, we introduce a new method based on ADF, called COAST. COAST extends the ADF so as to establish a library for acquiring effective subroutines with each generation. As the generations proceed, the subroutines acquired by COAST are shared among the population, whereby any individual in COAST can use this generalized knowledge by referring to the library. As a result, through these shared subroutines, COAST gives robustness to the system.

2. COAST

The main concept of COAST is to share some subroutines generated by ADF through the library. If the problem is changeable, COAST is able to cope with the change in order to arrange the subroutines or reuse them. The detail of COAST is referred to "Hondo et al 96". Figure 1 indicates the overview of COAST.

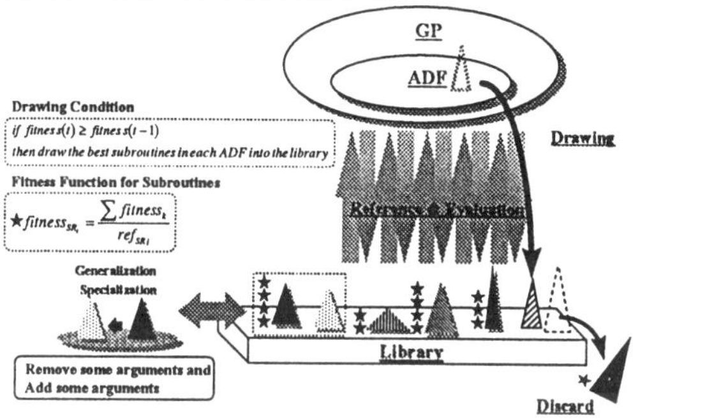

Figure 1 COAST.

3. Experimental Result

We make use of changeable Wall Following Problem to show the efficiency of COAST. The detail of configuration is shown in "Hondo et al 96". In Figure 2, COAST results in stable result comparison to others.

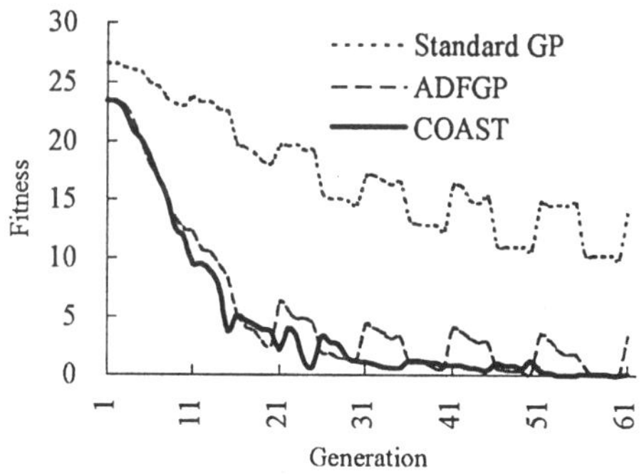

Figure 2 Result.

Bibliography

[Hondo et al 96] Hondo, N, Iba, H and Kakazu, Y: Sharing and Refinement for Reusable Subroutine of Genetic Programming, *IEEE International Conf. On Evolutionary Computing '96 (ICEC'96)*, 1996.

Recurrences with Fixed Base Cases in Genetic Programming*

Stefan J. Johansson
Dept. of Computer Science and Business Administration
University of Karlskrona/Ronneby, Soft Center, S-372 25 Ronneby, Sweden
E-mail: Stefan.Johansson@ide.hk-r.se
WWW: http://www.pt.hk-r.se/~sja/

* The work underlying this paper has been supported by the Dept. of Computer and Information Science at Linköping University, Sweden.

ABSTRACT

A new approach to how to evolve integer recurrences using Genetic Programming (GP) is taken where the recurrences fix their base cases in accordance with the set of test cases instead of evolving them. A number of experiments indicate that this approach outperforms the old one as well as random search in finding non-trivial recurrences.

1 Introduction

An integer series may have a shorter recursive description than the equivalent closed form and it may therefore be of interest to find such a form. John Koza made some experiments with the Fibonacci sequence [Koz. 92], but the results were probably poor (at least not described). This paper is an overview of the experiments [Joh. 96] where the proposed *fixed base cases* (fbc) approach is compared to the *evolved base cases* (ebc) approach used so far in the field.

2 Base cases approaches

In the ebc approach of finding integer recurrences, not only the body of the recurrence is evolved, but also the numeric values of the base cases. The approach *we* take is to make more use of the test cases given to the fitness function, than just measuring the fitness. This is done by fetching the base cases of the recurrences from the set of test cases instead of evolving them. By doing this, we reduce the search space *and* increase the probability of finding a correct solution, since no possible solutions are excluded in the reduction.

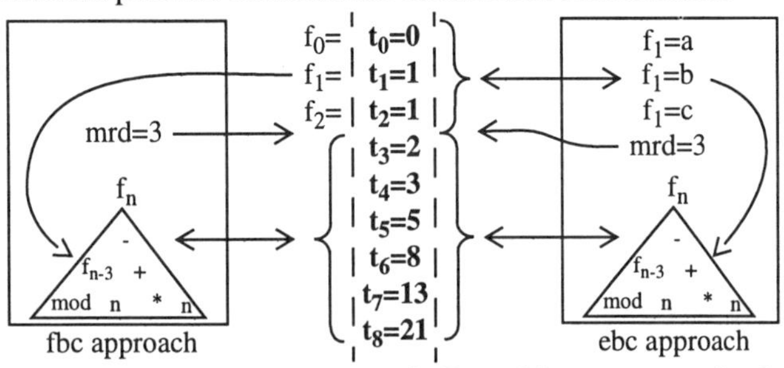

Figure 1: fbc vs. ebc approach. Boxed items are evolved; mrd is the maximum recursive depth and $t_0...t_8$ are test cases. $\rightarrow$ and $\leftrightarrow$ denote effect and comparison.

3 Experiments and results

A total of over 50 000 runs were made in the experiments where more than 720 different configurations with different population sizes, fitness measures and types of recurrences were tested. For a full description of the setup, see [Joh. 96].

In the base case experiments the approaches were compared under the same circumstances. The result was 461 - 1 in favor of the fbc approach. In the remaining 258 configurations none of the approaches found any solution. Probably the integer series were too hard to find or the population sizes were too small.

In the comparisons with random search (RS), individuals were randomly generated and evaluated under the same conditions as the evolved ones, i.e. the random individuals had an fbc or ebc approach. The results show that for easily generated recurrences RS is as good as, or even better than, GP on finding them, but the superiority of GP over RS increases with the complexity of the integer series. However this improvement is only valid for the fbc-runs. The results of the ebc-runs were too poor to draw any conclusions from.

Two more aspects of evolving integer recurrences were investigated. The comparison between different population sizes indicated that the larger sizes commonly used in GP are also optimal in the case of finding integer recurrences. The other aspect concerned different fitness functions, but unfortunately these runs did not give any conclusive results.

4 Conclusions

The fbc approach seems to be a promising way to find recurrences. In the work described, this observation applies to both GP and RS and there are reasons to suspect that this is the case for other search methods as well. Moreover, when the fbc approach is used, GP is better than RS on finding non-trivial recurrences. Also, a population size of 1000-4000 seems to be optimal for those recurrences.

Bibliography

Koza, John (1992), Genetic Programming: On the Programming of Computers by Means of Natural Selection. Cambridge, MA: The MIT Press

Johansson, Stefan J. (1996), Evolving Integer Recurrences using Genetic Programming, Technical Report, Faculteit Wiskunde en Informatica, Vrije Universiteit, Amsterdam

Evolutionary and Incremental Methods to Solve Hard Learning Problems

Ibrahim KUSCU
Cognitive and Computing Sciences
University of Sussex
Brighton BN1 9QH
E-mail: ibrahim@cogs.susx.ac.uk

Abstract

This paper briefly introduces the experiments and results of an evolutionary approach and an incremental approach to finding learning rules of several simple and hard supervised learning tasks.

In [Kuscu, forthcoming] a genetic based encoding schema has been presented as a potentially powerful tool to discover learning rules for several supervised tasks. The representation schema is very similar to the one used in Genetic Programming (GP). However, introducing prior knowledge into the representation of initial solutions using problem specific functions is minimal, if any at all. In this strategy potential learning rules are encoded as random mathematical expressions at variable lengths. The expressions are made up of random numbers and random variables. The variables are to be instantiated to input values of training set in a typical supervised learning. The expressions are evaluated to certain numbers and by the help of a squashing-function this value is mapped to a value in the range of output values. The fitness of an expression is determined by the number of correct mappings from the training set but the success of the expression in learning the task is determined by the degree of generalisation over the testing set. The model is tested on three Monks and Parity problems. The results showed that the model is useful in discovering problem-specific-functions describing the learning rules of supervised problems by using a relatively more general and fixed set of non-problem-specific functions. However, when the problems were larger and more complex (i.e. those supervised learning problems where the learning rule refers to a relationship *among* the input values rather than a direct correlation between value(s) of input(s) and output variable(s) [Clark and Thornton, Forthcoming]), the ability of the model in coding for good solutions and effectively generalising over the testing set was not found to be satisfactory.

In recent experiments [Kuscu, 1995] these issues are investigated further. The experiments involve an incremental encoding of an expression as a potential solution. Although the basic encoding strategy (forming potential solutions as random mathematical expressions) is the same as the one used in previous experiments, the searching strategy is different. Rather than using evolution with a population of expressions, a hill climbing strategy is employed by incrementally forming a single expression (i.e. by preserving the old and concatenating new expressions). In this way, it is hoped that problems requiring larger solutions might be effectively coded for. The model is tested on the Monks and Parity problems. An increased performance is observed in finding solutions to larger and more complex problems but a low performance is obtained when the model is tested over the unseen cases.

The forthcoming research efforts are concentrated on better clarifying the role of evolution not only in discovering hard learning rules but the degree of discovered solution's ability in generalising for the unseen cases.

Acknowledgements

This research is funded by the Middle East Technical University, Ankara, Turkiye.

References

[Clark and Thornton, Forthcoming] A. Clark and C. Thornton. Trading spaces: Computation, representation and the limits of uninformed learning. *Behavioral and Brain Sciences*, Forthcoming.

[Kuscu, 1995] I. Kuscu. Incrementally learning the rules for supervised tasks: Monk's problems. Technical Report CSRP-396, Uni. of Sussex, COGS, 1995.

[Kuscu, forthcoming] I. Kuscu. Evolution of learning rules for hard learning problems. In *the Proceedings of The Fifth Annual Conference on Evolutionary Programming*, forthcoming.

Detection of Patterns in Radiographs using ANN Designed and Trained with GA

Alejandro Pazos
RNASA-Lab, Faculty of Computer Science. University of A Coruña. 15071, A Coruña, Spain.
ciapazos@udc.es

Julian Dorado
RNASA-Lab, Faculty of Computer Science. University of A Coruña. 15071, A Coruña, Spain.
cijulian@udc.es

Antonino Santos
RNASA-Lab, Faculty of Computer Science. University of A Coruña. 15071, A Coruña, Spain.
nino@udc.es

ABSTRACT

This system explores the union between the hip joint and a prosthesis inserted in the femur. The objective is to make an early detection of the existence of any gapping between prosthesis and bone. To do this, we use an ANN built with a GA, that optimize not only the architecture of the ANN, but also the design of the training set. The output offered by the system is the probability of the existence of any spacing between prosthesis and bone.

1. Introduction

Taking advantage of the availability of digital X-rays at the Hospital of A Coruña, and working with Dr. Pernas, research has been done. In this case we're interested in femur prosthesis, in which any misfitting in the insertion on the bone can provoke pain in the patient. Post-operative X-rays were used to review the prosthesis-bone fitting but, it is not possible to carry out, by plain sight, an exact study owing to inflammation after the operation. The X-ray techniques pose the problem of variability of parameters, such as lighting. Owing to this, an Artificial Neural Network (ANN) has been used for avoiding the impact of changing conditions. The ANN used was built with a GA. We choose this option for the construction because of the great quantity of parameters which can possibly be fashioned.

2. Training Set and ANN Architecture

To carry out this work, we used seventy eight X-rays of hip prosthesis. To build a training file we selected a set of bone-prosthesis unions in which the degree of conviction of whether there existed a separation was significant. Starting from a set of neighboring pixels, there must be given a value of probability for a pixels correspondence to the bone or to a gap prosthesis-bone. The number of pixels needed to estimate what this pixel corresponds to, cannot be decided upon until an evaluation has begun of the initial results offered by the network and, it will be the beginning of a long process of testing configurations. Apart from this complexity there still remains the design process of the architecture itself. As a result of all this, to accelerate the construction of the ANN, a GA was used.

3. Genetic Algorithm and ANN

The tool used for applying the GA's was codified in C. In the evaluation function of the GA, the ANN is represented by the weights of its layers. The size of the population is configurative only at design time, while the mutation and crossover ratios may be varied throughout training. To substitute old individuals for newer generations, an algorithm selects the set of individuals of the population which are most similar to the new set it wishes to include. The crossover function of the GA is adapted to work with individuals which don't have the same number of layers. The crossing function choose the cutting point randomly upon the shortest length of the string and their weights are crossed at the chosen position. The mutation function is standard. The ANN resulting obtained the following form: an input layer of 15 neurons, one hidden layer with 5 neurons and, the output layer has just one single neuron. Before arriving at this configuration, the GA explored possibilities from 1 to 4 hidden layers, and from 4 to 25 neurons in the input layer.

4. Results

For the presentation of the results, we placed an overprint of false color over the initial X-ray. In a case like this one, there does not exist a quantitative valuation of the accuracy of the network's output, since it is impossible to make an objective measure. To verify the system we used X-ray evaluation sequences on each patient. Working with these sequences we can see how the prosthesis has evolved throughout the years. If a gap has developed following this period, we study the systems output on the first X-rays of the sequence to try to discover the formation of that gap. Once these parameters were defined, the clinicians chose twelve trial sequences, six which ended up with problems of separation. The doctor's opinion on the results obtained was good, and only in a few X-rays from three sequences the output of the system wasn't adequate, due to bad lighting in the region of interest. In the remaining X-rays there is a clear appreciation for how the system detects the presence of gaps from an early stage.

The Logic Grammars based Genetic Programming System

Man Leung Wong
Department of Systems Engineering and Engineering Management
The Chinese University of Hong Kong
mlwong@se.cuhk.hk

Kwong Sak Leung
Department of Computer Science and Computer Engineering
The Chinese University of Hong Kong
ksleung@cs.cuhk.hk

ABSTRACT

Inductive Logic Programming (ILP) and Genetic Programming (GP) are two approaches in program induction. However, they are restricted in the computer languages in which programs can be induced. Generic Genetic Programming (GGP) is a novel, powerful, and general approach that combines ILP and GP. Using GGP, programs in various programming languages such as Lisp, Prolog, and Fuzzy Prolog can be evolved. The approach is also powerful enough to handle context-sensitive information and domain-dependent knowledge. In this paper, we describe the details of an implementation of GGP called LOGENPRO (The LOgic grammars based GENetic PROgramming system).

1. Introduction

Inductive Logic Programming (Muggleton 1992) and Genetic Programming (Koza 1992, 1994; Kinnear 1994) are two approaches in program induction. However, they are restricted in the computer languages in which programs can be induced. Inductive Logic Programming (ILP) can only learn logic programs. On the other hand, Genetic Programming (GP) represents induced programs as parse trees. In theory, programs in any computer languages such as Prolog, C, and Fortran can be represented as parse trees. Hence, GP should be able to handle them as well. In fact, the process of translating a program in some languages to the corresponding parse tree is not trivial. Since the syntax of Lisp is so simple and uniform that the translation process can be achieved easily, programs evolved by GP are usually expressed in Lisp.

A novel approach called Generic Genetic Programming (GGP) that combines GP and ILP has been proposed (Wong and Leung 1995, 1996). Using GGP, programs in various programming languages such as Lisp, Prolog, and Fuzzy Prolog can be evolved. The approach is also powerful enough to handle context-sensitive information and domain-dependent knowledge which can be used to accelerate the learning speed and/or improve the quality of the programs induced. This paper describes the details of an implementation of GGP called LOGENPRO (The LOgic grammars based GENetic PROgramming system).

Since LOGENPRO can induce programs in various programming languages, it must be able to accept grammars of different languages and produce programs in them. LOGENPRO is based on logic grammars because they are expressive enough to represent context-sensitive information of some languages and domain-dependent knowledge of the target programs being induced.

Bibliography

Koza, J. R. (1992). *Genetic Programming: on the Programming of Computers by Means of Natural Selection.* Cambridge, MA: MIT Press.

Koza, J. R. (1994). *Genetic Programming II: Automatic Discovery of Reusable Programs.* Cambridge, MA: MIT Press.

Kinnear, K. E. Jr., editor (1994). *Advances in Genetic Programming.* Cambridge, MA: MIT Press.

Muggletion, S., editor (1992). *Inductive Logic Programming.* London: Academic Press.

Wong, M. L. and Leung, K. S. (1995). Combining Genetic Programming and Inductive Logic Programming using Logic Grammars. In *Proceedings of the 1995 IEEE International Conference on Evolutionary Computing.*

Wong, M. L. and Leung, K. S. (1996). Evolving recursive functions for the even-parity problem using genetic programming. In P. J. Angeline and K. E. Kinnear, Jr. (Eds.) *Advances in Genetic Programming 2.* MA: MIT Press.

Long Genetic Algorithms Papers

Genetic Algorithms with Analytical Solution

Erol Gelenbe
Department of Electrical and Computer Engineering
Duke University
Durham, N.C. 27708-0291
erol@ee.duke.edu

ABSTRACT

We study a class of genetic algorithms (GA) based on a population of chromosomes with mutation and crossover, as well as fitness. The chromosomes in these algorithms also have another property we call "activity" (which determines how fast a chromosome may mutate or interact with other chromosomes). We represent these algorithms by a multi-dimensional Markov process, and provide a new closed form expression for the probability distribution of the number of chromosomes of each type. Specifically we prove that in steady-state:
(i) the joint steady-state probability of the number of chromosomes of each type is the product of the marginal probabilities for each type of chromosome, and
(ii) the marginal probability of the number of chromosomes of each type has a geometric distribution. Another question we address is that of the choice of the GA parameters for a specific problem. We show that it is possible to choose parameters using a gradient based learning algorithm of complextity $O(n^3)$ and we apply this concept to the choice of the fitness parameter for our GA model.

1. Introduction

The area of Genetic Algorithms (GA) founded by Holland [2] was inspired by the manner in which a set of chromosomes of different types undergoes transformations (see for instance Muehlenbein [6] and Goldberg [7]).

The basic primitives associated with these algorithms are:

- An initialization of the population of chromosomes,

- Transformation operations, such as crossover -- which replaces a subset of the chromosomes by a new subset, mutation -- which simply replaces some chromosome by another, and fitness -- which determines how each type of chromosome will survive or be duplicated.

Most studies and applications of genetic algorithms use variants of these operations. For instance crossover is typically limited to replacing two chromosomes by two other chromosomes. Fitness may determine simply the probability that a chromosome within a particular population is allowed to appear in the next generation. There is a vast literature on genetic algorithms and their applications -- in particular to optimization problems. Let us just mention a few of the references where an analysis of various aspects of GAs has been presented [2, 3, 6, 7, 10, 11]. One aspect of GAs which has not been significantly developed concerns mathematical tractability, and GA applications are mainly based on Montecarlo simulation. Yet mathematical tools, if further developed, could make it easier to design GAs with a specific functional behavior, and also to test GAs more rapidly that what would be possible via a lengthy Montecarlo simulation. Analytical tractability would also avoid the major issue of the well-known issue of the statistical significance of a Montecarlo simulation.

The use of Markov processes in population models is very old (see for instance [1,15]) and recent applications of stochastic process models to GAs include [16, 17].

The novelty of our approach is that we derive closed form expressions for the stationary probability distribution of the state of any given GA formalized with a specific set of operations which we describe. This has the advantage of avoiding time consuming (and sometimes inaccurate) numerical methods needed to solve for the very large state-space of a GA model, or lengthy Montecarlo simulations which need to be undertaken when a mathematical solution is not used.

We also show (in Section 5) how the closed form solutions which we derive are useful in selecting the GA's parameters. Specifically we present an algorithm based on gradient descent for selecting the fitness parameter for a GA which is being designed to solve an optimization problem. This is similar to learning algorithms for recurrent random neural networks [14].

The unusual result we prove in Theorem 1 says that :

(i) The joint steady-state probability of the number of chromosomes of each type is the product of the marginal probabilities for each type of chromosome,
(ii) The marginal probability of the number of chromosomes of each type has a geometric distribution.

In this paper we will consider a set of GA primitives ((a) to (e) given below) which capture the essence of the basic ideas on GAs. We will then proceed to give an analytical solution to GAs operating under these primitive transformations based on continuous time and infinite discrete state-space Markov processes. Specifically, we will provide a closed-form analytical expression for the stationary solution of this class of models. Finally, we will show an application of our approach to a well known difficult optimization problem.

We will deal with a finite set of types of chromosomes simply denoted by the integers {1, ... , n}. At a given instant of time, our system will contain a finite but unbounded number of chromosomes of any type, and these numbers will be denoted by some vector $k = (k_1,...,k_n)$. Thus each k_i is some non-negative integer representing the number of chromosomes of type i which are currently present. In more precise terms, the state of the system at time t is the vector $k(t) = (k_1(t), ... ,k_n(t))$ whose value is $k = (k_1,...,k_n)$ with $k_i \geq 0$, i=1,...,n. In this model the total population size of chromosomes at any time t , $\Sigma^n{}_{i=1}\, k_i(t)$, is unbounded, while the total number of types of chromosomes is finite (n). This system evolves under the following set of elementary rules :

(a) A chromosome of type i can be spontaneously created (birth),
(b) A chromosome of type i can survive or it may not survive ; this relates to the fitness criterion of GA models,
(c) A chromosome of type i may be transformed into a chromosome of type j (mutation),
(d) A chromosome of type i may combine with a chromosome of type j (here j *need not necessarily be different* from i) to produce a chromosome of type k (simple crossover). In that case, the two originating chromosomes (of types i and j) disappear and are replaced by the type k chromosome. Note that we allow i and j to be the same, and indeed we *may also* have that k is the same as i or j (or both),
(e) A chromosome of type i may combine with a chromosome of type j, and both may be destroyed; we will call this the destruction (or joint destruction) operation.

Note that this corresponds to a model in which two individuals of any generation can only give rise to a single individual in the next generation (Rule (d)). While Rule (e) appears to be novel in genetic algorith models but is plausible if one considers combinations of chromosomes which are not viable.

Thus if we did not have Rule (a) the population of chromosomes would eventually become extinct, even if we did not have Rule (b), i.e. death. Therefore Rule (a) is needed to cover cases where we wish the population either to tend to some probabilistic steady-state or to grow indefinitiely. This paper deals principally with the former case, i.e. when there exists a stationary distribution of the size of the population of chromosomes as a whole.

2. The rates of change and the probabilities of events

Now that we have stated what is allowed to happen, we also have to indicate how these things happen and when. All events are considered over infinitesimal time periods Dt, and therefore occur within infinitesimally small intervals of time $[t , t + \Delta t)$.

Due to Rule (a), the probability that one chromosome of type i is created in the interval $[t , t + \Delta t)$ is denoted by $L_i \Delta t$. The probability that such an event will *not* occur is $(1 - L_i \Delta t)$.

Each type of chromosome has a specffic level of activity, describing how fast it gets involved in changes. We introduce this idea into our GA model, since we think it may be of use in applications and because it may also very well correspond to biological reality. This is quantified by an activity rate $a_i > 0$ which may be different for each type i of chromosome.

Mutation, fitness and crossover will are related to the activity rate as described below. Thus chromosomes with higher activity rates will tend to mutate or crossover more frequently. Clearly, this activity rate may be modulated by other effects such as the fitness, or the mutation probability. Thus the probability that the chromosome "is not fit", as well as the activity rate, will determine its capabilities with respect to survival, mutation and crossover.

The probability that a chromosome of type i is fit will be denoted by f_i . In the interval $[t , t + \Delta t)$ it will die and disappear from the system with probability $(1- f_i)a_i \Delta t$. On the other hand, a chromosome of type i will proceed to mutate or carry out a crossover only if it is fit and survives and this occurs with probability $f_i a_i \Delta t$. Notice that the activity rate also influences the destruction or death rate of chromosomes which are not fit.

Concerning Rule (c), the probability that a chromosome of type i mutates into one of type j is m_{ij} . Therefore in the interval $[t , t + \Delta t)$ the probability that it mutates into one of type j is is $f_i a_i m_{ij}$ Dt , while the probability that this does not happen is simply $(1 - f_i a_i m_{ij} \Delta t)$. Clearly, for this to occur at all, there must be some chromosome of type i in the system. In the next section we will show how this definitions of fitness relates to the usual definition [3, 4, 7, 8] which is a metric relative to the population size as a whole, rather than to a single individual.

Concerning Rule (d) for crossover, the probability that a chromosome of type i combines with one of type j, to yield a chromosome of type v is c_{ijv}. However for this to occur, the chromosome of type i must be activated and be fit as well. Thus, the probability that it will occur in the interval $[t, t + \Delta t)$ is $f_i a_i c_{ijv} \Delta t$, while the probability that this does not happen is $(1 - f_i a_i c_{ijv} \Delta t)$.

Finally, for Rule (e) for mutual destruction we have that the probability that a chromosome of type i combines with one of type j, to mutually destroy is d_{ij}. For this to occur, the chromosome of type i must be activated and be fit as well. The probability that it will occur in the interval $[t, t + \Delta t)$ is $f_i a_i d_{ij} \Delta t$, while the probability that this does not happen is $(1 - f_i a_i d_{ij} \Delta t)$.

Of course, a fit chromosome may also neither mutate nor take the initiative of combining with another chromosome for a crossover, and choose to remain for the time being in the same type with probability s_i:

$$\Sigma^{n}_{j=1} (m_{ij} + d_{ij}) + \Sigma^{n}_{j,v=1} c_{ijv} + s_i = 1.$$

The term $f_i a_i$ which will later appear in many expressions related to this model can be intuitively be interpreted as the mathematical counterpart of the type i chromosome being "fit and active". Note that because we introduce the notion of activation rate, and because we require a chromosome to be fit in order to mutate or do a crossover, our model distinguishes the case of a type i chromosome taking the initiative of combining with a type j chromosome, from the case where the initiative begins at a type i chromosome. These two cases can be made equivalent by setting $f_i a_i c_{ijv} = f_j a_j c_{jiv}$.

The parameters L_i, f_i, a_i, m_{ij}, d_{ij}, c_{ijv}, s_i characterize the manner in which creation, fitness, activity, mutation and crossover rules are being applied by the GA.

Note that if the crossover operation were not present in ths model, we would be merely dealing with a generalized birth and death process whose properties and solutions are well known [4,5]. The crossover operation, even in its limited present form, distinguishes the model discussed in this paper from these conventional multidimensional birth and death models.

3. Analytical expression for system state

Once a GA is set up with a given ensemble of parameters, it is then run as a Montecarlo (or sometimes deterministic) simulation to evaluate its behavior and effect. Though short term evolutions and behavior is of interest, statistical significance is often only available if one considers the long run or steady-state behavior of the model.

We will presently show that for the GA model we have defined above, the long run behavior need not be deduced from a simulation, since it is directly available from the Theorem we prove below.

Let us denote $p(k,t) = \text{Prob}\,[k(t) = k]$, i.e. the probability that the GA is in some state k at time t, assuming some arbitrary initial condition. Clearly, if we know p(k,t) for all t, then we have full knowledge of system behavior. For large enough t, however, the stationary or long term behavior can provide enough information. This is precisely what we will be dealing with here, in that we will provide formulas to easily calculate the steady-state probabilities for $k = (k_1,\dots,k_n)$:

$$(1) \qquad p(k) = \lim_{t \to \infty} \Pr[k(t) = k]$$

The main result we have provides an analytical exprssion for (1). We present it in this section, and give the rather lengthy proof in the next Section.

Theorem 1 The steady-state probability characterizing completely the GAs behavior is independent of the intial state configuration k(0). It is given by the composite geometric distribution:

$$(2) \qquad p(k_1,\dots,k_n) = \Pi^{n}_{i=1} (1 - r_i) r_i^{k_i},$$

where the r_i, i=1, ... , n are the unique solution to the following system of equations:

$$(3)\ r_i = \frac{\{L_i + S^{n}_{j=1} r_j f_j a_j m_{ji} + S^{n}_{u,v=1} r_u f_u a_u r_v c_{uv}\}}{\{a_i(1 - f_i s_i) + S^{n}_{j,v=1} r_j f_j a_j c_{jiv} + S^{n}_{j} r_j f_j a_j d_{ji}\}}$$

Expression (3) has an intuitive explanation. The denominator contains all those terms which result in the creation of chromosomes of type i, while the denominator contains the terms which result in the depletion of type i chromosomes. Note from (2) that is the probability that there is at least one chromosome of type i in the system at steady state.

On the other hand, (2) is a remarkable and unusual result since it states that the joint probability of system state can be written as the product of the marginal probabilities related to each individual population of chromosomes, and that the distribution of the number of chromsomes of each type obeys a geometric distribution.

Before entering into the proof we will make the following two remarks.

Remark 1 (Relative Fitness)

We will relate the parameter f_i to another measures of fitness: it is the relative fitness of the population of Type i relative to the fitness of the population as a whole and we represent it by F_i .

From (3) we can easily show that the average number of individuals of Type i is given by : $N_i = r_i /(1 - r_i)$, Similarly, the total average number of individuals can be calculated as:

$$N = \Sigma^{n}_{i=1} r_i /(1 - r_i) = r/(1 - r),$$

where $r = \Sigma^{n}_{i=1} r_i$.

Then the average fitness of the population is

$f = \Sigma^{n}_{j=1} f_j N_j/N$ and the relative fitness of the chromosomes of Type i is $F_i = f_i /f$.

Remark 2 (Fictitious chromosome types and Rule (e))

This second remark will simplify the subsequent analysis. Let us add a "fictitious" chromosome type, call it type I, so that I is not in the set {1, ... , n}. This will be a type of chromosome into which one can <u>enter</u> only by crossover so that :

$$c_{ijI} \text{ and } c_{jiI}$$

may be non-zero for i,j = 1, ... , n .

However we are not be able to <u>leave</u> this chromsome type in any way, i.e. :

$m_{iI} = 0$, $m_{Ii} = 0$, $c_{Iji} = 0$, $c_{jIi} = 0$, for all i ,j = 1 , ... , n and $L_I = 0$,

so that from (3) we necessarily have $r_I = 0$, since all terms in the numerator are null. However if we set $d_{ji} = c_{jiI}$ for each j,i we end up with :

$$r_i = \frac{\{ L_i + \Sigma^{n}_{j=1} r_j f_j a_j m_{ji} + \Sigma^{n}_{u,v=1} r_u f_u a_u r_v c_{uvi} \}}{\{ a_i(1 - f_i s_i) + \Sigma^{n}_{j,v=1} r_j f_j a_j c_{jiv} + \Sigma^{n}_{j} r_j f_j a_j c_{jiI} \}}$$

or

$$r_i = \frac{\{ L_i + \Sigma^{n}_{j=1} r_j f_j a_j m_{ji} + \Sigma^{n}_{u,v=1} r_u f_u a_u r_v c_{uvi} \}}{\{ a_i(1 - f_i s_i) + \Sigma^{n+1}_{j,v=1} r_j f_j a_j c_{jiv} \}}$$

where we use the (n+1)-st chromosome type to denote the type I chromosome. This simple development indicates that the model with an additional fictitious chromosome type but no chromosome destruction (Rule (e)) is equivalent to the model with a Rule (e). Therefore the proof of Theorem 1 will be carried out as if Rule (e) did not exist, without loss of generality.

4. Sketch of the Proof

Let us now briefly sketch the proof. Define the vectors of integers :

$$k^{+}_{i} = (k_1, \dots , k_i+1, \dots , k_n),$$
$$k^{-}_{i} = (k_1, \dots , k_i -1, \dots , k_n)$$
$$k^{+-}_{ij} = (k_1, \dots , k_i+1, \dots , k_j -1, \dots , k_n)$$
$$k^{++-}_{ijm} = (k_1, \dots , k_i+1, \dots , k_j+1, \dots , k_m -1, \dots , k_n)$$

We denote by $\{k(t): t \geq 0\}$ the continuous time Markov chain representing the state of the network. It satisfies a system of Chapman-Kolmogorov equations, whose steady-state representation is :

$$\begin{aligned}
(4)\quad & p(k) \Sigma_i [L(i) + a_i (1 - f_i s_i) \mathbf{1}[k_i>0]] \\
&= \Sigma_i [p(k_i^{+}) a_i (1 - f_i) + p(k^{-}_i) L(i) \mathbf{1}[k_i>0] \\
&\quad + \Sigma_j \{ p(k^{+-}_{ij}) a_i f_i m_{ij} \mathbf{1}[k_j>0] \\
&\quad + \Sigma_v p(k^{++-}_{ijv}) f_i a_i c_{ijv} \mathbf{1}[k_v>0] \\
&\quad + p(k_i^{+}) f_i a_i c_{ijv} \mathbf{1}[k_j=0] \}]
\end{aligned}$$

where the following notation is used: **1**[X] =1 if X is true and 0 otherwise.

The last term on the right-hand-side of (4) covers the case where a chromosome of type i attempts a crossover operation with a chromosome of type j, only to discover that such a chromosome does not exist, leading to an unsuccessful crossover operation.

We first have to verify that (3) indeed satisfies (4) in order to carry out the proof of the Theorem. In order to do so, we substitute the general geometric form (2) into (4) and obtain :

$$\begin{aligned}
(5)\quad & \Sigma_i [L_i + a_i (1 - f_i s_i) \mathbf{1}[k_i>0]] \\
&= \Sigma_i [r_i(1 - f_i) a_i + L_i \mathbf{1}[k_i>0] / r_i \\
&\quad + \Sigma_j \{ r_i f_i a_i m_{ij} \mathbf{1}[k_j>0] / r_j + \Sigma_v r_i r_j f_i a_i c_{ijv} \mathbf{1}[k_v>0] / r_v \\
&\quad + \Sigma_v r_i f_i a_i c_{ijv} \mathbf{1}[k_j=0] \}]
\end{aligned}$$

The proof follows from by verifying that (3) satisfies (5).

5. Designing a GA for a specific problem

In this section we turn to the issue of the choice of parameters for a specific optimization problem. Suppose that this problem's solution can be expressed by a set of binary variables X_i , i=1, ..., n which we encode by using the r_i so that $r_i = P[X_i = 1]$. Suppose also that we are seeking a solution $\{ X_1 , \dots , X_n \}$ which minimizes some cost function $C(X_1 , \dots , X_n)$. The coresponding cost function in terms of the GA will be $C(r_1 , \dots , r_n)$.

Let us limit first our attention to the choice of the fitness parameter f_k , k =1,...,n although the same principles can be used for any other of the parameters of the GA model we have introduced. Then we suggest that this choice be made by using the standard gradient algorithm because it will lead to a local minimum of C' with respect to the set of parameters under consideration:

$$(6) \qquad f_k \leftarrow f_k - h\, \partial C/ \partial f_k$$

for some positive constant h . We need to compute the derivative term, for which we use the chain rule:

$$(7) \qquad \partial C/ \partial f_k = S_i[\, \partial C/ \partial r_i \,]\; \partial r_i/\partial f_k$$

Clearly the issue then is to be able to compute $\partial r_i / \partial f_k$ for any i and k. We shall see below that this can be achieved simply by solving a linear system of equations (8) at each step of the iteration .

To simplify matters, assume that $m_{ii} = c_{iiv} = d_{ii} = 0$ so that we eliminate all unnecessary transformations of the set of chromosomes. Let us rewrite (3) as $r_i = N_i / D_i$, where

$$N_i = \{ L_i + \Sigma^n_{j=1} r_j f_j a_j m_{ji} + \Sigma^n_{u,v=1} r_u f_u a_u r_v c_{uvi} \}$$

$$D_i = \{ a_i (1 - f_i s_i) + \Sigma^n_{j,v=1} r_j f_j a_j c_{jiv} + \Sigma^n_{j} r_j f_j a_j d_{ji} \}$$

We can then write:

$$r'_i = \partial r_i / \partial f_k = D_i^{-1} \{ N'_i - r_i D'_i \},$$

where N'_i and D'_i are the partial derivatives of the numerator and denominator of (3) with respect to f_k . We then have :

$$N'_i = \Sigma^n_{j=1} r'_j f_j a_j m_{ji} + r_k a_k m_{ki} + \Sigma^n_{v=1} r_k a_k r_v c_{kvi} + \Sigma^n_{u,v=1} (r'_u r_v + r_u r'_v) f_u a_u c_{uvi}$$

$$D'_i = \Sigma^n_{j,v=1} r'_j f_j a_j c_{jiv} + \Sigma^n_{v=1} r_k a_k c_{kiv} + \Sigma^n_{j=1} r'_j f_j a_j d_{ji} + r_k a_k d_{ki} + a_i s_i 1[i=k]$$

As a consequence, we rearrange terms in the above expressions to write:

$$N'_i - r_i D'_i = \Sigma^n_{j=1} r'_j f_j a_j [m_{ji} + \Sigma^n_{v=1} r_v c_{jvi} - r_i d_{ji} - r_i \Sigma^n_{v=1} c_{jiv}] + \Sigma^n_{j=1} \Sigma^n_{u=1} r'_j r_u f_u a_u c_{uji} - r_i \Sigma^n_{v=1} r_k a_k c_{kiv} - r_i r_k a_k d_{ki} + r_i a_i s_i 1[i=k]$$

Let us denote :

$$M_{ij} = D_j^{-1} f_i a_i \{ m_{ij} + \Sigma^n_{v=1} [\, r_v c_{ivj} - r_j c_{ijv}] - r_j d_{ij} \} + D_j^{-1} \Sigma^n_{u=1} r_u f_u a_u c_{uij}$$

We then express the derivative r' of the vector $r = (r_1 , \dots , r_n)$ as

$$(8) \qquad r' = r' M + d(k) , \text{ or } r' = d(k)[I - M]^{-1}$$

where the matrix $M = [\, M_{ij}]$ is given above and $d(k) = (d_1(k), \dots , d_n(k))$ is given by :

$$(9)\ d_i(k) = - r_i D_i^{-1} \{ \Sigma^n_{v=1} r_k a_k c_{kiv} + r_k a_k d_{ki} - a_i s_i 1[i=k] \} .$$

It is interesting to see however that M <u>does not</u> depend on k, whereas d(k) does of course. Thus $[I - M]^{-1}$ needs to be computed only once for all the f_k at each step of the gradient algorithm.

For each value of r , M and hence $[I - M]^{-1}$ can be calculated just once to obtain the derivatives of r_i the with respect to all of the f_k .

We now summarize the proposed gradient algorithm for the fitness parameters. For each iteration step (6) :

1. First calculate the r_i from the previous values of the f_i , for i=1,...,n .

2. Calculate the matrix M from r and the previous values of the f_i. Obtain the inverse $[I - M]^{-1}$. Note that the same M will be used for updating all the f_k.

3. Calculate d^k for each k =1, ... , n . Use (8) to calculate $\partial r_i / \partial f_k$ for each k.

4. Use (6), (7) and $r' = d(k)[I - M]^{-1}$ to calculate the new values of the f_k.

5. Stop if the difference between the previous and the current value of C' is sufficiently small.

Once the GA's parameters have been selected in this manner, the GA's solution is considered to provide a solution for the optimization problem at hand.

The crossover parameters can also be derived using a gradient algorithm. As in the previous section, we are suggesting that the crossover parameters be calculated on the basis of a gradient descent:

$$c_{uvk} \leftarrow c_{uvk} - g\, \partial C / \partial c_{uvk}$$

where C is the cost function that the GA is supposed to optimize.

6. Conclusions

In this paper we have introduced a mathematical model of a GA and have shown that it has a closed form analytical solution. The model is based on a continuous time, infinite state-space Markov chain, and the stationary solution we obtain indicates that the joint probability distrbiution of the number of chromosmes of each type is the product of the marginal distributions of the chromosome types. Furthermore, the stationary distribution of each chromosome type is a geometric distribution whose parameter is derived from a simple non-linear equation.

This paper develops the theoretical basis for this model, and then addresses the issue of how model parameters may be derived.

We see that the closed form solution of the model can allow us to derive a gradient based algorithm for selecting parameters for a GA which will be used to solve a specific optimization problem. In particular, we derive a gradient based algorithm for selecting the fitness of each chromosome type which will yield a local minimium of the cost function. Each step of this algorithm of complexity $O(n^3)$ where n is the number of chromosme types.

These principles can be extended to select other parameters of the model and will be discussed in a subsequent paper where we will also prove mathematical issues of existence and uniqueness of the model's solution.

Bibliography

Bharucha-Reid, A.T. "Elements of the Theory of Markov Processes and their Applications," McGraw-Hill, New York, 1960.

Kernighan, B. and Lin, S. (1971) "An efficient heuristic procedure for partitioning graphs", *The Bell System Technical Journal*, February, 1970.

Holland, J. (1975) "Adaptation in Natural and Artificial Systems," The University of Michigan Press, Ann Arbor, Michigan, 1975.

De Jong, Kenneth A. (1975) "An Analysis of the Behaviour of a Class of Genetic Adaptive Systems", Doctoral Thesis, Department of Computer and Communication Sciences, University of Michigan, Ann Arbor, 1975.

Gelenbe, E. and Mitrani, I. (1980) "Analysis and Synthesis of Computer Systems", Academic Press, New York and London, 1980.

Gelenbe, E. and Pujolle, G. (1988) "Introduction to Networks of Queues", J. Wiley and Sons, New York and London, 1988.

Muehlenbein, H., Schleuter, G. and Kramm, D. (1988) "Evolution algorithms in combinatorial optimization", *Parallel Computing*, Vol 7, No 2, pp 65, 1988.

Goldberg, D. (1989) "Genetic Algorithms in Search, Optimization and Machine Learning", Addison Wesley, New York, 1989.

Gelenbe, E. (1989) "Multiprocessor Performance". J. Wiley and Sons, New York and London, 1989.

Talbi, E. and Bessière, P. (1991) "Un algorithme génétique massivement parallèle pour le problème de partitionement de graphes". *Rapport de recherche. Laboratoire de Génie Informatique de Grenoble,* 1991.

Vose, M.D. and Liepins, G.E. (1991) "Punctuated equilibria in genetic search", *Complex Systems,* Vol. 5, 31-44, 1991.

Vose, M.D. (1991) "Formalizing genetic algorithms", *Technical Report (CS-91-127), Department of Computer Science, The University of Tennessee,* 1991.

Vose, M.D. *(1993)* "Modeling simple genetic algorithms", in Whitley, L. Darrell ed.), "Foundations of Genetic Algorithms 2", Morgan Kaufmann, San Mateo, 1993.

Gelenbe, E. (1993) "Learning in the recurrent random neural network", *Neural Computation,* Vol. 5, No. 1, 154-164, 1993.

Medhi, J. (1994) "Stochastic Processes," 2nd Edition, Wiley Eastern Ltd, New Delhi, 1994.

Gunther, R. (1994) "Convergence analysis of canonical genetic algorithms", *IEEE Transactions on Neural Networks,* Vol. 5, No. 1, 96-101, 1994.

Xiaofeng Qui, F. Palmieri (1994) "Theoretical analysis of evolutionary algorithms with infinite population size, Parts I and II", *IEEE Transactions on Neural Networks,* Vol. 5, No. 1, 102-129, 1994.

Silicon Evolution

Adrian Thompson
School of Cognitive and Computing Sciences, University of Sussex, Brighton BN1 9QH, UK.
`adrianth@cogs.susx.ac.uk`, `http://www.cogs.susx.ac.uk/users/adrianth/`

Abstract

The advent of new families of reconfigurable integrated circuits makes it possible for artificial evolution to manipulate a real physical substrate to produce electronic circuits evaluated in the real world. This raises new issues about the potential nature of electronic circuits, because evolution uses no modelling, abstraction or analysis; only physical behaviour. The simplifying constraints of conventional design methodologies can be dropped, allowing evolution to exploit the full range of physical dynamics available from the silicon medium. This claim is investigated theoretically and in simulation, before presenting the first reported direct evolution of the configuration of a Field Programmable Gate Array (FPGA). Evolution is seen to harness its natural dynamics and exploit them in achieving a real-world task.

1 Introduction

There is a type of Very-Large Scale Integrated circuit (a VLSI chip) known as a Field-Programmable Gate Array (FPGA). These chips do not have a predetermined function, but provide uncommitted electronic resources which the user can configure *from software.* One can imagine the attributes of the components, and the way in which they are interconnected, to be under the control of a set of transistor switches. The states of these switches are determined by the contents of digital memory cells (similar to normal computer memory) situated next to them on the silicon. By writing to this "configuration store," the user can physically create new electronic circuits as easily as writing to computer memory.

This paper discusses the use of artificial evolution to generate these circuit configurations automatically. I argue that once evolution is instantiating individuals as real physical electronic circuits, and fitness evaluation is the process of observing the laws of physics unfolding within them in real time, a new perspective is called for. Standard views of how electronic circuits should be structured are no longer appropriate. The reconfigurable hardware is a continuous-time, analogue dynamical electronic system: a primordial soup of resources to be exploited. However, once the simplifying design constraints used by human designers are dropped, can artificial evolution effectively control the behaviour of a totally unconstrained, recurrent, arrangement of high-speed electronic components? The main body of the paper addresses this question, firstly through a simulation study and then using a real FPGA (the first reported direct evolution of an FPGA configuration). We shall see that evolution can indeed sculpt the dynamics of the system.

The next section gives a little more detail on what reconfigurable hardware is, how it can be evolved, and why. The implications of removing inappropriate design constraints are then considered, and the experiments shown to corroborate the new perspective.

2 Evolvable Hardware

The notion of *evolvable hardware,* i.e. reconfigurable hardware to which artificial evolution can be applied, has been debated to an increasing extent over the last six years (de Garis, 1991; Higuchi et al., 1993; de Garis, 1993; Mange, 1993; Hemmi et al., 1994; Marchal et al., 1994; Mizoguchi et al., 1994; Thompson, 1995a, 1995b; Hirst, 1996; Thompson et al., 1996), but only now are new FPGAs making it feasible. Previously (York, 1993), commercial off-the-shelf chips have been unsuitable because of a limit on the number of times they could be reconfigured, the presence of a layer of secret proprietary software between the user and the chip, or the possibility of easily damaging the device with an invalid configuration.

It is possible to build custom reconfigurable hardware systems out of separate components (Thompson, 1995a), and small analogue FPGAs are now available (IMP, 1994). However, it is FPGAs intended for the implementation of digital logic systems which hold the most promise in the near future. Figure 1 shows a simplified typical architecture: there is a large two-dimensional array of reconfigurable logic blocks, each with a collection of wires running past it. Reconfigurable switches (shown as dots in the figure) determine how the inputs and out-

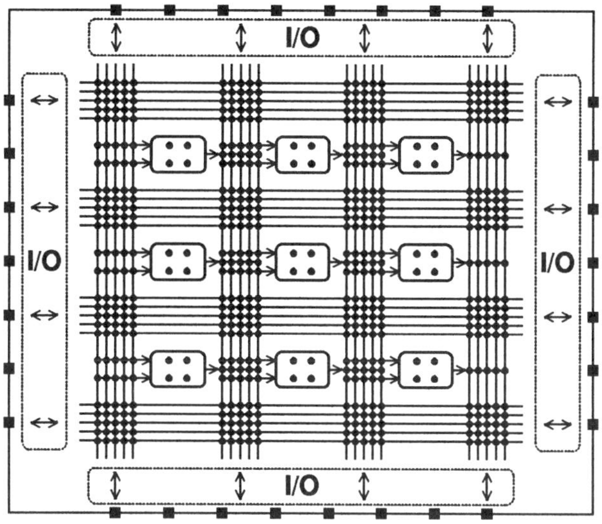

Figure 1: Part of a simplified typical FPGA architecture.

puts of the blocks are connected to the wires. Wires can also connect to other wires at some places. There are additional configuration switches within each block: in the figure, the four dots in each block represent the four bits of information required to specify which out of the set of all boolean functions of two inputs that block is to perform. There are special blocks around the periphery of the array, which interface it to the pins of the chip. Note that real commercial products are much more elaborate than this example.

The settings of the switches are stored in a configuration memory, which can be written to from software. By coding the contents of the configuration memory onto a genotype, an evolutionary algorithm can be used to automatically derive a circuit implemented on the FPGA to achieve a given task. To evaluate an individual's fitness, its genotype is expressed to produce a setting of the configuration memory; this is then applied to the chip, which instantiates the corresponding circuit. The behaviour of this real circuit is then evaluated to give a fitness score. This is sometimes called 'intrinsic' hardware evolution, as compared to the 'extrinsic' case where evolution is carried out using a simulation model, with only the final configuration being downloaded onto the chip. This paper addresses the issues associated with *intrinsic* hardware evolution.

Perhaps the most promising applications of evolved hardware are for high-speed pattern recognition and signal processing. These are domains where a hardware implementation is required because software is too slow, but yet conventional hardware design techniques are not well suited in some cases. Fitness evaluations could be extremely short, resulting in a rapid rate of evolution. Other possible applications in which traditional methods flounder are fault-tolerant design (Thompson, 1995b), design under low power/area constraints, and ill-defined tasks such as autonomous mobile robotics (Thompson, 1995a).

3 Embodiment versus Abstraction

In intrinsic hardware evolution, individuals of the evolving population receive fitness scores according to their performance when instantiated as real physical pieces of electronics. Evolution proceeds by taking note of the overall behavioural effect of variations made to the real circuits; this is very different from conventional design techniques, which proceed by manipulating abstract models.

The use of abstract models simplifies design by allowing some aspects of reality to be ignored, but the properties of the real hardware that have been 'abstracted away' must be suppressed: they must not be allowed to influence the final behaviour of the designed circuit. For example, a designer engaged in digital design does not need to think about the analogue behaviour of the transistors, but considers them as ON/OFF switches. To allow circuits designed at this level of abstraction to work in reality, the transistors must always be kept in either the ON state or the OFF state except during short transient periods while they are switching between them. Steps must be taken to ensure that these transients do not influence the overall behaviour of the system as predicted by the designer's digital model. In synchronous design, this is done by compartmentalising the system into modules which only communicate on the ticking of a clock: the transient dynamics are localised within each module and die away before the module is allowed to influence the rest of the system. Thus, both the spatial organisation and the dynamical behaviour of the circuit are constrained in order to support the designer's abstract model. This not only applies to digital design, but to all design methodologies: none can proceed far with detailed physical descriptions of the dynamical behaviour of the elementary components (e.g. transistors) before abstraction is necessary.

Hardware evolution, by observing the consequences of variations made to the real hardware, avoids the need for design abstractions and the accompanying constraints. Our view of the nature of electronic systems is heavily biased by our design methodologies and the constraints applied to facilitate their abstractions, so hardware evolution demands a radical rethink of what electronic circuits can be. Both the spatial structure (modularity) and the temporal structure (synchronisation and the rôle of phase in general) need to be considered.

3.1 Spatial Structure

As well as to support abstract models, modularity arises in designs according to the way in which the problem was decomposed. Humans typically use some sort of "divide and conquer" strategy; whether the decomposition is a functional one or a behavioural one (Brooks, 1991), the final structure arrived at usually has modules echoing that decomposition. The *evolutionary* process could also benefit from a kind of modularity, such that different phenotypic characteristics can be improved semi-independently by genetic mutations (Wagner, 1995). However, these 'modules' are not necessarily reflected in the phenotype circuit's spatial or topological structure: for example, they could correspond to *basins of attraction* in state-space. These are currently open questions, and are intimately tied to the study of *morphogenesis*: the expression of genotype by a growth process. (Thompson et al. (1996) considers these issues in more detail.) The important observation here is that any kind of modularity that *is* appropriate to evolution could be very different to that arising in design by humans. Conventional notions of modularity should not be imposed upon an evolving circuit; although human and evolutionary decomposition may sometimes coincide, evolution should ideally be allowed to explore the full range of possibilities without inheriting limitations from humans.

3.2 Temporal Structure

Real physical electronic circuits are continuous-time dynamical systems. They can display a broad range of dynamical behaviour, of which discrete-time systems, digital systems and even computational systems are but subsets. These subsets are much more amenable to non-evolutionary design techniques than dynamical electronic systems in general, because the restrictions to the dynamics brought by each subset support design abstractions, as described above. Evolution does not require abstract models, so there is no need to constrain artificially the dynamics of the reconfigurable hardware being used.

In particular, there no longer needs to be an *enforced* method of controlling the phase (temporal co-ordination) in reconfigurable hardware originally intended to implement digital designs. The phase of the system does not have to be advanced in lock-step by a global clock, nor even the local phase-controlling mechanisms of asynchronous digital design methodologies imposed. The success of pulse-stream neural networks (Murray et al., 1991; Murray, 1992), where *analogue* operations are performed using *binary* pulse-density signals, gives a clue that allowing the system's phase to unfold in real-time in a way useful to the problem at hand can add a powerful new dimension to electronic systems: time. Mead's highly successful analogue neural VLSI devices (e.g. the 'silicon retina'), exploiting the continuous-time dynamics of networks of analogue components (with the transistors mostly operating in the sub-threshold region), show how profitable an excursion into the space of general dynamical electronic systems can be (Mead, 1989).

Once unessential dynamical constraints have been removed, we are relying on evolution to arrange the dynamics of the system so that the overall behaviour is appropriate. Without evidence, it would be too much a leap of faith to expect this to work. For example, surely a complex continuous-time recurrent network of digital logic gates is doomed to fall into useless uncontrollable high-frequency oscillation? The remainder of the paper addresses exactly this issue. If evolution is able to control the dynamics of such seemingly unmanageable networks, then there is great potential for it to exploit usefully those very dynamics it has harnessed — dynamics normally precluded from contributing to the way an electronic system operates.

4 Experiment 1: Simulation

The task was to evolve a network of high-speed logic gates, in a simulation loosely based on the structure of an FPGA, to oscillate regularly at a much slower timescale than the gate delays. The task was simple, but success would demonstrate evolution's ability to manipulate dynamical time-scales.

The number of logic nodes available was fixed at 100, and the genotype determined which of the boolean functions of Table 1 was instantiated by each node, and how the nodes were connected. The nodes were analogous to the reconfigurable logic blocks of an FPGA, but an input could be connected to the output of any node without restriction. The linear bit-string genotype consisted of 101 segments (numbered 0..100 from left to right), each of which directly coded for the function of a node and the sources of its inputs, as shown in Table 2. (Node 0 was a special 'ground' node, the output of which was always clamped at logic zero.) This encoding is based on that used by Cliff, Harvey, and Husbands (1993). The source of each input was specified by counting forwards/backwards along the genotype (according to the 'Direction' bit) a certain number of segments (given by the 'Length' field), either starting from one end of the string, or starting from the current segment (dictated by the 'Addressing Mode' bit). When counting along the genotype, if one end was reached, then counting continued from the other.

At the start of the experiment, each node was assigned a real-valued propagation delay, selected uniformly randomly from the range 1.0 to 5.0 nanoseconds, and held to double precision accuracy. These delays were to be the input-output delays of the nodes during the entire experiment, no matter which functions the nodes performed. There were no delays on the interconnections. To commence a simulation of a network's behaviour, all of the outputs were set to logic zero. From that moment onwards, a standard asynchronous event-based logic simulation was performed (Miczo, 1987), with real-valued time being held to double precision accuracy. An equiv-

Name	Symbol
BUFFER	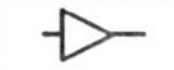
NOT	
AND	
OR	
XOR	
NAND	
NOR	
NOT-XOR	

Table 1: Node functions.

BITS	MEANING
0-4	Junk
5-7	Node Function
	POINTER TO FIRST INPUT
8	Direction
9	Addressing Mode
10-15	Length
	POINTER TO SECOND INPUT
16	Direction
17	Addressing Mode
18-23	Length

Table 2: Genotype segment for one node.

alent time-slicing simulation would have had a time-slice of 10^{-24} seconds, so the underlying synchrony of the simulating computer was only manifest at a time-scale 15 orders of magnitude smaller than the node delays, allowing the *asynchronous* dynamics of the network to be seen in the simulation. A low-pass filter mechanism meant that pulses shorter than 0.5ns never happened anywhere in the network.

The objective was for node number 100 to produce a square wave oscillation of 1kHz, which means alternately spending 0.5×10^{-3} seconds at logic **1** and at logic **0**. If k logic transitions were observed on the output of node 100 during the simulation, with the nth transition occurring at time t_n seconds, then the average error in the time spent at each level was calculated as :

$$\text{average error} = \frac{1}{k-1}\sum_{n=2}^{k} \mid (t_n - t_{n-1}) - 0.5 \times 10^{-3} \mid \tag{1}$$

For the purpose of this equation, transitions were also assumed to occur at the very beginning and end of the trial, which lasted for 10ms of simulated time. The fitness was simply the reciprocal of the average error. Networks that oscillated far too quickly or far too slowly (or not at all) had their evaluations aborted after less time than this, as soon as a good estimate of their fitness had been formed. The genetic algorithm used was a conventional generational one (Goldberg, 1989) with elitism, truncation of the five least-fit individuals, and linear rank-based selection. Population size was 30, single-point crossover probability 0.7 and mutation rate 6.0×10^{-4} per bit.

Figure 2 shows that the output of the best individual in the 40th generation (Figure 3) was approximately $4\frac{1}{2}$ thousand times slower than the best of the random initial population, and was six orders of magnitude slower than the propagation delays of the nodes. In fact, fitness was still rising at generation 40 when the experiment was stopped because of the excessive processor time needed to simulate this kind of network. This result suggests that it *is* possible for evolution to arrange for a network of high-speed components to generate much slower behaviour, without having to have constraints applied to the dynamics.

The evolved oscillators produced spike trains rather than the desired square wave. (A square wave could have been produced by the addition of a toggling flip-flop to the output, but this did not arise within the 40 generations.) Probing internal nodes indicated that *beating* between spike trains of slightly different frequencies was being used to generate a much lower frequency; beating only works for spikes, not for square waves. This does not mean that the task was easy: it is difficult for beats to reduce the frequency by the massive factor required and yet produce an output as regular as that seen in Figure 2. Beating does not just occur at a few nodes, but is distributed throughout the network: I was unable to attribute functions to particular subnetworks of components. Indeed, a heuristic graph-partitioning algorithm revealed that there are no significant 'modules' (cohesive subnetworks) in the topology of the circuit. The use of beats is essentially a continuous-time phenomenon: a highly effective solution that would not have been available via conventional methods. It is an example of evolution putting to use the complex dynamics it has harnessed.

5 Experiment 2: Reality

In the simulation experiment, we saw that evolution could control the dynamics of a noise-free purely digital network. However, real recurrent continuous-time logic networks are far from noise-free and digital. Even though the gates are nominally digital, they are essentially very high gain analogue amplifiers, and this can become a more appropriate description of their behaviour. Here, we repeat the previous experiment, but with all fitness evaluations taking place on a *real* FPGA. There is no simulation of the circuit, just real semiconductor physics.

The apparatus is shown in Figure 4. The FPGA is from the forthcoming XC6200 family of Xilinx, Inc. (Xilinx, 1995). The details of this device are confidential at

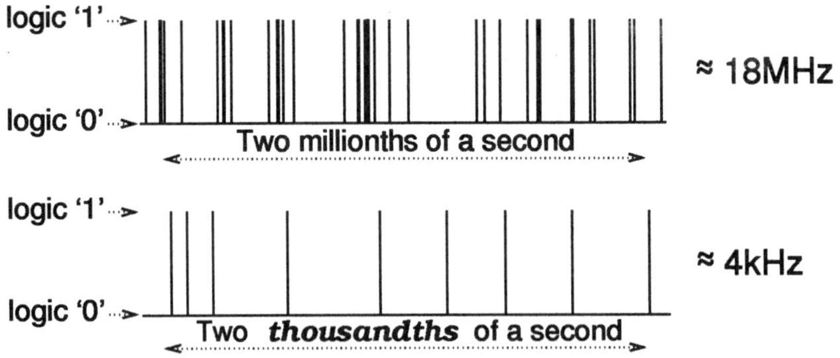

Figure 2: Output of the oscillator evolved in simulation. (Top) Best of the initial random population of 30 individuals, (Bottom) best of generation 40. Note the different time axes. A visible line is drawn for every output spike, and in the lower picture each line represents a single spike.

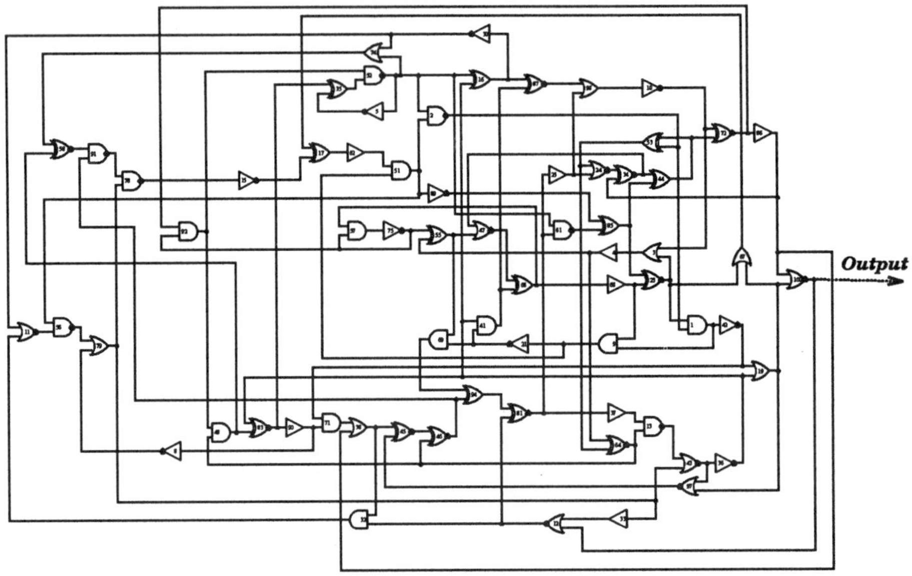

Figure 3: The 4kHz oscillator evolved in simulation (unconnected gates removed, leaving the 68 shown).

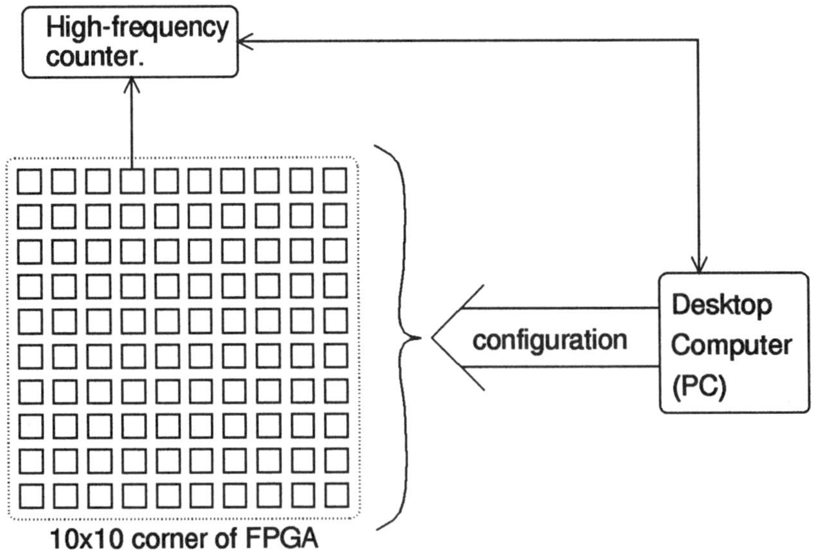

Figure 4: The experimental arrangement for oscillator evolution with the real FPGA.

the time of writing, but it is sufficient to say that a subset of its functionality was used such that it appeared quite similar to the simplified example given in Section 2. Only a 10×10 corner of a much larger chip was used, and the necessary configuration bits were directly encoded onto the linear bit-string genotype, of length 1800 bits. The GA was the same as above, except that the population size was 50 and there was no truncation in the rank selection. The bitwise mutation rate was set so as to give an expected mutation rate of 1.45 per genotype (see Thompson et al. (1996) for the theory behind this).

Each individual, once downloaded as a configuration onto the real hardware, was given a single one-second evaluation. A particular block at the edge of the array was designated to be the output (as shown), and monitored by a high-frequency counter. This counter (an HC11 micro-controller) is capable of counting positive-going logic transitions at a rate of up to 1MHz. Its count at the end of the one-second evaluation was taken as a direct measurement of frequency. Taking the fitness to be the inverse error as before, we have fitness $= 1/|f - n|$, where f is the desired frequency and n is the number of positive transitions counted over one second. If no transitions at all were counted, then the fitness was set to zero, and if $f - n$ was zero then the fitness was declared to be a million.

The initial population was hand-seeded by randomly generating a large number of circuits, inspecting their output by eye using an oscilloscope, and simply selecting those for which the output was not always constant. Starting from this same initial population, the GA was run for 40 generations for each of the target frequencies f = 10Hz, 1kHz, 100kHz, 1MHz. The measured frequency of oscillation of each individual over the course of each run is shown in Figure 5. As the diagram clearly shows, the population quickly converges on the desired frequency. The population is *not* just converging upon an individual which was already in the initial population: the maximum fitness in the population, as well as the mean, increases over time.

Although using the counter to measure frequency seems crude, there is evidence that the evolved individuals do indeed display the desired behaviour: if they are allowed to run for *two* seconds instead of one, then the number of counts measured during the second second is very close to that measured during the first. This implies a certain degree of regularity in the oscillation. Repeated evaluations over extended periods of time produce very similar readings.

How is the behaviour being achieved? Visual inspection of the output on an oscilloscope shows that *all* of the evolved solutions produce very high frequency waveforms of some kind, but that this high frequency component is not crossing the digital logic threshold, so is not being registered by the counter. Only at just the desired frequency is the signal making an excursion across the threshold long enough to be captured by the counter. All that can be seen on a cheap oscilloscope is a blur of high-frequency activity centered around an analogue voltage a short distance away from the logic threshold, with occasional faint traces of large-amplitude excursions.

Evolution really does seem to have tuned the circuit as a continuous-time analogue dynamical system, even though the manufacturers of the chip had discrete logic in mind. As an analogue system, the solution appears to be found more easily than in the equivalent purely digital noise-free simulation of the previous section. The physical characteristics of the FPGA have been exploited without any constraint whatsoever on the configurations allowed. The four runs show evolution manipulating the overall dynamics of the system on a timescale varying over five orders of magnitude.

6 Experiment 3: Application

The previous experiments show that *in principle* the evolution of unconstrained dynamical electronic circuits is possible, but the practicability of the approach in real applications is another matter. This section deals with a simple (but not 'toy') application intended as a first step into the pattern recognition domain identified earlier. Imagine a stream of binary data bits being sent down a telephone line coded as 10kHz and 1kHz tones: the task here is to evolve a circuit to decode the tones back to binary **0** and **1**. As for the oscillator experiment, a 10×10 corner of the real FPGA was used, but now with a single input as shown in Figure 6.

The GA and encoding were exactly as above, except that a random initial population was used, and the per-bit mutation rate was set such that the expected mutation rate was 2.7 per string (arrived at through trial and error). To evaluate an individual circuit, five half-second bursts of each tone were applied to the input in a random order (different each time). During each half-second interval, an analogue integrator integrated the voltage on the FPGA's output pin. At the end of each interval an analogue-to-digital conversion was performed on the integrator's output to give a measure of how high, on average, the voltage had been over the half-second. The integrator was then reset before the next tone. An individual's fitness was the absolute value of the difference between the mean integrator reading averaged over the cases when the input was 1kHz, and the mean reading when the input was 10kHz. Thus, the task was to maximise the difference in output voltage between the two cases, ideally producing a steady 0 Volts as soon as one tone is applied, and a steady 5 Volts immediately the other is present.

As shown in Figure 7, initially the circuits were as bad as possible, producing a steady 5V output in both

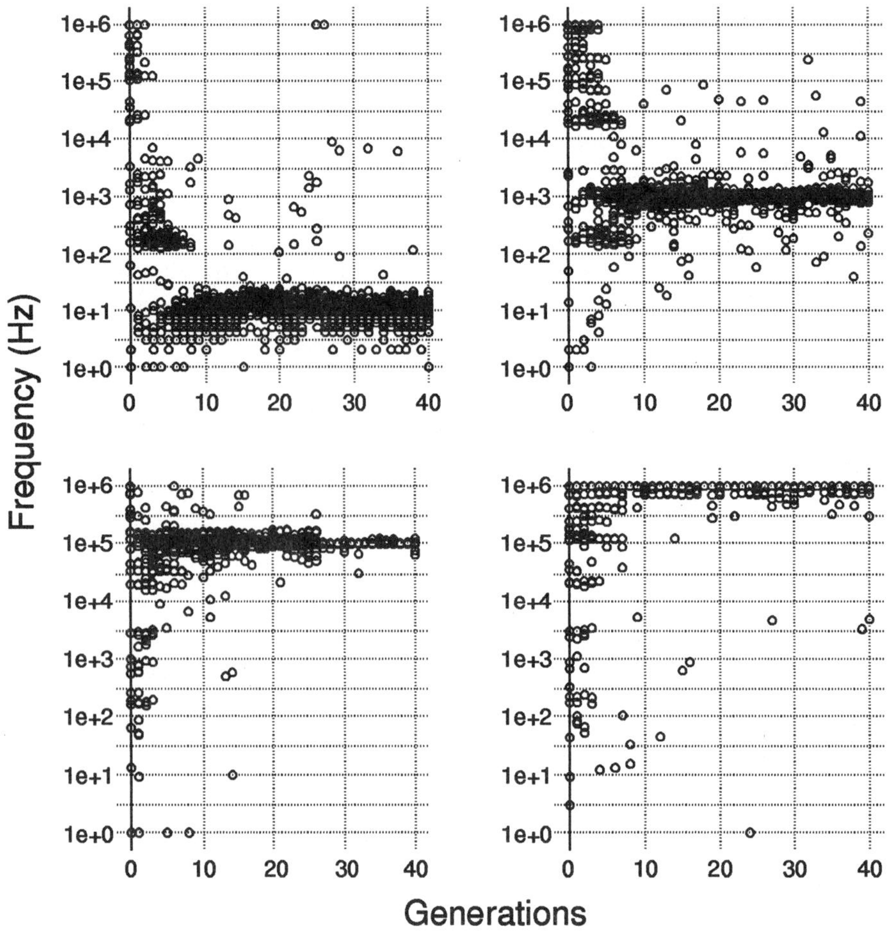

Figure 5: Frequency of oscillation of circuits in the population over the GA run (real FPGA hardware). The objectives were: Top left - 10Hz; Top right - 1kHz; Bottom left - 100kHz; Bottom right - 1MHz. Individuals with constant output (frequency = 0Hz) are not shown. Frequencies higher than 1MHz appear as exactly 1MHz due to the limited rate of the counter. Where many points are overlaid, they appear as one. Note that the frequencies are shown on a logarithmic scale. Each of these runs took just a few minutes to complete, due to the shortness of the evaluations.

cases. Over the generations, a wide range of partial solutions were seen (one is shown), usually involving complex waveforms with high-frequency components, and not respecting digital logic levels. After 3500 generations, the near-perfect solution shown was arrived at: eventually only tiny glitches (invisible in the figure) in the otherwise-perfect output were present, too small to be measurable by the integrator.

The task is simple, but it would be very difficult for a designer to solve this problem in such a small area, and with no external components. Normally, bulky external timing capacitors or an externally applied clock would be used: here evolution has arranged for the natural dynamics of the silicon to do everything on-chip. The complexity of the output seen for the intermediate solutions suggests that the circuit is exploiting rich and unusual dynamics even when producing the final orderly digital-looking output encouraged by the fitness function. This has been facilitated by the fitness evaluations being continuous-time analogue measurements.

The circuit operates over a 10 Celsius temperature range, but begins to gracefully degrade beyond this: many physical properties of VLSI chips are temperature sensitive. Ongoing research aims to tackle this problem by evaluating each individual circuit on multiple chips, each at different temperatures. This will not slow down evaluations, because the chips can be used in parallel. The objective is to evolve circuits that work over the full range of conditions (both external, like temperature, and internal properties peculiar to individual chips) which they could encounter in an industrial application.

7 Conclusion

Intrinsic hardware evolution has been presented as justifying a shift in the way electronic systems are conceptualised, towards an unconstrained exploitation of the physical dynamics. The first experiments evolv-

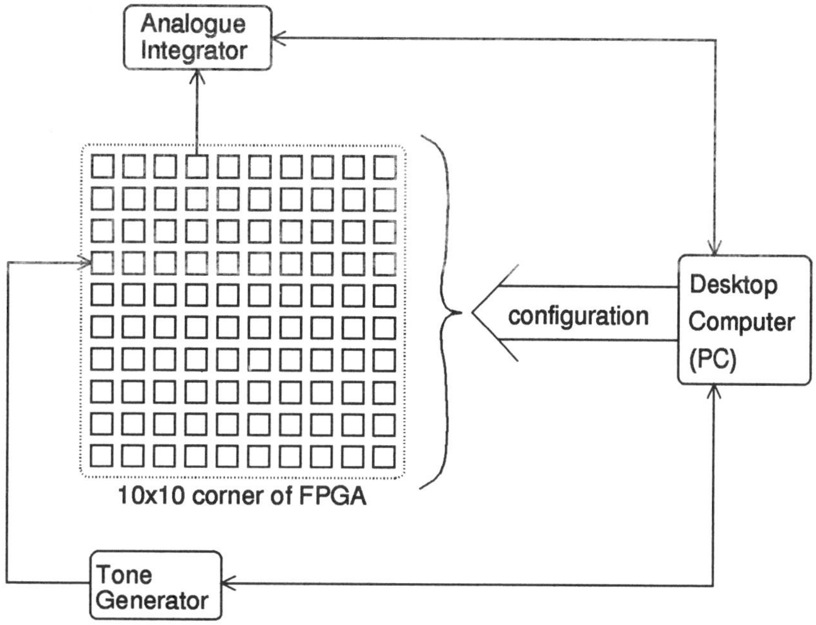

Figure 6: The experimental arrangement for evolution of the tone discriminator with the real FPGA.

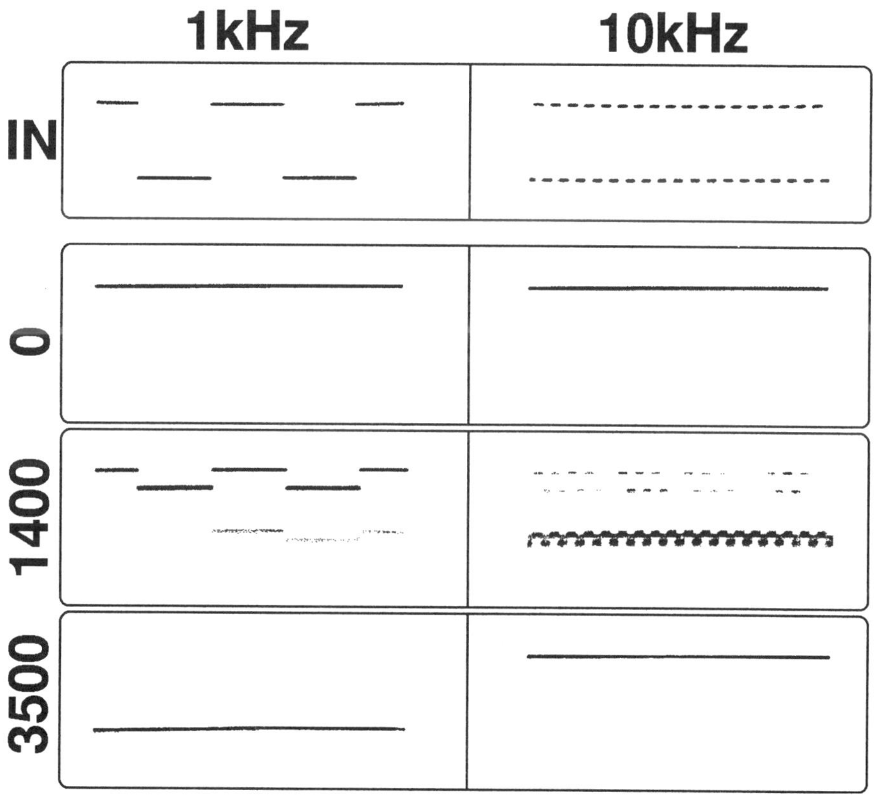

Figure 7: Photographs of the oscilloscope screen as the tone discriminator evolved on the real FPGA. (Top) the input waveforms; (Below) the corresponding output waveforms of the best of the population after 0, 1400 and 3500 generations. The probes were directly attached to the pins of the FPGA (the integrator is *not* part of the system).

ing FPGA configurations strongly reinforce the argument. High-speed pattern recognition and signal processing have been identified as particularly suitable applications; clearly there is much fundamental research still to be done.

Acknowledgements

Grateful thanks to Phil Husbands, Dave Cliff, Inman Harvey and Xilinx Scotland. Funding is by a D.Phil. scholarship from the School of Cognitive and Computing Sciences.

Bibliography

Hirst, A. J. (1996). (This paper gives an overview of the literature.) Notes on the evolution of adaptive hardware. In *Proc. of the 2nd Int. Conf. on Adaptive Computing in Engineering Design and Control (ACEDC96)*, Univ. of Plymouth UK. `http://kmi.open.ac.uk/~monty/evoladaphwpaper.html`

Brooks, R. A. (1991). Intelligence without representation. *Artificial Intelligence*, *47*, 139–159.

Cliff, Harvey, & Husbands. (1993). Explorations in evolutionary robotics. *Adaptive Behaviour*, *2*(1), 73–110.

de Garis, H. (1991). Genetic programming : Artificial nervous systems, artificial embryos and embryological electronics. In Schwefel & Männer (Eds.), *Parallel Problem Solving from Nature : Proc. of PPSN I*, pp. 117–123. Springer-Verlag.

de Garis, H. (1993). Evolvable hardware: Genetic programming of a Darwin Machine. In Reeves, Albrecht & Steele (Eds.), *Artificial Neural Nets and Genetic Algorithms - Proc. of the Int. Conf. in Innsbruck, Austria*, pp. 441–449. Springer-Verlag.

Goldberg, D. E. (1989). *Genetic Algorithms in Search, Optimisation & Machine Learning.* Addison Wesley.

Hemmi, H., Mizoguchi, J., & Shimohara, K. (1994). Development and evolution of hardware behaviours. In Brooks & Maes (Eds.), *Artificial Life IV*, pp. 371–376. MIT Press.

Higuchi, T., et al. (1993). Evolving hardware with genetic learning: A first step towards building a Darwin Machine. In *Proc. of the 2nd Int. Conf. on the Simulation of Adaptive Behaviour (SAB92).* MIT Press.

IMP, Inc. (1994). *IMP50E10 EPAC Electronically Programmable Analog Circuit: Preliminary product information sheet.*

Mange, D. (1993). Wetware as a bridge between computer engineering and biology. In *Proc. of the 2nd Eur. Conf. on Artificial Life (ECAL93)*, Brussels, pp. 658–667.

Marchal, P., et al. (1994). Embryological development on silicon. In Brooks & Maes (Eds.), *Artificial Life IV*, pp. 365–366. MIT Press.

Mead, C. A. (1989). *Analog VLSI and Neural Systems.* Addison Wesley.

Miczo, A. (1987). *Digital Logic Testing and Simulation.* Wiley New York.

Mizoguchi, J., Hemmi, H., & Shimohara, K. (1994). Production genetic algorithms for automated hardware design through an evolutionary process. In *IEEE Conference on Evolutionary Computation.*

Murray, A. F., et al. (1991). Pulsed silicon neural networks - following the biological leader. In Ramacher, & Rückert (Eds.), *VLSI Design of Neural Networks*, pp. 103–123. Kluwer Academic Publishers.

Murray, A. F. (1992). Analogue neural VLSI: Issues, trends and pulses. *Artificial Neural Networks*, *2*, 35–43.

Thompson, A. (1995a). Evolving electronic robot controllers that exploit hardware resources. In Morán et al. (Eds.), *Advances in Artificial Life: Proc. of the 3rd Eur. Conf. on Artificial Life (ECAL95)*, pp. 640–656. Springer-Verlag LNAI 929.

Thompson, A. (1995b). Evolving fault tolerant systems. In *Proc. of the 1st IEE/IEEE Int. Conf. on Genetic Algorithms in Engineering Systems (GALESIA'95), IEE Conference Publication No. 414*, pp. 524–529.

Thompson, A., Harvey, I., & Husbands, P. (1996). Unconstrained evolution and hard consequences. In Sanchez & Tomassini (Eds.), *Towards Evolvable Hardware.* Forthcoming: Springer-Verlag LNCS.

Wagner, G. P. (1995). Adaptation and the modular design of organisms. In Morán et al. (Eds.), *Advances in Artificial Life: Proc. of the 3rd Eur. Conf. on Artificial Life (ECAL'95)*, pp. 317–328. Springer-Verlag LNAI 929.

Xilinx, Inc. (1995). *The XC6200 FPGA Family Advance Product Information Sheet.*

York, T. A. (1993). Survey of field programmable logic devices. *Microprocessors and Microsystems*, *17*(7), 371–381.

Short Genetic Algorithms Papers

On Sensor Evolution in Robotics

Karthik Balakrishnan and Vasant Honavar
Artificial Intelligence Research Group
Iowa State University
Ames, IA - 50011.
balakris@cs.iastate.edu, honavar@cs.iastate.edu
http://www.cs.iastate.edu/~{balakris,honavar}/homepage.html

Abstract

In recent years, evolutionary algorithms (EAs) have been successfully used in the design of artificial neural networks for a variety of applications. The suitability of EAs for this design task stems from their ability to adaptively search large spaces in near-optimal ways. One direct application of this advance has been in the area of evolutionary robotics, where EAs are typically used for designing behavior controllers for robots and autonomous agents. While such designs have been found to work well in general, their performance is often limited by the number, placement, quality, efficacy, and reliability of the sensors that the robots are endowed with. In this paper we argue that designing the sensory systems of these robots, in addition to the usual practice of designing the controller, can lead to improvements in the performance of the robot. Our results indicate that the evolution of sensors is a useful enterprise, and can lead to efficient and often counterintuitive controller designs.

1 Introduction

Robots and autonomous agents are equipped with control mechanisms that allow them to produce specific behaviors in any given environment. Not only must these control mechanisms be efficient and robust, but also able to adapt to a changing environment. *Artificial neural networks* are therefore viable candidates for such behavioral control mechanisms [7, 5, 2].

However, realizing even the simplest of robotic behaviors requires fairly complex trade-offs among several, often conflicting, objectives. For example, to implement *wall-following* behavior, a robot has to optimize multiple criteria like efficient sensing, recognition of the wall, reasonably precise estimation of the distance to the wall, maintenance of a safe distance to avoid bumping into the wall (while keeping it within sensor range), appropriate control of speed, etc. Thus, design of appropriate neural network controllers (or neurocontrollers) to realize such behaviors is an instance of a difficult multi-criterion optimization problem.

Evolutionary Algorithms (EAs), simulated models of natural evolution, have been shown to be effective procedures for searching large, complex, multi-modal, and deceptive spaces [8, 6], and can thus be employed to solve multi-criterion optimization problems like the design of neurocontrollers for robots and autonomous agents [9, 12, 5, 10, 3]

Although a large body of work exists in this area, most of them assume a fixed sensory system architecture, while concentrating only on the design and development of the control mechanism. For example, the miniature robot Khepera [11] was used in the navigation studies in [5]. Since Khepera comes equipped with eight infra-red (IR) sensors, the neurocontrollers evolved in [5], assumed the availability of eight IR sensors. Fewer (or more) sensors might have resulted in better performance, but that dimension was not explored.

The wide variety of sensory systems represented in nature are suggestive of the power of evolution in determining good designs over many such dimensions, including — sensors of different *types* (e.g., vision, tactile, olfactory etc.), *characteristics* (e.g., range, sampling frequency, response function etc.), *numbers* (e.g., two eyes, one nose etc.), *placement* (e.g., front, rear, top etc.), modes of *sensory information integration* (e.g., through the use of appropriate neural circuits), etc.

Just as specific designs of the sensory system provide the endowed biological organisms with a survival advantage, it is natural to wonder if the performance of robots and autonomous agents would benefit from similar sensory system design considerations. This paper attempts to answer this question by using evolutionary techniques to discover good designs of sensory systems.

2 The Simulation Task

The robotic simulation that we consider in this paper is based on a task proposed by Teller [13]. Here the robot is placed in a square arena of $N \times N$ cells, in which M boxes are randomly scattered in the inner $(N-2) \times (N-2)$ grid. The arena is enclosed by *impenetrable* walls as shown in Figure 1.

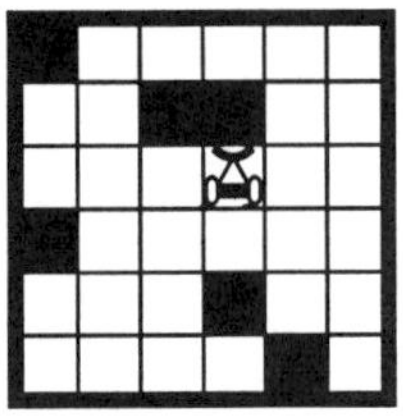

Figure 1: **The operating environment. Shaded squares represent boxes and unshaded ones, empty space.**

The robot is started off in a random position facing one of four directions — north, south, east, or west; with the task of clearing the arena by moving the boxes to the sides. As per the original specification of Teller, the robot has eight sensors which provide the robot with specific inputs so that it can detect the presence of boxes, walls, or empty spaces. The sensors are *fixed* to the robot (i.e., turn with the robot), and *sense* one square in each of the eight directions surrounding the robot. The robot determines its output action based on the information provided by the sensors, and possibly the history of its past actions.

The robot is capable of three possible actions — forward move, 90 degree anti-clockwise turn, and 90 degree clockwise turn. Turns are in-place, while move forward actions result in the robot moving ahead by one square. If there happens to be a box immediately in front of the robot when it moves forward, the box gets pushed ahead by one square. This is the primitive operation that the robot needs to clear the arena.

Attempts to push a box fail under two circumstances: the box may already be against a wall; or there may be another box adjacent to it in the direction in which the box is being pushed. The latter is tantamount to saying that the robot has the capacity to push *at most one* box at a time. The fitness of the robot is measured by the number of boxes it pushes to the walls within an allocated period of time. Our objective then, is to design a neurocontroller for this robotic bulldozer that allows it to attain a high fitness. We would also like to determine if the fitness might be improved upon by a judicious design of the number, placement, and range of sensors. For reasons mentioned in section 1, we use EAs to search for such designs.

3 Implementation Details

In our simulations we use *Genetic Algorithms* (GAs) [8, 6] to evolve sensory and neurocontroller designs. Our *genetic representation* had two parts, one encoding the placement and ranges of the sensors, and the other encoding the input connectivities of the units of the neurocontroller, as shown in Figure 2. The sensors were encoded using a 2-tuple (i, j), with the understanding that the corresponding sensor is placed to detect the square i cells *behind* and j cells to the *right* of the cell occupied by the robot. Thus, negative values of i mean the sensor is *frontal*, i.e., in front of the robot, while negative j's indicate that the sensor is positioned to the *left* of the robot. For example, a value of (-1,0) encodes a sensor that senses the square one cell to the *north* and zero cells to the *east* of the current position of a *north-facing* robot (labeled **N** in Figure 2). This sensor would detect the cell to the east if the robot turns clockwise through 90 degrees. (Henceforth, sensors will be labeled assuming a north-facing robot, with the understanding that the sensors turn with the robot.)

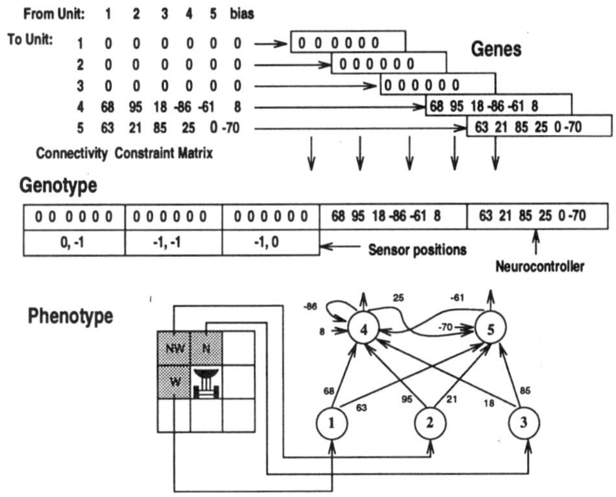

Figure 2: **Genetic representation used.**

The genetic operator *crossover* was defined to swap *logical blocks* (genes), where each block is the specification of either one sensor or one unit. *Mutation* was defined to randomly change values in the blocks, thereby producing changes in either the placement/range of sensors or in the input connections of a unit (depending on what the corresponding block encoded). Our simulations used populations of size 100, and the evolutionary runs lasted 100 generations. We used *uniform-crossover* with crossover probability of 0.5, and mutation with probability 0.1 per gene. The weights in the network were constrained to be *integers* in the range [-100$\cdots$+100], while sensor positions and ranges were bounded by the size of the arena.

Each sensor provides one of three input values to the corresponding input unit of the neurocontroller: 1 for a *box*, 0 for *empty space*, and -1 when it detects a *wall*. At each time-step, the network updates its inputs to reflect the sensor readings and each neuron in the network computes its output (by applying the threshold function to its summed input) and passes it on to each of the neurons to which it is connected. Thus, at each time step, signals are propagated through a depth of one link through the network. The output of the network at each time-step results in an (instantaneous) action by the robot. The neurocontrollers have two output units, one codes for the *action* (move vs turn), while the other determines the *direction of turn*. When the action unit produces a +1, the robot moves forward, while a -1 forces it to turn. On a turn action, the actual direction of turn is given by the output of the turn unit (-1 for clockwise and +1 for anti-clockwise).

The general framework for the simulation experiments was borrowed from [13]. Thus we used 6 × 6 arenas, with 6 boxes randomly distributed in the inner 4 × 4 grid. In each trial, the robot was placed in a randomly chosen empty cell, facing a random direction and allowed to perform $O(N^3) = 216$ actions (where each action takes one time step). For reasons mentioned in [3], Teller's simulation time estimate of $O(N^2) = 80$ was not used.

At the end of 216 time steps each box against the wall was awarded one point, while boxes in the corners fetched one extra point. This was deemed the fitness of the robot in the current trial. The overall fitness of the robot was evaluated by averaging its fitness over 100 random trials, where each trial involved placing it in a random environment and simulating it through 216 time steps as described above. This fitness value was used by evolution in the *fitness-proportionate selection* of parents.

4 Simulation Results

We report simulation results for the evolution of neural networks of the *initially feedforward* class. By this we mean that evolution starts off with feedforward networks, but through low probability mutation, retains the ability to add recurrent links, if need be. Our earlier work shows the empirical usefulness of such an approach in designing neural networks with near-minimal recurrence [3]. The absence of a link is represented by a zero value at the corresponding weight position (see, for example, Figure 2). For the purposes of keeping the analysis simple and the discussion focused, we have chosen to explore networks without any hidden units, although the addition of hidden units is seen to improve performance considerably [1]. In addition, only one kind of sensor is simulated in our experiments and the information acquisition by the sensors is assumed to be fault-free. Though this is an oversimplification, it can be addressed more rigorously by building realistic models of sensor noise and failures.

4.1 Experiment 1: Fixed Sensor Positions

In this set of experiments, evolution explores only the neurocontroller design space (while keeping the sensory system fixed). Thus, the robot had eight sensors placed to sense the eight immediate squares surrounding the robot, as shown in Figure 3.

We performed five evolutionary runs, each starting off with a different random seed (hence a different initial random population). The average fitness of the best network evolved in the five runs was 3.74, while the best network ever found had a *labeled fitness* of 5.03.

Since the fitness evaluations during evolution are based on averaging the fitness over 100 random environments, the fitness figures reported by different evolutionary runs are not truly comparable (as the 100 random environments might have been different in each run). Hence we took the best network found in each run, and reevaluated it over a chosen set of 500 environments. This fitness figure, which we call *corrected fitness*, is 4.07 for the best network. This also happens to be the network with the best labeled fitness (5.03), and is shown in Figure 3.

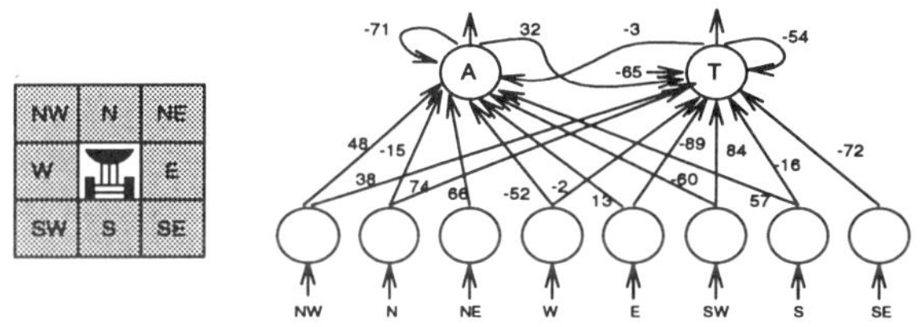

Figure 3: **Evolution of neurocontroller alone. The figure on the left shows the eight active sensors. Also notice the large negative self-loop at the action unit A.**

As observed in [3], the evolved neurocontrollers have a large negative self-loop at the output unit coding for action (labeled **A** in the figure). Such an arrangement biases the robots to interleave actions of moving forward and turning, thereby preventing them from getting into states in which they might be stuck till the end of simulation. This feature allows such robots to achieve higher fitnesses on this simulation task [3].

4.2 Experiment 2: Evolving Sensor Positions

In this experiment, evolution was allowed to operate not only on the structure of the neurocontroller, but also on the placement of the sensors. In order to determine if Teller's configuration of eight sensors [13], could be improved upon, we allowed evolution to start with random initial networks with the eight sensors placed as in Experiment 1. The sensor positions were incorporated as eight additional genes in the genotypic representation used in the above experiment. Thus, mutation could *move* sensors and crossover could exchange sensor positions between parents. We also incorporated a primitive *sensor switch* — where with *extremely low probability*, mutation could switch off sensors, thereby nullifying their effect on the behavior of the robot. The intuition here was that the evolutionary system might be forced to *co-optimize* the effective number of sensors in addition to their placement, although *no penalty was attached to the number of sensors used.*

The average fitness of the best network produced in five random runs was found to be 3.98. Even though there is little improvement in fitness over Experiment 1 (3.74), it is interesting to note that in each of the runs *at least one* sensor was switched off. In particular, one of the runs turned off *five sensors* yet attained among the highest fitnesses observed. The corresponding network and sensor placement are shown in Figure 4. This network, with a labeled fitness of 4.95, results in a corrected fitness of 4.26 (when evaluated on the *same* set of 500 random environments used in Experiment 1). One noticeable feature of this robot is that its active sensors are on its left. Thus, the robot should be more prone to clockwise turns, which would keep the object within sensor range even after a turn. This is visually confirmed by the robot runs.

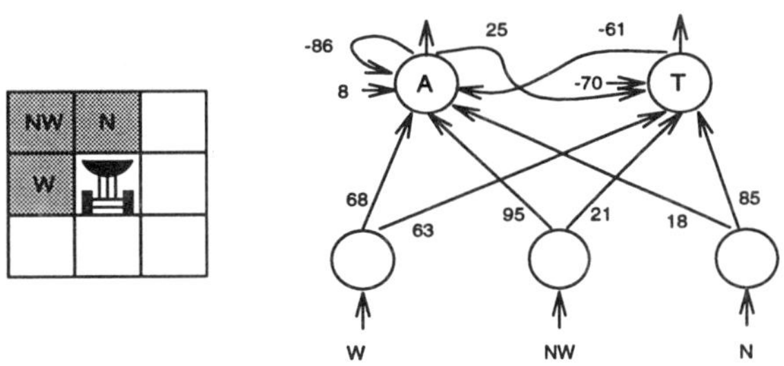

Figure 4: **Evolving controller and sensor positions. The most compact network found — only three sensors are used, the other five are discarded by evolution.**

The best network produced in this experiment had a labeled fitness of 5.41 and corrected fitness 4.74. This network, shown in Figure 5, uses only six of the eight sensors available. Further, evolution designs the network with two sensors *each* placed at the North-East (NE) and East (E) sides of the robot, thereby allowing the combined weights to exceed the [-100· · ·+100] *limit imposed by our simulation design* (in the case of the NE sensor).

The networks evolved in this experiment were observed to have the same basic design as those in Experiment

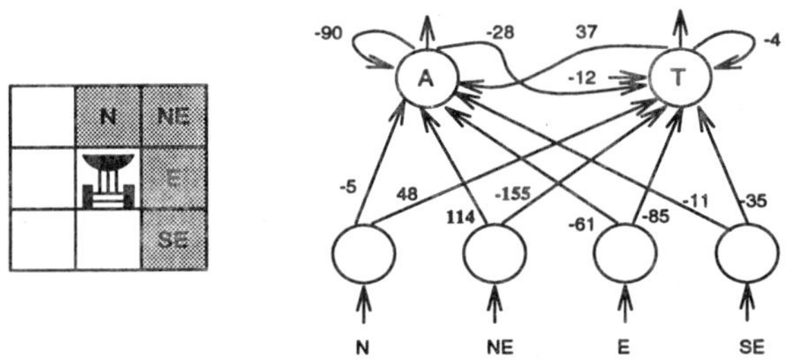

Figure 5: **Evolving controller and sensor positions. The best network discovered by evolution. Through the placement of two sensors at the NE position, evolution discovers a way of exceeding the [-100· · ·+100] limit on the weights.**

1, including the large negative self-loop at the output unit coding for action. Further, even though there was no pressure to optimize the number of sensors, evolution produced designs using fewer sensors than the eight available. It is interesting to consider possibilities of penalizing the number of sensors so as to bias evolutionary search in favor of designs with fewer sensors.

4.3 Experiment 3: Evolving Sensor Ranges

What would happen if evolution were allowed to tune the ranges of the sensors also? Would this lead to more effective designs? This experiment attempts to explore these questions.

Since all physical sensors have a maximum limit on their range, for the purposes of this study we fixed the maximum range of the simulated sensors to two cells. Note that we measure distances in terms of the number of intervening cells, not any physical metric like the Euclidean distance etc. For example, the sensor labeled 2NW in Figure 6 is considered to be two cells away from the position occupied by the robot.

In this experiment, evolution could not only rearrange the sensor placements, it could also tune their sensory ranges to zero, one, or two cells. A zero range is equivalent to a switched off sensor. Based on the evolutionary runs, the average fitness of the best networks evolved was found to be 4.28, while the best network produced by any one evolutionary run had a fitness of 5.35 (corrected fitness 4.66). The corresponding network architecture and sensor placement are shown in Figure 6.

Even in this case, the robot interleaves the move and turn motions (suggested by the weight of -100 at the action unit). In addition, this robot has two of its sensors placed behind it — a design that is highly counterintuitive considering the fact that the robot cannot move backwards. Observation of its actual behavior suggests that the robot uses its NE and SE sensors to move/orient itself such that the box is detected by the E sensor. At

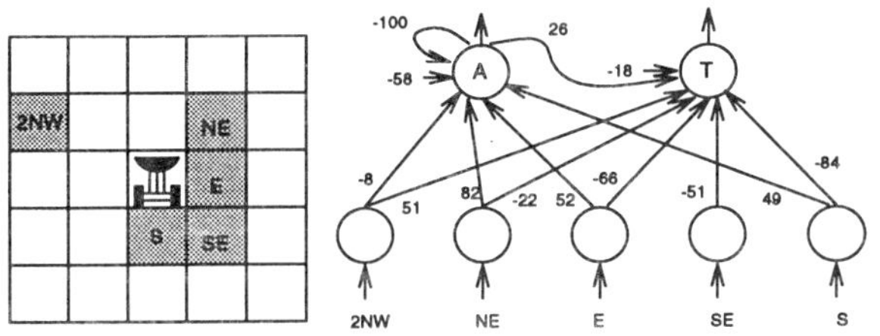

Figure 6: **Evolving controller, sensor positions, and range. The 2NW sensor allows the robot to *explore*.**

this point the robot turns towards the box, and even though it does not have a N sensor, manages to push the box given its tendency for interleaving turns and moves. The function of the 2NW sensor is twofold. When it detects the presence of a wall it suggests a *clockwise* turn to the neurocontroller, thereby moving the robot away from the wall. If it encounters a box, however, it helps the robot to turn in an *anti-clockwise* direction, i.e., towards the box. It is interesting to notice that this design too, uses fewer than the eight sensors initially provided.

5 Discussion

The results presented in this paper argue in favor of evolutionary design techniques that simultaneously explore the space of designs for sensory systems as well as controllers.

In our simulations, we observed that robots of higher fitness are produced if sensor placement and their ranges are also designed for the specific task. Further, the evolutionary system produced controllers utilizing far fewer sensors than the number actually made available. This is particularly interesting because these designs developed in the absence of any kind of penalty on the number of sensors used. It remains to be seen if explicit penalties on the number of sensors would lead to even more efficient designs.

The main limitation of the designs that were discovered in these experiments is the relatively low fitness of the robots — even the best robots are unable to clear the arena completely. The primary reason for this shortcoming is that the robots often get stuck in repetitive/cyclic paths. This can be handled by introducing noise in the system [4], thereby allowing the robots to break out of such fixed cycles. As some researchers have argued [4, 10], such noisy systems are of particular importance if controller designs evolved in simulation are to be successfully transferred on to real robots.

Sensory system design, in our view, has two main applications. First, it allows us to determine efficient, often counterintuitive designs for the control mechanism, as was seen in Experiments 2 and 3. Further, even though this paper assumed all sensors to be of the same kind, the EA-based design strategy could be easily extended to cope with sensors of different types. It could thus serve as a potentially useful approach to alleviate the problem of *sensor fusion* (how to combine/integrate information from multiple sensors, possibly of different kinds) in robotics.

Second, sensory system design is a necessary component in the development of robust, fault-tolerant systems. For example, many robots and autonomous agents are designed to operate in hazardous and unpredictable environments (nuclear waste cleanup, toxic chemical handling processes, space exploration etc.). In such applications, sensor failures might occur quite frequently. Hence there is a strong need to design systems that can operate reliably in such scenarios. This may be achieved by modeling the environment, including sensor failures, and performing studies as in Experiment 2 to evolve fault-tolerant sensor placements and appropriate controller designs.

These and other issues (e.g., fitness evaluation by placing the *real* robots directly in the physical environment of interest) are topics of ongoing research.

Acknowledgments

Vasant Honavar would like to thank the National Science Foundation for partial support of this research through grant NSF IRI-9409580.

References

[1] K. Balakrishnan and V. Honavar. Analysis of neurocontrollers designed by simulated evolution. Technical Report CS TR 95-25, Department of Computer Science, Iowa State University, Ames, IA - 50011, November 1995.

[2] K. Balakrishnan and V. Honavar. Evolutionary design of neural architectures — a preliminary taxonomy and guide to literature. Technical Report CS TR 95-01, Department of Computer Science, Iowa State University, Ames, IA - 50011, January 1995.

[3] K. Balakrishnan and V. Honavar. Analysis of neurocontrollers designed by simulated evolution. In *Proceedings of IEEE International Conference on Neural Networks ICNN'96*, 1996. To appear.

[4] D. Cliff, P. Husbands, and I. Harvey. Analysis of evolved sensory-motor controllers. In *Second European Conference on Artificial Life*, 1993.

[5] D. Floreano and F. Mondada. Automatic creation of an autonomous agent: Genetic evolution of a neural-network driven robot. In D. Cliff, P. Husbands, J-A Meyer, and S. Wilson, editors, *From Animals to Animats 3: Proceedings of the Third Inter-*

national Conference on Simulation of Adaptive Behavior, pages 421–430, Cambridge, MA, 1994. MIT Press.

[6] D. Goldberg. *Genetic Algorithms in Search, Optimization, and Machine Learning.* Addison-Wesley, Reading, MA, 1989.

[7] I. Harvey, P. Husbands, and D. Cliff. Issues in evolutionary robotics. In J. Meyer, H. Roitblat, and S. Wilson, editors, *From Animals to Animats II: Proceedings of the Second International Conference on Simulation of Adaptive Behavior*, pages 364–373, Cambridge, MA, 1992. MIT Press-Bradford Books.

[8] J. Holland. *Adaptation in Natural and Artificial Systems.* The University of Michigan Press, Ann Arbor, 1975.

[9] P. Husbands, I. Harvey, and D. Cliff. Analysing recurrent dynamical networks evolved for robot control. In *Proceedings of the 3rd IEE International Conference on Artificial Neural Networks.* IEE Press, 1993.

[10] O. Miglino, K. Nafasi, and C. Taylor. Selection for wandering behavior in a small robot. *Artificial Life*, 2(1):101–116, 1994.

[11] F. Mondada, E. Franzi, and P. Ienne. Mobile robot miniaturization: A tool for investigation in control algorithms. In *Proceedings of the Third International Symposium on Experimental Robotics*, Kyoto, Japan, 1993.

[12] S. Nolfi, J. Elman, and D. Parisi. Learning and evolution in neural networks. *Adaptive Behavior*, 3(1):5–28, 1994.

[13] A. Teller. The evolution of mental models. In Kim Kinnear, editor, *Advances in Genetic Programming*, pages 199–219. MIT Press, Cambridge, MA, 1994.

Testing software using Order-based Genetic Algorithms

Edward B. Boden and Gilford F. Martino
IBM AS/400 Endicott Programming
40DA, 257-3
1701 North Street
Endicott, NY 13760
607-752-6406
bodeneb@vnet.ibm.com and gilfordm@vnet.ibm.com

Abstract

A major strength of GAs is the ability to search large problem spaces, given a suitable fitness function. The feasibility of exploiting this strength of GAs to find software errors was explored by focusing on software APIs and commands that have completed development, hence are assumed to contain no, or very few, errors. A testing facility was developed in which a GA is used to generate API tests. The design of a fitness function which usefully guides the GA selection to find API errors, was the challenge. Initial results using the testing facility found two previously unreported exceptions in the test target API implementation, demonstrating this approach does appear to have potential.

1. Introduction

The cost to software vendors of errors in shipped products is high. Estimates of the cost to correct a single software bug range from $20,000 to $100,000. And although testing is extensive for function shipped as APIs or commands, it is far from exhaustive, due to the huge number of combinations. Hence, providers are left at the end of the development cycle with software that is likely to have some number of errors remaining (dependent of course, on size, complexity, and prior error history), and little choice but to ship anyway and treat the resulting errors as a cost of doing business (Deming, 1962). Our purpose in this study, was to determine the feasibility of using order-based genetic algorithms (GAs) to help find some of these software errors, as an adjunct to existing software test techniques and methods.

2. Problem Definition

Callable Application Programming Interfaces (APIs) are widely used by software vendors to provide function which customers and other vendors can use to build additional capability. The subroutine libraries shipped with compilers are an extremely common example of what is termed here a set of APIs. For example, the ANSI C programming language specifies a number of standard include (.h) files, which have function prototypes for well-known routines like printf(), malloc(), memcpy(), strtok() and many more. Contemporary operating systems ship many APIs (e.g. the AS/400 operating system ships over 800 (IBM 1995a)), and many middleware products and applications also ship APIs (e.g. Lotus Notes). Industry trends such as client/server, open systems, and now the internet, have all tended to drive APIs. Some examples are;

- IEEE POSIX 1001.3 (alone) specifies approximately 236 APIs (IEEE, 1990).
- Open Software Foundation's Distributed Computing Environment (OSF DCE) specifies about 455 APIs (OSF 1993).
- Microsoft's Windows 95, in such functional groups as MFC, OLE, MAPI, TAPI, etc. has over 400 API routines and object methods (SAMS 1995).

- The Taligent class summary document defines over 1100 classes, and well over 5000 methods (Taligent, 1994).

In addition to APIs, operating systems deliver function which is typically used directly by users, in the form of commands (e.g. DOS 'dir', Unix 'ls'). Even a relatively simple operating system such as PC DOS 7 has over 150 commands (IBM 1995c) and the AS/400 operating systems ships over 850 commands (IBM 1995b). Most software products (applications) also deliver function as commands. Even when the product will execute in some windowed environment (e.g. Motif or Windows 95), vendors have found that customers often want the function available by command, in addition to windows, so that the commands can be automated via scripts (e.g. Unix ksh, Rexx).

Hillis's (1990) investigations of using simulated evolution to solve an optimization problem, used co-evolving parasites which represented sorting procedures. The current work differs in that the GA is used to generate API or command sequences that cause errors in preexisting software, which is not subject to simulated evolution.

Though focused on errors in existing software, there is a relationship between the current work and genetic programming (GP) as described by Koza (1992). Rather than using programs trees as the genotype, and directly evolving them, the machine-language programs in memory (linear sequences of instructions) were the genotype studied by Nordin (1995). Genetic operators were appropriately defined: for example crossover was restricted to the machine instruction boundary, but worked on any of a selected instruction set. Nordin's 'unrestricted crossover' was 'crucial to the design', and allowed the operator to work blindly, without regard to what kind of instructions were moved. At a different level of abstraction, and hence function, the current work also employs unrestricted crossover on linear sequences of 'instructions'.

In addition to being functionally important, APIs and commands have the advantage (especially for a feasibility study) that writing test driver function is relatively simple, and requires no access to product or operating system 'internals'. This paper investigates using GAs to test APIs and commands with varying parameters to find new (unknown to development) software errors. We began with software that has already been shipped, to increase the likelihood that any error found is indeed a new error.

3. Approach

A testing facility was designed and implemented to test the feasibility of using GAs to find software bugs in API implementations. Three central design decisions for the test facility are these:

a) the chromosome encodes an ordered set of API calls or command invocations,

b) the chromosome also encodes parameter information for the API call or command invocation, and

c) the fitness function of the GA is a distributed application.

The chromosome uses order-based (Davis, 1991) encoding to represent an API call or command invocation, followed by ordered parameter codes. Encoding is at the byte level, allowing simple associative-array decoding of call and parameter codes. The genetic operations of order-based crossover and mutation (byte, non-order based) were used. (Other known order-based operations of swap and point-insertion, were not used.)

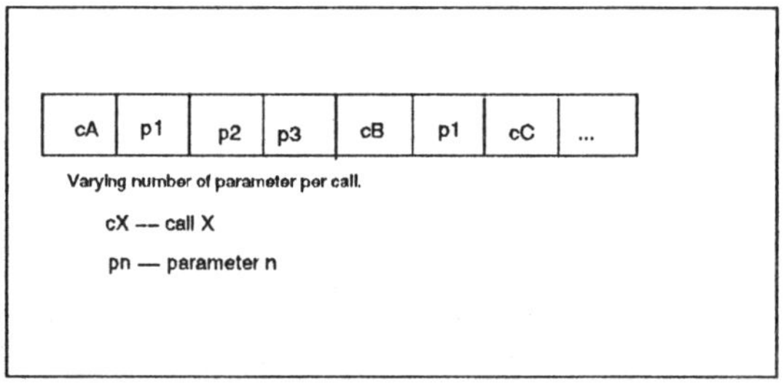

Figure 1. Order-based chromosome encoding

The GA uses standard techniques and parameters (Goldberg, 1989). For example, 80% single-point crossover, roulette wheel selection (dropping individuals with less than 10% of total fitness) and 2% (per individual) mutation, and a constant population size. The first generation is a population of randomly generated chromosomes. The cycle to process a single GA population consists of these steps;

1. On the GA-host system, each chromosome is used to derive an API test case (the phenotype)
2. each API test is sent to a test system where the API calls or command invocations are executed
3. The API test results are sent back to the GA-host system
4. The GA then evaluates the results
5. Chromosomes with higher fitness are proportionally selected for the next generation

In addition to roulette wheel selection, some initial exploration was done using two other selection techniques. Random selection was used to assist in developing a baseline for performance, in which every individual in each population was randomly chosen. And a 'top percent' selection was tried, in which the top 50% of the individuals in a population were copied into the next generation, with random individuals filling out the rest of the (constant size) population. This family of selection techniques relies on mutation and random individuals only, and does not employ crossover (adapted from Fogel, 1995).

Chromosome decoding uses various kinds of information about the API or commands being tested to create a specific API test case (phenotype). Associated with each API or command is a set of possible valid values. These value sets are used both in chromosome decoding and in the fitness evaluation function.

Using a distributed fitness function, which includes the execution of the API tests, has various advantages. The portion of the fitness function which executes on the test system has been kept to a minimum. A single GA-based test generator might drive multiple test systems. Individuals can also be sent to different (homogeneous) test systems, speeding up the evaluation function. In addition, this allows the generating system to be a different operating system than the test system, and minimizes the amount of function that is test-system specific. Further, it directly supports testing of APIs or commands that have multiple-system effects or interactions.

Despite the distributed nature of the fitness function, our experience is that 10,000s of API calls are easily generated, executed and evaluated in a few minutes. This is typically many times the size of the predefined test suite for a set of APIs.

4. Fitness function

The fitness landscape for locating actual program bugs is not well understood. It is perhaps very flat, with only rare, and very sudden, peaks. How can the GA do better than random search in such a space? We have made some hypotheses and implemented them in versions of the fitness function.

The fitness function is a weighted sum of various factors of a test response. Since each test may include many calls, each of these factors is itself a composite. The overall fitness is scaled so that the effect of numbers of calls is, by itself, not a factor. The following describes the fitness function:

```
single call fitness = W1*rcok + W2*rcnok +
                      W3*rcexcp +
                      W4*seq1 + W5*seq2  +
                      W6*seq3 + W7*seq4

test case fitness = scaled sum of single
                    call fitness's

where,
     W1-W7   weights (0-1.0)
     rcok    return code valid flag (0,1)
     rcnok   return code error flag (0,1)
     rcexcp  exception flag (0,1)
     seq1-2  sequence factors 1 to 4
```

The three "rc*" flags are mutually exclusive -- only one may have a '1' value per call or invocation. The sequence factors are explained in more detail below.

The main purpose of the study was to find previously undetected errors, so the key factor in weighing each test was the presence of any exceptions generated by the operating system. These were evaluated highly (by weight W3). (When such an individual is found, the test case (phenotype) is saved to a file, for later analysis and symptom re-creation.) Next in value are test cases with well-defined errors or return codes. Lowest are test cases with no errors. The value set associated with each API and command is used in the evaluation function, by rating more highly results outside the value set.

The rationale for evaluating defined error conditions more highly than no errors, is simply the intuition that error paths and states in the API implementation are likely to have been less exhaustively tested, as compared to the 'normal function' paths. Hence, if the GA population is steered toward individuals that exercise defined error conditions and states, the chances are better that a real exception might occur. If this hypothesis is incorrect, the fitness function weights for normal and error conditions results may be changed.

Another factor the fitness function exploits is the fact that each population individual is a sequence of API calls. Selected subsequences are identified as useful, based on expected or recommended usages for the APIs. (Note, however that this does not exclude tests that have other sequences.) As part of fitness evaluation, the response is analyzed to determine 'sequence-related' factors such as;

- the number of selected subsequences started
- the number of selected subsequences completed
- the length of the longest subsequence
- the length of the longest complete subsequence

These factors are weighted separately.

We use the term call-space for the set of permutations of N calls of APIs or command invocations, taken R at a time, with replacement. So the call-space has N**R elements. The motivation for using sequences as part of the evaluation is a) to reduce the amount of time spent searching highly unlikely areas of the huge (O(10**20)) space and b) the GA's search is steered to areas of the call-space that actual users of the API are likely to work in, hence improving the value of any previously unknown errors. (Not all errors are of equal cost to fix. It is standard practice in the industry to postpone fixing some problems until a later release, as a way of reducing costs. This may be done because the problem is extremely rare, there is simple work-around, or the consequences are trivial.)

5. Results

We implemented test driver code for 18 'Unix-style' APIs with an average number of parameters per API of 2.67, and assumed only 3 valid values per parameter. The maximum number of calls allowed in a single chromosome was 15. (Hence search space is over 18**15, without considering effect of varying parameter values.)

On numerous runs of the GA, the population would reach a plateau on average fitness and the maximum fitness individual. This apparently would persist indefinitely and suggests that the GA operators can be improved, perhaps with simply a higher mutation rate. The following figure shows summary population information for 10 generations of 18 test individuals ('inds') each. The 'avg eval seconds' is the average evaluation duration per individual. Notice that one individual in generation 1 was not completed, due to test system load.

pop#	avg f	highest f	avg Ncall	avg eval seconds	# inds complete
0	2.05198	8.82857	5.83	0	18
1	3.33577	8.84286	7	0.722	17
2	4.69483	8.82857	6.78	0	18
3	3.93402	8.82857	6.56	0	18
4	5.25653	8.82857	6	0.0556	18
5	5.92163	8.82857	5.17	0	18
6	5.97712	11.2286	4.56	0	18
7	8.16717	11.2286	5.89	0	18
8	7.48968	11.2286	5.67	0	18
9	8.33611	11.2286	6.33	0.0556	18
					179

Figure 2. Example population summary

After a few days of executing various test runs, with populations of 15-30 run over 10-50 generations, an operating system generated exception was found in one of the target APIs. Later, a second exception was found. The target APIs have been in the operating system for two releases, through two cycles of component and system test, in addition to usage by customer applications. These are both apparently new exceptions in the API implementation, not previously reported by customers and unknown to development.

Comparisons of the three selection techniques (roulette wheel, top percent and random) have shown these initial results;

- All three selection methods were able to find the same basic set of exceptions.
- The top percent (50%, without crossover) seemed to find exceptions faster than roulette wheel. This was expressed as average number of individual chromosomes, across populations. (Population size was constant per run and different runs were made with different population sizes.)
- Varying population sizes (4, 6, 20, 30) had the expected effect within a selection method and did not effect the relative results across selection methods.
- Roulette wheel selection could be run for as much 5 minutes without finding a exception. In contrast, Top percent and random selection rarely ran more than 3 minutes without finding a exception.

Typically, runs went 100-400 generations and lasted less than five minutes, until reaching a exception. During the testing, the exceptions were maintained so that additional results could be compared to the same target APIs and exceptions.

Preliminary analysis by development into the problems indicates that they are due to unexpected internal interactions within the API implementation, which maintains state information and relates to the high volume of calls generated. Essentially, the GA 'wandered' into untested portions of the API call-space. Our conclusion is that GAs may indeed be a viable addition to existing software testing practices.

6. Next steps

The utility of this application of GAs should be explored over a broader range of APIs, object methods and commands, including combinations. (Is it true, for example, that error conditions tend to have more bugs than normal paths?) The fitness function can also be changed to use additional information about the APIs or commands. (What are characteristics of the fitness landscape near by the major exceptions?) More research-oriented questions involve further analysis of the roulette wheel and top percent selection methods, and their relative performance. Additional variations on the GA operators such as tournament selection, including the mutation percentage in the chromosome, and multi-point crossover should be studied to determine if search time decreases. Other known order-based genetic operations (Davis, 1991) such as swap or point-insertion should be considered.

To readily accomplish this, additional controls need to be added to the portion of the system that acquires information about the target invocation set and the fitness function.

Aylor et al. (1991) studied the use of genetic algorithms for test vector generation for 'combinational' VLSI circuits. They first generated test sets using random methods, then used GAs to optimize and reduce overlapping coverage. They found consistently better test sets, in terms of test set size, with no loss of fault coverage, when run on two well-known benchmark circuits. The current work uses GAs to generate tests directly, and the tests are saved. Additional compacting of such saved tests may also improve performance.

Consideration can also be given to broader use of GAs in testing, perhaps to earlier phases of development, such as unit test.

Acknowledgements

To Karl Sims, for the glimmer of an idea: during his presentation "Evolving Virtual Creatures with directed graph genotypes" (ICGA '95), Karl spoke of creatures which accelerated off into infinity by hitting themselves with their own arm(!), due to a bug in his implementation of (the simulation of) conservation of momentum!

To Peter Angeline, for his insights into the literature, encouragement and incisive comments on an early draft.

References

Aylor, J.H., Cohoon, J.P., Feldhousen, E.L., Johnson, B.W., 1991. GATE--A Genetic Algorithm for Compacting Randomly Generated Test Sets. *International Journal of Computer Aided VLSI Design*, Vol 3, pp259-272.

Davis, L. ed. (1991), *Handbook of Genetic Algorithms*, New York, NY: Van Nostrand Reinhold.

Deming, W.E. (1962), *Out of the Crisis*, Cambridge, MA: MIT Press.

Fogel, D.B. (1995). *Evolutionary Computation.* New York, NY: IEEE.

Goldberg, D. (1989), *Genetic Algorithms in Search, Optimization & Machine Learning*, Reading, MA: Addison-Wesley Publishing Company, Inc.

Hillis, D. (1990), "Co-evolving Parasites Improve Simulated Evolution as an Optimization Procedure", in *Artificial Life II*, edited by C. Langton, C. Taylor, J. Farmer and S. Rasmussen. Reading, MA: Addison-Wesley Publishing Company, Inc.

IBM (1995a), *AS/400 Advanced Series System API Reference*, SC41-4801, Armonk NY: IBM Inc.

IBM (1995b), *AS/400 Advanced Series System Command Language Reference*, SC41-4722, Armonk NY: IBM Inc.

IBM (1995c), *PC DOS 7 User Guide*, S83G-9309, Armonk NY: IBM Inc.

IEEE (1990), *POSIX.1 ANSI/IEEE 1003.1*, Piscataway, NJ: IEEE Service Center.

Jones, T. (1995), "Crossover, Macromutation, and Population-based Search", in *Proceedings of the Sixth International Conference on Genetic Algorithms*, San Francisco, CA: Morgan Kaufmann Publishers Inc.

Koza, J.R. (1992), *Genetic Programming: On the Programming of Computers by Means of Natural Selection*, Cambridge, MA: MIT Press.

Nordin, P., Banzhaf, W. (1995), "Evolving Turing-Complete Programs for a Register Machine with Self-modifying Code", in *Proceedings of the Sixth International Conference on Genetic Algorithms*, San Francisco, CA: Morgan Kaufmann Publishers Inc.

SAMS (1995), *Programming Windows 95 Unleashed*, Indianapolis, IN: SAMS Publishing.

OSF (1993), *OSF DCE Application Development Reference*, Englewood Cliffs, NJ: Prentice Hall.

Taligent (1994), *Taligent Class Summary*, 10201 North De Anza Boulevard, Cupertino CA: Taligent Inc.

Optimizing Local Area Networks Using Genetic Algorithms

Andy Choi
Hewlett-Packard, Roseville Networks Division
8000 Foothills Blvd. M/S 5558
Roseville, CA 95747-5558
andyc@hprnd.rose.hp.com

ABSTRACT

This paper describes a genetic algorithm approach to the real-time optimization of a class of Local Area Network (LAN) topologies under measured traffic patterns. This approach consists of two parts: 1) a genetic algorithm approach to optimizing LAN topologies; and 2) a genetic programming approach to discovering good fitness functions for part (1). Results are given for applying the approach to a particular instance of the LAN optimization problem.

1. Introduction

Network management tools and techniques have become increasingly important in recent years due to society's increasing reliance on computer networks. Most network management tools currently available address problems of monitoring and trouble-shooting and follow the paradigm of correcting problems as they occur. In contrast, relatively few tools address the problem of network topology design, and virtually no tools exist that directly aid the network administrator in on-going network redesign. The topology of a network is the first factor in determining its efficiency and consequently its cost. Apart from the more costly solution of upgrading a network to a faster medium, the primary method of improving a network's performance is to divide or repartition its segments. Segmentation introduces the problem of deciding how to assign network nodes across the resulting segments, a problem whose optimal solution depends on the actual traffic on a network at any given time. This assignment problem is NP-complete. The real-time nature of this problem and its difficulty together help explain why so few network management tools exist to address it. The other explaining factor for the paucity of tools is that the enabling technologies needed to apply any technique have only recently been developed. In particular, traffic monitoring technologies with which to gather the real-time problem input have only recently become widespread, whereas dynamically reconfigurable interconnect devices that can be used to effect node re-assignments in real-time have only been introduced in the past year.

This paper examines a genetic algorithm approach to solving the node assignment problem and gives results for a particular instance of the problem. Sections 2 and 3 provide more background and formally define the problem. Section 4 describes the methodology, while sections 5 and 6 provide results of applying the approach to the chosen problem instance.

2. Background

Almost all local area networks are implemented using a shared-bandwidth technology such as CSMA/CD or token-ring. In these networks, multiple nodes are interconnected using devices called "hubs." A group of nodes interconnected only by hubs is collectively called a segment. Because nodes in a segment share the network bandwidth, only one node may transmit information (network packets) at a time. With more nodes and more information to transmit, LANs become increasingly "congested," introducing communication latencies and, in extreme cases, eventually ceasing to operate altogether. To increase network capacity and performance, nodes are divided into multiple segments, and the segments are interconnected using devices called "switches." A switch simply forwards packets between two segments when the transmitting and receiving nodes are on separate segments; packets whose source and destination reside on the same segment are isolated to the single segment. Traffic between segments is referred to as "intersegment traffic," while traffic isolated to one segment is called "intrasegment traffic." A new type of device called a "port-mobile hub" can be used to dynamically group nodes into segments. Instead of connecting nodes into a single segment, a port-mobile hub can group connected nodes into multiple segments; the specific groupings can be controlled programmatically from a remote management station.

Care must be taken when dividing nodes into multiple segments, as network traffic between segments incurs a small latency and consumes bandwidth in all segments that the traffic traverses. Also, higher bandwidth utilization can result in increased latency. A network administrator initially groups nodes into multiple segments based on knowledge of the expected node-to-node traffic patterns. For example, a network administrator might segment a LAN according to the functional divisions of a company and assign each project team to a different segment. When redesigning a LAN (regrouping segments), a network administrator can also make use of actual network statistics gathered during the previous configuration (e.g., Jedwab 1992).

Efficient LAN monitoring technologies and dynamically reconfigurable devices may enable LANs in some respects to become self-configuring and "self-healing," a goal which has traditionally been a holy grail of network management. This paradigm is shown in Figure 1.

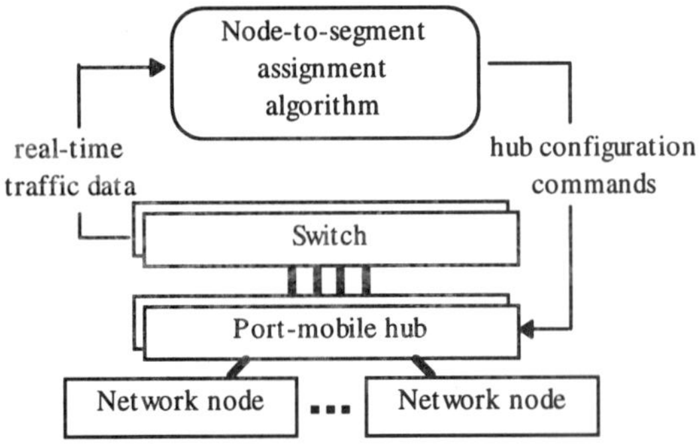

Figure 1. Self-configuring network paradigm.

3. Problem Statement

The network node assignment problem is a combinatorial optimization problem that can be defined as follows:

> Given a topology T of n nodes $\{N_1, N_2, ..., N_n\}$ and s segments $\{S_1, S_2, ..., S_s\}$ and an n-by-n matrix M of node-to-node traffic $N_i \rightarrow N_j$, find a mapping $N_i \rightarrow S_j$ such that the objective function f = (SA + SD) is minimized, where SA is the average intrasegment traffic and SD is the intrasegment traffic standard deviation.

Note that there is no limit on the number of nodes that may be mapped to a given segment; this reflects the functionality of port-mobile hubs. Ideally, the objective function would be a direct measure of end-to-end latency for the traffic matrix on a real-time simulation of the network configuration under evaluation. Since this is not feasable, the objective function (SA + SD) is chosen as an indirect approximation. The inclusion of SD in the objective function ensures that traffic is evenly distributed across the segments; without it, all nodes would be mapped to a single segment.

The network node problem is NP-complete (Garey 1979). Moreover, the objective function has many local minima when applied to traffic matrices collected from any real network (Fowler 1991).

The specific problem instance examined in this paper uses 4 segments and 48 nodes, corresponding to an actual network using a switch and a 48 port port-mobile hub with 4 internal segments (standard equipment).

Three traffic matrix test cases are used to evaluate the performance of a given node-assignment algorithm. Each of the first two matrices was created by monitoring a very congested Ethernet LAN for an 8 hour period and then extracting traffic data for the 48 nodes that communicated most with each other. The third traffic matrix was hand-crafted to create a difficult solution space for a genetic algorithm approach. The matrix represents 1 second of traffic for a network whose optimal node assignment divides the 48 nodes evenly among the 4 segments. Each node communicates 0.08Mb (megabits) with each of the other nodes in its own segment and communicates 0.01Mb with each node not in its own segment. The optimal node assignment results in a intrasegment traffic average of 9.6Mb with 0 deviation, corresponding to 1 very busy second of traffic on a standard 10Mb/s Ethernet network. The matrix has 4! = 24 optimal node assignments (out of a possible 4^{48}) due to mirror node assignments. The important aspect of this third matrix is that the difference between intersegment and intrasegment traffic between nodes is relatively small; the result of this is that single node assignment changes often lead the objective function towards non-global minima. Thus, the matrix is "deceptive" (though not fully deceptive) in the sense of (Goldberg 1989).

4. Methods

A combination of genetic algorithm (GA) and genetic programming (GP) techniques are used in the solution approach to the node assignment problem. The GA coding of the problem is fairly straightforward, with the exception that special measures are taken to overcome the typical problems encountered by GAs when applied to combinatorial optimization problems. The other interesting aspect of the GA coding is that instead of using the objective function as the fitness function, GP is used to attempt to evolve a fitness function that performs better. The complex relationship between SA and SD in the objective function when applied to a given traffic matrix suggests the possibility that GP can find a fitness function that, while still having the same global minimum as the objective function, causes the GA to converge more quickly. In addition, there are other related factors that might be useful in evolving such a fitness function, including intersegment traffic average and standard deviations as well as inter- and intrasegment traffic maximums. Changing the fitness function throughout a GA run may even prove useful, suggesting the current generation number as a possible factor in the fitness function.

The GA algorithms were written in C from scratch, whereas the GP algorithms make use of the DGPC (Dave's Genetic Programming in C) package. All experiments were run on an HP 715/50 workstation running HP-UX 9.05.

4.1 Genetic Algorithm Encoding for the Node Assignment Problem

A tableau for the GA coding is given in the following table.

Table 1. Tableau for the node assignment problem.

Objective:	Minimize the function f = (SA + SD).
Representation scheme:	· Structure: Fixed-length string, with $K = 4$, $L = 48$ · Mapping: Each of the 48 genes represents a node assignment. The gene loci correspond to indices in traffic matrix test cases, while alleles correspond to segment indices.
Fitness cases:	1 fitness case, chosen from the 3 traffic matrices described above.
Fitness (= raw fitness):	Fitness function is evolved through symbolic regression, as described in the next section. Lower values of the fitness function indicate better fitness (negative values are allowed).

Parameters:	$M = 250$, $G = 200$, $p_c = 0.6$, $p_m = 0.02$
Termination criteria:	The GA has run for G generations.
Result designation:	The best-so-far individual in the last generation population.

The representation is a fixed-length string of 48 base-4 digits. The string is interpreted as 48 node assignments across 4 segments, where string indices match node indices and digit values within the string represent segment indices. In the following example string, the nodes are mapped evenly across the four segments: nodes 1 to 12 are mapped to segment 1, nodes 13 to 24 map to segment 2, nodes 25 to 36 map to segment 3, and nodes 37 to 48 map to segment 4.

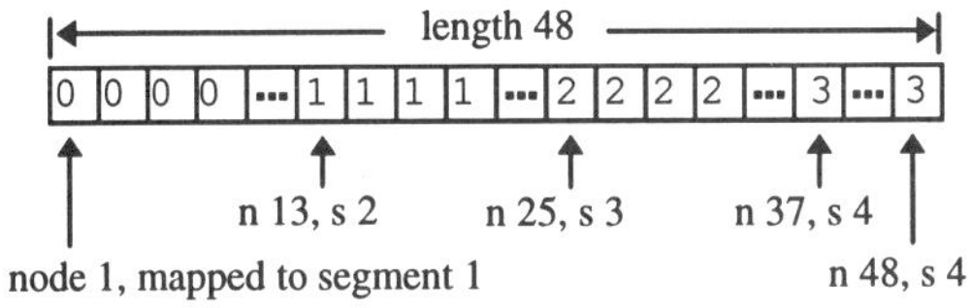

Figure 2. Example chromosome.

Note that each string is equivalent in fitness to 24 other strings, since only node groupings are important; which segment a given group of nodes is mapped to is unimportant. A base-4 encoding was chosen instead of a binary encoding, since the desired building blocks are composed of complete node-to-segment mappings, and individual allele bits have no meaning by themselves.

Node-to-segment assignments are traffic-dependent, so the GA is run using one traffic matrix at a time. To evaluate a fitness function, the traffic matrix fitness case is mapped onto the topology implied by the given chromosome. Then, each of the traffic statistics are computed. For example, for the chromosome in Figure 2, the first 12 genes map the first 12 node indices in a traffic matrix to segment 1; thus, the intrasegment traffic for segment 1 of traffic matrix M_1 is as follows:

$$\sum_{i=1}^{12}\sum_{j=1}^{12} M_1[i][j] + \sum_{i=1}^{12}\sum_{j=12}^{48} M_1[i][j] + \sum_{i=12}^{48}\sum_{j=1}^{12} M_1[i][j]$$

As stated, the deceptive matrix is designed to result in an optimal intrasegment traffic average of 9.6Mb. The other two matrices are normalized such that the lower bound of intrasegment traffic average is 10Mb (the optimal intrasegment average may not reach the lower bound).

Typical GA parameters were chosen: 200 generations of population size 250 with 0.6 probability of crossover and 0.02 probability of single-gene mutation. The seed population is composed of 125 randomly generated individuals and 125 individuals generated from the example chromosome in Figure 2 with 0.5 probability of single-gene mutation. As with other combinatorial optimization problems, the likelihood is high that chromosomes evolved from good chromosome parents will have lower fitness (more so than in non-combinatorial optimization GAs). Thus, both steady-state and elitist reproduction strategies are used. In the elitist strategy used, the best chromosome of the last generation replaces the worst chromosome of the current generation. Also, fitness scaling is used to emphasize fitness differences, as the difference in intrasegment traffic average between an optimal node allocation and a good one is unlikely to be greater than a few Mb. The scaling operator used simply offsets all fitness values by the best fitness, effectively basing the best fitness at 0. A two point crossover operator is used, with modifications described below.

Standard GAs typically perform poorly in combinatorial optimization, because they are not designed to operate directly on gene relationships (i.e., combinations). In the node assignment problem encoding, genes that are related may or may not be adjacent to each other in the chromosome encoding; i.e., the building blocks may be "disrupted." For example, two nodes that belong in the same segment may correspond to genes that are at opposite ends of the chromosome. Furthermore, even if a set of related genes are grouped together in a chromosome, their specific values may prevent their relation from being expressed during the crossover operation. For example, if the partial genes (000212) and (121333) are crossed between genes 3 and 4, the resulting children are (000333) and (121212). In this example, high fitness of the children results from the relationships expressed in the (000) and (333) sequences, yet the standard crossover function copies gene values without regard to relationships. Both these problems are addressed by a modified two-point crossover operator.

To partially address the second problem, parent chromosomes are first "aligned" before being crossed-over as follows:

```
Given parent chromosomes p1 and p2:
  - Find all allele tuples <a1_i, a2_i>, where:
       a1_i is the value of gene i of p1, and
       a2_i is the value of gene i of p2.
  - Choose the allele tuple <x, y> that occurs
    most frequently.
  - Change all occurrence of allele x in parent
    p1 to y and all occurrences of y in p1 to x.
```

In the example in Figure 3, the most frequent allele combination is <0, 3>, so each gene in parent 1 with value 0 is changed to value 3, and vice-versa. Crossover at the point between genes 4 and 5 then produces the children as shown.

parent-1:	00230000 ···	parent-2:	33330123 ···
parent-1':	3320	3333 ···	parent-2:	3333	0123 ···
child-1:	33200123 ···	child-2:	33333333 ···

Figure 3. Example partial genes undergoing alignment.

In effect, the alignment operator identifies a likely reason that explains the fitness of both of the parent genes. In terms of schema, the operation maximizes the specificity of the most specific schema (i.e., the one with fewest don't-care genes). In general, the aligned gene may actually have fewer genes in common with the other parent, in which case the original, unaligned gene is used in the crossover operation. While the effect of alignment is not usually as pronounced as in the example above, it should never be harmful, since alignment of a parent chromosome only changes gene values and not gene relationships.

To address the other problem posed by combinatorial optimization, building-block "disruption," chromosome encodings are co-evolved using the SYMBIOT method described in Paredis (1995). In SYMBIOT, parent chromosomes undergo a permutation prior to crossover. The specific permutation is chosen from a population of possible permutations that are co-evolved with the node-assignment chromosome population. After crossover, the resulting children undergo the reverse of the permutation applied to their parents. The children's fitnesses are then computed and the value ((child1-fitness + child2-fitness)/(parent1-fitness + parent2-fitness)) is "pushed" onto the "fitness history" of the selected permutation. During the evolution cycle of the permutation population, permutations are selected according to their fitness, which is computed as the average of the fitness values pushed onto them (10 fitness values are stored in each permutation chromosome's history); the history average is used to filter out noise. Permutation chromosomes are interpreted as mapping gene indices to gene values; for example, the chromosome (3210) implies the following permutations: (allele at gene index 0 replaces allele at gene index 3), (1→2), (2→1), (3→0). Permutation crossover is performed using cycle crossover (Goldberg 1989). If the permutation evolution converges, then by the end of the node-assignment problem evolution, the permutation selected most frequently should typically result in maximal building blocks in the node-assignment chromosomes. See Paredis (1995) for more details on the SYMBIOT algorithm.

4.2 Genetic Programming Encoding for Evolving Fitness Functions

A tableau for the GP coding is given in the following table.

Table 2. Tableau for evolving the fitness function for the node assignment problem.

Objective:	Find a good fitness function to use in the node assignment problem.
Terminal set:	SA (intrasegment traffic average), SD (intraseg. traf. standard deviation), SM (intraseg. traf. maximum), IA (interseg. traf. avg.), ID (interseg. traf. std. dev.), IM (interseg. traf. max.), GEN (generation), 250 random floating-point constants from the range [0, 1000].
Function set:	+, -, *, % (restricted division)
Fitness cases:	3 traffic matrices.
Fitness (= raw fitness):	The GA for the node assignment problem, as described above, is run using the current GP-evolved fitness function. The fitness is the average of the objective function (f = SD+SA) applied to the result of the GA when run on each of the three traffic matrices.
Parameters:	$M = 500$, $G = 20$, $p_{c\text{-}leaf} = 0.1$, $p_{c\text{-}node} = 0.8$, $p_m = 0$
Termination criteria:	The GP has run for G generations.
Result designation:	The best-so-far individual in the last generation population.

Evolution of the fitness function for the node assignment problem is coded in GP as a straightforward symbolic regression. As described previously, the terminal set includes not only intersegment traffic average (SA) and intersegment traffic standard deviation (SD), but also: intersegment traffic maximum (SM), intrasegment average (IA), intrasegment deviation (ID), intrasegment maximum (IM), and current generation (GEN), as well as 250 random floating-point constants. While SA, SD, SM, IA, ID, and IM are all inter-related, different combinations of the factors may lead to a better fitness function. GEN is included to allow for the discovery of GA evolution strategies in which pressure is applied to different factors throughout the evolution process. For example, the fitness function f = ((250-GEN)*(IA+ID) + GEN*(SA+SD)) would stress intersegment statistics at the beginning of a run but stress intrasegment statistics at the end; such a strategy might result in more node grouping in the beginning and then a gradual node redistribution towards the end. The [0, 1000] range of the floating-point constants was somewhat arbitrary.

The GP evolution uses one result-producing branch (RPB) of 50 nodes maximum and one automatically-defined-function (ADF) of 50 nodes maximum. The fitness of an evolved S-expression is computed by using it as the fitness function in one run of the node-assignment GA for each of the three traffic matrix fitness cases; the fitness is the average of the objective function (f = SA+SD) applied to each of the three GA-evolved genes and their respective traffic matrices. DGPC is seeded with the objective function, so that the best S-expression of generation 0 performs at least as well as the objective function.

Most of the default DGPC evolution parameters were used: MaxNewTreeDepth = 5, MaxDepthForMutation = 5, copy_fraction = 0.1, crossover_fraction_for_leaves = 0.1, crossover_fraction_for_node = 0.8, mutation_fraction = 0, GrowthMethod = 3. A non-default population size of 250 was used, with a maximum of 20 generations. The low population size is required, since the fitness evaluation is very time-intensive.

5. Results

The GP fitness function evolution was run once to completion, requiring over 6 days. Average and best-of-generation fitness are shown in Figure 5. GP fitness evolution.; for comparison, fitness is also shown for the GA when using the objective function as the fitness function. GA fitness with the objective function is 12.44. The best-of-generation 0 evolved function has fitness 12.25; fitness improves steadily until generation 7 and then reaches a minimum of 11.34 at generation 13. The difference of (12.25 - 11.34) corresponds in the network topology to a very significant ~0.9Mb/s bandwidth improvement. Although the lineages of the best-of-generation individuals were not recorded, the individuals themselves are listed in Figure 4 with a few possible gene inheritance paths highlighted.

All figures showing GA performance for the node-assignment problem graph the objective fitness of the best-of-generation gene; i.e., regardless of which fitness function is actually used, the graphed fitness shown corresponds to the objective function. Except when noted otherwise, the fitness shown is the average fitness over the three GA runs (for the three traffic matrices), using the best GP-evolved fitness function. Each run of the node-assignment GA required about 40 seconds.

Figure 7 compares the best-of-generation fitness for the GA using the GP-evolved function and the objective function as the fitness function. As shown, the evolved function performs significantly better. Performance of the GA when using the objective function as the fitness function did not improve even when the GA was allowed to run for 1000 generations.

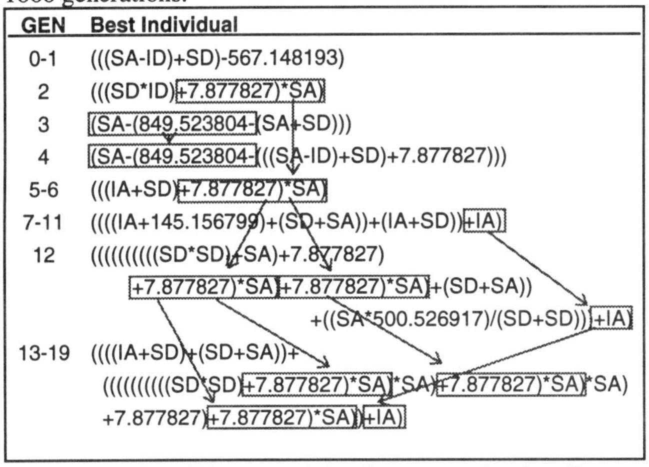

Figure 4. Possible gene inheritance paths for fitness function evolution.

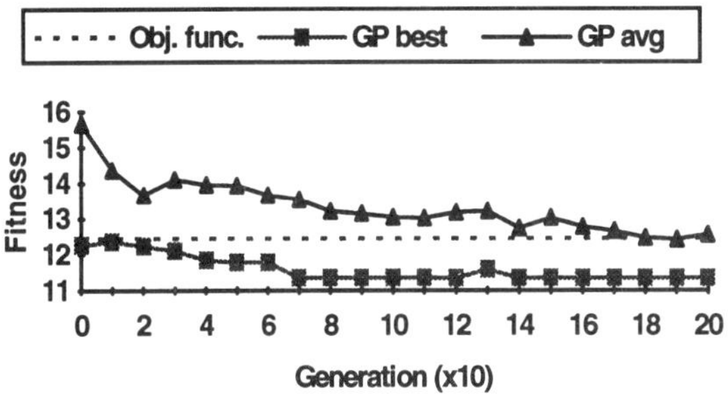

Figure 5. GP fitness evolution.

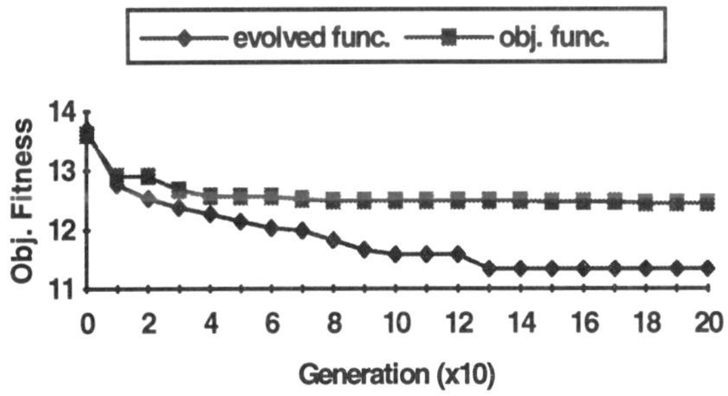

Figure 6. GA runs using objective function and evolved fitness function.

To validate the choice of population size and number of generations for the GA, the GA was run separately with the evolved fitness function for 400 generations of each of the population sizes M = {250, 500, 1000, 2000, 4000}. The results validate the initial choice of 200 generations of population 250, as shown in Figure 8. All runs converged before generation 200. Although larger population sizes reduce the number of generations required for convergence, the initial choice is efficient in terms of the number of fitness cases evaluated prior to convergence.

The GA was also run with all combinations of the crossover operation modifications (alignment and permutation co-evolution), as shown in Figure 8. Without either modification, the GA performs poorly. However, results are similar for combinations of either or both of the alignment and permutation co-evolution modifications; all three runs converge to the same minima. Still, using both modifications results in the quickest convergence. The alignment and permutation operations each add approximately 1 second of additional computation per run.

Additional sensitivity analysis of the GA parameters is given in Choi (1995).

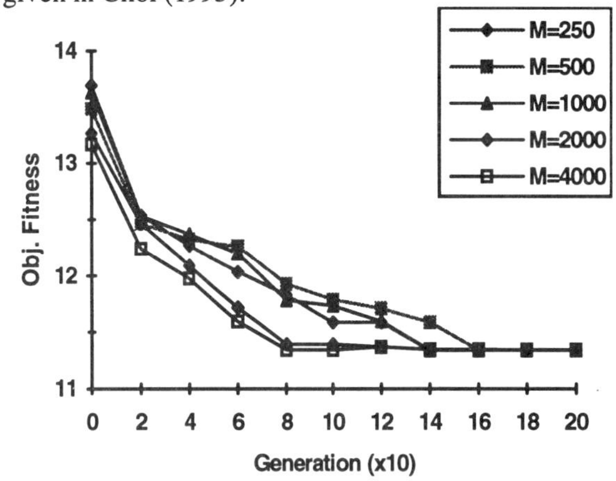

Figure 7. GA runs using various population sizes and numbers of generations.

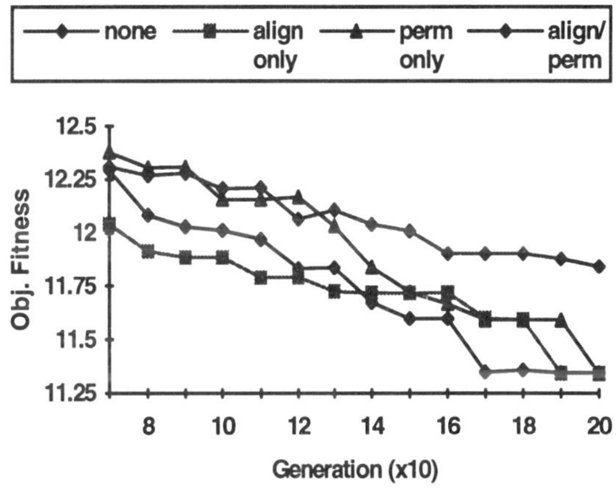

Figure 8. GA runs with various schema manipulations.

6. Discussion

A general concern with the approach used is that the GP-evolved fitness functions may favor the specific traffic matrix test cases used and perform less well on other test cases. To fully address this concern, the GA algorithm should be integrated into a real traffic environment, as diagrammed in Figure 1. Another approach to counteract the problem of over-fitting is to analyze the evolved function and to hand-write a fitness function that extracts only the features that are explainably important -- in effect, using GP as a means to discovering fitness function "strategies." On examination of the GP-evolved fitness function, it was found that only the first part of the expression, (IA+SD+SD+SA), is important when running the GA on the two traffic matrices that contain real data (i.e., if the remainder of the expression is removed, the GA still finds the same solution).

The rest of the expression is only important when running the GA on the deceptive matrix. Furthermore, the expression can be modified as (10*(SA+SD) + 10000*(SD+10)*SA) to obtain similar results. It appears that the second half of the expression works to optimize for traffic matrices whose optimal topologies result in near-zero intersegment standard deviation (SD ≅ 0).

Interestingly, the GP terminals SM, IM, and GEN were never expressed in any best-of-generation S-expression. However, inspection of the populations does show that they are being used. An example from generation 4 that yields slightly better GA performance than the objective function is ((SA / GEN) + SD). During initial generations, this function will group all nodes into a single segment to yield the lowest intrasegment traffic average; then, as the run progresses, the function stresses the importance of intrasegment traffic deviation, effectively dispersing the nodes across segments again. The success of such a strategy will likely depend on the traffic matrices used, but one implication of this strategy is that it may be helpful to use a seed chromosome for the GA that assigns all nodes to the same segment.

The approach of using GP to evolve fitness functions may have more general applicability than to just the problem described in this paper. The approach is an example of multi-level optimization, in which an upper-layer mechanism (in this case, GP) is used to optimize the parameters (in this case, the fitness function) of the lower-layer optimization. Previous work has focused on using upper-layer GAs to evolve lower-layer parameters such as mutation and crossover probabilities as well as the selection, mutation, and crossover operators themselves. Wang (1996) summarizes previous work and suggests a possible framework for such multi-level evolution. For the approach used in this paper, the usefulness of the upper-level GP optimization relies on the following lower-level problem characteristics: 1) the lower-level problem GA is to be run on a number of data sets in the future, and the lower-level problem has some number of data sets which can effectively be used beforehand as fitness cases for the upper-level GP; 2) the lower-level problem has a single-objective objective function that is composed of multiple related parameters (as opposed to a multi-objective function), and the relationships between those parameters and their effects on the GA evolution cannot be analyzed directly.

7. Conclusions

This paper described a genetic algorithm and genetic programming approach to optimizing local area networks in response to traffic conditions. In particular, a specific problem instance was chosen for its practical applicability to real-world LANs. The approach is shown to work well on three test case traffic matrices, including a hand-crafted "deceptive" one. Running time of the algorithm is small enough to allow it to be applied continually to a network at 1-minute intervals, resulting in a real-time, continual network topology optimization mechanism (that is based on current network traffic patterns). The next step in investigating this approach is to integrate it into an actual "live" LAN so as to obtain more realistic traffic data as well as to observe actual traffic dynamics as the algorithm is allowed to modify the LAN topology.

Finally, this paper briefly discused the more general use of GP in evolving lower-level fitness functions, based on the approach used in this paper.

Acknowledgements

Thanks to John Koza for his Stanford GA/GP class, for which much of this work was done as a final project. Thanks also to the GP-96 paper referees for their helpful comments.

Bibliography

Choi, A. 1995. Optimizing Local Area Networks Using Genetic Algorithms. In Koza, J. (editor). *Genetic Algorithms and Genetic Programming at Stanford 1995.* Stanford, CA: Stanford Bookstore. Pages 49-59.

Fowler, H.J. and Leland, W.E. 1991. Local Area Network Traffic Characteristics, with Implications for Broadband Network Congestion Management. *IEEE Journal on Selected Areas in Communications.* Volume 9..

Garey, M.R. and Johnson, D.S. 1979. *Computers and Intractability: A Guide to the Theory of NP-Completeness.* WH Freeman.

Goldberg, D.E. 1989. *Genetic Algorithms in Search, Optimization, and Machine Learning.* Reading, MA: Addison-Wesley.

Holland, J.H. 1975. *Adaptation in Natural and Artificial Systems.* Ann Arbor, MI: University of Michigan Press.

Jedwab, J. 1992. *Network Sampling for the HP Resource Manager.* Hewlett-Packard Laboratories Bristol technical report HPL-92-34. March 1992.

Koza, J.R. *Genetic Programming: On the Programming of Computers by Means of Natural Selection.* Cambridge, MA: The MIT Press.

Paredis, J. 1995. The Symbiotic Evolution of Solutions and their Representations. In Eshelman, L. (editor). *Proceedings of the Sixth International Conference on Genetic Algorithms.* San Francisco, CA: Morgan Kaufmann.

Phaal, P. 1994. *LAN Traffic Management.* New York: Prentice-Hall.

Wang, G., Goodman, E.D., and Punch, W.F. 1996. *Simultaneous Multi-Level Evolution.* Michigan State University Computer Science technical report GARAGe96-03-01. March 1996. http://isl.cps.msu.edu/GA/papers/GARAGe96-03-01.ps

A Genetic Algorithm for the Construction of Small and Highly Testable OKFDD-Circuits

Rolf Drechsler **Bernd Becker** **Nicole Göckel**

Institute of Computer Science
Albert-Ludwigs-University
79110 Freiburg im Breisgau, Germany
email: <name>@informatik.uni-freiburg.de

ABSTRACT

A Genetic Algorithm (GA) is applied to derive circuits that combine area efficiency and testability. These circuits are obtained from Ordered Kronecker Functional Decision Diagrams (OKFDDs). In "Becker and Drechsler (1995)" a heuristic approach has been presented. In this paper we show how these results can further be improved by GAs. Finally, we apply our minimization algorithm to technology mapping for FPGAs. We present experimental results to show the efficiency of our approach.

1 Introduction

Genetic Algorithms (GAs) are often used in optimization and machine learning (see "Goldberg (1989)" and "Davis (1991)"). In many applications they are superior to the classical optimization techniques, e.g. gradient-descent. Recently, GAs have succesfully been applied to several hard problems in *Computer Aided Design* (CAD) (see e.g. "Rahmani and Ono (1993)", "Esbensen and Mazumder (1994)" and "Drechsler, Becker, and Göckel (1995)").

In the following we briefly describe some of the problems occuring in testing: Even if circuits are correctly designed a fraction of them will behave faulty due to physical defects during the manufacturing process. This is one of the reasons why more and more approaches to circuit design try to incorporate testing aspects in the synthesis step (*Design For Testability* (DFT)).

In the area of DFT many different approaches have been presented in the last few years, one of them being DFT tools based on *Ordered Decision Diagrams* (ODDs), like OBDDs, OFDDs or OKFDDs "Ashar, Devadas, and Keutzer (1991)", "Becker (1992)", "Ashar, Devadas, and Keutzer (1993)" and "Becker and Drechsler (1995)".

OKFDDs are a superset of OBDDs and OFDDs and allow efficient representation of Boolean functions for which no efficient OBDDs and OFDDs exist. In "Becker and Drechsler (1995)" a heuristic approach to DFT using OKFDDs has been presented. As a result circuits derived from OKFDDs were obtained, that are smaller in size than circuits derived from OBDDs or OFDDs and have nice testability properties. But finding the minimal OKFDD is a difficult problem. In "Drechsler and Becker (1995)" a heuristic method for finding small OKFDDs was presented. But unfortunately the resulting circuits do not always have good testability properties, i.e. they are in general not 100% testable.

In this paper we apply *Genetic Algorithms* (GAs) to the problem of finding small OKFDD-circuits that have good testability properties, i.e. circuits with a high stuck-at fault coverage. We performed a large set of experiments and compared the GA-approach to the heuristic approach from "Becker and Drechsler (1995)". It is shown that often GAs can further improve the heuristic result. Finally, we apply the minimization algorithm to technology mapping for FPGAs and compare our results to the results from "Schaefer, Perkowski, and Wu (1993)", where OKFDDs have also been used for circuit design. The GA-approach can often improve on these results by a factor up to two. Obviously this simplifies the mapping process tremendously.

2 Preliminaries

We define OKFDDs and then describe how they can be transformed to combinational circuits.

2.1 OKFDDs

In the following we briefly review the essential definitions of OKFDDs from "Drechsler et al. (1994)" and "Drechsler and Becker (1995)".

A *Kronecker Functional Decision Diagram* (KFDD) is a graph-based representation of a Boolean function

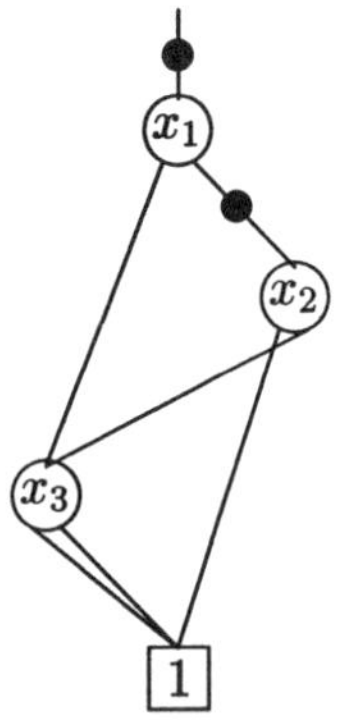

Figure 1: OKFDD for $f = x_1x_2 + x_3$

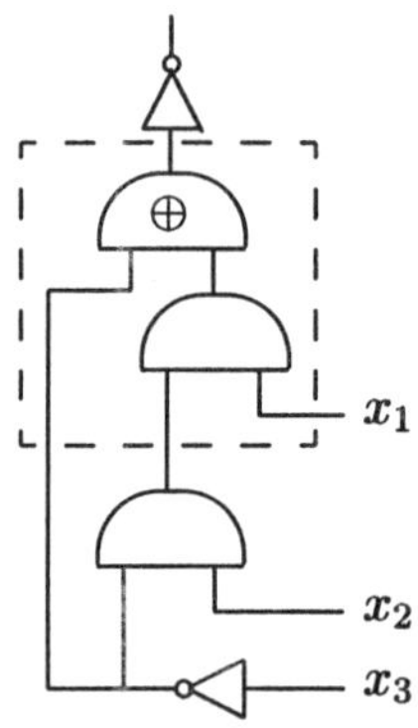

Figure 2: Circuit over EXORLIB for $f = x_1x_2 + x_3$

$f : \mathbf{B}^n \to \mathbf{B}$ over the variable set X_n (similar to BDDs), where one of the following three decompositions is carried out in each node:

$$f = \overline{x}_i f_i^0 + x_i f_i^1 \quad \textit{Shannon (S)} \quad (1)$$

$$f = f_i^0 \oplus x_i f_i^2 \quad \textit{positive Davio (pD)} \quad (2)$$

$$f = f_i^1 \oplus \overline{x}_i f_i^2 \quad \textit{negative Davio (nD)} \quad (3)$$

(f_i^0 defined by $f_i^0(x) := f(x_1, .., x_{i-1}, 0, x_{i+1}, .., x_n)$ for $x = (x_1, x_2, \ldots, x_n) \in \mathbf{B}^n$ denotes the *cofactor* of f with respect to $x_i = 0$, f_i^1 denotes the cofactor for $x_i = 1$ and f_i^2 is defined as $f_i^2 := f_i^0 \oplus f_i^1$, $\oplus$ being the Exclusive OR operation.)

If the underlying graph is *ordered* the corresponding KFDD is called an OKFDD.

Decomposition types are associated to the variables in X_n with the help of a *Decomposition Type List* (DTL) $d := (d_1, \ldots, d_n)$ where $d_i \in \{S, pD, nD\}$, i.e. for each variable one fixed decomposition is chosen.

On OKFDDs reductions can be defined. From "Drechsler et al. (1994)" it is known that reductions can be used to define canonical representations for not only OBDDs and OFDDs, but also for OKFDDs.

The size of an OKFDD can be further reduced, if *Complement Edges* (CEs), that are well known from OBDDs and OFDDs, are introduced. Then a node is used to represent a function and its complement at the same time. In the following it is described how these reduced OKFDDs can be transformed to a combinational logic circuit.

2.2 OKFDD-Circuits

In general, a *Combinational Logic Circuit* (CLC) is defined over a fixed library and modeled as a directed acyclic graph $C = (V, E)$ with some additional properties.

In the following we consider circuits over the *Standard library* (STD) consisting of *Primary Input* (PI) and *Output Ports* (PO), the 2-input, 1-output *AND, OR* gate and the 1-input, 1-output inverter *NOT*.

OKFDDs directly correspond to so called *OKFDD-circuits* defined over the library STD. An OKFDD-circuit can easily be obtained from an OKFDD by traversing the corresponding graph in topological order and replacing all nodes by the corresponding cells from the library.

Example 1 In Figure 1 an OKFDD for the function $f = x_1x_2 + x_3$ is shown (all nodes are of type pD). A dot on an edge symbolizes a CE. The corresponding circuit is shown in Figure 2. (For the interpretation over STD the EXOR-gate is substituted by a basic cell realization.)

2.3 Fault Model

We give a short description of the fault model considered in this paper.

A fault in the *Stuck-At Fault Model* (SAFM) causes exactly one input or output pin of a node in C to have a fixed constant value (0 or 1) independently of the values applied to the PIs of the circuit "Breuer and Friedman (1976)".

We finish with some general definitions and remarks. For this let C be any CLC over a fixed library.

Definition 1 An input t to C is a *test* for a fault f, iff the primary output values of C on applying t in the presence of f are different from the output values of C in the fault free case.

A fault is *testable*, iff there exists a test for this fault. The goal of any test pattern generation process is a *complete* test set for the circuit under test, i.e., a test set that contains a test for each testable fault. The construction of complete test sets requires the determination of the faults which are not testable (= *redundant*), even though it is easy to see that in general the detection of redundancies is *coNP-complete*. Redundancies have further unpleasant properties: They may e.g. invalidate tests for testable faults and often correspond to locations of the circuit where area is wasted "Breuer and Friedman (1976)". All in all, synthesis procedures which result in non-redundant circuits or at least circuits with known redundancies are desirable. A node v in C is called *fully testable*, if there does not exist a redundant fault with fault location v. If all nodes in C are fully testable, then C is called *fully testable*. In the following we try to synthesize circuits with a small number of redundant faults.

For OKFDD-circuits test pattern generation can be performed easily "Drechsler and Becker (1993)". Thus, the exact fault coverage can easily be determined. In the following we call it the *Maximal Fault Coverage* (MFC).

3 The Problem Domain

The size of OKFDDs is very sensitive to the variable ordering and the choice of the DTL. This size directly corresponds to the size of the resulting OKFDD-circuit. Additionally, the testability of the circuit should be considered.

Based on a result from "Becker and Drechsler (1995)" the MFC of an OKFDD-circuit can easily be determined by performing synthesis operations on the corresponding OKFDD.

In the following we consider the problem of the construction of small OKFDD-circuits that additionally have good testability properties.

4 Genetic Algorithm

In this section we describe the *Genetic Algorithm* (GA) as a whole and the genetic operators that are applied to the problem given above.

4.1 Representation

Most often GAs work on binary strings and the problem to be solved is encoded in a two-valued representation. In our application we have to store two informations per variable: the ordering position and the decomposition type.

1. **Ordering position:** We use integer strings to represent the orderings, since a binary encoding would require special repair algorithms to avoid the creation of invalid solutions. (The occuring problem is similar to the problem of tour encoding in the *traveling salesman problem*. Similar observations concerning encoding are reported in "Whitley, Starkweather, and Fuquay (1989)".) Each integer vector is a permutation of the variables and thus represents a valid ordering.

2. **Decomposition type:** For each variable one decomposition type out of three must be chosen. Thus, each element contains an n-dimensional three-valued vector. Using this three-valued encoding each string represents a valid solution.

A population is a set of elements. Each element in the population contains two arrays: one for the ordering position and one for the decomposition type of each variable. To the different arrays different genetic operators are applied in the following.

4.2 Objective Function and Selection

As an *objective function* f that measures the *fitness* of each element we use two fitness functions f_{size} and f_{test} that are combined with respect to the optimization goals:

1. f_{size}: This function measures the number of nodes of the OKFDD corresponding to the chosen variable ordering and decomposition types. This function has to be minimized to find small OKFDD-circuits. An OKFDD with a given variable ordering π_1 and a given DTL d_1 is transformed to an OKFDD with variable ordering π_2 and DTL d_2 by dynamic methods as described in "Drechsler and Becker (1995)".

2. f_{test}: This function measures the *Maximal Fault Coverage* (MFC) of the OKFDD-circuit corresponding to the chosen variable ordering and decomposition types with respect to the SAFM. This function has to be maximized to find an OKFDD-circuits with good testability properties. The fault coverage is determined by performing OKFDD synthesis operations as described in "Becker and Drechsler (1995)".

These two fitness functions are combined as follows to measure the fitness of each element:

$$f = a \cdot f_{size} + b \cdot (1 - f_{test})$$

Thus our problem is formulated as a minimization problem. Obviously, the choice of the parameters a and b largely influence the resulting fitness. If a is set to 0, then the algorithm only optimizes the testability without respecting the size of the circuit. If b is set to 0 only the size of the resulting circuit is minimized, but testability is not considered.

The selection is performed by *roulette wheel selection*, i.e. each string is chosen with a probability proportional to its fitness. Additionally, we also make use of *steady-state-reproduction* "Davis (1991)": Some of the best elements of the old population are included in the new one anyway. This strategy guarantees that the best element never gets lost and a fast convergency is obtained. GA practice has shown that this method is usually advantageous. (The choice of the number of elements to be copied in each iteration will be discussed in more detail later.)

4.3 Initialization

At the beginning of each GA-run an initial population is randomly generated. To each element a fitness is assigned using the method described above.

Often it is helpful to combine GAs with problem specific heuristics "Davis (1991)". The resulting GAs are called *Hybrid GAs* (HGAs). In the case that only minimization of size is of importance, but testability is not considered we combined our GA with *DTL-sifting* (DTL-sift) "Drechsler and Becker (1995)": a combination of dynamic variable ordering "Rudell (1993)" and dynamic choice of DTLs. (The details are left out due to page limitation.)

If testability is also considered DTL-sift often destroys the obtained structures and thus is left out.

4.4 Genetic Operators

We now introduce the genetic operators that are used by our GA. First the operators that are applied to the ordering positions are introduced. Then the operators for the DTLs are discussed.

4.4.1 Ordering Operators

In our application we use *Partially Matched Crossover* (PMX) "Goldberg and Lingle (1985)". This method creates two children from two parents. The parents are selected by the method described above. The operator chooses two cut positions at random.

Notice that a simple exchange of the parts between the cut positions (as often applied to binary coded GA-problems) is not possible, since this would often produce invalid solutions. The operator works as follows to *validate* the children after the exchange:

PMX: Construct the children by choosing the part between the cut positions from one parent and preserve the position and order of as many variables as possible form the second parent.

For more details about PMX see "Oliver, Smith, and Holland (1987)", "Goldberg (1989)" and "Michalewicz (1994)".

We also tested further crossover operators, like *ordered crossover* and *cycle crossover* "Davis (1985)", "Oliver, Smith, and Holland (1987)" and "Michalewicz (1994)", but they did not improve the results obtained. In contrast, often the quality of the solutions decreased.

Three different *mutation operators* are used:

Mutation (MUT): Select a parent element at random and choose one position. The value of the variable at this position is used to determine the position of the variable with which it is exchanged.

2-time Mutation: Perform MUT two times on the same parent.

Mutation with neighbour: Select a parent element at random and choose one position. Then perform MUT at this and the neighbouring position.

Notice, that the mutation operators are a generalization of exchanging neighbouring variables that is the basic operation for dynamic variable ordering "Fujita, Matsunga, and Kakuda (1991)" and "Rudell (1993)".

4.4.2 DTL Operators

In the following we briefly describe the genetic operators used for the manipulation of DTLs. All operators are directly applied to three-valued strings of finit length that represent elements in the population.

Crossover: Construction of two new elements x_1 and x_2 from two parents y_1 and y_2, where the first (second) part of x_1 (x_2) up to a cut position i is taken from y_1 and the second (first) part is taken from y_2.

2-time Crossover: Construction of two new elements x_1 and x_2 from two parents y_1 and y_2, where the first (second) part of x_1 (x_2) up to a cut position i is taken from y_1 (y_2), the second part up to a second cut position $j > i$ is taken from y_2 (y_1) and the last part is again taken from y_1 (y_2).

Uniform Crossover: Construction of two new elements x_1 and x_2 from two parents y_1 and y_2, where at each position the element is taken with a probability p from y_1 and y_2, respectively.

Mutation: Construction of a new element from a parent by copying the whole element and randomly changing its value at mutation position i.

2-time Mutation: Construction of a new element from a parent by copying the whole element and randomly changing its value at mutation positions i and j ($i \neq j$).

Mutation with neighbour: Construction of a new element from a parent by copying the whole element and randomly changing its value at mutation positions i and $i+1$.

Obviously, all genetic operators only generate valid solutions, if they are applied to the three-valued strings.

4.5 Algorithm

Using the operators introduced above our genetic algorithm works as follows:

- Initially a random population of finit strings is generated (see Subsection 4.3).
- The better half of the population is copied in each iteration without modification. Then the genetic operators are applied to another $\frac{pop}{2}$ elements. The elements are chosen according to their fitness as described in Subsection 4.2. The newly created elements are then mutated by the mutation operators

Table 1: Comparison of different fitness parameters

name	Heu		Test		Size		Mixed	
	size	mfc	size	mfc	size	mfc	size	mfc
add6	52	97.2%	183	99.8%	23	98.6%	23	98.7%
alu2	149	99.3%	222	99.8%	130	98.6%	170	99.6%
apex7	249	98.6%	1322	99.6%	264	97.5%	741	98.8%
dist	120	99.4%	159	100%	120	98.9%	136	99.0%
in2	221	99.0%	863	99.9%	175	97.8%	589	99.4%
rd53	13	96.5%	16	99.3%	13	99.1%	13	99.1%
root	59	99.7%	77	100%	59	100%	59	100%
Z5xp1	31	97.2%	60	99.5%	28	99.0%	49	99.9%

```
genetic_algorithm (benchmark) {
  generate_random_population () ;
  initialize_population () ;
  if (b==0) optimize_initial_population_with_DTL-sift () ;
  {
    apply_genetic_operators () ;
    calculate_fitness () ;
    update_population () ;
  } while (improvement obtained) ;
  return ;
}
```

Figure 3: Sketch of basic algorithm

with a given probability. After each iteration the size of the population is constant.

- The algorithm stops if no improvement is obtained for $50 \cdot log(best\_fitness)$ iterations, where $best\_fitness$ denotes the fitness of the best element in the population.

A sketch of the algorithm is given in Figure 3.

4.6 Parameter Settings

The size of the population is chosen equal to the number of variables of the considered Boolean function.

The genetic operators are iteratively applied corresponding to their probabilities:

1. *Reproduction* is performed with a probability of 20%.
2. *Crossover* is performed with a probability of 80%.
3. *Mutation* is carried out on the newly generated elements with a probability of 15%.

5 Experimental Results

In this section we present experimental results that were all carried out on the OKFDD-package from "Drechsler and Becker (1995)". All runtimes are given in CPU seconds on an *HP apollo 715/50* workstation.

In a first series of experiments we applied our GA to several *small* benchmark circuits from LGSynth91, i.e. circuits with less than 17 input variables. For these circuits we choose different values for the parameters that influence the objective function (see Subsection 4.2). The results are given in Figure 1. *size* and *mfc* denote the number of nodes of the OKFDD and the MFC, respectively, for benchmark *name*. Column *Heu* denotes the results from the heuristic approach presented in "Becker and Drechsler (1995)". In column *Size* (*Test*) the results are given for $b = 0$ ($a = 0$), i.e. if the fitness is only used with respect to size (testability). In column *Mixed* size and testability are considered in parallel.

As can be seen there exists a trade-off between representation size and testability, i.e. small circuits often have *bad* testability properties. In contrast the OKFDD-circuits that were optimized for testability always got more than 99% fault coverage.

Finally, to give an impression of the quality of the results we compare our method to an OKFDD approach to technology mapping for FPGAs. We compare our results for larger circuits with the most powerful heuristic from "Schaefer, Perkowski, and Wu (1993)". The results are given in Figure 2. *in* (*out*) denotes the number of inputs (outputs) of the corresponding circuit. The results from "Schaefer, Perkowski, and Wu (1993)" measured in number of nodes are given in column SPW. The results obtained by our GA and the execution times are given in the last two columns. (Notice, that the approach in "Schaefer, Perkowski, and Wu (1993)" does not consider testability.)

As can easily be seen our approach obtains much better results in all considered cases. For some benchmarks more than 60% improvement is obtained. Obviously this simplifies the mapping process tremendously.

Table 2: Results for FPGA design

name	in	out	SPW	GA	time
cc	21	20	117	39	219.0
c8	28	18	237	62	264.1
frg1	28	3	186	92	538.2
pdc	16	40	1118	564	216.5
spla	16	46	724	549	230.3
vg2	25	8	204	74	298.3
$\sum$			2586	1380	

6 Conclusions

We presented a new method for the construction of small and highly testable OKFDD-circuits using *Genetic Algorithms*. Using this method it is possible to optimize circuits with respect to size and testability in parallel.

The efficiency of our approach has been demonstrated by experimental results. A comparison to an FPGA-approach from "Schaefer, Perkowski, and Wu (1993)" has shown that our GA can construct much smaller circuits within reasonable running times.

Bibliography

Ashar, P., Devadas, S., and Keutzer, K., 1991. Gate-delay-fault testability properties of multiplexor-based networks. In *Int'l Test Conf.*, pages 887–896.

Ashar, P., Devadas, S., and Keutzer, K., 1993. Path-delay-fault testability properties of multiplexor-based networks. *Integration the VLSI Jour.*, 15(1):1–23.

Becker, B., 1992. *Synthesis for Testability: Binary Decision Diagrams*, volume 577 of *LNCS*. Symp. on Theoretical Aspects of Comp. Science.

Becker, B. and Drechsler, R., 1995. Synthesis for testability: Circuits derived from ordered kronecker functional decision diagrams. In *European Design & Test Conf.*, page 592.

Breuer, M.A. and Friedman, A.D., 1976. *Diagnosis & reliable design of digital systems.* Computer Science Press.

Davis, L., 1985. Applying adaptive algorithms to epistatic domains. In *Int'l Joint Conf. on Artificial Intelligence*, pages 162–164.

Davis, L., 1991. *Handbook of Genetic Algorithms.* van Nostrand Reinhold, New York.

Drechsler, R. and Becker, B., 1993. Rapid prototyping of fully testable multi-level AND/EXOR networks. *IFIP WG 10.5 Workshop on Applications of the Reed-Muller Expansion in Circuit Design*, pages 126–133.

Drechsler, R. and Becker, B., 1995. Dynamic minimization of OKFDDs. In *Int'l Conf. on Comp. Design*, pages 602–607.

Drechsler, R., Becker, B., and Göckel, N., 1995. A genetic algorithm for minimization of Fixed Polarity Reed-Muller expressions. In *Int'l Conf. on Artificial Neural Networks and Genetic Algorithms*, pages 392–395.

Drechsler, R., Sarabi, A., Theobald, M., Becker, B., and Perkowski, M.A., 1994. Efficient representation and manipulation of switching functions based on Ordered Kronecker Functional Decision Diagrams. In *Design Automation Conf.*, pages 415–419.

Esbensen, H. and Mazumder, P., 1994. SAGA: A unification of the genetic algorithm with simulated annealing and its application to macro-cell placement. In *VLSI Design Conf.*, pages 211–214.

Fujita, M., Matsunga, Y., and Kakuda, T., 1991. On variable ordering of binary decision diagrams for the application of multi-level synthesis. In *European Conf. on Design Automation*, pages 50–54.

Goldberg, D.E., 1989. *Genetic Algortithms in Search, Optimization & Machine Learning.* Addision-Wesley Publisher Company, Inc.

Goldberg, D.E. and Lingle, R., 1985. Alleles, loci, and the traveling salesman problem. In *Int'l Conference on Genetic Algorithms*, pages 154–159.

Michalewicz, Z., 1994. *Genetic Algorithms+Data Structures=Evolution Programs.* Springer-Verlag.

Oliver, I.M., Smith, D.J., and Holland, J.R.C., 1987. A study of permutation crossover operators on the traveling salesman problem. In *Int'l Conference on Genetic Algorithms*, pages 224–230.

Rahmani, A.T. and Ono, N., 1993. A genetic algorithm for channel routing problem. In *Int'l Conference on Genetic Algorithms*, pages 494–498.

Rudell, R., 1993. Dynamic variable ordering for ordered binary decision diagrams. In *Int'l Conf. on CAD*, pages 42–47.

Schaefer, I., Perkowski, M.A., and Wu, H., 1993. Multilevel logic synthesis for cellular FPGAs based on orthogonal expansions. *IFIP WG 10.5 Workshop on Applications of the Reed-Muller Expansion in Circuit Design*, pages 42–51.

Whitley, D., Starkweather, T., and Fuquay, D., 1989. Scheduling problems and traveling salesman: The genetic edge recombination operator. In *Int'l Conference on Genetic Algorithms*, pages 133–140.

Motion Planning and Design of Cam Mechanisms by means of a Genetic Algorithm

Rodolfo Faglia
Mech. Engin. Dept
Università di Brescia
Via Branze 38
25123 Brescia, ITALY
rfaglia@bsing.ing.unibs.it

David Vetturi
Mech. Engin. Dept
Università di Brescia
Via Branze 38
25123 Brescia, ITALY
vetturi@bsing.ing.unibs.it

ABSTRACT

The present work shows the application of a genetic algorithm for the design and the trajectory planning of a cam-follower system. After a brief description of the bases of the genetic procedure used, some results derived from a practical application of the method to the design of a cam profile are shown.

1. Nomenclature

n: number of the elements of a vector.

$x_1, x_2, \ldots, x_n$: elements of a vector ("genes")

$\mathbf{x}$: vector having $x_1, x_2, \ldots, x_n$ as elements.

$F = F(\mathbf{x}) = F(x_1, x_2, \ldots, x_n)$: generic functional depending on vector $\mathbf{x}$ (on variables $x_1, x_2, \ldots, x_n$).

F*: value of the functional to be reached (goal).

$\mathbf{x}^* = (x_1^*, x_2^*, \ldots, x_n^*)$: solution vector which satisfies $F(\mathbf{x}^*)=F^*$.

$\mathbf{x}_j^i$: j-th vector of the i-th generation.

m: number of vectors in a generation.

$\mathbf{X}^i = (\mathbf{x}_1^i, \mathbf{x}_2^i, \ldots, \mathbf{x}_m^i)$: set of vectors; i-th generation.

$\varepsilon_j^i = \left|F^* - F(\mathbf{x}_j^i)\right|$: error with regard to the goal when using vector j-th of generation i-th.

$A_j^i = \dfrac{b_j^i}{b^i}$

where $b_j^i = \dfrac{1}{\varepsilon_j^i}$ and $b^i = \sum_{j=1}^{m} b_j^i$: fitness index of vector $\mathbf{x}_j^i$.

S: total displacement.

T: total time to perform displacement S.

T*, S*: precision point coordinates.

$\delta_1, \delta_2, \delta_3, \delta_4, \delta_5, \delta_6, \delta_7$: coefficients to define the motion profile.

2. Introduction

Very often, during the motion planning of a cam follower, it becomes necessary to impose different constraints on the motion.

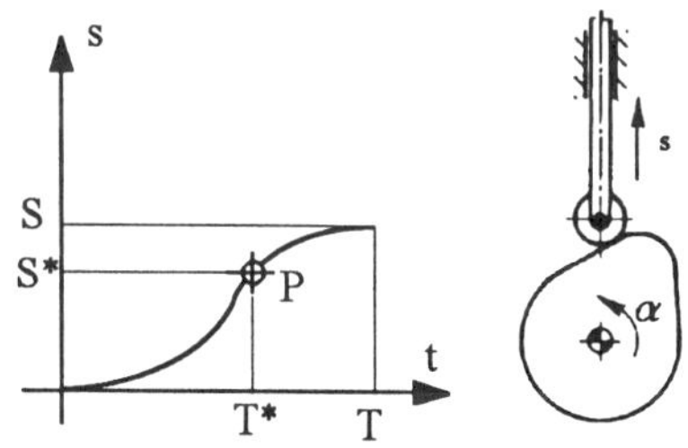

Figure 1. Motion of the follower of a cam mechanism with a precision point.

In particular, the problem refers to the definition of a motion profile which must accomplish, not only the typical constraints of the cam follower movements (null velocity at the beginning and at the end of the movement, obtaining a pre-fixed displacement S in a pre-fixed time T), but it must also be able to pass through a precision point P, at pre-fixed displacement S* and time T* (see figure 1). Very often the precision point is defined on the displacement graph, in other cases the precision point implies the passage with a certain velocity in correspondence to a particular displacement, and so on. In every case, in conclusion, near the constraints at the beginning and at the end of the motion, a further constraint is given.

There are many numerical and analytical methods, by which it is possible to obtain a function with the pre-fixed characteristics listed above; however, for practical applications, the results obtained from the majority of the methods must be discussed and further analysed. For instance, a good function in the displacement field could be very bad in the acceleration profile (high acceleration peaks, for example). Or, we could obtain a function which fulfils the imposed constraints, but it is not monotonic increasing (which is generally undesirable in cam mechanisms).

A genetic algorithm seems a right novel choice in this field to find the solution to the problem, obtaining motion profiles good for actual applications.

In fact: 1) the genetic algorithm offers not only one solution, but a whole "family" of solutions estimations which, with a more or less good approximation, are close to the goal, so that among them it is possible to select the one which appears the best for a practical application; 2) the iterations of the genetic algorithm have "memory" of the previous ones; this allows the final goal to be changed during some iterations, without driving the method into a chaotic situation as happens in other algorithms when the goal is not fixed; in this way, it is possible to approach different aims simultaneously; 3) from literature, the genetic algorithm seems very stable, easy to implement, and fairly quick to gain the solutions.

Besides, the authors were curious to discover how the genetic algorithm, successfully applied in other scientific fields (e.g., see Gupta (1993), Hajima et al. (1992), Holland (1992), Muscia et al. (1993), Yun Wei-min et al. (1994)), could work in a typical mechanical problem where the quality of the results can be very easily verified.

Another interesting problem linked with a cam mechanism design is to find the right geometry in order to obtain the best cam profiles.

Referring to the oscillating cam with the roller follower depicted in Fig.2, taken as an example, the main geometrical parameters to define a cam follower mechanism are: the length of the frame and of the follower with their angle as regard to a reference direction, and the roller radius. In general, these quantities are not totally fixed, but the designer admits small tolerances inside which it is possible to optimise the geometry of the system. In these cases small adjustments of the geometry allow to improve the dynamic behaviour of the cam profile, decreasing the pressure angle or / and increasing the curvature radius. The great difficulty of the procedure lies in the fact that it is very difficult to move inside the tolerances to find a right combination of all the parameters, so that an optimisation algorithm is needed.

In this case we have used a genetic algorithm again, in its typical role of optimisation method, obtaining good results.

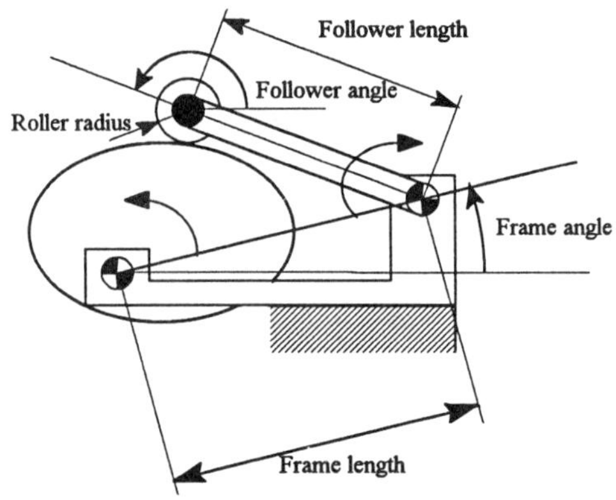

Figure 2. The geometrical parameters of a cam follower mechanism.

Summing up, the paper analyses the general properties of a genetic algorithm, shows its application to the trajectory planning problem briefly described above, particularly pointing out the good results obtained in designing the profile of a cam mechanism and foresees the application of a genetic algorithm as a method to optimise the geometrical parameters of a cam follower system.

3. Tne Genetic Algorithm used

The method discussed has been historically named "genetic algorithm" because it follows some procedures which appear in Darwin's doctrine of evolution, basically explaining that every creature must adapt to its environment; otherwise the creature and/or his progeny is doomed to extinction.

It is possible to say that the genetic process optimises the species because only the best elements can survive.

In the case of a numerical application, it is possible to build the following scheme: the "environment" is a functional where the "creature" is the functional variable (in general a vector), and the goal to be reached "to survive" is a pre fixed value of the functional itself. Mathematically, consider functional

$$F = F(\mathbf{x}) = F(x_1, x_2, \ldots, x_n).$$

The elements of the vector $\mathbf{x}$ are named "genes".

Let's consider a pre-fixed value F* of the functional; the aim is to find vector $\mathbf{x}^* = (x_1^*, x_2^*, \ldots, x_n^*)$, so that $F(\mathbf{x}^*)=F^*$.

To reach the goal, let's define a set of vectors $\mathbf{X}^1 = (\mathbf{x}^1_1, \mathbf{x}^1_2, \ldots, \mathbf{x}^1_m)$. $\mathbf{X}^1$ is the first family (generation) of vectors by means of which we try to find the solution.

For every vector forming the first generation, it is now possible to define an index by means of which we evaluate the closeness of the vector itself to the solution.

A possible index is the inverse of the error, so defined:

$$b^1_1 = \frac{1}{\varepsilon^1_1} \; ; \; b^1_2 = \frac{1}{\varepsilon^1_2} \; ; \; \ldots \; ; \; b^1_m = \frac{1}{\varepsilon^1_m}$$

where

$$\varepsilon^1_1 = \left|F^* - F(\mathbf{x}^1_1)\right| \; ; \; \varepsilon^1_2 = \left|F^* - F(\mathbf{x}^1_2)\right| \; ; \; \ldots \; ; \; \varepsilon^1_m = \left|F^* - F(\mathbf{x}^1_m)\right|$$

are the errors evaluated in correspondence to all the vectors of the first generation, as regard to the pre-fixed functional value F*.

It is useful to normalise the index above defined as follows:

$$A^1_j = \frac{b^1_j}{b^1} \quad \text{where} \quad b^1 = \sum_{j=1}^{m} b^1_j$$

So, at every vector $\mathbf{x}^1_j$ of the first generation we have linked a value A^1_j named "fitness index": a vector with a high fitness index will have good possibilities of surviving in the next generation, while a vector with a low fitness index will have poor possibilities of surviving in the next generation.

From the definition, it is easy to verify that

$$\sum_{1}^{m} A^1_j = 1$$

So, it is possible to build a diagram formed by a segment with unit length, and divide it into partitions proportional to A^1_j values, as suggested in Figure 3.

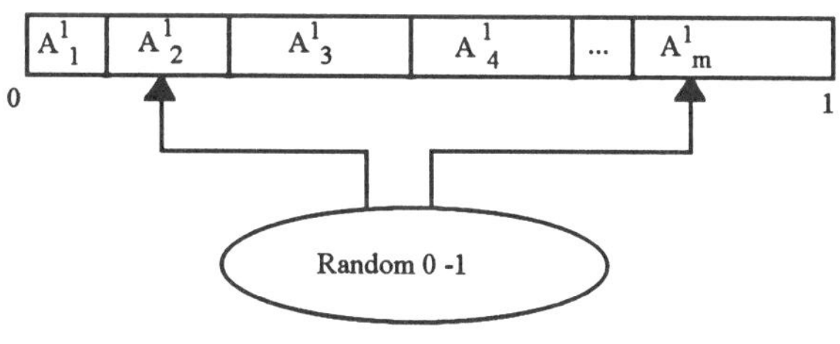

Figure 3. Random determination of the "parents" from a generation.

Now we can randomly draw two numbers inside the interval [0,1]; these two values will identify two fitness indexes and, consequently, two vectors.

These two selected vectors are named "parents" and form a couple of vectors from which a new vector originates, the "son". The procedure described above implies that the possibility of selecting vectors with a good fitness index is higher than drawing a vector with low fitness.

The son is generated from the parents by means of three basic mechanisms (see Figure 4) acting on the elements (genes) of the two vectors: the inheritance, the cross-over, and the mutation.

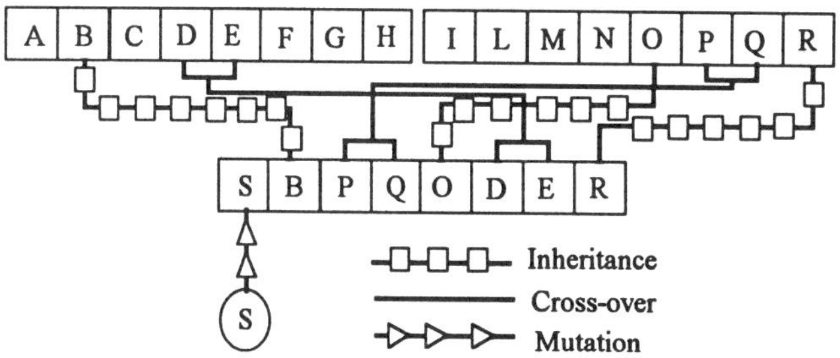

Figure 4. The mechanisms for the generation of a son vector.

The inherited gene directly passes from the parent to the son, without changing its value or its position in the vector; the cross-over mechanism determines the passage of a group of genes from one parent to the son, but in different positions; the mutation causes the presence of a random gene in the son. The three mechanisms are active under a random flag, so that the chances of the son turning out a perfect copy of one of the parents is as likely as it being completely different from them both.

The son's generation is iterated repeatedly in order to define a new generation $\mathbf{X}^2$. Starting from this new family, the phases described above for the first generation are repeated. The procedure stops when a vector with fitness near 1 (inside a tolerance pre-fixed by the user) is obtained.

In addition to the well known positive aspects of the method such as the very easy implementation (the more complex operation is the swap one!), and the negative ones (i.e. the right choice of the algorithm parameters (number of the vectors of a generation, maximum number of mutations, cross-over events, inherited genes) that can be optimised only after various attempts), here we want to focus attention upon a very important characteristic of the procedure, which will be very useful in the particular application of the algorithm on cam shaping, discussed in the introduction.

If the solution is one, the algorithm converges towards the solution. If the solution is multiple, then the vectors of a generation converge towards all the solutions until one is reached. Consider, for example, the scheme depicted in

Figure 5: the black points (A, B, C and D) are the solutions to an hypothetical problem; the numbered squares are the vectors of the various generations 1, 2, 3 and 4. At every generation we observe a closing in on the solutions. Imagine a platoon marching towards a series of objectives; it is not important which of them is gained first: every one has the same importance. At generation 4 solution A has been reached by one vector, so the algorithm stops.

Suppose now that, for some reason, among A, B, C and D, the best solution (best with regard to a "secondary goal") is B. (Imagine for example that B is the best because we are searching for the solution which is the nearest to point E). How can we get it ? How is it possible for the imaginary platoon to converge on objective B which is now more important than the others?

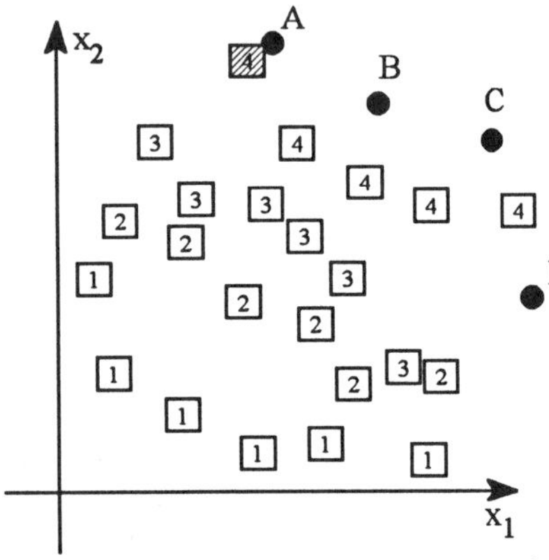

Figure 5. A schematic approach to the solution

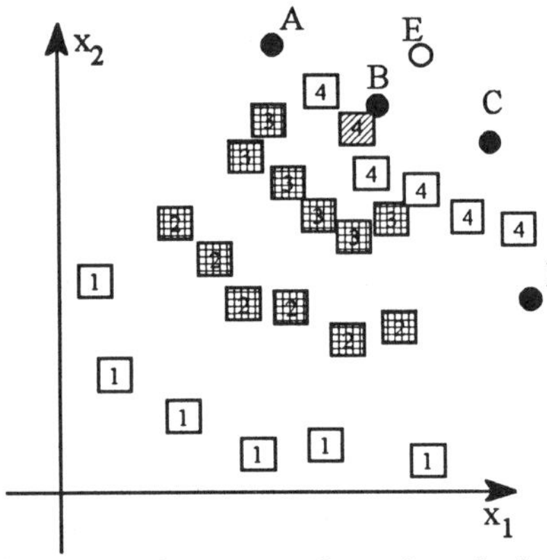

Figure 6. A schematic approach to the solution when iteration 2 and 3 are driven to the secondary goal.

It is possible, for some iterations, to change from the original goal to the secondary one (minimise the distance from E) so that the generation tends to converge on this new goal and the vectors "march" towards B. After some iterations we impose the primary goal again. Even if there is not a complete certitude, because all the algorithm is based upon random selections, there is a high possibility of going towards the solution of the primary goal which, of all the solutions, optimises also the secondary one (see Fig. 6).

The switching of the goals does not make the procedure go into a chaotic behaviour just because the genetic algorithm acts through families of vectors, and, inside a family there are vectors which fit the first goal well and drag the generation towards the primary one and others which fit the secondary objective better and lead the generation towards it.

4. The GA Applied to Trajectory Planning

Let's consider again, in a deeper way, the problem of the passing through a precision point of the displacement curve, explained in the introduction.

We have to satisfy:

A goal "zero": the displacement must start from 0 at time 0 with null velocity (first derivative null) and must end at a pre-fixed coordinate S at a given time T with null velocity again.

A goal "one" (primary goal): among all the possible solutions which fulfil goal "zero" we have to select the ones which pass through a pre-fixed precision point having coordinates (T*, S*) with pre-fixed tolerances ΔT and ΔS.

A goal "two" (secondary goal): among all the possible solutions which satisfy goal "one" we must select the one which optimises a pre-fixed parameter. (For example, in the particular field of cams mechanisms, a possible goal is to obtain the motion profile with the lowest negative acceleration peak; so the parameter to optimise is the acceleration of the system).

To fulfil goal "zero" it is possible to use the well known modified trapezoidal motion shape (Magnani et al., 1986). This kind of motion profile is defined by means of the total displacement S, the total motion time T, and 7 coefficients which represent the time percentages as regard the total time T of the acceleration diagram partitions, as depicted in Figure 7. In interval times $\delta_1, \delta_3, \delta_5$ and δ_7 the acceleration has a harmonic function profile, in δ_2, δ_6 it is constant, in δ_4 it assumes null value. Different combinations of the 7 coefficients affect only the motion profile, without changing the boundary conditions; in particular, velocity at the end points is always null.

Now, given values T* and S* which fix a precision point P, the problem is solved if the right combination of the seven coefficients which define a displacement profile passing through P is found.

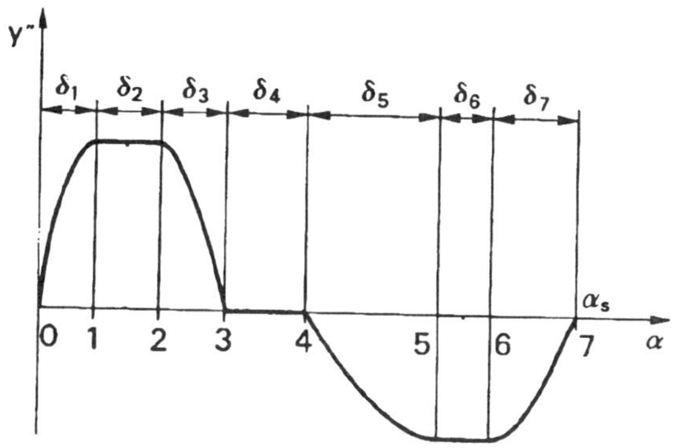

Figure 7. Acceleration profile defined by 7 coefficients

Goal "one" can be achieved by the use of the genetic algorithm. The fitness of the members of a generation (each one constituted by a vector with seven elements, the seven parameters which identify the motion shape) is evaluated as regard the distance from point (T*, S*).

Among all the possible solutions achieving goal "one", it is possible to select the ones which best fulfil goal "two" by changing the way of evaluating the fitness in some iterations of the algorithm: in these cases, the fitness is evaluated as regard goal "two", e.g. the minimum acceleration negative peak.

To judge the effectiveness of the procedure, a comparison between the genetic algorithm and the Newton-Raphson method has been carried out. Table 1 is a selection of some values of the secondary goal achieved by the two methods in several tests, when the primary goal has been reached by both. It is possible to note that the genetic algorithm offers the best solutions.

Table 1. Comparison between values of the secondary goal (negative peak acceleration) obtained by the genetic algorithm and the Newton-Raphson one. The genetic procedure has been used substituting two iterations of the secondary goal every ten iterations of the primary one.

	Method	
Test Number	Genetic algorithm	Newton-Raphson
1	76.65	113.60
2	65.33	100.90
3	73.79	132.56
4	94.46	119.17
5	92.08	123.53
6	79.02	126.18
7	92.60	100.60
8	67.37	99.40

We also tried to use Newton-Raphson algorithm by a cyclical switching of the goal, from the primary to the secondary one, but by doing so this method rarely converged.

5. Designing Cams with a GA based Software

This paragraph briefly compares the results (Table 2) obtained in profiling a cam by a very expert designer with the use of a typical cam software without any optimisation module, and by an inexpert student who used a software for cams design based on a genetic procedure written by the authors, having as input the length of the frame and the follower, their angles as regard to a reference direction, the roller radius with the respective tolerances and the seven coefficients describing the motion profile. The goal was to find the right combination of the geometrical parameters (inside given tolerances) and the seven coefficients of the motion to achieve the minimum pressure angle.

Note, form the comparison of the attempts, that the maximum pressure angles has been effectively improved.

Table 2. Comparison between the results obtained in profiling a cam by an expert designer and an inexpert student using a software endowed with a genetic algorithm procedure for optimisation.

	Expert designer	Genetic algorithm
Roller radius (mm)	10	10
Frame length (mm)	121	124
Frame angle (°)	0	0
Follower length (mm)	143	143.5
Follower initial angle (°)	136	137.2
Motion 7 parameters	125-250-125 0 125-250-125	125-230-120 512- 120-274-140
Max. pressure angle (°)	43.5°	40.8°
Min. Curvature radius (mm)	22.2	22.2

6. Other Results and Conclusions

The genetic algorithm procedure has been successfully utilised also for the trajectory planning with the passage through multiple precision points.

Moreover, particular genetic algorithms (named "Lamarckian") have been studied for use in trajectory planning: their characteristic is that the generation of the son from the parents is not so random but takes into account stricter pre-fixed rules (Davidor, 1991). These kinds of procedures have shown a lower iteration number to reach the solution, but a similar computing time to get it than the pure genetic algorithm.

The best result obtained has certainly been the implementation of the algorithm described in a software package actually effectively used for cams profile.

Bibliography

Gupta M.C. 1993 *Minimizing flow time variance in a single machine system using genetic algorithms.* European Journal of Operational Research, 70, Elsevier.

Hajima R., Takeda N., Ohashi H., Akiyama M. 1992. *Optimization of wiggler magnets ordering using a genethic algorithm*. Nuclear Instruments and Methods in Physics Research, A318, Elsevier.

Holland J.H. 1992. Genetic Algorithms. Scientific American.

Muscia R., Poloni C.1993. *Optimal kinematic synthesis of plane mechanisms by genetic algorithms.* Proceedings of the Int. Conf. Design to Manufactury in Modern Industry, Bled (Slovenia), June 1993.

Yun Wei-Min, Xi Yu-Geng. 1994. *Genetic optimum motion planning in koint space with dynamic controlo constraints.* Proceediings of the 3rd ICARCV '94 (International Conference on Automation, Robotics and Computer Vision), Singapore, Nov. 1994.

Magnani P.L. Ruggieri G, 1986. Mechanisms for automatic machines (in Italian). UTET Edition, Torino, Italy

Davidor Y. 1991 Genetic algorithms and robotics. *Series in Robotics and Automated Systems.* Vol.1. World Scientific.

Evolving Strategies Using the Nearest-Neighbor Rule and a Genetic Algorithm

Matthias Fuchs
Center for Learning Systems and Applications
Fachbereich Informatik, Universität Kaiserslautern
Postfach 3049, 67653 Kaiserslautern
Germany
E-mail: fuchs@informatik.uni-kl.de

ABSTRACT

We propose a method for evolving strategies based on the nearest-neighbor rule. A strategy corresponds to an action selection function which chooses an action depending on the current state of a (reactive) control problem to be solved. Given a set $\mathcal{I}$ of state/action pairs, our approach determines the action to be taken in state S' as the action A from $(S, A) \in \mathcal{I}$ for which S' and S are "nearest" among all states from members of $\mathcal{I}$. The set $\mathcal{I}$ is evolved using a genetic algorithm. Our approach exceeds "standard" condition/action rule based approaches regarding the ways the state space can be subdivided. This is achieved without confronting the genetic algorithm—unlike neural networks—with too hard search problems as preliminary experiments with variations of a pursuit (predator–prey) game demonstrate.

1 Introduction

The search for optimal or at least adequate strategies for solving (reactive) control problems is a central element of machine learning. Based on the current problem state, such a strategy determines some action which is then executed, transforming the current state into a successor state. (Intended and actual effect of an action possibly do not always agree, because an action *cannot* or *must not* be applied. Furthermore, interferences or disturbing effects "from the outside" may occur.) Given an initial state, the objective of a strategy is to eventually reach a goal state (e.g., a winning state as in game playing, [Sm83]) or to produce a (possibly infinite) sequence of states which satisfy certain conditions (e.g., pole balancing, [BSA83]) through its repetitive application resulting in continuous state transformations. Assuming the common representation of a state S as a vector of n *state variables* from the set $\Re$ of real numbers, a strategy corresponds to an *action-selection function* $\Psi : \Re^n \rightarrow \mathcal{A}$ which maps a state S from the *state space* $\Re^n$ to an action A from the set $\mathcal{A}$ of possible actions.

There are many ways to realize such a strategy Ψ. For many applications it is necessary that a strategy Ψ can be learned without being explicitly taught, e.g., by providing examples of its desired behavior. In this case, learning essentially amounts to a search problem. The search space of all strategies mostly is highly unstructured and usually huge, possibly infinite. The use of a genetic algorithm (GA, [Ho92], [Jo88], [Ko91]) has proven profitable under these unfavorable conditions where simple schemes to explore the search space like hill-climbing or (pure) random search have (in general) no chance to get sufficiently close to an optimum (cp. [Ko91]).

Despite the power of the GA, nevertheless a reasonable compromise between the intricacy of the search space and the expressiveness of the strategies should be sought. We propose here a design based on the *nearest-neighbor rule* (NNR, [CH67]). A strategy Ψ based on the NNR is simply given as a set $\mathcal{I}$ of $m_{\mathcal{I}}$ state/action pairs, i.e., $\mathcal{I} = \{(S_i, A_i) \in (\Re^n \times \mathcal{A}) \mid 1 \leq i \leq m_{\mathcal{I}}\}$. Given a state $S \in \Re^n$, the action A_j chosen by Ψ satisfies the condition that there is no S_i which is closer to S than S_j according to some distance measure.

Approaches based on the NNR fascinate with their simplicity. Nevertheless (or because of that) they have shown considerable success in connection with Instance-Based Learning ([AKA91]) and classification problems (e.g., [MST94]). To our knowledge, all applications of the NNR involve a set $\mathcal{I}$ which in some way derives from a set of *given* training examples. (More recent work in connection with classification, however, does not make an explicit use of training examples. See for instance [DK95].) We propose here to *evolve* the set $\mathcal{I}$ in order to optimize the performance of the associated strategy Ψ. The following section 2 will put our approach into relation to other approaches. After that, section 3 will introduce the GA tailored to the evolution of a set $\mathcal{I}$ of state/action pairs. Section 4 presents experimental results obtained in connection with a pursuit (predator–prey) game. Finally, section 5 concludes this paper.

2 Methods for Action Selection

The *search* for adequate strategies (action-selection functions) using the GA can be tackled in many ways. The basic design of strategies determines (a) how expressive a strategy we can expect to find, and (b) how difficult the search is going to be. Point (a) surely favors the use of neural nets, because a neural net can realize (almost) any mapping provided that it is equipped with an apt topology and the weights are chosen appropriately (cp. [Sp65]). But evolving weight configurations and possibly also evolving net topology is a strenuous task even for the GA (cp. [ZM93]). Considering the fact that neural nets are "over-sophisticated" for many applications, simpler designs entailing more manageable search spaces are called for. Similar considerations apply to genetic programming ([Ko91]).

Classifier Systems ([Ho86], [Ho92]) can—by design—profit from the GA much more easily. Also, they are a very powerful methodology. But the rather complex architecture of a Classifier System and the sophisticated techniques involved (e.g., credit assignment, bidding, etc.) still make it worthwhile thinking about simpler methods.

The use of condition/action rules is a very popular approach in this context. Commonly, such a rule is of the form "`if` $a_1 \leq x_1 \leq b_1 \wedge \ldots \wedge a_n \leq x_n \leq b_n$ `then` *execute* $A \in \mathcal{A}$", where each condition $a_i \leq x_i \leq b_i$ specifies the range of values for state variable x_i (w.r.t. this rule). In a set of rules, the rules are (implicitly) or-connected. Hence, such a set of rules can subdivide the state space $\Re^n$ into hyper-rectangles. These may be partially overlapping, which calls for disambiguation techniques, and there may be gaps which necessitate some kind of default rules. The major limitation of these condition/action rules is the restricted way they can subdivide the state space. An approach based on the NNR and a set of state/action pairs as sketched in section 1 allows for subdividing the state space without overlapping and gaps, where the possible sub-spaces include, but are not limited to hyper-rectangles. As a matter of fact, such an approach can piecewise-linearly approximate arbitrary sub-space boundaries.

In [SS95], both approaches are combined. But the GA is only applied to evolve sets of condition/action rules, which then provide the NNR based component with state/action pairs. We couple here the GA and the NNR approach tightly, meaning that the GA is immediately responsible for the set $\mathcal{I}$ of state/action pairs which determine the behavior of the associated strategy $\Psi_{\mathcal{I}}$.

3 Our Genetic Algorithm

The principle of our approach is as follows: Given a set $\mathcal{I} = \{(S_1, A_1), \ldots, (S_{m_{\mathcal{I}}}, A_{m_{\mathcal{I}}})\}$ of $m_{\mathcal{I}} \geq 1$ state/action pairs, the associated strategy $\Psi_{\mathcal{I}}$ determines the action to be chosen when presented with state S in the following way: Let $\mathcal{D} : \Re^n \times \Re^n \rightarrow \Re$ be a distance measure. (We employed the Euclidean distance for our experiments.) Then $\Psi_{\mathcal{I}}(S) = A_j$, if $\mathcal{D}(S, S_j) = \delta_{min}$, where $\delta_{min} = \min(\{\mathcal{D}(S_i, S) \mid 1 \leq i \leq m_{\mathcal{I}}\})$ is the minimal distance between states in $\mathcal{I}$ and S. (To resolve ambiguities, j is the smallest index satisfying the above condition.)

So, the search space the GA is confronted with is the set of all finite sets of state/action pairs. Consequently, an individual corresponds to such a set $\mathcal{I}$ of state/action pairs respectively the associated strategy $\Psi_{\mathcal{I}}$. (Our GA hence follows the Pittsburgh approach in that each individual encodes a complete "solution" of the problem.) We now define the genetic operators, namely *crossover* and *mutation*, adapted to the present structure of individuals. (We assume that the reader is familiar with the principles of GAs. See [Jo88] or [Ko91] for a comprehensive introduction.)

The **crossover operator** randomly selects two distinct parents from the pool of $r\%$ best (surviving) individuals. A subset of the state/action pairs of each parent individual is chosen at random, and the union of these two subsets yields the "child" individual (*offspring*). (Note that the surviving individuals here simply all have an equal chance to become reproductive.) Thus, this operator complies with the basic idea of crossover, namely providing the ability to combine good partial solutions. Furthermore, the GA is supported by the modularity of individuals: Subsets of individuals actually can often be good partial solutions, e.g., for early or final stages of the problem solving process, because a subset of state/action pairs that work well for certain stages of the problem solving process will also function for these stages when being isolated. (The lack of modularity is the major problem GAs encounter in connection with neural networks.)

The **mutation operator** modifies individuals stemming from crossover before they are admitted to the next generation. An individual $\mathcal{I}$ is subject to mutation with probability P_{mut}. *If* an individual $\mathcal{I}$ is selected for mutation, a state/action pair of $\mathcal{I}$ is—with probability P_{rnd}—replaced by a randomly generated state/action pair. This operator introduces—controlled by the probabilities P_{mut} and P_{rnd}—random influences which are an essential ingredient of the GA. (Note that there is a probability $1 - P_{mut}$ that offspring is not subject to mutation *at all*, meaning that P_{mut} controls the percentage of offspring resulting from "pure" crossover.)

The **initial population** of the GA is (as usual) generated completely at random, i.e., for each individual $\mathcal{I}$ of the initial population both its size $m_{\mathcal{I}}$ and the $m_{\mathcal{I}}$ state/action pairs are determined at random. The size $m_{\mathcal{I}}$ of an individual is limited to some arbitrary but fixed $M \in \mathbb{N}$, meaning $m_{\mathcal{I}} \leq M$ for all individuals $\mathcal{I}$. This way, we impose restrictions on the possible ways the state space can be subdivided, but we also simplify the search space. Introducing such an upper bound M is reasonable, because many problems do not necessitate cutting up the state space strongly; it mostly suffices to find a relatively parsimonious division. Nevertheless, an appropriate choice of M is crucial. (Condition/action rule based approaches have a very similar difficulty in limiting the number of rules in some way.)

Apart from the genetic operators, the **fitness measure** is crucial for the performance of the GA. In our case, the fitness of an individual $\mathcal{I}$ is assessed in terms of the problem solving expertise of its associated strategy $\Psi_{\mathcal{I}}$. Since the fitness measure is the only connection between the GA and the actual (optimization) problem, we can here of course not define a general fitness measure. We therefore have to postpone details until a concrete problem is specified in the subsequent section which presents experimental results we obtained with our approach. Despite the simplicity of our GA, it was nonetheless able to find strategies that can handle variations of a pursuit (predator–prey) game.

4 Experiments

4.1 Experimental Setting

We experimented with our NNR-based approach in connection with variations of a pursuit (predator–prey) game first introduced in [BJD85]. The predator–prey scenario has received a lot of attention regarding multi-agent systems (e.g., [Os95]), providing a test ground of scalable difficulty that is easy to define and to realize on a computer. The basic idea of the game consists in a pack of k predators trying to capture their prey. Here, the game area is a $N \times N$ grid. Predator i occupies square (x_i, y_i), and the prey occupies square (x, y). The southern boundary of the grid is considered as the x-axis, and the western boundary as the y-axis, as depicted by figure 1 for $N = 10$.

The game is played in discrete time steps. At each time step, both *all* the k predators and the prey simultaneously execute one of the following five moves: move one square to the north, south, east, west, or stay put. (There are hence no diagonal moves.) Movements that would cause a predator or the prey to cross a grid boundary are ignored (i.e., are considered as 'stay put'). (Predators cannot hinder each other. In particular, they may even occupy the same square.) For the predators, the objective of the game is to cooperate in order to capture the prey (as fast as possible). The prey is captured if at least one of the predators and the prey both occupy the same square on the grid at some point in time. Note that the predators cannot capture the prey by simply outrunning it.

The state $S^{(i)}$ of predator i is given by its relative position w.r.t. its $1 \leq j \leq k$ $(j \neq i)$ fellow predators and the prey. See annotations of figure 1 for an example. (The positions of the predators and the prey, and hence the state of a predator naturally depend on the current time step. But this dependency is made explicit only if it is necessary to avoid confusion.) Consequently, the state space is $\{-N+1, \ldots, N-1\}^{2k}$. The strategies $\Psi^{(1)}, \ldots, \Psi^{(k)}$ of the predators operating on these states and controlling their moves (actions) are to be learned by the GA.

Considering this state representation, predators decide which action to take based on so to speak "visual communication" represented by relative positions. There is no central control, i.e., there is no leader of the pack that can tell the others what to do. The k predators basically are independent, meaning that predator i proceeds according to its respective strategy $\Psi^{(i)}$, attempting to contribute to their common goal of capturing the prey. This individuality is restricted by each $\Psi^{(i)}$ utilizing the same set $\mathcal{I}$ of state/action pairs. (In [Ko91] (chapter 2), a similar constellation is referred to as "evolution of emergent behavior".) Consequently, an individual of the GA—representing the strategies of the members of the pack—does not consist of k sets of state/action pairs, but merely of one such set. This restriction simplifies the search space. However, it prevents individual predators to perform specialized tasks. But for this kind of pursuit game, it suffices that predators fulfill their duties according to their spatial position in the pack in order to be successful. Nonetheless, permitting (complete) individuality is an interesting subject for further studies.

The objective of the prey is to evade its predators (as long as possible). To achieve this goal, the prey moves according to one of the given *escape strategies* Φ_1 and Φ_2.

Escape Strategy Φ_1: The prey (located at square (x, y)) determines the (minimal) number of moves that separate it from each predator. This number respectively *distance* is given by $|x - x_i| + |y - y_i|$ for each predator i. The k distance values are sorted in ascending order yielding $(d_1, \ldots, d_k)$. The prey then "simulates" the four moves to the north, south, east and west, and determines the respective sorted distance values for each of these four moves (excluding those moves which cause the prey to cross a grid boundary) the same way as for its actual location. If none of the four moves results in a $(d'_1, \ldots, d'_k)$ which is "better" than $(d_1, \ldots, d_k)$, i.e., greater w.r.t. the usual lexicographic ordering $>_{lex}$ (left-

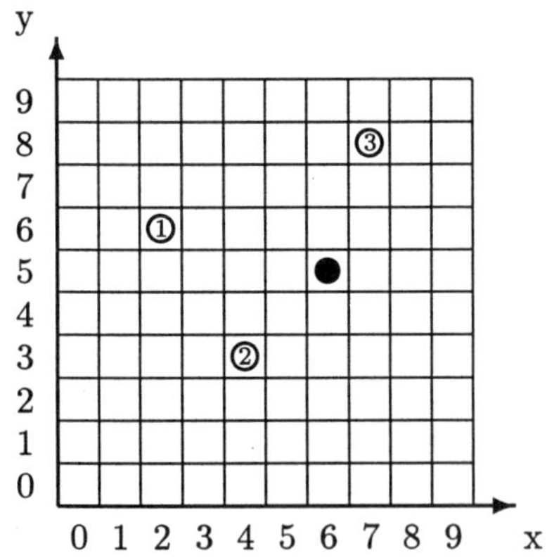

On the left, an example game situation is displayed. The prey (filled circle) occupies square $(x, y) = (6, 5)$. The $k = 3$ predators are located at squares $(x_1, y_1) = (2, 6)$, $(x_2, y_2) = (4, 3)$ and $(x_3, y_3) = (7, 8)$, respectively.
The state of predator 2 is

$$\begin{aligned} S^{(2)} &= (-2, 3, 3, 5, 2, 2) \\ &= (x_1 - x_2, y_1 - y_2, x_3 - x_2, \ldots, y - y_2) \end{aligned}$$

given by its relative position to predators 1, 3, and the prey.

Figure 1: The game area: a $N \times N$ grid, here shown for $N = 10$.

to-right comparison with >), then the prey stays put. Otherwise, the best move is chosen.

Escape Strategy Φ_2: Escape strategy Φ_2 is an extension of Φ_1. In addition to the distances from the predators, the prey here also takes into account its distances x, $N - x - 1$, y and $N - y - 1$ from the western, eastern, southern, and northern grid boundary, respectively. Consequently, there are $k + 4$ distance values which are treated uniformly as described for escape strategy Φ_1. By also trying to maximize the distance from grid boundaries, escape strategy Φ_2 alleviates a "flaw" of escape strategy Φ_1 which essentially makes the prey run away to some grid boundary where it can be captured much more easily than it can be when trying to stay in "open field" as long as possible.

We can now specify the missing component of the GA, namely the **fitness measure** ϑ. The fitness measure $\vartheta(\mathcal{I}) \in \mathbb{N}$ is to rate the performance of an individual $\mathcal{I}$, i.e., a set of state/action pairs. The performance of $\mathcal{I}$ is given by the capturing abilities of the associated strategy $\Psi_{\mathcal{I}}$ employed by all predators. ϑ should therefore reflect *if* the predators are successful, and in that case, how fast they captured the prey. In case of failure, ϑ should somehow express how close to success the predators came. Naturally, the predators cannot be granted an arbitrary amount of time. If they do not succeed in capturing the prey within T time steps, then their hunt is judged a failure. Hence

$$\vartheta(\mathcal{I}) = \begin{cases} t_s, & \text{success after } t_s \leq T \text{ steps} \\ \sum_{t=1}^{T} \sum_{i=1}^{k} \delta(i,t), & \text{in case of failure,} \end{cases}$$

where $\delta(i, t)$ is the distance $|x_i^{(t)} - x^{(t)}| + |y_i^{(t)} - y^{(t)}|$ of predator i from the prey at time step t. Hence the fitness measure $\vartheta(\mathcal{I})$ is the smaller the fitter $\mathcal{I}$ is considered to be. If two individuals have the same fitness measure, then the more "concise" and therefore in a way more "efficient" individual containing less state/action pairs is preferred.

4.2 Experimental Results

For our experiments, the side length N of the grid was set to 50. The maximal number M of state/action pairs of an individual was set to 30. The population size of the GA was 100, and the probabilities P_{mut} and P_{rnd} were set to 50% and 10%, respectively (cp. section 3). The survival rate was $r = 30\%$. (These settings were determined after a few preliminary tests.) At the start of a hunt, the prey is located in the middle of the grid, and the predators are lined up along the whole western boundary with equal distance from each other. The number of predators ranged from 2 to 5. (A single predator has no chance to capture a prey that uses escape strategy Φ_1 or Φ_2.) $T = 200$ time steps were permitted.

Table 1 displays the experimental results. The GA terminated as soon as a successful individual was found. The columns show the number of predators (k), the number of cycles performed (i.e., the number of generations generated) by the GA until termination, the run time (CPU time, obtained on a SPARCstation 10), the number of state/action pairs of the successful (best) individual (m), and the number of time steps necessary to capture the prey (t_s).

The number of cycles necessary to find a successful individual $\mathcal{I}$ demonstrate two things. First, escape strategy Φ_2 actually is significantly more difficult to cope with than escape strategy Φ_1. Second, the search naturally becomes the more difficult the more predators come into play. It is (intuitively) clear that more predators can capture the prey more easily. But this does not necessarily compensate for the (exponential) growth of the search space. This observation was sustained when we let the GA continue after it had found the first successful individual in connection with escape strategy Φ_1. The GA then terminated as soon as the best individual had not improved for 20 cycles. For $k = 3$, m was reduced to 5, and t_s to 60 after 49 cycles. (The GA terminated after $49 + 20 = 69$ cycles.) Similar reductions of m were

Table 1: Results of the GA regarding Φ_1 and Φ_2

	Escape Strategy Φ_1				Escape Strategy Φ_2			
k	cycles	run time	m	t_s	cycles	run time	m	t_s
2	7	16 sec.	28	105	301	273 sec.	7	111
3	3	14 sec.	25	80	93	342 sec.	16	80
4	8	49 sec.	22	84	1769	5319 sec.	10	82
5	16	168 sec.	9	96	2046	8544 sec.	13	72

possible for $k = 2$ and $k = 4$, although t_s diminished only marginally in these cases. For $k = 5$, the first successful individual did not improve for 20 cycles.

Furthermore, preliminary analyses of the predators' moves revealed that some interesting strategies had been developed, e.g., a phalanx-like attack pattern ($k = 4$), or three predators surrounding the prey on three sides and then "accompanying" it to the next grid boundary, where it then became trapped so that one of the predators had no trouble moving onto the prey's square. But these analyses also revealed that higher numbers of predators were not made use of properly. For instance, some of the predators simply "stood by". This kind of behavior is not very surprising, since basically two predators suffice to succeed, and "elegance" of the hunt is merely rewarded allusively by the fitness measure.

We have also run a series of experiments with an escape strategy which essentially performs random moves and hence introduces uncertainty. Due to space limitations, we can only report on results very briefly. The results show that strategies evolved which enabled even a single predator to capture the prey in all of 100 test trials (hunts). If the prey is employing a random escape strategy, a single predator can be successful. Essentially, the predator approaches the prey, and then continuously moves to some square next to the prey, "hoping" that this is also the square the prey picks at random for its next move. The average number of time steps necessary to capture the prey can—as expected—be reduced significantly when more predators come into play.

The results of our experiments naturally depend on the chosen generator of (pseudo-) random numbers, since our approach like any other application of a GA utilizes random effects. The results listed in table 1 were obtained using the built-in generator '`random`' provided by the GNU C-Compiler with the default seed 1. For these initial studies we refrained from a thorough test of the effects alternative sequences of random numbers have, but preliminary experiments on that score did not lead to the conclusion that the default sequence is especially favorable or unfavorable.

Finally, we would like to point out a possible improvement of the genetic search. In our experiments, random states were created by producing random numbers which were uniformly distributed in $\{-49, \dots, 49\}$, hence yielding states uniformly distributed in the respective state space $\{-49, \dots, 49\}^{2k}$. Since the hunt becomes the most complicated in its final crucial stage where predators and prey are very close to each other—that is, near the state $(0, \dots, 0)$—, the state space must be diversified more in the vicinity of $(0, \dots, 0)$ than in larger distances of $(0, \dots, 0)$ in order to account for the subtleties of the final phase of a hunt. Therefore, a sequence of random numbers where numbers close to 0 are more likely to occur appears to be profitable (which preliminary experiments have sustained).

5 Discussion

We have presented a method for evolving strategies for problem solving which essentially amount to action selection. The hub of our approach is the nearest-neighbor rule ([CH67]) which determines the action to be chosen based on the nearness of the current state of the problem solving process to the states in a set $\mathcal{I}$ of state/action pairs. This set $\mathcal{I}$ is evolved using a genetic algorithm. We have illustrated our approach in the light of a pursuit (predator–prey) game, where the objective of k predators is to capture a prey moving on a $N \times N$ grid.

Although this pursuit game provides an interesting experimental setting, there are further fields of application which are of bigger (practical) interest. For instance, learning how to steer some piece of machinery is surely within the scope of our method (readings of instruments represent the current state). Controlling the movements of limbs of an artificial animat (robot) in order to move or grasp something is a further interesting experimental environment where our approach can be of help (angles of joints represent the current state). But an adequate simulation of physical laws make these experiments too complicated for initial studies.

Furthermore, our approach has been successfully employed for obtaining optimized nearest-neighbor classifiers based on *generated* instances ([FA96]). Unlike common approaches based on the nearest-neighbor rule, the given set of training instances is only used to measure the fitness of a classifier that employs a set of instances evolved by our approach (states correspond to the attribute vectors describing the objects to be classified, and actions correspond to classes).

The genetic algorithm presented here is a "first generation" experimental (purely Darwinian) genetic search procedure which leaves much room for improvements. In particular the introduction of credit assignment to state/action pairs (which are the counterpart of condition/action rules), and the examination of Lamarckian operators capable of inducing more "intelligent" search behavior (cp. [Gr91]) are promising directions of future research.

We are convinced that our approach can be profitable for many (reactive) control problems and related problems. Its major advantage is the ability to subdivide the state space more flexibly than standard condition/action rule based approaches, which are merely able to single out a number of hyper-rectangles. Nevertheless, we believe that our approach—unlike neural networks—does not entail significantly more complex search spaces for the genetic algorithm. Naturally, it inherits the weaknesses of all methods based on the nearest-neighbor rule, e.g., difficulties to handle nominal values properly. Also, just like approaches based on condition/action rules, our approach is inappropriate if the state space must be cut up extremely. But it appears to be a suitable compromise between condition/action rule based approaches and neural networks which deserves further studies.

Acknowledgments

This work was supported by the *Deutsche Forschungsgemeinschaft (DFG)*. I thank the referees for valuable comments on the paper. I also thank Jörg Denzinger and Andreas Abecker for helpful discussions.

Bibliography

[AKA91] **Aha, D.W.; Kibler, D.; Albert, M.K.:** *Instance-Based Learning Algorithms*, Machine Learning 6, 1991

[BJD85] **Benda, M.; Jagannathan, V.; Dodhiawalla, R.:** *An Optimal Cooperation of Knowledge Sources*, Technical Report BCS-G2010-28, Boeing AI Center, 1985

[BSA83] **Barto, A.G.; Sutton, R.S.; Anderson, C.W.:** *Neuronlike elements that can solve difficult learning control problems*, IEEE Transactions on Systems, Man and Cybernetics **13**:835–846, 1983

[CH67] **Cover, T.M.; Hart, P.E.:** *Nearest Neighbor Pattern Classification*, IEEE Transactions on Information Theory, Vol. IT-13, Jan. 1967, pp. 21–27

[DK95] **Datta, P.; Kibler, D.:** *Learning Prototypical Concept Descriptions*, Proc. 12th International Conference on Machine Learning, 1995, Tahoe City, CA, USA, pp. 158–166

[FA96] **Fuchs, M.; Abecker, A.:** *Optimized Nearest-Neighbor Classifiers Using Generated Instances*, Technical Report LSA-96-02E, University of Kaiserslautern, 1996 [`http://www.uni-kl.de/AG-AvenhausMadlener/fuchs.html`]

[Gr91] **Grefenstette, J.J.:** *Lamarckian Learning in Multi-agent Environments*, Proc. 4th International Conference on Genetic Algorithms, 1991, San Diego, CA, USA, pp. 303–310

[Ho86] **Holland, J.H.:** *Escaping brittleness: The possibility of general-purpose learning algorithms applied to parallel rule-based systems* In R.S. Michalski, J.G. Carbonell, T.M. Mitchell (eds.), Machine Learning: An Artificial Intelligence Approach (Vol. 2), Los Altos, CA, Morgan Kaufmann, 1986

[Ho92] **Holland, J.H.:** *Adaptation in natural and artificial systems*, Ann Arbor: Univ. of Michigan Press, 2nd edition, 1992

[Jo88] **De Jong, K.:** *Learning with Genetic Algorithms: An Overview*, Machine Learning **3**:121–138, 1988

[Ko91] **Koza, J.R.:** *Genetic Programming: On the Programming of Computers by Means of Natural Selection*, MIT Press, Cambridge, MA, 1991

[MST94] **Michie, D.; Spiegelhalter, D.J.; Taylor, C.C.:** *Machine Learning, Neural and Statistical Classification*, Ellis Horwood, 1994

[Os95] **Osawa, E.-I.:** *A Metalevel Coordination Strategy for Reactive Cooperative Planning*, 1st International Conference on Multi-Agent Systems, 1995, San Francisco, CA, USA, pp. 297–303

[Sm83] **Smith, S.F.:** *Flexible Learning of Problem Solving Heuristics through Adaptive Search*, Proc. International Joint Conference On Artificial Intelligence, Karlsruhe, FRG, 1983, pp. 422–425

[Sp65] **Sprecher, D.A.:** *On the Structure of Continuous Functions of Several Variables*, Trans. Amer. Math. Soc., **115**:340-355, March 1965

[SS95] **Sheppard, J.W.; Salzberg, S.L.:** *Combining the Genetic Algorithm with Memory-Based Reasoning*, Proc. 6th International Conference on Genetic Algorithms, 1995, Pittsburgh, PA, USA

[ZM93] **Zhang, B.-T.; Mühlenbein, H.:** *Genetic Programming of Minimal Neural Nets Using Occam's Razor*, Proc. 5th International Conference on Genetic Algorithms, 1993, pp. 342–349

Recognition and Reconstruction of Visibility Graphs Using a Genetic Algorithm

Marshall S. Veach
Computer Science Dept.
Gates 1B Room 196
Stanford University
Stanford, California 94305-9015
mveach@cs.stanford.edu

ABSTRACT

Recognizing and reconstructing simple polygons from their combinatorial visibility graphs remains one of the open problems in computational geometry. Although much work has been completed on restricted versions of the problem, very little is known about the nature of the general solution. In this paper, I consider the use of a genetic algorithm (GA) as a means of solving the general form of the visibility graph reconstruction problem. The concept of a blocking vertex visibility graph is introduced; and a GA is described which makes use of this structure to construct polygons satisfying certain visibility constraints.

1. Introduction

For the purposes of this paper, the visibility graph of a simple polygon P, denoted $VG(P)$, is defined as a graph $G = (V,E)$, where there exists a one-to-one correspondence between the combinatorial vertices in V and the geometric vertices of P; an edge exists between two vertices in V iff there exists an internal line of sight (on the plane and unobstructed by polygonal edges) between the two corresponding vertices in the polygon, P. Figure 1 shows an example of a polygon and its associated visibility graph. This purely combinatorial structure captures the notion of which vertices can "see" each other within a polygon. Useful in such applications as robotics and pattern recognition (Prasad 1995) as well as having significant theoretical importance in the field of computational geometry (O'Rourke 1987), the visibility graph and related structures have been considered by several independent groups (Abello et al. 1992, Coullard 1991, Corneil et al. 1990, ElGindy 1985).

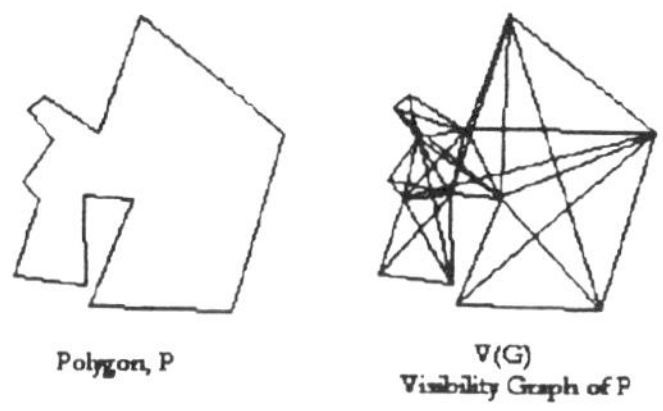

Figure 1

For the purposes of implementing a GA, a variant of the visibility graph was developed: the blocking vertex visibility graph. In a polygon, a blocking vertex is one whose internal angle is concave. These are vertices which are responsible for "blocking" one vertex from seeing another vertex in the polygon—they can be thought of as the cause of non-edges in the corresponding visibility graph. The blocking vertex visibility graph is similar to the visibility graph for a polygon except with the added information of associating a blocking vertex with all non-edges—this blocking vertex is chosen as being probably responsible for that non-edge.

Perhaps the canonical visibility graph problem, known as the visibility graph recognition problem, is deceptively easy to state: Given a graph $G = (V,E)$, determine whether or not there exists a simple polygon, P, for which $G = VG(P$; in other words, determine whether or not G is a visibility graph for some simple polygon on the plane. A closely related problem, the visibility graph reconstruction problem, has a similar description: Given a graph $G = (V,E)$, s.t. $G = VG(P)$ for some P, find P. In other words, given a graph which is a visibility graph, find its polygon. It should be clear that a solution to the reconstruction problem provides an implicit solution to the recognition problem.

This paper contributes a GA designed to recognize a visibility graph by reconstructing one of its associated polygons. The GA has the following notable properties: 1) the structures manipulated by the GA are interpreted as operations which modify geometric objects and 2) although there does not exist previous work to which one can compare the empirical performance of the GA, the results seem encouraging. The rest of this paper contains the following sections: section 2 provides a detailed description of the search space associated with the problem; the nature of the structures undergoing evolution; and the algorithm which computes the fitness measure for the individuals of the population; section 3 discusses the parameters chosen for the GA runs; section 4 details the results achieved by this work.; section 5 offers thoughts on how to extend these experiments; and in Section 6 the paper is concluded.

2. Methodology

This paper describes the design, implementation and testing of a genetic algorithm which, when given a valid visibility graph, finds a polygon whose visibility graph is equivalent to the input visibility graph. More precisely, this paper describes a set of genetic algorithms rather than a particular genetic algorithm. No single genetic algorithm was created which could solve the reconstruction problem for variable sized polygons. Rather, when solving any given instantiation of the reconstruction problem the parameters of the GA are a function of the size of the visibility graph. When the size of the visibility graph changes, the GA uses a different parameter set.

2.1. Description of Search Space

The genetic algorithm is searching for a polygon with a particular characteristic: namely, that it has a visibility graph which satisfies certain constraints. Thus, in general, the algorithm needs to search over the space of all possible polygons—which is, unfortunately, unboundedly large. So, in order to make the search feasible, the geometric space of the experiments described in this paper was reduced from the real coordinate system to an integer coordinate system of size 100x100. Although some restriction on real space must be made in order for the genetic algorithm to achieve success, it should be obvious that the particular restriction need only be a function of the computational resources available.

Estimating the number of polygons of any particular size (number of vertices), n, which are realizable on a given matrix turns out to be a non-trivial task in and of itself. One easily obtain a naive upper bound by counting the number of possible cells a vertex could occupy. For example, on a 100x100 matrix, this method yields an upper bound of:

$$\frac{100{,}000!}{(100{,}000 - n)!}$$

There are two problems with this estimate: first, it clearly counts polygons which are geometrically equivalent (but have different point configurations) more than once; and second, it counts point configurations which do not represent valid polygons. Unfortunately, it is not clear how to accurately count the occurrences of either one of these overestimations. Let it suffice to say that the search space is: a) a function of the size of the matrix and the size of the polygons being tested, and b) is no larger than the upper bound described above.

2.2. Description of String Structure

As described previously, the genetic algorithm must search over the set of all possible polygons in some geometric space. In order to do this, it must operate on chromosome string structures which somehow represent polygons. A straightforward design for the chromosome would be to interpret it as a set of x-y coordinate points on the plane. There are, however, two immediate problems with this design. First, an unknown (but, presumably, prohibitively large) percentage of random strings would not even be realizable as valid polygons--rather, some of the edges between the points would intersect resulting in a complex polygon. And more importantly, even when one was able to realize valid polygons from random chromosome strings, there would still be little chance that crossing two valid strings (in the reproduction phase of the GA) would, in turn, yield a new chromosome which was a valid polygon. Fortunately, a solution to the reconstruction problem does not depend on having a particular polygon vertex at any particular coordinate. In other words, it is the relative location of the polygon vertices (rather than the absolute location) which determine a polygon's visibility graph. As such, it seems much more intuitive for the chromosome to somehow represent this relational information.

The chromosome actually implemented with the GA was designed to address the above considerations. The chromosome does not (by itself) represent a polygon, per se. Rather, it represents operations which can be used to modify any given polygon. In the experiments described in this paper, all of the chromosomes were assumed to modify a regular polygon of size n which was centered in the geometric space—this polygon is the embryo polygon. The chromosome was designed so that every possible chromosome string represented a valid polygon. This was done by interpreting the string to consist of n objects, where each object holds information on how to modify one of the vertices of a regular n-gon without destroying its polygonal structure. The objects (which can be interpreted as vectors in two dimensional space) have two components: an integer direction, and a real numbered magnitude. So each object can be interpreted as holding operations which modify a vertex in a polygon by moving it in a particular direction for some distance. Actually, even this design will not have the desired affect if the magnitude is interpreted absolutely. If this was the case, then there would be no way to prevent the chromosome from pushing a polygon vertex through one of the existing polygonal edges. Instead, the distance portion of the object takes on a value in the range [0,1]. When evaluating a chromosome string, the GA first determines the maximum distance that a vertex can be pushed in a particular direction without creating a complex polygon. Then it moves that vertex the proportion of the possible distance as determined by the distance value in the object. In this manner, the GA is able to interpret all chromosome strings as valid polygons.

For example, consider an individual in the reconstruction system for a visibility graph with 4 vertices:

Object 1	Object 2	Object 3	Object 4
230 ; .73	172 ; .48	258 ; .89	0 ; 0

In the actual system, the above chromosome is stored and processed as a string of bits; but the above abstraction better illustrates the nature of the structure. Figure 2 shows the way that the above chromosome creates a polygon. Starting with a regular polygon of size 4, each of the objects are interpreted as operations for manipulating a vertex. The direction integer corresponds to one of the 400 points which make up the geometric space boundary; and, as described previously, the distance component moves the vertex some proportion of a "safe" distance. Notice that the fourth object's value of (0 ; 0) allows the vertex to remain in its original location.

2.3. THE FITNESS FUNCTION

Once the GA has translated a chromosome into a polygon, it performs calculations on that polygon to measure the fitness of the chromosome. Ultimately, the GA is searching for a chromosome which represents a polygon that has a visibility graph identical to the target graph. So the first step in evaluating a polygon is computing its visibility graph. This is a relatively straight forward procedure which can be accomplished in $O(n^2)$ (Hershberger 1989). For these experiments, the GA calculated the blocking vertex visibility graph of the polygon. A description of this structure can be found in Section 1 of this paper. The algorithm used in these experiments to compute the blocking vertex graph of a polygon is also straight forward and runs in $O(n^2)$. Once the visibility graph has been computed, it is matched against the target graph. If the two graphs are identical, then the GA has successfully reconstructed the polygon of the target graph. If, however, there are discrepancies between the two graphs, then the following method is used to calculate a relative fitness for the polygon:

Every discrepancy between the two graphs is caused by one of two possible problems: either a vertex, v_1, of the polygon is blocking an internal line of sight between vertices v_2 and v_3 in the polygon which is associated with an edge in the target graph, or v_1 is not blocking an internal line of sight between v_2 and v_3 which does not have an edge in the corresponding target graph. In either case, the fitness contribution made by such a discrepancy is equal to the area of the triangle made by v_1, v_2, and v_3. As v_1 moves closer to crossing the line of sight as dictated by the target graph, the fitness contribution made by the discrepancy becomes smaller and smaller. The total fitness for the entire polygon is found by summing the fitness contributions of every discrepancy between the two graphs.

Intuitively, the target graph dictates that certain internal lines of sight should be crossed by edges of the polygon and that the other internal lines of sight should be clear. The above fitness function approaches 0 as vertices in the polygons get geometrically closer to their appropriate regions. When all vertices are in their proper geometric regions, then the fitness function will be 0 and the problem will have been solved.

3. GENETIC ALGORITHM PARAMETERS

The following tableau summarizes the GA parameters used for the experiments described by this paper—it is important to note that the below tableau is actually a template for a GA which solves the problem for a fixed polygon size:

Objective	Solve the visibility graph reconstruction Problem. More specifically, given a visibility graph, find a polygon whose visibility graph matches the input graph.
Representation Scheme	Structure = fixed length string K = 2 L = n*19 (n = number of vertices in graph) Portions of the structures are interpreted as rules for building a polygon. A detailed description of this process can be found in Section 3.
Fitness Cases	One.
Raw Fitness	The sum of the area of all out of place vertices in the polygon.
Parameters	M = 500 , G = 1000
Termination Criteria	The GA has run for G generations or the best so far individual in the population has a fitness which is less than 1.
Result designation	The best-so-far individual in the population.

Table 1

Most of the choices for the GA parameters found in the above tableau are straight forward. Two, however, deserve further explanation: the population and the termination criteria. As described earlier in this paper, an individual of the population consists of n vectors. The direction component of the vector takes on a value in the range from 0 to 395 (the number of points on the border of the geometric space) and the distance component is

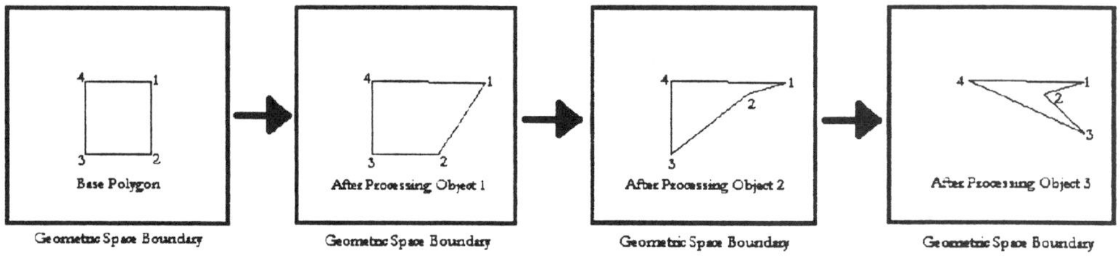

Figure 2

interpreted as a floating point number ranging between 0 and 1 precise to the 1/100th place. By using a population of 500 members, it increases the likelihood that all needed values will already exist in the initial random population—therefore, the GA will not need to generate any new numbers via mutation. Instead, it must merely find the best combination of the values in the initial population. It is important to note that these vector objects correspond to the concept of building blocks as described by Goldberg (Goldberg 1989). The GA operates on these objects (which are relatively short) by concatenating the highly fit ones together.

Another interesting parameter choice can be found in the termination criteria. The problem has not been solved unless the GA finds an individual with a fitness of 0. Since the current version of the GA terminates once the best individual has a fitness of less than 1, the system halts before the problem has been solved. This highlights one of the shortcomings of the GA approach. Genetic algorithms' strength lies in their ability to narrow a difficult search over a large space to the correct region of the solution. However, once the GA has found the correct region, it tends to have trouble actually finishing the job by finding the exact solution. This phenomena has been observed in a large number of previous GA applications including (now) the visibility graph reconstruction problem. Fortunately, the GA comes so close to the exact solution that a very simple hill climbing strategy finishes the job with a trivial amount of work. So once the GA has come within a unit of the exact solution, the best individual is passed to a hill climbing algorithm which finds the exact solution. It is important to note that there is nothing fundamental preventing the GA from finding the exact solution. The GA just tends to become increasingly inefficient as the changes necessary for better fitness become more exact. A strategy where the final answer provided by the GA is tweaked to obtain the exact solution is simply much more efficient than letting the GA churn on indefinitely.

4. Results & Discussion

In order to test the design described in this paper, a working implementation of the reconstruction system was developed on two separate machines: an HP 9000/755/99 with 512 MB RAM running HPUX 9.05 and a PC with the Intel Pentium chip, 16 MB RAM running MS-Windows95. On the HP, the compiler was GNU's gcc compiler. And on the PC, Borland Turbo C++ was used for compilation. On both machines, the genetic algorithm shell software, GENESIS, was used for the generic GA operations.

The first step in testing the system required the generation of test data. Since there does not exist an efficient method of generating random visibility graphs directly, a series of random polygons were created. The visibility graphs of these polygons were then generated to serve as test input for the reconstruction system. For the purposes of these experiments the size (in number of vertices) of the test graphs was limited to the range between 5 and 25. In particular, 20 random polygons (and their visibility graphs) were generated for each of these sizes; making for a total of 400 visibility graphs with which to test the system. Figures 3 and 4 summarize the results of running the GA on this data.

Figure 3 shows the average number of generations required for each input graph size. Notice how for smaller graph sizes, the GA either randomly generates a solving polygon, or finds a solution within one or two generations. This phenomena occurs due to the small number of distinct visibility graphs associated with small polygons. For example, there are only five distinct visibility graphs of size four. So, in a random initial population of size 500, the chances of generating a polygon with the target graph are extremely high. Unfortunately, the number of distinct visibility graphs associated with polygons of size n increases exponentially with n. However, despite this exponential increase in the number of possible polygons, the GA does an outstanding job of ignoring the extraneous information. This is reflected in the relationship between the size of the polygon and the number of generations required for solution. However, the number of generations required to solve any particular problem does not provide much information without knowing what kind of computational resources are required to process a generation. Figure 4 provides information which should allow readers to assess the feasibility of the reconstruction system.

Figure 4 provides the average wall clock time required to compute a solution for each of the input graph sizes. This information has been included to illustrate that the reconstruction system has potential for practical

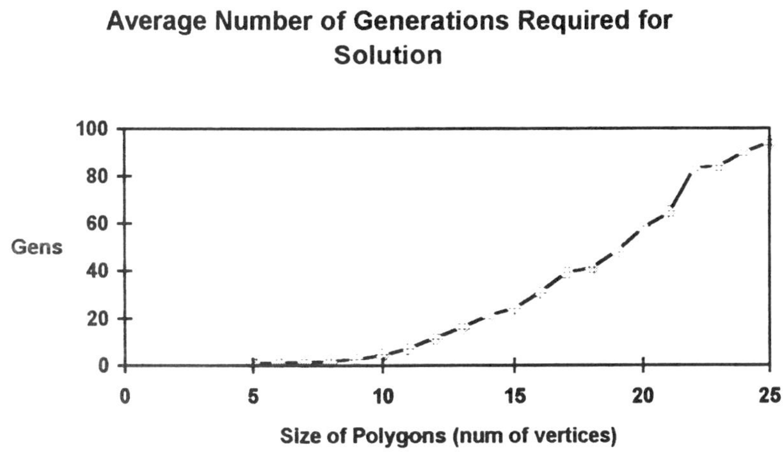

Figure 3

applications such as those mentioned in Section 1. Not only does the current implemented system have reasonable return times over this test data, but there exist many ways that the current system can be optimized to decrease the execution time required for a fitness calculation. The system described by this paper was developed to test only whether a genetic approach could achieve any success at solving the visibility graph reconstruction problem; it was not optimized for actual field use. Some specific thoughts on how the system can be optimized are included the next section.

5. FUTURE WORK

The work described in this paper can be extended in a number of ways:

The code which calculates the fitness of a particular individual in the population was not optimized. There are several changes which would significantly decrease the wall clock time required for a GA run: 1) the algorithm used to generate the visibility graph of a polygon can be improved by an order of magnitude, 2) a domain specific GA shell could be developed (rather than use GENESIS) which would eliminate any unnecessary operations, and 3) a table of values for trigonometric functions could be held in memory for use in geometric calculations. After optimization, the system should be tested with a larger number (and variety) of test cases. Also, one could determine a set of visibility graphs likely to occur in some practical application. By using this set as a test of the reconstruction system, one could make a stronger case for the use of the system in practical applications.

Also, a single GA which used variable length strings could replace the need to run multiple GAs with different parameter sets. Examining how well a variable length string GA handled this problem could prove interesting. It is possible that a variable length string GA would need fewer generations (on average) to achieve a solution.

6. CONCLUSION

In this paper, I have described a GA which attempts to solve the general form of the visibility graph reconstruction Problem. The results show that for the test cases provided to the GA, a correct answer can be returned in a reasonable amount of time. This provides encouragement not only for developing further solutions to visibility graph problems, but also for applying the GA paradigm to geometric structures in general.

Average Wall Clock Time Required for Calculation

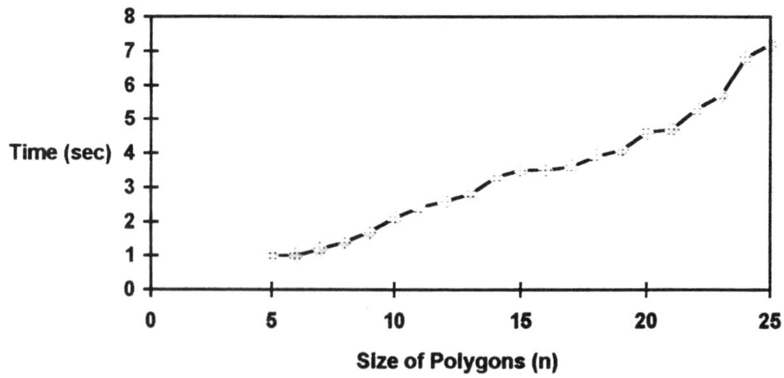

Figure 4

ACKNOWLEDGMENTS

I would like to thank James Abello and John Koza for their criticism and advice related to the work described in this paper.

BIBLIOGRAPHY

Abello, J., Egecioglu, O. and Kumar, K. 1992. Visibility graphs of staircase polygons and the weak Bruhat order I: from visibility graphs to maximal chains. Discrete and Computational Geometry.

Corneil, D. and Everett, H. 1990. Recognizing visibility graphs of spiral polygons. J. of Algorithms. (11), 1-26.

Coullard, C. 1991. Distance Visibility Graphs. Proceedings of the Seventh Annual Symposium on Computational Geometry. New York, New York: ACM Press. Pages 289- 296.

ElGindy, H. 1985. Hierarchical decomposition of polygons with applications, Ph.D. Thesis, McGill Univ.

Goldberg, D. 1989. Genetic Algorithms in Search, Optimization and Machine Learning. Addison-Wesley.

Hershberger, J. 1989. An optimal visibility graph algorithm for triangulated simple polygons. Algorithmica. (4), 141-155.

O'Rourke, J. 1987. Art Gallery Theorems and Algorithms. International Series of Monographs on Computer Science. Oxford: Oxford Univ. Press.

Prasad, L. and Iyengar, S. 1995. A note on the combinatorial structure of the visibility graph in simple polygons. Theoretical Computer Science. (140), 249 - 263.

Genetic Algorithms Poster Papers

The use of genetic algorithms in the optimization of competitive neural networks which resolve the stuck vectors problem

Tin Ilakovac[1], Željka Perković[2], Strahil Ristov[3]

[1,3] Institute "Ruđer Bošković", Bijenička 54, 1000 Zagreb, Croatia (tin@argus.irb.hr)

[2] Faculty For Elecrotechnical Engineering And Computer Science. Vukovarska avenija, Zagreb, Croatia

ABSTRACT

We show that for variable rates of mutation and crossover that depend on the global fitness of the population, and selection of reproduction pair that is asymmetric, solutions to problems with stringent constraints are found.

1. Neurons with a logic function

Here genetic algorithms (GA) are used for optimization of competitive artificial neural networks (ANN) to classification tasks. The learning problem called *stuck vectors* (Freeman, 1992) arises with competitive ANNs as there is a possibility of choosing a set of initial weight vectors which obstructs the learning process (Kohonen, 1990). Here a solution is proposed in form of a logical function that is attached to each neuron and manages the actions of a neuron, its learning and activation. Also, neurons change their sensitivity ranges S_R . ANNs of this type need no preadjustments of weights prior to the learning process. The simplified actions of the neuron (Fig 1.) are:

- neuron widens S_R until the first sample is sensed
- neuron learns if it senses the current sample and if no other neuron senses this sample
- neuron lessens S_R if it learns, otherwise it widens S_R; if S_R gets larger than a predefined maximum, the neuron shrinks it to zero

2. GA adaptation of ANN parameters

GA is used to optimize the above ANNs for better classification results obtained in less learning steps.

When ANNs in a population perform poorly on a classification task, GA increases the mutation rate, and decrease the crossover rate. Oppositely, when ANNs begin to obtain classification results, GA increase the crossover rate and decrease the mutation rate (Kitano, 1990). Mutations are distributed among the members of a population in favor of members with lower fitness value. A global fitness value F_G, which is a sum of fitness values of all members, defines the success of a population. We applied an exponential increase of mutation rate and a linear decrease of crossover rate with F_G.

The asymmetric selection of a reproduction pair provides a strict selection criteria for one member, and a less strict one for the other. This enable members with lower fitness values to mate, therefore parts of their genetic material can be expressed in the next generation. If these parts combine well with the parts of the best members, they can propagate to further generations.

3. Conclusion

We propose ANNs that work autonomously from the very beginning of the learning process. Our GA finds ANNs capable of learning the given task in 600 learning steps or less. This result could be reflected in the adaptation of complete ANN structures using GAs (Ilakovac, 1995).

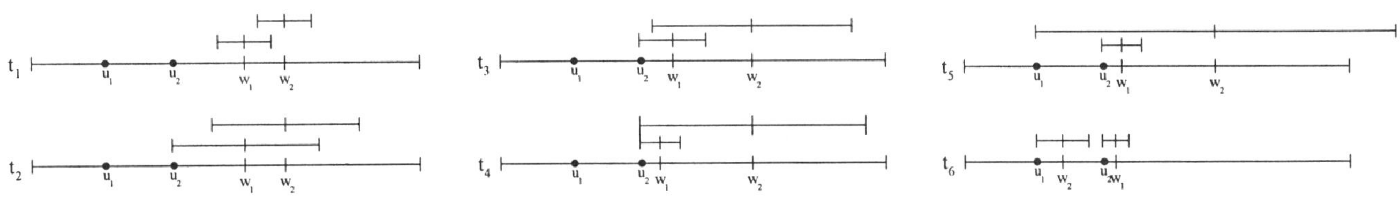

Figure 1. Two competitive neurons begin to learn in a *stuck vectors position* at t_1, and resolve the position at t_6. Two samples are presented by dots (samples alternate). The middle line of an elongated H denotes the position of the neuron's weight vector *w*, while left and right vertical lines denote its range of sensitivity.

Bibliography

T.Kohonen, "The Self-Organizing Map", *Proceedings of the IEEE*, Vol.78, No.9, 1990, pp. 1464-1480

J.A.Freeman and D.M.Skapura, *"Neural networks,"* Addison-Wesley, 1992

T.Ilakovac, "Adaptation of ANNs Using GAs", *Croatica Chemica Acta*, Vol. 68, No. 1, pp. 29-35, 1995

(this paper in 8-pages format) http://argus.irb.hr/~tin /gann-svp.tgz

An Extraction Method of a Car License Plate using a Distributed Genetic Algorithm

Dae Wook Kim, Sang Kyoon Kim and Hang Joon Kim
Dept. of Computer Engineering, Kyung Pook National University,
Taegu, 702-701, Korea
email: kimhj@bh.kyungpook.ac.kr

Extracting a license plate is an important stage in the automatic car identification. It is very difficult because car images are usually degraded and processing the images is computationally intensive.

In this paper, we proposed a new extraction method of a license plate using a distributed genetic algorithm. To overcome difficulties dealing with degraded images and requiring much processing time, we devised the method using a distributed genetic algorithm based on segmentation and some heuristic rules.

The overall processing order is as follows. After capturing an input image from a car, a Red-Green-Blue (RGB) color space of the image is transformed to a Hue-Lightness-Saturation (HLS) color space. In the segmentation stage, a label image is generated with a distributed genetic algorithm. From the label image, we extract the plate region by applying some heuristic rules.

Experimental results showed that the proposed method has the potential for extracting the plate region working well with real-world images. The advantages of the method is that: (1) it provides a robustness in dealing with degraded car images; (2) it extract more efficiently the plate region even from a image which has similar color between a car body and a plate region (3) it has inherent parallelism to improve processing time.

(a) Original image(500×300)

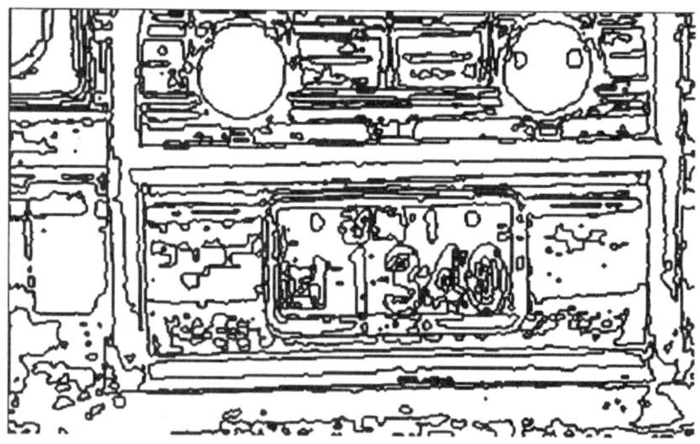

(b) Segmented image(region boundaries)

Fig 1. Segmentation result of an image.

Though the performance of the proposed method is very good, it could be improved further by using a more efficient representation model of color. Future works include also implementation of the method with parallel facilities.

Evolutionary Programming and Evolution Strategies Papers

Evolving Fractal Movies

Peter J. Angeline

Loral Federal Systems
1801 State Route 17C
Owego, NY 13823 USA
pja@lfs.loral.com

ABSTRACT

Interactive evolution of complex yet aesthetically pleasing images has generally concentrated on the acquisition of single images. In this paper, a method for evolving fractal movies is presented. Here, the interest is not only for each individual frame of the movie to be a fractal but that the progression of frames also have a fractal relationship. This paper defines a method for representing a fractal set of fractals, the Meta-IFS, and uses an interactive evolutionary program to create examples of this mathematical structure. Self-adaptive parameters are used in the evolving individuals to help identify and exploit trends in the interactive evolutionary process dynamically.

1.0 Introduction

Interactive evolution is generally used as a method for creating artifacts for environments where the specification of the needs or wants of the user are not able to be quantified in a fitness function. Often environments where the aesthetic concerns of the user are required as feedback on evolving structures are the object. Dawkins (1986) was the first to use such a system for evolving what he called "biomorphs," simple line drawings that resembled natural objects. Sims (1991) describes an interactive genetic program that creates visual images with significant aesthetic quality. Graf and Banzhaf (1995) describe a system where aesthetic considerations of the user are used to interactively evolve car body designs. Others have used interactive evolution for other artistic (Todd and Latham 1991) and non-artistic (Caldwell and Johnston 1991) applications.

In most systems using interactive evolution for artistic purposes, only a single image is evolved. To date, no work has been performed on the evolution of sequences of images that can be used to create interactively evolved movies. Ideally, these movies would carry some theme between the individual frames as well as provide each frame with interesting content.

Fractals (Mandelbrot 1977) are geometric objects that when rendered often display complex yet beautiful pictures. Much of the beauty of fractals result from their self-similar nature and their association with naturally occurring organic and non-organic structures. However, many methods for creating fractals are non-trivial and are inappropriate as representations for evolving desirable pictures.

Iterated Function Systems (IFS) (Barnsley 1988) represent one form of fractals that are easily defined and directly evolvable. An IFS is a structure that maps a given set of points in a space into a new set of points in that space. When iterated an infinite number of times, the set of points converges to a set of points that is a fractal. Methods exist for computationally approximating the fractal set of points in a reasonable amount of time. In addition, an IFS is easily represented by a matrix of real values and is an appropriate structure for an interactive evolutionary system. Unfortunately, a standard IFS still only defines a single frame rather than a sequence of frames that can be cast as a movie.

This paper defines a Meta-IFS to be an IFS that instead of mapping points in an image space to other points in the image space, maps a set of IFSes in an appropriately defined space into a fractal set of IFSes. Given that an IFS identifies a fractal in image space, a Meta-IFS identifies a fractal set of fractal images in a suitably defined space of IFSes. Sequences of IFSes created from the Meta-IFS then designate a sequence of fractals with a fractal relationship between them. This paper defines a *fractal movie* as such a sequence of images.

The objectives of this paper are two fold. The first objective is to introduce the concept of a Meta-IFS and how it can be interactively evolved to generate aesthetically interesting fractal movies. Second, this paper suggests that self-adaptive parameters added to interactively evolving individuals can form a model of the user's preferences and help to guide the interactive evolutionary process. This paper provides an example of a self-adaptive interactive evolutionary environment.

This paper begins with an introduction to fractals and iterated functions systems, then defines the Meta-IFS structure and describes s how it is used to generate fractal movies. Next, a description of the interactive evolutionary program investigated in this study is presented along with the usage of the self-adaptive parameters in the individuals. Finally an example fractal movie is generated and a discussion of the results and future work is presented.

2.0 Fractals and Iterated Function Systems

This section provides the rudimentary definitions for the mathematical objects of a metric space, a fractal, and an iterated function system. A number of details have been left out due to clarity and space restrictions. For a complete description of these topological concepts, refer to Barnsley (1988) or Peitgen, Jürgens and Saupe (1992).

2.1 Metric Spaces, Transformations and Fractals

Let X be a space of points, for instance $\Re^2$. Let d be a metric on that space such that $d:X \times X \to \Re$ measures the distance between points in X. The metric d must have the following properties:

$$\begin{array}{ll} d(x, y) = d(y, x) & \forall x, y \in X \\ 0 < d(x, y) < \infty & \forall x, y \in X, x \neq y \\ d(x, x) = 0 & \forall x \in X \\ d(x, y) \le d(x, z) + d(z, y) & \forall x, y, z \in X \end{array} \quad \text{(EQ 1)}$$

where the first property requires the metric to be reflexive, the second requires the metric to return a positive real value, the third forces the distance from a point to itself to always be zero and the last stipulates that the triangulation inequality holds. An example metric for $\Re^2$ is the standard Euclidean distance metric given by:

$$d(x, y) = \sqrt{(x_1 - y_1)^2 + (x_2 - y_2)^2} \quad \text{(EQ 2)}$$

which has all of the properties of Equation 1 in $\Re^2$. An association between a space X and a valid metric on that space d is called a *metric space* and is denoted (X, d).

A transformation $f:X \to X$ on a metric space (X, d) is said to be *contractive* if there is a constant $0 \le s < 1$ such that:

$$d(f(x), f(y)) \le s \cdot d(x, y) \qquad \forall x, y \in X \quad \text{(EQ 3)}$$

Thus a contractive transformation is one that maps a given metric space into the same space or a smaller space contained within the original space such that all of the points are closer together after the transformation. Any point $x \in X$ such that $f(x) = x$ is called a *fixed point* of the transformation.

The *Contraction Mapping Theorem* (see Barnsley 1988) states that given a complete metric space (X, d) and a contractive transformation $f:X \to X$ then there is a unique fixed point $x_f \in X$. Moreover, for any initial point $x \in X$, repeated applications of the contractive transformation will always converge to the fixed point. Contractive transformations are also called *contractive mappings*. This paper will use both interchangeably.

Formally, a *fractal* is a fixed point of a contraction mapping on the metric space $(H(X), h(d))$ where $H(X)$ is a special space called the *Hausdorff space* defined relative to the simpler space X and the metric $h(d)$ is the Hausdorff metric defined relative to the simpler metric d. The relationship between $H(X)$ and X is that points in $H(X)$ denote subsets of points in X with specific properties. Some fractal images are created by viewing the fractal set of points in X denoted by the fixed point of a contractive mapping on $(H(X), h(d))$. See Barnsley (1988) for additional details.

2.2 Iterated Function Systems and Fractals

An *iterated function system* (IFS) is defined as a complete metric space (X, d) and a finite set of transformations $w_n:X \to X$ and is denoted $\{X, w_n, n = 1 \ldots N\}$. Barnsley(1988, p. 82) proves that given an IFS $\{X, w_n, n = 1 \ldots N\}$ such that each transformation in the IFS is contractive then a contractive transformation on the metric space $(H(X), h(d))$ can be defined as the union of the transformations of the IFS. In other words, an IFS $\{X, w_n, n = 1 \ldots N\}$ which has only contractive transformations denotes a fractal set of points in the space X. This fractal set of points in X, determined by the fixed point in $H(X)$ is called the *attractor* of the IFS. See Barnsley (1988) for additional details and other properties of IFSes.

A special type of transformation is a *linear* or *affine* transformation which, using matrix notation and assuming the space is $\Re^3$, takes the form:

$$f\left(\begin{bmatrix} x \\ y \\ z \end{bmatrix}\right) = \begin{bmatrix} a & b & c \\ d & e & f \\ g & h & i \end{bmatrix} \begin{bmatrix} x \\ y \\ z \end{bmatrix} + \begin{bmatrix} j \\ k \\ l \end{bmatrix} \quad \text{(EQ 4)}$$

where all variables in the equation are numerical constants. Equation 4 essentially maps a given point to a new point such that each coordinate of the new point is a linear function of the coordinates of the original point. Such a transformation encodes a rotation, a scaling and a translation of the original space into a new space. Affine transformations have the property that they preserve angles between points in the space after transformation so that the resulting image is visually identical to the untransformed image, although rotated, scaled and translated. For the remainder of this paper, we concentrate only on IFSes that rely on contractive affine mappings.

2.3 The Chaos Game

One method of constructing the attractor for an IFS is to use a stochastic algorithm dubbed the Chaos Game (Barnsley 1988, p. 91-93). This method makes use of the fact that the points on the attractor of an IFS are closed under the transformations in the IFS. In other words, given any point on the attractor specified by an IFS, any transform in the IFS will give another point on the attractor of the IFS.

When plotting an attractor of an IFS with the Chaos Game, a probability of selection is associated with each transform of an IFS such that the sum of the probabilities associated with the transforms totals to one. To plot the attractor of the IFS, given a point in the space:

- randomly select a transform from the IFS according to their associated probabilities,
- apply that transform to the point and plot the result.

Obviously, transforms that have large probabilities of selection will have points associated with them drawn more quickly than those with smaller probabilities since those transforms will be applied more often to create the next plotted point. This is often useful when rendering images in that it allows a bias to be given to those portions of the attractor that are more complex or more interesting.

Barnsley (1988) proves that the set of points visited using the Chaos Game on a given IFS approaches the attractor of the IFS exponentially fast. In other words, after relatively few applications of the IFS the points visited will be arbitrarily close to points on the attractor of the IFS. Thus, when using the Chaos Game, it is sufficient to throw away some number of initial transition points before plotting the attractor of the IFS. In this study, the number of transition points generated before plotting any points from an IFS was 1000.

3.0 Generating Fractal Movies

This section defines a Meta-IFS, an IFS that designates a fractal set of IFSes and shows how to use the Chaos Game and another property of IFSes with a Meta-IFS to create fractal movies.

3.1 Meta-IFS Definition

Given the definition of an IFS, the coefficients of the transformations in the IFS can be taken to be coordinates in a larger continuous space. For instance, the IFS $\{\Re^3, w_n, n = 1\ldots5\}$ denotes an IFS on the space $\Re^3$ with five transformations. Given that each transformation in this IFS will have 12 coefficients, this IFS can be represented as a point in the space $\Re^{60}$ where each unique coordinate is a coefficient from one of the transformations in the IFS. For reference ease, we refer to a space used to represent IFSes in this manner as an *IFS space* and uses the notation $\Re^{\mathrm{IFS}}$ to denote such a space.

Given an IFS space, it is possible to define transformations that convert from a given IFS to a new IFS in IFS space. Once a transformation has been made, the new point in $\Re^{\mathrm{IFS}}$ can be "decoded" back into a set of transformations for an IFS and applied to the original space to show the attractor of the new IFS. Figure 1 shows the process of transforming a given IFS into a new IFS via an IFS space using a transformation.

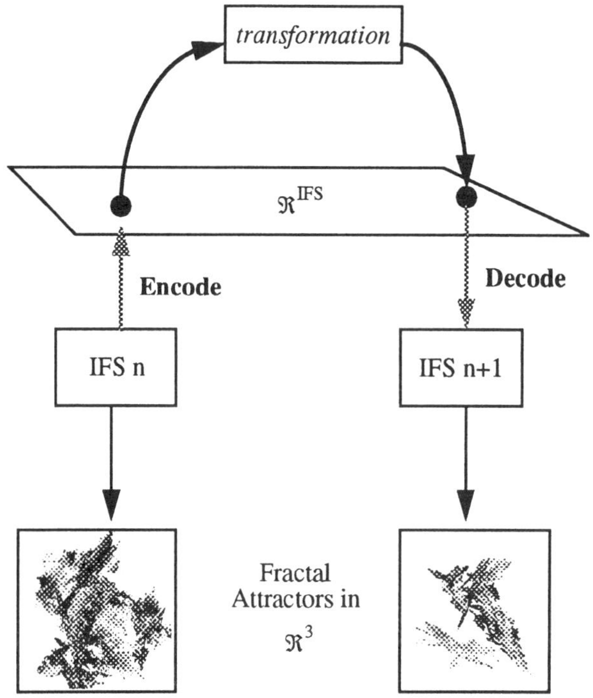

Figure 1: The relationship between two IFSes generated by a transformation in IFS space. Each IFS has an associated attractor here depicted in $\Re^3$ with the first two coordinates determining the position of the point and the third determining the color.

We define a *Meta-IFS* to be an iterated function system defined on an IFS space with an appropriate metric. In a Meta-IFS the individual points of the metric space are IFSes defined on a given metric space with a given number of transformations. When the transformations of the Meta-IFS are all contractive, the attractor of a Meta-IFS will be a fractal set of points in $\Re^{\mathrm{IFS}}$. It is unlikely that the attractor of some Meta-IFS is a set that includes only IFSes that have contractive transformations. Consequently, some of the IFSes created by any given Meta-IFS may not specify fractal sets in the original IFS space. However, the definition of a Meta-IFS does guarantee that the set of IFSes denoted by a Meta-IFS is a fractal set in IFS space. This is obvious from the fact that a Meta-IFS is an IFS and then applying the theorem proven in Barnsley (1988, p. 82) described in Section 2.2.

The number of parameters in a Meta-IFS M is given by the following relation:

$$\mathrm{size}(M) = mkn\left(kn^3 + 2kn^2 + kn + n + 1\right) \qquad \text{(EQ 5)}$$

where k is the dimensionality of the IFS, n is the number of transformations in the IFS and m is the number of transformations in the Meta-IFS. Thus, for a Meta-IFS with three transforms that creates three dimensional IFSes with five transforms the number of parameters in the Meta-IFS would be 3660.

3.2 Generating a Sequence of Fractals with a Meta-IFS

Given that a Meta-IFS identifies a fractal set of IFSes, to create a fractal movie any subset of IFSes designated by the Meta-IFS will have a fractal relationship. However, it is possible to select out only those individual IFSes generated by a Meta-IFS that use contractive mappings and therefore designate a fractal image. Thus it is possible to create a series of frames that consist of fractals where each successive frame is related to the previous frame in a fractal manner.

The difficulty comes in creating a Meta-IFS that produces a good sequence of interesting IFSes that when interpolated give an aesthetically pleasing fractal movie. The number of parameters in a Meta-IFS is rather large, as shown in Equation 5. Consequently, determining an interesting set of IFSes is a potentially daunting task of setting a very large number of non-linear parameters.

3.3 Fractal Morphing

The sequence of fractal images generated from a Meta-IFS will not necessarily be aesthetically pleasing. It should be expected that while the sequence of frames will have a fractal relationship they will not necessarily flow together in a visually obvious manner. Animation artists have a concept called "tweening" which is the creation of frames between key frames that depict a smooth transition between the two key frames.

A morphing property of IFSes can be exploited to transition between consecutively generated fractals from a Meta-IFS when creating a fractal movie. Peitgen, Jürgens and Saupe (1992, p. 257) states that an IFS constructed using a linear interpolation between two IFSes is proportionally visually similar to both of the generators. An interpolated IFS is created as follows:

$$I' = pI_1 + (1-p)I_2 \qquad \text{(EQ 6)}$$

where I_1 and I_2 are IFSes on the same space with the same number of transformations. In other words, a set of IFSes created through a progressive linear interpolation between two IFSes creates a series of fractals that "morph" from the first IFS into the second.

Given a Meta-IFS, M, and two IFSes on its attractor, A and B, it is probably not the case that an arbitrary linear interpolation between A and B is also on the attractor for M. Thus, while any interpolated IFS is a fractal, it may not be from the same fractal set of fractals as its generators.

3.4 Creating a Fractal Movie

Given a Meta-IFS, a fractal movie can be created in the following manner:

- Using the Chaos Game, generate a given number of IFSes from the attractor of the Meta-IFS.

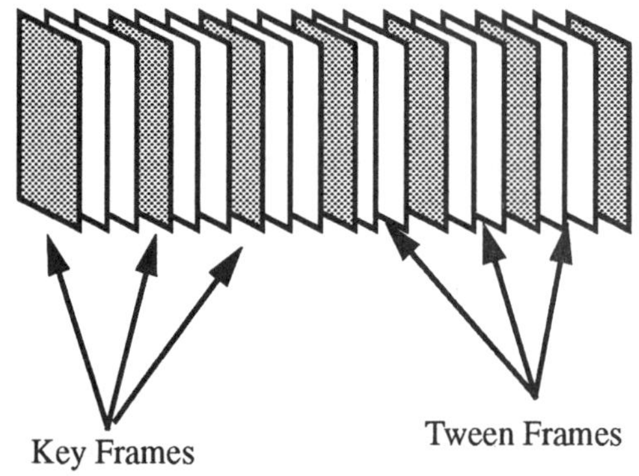

Figure 2: Organization of key frames and tween frames in a fractal movie. Key frames, shown in gray, are generated by successive applications of the Meta-IFS. Tween frames are created by a linear interpolation between the IFSes of the surrounding key frames.

- Arrange the generated IFSes into some order and designate them as the key frames of the movie,
- Generate a number of IFSes to be *tween frames* between each pair of successive key frames using the linear interpolation method shown in Equation 6.
- Generate the fractal attractors of all IFSes using the Chaos Game.

Figure 2 provides a pictorial representation of the organization of key frames and tween frames in a fully generated fractal movie.

4.0 A Self-Adaptive Interactive Evolutionary Program

Because of the large number of parameters in a Meta-IFS, creating this structure by hand so that it specifies an aesthetically pleasing set of IFSes is non-trivial. Rather than create a Meta-IFS by hand, the following section describes an interactive evolutionary program that is used to evolve Meta-IFSes that are aesthetically pleasing.

4.1 Evolutionary Programming

Evolutionary programming (EP) (Fogel, Owens and Walsh 1966; Fogel 1995) is a method of evolutionary computation that models evolution at the level of competing species. This is distinct from the genetic model used by *genetic algorithms* (Holland 1975; Goldberg 1987) and genetic programming (Koza 1992)[1] and the individual level model used by *evolution strategies* (Rechenberg; Schwefel). Because individuals

1. For additional information on the relationship between genetic programming and other evolutionary computations including evolutionary programming see Angeline (1996a).

from different species don't procreate, sexual recombination between individuals in the population to create offspring is not used. Instead, representation specific mutation operators create offspring from single parent structures.

Another distinction between EP and other evolutionary computations is the selection method. The type of selection typically performed in evolutionary programming is a form of tournament selection modified to minimize over-selection (i.e. multiple reproductions given to a single high-fitness parent) (Fogel 1995). First, a tournament score is associated with each member of the population based on its fitness relative to other population members. The algorithm assigns an individual's tournament score by randomly selecting *k* additional population members and summing the number with worse fitness. The population is then sorted by the tournament score and the best 50% is reserved as parents for the next generation. As a result, any given population member can be selected only once as a parent for the next generation.

Unlike the typical tournament selection algorithm used in genetic algorithms, this scheme forces both an individual and its offspring to have above average fitness in order to increase the representation of the lineage in the population. It thus places additional selection pressure on *mutable* fit population members rather than just those with high fitness. While this difference is subtle, it may hold advantages in complex environments.

4.2 Self-Adaptation During Interactive Evolution

Angeline (1995) defines a self-adaptive evolutionary computation as an evolutionary computation that uses the evolutionary dynamics to determine both the content of a population member and the parameters of the individual's reproduction. The parameters of an evolutionary computation acquire information about the space being searched by the evolutionary process and bias subsequent searches in directions that were fruitful in previous generations. Reproductive parameters manipulated by an evolutionary computation are typically called *strategy parameters*.

It is possible to view the strategy parameters in an evolutionary computation as a model of the reproductive process and its relation to the fitness landscape being traversed. In the case of self-adaptive evolutionary optimization, strategy parameters often acquire local gradient information for the fitness landscape and can be used to bias the creation of offspring along the gradient.

In interactive evolutionary systems, given that the user is the fitness function, strategy parameters can be used to acquire models of the user's preferences in a fashion analogous to acquiring the local gradient characteristics of a fitness landscape. Strategy parameters that more accurately reflect the preferences of the user will be more likely to produce variants that the user will consider improvements.

In this study, the Meta-IFS representation contains a set of strategy parameters, one parameter for each transform in the Meta-IFS. Usage of these parameters during offspring creation is described in the next section.

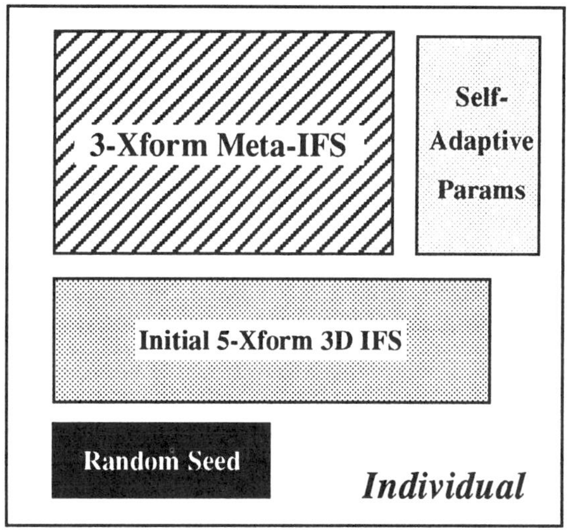

Figure 3: Components in a single interactively evolved individual. Total number of parameters in a single individual is 3724. See text for an explanation of the individual components.

It should be mentioned that a user that is more consistent in selecting preferred structures will be easier to model in the strategy parameters and will be able to find more interesting structures than a user that is inconsistent in assigning fitness. Angeline (1996b) shows that self-adaptive parameters are viable in environments with a moderate amount of noise. Consequently, small degrees of inconsistency in the rating of the sequences should be averaged out by the evolutionary process allowing the generation of offspring to become more consistent with user preferences on average as the interactive evolution progresses.

4.3 Meta-IFS Representation and Mutation

Figure 1 shows the representation for the individuals interactively evolved in this experiment. The structure contains four separate components. The first is the Meta-IFS associated with this individual. The Meta-IFS has three transformations and is defined to manipulate a three dimensional IFS with five transformations. Plugging this information into Equation 5 shows that the size of the represented Meta-IFS is 3660 parameters. Associated with each of the three transformations in the Meta-IFS are self-adaptive strategy parameters. The strategy parameter associated with a transform determines the severity of mutation applied to that transform in the Meta-IFS. It is typical in self-adaptive systems to associate a strategy parameter with each individual parameter in the representation. Due to the exceedingly large number of parameters in a Meta-IFS, it was decided that a single parameter per Meta-IFS transformation would be sufficient.

To manipulate a component in the Meta-IFS portion of the representation, Gaussian noise is applied to the parameter as follows:

$$x'_{ij} = x_{ij} + \sigma_i \cdot N(0, 1) \qquad \text{(EQ 7)}$$

where x_{ij} is the jth parameter for the ith transform in the Meta-IFS, σ_i is the strategy parameter for the ith transform of the Meta-IFS, and N(0,1) is a Gaussian random variable with mean 0 and variance 1. A strategy parameter is mutated in an analogous manner:

$$\sigma'_i = \sigma_i + \alpha \cdot \sigma_i \cdot N(0, 1) \qquad \text{(EQ 8)}$$

where α is a scaling factor set to 0.3 in the experiments.

The other components in the representation were introduced to make the process of creating fractal movies repeatable. While it is true that the set of points visited using the Chaos Game converges to the attractor of the IFS exponentially fast independent of the starting position, the sequence of points in the set visited using the Chaos Game is dependent on the initial point. The IFS structure in a population member's structure is the initial point in IFS space from which all other IFSes are constructed using the Meta-IFS. The parameters of the IFS were manipulated as in Equation 7 except the strategy parameter was set to a constant of 0.3.

The final component of the representation is a random seed. In order to recreate the exact sequence of IFSes visited by the Meta-IFS, the random seed used by the random number generators in this system was associated with the individual. Whenever the individual was evaluated, the individual's random seed was used as the initial value of the random number generation seed. Random seeds were manipulated also using Gaussian noise, although the specific mutation used on the random seed is unimportant given that any sufficient random number generator will have virtually no association between any two random seeds provided to it. It is sufficient to suggest that the random seed is simply altered in some fashion.

4.4 Statistical Feedback to the User

The user of this system receives statistical feedback from the presentation function. This is so the user can incorporate statistics about the distribution of the values plotted in the key frames into consideration while forming a fitness score for a sequence. The feedback provided includes the following:

- The range of x, y, and z values for each key frame.
- The average number of pixels per key frame colored by the Meta-IFS.
- The number of unique x, y, and z values generated for each key frame.

Runs that used arithmetic combinations of these features as a fitness function to a non-interactive evolutionary program generally did not produce sequences that were as aesthetically interesting as those produced interactively.

5.0 Additional Details and An Example Movie

In this run, a Meta-IFS with three transforms is interactively evolved to produce a series of three dimensional IFSes each with five transformations. When rendering an IFS, the first two coordinates were taken to be the position of the point in the plane and the third coordinate was taken to be the color of the point. The space for the fractal attractors was set to be the filled cube from -0.5 to 0.5 in all three dimensions. Any coordinate value that did not fall within the confines of the space were not plotted.

The points of the attractor were mapped to an image with a resolution of 256x256 pixels. The colormap used held 128 gray values. For each frame, 1000 transition points were generated prior to any points being plotted. After this, 500,000 points on the attractor of an IFS were plotted.

In order to evaluate a fractal movie, the key frames for a given Meta-IFS were shown to the user in the order they were generated and the statistics described in Section 4.4 were reported after the last key frame. The user was then asked to supply a score for only the series of key frames. Once a score was supplied, the set of key frames for the next child was presented. The population size was set at 20 and thus 10 child movies were shown each generation. The population of movies were evaluated in random order to prevent bias in scoring on the part of the user. At the end of a generation the complete movie including tween frames was generated for the individual that scored the highest in the previous generation. A single generation of rendering and scoring each of the 20 movies and then fully rendering the best of generation took approximately 9 minutes on a Pentium 120 personal computer running Linux and X Windows.

Figure 4 shows the key frames generated from a run of the described interactive evolutionary program. Each successive frame was generated by applying the evolved Meta-IFS to the preceding frame using the Chaos Game to select a transform from the Meta-IFS. Each IFS attractor is generated as above except the number of points plotted is 1,000,000 to more fully flesh out the attractor. This set of key frames took 27 generations to create using the described algorithm.

Figure 5 shows the interpolated frames between the first two key frames generated for the example fractal movie. The sequence shows a continuous deformation of the first key frame into the second.

6.0 Discussion and Conclusions

The sequence of key frames generated by the individuals in the early generations often produced points that were outside the defined space, which is to be expected. Over time, and through the constant vigilance of the user, the attractors of the IFSes generated by the Meta-IFS began to be confined to the designated space, although they often switched from sets that covered a reasonable portion of the space to sets that covered only minimal portions of the space. An emphasis was placed by the user to create key frames that covered rea-

Figure 4: Sequence of key frames generated from a Meta-IFS for an example fractal movie evolved using the interactive evolutionary program described in the paper. Each frame was rendered by plotting 500,000 points using the Chaos Game on a 256x256 image using 16 gray values. Sequence is shown in the order the key frames were generated using the Meta-IFS, the initial IFS and Random Seed in the evolved individual. Additional "tween" frames, as shown in Figure 5 are inserted between these frames to animate the sequence.

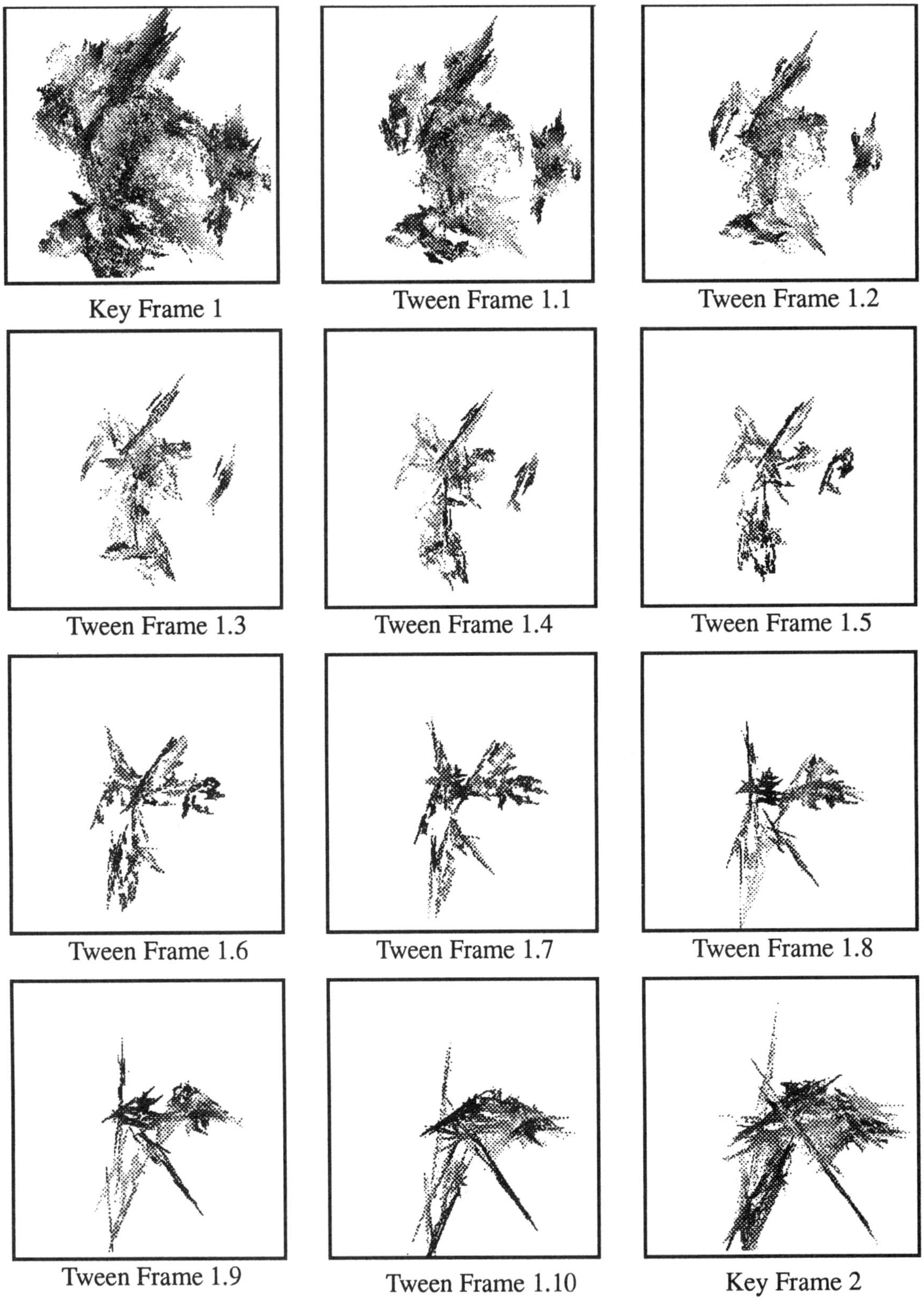

Figure 5: Sequence of tween frames between the first and second key frames generated for the example fractal movie. Each frame was rendered by plotting 500,000 points using the Chaos Game on a 256x256 image with 16 gray values. The sequence of percentages used in Equation 6 to generate the IFSes corresponding to the tween frames are given by *i*/12 where *i* is the tween frame number.

sonably similar portions of the intended space while capturing some aesthetic components based on structure and color.

It was relatively easy to create interesting, aesthetically pleasing, fractal movies with this system. It was also easy to create fractal movies that were limited to the confines of the space. Given that the resulting fractals were limited to the intended region, it can be assumed with some confidence that IFSes with contractive mappings were produced.

Few direct visual associations between parent and child were noted by the user during the experiments. This is not surprising since the change of random seed in an individual in itself can cause significant modifications to the sequence of key frames generated. Further experimentation should identify if any perceivable relationships between offspring and parents result from the mutation operations.

It is unclear from the experiments performed to date if the self-adaptive parameters aided in the discovery of the final fractal movies. While the concept may be useful, this environment is not an appropriate one to make any scientific claims about its applicability. A simpler environment should be designed to determine the usefulness of self-adaptive parameters in interactive evolutionary systems.

One effect of the linear interpolation between key frames is that the steady progression from one key frame to the other causes the transfer from one pair of key frames to the next to be rather abrupt. Once the interpolation between key frames reaches the next key frame and the new interpolation starts, the change in directions of perceived movement appear to occur in zero time. One solution to this is to use a non-linear interpolation method that uses a slower interpolation around the key frames and a faster one in the middle of the interpolation process. This should allow the transitions between the key frames to be smoother and more visually appealing.

Future work should also include the identification of representations for IFSes that guarantee that the interactively evolved transformations are contractive. Doug Hoskins has suggested the use of *singular value decompositions* as one potential representation.

Acknowledgments

Thanks to David Fogel for the invitation to the present this work. Thanks also to Doug Hoskins for in-depth discussions on fractals, IFSes and his verification that the concept of a Meta-IFS was worthwhile to explore. Thanks also to Steve Shafer for help with X Windows questions and being excited about this project.

References

Angeline, P. J. (1995) Adaptive and Self-Adaptive Evolutionary Computations. In *Computational Intelligence: A Dynamic System Perspective*, M. Palaniswami, Y. Attikiouzel, R. Marks, D. Fogel and T. Fukuda (eds.), IEEE Press, Piscataway, NJ, pp. 152-163

Angeline, P. J. (1996a) Genetic Programming's Continued Evolution. In *Advances in Genetic Programming: Volume II*, P. Angeline and K. Kinnear (eds.). Cambridge MA: MIT Press. To Appear.

Angeline, P. J. (1996b) The Effects of Noise on Self-Adaptive Evolutionary Optimization. In *The Proceedings of the Fifth Annual Conference on Evolutionary Programming*. L. Fogel, P. Angeline and T. Bäck (eds). Cambridge MA: MIT Press. To Appear.

Barnsley, M. (1988). *Fractals Everywhere*. Boston, MA: Academic Press, Inc.

Caldwell, C. and Johnston, V. (1991) Tracking a Criminal Suspect through Face-space with a Genetic Algorithm. In *Proceedings of the Fourth International Conference on Genetic Algorithms*. R. Belew and L. Booker (eds.), San Mateo CA: Morgan Kaufmann, pp. 416-422.

Dawkins, R. (1986) *The Blind Watchmaker.* Longman, Harlow.

Fogel, D. B. (1995) *Evolutionary Computation: Toward a New Philosophy of Machine Intelligence*, IEEE Press, Piscataway, NJ.

Fogel, L. J., Owens, A. J. and Walsh, M. J. (1966) *Artificial Intelligence through Simulated Evolution*, John Wiley, NY.

Goldberg, D. E. (1989) *Genetic Algorithms in Search, Optimization and Machine Learning*, Addison Wesley, Reading, MA.

Graf, J. and Banzhaf, W. (1995) Interactive Evolution of Images, In *The Proceedings of the Fourth Annual Conference on Evolutionary Programming*, J. McDonnell, R. Reynolds and D. Fogel (eds), Cambridge MA: MIT Press. pp. 56-65.

Holland, J. H. (1975) *Adaptation in Natural and Artificial Systems*, University of Michigan Press, Ann Arbor, MI.

Koza, J. R. (1992) *Genetic Programming: On the Programming of Computers by Means of Natural Selection*, MIT Press, Cambridge, MA.

Mandlebrot, B. B. (1977) The Fractal Geometry of Nature. New York: W. H. Freeman and Company.

Peitgen, H.-O., H. Jürgens and D. Saupe (1992). *Chaos and Fractals: New Frontiers of Science*. New York: Springer-Verlag.

Rechenberg, I. (1973) *Evolutionsstrategie: Optimierung Technischer Systeme nach Prinzipien der Biologischen Evolution*, Frommann-Holzberg Verlag, Stuttgart.

H.-P. Schwefel (1981) *Numerical Optimization of Computer Models*, John Wiley, Chichester, U.K.

Sims, K. (1991) Artificial Evolution for Computer Graphics. *Computer Graphics*, **25**, pp. 319-328.

Todd, S. and W. Latham (1991). *Mutator: A Subjective Human Interface for Evolution of Computer Sculptures*. IBM United Kingdom Scientific Center Report.

Preliminary Experiments on Discriminating between Chaotic Signals and Noise Using Evolutionary Programming

David B. Fogel
Lawrence J. Fogel
Natural Selection, Inc.
3333 N. Torrey Pines Ct., Suite 200
La Jolla, CA 92037
dfogel@natural-selection.com
lfogel@natural-selection.com

ABSTRACT

Evolutionary programming is a stochastic optimization algorithm that can be used for system identification. This paper focuses on the use of evolutionary programming for optimizing models of chaotic signals, both with and without additive noise. Preliminary results indicate that the method may be useful for estimating parameters of nonlinear chaotic sequences and can assist in detecting the presence or absence of a chaotic signal in an observed time series.

1 Introduction

Signal detection is a difficult and recurring problem in engineering. A time series is observed and it must be determined whether or not some regularity exists within the data. Typical procedures for accomplishing such detection include the use of matched filtering and other similar methods that involve exemplars of the signals to be recognized. But these techniques require *a priori* knowledge of the specific signals to be detected. They become much less valuable when the signals take on unknown characteristics, and even less useful when the signals are chaotic and may appear much like the random background noise in which they reside.

As an alternative to matched filtering methods, system identification techniques can be used to generate mathematical models of observed data. These models can be evaluated in light of information criteria (e.g., Akaike, 1974) that trade off their goodness-of-fit against the associated number of degrees of freedom. Models that explain observed data solely on the basis of random noise can also be evaluated and compared with other models to determine whether or not there is objective evidence to support the belief that signals other than random noise are embedded in the data.

Unfortunately, most common system identification techniques rely on gradient-based optimization of parameters of a predefined linear model (Ljung, 1987). Yet the function relating the worth of various parameters to the overall predictive error is often nonconvex and multimodal. The result of such identification methods is convergence to only locally optimal models. In some cases, such models may be sufficient. But when attention is devoted to modeling chaotic signals in noise, success appears less than likely.

Evolutionary computation has been proposed as an alternative method for optimizing linear and nonlinear models of data (Fogel, 1991; McDonnell and Waagen, 1994; and others). These stochastic optimization procedures are less likely to become trapped in locally optimal solutions than classic methods and can be used to simultaneously determine both the parameters and structure of general models. The current effort explores preliminary parameter estimation experiments with evolutionary programming for detecting chaotic signals in noise. Attention is focused on a Lorenz system and the logistic function.

2 Modeling with Evolutionary Programming

The fundamental approach to modeling begins with an observed time series, $\{y\}$, where y is a vector of observations. Any observation made at time t is described as $y[t]$. It may be that y is multivariate, that is, each observation may comprise many data. The modeling process is often conducted with a specific model form chosen *a priori* and the task is to find the optimum setting for parameters associated with the model such that the residual error to the observed data is minimized.

For example, suppose observations were generated from the system

$$y[t+1] = 0.5\, y[t] \qquad (1)$$

where each next observation is half of the previous observation. A model form incorporating this system is

$$y[t+1] = a\,y[t] \quad (2)$$

and the problem is then to find the value of a that best fits any observed sequence. When the system does not incorporate random perturbations, such parameter estimation is relatively easy and can be accomplished with complete accuracy by classical regression methods. Indeed, even when random perturbations do affect the observed data, if they are of the form

$$y[t+1] = a_0\,y[t] + \ldots + a_k\,y[t-k] + e[t+1] + c_0\,e[t] + \ldots + c_j\,e[t-j] \quad (3)$$

where $e[t]$ is the tth sample from an independent identically distributed zero mean random variable representing background noise, standard methods of least square parameter estimation can be used to find optimal values of $a_0, \ldots, a_k, c_0, \ldots, c_j$.

But when the system is nonlinear, where

$$y[t+1] = f(y,e) \quad (4)$$

and f is a nonlinear function on the previous values of y and e, classic methods of optimization, such as sequential least-squares, cannot be guaranteed to provide globally optimal parameter estimates.

Rather than rely on local convergence, stochastic optimization algorithms such as evolutionary programming can be used to generate parameter estimates. Given a sequence of observations $\{y\}$ and a model form, a population of potential parameter estimates is initialized at random and each set of parameter estimates is scored in light of the goodness-of-fit it yields to the data. Each of these initial trial solutions, termed *parents*, are mutated to create *offspring*. Often, only one offspring is created from each parent for convenience, but there is no restriction on the number of progeny. When searching for optimal estimates of real-valued parameters, mutation is often accomplished by adding a Gaussian random variable (other probabilistic mutations can be considered) with mean zero and selected variance to each parameter of each parent. The offspring are scored in the same manner as their parents and a stochastic competition is conducted between all parents and their offspring so as to allow the population to walk up hills on the energy landscape. This takes the form of a round-robin tournament in which each solution competes directly with a specified number of the randomly selected opponents (for the experiments described here, 10 competitions were used). In each encounter, if the solution is at least as good as its random opponent, it receives a "win." Those solutions with the most wins are selected to become parents for the next generation and the process iterates until a set of parameters with sufficient quality is attained or the available computing time is exhausted.

In pseudocode, the algorithm is written as:

```
program evolve;
begin
    initialize parents;
    score parents;
    repeat
            mutate parents;
            score offspring;
            select parents;  /* new parents */
    until (termination criterion);
end;
```

3 Method, Materials, Results

3.1 Lorenz System

The Lorenz system was used to provide an initial examination of the ability of evolutionary programming to address parameter estimation of chaotic signals. The studied system involved three variables, x, y, and z:

$$dx/dt = -\sigma x + \sigma y$$
$$dy/dt = -xz + rx - y$$
$$dz/dt = xy - bz$$

where $\sigma = 10$, $r = 28$, and $b = 8/3$. To facilitate experiments, this set of differential equations was simulated as a set of difference equations using the Euler method and a time step of 0.01 seconds. It was recognized that superior fidelity could be achieved using a Runge-Kutta expansion of higher order. Implementing such an expansion was not viewed necessary for two reasons:

(1) The sensitivity to initial conditions of the set of difference equations persists, albeit in a different manner from the studied Lorenz system, and

(2) The statistical question to be answered was the effectiveness of evolutionary programming for providing a method for discriminating between chaotic signals and noise, not necessarily between a particular chaotic system (e.g., the continuous time system) and noise.

Figure 1 indicates the phase space trajectory of the simulated system when given the initial conditions of $x[0] = 0.02$, $y[0] = 0.01$, $z[0] = 0.03$. These initial conditions were arbitrarily chosen and other small perturbations away from (0,0,0) were expected to yield similar results.

3.1.1 Experiment #1

The simulated system was driven without any random perturbation (noise) for 1,000 time steps (Figure 2). Evo-

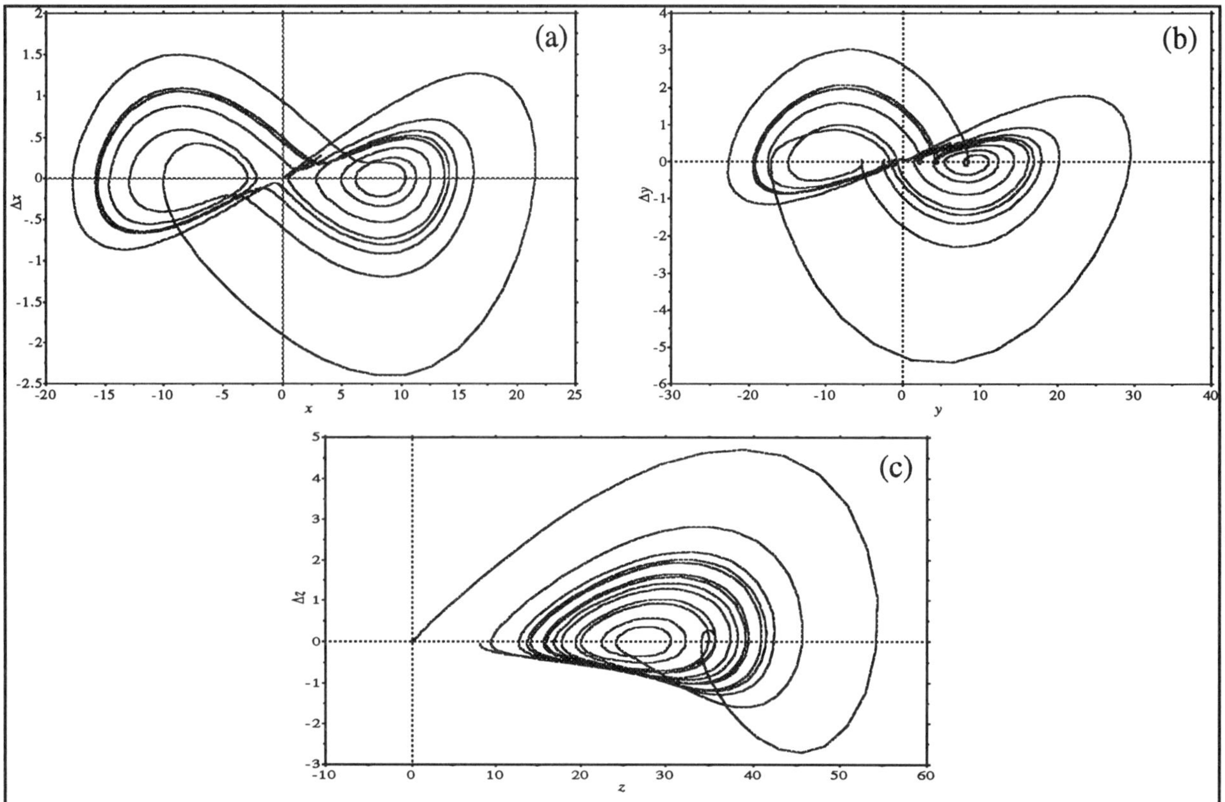

Figure 1. The discrete time phase space trajectory of the simulated Lorenz system for each of the three dimensions (a) x, (b) y, and (c) z.

lutionary programming was implemented to estimate σ, r, and b, given the general form of the Euler expansion of the true system. The initial population consisted of 50 parents. Each parent was a vector of three real-values corresponding to the parameters to be estimated, with components distributed uniformly over the range [0,100]. One offspring was generated from each parent by adding to each component a Gaussian random variable with mean zero and variance equal to the mean squared one-step-ahead prediction error of the parent. Competition and selection were handled as described in the previous section. Five trials were conducted and evolution was halted after 500 generations (arbitrarily chosen).

The results of each trial were remarkably similar. Table 1 indicates the estimated parameter values for σ, r, and b in each trial. Typically, the Euclidean distance to the true parameter vector was less than 10^{-9}. The quality of fit to the time series precludes graphing both the actual and predicted values because there is no perceivable difference. Figure 3 indicates the rate of improvement in the mean squared predictive error for the best parameter set in the population as a function of the number of generations in trial #2. These results are typical of other trials.

While it is quite obvious that the evolutionary pro-

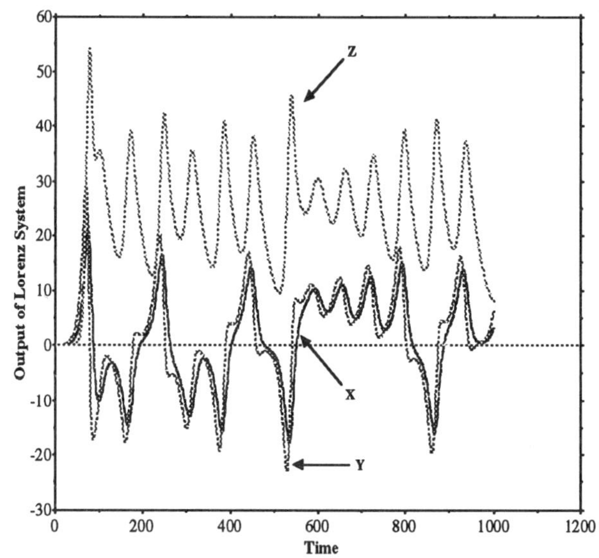

Figure 2. The simulated Lorenz system driven for 1,000 samples.

gramming was able to estimate parameters of the given model such that there is a high degree of confidence that the data is not purely random, this must be stated with statistical rigor. Many methods have been proposed for such discrimination, including Akaike's information cri-

Table 1. The estimated values for σ, r, and b, in the Lorenz system using evolutionary programming in each of the five trials with no random perturbations to the observations. The estimates are consistent across all trials and are very close to the true values.

Trial #	σ	r	b
1	10.000000003234	27.999999999573	2.666666666572
2	10.000000000949	28.000000000612	2.666666666601
3	10.000000004087	27.999999998602	2.666666666660
4	9.999999999743	28.000000000038	2.666666666669
5	10.000000003584	28.000000001177	2.666666666669

terion (AIC) (Akaike, 1974), the minimum description length principle (MDL) (Rissanen, 1978), C_p (Mallows, 1973), predicted squared error (PSE) (Barron, 1984), and others. The theoretical foundations of these criteria are beyond the scope of this paper. For convenience, the AIC will be used here. Broadly, the AIC is a weighted sum of the model's goodness-of-fit (measured by residual squared error) and the number of degrees of freedom allowed. For comparison, models with lower values of the AIC are favored as being better explanations of the data. For more details on this procedure, see Akaike (1974), Fogel (1991).

Under the assumption of Gaussian residuals, the calculation of the AIC can be reduced to:

$$AIC(\theta) = N \ln(SSE/N) + 2p$$

where there are p degrees of freedom, SSE represents the total sum of squared errors, N is the sample size, and θ is the parameter vector being estimated. An AIC over the three dimensions of concern (x,y,z) may be calculated as the sum of the individual AIC measures for each independent dimension with the appropriate degrees of freedom (in this case, three).

The AIC for each fitted model appears in Table 2. In contrast, the AIC for the model that explains the data as purely Gaussian noise with constant mean and variance is:

$$1000 \times (\ln(63.128) + \ln(85.507) + \ln(103.785)) + 2(3) = 13{,}242.08$$

Therefore, the evolved models with parameter estimates for σ, r, and b can be said to provide a better explanation of the data than the constant mean pure noise model.

3.1.2 Experiment #2

A second set of trials involved the incorporation of measurement errors added to the signal. For this test, white Gaussian noise was added to each data point such that the signal-to-noise ratio (SNR) was 0 dB, calculated by using the formula $10 \times \log_{10}(\sigma^2_s/\sigma^2_n)$, where σ^2_s is the variance of the signal and σ^2_n is the variance of the noise. All other conditions were held constant from the previous set of trials.

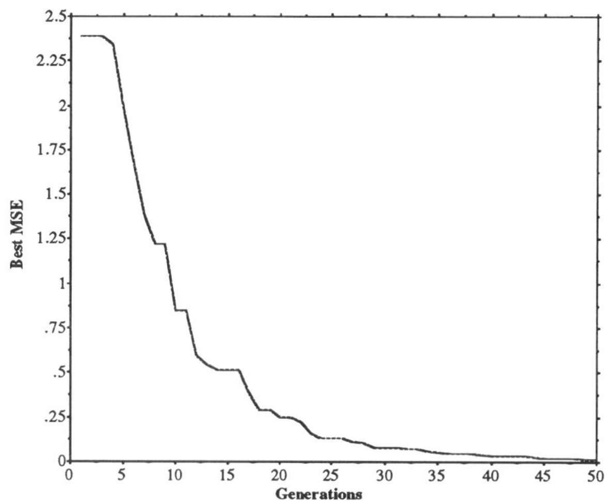

Figure 3. The rate of improvement in the mean squared predictive errors for the best parameter set in the population as a function of the number of generations in trial #2 with the Lorenz system (no noise).

Table 2. The AIC for each fitted model of the five trials with evolutionary programming to estimate the parameters of the Lorenz system in the absense of noise. Each of the AIC values for the modeled system is less than the AIC for a model that explains the data as purely Gaussian noise (AIC = 13,242.08). Therefore, there is evidence that the models with the evolved parameters provide a better explanation of the observations.

Trial #	AIC Score for Best Evolved Model
1	-24,614.54
2	-22,164.49
3	-21,992.79
4	-26,823.34
5	-20,129.57

Five trials were conducted. The final estimates of the three parameters are shown in Table 3. The additive noise adversely affects the ability of the evolutionary program to accurately estimate the parameters. The goodness-of-fit to each of the three variables (x,y,z) for trial #1 is shown

Table 3. The estimated values for σ, r, and b using evolutionary programming in each of the five trials at a signal to noise ratio of 0.0. The additive noise adversely affects the ability of the evolutionary program to accurately estimate the true parameters ($\sigma = 10, r = 28, b = 8/3$).

Trial #	σ	r	b
1	37.575103	25.082325	13.687363
2	33.422355	25.979094	16.052874
3	43.576730	25.389562	18.871481
4	36.266158	23.168835	14.975856
5	39.147237	25.326023	14.793896

in Figure 4. This is representative of the performance on the other trials. The rate of error reduction for the best parameter estimates in trial #1 is shown in Figure 5.

The previously described AIC technique was used to assess whether or not the evolved models explain a sufficient amount of the variation in the data to claim that a signal is present. The AIC for each trial, along with the residual mean squared error for each variable and the AIC value for the random noise model, is indicated in Table 4. In each case, the AIC for the chaotic model is lower than the AIC value for the noise-only model, and thus can be said to provide a better explanation of the data.

3.2 Logistic System

The logistic difference equation

$$x[t+1] = \lambda \times x[t] \times (1 - x[t])$$

is capable of diverse behavior (May, 1976) and depends on the value of λ. If $1 < \lambda < 3$, the fixed point $x = 1 - \lambda^{-1}$ is an attractor and the system will settle at the stable point. At $\lambda = 3$, the system bifurcates to give a cycle of period two, which is stable for $3 < 1 < 1+6^{0.5}$. As λ increases beyond this value, successive bifurcations give rise to a cascade of period doublings which lead to an apparently chaotic sequence for $3.57 < \lambda \leq 4$. The experiments presented here involved conditions where $\lambda = 4$. One thousand samples were generated from this system.

A more general structural form was used in experiments involving the logistic equation. A multilayer feed forward neural net possessing two input units, two hidden units, and a single output unit (2-2-1) was chosen as the nonlinear functional form (see Figure 6). The two input units corresponded to the values of $x[t]$ and $x[t-1]$, respectively, while the output unit provided a prediction of $x[t+1]$. The values of $x[t-1]$ do not enter into the governing equation, yet were provided to the evolutionary program in order to examine the manner in which these uncorrelated data would be handled. The nonlinear functions at both hidden nodes were the familiar sigmoid $f(d) = (1 + e^{-d})^{-1}$, where d is the dot product of the appropriate weights and the inputs to the node, offset by a given bias term. The output of the net was calculated as the simple dot product of the weights and inputs to the final node, offset by the associated bias term. No sigmoid function was applied to this value. Fifty parents were maintained at each iteration, with each parent containing each of the real-valued weights and bias terms of the network. Offspring were created as with the Lorenz system experiments and evolution was again halted after 500 generations in each trial.

3.2.1 Experiment #1

Five trials were conducted using an initial point of $x[0] = 0.2$ and no additive noise. Figure 7 indicates the predicted and actual values of the sequence using the final best evolved network for trial #1. The rate of improvement for the best network in the population in trial #1 is shown in Figure 8. The goodness-of-fit typical of Figure

Table 4. The AIC for each trial, the residual mean squared error for each variable, and the AIC value for the random noise model of the Lorenz system at a signal-to-noise ratio of 0.0. In each case, the AIC for the chaotic model is lower than the AIC value for the noise-only model, providing a better explanation of the data.

Trial	AIC (Lorenz)	Res(x)	Res(y)	Res(z)	AIC (Noise Only)	Res(x)	Res(y)	Res(z)
1	14729.2	101.90	154.95	155.57	15218.37	128.76	159.34	197.43
2	14733.97	83.56	159.67	186.67	15247.59	114.08	170.08	214.95
3	15335.36	109.15	176.49	223.48	15516.52	132.73	169.28	242.90
4	15152.89	107.68	194.99	178.17	15288.17	121.69	180.44	196.23
5	14842.78	104.22	157.26	167.41	15345.92	125.93	171.51	211.36

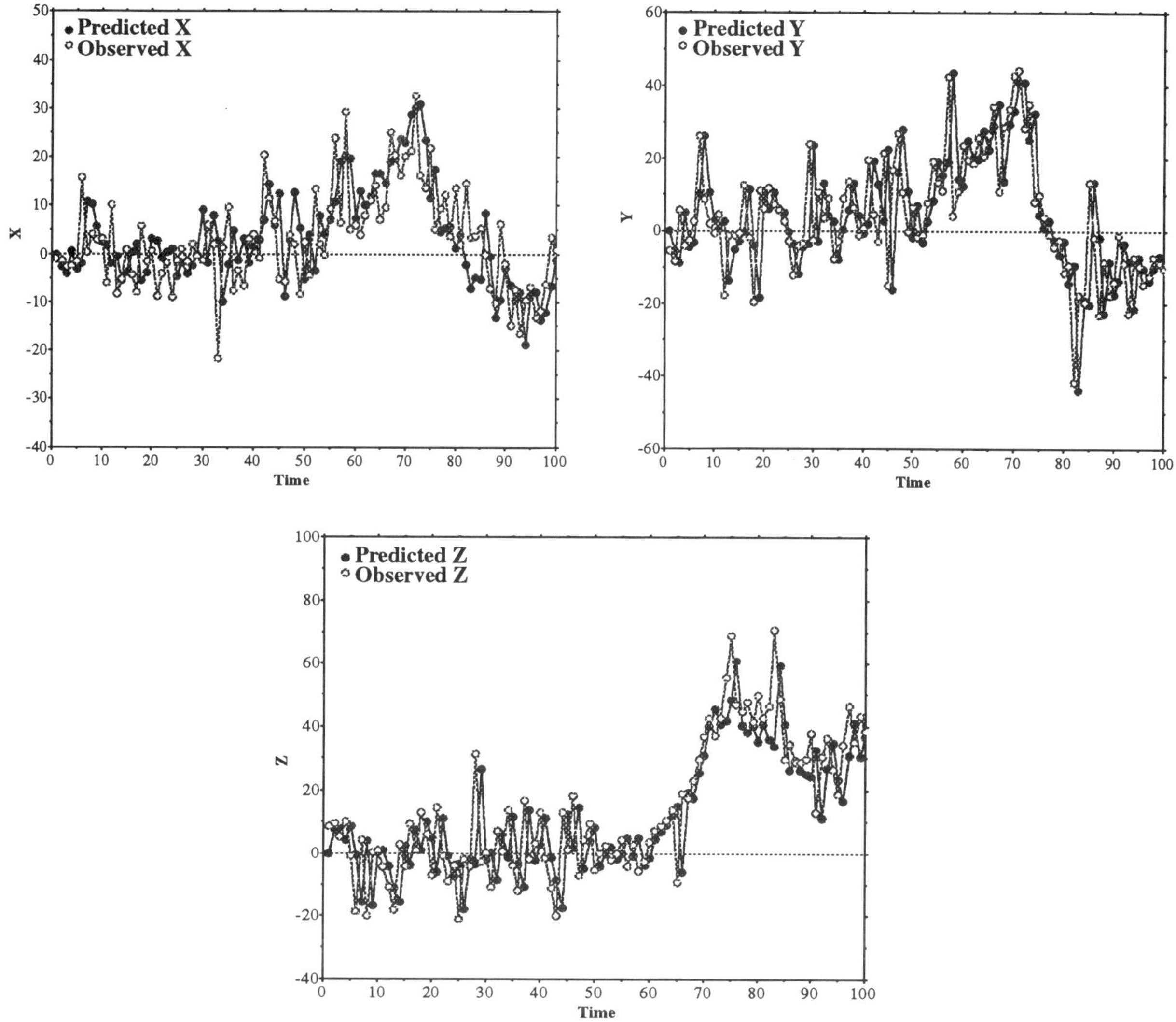

Figure 4. The goodness-of-fit to each of the three observed variables (a) *x*, (b) *y*, (c) *z*, for trial #1 with the Lorenz system under a 0.0 signal-to-noise ratio. Only the first 100 samples are presented for clarity.

---

7 provides clear evidence that the signal is not purely random noise, and it was not deemed necessary to calculate AIC values to confirm this conclusion. Table 5 shows the weights from the two input nodes to each of the two hidden nodes for each of the five trials. Note that the weights associated with the input node for the sample $x[t-1]$ were set very close to zero, as would be expected because $x[t-1]$ does not enter into the generation of $x[t+1]$.

3.2.2 Experiment #2

Five additional trials were conducted with white Gaussian noise being added to each measured point. The SNR was reduced to 0 dB (relative to the background noise). Figure 9 indicates the predicted and actual values of the sequence for trial #1 using the final best evolved network, while Figure 10 shows the rate of improvement of the best network as a function of the number of generations.

The standard deviation of the logistic function over 1,000 samples is 0.353933. With the SNR at 0 dB, this standard deviation is the approximate standard deviation of the additive noise as well. Figure 11 indicates the histogram of the residual errors for trial #1 when using the best evolved neural network predictor. For Gaussian distributions, the point of inflection on the density curve corresponds with a distance of one standard deviation from the mean. It appears that the residuals are approximately Gaussian and possess a standard deviation (0.45787) that is fairly close to optimum.

To make a statistical determination of the ability of

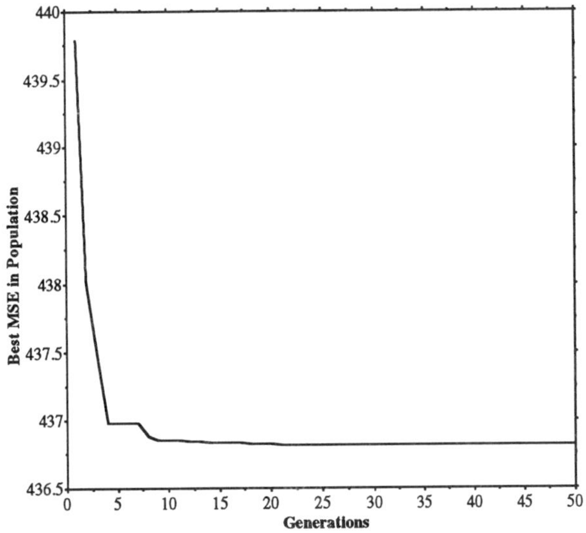

Figure 5. The rate of improvement in the mean squared predictive error for the best parameter estimates in trial #1 with the Lorenz system under a 0.0 signal-to-noise ratio (first 50 generations).

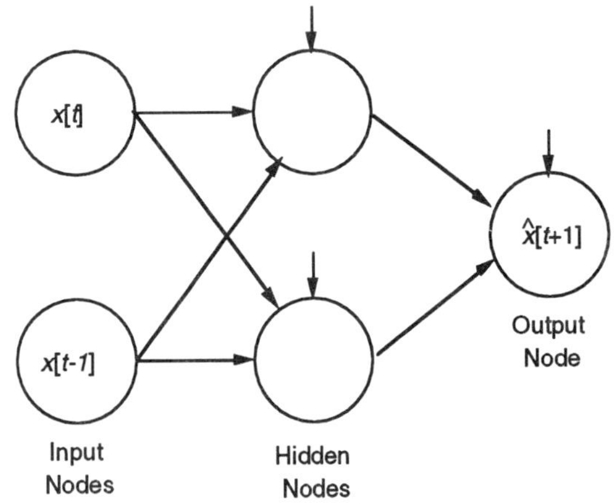

Figure 6. The multilayer neural network used to model the logistic equation. The network includes inputs from the previous two time steps even though the actual system depends only on the previous time step. Each hidden node performs a weighted sum of the inputs, offset by a bias term, and passes the result through a sigmoid filter. The output node performs the same operation without passing the result through the filter.

the evolutionary modeling to discriminate between the logistic sequence and pure noise, the AIC value for each trial is indicated in Table 6, along with the expected AIC score for the pure noise model. The AIC score, which is lower in every case for the evolved neural network, indicates the successful identification of the process as not being purely random noise.

4 Conclusions

The results suggest the potential for evolving models of chaotic data, even in background noise. Evolutionary programming can be used to optimize parameter estimates associated with models of chaotic time series in light of observed data. These preliminary results support other recent efforts that address the evolutionary modeling of chaotic data by simultaneously adjusting the model form and parameters (McDonnell and Waagen, 1994; May, 1976; Breeden, 1994; English, 1994; and others). Assessing the degradation of signal detection as a function of SNR, both under measurement and process noise, remains for future investigation.

Acknowledgments

This research was supported in part under contract number N66001-91-D-0272, Naval Command Control and Ocean Surveillance Center. The authors appreciate the assistance of T. Passamante and C. Persons, and the technical comments of J.B. Golden and V.W. Porto.

Bibliography

H. Akaike, "A new look at the statistical model identification," *IEEE Trans. Auto. Control*, Vol. AC-19:6, pp. 716-723, 1974.

A.R. Barron, "Predicted squared error: a criterion of automatic model selection," *Self-Organizing Methods in Modeling*, S.J. Farlow (ed.), Marcel Dekker, NY, 1984.

J. Breeden, "An EP/GA Synthesis for Optimal State Space Representations," *Proc. of the Third Annual Conference on Evolutionary Programming*, A.V. Sebald and L.J. Fogel (eds.), World Scientific, River Edge, NJ,pp. 216-223, 1994.

T.M. English, "Generalization in Populations of Recurrent Neural Networks," *Proc. of the Third Annual Conference on Evolutionary Programming*, A.V. Sebald and L.J. Fogel (eds.), World Scientific, River Edge, NJ, pp. 26-33, 1994.

D.B. Fogel, *System Identification through Simulated Evolution: A Machine Learning Approach to Modeling*, Ginn Press, Needham, MA, 1991.

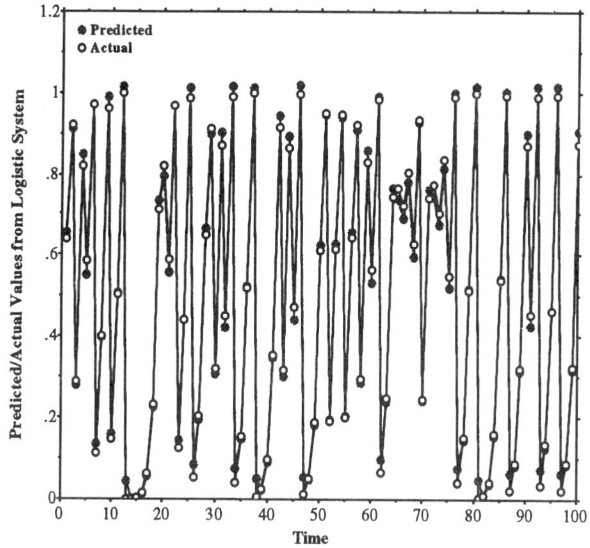

Figure 7. The predicted and actual values of the sequence using the final best evolved network for trial #1 with the logistic equation without additive noise (samples 1-100).

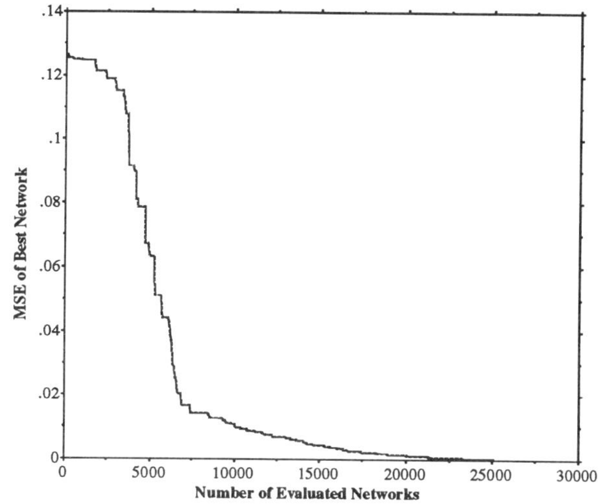

Figure 8. The rate of improvement for the best network in the population in trial #1 with the logistic equation without additive noise as a function of the number of evaluated networks.

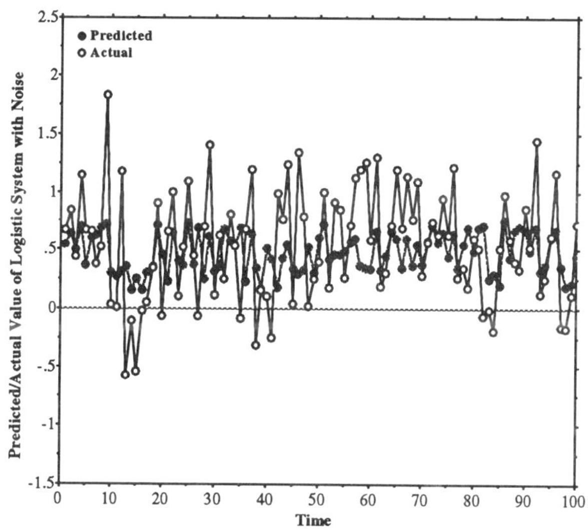

Figure 9. The predicted and actual values of the sequence using the final best evolved network for trial #1 with the logistic equation with additive noise at 0 SNR (samples 1-100).

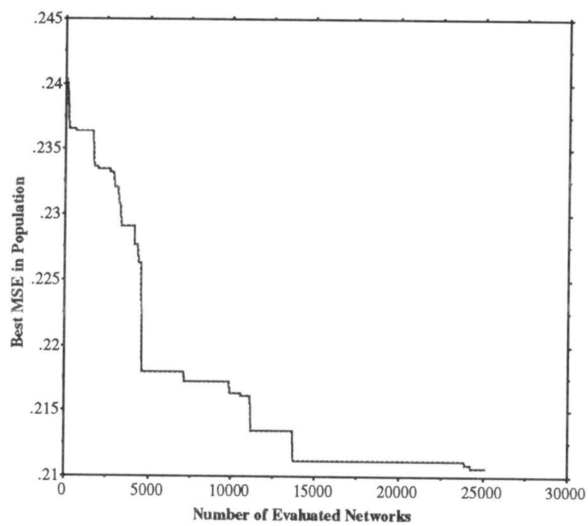

Figure 10. The rate of improvement for the best network in the population in trial #1 with the logistic equation with additive noise at 0 SNR as a function of the number of evaluated networks.

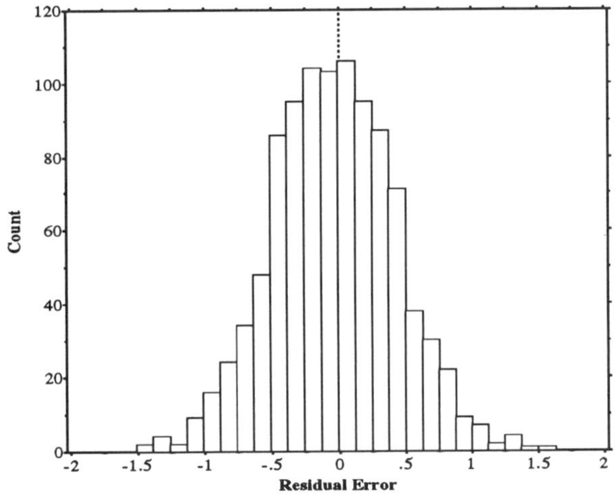

Figure 11. Histogram of the residual errors when using the best evolved neural network predictor for trial #1 with the logistic equation with additive noise at 0 SNR. The histogram appears to closely resemble a Gaussian and the error appears to be approximately as low as could be achieved.

L. Ljung, *System Identification: Theory for the User*, Prentice-Hall, NY, 1987.

C.L. Mallows, "Some comments on C_p," *Technometrics*, Vol. 15, pp. 661-675, 1973.

R.M. May, "Simple mathematical models with very complicated dynamics," *Nature*, Vol. 261, pp. 459-467, 1976.

J.R. McDonnell and D. Waagen, "Evolving recurrent perceptrons for time-series modeling," *IEEE Trans. Neural Networks*, Vol. 5:1, pp. 24-38, 1994.

J. Rissanen, "Modeling by shortest data description," *Automatica*, Vol. 14, pp. 465-471, 1978.

Table 5: The weights from the two input nodes to each of the two hidden nodes for each of the five trials with the logistic equation in the absense of noise. The weights associated with the input node for the sample $x[t\text{-}1]$ were set very close to zero, as would be expected because $x[t\text{-}1]$ does not enter into the generation of $x[t+1]$.

Trial	Weight from Input Node	Weight to 1st Hidden Node	Weight to 2nd Hidden Node
1	$x[t]$	-5.7782	-2.4879
	$x[t\text{-}1]$	0.0344	0.0154
2	$x[t]$	-7.0248	3.7479
	$x[t\text{-}1]$	-0.0181	0.0369
3	$x[t]$	-5.5608	3.7668
	$x[t\text{-}1]$	0.0106	-0.0182
4	$x[t]$	-4.3576	3.4727
	$x[t\text{-}1]$	-0.0064	-0.0087
5	$x[t]$	-4.1234	-4.7058
	$x[t\text{-}1]$	0.0059	0.0006

Table 6: The AIC value for each trial with the logistic equation at 0 SNR. The AIC score is lower in every case for the evolved neural network, indicating the successful identification of the process as not purely random noise.

Trial	AIC for Evolved Neural Network	Expected AIC Noise Only Hypothesis
1	-1551.7947	-1380.1527
2	-1429.7115	-1380.1527
3	-1515.4545	-1380.1257
4	-1453.1662	-1380.1257
5	-1556.4303	-1380.1527

Discovering Patterns in Spatial Data Using Evolutionary Programming

Adam Ghozeil
David B. Fogel
Natural Selection, Inc.
3333 N. Torrey Pines Ct., Suite 200
La Jolla, CA 92037
dfogel@natural-selection.com

ABSTRACT

The problem of unsupervised learning of patterns in spatial data is addressed using evolutionary programming. Hyperbox clusters of two-dimensional data are evolved in light of a minimum description length (MDL) principle. A series of experiments of increasing complexity indicates that evolutionary programming can be useful in generating clusters for data when no a priori classification information is available. In some cases, the best evolved clusters have lower MDL scores than the clusters that would correspond to the true probability distribution of the data. Brief consideration is given to using other functions for clustering, such as hyperellipses.

1 Introduction

Intelligence may be defined as the capability of a system to adapt its behavior to meet its goals in a range of environments (Fogel 1995, p. xiii). If the range of environments includes potentially novel conditions, then successful adaptation requires pattern *discovery*. In contrast, pattern recognition, the identification of patterns already known (thus re-cognized) or data that is classified a priori, is inherently limited. An individual that can discover new patterns in its environment will possess a selective advantage over an individual that can only respond to a fixed set of stimuli. This effect is amplified when the environment is considered to include all other individuals. Those that can only anticipate previously recognized symbols will be continually surprised by others that can vary their behavior to solve new problems in new ways. Surprise (i.e., making an incorrect prediction) can be lethal. The process of mutation and selection on a population of individual predictive algorithms (either in carbon or silicon) can be expected to lead to individuals that minimize the costs of predictive error by successively unfolding the primary patterns that are associated with important future events, then the secondary patterns, the tertiary patterns, and so forth (Atmar 1994).

Pattern discovery (or *unsupervised learning*) in spatial and temporal or otherwise indexed data has been described as a key element of intelligence, both in humans (Hofstadter, 1995), and in general evolutionary systems (Fogel et al. 1966; Atmar 1976). The current effort focuses solely on discovering clusters in spatial data. Evolutionary programming is used to optimize the clustering of two-dimensional data in light of a minimum coding criterion. The data is assumed to be drawn from a finite range of values and clusters are presumed to come from a multivariate uniform density function. The rationale for these assumptions is derived from the desire to use the spatial pattern discovery algorithm for subsequent fuzzy classification, and is discussed further in the next section (see Figure 1). Preliminary experiments indicate that the evolutionary algorithm provides reasonable clusters for known spatial patterns.

2 Background

Simpson (1992, 1993) introduced the idea of using hyperboxes to cluster or classify spatial data[1]. Each hyperbox is viewed as a fuzzy cluster, a fuzzy set in which all of the elements within the hyperbox have membership 1.0 for being in that set, and elements outside the hyperbox can have positive membership in the set depending on a fuzzy membership rule for that set (Figure 1). A universe

[1] Others (e.g., Buckles et al. 1994) have employed a similar idea that uses hyperellipses to cluster data, rather than hyperboxes.

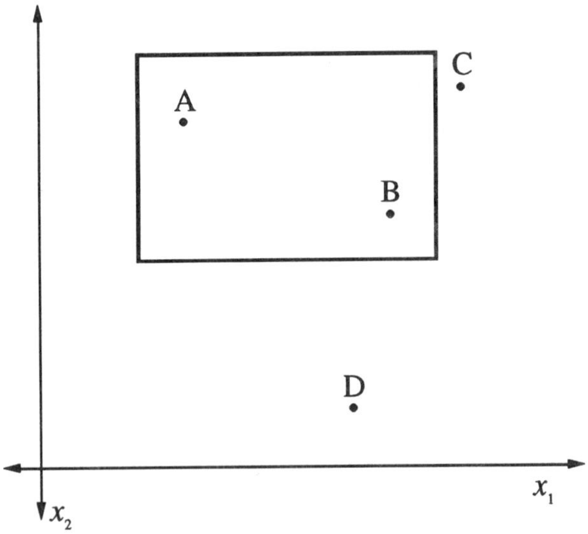

Figure 1. A hyperbox defines a fuzzy cluster. In the diagram, there are two measurable parameters, x_1 and x_2. A box is drawn that defines a cluster. Points contained within the boundaries of the box (e.g., points A and B) are said to be complete members of that cluster, or possess membership 1.0. Points outside the box (e.g., C and D) are still members of the cluster defined by the box, but possess membership less than 1.0. Typically, membership is based on a metric such that D is given less membership than C.

of discourse defines the minimum and maximum values for each dimension of the data. Simpson (1992, 1993) used a deterministic procedure to place and appropriately size hyperboxes to describe data. Hyperboxes were created and sized by considering the data in an ordered sequence. A hyperbox was placed around preliminary data. As subsequent data was added, either the present hyperbox was grown to include the new data, or a new hyperbox was added and the process continued. This procedure was of limited efficacy because it required trial-and-error setting of operator parameters (e.g., maximum box size) and the final solution depended on the order of presentation of the data, even when the data possessed only spatial and not sequential characteristics.

Fogel and Simpson (1993a, 1993b) used evolutionary programming (population-based mutation and selection; readers who are unfamiliar with evolutionary programming are referred to Fogel, 1995, pp. 75-84) to optimize the position of hyperboxes to cluster data in light of a minimum description length criterion (Rissanen 1984). The criterion trades off the goodness-of-fit to the data with the number of degrees of freedom used (see further discussion below). A population of sets of clusters underwent random mutation and selection with the result being convergence toward parsimonious descriptions of the available data. The experiments in Fogel and Simpson (1993a, 1993b) were restricted to evolving hyperboxes that were aligned with coordinate axes. The current work extends these previous efforts by including the capability to rotate the hyperboxes. It also employs a self-adaptive procedure such that the evolutionary program can adapt its search parameters during the course of the evolution.

3 Method & Materials

3.1 Evolutionary Programming Procedure

The population was initialized with 50 parent solutions. Each solution encoded a complete clustering. The experiments involved only two-dimensional data and the representation chosen for each hyperbox was the five-tuple:

$$(x, y, \theta, w, h)$$

where x and y define the (x,y) position of the center of the hyperbox, θ is the rotation (in radians) anti-clockwise of the hyperbox around its center, w is the length of the sides of hyperbox in the direction of θ (i.e., width), and h is the length of the sides of the hyperbox in the direction perpendicular to θ (i.e., height). Each solution also incorporated a five-tuple of self-adaptive parameters (see Fogel, 1995, p. 146) that were used in the generation of offspring:

$$(\sigma_x, \sigma_y, \sigma_\theta, \sigma_w, \sigma_h)$$

where each self-adaptive parameter was a standard deviation used in the perturbation of the parent parameters. Each parent also used a parameter *NBox* to define the number of hyperboxes in the solution, and two self-adaptive mutation factors controlling how often a box is added or deleted in the creation of an offspring, σ_{AddBox} and σ_{DelBox}. All parents were initialized to possess five hyperboxes. The parameters σ_{AddBox} and σ_{DelBox} were initialized to 0.25. For each hyperbox, the values of x, y, w, and h were distributed in accordance with a uniform random variable over the range (0,100) (noted as $U(0,100)$). The parameter θ for each hyperbox was distributed as $U(0,2\pi)$. The self-adaptive parameters σ_x, σ_y, σ_w, and σ_h were initialized to 5.0, while the parameter σ_θ was initialized to 0.5.

Each parent produced a single offspring. For each hyperbox in the parent, a corresponding hyperbox for the offspring (denoted by the prime symbols below) was created by adding a Gaussian random variable to the parent's hyperbox parameters:

$$x' = x + N(0,\sigma_x)$$
$$y' = y + N(0,\sigma_y)$$
$$\theta' = \theta + N(0,\sigma_\theta)$$
$$w' = w + N(0,\sigma_w)$$
$$h' = h + N(0,\sigma_h)$$

where $N(0,\sigma)$ is a zero mean Gaussian random variable with standard deviation σ.

The number of hyperboxes (*NBox*) was mutated by first possibly deleting a number of hyperboxes determined by multiplying the maximum number of hyperboxes (in this case, five), by a random variable distributed as $U(0,1)^{\sigma_{DelBox}}$, and truncating any fractional portion. Hyperboxes were selected for removal at random over all hyperboxes remaining until the number to be deleted has been reached (which could be zero) or the minimum number of boxes remain (in this case, one). Hyperboxes were then added in a similar procedure as they were deleted, using σ_{AddBox}, up to the maximum number of hyperboxes. New hyperboxes were parameterized in a similar manner as for initial parents.

The self-adaptive parameters were updated after adjusting the primary hyperbox parameters. The parameters σ_{AddBox} and σ_{DelBox} were modified as:

$$\sigma'_{AddBox} = \sigma_{AddBox} \cdot \exp(N(0.3,\tau_1) + N(0.3,\tau_2))$$
$$\sigma'_{DelBox} = \sigma_{DelBox} \cdot \exp(N(0.3,\tau_1) + N(0.3,\tau_2))$$

where $\tau_1 = [2(NBox)p]^{-0.5}$ and $\tau_2 = [2(NBox)p^{0.5}]^{-0.5}$, following Bäck and Schwefel (1993)[2]. All other self-adaptive parameters were determined by:

$$\sigma'_x = \sigma_x \cdot \exp(N(0.3,\tau_1) + N_i(0.3,\tau_2))$$
$$\sigma'_y = \sigma_y \cdot \exp(N(0.3,\tau_1) + N_i(0.3,\tau_2))$$
$$\sigma'_\theta = \sigma_\theta \cdot \exp(N(0.3,\tau_1) + N_i(0.3,\tau_2))$$
$$\sigma'_w = \sigma_w \cdot \exp(N(0.3,\tau_1) + N_i(0.3,\tau_2))$$
$$\sigma'_h = \sigma_h \cdot \exp(N(0.3,\tau_1) + N_i(0.3,\tau_2))$$

where the value of $N(0.3,\tau_1)$ was held fixed across all self-adaptive parameters, and the notation $N_i(0.3,\tau_2)$ indicates that the Gaussian random variable was resampled for each self-adaptive parameter. The constant term of 0.3 was included based on experimentation, and appeared to offer a bias toward increasing the mutation step sizes in balance to indirect selective pressures to reduce the step sizes.

All parents and offspring were evaluated with respect to their minimum description length (MDL) scores (explained fully below) and then faced competition for survival in a stochastic tournament. Each solution in the population was compared to 10 randomly chosen opponents. Each solution was assigned a number of "wins" corresponding to the number of chosen opponents with equivalent or larger (i.e., worse) MDL scores. Selection then retained the 50 solutions with the greatest number of wins, thus allowing for the probabilistic survival of solutions that otherwise would rank in the lower half of the population.

The creation of offspring, evaluation, and selection constituted a single generation. The current experiments were executed over 2000 generations.

3.2 Evaluation

Scoring was accomplished using an MDL principle (Rissanen 1984). The MDL was derived such that a minimum coding for a given set of data would be obtained. The specific criterion is:

$$\mathrm{MDL}(\omega) = -\log_2(f(\mathbf{x}|\omega)) + 0.5p \log_2 n$$

where $\mathbf{x}$ is the observed data, $f(\mathbf{x}|\omega)$ is the conditional likelihood function given a parameter vector ω, p is the number of independently adjusted parameters in ω, and n is the data size. The first term indicates the goodness-of-fit of the clustering while the second term provides a penalty function for having to use a given number of clusters. The optimum ω minimizes the MDL.

The application of the MDL requires the assumption of a parametric distribution of points in a cluster defined as the interior of a positioned hyperbox. This was assumed to be uniform over the volume of the hyperbox. The MDL for any hyperbox was then determined simply by counting the number of data points contained in that hyperbox (p was the same for any hyperbox in d dimensions under the current representation). After all assigned hyperboxes were assessed, consideration was given to points that were not contained in any hyperbox. These points were treated as outliers drawn form a uniform distribution over the entire available range of the data [0,100] in each dimension. The total MDL for a given solution was determined by summing each of the individual MDL scores for each hyperbox in that solution including the hyperbox containing any outliers.

There are specific considerations in the use of the MDL that deserve greater attention (Fogel and Simpson 1993a):

(1) Under the assumed uniform distribution, the likelihood for any point in a hyperbox is the inverse of the volume of the hyperbox. Thus the MDL will be minimized at negative infinity should any hyperbox converge to bound even a single data point and have zero width in any dimension. The penalty term associated with hyperboxes that contain only one datum ($n = 1$) vanishes. To relieve this problem, the MDL of a hyperbox was only scored when it contained at least $d+1$ points for data in d dimensions. This is intuitively reasonable as it makes little sense to speak of a "cluster" of a single point, or even d points in d dimensions. In addition, the minimum values for w and h were set to 1.0.

[2] The product of *NBox* and p yields the total dimensions involved in that particular solution. Bäck and Schwefel suggest using the total dimensions as a factor to set the values of τ_1 and τ_2.

(2) If a hyperbox contains no data ($n = 0$), the penalty term becomes negative infinity and again the MDL score is minimized inappropriately. Requiring a hyperbox to contain at least d+1 points before it will be scored removes this problem, but then extraneous hyperboxes that contain no data incur no penalty. This can be handled in two ways. A small penalty term can be applied to the overall MDL of a solution for each hyperbox regardless of the number of data points each contains, or hyperboxes that contain no data can be pruned at the completion of a clustering run. The latter method was used in the current experiments.

(3) The MDL for any outliers can be assessed even if there is only one such datum. As the size of the outlier hyperbox is constrained to be the entire available range of the data in each dimension, there is no possibility of driving the MDL score to negative infinity by perfectly fitting a hyperbox to the outlier. The choice of the value of p for the outlier hyperbox may at first appear problematic as it is constrained to always be the range of the data. Its location is not free to vary. Yet, just like any other hyperbox, it is defined by $(d^2 + 3d)/2$ free parameters. As the MDL is used to assess a fixed clustering, it is reasonable to impose a penalty term for the outlier hyperbox just as for any other hyperbox.

(4) The best coding in terms of the MDL depends on the range of the data values. The MDL is composed of two terms: the first assesses the likelihood of the data given a parametric model describing their distribution, while the second imposes a penalty for the degrees of freedom of the model weighted by the length of the data. If the data is scaled to successively smaller values, the penalty term remains the same, but the likelihood function will grow larger. Therefore, the smaller the range of the data, the more hyperboxes will be allowed by the MDL. For the current experiments, the data was arbitrarily chosen to be in the range [0,100]. The variability of clustering based on different scalings remains for future investigation.

4 Problems Investigated

Four specific clustering problems were considered. The first involved a single set of 100 data distributed in a hyperbox rotated by 30°. The second involved two non-overlapping hyperboxes of 50 data in each set, with the hyperboxes being rotated so as not to be aligned with coordinate axes. The third problem involved two overlapping hyperboxes of 50 data in each set, with the two hyperboxes placed perpendicular to each other. The final problem involved three overlapping hyperboxes arranged in a slanted H pattern. Fifty trials were conducted on each problem in which a new random seed was used to initialize the evolutionary program, but the data was held fixed across trials.

5 Results

Figure 2 indicates the improvement of the best set of clusters in the population and the mean of all parents in the population as a function of the number of generations averaged over all 50 trials on the first problem. The figure also depicts the data and a typical final best solution, which appropriately yielded a single cluster that fit the data tightly. Figures 3-5 show similar optimization curves and typical best solutions on the second, third, and fourth problems. Again, the best evolved clusterings are reasonable; the solutions depicted for cases 3 and 4 possess more hyperboxes than were used to generate the data, but overlapping clusters are discouraged by the MDL so such results should be anticipated.

To assess the quality of the evolved solutions in each case, the MDL score for the set of hyperboxes (or the single hyperbox) that defined the range of the distribution from which the data were drawn was calculated. In the case of the overlapping boxes, the total areas for the overlapping boxes were used. Table 1 shows the MDL scores for the best evolved clusterings and the MDL scores for the "true" clusterings. In cases 1 and 3, the evolutionary program was able to discover clusterings with lower MDL scores than the so-called truth. This reflects the ability of the algorithm to adapt to variability in the actual sampled data. The sampled data did not arise all along on the boundaries of the defining hyperboxes, thus the evolved hyperboxes could be smaller than the "true" hyperboxes and generate lower MDL scores. In cases 2 and 4, the mean best evolved clustering generated MDL scores that were close to but above "truth."

6 Discussion

Many traditional clustering techniques rely on local heuristics applied to individual data points (e.g., nearest-neighbor methods). In contrast, information criteria, such as the MDL, can provide a global metric for evaluating alternative entire clusterings. Determining the optimum number of clusters and the locations of those clusters is a combinatorial optimization problem that appears well suited for evolutionary programming. The current experiments generated clusterings that appeared appropriate in each case and the best evolved MDL score was comparable with the MDL score that would have been assigned to the original generating distributions. Assessing the statistical fluctuation of the evolved best MDL as compared to the supposed true best MDL score in these experiments remains for future investigation.

The choice of hyperboxes for clustering is somewhat arbitrary, and other shapes could be used, notably hyper-

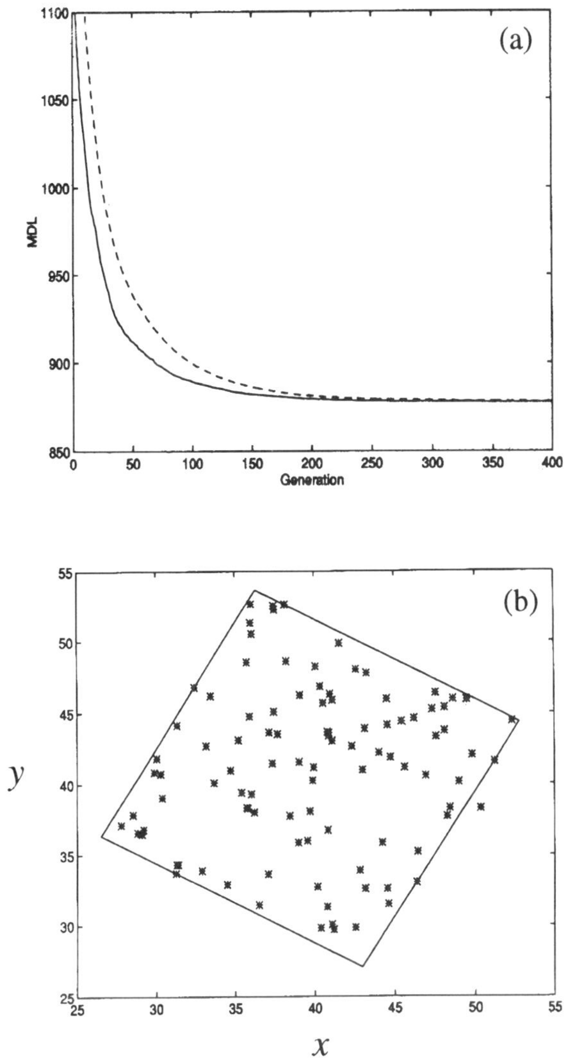

Figure 2. (a) The mean best score in the population (solid line) and the mean of the average score of all parents in the population (dashed line) as a function of the number of generations averaged over 50 trials for the case where the data are distributed in a single slanted box. (b) A typical final best evolved clustering, which consisted of a single box that covered all but a single data point (data indicated by asterisks).

ellipses. Hyperellipses have received more attention in the literature (e.g., radial basis function neural networks) because a network possessing a single hidden layer of a sufficient number of such functions and linear output node can be used for universal function approximation (Hornik et al. 1989). No similar statement has been offered in general for hyperboxes, at least to the authors' knowledge, although such a result does not appear unreasonable in light of Riemann integrals. Future efforts involve comparisons between the use of hyperboxes and hyperellipses,

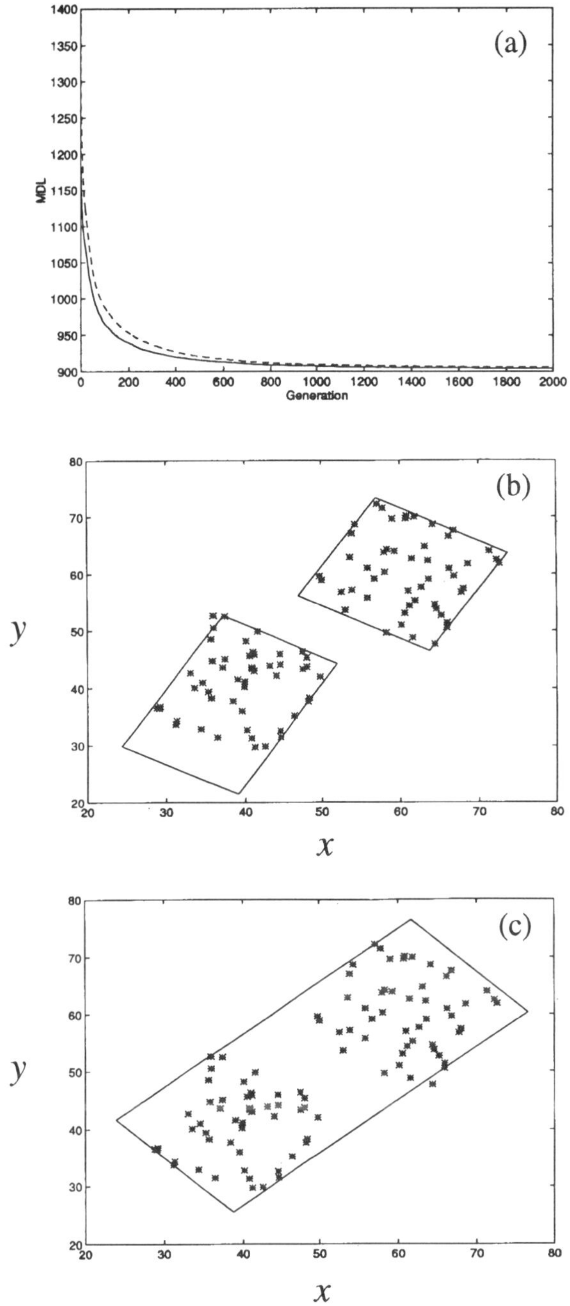

Figure 3. (a) The mean best score in the population (solid line) and the mean of the average score of all parents in the population (dashed line) as a function of the number of generations averaged over 50 trials for the case where the data are distributed in two separated slanted boxes. (b) A typical final best evolved clustering. The lower left-hand hyperbox contains some remaining empty space. (c) The final best evolved clustering in the 50th trial. Although this clustering consists of only a single hyperbox, it scored within one standard deviation above the mean of all best clusterings.

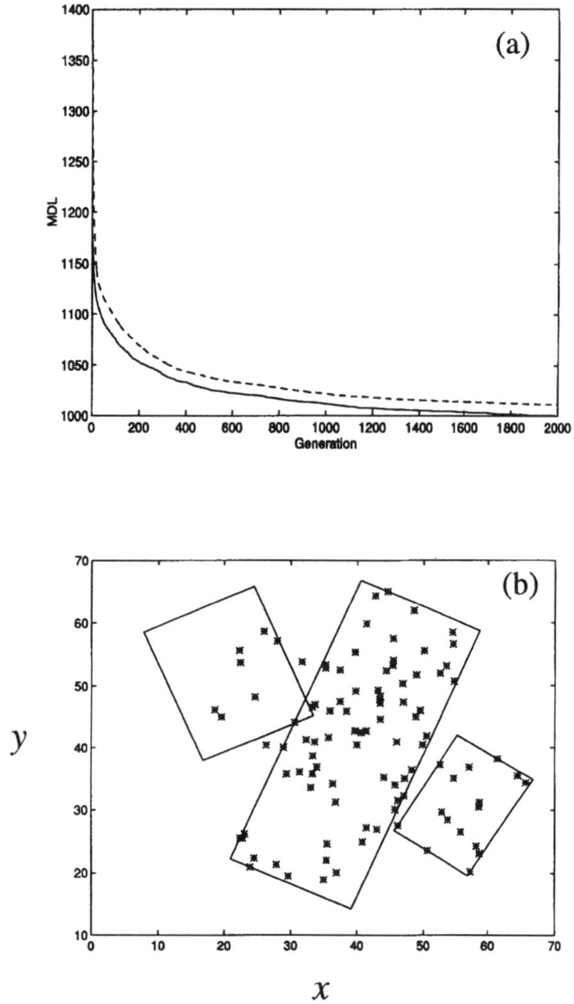

Figure 4. (a) The mean best score in the population (solid line) and the mean of the average score of all parents in the population (dashed line) as a function of the number of generations averaged over 50 trials for the case where the data are distributed in two overlapping perpendicular boxes. (b) A typical final best evolved clustering. The MDL penalizes solutions that place data in two or more hyperboxes, thus the clustering that would correspond to two overlapping hyperboxes is discouraged. The depicted final clustering has three hyperboxes, although the placement and size of the upper left-hand hyperbox could be improved.

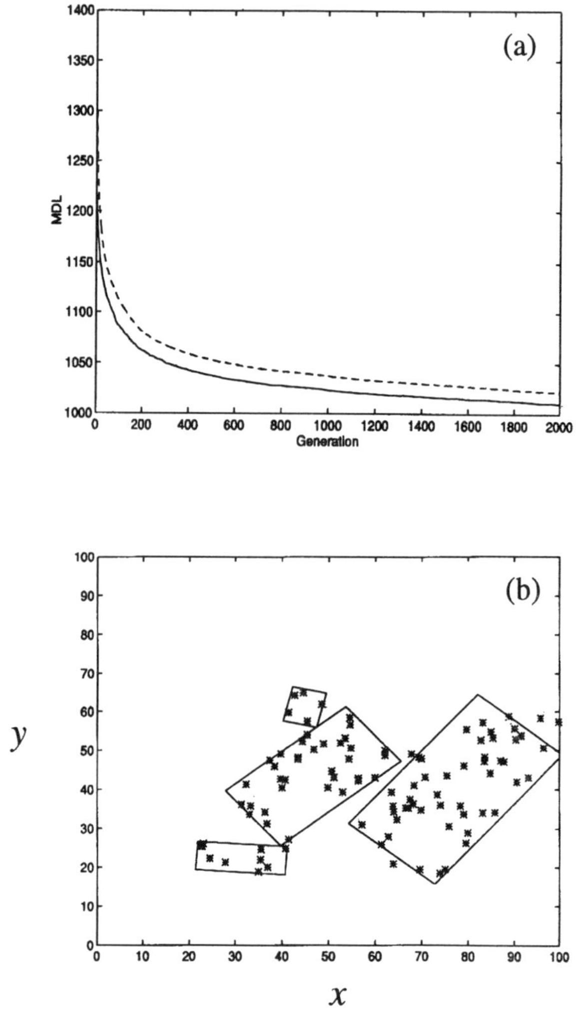

Figure 5. (a) The mean best score in the population (solid line) and the mean of the average score of all parents in the population (dashed line) as a function of the number of generations averaged over 50 trials for the case where the data are distributed in three boxes forming a slanted H. (b) A typical final best evolved clustering. Again, due to the penalty for placing points in more than one hyperbox, the depicted best solution possesses four nonoverlapping hyperboxes. Empty spaces that result from randomly generating the data lead to the observed partitions between clusters.

an assessment of the potential for combining both hyperboxes and hyperellipses, and consideration of fuzzy measures within the framework of the minimum description length principle.

Acknowledgments

The authors would like to thank J.B. Golden, V.W. Porto, and L.J. Fogel for their comments and criticisms.

Bibliography

J.W. Atmar (1976) "Speculation on the Evolution of Intelligence and Its Possible Realization in Machine Form," Doctoral Dissertation, New Mexico State University, Las Cruces.

W. Atmar (1994) "Notes on the Simulation of Evolution," *IEEE Trans. Neural Networks*, Vol. 5:1, pp. 130-147.

Table 1. Summary of the results for 50 trials on each of the four cases. The true box dimensions used for generating the data are listed, along with the mean best minimum description length (MDL) value after 2000 generations, the average of the mean MDL score in the population after 2000 generations, and the calculated MDL score for a set of boxes (or single box) that would match the dimensions of the true boxes being used to generate the data. Scores lower than (i.e., superior to) "truth" can be obtained because randomly generated data requires less area than is defined by the boundaries of the generating boxes (i.e., to require the entire area, the generated data would have to fall on the boundaries).

Case	Box Dimensions	Mean Best MDL	Avg. Mean MDL	"Truth"
1	Box 1: 20 × 20	877.7	877.8	887
2	Box 1: 20 × 20 Box 2: 20 × 20	909.1	912.2	893
3	Box 1: 50 × 20 Box 2: 50 × 20	1004.6	1014.5	1025
4	Box 1: 50 × 20 Box 2: 50 × 20 Box 3: 20 × 2	1014.0	1026.4	1008

B.P. Buckles, F.E. Petry, D. Prabhu, R. George, and R. Srikanth (1994) "Fuzzy Clustering with Genetic Search," *Proceedings of the First IEEE Conference on Evolutionary Computation*, IEEE Press, Piscataway, NJ, pp. 46-50.

D.B. Fogel (1995) *Evolutionary Computation: Toward a New Philosophy of Machine Intelligence*, IEEE Press, Piscataway, NJ.

D.B. Fogel and P.K. Simpson (1993a) "Evolving Fuzzy Clusters," *Proceedings of the 1993 IEEE International Conference on Neural Networks*, IEEE Press, Piscataway, NJ.

D.B. Fogel and P.K. Simpson (1993b) "Experiments with Evolving Fuzzy Clusters," *Proceedings of the Second Annual Conference on Evolutionary Programming*, D.B. Fogel and W. Atmar (eds.), Evolutionary Programming Society, La Jolla, CA, pp. 90-97.

L.J. Fogel, A.J. Owens, and M.J. Walsh (1966) *Artificial Intelligence through Simulated Evolution*, John Wiley, NY.

D. Hofstadter (1995) *Fluid Concepts and Creative Analogies: Computer Models of the Fundamental Mechanisms of Thought*, Basic Books, NY.

K. Hornik, M. Stinchcombe, and H. White (1989) "Multilayer Feedforward Networks are Universal Approximators," *Neural Networks*, Vol. 2, pp. 359-366.

J. Rissanen (1984) "Universal Coding, Information, Prediction and Estimation," *IEEE Transactions on Information Theory*, Vol. 30, pp. 629-636.

P.K. Simpson (1992) "Fuzzy Min Max Neural Networks. I. Classification," *IEEE Trans. on Neural Networks*, Vol. 3:5, 776-786.

P.K. Simpson (1993) "Fuzzy Min-Max Neural Networks. II. Clustering," *IEEE Trans. Fuzzy Systems*, Vol. 1:1, 32-45.

Evolving Reduced Parameter Bilinear Models for Time Series Prediction using Fast Evolutionary Programming

Sathyanarayan S. Rao, Kumar Chellapilla
Department of Electrical and Computer Engineering
Villanova University
Villanova, U.S.A
rao@ece.vill.edu, kumarc@ece.vill.edu
HTTP://www.ece.vill.edu/user/kumarc

Abstract

We propose fast evolutionary programming (FEP) for optimizing the parameters of a reduced parameter bilinear model (RPBL) used for predicting nonlinear and chaotic time series. The RPBL model is capable of effectively representing nonlinearity and multiperiodicity with the additional advantage of using fewer parameters than a conventional bilinear model. FEP has been shown to have reasonable optimization performance and can be coupled with the ability to determine model structure. Unlike conventional methods where the model order is chosen first and the parameters of the model are determined subsequently, both the model order and the parameters are evolved simultaneously in light of the minimum description length (MDL) criterion. The effectiveness of FEP is first tested on the Bohachevsky and Rosenbrock functions and then used for prediction of the sunspot series, laser generated data, Mackey-Glass series, and astrophysical data. Experimental results show that FEP, in comparison with conventional evolutionary programming, evolves RPBL models with lower normalized mean squared error (NMSE) and also lower model order.

1 Introduction

Bilinear time series models constitute a special class of nonlinear models for time series prediction. Advantages include computational simplicity due to its nearly linear form and the ability to approximate to an arbitrary degree of accuracy any well-behaved Volterra series relationship over a finite interval. One of the disadvantages of the bilinear model is that the number of parameters to be estimated is typically large. Recently Zhang and Hagan (1994) have proposed a reduced parameter bilinear (RPBL) model that can represent a bilinear structure with fewer parameters and with the additional capacity to incorporate seasonal and periodic variations in the model.

Simultaneously determining both the RPBL model order and the parameters requires a technique that is amenable to combinatorial optimization problems. Stochastic search techniques such as simulated annealing, genetic algorithms, and evolutionary programming have been found to be robust and effective in solving these types of problems. Evolutionary programming (EP) has been shown to be useful in simultaneous identification and parameter estimation (Fogel et al. 1966). It is particularly suitable for application to problems typified by chaos, chance, temporality, and nonlinear interactivity that are not tractable to classical optimization methods.

A new modified simulated evolutionary optimization (MSEO) algorithm, by Rao and Ramasubrahmanyam (1996) has been used for the evolution of bilinear models for time series prediction. The modeling of time series is carried out as a two step process. In the first step, the model order and the parameters are evolved simultaneously using the minimum description length (MDL) criterion. In the second step, the model order of the winner is held constant and the parameters of the winner are evolved, using the normalized mean squared error criterion.

In this paper we enhance the modeling process by performing both steps simultaneously and accelerate the evolution by the use of fast evolutionary programming (FEP). We present FEP in section 2. In section 3, we test the effectiveness of FEP in view of the meta-EP and MSEO algorithms on the Rosenbrock and Bohachevsky functions. We present the RPBL model in section 4 and the model identification criterion in section 5. We evolve the bilinear models for time series prediction in sections 6. The prediction results are given in section 7, and conclusions are offered in section 8.

2 Fast Evolutionary Programming

Fogel *et al.* (1966) proposed a population-based random search strategy termed evolutionary programming, where instead of keeping the single best point in each generation, a population of search points is maintained. EP has been extended (Rao and Ramasubrahmanyam 1996) into modified simulated evolutionary optimization (MSEO) for the modeling of time series. Fogel *et al.* (1991) and Bäck *et al.* (1993) have indicated that EP with self-adaptive mutation (meta-EP) usually performs better than EP without adaptive mutation.

Yao and Liu (1996) have shown empirically that FEP, which uses a Cauchy mutation operator performs much better than EP using the Gaussian mutations for several multimodal functions with many local minima. Further it is comparable to EP in performance for unimodal and multimodal functions with only a few local minima. FEP is implemented as follows (Yao and Liu 1996):

1. *Initialization:* Generate the initial population of μ individuals, and set the generation number, k to *1*. Each individual is taken as a pair of real-valued vectors, $(x_i, \eta_i), \forall i \in \{1, \ldots, \mu\}$.

2. Evaluate the error score for each individual, $(x_i, \eta_i), \forall i \in \{1, \ldots, \mu\}$, in terms of the objective function, $f(x_i)$.

3. *Mutation:* Each parent $(x_i, \eta_i), \forall i \in 1, \ldots, \mu$, creates a single offspring (x_i', η_i') by:

$$x_i'(j) = x_i(j) + \eta_i(j)C(0, 1) \tag{1}$$

$$\eta_i'(j) = \eta_i(j)\exp(\tau' N(0, 1) + \tau N_j(0, 1)) \tag{2}$$

for $j = 1, \ldots, n$, where $x_i(j), x_i'(j), \eta_i(j)$ and $\eta_i'(j)$ denote the jth component of the vectors x_i, x_i', η_i, and η_i', respectively. $C(0,1)$ denotes a one-dimensional random variable with a Cauchy distribution centered at the origin and with the scale parameter t equal to 1. The one dimensional Cauchy density function centered at the origin is defined by:

$$g_t(x) = \frac{1}{\pi}\frac{t}{t^2 + x^2}, \quad -\infty < x < \infty, \tag{3}$$

where $t > 0$ is the scale parameter. The corresponding distribution function is

$$G_t(x) = \frac{1}{2} + \frac{1}{\pi}\operatorname{atan}\left(\frac{x}{t}\right). \tag{4}$$

The shape of $g_t(x)$ resembles that of the Gaussian density function but approaches the axis so slowly that an expectation does not exist. $N(0,1)$ denotes a normally distributed one-dimensional random number with mean zero and standard deviation one. $N_j(0,1)$ indicates that the random number is generated anew for each value of j. The factors τ and τ' are commonly set to $(\sqrt{2\sqrt{n}})^{-1}$ and $(\sqrt{2n})^{-1}$, respectively.

4. *Fitness:* Calculate the error of each offspring $(x_i', \eta_i'), \forall i \in \{1, \ldots, \mu\}$.

5. Conduct pairwise comparison over the union of parents (x_i, η_i) and offspring (x_i', η_i'), $\forall i \in \{1, \ldots, \mu\}$. For each individual, q opponents are chosen randomly from all the parents and offspring with equal probability. For each comparison, if the individual's error is no greater than the opponent's, it receives a "win."

6. *Selection:* Select the μ individuals out of (x_i, η_i) and (x_i', η_i'), $\forall i \in \{1, \ldots, \mu\}$, that have the most wins to be parents of the next generation.

7. Stop if the halting criterion is satisfied; otherwise, increment the generation number, that is $k = k + 1$ and go to step 3.

In our simulations, the halting criterion was taken as a certain maximum number of generations (*kmax).*

3 Experiments with FEP on Test Problems

To provide a comparison between the EP variations presented, the two-dimensional Rosenbrock and Bohachevsky functions were chosen. The task was to find the global minima of these functions. A population size of 100 was taken with the initial values uniformly distributed in the range of [-5,5]. The initial perturbation variances for the meta-EP and FEP were taken from a uniform random distribution in [0,1]. The number of opponents was arbitrarily set to 10.

The Bohachevsky function is:

$$F(x, y) = x^2 + 2y^2 - 0.3\cos(3\pi x) - 0.4\cos(4\pi y) + 0.7 \tag{5}$$

It has numerous local minima and a global minimum of zero at the origin. The function is illustrated in Fig. 1 with *x,y* ranging over [-1,1]. The FEP algorithm was compared with meta-EP and MSEO. Fifty Monte Carlo simulations were performed. The mean error trajectory of the best population member at each generation during the search on the Bohachevsky surface is given in Fig. 3.

The Rosenbrock function is:

$$F(x, y) = 100(x^2 - y)^2 + (1 - x)^2 \tag{6}$$

It has a parabolic valley along the curve $y = x^2$, with a unique minimum of zero at the point (1,1). There are no local minima associated with the function. The function is illustrated in Fig. 2 with *x,y* ranging over [-4,4]. Once again the FEP algorithm was compared with meta-EP and MSEO. Fifty Monte Carlo simulations were performed. The mean error trajectory of the best population member at each generation during the search on the Rosenbrock surface is given in Fig. 4.

From the plots and the convergence trajectories we observe that FEP performs better than MSEO and meta-EP in terms of convergence rate. This is likely due to the fatter tails of the Cauchy distribution.

4 Reduced Parameter Bilinear Model

The reduced parameter bilinear model is defined as

$$\phi(z^{-1})Z_t = \theta(z^{-1})a_t + (\xi(z^{-1})Z_t)(\zeta(z^{-1})a_t) \tag{7}$$

where $\{a_t\}$ is a sequence of independent zero mean random variables and

Fig 1. Bohachevsky function surface

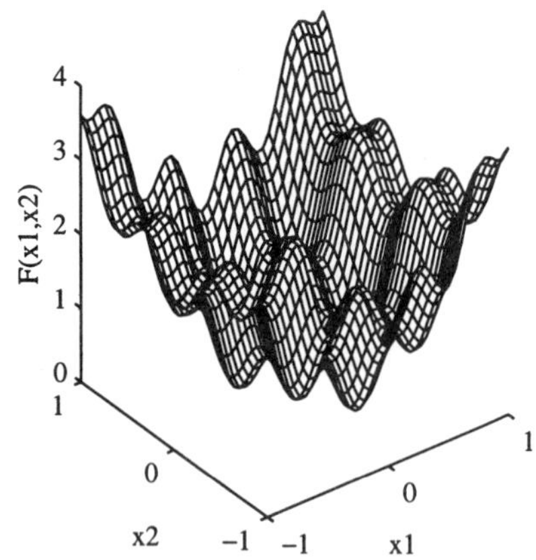

Fig 2. Rosenbrock function surface

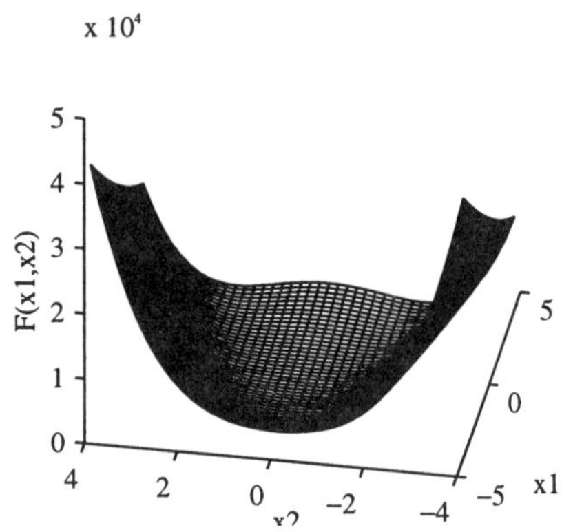

Fig 3. Optimization on Bohachevsky function. 50 Monte Carlo Trials. Best Scores.

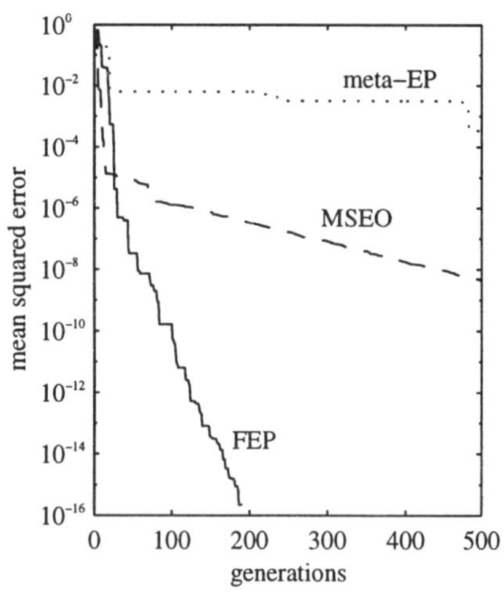

Fig 4. Optimization on Rosenbrock function. 50 Monte Carlo Trials. Best Scores.

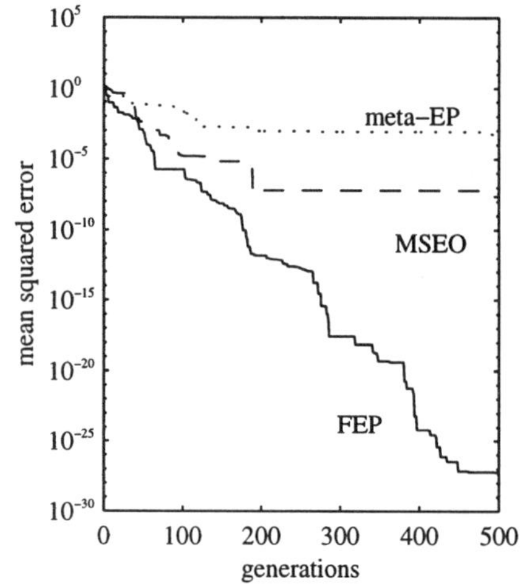

$$\begin{aligned}\phi(z^{-1}) &= 1-\phi_1 z^{-1}-\phi_2 z^{-2}-\ldots-\phi_p z^{-p}\\ \theta(z^{-1}) &= 1-\theta_1 z^{-1}-\theta_2 z^{-2}-\ldots-\theta_q z^{-q}\end{aligned} \tag{8}$$

$$\begin{aligned}\xi(z^{-1}) &= z^{-1}+\xi_2 z^{-2}+\ldots+\xi_m z^{-m}\\ \zeta(z^{-1}) &= \zeta_1 z^{-1}+\zeta_2 z^{-2}+\ldots+\zeta_k z^{-k}\end{aligned} \tag{9}$$

5 Model Identification

The identification procedure consists of determination of orders p, q, m, k and for the determined model order, estimating the parameters $\{\phi_j\}, \{\theta_j\}, \{\xi_j\}, \{\zeta_j\}$. The model order is determined as that which minimizes the Minimum Description Length (MDL) criterion defined in Candy (1988) as

$$(N-\gamma)\log(\sigma_e^2) \quad + \\ \left(\frac{1}{2}\right)(\text{number of independent parameters})\log(N-\gamma) \tag{10}$$

where N is the number of observations of the time-series, $\gamma = \max(p, q, m, k)$ and

$$\sigma_e^2 = \left(\frac{1}{N-\gamma}\right)\sum_{t=\gamma+1}^{N}(Z_t-\hat{Z}_t)^2 \tag{11}$$

$\hat{Z}_t$ is the predicted output at time t obtained using the model with order (p, q, m, k). Using evolutionary programming, the model order is estimated as follows: Each organism in the population is a vector which contains the model order (p, q, m, k) followed by the parameters themselves. Gaussian mutations were performed, simultaneously, on the model order and the parameters. In each generation, the model order parameters were perturbed with continuous gaussian random numbers, with corresponding model order variances, and rounded to the nearest integer to obtain the new model order. The model order variances were also evolved, in the same manner as the variances of the model parameters. The model order evolved with each generation according to the MDL fitness criterion and the fittest vector after a certain number of generations (after which there is no more improvement in the fitness) contained the desired model order and the model parameters that could be used for time-series prediction.

6 Evolving Bilinear Time Series Prediction Models

To verify the capability of the evolutionary algorithms in evolving bilinear time series models, a bilinear process Z_t was simulated by the following difference equation:

$$Z_t = 0.4Z_{t-1} + 0.5Z_{t-2} + a_t + 0.4Z_{t-1}a_{t-1} \qquad (12)$$

The models evolved by the various evolutionary algorithms are given in table 1. From the results it can be observed that the fitted models were an accurate representation of the original bilinear model.

Next the evolutionary optimization techniques were applied for the modeling of four time series namely, the sunspot series, the Mackey-Glass series, the laser generated data series, and astrophysical data. The performance of the final evolved models were evaluated using the normalized mean square error (NMSE) which is defined as

$$NMSE = \frac{\sum_k (Z_k - \hat{Z}_k)^2}{\sum_k (Z_k - \bar{Z})^2} \qquad (13)$$

where Z_k, $\hat{Z}_k$, $\bar{Z}$ correspond to the time series, its prediction, and its mean, respectively.

7 Prediction Results

A. The Sunspot series

The first set of experiments was conducted on Wolf's sunspot series for the years 1700-1988. These numbers are indicative of the average relative number of sunspots observed each day of the year and serve as a standard bench mark for time-series modeling, where the objective is to generate a single-step prediction based on past observations. Consistent with McDonnell *et al.* (1994), the data set was partitioned into a training set over years 1700-1920 and test sets for years 1921-1955 and 1956-1979, respectively. The resulting bilinear models evolved by the three evolutionary algorithms in consideration are given in table 2. The actual and predicted outputs and prediction error of the model evolved by FEP are given in Figs. 5(a) and 5(b).

Fig. 5(a) Predicted and actual sunspot series

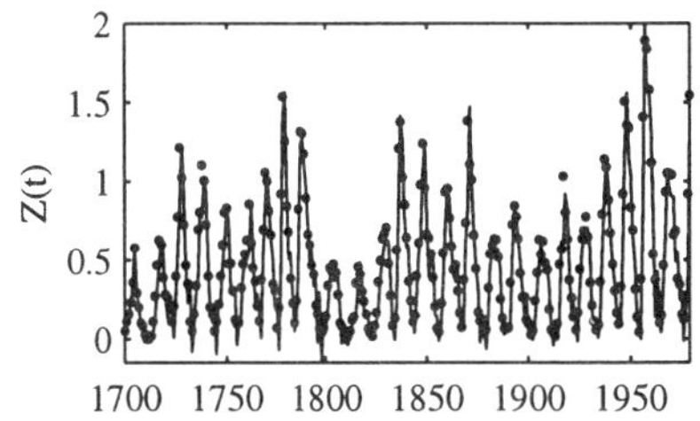

__: Predicted series and ...: Actual series

Fig. 5(b) Prediction error of the model evolved by FEP

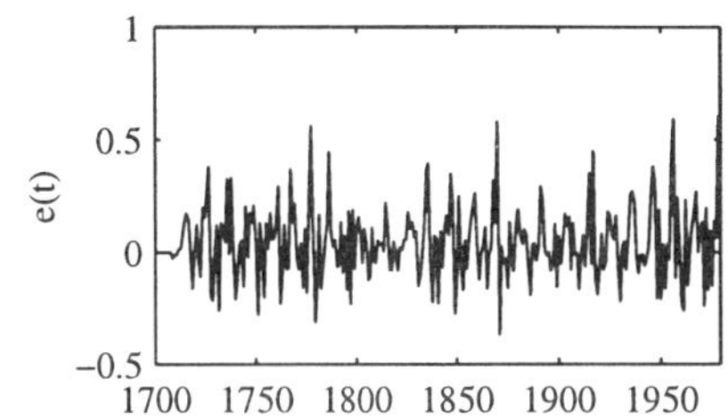

B. The Mackey-Glass equation.

The Mackey-Glass equation represents a model for white blood cell production in leukemia patients (Mackey and Glass 1977). This model is complicated by the addition of a time delay τ in the nonlinear differential equation

$$\dot{x}(t) = \frac{ax(t-\tau)}{1 + x^c(t-\tau)} - bx(t) \qquad (14)$$

where the free parameters are selected $a = 0.2$, $b = 0.1$, $c = 10$, and $\tau = 1$ as chosen in (McDonnell and Waagen 1994). The training data consisted of 500 data points, extracted from a longer run so that the initial transients would have no effect. The resulting bilinear models evolved are given in table 3. The actual and predicted outputs and prediction error of the model evolved by FEP are given in Figs. 6(a) and 6(b), respectively.

C. Laser Generated Data

The laser generated data consists of measurements made on an 81.5 micron 14NH3 cw (FIR) laser, pumped optically by the P(13) line of an N20 laser via the vibrational aQ(8,7) NH3 transition. The intensity data was recorded by a LeCroy oscilloscope. The experimental signal-to-noise ratio was about 300. The data are a cross-cut through periodic to chaotic intensity pulsations of the laser. The chaotic pulsations follow the theoretical Lorenz model of a two-level system. The data are a good example of the complicated behavior that can be seen in a clean, stationary, low-dimensional, nontrivial physical system for which the underlying governing equations and dynamics are well understood. A training data sequence of 350 data points was chosen. The evolved bilinear models and their performance measures are given in table 4. The actual and predicted outputs and prediction error of the model evolved by FEP are given in Figs. 7(a) and 7(b), respectively.

Fig. 6(a) Predicted and actual Mackey-Glass series

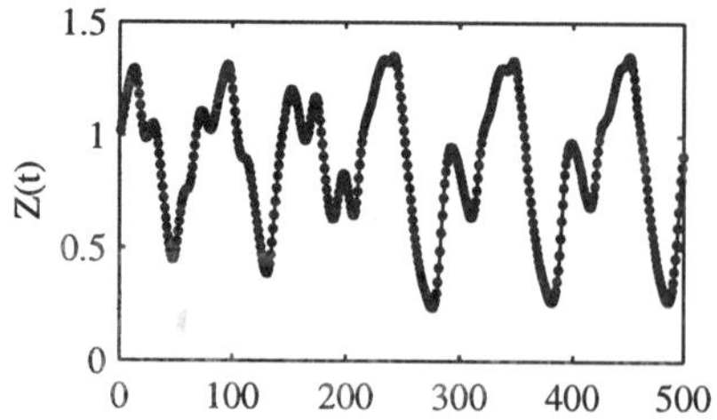

__: Predicted series and ...: Actual series

Fig. 6(b) Prediction error of the model evolved by FEP

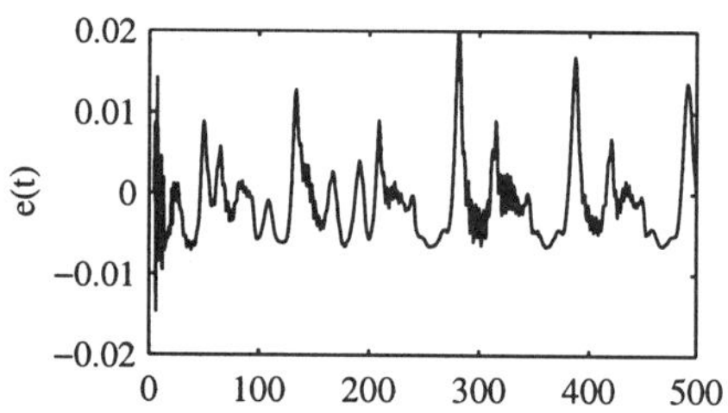

Fig. 7(a) Predicted and actual laser generated data series

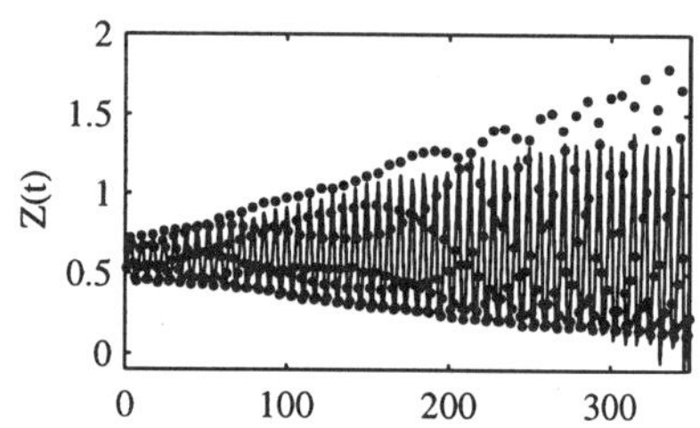

__: Predicted series and ...: Actual series

Fig. 7(b) Prediction error of the model evolved by FEP

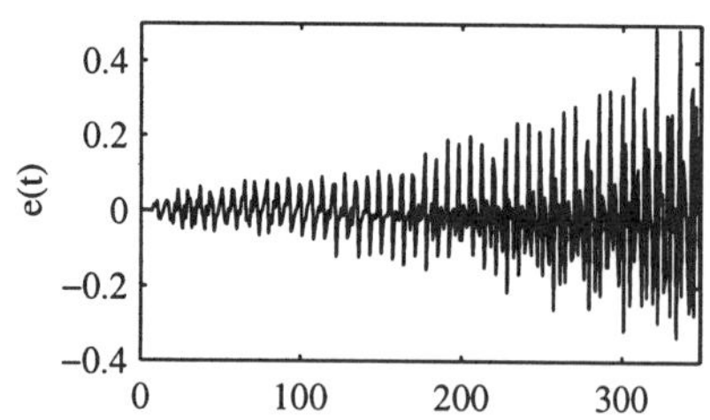

D. Astrophysical Data

The astrophysical data consists of measurements of the light curve (time variation on intensity) of the variable white dwarf star PG1159-035 during March 1989. It was recorded by the Whole Earth Telescope (a coordinated group of telescopes distributed around the Earth that permits the continuous observation of an astronomical object). The intensity variations of the star arise from the excited modes, which are spherical harmonics. This data set represents the optical oscillations of a physical system, so unlike the laser data set, which has very little noise, the measurement process for these data is much noisier. The evolved models are given in table 5. The actual and predicted outputs and prediction error of the model evolved by FEP are given in Figs. 8(a) and 8(b), respectively.

Fig. 8(a) Predicted and actual astrophysical data series

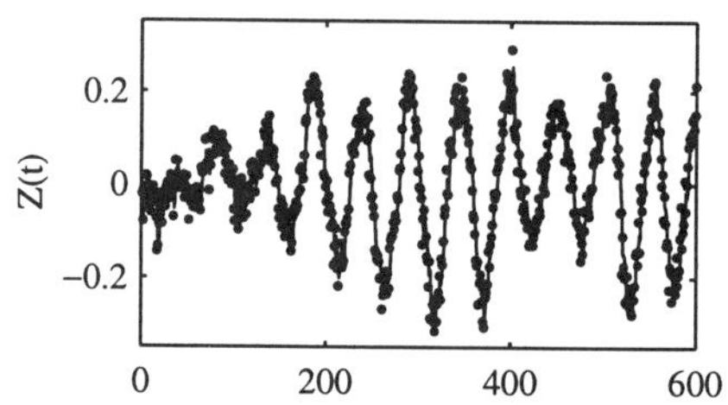

__: Predicted series and ...: Actual series

Fig. 8(b) Prediction error of the model evolved by FEP

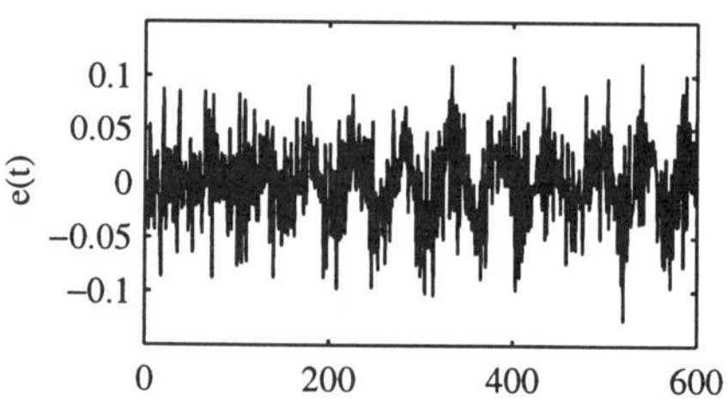

8 Discussion

The FEP technique has been successfully applied to optimize the parameters of a reduced parameter bilinear model used for predicting nonlinear chaotic time series. The performance of the FEP algorithm has been shown to be superior to those of meta-EP (Fogel et al. 1991) and MSEO (Rao and Ramasubrahmanyam 1996) in terms of faster convergence and more parsimonious solutions that possess less error. This better performance may be attributed to the ability of the Cauchy random variable to generate occasionally useful large jumps. This ability of the Cauchy random variable was found to be useful in accelerating the evolution of the parameters of reduced parameter bilinear models for time series modeling. The usefulness of such a property in optimization problems in the identification of control systems, and modeling of nonlinear signals is an open area for research.

REFERENCES

T.Bäck and H.-P. Schwefel (1993), "An overview of evolutionary algorithms for parameter optimization," *Evolutionary Computation,* 1(1):pp. 1-23.

J. V. Candy (1988), *Signal Processing: the model-based approach*, McGraw-Hill Book Company.

L. J. Fogel, A. J. Owens and M. J. Walsh (1966), *Artificial Intelligence Through Simulated Evolution*, New York: John Wiley & Sons.

D. B. Fogel, L. J. Fogel, J. W. Atmar (1991), "Meta-evolutionary programming," in *Proc. of the 25th Asilomar Conf. on Signals, Systems and Computers*, R. R. Chen, Ed. IEEE computer Society, pp. 540-545.

M. C. Mackey and L. Glass (1977), "Oscillations and chaos in physiological control systems," *Science,* 197.

J. R. McDonnell, and D. Waagen (1994), "Evolving Recurrent perceptrons for time-series modeling," *IEEE Trans. Neural Networks*, vol. 5, no. 1.

S. S. Rao, and A. Ramasubrahmanyam (1996), "Evolving bilinear models for Time Series Prediction," submitted to *IEEE Trans. Neural Networks*.

X. Yao and Y. Liu (1996), "Fast evolutionary programming," *Evolutionary Programming V: Proc. of 5th Annual Conf. on Evol. Prog.,* L. J. Fogel, P. J. Angeline, T. Bäck (editors), MIT Press, Cambridge, MA, forthcoming.

Y. Zhang and M. T. Hagan (1994), "A Reduced Parameter Bilinear Time Series Model," *IEEE Trans. Signal Processing*, vol. 42, no. 7, pp. 1867-1870.

Table 1: The evolved models for the test problem and the corresponding NMSE

ALGORITHM	MODEL ORDER	MODEL	NMSE
FEP	**(2,1,2,2)**	$Z_t = 0.3925Z_{t-1} + 0.5068Z_{t-2} + a_t + 0.0070a_{t-1} + (-0.7338Z_{t-1} - 0.0123Z_{t-2})(-0.5595a_{t-1} + 0.0188a_{t-2})$	**9.8993E-6**
MSEO	**(3,2,2,1)**	$Z_t = 0.3909Z_{t-1} + 0.5231Z_{t-2} - 0.0061Z_{t-3} + a_t + 0.0076a_{t-1} - 0.0187a_{t-2} + (-0.8547Z_{t-1} - 0.0322Z_{t-2})(0.0016a_t - 0.4574a_{t-1})$	**8.2235E-4**
META-EP	**(2,1,2,3)**	$Z_t = 0.3016Z_{t-1} + 0.5859Z_{t-2} + a_t + 0.0102a_{t-1} + (-0.1764Z_{t-1} + 0.2673Z_{t-2})(-0.8036a_{t-1} + 0.8687a_{t-2} - 0.2712a_{t-3})$	**0.0055**

Table 2: The evolved models for the sunspot series and the corresponding NMSE

ALGORITHM	MODEL ORDER	MODEL	NMSE
FEP	**(2,3,9,2)**	$Z_t = 0.3762Z_{t-1} + 0.3386Z_{t-2} + a_t + 0.9993a_{t-1} + 0.9032a_{t-2} + 0.2502a_{t-3} + \left(\begin{array}{l} 0.1370Z_{t-1} - 0.9997Z_{t-2} + 0.1085Z_{t-3} \\ + 0.9088Z_{t-4} + 0.1133Z_{t-5} + 0.9464Z_{t-6} \\ + 0.9998Z_{t-7} + 0.1353Z_{t-8} + 0.9745Z_{t-9} \end{array}\right) X(0.2989a_{t-1} + 0.1007a_{t-2})$	**1700-1920** **0.0680** **1921-1955** **0.1130** **1956-1979** **0.1895**

Table 2: The evolved models for the sunspot series and the corresponding NMSE

ALGORITHM	MODEL ORDER	MODEL	NMSE
MSEO	(4,2,4,4)	$Z_t = 0.2740Z_{t-1} + 0.6688Z_{t-2} - 0.1958Z_{t-3} - 0.0463Z_{t-4}$ $+ a_t + 1.5533a_{t-1} + 1.2948a_{t-2}$ $(-0.3649Z_{t-1} + 0.4854Z_{t-2} - 0.2490Z_{t-3} + 0.1892Z_{t-4})$ $X(-0.3888a_{t-1} + 0.0094a_{t-2} + 0.0709a_{t-3} - 0.4785a_{t-4})$	1700-1920 0.0896 1921-1955 0.1328 1956-1979 0.2682
META-EP	(2,1,7,4)	$Z_t = 0.9999Z_{t-1} - 0.1940Z_{t-2} + a_t + 0.9999a_{t-1}$ $+\begin{pmatrix} 0.1721Z_{t-1} + 0.0868Z_{t-2} + 0.1216Z_{t-3} - 0.1908Z_{t-4} \\ + 0.1601Z_{t-5} + 0.1263Z_{t-6} - 0.9999Z_{t-7} \end{pmatrix}$ $X(-0.9999a_{t-1} - 0.4381a_{t-2} + 0.4768a_{t-3} + 0.2280a_{t-4})$	1700-1920 0.1065 1921-1955 0.1571 1956-1979 0.1930

Table 3: The evolved models for the Mackey-Glass series ($\tau = 1$) and the corresponding NMSE

ALGORITHM	MODEL ORDER	MODEL	NMSE
FEP	(3,1,3,4)	$Z_t = 0.1000Z_{t-1} + 0.3053Z_{t-2} - 0.2940Z_{t-3} + a_t + 0.8740a_{t-1}$ $+ (0.9996Z_{t-1} + 0.7703Z_{t-2} + 0.9994Z_{t-3})$ $X(0.8071a_{t-1} + 0.1000a_{t-2} + 0.5317a_{t-3} + 0.1963a_{t-4})$	2.6312E-4
MSEO	(1,1,7,5)	$Z_{t-1} + a_t + 0.9999a_{t-1}$ $\begin{pmatrix} 0.9998Z_{t-1} - 0.9987Z_{t-2} + 0.8823Z_{t-3} - 0.0503Z_{t-4} \\ - 0.9993Z_{t-5} + 0.7435Z_{t-6} + 0.9997Z_{t-7} \end{pmatrix}$ $X(0.9996a_{t-1} + 0.8030a_{t-2} + 0.8191a_{t-3} + 0.9984a_{t-4} + 0.5037a_{t-5})$	6.1188E-4
META-EP	(2,5,2,2)	$Z_t = 0.9999Z_{t-1} + 0.0036Z_{t-2} + a_t + 0.8776a_{t-1} + 0.0944a_{t-2}$ $+ 0.8866a_{t-3} + 0.1000a_{t-4} + 0.4386a_{t-5}$ $+ (-0.8299Z_{t-1} - 0.5765Z_{t-2})(-0.9511a_{t-1} - 0.5589a_{t-2})$	5.8345E-4

Table 4: The evolved models for the laser generated data and the corresponding NMSE

ALGORITHM	MODEL ORDER	MODEL	NMSE
FEP	(6,2,2,2)	$Z_t = 0.9202Z_{t-1} - 0.2635Z_{t-2} - 0.0673Z_{t-3} - 0.0244Z_{t-4}$ $- 0.0190Z_{t-5} + 0.4574Z_{t-6} + a_t + 0.7473a_{t-1} - 0.7797a_{t-2}$ $+ (0.2217Z_{t-1} - 0.0532Z_{t-2})(0.1684a_{t-1} - 0.7376a_{t-2})$	0.1026
MSEO	(4,1,10,2)	$Z_t = 0.7980Z_{t-1} - 0.2338Z_{t-2} + 0.0972Z_{t-3} + 0.0735Z_{t-4} + a_t + a_{t-1}$ $+\begin{pmatrix} -0.6607Z_{t-1} - 0.9991Z_{t-2} + 0.2393Z_{t-3}(-0.4316Z_{t-4} + 0.9882Z_{t-5}) \\ 0.4042Z_{t-6} + Z_{t-7} - 0.3240Z_{t-8} + 0.4655Z_{t-9} + 0.4688Z_{t-10} \end{pmatrix}$ $X(a_{t-1} + 0.6854a_{t-2})$	0.1437

Table 4: The evolved models for the laser generated data and the corresponding NMSE

ALGORITHM	MODEL ORDER	MODEL	NMSE
META-EP	**(7,1,5,2)**	$Z_t = 0.5468Z_{t-1} - 0.2103Z_{t-2} + 0.2016Z_{t-3} - 0.1967Z_{t-4}$ $- 0.0470Z_{t-5} + 0.3397Z_{t-6} + 0.4100Z_{t-7} + a_t + 0.8302a_{t-1}$ $+ (-0.1015Z_{t-1} + 0.7673Z_{t-2} - 0.0957Z_{t-3} + 0.8783Z_{t-4} + 0.4383Z_{t-5})$ $\mathrm{X}(-0.0441a_{t-1} + 0.0039a_{t-2})$	**0.1203**

Table 5: The evolved models for the astrophysical data and the corresponding NMSE

ALGORITHM	MODEL ORDER	MODEL	NMSE
FEP	**(2,2,2,2)**	$Z_t = 0.9999Z_{t-1} - 0.03634Z_{t-2} + a_t + 0.6322a_{t-1} + 0.1661a_{t-2}$ $+ (-0.3092Z_{t-1} + 0.2401Z_{t-2})(0.3902a_{t-1} + 0.6964a_{t-2})$	**0.0967**
MSEO	**(2,3,9,6)**	$Z_t = 0.2823Z_{t-1} + 0.6568Z_{t-2} + a_t + 0.6572a_{t-1} + 0.6253a_{t-2}$ $- 0.1211a_{t-3} + \begin{pmatrix} 0.5788Z_{t-1} + 0.6031Z_{t-2} - 0.5078Z_{t-3} \\ -0.2970Z_{t-4} - 0.4731Z_{t-5} - 0.7745Z_{t-6} \\ + 0.5982Z_{t-7} - 0.2465Z_{t-8} + 0.5460Z_{t-9} \end{pmatrix}$ $X\begin{pmatrix} -0.0946a_{t-1} - 0.6009a_{t-2} - 0.5744a_{t-3} \\ - 0.8693a_{t-4} - 0.0526a_{t-5} - 0.3846a_{t-6} \end{pmatrix}$	**0.0968**
META-EP	**(2,2,2,2)**	$Z_t = 0.9999Z_{t-1} - 0.037Z_{t-2} + a_t + 0.6524a_{t-1} + 0.1574a_{t-2}$ $+ (0.9999Z_{t-1} + 0.7959Z_{t-2})(-0.9999a_{t-1} + 0.6563a_{t-2})$	**0.0970**

Classifier Systems Papers

Three-dimensional Shape Optimization Utilizing a Learning Classifier System

Robert A. Richards
Mechaniucal Design Division: SLAC
PO Box 9723
Stanford University
Stanford, CA 94309-9723
buc@slac.stanford.edu

Sheri D. Sheppard
Mechanical Engineering Dept.
Terman 503
Stanford University
Stanford, CA 94305
sheppard@cdr.stanford.edu

ABSTRACT

A methodology to perform generalized zeroth-order two- and three-dimensional shape optimization utilizing a learning classifier system is presented and applied. Specifically, the methodology has the objective of determining the optimal boundary to minimize mass while satisfying constraints on stress and geometry. Even with the enormous advances in shape optimization no method has proven to be satisfactory across the broad spectrum of optimization problems facing the modern engineer. The shape optimization via hypothesizing inductive classifier system complex (SPHINcsX) instantiates the methodology in a software package overcoming many of the limitations of today's conventional shape optimization techniques. From environmental input, and a population of initially randomly generated rules, SPHINcsX is expected to learn to make the appropriate component shape modifications to reach a minimum mass design while satisfying all stress constraints. SPHINcsX not only learns from a clean slate, but confronts the additional challenge of learning without sensitivity information that most other shape optimization algorithms deem essential. SPHINcsX has proven adept at solving both two- and three-dimensional shape optimization problems.

1. Introduction

This paper surveys the Shape oPtimization via Hypothesizing Inductive classifier system compleX (SPHINcsX, pronounced *sfinks*). The complex is a software package which embodies a novel methodology for performing zeroth-order stress constrained shape optimization by exploiting an inductive classifier system. The complex starts with a design component and modifies boundaries to optimize the shape (minimizing its mass) while meeting all design constraints.

The broadest objectives of the research, which drive the development of SPHINcsX, are:

- to develop a *general* scaleable methodology for component design optimization, from simple two-dimensional sizing designs to three-dimensional shape designs,
- to investigate the classifier system's applicability as a tool for mechanical engineering design.

Some of SPHINcsX's sub-objectives, to make it *general*, include:

- perform without the benefit of sensitivity information (derivatives of the objective function and constraints with respect to the design variables), that is; a zeroth-order method,
- be independent of the analysis tool,
- be virtually independent of the boundary representation.

The scope of SPHINcsX is limited as follows:

- optimization occurs on isotropic components acted upon by loads, and in static equilibrium due to a set of fixtures, or constraints,
- the design modifications occur by modifying the boundaries of the component; no new boundaries, such as voids, are introduced,
- the objective is to minimize the component's mass,
- constraints are placed on the maximum stress allowed.

The overall approach is to take the problem of shape optimization (see Zienkiewicz & Campbell (1973) and Haftka & Grandhi (1986)) and interface it with a relatively simple classifier system. The resulting complex takes an initial design, calculates the stress properties and applies a set of rules (classifiers) to generate a new design. This iterative process continues until an optimum design is created. The evolving design's changes in merit are used as feedback to the complex and its underlying classifier system. Utilizing the feedback and a genetic algorithm, good rules will gain strength and breed with other good rules in an attempt to evolve better rules.

There are three distinct classes of shape optimization problems. In order of complexity, these are: size, shape, and topological optimization. The current investigation considers shape optimization, which subsumes size optimization, but does not deal with topological optimization

The objective is to minimize the mass of the design by manipulation of non-constrained boundaries. The explicit constraints consist of the bounds placed on the design variables, if any. The design points defining the boundary are the design variables. Since the optimization performed in this study is relative to stress, the major implicit constraint is the maximum allowable von Mises stress,. The design is assumed optimized when all the design's stresses are within a certain percentage of the maximum allowable, or the incremental improvement falls below a threshold.

2. Process of Shape Optimization

The initial condition starts with a (normally feasible) design. A *feasible design* is one where none of the constraints are violated. Figure 1 shows that there are four major modules in the shape optimization process: *boundary representation*, *analysis*, *optimization*, *termination*.

The boundary representation must be defined as a part of the design's initial condition. The optimal shape depends on the boundary representation and the design variables selected to represent the modifiable boundaries. SPHINcsX attempts to treat the boundary representation module as a black box; but since the optimization module must interface with the boundary representation module it must be cognizant of the representation used. Boundary representations used with SPHINcsX are restricted so that the design variables are actually on the design boundary. This restriction still allows for most boundary representations, including lines, arcs, circles, ellipses, cubic splines, and surface patches.

The analysis module's purpose is to determine the values of the: *objective function*, *implicit constraints*, *auxiliary information* (needed by the optimization module). The analysis module utilized in conjunction with SPHINcsX is envisioned as a black box which will provide the stress and mass information to the optimization module allowing for analysis technique independence. There is no auxiliary information (such as sensitivities) needed by the SPHINcsX optimization module.

The termination module checks to see if any of the termination criteria are met. From here the system branches, stopping if the termination module determines the termination branch should be taken, otherwise the branch is made to the optimization module.

The optimization module determines which design variables to modify and the magnitude of the modifications. The optimization module then changes the design variable values. These new design variable values are subsequently passed to the boundary representation module, completing the optimization loop (iteration).

3. Classifier System & Genetic Algorithm

A classifier system (CS) is a machine learning system that learns rules, called classifiers. These classifiers guide the system's performance in an arbitrary environment. Using a genetic algorithm, the CS potentially evolves better populations of classifiers. For a complete description of classifier systems see the founding work by Holland and Reitman (1978), as well as later works by Riolo (1988), Booker et al. (1989), Goldberg (1989), Dorigo & Sirtori

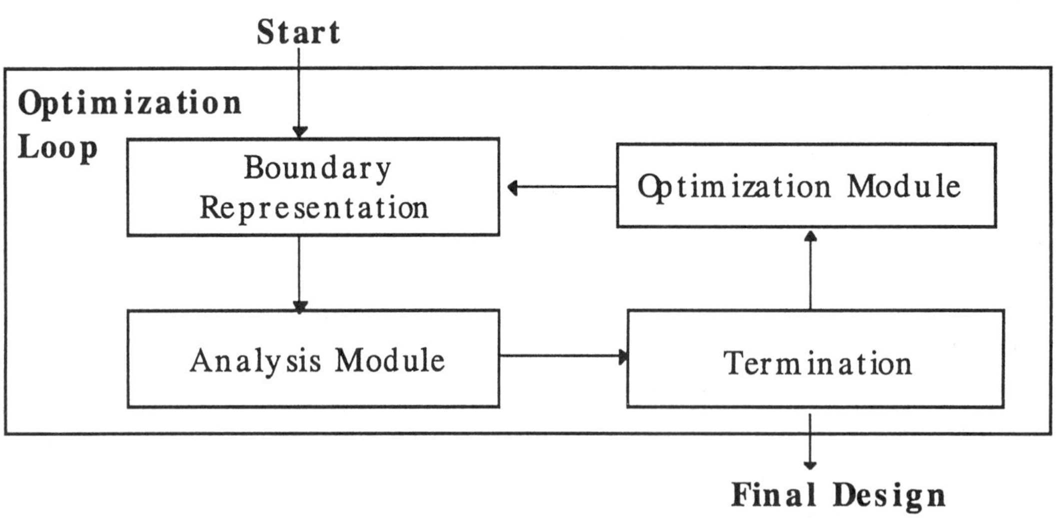

Figure 1. Shape Optimization Process Overview

(1991), and Wilson (1994).

The following sub-sections will describe research specific aspects of the Michigan approach stimulus-response classifier system used in SPHINcsX. SPHINcsX uses both a genetic algorithm and a covering operator.

3.1 Detector Interface

The detectors filter all the information produced by the environment, sensing only the subset deemed important by the classifier system. For this study von Mises stresses are of greatest concern. Patterns of the stress state in zones around each design point are presented to the CS.

Figure 2 illustrates the locations for stress sensing for a two-dimensional boundary. The sensing occurs at each design point (the box in Figure 2), two straddle points (represented as circles on the surface in Figure 2), and an interior point (represented as a circle interior to the design point in Figure 2). The straddle points are two points, one on each side of the design point and lying on the boundary. The interior point is a point interior to the design point, lying on the line, defined by the surface normal at the design point. Each (non-design) point is one *global element length* (EL) from the design point. For the interior point the distance is measured along the surface normal at the design point. For the straddle points, the distance is measured along the actual boundary geometry. The *global element length* is a user specified parameter, if the finite-element method, see Zienkiewicz (1980), is used in the analysis module then a global element length is equivalent to the length of the elements defined along the boundary. However, if closed form solutions are used, a global element length is still needed, and should be set to a value appropriate for the model if the finite-element method was being applied.

The classifiers' format is driven by the input to which the classifiers are to match. The classifiers will now see a maximum of four distinct stresses per design point (in the two-dimensional case), as well as the mass.

For scalability from two-dimensional to three-dimensional problems, the straddle point stresses will be compared by the detectors and only the most *prominent* will be passed as a message to the classifier population. *Prominent* in this context means the stress whose value differs greatest from the maximum allowable von Mises stress. From this information the detectors process the three stresses into a *detector message*; one detector message is created per design point. Studies (Richards, 1995) have revealed that 6 bits per detector message provide adequate granularity while not requiring exorbitant processing overhead.

3.2 Effector Interface

Modifications to a design occur after all classifiers that match an environmental message bid in an *auction* for the right to execute their action (i.e., consequent). When the victorious classifier invokes its action it is applied to each design point that produced an environmental message, which matched the victorious classifier's antecedent.

The design point's modification consists of the distance to move and the direction of the move. In the case, where a design point is not constrained in its directional freedom, the direction of movement occurs collinear to the surface normal. The effectors work in conjunction with the boundary representation to determine the direction of movement and therefore the classifier consequent does not factor into the direction determination

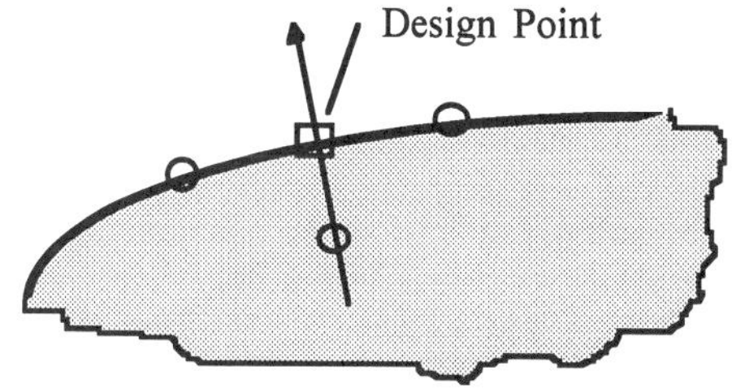

Figure 2. Design Point with Associated Stress Sensing Locations

The range of possible movements is extensive — the complex should be able to optimize components used in structures ranging from bridges to micro-machinery. The initial move limit is based on the user defined, global element length. After a design point has been moved three times, approximate functions of stress versus design point location are generated.

For SPHINcsX, the number of gradations for the point's move is 64. The mapping of the relative move magnitude onto the range defined by the maximum move is accomplished as follows:

physical magnitude =
(max. move) * (relative move magnitude - 31.5) / 32

As the system nears the optimum, the range defined by the move limits also decreases, resulting in ever finer design point manipulations.

3.3 Feedback Interface

The feedback interface collects the mass information of the component. In many cases the change in mass provides a viable ranking of the change when no stress constraint violations occurred pre- or post-modification. For stress constrained optimization problems, the *Total Normalized Stress Error* (TNSE), as defined by Hsu (1992), complements the mass as an evaluator. Therefore, the change in mass and the change in TNSE per optimization loop are the primary feedback components. With knowledge of the changes in stress, mass, and TNSE, modifications are determined to be beneficial or detrimental.

4. Learning Regime

The SPHINcsX's learning regime algorithm begins with the initialization of the classifier system. A *tabula rasa* is used for the initial population. The population is created by randomly picking a *0*, *1*, or # for each position for each classifier's antecedent. The probability of picking a # for each position is determined by the, *# probability parameter* (0.6). All the classifier system parameter settings used in SPHINcsX may be found in Richards (1995).

In order to teach SPHINcsX in learning mode, a suite of design optimization problems is needed. The educational process consists of a progression through a learning suite of designs on which optimization is attempted. The learning suite contains a range of designs covering the spectrum from simple, one variable sizing optimization designs to multiple variable sizing optimization designs. The learning suite utilizes closed form designs only.

The learning suite is broken down into *clusters*; *clusters* are groups of similar designs, that have a common unique characteristic. The design *clusters* composing the learning suite are named as follows ranging from simple to more complex: *Tension Rod (TRg)*, *Pressure Vessel (PV)*, *Torsion Solid Rod (TSR)*, *Cantilever Beam (CB)*, *Torsion Elliptical Solid Rod (TESR)*, and *Torsion Elliptical Hollow Rod (TEHR)*.

Next, the initialization of the design to be optimized occurs. A list of the necessary design initializations consists of setting up the parameters defined by the learning suite. The educational process consists of a progression through a learning suite of designs on which optimization is attempted. The selected design has its design variables stochastically set to their initial values, the resulting initial design is then used in the SPHINcsX algorithm where an attempted optimization proceeds. After the attempted optimization, the process continues with the selection of another initial design. *Attempted* is used because a maximum of 100 iterations is set, so that if 100 iterations occur and the optimum has not been reached, the optimization is ceased and a new design is stochastically generated.

4.1 Learning Performance

The progress of the learning regime is evaluated by monitoring the performance metrics, *P1*, *P2*, *P3*. *P1* is the ratio of the number of correct responses to the total number of responses. *P2* is the total number of correct responses during an epoch to the epoch length. *P3* is the number of iterations needed to optimize the design. The performance metrics, *P1* and *P2* should show asymptotic improvement with increasing learning iterations, and thus reveal not only the learning progress but also a point where the diminishing returns are so small that continued learning is not justified. The genetic algorithm's purpose is exploration, so the populace after the GA application will most likely initially perform worse because some new members will be provided with artificially high strengths. Therefore, the asymptotic improvements in *P1* and *P2* may not be monotonic.

The learning iterations, t_I, t_{II}, t_{III}, t_{IV}, t_V, and t_{VI}, when each of the design clusters are introduced into the learning regime are not set *a priori*, but are determined via the monitoring of the performance metrics. Each introduction of a new design cluster is expected to cause an initial decrease in all the performance metrics.

4.2 Evolution

Learning commences by selecting a design from the general tension rod cluster, TRg, using 5 different radii, which are set stochastically. The information provided by the performance metrics must be examined in tandem (with each other) to assess when learning has occurred to the level where continued learning with the same cluster of designs is inefficient.

After six epochs, *P2* has reached nearly an 80% correct response rate. *P2* rate is not expected to ever reach 100% because the GA is applied every epoch and introduces new classifiers, some of which will invariably make incorrect decisions. The performance plateaued approaching 1,500 iterations, at this point the PV cluster was introduced.

After 5,000 learning iterations, the initial design selection scheme had progressed through the TRg cluster, PV cluster, TSR cluster, CB cluster, TESR cluster, and TEHR cluster. Each of the clusters went through an evolution similar to that described for the tension rod cluster.

Figure 3 depicts *P3* performance throughout the learning regime. Note blank regions of more than 100 iterations occur when SPHINcsX failed to converge within the maximum allowed 100 iterations. Recall also that *P3* was not expected to decrease monotonically through a design cluster for the initial design was set stochastically. Table 1 summarizes some of the aspects of the learning process, showing the ranges of learning iterations (defined empirically, as stated above) applied to each design cluster as well as presenting the level of the performance metrics *P1* and *P2* at the end of each cluster's learning phase.

Table 1. Learning Summary

Design Cluster	Iteration (Start)	Iteration (End)	P1	P2
TRg (n=5)	1	1,500 (t_I)	69	82
PV	1,501	2,000 (t_{II})	70	76
TSR	2,001	2,800 (t_{III})	69	71
CB	2,801	3,500 (t_{IV})	70	70
TESR	3,501	4,400 (t_V)	69	71
TEHR	4,401	5,000 (t_{VI})	70	74

5. Applications

To test SPHINcsX, it is applied to design optimization problems using its learned rules in application mode. Two design optimization problems with increasing difficulty presented below provide a test of SPHINcsX's shape optimization ability.

5.1 Shape Optimization: Torque Arm

The goal of the torque arm design problem is to find the shape that minimizes the mass of the torque arm by varying only the shape of the boundary between points A and B as shown in Figure 4, without exceeding the maximum allowable von Mises stress. The torque arm's shape is invariant through its thickness, t, and is loaded by both a transverse load, F_b, and a tensile load, F_t, on the inside of the small hole. In addition, the torque arm is constrained around the circumference of the larger hole located at x equal to 0.0416 m. Figure 4 depicts the initial design, displaying the fixed dimensions, loading and constraints.

SPHINcsX converged to within 2% of a fully stressed boundary after 36 iterations. Only one finite element analysis was required in each iteration. The initial mass was, 0.824 kg. The mass of the final design was 0.540 kg, a 34.5% decrease in mass. This example demonstrates the capability of SPHINcsX in solving a realistic design problem. Figure 5 depicts the shape of the optimized design.

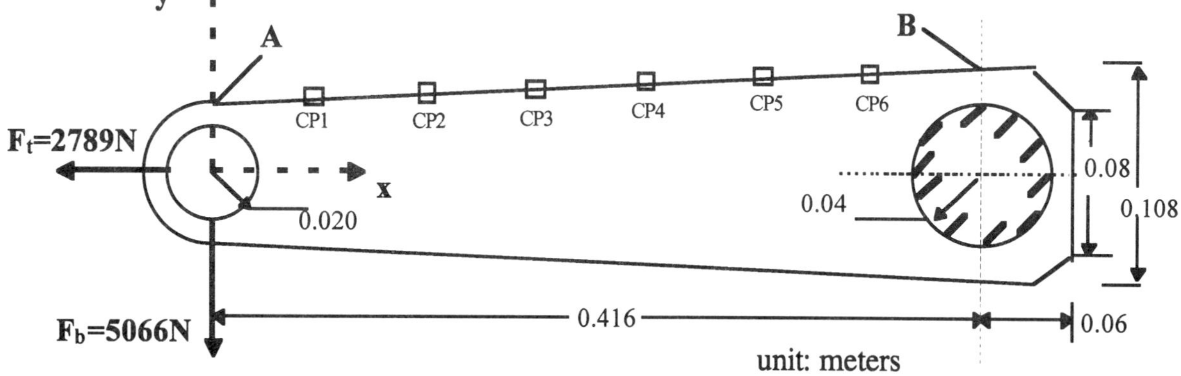

Figure 4. Initial Configuration and Loading Conditions of the Torque Arm

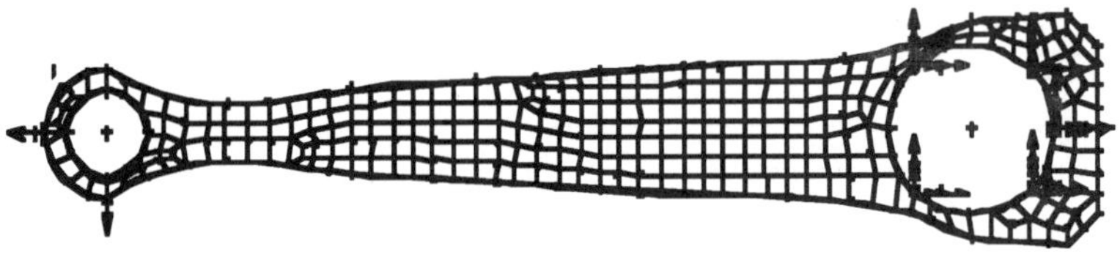

Figure 5. Optimized Analysis Model of the Torque Arm

5.2 Three Dimensional Shape Optimization: Pressure Vessel

The prominent straddle point definition included provisions to seamlessly allow for three-dimensional boundaries. In either the two-dimensional (2-D) case or the three-dimensional (3-D) case, each design point has one associated interior point that is one element length interior to the design point along the surface normal. The number of straddle points per design point varies depending on the boundary representation and the analysis model; however, the concept of prominent straddle point leads to the same environmental message format being constructed by the detectors in both the 2-D and 3-D scenarios. Thus, the structure of the individual classifiers remains the same, as does all of the internal operations of SPHINcsX. Therefore, the only extension to SPHINcsX necessary to move into the 3-D realm is the ability to read multiple straddle point stresses from three-dimensional analysis models.

This application utilizes a spherical pressure vessel, with a fixed spherical exterior loaded by a constant pressure. The initial interior cavity is an irregular non-optimal void. The goal of this problem is to find the shape of the cavity that minimizes the mass of the pressure

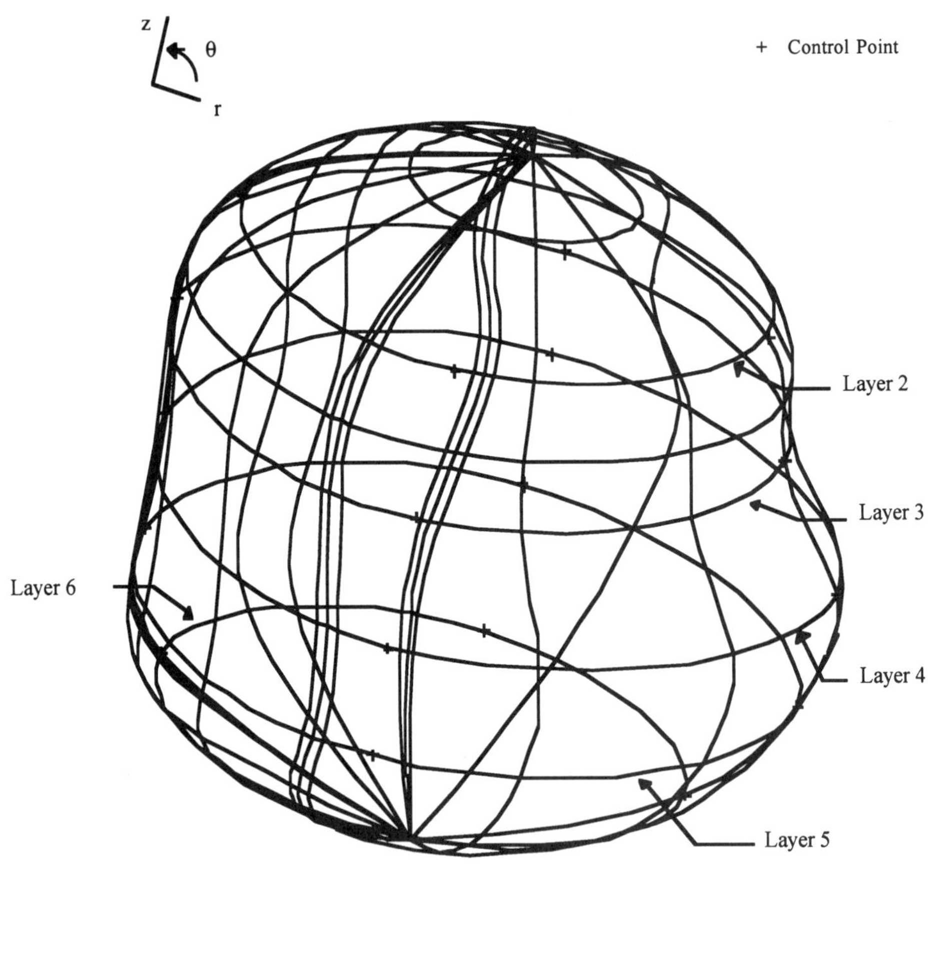

Figure 6. Initial Cavity of the Spherical Pressure Vessel

vessel without exceeding the maximum allowable von Mises stress. The void is modeled by seven layers of design variables, as shown in Figure 6, the first and seventh (top and bottom) layers each only have a single point representation, each of the other layers is represented by four design points. Therefore, there are a total of 22 design variables. The non-end layers' four design points are distributed evenly about the vertical axis but their radial distances from the vertical axis are not equal or interdependent. So a closed loop fitted to an internal layer's four design points does not result in a circle, but an irregular curve.

The initial design has both under stressed regions and over-stressed regions. Note that no supplemental education is provided before attempting to apply SPHINcsX to a three dimensional problem. The classifier population that learned via a suite of sizing problems was expected to handle this challenge.

SPHINcsX converged to within 1% of the maximum allowable stress over each curve after 51 iterations. Verification of this thick walled spherical pressure vessel solution is accomplished by comparing it to an analytical solution. The SPHINcsX solution closely approached the spherical optimal shape.

5.3 Performance Analysis

Considering, first, other methods which cover SPHINcsX's scope, few exist and those that do suffer debilitating limitations. For example, genetic algorithm based systems (see Jensen (1992)) can handle all problems in SPHINcsX's scope also without auxiliary information but the solution times would be (conservatively) in the thousands of iterations. SPHINcsX's computational requirements should be drastically less in 3-D than techniques which depend on sensitivity information, but successful applications in the literature are scarce (Kodiyalam et al. (1992)) and applying commercial solutions is restricted by the fact they can not work with general shape representations as input to their optimization modules. The literature does provide some 3-D sizing optimization examples, Yang (1989), Botkin & Yang (1991), and Noel et al. (1995). An illustration of the considerable cost of performing 3-D optimization utilizing sensitivities is provided by Chargin et al. (1991) where, with 23 design variables, the sensitivity calculation per analysis required over six times the computational overhead as the underlying analysis.

Modern optimization methods provide more competition to SPHINcsX in the realm of two-dimensional shape optimization. Recall that for the torque arm SPHINcsX determined the optimum in 36 iterations, this is competitive with other methods, however, many specialized methods can perform better. Belegundu (1993) found an optimum shape of a similar torque arm design in 17 analyses, while Hsu (1992) took less than 10 analyses.

6. Summary and Conclusions

SPHINcsX resulted from the need to discover an efficient generally applicable scaleable shape optimization methodology. In the search for such a methodology, non-calculus-based search and optimization techniques were explored to determine if any could provide results superior to the state of the art. Machine learning in general and the classifier system in particular appeared to possess the desired features.

SPHINcsX learned to perform general shape optimization and did so without the benefit of sensitivity information or any other auxiliary information. Furthermore, *generalized learning was achieved via sizing design problems only.* It is a further credit to the methodology learned without knowing anything about the initial shape, the representation of the curve under modification, the concept of stress and strain or optimization. Essentially, SPHINcsX learned to recognize patterns of stress and execute appropriate action. That is, the mapping between a control point's stress pattern and beneficial control point modification has been discovered.

The results illustrate the computational efficiency of SPHINcsX. The computational effort, after learning, is small relative to that of methods using sensitivity analysis. The pressure vessel application demonstrated how the computational effort increased less than linearly with the increase in design variables. SPHINcsX demonstrates that the classifier system indeed can be a valuable tool in helping to solve shape optimization problems, and its generality portents its value in other mechanical engineering domains.

Bibliography

Belegundu, Ashok D., "Optimizing the Shapes of Mechanical Components", *Mechanical Engineering*, January 1993, pp. 46-48.

Booker, L.B., D.E. Goldberg, and J.H. Holland, "Classifier Systems and Genetic Algorithms", *Artificial Intelligence*, Vol. 40, 1989, pp. 235-282.

Botkin, M.E. and R. J. Yang, "Three-Dimensional Shape Optimization with Substructuring", *AIAA Journal*, March 1991, pp. 486-488.

Chargin, Mladin K., Ingo Raasch, Ralph Bruns, and Dawson Deuermeyer. "General Shape Optimization Capability", *Finite Elements in Analysis and Design*, Vol. 7, 1991, pp. 343-354.

Dorigo, Marco and Enrico Sirtori, "Alecsys: A Parallel Laboratory for Learning Classifier Systems", *Proceedings of Fourth International Conference on Genetic Algorithms* - July 13-16, 1991 - San Diego - California, Morgan Kaufmann , San Mateo, CA.

Goldberg, David E.., *Genetic Algorithms in Search, Optimization & Machine Learning*, Addison-Wesley, NY, 1989.

Haftka, R.T., and R.V. Grandhi, "Structural Shape Optimization - A Survey", *Computer Methods in*

Applied Mechanics and Engineering, Elsevier Science Publishers B.V., North-Holland, 1986, pp. 91-106.

Holland, J.H. and J.S. Reitman, "Cognitive Systems Based on Adaptive Algorithms", in *D.A. Waterman and F. Hayes-Roth (eds.), Pattern-Directed Inference Systems*, Academic Press, NY, 1978.

Hsu, Y.L, *Zeroth Order Optimization Methods of Two Dimensional Shape Optimization*, Ph.D. Dissertation, Department of Mechanical Engineering , Stanford University, 1992.

Jensen, Eric Dean, *Topological Structural Design Using Genetic Algorithms*, Ph.D. Dissertation, Purdue University, Lafayette, IN, 1992.

Kodiyalam, Srinivas, Virendra Kumar and Peter M. Finnigan, "Constructive Solid Geometry Approach to Three-Dimensional Structural Shape Optimization", *AIAA Journal*, Vol. 30, No. 5, May 1992, pp. 1408-1415.

Noel, F., J.C. Leon and P. Trompette, "Shape Optimization of Three-dimensional Parts based on a Closed Loop between Structure Analysis and Geometric Modeling", *Engineering with Computers*, Vol. 11, 1995, pp. 114-121.

Richards, Robert A., *Zeroth-Order Shape Optimization Utilizing a Learning Classifier System*, Ph.D. Dissertation, Stanford University, Stanford, CA, 1995.

Riolo, Rick L., *Empirical Studies of Default Hierarchies and Sequences of Rules in Learning Classifier Systems*, Ph.D. Dissertation, Computer Science and Engineering Department, University of Michigan, 1988.

Wilson, Stewart W., "ZCS: A Zeroth Level Classifier System", *Evolutionary Computation*, Vol. 2, No. 1, Spring 1994, pp. 1-18.

Yang, R. J., "A Three-Dimensional Shape Optimization System — SHOP3D", *Computers & Structures*, Vol. 31. No. 6, 1989, pp. 881-890.

Zienkiewicz, O. C., and J.S. Campbell, "Shape Optimization and Sequential Linear Programming", Gallagher, R.H. and Zienkiewicz, O.C. (eds.), *Optimal Structural Design*, Wiley, NY, 1973.

Zienkiewicz, O. C., *The Finite Element Method in Structural and Continuum Mechanics*, McGraw Hill, NY, 1980.

Classifier System Renaissance: New Analogies, New Directions

H. Brown Cribbs, III
Engineering Science and Mechanics
The University of Alabama
P.O. Box 870278
Tuscaloosa, Alabama 35487-0278
brown@galab3.mh.ua.edu

Robert E. Smith
Engineering Science and Mechanics
The University of Alabama
Tuscaloosa, Alabama
rob@comec4.mh.ua.edu
phone: (205) 348-1618

ABSTRACT

Learning classifier systems (LCSs) have existed for nearly twenty years (Holland & Reitman, 1978). Research efforts in reinforcement learning (RL), evolutionary computation (EC), and neural networks have enhanced the original LCS paradigm. New thoughts from these areas have created a "renaissance" period for the LCS. This paper highlights some key LCS advancements and the fields that inspired them. One inspiration, from neural networks, is examined for a novel LCS approach to autonomous mobile robots. A simple, LCS-controlled robot simulation is presented. This simulation shows the potential benefits of combined biological paradigms and the hybridization of ideas in the LCS. Future directions for LCS research are discussed.

KEYWORDS: learning classifier system, machine learning, reinforcement learning, genetic algorithm, genetic programming, evolutionary computation, neural networks.

1 Introduction

The learning classifier system concept has existed for nearly twenty years. Recently, inspirations from other fields have led to something of an LCS "renaissance." Advances in reinforcement learning have led to more powerful and theoretically-backed LCS credit assignment schemes. Neural, fuzzy, and GP systems are inspiring ways to re-think the traditional LCS representation scheme. The separation of conflict resolution parameters (like strength) and fitness has allowed for more powerful LCS implementations. These new LCS inspirations and some of their effects are discussed in the following sections. One of many new LCS variants (a neural network LCS for Q-learning) is presented with preliminary results. Future research directions are discussed in light of the new inspirations presented.

2 Reinforcement Learning

The LCS has been traditionally applied to reinforcement learning (RL) problems. In a RL problem, a control system is provided performance feedback which does not directly indicate optimal actions. This feedback is often called *reward*. In contrast, supervised learning problems provide error in terms of optimal actions.

In recent years, analytical and empirical research on RL has grown drastically. Many new methods have been developed and investigated. These include Watkin's Q-*learning* algorithm (1989) and Sutton's *TD-λ* schemes (1987). New RL schemes are naturally extensible to LCS strength update. In fact, the bidding and bucket brigade procedures (Holland, 1986) typically used in the LCS are reinforcement learning update schemes.

The bucket brigade is remarkably similar to Q-learning, but it has one important difference. This can be seen in terms of a general expression for updating the strength (the traditional LCS measure of expected payoff) of a classifier i, S_i^t, with an update term, dS_i:

$$S_i^{t+1} = S_i^t + \eta \cdot dS_i \qquad (1)$$

In the bucket brigade, the update term dS_i is

$$dS_i = R_i + \lambda S_j^t - S_i^t \qquad (2)$$

where S_j^t is the strength of classifier j that acts at time $t+1$, R_i is the reward received when classifier i acts at time t, and λ is the *discount parameter*. It is interesting to note that the discount parameter was not included in many past LCS studies. However, its inclusion is now standard practice in RL methods and the LCS. In many RL problems, a discount factor $0 < \lambda \leq 1$ is necessary for stable, bounded growth of estimated payoff.

In Q-learning, the update term dS_i is similar to that given above, except that S_j^t is always the highest strength of the classifiers matched at time $t+1$. This is not necessarily the classifier picked to act by a (typically stochastic) conflict resolution scheme. This seemingly subtle difference leads to Watkin's proof that Q-learning converges to an optimal strategy given infinite exploration. The bucket brigade procedure does not lead to this optimal convergence. However, it is interesting to note that a recent RL study (Rummery & Niranjan, 1994) and a recent LCS study (Twardowski, 1993) both indicate that procedures like the bucket brigade may be superior to Q-learning for systems with realizable amounts of exploration.

Wilson points out the ties between the LCS and Q-learning in two of his recent papers (Wilson, 1994; Wilson, 1995). Smith and Cribbs also use Q-learning in the LCS presented later in this paper. Other elements from the growing body of RL research are bearing on LCS investigation as well, including examinations of exploration versus exploitation in parameter updating (Thrun, 1992) and alternate updating schemes (Schwartz, 1993). As general understanding of RL as a discipline expands, the mathematically foundations of RL can naturally supplant the analogies typically used as a foundation for LCS credit assignment schemes.

3 Separating Action Selection and Fitness

Due to reward's relationship to the system performance, rule updates rely directly on the actions taken. As the rule-set is updated and changed by evolution the worth of each action may become skewed by newer classifiers that are still in training with respect to the update term, dS.

As the rule-set is tuned, the updates diminish, i.e., $dS \rightarrow 0$ as $t \rightarrow \infty$. This phenomena can be viewed as the settling of the strength relative to the utility of the classifier. Once the strength has settled, the classifier's expected payoff is effectively fixed; hence, behavioral utility can be assessed without update noise. However, throughout this process it is important to ask whether expected payoff (in the form of strength or a Q-value) is an appropriate measure of classifier fitness.

Traditionally, the strength parameter is used as both a basis for action selection and as fitness of LCS rules. Strength is an estimator of predicted payoff. Wilson (1995) points out that such an estimator may be an inappropriate measure of fitness. It may be necessary to maintain classifiers with a variety of expected payoffs to achieve good performance. If expected payoff is used as fitness, this may run counter to the action of the GA. Therefore, Wilson recommends a fitness measure based on the *accuracy* of the classifier's payoff estimate, as opposed to its value. Smith and Cribbs (1994) show a similar, variance based fitness measure, which was partially inspired by Fahlman's (1990) cascade correlation algorithm for neural networks.

These new fitness measures, and the inspirations of RL for credit assignment, reduce coupling between the action selection parameters and fitness parameters in the LCS. In the future, this may allow for more isolated, detailed examination of the co-evolution of classifiers under various fitness measures.

4 Representation

Although the {1, 0, #} representation of rules with general conditions is traditional in LCSs, current research tends to indicate that it is not a required part of the LCS paradigm. Genetic programming (GP), fuzzy, and neural network based LCSs have been implemented. This section outlines some of these new LCS representations.

Note that there are two distinct forms of LCS representation. These are the so-called "Michigan" and "Pitt" approaches. These forms differ in what is actually being evolved. In the "Michigan" approach, each GA individual is a single rule, and the population is the rule set. The rule-set is incrementally changed by evolving individual rules. "Michigan" selects rules to be operated on genetically; hence, segments of existing rules are recombined. In the "Pitt" approach, each GA individual is a complete rule set. The GA operates on a *population of rule-sets*. "Pitt" attempts to evolve whole rule-sets by crossing existing rule sets. In the following discussion, the "Michigan" approach is assumed. However, note that hybrids of "Michigan" and "Pitt", perhaps with incremental expansion of the number of rules in a GA individual, are often discussed in the LCS community (Smith & Valenzuela-Rendón, 1992).

Wilson (1994) has suggested the use of a GP-like representation in the LCS in his discussion of *s-classifiers*. Encoding a classifier's conditions with the logical operators `AND`, `OR`, and `NOT` is straight-forward. Evolution of s-classifiers may be performed by the standard cut-and-splice operations of GP. While initial trials of this concept have used binary actions (Tufts, 1995), actions could also be encoded in a GP-like format (Wilson, 1994). Given its representational structure, GP may be the form of EC that most readily extensible to an LCS. Evolutionary programming (Fogel et al., 1966) and evolutionary strategies (Schwefel, 1965) may also be feasible, although none are currently discussed in the literature.

Given the growing popularity of fuzzy control systems, and their rule-based structure, fuzzy representations are a natural extension of the LCS. A number of studies have investigated fuzzy LCS representations (Valenzuela-Rendón, 1991; Parodi & Bonelli, 1993)

Neural networks offer another LCS representation. Many studies compare neural networks and the LCS. Smith and Cribbs (1994) offer a concise representational analogy, which is revisited here.

The basis for a neural network LCS lies in the net's ability to fire discriminantly for various input cases. This is similar to the selectivity of a rule's condition. In addition to selective firing, the ability to ignore certain input fields allows neurons to generalize over the inputs. An LCS generalizes through use of the #, while the neural network uses zero valued weights or null connections to achieve the same "don't care" effect. In this sense, a # is equivalent to a zero valued weight in a neural network. One can complete this analogy by moving from the LCS's *unipolar binary*, {1, 0}, representation to a *bipolar binary*, {1, −1}, representation. The resulting neural net representation for the LCS zero condition could be −1. The −1 acts as a analog complement operator, similar to the zero in the LCS. With the zero and # conditions redefined, the alphabet of neuron input weights $\{1, -1, 0\}$, stands as an equivalent to the LCS's $\{1, 0, \#\}$, alphabet.

With the LCS alphabet redefined, generalization can be re-addressed from a neural network standpoint. The input vector for both neural nets and LCSs contain multiple fields or variables. If one were to construct a space dimensioned by the number of variables in the input vector, an input hyper-space would exist. Generalization is evident in the input hyper-space partitioning, where hyper-planes divide the input hyper-space. Partitioning helps the network to classify inputs. Booker (personal communication) has also considered this notion, using hyperplane coefficients in an LCS for function approximation.

Albus' (1975) Cerebellar Model Articulation Controller (CMAC) neural network has a similar notion of limited receptive fields that partition a hyper-space. A neuron's receptive field is a sparsely connected set of inputs. Absence of full connectivity in the neuron's inputs creates a selective firing neuron. Much as the LCS uses conditions to control rule activation, the receptive field governs when a neuron fires. A receptive field is defined in the input connections to the neuron. For an neural network LCS, the receptive field's weights are constrained to the neural alphabet discussed. A neuron, with receptive field input connectivity, fires only when the input vector matches the receptive field. Figure 1 depicts a simple LCS rule and its receptive field neuron counterpart. For simplicity, Figure 1 does not show the zero connection weights, just the non-zero connection weights.

A neural-representation LCS is able to classify inputs by assessing the receptive fields the input excites. Neurons with receptive fields can be viewed as rules in an LCS. The activation function, $f(x)$, in Figure 1 is defined for the neural LCS as a *threshold logic unit* (TLU). TLUs, coupled with a receptive field, fire only when the input exactly matches the non-zero weights of the input to hidden layer connections. A TLU's excited state corresponds to an output of one.

Receptive fields as classifiers were first addressed by Wilson (1990). Inspired by this work, Smith and Cribbs mapped an LCS onto a neural network (Smith & Cribbs, 1994; Cribbs, 1995). Such a mapping shows a combination of biological paradigms—evolutionary neural networks. It also allows for future LCS studies to borrow ideas and techniques from neural networks.

5 Proving Grounds

Given the new inspirations discussed above, new variations of the LCS are beginning to appear. This section discusses one such variant. The new LCS is tested in a novel robot learning environment inspired by that of Mahadevan and Connell (1992). The LCS implementation applied to this problem used the described neural network mapping. In addition to the neural network representation, Q-learning was used as the basis for the feedback. Negative gradient descent (Rumelhart et al., 1986) was performed for the hidden to output layer weight updates. The weight update used a transformation of the Q-learning equation:

$$dQ = \alpha \cdot [r + \gamma \cdot \max(Q_j) - Q_i^t] \quad (3)$$

where Q_i^t is the Q-value being updated, Q_j are the Q-values for the current state, and r is the reward for the last action. The terms α and γ act as an update rate and a discount factor respectively. Using Equation 3 as the basis for negative gradient descent, weight updates can be made to the hidden to output layer weights. This system was tested on a problem called, *Wanderfree*, which is discussed below.

5.1 The Wanderfree Robot Simulation

As the name suggests, the goal of the robot in Wanderfree is to "roam freely" about a room, without bumping into any objects. The room is divided into a grid of 12 inch cells.

The robot is equipped with four sensors:

- two at 30 degrees, left and right of center, and
- two at 15 degrees, left and right of center.

These sensors can detect objects and classify the object's distance as either *near* (within 18 inches) or *far* (from 18 to 36 inches away). The robot's possible actions are:

- moving forward, one step (step=radius of robot, 6 inches),
- turning left 22 degrees,
- turning left 45 degrees,
- turning right 22 degrees, and
- turning right 45 degrees.

An LCS based on the neural network representation presented was used. Q-learning was used as feedback for the negative gradient descent weight update. The reward reported to the robot about its actions were:

- 500 for movement into a new cell,
- -100 for turning in any direction, and
- $-10,000$ for bumping an object.

To facilitate a wider roaming robot the 500 new cell reward is awarded upon entrance into a new cell only. After the robot leaves a cell, that cell's reward regenerates at a rate of 5 reward points per time step, up to the maximum cell reward of 500.

A block-style GA was used for rule-discovery. The fitness function correlates firing to whether a positive reward was received. This is effectively a variance-based method similar to that used in the *cascade correlation* neural network architecture (Fahlman & Lebiere, 1990). The fitness function:

$$base = Fire + Reward \quad (4)$$

$$extra = Fire \cdot Reward \quad (5)$$

$$fitness = (base + extra) \cdot decisiveness \quad (6)$$

is separate from LCS strengths, or from Q-values. Instead, neuron behavior is correlated to performance. The *base* fitness depicts the desire to either have neurons firing OR (denoted as +) have a positive, non-zero reward from the system. The *extra* fitness portion gives extra fitness points for neuron that fire when there is a reward ($\cdot$ denotes a logical AND). Equation 6 shows the *base* and

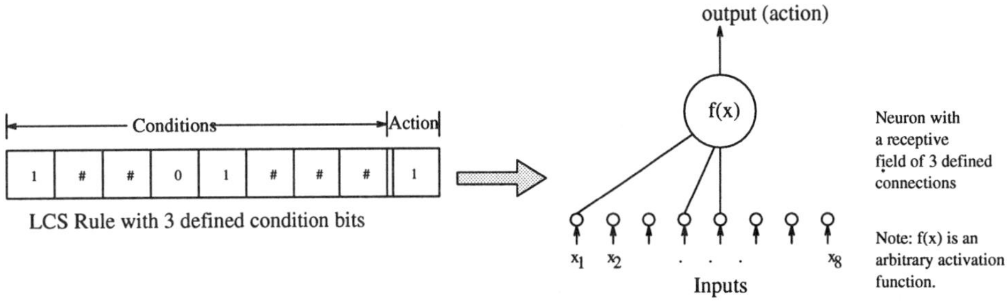

Figure 1: A simple LCS rule and its receptive field equivalent.

extra fitness components added together and weighted by the classifier's *decisiveness*. Classifier *decisiveness* is the *variance of the classifier's output layer weights* (from a hidden layer neuron to the output layer neurons). Decisiveness is normalized (between zero and one) by the maximum variance in output layer weights. The *decisiveness* factor promotes diversity in the hidden layer neurons' output weights. Decisiveness emphasizes greater variance in a classifier's output weights, and detracts from classifiers with smaller variances, encouraging decisive action classifiers in the rule-set.

Table 1 summarizes the values used for the major components of the neural LCS. The GA parameter $p_{\#}$ is

Table 1: LCS Parameters.

Parameter	Value
Q-Learning	
α	0.2
γ	0.75
Neural Net	
Hidden Layer Size	100
η	0.016
η tolerance	1.04
$\zeta_{increase}$	1.05
$\zeta_{decrease}$	0.70
GA	
$p_{crossover}$	0.80
$p_{mutation}$	0.01
$p_{\#}$	0.10
block	10
trigger	30

the probability that a given mutation will mutate to the # condition, i.e., a zero weight connection. In order to avoid problems in selecting the proper learning rate an *adaptive learning rate* method was used (Vogl et al., 1988). This technique uses a windowed average to either increase or decrease the learning rate, η. The neural network parameters $\zeta_{increase}$ and $\zeta_{decrease}$ are the learning rate multipliers for adaptive learning rate.

5.2 Results of Wanderfree

The neural network-based LCS performed admirably in the Wanderfree environment. Figure 2 shows a trail left by the robot. The robot is represented by large circle with a small circle inside (denoting the front of the robot). The figure's coordinates are in inches. At each directional change the robot has successfully detected and avoided the objects. Note that the room is closed; therefore walls must also be avoided.

The robot made approximately 400 moves before grazing (bumping) the triangular object in the upper left corner of the room. Given the robot's limited sensors, it may be impossible to avoid eventual grazing. The rule-set, contained in the hidden neuron layer, used 43 classifiers (neurons) with a total connectivity of 32% to achieve this behavior. Random networks with similar connectivity fail to learn good behaviors in this environment.

6 Conclusions and Comments

Several growing lines of thought are now beginning to influence LCS research. These include

- reinforcement learning techniques,
- the separation of expected payoff statistics (like strength or Q-values) from fitness, and
- alternate classifier representations from fuzzy, neural, GP, and other sources.

Each of these areas promises to lead to a firmer understanding of the complex dynamics of LCSs, and the development of new LCS-like systems. One such system, a neural LCS for Q-learning, was presented with promising preliminary results.

Future LCS developments may include newer RL methods, better fitness methods, diversity in rule-discovery methods, and various representation options. As the developments are pursued, it is hoped that the LCS will help make evolutionary computation a more powerful part of machine learning.

Acknowledgements

The authors wish to thank Lashon Booker and Stewart Wilson for their interaction through electronic mail. The authors further acknowledge support for this work by the National Science Foundation (grant ECS-9212066).

Bibliography

Albus, J. S. (1975). A new approach to manipulator control: The cerebellar model articulation controller. *Transactions of the ASME Journal of Dynamic Systems, Measurement, and Control*, 220–227.

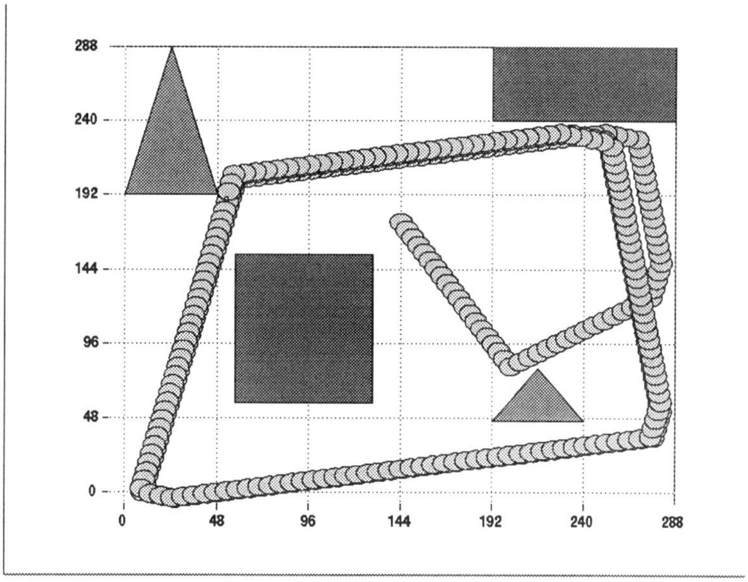

Figure 2: Behavioral trace for a robot in a novel environment.

Cribbs, H. B. (1995). Cooperative learning classifier systems: A neural network approach. Master's thesis, The University of Alabama, Tuscaloosa.

Fahlman, S. E., & Lebiere, C. (1990, February). *The cascade-correlation learning architecture* (Tech. Rep. CMU-CS-90-100). Pittsburgh, PA: Carnegie Mellon University.

Fogel, L. J., Owens, A. J., & Walsh, M. J. (1966). *Artificial intelligence through simulated evolution.* New York: John Wiley.

Holland, J. H. (1986). *Induction.* Cambridge, MA: MIT Press.

Holland, J. H., & Reitman, J. S. (1978). Cognitive systems bases on adaptive systems. In D. A. Waterman, & F. Hayes-Roth (Eds.), *Pattern directed inference systems* (313–329). New York: Academic Press.

Mahadevan, S., & Connell, J. (1992). Automatic programming of behavior-based robots using reinforcement learning. In *Artificial Intelligence 55*, 311–365. Elsevier Science Publishers.

Parodi, A., & Bonelli, P. (1993). A new approach to fuzzy classifier systems. In S. Forrest (Ed.), *Proceedings of the Fifth Annual International Conference on Genetic Algorithms*, 223–230, San Mateo, California. Morgan Kaufmann.

Rumelhart, D. E., Hinton, G. E., & Williams, R. J. (1986). Learning representation by backpropagating errors. *Nature*, *1986*(323), 533–536.

Rummery, G. A., & Niranjan, M. (1994). *On-line Q-learning using connectionist systems* (Tech. Rep. CUED/F-INFENG/TR 166). Cambridge CB2 1PZ, England: Cambridge University Engineering Department.

Schwartz, A. (1993). A reinforcement learning method for maximizing undiscounted rewards. In *Machine Learning: Proceedings of the Tenth International Conference*, San Mateo, CA. Morgan Kaufmann.

Schwefel, H. P. (1965). *Kybernetische evolution als strategie der experimentellen forschung in der Strömungstechnik* (Diplomarbeit). Berlin: Hermann-Föttinger-Institut für Strömungstechnik der Technische Universitat.

Smith, R. E., & Cribbs, H. B. (1994). Is a learning classifier system a type of neural network? *Evolutionary Computation*, *2*(1), 19–36.

Smith, R. E., & Valenzuela-Rendón, M. (Eds.) (1992). *Collected abstracts for the first international workshop on learning classifier systems.* Unpublished Manuscript.

Sutton, R. S. (1987). *Learning to predict by the methods of temporal differences* (Tech. Rep. TR87-509.1). Waltham, MA: GTE Laboratories, Inc.

Thrun, S. B. (1992). The role of exploration in learning control. In *Handbook of intelligent control: Neural, fuzzy, and adaptive approaches* (527–560). New York: Van Nostrand Reinhold.

Tufts, P. (1995). Dynamic classifiers: Genetic programming and classifier systems. In *AAAI Fall Symposium.* AAAI.

Twardowski, K. (1993). Credit assignment for pole balancing with learning classifier systems. In S. Forrest (Ed.), *Proceedings of the Fifth Annual International Conference on Genetic Algorithms*, 238–245, San Mateo, California. Morgan Kaufmann.

Valenzuela-Rendón, M. (1991). The fuzzy classifier system: A classifier system for continuously varying variables. In R. K. Belew, & L. B. Booker (Eds.), *Proceedings of the Fourth International Conference on Genetic Algorithms*, 346–353.

Vogl, T. P., Mangis, J. K., Rigler, A. K., Zink, W. T., & Alkon, D. L. (1988). Accelerating the convergence of the backpropagation method. *Biologial Cybernetics*, *59*, 257–263.

Watkins, J. C. H. (1989). *Learning with delayed rewards.* Unpublished doctoral dissertation, King's College, London.

Wilson, S. W. (1990). Perceptron redux: Emergence of structure. In S. Forrest (Ed.), *Emergent Computation: Proceedings of the Ninth Annual International Conference of the Center for Nonlinear Studies on Self-Organization, Collective, and Cooperative Phenomena in Natural and Artificial Computing Networks*, 249–256, Amsterdam. North-Holland.

Wilson, S. W. (1994). ZCS: A zeroth level classifier system. *Evolutionary Computation, 2*(1), 1–18.

Wilson, S. W. (1995). Classifier fitness based on accuracy. *Evolutionary Computation, 3*(2), 149–175.

Natural Niching for Evolving Cooperative Classifiers

Jeffrey Horn
Department of Computer Science
and the Illinois Genetic Algorithms Laboratory
University of Illinois at Urbana-Champaign
117 Transportation Building
104 South Mathews Avenue
Urbana, IL 61801-2996
Phone: 217/333-2346
Fax: 217/244-5705
Email: jeffhorn@uiuc.edu
WWW: http://www.uiuc.edu/ph/www/illigal

David E. Goldberg
Department of General Engineering
and the Illinois Genetic Algorithms Laboratory
University of Illinois at Urbana-Champaign
117 Transportation Building
104 South Mathews Avenue
Urbana, IL 61801-2996
Phone: 217/333-0897
Fax: 217/244-5705
Email: deg@uiuc.edu
WWW: http://www.uiuc.edu/ph/www/illigal

ABSTRACT

An *evolutionary classifier*, such as a learning classifier system (LCS) or a genetic programming boolean concept learner, must maintain a population of diverse rules that together solve a problem (e.g., classify examples). To maintain "cooperative diversity" while applying a selection operator to the population of rules, as in the *Michigan-style* LCS, the evolutionary algorithm must incorporate some kind of *niching* mechanism. The natural way to accomplish niching in an LCS is to force competing rules to *share* resources (i.e., rewards). The *implicit* or "natural" niching and speciation induced by such resource sharing, is shown to be robust in the face of severe selective pressure, low population sizes, and overlapping rule coverage. Specifically in this paper we analyze the two-niche (two competing/cooperating rules) case. We find closed form approximations for niche maintenance and niche convergence times, giving us the beginnings of a first predictive model for interacting (cooperating) rules in an evolving population. Finally, we make the case for niching/speciation as a basic, indirect form of cooperation that is fundamental to, and underlying, all other types of more direct cooperation, and which the LCS must therefore promote. Although we focus on the LCS as an example of a specific and well-known evolutionary classifier, all of our results are general enough to apply to any evolutionary algorithm, such as genetic programming (GP), that applies selection to a population of diverse classifiers.

1 Introduction

The unique power of the learning classifier system, or LCS (see (Holland, 1971, 1975, 1985, 1992; Booker, 1982; Goldberg, 1983, 1989a; Wilson, 1986; Wilson & Goldberg, 1989)), is its ability to discover, through artificial genetic search, a *group* of cooperating rules that together represent a concept, or implement a set of behaviors, that solve the problem at hand. In the case of the so-called *Michigan-style* LCS, which is the subject algorithm of the following analysis, the search for cooperative rule sets takes place within a single population of competing and cooperating rules. Thus cooperation and competition occur simultaneously among different rules (individuals) and sets of rules (subpopulations). This model of intertwined, multi-level cooperative and competitive interactions seems natural, based on real-world examples of ecologies, economies, and societies. Yet its potential power of expressing complex concepts and behaviors is offset by the tremendous complexity of its population dynamics.

Although the subject algorithm of our analysis is the Michigan-style LCS, our results are generally applicable to all types of evolutionary (i.e., selection-based) algorithms which adapt a population of diverse classifiers (e.g., rules). Because of our simplifying assumptions, the analysis is not specific to any type of recombination, mutation, or encoding. The classifiers in the population can be string encoded rules (as in an LCS), decision trees (as in GP), fuzzy logic rules, or neural networks. The analysis here assumes only fitness proportionate reproduction, and a fitness assignment that is proportionate to the *reward* (credit) for correct classifications.

In this paper we continue earlier work (Horn, Goldberg, & Deb, 1994) that attempts to lay a foundation for a complete analytical model of cooperative-competitive evolution by isolating what is arguably the lowest but most fundamental type of cooperation: problem decomposition via *niching*. That is, various rules, or *species*, solve different parts of the problem, earning separate, distinct rewards (i.e., *niches*). Rules that compete to perform the same task (e.g., classify the same examples) compete for the same rewards, while rules that do not overlap in their tasks/rewards are cooperating, in an indirect manner. As we shall show, such indirect or implicit cooperation is far from weak in its ability to survive strong selective pressure for many generations.

In this paper we focus exclusively on one particular method of niching: reward *sharing*. The *implicit niching* induced by the natural and simple method of dividing up reward/credit/fitness among all individuals who have earned it, is a powerful promoter of cooperation. It is capable of maintaining cooperative (i.e., non-competing) rule sets even under constant and rigorous application of genetic algorithm selection and search.

We first describe briefly the existing work on the use and analysis of sharing in the LCS. We then proceed to make a strong case for the critical need for sharing-induced niching in the Michigan-style LCS. Finally, we expand our understanding of the nature, capabilities, and limitations of such niching-induced cooperation by extending the results of Horn, Goldberg, and Deb (1994). Specifically we perform a critical timing analysis of niche convergence and maintenance. The ultimate goal of this line of research is a complete predictive model of niching in evolutionary computation systems. The results to date, and the expected model, are generally applicable to all types of evolutionary computation where the goal is successful search for emergent cooperation via selective competition.

2 Background

In this section we briefly review the nature of niching, the need for niching in an LCS, existing work on niched GAs, and some previous work using niching in classifier systems.

2.1 The Need for Niching in the LCS

In a simple GA (or GP) each individual is evaluated according to a single scalar fitness function independent of other members in the population. We can thus view the task of the simple GA as optimization of the sum, or average, of the population's fitnesses. The optimum population consists entirely of copies of the best individual. But when individuals influence each other's fitness evaluation, the task of the GA can no longer be modeled as an optimization of total population fitness. Such evolution has gone by the name of *co-evolution*, *coadaptation*, and *context-dependent function optimization*. Co-evolution is more natural, more realistic, potentially more powerful, and certainly more complex than the convergence of a simple GA. In the LCS, we ask the GA to search through the space of all possible rules to find and maintain a diverse, cooperative subpopulation.

In the LCS community it is commonly believed that the GA is used so sparingly, and with so many other specialized operators at work, that it is virtually impossible for a single rule to take over the population. And this is often the case. Specialized selection operators such as elitist selection, or protective group measures such as the formation of classifier corporations, or any kind of cooperative-reinforcement fitness function such as the bucket brigade or epochal credit assignment, all help counteract convergence. But this observation does not help us understand or predict the behavior of the LCS.

The LCS in its simplest form, arguably a stimulus-response LCS for binary classification as we outline below, should be able to maintain a diverse group of high-quality rules that together classify the examples at hand. Furthermore, if we try to maintain diversity by lowering selection pressure, we deprive the LCS of the power of GA search. Therefore, maintenance of rule sets must take place over a time frame that is at least an order of magnitude greater than the convergence time of the GA. The only way to maintain such high-quality diversity in the face of high selection pressure is to balance convergence with a restorative force, such as "niching pressure" (Horn, 1993).

2.2 Previous work

The important role of niching in evolutionary classification systems has been recognized by many researchers. Perhaps the earliest and most consistent advocate of the critical role of some type of niching in an LCS was Booker (1982, 1989). He implemented niching using an indirect form of sharing based on partial match scores of individuals with rewards, and introduced mating restrictions to limit mixing between niches. Wilson also demonstrated the ability of niching, via sharing of reward/credit, to maintain diverse disjuncts in a population representing a disjunctive boolean concept (Wilson, 1986) or a survival strategy (Wilson, 1987). In his more recent work (Wilson, 1994, 1995), he simplifies the LCS algorithm and isolates key dynamics, such as niching.

Smith and Valenzuela-Rendón (1989) identified the general need for niching in the LCS, but rather than use the *implicit niching*, via reward sharing, suggested by Booker, Wilson, and others, they applied *explicit niching*, via *fitness sharing*, to a small LCS using Hamming distance as a metric on the space of rules. They were successful in maintaining a set of diverse rules that together covered the examples and solved the problem. However, they apparently ran into the *separation* problem inherent in fitness sharing and identified in (Goldberg, Deb, & Horn, 1993). Briefly, the separation of desirable and undesirable individuals can be a problem for fitness sharing because of the fixed niche radius σ_{sh}. We conjecture that implicit niching using reward sharing should avoid this particular type of separation problem, but this remains to be shown. Mahfoud (1995a, 1995b) has recently had success also using explicit niching for GA-based classification of boolean concepts, although he does not call his algorithm an LCS. There is some evidence in his results that the separation problem crops up, but again this remains to be shown.

Perhaps the clearest demonstration of explicit niching for GA-based disjunctive concept learning is by McCallum and Spackman (1990). Their "sharing examples" (or "splitting examples") method is identical to the natural, simple method of equaling sharing reward/credit that is the subject of our analysis. They demonstrate stable steady state populations under GA selection. Unfortunately, Neri and Saitta (1995) appear to have misinterpreted McCallum and Spackman's description. Neri and Saitta show that their interpretation of splitting examples does not arrive at a diverse steady state but rather converges to a uniform population. However, Neri and Saitta go on to show implicit niching, correctly interpreted from (Horn, Goldberg, & Deb, 1994) does maintain a stable, diverse steady-state population of concept disjuncts. They go on to show how their *universal suffrage* niching method also maintains the desirable steady-state distributed concept representation. Furthermore, their analysis introduces some new techniques for modeling niching.

Most recently, Smith, Forrest, and Perelson (1993) analyzed implicit niching in the immune system model. They noted that a similar niching process takes place in the LCS. They went on to show that the immune system model does indeed exhibit "emergent fitness sharing" with many of the properties of GA explicit fitness sharing. The immune system model is very similar to the S-R LCS, with the exception that rules cannot use "don't care" wildcard symbols (i.e., "#"). Generalization is induced by taking a limited size random sample of rules (a.k.a. antibodies) to compete to cover an example (antigen). Much of their analysis was thus focused on the sampling process and its effects.

In this paper we avoid the issue of rule generality and concentrate on the implicit niching induced by arbitrarily breaking ties between rules competing for a resource. This arbitrary tie-breaking is used in the immune system model as well as the GA. It leads to an expected even division of a resource among all individuals competing for it, and is the key to sharing, implicit or explicit.

3 The Problem at Hand

In this section we describe the critical process of isolating one LCS mechanism, implicit niching, and its effect on LCS behavior. To do so, we idealize the LCS by eliminating or simplifying operators. At the same time, we must simplify the corresponding classification task which the resulting LCS, with its remaining idealized mechanisms, should accomplish.

To understand the complex relationships that can and do emerge among rules in the LCS, and among individuals in coadaptive systems in general, we believe it is necessary to analyze each type of rule interaction in isolation. Here we concentrate on the *weak cooperation* induced by rule competition for limited resources (i.e., finite rewards). The sub-goal of weak cooperation is to *cover* (exploit) as much of the resources as possible. The only type of rule interaction is competition for the same resource, and the natural mechanism for handling such competition (and encouraging search for uncovered resources) is *sharing* of contested resources. Thus similar rules share common resources by dividing them up among themselves. This simple method induces *niching* or *speciation*, an emergent phenomenon that we suggest is prerequisite to all other types of cooperation (i.e., *strong* cooperation).

3.1 The Assumptions, in Brief

We begin our process of functional decomposition by making the following assumptions about the LCS:

- Ternary alphabet
- Stimulus-response classifiers
- Binary classification
- Equal specificity of all rules

Our first simplification is a common one in LCS analysis as well as LCS applications: we use a ternary alphabet. The classifier conditions are coded from the set $\{0, 1\#\}$, where # is the "don't care", or wildcard, character. Responses are limited to the alphabet $\{0, 1\}$. We assume k attributes to describe the environmental input. Each attribute is encoded by ℓ_i bits, $i \in 1..k$. So our environmental input vectors will have $\ell = \sum_{i=1}^{k} \ell_i$ bits total. We can concatenate all our attributes into a single bit string representing an individual example (or more generally, an environmental input vector). Equivalently, we can imagine that we have ℓ binary attributes in our problem. Thus each rule's condition is a string of length ℓ taken from $\{0, 1, \#\}$ (e.g., ###00#1011##0#11).

Our second simplification is also a common one in LCS analysis: we limit ourselves to a *stimulus-response* (S-R) classifier system. In a S-R LCS there is no internal message list and hence no passing of messages between classifiers. The output of each classifier is a single response to a single input from the environment. With no message list, the LCS cannot evolve "strong" cooperation. That is, rules cannot communicate directly with each other, nor can they pay or charge each other credit. We thus avoid the question of delayed reward, eliminating the need for complex credit assignment mechanisms, such as the bucket-brigade algorithm. What we are left with is *weak* cooperation, in which rules cooperate indirectly to *cover* the different reward situations (e.g., correct classifications of examples).

We further restrict our classification task to binary classification (i.e., single class membership). The goal of the LCS is to learn to classify instances as either members of a class (or concept) or not members. The output of each classifier is either a "1" (member of class) or a "0" (not a member). If we also assume a default rule (###...# ⇒ 0), then the task of the LCS is to find accurate rules that classify the exceptions (i.e., the positive examples). Therefore, we can assume that all rules besides the default have an output of "1". We can then leave the response (output) out of the encoding, and search only among the possible condition vectors (e.g., 110###0#0###101#).

Finally, we make a very important simplification to eliminate the issue of specificity versus generality. We assume that all rules in the population have the same number of "don't care" characters #. With each rule applying to an equal volume of the total classification space[1], there can be no preference for specific or general rules. All reasonable fitness functions for individual rules reduce to a function of accuracy only. And since all rules have equal volumes of coverage (applicability), any reasonable fitness function based on accuracy must be a monotonically increasing function of the number of examples covered by the rule. So we can simply use the number of examples covered (again, assuming all examples are positive ones) to order the rules according to fitness, even though we don't know the exact, possibly nonlinear, fitness function.

3.2 What's Left? A Hard Problem: Covering

After all the above simplifications and assumptions, it is essential that we verify that we have still have a difficult, interesting, albeit abstract, problem. When we limit ourselves to binary classification and assume a default class (i.e., we are really only trying to identify one class), we are left with an instance of the set covering problem. We are trying to cover the positive examples with a small set of accurate rules. Figure 1 illustrates our version of set covering. The large rectangle represents all possible 2^ℓ instances, where ℓ is the number of bits used to encode each instance. In Figure 1, we use $\ell = 10$ bits. The entire instance space is thus described by the rule (########## ⇒ 1) or simply ##########. The actual concept to be learned is shown by the shaded regions. The LCS is given only a subset of these instances as (positive) examples. The LCS is constrained to a rules

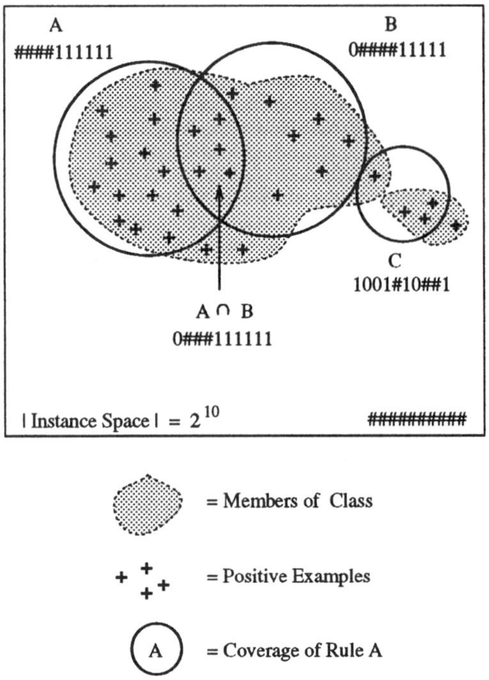

Figure 1: Binary classification as a covering problem. The task of the LCS is to find classifiers (rules, represented as circles) to cover the examples (positive only) of the unknown concept (single class, represented as shaded regions).

syntax of $\{0, 1, \#\}^{10}$. We represent this constraint by a circle for the "coverage" of a rule. Rules with more # characters cover more of the instance space, and are represented by larger diameter circles. As Figure 1 shows, rules can vary in specificity (coverage of instance space) as well as in number of examples[2] covered. Furthermore, rules can overlap to almost any degree in coverage of instances and/or examples.

In general, the binary classification task of the LCS is to find a *small* subset of *accurate* rules that *cover* the examples. These are three separate objectives. However, the second objective, individual rule accuracy, is problematic. How, exactly, do we measure accuracy? Clearly accuracy increases with the number of examples correctly classified and decreases with the number classified incorrectly. But these two numbers represent two conflicting objectives themselves. One rule might have both a higher number of correct classifications and a higher number of incorrect classifications than another rule. Which is preferred? In addition, we often want rules with greater predictive power, that is, more generality because of their greater coverage of instance space. We might prefer less accurate but more general rules to more accurate, more specific rules. So the objective of rule accuracy is itself

[1]The entire classification space is of size 2^ℓ, and is represented by ##...#.

[2]For the rest of this paper we omit the word "positive", and assume all examples are positive ones.

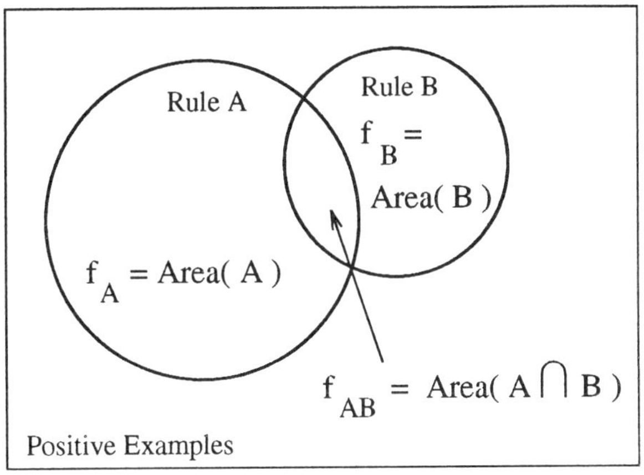

Figure 2: By assuming rules of equal specificity (number of #'s), we can use diameter in the space of examples to indicate a rule's accuracy as "coverage of *examples*". We can then visualize our definitions for the objective fitness of rules, f_A and f_B, and the "fitness" of the overlap in rule coverage, f_{AB} as areas. (Fitness might be measured as number of examples covered.)

multiobjective. It is tied up in the issues of generality-specificity versus accuracy. These are critical issues, but we believe they are separable from the issues of rule set size (conciseness of concept description) and coverage.

To avoid considerations of generality versus specificity, and hence reduce accuracy to a single objective that can be measured by a scalar, we consider only rules of the same order (number of don't cares). All such rules have the same specificity. Since each rule has the same coverage of instance space, their accuracies can be computed as simply the number of examples covered. Maximizing the number of examples covered maximizes accuracy. In terms of Figure 1, the LCS is further constrained to using circles of a specific, constant size. This observation prompts us to redefine our Venn diagram terms and use the diameter of the circles to represent accuracy. In Figure 2, the large rectangle represents the space of all *examples* given to the LCS for learning. The size of a circle represents the number of *examples* covered by the corresponding rule, and hence its accuracy. The overlaps of circles represent overlaps of coverage among rules, and thus contain the examples "shared" by two or more rules.

3.3 Definitions

Let f_A and f_B be the objective fitnesses for rules **A** and **B** respectively. The objective fitness could be taken as the number of examples covered by that rule, in the case of binary classification. Let f_{AB} be the amount of resources in the overlapping coverage of rules **A** and **B**. That is, f_{AB} is the amount of resources shared by **A** and **B** (e.g., the number of examples covered by both). We also define the fitness ratio $r_f \equiv \frac{f_A}{f_B}$ and the ratio of overlap $r_o \equiv \frac{f_{AB}}{f_A}$. Figure 2 illustrates some of these definitions. Note that $0 \leq r_o \leq 1/r_f$. (For example, if both rules cover the same number of examples, $f_A = f_B$ and therefore $r_f = 1$. Then the ratio r_o can vary between 0 and 1, as we increase the overlap in coverage.)

Let n_A, n_B be the number of copies of rules **A** and **B**, respectively, in our population. Then we can calculate the shared (expected) fitness of rule **A**:

$$f_{sh,A} = \frac{f_A - f_{AB}}{n_A} + \frac{f_{AB}}{n_A + n_B}. \tag{1}$$

Similarly for rule **B**,

$$f_{sh,B} = \frac{f_B - f_{AB}}{n_B} + \frac{f_{AB}}{n_A + n_B}. \tag{2}$$

4 Time Scale Analysis of Niching

In this section we look for two bounds on the expected times of niche maintenance for the two niche case. Specifically, we first determine the expected steady-state equilibrium point (population distribution) for two distinct, but possibly overlapping niches (rules) **A** and **B**. We then seek the expected time until *niche extinction*: the time until one or the other species is eliminated from the population (and hence cannot return). We then attempt to determine the *niche convergence time*: the expected number of generations to reach steady-state. When these two time bounds are close, we can expect poor niche maintenance. When they are far apart, we should expect long-term, steady-state niche maintenance. Under what conditions of niche overlap, fitness ratio, and population size can we expect to find the *phase transition* from poor to good niche maintenance?

4.1 Steady-states and equilibrium points

For niched GAs that use any kind of *resource sharing*, equilibrium (under selection only) will occur when the shared fitnesses of all population members are equal (Goldberg & Richardson, 1987; Deb, 1989). In the two niche case, the general equilibrium condition is:

$$f_{sh,A} = f_{sh,B}. \tag{3}$$

Substituting the formulae for shared fitness in an LCS, Equations 1 and 2,

$$\frac{f_A - f_{AB}}{n_A} + \frac{f_{AB}}{n_A + n_B} = \frac{f_B - f_{AB}}{n_B} + \frac{f_{AB}}{n_A + n_B}. \tag{4}$$

Solving for $\frac{n_B}{n_A}$,

$$\frac{n_B}{n_A} = \frac{f_B - f_{AB}}{f_A - f_{AB}} = \frac{\frac{f_B}{f_A} - \frac{f_{AB}}{f_A}}{1 - \frac{f_{AB}}{f_A}}. \tag{5}$$

Substituting our ratios r_f and r_o, rearranging, and naming the result $r_{eq,n} \equiv \frac{n_B}{n_A}$ (at equilibrium),

$$r_{eq,n} = \frac{\frac{1}{r_f} - r_o}{1 - r_o}. \tag{6}$$

We see that the equilibrium point changes with overlap. With no overlap, that is when $r_o = 0$, then $r_{eq,n} = 1/r_f$. While at complete overlap ($r_o = 1/r_f$) the equilibrium point is 0 (i.e., the population will converge to a uniform one, consisting of all **A**s).

Finally, let us extract from the equation for $r_{eq,n}$ the actual *proportion* of **A**'s in the population, $P_A = n_A/N$, which we will need later in the paper:

$$r_{eq,n} = \frac{n_B}{n_A} = \frac{N - n_A}{n_A} = \frac{1 - P_A}{P_A} = \frac{\frac{1}{r_f} - r_o}{1 - r_o}.$$

Solving for P_A and calling it $P_{A,eq}$:

$$P_{A,eq} = \frac{1 - r_o}{1 - 2r_o + \frac{1}{r_f}}. \tag{7}$$

We note that in general $P_A \neq p_A$; that is, the *probability* p_A of choosing **A** from the current population using fitness proportionate selection is **not** the same as the *proportion* P_A of **A** in the current population. However, at equilibrium (i.e., any steady-state, such as uniform populations) this IS true (i.e., $p_A = P_A$).

4.2 Rule maintenance times

In (Horn, Goldberg, & Deb, 1994), the expected time to loss of one of the two niches (i.e., classes of individuals) was calculated exactly by a Markov chain analysis, just as Horn (1993) calculated for *fitness sharing* in a niched simple GA . This quantity is simply the expected time to absorption by either of the two "converged" states (i.e., uniform populations of one or the other class of individual). In this subsection we review their exact results from Markov chain models, and then propose an approximate closed form expression.

4.2.1 Exact Model

Calculating the exact expected absorption time for an absorbing Markov chain is straightforward, although computationally expensive for large matrices (i.e., large population size N and/or many different niches). A random initial population, which has little chance of being in either absorbing state, is assumed. Horn, et. al. (1994) found that the expected absorption time grows exponentially with population size N. The exponent in the growth decreases as r_f moves away from one, just as in the case of the niched GA (Horn, 1993). Finally, holding r_f constant, they found that the exponent of growth (in N) also decreases with increasing niche overlap, just as in the niched GA (Horn, 1993)[3].

The time to absorption (t_{abs}) is also called *niche* (or *rule*) *maintenance* time or niche *extinction*, *failure* or *loss* time (e.g., Mahfoud, 1995a, 1995b). Here we look at how t_{abs} varies with overlap. That is, we vary $0 < r_o \leq 1/r_f$ while holding r_f constant. In Figure 3 we show the growth in expected absorption time as a function of N for $r_o = 0.0$, $r_o = 0.1$, $r_o = 0.3$, and $r_o = 0.5$, while $r_f = 2$ (the solid dots represent the output of the exact (Markov) models).

This growth (Figure 3) appears to be exponential. Indeed, for the case of "perfect sharing" (Horn, 1993), when $r_o = 0$, we can find the exact exponent.

Let p_A be the probability of selecting a single individual copy of rule **A** for reproduction from a particular population with n_A copies of rule **A** and n_B copies of rule **B** (with $n_A + n_B = N$). Under proportionate selection, $p_A = \frac{n_A f_{sh,A}}{n_A f_{sh,A} + n_B f_{sh,B}}$, where $f_{sh,A}$ and $f_{sh,B}$ are the *shared* fitnesses of rules **A** and **B** respectively, in the current population. Substituting in the shared fitnesses $f_{sh,A}$ and $f_{sh,B}$ from Equations 1 and 2, and rearranging, yields

$$p_A = \frac{f_A - f_{AB} + f_{AB}\frac{n_A}{N}}{f_A - f_{AB} + f_{AB}\frac{n_A}{N} + f_B - f_{AB} + f_{AB}\frac{n_B}{N}}. \tag{8}$$

Dividing numerator and denominator by f_A, the above becomes

$$p_A = \frac{1 - \frac{f_{AB}}{f_A} + \frac{f_{AB}}{f_A}\frac{n_A}{N}}{1 - \frac{f_{AB}}{f_A} + \frac{f_{AB}}{f_A}\frac{n_A}{N} + \frac{f_B}{f_A} - \frac{f_{AB}}{f_A} + \frac{f_{AB}}{f_A}\frac{n_B}{N}}. \tag{9}$$

Remembering our defined ratios $r_f = f_A/f_B$ for fitness, and $r_o = f_{AB}/f_A$ for overlap, we find that

$$p_A = \frac{1 + r_o(\frac{n_A}{N} - 1)}{1 - r_o + r_o\frac{n_A}{N} + \frac{1}{r_f} - r_o + r_o\frac{n_B}{N}}. \tag{10}$$

Finally, simplification yields

$$p_A = \frac{1 - r_o + r_o\frac{n_A}{N}}{1 - r_o + \frac{1}{r_f}}. \tag{11}$$

With no overlap ($r_o = 0$), the probability of choosing **A** is constant: $p_A = \frac{1}{1+1/r_f} = \frac{r_f}{r_f+1}$. Therefore in the Markov chain model for no overlap, the transition probabilities for all transient states are identical. It follows that prior to absorption we always have the same probability of absorption, namely $p_{absorb} = (p_A)^N + (p_B)^N$, where $(p_A)^N$ is simply the probability of choosing N **A**s given that p_A is the probability of choosing a single copy of **A** (under proportionate selection). With no overlap,

[3]Note that the case of no overlap ($r_o = 0$) in implicit niching is exactly the same as for the niched GA with no overlap (i.e., "perfect sharing" (Horn, 1993)).

$(p_A)^N = (\frac{r_f}{r_f+1})^N$. Similarly, $(p_B)^N = (\frac{1}{r_f+1})^N$, when $r_o = 0$. The expected time to absorption, assuming that the chain is started in a transient state, is the inverse of the probability of absorption, $\frac{1}{(p_A)^N+(p_B)^N}$:

$$E[t_{abs}] = \frac{1}{(\frac{1}{r_f+1})^N + (\frac{r_f}{r_f+1})^N}. \quad (12)$$

After some rearranging:

$$E[t_{abs}] = \frac{(r_f+1)^N}{r_f^N + 1}. \quad (13)$$

The expected absorption time above grows exponentially in N for any $r_f > 0$. In particular, when $r_f = 1$, the growth is $O(2^N)$.

When $r_o \neq 0$ (i.e., niche overlap), it might be impossible to calculate a closed form expression for the *exact* expected absorption time. However, in Figure 3 it certainly appears that all such growth is exponential, at least up until $r_o = 1/r_f$, at which time the growth should become logarithmic (linear) in N, since this is the case of selection (drift)[4].

4.2.2 Closed Form Approximation

A closed form expression for $E[t_{abs}]$ with overlap ($r_o > 0$) is desirable for our timing analysis however, and we define an approximate one here. We use the approach above (for $r_o = 0$) by simply assuming that the population is *always* at the equilibrium point until the actual absorption event. That is, we assume only three possible states: equilibrium, and the two absorbing states (uniform populations). We do not try to justify this assumption here, nor do we try to bound the error, leaving this to future work. Rather, we merely observe that under the assumptions of very stable steady states (e.g., large populations, low-variance selection), our approximation might be justified.

Justification aside, we go ahead and plug into the general equation

$$E[t_{abs}] = \frac{1}{(p_A)^N + (p_B)^N} = \frac{1}{(p_A)^N + (1-p_A)^N},$$

where p_A and p_B are the probabilities of selecting **A** and **B** respectively, from the current population under proportionate selection. The above is simply the expected waiting time until selection chooses N copies of **A** or N copies of **B**. Our assumption above translates here into the assumption that p_A is constant, at or near equilibrium, so that we can just substitute $p_{A,eq}$ from Equation 7 above:

$$E[t_{abs}] = \frac{1}{(\frac{1-r_o}{1-2r_o+\frac{1}{r_f}})^N + (1 - \frac{1-r_o}{1-2r_o+\frac{1}{r_f}})^N}, \quad (14)$$

which does not simplify much.

To test our closed form approximation, we plot both the exact results (from the Markov chain model) and the approximations in Figure 3. In Figure 3, the solid dots are the plotted points for the exact (Markov) model, while the dashed lines are defined by Equation 14. In general, we see close agreement over the range plotted, except for high overlap (e.g., complete overlap when $r_o = 1/r_f = 0.5$). As expected, our simple closed form expression modeling absorption time loses accuracy as for high overlap cases. Intuitively, with high overlap the variance of the steady-state distribution will be high, and the population will not spend as much near the steady-state. Since the GA is spending less time at steady-state, the use of the steady-state absorption probability as an estimate of the overall absorption probability introduces increasing error, as overlap increases.

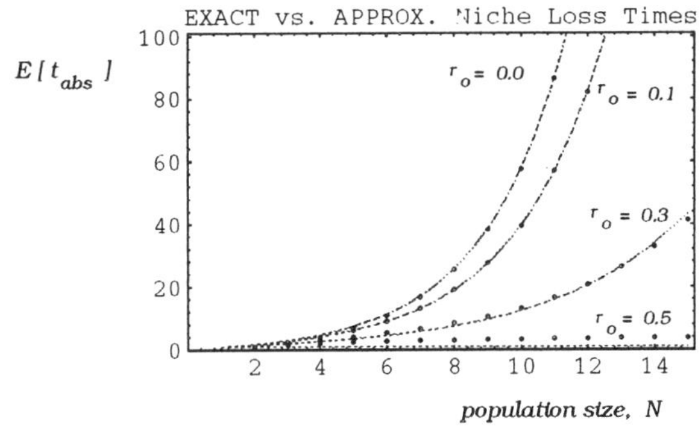

Figure 3: A comparison of *exact* expected niche loss times to the approximated times, as a function of population size. The exact results (from the Markov models) are shown as solid dots. The approximations, from our closed form expressed model, are shown as dashed lines. The plots indicate general agreement for small niche overlap (r_o). For all plots shown $r_f = 2$.

4.3 Steady-state Convergence Times

Having found an upper bound on expected lifetimes of niches (rules), and convinced ourselves that steady-state populations of rules can be maintained, we now turn to the question of how such steady-states are reached. Under normal selection (e.g., in a simple GA) it has been shown that convergence to the "equilibrium point" of a uniform population takes place very quickly. In the case of selective preference (i.e., $r_f \neq 1$) the convergence time grows logarithmically in population size N, while under genetic drift (i.e., $r_f = 0$), the expected convergence

[4]It has been suggested (Horn, 1993) that exponential growth in niche maintenance time, as N grows, is entailed by any restorative pressure, and thus could be used as an indicator of a "true nicher". If that is the case, then LCS implicit niching clearly qualifies.

time grows linearly in N. Can we expect similarly quick convergence to "niching equilibrium"?

To answer such questions, we use the simple, well-known method of *expected proportion equations* to model niche convergence. This method of tracking only the expected next generation *population*, rather than tracking the entire *distribution* over all possible populations, has been put to good use many times in the GA literature (e.g., Smith & Valenzuela-Rendón, 1989; Goldberg & Deb, 1991; Neri & Saitta, 1995). The methodology has acquired several names. Vose (1993), noting that the accuracy of the proportion equations asymptotically approaches one with increasing population size, calls it the *infinite population model.* Recently, Neri and Saitta (1995) coined the term *virtual average population.* For an in-depth consideration of the accuracy of these models for finite populations, see Ben Goertzel's book (Goertzel, 1993), in which he calls the trace given by the recurrent proportion equations the *local expected path*, as opposed to the "global" (exact) *expected path* given by the full Markov model.

4.3.1 Niching Convergence Behavior

Under proportionate selection, the expected proportion of the next generation's population given to an individual is *equal* to the probability of selecting that individual for reproduction in the current generation's population. Thus if $P_{A,t}$ is the proportion of the current population, at time t, consisting of copies of rule **A**, then $E[P_{A,t+1}] = p_A$, where p_A is calculated at time (generation) t. Substituting for p_A from Equation 11 above, and noting that $P_{A,t} = \frac{n_A}{N}$ at time t, we find that

$$E[P_{A,t+1}] = \frac{1 - r_o + r_o\frac{n_A}{N}}{1 - r_o + \frac{1}{r_f}} = \frac{1 - r_o + r_o P_{A,t}}{1 - r_o + \frac{1}{r_f}}.$$

Rearranging, we find that

$$E[P_{A,t+1}] = \frac{1 - r_o}{1 - r_o + \frac{1}{r_f}} + \frac{r_o}{1 - r_o + \frac{1}{r_f}} P_{A,t}.$$

Now if we make the major assumption that $E[P_{A,t+1}] \approx P_{A,t+1}$ (which is a strict equality only for deterministic selection operators or for infinite populations (Vose, 1993)) then the above is a simple linear recurrence relation on $P_{A,t}$. Defining $P_{A,0}$ to be the initial population, we solve the difference equation and rearrange to get:

$$P_{A,t} = \frac{1 - r_o}{1 - 2r_o + \frac{1}{r_f}} + \left(P_{A,0} - \frac{1 - r_o}{1 - 2r_o + \frac{1}{r_f}}\right)\left(\frac{r_o}{\frac{1}{r_f} - r_o + 1}\right)^t. \quad (15)$$

Remembering that at equilibrium (steady-state) $P_{A,eq} = \frac{1-r_o}{1-2r_o+\frac{1}{r_f}}$ (Equation 7), and introducing $\beta \equiv \frac{r_o}{\frac{1}{r_f}-r_o+1}$, we simplify to

$$P_{A,t} = P_{A,eq} - (P_{A,eq} - P_{A,0})\beta^t$$

or

$$P_{A,t} = P_{A,eq}(1 - \beta^t) + P_{A,0}\beta^t. \quad (16)$$

We note that $\beta \leq 1$. This must be so as the numerator of β, $r_o \equiv \frac{f_{AB}}{f_A}$ must always be ≤ 1, while the denominator must always be ≥ 1, since $f_{AB} \leq f_B \Rightarrow \frac{f_{AB}}{f_A} \leq \frac{f_B}{f_A} \Rightarrow r_o \leq \frac{1}{r_f}$, and so $1 \leq \frac{1}{r_f} - r_o + 1$. In general, $\beta < 1$ and Equation 16 illustrates the exponential decay of the initial state's effect on the current population, and the corresponding exponential growth of the "long term" steady state's influence.

The thick line plotted in Figure 4 illustrates a typical convergence graph. With fitness ratio $r_f = 2$ and overlap ratio $r_o = 1/3$, the equilibrium proportion $P_{A,eq} = \frac{1-r_o}{1-2r_o+\frac{1}{r_f}} = 4/5 = 0.8$. In Figure 4 we see very fast convergence to this equilibrium.

To see the effect of varying the degree of overlap, Figure 4 shows convergence curves for a number of different r_o values. The lowest curve corresponds to $r_o = 1/10$, very little overlap, in which we expect to see almost instantaneous (i.e., one generation) convergence to steady-state ($P_{A,eq}$). The other curves show how increasing overlap increases the convergence time but also changes the equilibrium point. At maximum overlap, $r_o = 1/r_f = 1/2$, we see convergence to a population of all **A**s.

In the degenerate case of $f_A = f_B = f_{AB}$, the two species/rules **A** and **B** are identical, completely overlapping, and Equation 16 reduces to a constant, $P_{A,0}$, indicating no selective pressure away from the initial population distribution. This is the case of *genetic drift*, in which equations of expectation (e.g., Equation 16) are least useful, failing to describe the variance of the selection operator. In the case of genetic drift, this variance dominates the dynamics, and leads to eventual loss of one or the other species.

4.3.2 Convergence Times

Here we derive a useful expression for (two-)niche convergence times by solving Equation 16 for time t (in generations). Rearranging Equation 16, we move t to one side:

$$\beta^t = \frac{P_{A,t} - P_{A,eq}}{P_{A,0} - P_{A,eq}}.$$

Taking the logarithm of both sides and solving for t, yields:

$$t = \frac{Log[\frac{P_{A,eq}-P_{A,t}}{P_{A,eq}-P_{A,0}}]}{Log[\beta]}. \quad (17)$$

In general $P_{A,t}$ approaches $P_{A,eq}$ asymptotically with t. More practically, we introduce the discrete nature of a

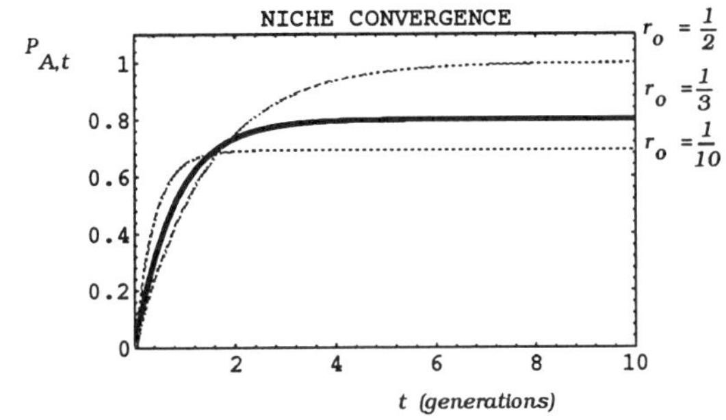

Figure 4: Expected convergence with varying overlap: $r_o = 1/10$, $r_o = 1/3$, and $r_o = 1/2$ (complete overlap). Fitness ($r_f = 2$) is the same for all plots.

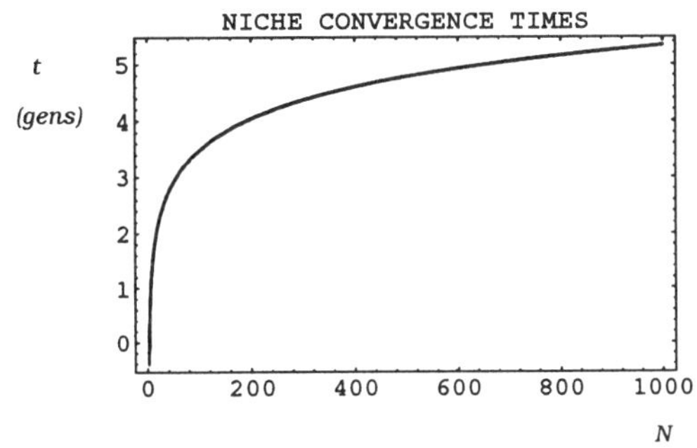

Figure 5: Expected niche convergence time grows logarithmically in population size.

finite population of size N, by asking for the time it takes to go from one copy of **A** ($P_{A,0} = 1/N$) to within at most one individual ($1/N$) of the equilibrium population ($P_{A,t} = P_{A,eq} - 1/N$):

$$t = \frac{Log[\frac{1/N}{P_{A,eq}-1/N}]}{Log[\beta]}.$$

Simplifying yields:

$$t = \frac{-Log[P_{A,eq}N - 1]}{Log[\beta]}. \tag{18}$$

We can see immediately that expected convergence time grows logarithmically in population size N, as we suspected. Figure 5 illustrates a typical expected growth in convergence time with increasing N. Here $r_f = 2$ and $r_o = 1/3$ (remember that β is a function of r_f and r_o).

4.4 Putting It All Together: Critical Population Sizes

In this subsection we make an initial attempt to bring the above two timing results together in a rather crude fashion. We hope to refine and make rigorous some *niching control maps* in future.

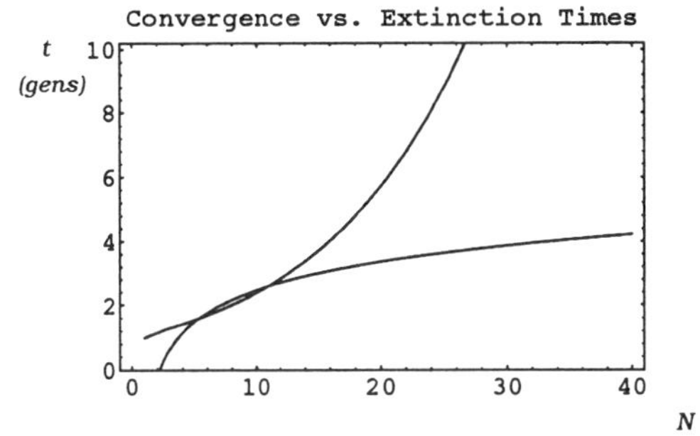

Figure 6: Expected niche extinction times (upper curve) versus "expected" niche convergence times (lower curve). Fitness ratio $r_f = 2$ with very high overlap $r_o = 0.45$ (near maximum).

In Figure 6 we illustrate a simple way of using these two results together. The upper curve plots the expected niche extinction times (using the approximations) as a function of population size. The lower curve is the expected niche convergence time. Here again the fitness ratio $r_f = 2$, but now the overlap ratio is relatively high, $r_o = 0.45$. This overlap means that $r_f * r_o = 90\%$ of **B**'s covered examples are also covered by **A**. This relatively high ratio of overlap was chosen to illustrate clearly the predicted niching failure at low population sizes (e.g., $N < 20$). At high enough population sizes (e.g., $30 < N$), the difference between the niche convergence and niche extinction times is quite high.

Characterizing the critical phase transition from poor niche maintenance to robust niche maintenance as population size increases, is tricky. The niche convergence model (lower curve) breaks down for small population sizes, while the niche extinction time approximation (upper curve) loses accuracy with increasing overlap. Yet our initial models above give us clear indication that implicit niching, in a learning classifier system or in any GA or GP with distinct and finite resources/rewards, works well *within bounds*. And apparently we have some hope of defining those boundaries.

5 Limitations

Before discussing the implications of the above results, we address the limitations on our analysis imposed by our simplifications and assumptions described in Section 3.1. Again, briefly, those assumptions are (1) a ternary alphabet $\{0, 1, \#\}$, (2) stimulus-response (S-R) classifiers, (3) binary classification task, and (4) equal specificity of all rules.

Ternary alphabets are common in both practical and theoretical LCS work, and are equal in expressive power to higher cardinality alphabets. So the assumption of a ternary alphabet is not a simplification and hence does not limit the applicability of our results.

The restriction to S-R classifiers is a real simplification, however, and does place some limits on the extension of our results. Still, the S-R LCS is quite powerful (Wilson, 1994). We could argue that before adding the complexities of message-passing, we should first understand and master the power of the S-R LCS. We could also look at the additional rule interactions possible with message-passing in terms of their effects on the steady-states of the S-R LCS, thus extending our results to include non-S-R classifiers and strong cooperation.

Our assumption of binary classification is a simplification that allowed us to graphically depict rule coverage and overlap, but which actually has no effect on the mathematical analysis in this paper. The concepts of rule coverage and overlap are just as valid in the case of k-ary classification, but they can no longer be measured and ordered simply by numbers of examples.

The fourth assumption is the only one that places any significant limitations on the applicability of our analytical results. Specifically, LCS implicit fitness sharing describes the effects of rules on the "coverage" component of each other's fitness function (where coverage is based on the number of correct classifications). Rules of the same order of specificity can vary only in this component, and so implicit fitness sharing captures the entirety of rule interaction within a single level of generality. Rules of different orders, however, can vary in a second component of fitness: accuracy (which is based on the number of errors). But coverage and accuracy are conflicting objectives, with no general way to combine them into a single scalar fitness measure. So the extension of our results to include rules of all orders of specificity will depend on the user-specified combination of the two fitness components. However, the effect of a more general rule on a more specific rule, with overlapping coverage, will probably be some kind of sharing of the overlapped resources.

6 Conclusions

The specific focus of this paper was on niching in the LCS. But the results are generally applicable to any evolutionary algorithm maintaining useful diversity under selective pressure on a task with separate rewards/resources.

It has long been known that some kind of natural and implicit speciation takes place in the LCS if we break tied competitions for resources arbitrarily. In (Horn, Goldberg, & Deb, 1994), our goal was to demonstrate a meaningful LCS *dynamic equilibrium point.* By meaningful we mean one that will be reached quickly (i.e., fast convergence rate and high *restorative pressure*), and last long (i.e., long niche extinction times). In 1994, we showed the ability of the LCS to balance selection pressure with restorative force, and thus maintain high quality, diverse niches virtually indefinitely. This balance in turn allows us to apply the GA more vigorously in the LCS. We can then hope to improve LCS rule discovery. But for successful exploration and exploitation of niches, the LCS must have a large enough population to maintain subpopulations at all desirable niches.

In this study we extended our model of implicit niching by finding a closed form *approximation* for expected absorption times in the Markov chain models, and also by deriving a closed form estimation of niching convergence rates using the expected proportion models. We now have upper and lower bounds, however crude, on expected "niche existence" times, as functions of population size, fitness, and niche overlap. These are descriptive models for the moment, but by carefully adjusting population size, scaling fitness, and by introducing a tunable parameter for tolerating niche overlap, we can eventually use the models to *prescribe* proper parameter settings to maintain exactly the cooperative population distributions we desire. That is, we should be able to control selection's distinction between cooperation and competition, and the inherent tradeoffs of accuracy (objective fitness) versus generality versus coverage (overlap), etc.

We hope we have demonstrated that niching in an LCS is absolutely necessary to any kind of cooperation. By definition a cooperative group of rules is a set of *diverse* rules, and without some kind of niching the GA in the LCS will not maintain diversity. If we want serious GA search in our LCS, we must have significant selection pressure. If we want to maintain cooperative rules, we must balance that selection pressure with a restorative pressure. If we can't do that, we won't be able to induce any type of *strong cooperation*, such as that necessary for temporal rule chains and default hierarchies. All such strong cooperation implies successful weak cooperation (i.e., covering). Because of such entailment, niching is fundamental to the success of rule search *and* maintenance in the LCS, and perhaps to any cooperative evolutionary algorithm.

The most important point of this paper, however, is the demonstration that the dynamics of an interacting, coevolving, cooperative/competitive population *can* be analyzed, described, predicted, and controlled.

Acknowledgments

The first author is grateful for early support provided by NASA under contract number NGT-50873. Both authors acknowledge the support of the Air Force Office of Scientific Research (AFOSR), Air Force Materiel Command, USAF, under grants numbered F4960-94-1-0103 and F49620-95-1-0338. The U.S. Government is authorized to reproduce and distribute reprints for Governmental purposes notwithstanding any copyright notation thereon. The views and conclusions contained herein are those of the authors and should not be interpreted as necessarily representing the official policies or endorsements, either expressed or implied, of the AFOSR or the U.S. Government.

Bibliography

Booker, L. B. 1982. Intelligent behavior as an adaptation to the task environment. *Dissertation Abstracts International, 43*(2), 469B. (University Microfilms No. 8214966)

Booker, L. B. 1989. Triggered rule discovery in classifier systems. In J. D. Schaffer, (Ed.), *Proceedings of the Third International Conference on Genetic Algorithms (ICGA 3).* San Mateo, CA: Morgan Kaufmann. Pages 265–274.

De Jong, K. A. 1975. An analysis of the behavior of a class of genetic adaptive systems (Ph.D. thesis, University of Michigan, Ann Arbor). *Dissertation Abstracts International, 36*(10), 5140B. (University Microfilms No. 76-9381)

Deb, K. 1989. *Genetic algorithms in multimodal function optimization.* Masters thesis and TCGA Report No. 89002. Tuscaloosa, AL: Department of Engineering Mechanics, University of Alabama.

Deb, K., & Goldberg, D. E. 1989. An investigation of niche and species formation in genetic function optimization. In J. D. Schaffer (Ed.), *Proceedings of the Third International Conference on Genetic Algorithms (ICGA 3).* San Mateo, CA: Morgan Kaufmann. Pages 42–50.

Goertzel, B. 1993. *The Evolving Mind.* Langhorne, PA: Gordon and Breach.

Goldberg, D. E. 1983. *Computer-aided gas pipeline operation using genetic algorithms and rule learning* (Ph.D. thesis, University of Michigan).

Goldberg, D. E. 1989. *Genetic algorithms in search, optimization, and machine learning.* Reading, MA: Addison-Wesley.

Goldberg, D. E., & Deb, K. 1991. A comparative analysis of selection schemes used in genetic algorithms. In G. J. E. Rawlins, (Ed.), *Foundations of Genetic Algorithms (FOGA).* San Mateo, CA: Morgan Kaufmann. Pages 69–93.

Goldberg, D. E., Deb, K., & Horn, J. 1992. Massive multimodality, deception, and genetic algorithms. In R. Männer & B. Manderick, (Eds.), *Parallel Problem Solving From Nature, 2.* North-Holland. Pages 37–46.

Goldberg, D. E., Horn, J., & Deb, K. 1992. *What makes a problem hard for a classifier system?* (IlliGAL Report No. 92007). Urbana, IL: Department of General Engineering, University of Illinois at Urbana-Champaign. Presented as an extended abstract at *The First International Workshop on Learning Classifier Systems.* Houston, Texas, USA, October 1992.

Goldberg, D. E., & Richardson, J. 1987. Genetic algorithms with sharing for multimodal function optimization. In J. Grefenstette, (Ed.), *Proceedings of the Second International Conference on Genetic Algorithms (ICGA 2).* Hillsdale, NJ: Lawrence Erlbaum Associates. Pages 41–49.

Goldberg, D. E., & Segrest, P. 1987. Finite Markov chain analysis of genetic algorithms. In J. Grefenstette, (Ed.), *Proceedings of the Second International Conference on Genetic Algorithms (ICGA 2).* Hillsdale, NJ: Lawrence Erlbaum Associates. Pages 1–8.

Holland, J. H. 1971. Processing and processors for schemata. *Proceedings of the Symposium on Associative Information Techniques.* Pages 127–146.

Holland, J. H. 1975. *Adaptation in natural and artificial systems.* Ann Arbor, MI: University of Michigan Press.

Holland, J. H. 1985. Properties of the bucket-brigade algorithm. In Grefenstette, J. J. (Ed.), *Proceedings of an International Conference on Genetic Algorithms and Their Applications (ICGA 1).* Pittsburgh, PA: Carnegie-Mellon University. Pages 1–7.

Holland, J. H. 1992. *Adaptation in natural and artificial systems.* Cambridge, MA: MIT Press.

Horn, J. 1993. Finite Markov chain analysis of genetic algorithms with niching. In S. Forrest, (Ed.), *Proceedings of the Fifth International Conference on Genetic Algorithms (ICGA 5).* San Mateo, CA: Morgan Kaufmann. Pages 110–117.

Horn, J., Goldberg, D. E., & Deb, K. 1994. Implicit Niching in a Learning Classifier System: Nature's Way. *Evolutionary Computation, 2*(1). Pages 37–66.

Mahfoud, S. W. 1992. Crowding and preselection revisited. In R. Männer & B. Manderick (Eds.), *Parallel Problem Solving From Nature, 2.* North-Holland. Pages 27–36.

Mahfoud, S. W. 1993. Simple analytical models of genetic algorithms for multimodal function optimization. One-page summary in S. Forrest, (Ed.), *Proceedings of the Fifth International Conference on*

Genetic Algorithms (ICGA 5). San Mateo, CA: Morgan Kaufmann, Page 643. Full paper available as IlliGAL Report No. 93001. Urbana: Department of General Engineering, University of Illinois at Urbana-Champaign.

Mahfoud, S. W. 1995a. Population size and genetic drift in fitness sharing. In L. D. Whitley & M. D. Vose, (Ed.s), *Foundations of Genetic Algorithms, 3 (FOGA 3)*. San Francisco, CA: Morgan Kaufmann. Pages 185–224.

Mahfoud, S. W. 1995b. *Niching Methods for Genetic Algorithms* (Ph.D. thesis, University of Illinois at Urbana-Champaign). Available as IlliGAL Report No. 95001. Urbana: Department of General Engineering, University of Illinois at Urbana-Champaign.

McCallum, R. A., & Spackman, K. A. 1990. Using genetic algorithms to learn disjunctive rules from examples. In B. W. Porter & R. J. Mooney, (Ed.s), *Machine Learning: Proceedings of the Seventh International Conference*. Palo Alto, CA: Morgan Kaufmann. Pages 149–152.

Neri, F., & Saitta, L. 1995. Analysis of genetic algorithms evolution under pure selection. In L. J. Eshelman, (Ed.), *Proceedings of the Sixth International Conference on Genetic Algorithms*. San Francisco, CA: Morgan Kaufmann. Pages 32–39.

Smith, R. E., Forrest, S., & Perelson, A. S. 1993. Searching for diverse, cooperative subpopulations with genetic algorithms. *Evolutionary Computation, 1*(2). Pages 127–149.

Smith, R. E., & Valenzuela-Rendón, M. 1989. A study of rule set development in a learning classifier system. In J. D. Schaffer, (Ed.), *Proceedings of the Third International Conference on Genetic Algorithms (ICGA 3)*. San Mateo: Morgan Kaufmann. Pages 340–346.

Vose, M. D. 1993. Modeling simple genetic algorithms. In L. D. Whitley, (Ed.), *Foundations of Genetic Algorithms, 2 (FOGA 2)*. San Mateo, CA: Morgan Kaufmann. Pages 63–73.

Wilson, S. W. 1986. *Classifier learning of a boolean function*. (Research Memo RIS No. 27r). Cambridge: The Rowland Institute for Science.

Wilson, S. W. 1987. Classifier systems and the animat problem. *Machine Learning, 2*. Pages 199–228.

Wilson, S. W. 1994. ZCS: A zeroth level classifier system. *Evolutionary Computation, 2*(1). Pages 1–18.

Wilson, S. W. 1995. A classifier system based on accuracy. *Evolutionary Computation, 3*(2). Pages 149–175.

Wilson, S. W. & Goldberg, D. E. 1989. A critical review of classifier systems. In J. D. Schaffer, (Ed.), *Proceedings of the Third International Conference on Genetic Algorithms*, San Mateo, CA: Morgan Kaufmann. Pages 244–255.

AUTHOR INDEX

SUBJECT INDEX